ALGEBRA FOR JEE Main and Advanced

Man Mohan Jha

Pawan Khandelwal

ALGEBRA FOR JEE MAIN AND ADVANCED

UNIVERSITIES PRESS (INDIA) PRIVATE LIMITED

Registered Office
3-6-747/1/A & 3-6-754/1, Himayatnagar, Hyderabad 500 029, Telangana, India
info@universitiespress.com; www.universitiespress.com

Distributed by
Orient Blackswan Private Limited

Registered Office
3-6-752 Himayatnagar, Hyderabad 500 029, Telangana, India

Other Offices
Bengaluru / Chennai / Guwahati / Hyderabad / Kolkata / Mumbai / New Delhi / Noida / Patna

First published 2025

ISBN: 978-93-93330-87-1

502259

Typeset in Times NR MT Std 10pt *by*
SRS Publishing Services, Puducherry

Printed at
B.B Press, Tronica City, Ghaziabad, U.P. 201 103

Published by
Universities Press (India) Private Limited
3-6-747/1/A & 3-6-754/1, Himayatnagar, Hyderabad 500 029, Telangana, India

Preface

This book was conceptualised and subsequently designed as a textbook on Algebra, with the aim of simplifying students' understanding of their coursework while also supporting their preparation for competitive entrance examinations.

We have endeavoured to explain the fundamental concepts in a clear and simple manner, with supplementary elaborations provided wherever necessary. Our methodology in structuring the book has been as follows:

- Each concept is explained in detail, accompanied by suitable examples that illustrate its application.
- Subjective solved examples are included throughout, aimed at enhancing both the understanding and problem-solving skills of students.
- A variety of objective and subjective assignments are provided, including MCQs (with one or more correct options), integer-based, comprehension-based, and match matrix-type questions.
- The assignments have been deliberately designed with varying levels of difficulty, ranging from basic to advanced. The questions have also been curated to test the lateral application of concepts wherever possible.

In this book, we have sought to deepen students' understanding of Algebra and equip them with a comprehensive toolkit to confidently approach any algebraic problem they may encounter. Working on this book has been personally challenging yet immensely rewarding, and we are truly pleased to present it to our readers.

We have made every effort to ensure the text is free of errors and would be grateful to readers who point out any oversights or offer constructive suggestions.

Man Mohan Jha
Pawan Khandelwal

Acknowledgements

We would like to express our sincere gratitude to Universities Press Private Limited for believing in our vision, granting the opportunity to channel our years of teaching experience and accumulated knowledge into this book, and for their support throughout the publication process. In particular, we thank Dr Gita S Dattatri for her patient editing and suggestions, Mr Kallol Das, Mr Thomas Mathew Rajesh and Mr Dushyant Singh for their thoughtful guidance and valuable input in shaping the direction and structure of this work.

We are deeply thankful to our fellow educators – Mr Manoj Aggarwal, Mr Rajesh Pandey and Mr Manohar Bhiryani – for their insightful suggestions, which have added clarity and depth to the book.

Man Mohan Jha would like to dedicate this book to his late mother, Dhanmaya Jha, whose perseverance in the face of adversity has been a lifelong source of inspiration. He extends heartfelt thanks to his daughter, Dr Supriya Rajhans; his son-in-law, Dr Bhaswar Dutta Gupta; his daughter-in-law, Jyoti Jha; and his son, Sumit Rajhans. Special thanks go to his wife, Punita Jha, for her unwavering support and encouragement throughout the journey.

Pawan Khandelwal pays tribute to his forefather, the late Mr Jagdish Prasad Khandelwal, whose values continue to guide him. He expresses his deepest appreciation to his parents and extended family. Most especially, he thanks his wife, Neha, and his two sons, Sadhil and Siddhant, for their love, patience and constant encouragement, without which this book would not have been possible.

We are both grateful to one another for bringing complementary strengths to this collaboration and for enriching the work through shared experience and mutual respect.

Man Mohan Jha
Pawan Khandelwal

Contents

Preface . . . . . . *iii*
Acknowledgements . . . . . . *iv*

Chapter 1: Quadratic Equations and Expressions **1.1**

1.1 Introduction 1.1
1.2 Quadratic Equations 1.1
1.3 Quadratic Expressions 1.7
1.4 Know Your Concepts 1.11
1.5 JEE Main: Sample Questions 1.12
1.6 JEE Advanced: Sample Questions 1.16
1.7 JEE Main: Archive 1.23
1.8 Answer Key 1.25
1.9 Solutions 1.27

Chapter 2: Progression and Series **2.1**

2.1 Introduction 2.1
2.2 Arithmetic Progression 2.1
2.3 Geometric Progression 2.2
2.4 Harmonic Progression 2.3
2.5 Relation between A.M., G.M. and H.M. 2.4
2.6 Sum of Natural Number Expressions 2.4
2.7 Arithmetico–Geometric Series 2.5
2.8 Know Your Concepts 2.14
2.9 JEE Main: Sample Questions 2.14
2.10 JEE Advanced: Sample Questions 2.20
2.11 JEE Main: Archive 2.29
2.12 Answer Key 2.36
2.13 Solutions 2.38

Chapter 3: Complex Numbers **3.1**

3.1 Introduction 3.1
3.2 Conjugate of a Complex Number 3.1
3.3 Modulus of a Complex Number 3.1
3.4 Argument or Amplitude of a Complex Number 3.2
3.5 Polar Form of a Complex Number 3.2
3.6 Euler's Formula 3.3
3.7 Demoivre's Theorem 3.3
3.8 Cube Roots of Unity 3.3
3.9 n^{th} Roots of Unity 3.3
3.10 Rotation 3.4
3.11 Equation of a Straight Line 3.4
3.12 Equation of Circle 3.5
3.13 Know Your Concepts 3.11
3.14 JEE Main: Sample Questions 3.12
3.15 JEE Advanced: Sample Questions 3.17
3.16 JEE Main: Archive 3.26
3.17 Answer Key 3.32
3.18 Solutions 3.34

Chapter 4: Permutation and Combination **4.1**

4.1 Introduction 4.1
4.2 Permutation 4.1
4.3 Fundamental Principle of Counting 4.2
4.4 Circular Permutation 4.3
4.5 Sum of Numbers Formed by Using the Digits 4.3
4.6 Tests of Divisibility 4.4
4.7 Rank of Words 4.4
4.8 Combination 4.5
4.9 De-arrangement 4.7
4.10 Multinomial Theorem 4.7
4.11 Non-negative Integral Solution Method 4.8

4.12 Principle of Inclusion and Exclusion . . . 4.8
4.13 Formation on Geometrical Figures . . . 4.9
4.14 Know Your Concepts . . . 4.17
4.15 JEE Main: Sample Questions . . . 4.18
4.16 JEE Advanced: Sample Questions . . . 4.23
4.17 JEE Main: Archive . . . 4.31
4.18 Answer Key . . . 4.35
4.19 Solutions . . . 4.37
4.20 Summary . . . 4.72

Chapter 5: Binomial Theorem 5.1

5.1 Introduction . . . 5.1
5.2 Properties of Binomial Theorem . . . 5.1
5.3 Greatest Term . . . 5.2
5.4 Properties of Binomial Coefficients . . . 5.2
5.5 Know Your Concepts . . . 5.9
5.6 JEE Main: Sample Questions . . . 5.10
5.7 JEE Advanced: Sample Questions . . . 5.15
5.8 JEE Main: Archive . . . 5.21
5.9 Answer Key . . . 5.28
5.10 Solutions . . . 5.29

Chapter 6: Determinants and Matrices 6.1

6.1 Determinants . . . 6.1
6.2 Matrices . . . 6.10
6.3 Know Your Concepts . . . 6.21
6.4 JEE Main: Sample Questions . . . 6.23
6.5 JEE Advanced: Sample Questions . . . 6.29
6.6 JEE Main: Archive . . . 6.40
6.7 Answer Key . . . 6.52
6.8 Solutions . . . 6.53

Chapter 7: Probability 7.1

7.1 Introduction . . . 7.1
7.2 Probability of Occurrence of an Event 7.2
7.3 Odds in Favour and Odds Against an Event . . . 7.2
7.4 Conditional Probability . . . 7.3
7.5 Theorem of Total Probability . . . 7.4
7.6 Bayes' Theorem (Inverse Probability Theorem) . . . 7.4
7.7 Binomial Theorem on Probability . . . 7.6
7.8 Probability Distribution of a Random Variable . . . 7.6
7.9 Know Your Concepts . . . 7.17
7.10 JEE Main: Sample Questions . . . 7.18
7.11 JEE Advanced: Sample Questions . 7.24
7.12 JEE Main: Archive . . . 7.33
7.13 Answer Key . . . 7.40
7.14 Solutions . . . 7.41

Chapter 8: Statistics 8.1

8.1 Introduction . . . 8.1
8.2 Measures of Central Value . . . 8.1
8.3 Dispersion . . . 8.2
8.4 Short Cut Method . . . 8.4
8.5 Step Deviation Method for Finding S.D. . . . 8.5
8.6 Know Your Concepts . . . 8.10
8.7 JEE Main: Sample Questions . . . 8.11
8.8 JEE Main: Archive . . . 8.13
8.9 Answer Key . . . 8.17
8.10 Solutions . . . 8.17
8.12 Summary . . . 8.36

Chapter 9: Set Theory and Real Number System 9.1

9.1 Sets . . . 9.1
9.2 Types of Sets . . . 9.1
9.3 Set Operations . . . 9.2
9.4 Cardinal Number of a Set . . . 9.3
9.5 Real Number System . . . 9.5
9.6 Concept of Intervals . . . 9.6
9.7 Wavy Curve Method . . . 9.6
9.8 Absolute value . . . 9.6
9.9 Greatest Integer and Fractional Part . 9.7
9.10 Know Your Concepts . . . 9.10
9.11 JEE Main: Sample Questions . . . 9.11
9.12 Answer Key . . . 9.12
9.13 Summary . . . 9.12

CHAPTER 1

Quadratic Equations and Expressions

OBJECTIVES

1.1 Introduction
1.2 Quadratic Equations
1.3 Quadratic Expressions
1.4 Know Your Concepts
1.5 JEE Main: Sample Questions
1.6 JEE Advanced: Sample Questions
1.7 JEE Main: Archive
1.8 Answer Key
1.9 Solutions
1.10 Summary

1.1 Introduction

'Quadratic equations' is considered to be one of the most important and useful topics in Algebra. Quadratic equations do appear in varied form in most of the topics in Mathematics.

This chapter covers the concepts and methods required to find solutions of a quadratic equation along with the relation of the roots of a quadratic equation with its coefficients. Each theoretical concept has been explained through multiple solved examples. The chapter covers the analytical and qualitative analysis of the quadratic expressions in detail. The chapter has been constructed in such a way that the topic is punctuated and clarified through a number of solved and unsolved problems.

1.2 Quadratic Equations

An equation of the form $ax^2+bx+c=0$ where $a \neq 0$ and a, b, c are real numbers, is called a *quadratic equation*. The roots of the quadratic equation are those values of 'x' at which the quadratic equation is satisfied.

Since $ax^2+bx+c=0$

$$\Rightarrow \quad x^2+\frac{b}{a}x+\frac{c}{a}=0$$

$$\Rightarrow \quad x^2+\frac{b}{a}x+\frac{b^2}{4a^2}-\frac{b^2}{4a^2}+\frac{c}{a}=0$$

$$\Rightarrow \quad \left(x+\frac{b}{2a}\right)^2-\frac{b^2-4ac}{4a^2}=0$$

$$\Rightarrow \quad \left(x+\frac{b}{2a}\right)^2=\frac{b^2-4ac}{4a^2}$$

$$\Rightarrow \quad x+\frac{b}{2a}=\pm\frac{\sqrt{b^2-4ac}}{2a}$$

$$\Rightarrow \quad x=-\frac{b}{2a}\pm\frac{\sqrt{b^2-4ac}}{2a}$$

$$\Rightarrow \quad x=\frac{-b\pm\sqrt{b^2-4ac}}{2a}$$

Let, $\alpha=\frac{-b+\sqrt{b^2-4ac}}{2a}$ and $\beta=\frac{-b-\sqrt{b^2-4ac}}{2a}$ be the two roots of the quadratic equation, then $\alpha+\beta=-\frac{b}{a}$ and $\alpha\beta=\frac{c}{a}$.

As the nature of roots of the quadratic equation depend upon $\left(b^2-4ac\right)$, $\left(b^2-4ac\right)$ is called the discriminant of the quadratic equation $ax^2+bx+c=0$ and is denoted by D.

> If α and β are the roots of quadratic equation $ax^2+bx+c=0$, then it follows that
>
> $$ax^2+bx+c=a\left(x^2+\frac{b}{a}x+\frac{c}{a}\right)$$
>
> $$=a\left(x^2-(\alpha+\beta)x+\alpha\beta\right)$$
>
> $$ax^2+bx+c=a\,(x-\alpha)(x-\beta)$$

1.2.1 Nature of Roots

If $a, b, c \in R$ and $a \neq 0$, then

1. The quadratic equation has real and equal roots if and only if $D = 0$.
2. The quadratic equation has complex roots with non-zero imaginary parts if and only if $D < 0$.
3. The quadratic equation has real and distinct roots if and only if $D > 0$.
4. If D_1 and D_2 are the discriminants of two quadratic equations, then
   (a) $D_1 + D_2 \geq 0$, then at least one of D_1 and $D_2 \geq 0$.
   (b) $D_1 + D_2 < 0$, then at least one of D_1 and $D_2 < 0$.
5. If $a, b, c \in Q$ where Q is a rational number and D is a perfect square of a rational number, the roots are rational and in case it is not a perfect square, the roots are irrational.
6. If $a, b, c \in R$ and $(p + iq)$ is one of roots of the quadratic equation, then the other root must be $(p - iq)$ (where $q \neq 0; p, q \in R$ and $i = \sqrt{-1}$).
7. If $a, b, c \in Q$ (rational number) and $p + \sqrt{q}$ is one root of quadratic equation where $q \neq 0$, then the other root must be $p - \sqrt{q}$.
8. If $a = 1$ and $b, c \in I$ and the roots of the quadratic equation are rational numbers, then the roots must be integers.
9. If in a quadratic equation $ax^2 + bx + c = 0$, $a + b + c = 0$ and a, b, c are rational, then 1 is the root of the equation.
10. If the equation $ax^2 + bx + c = 0$ has more than two roots, then it will become an identity and this implies that $a = b = c = 0$.

SOLVED EXAMPLES

Problems based on the nature of roots of a quadratic equation.

Example 1.1 Find the value of m for which the equation $(1 + m)\, x^2 - 2(1 + 3m)x + (1 + 8m) = 0$ has equal roots.

Solution As roots of the equation

$(1 + m)\, x^2 - 2(1 + 3m)\, x + (1 + 8m) = 0$ are equal,

Hence, $D = 0$

$\Rightarrow \quad 4(1+3m)^2 - 4\,(1 + m)\,(1 + 8m) = 0$

$\Rightarrow \quad m^2 - 3m = 0$

$\Rightarrow \quad m = 0, 3$

Example 1.2 Show that if the roots of the equation $(a^2 + b^2)\, x^2 + 2x\,(ac + bd) + c^2 + d^2 = 0$ are real, they will be equal.

Solution As the roots of the equation

$\left(a^2 + b^2\right)x^2 + 2x\left(ac + bd\right) + c^2 + d^2 = 0$ are real,

Hence, $D \geq 0$

$\Rightarrow \quad 4\left(ac + bd\right)^2 - 4\left(a^2 + b^2\right)\left(c^2 + d^2\right) \geq 0$

$\Rightarrow \quad \left(a^2c^2 + b^2d^2 + 2abcd\right) - \left(a^2c^2 + a^2d^2 + b^2c^2 + b^2d^2\right) \geq 0$

$\Rightarrow \quad -\left(a^2d^2 + b^2c^2 - 2abcd\right) \geq 0$

$\Rightarrow \quad -\left(ad - bc\right)^2 \geq 0$

$\Rightarrow \quad ad - bc = 0$

$\Rightarrow$ Roots are equal.

Example 1.3 If the roots of the equation, $(b - c)\, x^2 + (c - a)\, x + (a - b) = 0$ are equal, then prove that a, b, c are in A.P.

Solution It is given that the roots of the equation $(b - c)x^2 + (c - a)x + (a - b) = 0$ are equal

As $(b - c) + (c - a) + (a - b) = 0$

$\Rightarrow$ One root is 1.

Hence both roots are 1.

Product of roots $= \dfrac{a - b}{b - c}$

$\Rightarrow \quad 1 = \dfrac{a - b}{b - c}$

$\Rightarrow \quad a - b = b - c$

$\Rightarrow \quad a, b, c$ is in A.P.

Example 1.4 If the roots of the equation $x^2 - 8x + a^2 - 6a = 0$ are real and distinct, then find all possible values of a.

Solution Since the roots of the given equation are real and distinct, we must have $D > 0$.

$\Rightarrow \quad 64 - 4\left(a^2 - 6a\right) > 0$

$\Rightarrow\ 16-a^2+6a>0$

$\Rightarrow\ a^2-6a-16<0$

$\Rightarrow\ (a-8)(a+2)<0$

$\Rightarrow\ -2<a<8$

$\Rightarrow$ Hence, $a\in(-2,8)$.

Example 1.5 If $a+b+c=0$ and a, b, c are rational, prove that the roots of the equation

$(b+c-a)x^2+(c+a-b)x+(a+b-c)=0$,

are rational.

Solution Since the given equation has rational roots, D must be a perfect square.

Now $D=(c+a-b)^2-4(b+c-a)(a+b-c)$

As $a+b+c=0$

$\Rightarrow\ b=-(c+a)$

Hence,

$$D=4(c+a)^2-4[(-c-a+c-a)(-c-a+a-c)]$$

$$\Rightarrow D=4\left[(c+a)^2-4ac\right]=4(c-a)^2$$

which is a perfect square.
Hence, roots are rational.

Example 1.6 If one root of the equation $(l-m)x^2+lx+1=0$ be the double of the other and if l be real, find the greatest value of m.

Solution Let α and 2α be the roots of the equation $(l-m)x^2+lx+1=0$

Then,

$$\alpha+2\alpha=-\frac{l}{l-m}$$

$$\Rightarrow\ \alpha=-\frac{l}{3(l-m)}$$

and, $\alpha\cdot 2\alpha=\dfrac{1}{l-m}$

$$\Rightarrow\ 2\alpha^2=\frac{1}{l-m}$$

$$\Rightarrow\ 2\frac{l^2}{9(l-m)^2}=\frac{1}{l-m}$$

$$\Rightarrow\ 2l^2=9(l-m)$$

$$\Rightarrow\ 2l^2-9l+9m=0$$

As this is quadratic in l and l is real,

$D\ge 0$

$\Rightarrow\ 81-72m\ge 0$

$\Rightarrow\ 72m\le 81$

$$\Rightarrow\ m\le\frac{81}{72}$$

$$\Rightarrow\ m\le\frac{9}{8}$$

Hence greatest value of $m=\dfrac{9}{8}$.

Example 1.7 If α and β are the roots of the equation $x^2-a(x+1)-b=0$, then find the value of

$$\frac{\alpha^2+2\alpha+1}{\alpha^2+2\alpha+b}+\frac{\beta^2+2\beta+1}{\beta^2+2\beta+b}.$$

Solution The given equation is $x^2-ax-(a+b)=0$.
As α and β are the roots of the equation,

$\alpha+\beta=a,\ \alpha\beta=-(a+b)$

$\Rightarrow\ \alpha\beta=-a-b$

$\Rightarrow\ b=-(\alpha+\beta)-\alpha\beta$

$\Rightarrow\ b=-\alpha-\beta-\alpha\beta$

$\Rightarrow\ b-1=-\alpha-1-\beta-\alpha\beta$

$\Rightarrow\ b-1=-1(\alpha+1)-\beta(\alpha+1)$

$\Rightarrow\ (b-1)=-(\alpha+1)(\beta+1)$

Now,

$$\frac{\alpha^2+2\alpha+1}{\alpha^2+2\alpha+b}+\frac{\beta^2+2\beta+1}{\beta^2+2\beta+b}$$

$$=\frac{(\alpha+1)^2}{(\alpha+1)^2+(b-1)}+\frac{(\beta+1)^2}{(\beta+1)^2+(b-1)}$$

$$=\frac{(\alpha+1)^2}{(\alpha+1)^2-(\alpha+1)(\beta+1)}+\frac{(\beta+1)^2}{(\beta+1)^2-(\alpha+1)(\beta+1)}$$

$$=\frac{(\alpha+1)^2}{(\alpha+1)(\alpha+1-\beta-1)}+\frac{(\beta+1)^2}{(\beta+1)(\beta+1-\alpha-1)}$$

$$=\frac{(\alpha+1)}{(\alpha-\beta)}+\frac{(\beta+1)}{(\beta-\alpha)}$$

$$=\frac{\alpha+1-\beta-1}{\alpha-\beta}=1$$

Example 1.8 If the roots of the quadratic equation $x^2-(2n+18)x-n-11=0, n \in I$ are rational, then find the value of n.

Solution As the roots of the given equation $x^2-(2n+18)x-n-11=0$ are rational, D must be a perfect square.

$n^2+19n+92$ must be a perfect square

Let, $n^2+19n+92=m^2$

$\Rightarrow \quad n^2+19n+92-m^2=0$

$\Rightarrow \quad n=\dfrac{-19\pm\sqrt{4m^2-7}}{2}$

$\Rightarrow \quad 4m^2-7$ is a perfect square

Let $4m^2-7=\beta^2$

$\Rightarrow \quad 4m^2-\beta^2=7$

$\Rightarrow \quad (2m+\beta)(2m-\beta)=7$

Either $(2m+\beta)=\pm1$ or $(2m-\beta)=\pm7$

$\Rightarrow \quad (2m+\beta)=\pm7$ or $(2m-\beta)=\pm1$

$\Rightarrow \quad 2m=\pm4$

$\Rightarrow \quad m=\pm2$

$\Rightarrow \quad n^2+19n+92=4$

$\Rightarrow \quad (n+8)(n+11)=0$

$\Rightarrow \quad n=-8,-11$

Alternative

$n^2+19n+92$ should be a perfect square.

$n^2+19n+88+4$ should be a perfect square.

$(n+8)(n+11)+4$ should be a perfect square.

Which holds for $n=-8$ and $n=-11$.

Example 1.9 Find the value of n for which the expression $9x^4-12x^3+nx^2-8x+4$ is a perfect square.

Solution As $9x^4-12x^3+nx^2-8x+4$ is a perfect square,

$\Rightarrow \quad x^2\left[9x^2-12x+n-\dfrac{8}{x}+\dfrac{4}{x^2}\right]$

is a perfect square

$\Rightarrow \quad \left(3x+\dfrac{2}{x}\right)^2-12+n-4\left(3x+\dfrac{2}{x}\right)$

should be a perfect square

$\Rightarrow \quad \left(3x+\dfrac{2}{x}\right)^2-4\left(3x+\dfrac{2}{x}\right)+(n-12)$

should be a perfect square

$\Rightarrow \quad n-12=4$

$\Rightarrow \quad n=16$

Example 1.10 If p, q, r and s are real numbers and $pr = 2(q + s)$, then prove that at least one of the equations $x^2+px+q=0$ and $x^2+rx+s=0$ has real roots.

Solution Let D_1 and D_2 be the discriminants of the given equation $x^2+px+q=0$ and $x^2+rx+s=0$ respectively. Now,

$\Rightarrow D_1+D_2=p^2-4q+r^2-4s=p^2+r^2-4(q+s)$

$=p^2+r^2-2pr \qquad [pr=2(q+s)]$

$=(p-r)^2\geq 0$

$\Rightarrow \quad D_1+D_2\geq 0$

Hence, at least one of the equations $x^2+px+q=0$ and $x^2+rx+s=0$ has real roots.

1.2.2 Condition for Common Roots

- Find the condition that the quadratic equation $a_1x^2+b_1x+c_1=0$ and $a_2x^2+b_2x+c_2=0$ have a common root.

  Let α be the common root of the quadratic equation

  $a_1x^2+b_1x+c_1=0$ and $a_2x^2+b_2x+c_2=0$

  $\Rightarrow \quad a_1\alpha^2+b_1\alpha+c_1=0$ and $a_2\alpha^2+b_2\alpha+c_2=0$

  Using cross multiplication

  $$\frac{\alpha^2}{b_1c_2-b_2c_1}=\frac{\alpha}{c_1a_2-c_2a_1}=\frac{1}{a_1b_2-a_2b_1}$$

  $\Rightarrow \alpha=\dfrac{b_1c_2-b_2c_1}{c_1a_2-c_2a_1}$ and $\alpha=\dfrac{c_1a_2-c_2a_1}{a_1b_2-a_2b_1}$

  $\Rightarrow \dfrac{b_1c_2-b_2c_1}{c_1a_2-c_2a_1}=\dfrac{c_1a_2-c_2a_1}{a_1b_2-a_2b_1}$

  which is the required condition.

- Both roots are common.

Let α and β be the roots of the equation

$a_1x^2+b_1x+c_1=0$ and $a_2x^2+b_2x+c_2=0$

$$\Rightarrow \alpha+\beta=\frac{-b_1}{a_1}=\frac{-b_2}{a_2}$$

$$\Rightarrow \frac{a_1}{a_2}=\frac{b_1}{b_2}$$

$$\Rightarrow \alpha\beta=\frac{c_1}{a_1}=\frac{c_2}{a_2}$$

$$\Rightarrow \frac{a_1}{a_2}=\frac{c_1}{c_2}$$

Hence, $\frac{a_1}{a_2}=\frac{b_1}{b_2}=\frac{c_1}{c_2}$.

SOLVED EXAMPLES

Problems based on common roots of a quadratic equation.

Example 1.11 Prove that the equations

$(q-r)x^2+(r-p)x+(p-q)=0$ and

$(r-p)x^2+(p-q)x+(q-r)=0$

have a common root.

Solution Since the sum of the coefficients of the quadratic equation

$$(q-r)x^2+(r-p)x+(p-q)=0 \text{ is}$$

$$q-r+r-p+p-q=0$$

Also, the sum of the coefficients of the quadratic equation

$$(r-p)x^2+(p-q)x+(q-r)=0 \text{ is}$$

$$r-p+p-q+q-r=0$$

Hence 1 is a common root of both the equations.

Example 1.12 Find the value of p if the equations $3x^2-2x+p=0$ and $6x^2-17x+12=0$ have a common root.

Solution Let α be the common root of the equations $3x^2-2x+p=0$ and $6x^2-17x+12=0$

Hence, $3\alpha^2-2\alpha+p=0$

$$6\alpha^2-17\alpha+12=0$$

By cross multiplication

$$\frac{\alpha^2}{-24+17p}=\frac{\alpha}{6p-36}=\frac{1}{-51+12}$$

$$\Rightarrow \alpha=\frac{17p-24}{6p-36}=\frac{6p-36}{-39}$$

$$\Rightarrow (6p-36)^2=39(24-17p)$$

$$\Rightarrow 12p^2+77p+120=0$$

$$\Rightarrow (4p+15)(3p+8)=0$$

Hence, $p=\frac{-15}{4},\frac{-8}{3}$

Example 1.13 If the equations

$x^2+3x+5=0$ and $ax^2+bx+c=0$

have a common root and $a, b, c \in N$, then find the minimum value of $a+b+c$.

Solution The discriminant of the quadratic equation $x^2+3x+5=0$ is negative.

$\Rightarrow$ Its roots are imaginary.

$\Rightarrow$ Hence both roots are common.

Therefore, $\frac{a}{1}=\frac{b}{3}=\frac{c}{5}=k(\text{say})$

$\Rightarrow a=k, b=3k, c=5k$

$\Rightarrow a+b+c=9k$

Hence the minimum value of $a+b+c=9$.

Example 1.14 If $x^2+ax+bc=0$ and $x^2+bx+ca=0$ have a common root, then show that their other roots satisfy the equation $x^2+cx+ab=0$.

Solution Let α, β and α_1, β_1 be the roots of the equations $x^2+ax+bc=0$ and $x^2+bx+ca=0$ respectively.

Now as α is common root,

$$\Rightarrow \alpha^2+a\alpha+bc=0$$

$$\alpha^2+b\alpha+ca=0$$

$$=\frac{\alpha^2}{a^2c-b^2c}=\frac{\alpha}{bc-ca}=\frac{1}{b-a}$$

$$\text{Hence, } \alpha=\frac{c(a^2-b^2)}{c(b-a)}=-(a+b)$$

$$\text{and } \alpha=\frac{c(b-a)}{(b-a)}=c$$

Hence, $c=-(a+b)$

Now,

$$\alpha \cdot \beta = bc$$

$$\Rightarrow \quad c\beta = bc$$

$$\Rightarrow \quad \beta = b$$

Also, $\alpha \cdot \beta_1 = ca$

$$\Rightarrow \quad c\beta_1 = ca$$

$$\Rightarrow \quad \beta_1 = a$$

Now the equation whose roots are β and β_1 is $x^2 - (\beta + \beta_1)x + \beta\beta_1 = 0$

$$\Rightarrow \quad x^2 - (b+a)x + ba = 0$$

$$\Rightarrow \quad x^2 + cx + ab = 0$$

1.2.3 Relation Between Roots and Coefficient of a Polynomial Equation

If α and β are the roots of the quadratic equation $ax^2 + bx + c = 0$. we write

$$ax^2 + bx + c = a((x-\alpha)(x-\beta))$$

$$\Rightarrow \quad ax^2 + bx + c = a(x^2 - (\alpha + \beta)x + \alpha\beta)$$

Comparing coefficients of like powers of x from both sides, we get

$$\alpha + \beta = -\frac{b}{a} \text{ and } \alpha\beta = \frac{c}{a}$$

Similarly, if α, β, γ are the roots of the equation $ax^3 + bx^2 + cx + d = 0$, we write

$$\Rightarrow ax^3 + bx^2 + cx + d = a(x-\alpha)(x-\beta)(x-\gamma)$$

$$\Rightarrow ax^3 + bx^2 + cx + d$$

$$= a\left[x^3 - (\alpha+\beta+\gamma)x^2 + (\alpha\beta + \beta\gamma + \gamma\alpha)x - \alpha\beta\gamma\right]$$

Comparing coefficients of like powers of x from both sides, we get

$$\alpha + \beta + \gamma = -\frac{b}{a}, \alpha\beta + \beta\gamma + \gamma\alpha = \frac{c}{a}, \text{ and } \alpha\beta\gamma = -\frac{d}{a}$$

Similarly, if α, β, γ, δ are the roots of the equation $ax^4 + bx^3 + cx^2 + dx + e = 0$, then we can write

$$\Rightarrow \quad ax^4 + bx^3 + cx^2 + dx + e$$

$$= a(x-\alpha)(x-\beta)(x-\gamma)(x-\delta)$$

$$\Rightarrow ax^4 + bx^3 + cx^2 + dx + e$$

$$= a[x^4 - (\alpha+\beta+\gamma+\delta)x^3 + (\alpha\beta + \alpha\gamma + \alpha\delta + \beta\gamma + \beta\delta + \gamma\delta)x^2 - (\alpha\beta\gamma + \alpha\beta\delta + \alpha\gamma\delta + \beta\gamma\delta)x + \alpha\beta\gamma\delta]$$

Comparing coefficients of like powers of x from both sides, we get

$$\alpha + \beta + \gamma + \delta = -\frac{b}{a},$$

$$\alpha\beta + \alpha\gamma + \alpha\delta + \beta\gamma + \beta\delta + \gamma\delta = \frac{c}{a},$$

$$\alpha\beta\gamma + \alpha\beta\delta + \alpha\gamma\delta + \beta\gamma\delta = -\frac{d}{a},$$

and $\alpha\beta\gamma\delta = \frac{e}{a}$

SOLVED EXAMPLES

Problems based on relation between roots and coefficients of a quadratic equation.

Example 1.15 If one root of the equation $x^2 - x - K = 0$ is the square of the other, then find the value of K.

Solution Let α and α^2 be the roots of the equation $x^2 - x - K = 0$

Hence, $\quad \alpha + \alpha^2 = 1 \quad (1)$

and $\quad \alpha^3 = -K \quad (2)$

From (1)

$$\Rightarrow \quad \alpha(1+\alpha) = 1$$

$$\Rightarrow \quad \alpha^3(1+\alpha)^3 = 1$$

$$\Rightarrow \quad \alpha^3\left[1 + \alpha^3 + 3\alpha(1+\alpha)\right] = 1$$

$$\Rightarrow \quad -K[1 - K + 3] = 1$$

$$\Rightarrow \quad K^2 - 4K - 1 = 0$$

$$\Rightarrow \quad K = \frac{4 \pm \sqrt{16+4}}{2}$$

$$\Rightarrow \quad K = 2 \pm \sqrt{5}$$

Example 1.16 If the roots α, β, γ of the equation $x^3 + ax + a = 0$ (where $a \in R - \{0\}$) satisfy $\frac{\alpha^2}{\beta} + \frac{\beta^2}{\gamma} + \frac{\gamma^2}{\alpha} = -8$, then find the value of a and the roots of the cubic equation.

Solution As α, β, γ are the roots of the equation $x^3 + ax + a = 0$

$$\Rightarrow \quad \alpha + \beta + \gamma = 0$$

$$\Rightarrow \quad \alpha\beta + \beta\gamma + \gamma\alpha = a$$

$$\Rightarrow \quad \alpha\beta\gamma = -a$$

Also, $\alpha^3 = -a\alpha - a$

$\Rightarrow \quad \beta^3 = -a\beta - a$

$\Rightarrow \quad \gamma^3 = -a\gamma - a$

$$\text{Now, } \frac{\alpha^3\gamma + \beta^3\alpha + \gamma^3\beta}{\alpha\beta\gamma} = -8$$

$$\Rightarrow \quad \frac{(-a\alpha - a)\gamma + (-a\beta - a)\alpha + (-a\gamma - a)\beta}{\alpha\beta\gamma} = -8$$

$$\Rightarrow \quad \frac{-a(\alpha + \beta + \gamma) - a(\alpha\gamma + \beta\alpha + \gamma\beta)}{\alpha\beta\gamma} = -8$$

$$\Rightarrow \quad \frac{-a^2}{-a} = -8$$

$\Rightarrow \quad a = -8$

Hence, the given equation is $x^3 - 8x - 8 = 0$

$\Rightarrow \quad x = -2, 1 \pm \sqrt{5}$

Example 1.17 If $\alpha, \beta, \gamma, \delta$ are the roots of the equation $x^4 + px^3 + qx^2 + rx + s = 0$, then find the value of $(\alpha^2+1)(\beta^2+1)(\gamma^2+1)(\delta^2+1)$.

Solution As $\alpha, \beta, \gamma, \delta$ are the roots of the equation $x^4 + px^3 + qx^2 + rx + s = 0$

$$\Rightarrow \quad x^4 + px^3 + qx^2 + rx + s = (x-\alpha)(x-\beta)(x-\gamma)(x-\delta)$$

$$\Rightarrow \quad x^4 + px^3 + qx^2 + rx + s = (\alpha-x)(\beta-x)(\gamma-x)(\delta-x)$$

Put $x = i$ and $-i$ respectively

We get,

$$(1-q+s) + i(r-p) = (\alpha-i)(\beta-i)(\gamma-i)(\delta-i)$$

and $(1-q+s) - i(r-p) = (\alpha+i)(\beta+i)(\gamma+i)(\delta+i)$

Multiplying them, we get

$$(\alpha^2+1)(\beta^2+1)(\gamma^2+1)(\delta^2+1) = (1-q+s)^2 + (r-q)^2$$

1.3 Quadratic Expressions

An expression of the form $ax^2 + bx + c$, where a, b, $c \in R$ and $a \neq 0$ is called a quadratic expression in x. So, in general, a quadratic expression is represented by

$$f(x) = ax^2 + bx + c \text{ or } y = ax^2 + bx + c$$

1.3.1 Graph of Quadratic Expressions

Let $y = ax^2 + bx + c$

$$\Rightarrow \quad y = a\left[x^2 + \frac{b}{a}x + \frac{c}{a}\right]$$

$$\Rightarrow \quad y = a\left[x^2 + \frac{b}{a}x + \frac{b^2}{4a^2} - \frac{b^2}{4a^2} + \frac{c}{a}\right]$$

$$\Rightarrow \quad y = a\left[\left(x + \frac{b}{2a}\right)^2 - \frac{b^2 - 4ac}{4a^2}\right]$$

$$\Rightarrow \quad y = a\left[\left(x + \frac{b}{2a}\right)^2 - \frac{D}{4a^2}\right] \quad (1.1)$$

$$\Rightarrow \quad y + \frac{D}{4a} = a\left(x + \frac{b}{2a}\right)^2 \quad (1.2)$$

Which represent a parabola whose vertex is $\left(-\frac{b}{2a}, -\frac{D}{4a}\right)$, if $a > 0$, then it represents an upward parabola and if $a < 0$, then it represents a downward parabola.

Case 1. Consider $a > 0$

(I) From equation (1) if $D < 0$ then $y > 0$ for all real value of x, that is, if $a > 0$ and $D < 0$, then $ax^2 + bx + c$ is positive for all real value of x.

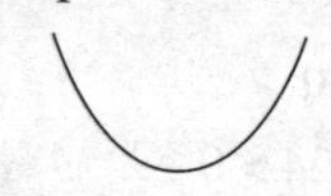

(II) $D = 0$, then $y \geq 0$.

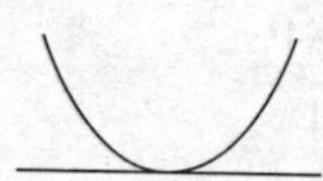

(III) $D > 0$, then $ax^2 + bx + c = 0$ has two distinct real roots.

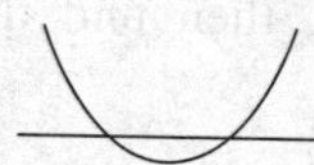

Case 2. If $a < 0$

(I) If $D < 0$, then $y < 0$ for all real values of x.

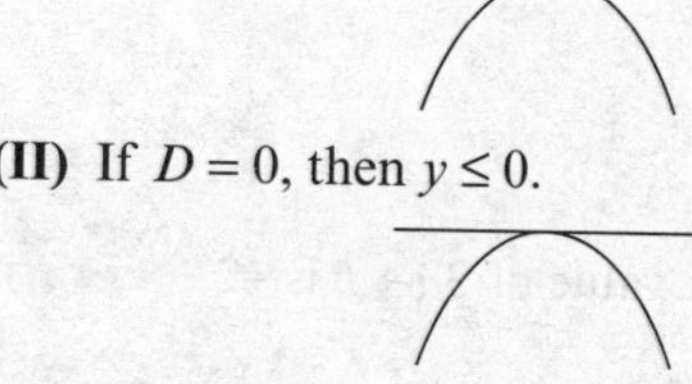

(II) If $D = 0$, then $y \leq 0$.

(III) $D > 0$, then two distinct real roots.

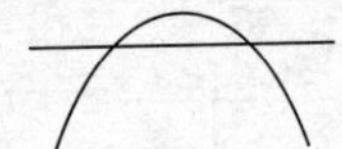

Note:

(a) For an upward parabola, minimum value will be at its vertex.

(b) For a downward parabola, maximum value exists at its vertex.

SOLVED EXAMPLES

Problems based on the graph of a quadratic expression.

Example 1.18 Find the least value of n such that $(n-2)x^2+8x+(n+4)>0$, for all $x \in R$, where $n \in N$.

Solution $(n-2)x^2+8x+(n+4)>0$, for all $x \in R$

$\Rightarrow\ a>0$ and $D<0$

$\Rightarrow\ n-2>0$

$\Rightarrow\ n>2$

and $D<0$

$\Rightarrow\ 64-4(n-2)(n+4)<0$

$\Rightarrow\ 16-(n^2+2n-8)<0$

$\Rightarrow\ -n^2-2n+24<0$

$\Rightarrow\ n^2+2n-24>0$

$\Rightarrow\ n^2+6n-4n-24>0$

$\Rightarrow\ n(n+6)-4(n+6)>0$

$\Rightarrow\ (n+6)(n-4)>0$

$\Rightarrow\ n<-6$ or $n>4$

Hence $n>4$, so the least integral value of $n=5$.

Example 1.19 If $ax^2+bx+6=0$ does not have distinct real roots, then find the least value of $3a+b$.

Solution Since, $f(x)=ax^2+bx+6=0$ does not have distinct real roots and $f(0)=6$

$\Rightarrow\ f(x)\geq 0,\ x\in R$

$\Rightarrow\ f(3)\geq 0$

$\Rightarrow\ 9a+3b+6\geq 0$

$\Rightarrow\ 3a+b\geq -2$

Therefore, the least value of $3a+b$ is -2.

Example 1.20 If the roots of the equation $ax^2+bx+c=0$, $a\neq 0$ (a, b, c are real numbers), are imaginary and $a+c<b$, then prove that $4a-2b+c<0$.

Solution Let $f(x)=ax^2+bx+c=0$

Since roots are imaginary, so either $f(x)>0$ or $f(x)<0$.

Since $f(-1)=a-b+c<0$ $\quad$ [as $a+c<b$]

$\Rightarrow\ f(x)<0$ where $x\in R$

$\Rightarrow\ f(-2)<0$

$\Rightarrow\ 4a-2b+c<0$

1.3.2 Location of Roots (Interval in Which Roots Lie)

(I) When both roots are greater than a real number K.

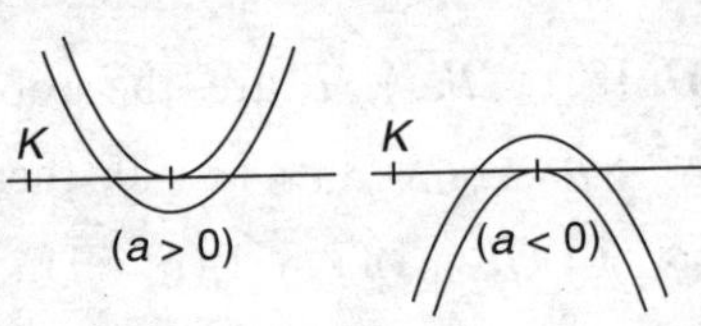

(a) $D\geq 0$

(b) $\frac{-b}{2a}>K$

(c) $af(K)>0$

(II) When both roots are less than a real number K.

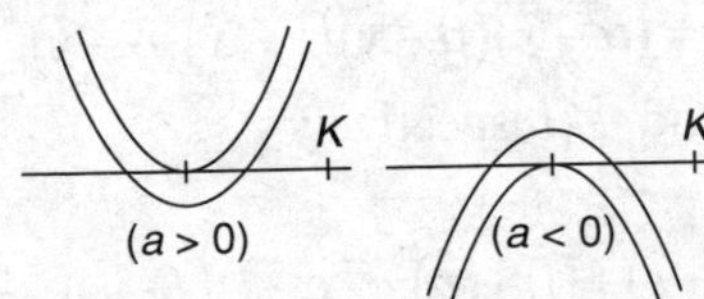

(a) $D\geq 0$

(b) $\frac{-b}{2a}<K$

(c) $af(K)>0$

(III) When a real number K lies between the roots.

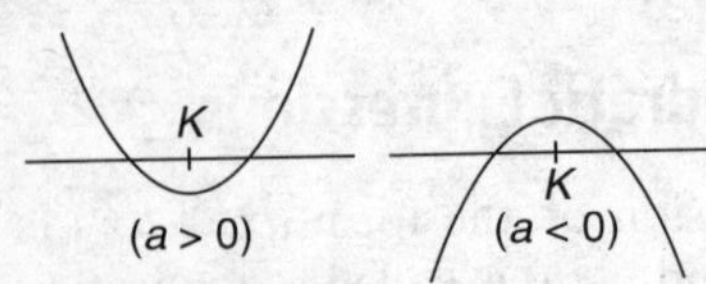

(a) $D>0$

(b) $af(K)<0$

(IV) If exactly one root of $f(x)=0$ lies in the interval (K_1, K_2).

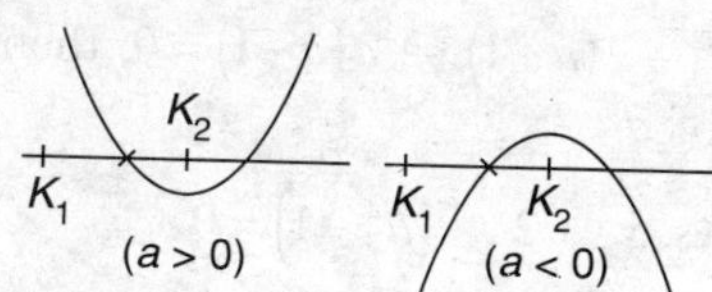

(a) $D>0$

(b) $f(K_1)f(K_2)<0$

(V) If both roots of $f(x)=0$ lies between K_1 and K_2.

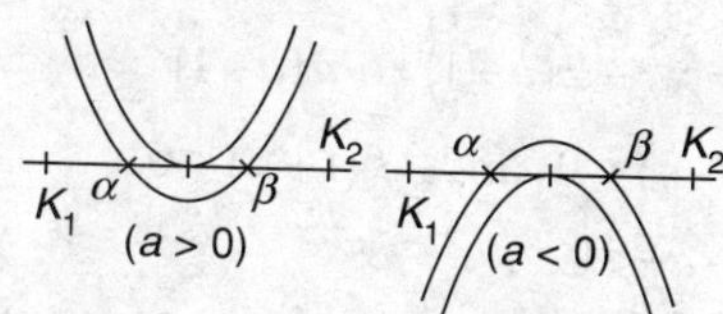

(a) $D\geq 0$

(b) $K_1<-\dfrac{b}{2a}<K_2$, where $\alpha\leq\beta$ and $K_1<K_2$.

(c) $af(K_1)>0$

(d) $af(K_2)>0$

(VI) If K_1 and K_2 lie between the roots of $f(x)=0$

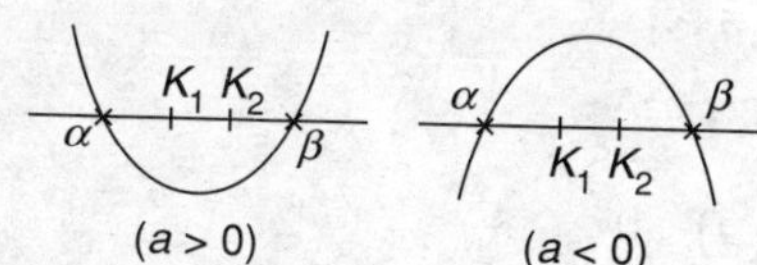

(a) $D>0$

(b) $af(K_1)<0$

c) $af(K_2)<0$, where $\alpha<\beta$

SOLVED EXAMPLES

Problems based on the location of roots

Example 1.21 Find the range of values of a for which the equation $x^2-(a-5)x+\left(a-\dfrac{15}{4}\right)=0$, has at least one positive root.

Solution

Case 1: When both roots are positive,

$D\geq 0, a-5>0, a-\dfrac{15}{4}>0$

Now, $D\geq 0$

$\Rightarrow\ (a-5)^2-4\dfrac{(4a-15)}{4}\geq 0$

$\Rightarrow\ a^2-10a+25-4a+15\geq 0$

$\Rightarrow\ a^2-14a+40\geq 0$

$\Rightarrow\ a^2-10a-4a+40\geq 0$

$\Rightarrow\ a(a-10)-4(a-10)\geq 0$

$\Rightarrow\ (a-4)(a-10)\geq 0$

$\Rightarrow\ a\leq 4 or\ \ a\geq 10$

Again, $a-5>0$

$\Rightarrow\ a>5$

$\Rightarrow\ a-\dfrac{15}{4}>0$

$\Rightarrow\ a>\dfrac{15}{4}$

Thus, $a\geq 10$

$\Rightarrow\ a\in[10,\infty]$

Case 2: When exactly one positive,

$$a-\frac{15}{4}<0$$

$\Rightarrow\ a\subset\left(-\infty,\dfrac{15}{4}\right)$

Thus $a\in\left(-\infty,\dfrac{15}{4}\right)\cup[10,\infty]$.

Example 1.22 If both roots of the quadratic equation $x^2-(m-3)x+m=0, (m\in R)$ are greater than 2, then find the value of m.

Solution As both roots are greater than 2

(a) $D\geq 0$ **(b)** $\dfrac{-b}{2a}>2$ **(c)** $af(2)>0$

(a) $(m-3)^2-4m\geq 0$

$\Rightarrow m^2-6m+9-4m\geq 0$

$\Rightarrow m^2-10m+9\geq 0$

$\Rightarrow m^2-9m-m+9\geq 0$

$\Rightarrow m(m-9)-1(m-9)\geq 0$

$\Rightarrow m\leq 1$ or $m\geq 9$

(b) $\dfrac{m-3}{2}>2$

$\Rightarrow m>7$

(c) $4-2(m-3)+m>0$

$\Rightarrow 4-2m+6+m>0$

$\Rightarrow\ 10>m$

$\Rightarrow m<10$

Hence, $m \in [9,10)$.

Example 1.23 If α is a root of equation $ax^2+bx+c=0$ and β is a root of equation $-ax^2+bx+c=0$, then prove that there will be a root of the equation $\frac{a}{2}x^2+bx+c=0$ lying between α and β.

Solution Let $\quad f(x)=\frac{a}{2}x^2+bx+c$

$$\Rightarrow f(\alpha)=\frac{a}{2}\alpha^2+b\alpha+c$$

$$\Rightarrow f(\alpha)=a\alpha^2+b\alpha+c-\frac{a}{2}\alpha^2$$

$$\Rightarrow f(\alpha)=-\frac{a}{2}\alpha^2 \quad (1)$$

$$\left(\text{as, } a\alpha^2+b\alpha+c=0\right)$$

and $\quad f(\beta)=\frac{a}{2}\beta^2+b\beta+c$

$$\Rightarrow f(\beta)=-a\beta^2+b\beta+c+\frac{3}{2}a\beta^2$$

$$\Rightarrow f(\beta)=\frac{3}{2}a\beta^2 \quad (2)$$

$$\left(\text{as, } -a\beta^2+b\beta+c=0\right)$$

$$f(\alpha)f(\beta)=-\frac{3}{4}a^2\alpha^2\beta^2<0$$

Hence, $f(x)=0$ has one real root between α and β.

Example 1.24 If $a<b<c<d$, then prove that for any real λ, the quadratic equation $(x-a)(x-c)+\lambda(x-b)(x-d)=0$ has real roots.

Solution Let,

$$f(x)=(x-a)(x-c)+\lambda(x-b)(x-d)$$

Now, $\quad f(a)=\lambda(a-b)(a-d)$

and, $\quad f(c)=\lambda(c-b)(c-d)$

So, $f(a)\cdot f(c)=\lambda^2(a-b)(a-d)(c-b)(c-d)<0$

(as $a<b<c<d$)

Hence $f(x)=0$ has one real root between a and c.

Since, $f(x)=0$ is a quadratic equation with real coefficients, both the roots of $f(x)=0$ will be real.

Example 1.25 If the roots of the equation $x^2-2x-(a^2-1)=0$ lie between the roots of the equation $x^2-2(a+1)x+a(a-1)=0$, then find the real value of a.

Solution As, $x^2-2x-(a^2-1)=0$

$$\Rightarrow x=\frac{2+\sqrt{4+4(a^2-1)}}{2}$$

$$\Rightarrow x=1\pm\sqrt{a^2}$$

$$\Rightarrow x=1-a, 1+a$$

Let $f(x)=x^2-2(a+1)x+a(a-1)$

Since, $f(1-a)<0$ and $f(1+a)<0$

Now, $f(1-a)<0$

$$\Rightarrow (1-a)^2-2(a+1)(1-a)+a(a-1)<0$$

$$\Rightarrow (1-a)\left[(1-a)-2(a+1)-a\right]<0$$

$$\Rightarrow (1-a)[1-a-2a-2-a]<0$$

$$\Rightarrow (1-a)(-1-4a)<0$$

$$\Rightarrow (a-1)(4a+1)<0$$

$$\Rightarrow -\frac{1}{4}<a<1 \quad (1)$$

Also,

$$\Rightarrow f(1+a)<0$$

$$\Rightarrow (1+a)^2-2(a+1)(1+a)+a(a-1)<0$$

$$\Rightarrow -(a+1)^2+a(a-1)<0$$

$$\Rightarrow -a^2-2a-1+a^2-a<0$$

$$\Rightarrow -3a-1<0$$

$$\Rightarrow 3a+1>0$$

$$\Rightarrow a>-\frac{1}{3} \quad (2)$$

From (1) and (2) $a\in\left(\frac{-1}{4},1\right)$

SOLVED EXAMPLES

Problems based on Range of Quadratic Expression

Example 1.26 Show that the expression $\frac{x^2-3x+4}{x^2+3x+4}$, lies between $\frac{1}{7}$ and 7 for real values of x.

Solution Let, $y=\dfrac{x^2-3x+4}{x^2+3x+4}$

$\Rightarrow\ yx^2+3xy+4y=x^2-3x+4$

$\Rightarrow\ (y-1)x^2+3(y+1)x+4(y-1)=0$

As x is real, $D\geq 0$

$\Rightarrow\ 9(y+1)^2-16(y-1)^2\geq 0$

$\Rightarrow\ [3(y+1)+4(y-1)][3(y+1)-4(y-1)]\geq 0$

$\Rightarrow\ (7y-1)(-y+7)\geq 0$

$\Rightarrow\ (7y-1)(y-7)\leq 0$

$\Rightarrow\ \dfrac{1}{7}\leq y\leq 7$

$\dfrac{x^2-3x+4}{x^2+3x+4}$ lies between $\dfrac{1}{7}$ and 7.

Example 1.27 If $\dfrac{x^2+ax+3}{x^2+x+a}$ takes all real values for possible real values of x, then prove that $4a^3+39<0$.

Solution Let, $y=\dfrac{x^2+ax+3}{x^2+x+a}$

$\Rightarrow\ yx^2+yx+ya=x^2+ax+3$

$\Rightarrow\ (y-1)x^2+(y-a)x+(ya-3)=0$

As x is real, $D\geq 0$

$\Rightarrow\ (y-a)^2-4(y-1)(ya-3)\geq 0$

$\Rightarrow\ (y^2-2ay+a^2)-4(y^2a-3y-ay+3)\geq 0$

$\Rightarrow\ (1-4a)y^2+y(4a-2a+12)+(a^2-12)\geq 0$

$\Rightarrow\ (1-4a)y^2+(2a+12)y+(a^2-12)\geq 0$

for all real y

Hence,

$1-4a>0$ and $(2a+12)^2-4(1-4a)(a^2-12)\leq 0$

$\Rightarrow\ a<\dfrac{1}{4}$ and $(a+6)^2-(a^2-12-4a^3+48a)\leq 0$

$\Rightarrow\ a^2+12a+36-a^2+12+4a^3-48a\leq 0$

$\Rightarrow\ 4a^3-36a+48\leq 0$

$\Rightarrow\ 4a^3\leq 36a-48<36\cdot\dfrac{1}{4}-48\qquad\left[\text{as, } a<\dfrac{1}{4}\right]$

$\Rightarrow\ 4a^3<9-48$

$\Rightarrow\ 4a^3<-39$

$\Rightarrow\ 4a^3+39<0$

1.4 Know Your Concepts

1. Find the nature of roots of the quadratic equation $(l-m)x^2-5(l+m)x-2(l-m)=0$, where l, m and n are real and $l\neq m$.
2. If $x^2-ax+b=0$ and $x^2-px+q=0$ have a root in common and the second equation has equal roots, find the value of $\dfrac{ap}{b+q}$.
3. If $(\lambda^2+\lambda-2)x^2+(\lambda+2)x<1$, for all $x\in R$, then find the interval in which λ lies.
4. Find the total number of quadratic equations of the form $px^2+qx+1=0$, having real roots where $p, q\in\{1, 2, 3, 4\}$.
5. If the ratio of the roots of the equation $ax^2+bx+c=0$ be equal to that of the roots of the equation $a_1x^2+b_1x+c_1=0$, then prove that $\left(\dfrac{b}{b_1}\right)^2=\dfrac{ca}{c_1a_1}$.
6. If one root of the equation $x^2+px+q=0$ is the square of the other, then find the value of $p^3-q(3p-1)+q^2$.
7. If c, d are the roots of the equation $(x-a)(x-b)-k=0$, find the equation whose roots are a, b.
8. If α, β are the roots of the equation $3x^2+2x+1=0$, then find the equation whose roots are $\dfrac{1-\alpha}{1+\alpha}, \dfrac{1-\beta}{1+\beta}$.
9. Prove that the roots of the equation $(a+b-c)x^2+2bx+(b+c-a)=0$ are rational, provided a, b, c are rational.
10. If a, b, c are odd integers, then prove that roots of $ax^2+bx+c=0$ cannot be rational.
11. If α, β are the roots of $ax^2+bx+c=0\,(a\neq 0)$ and $\alpha+\delta$, $\beta+\delta$ are the roots of $Ax^2+Bx+C=0\,(A\neq 0)$ for some constant δ, then prove that $\dfrac{b^2-4ac}{a^2}=\dfrac{B^2-4AC}{A^2}$.
12. Let α and β be the roots of $x^2-6x-2=0$ with $\alpha>\beta$. If $a_n=\alpha^n-\beta^n$ for $n\geq 1$, then find the value of $\dfrac{a_{10}-2a_8}{2a_9}$.

13. Let α, β be the roots of the equation $px^2+qx+r=0, p\neq 0$. If p,q,r are in A.P. and $\frac{1}{\alpha}+\frac{1}{\beta}=4$, then find the value of $|\alpha-\beta|$.

14. Find the value of m for which the equation $\frac{a}{x+a+m}+\frac{b}{x+b+m}=1$ has roots equal in magnitude but opposite in sign.

15. Find the set of all solutions of the equation $2^{|x|}-|2^{x-1}-1|=2^{x-1}+1$.

16. Prove that the equation
$$\frac{A^2}{x-a}+\frac{B^2}{x-b}+\frac{C^2}{x-c}+\ldots+\frac{K^2}{x-k}=x+l$$
has no imaginary root, where $A, B, C, \ldots, K, a, b, c, \ldots, k$ and l are real.

17. Find all the integral values of 'a' for which the quadratic equation $(x-a)(x-10)+1=0$ has integral roots.

18. Let a and b be two roots of the equation $x^3+px^2+qx+r=0$ satisfying the relation $ab+1=0$. Prove that $r^2+pr+q+1=0$.

19. If the three equations $x^2+ax+12=0$ and $x^2+bx+15=0$ and $x^2+(a+b)x+36=0$ have a common positive root, find a and b and the roots of the equations.

20. If $(y^2-5y+3)(x^2+x+1)<2x$ for all $x\in R$, then find the interval in which y lies.

1.5 JEE Main: Sample Questions

Single Answer Type

1. Let α, β be two real roots of equation $x^2+ax+b=0$ ($a, b\in R$ and $b\neq 0$). If the quadratic equation $g(x)=0$ has two roots $\alpha+\frac{1}{\alpha}, \beta+\frac{1}{\beta}$ such that sum of roots is equal to product of roots, then the complete range of b is

(A) $\left[\frac{1}{3}, 3\right]$ (B) $\left(\frac{1}{3}, 3\right]$

(C) $\left[\frac{1}{3}, 3\right)$ (D) $\left(-\infty, \frac{1}{3}\right)\cup(3,\infty)$

2. If the equations $x^3+x^2-4x-4=0$ and $x^2+px+2p=0$ ($p\in R$) have two roots common then the value of p is

(A) -2 (B) -1 (C) 1 (D) 3

3. For real x, the value of the rational function $f(x)=\frac{x^2-2x+2}{2x-2}$ cannot lie in the interval

(A) $(-\infty, -1)$ (B) $(-1, 1)$
(C) $(1, 2)$ (D) $(2, \infty)$

4. If the equation $x^2+4x\sin\theta+\tan\theta=0$ $\left(0<\theta<\frac{\pi}{2}\right)$ has repeated roots, then θ equals

(A) $\frac{\pi}{12}$ (B) $\frac{\pi}{6}$

(C) $\frac{\pi}{12}$ or $\frac{5\pi}{12}$ (D) $\frac{\pi}{6}$ or $\frac{\pi}{12}$

5. If $\tan^2\frac{3\pi}{8}$ is a root of the equation $2x^2-3ax+4b=0$, where $a, b\in Q$, then the value of $a+4b$ can be equal to

(A) 5 (B) 8 (C) 6 (D) 4

6. If e^{λ} and $e^{-\lambda}$ are the roots of equation $3x^2-(a+b)x+2a=0$, $a, b, \lambda\in R, \lambda\neq 0$ then least integral value of b is

(A) 4 (B) 5 (C) 9 (D) 10

7. Which one of the following can best represent the graph of quadratic polynomial $y=ax^2+bx+c$ (where $a<0, b<0$ and $c>0$)?

(A) (B)

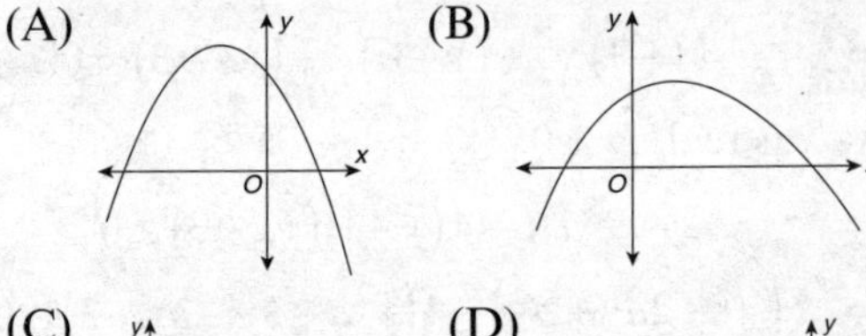

(C) (D)

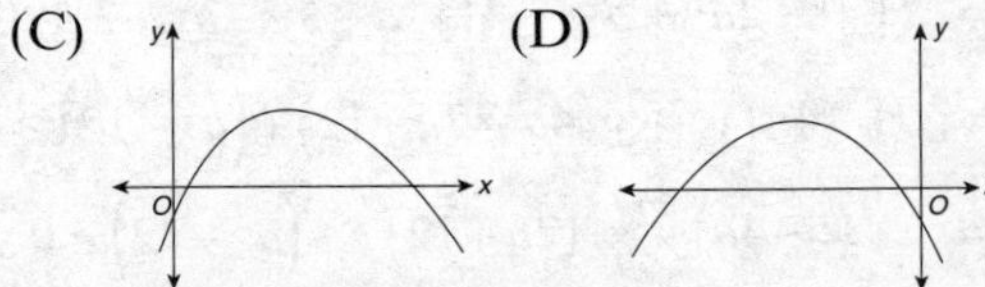

8. Let α, β, γ and δ be the roots (real or non-real) of equation $x^4-3x+1=0$. The value of $\alpha^3+\beta^3+\gamma^3+\delta^3$ is equal to

(A) 6 (B) 9 (C) 12 (D) 15

9. If α, β are roots of the equation $x^2-2mx+m^2-1=0$ then the number of integral values of m for which $\alpha, \beta\in(-2, 4)$ is

(A) 0 (B) 1 (C) 2 (D) 3

10. If the quadratic equations $3x^2+ax+1=0$ and $2x^2+bx+1=0$ have a common root, then the value of the expression $5ab-2a^2-3b^2$ is

(A) 0 (B) 1 (C) -1 (D) 2

11. If both roots of the quadratic equation $x^2 - 2kx + k^2 + 2k - 5 = 0$ are less than 4, then exhaustive set of values of k is
(A) $k \in R$ (B) $k < 4$
(C) $k \leq \frac{5}{2}$ (D) $k > 0$

12. The number of non-negative integral values of k for which the equation $5x^2 + (13 - k)x - 3k - 6 = 0$ has at least one real solution in $(-2, 2)$ are
(A) 21 (B) 20 (C) 7 (D) 8

13. If the roots of the quadratic polynomial $f(x) = 2x^2 - 3x + k^2 - 3k + 4$ are of opposite sign, then the set of the values of k is
(A) ϕ (B) $(-4, 1)$
(C) $(4, -1)$ (D) $(-\infty, -4) \cup (1, \infty)$

14. If one of the roots of $ax^2 + bx + c = 0$ is greater than 2 and other is less than -1 and if the roots of $cx^2 + bx + a = 0$ are α and β, then
(A) $\alpha < \frac{1}{2}$ and $\beta > -1$
(B) $\alpha < \frac{1}{2}$ and $\beta < -1$
(C) $\alpha > \frac{1}{2}$ and $\beta > -1$
(D) $\alpha < 2$ and $\beta > -1$

15. If $x \in R$ then range of $f(x) = \frac{x^2 + 2x - 3}{2x^2 + 3x - 9}$ is
(A) $(-\infty, \infty)$ (B) $R - \left\{\frac{1}{2}\right\}$
(C) $R - \left\{\frac{4}{9}, \frac{1}{2}\right\}$ (D) $R - \left\{\frac{3}{2}\right\}$

16. If the vertex of parabola $y = 2x^2 - 4x + 6$ is (m, n), then the parabola whose vertex $\left(2m, \frac{n}{2}\right)$ is
(A) $y = x^2 - 4x + 8$
(B) $y = x^2 + 4x + 6$
(C) $y = -2x^2 + 8x - 6$
(D) $y = -2x^2 + 8x - 8$

17. Let α, β, γ be real numbers which satisfy the relations $\alpha + \frac{1}{\beta\gamma} = \frac{1}{2}$, $\beta + \frac{1}{\alpha\gamma} = \frac{8\sqrt{3}}{4}$, $\gamma + \frac{1}{\alpha\beta} = \frac{1}{18}$ then the value of $\frac{\beta - \gamma}{\alpha - \gamma}$ is equal to
(A) $\frac{-7}{20}$ (B) $\frac{-11}{40}$ (C) $\frac{11}{20}$ (D) $\frac{-1}{40}$

18. If α, β are the roots of $x^2 - p(x + 1) - c = 0$, $c \neq 1$, then the value of $(\alpha + 1)(\beta + 1)$ is equal to
(A) $1 - c$ (B) $p + c$
(C) $1 + c$ (D) $1 - 2p$

19. Let α and β are the roots of the equation $x^2 - (p - 2)x - p - 1 = 0$ $(p \in R)$. If $\alpha^2 + \beta^2$ is least, then p equals
(A) 0 (B) 1 (C) 2 (D) 5

20. Number of integral values of x satisfying the inequality $\frac{x^2 + 6x - 7}{|x + 4|} < 0$ is
(A) 5 (B) 6 (C) 7 (D) 8

21. If α, β, γ are roots of equation $x^3 - 2x^2 - 1 = 0$ and $T_n = \alpha^n + \beta^n + \gamma^n$, then value of $\frac{T_{11} - T_8}{T_{10}}$ is equal to
(A) 1 (B) 2 (C) -1 (D) 3

22. The zeroes of the quadratic polynomial $f(x) = 2x^2 - 3x + k^2 - 3k + 4$ lie on either side of the origin then the set of the values of k is
(A) ϕ (B) $(-4, 1)$
(C) $(4, -1)$ (D) $(-\infty, -4) \cup (1, \infty)$

23. If the quadratic polynomial $P(x) = 4x^2 + kx + 8x + 9$ where $k \in I$, is positive for every real 'x' then the greatest absolute values of 'k' is
(A) 20 (B) 19 (C) 4 (D) 3

24. The number of integral value(s) of 'a' so that the graph of $y = 16x^2 + 8(a + 5)x - 7a - 5$ is always above the x-axis is
(A) 12 (B) 13 (C) 14 (D) 15

25. If α and β are the roots of the equation $ax^2 + bx + c = 0$, then $\frac{1}{a\alpha + b} + \frac{1}{a\beta + b}$ is equal to
(A) $\frac{b}{ac}$ (B) $\frac{-b}{ac}$ (C) $\frac{ac}{b}$ (D) $\frac{-ac}{b}$

26. The sum of value(s) of x satisfying the equation $\sqrt{x^2 + 8x + 16} = 2$ is
(A) -6 (B) -2 (C) -8 (D) 0

27. If α, β are the roots of the quadratic equation $x^2 + 2(1 - \cos 3\theta)x - 2\sin^2 3\theta = 0$ $(\theta \in R)$, then the maximum value of $\alpha^2 + \beta^2$ is equal to
(A) 0 (B) 4 (C) 8 (D) 16

28. Let $P(x) = x^2 - (2 - p)\,x + p - 2$. If $P(x)$ assumes both positive and negative values $\forall\, x \in R$, then the range of p is

(A) $(-\infty, 2) \cup (6, \infty)$ (B) $(2, 6)$
(C) $(-\infty, 2)$ (D) $(6, \infty)$

29. If α, β, γ are roots of equation $111x^3 - 11x - 1 = 0$, then $(\alpha\beta)^{-2} + (\beta\gamma)^{-2} + (\gamma\alpha)^{-2}$ equals

(A) 2332 (B) 1331
(C) 1210 (D) 2442

30. If the general expression of degree 2 given by $3x^2 + xy + ky^2 + 10x - 3y + 7$ can be factorised into two linear factors then value of k is

(A) 4 (B) 2 (C) –2 (D) –4

31. If the minimum value of $P(x) = x^2 - 2x + (a^2 - 11a + 11)$ where $a \in R$ is greater than zero then the number of integers which are not in the range of 'a' are

(A) 9 (B) 10 (C) 11 (D) 8

32. The largest integral value of k for which the quadratic trinomial $P(x) = (k - 2)\,x^2 + 8x + k + 4$ is non-positive for all real values of x is

(A) 2 (B) 4 (C) –6 (D) –2

33. The possible set of values of a for which one roots of equation $x^2 + 2(a - 1)x + a + 5 = 0$ is smaller than 1 and other root is greater than 3 is

(A) $\left(\frac{-4}{3}, \frac{-8}{7}\right)$ (B) $\left(-\infty, \frac{-8}{7}\right)$

(C) $\left(-\infty, \frac{-4}{3}\right)$ (D) $\left(\frac{8}{7}, \infty\right)$

34. The complete solutions set of the inequality $\frac{3^x(2x-5)(x^2+x+2)}{(\cos x - 2)(x^2+x)} \le 0$ is

(A) $(-\infty, -1)$ (B) $\left(\frac{5}{2}, \infty\right)$

(C) $\left(-1, \frac{5}{2}\right]$ (D) $(-1, 0) \cup \left[\frac{5}{2}, \infty\right)$

35. If the graph of $f(x) = x^2 + (3 - k)\,x + k$, (where $k \in R$) lies above and below x-axis, then k cannot be

(A) –1 (B) 0 (C) 1 (D) 10

36. If roots of the equation, $a\,(x - \ell)\,(x - m) + n = 0$ are α, β then roots of the equation, $a\left(\frac{x}{x-k} - \alpha\right)\left(\frac{x}{x-k} - \beta\right) - n = 0$ are

(A) $\frac{\ell}{\ell+1}, \frac{m}{m+1}$ (B) $\frac{k(\ell+1)}{\ell}, \frac{k(m+1)}{m}$

(C) $\frac{k\ell}{\ell-1}, \frac{km}{m-1}$ (D) $\frac{\ell-1}{\ell}, \frac{m-1}{m}$

37. The quadratic equation $x^2 - 8x + a = 0$ and $x^2 - bx + 16 = 0$ have one root in common and the other roots are in the ratio 3:4.

(A) the common root is a prime number
(B) the value of a is 12
(C) the value of b is 10
(D) the sum of $(a + b)$ is a rational number

38. Least integral value of k for the quadratic equation $x^2 - 2(k + 2) + k^2 + 12 = 0$ has two distinct solution, is

(A) 1 (B) 2 (C) 3 (D) 4

39. If exactly one root of equation $(k^2 + 3k + 2)\,x^2 + (k^2 + k - 2)x + k + 5 = 0$ is infinity then possible value of 'k' is/are

(A) –2 (B) –1 (C) 1 (D) –5

40. If x is real, then the maximum value of $\frac{3x^2+9x+17}{3x^2+9x+7}$ is

(A) 41 (B) 1 (C) $\frac{1}{4}$ (D) $\frac{17}{7}$

41. Sum of all the roots of the equation $\log_7(2^x - 1) + \log_7(2^x - 7) = 1$, is

(A) 8 (B) 3 (C) 4 (D) 1

42. If the quadratic polynomial $f(x) = (a - 3)\,x^2 - 2ax + 3a - 7$ ranges from $[-1, \infty)$ for every $x \in R$, then the value of a lies in

(A) [0, 2] (B) [3, 5] (C) [4, 6) (D) [5, 7]

43. If quadratic equation $x^2 - 3ax + a^2 = 9$ has roots of the opposite sign, then the sum of all possible values of a equals

(A) 0 (B) –2 (C) 4 (D) 9

44. Number of natural numbers for which the number $\log_{4-x}(x^2 - 14x + 45)$ is defined is

(A) 4 (B) 3 (C) 2 (D) 1

45. The graph of $y = ax^2 + bx + c$ is as shown. Which one of the following is not correct?

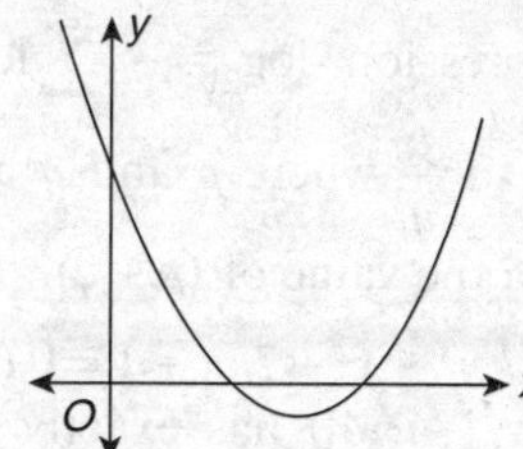

(A) $a > 0$ (B) $b^2 - 4ac > 0$
(C) $b < 0$ (D) $abc > 0$

46. The number of integral value of x if $5x - 1 < (x + 1)^2 < 7x - 3$, is
(A) 0 (B) 1 (C) 2 (D) 3

47. A quadratic equation having rational coefficients whose one root is $2\cos^2\frac{\pi}{10}$ is
(A) $4x^2 - 10x + 5 = 0$
(B) $2x^2 - 50x + 10 = 0$
(C) $4x^2 + 10x + 5 = 0$
(D) $2x^2 + 50x - 10 = 0$

48. The value of x if $\sqrt{x^2 + \sqrt{x^2 + \sqrt{x^2 + \ldots}}} = 9$, is
(A) $6\sqrt{2}$ (B) $3\sqrt{10}$ (C) $3\sqrt{6}$ (D) $2\sqrt{6}$

49. If the domain of $g(x)$ is [3, 4], then the domain of $g(\log_2(x^2 + 3x - 2))$ is
(A) $[-4, -1] \cup [2, 7]$ (B) $[-3, 2]$
(C) $[-6, -5] \cup [2, 3]$ (D) $\left[\frac{3}{2}, 5\right]$

50. The least positive integral value of 'k' for which $(x + k - 1)(x - 2k - 7) < 0 \; \forall \; x \in [-1, 2]$ is
(A) 1 (B) 2 (C) 3 (D) 4

51. Let u, v, w satisfy the equations $uvw = -6$, $uv + vw + wu = -5$, $u + v + w = 2$, where $u > v > w$. If $\vec{V}_1 = u\hat{i} + v\hat{j} + w\hat{k}$ and $\vec{V}_2 = 2\hat{i} + 3\hat{j} - 5\hat{k}$ then the scalar product $\vec{V}_1 \cdot \vec{V}_2$ equals
(A) 9 (B) 19 (C) 21 (D) 25

52. If m and n are real and different satisfying the relation $m^2 = 4m + 4$ and $n^2 = 4n + 4$, then $(m^2 + n^2 + 4mn)$ is equal to
(A) 2 (B) 4 (C) 8 (D) 16

53. The smallest integral value of α for which the inequality $1 + \log_5(x^2 + 1) \leq \log_5(\alpha x^2 + 4x + \alpha)$ is true for all $x \in R$, is
(A) 4 (B) 5 (C) 6 (D) 7

54. Suppose that 'a' and 'b' are non-zero real numbers and the equation $x^2 + ax + b = 0$ has solution a and b. Then the pair (a, b), is
(A) (–2, 1) (B) (–1, 2)
(C) (1, –2) (D) (2, –1)

55. If quadratic equation $ax^2 + bx + c = 0$ ($a, b, c, \in R$ and $a \neq 0$) has non-real roots, such that $a + b + c > 0$, then consider the statements
(I) $a - b + c < 0$
(II) $4a + 2b + c > 0$
(III) $a - 2b + 4c > 0$
The true statements are
(A) I, II (B) II, III
(C) III, I (D) I, II, III

56. The sum of all values of p for which one root of the quadratic equation $x^2 + 3(1 - 3p)x + 2 = 0$ is twice as large as the other root, is
(A) 1 (B) –1 (C) $\frac{-2}{3}$ (D) $\frac{2}{3}$

57. If one solution of $x^3 - 2x^2 + ax + 10 = 0$ is the additive inverse of another, then which one of the following inequalities is true?
(A) $-40 < a < -30$ (B) $-30 < a < -20$
(C) $-20 < a < -10$ (D) $-10 < a < 0$

58. If the roots of the equation $x^2 - 2ax + a^2 + a - 3 = 0$ are real and less than 3, then
(A) $a < 2$ (B) $2 \leq a \leq 3$
(C) $3 < a \leq 4$ (D) $a > 4$

Integer Answer Type

1. If $x = 2 + i$ and $x = 1 - 3i$ are the roots of the equation $x^4 + ax^3 + 23x^2 + bx + c = 0$ ($a, b, c \in R$) then find the value of $|a + b + c|$.

2. Find the number of integral values of a for which the numbers 1 and a lie between the numbers x_0 and y_0, where (x_0, y_0) is a solution of the system of equations $x + y - 1 = 2a$, $2xy = a^2 - a$.

3. If α, β and γ are the roots of the equation $5x^3 - qx - 1 = 0$, $(q \in R)$ then find the value of $\frac{\alpha^2-3}{\beta\gamma} + \frac{\beta^2-3}{\gamma\alpha} + \frac{\gamma^2-3}{\alpha\beta}$.

4. If the quadratic equation $mx^2 - nx + 12 = 0$, where m and n are positive integers not exceeding 10, has both roots greater than 2, then find the number of possible ordered pairs (m, n).

5. Let $f(x) = px^2 + px - 2$, $g(x) = x^2 + (2p + 1)x + (p + 2)$ $\forall\ x \in R$, where p is a real constant. Find the number of integral values of p so that $f(x) > g(x)$ $\forall\ x \in R$

6. Let A denote the number of integers satisfying the inequality $\frac{3}{x^2 - x + 1} > 1$.
B denotes the value of k for which the equation $(2k^2 - 5k - 3)x^2 + (k^2 - 5k + 6)x + (k^2 - 9) = 0$ has more than 2 roots and X denotes the value of the expression $2x^4 - 30x^2 + 8x + 10$ where $x = \tan\left(\frac{5\pi}{12}\right)$. Find the value of $(X \div A + B)$.

7. The roots α and β of a quadratic equation are the square of two consecutive natural numbers. The geometric means of the two roots is 1 greater than the positive difference of the roots. If exactly one root of $x^2 - kx + 32 = 0$ lies between α and β then find the number of integral value(s) of k.

8. If exactly one root of the quadratic equation $x^2 + (2a - 1)x + (a + 3) = 0$ lies in the interval $[0, 2]$ then find the square of the least integral value of a.

9. Find the number of integral value(s) of $a \in (8, \infty)$ for which the expression $\log_{10}(4x^2 - ax + 3)$ is always non-negative for all $x \in [1, 3]$.

10. If the inequality $\cot^2 x + (k + 1)\cot x - (k - 3) < 0$ is true for at least one $x \in \left(0, \frac{\pi}{2}\right)$, then find the least positive integral value of k.

11. The quadratic polynomial $P(x)$ satisfies the equation $P(x) - P(x + 1) = 2x - 1$ $\forall\ x \in R$. If the greatest value of $P(x)$ in $(-\infty, 0]$ is 2, then find the value of $P(-1) + P(1)$.

12. Let $x^2 + 2y^2 - 2xy - 2 \geq k(x + 2y)$ $\forall\ x, y \in R$ then find the number of integral values of k.

13. The expression $\log_2 5 - \sum_{k=1}^{4} \log_2\left(\sin\frac{k\pi}{5}\right)$ reduces to $\frac{p}{q}$ where p and q are co-prime then find the value of $(p + q)$.

14. If $x^4 + 2kx^3 + x^2 + 2kx + 1 = 0$ (where $k \in R$ and $k \in [-4, 4]$) has exactly two distinct positive and two distinct negative roots, then find the number of integral values of k.

15. Let $f(x) = 3\sin x + 2\cos x - 4(\sin^3 x + \cos^3 x - \sin^2 x\cos x)$ $\forall\ x \in R$ and m and M are the minimum and maximum value of $f(x)$ and roots of $g(x) = x^2 + (2k - 1)x + k^2 - k$ are real and unequal which does not lie in $\left[\frac{m}{\sqrt{5}}, \frac{M}{\sqrt{5}}\right]$ then find the number of integral value(s) of k.

16. If the roots of the equation $x^2 + (p + 1)x + 2q - q^2 + 3 = 0$ are real and unequal $\forall\ p \in R$ then find the minimum integral value of $(q^2 - 2q)$.

17. Find the smallest integral value of 'a' for which $(a - 1)x^2 - (a + 1)x + (a - 1) \geq 0$ for all $x \geq 2$.

18. If the range of the value of a for which the roots of the equation $x^2 - 2x - a^2 + 1 = 0$ lie between the roots of the equation $x^2 - 2(a + 1)x + a(a - 1) = 0$ is (p, q), find the value of $\left(q + \frac{1}{p^2}\right)$.

19. Find the largest natural number 'a' for which the maximum value of $f(x) = a - 1 + 2x - x^2$ is smaller than the minimum value of $g(x) = x^2 - 2ax + 10 - 2a$.

1.6 JEE Advanced: Sample Questions

Single Answer Type

1. Number of integral values of m for which $x^2 - 2mx + 6m - 41 < 0$, $\forall\ x \in (1, 6]$, is
(A) 8 (B) 9 (C) 10 (D) 11

2. If the equations $x^3 - x - 6 = 0$ and $x^2 - bx - c = 0$ $(b, c \in R)$ have exactly one root in common, then the value of $(2b + c)$ is
(A) 1 (B) 2 (C) 3 (D) 4

3. Let $f(x) = x^2 + ax + b$. If the maximum and the minimum values of $f(x)$ are 3 and 2 respectively for $0 \le x \le 2$ then the possible ordered pair(s) of (a, b) is/are

(A) $(-2, 3)$ (B) $\left(\frac{-3}{2}, 2\right)$

(C) $\left(\frac{-5}{2}, 3\right)$ (D) $\left(\frac{-5}{2}, 2\right)$

4. Let α, β are the roots of the equation $x^2 + ax + b = 0$ such that $\frac{3\alpha - 5 - \alpha^2}{(\beta - 3)} = \beta$, then value of $|a^2 - 2b + 3a|$ is (where, $\beta \ne 3$)

(A) 0 (B) 3 (C) 5 (D) 7

5. If one of the roots of $ax^2 + ax + a + 1 = 0$ is less than 1 and the other is more than 1, then the complete set of values of a is given by

(A) $a \in \left(\frac{-1}{2}, 0\right)$ (B) $\left(0, \frac{1}{2}\right)$

(C) $\left(\frac{-1}{3}, 0\right)$ (D) $\left(0, \frac{1}{3}\right)$

6. If α, β are roots of the equation $x^2 + ax + 2 = 0$ such that $\alpha^6\beta^5 = 32$, then the value of 'a' is

(A) -2 (B) 2 (C) -3 (D) 3

7. Let $f(x) = x^{10} - 2x^9 + 4x^8 - 8x^7 + \ldots + 1024$. When $f(x)$ is divided by $x^3 - (a + 2)x^2 + 2ax$, then the remainder is $341x^2 - 682x + 1024$. If 'a' is any arbitrary constant, then 'a' may be equal to

(A) -1 (B) 1 (C) 2 (D) -2

8. The number of integral values of 'b' for which inequality $(a^2 + 1)x^2 + 4(a + b)x + 2 < 0$ is true for atleast one 'x' and $\forall\, a \in R$, is

(A) 0 (B) 1 (C) 2 (D) 3

9. If a, b, c, d are non-zero real number such that c and d are the roots of $x^2 + ax + b = 0$ and a and b are the roots of $x^2 + cx + d = 0$, then the absolute value of $(a + 2b + 3c + 4d)$ is

(A) 2 (B) 4 (C) 6 (D) 8

10. If graph of $f(x) = x^2 + bx + c$ is drawn (in adjacent diagram), where $b, c \in I$, then the number of such quadratic equations $f(x) = 0$ is

(A) 1 (B) 2 (C) 3 (D) 4

11. If α, β, γ be the roots of the equation $x^3 + ax + a = 0$ (where $a \in R, a \ne 0$) satisfy $\frac{\alpha^2}{\beta} + \frac{\beta^2}{\gamma} + \frac{\gamma^2}{\alpha} = -8$, then a is

(A) -8 (B) -2 (C) 2 (D) 8

12. Sum of all the possible values of k for which $x^3 + 3x^2 + 3 - k = 0$ has all the roots as integers is equal to

(A) 10 (B) 9 (C) 8 (D) 7

13. Let α and β be the roots of $x^2 - 10x + 2k^2 + 2k = 0$ and α, γ be the roots of $x^2 - (3k + 2)x + k^3 + 3 = 0$ (where $k \in I$). If α, γ and β are in A.P., then the value of $\alpha + \beta^2 + \gamma^3$ is

(A) 135 (B) 147 (C) 165 (D) 178

14. $f(x)$ is a quadratic polynomial (leading coefficient > 0) and if $f(x) = 0$ has two real roots α and β such that $-1 < \alpha < 1 < \beta < 3$ and $f(0) < f(2) < 0$, then

(A) $f(3) > f(-1)$ (B) $f(1) < f(2)$

(C) $f(3) < f(-1)$ (D) $f(1) > f(2)$

15. If α, β are the roots of $cx^2 + bx + a = 0$, then the roots of equation $a(x + 1)^2 + b(x + 1) + c = 0$ are

(A) $\alpha - 1, \beta - 1$ (B) $\frac{1-\alpha}{\alpha}, \frac{1-\beta}{\beta}$

(C) $\alpha + 1, \beta + 1$ (D) $\frac{\alpha+1}{\alpha}, \frac{1+\beta}{\beta}$

16. If $\alpha + \frac{1}{\alpha}, \beta + \frac{1}{\beta}$ are zeroes of $f(x) = x^2 - 5x - a$ where $\alpha, \beta \in (0, \infty)\ \forall\, x \in R$, then the set of values of a is

(A) $\left[-6, \frac{25}{4}\right]$ (B) $\left[\frac{-25}{4}, 6\right]$

(C) $\left[-\frac{25}{4}, -6\right]$ (D) $\left(-\frac{25}{4}, -6\right)$

17. The number of integral values of x which satisfies the inequality

$\frac{(x+6)^2(x+5)(x+1)^2}{(x-3)} \le 0$, is

(A) 10 (B) 6 (C) 8 (D) 9

18. The number of real roots of the equation $(x^2 + x + 1)(x^2 + x + 3) = 8$ is

(A) 0 (B) 2

(C) 4 (D) cannot be determined

19. If $\alpha^2 = 10\alpha - 12$, $\beta^2 = 15\beta - 27$ and $\alpha > \beta$, then the value of $\frac{2\beta+3\alpha}{\alpha\beta}$ is

(A) $\frac{5}{3}$ (B) $\frac{8}{5}$ (C) $\frac{7}{4}$ (D) $\frac{11}{6}$

20. If α and β are the roots of the equation $x^2 - x + 11 = 0$, then the value of $3\alpha^3 - 3\alpha^2 + 2\beta^3 - 2\beta^2 + 11\alpha$ is

(A) 33 (B) −33 (C) 22 (D) −22

21. If α and β are the roots of the equation $x^2 - 2x - 5 = 0$, then the equation whose roots are $\frac{\alpha}{\alpha-1}$ and $\frac{\beta}{\beta-1}$ is

(A) $6x^2 - 12x + 5 = 0$
(B) $2x^2 + 8x - 5 = 0$
(C) $-2x^2 + 8x - 5 = 0$
(D) $6x^2 + 12x - 5 = 0$

22. Given $f(x) = x^2 - ax + \beta$ (where $\alpha, \beta \in R$). If the equation $f(f(x)) = 0$ has two roots 1 and 2, then f(0) is equal to

(A) $\frac{-3}{2}$ (B) $\frac{-1}{2}$ (C) $\frac{3}{2}$ (D) $\frac{1}{2}$

23. If a, b, c, d are roots of the equation $x^4 - 2px - \sqrt{2007} = 0$, then the product of $\frac{a+b+c}{abc}$, $\frac{b+c+d}{bcd}$, $\frac{c+d+a}{cda}$, $\frac{d+a+b}{abd}$ is equal to

(A) $-\sqrt{2007}$ (B) $\sqrt{2007}$
(C) $\frac{1}{2007}$ (D) 2007

24. If $ax^{17} + bx^{16} + 1$ is divisible by $x^2 - x - 1$, then integral value of a is

(A) 1 (B) 987 (C) −1597 (D) 1597

25. Number of solutions of the equation $|\sin(\pi x)| + \log_5(6x - x^2 - 4) + \sqrt{6x - x^2 - 5} = 0$ is

(A) 4 (B) 2
(C) 1 (D) no solution

Multiple Answer Type

1. Let $P(x)$ and $Q(x)$ be two polynomials. If $f(x) = P(x^4) + x\,Q(x^4)$ is divisible by $x^2 + 1$, then

(A) $P(x)$ is divisible by $(x - 1)$
(B) $Q(x)$ is divisible by $(x - 1)$
(C) $f(x)$ is divisible by $(x - 1)$
(D) none of these

2. Consider the equation $x^2 + x - a = 0$; $a \in N$, if equation has integral roots, then

(A) $a = 2$ (B) $a = 6$
(C) $a = 12$ (D) $a = 20$

3. If $ax^2 + bx + c = 0$ and $cx^2 + bx + a = 0$ ($a, b, c \in R$) have a common non-real root, then

(A) $-2|a| < b < 2|a|$ (B) $-2|c| < |b| < 2|c|$
(C) $a = \pm c$ (C) none of these

4. Consider the graph of $f(x) = ax^2 + bx + c$ in the figure below, we can conclude that

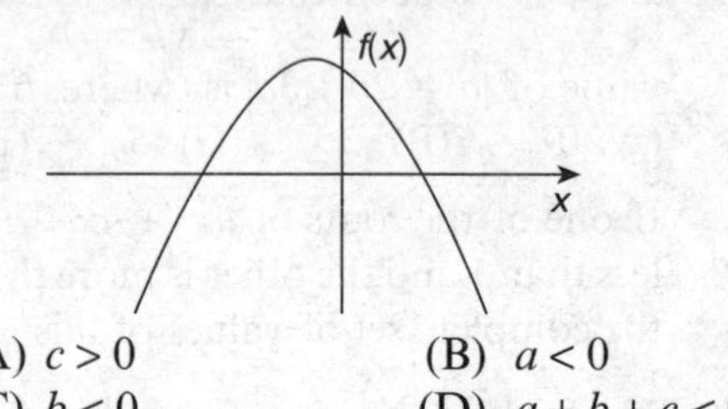

(A) $c > 0$ (B) $a < 0$
(C) $b < 0$ (D) $a + b + c < 0$

5. For the quadratic equation $ax^2 + bx + c = 0$, if 'a' and $4a + 2b + c$ are of same sign, then

(A) one root greater than 2 and the other root less than 2
(B) both roots greater than 2
(C) both roots less than 2
(D) none of these

6. If the equation $cx^2 + bx - 2a = 0$ has no real roots and $a < (b + c)/2$, then

(A) $ac < 0$ (B) $a < 0$
(C) $(c - a)/2 > a$ (D) $(c + 2b)/8 > a$

7. The value(s) of x which satisfy the equation $|x^2 + 3x| + x^2 - 2 = 0$ is (are)

(A) −2/3 (B) 2/3 (C) −1/2 (D) 1/2

8. If the equation $ax^2 + bx + c = 0$ ($a < 0$) has two roots α and β such that $\alpha < -3$ and $\beta > 3$, then

(A) $9a + 3|b| + c > 0$ (B) $c > 0$
(C) $4a + 2|b| + c > 0$ (D) none of these

9. For $m \in R$, let $f(x) = x^3 + 3mx^2 - 3x - 3m + 2$ has 3 real roots x_1, x_2 and x_3, also k denotes the value of m for which $(x_1^2 + x_2^2 + x_3^2)$ is minimum. Which of the following statement(s) is/are correct?

(A) $f(x)$ has all 3 roots real and distinct corresponding to m for which is $(x_1^2 + x_2^2 + x_3^2)$ minimum.
(B) The value of k lies in [−1, 1].

(C) The minimum value of $\sum_{i=1}^{3} x_i^2$ equals 6.

(D) The number of roots of $\sin x = k$ in $[0\ 2\pi]$ equals 3.

10. If the expression $ax^4 + bx^3 - x^2 + 2x + 3$ leaves a remainder $4x + 3$ when divided by $(x^2 + x - 2)$ then

(A) $(a^2 + b^2) = 5$ (B) $(2b - a) = 3$

(C) $(4a + b) = 9$ (D) $(a - b) = 1$

11. The integral value(s) of 'a' for which the roots of $ax(x + 2) + a - (x + 2) = 0$ are rational is given by

(A) 2 (B) 5 (C) 12 (D) 20

12. If $1 + \log_5(x^2 + 1) \geq \log_5(ax^2 + 4x + a)$, $\forall\, x \in R$ then 'a' can be equal to

(A) 3 (B) $\frac{5}{2}$ (C) 2 (D) $\frac{3}{2}$

13. If the equation $\sin^2\theta - a \sin\theta + b = 0$ has only one solution in $(0, \pi)$, then

(A) $a \in (-\infty, 1] \cup [2, \infty)$

(B) $b \in (-\infty, 0] \cup [1, \infty)$

(C) $a = 1 + b$

(D) $a + b = 1$

14. If the expression $y = \dfrac{x - k}{x^2 - 4x + k}$ can take all real values for permissible real values of x, then the possible integral value of k can be

(A) 0 (B) 1 (C) 2 (D) 3

15. The value of 'k' for which both roots of the equation $4x^2 - 2x + k = 0$ are completely in $(-1, 1)$, then k may be equal to

(A) –1 (B) 0 (C) 2 (D) –3

16. For which of the following graphs of the quadratic expression $y = ax^2 + bx + c$, is the product abc positive?

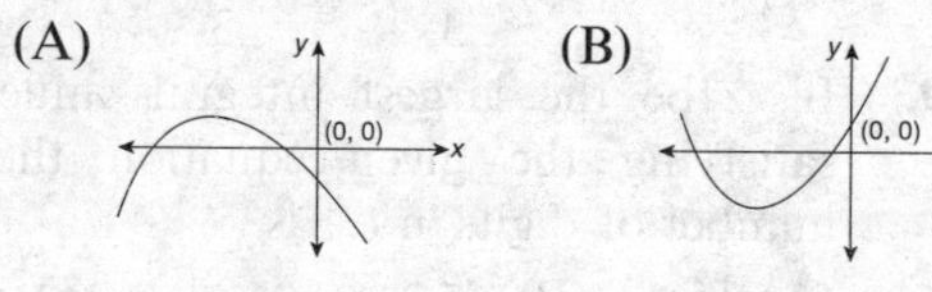

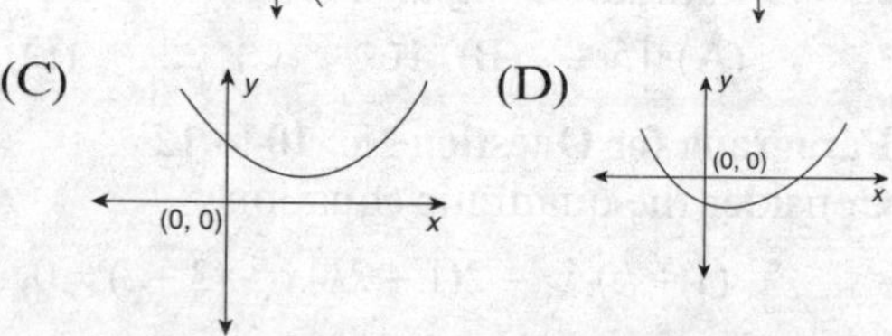

17. Let $f(x) = Ax^2 + Bx + C$ where $f(x)$ is an integer when x is an integer, then

(A) $A + B + C$ is an integer

(B) $f'''(x)$ is an integer

(C) C is an integer

(D) $A - B + C$ is an integer

18. Let $\alpha\beta\gamma\delta < 0$, $\alpha + \beta + \gamma + \delta > 0$ and $\frac{\alpha}{|\alpha|} + \frac{\beta}{|\beta|} + \frac{\gamma}{|\gamma|} + \frac{\delta}{|\delta|} = x$, then the possible value(s) of $x^4 - 32x + 25$ is/are

(A) – 23 (B) 105 (C) 41 (D) 64

19. If A, B, C are angles of ΔABC and $\sin 2A$, $\sin 2B$, $\sin 2C$ are roots of equation $625x^3 - 400x^2 - 560x + 384 = 0$, then select the correct option(s).

(A) ΔABC is an obtuse angled triangle.

(B) ΔABC is an acute angled triangle.

(C) $\sum \tan A = \frac{-1}{3}$

(D) $\sum \tan A = \frac{1}{3}$

20. If α, β^2 are roots of $5x^2 - 20x + 3k = 0$ and β, α^2 are roots of $kx^2 - 50x + 45 = 0$ (a, b, $\alpha, \beta \in R$ and $k \neq 0$), then

(A) $a\beta = 3$ (B) $k = 5$

(C) $\alpha^3 - 10\alpha + 3\beta = 0$ (D) $\beta^3 - 4\beta + \alpha = 0$

21. If $(x - p)(x - q) + 5(x - r)(x - s) = 0$, where $p < r < q < s$, then

(A) one root lies in (p, r) and the other root lies in (q, s)

(B) one root lies in (r, s) and the other root lies in (s, ∞)

(C) one root lies in (p, q) and the other root lies in (q, s)

(D) one root lies in $(-\infty, p)$ and the other root lies in (s, ∞)

22. If $f(x) = 2x^2 + kx + 50$, $k \geq 20$ and magnitude of difference of the roots of the equation $f(x) = 0$ is 8, then which of the following can be true?

(A) minimum value of $f(x)$ is –32

(B) minimum value of $f(x)$ is –64

(C) graph of $f(x)$ cuts positive y-axis

(D) graph of $f(x)$ cuts negative y-axis

23. Let $f(x)$ be an even quadratic polynomial such that $-4 \le f(1) \le -1$ and $-1 \le f(2) \le 5$. Which of the following statements is true?

(A) Leading coefficient of $f(x)$ is $\frac{f(2)-f(1)}{3}$

(B) $f(0) = \frac{f(2)-4f(1)}{3}$

(C) Maximum value of $f(3)$ is 20

(D) Maximum value of $f(3)$ is 12

24. The equation $x^3 + 4x^2 + px + q = 0$ and $x^3 + 6x^2 + px + r = 0$ have two roots in common (where $p, q, r \in R$). If their uncommon roots are the roots of the equation $x^2 + 2ax + 8c = 0$, then

(A) $a + c = 8$ (B) $3q = 2r$

(C) $a + c = 2$ (D) $3r = 2q$

25. Graph of parabola $y = ax^2 + bx + c$ is shown in the adjacent figure where V is the vertex and ΔAVB is a right angled triangle, right angled at V. If $a, b, c \in R$ and $D = b^2 - 4ac$, then

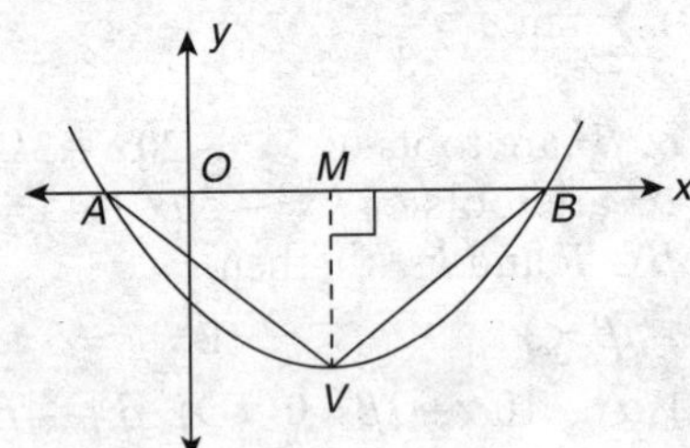

(A) $D = 2$ (B) $D = 4$

(C) $b \in (-2, 2)$ (D) $b \in (-2, 0)$

Comprehension Based Questions

Paragraph for Question No. 1 to 3

Let us consider a rational function $f(x) = \frac{x^2 - 3x - 4}{x^2 - 3x + 4}$ and a quadratic function $g(x) = x^2 - (b+1)x + (b-1)$, where b is a parameter.

1. The sum of integers in the range of $f(x)$, is

(A) −10 (B) −11 (C) −5 (D) −6

2. If both roots of the equation $g(x) = 0$ are greater than −1, then b lies in the interval

(A) $\left(-\frac{1}{2}, \infty\right)$ (B) $(-2, \infty)$

(C) $\left(-\infty, -\frac{1}{4}\right)$ (D) $(-\infty, -2)$

3. The largest natural number b satisfying $g(x) > -2$ for all $x \in R$, is

(A) 1 (B) 2 (C) 3 (D) 4

Paragraph for Question No. 4 to 6

The vertices of a triangle ABC lie on the curve $f(x) = ax^2 + bx + c$ as shown in the figure. Let the triangle ABC be a right angled isosceles whose hypotenuse $AC = 6\sqrt{3}$ units, then

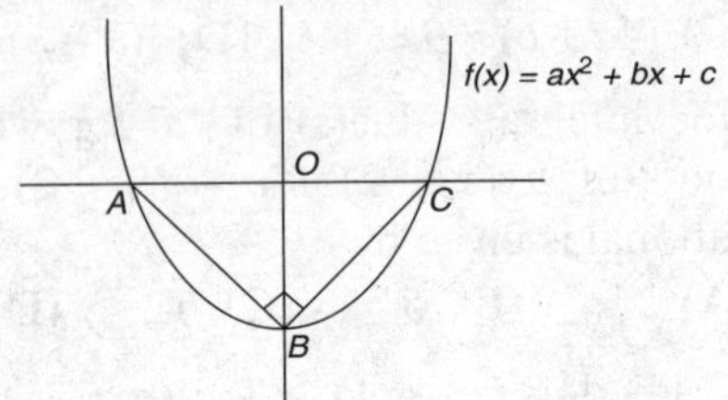

4. The value of a is

(A) $\frac{1}{\sqrt{3}}$ (B) $\frac{1}{3}$ (C) $\frac{1}{2\sqrt{3}}$ (D) $\frac{1}{3\sqrt{3}}$

5. The minimum value of $f(x)$ is

(A) $-3\sqrt{3}$ (B) −3

(C) 3 (D) $3\sqrt{3}$

6. Number of integral value of K for which $K/2$ lies between the roots of $f(x) = 0$, is

(A) 22 (B) 21

(C) 20 (D) 19

Paragraph for Question No. 7 to 9

Consider the equation

$\sqrt{x^2 + 2x + 1} - \sqrt{4x^2 + 4x + 1} + \sqrt{4x^2} = 2$

7. The number of integral solution(s) of equation is/are

(A) 0 (B) 2 (C) 3 (D) 4

8. The sum of all the solutions of the equation is

(A) 0 (B) $\frac{-2}{3}$ (C) $\frac{2}{3}$ (D) 1

9. If α be the largest integral value of x satisfying the given equation then the number of digits in α^{50} is

(A) 15 (B) 16 (C) 23 (D) 24

Paragraph for Question No. 10 to 12

Consider the quadratic equation

$$(1 + k)\, x^2 - 2(1 + 2k)x + (3 + i) = 0,$$

where $k \in R - \{-1\}$.

10. The number of integral values of k such that the given quadratic equation has imaginary roots is
(A) 0 (B) 1 (C) 2 (D) 3

11. The set of values of k such that the given quadratic equation has both the roots positive is
(A) $k \in R$
(B) $k \in (-\infty, -3) \cup \left[\sqrt{\frac{2}{3}}, \infty\right)$
(C) $k \in (-\infty, -3) \cup (-1, \infty)$
(D) $k \in \left(-\infty, -\sqrt{\frac{2}{3}}\right] \cup \left[\frac{2}{3}, \infty\right)$

12. The number of real values of k such that the given quadratic equation has roots in the ratio 1 : 2 is
(A) 0 (B) 1 (C) 2 (D) ∞

Paragraph for Question No. 13 to 15
Let $f(x)$ and $g(x)$ be two polynomials of degree 2 defined as $f(x) = x^2 - 2(k-1)x + 1$ and $g(x) = kax^2 + bx + c$. Given that $a^2 + b^2 + c^2 - 2a + 6b - 4c + 14 = 0$, $a, b, c \in R$ then

13. The value of $f(0) + g(0)$ is
(A) 1 (B) 2 (C) 3 (D) 4

14. The graph of $f(x)$ and $g(x)$ touch each other; then, sum of all the real values of k is
(A) 0 (B) 4 (C) 6 (D) 8

15. The number of values of k for which $f(x) = 0$ and $g(x) = 0$ has 3 real and distinct roots provided that there is no common solution among the two equations is/are
(A) 0 (B) 1 (C) 2 (D) 3

Match Matrix

1. One or more than one option may match

Column I	Column II
(A) If the expression $ax^4 + bx^3 - x^2 + 2x + 3$ leaves a remainder $4x + 3$ when divided by $(x^2 + x - 2)$, then $(2b - a)$ is	(P) 2
(B) If the vertex of $y = ax^2 + bx + c$ is $(4, -5)$ and two x-intercepts one positive and one negative, then $8a + b$ is	(Q) 0
(C) The integral value of 'a' for which the roots of $ax(x+2) + a - (x+2) = 0$ are rational is	(R) 3
(D) If α, β^2 are roots of $5x^2 - 20x + 3k = 0$ and β, α^2 are roots of $kx^2 - 50x + 45 = 0$ $(a, b, \alpha, \beta \in R$ and $k \neq 0)$, then $\beta^3 - 4\beta + \alpha$ is	(S) 6

2. Only one option is matching
Let $f(x) = x^2 - (2a+3)x + a^2 + 3a$

Column I	Column II
(A) If $f(x) = 0$ has both positive roots, then complete set of values of a is	(P) (1, 2)
(B) If both roots of $f(x) = 0$ belong to interval (1, 5), then complete set of values of a is	(Q) (−1, 1)
(C) If one root of $f(x) = 0$ is less than 1 whereas other root exceeds 2, then the complete set of values of a is	(R) $(0, \infty)$
(D) If the inequality $f(x) < 0$ has at least one solution, then complete set of values of a is	(S) $(-\infty, \infty)$

3. Let $P(x) = 2x^2 - 12x + c\ \forall\ x \in R$ where c is a real constant, then

Column I	Column II
(A) If greatest value of $p(x)$ for $x \in [1, 2]$ is 1, then c equals	(P) 8
(B) If smallest value of $P(x)$ for $x \in [1, 5]$ is -1, then c equals	(Q) 11
(C) If the greatest value of $P(x)$ for $x \in [1, 4]$ is 2, then c equals	(R) 12
	(S) 17

4. Consider the quadratic trinomial $P(x) = 3x^2 - 7kx + (5k - 3)$ where k is the parameter. Find the range of k in the following conditions given in column I.

Column I	Column II
(A) One root is greater and other root is less than 2	(P) $(1, \infty)$
(B) One root lies in $(-2, 0)$ and other root lies in $(0, 1)$	(Q) $(-\infty, 0)$
(C) One root is less than -1 and other root is greater than 1	(R) ϕ
	(S) $\left(\frac{-9}{10}, 0\right)$

5.

Column I	Column II
(A) If $\log_3 x + \log_3 y \geq 4$, then the least value of $(x + y)$ is	(P) 1
(B) In ΔABC, if $\sum \cos^2 A = m + n(\prod \cos A)$ (where $m, n \in I$) then the value of $\lvert m - n\rvert$	(Q) 4/3
(C) The number of integral value(s) of x satisfying both the inequalities $\log_{\sqrt{5}}(4 - x) \geq 0$ and $\log_{(x-1)}(x^2 + 2) \geq 0$ is	(R) 3
	(S) 18

Subjective Answer Type (Hard Nut)

1. If $1, a_1, a_2, \ldots, a_{n-1}$ are the n roots of the equation $x^n - 1 = 0$, find the value of $(1-a_1)(1-a_2)\ldots(1-a_{n-1})$.

2. If $ax^2 - bx + c = 0$ have two distinct real roots lying in the interval $(0, 1)$, $a, b, c \in N$, then prove that $\log_5(abc) \geq 2$.

3. If $A = \begin{vmatrix} a & 1 & 1 \\ 1 & b & 1 \\ 1 & 1 & c \end{vmatrix}$ where $a, b, c > 0$ and $A > 0$ and there exist two real roots of $A(x^2 + x) + 1 = 0$, then find the minimum numerical value of $a\,b\,c$.

4. If the roots of the equation $ax^2 - 2bx + c = 0$ are imaginary, find the number of real roots of $4e^x + (a+c)^2(x^3 + x) = 4b^2x$.

5. If $ax + by = 1$ and $cx^2 + dy^2 = 1$ have only one solution, then find the value of $\frac{a^2}{c} + \frac{b^2}{d}$.

6. Find the value of parameter 'a' for which the equation $x^4 - (a+1)x^3 + x^2 + (a+1)x - 2 = 0$ have at least two distinct positive real roots.

7. Find 'a', if $ax^2 - 4x + 9 = 0$ has integral roots.

8. Let $f(x) = ax^2 + bx + c\,(abc > 0)$ and $f(x) = 0$ does not have two distinct real roots. If the point $(1, 3)$ lies on the curve $y = f(x)$, if $b \in (\alpha, \beta)$, then find the value of $\alpha + \beta^3$.

9. If a,b,c,d are four distinct real numbers and they are in arithmetic progression. If

$$2(a-b)+x(b-c)^2+(c-a)^3$$
$$=2(a-d)+(b-d)^2+(c-d)^3.$$

Then find the value of x.

10. Find all real values of parameter 'K' for which the equation $\sin^2 x - |K-2|\sin x = \cos^2 x + |K-2| - 1$ may have a solution.

11. Let $f(x) = ax^2 + bx + c$, $a \neq 0$, $a, b, c \in I$. Suppose that $f(1) = 0$, $50 < f(7) < 60$ and $70 < f(8) < 80$. If $13K < f(10) < 13(K+1)$ for some integer K, then find the value of K.

12. If $p(x) = ax^2 + bx + c$ leaves a remainder of 4 when divided by x, a remainder of 3 when divided by $(x+1)$ and a remainder of 1 when divided by $(x-1)$, then find $p(2)$.

13. Let $f(x) = x^3 + 3x^2 + 6x + 9\sin x$, then prove that equation $\frac{1}{y-f(1)} + \frac{2}{y-f(2)} + \frac{3}{y-f(3)} = 0$ has real roots.

14. If α and $\alpha+1$, for some values of α are the roots of the equation

$$\sum_{k=1}^{n}\left[(x+k-1)(x+k)\right]-10=0,$$ find n.

15. If $2x^3 + ax^2 + bx + 4 = 0$ (a and b are positive real number) has 3 real roots, then prove that $a + b \geq 6(2^{1/3} + 4^{1/3})$.

16. Find out minimum non-negative real values of a, b and c given that the equation $x^4 + ax^3 + bx^2 + cx + 1 = 0$ has real roots.

17. If 'c' is positive and $2ax^2 + 3bx + 5c = 0$ does not have any real roots, then prove that $2a - 3b + 5c > 0$.

18. If the equation $x^4 + px^3 + qx^2 + rx + 5 = 0$ has four positive real roots, then find the minimum value of pr.

19. If $(a^2 - 14a + 13)x^2 + (a+2)x - 2 = 0$ does not have two distinct real roots, then find the maximum value of $(a^2 - 15a)$.

20. Let α, β be the roots of the equation $ax^2 + bx + c = 0$ and $S_n = \alpha^n + \beta^n$, prove that a $S_{n+1} + b\,S_n + c\,S_{n-1} = 0$.

1.7 JEE Main: Archive

1. If the sum of the squares of the reciprocals of the roots α and β of the equation $3x^2 + \lambda x - 1 = 0$ is 15, then $6\left(\alpha^3+\beta^3\right)^2$ is equal to

(A) 18 (B) 24 (C) 36 (D) 96

[Main 2022, 24 June 1st Shift]

2. The sum of all the real roots of the equation $\left(e^{2x}-4\right)\left(6e^{2x}-5e^{x}+1\right)=0$ is

(A) $\log_e 3$ (B) $-\log_e 3$

(C) $\log_e 6$ (D) $-\log_e 6$

[Main 2022, 24 June 2nd Shift]

3. For a natural number n, let $a_n = 19^n - 12^n$. Then, the value of $\frac{31\alpha_9 - \alpha_{10}}{57\alpha_8}$ is

[Main 2022, 25 June 1st Shift]

4. Let $a, b \in R$ be such that the equation $ax^2 - 2bx + 15 = 0$ has a repeated root α. If α and β are the roots of the equation $x^2 - 2bx + 21 = 0$, then $\alpha^2 + \beta^2$ is equal to

(A) 37 (B) 58 (C) 68 (D) 92

[Main 2022, 25 June 2nd Shift]

5. The sum of the cubes of all the roots of the equation $x^4 - 3x^3 - 2x^2 + 3x + 1 = 10$ is ______. **[Main 2022, 26 June 1st Shift]**

6. If the sum of all the roots of the equation $e^{2x} - 11e^{x} - 45e^{-x} + \frac{81}{2} = 0$ is $\log_e P$, then P is equal to ______.

[Main 2022, 27 June 1st Shift]

7. Let α, β be the roots of the equation $x^2 - 4\lambda x + 5 = 0$ and α, γ be the roots of the equation $x^2 - \left(3\sqrt{2}+2\sqrt{3}\right)x + 7 + 3\lambda\sqrt{3} = 0$. If $\beta + \gamma = 3\sqrt{2}$, then $\left(\alpha + 2\beta + \gamma\right)^2$ is equal to

[Main 2022, 27 June 2nd Shift]

8. The number of real solutions of the equation $e^{4x} + 4e^{3x} - 58e^{2x} + 4e^{x} + 1 = 0$ is ______.

[Main 2022, 28 June 1st Shift]

9. Let $f(x)$ be a quadratic polynomial such that $f(-2)+f(3)=0$. If one of the roots of $f(x)=0$ is -1, then the sum of the roots of $f(x)=0$ is equal to

(A) $\frac{11}{3}$ (B) $\frac{7}{3}$ (C) $\frac{13}{3}$ (D) $\frac{14}{3}$

[Main 2022, 28 June 2nd Shift]

10. Let α be a root of the equation $1+x^2+x^4=0$. Then the value of $\alpha^{1011}+\alpha^{2022}-\alpha^{3033}$ is equal to

(A) 1 (B) α
(C) $1+\alpha$ (D) $1+2\alpha$

[Main 2022, 29 June 2nd Shift]

11. Let α be a root of the equation $(a-c)x^2+(b-a)x+(c-b)=0$ where a,b,c are distinct real numbers such that the matrix $\begin{bmatrix} \alpha^2 & \alpha & 1 \\ 1 & 1 & 1 \\ a & b & c \end{bmatrix}$ is singular. Then the value of

$$\frac{(a-c)^2}{(b-a)(c-b)}+\frac{(b-a)^2}{(a-c)(c-b)}+\frac{(c-b)^2}{(a-c)(b-a)}$$

is

(A) 6 (B) 3 (C) 9 (D) 12

[Main 2023, 24 Jan 1st Shift]

12. Let $\lambda \in R$ and let the equation E be $|x^2|-2|x|+|\lambda-3|=0$. Then the largest element in the set $S=\{x+\lambda : x \text{ is an integer solution of } E\}$ is ______.

[Main 2023, 24 Jan 1st Shift]

13. The number of real solutions of the equation $3\left(x^2+\frac{1}{x^2}\right)-2\left(x+\frac{1}{x}\right)+5=0$, is

(A) 4 (B) 0 (C) 3 (D) 2

[Main 2023, 24 Jan 2nd Shift]

14. Let $S=\left\{\alpha : \log_2\left(9^{2\alpha-4}+13\right)-\log_2\left(\frac{5}{2}\cdot 3^{2\alpha-4}+1\right)=2\right\}$.

Then the maximum value of β for which the equation

$$x^2-2\left(\sum_{\alpha\in S}\alpha\right)^2 x+\sum_{\alpha\in S}(\alpha+1)^2\beta=0$$

has real roots, is ______.

[Main 2023, 25 Jan 1st Shift]

15. Let $a \in R$ and let α, β be the roots of the equation $x^2+60^{\frac{1}{4}}x+a=0$. If $\alpha^4+\beta^4=-30$, then the product of all possible values of a is ______. **[Main 2023, 25 Jan 2nd Shift]**

16. Let $\lambda \neq 0$ be a real number. Let α, β be the roots of the equation $14x^2-31x+3\lambda=0$ and α, γ be the roots of the equation $35x^2-53x+4\lambda=0$. Then $\frac{3\alpha}{\beta}$ and $\frac{4\alpha}{\gamma}$ are the roots of the equation:

(A) $7x^2+245x-250=0$
(B) $7x^2-245x+250=0$
(C) $49x^2-245x+250=0$
(D) $49x^2+245x+250=0$

[Main 2023, 29 Jan 1st Shift]

17. If the value of real number $a>0$ for which $x^2-5ax+1=0$ and $x^2-ax-5=0$ have a common real roots is $\frac{3}{\sqrt{2\beta}}$ then β is equal to ______. **[Main 2023, 30 Jan 2nd Shift]**

18. The number of real roots of the equation $\sqrt{x^2-4x+3}+\sqrt{x^2-9}=\sqrt{4x^2-14x+6}$, is:

(A) 0 (B) 1 (C) 3 (D) 2

[Main 2023, 31 Jan 1st Shift]

19. The equation $e^{4x}+8e^{3x}+13e^{2x}-8e^x+1=0, x\in R$ has:

(A) two solutions and both are negative
(B) no solution
(C) four solutions two of which are negative
(D) two solutions and only one of them is negative

20. Let

$$S=\left\{x\in R : \left(\sqrt{3}+\sqrt{2}\right)^x+\left(\sqrt{3}-\sqrt{2}\right)^x=10\right\}$$

Then the number of elements in S is:

(A) 4 (B) 0 (C) 2 (D) 1

[Main 2024, 1 Feb 1st Shift]

21. Let α and β be the roots of the equation $px^2 + qx - r = 0$, where $p \neq 0$. If p, q and r be the consecutive terms of a non-constant G.P. and $\frac{1}{\alpha}+\frac{1}{\beta}=\frac{3}{4}$, then the value of $(\alpha-\beta)^2$ is:

(A) $\frac{80}{9}$ (B) 9 (C) $\frac{20}{3}$ (D) 8

[Main 2024, 1 Feb 2nd Shift]

22. If α, β are the roots of the equation, $x^2 - x - 1 = 0$ and $S_n = 2023\alpha^n + 2024\beta$, then

(A) $2S_{12} = S_{11} + S_{10}$ (B) $S_{12} = S_{11} + S_{10}$

(C) $2S_{11} = S_{12} + S_{10}$ (D) $S_{11} = S_{10} + S_{12}$

[Main 2024, 27 Jan 2nd Shift]

23. Let the set $C = \{(x, y) | x^2 - 2^y = 2023, x, y \in N\}$. Then $\sum_{(x,y)=C}(x+y)$ is equal to

[Main 2024, 29 Jan 2nd Shift]

24. Let $\alpha, \beta \in N$ be roots of equation $x^2 - 70x + \lambda = 0$, where $\frac{\lambda}{2}, \frac{\lambda}{3} \notin N$. If λ assumes the minimum possible value, then $\frac{(\sqrt{\alpha-1}+\sqrt{\beta-1})(\lambda+35)}{|\alpha-\beta|}$ is equal to:

[Main 2024, 30 Jan 1st Shift]

25. The number of real solutions of the equation $x(x^2 + 3|x| + 5|x-1| + 6|x-2|) = 0$ is ______. **[Main 2024, 30 Jan 2nd Shift]**

26. For $0 < x < b < a$, let

$(a+b-2c)x^2 + (b+c-2a)x + (c+a-2b) = 0$

and $\alpha \neq 1$ be one of its roots. Then, among the two statements

(I) If $\alpha \in (-1,0)$, then b cannot be the geometric mean and a and c.

(II) If $\alpha \in (0,1)$, then b may be the geometric mean of a and c.

(A) both (I) and (II) are true

(B) neither (I) nor (II) is true

(C) only (II) is true

(D) only (I) is true

[Main 2024, 31 Jan 1st Shift]

27. Let S be the set of positive integral values of a for which

$$\frac{ax^2 + 2(a+1)x + 9a + 4}{x^2 - 8x + 32} < 0, \forall x \in R.$$

Then, the number of elements in S is:

(A) 1 (B) 0 (C) ∞ (D) 3

[Main 2024, 31 Jan 1st Shift]

28. The number of solutions, of the equation $e^{\sin x} - 2e^{-\sin x} = 2$ is

(A) 2 (B) more than 2

(C) 1 (D) 0

[Main 2024, 31 Jan 2nd Shift]

29. Let a, b, c be the length of three sides of a triangle satisfying the condition $(a^2 + b^2)x^2 - 2b(a+c).x + (b^2 + c^2) = 0$. If the set of all possible values of x is the interval (α, β), then $12(\alpha^2 + \beta^2)$ is equal to ______. **[Main 2024, 31 Jan 2nd Shift]**

1.8 Answer Key

1.8.1 Know Your Concepts

1. Real and Distinct **2.** (2) **3.** $\lambda \in \left(-2, \frac{2}{5}\right)$ **4.** (7) **6.** (0)

7. $(x-c)(x-d) + K = 0$ **8.** $(x^2 - 2x + 3 = 0)$ **12.** (3) **13.** $\frac{2\sqrt{13}}{9}$

14. (0) **15.** $[1, \infty) \cup \{-1\}$ **17.** (8, 12)

19. $a = -7, b = -8$, (3, 4), (3, 5), (3, 12) **20.** $\left(\frac{5-\sqrt{5}}{2}, \frac{5+\sqrt{5}}{2}\right)$

1.8.2 JEE Main: Sample Questions

Single Answer Type

1. (C)	**2.** (B)	**3.** (B)	**4.** (C)	**5.** (C)	**6.** (B)	**7.** (A)	**8.** (B)
9. (D)	**10.** (B)	**11.** (C)	**12.** (D)	**13.** (A)	**14.** (A)	**15.** (C)	**16.** (C)
17. (A)	**18.** (A)	**19.** (B)	**20.** (B)	**21.** (B)	**22.** (A)	**23.** (B)	**24.** (A)
25. (A)	**26.** (C)	**27.** (D)	**28.** (A)	**29.** (D)	**30.** (D)	**31.** (B)	**32.** (C)
33. (C)	**34.** (D)	**35.** (C)	**36.** (C)	**37.** (A)	**38.** (C)	**39.** (B)	**40.** (A)
41. (B)	**42.** (D)	**43.** (A)	**44.** (C)	**45.** (D)	**46.** (B)	**47.** (A)	**48.** (A)
49. (C)	**50.** (C)	**51.** (B)	**52.** (C)	**53.** (D)	**54.** (C)	**55.** (B)	**56.** (D)
57. (D)	**58.** (A)						

Integer Answer Type

1. (6)	**2.** (4)	**3.** (3)	**4.** (1)	**5.** (0)	**6.** (8)	**7.** (1)	**8.** (9)
9. (0)	**10.** (4)	**11.** (2)	**12.** (0)	**13.** (0005)	**14.** (8)	**15.** (0)	**16.** (4)
17. (3)	**18.** (17)	**19.** (1)					

1.8.3 JEE Advanced: Sample Questions

Single Answer Type

1. (D)	**2.** (D)	**3.** (A)	**4.** (C)	**5.** (C)	**6.** (C)	**7.** (A)	**8.** (A)
9. (D)	**10.** (A)	**11.** (A)	**12.** (A)	**13.** (B)	**14.** (B)	**15.** (B)	**16.** (C)
17. (D)	**18.** (B)	**19.** (A)	**20.** (D)	**21.** (A)	**22.** (A)	**23.** (C)	**24.** (B)
25. (B)							

Multiple Answer Type

1. (A, B, C)	**2.** (A, B, C, D)	**3.** (A, B, C)	**4.** (A, B, C)
5. (B, C)	**6.** (A, B, C, D)	**7.** (A, D)	**8.** (A, B, C)
9. (B, C, D)	**10.** (A, B)	**11.** (A, C, D)	**12.** (A, B)
13. (A,B,C)	**14.** (B, C)	**15.** (A, B)	**16.** (B, D)
17. (A, B, C, D)	**18.** (A, B)	**19.** (A, C)	**20.** (A, B, D)
21. (A, C)	**22.** (A, C)	**23.** (A, C)	**24.** (A, B)
25. (B, D)			

Comprehension Based Questions

1. (D)	**2.** (A)	**3.** (B)	**4.** (D)	**5.** (A)	**6.** (C)	**7.** (B)	**8.** (A)
9. (B)	**10.** (B)	**11.** (B)	**12.** (C)	**13.** (C)	**14.** (C)	**15.** (A)	

Match Matrix

1. (A – R; B – Q; C – P, Q, S; D – Q)
2. (A – R; B – P; C – Q; D – S)
3. (A – Q; B – S; C – R)
4. (A – P; B – S; C – R)
5. (A – S; B – R; C – P)

Subjective Answer Type (Hard Nut)

1. n	**3.** (8)	**4.** (one)	**5.** (1)	**6.** $a \in \left[2\sqrt{2}-1, \infty\right)$	**7.** $\frac{1}{3}$	**8.** (4)	
9. $x \in (-\infty, -8] \cup [16, \infty)$	**10.** $K \in [1, 3]$	**11.** (10)	**12.** (–6)	**14.** (5)	**16.** (4, 6, 4)		
18. (80)	**19.** (–9)						

1.8.4 JEE Main: Archive

1. (B)	**2.** (B)	**3.** (4)	**4.** (B)	**5.** (36)	**6.** (45)	**7.** (98)	**8.** (2)
9. (A)	**10.** (A)	**11.** (B)	**12.** (5)	**13.** (B)	**14.** (25)	**15.** (45)	**16.** (C)
17. (13)	**18.** (B)	**19.** (A)	**20.** (C)	**21.** (A)	**22.** (B)	**23.** (46)	**24.** (60)
25. (1)	**26.** (A)	**27.** (B)	**28.** (D)	**29.** (36)			

1.9 Solutions

1.9.1 Know Your Concepts

1. $D=25(l+m)^2+8(l-m)^2>0 \Rightarrow$ Roots are real and distinct.

2. Since, $x^2-ax+b=0$ and $x^2-px+q=0$ have a common root, but $x^2-px+q=0$ has equal roots $\Rightarrow$ both roots are common and $p^2=4q$.

Therefore $1=\dfrac{q}{p}=\dfrac{b}{q} \quad \Rightarrow a=p, b=q$

Now $\dfrac{ap}{b+q}=\dfrac{p^2}{2q}=\dfrac{4q}{2q}=2.$

3. $(\lambda^2+\lambda-2)x^2+(\lambda+2)x-1<0$

$\Rightarrow(\lambda^2+\lambda-2)<0$ and

$(\lambda+2)^2+4(\lambda^2+\lambda-2)<0$

Now, $\lambda^2+\lambda-2<0 \Rightarrow -2<\lambda<1$ (1)

and $(\lambda+2)^2+4(\lambda^2+\lambda-2)<0$

$\Rightarrow 5\lambda^2+8\lambda-4<0 \Rightarrow -2<\lambda<\dfrac{2}{5}$ (2)

from (1) and (2) $\lambda\in\left(-2,\dfrac{2}{5}\right)$

4. $D=q^2-4p\geq 0 \Rightarrow q^2\geq 4q$

q	p	4p	$q^2-4p\geq 0$	Possible value
1	1	4	$q^2-4p<0$	0
2	1	4	$q^2-4p=0$	1
3	1, 2	4, 8	$q^2-4p>0$	2
4	1, 2, 3, 4	4, 8, 12, 16	$q^2-4p\geq 0$	4

Hence the total number of quadratic equations = 7.

5. Let α, β be roots of $ax^2+bx+c=0$

$\Rightarrow \alpha+\beta=-\dfrac{b}{a}, \alpha\beta=\dfrac{c}{a}$

Also let γ, δ be roots of $a_1x^2+b_1x+c_1=0$

$\Rightarrow \gamma+\delta=-\dfrac{b_1}{a_1}, \gamma\delta=\dfrac{c_1}{a_1}$

Since, $\dfrac{\alpha}{\beta}=\dfrac{\gamma}{\delta}\Rightarrow\dfrac{\alpha}{\beta}+1=\dfrac{y}{\delta}+1$

$\Rightarrow\dfrac{\alpha+\beta}{\gamma+\delta}=\dfrac{\beta}{\delta}\Rightarrow\dfrac{(\alpha+\beta)^2}{(\gamma+\delta)^2}=\dfrac{\alpha}{\gamma}\cdot\dfrac{\beta}{\delta}$

$\Rightarrow\dfrac{\frac{b^2}{a^2}}{\frac{b_1^2}{a_1^2}}=\dfrac{\frac{c}{a}}{\frac{c_1}{a_1}}\Rightarrow\left(\dfrac{b}{b_1}\right)^2=\dfrac{ca}{c_1a_1}$

6. Let α, α^2 be roots

$\alpha+\alpha^2=-p; \alpha^3=q$

Now, $\alpha(1+\alpha)=-p\Rightarrow\alpha^3(1+\alpha)^3=-p^3$

$\Rightarrow\alpha^3\{1+\alpha^3+3\alpha(1+\alpha)\}=-p^3$

$\Rightarrow q(1+q-3p)=-p^3$

$\Rightarrow p^3-q(3p-1)+q^2=0$

7. Since, c and d are the roots of

$(x-a)(x-b)-K=0$

$\Rightarrow(x-a)(x-b)-K=(x-c)(x-d)$

$\Rightarrow(x-c)(x-d)+K=(x-a)(x-b)$

Therefore, a, b are roots of $(x-c)(x-d)+K=0$

8. Let $\dfrac{1-\alpha}{1+\alpha}=y$ and $\alpha=\dfrac{1-y}{1+y}$

As α is a root of $3x^2+2x+1=0$

$\Rightarrow 3\alpha^2+2\alpha+1=0$

$\Rightarrow 3\left(\dfrac{1-y}{1+y}\right)^2+2\left(\dfrac{1-y}{1+y}\right)+1=0$

$\Rightarrow 3(1-y)^2+2(1-y^2)+(1+y)^2=0$

$\Rightarrow 3(1-2y+y^2)+2-2y^2+1+2y+y^2=0$

$\Rightarrow 2y^2-4y+6=0\Rightarrow x^2-2x+3=0$

9. $D=4b^2-4(a+b-c)(b+c-a)$

$=4\left[b^2-\{(b+(a-c))(b-(a-c))\}\right]$

$=4\left[b^2-\{b^2-(a-c)^2\}\right]$

$= 4(a-c)^2$ which is a perfect square

As a, b, c are rational $\Rightarrow$ Roots are rational.

10. Suppose the roots are rational, then $D = b^2 - 4ac$ must be a perfect square, as a, b, c are rational.

Let $b^2 - 4ac = \lambda^2$, as a, b, c are odd integers, therefore λ will be odd.

Let $b = 2m+1$, $\lambda = 2n+1$

$\therefore b^2 - \lambda^2 = (b+\lambda)(b-\lambda)$

$= 2(m+n+1).2(m-n)$

Since either $(m+n+1)$ or $(m-n)$ is always even, hence $b^2 - \lambda^2$ is a multiple of 8 but $4ac$ is only a multiple of 4 (not of 8) which is a contradiction. Hence roots cannot be rational.

11. $\alpha+\beta = -\dfrac{b}{a}, \alpha\beta = \dfrac{c}{a}$

$(\alpha+\delta)+(\beta+\delta) = -\dfrac{B}{A}, \ (\alpha+\delta)(\beta+\delta) = \dfrac{C}{A}$

Now, $\left[(\alpha+\delta)-(\beta+\delta)\right]^2$

$= \left[(\alpha+\delta)+(\beta+\delta)\right]^2 - 4(\alpha+\delta)(\beta+\delta)$

$\Rightarrow (\alpha+\beta)^2 - 4\alpha\beta = \dfrac{B^2}{A^2} - 4\dfrac{C}{A}$

$\Rightarrow \dfrac{b^2}{a^2} - \dfrac{4c}{a} = \dfrac{B^2}{A^2} - \dfrac{4C}{A}$

$\Rightarrow \dfrac{b^2-4ac}{a^2} = \dfrac{B^2-4AC}{A^2}$

12. Since, $\alpha^2 - 2 = 6\alpha \Rightarrow \alpha^{10} - 2\alpha^8 = 6\alpha^9$

Similarly, $\beta^{10} - 2\beta^8 = 6\beta^9$

$\Rightarrow (\alpha^{10}-\beta^{10}) - 2(\alpha^8-\beta^8) = 6(\alpha^9-\beta^9)$

$\Rightarrow \dfrac{a_{10}-2a_8}{2a_9} = 3$

13. $\alpha+\beta = \dfrac{-q}{p}, \ \alpha\beta = \dfrac{r}{p}$

$q = \dfrac{p+r}{2}$ and $\dfrac{\alpha+\beta}{\alpha\beta} = 4 \Rightarrow \dfrac{-q}{r} = 4$

$\Rightarrow q = -4r$

$\Rightarrow p = -9r$

Now, $|\alpha-\beta| = \sqrt{(\alpha+\beta)^2 - 4\alpha\beta}$

$= \sqrt{\dfrac{q^2}{p^2} - \dfrac{4r}{p}} = \sqrt{\dfrac{q^2-4pr}{p^2}}$

$= \sqrt{\dfrac{16r^2+36r^2}{81r^2}} = \dfrac{2\sqrt{13}}{9}$

14. (0)

15. $2^{|x|} - \left|2^{x-1}-1\right| = 2^{x-1}+1$

When $x < 0$

$2^{-x} + 2^{x-1} - 1 = 2^{x-1} + 1 \Rightarrow 2^{-x} = 2 \Rightarrow x = -1$

When $0 \le x < 1$

$2^x + 2^{x-1} - 1 = 2^{x-1} + 1$

$\Rightarrow 2^x = 2; \Rightarrow x = 1$

When $x \ge 1$

$\Rightarrow 2^x = 2^x$. Thus, $x \in [1, \infty) \cup \{-1\}$

16. Let $\alpha + i\beta$ be a root of

$\dfrac{A^2}{x-a} + \dfrac{B^2}{x-b} + \dfrac{C^2}{x-c} + \ldots + \dfrac{K^2}{x-K} = x + \ell$

$\Rightarrow \dfrac{A^2}{(\alpha-a)+i\beta} + \dfrac{B^2}{(\alpha-b)+i\beta} + \dfrac{C^2}{(\alpha-c)+i\beta} +$

$\ldots + \dfrac{K^2}{(\alpha-K)+i\beta} = (\alpha+\ell)+i\beta \quad (1)$

As imaginary roots occur in conjugate pairs

$\therefore \dfrac{A^2}{(\alpha-a)-i\beta} + \dfrac{B^2}{(\alpha-b)-i\beta} + \dfrac{C^2}{(\alpha-c)-i\beta} +$

$\ldots + \dfrac{K^2}{(\alpha-K)-i\beta} = (\alpha+\ell)-i\beta \quad (2)$

(1) – (2)

$-2i\beta\left[\dfrac{A^2}{(\alpha-a)^2+\beta^2} + \dfrac{B^2}{(\alpha-b)^2+\beta^2} + \right.$

$\left.\dfrac{C^2}{(\alpha-c)^2+\beta^2} + \ldots + \dfrac{K^2}{(\alpha-K)^2+\beta^2} + 1\right] = 0$

$\Rightarrow \beta = 0.$

Hence the given equation has no imaginary root.

17. The given equation is

$(x-a)(x-10)+1 = 0$

$\Rightarrow x^2 - (10+a)x + (10a+1) = 0$

$\Rightarrow (10+a)^2 - 4(10a+1)$ is a perfect square

$\Rightarrow a^2 - 20a + 96 = \lambda^2$ where $\lambda \in I$

$\Rightarrow (a-10)^2 - \lambda^2 = 4$

$\Rightarrow (a+\lambda-10)(a-\lambda-10) = 4$

When $\begin{array}{r} a+\lambda-10=-2 \\ \underline{a-\lambda-10=-2} \\ 2a-20=-4 \end{array}$

$\Rightarrow a = 8$

When $\begin{array}{r} a+\lambda-10=2 \\ \underline{a-\lambda-10=2} \\ 2a-20=4 \end{array}$

$\Rightarrow a = 12$. Hence, $a = 8, 12$

18. Let the third root be α

Now $ab\alpha = -r \Rightarrow \alpha = r$

As α is a root of $x^3 + px^2 + qx + r = 0$

$\Rightarrow r^3 + pr^2 + qr + r = 0$

$\Rightarrow r^2 + pr + q + 1 = 0$

19. Let α be a common positive root

$\therefore \alpha^2 + a\alpha + 12 = 0,\ \alpha^2 + b\alpha + 15 = 0$

and $\alpha^2 + (a+b)\alpha + 36 = 0$

Now, $\alpha^2 + \alpha^2 + (a+b)\alpha + 27 = 0$

$\Rightarrow \alpha^2 + \alpha^2 + (a+b)\alpha + 36 - 9 = 0$

$\Rightarrow \alpha = 3$

Therefore, $9 + 3a + 12 = 0 \Rightarrow a = -7$

$9 + 3b + 15 = 0 \Rightarrow b = -8$

Now, $x^2 - 7x + 12 = 0 \Rightarrow x = 3, 4$

$x^2 - 8x + 15 = 0 \Rightarrow x = 3, 5$

$x^2 - 15x + 36 = 0 \Rightarrow x = 3, 12$

20. $y^2 - 5y + 3 < \dfrac{2x}{x^2 + x + 1}$ (1)

Let, $P = \dfrac{2x}{x^2 + x + 1}$

$\Rightarrow Px^2 + (P-2)x + P = 0$

$\Rightarrow (P-2)^2 - 4P^2 \geq 0$

$\Rightarrow (P-2+2P)(P-2-2P) \geq 0$

$\Rightarrow (3P-2)(-P-2) \geq 0$

$\Rightarrow (3P-2)(P+2) \leq 0 \Rightarrow -2 \leq P \leq \dfrac{2}{3}$

$\therefore$ from (1), $y^2 - 5y + 3 < -2$

$\Rightarrow y^2 - 5y + 5 < 0 \Rightarrow y \in \left(\dfrac{5-\sqrt{5}}{2}, \dfrac{5+\sqrt{5}}{2}\right)$

1.9.2 JEE Main: Sample Questions

Single Answer Type

1. **(C)**

$x^2 + ax + b = 0 < \begin{matrix} \alpha \\ \beta \end{matrix}$

$\alpha + \beta = -a;\ \alpha\beta = b$ (1)

$g(x) = 0 < \begin{matrix} \alpha + \frac{1}{\alpha} \\ \beta + \frac{1}{\beta} \end{matrix}$

sum of roots = product of roots

$\alpha + \beta + \left\{1, \dfrac{2}{3}\right\} + \dfrac{1}{\beta} = \left(\alpha + \dfrac{1}{\alpha}\right)\left(\beta + \dfrac{1}{\beta}\right)$

$\Rightarrow \alpha + \beta + \dfrac{(\alpha+\beta)}{\alpha\beta} = a\beta + \dfrac{\alpha}{\beta} + \dfrac{\beta}{\alpha} + \dfrac{1}{\alpha\beta}$

Using equation (1)

$\Rightarrow -a - \left\{\dfrac{2}{3}, 6\right\} = b + \dfrac{a^2 - 2b}{b} + \dfrac{1}{b}$

$\Rightarrow a^2 + a(b+1) + (b^2 - 2b + 1) = 0$

$D \geq 0$ (Quadratic in a)

$(b+1)^2 - 4(b^2 - 2b + 1) \geq 0$

$\Rightarrow (3b-1)(b-3) \leq 0 \Rightarrow b \in \left[\dfrac{1}{3}, 3\right]$

2. **(B)**

$x^3 + x^2 - 4x - 4 \equiv (x+1)(x-2)(x+2)$

$\equiv (x^2 - x - 2)(x+2)$

$\Rightarrow p = -1$

3. **(B)**

$y = \dfrac{x^2 - 2x + 2}{2x - 2}$

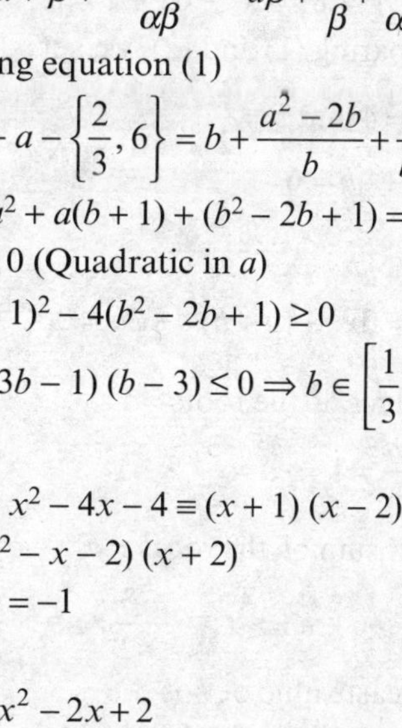

$\therefore x^2 - 2x + 2 = 2xy - 2y$

$\Rightarrow x^2 - 2x(y+1) + 2(y+1) = 0$

As $x \in R$, so $D \geq 0$

$\Rightarrow (y+1)^2 - 2(y+1) \geq 0$

$\Rightarrow (y+1)(y-1) \geq 0$

$\Rightarrow y \geq 1$ or $y \leq -1$

Hence, y cannot lie in $(-1, 1)$

4. (C)

$D = 0 \Rightarrow 16 \sin^2\theta = 4 \tan\theta$

$\Rightarrow 4 \sin^2\theta = \tan q \Rightarrow 2 \sin 2\theta = 1$

$\sin 2\theta = \frac{1}{2} \Rightarrow 2\theta = \frac{\pi}{6}$ or $\frac{5\pi}{6}$

$\Rightarrow \theta = \frac{\pi}{12}$ or $\frac{5\pi}{12}$

5. (C)

$\tan^2 \frac{3\pi}{8} = \left(\sqrt{2}+1\right)^2 = 3 + 2\sqrt{2}$

$\therefore x = 3 + 2\sqrt{2}$ satisfies the equation

$2x^2 - 3ax + 4b = 0$ (1)

$(x-3)^2 = 8 \Rightarrow x^2 - 6x + 1 = 0$ (2)

comparing (1) and (2), we get

$\therefore a = 4$ and $b = \frac{1}{2}$

$\Rightarrow a + 4b = 6$

6. (B)

$f(x) = 3x^2 - (a+b)x + 2a = 0 \begin{cases} e^{-\lambda} \\ e^{\lambda} \end{cases}$

Product of the roots = 1

$\Rightarrow \frac{2a}{3} = 1 \Rightarrow a = \frac{3}{2}$

Now, sum of the roots $= e^{-\lambda} + e^{\lambda} = \frac{a+b}{3} > 2$

$\Rightarrow b > 6 - a \Rightarrow b > 6 - \frac{3}{2} \Rightarrow b > \frac{9}{2}$

$\therefore$ Least value of b is 5.

7. (A)

Clearly $\frac{4R}{\sqrt{2}}$ and $\frac{c}{a} < 0$

8. (B)

$x^4 = 3x - 1$

$\therefore x^3 = 3 - \frac{1}{x} \begin{cases} \alpha \\ \beta \\ \gamma \\ \delta \end{cases}$

$\alpha^3 = 3 - \frac{1}{\alpha}; \beta^3 = 3 - \frac{1}{\beta}; \gamma^3 = 3 - \frac{1}{\gamma}; \delta^3 = 3 - \frac{1}{\delta}$

on adding $\sum \alpha^3 = 12 - \left(\frac{1}{\alpha} + \frac{1}{\beta} + \frac{1}{\gamma} + \frac{1}{\delta}\right)$

$= 12 - \frac{\sum \alpha\beta\gamma}{\alpha\beta\gamma\delta} = 12 - 3 = 9$

9. (D)

$x = \frac{2m \pm \sqrt{4m^2 - 4(m^2-1)}}{2}$

$= \frac{2m \pm 2}{2} = m + 1$ or $m - 1$

Hence $-2 < m + 1 < 4 \Rightarrow -3 < m < 3$

$-2 < m, -1 < 4 \Rightarrow -1 < m < 5$

Hence $m \in (-1, 3) \Rightarrow m = 0, 1, 2$

10. (B)

$6x^2 + 2ax + 2 = 0$ and $6x^2 + 3bx + 3 = 0$

subtracting $x(2a - 3b) - 1 = 0$

$\Rightarrow x = \frac{1}{2a - 3b}$ (put in any equation)

$\therefore 2\frac{1}{(2a-3b)^2} + \frac{b}{2a-3b} + 1 = 0$

$2 + b(2a - 3b) + (2a - 3b)^2 = 0$

$4a^2 + 9b^2 - 12ab + 2ab - 3b^2 + 2 = 0$

$-10ab + 6b^2 + 4a^2 + 1 = 0$

$\Rightarrow 5ab - 3b^2 - 2a^2 = 1 \Rightarrow$ (B)

11. (C)

$f(x) = x^2 - 2kx + k^2 + 2k - 5$

(i) $f(4) > 0$

$16 - 8k + k^2 + 2k - 5 > 0$

$k^2 - 6k + 11 > 0$

$k \in R$

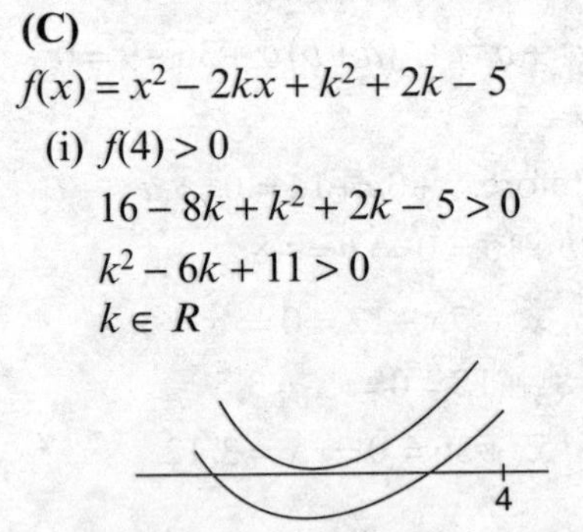

(ii) $D \geq 0$; $4k^2 - 4(k^2 + 2k - 5) \geq 0$

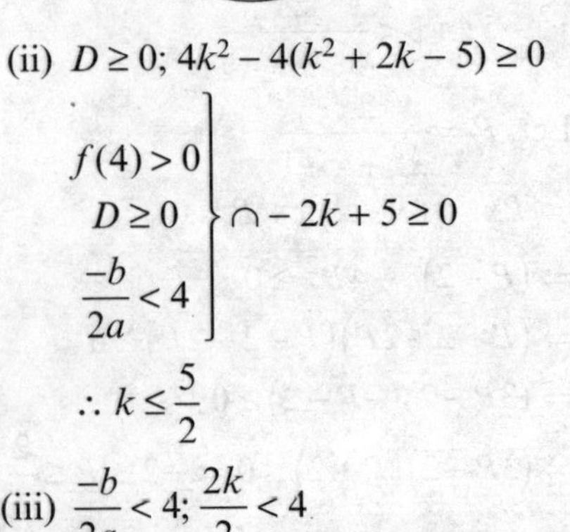

$\left.\begin{array}{l} f(4) > 0 \\ D \geq 0 \\ \frac{-b}{2a} < 4 \end{array}\right\} \cap -2k + 5 \geq 0$

$\therefore k \leq \frac{5}{2}$

(iii) $\frac{-b}{2a} < 4$; $\frac{2k}{2} < 4$

$\therefore k < 4$

Hence, from (i) ∩ (ii) ∩ (iii); $k \in \left(-\infty, \frac{5}{2}\right]$

12. **(D)**

$5x^2 + (13 - k)x - (3k + 6) = 0$

$$x = \frac{(k-13) \pm \sqrt{(13-k)^2 + 20(3k+6)}}{10}$$

$$x = \frac{(k-13) \pm \sqrt{k^2 - 26k + 169 + 60k + 120}}{10}$$

$$x = \frac{(k-13) \pm (k+17)}{10} \Rightarrow x = -3, \frac{2k+4}{10}$$

Since, $x \in (-2, 2)$

$\therefore -2 < \frac{2k+4}{10} < 2; -20 < 2k + 4 < 20$

$-24 < 2k < 16; -12 < k < 8.$

13. **(A)**

$f(0) < 0$ not possible.

14. **(A)**

Roots of $ax^2 + bx + c = 0$ and $cx^2 + bx + a = 0$ are reciprocals of each other.

Hence, $\alpha < \frac{1}{2}$ and $\beta > -1$.

15. **(C)**

$$f(x) = \frac{(x+3)(x-1)}{(2x-3)(x+3)}$$

$x \neq -3, \frac{3}{2}$

$f(x) = y = \frac{x-1}{2x-3}; 2xy - 3y = x - 1$

$$x = \frac{3y-1}{2y-1}$$

$\therefore y \neq \frac{1}{2}$

but since $x \neq -3$ hence $y \neq \frac{4}{9}$

$\therefore y \in R - \left\{\frac{4}{9}, \frac{1}{2}\right\}$

16. **(C)**

Vertex of $y = 2x^2 - 4x + 6$

$m = x = \frac{-b}{2a} = 1$

$n = y = \frac{-D}{4a} = 4$

$\therefore$ Now check the alternatives for parabola having vertex. $\left(2m, \frac{\pi}{2}\right)$, that is, (2, 2).

17. **(A)**

$$\alpha\beta\gamma + 1 = \frac{\beta\gamma}{2} = \frac{-\alpha\gamma}{10} = \frac{\alpha\beta}{18} = \lambda$$

$\frac{\beta}{\alpha} = \frac{-1}{5}; \frac{\gamma}{\alpha} = \frac{1}{9}$

$$\frac{\beta - \gamma}{\alpha - \gamma} = \frac{\frac{\beta}{\alpha} - \frac{\gamma}{\alpha}}{1 - \frac{\gamma}{\alpha}} = \frac{-7}{20}$$

18. **(A)**

The given equation may be written as

$x^2 - px - (p + c) = 0$

$\therefore \alpha + \beta = p$ and $\alpha\beta = -(p + c)$

Now $(\alpha + 1)(\beta + 1) = a\beta + \alpha + \beta + 1$

$= -p - c + p + 1 = 1 - c$

19. **(B)**

$\alpha^2 + \beta^2 = (\alpha + \beta)^2 - 2a\beta = (p - 1)^2 + 5$

20. **(B)**

$x \neq -4$, solve $x^2 + 6x - 7 < 0$

$\Rightarrow (x + 7)(x - 1) < 0$

$\Rightarrow x \in (-7, -4) \cup (-4, 1)$

$x \in \{-6, -5, -3, -2, -1, 0\}$

21. **(B)**

$x^3 - 2x^2 - 1 = 0$ or $a^3 - 1 = 2a^2$

$$\therefore \frac{T_{11} - T_8}{T_{10}} = \frac{\alpha^{11} + \beta^{11} + \gamma^{11} - [\alpha^8 + \beta^8 + \gamma^8]}{\alpha^{10} + \beta^{10} + \gamma^{10}}$$

$$= \frac{\alpha^8(\alpha^3 - 1) + \beta^8(\beta^3 - 1) + \gamma^8(\gamma^3 - 1)}{\alpha^{10} + \beta^{10} + \gamma^{10}}$$

$$= \frac{2(\alpha^{10} + \beta^{10} + \gamma^{10})}{\alpha^{10} + \beta^{10} + \gamma^{10}} = 2$$

22. **(A)**

$f(0) < 0$ not possible

23. **(B)**

$P(x) = 4x^2 + (k + 8)x + 9, x \in R$

$D < 0 \Rightarrow (k + 8)^2 - 144 < 0$

i.e., $(k + 20)(k - 4) < 0$

$-20 < k < 4 \Rightarrow |k|_{\text{largest}} = 19$

24. **(A)**

We have $16x^2 + 8(a + 5)x - (7a + 5) > 0$, $\forall x \in R$

$\therefore$ Discriminant must be negative.

Now, Discriminant $= D = 64\,((a+5)^2 + 7a + 5)$
As $D < 0$
$\Rightarrow a^2 + 17a + 30 < 0 \Rightarrow (a+15)\,(a+2) < 0$
$\therefore -15 < a < -2$
Hence number of integral value(s) = 12

25. **(A)**
$\alpha + \beta = \dfrac{-b}{a}$
$\therefore a\alpha + b = -a\beta$ and $a\beta + b = -a\alpha$
$\therefore S = \dfrac{-1}{a}\left(\dfrac{\alpha+\beta}{\alpha\beta}\right) = \dfrac{-1}{a}\left(\dfrac{-\frac{b}{a}}{\frac{c}{a}}\right) = \dfrac{b}{ac}$

26. **(C)**
$|x+4| = 4$
Case-I: If $x + 4 \geq 0$ then $x + 4 = 2$
$x = -2$
Case-II: If $x + 4 < 0$ then $-(x+4) = 2$
$x + 4 = -2 \Rightarrow x = -6$
sum of value(s) of x is -8

27. **(D)**
We have $x^2 + 2(1 - \cos 3\theta)\,x - 2\sin^2 3\theta = 0$ (roots α, β)
$\alpha + \beta = -2(1 - \cos 3\theta)$; $\alpha\beta = -2\sin^2 3\theta$
Now, $\alpha^2 + \beta^2 = (\alpha+\beta)^2 - 2\alpha\beta$
$= 4(1 - \cos 3\theta)^2 + 4\sin^2 3\theta$
$= 4(1 - 2\cos 3\theta + \cos^2 3\theta + \sin^2 3\theta)$
$\Rightarrow \alpha^2 + \beta^2 = 4\,(2 - 2\cos 3\theta)$
Clearly, maximum value of $\alpha^2 + \beta^2$ is 16 when $\cos 3\theta = -1$

28. **(A)**
Let $f(x) = x^2 - (2-p)x + (p-2)$
We must have $D > 0$

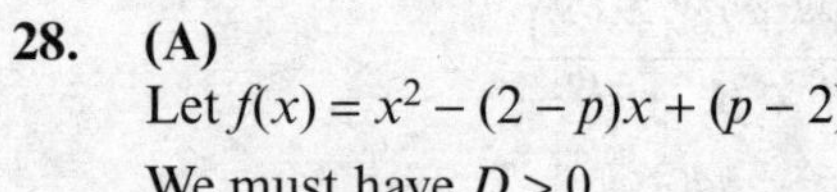
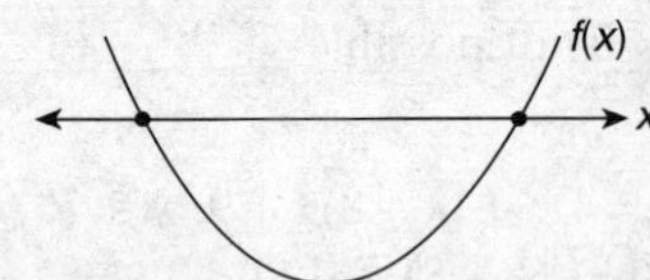

$\Rightarrow (p-2)\,(p-6) > 0$
$\therefore p \in (-\infty, 2) \cup (6, \infty)$

29. **(D)**
$111x^3 - 11x - 1 = 0$
$\therefore \alpha + \beta + \gamma = 0$
$\Sigma\alpha\beta = \dfrac{-11}{111}$
$\alpha\beta\gamma = \dfrac{1}{111}$
$\Rightarrow \dfrac{1}{(\alpha\beta)^2} + \dfrac{1}{(\beta\gamma)^2} + \dfrac{1}{(\gamma\alpha)^2}$
$= \dfrac{\alpha^2 + \beta^2 + \gamma^2}{(\alpha\beta\gamma)^2} = \dfrac{(\alpha+\beta+\gamma)^2 - 2\Sigma\alpha\beta}{(\alpha\beta\gamma)^2}$
$= 2442$

30. **(D)**
$\Delta = abc + 2fgh + af^2 - bg^2 - ch^2 = 0$

31. **(B)**
$D < 0$; $a \in (-\infty, 1) \cup (10, \infty)$

32. **(C)**
$k - 2 < 0$ and $D \leq 0$
$K \in (-\infty, 6)$

33. **(C)**
$f(1) < 0$ and $f(3) < 0$; $a \in \left(-\infty, \dfrac{-4}{3}\right)$

34. **(D)**
$\dfrac{(2x-5)}{x(x+1)} \geq 0$

35. **(C)**
$D > 0$; $k \in (-\infty, 1) \cup (9, \infty)$

36. **(C)**
$a(x - \ell)\,(x - m) + n = 0$
If roots are α and β
then $a(x-\ell)(x-m) + n = a(x-\alpha)\,(x-\beta)$
$\Rightarrow a(x-\ell)\,(x-m) = a(x-\alpha)\,(x-\beta) - n$
or we can say roots of $a(x-\alpha)\,(x-\beta) - n = 0$ are ℓ, m
Now by transformation for
$a\left(\dfrac{x}{x-k} - \alpha\right)\left(\dfrac{x}{x-k} - \beta\right) - n = 0$, we have
$\dfrac{x}{x-k} = \ell, m \Rightarrow x = \dfrac{k\ell}{\ell-1}, \dfrac{km}{m-1}$

37. **(A)**

$x^2 - 8x + a = 0$ with roots α and 3β and

$x^2 - bx + 16 = 0$ with roots α and 4β

$\alpha + 3\beta = 8$ and $3\alpha\beta = a\alpha + 4\beta = b$
and $4\alpha\beta = 16$

$\Rightarrow \alpha\beta = 4$

$\therefore \alpha = 12$

Hence, $x^2 - 8x + 12 = 0$

$x = 2$ or $x = 6$

If common root is 2 then $\beta = 2$ for which we get 3β and 4β both integers but if common root is 6 then $\beta = \frac{2}{3}$ for which 4β for which 4β is not integer. Hence, common roots is 2.

38. **(C)**

Put $D > 0 \Rightarrow k > 2 \Rightarrow k = 3$

39. **(B)**

If exactly one root of $ax^2 + bx + c = 0$ is infinite then $a = 0$ but $b \neq 0$

$\therefore k = -1$ only

40. **(A)**

Let $y = \dfrac{3x^2 + 9x + 17}{3x^2 + 9x + 7}$

$\Rightarrow 3x^2 (y - 1) + 9x(y - 1) + (7y - 17) = 0$

As x is real, so $D \geq 0$

$\Rightarrow 81(y - 1)^2 - 12(y - 1)(7y - 17) \geq 0$

$\Rightarrow (y - 1)(y - 41) \leq 0 \Rightarrow 1 \leq y \leq 41$

But $y \neq 1$, so $1 < y \leq 41 \Rightarrow y_{max} = 41$

Aliter: $y = \dfrac{3x^2 + 9x + 17}{3x^2 + 9x + 7}$

$= \dfrac{(3x^2 + 9x + 7) + 10}{3x^2 + 9x + 7} = 1 + \dfrac{10}{3x^2 + 9x + 7}$

$y_{max} = 1 + \dfrac{10}{(3x^2 + 9x + 7)_{min.}}$

$y_{max} = 1 + \dfrac{10}{\frac{3}{12}\left(-\frac{d}{4a}\right)}$

$y_{max} = 41$

41. **(B)**

$(2^x - 1)(2^x - 7) = 7 \Rightarrow x = 3$ is the only solutions

42. **(D)**

$f(x) = (a - 3)x^2 - 2ax + 3a - 7$

Since range of $f(x) \in [-1, \infty)\ \forall\ x \in R$

Hence range of $(a - 3)x^2 - 2ax + 3a - 6 \in [0, \infty)$

$\therefore a - 3 > 0$ and $D = 0$

On solving we get $a = 6$ only

43. **(A)**

$\because$ Roots of quadratic $x^2 - 3ax + a^2 = 9$ are opposite sign

$\therefore a^2 - 9 < 0 \Rightarrow -3 < a < 3$

$\therefore$ Sum of all possible values of $a = 0$

44. **(C)**

$x^2 - 14x + 45 > 0$; $(x - 9)(x - 5) > 0$

$\Rightarrow x > 9$ or $x < 5$ and $4 - x > 0$

$\Rightarrow 4 > x \Rightarrow x < 4$

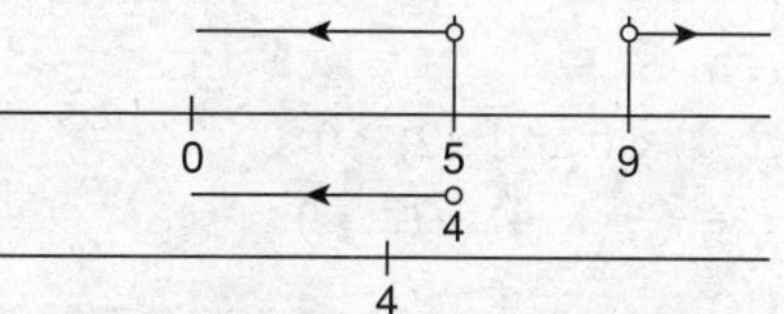

$\therefore 0 < x < 4 \Rightarrow x \in \{1, 2, 3\}$ but $4 - x \neq 1$

$\Rightarrow x \neq 3$; $x \in \{1, 2\}$

$\therefore$ There are two natural numbers.

45. **(D)**

$a > 0$; $f(0) = c > 0$; $\dfrac{-b}{a} > 0$

$\therefore -b > 0 \Rightarrow b < 0$

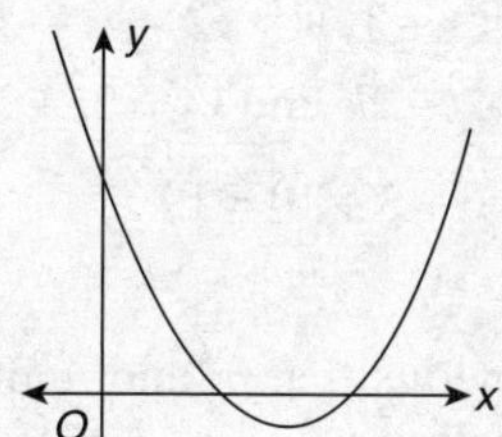

Hence, $abc < 0$ and $D = b^2 - 4ac > 0$.

46. **(B)**

We have $(x + 1)^2 > 5x - 1 \Rightarrow x^2 - 3x + 2 > 0$

$(x - 1)(x - 2) > 0 \Rightarrow x < 1$ or $x > 2$ (1)

And $(x + 1)^2 < 7x - 3 \Rightarrow x^2 - 5x + 4 < 0$

$(x - 1)(x - 4) < 0 \Rightarrow 1 < x < 4$ (2)

$\therefore (1) \cap (2) \Rightarrow x \in (2, 4)$

$\therefore$ Only one integral value of x which is 3.

47. (A)

Let one root be α

$\therefore \alpha = 2\cos^2 18° = 1 + \cos 36°$

$= 1 + \frac{(\sqrt{5}+1)}{4} = \frac{5+\sqrt{5}}{4}$

$\therefore$ other root $= \beta = \frac{5-\sqrt{5}}{4}$

$\therefore \alpha + \beta = \frac{5}{2}$ and $\alpha\beta = \frac{20}{16} = \frac{5}{4}$

$\therefore$ Quadratic equation is $4x^2 - 10x + 5 = 0$

48. (A)

$\sqrt{x^2+9} = 9 \Rightarrow x^2 + 9 = 81$

$\Rightarrow x^2 = 72 \Rightarrow x = \pm 6\sqrt{2}$

49. (C)

$3 \le \log_2(x^2 + 3x - 2) \le 4$

50. (C)

$f(-1) < 0$ and $f(2) < 0$

$K \in (-\infty, -4) \cup (2, \infty)$

51. (B)

The equation whose roots are u, v, w is

$x^3 - (u+v+w)x^2 + (uv+vw+wu)x - uvw = 0$

or $x^3 - 2x^2 - 5x + 6 = 0$

by trial method, $x = 1$ satisfies it.

$\therefore x^3 - 2x^2 - 5x + 6 = (x-1)(x^2 - x - 6)$

Now $x^2 - x - 6 = 0$

$\Rightarrow (x-3)(x+2) = 0$

$\therefore x = -2, 3$

$\therefore u = 3, v = 1, w = -2$ (As $u > v > w$)

$\vec{V_1} = 3\hat{i} + \hat{j} - 2\hat{k}$ and $\vec{V_2} = 2\hat{i} + 3\hat{j} - 5\hat{k}$

$\vec{V_1}\cdot\vec{V_2} = 6 + 3 + 10 = 19$

52. (C)

It is obvious that m and n are the roots of the equation $x^2 - 4x - 4 = 0$

$\Rightarrow m + n = 4$ and $mn = -4$

Now $m^2 + n^2 + 4mn = (m+n)^2 + 2mn$

$= (4)^2 + 2(-4) = 16 - 8 = 8$

53. (D)

We have, $(5-\alpha)x^2 - 4x + (5-\alpha) \le 0 \;\forall\; x \in R$

$\therefore 5 - \alpha < 0$ and Disc. ≤ 0

$\Rightarrow \alpha > 5$ and $16 \le 4(\alpha - 5)^2$

$\Rightarrow \alpha > 5$ and $(5-\alpha)^2 - (2)^2 \ge 0$

$\Rightarrow \alpha > 5$ and $(\alpha - 7)(\alpha - 3) \ge 0$

So, $\alpha \in [7, \infty)$

54. (C)

$x^2 + ax + b = 0$

$a + b = -a \Rightarrow 2a + b = 0$

and $ab = b \Rightarrow b = 0$ (rejected)

$\therefore a = 1; b = -2 \Rightarrow (1, -2)$

55. (B)

Let $f(x) = ax^2 + bx + c$ and $f(1) = a + b + c > 0$ and roots are imaginary.

It implies that graph (parabola) of $f(x)$ will be completely above x-axis.

So $f(-1) > 0$, $f(2) > 0$ and $f\left(\frac{-1}{2}\right) > 0$

$\Rightarrow a - b + c > 0$, $4a + 2b + c > 0$,

$a - 2b + 4c > 0$

56. (D)

Given,

$x^2 + 3(1-3p)x + 2 = 0$

$\alpha \quad 2\alpha$

$\therefore 3\alpha = 3(3p-1) \Rightarrow \alpha = (3p-1)$

Also, $2\alpha^2 = 2 \Rightarrow \alpha^2 = 1 \Rightarrow \alpha = \pm 1$

So, for $\alpha = -1$, $p = 0$ and for $\alpha = 1$, $p = \frac{2}{3}$

$\Rightarrow$ sum of all possible values of

$p = 0 + \frac{2}{3} = \frac{2}{3}$

57. (D)

Let x_1, x_2, x_3 are the roots of

$x^3 - 2x^2 + ax + 10 = 0$

$\therefore \underbrace{x_1 + x_2}_{\substack{\text{add. inverse} \\ \Rightarrow 0}} + x_3 = 2 \Rightarrow x_3 = 2$

$\therefore 8 - 8 + 2a + 10 = 0 \Rightarrow a = -5$

58. (A)

$D \ge 0$ (1)

$f(3) > 0$ (2)

$\frac{-b}{2a} < 3$ (3)

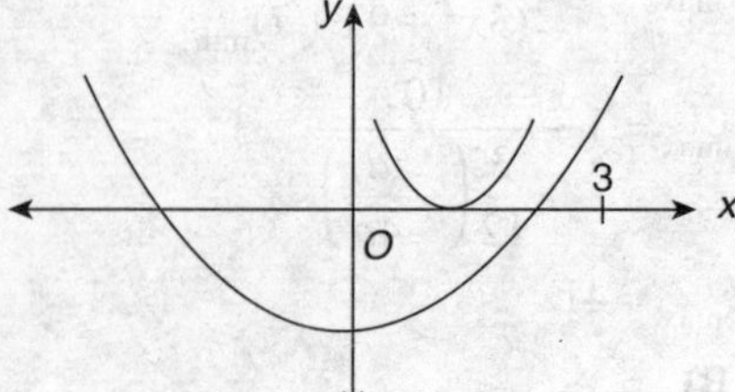

$(1) \Rightarrow 4a^2 - 4(a^2 + a - 3) \ge 0$

$\Rightarrow 12 - 4a \ge 0 \Rightarrow a \le 3$ (4)

$(2) \Rightarrow 9 - 6a + a^2 + a - 3 > 0$

$\Rightarrow a^2 - 5a + 6 > 0$

$\Rightarrow (a-3)(a-2) > 0$

$\Rightarrow a < 2$ or $a > 3$ (5)

$(3) \Rightarrow \dfrac{2a}{2} < 3 \Rightarrow a < 3$ (6)

Hence, $a < 2$

Integer Answer Type

1. (6)

$x^4 + ax^3 + 23x^2 + bx + c = 0$; roots $2+i$, $2-i$, $1-3i$, $1+3i$

$\therefore x^4 + ax^3 + 23x^2 + bx + c$

$\equiv (x^2 - 4x + 5)(x^2 - 2x + 10)$ (1)

Put $x = 1$ on both sides of above equation (1), we get.

$a + b + c + 24 = (2)\cdot(9)$

$\Rightarrow a + b + c = -6$

$\Rightarrow |a + b + c| = 18 - 24 = 6$

2. (4)

Given, $x + y - 1 = 2a$ (1)

$2xy = a^2 - a$ (2)

$\therefore$ From equation (1), put $y = (1 + 2a - x)$ in equation (2), we get

$2x(1 + 2a - x) = a^2 - a$

$\Rightarrow 2x^2 - 4ax - 2x + a^2 - a = 0$

Let $f(x) = 2x^2 - 4ax - 2x + a^2 - a$

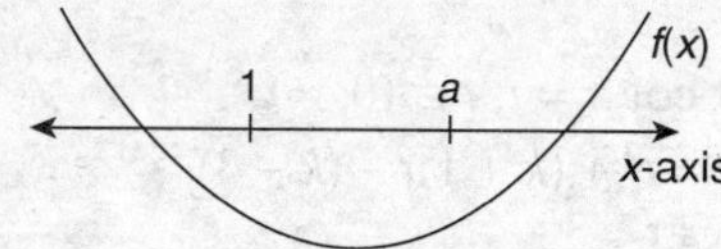

Clearly, $f(1) < 0$ and $f(a) < 0$ must be satisfied simultaneously.

Now $f(1) < 0 \Rightarrow 0 < a < 5$ (1)

and $f(a) < 0 \Rightarrow a < -3$ or $a > 0$ (2)

$\therefore (3) \cap (4) \Rightarrow a \in (0, 5)$

Hence, number of integral values of a are four, that is, $a = 1, 2, 3, 4$.

3. (3)

We have $5x^3 - qx - 1 = 0$; roots α, β, γ

Now, $\dfrac{\alpha^2 - 3}{\beta\gamma} + \dfrac{\beta^2 - 3}{\gamma\alpha} + \dfrac{\gamma^2 - 3}{\alpha\beta}$

$= \dfrac{\alpha^3 - 3\alpha + \beta^3 - 3\beta + \gamma^3 - 3\gamma}{\alpha\beta\gamma}$

$= \dfrac{\alpha^3 + \beta^3 + \gamma^3 - 3(\alpha + \beta + \gamma)}{\alpha\beta\gamma}$

$\because \alpha + \beta + \gamma = 0$

$\Rightarrow \alpha^3 + \beta^3 + \gamma^3 = 3\alpha\beta\gamma = \dfrac{3\alpha\beta\gamma}{\alpha\beta\gamma} = 3$

Aliter: $\alpha\beta\gamma = \dfrac{1}{5}$ and $\alpha + \beta + \gamma = 0$

$\therefore \left(\dfrac{\alpha^2 - 3}{\beta\gamma} + \dfrac{\beta^2 - 3}{\gamma\alpha} + \dfrac{\gamma^2 - 3}{\alpha\beta}\right)$

$= 5\alpha(\alpha^2 - 3) + 5\beta(\beta^2 - 3) + 5\gamma(\gamma^2 - 3)$

$= 5(\alpha^3 + \beta^3 + \gamma^3) - 15(\alpha + \beta + \gamma)$

$= 5(3\alpha\beta\gamma) = 15 \times \dfrac{1}{5} = 0003$ [As $\alpha + \beta + \gamma = 0$]

4. (1)

$1 \le m, n \le 10$

(i) $D \ge 0$

(ii) $\dfrac{-b}{2a} > 2$

(iii) $f(2) > 0$

$\cap$ 2

(i) $D \ge 0 \Rightarrow n^2 - 48m \ge 0$ (1)

(ii) $\dfrac{-b}{2a} > 2 \Rightarrow n > 4m$ (2)

(iii) $f(2) > 0 \Rightarrow 2m - n + 6 > 0$ (3)

$\therefore$ From $(1) \cap (2) \cap (3)$

Since m, n are positive integers which are less than or equal to 10

Hence, $m = 1$ and $n = 7$ only are possible.

5. (0)

Given $f(x) > g(x)\ \forall\ x \in R$

$\Rightarrow px^2 + px - 2 > x^2 + (2p + 1)x + p + 2\ \forall\ x \in R$

$\Rightarrow (p - 1)x^2 - (p + 1)x - p - 4 > 0\ \forall\ x \in R$

So, $p - 1 > 0 \Rightarrow p > 1$

and, $D < 0 \Rightarrow (p + 1)^2 + 4(p - 1)(p + 4) < 0$

$\Rightarrow p^2 + 2p + 1 + 4(p^2 + 3p - 4) < 0$

$\Rightarrow 5p^2 + 14p - 15 < 0$

$\Rightarrow \left(p - \frac{-14 - \sqrt{196+300}}{10}\right)$

$\left(p - \frac{-14 + \sqrt{196+300}}{10}\right) < 0$

$\Rightarrow \quad (p - (-3.6))(p - 0.8) < 0$

$p \in (-3.6, 0.8)$ but $p > 1$

Hence, there is no value of p for which $f(x) > g(x)\ \forall\ x \in R$.

6. **(8)**

$A = \frac{3}{x^2 - x + 1} - 1 > 0$

$\Rightarrow \frac{3 - x^2 + x - 1}{x^2 - x + 1} > 0 \Rightarrow x^2 - x - 2 < 0$

$\Rightarrow (x - 2)(x + 1) < 0 \Rightarrow x \in (-1, 2)$

Number of integers $\{0, 1\} \Rightarrow A = 2$

$B =$ coefficient of $x^2 = (k - 3)(2k + 1) = 0$

Coefficient of $x(k - 3)(k - 2) = 0$

Absolute term $= (k - 3)(k + 3) = 0$

$B = 3$

$X = 2x^4 - 30x^2 + 8x + 10$

given $x = 2 + \sqrt{3} \Rightarrow (x - 2)^2 = 3$

$\Rightarrow x^2 - 4x + 1 = 0$

$\underbrace{2x^2(x^2 - 4x + 1)}_{\text{zero}} + 8x^3 - 32x^2 + 8x + 10$

$\Rightarrow \underbrace{8x(x^2 - 4x + 1)}_{\text{zero}} + 10 = 10 \Rightarrow X = 10$

$(X \div A + B) = (10 \div 2 + 3) = 5 + 3 = 8$

7. **(1)**

Let $\alpha = n^2, \beta = (n + 1)^2$

$\sqrt{\alpha\beta} = |\alpha - \beta| + 1$

Hence, $n = 2$

$\therefore \alpha = 4, \beta = 9$

Now, $f(4) \cdot f(9) < 0$ and also checking boundary points.

We get $k \in \left[12, \frac{113}{9}\right)$

8. **(9)**

$f(x) = x^2 + (2a - 1)x + (a + 3)$

$f(0) = (a + 3)$

$f(2) = 5(a + 1)$

$f(0) \cdot f(2) < 0 \Rightarrow a \in (-3, -1)$

Now, check when $f(0) = 0$ or $f(2) = 0$

$f(0) = 0 \Rightarrow a = -3$

when $a = -3$, the quadratic equation becomes

$x^2 - 7x = 0 \Rightarrow x = 0$ and $x = 7$

Hence, $a = -3$ is acceptable

when $f(2) = 0 \Rightarrow a = -1$

the quadratic equation becomes

$x^2 - 3x + 2 = 0$

$\Rightarrow x = 1$ or $x = 2$

Hence, $a = -1$ is not acceptable

$\therefore a \in [-3, -1]$

9. **0**

Let $y = \log_{10}(4x^2 - ax + 3)$

Since y is non-negative for all $x \in [1, 3]$

$\therefore y \geq 0$ for all $x \in [1, 3]$

$\Rightarrow (4x^2 - ax + 3) \geq 1$ for all $x \in [1, 3]$

$4x^2 - ax + 2 \geq 0$ for all $x \in [1, 3]$

Case I: When $\frac{a}{8} \leq 1$

This case is not possible

Case II: $1 < \frac{a}{8} < 3 \Rightarrow a \in (8, 24)$

now, $f\left(\frac{a}{8}\right) \geq 0 \Rightarrow a \in \left[-4\sqrt{2}, 4\sqrt{2}\right]$

This is not possible.

Case III: $\frac{a}{8} \geq 3 \Rightarrow a \geq 24$

$f(3) \geq 0$

$a \leq \frac{38}{3}$, which is not possible

Hence, from the above cases $a \in \phi$.

10. **(4)**

Let $\cot x = t;\ t \in (0, \infty)$

$f(t) = t^2 + (k + 1)t - (k - 3)$

Case I:

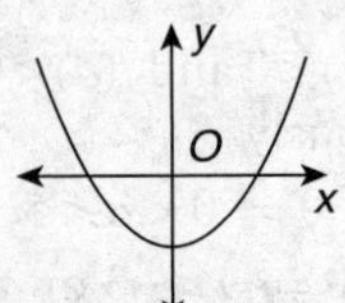

$f(0) < 0 \Rightarrow k > 3$

Case II:

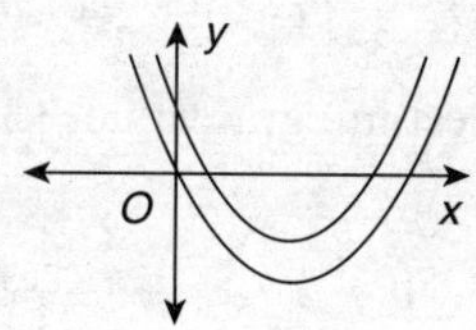

$$\left.\begin{array}{l} D>0 \\ \dfrac{-b}{2a}>0 \\ f(0)\geq 0 \end{array}\right\} \cap k\in\left(-\infty, -3-2\sqrt{5}\right)$$

From case I and case II

$\therefore\ k\in\left(-\infty, 3-2\sqrt{5}\right)\cup(3, \infty)$

$\therefore\ k_{\text{least positive integer}} = 4$

11. **(2)**

$P(x) - P(x + 1) = 2x - 1\ \forall\ x\in R$

$P(x) = ax^2 + bx + c$

$ax^2+bx+c-[a(x+1)^2+b(x+1)+c]=2x-1$

$-[ax^2 + 2ax + a + bx + b]$

$-2ax - (a + b) = 2x - 1$

$-2a = 2 \qquad a + b = 1$

$a = -1 \qquad b = 2$

$P(x) = -x^2 + 2x + c$

$\because$ Greatest value of $P(x)$ in $(-\infty, 0]$ is 2

$\therefore\ P(0) = c = 2;\ P(x) = -x^2 + 2x + 2$

$P(1) = -1 + 2 + 2 = 3;\ P(-1) = -1 - 2 + 2 = -1$

$\therefore\ P(1) + P(-1) = 2$

12. **(0)**

$x^2 - (2y + k)x + 2y^2 - 2ky - 2 \geq 0\ \forall\ x\in R$

$D\leq 0 \Rightarrow (2y + k)^2 - 4(2y^2 - 2ky - 2) \leq 0$

$\Rightarrow 4y^2 + k^2 + 4ky - 8y^2 + 8ky + 8 \leq 0$

$\Rightarrow -4y^2 + 12ky + k^2 + 8 \leq 0$

$\Rightarrow 4y^2 - 12ky - (k^2 + 8) \geq 0\ \forall\ y\in R$

$D\leq 0 \Rightarrow 144k^2 + 4.4.(k^2 + 8) \leq 0$

$\therefore\ k\in\phi$

13. **(0005)**

$$E = \log_2 5 - \sum_{k=1}^{4}\log_2\left(\sin\frac{k\pi}{5}\right)$$

$$E = \log_2 5 - \left(\log_2\sin\frac{\pi}{5} + \log_2\sin\frac{2\pi}{5} + \log_2\sin\frac{3\pi}{5} + \log\frac{4\pi}{5}\right)$$

$$E = \log_2 5 - \left(\log_2\left(\sin^2\frac{\pi}{5}\sin^2\frac{2\pi}{5}\right)\right)$$

$$E = \log_2 5 - \log_2\left(\frac{1}{4}\left(2\sin^2\frac{\pi}{5}\cdot 2\sin^2\frac{2\pi}{5}\right)\right)$$

$$E = \log_2 5 - \log_2\left(\frac{1}{4}(1-\cos 72°)(1-\cos 144°)\right)$$

$$E = \log_2 5 - \log_2\left(\frac{1}{4}\left(1-\frac{\sqrt{5}-1}{4}\right)\left(1+\frac{\sqrt{5}+1}{4}\right)\right)$$

$$E = \log_2 5 - \log_2\left(\frac{1}{4}\left(\frac{20}{16}\right)\right)$$

$$E = \log_2 5 - \log_2\left(\frac{5}{16}\right) = \log_2 5 - \log_2 5 + 4$$

$$\therefore\ E = \frac{4}{1} = \frac{p}{q}$$

14. **(8)**

$$\left(x^2+\frac{1}{x^2}\right) + 2k\left(x+\frac{1}{x}\right) + 1 = 0$$

$$\left(x+\frac{1}{x}\right)^2 + 2k\left(x+\frac{1}{x}\right) - 1 = 0$$

Put $\left(x+\dfrac{1}{x}\right) = t$

$t^2 + 2kt - 1 = 0$

Now, one root of $t^2 + 2kt - 1 = 0$ will lie in $(-\infty, -2)$ and other root in $(2, \infty)$

$$\left.\begin{array}{l} f(-2)<0 \\ f(2)<0 \end{array}\right\}\cap$$

Hence, $k\in\left(-\infty, \dfrac{-3}{4}\right)\cap\left(\dfrac{3}{4}, \infty\right)$

15. **(0)**

$f(x) = 3\sin x + 2\cos x - 4(\sin^3 x + \cos^3 x - (1 - \cos^2 x)\cos x)$

$f(x) = (3\sin x - 4\sin^3 x) - 2(4\cos^3 x - 3\cos x)$

$f(x) = \sin 3x - 2\cos 3x$

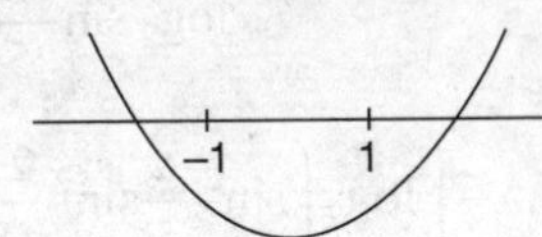

$\therefore f(x) \in \left[-\sqrt{5}, \sqrt{5}\right]$

Now, roots of $g(x) = x^2 + (2k - 1)x + k^2 - k$ does not lie in $[-1, 1]$

Hence, $g(-1) < 0$ and $g(1) < 0$

$\therefore g(-1) = 1 - 2k + 1 + k^2 - k < 0$

$\Rightarrow k^2 - 3k + 2 < 0 \Rightarrow k \in (1, 2)$ (1)

$\Rightarrow g(1) = 1 + (2k - 1) + k^2 - k < 0$

$\therefore k^2 + k < 0$

$\Rightarrow k(k + 1) < 0 \Rightarrow k \in (-1, 0)$ (2)

$\therefore$ From (1) $\cap$ (2), we get $k \in \phi$

16. **(4)**

$x^2 + (p + 1)x + 2q - q^2 + 3 = 0$

$\therefore D > 0$

$\Rightarrow (p + 1)^2 - 4(2q - q^2 + 3) > 0$

$\Rightarrow p^2 + 2p + 1 - 8q + 4q^2 - 12 = 0$

$\Rightarrow p^2 + 2q + (4q^2 - 8q - 11) > 0 \;\forall\; p \in R$

$\Rightarrow 4 - 4(4q^2 - 8q - 11) < 0$

$\Rightarrow 1 - 4q^2 + 8q + 12 < 0$

$\Rightarrow 4q^2 - 8q - 12 > 0 \Rightarrow q^2 - 2q - 3 > 0$

$\Rightarrow q^2 - 2q > 3$

$\therefore$ Minimum integral value of $(q^2 - 2q)$ is 4.

17. **(3)**

$a(x^2 - x + 1) - (x^2 + x + 1) \geq 0$

$$\Rightarrow a \geq \frac{x^2 + x + 1}{x^2 - x + 1};\; a \geq \frac{(x^2 - x + 1) + 2x}{x^2 - x + 1}$$

$$\Rightarrow a \geq 1 + \frac{2x}{x^2 - x + 1};\; a \geq 1 + \frac{2}{\left(x + \frac{1}{x}\right) - 1}$$

$$a \geq 1 + \left.\frac{2}{\left(x + \frac{1}{x}\right) - 1}\right]_{\max}$$

Now for $x \geq 2$

$$\therefore \left. x + \frac{1}{x}\right]_{\min} = \frac{5}{2}$$

$$\Rightarrow a \geq 1 + \frac{2 \cdot 2}{3};\; a \geq 1 + \frac{4}{3} \text{ or } a \geq \frac{7}{3}$$

$\therefore$ Smallest integral value = 3.

18. **(17)**

We have

$x^2 - 2x - (a^2 - 1) = 0$ with roots α, β

$$\therefore \alpha, \beta = \frac{2 \pm \sqrt{4 + 4(a^2 - 1)}}{2} = 1 \pm a$$

Let $f(x) = x^2 - 2(a + 1)x + a(a - 1)$

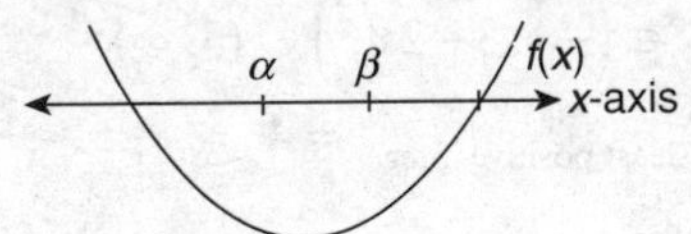

Now, $f(\alpha) < 0$ and $f(\beta) < 0$ must hold simultaneously.

So, $f(\alpha) < 0 \Rightarrow a > \frac{-1}{3}$ (1)

and $f(\beta) < 0 \Rightarrow \frac{-1}{4} < a < 1$ (2)

$\therefore$ From (1) and (2), we get

$$a \in \left(-\frac{1}{4}, 1\right) \Rightarrow p = \frac{-1}{4} \text{ and } q = 1$$

$$\Rightarrow \left(q + \frac{1}{p^2}\right) = 1 + 16 = 17$$

19. **(1)**

Let V_1 and V_2 be the vertices of the parabola $f(x)$ and $g(x)$ respectively.

[Hence max. value of $f(x) = a$ and min. value of $g(x) = b - 2a - a^2$]

We have $V_1 \equiv (1, a)$ and

$V_2 \equiv (a, -a^2 + 10 - 2a)$

Now, $a < -a^2 + 10 - 2a$

$\Rightarrow a^2 + 3a - 10 < 0$

$\therefore (a + 5)(a - 2) < 0$

$\Rightarrow -5 < a < 2$

$\therefore$ Largest natural number $a = 1$

1.9.3 JEE Advanced: Sample Questions

Single Answer Type

1. **(D)**

Let $Q(x) = x^2 - 2mx + 6m - 41$

As, $Q(x) < 0$ for all values of $x \in (1, 6]$. So

$\left.\begin{array}{l}(1)\quad Q(1) \le 0 \\ (2)\quad Q(6) < 0\end{array}\right\}$ must be satisfied simultaneously.

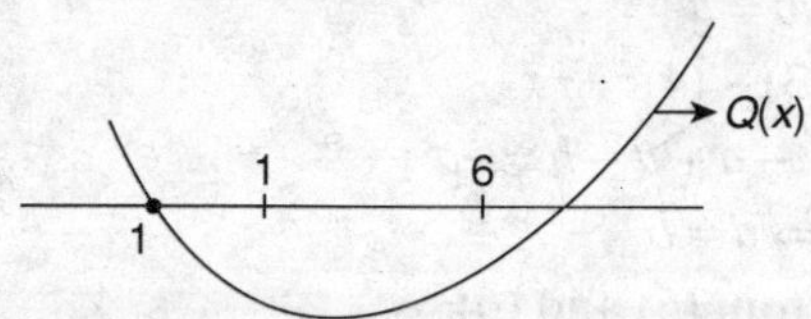

Now, $Q(1) \le 0 \Rightarrow 1 - 2m + 6m - 41 \le 0$

$\Rightarrow m \le 10$ (1)

and $Q(6) < 0 \Rightarrow 36 - 12m + 6m - 41 < 0$

$\Rightarrow m > \dfrac{-5}{6}$ (2)

Now, $(1) \cap (2) \Rightarrow m \in \left(\dfrac{-5}{6}, 10\right]$

2. (D)

As, $x^3 - x - 6 = 0$

$\Rightarrow (x-2)\underbrace{(x^3 + 2x + 3)}_{\text{Non-real roots}} = 0$ (1)

and $x^2 - bx - c = 0$ $(b, c \in R)$ (2)

As, exactly one root is common, so common roots can't be non-real, otherwise two roots will be common.

So, common root = 2

Hence, $(2)^2 - b(2) - c = 0 \Rightarrow (2b + c) = 4$.

3. (A)

$f(x) = x^2 + ax + b$

Vertex of $f(x)$ is at $\dfrac{-a}{2}$

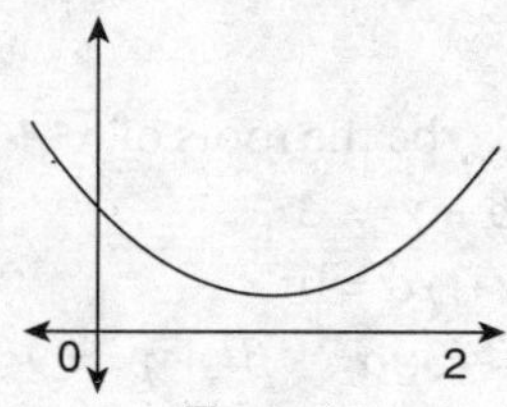

Figure-1

Case I: If $0 \le -\dfrac{a}{2} \le 2$

$\Rightarrow 0 \le -a \le 4; -4 \le a \le 0$

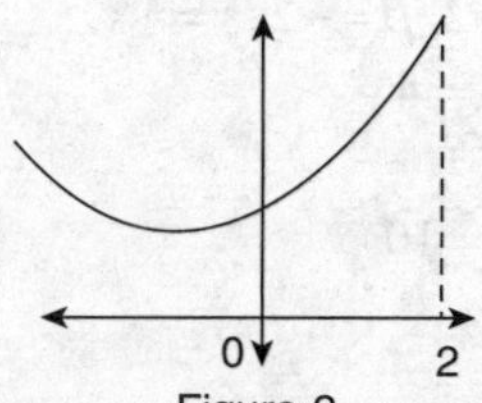

Figure-2

$f(x)]_{\min} = f\left(-\dfrac{a}{2}\right) = \dfrac{a^2}{4} - \dfrac{a^2}{2} + b = b - \dfrac{a^2}{4} = 2$

(given) (1)

and $f(x)]_{\max} = f(0)$ or $f(2)$

But $f(0) = b$ and $f(2) = 4 + 2a + b$

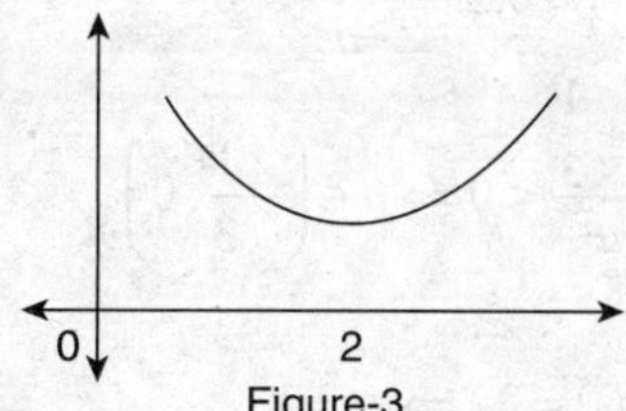

Figure-3

Let $b = 3$

From (1) $3 - \dfrac{a^2}{4} = 2; \dfrac{a^2}{4} = 1 \Rightarrow a = 2$ or -2

But $a = 2$ is rejected hence $a = -2$

$\therefore (-2, 3)$

If $f(2)$ is maximum

$4 + 2a + b = 3 \Rightarrow 2a + b = -1$ (2)

(2) – (1) gives, $2a + \dfrac{a^2}{4} = -3$

$\Rightarrow a^2 + 8a + 12 = 0$

$\Rightarrow (a + 6)(a + 2) = 0$

$\therefore a = -2$ or $a = -6$ (rejected)

If vertex is < 0

$\Rightarrow f(0)$ is minimum and $f(2)$ is maximum (Figure 3)

if vertex is > 0

$\Rightarrow f(0)$ is maximum and $f(2)$ is minimum (Figure 3)

In both these cases no accepted order pairs.

Hence only (A) is correct.

4. (C)

$\alpha + \beta = -a$ and $\alpha\beta = b$

$\therefore 3\alpha - 5 - \alpha^2 = \beta^2 - 3\beta$

$\Rightarrow \beta^2 + \alpha^2 = 3(\alpha + \beta) - 5$

$= (\alpha + \beta)^2 = 3(\alpha + \beta) + 2\alpha\beta - 5$

$a^2 = -3a + 2b - 5$

$\Rightarrow |a^2 + 3a - 2b| = 5$

5. (C)

$\because a \ne 0$

$f(x) = x^2 + x + \dfrac{a+1}{a} = 0$

$\Rightarrow f(1) < 0;\ 2 + \dfrac{a+1}{a} < 0$

$\Rightarrow \dfrac{a+\frac{1}{3}}{a} < 0 \Rightarrow a \in \left(-\frac{1}{3}, 0\right)$

6. (C)

$\alpha(\alpha\beta)^5 = 32 \Rightarrow \alpha = 1$

The value of α will satisfy the given equation

$1 + a + 2 = 0 \Rightarrow a = -3$

7. (A)

$f(x) = x^{10} - 2x^9 + 4x^8 - \ldots + 1024$

$x^3 - (a+2)x^2 + 2ax = x(x-2)(x-a)$

$\because f(x) = x(x-2)(x-a) + 341x^2 - 682x + 1024$

$\Rightarrow f(a) = 341a^2 - 682a + 1024$

$= \dfrac{a^{10}\left(1-\left(-\frac{2}{a}\right)^{11}\right)}{1+\frac{2}{a}}$

$\Rightarrow 341a^2 - 682a + 1024 = \dfrac{a^{11}+2^{11}}{a+2}$

$a = -1$

8. (A)

$D > 0\ \forall\ x \in R$

$8(2(a+b)^2 - (a^2+1)) > 0$

$\Rightarrow a^2 - (4b)a + 2b^2 - 1 > 0\ \forall\ a \in R$

$D < 0$

$\Rightarrow 16b^2 - 4(2b^2 - 1) < 0$

$8b^2 + 4 < 0 \Rightarrow$ no solution

9. (D)

$x^2 + ax + b = 0$ with roots c, d

$x^2 + cx + d = 0$ with roots a, b

$c^2 + ac + b = 0$ (1)

$a^2 + ac + d = 0$ (2)

(1) – (2) $c^2 - a^2 = d - b$

Also $c + d = -a$ (3)

$a + b = -c$ (4)

$cd = b$ (5)

$ab = d$ (6)

(3) – (4)

$c - a + d - b = -a + c$

$\Rightarrow d = b$

from (5) and (6)

$a = c = 1$

$b = -2 = d$

10. (A)

From diagram $\alpha - \beta = 3$, $\alpha > 0$, $\beta < 0$, $|\beta| > \alpha$

$\Rightarrow b^2 - 4c = 9$, $c < 0$ and $b > 0$

$\Rightarrow c = \dfrac{b^2-9}{4} < 0 \Rightarrow b = (0, 3)$

$\therefore b \in I \Rightarrow b = 1, 2$

but for $b = 2$, $c \notin I$

$\Rightarrow b = 1 \Rightarrow c = -2$

$\Rightarrow$ Only one quadratic equation $x^2 + x - 2 = 0$

11. (A)

$\dfrac{\alpha^3\gamma + \beta^3\alpha + \gamma^3\beta}{\alpha\beta\gamma} = -8$

$\Rightarrow \dfrac{(-a\alpha - a)\gamma + (-a\beta - a)\alpha + (-a\gamma - a)\beta}{\alpha\beta\gamma} = -8$

$\Rightarrow \dfrac{-a\Sigma\alpha\beta - a\Sigma\alpha}{\alpha\beta\gamma} = -8$

$\Rightarrow \dfrac{-a.a}{-a} = -8$

$\Rightarrow a = -8$

12. (A)

Let α, β, γ be the roots of $x^3 + 3x^2 + 3 - k = 0$

$\Rightarrow \alpha + \beta + \gamma = -3$

$\alpha\beta + \beta\gamma + \gamma\alpha = 0$

$\alpha\beta\gamma = k - 3;\ \alpha^2 + \beta^2 + \gamma^2 = 9$

$\because \alpha, \beta, \gamma \in I$

Case-I: $\alpha^2 = 9$, $\beta^2 = \gamma^2 = 0$

$\Rightarrow \alpha = -3, \beta = \gamma = 3 \Rightarrow k = 3$

Case-II: $\alpha^2 = 4$, $\beta^2 = 4$, $\gamma^2 = 1$

$\Rightarrow \alpha = -2, \beta = -2, \gamma = 1$

$\Rightarrow k - 3 = 4$

$\Rightarrow k = 7$

$\Rightarrow$ sum = 10

13. **(B)**

$\alpha+\beta=10 \Rightarrow \gamma=5$

$\therefore$ Put value of γ in $x^2-(3k+2)x+k^3+3=0$

$\Rightarrow k^3-15k+18=0$

$\Rightarrow k=3$

$\therefore\ \alpha=6, \beta=4, \gamma=5$

14. **(B)**

$\because f(0)<f(2)$

$\Rightarrow f(1)<f(2)$ and $f(3)>f(-1)$

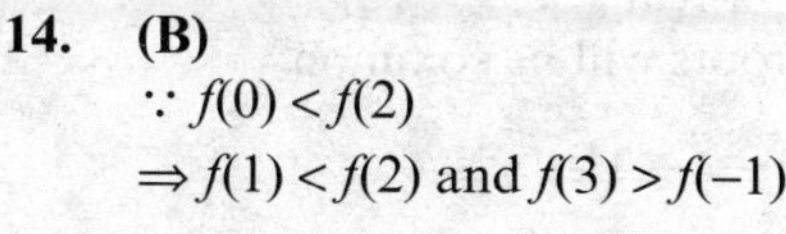

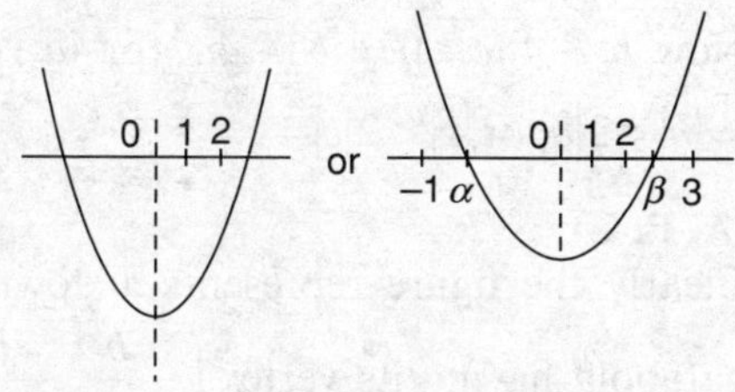

15. **(B)**

Roots of the equation $cx^2+bx+a=0$ are α, β then roots of the equation $ax^2+bx+c=0$ are $\dfrac{1}{\alpha}, \dfrac{1}{\beta}$ then roots of the equation

$a(x+1)^2+b(x+1)+c=0$ are $\dfrac{1}{\alpha}-1, \dfrac{1}{\beta}-1$

$=\dfrac{1-\alpha}{\alpha}, \dfrac{1-\beta}{\beta}$

16. **(C)**

$\alpha+\dfrac{1}{\alpha}>2, \beta+\dfrac{1}{\beta}>2$

$\Rightarrow$ both the roots are real and greater than 2.

$\Rightarrow$ (graph, 2) $\begin{cases} f(2)\ge 0 \\ D\ge 0 \\ -\dfrac{b}{2a}>2 \end{cases}$

$$\Rightarrow \begin{cases} 2-10-a\ge 0 \\ 25+4a\ge 0 \\ \dfrac{5}{2}>2 \end{cases} \Rightarrow a\in\left[\dfrac{-25}{4}, -6\right]$$

17. **(D)**

$x\in[-5, 3)\cup\{-6\}$

+   +   −   −   +

−6   −5   −1   3

$\therefore$ 9 integral values of x.

18. **(B)**

Let $x^2+x=t$

$\Rightarrow (t+1)(t+3)=8 \Rightarrow t=1$ or -5

$\therefore\ x^2+x-1=0$ or $x^2+x+5=0$

$\Downarrow \qquad \Downarrow$

Two real solutions; two imaginary solutions.

19. **(A)**

From equations $\dfrac{\alpha}{2}$ and $\dfrac{\beta}{3}$ are roots of

$x^2-5x+3=0$

$\therefore\ 3\alpha+2\beta=30$ and $\alpha\beta=18$

20. **(D)**

$\alpha^2-\alpha+11=0 \Rightarrow 3\alpha^3-3\alpha^2+33\alpha=0$ (1)

$\beta^2-\beta+11=0$

$2\beta^3-2\beta^2+22\beta=0$ (2)

(1) + (2)

$\Rightarrow 3\alpha^3-3\alpha^2+2\beta^3-2\beta^2+33\alpha+22\beta=0$

$\Rightarrow 3\alpha^3-3\alpha^2+2\beta^3-2\beta^2+11\alpha$

$=-22(\alpha+\beta)=-22$

21. **(A)**

Let $y=\dfrac{x}{x-1} \Rightarrow x=\dfrac{y}{y-1}$

$\therefore\ \dfrac{y^2}{(y-1)^2}-\dfrac{2y}{y-1}-5=0$

$\Rightarrow y^2-2y(y-1)-5(y-1)^2=0$

$\Rightarrow 6y^2-12y+5=0$

22. **(A)**

Given $f(f(1))=0$ and $f(f(2))=0$

That is, equation $f(x)=0$ has two roots $f(1)$ and $f(2)$

$\therefore\ f(1)+f(2)=-\alpha$ and $f(1)\cdot f(2)=\beta$

$\Rightarrow 5+3\alpha+2\beta=-\alpha$ and

$(1+\alpha+\beta)(4+2\alpha+\beta)=\beta$

$\Rightarrow 2\alpha+\beta=-\dfrac{5}{2}$ and

$\left(1-\dfrac{5}{4}-\dfrac{\beta}{2}+\beta\right)\cdot\left(4-\dfrac{5}{2}\right)=\beta$

$\Rightarrow \beta=-\dfrac{3}{2}$

$\therefore\ f(0)=\beta=-\dfrac{3}{2}$

23. **(C)**
As $a + b + c + d = 0$

$$\therefore \left(\frac{-d}{abc}\right)\left(\frac{-a}{bcd}\right)\left(\frac{-b}{cda}\right)\left(\frac{-c}{abd}\right)$$

$$= \frac{1}{(abcd)^2} = \frac{1}{2007}$$

24. **(B)**
Let roots of $x^2 - x - 1 = 0$ are α, β

$\Rightarrow \alpha + \beta = 1,\ \alpha\beta = -1$

$$\text{also} \begin{cases} a\alpha^{17} + b\alpha^{16} + 1 = 0 \\ a\beta^{17} + b\beta^{16} + 1 = 0 \end{cases}$$

$$\Rightarrow a\alpha + b = -\frac{1}{\alpha^{16}}$$

$$\alpha\beta + b = -\frac{1}{\beta^{16}}$$

$$\Rightarrow a(\alpha - \beta) = \frac{\alpha^{16} - \beta^{16}}{(\alpha\beta)^{16}}$$

$$\Rightarrow a = (\beta^8 + \alpha^8)(\beta^4 + \alpha^4)(\beta^2 + \alpha^2)(\beta + \alpha)$$

$$= 47 \times 7 \times 3 \times 1 = 987$$

25. **(B)**
$6x - x^2 - 5 \geq 0$

$\Rightarrow x^2 - 6x + 5 \leq 0 \Rightarrow x \in [1, 5]$

$\therefore |\sin \pi x| + \log_5(6x - x^2 - 4) + \sqrt{6x - x^2 - 5} = 0$

$\Rightarrow \sin \pi x = 0;\ 6x - x^2 - 4 = 1;\ 6x - x^2 - 5 = 0$

$\Rightarrow x = 1, 5.$

Multiple Answer Type

1. **(A, B, C)**
$f(x) = P(x^4) + xQ(x^4)$
is divisible by $(x + i)(x - i)$

$\Rightarrow f(i) = P(1) + iQ(1) = 0$

$\Rightarrow f(-i) = P(1) - iQ(1) = 0$

$\Rightarrow P(1) = Q(1) = f(1) = 0 \Rightarrow P(x), Q(x)$
and $f(x)$ are divisibly by $(x-1)$

2. **(A, B, C, D)**
$1 + 4a$ should be perfect square
As $1 + 4a$ is always odd

$\Rightarrow 1 + 4a = (2\lambda + 1)^2,\ \lambda \in I$

$\Rightarrow a = \lambda(\lambda + 1$; Hence (A), (B), (C), (D)

3. **(A, B, C)**
$D_1 = b^2 - 4ac < 0,\ D_2 = b^2 - 4ac < 0$ as the root is non-real

$\Rightarrow$ both roots will be common

$$\Rightarrow \frac{a}{c} = \frac{b}{b} = \frac{c}{a} = 1 \Rightarrow a = c$$

Now, $b^2 - 4ac < 0 \Rightarrow b^2 - 4a^2$ (or $4c^2$) < 0

$\Rightarrow |b| < 2|a|$ or $|b| < 2|c|$

4. **(A, B, C)**
Clearly the figure represents a downward parabola having its vertex $\left(-\frac{b}{2a}, -\frac{D}{4a}\right)$ in the second quadrant

$$\Rightarrow a < 0, -\frac{b}{2a} < 0 \Rightarrow b < 0$$

Also, $ax^2 - 4ac < 0$ $ax^2 + bx + c = 0$ has roots of opposite signs

$$\Rightarrow \frac{c}{a} < 0 \Rightarrow c > 0$$

5. **(B, C)**
a and $(4a + 2b + c)$ are of same sign

$$\Rightarrow a(4a + 2b + c) > 0 \Rightarrow a^2\left(4 + 2\frac{b}{a} + \frac{c}{a}\right) > 0$$

$\Rightarrow 4 - 2(\alpha + \beta) + \alpha\beta > 0$ where α, β are the roots of the given equation

$\Rightarrow (\alpha - 2)(\beta - 2) > 0$

$\Rightarrow$ Either $\alpha > 2, \beta > 2$ or $\alpha < 2, \beta < 2$

6. **(A, B, C, D)**
Let $f(x) = cx^2 + bx - 2a$

$$\frac{b+c}{2} > a \Rightarrow b + c - 2a > 0 \Rightarrow f(1) > 0$$

Now, $f(x) = 0$ has no real root

$\Rightarrow f(x) > 0 \forall x \Rightarrow f(0) > 0 \Rightarrow a < 0$

Clearly $c > 0 \Rightarrow ac < 0$;

Also, $f(-1) > 0 \Rightarrow c - b > 2a$

$$\text{Similarly, } f\left(\frac{1}{2}\right) > 0 \Rightarrow \frac{c + 2b}{8} > a$$

7. **(A, D)**

8. **(A, B, C)**

Let, $f(x) = ax^2 + bx + c$

$\Rightarrow -3$ and 3 lying between the roots of $f(x) = 0$

$\Rightarrow f(-3) > 0$ and $f(3) > 0$

$\Rightarrow 9a + 3|b| + c > 0$

Also, 0, – 2, 2 are lying between the roots

Therefore, $c > 0$ and $4a + 2|b| + c > 0$

9. **(B, C, D)**

$$E = \sum_{i=1}^{3} x_1^2 = 9m^2 - 2(-3) = 9m^2 + 6$$

which is minimum when $m = 0$ and minimum value is 6.

$\therefore k = 0$

For $m = 0$ $f(x) = x^3 - 3x + 2$

$\therefore f(x) = (x-1)^2 (x+2)$

$\therefore$ Roots are all real but not all 3 distinct.

10. **(A, B)**

$ax^4 + bx^3 - x^2 + 2x + 3$

$= Q(x-1)(x+2) + (4x+3)$

Put $x = 1$ to get $a + b = 3$

Put $x = -2$ to get $2a - b = 0$

$\therefore a = 1$ and $b = 2$

11. **(A, C, D)**

$ax^2 + (2a-1)x + (a-2) = 0$

$D = (2a-1)^2 - 4a(a-2) = 4a + 1$ must be a perfect square.

Now, check from the options.

12. **(A, B)**

We have $\log_5(5)(x^2+1) \geq \log_5(ax^2 + 4x + a)$

$\Rightarrow 5(x^2+1) \geq ax^2 + 4x + a \;\forall\; x \in R$

$\Rightarrow x^2(5-a) - 4x + (5-a) \geq 0 \;\forall\; x \in R$

As $5 > a$ (1)

and $D \leq 0$

$\Rightarrow 16 - 4(5-a)^2 \leq 0$

$\Rightarrow (2 - (5-a))(2 + 5 - a) \leq 0$

$\Rightarrow (a-3)(a-7) \geq 0 \Rightarrow (-\infty, 3] \cap [7, \infty)$ (2)

$\therefore (1) \cap (2) \Rightarrow a \in (-\infty, 3]$ (3)

But for the domain of $\log_5(ax^2 + 4x + a)$

We must have $ax^2\ 4x + a > 0 \;\forall\; x \in R$

$\Rightarrow a > 0$ and $16 - 4a^2 < 0$

$\Rightarrow (-\infty, -2) \cap (2, \infty)$

$\therefore$ domain of 'a' is $(2, \infty)$ (4)

$\Rightarrow (3) \cap (4) \Rightarrow a \in (2, 3]$

13. **(A,B,C)**

$\sin^2\theta - a \sin\theta + b = 0$ has only one solution in $(0, \pi)$,

$\Rightarrow \sin\theta = 1$ gives one solution and $\sin\theta = \alpha$ gives other solution.

such that $\alpha \geq 1$ or $\alpha \leq 0$ as $\theta \in (0, \pi)$

$\Rightarrow (\sin\theta - 1)(\sin\theta - \alpha) = 0$ is the same equation as $\sin^2\theta - a\sin\theta + b = 0$

$\Rightarrow 1 + \alpha = a$ and $\alpha = b$

$\Rightarrow 1 + b = a$ and $b \geq 1$ or $b \leq 0$

$\Rightarrow b \in (-\infty, 0] \cup [1, \infty)$ and

$a \in (-\infty, 1] \cup [2, \infty)$

14. **(B, C)**

We have, $yx^2 - 4yx + ky = x - k$

$\Rightarrow yx^2 - (4y+1)x + (y+1)k = 0,$

as $x \in R$ so $D \geq 0$

$\Rightarrow (4y+1)^2 - 4y(y+1)k \geq 0$

$\Rightarrow 16y^2 + 8y + 1 - 4ky^2 - 4ky \geq 0$

$\Rightarrow (16-4k)y^2 + (8-4k)y + 1 > 0 \;\forall\; y \in R$

This will be true if

$\Rightarrow 16 - 4k > 0 \Rightarrow 16 > 4k$ $(k < 4)$ and

$\Rightarrow (8-4k)^2 - 4(16-4k) \leq 0$

$\Rightarrow 64 + 16k^2 - 64k - 64 + 16k \leq 0$

$\Rightarrow k^2 - 3k \leq 0 \Rightarrow k(k-3) \leq 0 \Rightarrow k \in [0, 3]$

If $k = 0$ then $y = \dfrac{x}{x(x-4)}$

Hence, y cannot be equal to $\left(\dfrac{-1}{4}\right)$. (Think!)

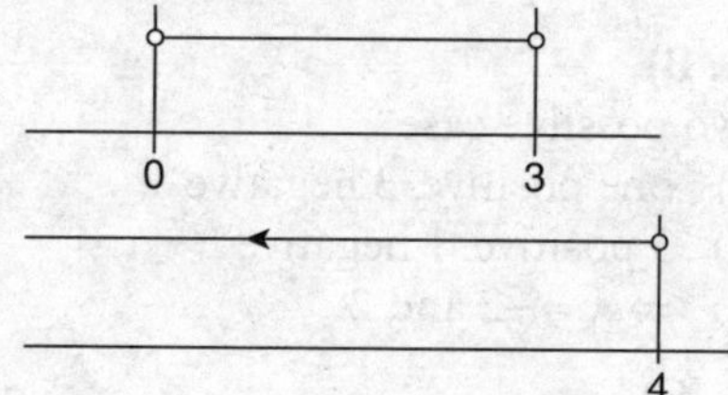

If $k = 3$ then, $y = \dfrac{x-3}{x^2 - 4x + 3} = \dfrac{x-3}{(x-3)(x-1)}$

$\Rightarrow y$ cannot be equal to $\left(\dfrac{1}{2}\right)$. (Think!)

$\therefore k \in (0, 3)$. Hence, $k \in \{1, 2\}$

15. **(A, B)**

Let $f(x) = 4x^2 - 2x + k$

OR

Here $D \geq 0 \Rightarrow k \leq \frac{1}{4}$ (1)

$-1 < \frac{-B}{2A} < 1 \Rightarrow -1 < \frac{1}{4} < 1$ (2)

$f(-1) > 0 \Rightarrow k > -6$ (3)

$f(1) > 0 \Rightarrow k > -2$ (4)

So, $(1) \cap (2) \cap (3) \cap (4) \Rightarrow k \in \left(-2, \frac{1}{4}\right]$

16. **(B, D)**

(A) $a < 0; \frac{-b}{2a} < 0 \Rightarrow b < 0$

$c < 0 \Rightarrow abc < 0$

(B) $a > 0, b > 0, c > 0 \Rightarrow abc > 0$

(C) $a > 0, b < 0, c > 0 \Rightarrow abc < 0$

(D) $a > 0, b < 0, c < 0 \Rightarrow abc > 0$

17. **(A, B, C, D)**

$f(0) = C = I_1$

$f(1) = A + B + C = I_2$ (1)

$f(1) = A + B + C$ is an integer (2)

$f''(x) = 0$ is an integer

$f(-1) = A - B + C$ is an integer

18. **(A, B)**

Two possible case

(a) one positive, 3 negative

(b) 3 positive, 1 negative

$\Rightarrow x = -2$ and 2

19. **(A, C)**

$\sin 2A + \sin 2B + \sin 2C = \frac{400}{625} = \frac{16}{25}$

$\sin A \cdot \sin B \sin C = \frac{4}{25}$

$\sin 2A \cdot \sin 2B \cdot \sin 2C = \frac{-384}{625}$

$8 \times \frac{4}{25} \times \cos A \cos B \cos C = \frac{-384}{625}$

$\Rightarrow \cos A \cos B \cos C = \frac{-12}{25}$

$\Sigma \tan A = \tan A \tan B \tan C = \frac{-1}{3}$

20. **(A, B, D)**

$\alpha + \beta^2 = 4, \alpha\beta^2 = \frac{3k}{5}, \beta\alpha^2 = \frac{45}{k}$

and $\beta + \alpha^2 = \frac{50}{k}$

$\Rightarrow \alpha^3\beta^3 = 27 \Rightarrow \alpha\beta = 3$

$\Rightarrow \alpha = 3, \beta = 1$ and $k = 5$

$\because \beta^2$ is root of $5x^2 - 20x + 3k = 0$

$\Rightarrow \beta^4 - 4\beta^2 + 3 = 0$

$\Rightarrow \beta^3 - 4\beta + \frac{3}{\beta} = 0$ or $\beta^3 - 4\beta + \alpha = 0$

Similarly $\alpha^4 - 10\alpha^2 + 9 = 0$

$\Rightarrow \alpha^3 - 10\alpha + 3\beta = 0$

21. **(A, C)**

Let $f(x) = (x - p)(x - q) + 5(x - r)(x - s)$

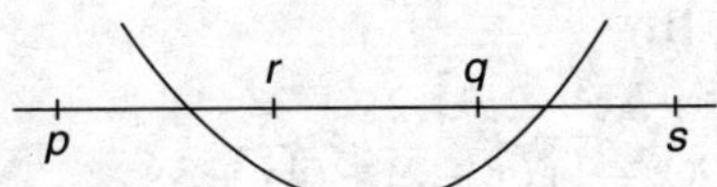

$\because f(p) = +\text{ve}$

$f(r) = -\text{ve}$

$f(q) = -\text{ve}$ and $f(s) = +\text{ve}$

$\therefore$ One root lies in (p, r) and other lies in (q, s)

22. **(A, C)**

$f(x) = 2x^2 + kx + 50$

$|\alpha - \beta| = 8 \Rightarrow \frac{\sqrt{D}}{|a|} = 8 \Rightarrow D = 256$

$\therefore$ Minimum value $= -\frac{D}{4a} = -\frac{256}{8} = -32$

$\because$ at $x = 0, f(x) = 50$

$\Rightarrow$ graph cuts positive y-axis.

23. **(A, C)**

Let $f(x) = ax^2 + c$

$f(1) = a + c, -4 \leq a + c \leq -1$

$f(2) = 4a + c, -1 \leq 4a + c \leq 5$

$\Rightarrow a = \frac{f(2) - f(1)}{3}, f(0) = c \Rightarrow \frac{4f(1) - f(2)}{3}$

$f(3) = 9a + c$

$= 3f(2) - 3f(1) + \frac{4f(1)}{3} - \frac{f(2)}{3}$

$= \frac{8f(2) - 5f(1)}{3}$

$\therefore f(3)_{max} = \frac{8.5 - 5(-4)}{3} = 20$

24. **(A, B)**

Let $x^3 + 4x^2 + px + q = 0$ has α, β, γ as roots and $x^3 + 6x^2 + px + r = 0$ has α, β, δ as roots

$\therefore \alpha + \beta + \gamma = -4$ (1)

$\alpha\beta + \alpha\gamma + \beta\gamma = p$ (2)

$\alpha\beta\gamma = -q$ (3)

$\therefore \alpha + \beta + \delta = -6$ (4)

$\alpha\beta + \beta\delta + \alpha\delta = p$ (5)

$\alpha\beta\delta = -r$ (6)

$\Rightarrow (1) - (4) \Rightarrow \gamma - \delta = 2$

and $(2) - (5) \Rightarrow (\alpha + \beta)(\gamma - \delta) = 0$

$\Rightarrow \alpha + \beta = 0$

$\Rightarrow \gamma = -4$ and $\delta = -6$

Equation having γ and δ as roots

$x^2 + 10x + 24 = 0$

$a = 5, c = 3 \Rightarrow a + c = 8$

$\frac{(3)}{(6)} \Rightarrow \frac{\gamma}{\delta} = \frac{q}{r} \Rightarrow 3q - 2r = 0$

25. **(B, D)**

$\because MA = MB \Rightarrow \Delta$ is right angled isosceles

$\Rightarrow MV = MA = MB = \left|-\frac{D}{4a}\right|$

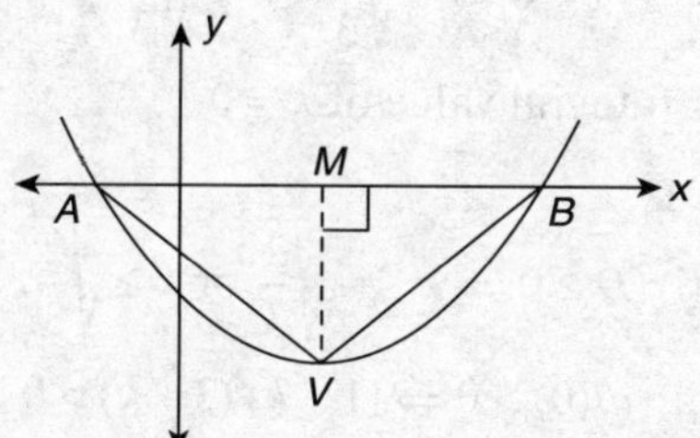

$\Rightarrow$ difference of roots $= |\alpha - \beta| = \frac{\sqrt{D}}{|a|}$

$= 2\left|\frac{D}{4a}\right|$

$\Rightarrow \frac{D}{a^2} = \frac{D^2}{4a^2} \Rightarrow D = 4$

also $b^2 - 4ac = D = 4\{a > 0$ and $c < 0\}$

$\Rightarrow 4ac = b^2 - 4 < 0 \Rightarrow b \in (-2, 2)$

and vertex $-\frac{b}{2a} > 0 \Rightarrow b < 0$

Comprehension Based Questions

1. **(D)**

$y = \frac{x^2 - 3x - 4}{x^2 - 3x + 4}$

$\Rightarrow (y-1)x^2 - 3(y-1)x + 4(y+1) = 0$

As x is real $\Rightarrow 9(y-1)^2 \geq 16(y^2 - 1)$

$\Rightarrow 7y^2 + 18y - 25 \leq 0 \Rightarrow (y-1)(7y+25) \leq 0$

$\Rightarrow -\frac{25}{7} \leq y \leq 1$

But $y \neq 1$; So, range of $f(x) = \left[-\frac{25}{7}, 1\right)$

Clearly integers in the range of $f(x)$ are $-3, -2, -1,$; sum of integers $= -6$

2. **(A)**

Since both roots of

$g(x) = x^2 - (b+1)x + (b-1) = 0$ are greater than -1

(i) $D > 0$

(ii) $-\frac{b}{2a} > -1$ and

(iii) $g(-1) > 0$

Hence $b \in \left(-\frac{1}{2}, \infty\right)$

3. **(B)**

$g(x) > -2 \ \forall \ x \in R$

$\Rightarrow x^2 - (b+1)x + (b+1) > 0 \forall x \in R$

$\Rightarrow (b+1)^2 < 4(b+1) \Rightarrow -1 < b < 3$

$\Rightarrow$ largest natural number $b = 2$

4. **(D)**

$AC = 6\sqrt{3}$

and $(AB)^2 + (BC)^2 = (AC)^2$

$\Rightarrow 2(AB)^2 = (6\sqrt{3})^2 \ [AB = BC]$

$\Rightarrow AB = BC = \frac{6\sqrt{3}}{\sqrt{2}}$

and $(OB) = \sqrt{\left(\frac{6\sqrt{3}}{\sqrt{2}}\right)^2 - (3\sqrt{3})^2}$

$$= \sqrt{(3\sqrt{3})^2\left(\left(\frac{2}{\sqrt{2}}\right)^2 - 1\right)} = 3\sqrt{3}$$

$\therefore A = (-3\sqrt{3}, 0), B(0, -3\sqrt{3}), C(3\sqrt{3}, 0)$

Since, $y = ax^2 + bx + c$ passes through A, B and C

So, we get $a = \frac{1}{3\sqrt{3}}, b = 0, c = -3\sqrt{3}$

5. **(A)**
Minimum value of

$y = \frac{x^2}{3\sqrt{3}} - 3\sqrt{3}$ is at $x = 0$, that is, $-3\sqrt{3}$

6. **(C)**
Roots of $f(x) = 0$

That is, $\frac{x^2}{3\sqrt{3}} - 3\sqrt{3} = 0$ are, $x = \pm 3\sqrt{3} = 0$

Number of integral values of K for which $\frac{K}{2}$ lies in $(-3\sqrt{3}, 3\sqrt{3})$ is 20

7. **(B)**

$\sqrt{x^2 + 2x + 1} - \sqrt{4x^2 + 4x + 1} + \sqrt{4x^2} = 2$

$\sqrt{(x+1)^2} - \sqrt{(2x+1)^2} + \sqrt{(2x)^2} = 2$

$|x+1| - |2x+1| + |2x| = 2$

Case (i): When $x < -1$

$-x - 1 + 2x + 1 - 2x = 2$

$x = -2$

Case (ii): When $-1 \le x < \frac{-1}{2}$

$x + 1 + 2x + 1 - 2x = 2$

$x = 0$

Case (iii): When $\frac{-1}{2} \le x < 0$

$x + 1 - 2x - 1 - 2x = 2$

$x = \frac{-2}{3}$

Case (iv): When $x \ge 0$

$x + 1 - 2x - 1 - 2x = 2$

$x = 2$

So, integral solutions are $x = 2$ and $x = -2$

No. of integral solution is '2'. Correct option is (B)

8. **(A)**
Sum of all the solutions is $-2 + 2 = 0$
Correct option is (A)

9. **(B)**
$|x + 1| - |2x + 1| + |2x| = 2$
Case I: $x < -1$; $-x - 1 + 2x + 1 - 2x = 2$*
$x = -2$ (accepted)
Case II: $-1 \le x < \frac{-1}{2}$; $x + 1 + 2x + 1 - 2x = 2$
$x = 0$ (rejected)
Case III: $\frac{-1}{2} \le x < 0$
$x + 1 - (2x + 1) - 2x = 2$;
$x + 1 - 2x - 1 - 2x = 2$
$-3x = 2$
$x = \frac{-2}{3}$ (rejected)
Case IV: $x \ge 0$
$(x + 1) - (2x + 1) + 2x = 2$;
$x + 1 - 2x - 1 + 2x = 2$
$x = 2$ (accepted)
Hence, number of integral solutions is 2.
Now, $\alpha = 2$
Number of digits in 2^{50} is 16.
$\log_{10}(2)^{50} = 50 \log_{10}2 = 15.05$

10. **(B)**
$D = 4(1 + 2k)^2 - 4(1 + k)(3 + k) = 4(3k^2 - 2)$
$3k^2 - 2 < 0 \Rightarrow -\sqrt{\frac{2}{3}} < k < \sqrt{\frac{2}{3}}$
$\therefore$ Integral value of $k = 0$

11. **(B)**
(i) $D \ge 0 \Rightarrow k \le -\sqrt{\frac{2}{3}}$ or $k \ge \sqrt{\frac{2}{3}}$
(ii) $af(0) > 0 \Rightarrow (1 + k)(3 + k) > 0$
$\Rightarrow k < -3$ or $k > -1$
(iii) $\frac{-b}{2a} > 0 \Rightarrow \frac{1 + 2k}{1 + k} > 0$
$\Rightarrow k < -1$ or $k > -\frac{1}{2}$

$\therefore$ Common values of (i), (ii) and (iii) are $k < -3$

or $k \ge \sqrt{\frac{2}{3}}$

12. (C)
Let the roots be α and 2α

$\therefore\ \alpha+2\alpha=\frac{2(1+2k)}{1+k} \Rightarrow \alpha=\frac{2(1+2k)}{3(1+k)}$

and $\alpha\cdot 2\alpha=\frac{3+k}{1+k}$

$\Rightarrow 2\times\left(\frac{2(1+2k)}{3(1+k)}\right)^2=\frac{3+k}{1+k}$

$\Rightarrow 8(1+2k)^2=9(3+k)(1+k)$

$\Rightarrow 8(4k^2=4k+1)=9(k^2+4k+3)$

$\Rightarrow 23k^2-4k-19=0$

$\Rightarrow 23x^2-23k+19k-19=0$

$\Rightarrow 23k(k-1)+19(k-1)=0$

$\Rightarrow (k-1)(23k+19)=0$

$\therefore\ k=1$ or $k=\frac{-19}{23}$

$\therefore$ Two real values of k.

13. (C)

14. (C)

15. (A)
$(a-1)^2+(b+3)^2+(c-2)^2=0$
Hence, $a=1$; $b=-3$; $c=2$
$f(x)=x^2-2(k-1)x+1$
$g(x)=kx^2-3x+2$

(i) $f(0)+g(0)=3$

(ii) $f(x)=g(x)$
$(1-k)x^2-(2k-5)x-1=0$
Now, $D=0\Rightarrow 4k^2-24k+29=0$
Sum of all real values of $k=6$

(iii) Let D_1 be the discriminant of $f(x)=0$ and D_2 be the discriminant of $g(x)=0$
$D_1=4(k-1)^2-4$; $D_2=(9-8k)$
Now,
Case I: $D_1>0$ and $D_2=0$
$(k-1)^2-1>0$ and $k=\frac{9}{8}$
Hence, no value of k from this case.
Case II: $D_2>0$ and $D_1=0$
$9-8k>0$ and $(k-1)^2-1=0$
From this case $k=0$ which is rejected.
Hence, no value of k from this case.
Therefore, no value of k is possible.

Match Matrix

1. (A – R; B – Q; C – P, Q, S; D – Q)

(A) $ax^4+bx^3-x^2+2x+3$
$=Q(x-1)(x+2)+(4x+3)$
Put $x=1$, we get $a+b=3$
Put $x=-2$, we get $2a-b=0$
$\therefore a=1$ and $b=2\Rightarrow 2b-a=3$

(B) From figure

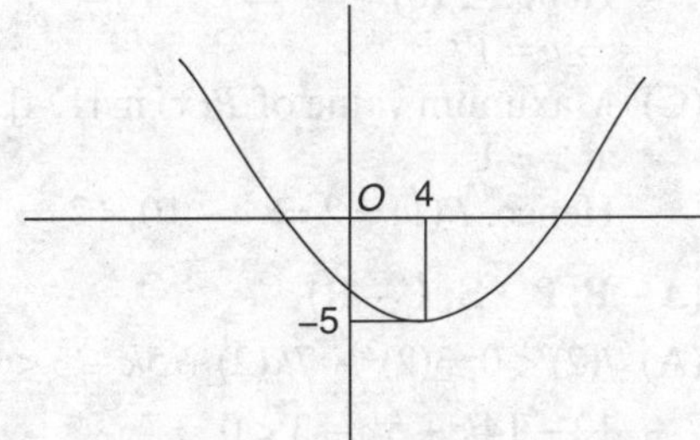

$a>0$ and $-\frac{b}{2a}=4\Rightarrow -\frac{b}{2a}>0\ \therefore b<0$

$f(0)=c<0$; Also, $8a+b=0$

(C) $ax^2+(2a-1)x+(a-2)=0$

$D=(2a-1)^2-4a(a-2)=4a+1$

must be a perfect square

(D) $\alpha+\beta^2=4, \alpha\beta^2=\frac{3K}{5}$,

$\beta\alpha^2=\frac{45}{K}, \beta+\alpha^2=\frac{50}{K}$

$\Rightarrow \alpha^3\beta^3=27\Rightarrow \alpha\beta=3$

$\Rightarrow \alpha=3, \beta=1$ and $K=5$

As β^2 is a root of $5x^2-20x+3K=0$

$\Rightarrow \beta^4-4\beta^2+3=0\Rightarrow \beta^3-4\beta+\frac{3}{\beta}=0$

$\Rightarrow \beta^3-4\beta+\alpha=0$

2. (A – R; B – P; C – Q; D – S)

$x^2-(2a+3)x+a(a+3)=0$

$\Rightarrow (x^2-2ax+a^2)-3(x-a)=0$

$\Rightarrow (x-a)^2-3(x-a)=0$

$\Rightarrow (x-a)(x-a-3)=0$

$\Rightarrow$ roots are a and $a+3$

Now verify the options.

3. **(A – Q; B – S; C – R)**

(A) Maximum value of $P(x)$ in [1, 2] occurs at $x = 1$.
Hence, $P(1) = 1 \Rightarrow c - 10 = 1 \Rightarrow c = 11$

(B) Minimum value of $P(x)$ in [1, 5] occurs at $x = 3$.
Hence, $P(3) = -1 \Rightarrow c - 18 = -1$
$\Rightarrow c = 17$.

(C) Maximum value of $P(x)$ in [1, 4] occurs at $x = 1$
Hence, $P(1) = 2 \Rightarrow c - 10 = 2 \Rightarrow c = 12$.

4. **(A – P; B – S; C – R)**

(A) $f(2) < 0$; $3(2)^2 - 7k(2) + 5k - 3 < 0$

$12 - 14k + 5k - 3 < 0$

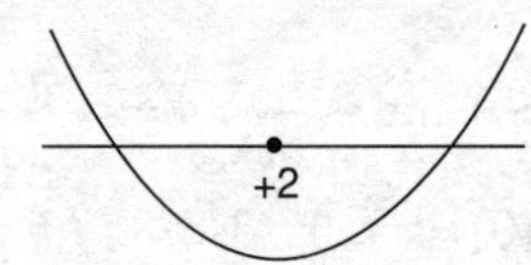

$9 - 9k < 0$

$\therefore k > 1$

(B) $\left.\begin{matrix} f(-2) > 0 \\ f(0) < 0 \\ f(1) > 0 \end{matrix}\right\} \cap f(-2) > 0$

$\Rightarrow 3(4) + 14k + 5k - 3 > 0$

$\Rightarrow 9 + 19k > 0$

$\Rightarrow k > \dfrac{-9}{19}$ (1)

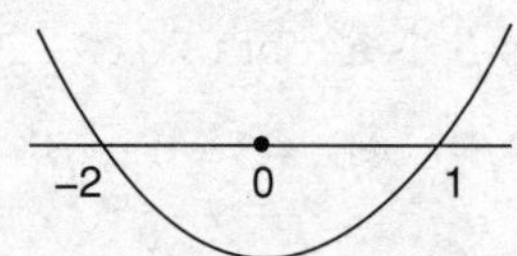

$f(0) < 0 \Rightarrow 5k - 3 < 0 \Rightarrow k < \dfrac{3}{5}$ (2)

$f(1) > 0 \Rightarrow 3 - 7k + 5k - 3 > 0 \Rightarrow k < 0$ (3)

$(1) \cap (2) \cap (3)$, $k \in \left(\dfrac{-9}{19}, 0\right)$

(C) $\left.\begin{matrix} f(-1) < 0 \\ f(1) < 0 \end{matrix}\right\} \cap$

$f(-1) = 3 + 7k + 5k - 3 < 0 \Rightarrow k < 0$ (1)

$f(1) = 3 - 7k + 5k - 3 < \Rightarrow k > 0$ (2)

$(1) \cap (2)$

$k \in \phi$

5. **(A – S; B – R; C – P)**

(A) $\log_3(xy) \geq 4 \Rightarrow xy \geq 81$

Now, $(\sqrt{x} - \sqrt{y})^2 \geq 0$

$x + y \geq 2\sqrt{xy} \Rightarrow x + y \geq 18$

(B) In ΔABC, $\Sigma \cos^2 A = 1 - 2\cos A \cos B \cos C$

$\therefore m = 1$ and $n = -2$

(C) $\log_{\sqrt{5}}(4-x) \geq 0$

$4 - x > 0$ and $4 - x \geq 1$

$\therefore x \in (-\infty, 3]$ (1)

Case I: $x - 1 > 1 \Rightarrow x > 2$

$x^2 + 2 \geq 1 \Rightarrow x^2 \geq -1$

$\therefore x \in (2, \infty)$

Case II: $0 < x - 1 < 1$

This case is not possible.

Therefore from case-I and case-II

$\therefore x \in (2, \infty)$ (2)

Hence, from (1) and (2)

$\therefore x \in (2, 3]$

Subjective Answer Type (Hard Nut)

1. $x^n - 1 = (x-1)(x-a_1)(x-a_2)\ldots(x-a_{n-1})$

$\Rightarrow \dfrac{x^n - 1}{x-1} = (x-a_1)(x-a_2)\ldots(x-a_{n-1})$

$\Rightarrow 1 + x + x^2 + \ldots + x^{n-1}$

$= (x-a_1)(x-a_2)\ldots(x-a_{n-1})$

Put $x = 1$

$(1-a_1)(1-a_2)\ldots(1-a_{n-1}) = n$

2. $\alpha + \beta = \dfrac{b}{a}$, $\alpha\beta = \dfrac{c}{a}$

As, $0 < \alpha < 1$ and

$0 < \beta < 1 \Rightarrow 0 < 1 - \alpha < 1, 0 < 1 - \beta < 1$

Now, $\dfrac{\alpha + (1-\alpha)}{2} \geq \sqrt{\alpha(1-\alpha)}$

$\Rightarrow 0 < \alpha(1-\alpha) \le \frac{1}{4}$

Similarly, $0 < \beta(1-\beta) \le \frac{1}{4}$

Therefore, $0 < \alpha\beta(1-\alpha)(1-\beta) < \frac{1}{16}$

[As α, β are distinct]

$\Rightarrow 0 < \alpha\beta\{1-(\alpha+\beta)+\alpha\beta\} < \frac{1}{16}$

$\Rightarrow 0 < \frac{c}{a}\left(1-\frac{b}{a}+\frac{c}{a}\right) < \frac{1}{16}$

$\Rightarrow 0 < c(a-b+c) < \frac{a^2}{16}$

As, $a, b, c \in N$

$\Rightarrow$ minimum value of $c(a-b+c)=1$ and $c=1$

$\Rightarrow \frac{a^2}{16} > 1 \Rightarrow a \ge 5$

Also, $b^2 > 4ac \Rightarrow b^2 > 20 \Rightarrow b \ge 5$

Hence, $abc \ge 25$

$\Rightarrow \log_5(abc) \ge \log_5 5^2 \Rightarrow \log_5(abc) \ge 2$

3. Since, $A > 0$ and $Ax^2 + Ax + 1 = 0$ has two real roots $\Rightarrow D \ge 0$

$\Rightarrow A^2 - 4A \ge 0 \Rightarrow A(A-4) \ge 0$

$\because A > 0 \Rightarrow A - 4 \ge 0$

$\Rightarrow A \ge 4$

Now, $A = \begin{vmatrix} a & 1 & 1 \\ 1 & b & 1 \\ 1 & 1 & c \end{vmatrix} = abc-(a+b+c)+2$

$\Rightarrow abc-(a+b+c)+2 \ge 4$

$\Rightarrow abc \ge (a+b+c)+2$ (1)

Now $\frac{a+b+c}{3} \ge (abc)^{\frac{1}{3}}$

$\Rightarrow (a+b+c)+2 \ge 3(abc)^{\frac{1}{3}}+2$

$\Rightarrow abc \ge 3(abc)^{\frac{1}{3}}+2$

Put $(abc)^{\frac{1}{3}} = y \Rightarrow y^3 \ge 3y+2$

$\Rightarrow y^3-3y-2 \ge 0 \Rightarrow (y+1)^2(y-2) \ge 0$

Hence, $y \ge 2 \Rightarrow (abc)^{\frac{1}{3}} \ge 2$

$\Rightarrow abc \ge 8 \Rightarrow$ minimum value of $abc = 8$

4. Let $f(x) = 4e^x + (a+c)^2(x^3+x) - 4b^2x$

$f'(x) = 4e^x + (a+c)^2 - 4b^2 + 3x^2(a+c)^2$

It is given, $ax^2 - 2bx + c = 0$ has imaginary roots

$\Rightarrow b^2 - ac < 0$

$\Rightarrow -4b^2 + 4ac > 0$

$\Rightarrow -4b^2 + (a-c)^2 + 4ac > 0$

$\Rightarrow -4b^2 + (a+c)^2 > 0$

So, $f'(x) > 0$

$\Rightarrow f(x)$ is a monotonically increasing function

Also, $f(0) = 4$ and $\lim_{x\to-\infty} f(x) = -\infty$

Hence, $f(x)$ has one and only one real root

5. $y - \frac{1-ax}{b}$

$\therefore cx^2 + d\left(\frac{1-ax}{b}\right)^2 = 1$

$\Rightarrow (b^2c + a^2d)x^2 - 2adx + d - b^2 = 0$

As roots are equal

$\therefore 4a^2d^2 - 4(b^2c + a^2d)(d - b^2) = 0$

$\Rightarrow b^2c + a^2d = cd \Rightarrow \frac{b^2}{d} + \frac{a^2}{c} = 1$

6. Since,

$x^4 - (a+1)x^3 + x^2 + (a+1)x - 2$

$= (x^4-1) - (a+1)x^3 + x^2 + (a+1)x - 1$

$= (x^2-1)(x^2+1) - (a+1)x(x^2-1) + (x^2-1)$

$= (x^2-1)\left[x^2 - (a+1)x + 2\right] = 0$

Hence, $x^2 - (a+1)x + 2 = 0$ should have at least one positive real root

Let α, β be its roots

$\alpha\beta = 2 \Rightarrow$ both roots positive

$\Rightarrow \alpha + \beta > 0 \Rightarrow a+1 > 0 \Rightarrow a > -1$ (1)

Also, $D \ge 0 \Rightarrow (a+1)^2 - 8 \ge 0$

$\Rightarrow (a+1)^2 \ge 8 \Rightarrow a+1 \le -2\sqrt{2}$ or $a+1 \ge 2\sqrt{2}$

$\Rightarrow a \le -2\sqrt{2} - 1$ or $a \ge 2\sqrt{2} - 1$ (2)

From (1) and (2), $a \in \left[2\sqrt{2}-1, \infty\right)$

7. Put $a=\frac{1}{b}$, therefore given equation becomes

$x^2-4bx+9b=0$

This equation has integral roots if b is an integer and $16b^2-36b$ is a perfect square

Let $b(4b-9)=K^2$, where K is an integer

$\Rightarrow 4b^2-9b-K^2=0$

$\Rightarrow \left(2b-\frac{9}{4}\right)^2-K^2=\frac{81}{16}$

$\Rightarrow (8b-9)^2-16K^2=81$

$\Rightarrow (8b-9-4K)(8b-9+4K)=81=3\times 27$

$8b-9-4K=3$

and $8b-9+4K=27$

$\Rightarrow 16b-18=30 \Rightarrow b=3 \Rightarrow a=\frac{1}{3}$

8. As $ax^2+bx+c=0$ does not have two distinct real roots, so either $f(x)\ge 0 \forall x\in R$ or $f(x)\le 0 \forall x\in R$

Now as $f(1)=3$, so $f(x)\ge 0 \forall x\in R$

$\Rightarrow a>0$ and $f(0)=c>0$. As $abc>0 \Rightarrow b>0$

Therefore, a, b, c are positive

Now, $f(1)=a+b+c=3$

Also, $\frac{a+b+c}{3}\ge (abc)^{\frac{1}{3}}$

$\Rightarrow abc\le 1$

From, $D<0$, $b^2<4ac \Rightarrow b^3<4abc$

$\Rightarrow b^3<4$

$\Rightarrow 0<b<4^{\frac{1}{3}}$

$\Rightarrow \alpha=0,\ \beta=4^{\frac{1}{3}}$

$\Rightarrow \alpha+\beta^3=4$

9. Since, a, b, c, d are in A.P.

$\Rightarrow b-a=c-b=d-c=K$ (say)

$\therefore b=a+K$

$c=a+2K$

$d=a+3K$

$\therefore 2(a-b)+x(b-c)^2+(c-a)^3$

$=2(a-d)+(b-d)^2+(c-d)^3$

$=-2K+xK^2+(2K)^3=-6K+4K^2-K^3$

$\Rightarrow K\left[9K^2+(x-4)K+4\right]=0$

$\Rightarrow 9K^2+(x-4)K+4=0$

[$K\ne 0$ as a, b, c, d are distinct]

As K is real

$(x-4)^2-144\ge 0$

$\Rightarrow x^2-8x-128\ge 0$

$\Rightarrow (x-16)(x+8)\ge 0$

$\Rightarrow x\le -8$ or $x\ge 16$

$\Rightarrow x\in(-\infty,-8]\cup[16,\infty)$

10. Put $\sin x=t \Rightarrow t\in[-1,1]$

$\therefore 2t^2-|K-2|t-|K-2|=0$

Let, $f(t)=2t^2-|K-2|t-|K-2|$

$D=|K-2|^2+8|K-2|\ge 0\ \forall\ K\in R$ (1)

For the given equation to have real roots

$f(-1)\ge 0$, $f(1)\ge 0$ and $-1\le -\frac{b}{2a}\le 1$

$f(-1)=2+|K-2|-|K-2|>0$

$f(1)=2-2|K-2|\ge 0 \Rightarrow |K-2|\le 1$

$\Rightarrow 1\le K\le 3$ (2)

and $\frac{|K-2|}{4}\le 1 \Rightarrow -2\le K\le 6$ (3)

From (2) and (3)

$K\in[1,3]$

11. $f(1)=a+b+c=0$

$50<f(7)<60 \Rightarrow 50<49a+7b+c<60$

$\Rightarrow 50<48a+6b<60$

$\Rightarrow \frac{50}{6}<8a+b<10$

$\Rightarrow 8a+b=9$

Similarly, $10<9a+b<\frac{80}{7} \Rightarrow 9a+b=11$

$\Rightarrow a=2,\ b=-7,\ c=5$

$\Rightarrow f(x)=2x^2-7x+5$

$f(10)=200-70+5=135$

As, $13K<135<13(K+1) \Rightarrow K=10$

12. $P(x)=Q_1x+4 \Rightarrow P(0)=4 \Rightarrow c=4$

$P(x)=Q_2(x+1)+3$

$\Rightarrow P(-1)=3 \Rightarrow a-b+c=3$

$P(x) = Q_3(x-1)+1 \Rightarrow P(1) = 1$

$\Rightarrow a+b+c=1$

$\therefore a-b=-1$

$a+b=-3 \Rightarrow 2a=-4 \Rightarrow a=-2$

and $b=-1$

Thus, $P(x) = -2x^2 - x + 4$

$\Rightarrow P(2) = -6$

13. Since, $f(x) = x^3 + 3x^2 + 6x + 9\sin x$

$\Rightarrow f'(x) = 3x^2 + 6x + 6 + 9\cos x$

$\Rightarrow f'(x) \geq 0$

$\Rightarrow f(x)$ is an increasing function

$\Rightarrow f(1) < f(2) < f(3)$

Let $f(1) = a, f(2) = b, f(3) = c$

$\Rightarrow a < b < c$

Let $\phi(y) = (y-b)(y-c) + 2(y-a)(y-c) + 3(y-a)(y-b)$

$\phi(a) = (a-b)(a-c) > 0$

$\phi(b) = 2(b-a)(b-c) < 0$

$\phi(c) = 3(c-a)(c-b) > 0$

As $\phi(a)\phi(b) < 0$

$\Rightarrow$ one root between a and b

and the other root between b and c

Hence they are real roots.

14. The given equation is

$$\sum_{K=1}^{n}\left[x^2 + (2K-1)x + \left(K^2 - K\right)\right] - 10 = 0$$

$$\Rightarrow nx^2 + \left(\frac{2n(n+1)}{2} - n\right)x + \frac{n(n+1)(2n+1)}{6} - \frac{n(n+1)}{2} - 10 = 0$$

$$\Rightarrow nx^2 + n^2x + \frac{n\left(n^2-1\right)}{3} - 10 = 0$$

As α and $\alpha+1$ are roots

$$2\alpha + 1 = -n \Rightarrow \alpha = -\frac{(n+1)}{2}$$

$$\alpha(\alpha+1) = \frac{n^2-1}{3} - \frac{10}{n}$$

$$\Rightarrow -\frac{(n+1)}{2}\left(\frac{1-n}{2}\right) = \frac{n^2-1}{3} - \frac{10}{n}$$

$$\Rightarrow \frac{n^2-1}{3} - \frac{\left(n^2-1\right)}{4} = \frac{10}{n}$$

$\Rightarrow \left(n^2-1\right)n = 120$

$\Rightarrow (n-1)n(n+1) = 4 \times 5 \times 6$

$\Rightarrow n = 5$

15. Let $-\alpha, -\beta, -\gamma$ be roots of

$2x^3 + ax^2 + bx + 4 = 0$

$\alpha + \beta + \gamma = \frac{a}{2}$

$\alpha\beta + \alpha\gamma + \beta\gamma = \frac{b}{2}$

$\alpha\beta\gamma = 2$

Now, $\frac{\alpha+\beta+\gamma}{3} \geq (\alpha\beta\gamma)^{\frac{1}{3}} \Rightarrow \frac{a}{6} \geq 2^{\frac{1}{3}}$

$\Rightarrow a \geq 6\left(2^{\frac{1}{3}}\right)$

Again, $\frac{\alpha\beta+\alpha\gamma+\beta\gamma}{3} \geq (\alpha\beta\gamma)^{\frac{2}{3}}$

$\Rightarrow \frac{b}{6} \geq 2^{\frac{2}{3}} \Rightarrow b \geq 6\left(4^{\frac{1}{3}}\right)$

$\therefore a+b \geq 6\left(2^{\frac{1}{3}} + 4^{\frac{1}{3}}\right)$

16. Let $-\alpha, -\beta, -\gamma, -\delta$ be roots of

$x^4 + ax^3 + bx^2 + cx + 1 = 0$

$\alpha + \beta + \gamma + \delta = a$

$\alpha\beta + \alpha\gamma + \alpha\delta + \beta\gamma + \beta\delta + \gamma\delta = b$

$\alpha\beta\gamma + \alpha\beta\delta + \alpha\gamma\delta + \beta\gamma\delta = c$

$\alpha\beta\gamma\delta = 1$

Now, $\frac{\alpha+\beta+\gamma+\delta}{4} \geq (\alpha\beta\gamma\delta)^{\frac{1}{4}}$

$\Rightarrow a \geq 4$, similarly, $b \geq 6$ and $c \geq 4$

$\therefore a = 4, b = 6, c = 4$

17. Let $f(x) = 2ax^2 + 3bx + 5c$

As $f(x) = 0$, does not have real roots

$\Rightarrow f(x) > 0$ or $f(x) < 0 \forall x \in R$

But $f(0)=5c>0$ [As $c>0$]

$\Rightarrow f(x)>0, \forall x \in R$

$\Rightarrow f(-1)>0 \Rightarrow 2a-3b+5c>0$

18. Let $\alpha, \beta, \gamma, \delta$ be the four positive real roots of the given equation. Then

$\alpha+\beta+\gamma+\delta=-p$

$\alpha\beta+\alpha\gamma+\alpha\delta+\beta\gamma+\beta\delta+\gamma\delta=q$

$\alpha\beta\gamma+\alpha\beta\delta+\alpha\gamma\delta+\beta\gamma\delta=-r$

$\alpha\beta\gamma\delta=5$

Using A.M. $\geq$ G.M.

$$\Rightarrow \frac{\alpha+\beta+\gamma+\delta}{4}\cdot\frac{\alpha\beta\gamma+\alpha\beta\delta+\alpha\gamma\delta+\beta\gamma\delta}{4} \geq (\alpha\beta\gamma\delta)^{\frac{1}{4}}(\alpha\beta\gamma\delta)^{\frac{3}{4}}$$

$$\Rightarrow \left(-\frac{p}{4}\right)\left(-\frac{r}{4}\right) \geq 5 \Rightarrow pr \geq 80$$

$\Rightarrow$ Minimum value of $pr=80$

19. Let, $f(x)=\left(a^2-14a+13\right)x^2+(a+2)x-2$

As $f(x)=0$ does not have two distinct real roots, so either

$f(x)\geq 0$ or $f(x)\leq 0 \forall x \in R$

But $f(0)=-2 \Rightarrow f(x)\leq 0 \forall x \in R$

$\Rightarrow f(-1)\leq 0$

$\Rightarrow \left(a^2-14a+13\right)-(a+2)-2\leq 0$

$\Rightarrow a^2-15a\leq -9$

So, maximum value of $a^2-15a=-9$

20. Since, α and β are roots of $ax^2+bx+c=0$

Therefore, $a\alpha^2+b\alpha+c=0$ and

$a\beta^2+b\beta+c=0$

Now, $aS_{n+1}+bS_n+cS_{n-1}$

$=a\left(\alpha^{n+1}+\beta^{n+1}\right)+b\left(\alpha^n+\beta^n\right)+c\left(\alpha^{n-1}+\beta^{n-1}\right)$

$=\alpha^{n-1}\left(a\alpha^2+b\alpha+c\right)+\beta^{n-1}\left(a\beta^2+b\beta+c\right)$

$=0+0=0$

1.9.4 JEE Main: Archive

1. **(B)**

α, β roots of equation $3x^2+\lambda x-1=0$

$$\alpha+\beta=\frac{-\lambda}{3}, \alpha\beta=\frac{-1}{3}$$

$$\frac{1}{\alpha^2}+\frac{1}{\beta^2}=\frac{(\alpha+\beta)^2-2\alpha\beta}{\alpha^2\beta^2}=15$$

$\lambda^2=9$

Now

$$6\left(\alpha^3+\beta^3\right)^2=6\left((\alpha+\beta)\left((\alpha+\beta)^2-3\alpha\beta\right)\right)^2$$

$$=6\left(\frac{\lambda^2}{9}\right)\left\{\frac{\lambda^2}{9}+1\right\}^2=24$$

2. **(B)**

$\left(e^{2x}-4\right)\left(6e^{2x}-3e^x-2e^x+1\right)=0$

$\left(e^{2x}-4\right)\left(3e^x-1\right)\left(2e^x-1\right)=0$

$e^{2x}=4$ or $e^x=\frac{1}{3}$ or $e^x=\frac{1}{2}$

$\Rightarrow$ sum of real roots $=\frac{1}{2}\ln 4+\ln\frac{1}{3}+\ln\frac{1}{2}$

$=-\ln 3$

3. **(4)**

$a_n=19^n-12^n$

$$\frac{31\alpha_9-\alpha_{10}}{57\alpha_8}=\frac{31\left(19^9-12^9\right)-\left(19^{10}-12^{10}\right)}{57\alpha_8}$$

$$\frac{19(31-19)-12^9(31-12)}{57\alpha_8}$$

$$=\frac{19^9\cdot 12-12^{19}\cdot 19}{57\alpha_8}$$

$$=\frac{12\cdot 19\left(19^8-12^8\right)}{57\alpha_8}=4$$

4. **(B)**

$ax^2-2bx+15=0$

$2\alpha=\frac{2b}{a}, \alpha^2=\frac{15}{a}$

$\frac{\alpha}{2}=\frac{15}{2b}$

$\alpha=\frac{15}{b}$

$x^2-2bx+21=0$

$$\left(\frac{15}{b}\right)^2-2b\left(\frac{15}{b}\right)+21=0$$

$b^2=25$

$\alpha+b=2b, \alpha\beta=21$

$\alpha^2+\beta^2=4b^2-42$

$=58$

5. **(36)**

$x^4-3x^3-2x^2+3x+1=10$

$x=0$ is not the root of this equation so divide it by x^2

$x^2-3x-2+\frac{3}{x}+\frac{1}{x^2}=0$

$x^2+\frac{1}{x^2}-2+2-3\left(x-\frac{1}{x}\right)-2=0$

$\left(x-\frac{1}{x}\right)^2-3\left(x-\frac{1}{x}\right)=0$

$x-\frac{1}{x}=0, x-\frac{1}{x}=3$

$x^2-1=0, x^2-3x-1=0$

$x=\pm1, \gamma+\delta=3$

$\alpha=1, \beta=-1, \gamma\delta=-1$

$\alpha^3+\beta^3+\gamma^3+\delta^3$

$1-1+(\gamma+\delta)\left((\gamma+\delta)^2-3\gamma\delta\right)$

$0+3(9-3(-1))+3(12)=36$

6. **(45)**

$e^{2x}-11e^x-45e^{-x}+\frac{81}{2}=0]$

$\left(e^x\right)^3-11\left(e^x\right)^2-45+\frac{81e^x}{2}=0]$

$e^x=t$

$2t^3-22t^2+81t-90=0$

$t_1t_2t_3=45$

$e^{x_1}\cdot e^{x_2}\cdot e^{x_3}=45$

$e^{x_1+x_2+x_3}=45$

$\log_e e^{x_1+x_2+x_3}=\log_e 45$

$x_1+x_2+x_3=\log_e 45$

$\log_e P=\log_e 45$

$P=45$

7. **(98)**

$x^2-4\lambda x+5=0<^{\alpha}_{\beta}$

$x^2-\left(3\sqrt{2}+2\sqrt{3}\right)x+\left(7+3\lambda\sqrt{3}\right)=0<^{\alpha}_{\gamma}$

$\alpha+\beta=4\lambda$

$\alpha+\gamma=3\sqrt{2}+2\sqrt{3}$

$\beta+\lambda=3\sqrt{2}$ $\qquad$ $\alpha\gamma=7+3\lambda\sqrt{3}$

$\therefore \alpha=2\lambda+\sqrt{3}$ $\qquad$ $\alpha\beta=5$

$\beta=2\lambda-\sqrt{3}$ $\qquad$ $4\lambda^2=8\Rightarrow\lambda=\sqrt{2}$

$\therefore (\alpha+2\beta+\lambda)^2=\left(4\alpha+3\sqrt{2}\right)^2=\left(7\sqrt{2}\right)^2=98$

8. **(2)**

$e^{4x}+4e^{3x}-58e^{2x}+4e^x+1=0$

Let

$f(x)=e^{2x}\left(e^{2x}+\frac{1}{e^{2x}}+4\left(e^x+\frac{1}{e^x}\right)-58\right)$

$e^x+\frac{1}{e^x}$

Let $h(t)=t^2+4t-58=0$

$t=\frac{-4\pm\sqrt{16+4.58}}{2}$

$=\frac{-4\pm2\sqrt{62}}{2}$

$t_1=-2+2\sqrt{62}$

$t_2=-2-2\sqrt{62}$ (not possible)

$t\geq2$

$e^x+\frac{1}{e^x}=-2+2\sqrt{62}$

$e^{2x}-\left(-2+2\sqrt{62}\right)e^x+1=0$

$\left(-2+2\sqrt{62}\right)-4$

$4+4.62-8\sqrt{62}-4$

$248-8\sqrt{62}>0$

$-\frac{b}{2a}>0$

Both roots are positive; 2 real roots

9. **(A)**

$f(-2)+f(3)=0$

$f(x)=(x+1)(ax+b)$

$f(-2)+f(3)=-1(-2a+b)+4(3a+b)=0$

$2a-b+12a+4b=0$

$14a+3b=0$

$-\frac{b}{a} = \frac{14}{3}$

Sum of roots $= \left(-1 + \frac{-b}{a}\right) = -1 + \frac{14}{3} = \frac{11}{3}$

10. (A)

$x^4 + x^2 + 1 = 0$

$\Rightarrow (x^2 + x + 1)(x^2 - x + 1) = 0$

$\Rightarrow x = \pm\omega, \pm\omega^2$ where $\omega = 1^{\frac{1}{3}}$ and imaginary.

So $\alpha^{1011} + \alpha^{2022} - \alpha^{3033} = 1 + 1 - 1 = 1$

11. (B)

$\Delta = 0 = \begin{vmatrix} 1 & 1 & 1 \\ a & b & c \end{vmatrix}$

$\Rightarrow \alpha^2(c-b) - \alpha(c-a) + (b-a) = 0$

It is singular when $\alpha = 1$

$\frac{(a-c)^2}{(b-a)(c-b)} + \frac{(b-a)^2}{(a-c)(c-b)} + \frac{(c-b)^2}{(a-c)(b-a)}$

$\frac{(a-b)^3 + (b-c)^3 + (c-a)^3}{(a-b)(b-c)(c-a)}$

$= 3\frac{(a-b)(b-c)(c-a)}{(a-b)(b-c)(c-a)} = 3$

12. (5)

$|x|^2 - 2|x| + |\lambda - 3| = 0$

$|x|^2 - 2|x| + |\lambda - 3| - 1 = 0$

$(|x| - 1)^2 + |\lambda - 3| = 1$

At $\lambda = 3$, $x = 0$ and 2,

at $\lambda = 4$ or 2, then

$x = 1$ or -1

So maximum value of $x + \lambda = 5$

13. (B)

$3\left(x^2 + \frac{1}{x^2}\right) - 2\left(x + \frac{1}{x}\right) + 5 = 0$

$3\left[\left(x + \frac{1}{x}\right)^2 - 2\right] - 2\left(x + \frac{1}{x}\right) + 5 = 0$

Let $x + \frac{1}{x} = t$

$3t^2 - 2t - 1 = 0$

$3t^2 - 3t + t - 1 = 0$

$3t(t-1) + 1(t-1) = 0$

$(t-1)(3t+1) = 0$

$t = 1, -\frac{1}{3}$

$x + \frac{1}{x} = 1, -\frac{1}{3} \Rightarrow$ No solution.

14. (25)

$\log_2\left(9^{2\alpha-4} + 13\right) - \log_2\left(\frac{5}{2} \cdot 3^{2\alpha-4} + 1\right) = 2$

$\Rightarrow \frac{9^{2\alpha-4} + 13}{\frac{5}{2}3^{2\alpha-4} + 1} = 4$

$\Rightarrow \alpha = 2$ or 3

$\sum_{\alpha \in S} \alpha = 5$ and $\sum_{\alpha \in S} (\alpha + 1)^2 = 25$

$\Rightarrow x^2 - 50x + 25\beta = 0$ has real roots

$\Rightarrow \beta \le 25$

$\Rightarrow \beta_{\max} = 25$

15. (45)

Let α and β be roots of $x^2 + 60^{\frac{1}{4}} x + a = 0$

$\alpha + \beta = -60^{\frac{1}{4}}$ and $\alpha\beta = a$

Given $\alpha^4 + \beta^4 = -30$

$\Rightarrow (\alpha^2 + \beta^2)^2 - 2\alpha^2\beta^2 = -30$

$\Rightarrow \left\{(\alpha + \beta)^2 - 2\alpha\beta\right\}^2 - 2a^2 = -30$

$\Rightarrow \left\{60^{\frac{1}{2}} - 2a\right\}^2 - 2a^2 = -30$

$\Rightarrow 60 + 4a^2 - 4a \times 60^{\frac{1}{2}} - 2a^2 = -30$

$\Rightarrow 2a^2 - 4 \cdot 60^{\frac{1}{2}} a + 90 = 0$

Product $= \frac{90}{2} = 45$

16. (C)

$14x^2 - 31x + 3\lambda = 0$

$\alpha + \beta = \frac{31}{14}$ (1)

and $\alpha\beta = \dfrac{3\lambda}{14}$ (2)

$35x^2 - 53x + 4\lambda = 0$

$\alpha + \gamma = \dfrac{53}{35}$ (3)

and $\alpha\gamma = \dfrac{4\lambda}{35}$ (4)

$\dfrac{(2)}{(4)} \Rightarrow \dfrac{\beta}{\gamma} = \dfrac{3\times 35}{4\times 14} = \dfrac{15}{8} \Rightarrow \beta = \dfrac{15}{8}\gamma$

$(1)-(3) \Rightarrow \beta - \gamma = \dfrac{31}{14} - \dfrac{53}{35} = \dfrac{155 = 106}{70} = \dfrac{7}{10}$

$\dfrac{15}{8}\gamma - \gamma = \dfrac{7}{10} \Rightarrow \gamma = \dfrac{4}{5}$

$\Rightarrow \beta = \dfrac{15}{8} \times \dfrac{4}{5} = \dfrac{3}{2}$

$\Rightarrow \alpha = \dfrac{31}{14} - \beta = \dfrac{31}{14} - \dfrac{3}{2} = \dfrac{5}{7}$

$\Rightarrow \lambda = \dfrac{14}{3}\alpha\beta = \dfrac{14}{3} \times \dfrac{5}{7} \times \dfrac{3}{2} = 5$

So, sum of roots $\dfrac{3\alpha}{\beta} + \dfrac{4\alpha}{\gamma} = \left(\dfrac{3\alpha\gamma + 4\alpha\beta}{\beta\gamma}\right)$

$= \dfrac{\left(3\times\dfrac{4\lambda}{35} + 4\times\dfrac{3\lambda}{14}\right)}{\beta\gamma} = \dfrac{12\lambda(14+35)}{14\times 35\beta\gamma}$

$\dfrac{49\times 12\times 5}{490\times\dfrac{3}{2}\times\dfrac{4}{5}} = 5$

17. **(13)**

Two equations have common root

$\therefore (4a)(26a) = (-6)^2 = 36$

$\Rightarrow a^2 = \dfrac{9}{26}$

$\therefore a = \dfrac{3}{\sqrt{26}} \Rightarrow \beta = 13$

18. **(B)**

$\sqrt{(x-1)(x-3)} + \sqrt{(x-3)(x+3)}$

$= \sqrt{4\left(x - \dfrac{12}{4}\right)\left(x - \dfrac{2}{4}\right)}$

$\Rightarrow \sqrt{x-3} = 0 \Rightarrow x = 3$ which is in domain

or

$\sqrt{x-1} + \sqrt{x+3} = \sqrt{4x-2}$

$2\sqrt{(x-1)(x+3)} = 2x - 4$

$x^2 + 2x - 3 = x^2 - 4x + 4$

$6x = 7$

$x = \dfrac{7}{6}$ (rejected)

19. **(A)**

$e^{4x} + 8e^{3x} + 13e^{2x} - 8e^x + 1 = 0$

Let $e^x = t$

Now, $t^4 + 8t^3 + 13t^2 - 8t + 1 = 0$

Dividing equation by t^2,

$t^2 + 8t + 13 - \dfrac{8}{t} + \dfrac{1}{t^2} = 0$

$t^2 + \dfrac{1}{t^2} + 8\left(t - \dfrac{1}{t}\right) + 13 = 0$

$\left(t - \dfrac{1}{t}\right)^2 + 2 + 8\left(t - \dfrac{1}{t}\right) + 13 = 0$

Let $t - \dfrac{1}{t} = z$

$z^2 + 8z + 15 = 0$

$(z+3)(z+5) = 0$

$z = -3$ or $z = -5$

So, $t - \dfrac{1}{t} = -3$ or $t - \dfrac{1}{t} = -5$

$t^2 + 3t - 1 = 0$ or $t^2 + 5t - 1 = 0$

$t = \dfrac{-3 \pm \sqrt{13}}{2}$ or $t = \dfrac{-5 \pm \sqrt{29}}{2}$

as $t = e^x$ so t must be positive,

$t = \dfrac{\sqrt{13} - 3}{2}$ or $\dfrac{\sqrt{29} - 5}{2}$

20. **(C)**

$\left(\sqrt{3} + \sqrt{2}\right)^x + \left(\sqrt{3} - \sqrt{2}\right)^x = 10$

Let $\left(\sqrt{3} + \sqrt{2}\right)^x = t$

$t + \dfrac{1}{t} = 10$

$t^2 - 10t + 1 = 0$

$t=\frac{10\pm\sqrt{100-4}}{2}=5\pm2\sqrt{6}$

$\left(\sqrt{3}+\sqrt{2}\right)^x=\left(\sqrt{3}\pm\sqrt{2}\right)^2$

$x=2$ or $x=-2$

Number of solutions = 2

21. (A)

$px^2+qx-r=0<^{\alpha}_{\beta}$

$p=A, q=AR, r=AR^2$

$Ax^2+ARx-AR^2=0$

$x^2+Rx-R^2=0<^{\alpha}_{\beta}$

$\because \frac{1}{\alpha}+\frac{1}{\beta}=\frac{3}{4}$

$\therefore \frac{\alpha+\beta}{\alpha\beta}=\frac{3}{4}\Rightarrow\frac{-R}{-R^2}=\frac{3}{4}\Rightarrow R=\frac{4}{3}$

$(\alpha-\beta)^2=(\alpha+\beta)^2-4\alpha\beta$

$=R^2-4\left(-R^2\right)=5\left(\frac{16}{9}\right)$

$=\frac{80}{9}$

22. (B)

$x^2-x-1=0$

$S_n=2023\alpha^n+2024\beta^n$

$S_{n-1}+S_{n-2}=2023\alpha^{n-1}+2024\beta^{n-1}+$
$2023\alpha^{n-2}+2024\beta^{n-2}$

$=2023\alpha^{n-2}[1+\alpha]+2024\beta^{n-2}[1+\beta]$

$=2023\alpha^{n-2}\left[\alpha^2\right]+2024\beta^{n-2}\left[\beta^2\right]$

$=2023\alpha^n+2024\beta^n$

$S_{n-1}+S_{n-2}=S_n$

Put $n=12$

$S_{11}+S_{10}=S_{12}$

23. (46)

$x^2-2^y=2023$

$\Rightarrow x=45, y=1$

$\sum_{(x,y)=C}(x+y)=46$

24. (60)

$x^2-70x+\lambda=0$

$\alpha+\beta=70$

$\alpha\beta=\lambda$

$\therefore \alpha(70-\alpha)=\lambda$

Since, 2 and 3 does not divide λ

$\therefore \alpha=5, gb=65, \lambda=325$

By putting values of α, β, λ we get the required value 60.

25. (1)

$x=0$ and $x^2+3|x|+5|x-1|+6|x-2|=0$

Here all the terms are +ve except at $x=0$

So, there is no value of x that satisfies this equation

Only solution $x=0$

Number of solutions is 1.

26. (A)

$f(x)=(a+b-2c)x^2+$
$(b+c-2a)x+(c+a-2b)$

$f(x)=a+b-2c+b+c-2a+c+a-2b=0$

$f(1)=0$

$\therefore \alpha\cdot1=\frac{c+a-2b}{a+b-2c}$

$\alpha=\frac{c+a-2b}{a+b-2c}$

If $-1<\alpha<0$

$-1<\frac{c+a-2b}{a+b-2c}<0$

$b+c<2a$ and $b>\frac{a+c}{2}$

therefore, b cannot be G.M. between a and c.

If $0<a<1$

$0<\frac{c+a-2b}{a+b-2c}<1$

$b>c$ and $b<\frac{a+c}{2}$

Therefore b may be the G.M. between a and c.

27. (B)

$ax^2+2(a+1)x+9a+4<0\forall x\in R$

$\therefore a<0$

28. **(D)**

Take $e^{\sin x} = t\,(t > 0)$

$\Rightarrow t - \dfrac{2}{t} = 2$

$\Rightarrow \dfrac{t^2 - 2}{t} = 2$

$\Rightarrow t^2 - 2t - 2 = 0$

$\Rightarrow t^2 - 2t + 1 = 3$

$\Rightarrow (t-1)^2 = 3$

$\Rightarrow t = 1 \pm \sqrt{3}$

$\Rightarrow t = 1 \pm 1.73$

$\Rightarrow t = 2.73$ or -0.73 (rejected as $t > 0$)

$\Rightarrow e^{\sin x} = 2.73$

$\Rightarrow \log_e e^{\sin x} = \log_e 2.73$

$\Rightarrow \sin x = \log_e 2.73 > 1$

So no solution.

29. **(36)**

$(a^2 + b^2)x^2 - 2b(a+c)x + b^2 + c^2 = 0$

$\Rightarrow a^2x^2 - 2abx + b^2 + b^2x^2 - 2bcx + c^2 = 0$

$\Rightarrow (ax-b)^2 + (bx-c)^2 = 0$

$\Rightarrow ax - b = 0, bx - c = 0$

$\Rightarrow a + b > c \qquad b + c > a \; c + a > b$

$\Rightarrow a + ax > bx \; \left| ax + bx > a \right| ax^2 + a > ax$

$a + ax > ax^2 \left| \begin{array}{l} ax + ax^2 > 0 \\ x^2 - x + 1 > 0 \end{array} \right.$

$x^2 - x - 1 < 0 \left| x^2 + x - 1 > 0 \right|$ always true

$\dfrac{1-\sqrt{5}}{2} < x < \dfrac{1+\sqrt{5}}{2}$

$x < \dfrac{-1-\sqrt{5}}{2}$, or $x > \dfrac{-1+\sqrt{5}}{2}$

$\Rightarrow \dfrac{\sqrt{5}-1}{2} < x < \dfrac{\sqrt{5}+1}{2}$

$\Rightarrow \alpha = \dfrac{\sqrt{5}-1}{2}, \beta = \dfrac{\sqrt{5}+1}{2}$

$$2(\alpha^2 + \beta^2) = 12\left(\frac{(\sqrt{5}-1)^2 + (\sqrt{5}+1)^2}{4}\right) = 36$$

1.10 Summary

- Relation between the roots and the coefficient of a quadratic equation $ax^2 + bx + c = 0$ whose roots are α and β is
  $ax^2 + bx + c = a(x-\alpha)(x-\beta)$.
  Similarly, if α, β, γ are the roots of the cubic equation $ax^3 + bx^2 + cx + d = 0$, then
  $ax^3 + bx^2 + cx + d = a(x-\alpha)(x-\beta)(x-\gamma)$
  and if $\alpha, \beta, \gamma, \delta$ are the roots of the equation $ax^4 + bx^3 + cx^2 + dx + e = 0$, then
  $ax^4 + bx^3 + cx^2 + dx + e$
  $= a(x-\alpha)(x-\beta)(x-\gamma)(x-\delta)$.
- Nature of roots of the quadratic equation:
  - $D = 0$: Roots are real and equal
  - $D > 0$: Roots are real and distinct
  - $D < 0$: Roots are imaginary
  - If D is a perfect square, roots are rational
  - If $a = 1$ and $b, c \in I$ and roots are rational, then roots must be integers
  - If $p + iq$ is one root, then the other root will be $p - iq$
  - If $p + \sqrt{q}$ is one root, then the other root will be $p - \sqrt{q}$
  - If D_1 and D_2 are discriminants of two quadratic equations and $D_1 + D_2 \geq 0$, then at least one of D_1 and $D_2 \geq 0$. If $D_1 + D_2 < 0$ then at least one of D_1 and $D_2 < 0$.
  - If $ax^2 + bx + c = 0$ has more than two roots, then it will be an identity and this implies that $a = b = c = 0$
- If $a_1x^2 + b_1x + c_1 = 0$ and $a_2x^2 + b_2x + c_2 = 0$
  - have one root in common, then
    $\dfrac{b_1c_2 - b_2c_1}{c_1a_2 - c_2a_1} = \dfrac{c_1a_2 - c_2a_1}{a_1b_2 - a_2b_1}$
  - If both roots are common, then
    $\dfrac{a_1}{a_2} = \dfrac{b_1}{b_2} = \dfrac{c_1}{c_2}$
- If in a quadratic equation $ax^2 + bx + c = 0$:
  - $a > 0$, it represents an upward parabola having vertex $\left(\dfrac{-b}{2a}, \dfrac{-D}{4a}\right)$

- $a < 0$, it represents a downward parabola having vertex $\left(\frac{-b}{2a}, \frac{-D}{4a}\right)$

- Some more cases:
  - When both roots are greater than a real number k

    (I) $D \geq 0$ (II) $\frac{-b}{2a} > k$ (III) $af(k) > 0$
  - When both roots are less than a real number k

    (I) $D \geq 0$ (II) $\frac{-b}{2a} < k$ (III) $af(k) > 0$
  - If a real number k lies between the roots

    (I) $D > 0$ (II) $af(k) < 0$
  - If exactly one root lies in the interval (k_1, k_2), then

    (I) $D > 0$ (II) $f(k_1)f(k_2) < 0$.

CHAPTER 2

Progression and Series

OBJECTIVES

2.1 Introduction
2.2 Arithmetic Progression
2.3 Geometric Progression
2.4 Harmonic Progression
2.5 Relation between A.M. G.M. and H.M.
2.6 Sum of Natural Number Expressions
2.7 Arithmetico–Geometric Series
2.8 Know Your Concepts
2.9 JEE Main: Sample Questions
2.10 JEE Advanced: Sample Questions
2.11 JEE Main: Archive
2.12 Answer Key
2.13 Solutions
2.14 Summary

2.1 Introduction

A sequence is a list of objects that have been ordered in a sequential fashion, such that each member either comes before or after every other member. In this topic the objects are usually numbers.

If the consecutive terms bear a fixed relation, then the sequence is known as a progression. If the two consecutive numbers of the group have a constant difference, then such numbers are said to be in 'Arithmetic Progression'. If the numbers are such that the square of one number is equal to the product of the number before it and after it, then such numbers are said to be in 'Geometric Progression'. If the characteristic is that the reciprocals of the given numbers are in arithmetic progression, then such numbers are said to be in 'Harmonic Progression'.

2.1.1 Sequence

A succession of numbers $t_1, t_2, t_3, \ldots, t_n, \ldots$ formed according to some definite rule is called a *sequence*. $t_1, t_2, t_3, \ldots, t_n$ are respectively called the first, second, ..., n^{th} term of the sequence. In other words, a sequence is a function whose domain is the set of natural numbers N, or a subset of N, and the range is a set of real numbers.

2.1.2 Progression

If the terms of a sequence are written under specific conditions, then the sequence is called a *progression*.

2.2 Arithmetic Progression

A succession of numbers is said to be an *arithmetic progression* (A.P.) if the difference between any two consecutive terms is constant. This constant is called the *common difference*.

2.2.1 Find the nth term of an A.P.

Let a be the first term and d be the common difference of an A.P.

Now, first term $= a$

Second term $= a + d$

Third term $= a + 2d$

---

n^{th} term $= a + (n-1)d$

Thus, the nth term of an A.P. is $a + (n-1)d$ and it is denoted by t_n.

Hence, $t_n = a + (n-1)d$.

Note: The last term of an A.P. is denoted by l and is defined as the nth term of an A.P.

Hence, $l = a + (n-1)d$.

2.2.2 Find the Sum of *n* Terms of an A.P.

Let a be the first term, d be the common difference and l be the last term of an A.P. Also let S_n denote the sum of n terms of the A.P.

Then,

$$S_n = a+(a+d)+(a+2d)+\ldots+(l-2d)+(l-d)+l \quad (2.1)$$

$$S_n = l+(l-d)+(l-2d)\ldots+(a+2d)+(a+d)+a \quad (2.2)$$

Adding (2.1) and (2.2)

$$2S_n = (a+l)+(a+l)+(a+l)+\ldots+(a+l)$$

$$2S_n = n(a+l)$$

$$\Rightarrow \quad S_n = \frac{n}{2}(a+l) \quad (2.3)$$

$$\Rightarrow \quad S_n = \frac{n}{2}[a+a+(n-1)d]$$

$$\Rightarrow \quad S_n = \frac{n}{2}[2a+(n-1)d](2.4) \quad (2.4)$$

Note:

(i) Three numbers in A.P. are $a-d, a, a+d$

(ii) Four numbers in A.P. are $a-3d, a-d, a+d, a+3d$

(iii) Five numbers in A.P. are $a-2d, a-d, a, a+d, a+2d$

2.2.3 Find the Single Arithmetic Mean Between Any Two Given Numbers

Let A be the A.M. between any two given numbers a and b.

Then a, A, b will be in A.P.

$$\Rightarrow \quad A-a=b-A$$

$$\Rightarrow \quad 2A=a+b$$

$$\Rightarrow \quad A=\frac{a+b}{2}$$

2.2.4 Insert *n* A.M.s Between Any Two Given Numbers

Let $A_1, A_2, \ldots, A_n$ be n A.M.s inserted between any two given numbers a and b.

Hence, $a, A_1, A_2, \ldots, A_n, b$ are in A.P.

Now, $b=(n+2)^{\text{th}}$ term

$$=a+(n+2-1)d$$

$$\Rightarrow \quad d=\frac{(b-a)}{(n+1)}$$

Now,

$$A_1 = a+d = a+\frac{b-a}{n+1} = \frac{na+a+b-a}{n+1}$$

$$\Rightarrow \quad A_1 = \frac{na+b}{n+1}$$

$$A_2 = a+2d = a+\frac{2(b-a)}{n+1} = \frac{na+a+2b-2a}{(n+1)}$$

$$\Rightarrow \quad A_2 = \frac{(n-1)a+2b}{n+1}$$

$$A_n = a+nd = a+\frac{n(b-a)}{n+1} = \frac{na+a+nb-na}{(n+1)}$$

$$\Rightarrow \quad A_n = \frac{a+nb}{n+1}$$

Thus,

$$A_1 = \frac{na+b}{n+1}$$

$$A_2 = \frac{(n-1)a+2b}{n+1}$$

$$A_n = \frac{a+nb}{n+1}$$

2.2.5 Find the Sum of *n* A.M. Inserted Between *a* and *b*

$$A_1 + A_2 + \ldots + A_n$$

$$= \frac{1}{(n+1)}\left[(n+(n-1)+\ldots+2+1)a+(1+2+\ldots+n)b\right]$$

$$= \frac{1}{(n+1)}\frac{n(n+1)}{2}(a+b)$$

$$A_1 + A_2 + \ldots + A_n = n\frac{(a+b)}{2}$$

Note: In an A.P., the sum of the terms equidistant from beginning and end is equal to the sum of first and last term.

2.3 Geometric Progression

A succession of numbers is said to be a *geometric progression* (G.P.), if the ratio of any two consecutive terms is constant, and this constant is called common ratio (c.r.).

2.3.1 Find the nth term of a G.P.

Let a be the first term and r be the c.r. of a G.P.

Now, first term = a

Second term = ar

Third term = ar^2

---

n^{th} term = ar^{n-1}

$\Rightarrow \quad t_n = ar^{n-1}$

2.3.2 Find the Sum of n Terms of a G.P.

Let a be the first term and r be the c.r. of a G.P. Also, let S_n denote the sum of n terms of a G.P. Now,

$$S_n = a + ar + ar^2 + \ldots + ar^{n-2} + ar^{n-1} \quad (2.5)$$

$$\Rightarrow \quad rS_n = ar + ar^2 + \ldots + ar^{n-1} + ar^n \quad (2.6)$$

Subtracting (2.6) from (2.5)

$$(1-r)S_n = a - ar^n$$

$$\Rightarrow \quad S_n = \frac{a(1-r^n)}{(1-r)}(r<1)$$

$$\Rightarrow \quad S_n = \frac{a(r^n-1)}{(r-1)}(r>1)$$

Sum of infinite G.P. if $|r| < 1 = \frac{a}{1-r}$.

2.3.3 Find the Single G.M. Inserted Between Any Two Given Positive Numbers

Let G be the G.M. between any two given positive numbers a and b.

$\therefore \quad a, G, b$ are in G.P.

$$\Rightarrow \quad \frac{G}{a} = \frac{b}{G}$$

$$\Rightarrow \quad G^2 = ab$$

$$\Rightarrow \quad G = \sqrt{ab}$$

2.3.4 Insert n Geometric Means Between Two Given Positive Numbers

Let $G_1, G_2, \ldots, G_n$ be n G.M.s inserted between any two given positive numbers a and b.

Then, $a, G_1, G_2, \ldots, G_n, b$ are in G.P.

Now, $b = (n+2)$ th term $= ar^{n+2-1}$

$$\Rightarrow \quad r = \left(\frac{b}{a}\right)^{\frac{1}{n+1}}$$

Now,
$$G_1 = ar = a\left(\frac{b}{a}\right)^{\frac{1}{n+1}}$$

$$G_2 = ar^2 = a\left(\frac{b}{a}\right)^{\frac{2}{n+1}}$$

$$G_n = ar^n = a\left(\frac{b}{a}\right)^{\frac{n}{n+1}}$$

2.3.5 Find the Product of n G.M.s Inserted Between Two Positive Numbers a and b

$$G_1 \cdot G_2 \cdot G_3 \ldots G_n = a^n \left(\frac{b}{a}\right)^{\frac{1+2+\ldots+n}{n+1}}$$

$$= a^n \left(\frac{b}{a}\right)^{\frac{n(n+1)}{2(n+1)}}$$

$$= a^n \frac{b^{\frac{n}{2}}}{a^{\frac{n}{2}}}$$

$$= a^{\frac{n}{2}} b^{\frac{n}{2}}$$

$$= \left(\sqrt{ab}\right)^n$$

Note:

(i) Three numbers in G.P. are $\frac{a}{r}, a, ar$

(ii) Four numbers in G.P. are $\frac{a}{r^3}, \frac{a}{r}, ar, ar^3$

2.4 Harmonic Progression

A *Harmonic Progression* (H.P.) is a sequence, if the reciprocals of its terms are in A.P.

2.4.1 Find the Single H.M. Between Any Two Given Numbers

Let H be the H.M. between any two given numbers a and b.

Then, a, H, b are in H.P.

$\Rightarrow \quad \frac{1}{a}, \frac{1}{H}, \frac{1}{b}$ are in A.P.

$$\Rightarrow \quad \frac{1}{H}-\frac{1}{a}=\frac{1}{b}-\frac{1}{H}=\frac{2}{H}=\frac{1}{a}+\frac{1}{b}$$

$$\Rightarrow \quad H=\frac{2}{\frac{1}{a}+\frac{1}{b}}$$

2.4.2 Insert *n* H.M.s Between Any Two Given Numbers

Let $H_1, H_2, H_3 \ldots H_n$ be n H.M.s inserted between any two given numbers a and b.

Therefore, $a, H_1, H_2, \ldots, H_n, b$ are in H.P.

$\Rightarrow \quad \frac{1}{a}, \frac{1}{H_1}, \frac{1}{H_2}, \ldots, \frac{1}{H_n}, \frac{1}{b}$ are in A.P.

$$\text{Now,} \quad \frac{1}{b}=(n+2)^{\text{th}} \text{ term}=\frac{1}{a}+(n+2-1)d$$

$$\Rightarrow \quad \frac{1}{b}-\frac{1}{a}=(n+1)d$$

$$\Rightarrow \quad d=\frac{(a-b)}{(n+1)ab}$$

$$\text{Now,} \quad \frac{1}{H_1}=\frac{1}{a}+d=\frac{1}{a}+\frac{(a-b)}{(n+1)ab}$$

$$=\frac{(n+1)b+a-b}{(n+1)ab}$$

$$=\frac{a+nb}{(n+1)ab}$$

$$\frac{1}{H_n}=\frac{1}{a}+nd=\frac{1}{a}+\frac{n(a-b)}{(n+1)ab}$$

$$=\frac{(n+1)b+n(a-b)}{(n+1)ab}$$

$$=\frac{na+b}{(n+1)ab}$$

$$\text{Thus,} \quad H_1=\frac{(n+1)ab}{a+nb}$$

$$H_n=\frac{(n+1)ab}{na+b}$$

Note:

(i) The A.M. of n numbers

$$a_1, a_2, \ldots, a_n=\frac{a_1+a_2+\ldots+a_n}{n}$$

(ii) The G.M. of n numbers

$$a_1, a_2, \ldots, a_n=(a_1 \cdot a_2 \ldots a_n)^{\frac{1}{n}}$$

(iii) The H.M. of n numbers

$$a_1, a_2, \ldots, a_n=\frac{n}{\frac{1}{a_1}+\frac{1}{a_2}+\ldots+\frac{1}{a_n}}$$

2.5 Relation between A.M., G.M. and H.M.

Let A, G, H respectively denote A.M., G.M. and H.M. between any two given positive numbers a and b.

$$\text{Then, } A=\frac{a+b}{2}, G=\sqrt{ab}, H=\frac{2ab}{a+b}$$

$$\text{Now, } AH=\frac{a+b}{2}\cdot\frac{2ab}{a+b}=ab=G^2$$

$\Rightarrow \quad A, G, H$ are in G.P.

$$\text{Again, } A-G=\frac{a+b}{2}-\sqrt{ab}=\frac{1}{2}\left(\sqrt{a}-\sqrt{b}\right)^2 \geq 0$$

$\Rightarrow \quad A \geq G$

As, $\frac{A}{G}=\frac{G}{H}$

$\Rightarrow \quad G \geq H$

Thus, $A \geq G \geq H$.

Equality sign exists if the numbers are equal.

2.6 Sum of Natural Number Expressions

I. Sum of the first n natural numbers

$$1+2+3+\ldots+n=\sum_{r=1}^{n} r=\frac{n(n+1)}{2}$$

II. Sum of the squares of first n natural numbers.

$$1^2+2^2+3^2+\ldots+n^2=\sum_{r=1}^{n} r^2=\frac{n(n+1)(2n+1)}{6}$$

Proof

Since, $r^3-(r-1)^3=3r^2-3r+1$

Put $r=1, 2, 3, \ldots, n$ respectively, we get

$$1^3 - 0^3 = 3.1^2 - 3.1 + 1$$
$$2^3 - 1^3 = 3.2^2 - 3.2 + 1$$
$$3^3 - 2^3 = 3.3^2 - 3.3 + 1$$
$$n^3 - (n-1)^3 = 3.n^2 - 3.n + 1$$

Adding, we get,

$$n^3 = 3\left(1^2 + 2^2 + 3^2 + \ldots + n^2\right) - 3\left(\frac{n(n+1)}{2}\right) + n$$

$$\Rightarrow 3\left(1^2 + 2^2 + 3^2 + \ldots + n^2\right)$$

$$= n^3 + 3\left(\frac{n(n+1)}{2}\right) - n$$

$$= n\left[n^2 + \frac{3(n+1)}{2} - 1\right]$$

$$= \frac{n}{2}\left[2n^2 + 3n + 1\right]$$

$$= \frac{n(n+1)(2n+1)}{2}$$

$$\Rightarrow 1^2 + 2^2 + 3^2 + \ldots + n^2 = \frac{n(n+1)(2n+1)}{6}$$

III. Sum of the cubes of first n natural numbers.

$$1^3 + 2^3 + 3^3 + \ldots + n^3 = \sum_{r=1}^{n} r^3 = \left(\frac{n(n+1)}{2}\right)^2$$

IV. Sum of the powers four of the first n natural numbers.

$$1^4 + 2^4 + 3^4 + \ldots + n^4$$

$$= \sum_{r=1}^{n} r^4 = \frac{n(n+1)(2n+1)\left(3n^2 + 3n - 1\right)}{30}$$

2.7 Arithmetico–Geometric Series

A series formed by multiplying the corresponding terms of an A.P. and a G.P. is called an *Arithmetico – Geometric Series.*

2.7.1 Sum of *n* Terms of an Arithmetico–Geometric Series

Let

$$S_n = a + (a+d)r + (a+2d)r^2 + \ldots + \left[a + (n-2)d\right]r^{n-2} + \left[a + (n-1)d\right]r^{n-1} \quad (2.7)$$

Multiplying both side by r

$$rS_n = ar + (a+d)r^2 + \ldots + \left[a + (n-2)d\right]r^{n-1} + \left[a + (n-1)d\right]r^n \quad (2.8)$$

Subtracting (2.8) from (2.7), we get

$$(1-r)S_n$$

$$= a + \left(dr + dr^2 + \ldots + dr^{n-1}\right) - \left[a + (n-1)d\right]r^n$$

$$= a + \frac{dr\left(1 - r^{n-1}\right)}{(1-r)} - \left[a + (n-1)d\right]r^n$$

$$\Rightarrow S_n = \frac{a}{1-r} + \frac{dr\left(1 - r^{n-1}\right)}{(1-r)2} - \frac{\left[a + (n-1)d\right]r^n}{(1-r)}$$

2.7.2 Method of Differences

I. Let $S_n = t_1 + t_2 + t_3 + \ldots + t_n$

II. Rewrite the given series with each term shifted by one place to the right.

III. Then, subtract the second expression of S_n from the first expression to obtain t_n.

2.7.2.1 V_n Method

To find the sum of the series of the forms

i. $a_1 a_2 \ldots a_r + a_2 a_3 \ldots a_{r+1} + \ldots + a_n a_{n+1} \ldots a_{n+r-1}$

ii. $\frac{1}{a_1 a_2 \ldots a_r} + \frac{1}{a_2 a_3 \ldots a_{r+1}} + \ldots + \frac{1}{a_n a_{n+1} \ldots a_{n+r-1}}$

where $a_1, a_2, \ldots, a_n, \ldots$ are in A.P.

For the form (i),

Suppose $t_n = \leftarrow n(n+1)(n+2)(n+3) \rightarrow$

Write them

$$t_n = \frac{1}{5}\left[(n+4) - (n-1)\right]n(n+1)(n+2)(n+3)$$

$$= \frac{1}{5}\left[n(n+1)(n+2)(n+3)(n+4) - (n-1)n(n+1)(n+2)(n+3)\right]$$

$$= \frac{1}{5}\left[V_n - V_{n-1}\right]$$

Put $n = 1, 2, 3, \ldots, n$ and add we get the required sum.

For the form (ii),

$$\text{Suppose, } t_n = \frac{1}{n(n+1)(n+2)(n+3)}$$

$$\Rightarrow \quad t_n = \frac{1}{3}\left[\frac{(n+3)-n}{n(n+1)(n+2)(n+3)}\right]$$

$$\Rightarrow \quad t_n = \frac{1}{3}\left[\frac{1}{n(n+1)(n+2)} - \frac{1}{(n+1)(n+2)(n+3)}\right]$$

$$= \frac{1}{3}\left[V_{n-1} - V_n\right]$$

Put n = 1, 2, 3, …, n and add we get the required sum.

Note:

I. Weighted Means: Let $a_1, a_2, \ldots, a_n$ be n positive real numbers and $m_1, m_2, \ldots, m_n$ be n positive rational numbers. Then we define weighted Arithmetic Mean (A^*), weighted Geometric Mean (G^*) and weighted Harmonic Mean (H^*) as

$$A^* = \frac{m_1a_1 + m_2a_2 + \ldots + m_na_n}{m_1 + m_2 + \ldots + m_n}$$

$$G^* = \left(a_1^{m_1} a_2^{m_2}, \ldots, a_n^{m_n}\right)^{\frac{1}{(m_1+m_2+\ldots+m_n)}}$$

$$\text{and } H^* = \frac{m_1 + m_2 + \ldots + m_n}{\left(\frac{m_1}{a_1} + \frac{m_2}{a_2} + \ldots + \frac{m_n}{a_n}\right)}$$

II. Arithmetic Mean of mth power: Let $a_1, a_2, \ldots, a_n$ be n positive real numbers (not all equal) and let m be a real number. Then,

$$\frac{a_1^m + a_2^m + \ldots + a_n^m}{n} > \left(\frac{a_1 + a_2 + \ldots + a_n}{n}\right)^m$$

if $m \in R - [0, 1]$

However if $m \in (0, 1)$, then

$$\frac{a_1^m + a_2^m + \ldots + a_n^m}{n} < \left(\frac{a_1 + a_2 + \ldots + a_n}{n}\right)^m$$

If $m \in \{0, 1\}$, then

$$\frac{a_1^m + a_2^m + \ldots + a_n^m}{n} = \left(\frac{a_1 + a_2 + \ldots + a_n}{n}\right)^m$$

SOLVED EXAMPLES

Example 2.1 If m times the mth term of an A.P. is equal to n times the nth term, find its $(m + n)^{\text{th}}$ term

Solution Let a be the first term and d be the c.d. of an A.P.

From question,

$$mt_m = nt_n$$

$$\Rightarrow \quad m\{a+(m-1)d\} = n\{a+(n-1)d\}$$

$$\Rightarrow \quad a(m-n) = (n^2 - n - m^2 + m)d$$

$$\Rightarrow \quad a(m-n) = \{(m-n) - (m+n)(m-n)\}d$$

$$\Rightarrow \quad a = (1-m-n)d \qquad (1)$$

Now, $t_{m+n} = a + (m+n-1)d$

$$= -(m+n-1)d + (m+n-1)d = 0$$

Example 2.2 How many terms are identical in the two arithmetic progressions 2, 4, 6, 8, … up to 100 terms and 3, 6, 9, … up to 80 terms?

Solution Let r terms be identical

Now, identical terms are 6, 12, 24, …

Its rth term $= 6 + (r-1)6 = 6r$ (1)

Now, 100th term of 2, 4, 6, 8, …

$$= 2 + (100-1).2 = 2 + 198 = 200$$

And, 80th term of 3, 6, 9, …

$$= 3 + (80-1).3 = 240$$

Clearly the rth term of the sequence of identical terms cannot be greater than 200.

$$\therefore \quad 6r \le 200$$

$$\Rightarrow \quad r \le 33\frac{1}{3}$$

$$\Rightarrow \quad r = 33$$

Hence 33 terms are identical.

Example 2.3 Find the number of integers between 100 and 1000 that are

(i) divisible by 7.

(ii) not divisible by 7.

Solution

(i) Here $a = 105$, $d = 7$, $l = 994$.

We know, $l = a + (n-1)d$

$\Rightarrow 994 = 105 + (n-1).7$

$\Rightarrow n = 128$

Hence, number of integers between 100 and 1000 divisible by 7 = 128.

(ii) Number of integers not divisible by 7

$= 899 - 128 = 771$

Example 2.4 If a, b, c be the pth, q^{th} and r^{th} terms respectively of an A.P., then find the value of

$$a(q-r)+b(r-p)+c(p-q).$$

Solution Let x be the first term and y be the c.d. of an A.P.

Now,
$$a = x+(p-1)y$$
$$b = x+(q-1)y$$
$$c = x+(r-1)y$$
$$\Rightarrow \quad a(q-r)+b(r-p)+c(p-q)=0$$

Example 2.5 Divide 20 into four parts which are in A.P. such that the ratio of the product of the first and fourth to the product of the second and third is 2 : 3.

Solution Four numbers in A.P. are

$$a-3d,\ a-d,\ a+d,\ a+3d$$

From question,

$$a-3d+a-d+a+d+a+3d=20$$
$$\Rightarrow 4a=20$$
$$\Rightarrow a=5$$

Now,
$$\frac{(a-3d)(a+3d)}{(a-d)(a+d)}=\frac{2}{3}$$
$$\Rightarrow \frac{25-9d^2}{25-d^2}=\frac{2}{3}$$
$$\Rightarrow 75-27d^2=50-2d^2$$
$$\Rightarrow 25d^2=25$$
$$\Rightarrow d=\pm 1$$

Take $d = 1$,	Take $d = -1$
$\Rightarrow \quad a-3d=5-3=2$	$a-3d=8$
$\Rightarrow \quad a-d=5-1=4$	$a-d=6$
$\Rightarrow \quad a+d=5+1=6$	$a+d=4$
$\Rightarrow \quad a+3d=5+3=8$	$a+3d=2$

Hence four numbers are 2, 4, 6, 8 or 8, 6, 4, 2.

Example 2.6 If the sum of n terms of a series is $5n^2+3n$, find its nth term. Are the terms of this series in A.P.?

Solution As we know,

$$t_n = S_n - S_{n-1}$$
$$\Rightarrow \quad t_n=\left(5n^2+3n\right)-\left(5(n-1)^2+3(n-1)\right)$$
$$=5(n^2-(n-1)^2+3(n-n+1)$$
$$=5\left(n^2-n^2+2n-1\right)+3$$
$$=10n-5+3$$
$$=10n-2$$

Put, $n = 1, 2, 3, \ldots$

$$t_1=8,\ t_2=18,\ t_3=28,\ldots$$

Obviously, they are in A.P.

Example 2.7 Let the sum of n, $2n$, $3n$ terms of an A.P. be S_1, S_2 and S_3 respectively, prove that $S_3=3(S_2-S_1)$.

Solution Let a be the first term and d be the c.d. of an A.P.

$$S_1=\frac{n}{2}\left[2a+(n-1)d\right]$$
$$S_2=\frac{2n}{2}\left[2a+(2n-1)d\right]$$
$$S_3=\frac{3n}{2}\left[2a+(3n-1)d\right]$$

Now, $S_2-S_1=\frac{n}{2}\left[4a+(4n-2)d-2a-(n-1)d\right]$

$$=\frac{n}{2}\left[2a+(3n-1)d\right]$$
$$\Rightarrow \quad 3(S_2-S_1)=\frac{3n}{2}\left[2a+(3n-1)d\right]=S_3$$

Example 2.8 If $\frac{b+c-a}{a}, \frac{c+a-b}{b}, \frac{a+b-c}{c}$ are in A.P., show that $\frac{1}{a}, \frac{1}{b}, \frac{1}{c}$ are in A.P. provided $a+b+c \neq 0$.

Solution Since $\frac{b+c-a}{a}, \frac{c+a-b}{b}, \frac{a+b-c}{c}$ are in A.P.,

$$\Rightarrow \quad \frac{b+c-a}{a}+2, \frac{c+a-b}{b}+2, \frac{a+b-c}{c}+2 \text{ are in A.P.}$$
$$\Rightarrow \quad \frac{a+b+c}{a}, \frac{a+b+c}{b}, \frac{a+b+c}{c} \text{ are in A.P.}$$
$$\Rightarrow \quad \frac{1}{a}, \frac{1}{b}, \frac{1}{c} \text{ are in A.P.}$$

Example 2.9 If a, b, c are in A.P., prove that $a\left(\frac{1}{b}+\frac{1}{c}\right), b\left(\frac{1}{c}+\frac{1}{a}\right), c\left(\frac{1}{a}+\frac{1}{b}\right)$ are in A.P.

Solution Since, a, b, c are in A.P.

$$\Rightarrow \frac{a}{abc}, \frac{b}{abc}, \frac{c}{abc} \text{ are in A.P.}$$

$$\Rightarrow \frac{1}{bc}, \frac{1}{ca}, \frac{1}{ab} \text{ are in A.P.}$$

$$\Rightarrow \frac{ab+bc+ca}{bc}, \frac{ab+bc+ca}{ca}, \frac{ab+bc+ca}{ab} \text{ are in A.P.}$$

$$\Rightarrow \frac{a(b+c)}{bc}+1, \frac{b(a+c)}{ca}+1, \frac{c(b+a)}{ab}+1 \text{ are in A.P.}$$

$$\Rightarrow a\left(\frac{1}{b}+\frac{1}{c}\right), b\left(\frac{1}{c}+\frac{1}{a}\right), c\left(\frac{1}{a}+\frac{1}{b}\right) \text{ are in A.P.}$$

Example 2.10 If n A.M.s are inserted between 1 and 31 such that the ratio of the 7th and $(n-1)^{\text{th}}$ means is 5 : 9, find n.

Solution Let $A_1, A_2, \ldots, A_n$ be n A.M.s inserted between 1 and 31.

Hence, $1, A_1, A_2, \ldots, A_n, 31$ are in A.P.

Now, $31 = (n+2)^{\text{th}}$ term $= 1+(n+2-1)d$

$$\Rightarrow d = \frac{30}{n+1}$$

Now, $A_1 = 1+\frac{30}{n+1} = \frac{n+31}{n+1}$

$$A_2 = 1+\frac{60}{n+1} = \frac{n+61}{n+1}$$

$$A_7 = 1+\frac{210}{n+1} = \frac{n+211}{n+1}$$

$$A_{n-1} = 1+\frac{(n-1).30}{n+1} = \frac{n+1+30n-30}{n+1}$$

$$= \frac{31n-29}{n+1}$$

Now, $\frac{n+211}{31n-29} = \frac{5}{9}$

$$\Rightarrow 9n+1899 = 155n-145$$

$$\Rightarrow 146n = 2044$$

$$\Rightarrow n = 14$$

Example 2.11 If the $(p+q)^{\text{th}}$ term of a G.P. is a and the $(p-q)^{\text{th}}$ term is b, prove that its pth term is $\sqrt{ab}$.

Solution Let x be the first term and y be the c.r. of a G.P.

Now, $a = xy^{p+q-1}$

And $b = xy^{p-q-1}$

$$\Rightarrow \frac{a}{b} = y^{p+q-1-p+q+1} = y^{2q}$$

$$\Rightarrow y = \left(\frac{a}{b}\right)^{\frac{1}{2q}}$$

$$\Rightarrow a = x\left(\frac{a}{b}\right)^{\frac{p+q-1}{2q}}$$

$$\Rightarrow x = a\left(\frac{b}{a}\right)^{\frac{p+q-1}{2q}}$$

$$\Rightarrow x = a^{\frac{2q-p-q+1}{2q}} b^{\frac{p+q-1}{2q}}$$

$$\Rightarrow x = a^{\frac{q-p+1}{2q}} b^{\frac{p+q-1}{2q}}$$

Now, p^{th} term $= xy^{p-1}$

$$= a^{\frac{q-p+1}{2q}} b^{\frac{p+q-1}{2q}} \left(\frac{a}{b}\right)^{\frac{p-1}{2q}}$$

$$= a^{\frac{q-p+1+p-1}{2q}} b^{\frac{p+q-1-p+1}{2q}}$$

$$= a^{\frac{1}{2}} b^{\frac{1}{2}}$$

$$= \sqrt{ab}$$

Example 2.12 In a set of four numbers, the first three are in G.P. and the last three are in A.P. with a common difference of 6. If the first number is same as the fourth, find the four numbers.

Solution Let the last three numbers which are in A.P. be $\beta, \beta+6, \beta+12$ and first number be α.

From question, $\alpha = \beta+12$

Also, $\alpha, \beta, \beta+6$ are in G.P.

$$\Rightarrow \beta^2 = \alpha(\beta+6)$$

$$\Rightarrow \beta^2 = (\beta+12)(\beta+6)$$

$$\Rightarrow 18\beta+72 = 0$$

$$\Rightarrow \beta = -4$$

$$\Rightarrow \alpha = -4+12 = 8$$

Hence, the four numbers are 8, –4, 2 and 8.

Example 2.13 Find the sum of n terms of the series $8+88+888+\ldots$

Solution Let, $S_n = 8+88+888+\ldots$ to n terms

$$= \frac{8}{9}[9+99+999+\ldots \text{to } n \text{ terms}]$$

$$= \frac{8}{9}\left[(10-1)+(10^2-1)+(10^3-1)+\ldots \text{to } n \text{ terms}\right]$$

$$S_n = \frac{8}{9}\left[\frac{10(10^n-1)}{10-1} - n\right]$$

$$= \frac{8}{81}\left[10^{n+1}-10-9n\right]$$

Example 2.14 If a function $f(x)$ satisfying $f(x+y) = f(x)\cdot f(y)$ for all $x, y \in N$ such that $f(1) = 3$ and $\sum_{r=1}^{n} f(r) = 120$, then find the value of n.

Solution Putting $x = y = 1$, we get $f(2) = 3^2$

$x = 1, y = 2$, we get $f(3) = 3^3$

and so on.

Now, $\quad f(1)+f(2)+\ldots+f(n) = 120$

$\Rightarrow \quad 3+3^2+\ldots+3^n = 120$

$\Rightarrow \quad \dfrac{3(3^n-1)}{(3-1)} = 120$

$\Rightarrow \quad 3^n - 1 = 80$

$\Rightarrow \quad 3^n = 81$

$\Rightarrow \quad n = 4$

Example 2.15 If $a^x = b^y = c^z$ and x, y, z are in G.P., Prove that $\log_b a = \log_c b$.

Solution Let $a^x = b^y = c^z = K$

$\Rightarrow \quad x\log a = y\log b = z\log c = \log K$

$\Rightarrow \quad x = \dfrac{\log K}{\log a}, y = \dfrac{\log K}{\log b}, z = \dfrac{\log K}{\log c}$

Now, x, y, z are in G.P.

$\Rightarrow \quad \dfrac{y}{x} = \dfrac{z}{y}$

$\Rightarrow \quad \dfrac{\log a}{\log b} = \dfrac{\log b}{\log c}$

$\Rightarrow \quad \log_b a = \log_c b$

Example 2.16 The arithmetic mean between two positive numbers a and b, where $a > b$, is two times their geometric mean, prove that $a:b = (2+\sqrt{3}):(2-\sqrt{3})$.

Solution From question, $\dfrac{a+b}{2} = 2\sqrt{ab}$

$\Rightarrow \quad \dfrac{a+b}{2\sqrt{ab}} = 2$

Applying componendo and dividendo, we get

$$\frac{a+b+2\sqrt{ab}}{a+b-2\sqrt{ab}} = \frac{2+1}{2-1}$$

$\Rightarrow \quad \dfrac{(\sqrt{a}+\sqrt{b})^2}{(\sqrt{a}-\sqrt{b})^2} = \dfrac{3}{1}$

$\Rightarrow \quad \dfrac{(\sqrt{a}+\sqrt{b})}{(\sqrt{a}-\sqrt{b})} = \dfrac{\sqrt{3}}{1}$

Again, applying componendo and dividendo, we get

$$\frac{\sqrt{a}+\sqrt{b}+\sqrt{a}-\sqrt{b}}{\sqrt{a}+\sqrt{b}-\sqrt{a}+\sqrt{b}} = \frac{\sqrt{3}+1}{\sqrt{3}-1}$$

$\Rightarrow \quad \dfrac{\sqrt{a}}{\sqrt{b}} = \dfrac{\sqrt{3}+1}{\sqrt{3}-1}$

Squaring, we get $\dfrac{a}{b} = \dfrac{4+2\sqrt{3}}{4-2\sqrt{3}}$

$\Rightarrow \quad a:b = (2+\sqrt{3}):(2-\sqrt{3})$

Example 2.17 If one A.M. 'A' and two G.M.s 'p' and 'q' be inserted between two given numbers, Show that $\dfrac{p^2}{q} + \dfrac{q^2}{p} = 2A$.

Solution Let two given numbers be a and b. From question, a, A, b are in A.P.

$\Rightarrow \quad A = \dfrac{a+b}{2}$

a, p, q, b are in G.P.

$\Rightarrow \quad \dfrac{p}{a} = \dfrac{q}{p} = \dfrac{b}{q}$

$\Rightarrow \quad p^2 = aq$

$\Rightarrow \quad q^2 = bp$

$\Rightarrow \quad \dfrac{p^2}{q} + \dfrac{q^2}{p} = a+b = 2A$

Example 2.18 If $\sum_{r=1}^{\infty}\frac{1}{(2r-1)^2}=\frac{\pi^2}{8}$, then find the value of $\sum_{r=1}^{\infty}\frac{1}{r^2}$.

Solution Let

$$\frac{1}{1^2}+\frac{1}{2^2}+\frac{1}{3^2}+\ldots+\infty=P \quad (1)$$

It is given that

$$\frac{1}{1^2}+\frac{1}{3^2}+\frac{1}{5^2}+\ldots+\infty=\frac{\pi^2}{8} \quad (2)$$

(1) can be written as,

$$P=\left(\frac{1}{1^2}+\frac{1}{3^2}+\frac{1}{5^2}+\ldots+\infty\right)+\frac{1}{2^2}\left(\frac{1}{1^2}+\frac{1}{2^2}+\frac{1}{3^2}+\ldots+\infty\right)$$

$$\Rightarrow\ P=\frac{\pi^2}{8}+\frac{P}{4}$$

$$\Rightarrow\ \frac{3P}{4}=\frac{\pi^2}{8}$$

$$\Rightarrow\ P=\frac{\pi^2}{6}$$

Example 2.19 If $x^2+9y^2+25z^2=xyz\left(\frac{15}{x}+\frac{5}{y}+\frac{3}{z}\right)$, then prove that x, y, z are in H.P.

Solution $x^2+9y^2+25z^2=15yz+5xz+3xy$

$$\Rightarrow\ (x-3y)^2+(3y-5z)^2+(5z-x)^2=0$$

$$\Rightarrow\ x=3y=5z=\lambda \quad \text{(say)}$$

$$\Rightarrow\ x=\lambda,\ y=\frac{\lambda}{3},\ z=\frac{\lambda}{5}$$

$\Rightarrow$ x, y, z are in H.P.

Example 2.20 If $x_1, x_2, x_3 \ldots, x_n$ are in H.P, then prove that

$$x_1x_2+x_2x_3+x_3x_4+\ldots+x_{n-1}x_n=(n-1)x_1x_n.$$

Solution Since, $x_1, x_2, x_3 \ldots, x_{n-1}, x_n$ are in H.P.,

$\Rightarrow$ $\frac{1}{x_1},\frac{1}{x_2},\frac{1}{x_3},\ldots,\frac{1}{x_{n-1}},\frac{1}{x_n}$ are in A.P.

$$\Rightarrow\ \frac{1}{x_2}-\frac{1}{x_1}=\frac{1}{x_3}-\frac{1}{x_2}=\ldots=\frac{1}{x_n}-\frac{1}{x_{n-1}}=d \quad \text{(say)}$$

Now,

$$x_1-x_2=dx_1x_2$$
$$x_2-x_3=dx_2x_3$$
$$x_{n-1}-x_n=dx_{n-1}x_n$$

Adding we get

$$x_1-x_n=d\left(x_1x_2+x_2x_3+\ldots+x_{n-1}x_n\right) \quad (1)$$

Also,

$$\frac{1}{x_2}-\frac{1}{x_1}=d$$
$$\frac{1}{x_3}-\frac{1}{x_2}=d$$
$$\frac{1}{x_n}-\frac{1}{x_{n-1}}=d$$

Adding we get

$$\frac{1}{x_n}-\frac{1}{x_1}=(n-1)d$$

$$\Rightarrow\ x_1-x_n=(n-1)d\,x_1x_n \quad (2)$$

From (i) and (ii), we get

$$x_1x_2+x_2x_3+x_3x_4+\ldots+x_{n-1}x_n=(n-1)x_1x_n.$$

Example 2.21 If a, b, c are in H.P., prove that

$$\left(\frac{1}{a}+\frac{1}{b}-\frac{1}{c}\right)\left(\frac{1}{b}+\frac{1}{c}-\frac{1}{a}\right)=\frac{4}{ac}-\frac{3}{b^2}.$$

Solution As a, b, c are in H.P.,

$$\Rightarrow\ b=\frac{2}{\frac{1}{a}+\frac{1}{c}}$$

$$\Rightarrow\ b=\frac{2ac}{a+c}$$

Now,

$$\left(\frac{1}{a}+\frac{1}{b}-\frac{1}{c}\right)\left(\frac{1}{b}+\frac{1}{c}-\frac{1}{a}\right)$$
$$=\left[\frac{1}{b}+\left(\frac{1}{a}-\frac{1}{c}\right)\right]\left[\frac{1}{b}-\left(\frac{1}{a}-\frac{1}{c}\right)\right]$$
$$=\frac{1}{b^2}-\left(\frac{1}{a}-\frac{1}{c}\right)^2$$
$$=\frac{1}{b^2}-\left\{\left(\frac{1}{a}+\frac{1}{c}\right)^2-4\frac{1}{a}\cdot\frac{1}{c}\right\}$$
$$=\frac{1}{b^2}-\frac{4}{b^2}+\frac{4}{ac}$$
$$=\frac{4}{ac}-\frac{3}{b^2}$$

Example 2.22 Find the sum of n terms of the series $1^2+3^2+5^2+7^2+\ldots$ to n terms.

Solution Since, r^{th} term of

$$1, 3, 5, 7\ldots = 1+(r-1).2 = 2r-1$$

If t_r denote the rth term of the given series,

Then $\quad t_r = (2r-1)^2$

$\Rightarrow \quad t_r = 4r^2 - 4r + 1$

Putting $r = 1, 2, 3, \ldots, n$ respectively, we get

$$t_1 = 4\times 1^2 - 4\times 1 + 1$$

$$t_2 = 4\times 2^2 - 4\times 2 + 1$$

$$t_n = 4\times n^2 - 4\times n + 1$$

Adding we get

$$t_1 + t_2 + \ldots + t_n = 4\frac{n(n+1)(2n+1)}{6} - 4\frac{n(n+1)}{2} + n$$

$$= n\left[\frac{2}{3}\left(2n^2+3n+1\right) - 2(n+1) + 1\right]$$

$$= n\left[\frac{4n^2+6n+2-6n-6+3}{3}\right]$$

$$= \frac{1}{3}n\left(4n^2-1\right)$$

Example 2.23 Find the sum of n terms of the series $5+7+11+17+25+\ldots$

Solution Let, $S_n = 5+7+11+17+25+\ldots+t_n$

$\Rightarrow \quad S_n = 5+7+11+17+\ldots+t_{n-1}+t_n$

Subtracting we get

$$0 = 5+2+4+6+8+\ldots\text{to } n \text{ terms} - t_n$$

$\Rightarrow \quad t_n = 5 + (2+4+6+8+\ldots\text{to}(n-1)\text{terms})$

$\Rightarrow \quad t_n = 5 + \frac{n-1}{2}\left[2\times 2 + (n-1-1)\times 2\right]$

$$= 5 + \frac{n-1}{2}\times 2(2+n-2)$$

$$= 5 + n^2 - n$$

$\Rightarrow \quad t_n = n^2 - n + 5$

$\Rightarrow \quad \sum_{n=1}^{n} t_n = \sum_{n=1}^{n} n^2 - \sum_{n=1}^{n} n + 5\sum_{n=1}^{n} 1$

$\Rightarrow \quad S_n = \frac{n(n+1)(2n+1)}{6} - \frac{n(n+1)}{2} + 5n$

$$= n\left[\frac{2n^2+3n+1}{6} - \frac{n+1}{2} + 5\right]$$

$$= \frac{n}{6}\left[2n^2+3n+1-3n-3+30\right]$$

$$= \frac{n}{6}\left(2n^2+28\right) = \frac{1}{3}n\left(n^2+14\right)$$

Example 2.24 Find the sum of n terms of the series $1+2x+3x^2+4x^3+\ldots$

Solution Let, S_n denote the sum of n terms.

$\Rightarrow S_n = 1+2x+3x^2+4x^3+\ldots+(n-1)x^{n-2}+nx^{n-1}$

$\Rightarrow \quad xS_n = x+2x^2+3x^3+\ldots+(n-1)x^{n-1}+nx^n$

Subtracting, we get

$\Rightarrow \quad (1-x)S_n = 1+x+x^2+x^3+\ldots+x^{n-1}-nx^n$

$\Rightarrow \quad (1-x)S_n = \frac{1\left(1-x^n\right)}{(1-x)} - nx^n$

$\Rightarrow \quad S_n = \frac{1\left(1-x^n\right)}{(1-x)^2} - \frac{nx^n}{(1-x)}$

Example 2.25 If $\sum_{r=1}^{n} T_r = \frac{n}{8}(n+1)(n+2)(n+3)$, then find $\sum_{r=1}^{n}\frac{1}{T_r}$.

Solution Since, $S_n = \frac{n}{8}(n+1)(n+2)(n+3)$

$$S_{n-1} = \frac{(n-1)}{8}n(n+1)(n+2)$$

$$T_n = S_n - S_{n-1} = \frac{n}{8}(n+1)(n+2)\left[(n+3)-(n-1)\right]$$

$$= \frac{n}{8}(n+1)(n+2)\times 4$$

$$= \frac{1}{2}n(n+1)(n+2)$$

$\Rightarrow \quad T_r = \frac{1}{2}r(r+1)(r+2)$

$\Rightarrow \quad \frac{1}{T_r} = \frac{2}{r(r+1)(r+2)}$

$\Rightarrow \quad \frac{1}{T_r} = \frac{(r+2)-r}{r(r+1)(r+2)}$

$\Rightarrow \quad \frac{1}{T_r} = \left[\frac{1}{r(r+1)} - \frac{1}{(r+1)(r+2)}\right]$

$\Rightarrow \frac{1}{T_r} = [V_{r-1} - V_r]$ where $V_r = \frac{1}{(r+1)(r+2)}$

Putting $r = 1, 2, 3, \ldots, n$ respectively, we get

$$\frac{1}{T_1} = [V_0 - V_1]$$
$$\frac{1}{T_2} = [V_1 - V_2]$$
$$\frac{1}{T_3} = [V_2 - V_3]$$
$$\frac{1}{T_n} = [V_{n-1} - V_n]$$

Adding, we get

$$\sum_{r=1}^{n} \frac{1}{T_r} = V_0 - V_n = \frac{1}{2} - \frac{1}{(n+1)(n+2)}$$
$$= \frac{n^2+3n+2-2}{2(n+1)(n+2)} = \frac{n(n+3)}{2(n+1)(n+2)}$$

Example 2.26 If $x+y+z=1$ and x, y, z are positive numbers, Show that

$$(1-x)(1-y)(1-z) \ge 8xyz.$$

Solution $\frac{y+z}{2} \ge \sqrt{yz}$

$\Rightarrow \quad y+z \ge 2\sqrt{yz}$

$$\frac{z+x}{2} \ge \sqrt{zx}$$

$\Rightarrow \quad z+x \ge 2\sqrt{zx}$

$$\frac{x+y}{2} \ge \sqrt{xy}$$

$\Rightarrow \quad x+y \ge 2\sqrt{xy}$

Multiplying $(y+z)(z+x)(x+y) \ge 8xyz$

$\Rightarrow \quad (1-x)(1-y)(1-z) \ge 8xyz \quad (\text{as } x+y+z=1)$

Example 2.31 If a, b, c are distinct positive real numbers, prove that $\frac{ab}{c^3}+\frac{bc}{a^3}+\frac{ca}{b^3} > \frac{1}{a}+\frac{1}{b}+\frac{1}{c}$.

Solution $\dfrac{\frac{ab}{c^3}+\frac{bc}{a^3}}{2} > \sqrt{\frac{ab}{c^3}\cdot\frac{bc}{a^3}}$

$\Rightarrow \quad \dfrac{\frac{ab}{c^3}+\frac{bc}{a^3}}{2} > \frac{b}{ac}$

Similarly, $\quad \dfrac{\frac{bc}{a^3}+\frac{ca}{b^3}}{2} > \frac{c}{ab}$

and $\quad \dfrac{\frac{ca}{b^3}+\frac{ab}{c^3}}{2} > \frac{a}{bc}$

Adding we get

$$\frac{ab}{c^3}+\frac{bc}{a^3}+\frac{ca}{b^3} > \frac{b}{ac}+\frac{c}{ab}+\frac{a}{bc} \quad (1)$$

Now,

$$\dfrac{\frac{b}{ac}+\frac{c}{ab}}{2} > \sqrt{\frac{b}{ac}\cdot\frac{c}{ab}}$$

$\Rightarrow \quad \dfrac{\frac{b}{ac}+\frac{c}{ab}}{2} > \frac{1}{a}$

Similarly,

$$\dfrac{\frac{c}{ab}+\frac{a}{bc}}{2} > \frac{1}{b}$$

and $\quad \dfrac{\frac{a}{bc}+\frac{b}{ac}}{2} > \frac{1}{c}$

Adding, we get

$$\frac{b}{ac}+\frac{c}{ab}+\frac{a}{bc} > \frac{1}{a}+\frac{1}{b}+\frac{1}{c} \quad (2)$$

From (1) and (2), we get

$$\frac{ab}{c^3}+\frac{bc}{a^3}+\frac{ca}{b^3} > \frac{1}{a}+\frac{1}{b}+\frac{1}{c}$$

Example 2.32 If a, b, c are the sides of triangle, then find the minimum value of

$$\frac{a}{b+c-a}+\frac{b}{c+a-b}+\frac{c}{a+b-c}.$$

Solution Let $b+c-a = x$

$$c+a-b = y$$
$$a+b-c = z$$

$\Rightarrow \quad y+z = 2a,\ z+x = 2b,\ x+y = 2c$

Hence,

$$\frac{a}{b+c-a}+\frac{b}{c+a-b}+\frac{c}{a+b-c} = \frac{y+z}{2x}+\frac{z+x}{2y}+\frac{x+y}{2z}$$
$$= \frac{1}{2}\left[\frac{y}{x}+\frac{z}{x}+\frac{z}{y}+\frac{x}{y}+\frac{x}{z}+\frac{y}{z}\right]$$

Now,

$$\frac{\frac{y}{x}+\frac{z}{x}+\frac{z}{y}+\frac{x}{y}+\frac{x}{z}+\frac{y}{z}}{6} \geq \left(\frac{y}{x}\cdot\frac{z}{x}\cdot\frac{z}{y}\cdot\frac{x}{y}\cdot\frac{x}{z}\cdot\frac{y}{z}\right)^{\frac{1}{6}}$$

$$\Rightarrow \quad \frac{1}{2}\left[\frac{y+z}{x}+\frac{z+x}{y}+\frac{x+y}{z}\right] \geq 3$$

$\Rightarrow$ Minimum value = 3

Example 2.33 Let $a_1, a_2, a_3, \ldots$ be in H.P. with $a_1 = 5$ and $a_{20} = 25$, then find the least positive integer n for which $a_n < 0$.

Solution Since, $a_1, a_2, a_3, \ldots$ are in H.P.,

$$\Rightarrow \quad \frac{1}{a_1}, \frac{1}{a_2}, \frac{1}{a_3}, \ldots \text{ are in A.P.}$$

Now, $$\frac{1}{a_n} = \frac{1}{a_1} + (n-1)d < 0 \qquad (1)$$

Also, $$\frac{1}{a_{20}} = \frac{1}{a_1} + (20-1)d$$

$$\Rightarrow \quad \frac{1}{25} = \frac{1}{5} + 19d$$

$$\Rightarrow \quad 19d = \frac{-4}{25}$$

$$\Rightarrow \quad d = \frac{-4}{25\times 19}$$

From (1)

$$\frac{1}{5} - \frac{4(n-1)}{25\times 19} < 0$$

$$\Rightarrow \quad 4(n-1) > \frac{25\times 19}{5}$$

$$\Rightarrow \quad n \geq \frac{99}{4}$$

$\Rightarrow$ Least positive interger = 25

Example 2.34 If the nth term of a series is given by $t_n = \frac{n^5+n^3}{n^4+n^2+1}$ and if the sum of its n terms is given by $S_n = an^2 + a + \frac{1}{bn^2+b}$ find the value of $\frac{b_n}{a_n}$.

Solution $t_n = \frac{n^5+n^3}{n^4+n^2+1} = n - \frac{n}{n^4+n^2+1}$

$$\Rightarrow \quad t_n = n - \frac{n}{(n^2+1)^2 - n^2}$$

$$= n + \frac{1}{2}\left[\frac{1}{n^2+n+1} - \frac{1}{n^2-n+1}\right]$$

$$\Rightarrow \quad S_n = \sum t_n = \frac{n(n+1)}{2} + \frac{1}{2}\left[\frac{1}{n^2+n+1} - 1\right]$$

$$= \left(\frac{n^2}{2}+\frac{n}{2}+\frac{1}{8}\right) - \frac{5}{8} + \frac{1}{\left(n\sqrt{2}+\frac{1}{\sqrt{2}}\right)^2 + \frac{3}{2}}$$

$$= \left(\frac{n}{\sqrt{2}}+\frac{1}{2\sqrt{2}}\right)^2 - \frac{5}{8} + \frac{1}{\left(n\sqrt{2}+\frac{1}{\sqrt{2}}\right)^2 + \frac{3}{2}}$$

Clearly,

$$a_n = \frac{n}{\sqrt{2}} + \frac{1}{2\sqrt{2}}, a = -\frac{5}{8}, b_n = n\sqrt{2} + \frac{1}{\sqrt{2}}, b = \frac{3}{2}$$

$$\Rightarrow \quad \frac{b_n}{a_n} = \frac{\frac{2n+1}{\sqrt{2}}}{\frac{2n+1}{2\sqrt{2}}} = 2$$

Example 2.35 If the equation

$$x^4 - 4x^3 + ax^2 + bx + 1 = 0$$

has four positive roots. Find the values of a and b.

Solution Let x_1, x_2, x_3, x_4 be the roots of the equation $x^4 - 4x^3 + ax^2 + bx + 1 = 0$.

Then, $x_1 + x_2 + x_3 + x_4 = 4$ and $x_1 \cdot x_2 \cdot x_3 \cdot x_4 = 1$

Now, A.M. of $x_1, x_2, x_3, x_4 = \frac{x_1+x_2+x_3+x_4}{4} = 1$

and G.M. $= (x_1x_2x_3x_4)^{\frac{1}{4}} = 1$

As A.M. = G.M.

$\Rightarrow x_1 = x_2 = x_3 = x_4 = 1$

Hence,

$$x^4 - 4x^3 + ax^2 + bx + 1 = (x-1)^4$$

$$= x^4 - 4x^3 + 6x^2 - 4x + 1$$

$$\Rightarrow a - 6, b = 4$$

Example 2.36 Find the sum of n terms of the series

$$\frac{4}{1.2.3} + \frac{5}{2.3.4} + \frac{6}{3.4.5} + \ldots$$

Solution $t_r = \frac{(r+3)}{r(r+1)(r+2)}$

$$=\frac{1}{(r+1)(r+2)}+\frac{3}{r(r+1)(r+2)}$$

$$t_r=\left[\frac{1}{r+1}-\frac{1}{r+2}\right]+\frac{\frac{3}{2}((r+2)-r)}{r(r+1)(r+2)}$$

$$t_r=\left[\frac{1}{r+1}-\frac{1}{r+2}\right]+\frac{3}{2}\left[\frac{1}{r(r+1)}-\frac{1}{(r+1)(r+2)}\right]$$

$$\Rightarrow \sum_{r=1}^{n} t_r=\left[\frac{1}{2}-\frac{1}{n+2}\right]+\frac{3}{2}\left[\frac{1}{2}-\frac{1}{(n+1)(n+2)}\right]$$

$$=\frac{5}{4}-\frac{1}{(n+2)}\left[1+\frac{3}{2(n+1)}\right]$$

$$=\frac{5}{4}-\frac{1}{2(n+1)(n+2)}(2n+5)$$

2.8 Know Your Concepts

1. The sum of n terms of two series in A.P. are in the ratio $(3n-13):(5n+21)$. Find the ratio of their 24th term.
2. If n arithmetic means are inserted between 20 and 80 such that the ratio of first mean to that of the last mean is 1 : 3, find the value of n.
3. Three numbers whose sum is 15 are in A.P. If 1, 4, 19 be added to them respectively the resulting numbers are in G.P., find the numbers.
4. If the product of three consecutive numbers in G.P. is 216 and the sum of the product taken in pairs is 156, then find the numbers.
5. The sum of three numbers in H.P. is 11 and the sum of their reciprocals is 1, find the numbers.
6. If a^2, b^2, c^2 are in A.P., then prove that $b+c,\ c+a,\ a+b$ are in H.P.
7. The interior angles of a polygon are in A.P. The smallest angle is 120° and the common difference is 5°. Find the number of sides of the polygon.
8. If the terms of the A.P. $\sqrt{a-x}, \sqrt{x}, \sqrt{a+x}, \ldots$ are all integers where $a, x > 0$, then find the least composite value of a.
9. Find the coefficient of x^{99} in the polynomial $(x-1)(x-2)\ldots(x-100)$.
10. The H.M. of two numbers is 4, and their A.M., A and G.M., G satisfy the relation $2A+G^2=27$. Find the two numbers.
11. The sum of three consecutive terms in H.P. is 37 and the sum of their reciprocal is $\frac{1}{4}$. Find the numbers.
12. In an increasing G.P. the sum of the first and the last term is 66, the product of the second and the last but one is 128 and the sum of the terms is 126. How many terms are in the progressions?
13. Find the four numbers in G.P. whose sum is 85 and product is 4096.
14. If $\log_3 2, \log_3(2^x-5)$, and $\log_3\left(2^x-\frac{7}{2}\right)$ are in A.P., Find the value of x.
15. If a, b, c are in A.P., b, c, d are in G.P. and c, d, e are in H.P., prove that a, c, e are in G.P.
16. The series of natural numbers is divided into groups (1), (2, 3, 4), (5, 6, 7, 8, 9), (10, 11, 12, 13, 14, 15, 16) and so on. Prove that the sum of the numbers in the nth group is $(n-1)^3+n^3$.
17. Let a_n be the nth term of an A.P. Show that $a_1^2-a_2^2+a_3^2-a_4^2+\ldots+a_{2n-1}^2-a_{2n}^2$ $=\frac{n}{(2n-1)}\left(a_1^2-a_{2n}^2\right)$.
18. The sum of three numbers in G.P. is 21 and the sum of their squares is 189, Find the numbers.
19. If $x+y+z=15;\ a, x, y, z, b$ are in A.P. and $\frac{1}{x}+\frac{1}{y}+\frac{1}{z}=\frac{5}{3}$.

    Further $\frac{1}{a}, \frac{1}{x}, \frac{1}{y}, \frac{1}{z}, \frac{1}{b}$ are in A.P., then find the G.M. of a and b.
20. Find the sum of the n terms of the series $1^2+(1^2+2^2)+(1^2+2^2+3^2)+\ldots$

2.9 JEE Main: Sample Questions

Single Answer Type

1. If the sum of the first 11 terms of an arithmetic progression equals that of the first 19 terms, then the sum of its first 30 terms, is
   (A) equal to 0
   (B) equal to –1

(C) equal to 1
(D) not possible to determine

2. If the roots of the cubic equation $x^3 - 5x^2 + kx - 27 = 0$ are in G.P. then the value of k is

(A) 15 (B) 12 (C) 9 (D) 6

3. If a, b, c are non-zero real numbers then the minimum value of the expression

$$\left(\frac{(a^8+4a^4+1)(b^4+3b^2+1)(c^2+2c+2)}{a^4b^2}\right)$$

equals

(A) 24 (B) 30 (C) 48 (D) 12

4. If the sum of three numbers of an infinite G.P. is 21 and the sum of their squares is 189 then the common ratio of G.P. is

(A) 2 (B) $\frac{18}{7}$ (C) $\frac{1}{3}$ (D) $\frac{1}{4}$

5. The sum of 51 + 52 + 53 + … + 100 is equal to

(A) 2775 (B) 3775
(C) 2275 (D) 1275

6. If the sum of third and ninth term of an arithmetic progression is equal to 8 and sum of its first seven terms is 14, then which of the following statements is not correct?

(A) Ratio of the first term to common difference is one.
(B) Sum of the first term and common difference is zero.
(C) Sum of the first eleven terms is 44.
(D) Absolute value of the difference of the first term and common difference is two.

7. Let $a_1, a_2, \dots a_{10}$ be in A.P. and $h_1, h_2, \dots, h_{10}$ be in H.P. If $a_1 = h_1 = 2$ and $a_{10} = h_{10} = 3$ then a_4h_7 is:

(A) 2 (B) 3 (C) 5 (D) 6

8. If the arithmetic mean of the roots of a quadratic equation is 7 and the arithmetic mean of their reciprocals is $\frac{7}{2}$, then the quadratic equation is

(A) $x^2 - 7x + 2 = 0$
(B) $x^2 - 14x + 2 = 0$
(C) $x^2 - 14x + 4 = 0$
(D) $x^2 - 7x + 4 = 0$

9. The geometric series $a + ar + ar^2 + ar^3 + \dots \infty$ has a sum of 7 and the terms involving odd powers of r has a sum of 3. The value of $(a + r)$ equals

(A) $\frac{4}{3}$ (B) $\frac{10}{7}$ (C) $\frac{5}{2}$ (D) $\frac{7}{3}$

10. The roots of $64x^3 - 144x^2 + 92x - 15 = 0$ are in A.P. The magnitude of the difference between the largest and smallest roots is.

(A) 2 (B) 1 (C) $\frac{1}{2}$ (D) $\frac{3}{8}$

11. If a, b, c are positive real numbers such that $a + b + c = 3$, then the minimum value of $\frac{a}{6-a}+\frac{b}{6-b}+\frac{c}{6-c}$ is equal to

(A) $\frac{1}{5}$ (B) $\frac{1}{3}$ (C) $\frac{3}{5}$ (D) $\frac{4}{5}$

12. Consider three sequences an A.P., a G.P. and a H.P. as follows

$a, A_1, A_2, A_3, \dots, A_9, b$ (A.P.)
$a, G_1, G_2, G_3, \dots, G_9, b$ (G.P.)
$a, H_1, H_2, H_3, \dots, H_9, b$ (H.P.)

If a, b are the roots of the equation $x^2 - (p + 1)x + 16 = 0, p \in R$, then the value of $\prod_{r=1}^{4} A_r H_{10-r} G_{10-2r}$ is

(A) 16 (B) 64 (C) 4^9 (D) 4^{12}

13. The limiting sum of the infinite series, $\frac{1}{10}+\frac{2}{10^2}+\frac{3}{10^3}$ … whose n^{th} term is $\frac{n}{10^n}$ is equal to

(A) $\frac{1}{9}$ (B) $\frac{10}{81}$ (C) $\frac{1}{8}$ (D) $\frac{17}{72}$

14. If a, b and c are three consecutive positive terms of a G.P. then the graph of $y = ax^2 + bx + c$ is

(A) a curve that intersects the x-axis at two distinct points
(B) entirely below the x-axis
(C) entirely above the x-axis
(D) tangent to the x-axis

15. If $S(n) = 1 + \frac{1}{2} + \frac{1}{3} + \dots + \frac{1}{n}$,

then $\dfrac{(x^2+1)+\frac{100}{(x^2+1)}}{2} \geq 10$ is equal to

(A) $(n+1)^2 - \frac{n(n+1)}{2}S(n)$

(B) $(n+1)^2 S(n) - \frac{n(n+1)}{2}$

(C) $(n+1)^2 - \frac{n(n-1)}{2}S(n)$

(D) $n^2 S(n) - \frac{n(n+1)}{2}$

16. The expression $\frac{1+3+5+\ldots+99}{2+4+6+\ldots+100}$ when simplified, reduces to the fraction

(A) $\frac{49}{50}$ (B) $\frac{50}{51}$ (C) $\frac{99}{100}$ (D) $\frac{100}{101}$

17. Let α, β be the roots of $x^2 - x + p = 0$ and γ, δ be the roots of $x^2 - 4x + q = 0$. If α, β, γ, δ (in that order) are in arithmetic progression, then

(A) $p = \frac{15}{64}$ (B) $p = \frac{37}{64}$

(C) $q = \frac{274}{64}$ (D) $q = \frac{247}{64}$

18. Consider a decreasing G.P.: $g_1, g_2, g_3, \ldots, g_n$, … such that $g_1 + g_2 + g_3 = 13$ and $g_1^2 + g_2^2 + g_3^2 = 91$ then which of the following does not hold?

(A) the greatest term of the G.P. is 9

(B) $3g_4 = g_3$

(C) $g_1 = 1$

(D) $g_2 = 3$

19. The sum of the infinite series

$$1 + \left(1+\frac{1}{5}\right)\left(\frac{1}{2}\right) + \left(1+\frac{1}{5}+\frac{1}{5^2}\right)\left(\frac{1}{2^2}\right) + \ldots \text{ is}$$

(A) $\frac{20}{9}$ (B) $\frac{10}{9}$ (C) $\frac{5}{9}$ (D) $\frac{5}{3}$

20. If $S_n = \sum_{r=1}^{n} \frac{2r+1}{r^4 + 2r^3 + r^2}$, then S_{10} is equal to

(A) 1 (B) $\frac{121}{120}$ (C) $\frac{120}{121}$ (D) $\frac{60}{61}$

21. Let $S = \frac{3}{2} + \frac{3}{6} + \frac{3}{12} + \frac{3}{20} + \ldots \infty$ then S is equal to

(A) 1 (B) 2 (C) $\frac{3}{2}$ (D) 3

22. The value of x satisfying the equation $(2011)^{\log_{10} x} = 11^{\log_{10}(1+3+5+\ldots+4021)}$ lies in the interval

(A) (95, 100) (B) (100, 105)

(C) (110, 115) (D) (120, 125)

23. Let $a_1, a_2, a_3, \ldots, a_{10}$ are in G.P. with $a_{51} = 25$ and $\sum_{i=1}^{101} a_i = 125$, then the value of $\sum_{i=1}^{101}\left(\frac{1}{a_i}\right)$ equals

(A) 5 (B) $\frac{1}{5}$ (C) $\frac{1}{25}$ (D) $\frac{1}{125}$

24. If 'b' be the pth term of a G.P., whose $(p+q)^{th}$ and $(p-q)^{th}$ terms are $a(a > 0)$ and c respectively and also $f(x) = ax^2 - 2bx + c$, then for all $x \in R$

(A) $af(x) \geq 0$ (B) $af(x) \leq 0$

(C) $af(x) > 0$ (D) $af(x) < 0$

25. The harmonic mean of the roots of the equation

$$\left(5+\sqrt{2}\right)x^2 - \left(4+\sqrt{5}\right)x + 8 + 2\sqrt{5} = 0 \text{ is}$$

(A) 2 (B) 4 (C) 6 (D) 8

26. If the positive numbers a, b, c are in G.P., then $\log_a 2, \log_b 2, \log_c 2$ are in

(A) A.P. (B) G.P.

(C) H.P. (D) none of these

27. If $x = 30°$, then the sum of the series $1 + \cos(2x) + \cos^2(2x) + \cos^3(2x) + \ldots \infty$, is

(A) ∞ (B) 4 (C) 2 (D) 1

28. Let p, q, r and s be four positive numbers such that $pqrs = 1$ then the least value of $(1+p)(1+q)(1+r)(1+s)$ is equal to

(A) 4 (B) 16 (C) 64 (D) 256

29. Let $a_1, a_2, a_3, \ldots, a_{49}, a_{50}$ be in arithmetic progression. If $a_1 = 4$ and $a_{50} = 144$, then the value of

$$\frac{1}{\sqrt{a_1}+\sqrt{a_2}} + \frac{1}{\sqrt{a_2}+\sqrt{a_3}} + \frac{1}{\sqrt{a_3}+\sqrt{a_4}} \ldots +$$

is equal to

(A) $\frac{14}{13}$ (B) $\frac{49}{14}$ (C) $\frac{50}{13}$ (D) $\frac{48}{17}$

30. If 100 times the 100th term of an A.P. with non-zero common difference equals the 50 times its 50th term, then the 150th term of this A.P is
(A) 150
(B) zero
(C) –150
(D) 150 times its 50th term

31. Let S_n denote the sum of first n terms of an arithmetic progression whose first term is –4 and the common difference is 1.
If $V_n = 2S_{n+2} - 2S_{n+1} + S_n$ where $n \in N$, then
(A) $\sum_{n=1}^{10} V_n = -5$ (B) $\sum_{n=1}^{10} V_n = 10$
(C) minimum value of V_n is –9
(D) minimum value of V_n is $\frac{-73}{8}$

32. The first, third and eighth terms of an arithmetic sequence with common difference 9 (taken in that order) are in geometric progression. The seventh term of arithmetic sequence is
(A) –42 (B) 66 (C) –21 (D) 34

33. Consider a sequence $\{a_n\}$ with $a_1 = 2$ and $a_n = \frac{a_{n-1}^2}{a_{n-2}}$ $\forall\ n \geq 3$. If a_2 and a_5 are +ve integers, and $a_5 \leq 512$, then number of possible value of a_5 are
(A) 1 (B) 2 (C) 3 (D) 4

34. Consider the ten numbers $ar, ar^2, ar^3, \ldots ar^{10}$. If their sum is 18 and the sum of their reciprocals is 6 then the product of these ten numbers, is
(A) 324 (B) 343 (C) 243 (D) 729

35. Number of terms common to the two sequences 17, 21, 25, …, 417 and 16, 21, 26, …, 466 is
(A) 19 (B) 20 (C) 21 (D) 22

36. If p, q, r are in H.P., and p and r are different having same sign then the roots of the equation $px^2 + qx + r = 0$ are
(A) real and equal
(B) real and distinct
(C) irrational
(D) imaginary

37. Let a_n, $n \in N$ be an A.P. with common difference 'd' and all whose terms are non-zero. If n approaches infinity, then the sum $\frac{1}{a_1a_2} + \frac{1}{a_2a_3} + \ldots + \frac{1}{a_na_{n+1}}$ will approach
(A) $\frac{1}{a_1d}$ (B) $\frac{2}{a_1d}$ (C) $\frac{2}{2a_1d}$ (D) a_1d

38. The sixth term of an A.P. $a_1, a_2, a_3, \ldots, a_n$ is 2. If the quantity $a_1a_4a_5$ is minimum then the common difference of the A.P., is
(A) $\frac{2}{3}$ (B) $\frac{8}{5}$ (C) $\frac{1}{3}$ (D) $\frac{2}{9}$

39. An infinite number of triangles are formed as shown in figure. If the total area of these triangles is A then $8A$ is equal to

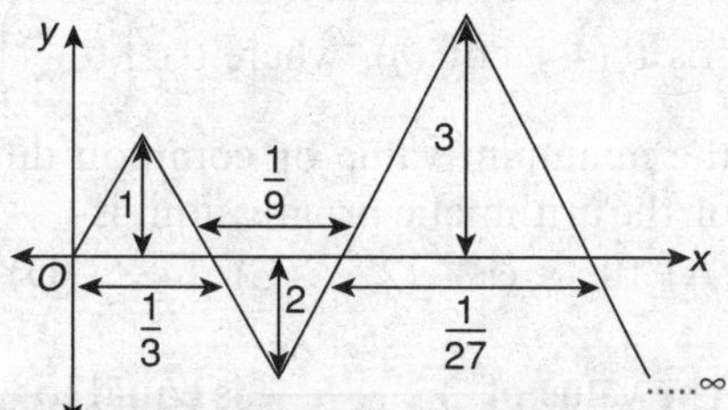

(A) 3 (B) 4 (C) 1 (D) 2

40. Sides a, b, c of a triangle are in G.P. if $\ln\left(\frac{5c}{a}\right), \ln\left(\frac{3b}{5c}\right)$ and $\ln\left(\frac{5c}{a}\right)$ are in A.P., then triangle must be
(A) isosceles (B) equilateral
(C) obtuse angled (D) none of these

41. For $x > 0$, the sum of the series $\frac{1}{1+x} - \frac{(1-x)}{(1+x)^2} + \frac{(1-x)^2}{(1+x)^3} - \ldots\infty$ is equal to
(A) $\frac{1}{4}$ (B) $\frac{1}{2}$ (C) $\frac{3}{4}$ (D) 1

42. In an arithmetic progression, if $S_n = n(5 + 3n)$ and $t_n = 32$, then the value of n is
[**Note:** S_n and t_n denote the sum of the first n terms and the nth term of the arithmetic progression respectively.]
(A) 4 (B) 5 (C) 6 (D) 7

43. Let $a_1, a_2, a_3, \ldots$ be an arithmetic progression with common difference 2. Let S_n be the sum of the first n terms of the sequence.

If $\frac{S_{3n}}{S_n}$ does not depend on n then the sum of the first 10 terms of the sequence is

(A) 100 (B) 125 (C) 150 (D) 225

44. The sum of a certain infinite geometric series is 20. When all the terms in the series are squared, the sum of the resulting series is 80. If the first term of the original series is expressed in lowest terms as $\frac{p}{q}, (p, q \in N)$ then the value of $(p + q)$ is

(A) 21 (B) 22 (C) 23 (D) 24

45. If the sum of the first n terms of an arithmetic progression is denoted by S_n and $S_n = 6n \sec^2\theta + n(n-1)(\sin^2\theta(4+\tan^2\theta) + \cos^2\theta(4+\cot^2\theta))$, where $\theta \in \left(0, \frac{\pi}{2}\right)$. Then the minimum value of common difference of the arithmetic progression, is

(A) 10 (B) 12 (C) 14 (D) 16

46. The value of $\sum_{n=1}^{\infty} \frac{5^n - 2^n}{10^n}$ is equal to

(A) 1 (B) $\frac{1}{4}$ (C) $\frac{1}{3}$ (D) $\frac{3}{4}$

47. If the roots of the equation $72x^3 - 108x^2 + 46x - 5 = 0$ are in arithmetic progression then the difference between largest and smallest root lies in the interval

(A) $\left(0, \frac{1}{2}\right)$ (B) $\left(\frac{1}{2}, \frac{1}{\sqrt{2}}\right)$

(C) $\left(\frac{1}{\sqrt{2}}, 1\right)$ (D) (1, 2)

48. If $\frac{1}{x} + \frac{1}{2x^2} + \frac{1}{4x^3} + \frac{1}{8x^4} + \frac{1}{16x^5} + \ldots\infty = \frac{1}{64}$, then the value of x is

(A) $\frac{125}{2}$ (B) $\frac{127}{2}$ (C) $\frac{129}{2}$ (D) $\frac{131}{2}$

49. The first term of an A.P. is 5, the last is 45 and their sum is 400. If the number of terms is n and d is the common difference, then $\left(\frac{n}{d}\right)$ is equal to

(A) 9 (B) 8 (C) 7 (D) 6

50. Let a_n denote the nth term of a G.P. given that $S = \sum_{n=1}^{10} a_n$ and $T = \sum_{n=1}^{10} \frac{1}{a_n}$, then the value of $\prod_{n=1}^{10} a_n$ is equal to

(A) $\left(\frac{S}{T}\right)^{10}$ (B) $\left(\frac{S}{T}\right)^{5}$

(C) $\left(\frac{S}{T}\right)^{9}$ (D) $\left(\frac{S}{T}\right)^{4}$

51. Let $a_1, a_2, a_3, \ldots, a_n$ be in geometric progression. If $a_1 + a_3 + a_5 = 455$ and $a_2 + a_4 + a_6 = 1365$, then the ratio between each consecutive term is

(A) 2 (B) 3

(C) 4 (D) any other number

52. A geometric sequence has four positive terms a_1, a_2, a_3, a_4. If $\frac{a_3}{a_1} = 9$ and $a_1 + a_2 = \frac{4}{3}$, then a_4 equals

(A) 3 (B) 9 (C) 27 (D) $3\sqrt{3}$

53. The sum to infinity of the series $1 + \frac{3}{4} + \frac{6}{4^2} + \frac{10}{4^3} + \frac{16}{4^4} + \ldots$ is

(A) $\frac{16}{9}$ (B) $\frac{64}{27}$ (C) $\frac{4}{3}$ (D) $\frac{32}{27}$

54. Two consecutive numbers k, $k + 1$ are removed from natural numbers 1, 2, 3, … n such that the arithmetic mean reduces by 1, then $(n - k)$ is equal to

(A) 1 (B) 2 (C) 3 (D) 4

55. In a geometric progression consisting of positive terms if $a_n = a_{n-1} + a_{n-2}$, $n \geq 3$, (where $n \in N$ and a_n denotes n^{th} term of G.P.) and so on, then the common ratio of this progression equals

(A) $\frac{\sqrt{5}+1}{2}$ (B) $\frac{\sqrt{5}-1}{2}$

(C) $\frac{1-\sqrt{5}}{2}$ (D) $\frac{\sqrt{5}}{2}$

56. If $a_1, a_2, \ldots, a_n$ are in H.P. then the expression $a_1a_2 + a_2a_3 + \ldots + a_{n-1}a_n$ is equal to

(A) na_1a_n (B) $(n-1)a_1a_n$

(C) $n(a_1 - a_n)$ (D) $(n-1)(a_1 - a_n)$

57. If $a = 2^b$, $b = \prod_{c=2}^{15} \log_e(c+1)$, then the value of $\log_2(\log_{16} a)$ is

(A) 0 (B) 2 (C) 4 (D) 16

58. Let 9 G.M.s be inserted between 100 and 3200. Then number of rational G.M.s are

(A) 1 (B) 2 (C) 3 (D) 4

59. If the first term of an A.P. is an integer with common difference 2, then the number of possible values of k for which the sum of first k terms is 153 is $(k > 1)$

(A) 3 (B) 4 (C) 5 (D) 6

60. If $S_1, S_2, S_3, \ldots, S_n$ are the sums of infinite geometric series whose first terms are 2, 4, 6, …, $2n$ and whose common ratios are $\frac{1}{3}, \frac{1}{5}, \frac{1}{7}, \ldots, \frac{1}{2n+1}$

$$T_n = \frac{1}{S_1S_2S_3} + \frac{1}{S_2S_3S_4} + \ldots + \frac{1}{S_nS_{n+1}S_{n+2}}$$

Then value of T_n when $n \to \infty$ is

(A) $\frac{1}{15}$ (B) $\frac{1}{60}$ (C) $\frac{1}{12}$ (D) $\frac{1}{3}$

Integer Answer Type

1. If $\alpha = \log_e \sqrt{\cot \frac{\pi}{12}}$, then find the value of

$$\left(\frac{\sum_{k=0}^{\infty} e^{-2k\alpha}}{\sum_{k=0}^{\infty} (-1)^k e^{-2k\alpha}}\right)^2.$$

2. Let p be the greatest integer for which $5m^2 - 16m$, $2pm$, p^2 are distinct consecutive terms of an A.P. where $m \in R$. If common difference of the A.P. is $\left(\frac{A}{B}\right)$, $A, B \in N$ then find the least value of $(26B - A)$.

3. Let $a_1a_2a_3a_4$ be *a* four-digit natural number such that a_1, a_2, a_3 are in arithmetic progression and a_2, a_3, a_4 are in geometric progression. If $a_4 = a_1 - 5$ and $d + 4r = 0$ then find $\sum_{i=1}^{4} a_i$.

[**Note:** d represents the common difference of arithmetic progression and r represents the common ratio of geometric progression.]

4. Let $\alpha, \beta, \gamma, \delta$ be the zeroes of $P(x) = 5x^4 + px^3 + qx^2 + rx + s$ $(p, q, r, s \in R)$ and α, γ, δ be the zeroes of $Q(x) = x^3 - 9x^2 + ax - 24$ $(\alpha < \beta < \gamma < \delta)$. If α, γ, δ (taken in that order) are in arithmetic progression and $\alpha, \beta, \gamma, \delta$ (taken in that order) are in harmonic progression, then find the value of $\left|\frac{P(1)}{Q(1)}\right|$.

5. Let $<a_n>$ be an arithmetic sequence such that arithmetic mean of $a_1, a_3, a_5, \ldots, a_{97}, a_{99}$ is 1.

Find the value of $\left|\sum_{r=1}^{50} (-1)^{\frac{r(r+1)}{2}} \cdot a_{2r-1}\right|$.

6. If the A.M., G.M. and H.M. of the first and the last terms of the series 100, 101, 102, …, $n - 2$, $n - 1$, n are the terms of the series itself, then find the value of n $(100 < n \leq 500)$.

7. Let r_1, r_2, r_3 be the three (not necessarily distinct) solutions to the equation $x^3 + 4x^2 - ax + 1 = 0$. If a can be any real number, then find the minimum value of

$$\left(r_1 + \frac{1}{r_1}\right)^2 + \left(r_2 + \frac{1}{r_2}\right)^2 + \left(r_3 + \frac{1}{r_3}\right)^2.$$

8. If the sum of the series

$$\sigma = 0.9\left(99 + 999 + 9999 + \ldots + \underbrace{999\ldots9}_{\text{string of 101 digits all equal to 9}}\right)$$

can be expressed in the form of $(10^a - b)$ where $a, b \in N$, then find $(a + b)$.

9. Let $\langle a_n \rangle$ be an arithmetic sequence. If $\sum_{r=1}^{10^{99}} a_{2r} = 10^{100}$ and $\sum_{r=1}^{10^{99}} a_{2r-1} = 10^{99}$, then find the common difference of the arithmetic sequence.

10. Let x, y, z, w be four non-zero real numbers such that x, y, z (in order) are in arithmetic progression and y, z, w (in order) are in geometric progression. If $x + w = 16$ and $y + z = 8$, then find the absolute value of $(x^2 - y^2 + 2z - w)$.

11. Let $T_n = 1+2+3+\ldots+n$. If $\sum_{i=3}^{\infty}\sum_{k=1}^{\infty}\left(\frac{3}{T_i}\right)^k = \frac{p}{q}$ where p and q are co-prime, then find the value of $(3q - p - 8)$.

12. The sum of the first 10 terms of the series $\frac{7}{2^2 5^2} + \frac{13}{5^2 8^2} + \frac{13}{8^2 11^2} + \ldots$ is $\frac{m}{n}$ then find the value of $(n - 12m)$.

13. If $x, y \in R$ satisfies the equation $3x^2 + 2y^2 - 6y - 3x + \frac{21}{4} = 0$, then the value of $\sum_{r=0}^{10}(2y + rx) = k$ then find the value of $\left[\frac{k}{11}\right]$.
(**Note:** $[x]$ denotes greatest integer less than or equal to x.)

14. If the equation $x^4 - (12k + 5)x^2 + 16k^2 = 0$ $(k > 0)$ has four real solutions which are in A.P. then the value of $k = \frac{a}{b}$ where a and b are co-prime, then find the value of $(a - 2b)$.

15. If a, b, c are in A.P. and $(a + 2b - c)(4b + 2c - 2a)(c + a - b) = \lambda, abc$, then find the value of λ.

16. Let $a_1, a_2, \ldots$ be in A.P., $g_1, g_2, \ldots$ be in G.P. and $h_1, h_2, \ldots$ be in H.P. If $a_1 = g_1 = h_1 = 1$, $a_{10} = g_{10} = 2$ and $h_{10} = \frac{1}{5}$, then find the value of $[a_{19}g_{10}h_{28} + a_{28}g_{19}h_{19}]$.
(**Note:** $[x]$ denotes greatest integer less than or equal to x.)

17. If the roots of cubic $x^3 - 6x^2 + kx - 8 = 0$ are positive real numbers, then find the value of $(k - 7)$.

18. A set of consecutive natural numbers from 1 to n are written and one number is removed from it. The arithmetic mean of the remaining numbers is found to be $\frac{71}{2}$, then the sum of all possible value(s) of the number removed from the set is λ then find the value of $\sqrt{\lambda}$.
(**Note:** $[x]$ denotes greatest integer less than or equal to x.)

19. The sum of a certain infinite geometric series is 20. When all the terms in the series are squared, the sum of the resulting series is 80. If the first term of the original series is expressed in lowest terms as (p/q) where $p, q \in N$, then find the value of $(p + q)$.

20. For $a, b > 0$, let $5a - b, 2a + b, a + 2b$ be in A.P. and $(b + 1)^2, ab + 1, (a - 1)^2$ are in G.P., then compute $(a^{-1} + b^{-1})$.

2.10 JEE Advanced: Sample Questions

Single Answer Type

1. If $S = 1.2(2^2 - 1^2) + 2.3(3^2 - 2^2) + 3.4(4^2 - 3^2) + \ldots$ up to 50 terms, then S is
(A) $50(52)^3$ (B) $50 \times (51)^2 \times 26$
(C) $49(50)^3$ (D) $50 \times (51)^2 \times 13$

2. If the roots of the equation $ax^2 + bx + c = 0$ are of the form $\frac{\alpha}{\alpha-1}$ and $\frac{\alpha+1}{\alpha}$ $(\alpha \neq \pm 1, 0)$, then the value of $(a + b + c)^2$ is
(A) $2b^2 - ac$ (B) $b^2 - 2ac$
(C) $b^2 - 4ac$ (D) $4b^2 - 2ac$

3. Let g_n be the nth term of a G.P. If $\prod_{r=1}^{100} g_{2r} = \alpha$ and $\prod_{r=1}^{100} g_{2r-1} = \beta$, then the common ratio of G.P is
(A) $\left(\frac{\beta}{\alpha}\right)^{\frac{1}{100}}$ (B) $\left(\frac{\alpha}{\beta}\right)^{\frac{1}{100}}$
(C) $\left(\frac{\beta}{\alpha}\right)^{\frac{1}{200}}$ (D) $\frac{\beta}{\alpha}$

4. If $(1 + x)(1 + x^3)(1 + x^5) \ldots (1 + x^{21})$ is a polynomial of 'n' degree, then the value of 'n' is
(A) 121 (B) 231
(C) 85 (D) 21

5. The minimum value of $\left(16x+\frac{4}{x}+1\right)$, $x > 0$, is

(A) 13 (B) 12 (C) 17 (D) 16

6. If sum $S=\sum_{r=1}^{2015}\frac{(-1)^{r+1}}{r}$, then S is also equal to

(A) $\sum_{r=1008}^{2015}\frac{(-1)^r}{r}$ (B) $\sum_{r=1008}^{2015}\frac{1}{r}$

(C) $\sum_{r=1008}^{2015}\frac{(-1)^{r+1}}{r}$ (D) $\sum_{r=1007}^{2015}\frac{1}{r}$

7. If $t_n=\frac{n^3-n}{2}$, then $\sum_{n=2}^{\infty}\frac{1}{t_n}$ is

(A) $\frac{1}{8}$ (B) $\frac{1}{4}$ (C) $\frac{1}{2}$ (D) 1

8. If a, b, c (all distinct positive) are in H.P., then the equation

$x^2 - kx + 2b^{101} - a^{101} - c^{101} = 0$ ($k \in R$) has

(A) both roots positive
(B) both roots negative
(C) one positive and one negative root
(D) both roots imaginary

9. If $x, y, z \in R^+$ such that $x + y + z = 4$, then the maximum possible value of xyz^2 is

(A) 1 (B) 2 (C) 3 (D) 4

10. The general solution of the equation cosec $x = 1 + \sin x + \sin^2 x + \sin^3 x + \ldots$ up to ∞ terms is

(A) $n\pi+(-1)^n\frac{\pi}{6}, n\in I$

(B) $n\pi+(-1)^n\frac{\pi}{3}, n\in I$

(C) $2n\pi+\frac{\pi}{6}, n\in I$ (D) $2n\pi-\frac{\pi}{3}, n\in I$

11. If S_1, S_2, S_3 is the sum of the first n terms of 3 A.P.s, the first terms of each being 1 and respective common differences are 1, 2 and 3, then

(A) $S_1 + S_2 = S_3$ (B) $S_1 + S_3 = 2S_2$

(C) $S_1 + S_3 = S_2$ (D) $S_1 + 2S_2 = 3S_3$

12. If $x_1, x_2, x_3, \ldots$ up to $2n$ terms are in A.P., then the value of $x_1^2 - x_2^2 + x_3^2 - x_4^2 + \ldots$ is

(A) $\frac{n}{2n-1}\left(x_1^2-x_{2n}^2\right)$ (B) $\frac{2n}{n-1}\left(x_{2n}^2-x_1^2\right)$

(C) $\frac{n}{n+1}\left(x_1^2+x_2^2\right)$ (D) $\frac{2n}{n+1}\left(x_1^2-x_2^2\right)$

13. Let $2p + 3q + 4r = 15$ and if the expression $p^3q^5r^7$ takes the maximum value, then $p : q : r$ is equal to (where $p, q, r \in R^+$)

(A) 18 : 20 : 21 (B) 3 : 5 : 7

(C) 2 : 3 : 4 (D) 6 : 15 : 28

14. $a_1, a_2, a_3, \ldots, a_n$ are in A.P. such that $a_n = 100$, $a_{50} - a_{49} = \frac{3}{5}$, then the 15th term of the A.P. from the end is

(A) $\frac{454}{5}$ (B) $\frac{448}{5}$ (C) $\frac{452}{5}$ (D) $\frac{458}{5}$

15. If x, y and z are three distinct real numbers in geometric progression and $ay = x + y + z$ then the complete interval of 'a' is

(A) $(-1, 3)$
(B) $[-1, 3]$
(C) $(-\infty, -1) \cup (3, \infty)$
(D) $(-\infty, -1] \cup [3, \infty)$

16. Let $b_i > 1$ for $i = 1, 2, \ldots, 101$. Suppose $\log_e b_1, \log_e b_2 \ldots, \log_e b_{101}$ are in arithmetic progression (A.P.) with the common difference $\log_e 2$. Suppose $a_1, a_2, \ldots, a_{101}$ are in A.P. such that $a_1 = b_1$ and $a_{51} = b_{51}$.

If $t = b_1 + b_2 + \ldots + b_{51}$ and $s = a_1 + a_2 + \ldots + a_{51}$ then

(A) $s > t$ and $a_{101} > b_{101}$
(B) $s > t$ and $a_{101} < b_{101}$
(C) $s < t$ and $a_{101} > b_{101}$
(D) $s < t$ and $a_{101} < b_{101}$

17. If p, q, r are in H.P., and p and r are different, having the same sign, then the roots of the equation $px^2 + qx + r = 0$ are

(A) real and equal
(B) real and distinct
(C) irrational
(D) imaginary

18. If α, β, γ are the three positive roots of cubic equation $x^3 - ax^2 + bx - 16\sqrt{6} = 0$ then the minimum value of $\left(\frac{4}{\alpha\beta}+\frac{3}{\beta\gamma}+\frac{2}{\gamma\alpha}\right)$ is

(A) 1 (B) $\frac{3}{2}$ (C) $\frac{2}{3}$ (D) $\frac{3}{4}$

19. $\sqrt{1+2(0.4)+3(0.4)^2+4(0.4)^3+\ldots\infty}$ has the value equal to
(A) $\frac{5}{3}$ (B) $\frac{5}{4}$ (C) 1 (D) $\frac{6}{5}$

20. The value which is not possible for $\sum_{k=1}^{n} k^3$ is
(A) 25 (B) 36 (C) 225 (D) 441

21. The minimum value of $\frac{x^4+2x^2+101}{x^2+1}$ for $x \in R$, is equal to
(A) $\frac{4}{5}$ (B) 10 (C) $\frac{8}{3}$ (D) 20

22. In a convex polygon, the degree measures of the interior angles form an arithmetic progression. If the smallest angle is 159° and the largest angle is 177°, then the number of sides in the polygon, is
(A) 21 (B) 27 (C) 30 (D) 31

23. The value of the sum $\sum_{k=1}^{\infty}\sum_{n=1}^{\infty}\frac{k}{2^{n+k}}$ is equal to
(A) 5 (B) 4 (C) 3 (D) 2

24. If the first term of a G.P. $a_1, a_2, a_3, \ldots$ is unity such that the quantity $4a_2 + 5a_3$ has the least value, then the sum to infinite number of terms of the sequence is
(A) cannot be found out as it diverges
(B) $\frac{5}{3}$ (C) $\frac{7}{5}$ (D) $\frac{5}{7}$

25. If the roots of the cubic equation $x^3 - 7x^2 + cx - 8 = 0$ are in G.P., then the value of 'c', is
(A) 10 (B) 12 (C) 14 (D) 16

Multiple Answer Type

1. If a, b, c are three terms of an A.P. such that $a \neq b$, then $\frac{b-c}{a-b}$ may be equal to
(A) $\sqrt{2}$ (B) 1 (C) $\sqrt{3}$ (D) 3

2. If three distinct positive real numbers a, b, c are in H.P., then
(A) $a+c>2b$ (B) $a^2+c^2>2ac$
(C) $a^2+c^2>2b^2$ (D) none of these

3. The number 111 … 11 (up to 91 digits) is a
(A) prime number
(B) composite number
(C) divisible by $\frac{10^7-1}{10-1}$
(D) none of these

4. If a, b, c, d are distinct positive numbers in A.P., then
(A) $ad < bc$ (B) $a+c<b+d$
(C) $a+d=b+c$
(D) $(a+1)(d+1)<(b+1)(c+1)$

5. If $b_1, b_2, b_3 (b_1 > 0)$ are three consecutive terms of a G.P. with common ratio r, the value of r for the inequality $b_3 > 4b_2 - 3b_1$ holds is given by
(A) $r<1$ (B) $r>3$
(C) $r=0.5$ (D) $r=5.2$

6. If a, b, c are in H.P., then
(A) $\frac{a}{b+c-a}, \frac{b}{c+a-b}, \frac{c}{a+b-c}$ are in H.P.
(B) $\frac{2}{b}=\frac{1}{b-a}+\frac{1}{b-c}$
(C) $a-\frac{b}{2}, \frac{b}{2}, c-\frac{b}{2}$ are in G.P.
(D) $\frac{a}{b+c}, \frac{b}{c+a}, \frac{c}{a+b}$ are in H.P.

7. For a positive integer n,
let $a_n = 1+\frac{1}{2}+\frac{1}{3}+\frac{1}{4}+\ldots+\frac{1}{2^n-1}$, then
(A) $a(n)<n$ (B) $a(n)>\frac{n}{2}$
(C) $a(2n)>n$ (D) $a(2n)<2n$

8. If the first and the $(2n+1)^{\text{th}}$ terms of an A.P., a G.P. and a H.P. of positive terms are equal and their $(n+1)^{\text{th}}$ terms are a, b and c respectively, then
(A) $a=b=c$ (B) $a \geq b \geq c$
(C) $a+c=2b$ (D) $ac=b^2$

9. If $a, b, c \in R^+$, then $\frac{a^2+1}{b+c}+\frac{b^2+1}{c+a}+\frac{c^2+1}{a+b}$ can never be equal to
(A) 1 (B) 2 (C) $\frac{8}{3}$ (D) 3

10. If a, b, c, d are four distinct positive real number in A.P., then

(A) $\frac{1}{a}+\frac{1}{d}>\frac{1}{b}+\frac{1}{c}$ (B) $\frac{1}{a}+\frac{1}{d}<\frac{1}{b}+\frac{1}{c}$

(C) $ad < bc$ (D) $\frac{1}{b}+\frac{1}{c}>\frac{4}{a+d}$

11. If a, b, c are in A.P. and a^2, b^2, c^2 are in H.P., then which of the following is/are true?

(A) $\frac{a}{2}$, b, c are in G.P.

(B) $a, -\frac{b}{2}, c$ are in G.P.

(C) $-\frac{a}{2}, b, c$ are in G.P.

(D) $a = b = c$

12. For any natural number $n > 1$, consider

$$S=\frac{1}{n^2+1}+\frac{2}{n^2+2}+\frac{3}{n^2+3}+\ldots+\frac{n}{n^2+n},$$

then

(A) $S<\frac{1}{2}+\frac{1}{2n}$ (B) $S>\frac{1}{2}+\frac{1}{2n}$

(C) $S>\frac{1}{2}$ (D) $S<1$

13. For a positive integer 'n',

let $S_n=1+\frac{1}{2}+\frac{1}{3}+\frac{1}{4}+\ldots+\frac{1}{2^n-1}$, then

(A) $S(200)>100$ (B) $S(200)<100$

(C) $S(100)<100$ (D) $S(100)>100$

14. If the roots of cubic $x^3-6x^2-4x+c=0$ are the first 3 terms of an A.P. then the sum to the first 10 terms of the A.P. is

(A) −250 (B) −190

(C) 230 (D) 290

15. Let A_1 denote the arithmetic mean of two numbers $a = 1$ and $b = 2$. A_2 denotes the arithmetic mean of A_1 and b. For $n \geq 3$, let A_n be the arithmetic mean of A_{n-1} and b, then

(A) $A_{10}=\frac{1023}{512}$ (B) $A_{10}=\frac{2047}{1024}$

(C) $\sum_{n=1}^{10} A_n > 19$ (D) $\sum_{n=1}^{10} A_n < 19$

16. If a and b are positive real numbers such that $a^2b^3=\frac{4}{9}$, then $(a + b)$ can be equal to

(A) $\frac{5}{3}$ (B) $\frac{4}{3}$ (C) 2 (D) $\frac{5}{2}$

17. If $\frac{1}{a}, \frac{1}{b}, \frac{1}{c}$ are in A.P. and a, b, $-2c$ are in G.P. where a, b, c are non-zero, then

(A) $a^3 + b^3 + c^3 = 3abc$

(B) $-2a$, b, $-2c$ are in A.P.

(C) $-2a$, b, $-2c$ are in G.P.

(D) a^2, b^2, $4c^2$ are in G.P.

18. If $(1 + 3 + 5 + \ldots + a) + (1 + 3 + 5 + \ldots + b) = (1 + 3 + 5 + \ldots + c)$ where each set contains the sum of consecutive odd integers then the smallest possible value of $(a + b + c)$ is less than or equal to

(A) 19 (B) 21 (C) 29 (D) 31

19. Let $a_1, a_2, a_3 \ldots$ and $b_1, b_2, b_3 \ldots$ be arithmetic progressions such that $a_1 = 25$, $b_1 = 75$ and $a_{100} + b_{100} = 100$. Then

(A) the difference between successive terms in progression 'a' is opposite of the difference in progression 'b'

(B) $a_n + b_n = 100$ for any n

(C) $(a_1+b_1), (a_2+b_2), (a_3+b_3), \ldots$ are in A.P.

(D) $\sum_{r=1}^{100}(a_r+b_r)=10000$

20. If the roots of the equation, $x^3 + px^2 + qx - 1 = 0$ form an increasing G.P. where p and q are real, then

(A) $p + q = 0$

(B) $p \in (-3, \infty)$

(C) one of the roots is unity

(D) one root is smaller then 1 and one root is greater than 1

21. If S_n denotes the sum of the first n terms of an arithmetic progression and a_n denotes the nth term of the same A.P., given $S_n = n^2p$; $S_k = k^2p$ where $k, p, n \in N$ and $k \neq n$ then,

(A) $a_1 = p$

(B) common difference = $2p$

(C) $S_p = p^3$

(D) $a_p = 2p^2 - p$

22. The first term of an infinite geometric series is 21. The second term and the sum of the series are both positive integers. The possible value(s) of the second term can be
(A) 12 (B) 14 (C) 18 (D) 20

23. Which of the following statement(s) is(are) correct?
(A) Sum of the reciprocal of all the n harmonic means inserted between a and b is equal to n times the harmonic mean between two given numbers a and b.
(B) Sum of the cubes of first n natural number is equal to square of the sum of the first n natural numbers.
(C) If $a, A_1, A_2, A_3, \ldots A_{2n}, b$ are in A.P. then $\sum_{i=1}^{2n} A_i = n(a+b)$.
(D) If the first term of the geometric progression $g_1, g_2, g_3, \ldots \infty$ is unity, then the value of the common ratio of the progression such that $(4g_2 + 5g_3)$ is minimum equals $\frac{2}{5}$.

24. Positive integers $a_1, a_2, a_3, \ldots$ form an A.P. If $a_1 = 10$ and $a_{a_2} = 100$, then
(A) common difference of A.P. is equal to 6
(B) the value of a_3 is equal to 20
(C) the value of a_{22} is equal to 136
(D) the value of a_{a_3} is equal to 820

25. If a, b, c, d are distinct positive numbers in H.P., then
(A) $\sqrt{\frac{a+d}{b+c}} + \sqrt{\frac{b+c}{a+d}} = \sqrt{\frac{ad}{bc}} + \sqrt{\frac{bc}{ad}}$
(B) $abc + bcd = abd + acd$
(C) $\frac{1}{a} + \frac{1}{c} = \frac{1}{b} + \frac{1}{d}$
(D) $\left(\frac{1}{a}+1\right)\left(\frac{1}{d}+1\right) < \left(\frac{1}{b}+1\right)\left(\frac{1}{c}+1\right)$

Comprehension Based Questions

Paragraph for Question No. 1 to 3

Four different integers form an increasing A.P. One of these numbers is equal to the sum of the squares of the other three numbers, then

1. The smallest number is
(A) −2 (B) 0 (C) −1 (D) 2

2. The common difference of the four numbers is
(A) 2 (B) 1 (C) 3 (D) 4

3. The sum of all the four numbers is
(A) 10 (B) 8 (C) 2 (D) 6

Paragraph for Question No. 4 to 6

Let $A_1, A_2, A_3, \ldots, A_m$ be the arithmetic means between −2 and 1027 and $G_1, G_2, G_3, \ldots, G_n$ be geometric means between 1 and 1024. The product of the geometric means is 2^{45} and the sum of the arithmetic means is 1025×171.

4. The value of n is
(A) 7 (B) 9
(C) 11 (D) none of these

5. The value of m is
(A) 340 (B) 342 (C) 344 (D) 346

6. The value of $G_1 + G_2 + G_3 + \ldots + G_n$ is
(A) 1022 (B) 2044
(C) 512 (D) none of these

Paragraph for Question No. 7 to 9

The sum of an infinitely decreasing geometric progression whose first term is a and common ratio is r, is equal to the least value of the quadratic trinomial $P(x) = 3x^2 - x + \frac{25}{12}$, in [0, 2]. Also the first term of the geometric progression is equal to the square of its common ratio.

7. The value of $(r^2 + 2r)$ is
(A) 1 (B) 2 (C) $\frac{3}{2}$ (D) $\frac{1}{2}$

8. If the minimum value of $P(x)$ lies between roots of equation $x^2 + (k+1)x + a + 2\sqrt{3} = 0$, then the maximum integral value of k is
(A) −6 (B) −1 (C) −7 (D) −10

9. If the range of function $y = \frac{P(x)}{P'(x)}$ is $(-\infty, p] \cup [q, \infty)$, then $(p + q)$ is

[**Note:** $P'(x)$ denotes derivative of $P(x)$ with respect to x.]

(A) $\frac{1}{2}$ (B) $\frac{1}{3}$ (C) $\frac{1}{6}$ (D) 0

Paragraph for Question No. 10 to 12

Let equation $x^3 + px^2 + qx - q = 0$ where $p, q \in R - \{0\}$ has 3 real roots α, β, γ in H.P., then

10. $(9p + 2q)$ has the value of equal to

(A) 9 (B) −18 (C) −27 (D) 1

11. The minimum value of $\frac{1}{\alpha^2}+\frac{1}{\beta^2}+\frac{1}{\gamma^2}$ is (you may use the inequality $a^2 + b^2 + c^2 \geq ab + bc + ca$ for any $a, b, c \in R$)

(A) $\frac{1}{3}$ (B) 1 (C) $\frac{4}{3}$ (D) 3

12. $\frac{p}{q}$ has the minimum value equal to

(A) $\frac{-1}{2}$ (B) $\frac{-1}{3}$ (C) $\frac{-1}{4}$ (D) −1

Paragraph for Question No. 13 to 15

Let the quantities 1, $\log_y x$, $\log_z y$ and $-15 \log_x z$ be the first four terms of an arithmetic progression with common difference d, all terms of the A.P. being real.

13. Common difference d of the A.P. satisfies the equation

(A) $6d^3 + 11d^2 + 6d - 16 = 0$

(B) $6d^3 + 11d^2 - 6d - 16 = 0$

(C) $6d^3 + 11d^2 + 6d + 16 = 0$

(D) $6d^3 + 11d^2 - 6d + 16 = 0$

14. The value of $(xz^{-3} + xy + yz^3)$ is equal to

(A) 0 (B) 1 (C) 2 (D) 3

15. Magnitude of the sum of the first 25 terms of the A.P. is equal to

(A) 575 (B) 675 (C) 625 (D) 552

Match Matrix

1. Only one matching option is possible

Column I	Column II
(A) If $a > 1$, the least value of $2\log_{100} a - \log_a 0.0001$ is	(P) 6
(B) If $a, b, c > 0$ and the minimum value of $a(b^2+c^2)+b(c^2+a^2)+c(a^2+b^2)$ is λabc, then λ is	(Q) 3
(C) If $x, y, z > 0$, then the minimum value of $x^{(\log y - \log z)} + y^{(\log z - \log x)} + z^{(\log x - \log y)}$ is	(R) 4
(D) If $a, b, c \in R^+$ such that $a+b+c = 18$ and the maximum value of $a^2b^3c^4$ is $\lambda_1^{19}\lambda_2^3$, then the value of $\lambda_1 \cdot \lambda_2$ is	(S) 7

2. Only one matching option is possible

Column I	Column II
(A) If $F(n+1) = \frac{2F(n)+1}{2}$ for $n = 1, 2, 3 \ldots$ and $F(1) = 2$, then $F(101)$ is equal to	(P) 42
(B) If $a_1, a_2, a_3, \ldots, a_{21}$ are in A.P. and $a_3 + a_5 + a_{11} + a_{17} + a_{19} = 10$, then $\sum_{i=1}^{21} a_i$ is equal to	(Q) 1620
(C) 10th term of the sequence $S = 1+5+13+29+\ldots$ is	(R) 52
(D) The sum of all two-digit numbers which are not divisible by 2 or 3 is	(S) 2045

3.

Column I	Column II
(A) For an A.P. $a_1, a_2, a_3, \ldots, a_n, \ldots$ $a_1 = \frac{5}{2}$; $a_{10} = 16$. If $\sum_{i=1}^{n} a_i = 110$ then 'n' equals	(P) 9
(B) The interior angles of a convex non-equiangular polygon of 9 sides are in A.P. The least positive integer that limits the upper value of the common difference between the measures of the angles in degree is	(Q) 10
(C) For an increasing G.P. $a_1, a_2, a_3, \ldots, a_n, \ldots$ $a_6 = 4a_4$; $a_9 - a_7 = 192$. If $\sum_{i=4}^{n} a_i = 1016$ then 'n' equals	(R) 11
	(S) 12

4.

Column I	Column II
(A) Let S_n, S_{2n}, S_{3n} are the sums of n, $2n$, $3n$ terms of an arithmetic progression, then $\frac{S_{3n}}{S_{2n} - S_n}$ is equal to	(P) 3
(B) The value of $\sum_{i=1}^{n}\sum_{j=1}^{i}\sum_{k=1}^{j} 1 = 220$, then the value of n equals	(Q) 4
(C) Let g_n be the nth term of the geometric progression of positive numbers. If $\sum_{n=1}^{100} g_{2n} = \frac{10}{3}$ and $\sum_{n=1}^{100} g_{2n-1} = \frac{5}{9}$, then the common ratio of geometric progression, is	(R) 6
(D) A geometric progression consists of an even number of terms. If the sum of all the terms is 5 times the sum of terms occupying the odd place, then its common ratio is	(S) 8
	(T) 10

5.

Column I	Column II
(A) If x and y are positive real numbers and p, q are any positive integers then the possible value which the expression $\frac{(1+x^{2p})(1+y^{2q})}{x^p y^q}$ can take is	(P) 3
(B) Given a and b are positive numbers not equal to 1 such that $\log_b(a^{\log_2 b}) = \log_a(a^{\log_4 a})$ and $\log_a(p-(b-9)^2) = 2$ then the minimum integral value for p is	(Q) 5

(Continued)

Column I	Column II
(C) Possible values of x simultaneously satisfying the system of inequalities $\frac{(x-6)(x-3)}{x+2} \geq 0$ and $\frac{x-5}{x+1} \leq 3$ is	(R) 9
(D) If the numbers, $(3^{1+x}\ 3^{1-x})$, $(a/2)$, $(9^x + 9^{-x})$ form an A.P., then possible value(s) of 'a' is/are ($x \in R$)	(S) 10

Subjective Answer Type (Hard Nut)

1. Let α, β be the roots of $x^2 - x + p = 0$ and γ, δ be the roots of $x^2 - 4x + q = 0$. If α, β, γ, δ are in G.P., then find the integral values of p and q.

2. The sum of the squares of three distinct real numbers which are in G.P. is s^2. If their sum is 'as'. Show that $a^2 \in \left[\frac{1}{3}, 1\right) \cup (1, 3]$.

3. Let a, b, c be positive integers such that $\frac{b}{a}$ is an integer. If a, b, c are in G.P. and the A.M. of a, b, c is $(b+2)$, then find the value of $\frac{a^2+a-14}{a+1}$.

4. A cricketer plays n matches $(n \geq 1)$, total number of runs scored by him is $\frac{(n+1)}{4}\left(2^{n+1} - n - 2\right)$. If he scores $K.2^{n-k+1}$ runs in K^{th} match $(1 \leq K \leq n)$. Find the value of n.

5. Find the sum of n terms of the series $\frac{3}{2^3} + \frac{4}{2^4.3} + \frac{5}{2^6.3} + \frac{6}{2^7.5} + \ldots$

6. Find the sum of n terms of the series $\frac{7}{4.1.2} + \frac{10}{4^2.2.3} + \frac{13}{4^3.3.4} + \ldots$

7. Let $a_1, a_2, \ldots, a_{11}$ be real numbers satisfying $a_1 = 15, 27 - 2a_2 > 0$ and $a_K = 2a_{K-1} - a_{K-2}$ for $K = 3, 4, \ldots, 11$.

   If $\frac{a_1^2 + a_2^2 + \ldots + a_{11}^2}{11} = 90$.

   Find the value of $\frac{a_1 + a_2 + \ldots + a_{11}}{11}$.

8. Determine whether the number 111 … 11 (up to 91 digits) is prime or composite.

9. If n is a positive integer, prove that $2^n > 1 + n\sqrt{2^{n-1}}, n > 1$.

10. If a, b, c, d are four distinct positive quantities in H.P. Show that
    (i) $a + d > b + c$
    (ii) $ad > bc$

11. Find the sum of the first 100 terms of the series whose general term is $a_K = \left(K^2 + 1\right)K!$

12. If $\sqrt{1+\frac{1}{1^2}+\frac{1}{2^2}} + \sqrt{1+\frac{1}{2^2}+\frac{1}{3^2}} + \ldots + \sqrt{1+\frac{1}{1999^2}+\frac{1}{2000^2}} = p + \frac{p}{q}$, then find the value of $|p - q|$.

13. If $\sum_{r=1}^{\infty} \frac{1}{r^4} = 1$, find the value of $\sum_{r=1}^{\infty} \frac{1}{(2r-1)^4}$.

14. Find the sum of the series $\frac{(1^2+1-1)1!}{1^2+3.1+2} + \frac{(2^2+2-1)2!}{2^2+3.2+2} + \ldots + \frac{(n^2+n-1)n!}{n^2+3.n+2}$.

15. Find the minimum value of $\left(a+\frac{1}{a}\right)^2 + \left(b+\frac{1}{b}\right)^2$ where $a > 0$, $b > 0$ and $a + b = \sqrt{15}$.

16. Find the sum of the series $\sum_{r=0}^{n} \frac{(n-3r+1)}{(n-r+1)} \frac{nc_r}{2^r}$.

17. Let P be any point inside the triangle ABC of side length 6, 5, 5 units and p_1, p_2, p_3 be the lengths of perpendiculars drawn from P

to the sides of triangle. Find the maximum value of $p_1p_2p_3$.

18. Three numbers in A.P. are removed from the first n consecutive natural numbers and average of the remaining numbers is found to be $\frac{43}{4}$. Find n as well as the removed numbers if one of the removed numbers is a perfect square.

19. Find the sum of n terms of the series $\frac{8}{5}+\frac{16}{65}+\frac{24}{325}+\ldots$ Also find S_∞.

20. If a, b, c are in G.P. and $\log_c a, \log_b c, \log_a b$ are in A.P., then find the common difference of the A.P.

21. Let a, b, c, d be four distinct real numbers in A.P., then find the smallest positive value of K satisfying

$$2(a-b)+K(b-c)^2+(c-a)^3$$
$$=2(a-d)+(b-d)^2+(c-d)^3.$$

22. If the sum of the terms of an infinitely decreasing G.P. is equal to the greatest value of the function $f(x)=x^3+3x-9$ on the interval $[-5,3]$ and the difference between the first and second terms is $f'(0)$, then find the common ratio.

23. Let r be the positive real zero of

$$P(x)=9x^5+7x^2-9;$$

if $s=r^4+2r^9+3r^{14}+4r^{19}+\ldots$ to ∞ can be expressed as the rational number $\left(\frac{a}{b}\right)$ in the lowest term, then find sum of the digits in $(a+b)$.

24. The roots α and β of a quadratic equation are the square of two consecutive natural numbers. The geometric mean of the two roots is 1 greater than the difference of the roots. If α and β lies between the roots of $x^2-cx+1=0$, then find the minimum integral value of c.

25. Let

$$S=\frac{1}{\sin 8^\circ}+\frac{1}{\sin 16^\circ}+\frac{1}{\sin 32^\circ}+\ldots+$$
$$\frac{1}{\sin 4096^\circ}+\frac{1}{\sin 8192^\circ}$$

If $S=\frac{1}{\sin\alpha}$ where $\alpha\in(0,90^\circ)$, then find α in degree.

26. If the sum of first 10 terms of the series $\frac{7}{2^2 5^2}+\frac{13}{5^2 8^2}+\frac{19}{8^2 11^2}+\ldots$ is $\frac{\alpha}{\beta^n}$, find the value of n.

27. Let

$$T_n=1+2+3+\ldots+n. \text{ If } \sum_{i=3}^{\infty}\sum_{K=1}^{\infty}\left(\frac{3}{T_i}\right)^K=\frac{p}{q}$$

where p and q are co-prime, then find the value of $(3q-p-8)$.

28. Let x, y, z, w be four non-zero real numbers such that x, y, z are in A.P. and y, z, w are in G.P. If $x+w=16$ and $y+z=8$, then find the absolute value of $\left(x^2-y^2+2z-w\right)$.

29. Let α, β, γ be the roots of the equation $x^3+4x^2-ax+1=0$. Then find the minimum value of

$$\left(\alpha+\frac{1}{\alpha}\right)^2+\left(\beta+\frac{1}{\beta}\right)^2+\left(\gamma+\frac{1}{\gamma}\right)^2.$$

30. Let $a_1\,a_2\,a_3\,a_4$ be a four-digit natural number such that a_1, a_2, a_3 are in A.P. and a_2, a_3, a_4 are in G.P. If $a_4=a_1-5$ and $d+4r=0$ where d denotes c.d. and r denote c.r., then find the value of $\sum_{i=1}^{4}a_i$.

31. If a, b, c, d are distinct positive real numbers such that $3s=a+b+c+d$ and

$$\frac{abcd}{(s-a)(s-b)(s-c)(s-d)}>\alpha^2,$$

find the value of α.

32. Find the HCF of the minimum non-negative values of a, b and c, given that the equation $x^4 + ax^3 + bx^2 + cx + 1 = 0$ has only real roots.

33. A sequence $a_1, a_2, a_3, \ldots, a_n$ of real numbers is such that $a_1 = 0$,
$|a_2| = |a_1 + 1|, |a_3| = |a_2 + 1| \ldots |a_n| = |a_{n-1} + 1|$,
where the A.M. of $a_1, a_2, \ldots, a_n$ cannot be less than $-\frac{\lambda}{\mu}$, then find the value of $\lambda + \mu$.

34. If a, b, c are three distinct positive numbers in HP. Prove that $\frac{a+b}{2a-b} + \frac{c+b}{2c-b} > 4$.

35. Let a, b, c, d, e be positive numbers such that $abcd = e^4$ and
$(1+a)(1+b)(1+c)(1+d) \geq (1+e)^{\lambda}$,
find the value of λ.

2.11 JEE Main: Archive

1. If $\{u_i\}_{i=1}^{n}$ where n is an even integer, is an arithmetic progression with common difference 1, and
$\sum_{i=1}^{n} a_i = 192, \sum_{i=1}^{n/2} a_{2i} = 120$, then n is equal to
(A) 48 (B) 96 (C) 92 (D) 104
[Main 2022 24th June 1st Shift]

2. If $\frac{1}{2 \cdot 3^{10}} + \frac{1}{2^2 \cdot 3^9} + \ldots + \frac{1}{2^{10} \cdot 3} = \frac{K}{2^{10} \cdot 3^{10}}$,
then the remainder when K is divided by 6 is
(A) 1 (B) 2 (C) 3 (D) 5
[Main 2022 25th June 1st Shift]

3. The greatest integer less than or equal to the sum of first 100 terms of the sequence $\frac{1}{3}, \frac{5}{9}, \frac{19}{27}, \frac{65}{81}, \ldots$ is equal to ______.
[Main 2022 25th June 1st Shift]

4. The sum $1 + 2 \cdot 3 + 3 \cdot 3^2 + \ldots + 10 \cdot 3^9$ is equal to
(A) $\frac{2 \cdot 3^{12} + 10}{4}$ (B) $\frac{19 \cdot 3^{10} + 1}{4}$
(C) $5 \cdot 3^{10} - 2$ (D) $\frac{9 \cdot 3^{10} + 1}{2}$
[Main 2022 25th June 2nd Shift]

5. Let $A = \sum_{i=1}^{10} \sum_{j=1}^{10} \min\{i, j\}$ and
$B = \sum_{i=1}^{10} \sum_{j=1}^{10} \max(i, j)$. Then $A + B$ is equal to ______.
[Main 2022 26th June 1st Shift]

6. If $A = \sum_{n=1}^{\infty} \frac{1}{\left(3 + (-1)^n\right)^n}$ and
$B = \sum_{n=1}^{\infty} \frac{(-1)^n}{\left(3 + (-1)^n\right)^n}$, then $\frac{A}{B}$ is equal to:
(A) $\frac{11}{9}$ (B) 1 (C) $-\frac{11}{9}$ (D) $-\frac{11}{3}$
[Main 2022 26th June 2nd Shift]

7. If $a_1 (> 0), a_2, a_3, a_4, a_5$ are in a G.P., $a_2 + a_4 = 2a_3 + 1$ and $3a_2 + a_3 = 2a_4$, then $a_2 + a_4 + 2a_5$ is equal to ______.
[Main 2022 26th June 2nd Shift]

8. If $x = \sum_{n=0}^{\infty} a^n, y = \sum_{n=0}^{\infty} b^n, z = \sum_{n=0}^{\infty} c^n$, where a, b, c are in A.P. and $|a| < 1, |c| < 1$, and $abc \neq 0$, then
(A) x, y, z are in A.P.
(B) x, y, z are in G.P.
(C) $\frac{1}{x}, \frac{1}{y}, \frac{1}{z}$ are in A.P.
(D) $\frac{1}{x} + \frac{1}{y} + \frac{1}{z} = 1 - (a + b + c)$
[Main 2022 27th June 1st Shift]

9. If the sum of the first ten terms of the series $\frac{1}{5} + \frac{2}{65} + \frac{3}{325} + \frac{4}{1025} + \frac{5}{2501} + \ldots$ is $\frac{m}{n}$, where m and n are co-prime numbers, then $m + n$ is equal to ______.
[Main 2022 27th June 1st Shift]

10. Let $S = 2 + \frac{6}{7} + \frac{12}{7^2} + \frac{20}{7^3} + \frac{30}{7^4} + \ldots$ then $4S$ is equal to
(A) $\left(\frac{7}{3}\right)^2$ (B) $\frac{7^3}{3^2}$ (C) $\left(\frac{7}{3}\right)^3$ (D) $\frac{7^2}{3^3}$
[Main 2022 27th June 2nd Shift]

11. If $a_1, a_2, a_3, \ldots$ and $b_1, b_2, b_3, \ldots$ are in A.P. and $a_1 = 2, a_{10} = 3, a_1 b_1 = 1 = a_{10} b_{10}$ then $a_4 b_4$ is equal to

(A) $\frac{35}{27}$ (B) 1 (C) $\frac{27}{28}$ (D) $\frac{28}{27}$

[Main 2022 27th June 2nd Shift]

12. Let $A_1, A_2, A_3, \ldots$ be an increasing geometric progression of positive real numbers. If $A_1 A_3 A_5 A_7 = \frac{1}{1296}$ and $A_2 + A_4 = \frac{7}{36}$, then, the value of $A_6 + A_7 + A_{10}$ is equal to

(A) 33 (B) 37 (C) 43 (D) 48

[Main 2022 28th June 1st Shift]

13. If n arithmetic means are inserted between 1and 100 such that the ratio of the first mean to the last mean is 1: 7 and $a + n = 33$, then the value of n is

(A) 21 (B) 22 (C) 23 (D) 24

[Main 2022 28th June 2nd Shift]

14. Let for $n = 1, 2, \ldots, 50$, S_n be the sum of the infinite geometric progression whose first term is n^2 and whose common ratio is $\frac{1}{(n+1)^2}$.

Then the value of $\frac{1}{26} + \sum_{n=1}^{50}\left(S_n + \frac{2}{n+1} - n - 1\right)$ is equal to

[Main 2022 28th June 2nd Shift]

15. Let $\{a_n\}_{n=0}^{\infty}$ be a sequence such that $a_0 = a_1 = 0$ and $a_{n+2} = 2a_{n+1} - a_n + 1$ for all $n \geq 0$. Then, $\sum_{n=2}^{\infty} \frac{a_n}{7^n}$ is equal to

(A) $\frac{6}{343}$ (B) $\frac{7}{216}$ (C) $\frac{8}{343}$ (D) $\frac{49}{216}$

[Main 2022 29th June 1st Shift]

16. The sum of the infinite series $1 + \frac{5}{6} + \frac{12}{6^2} + \frac{22}{6^3} + \frac{35}{6^4} + \frac{51}{6^5} + \frac{70}{6^6} + \ldots$ is equal to

(A) $\frac{425}{216}$ (B) $\frac{429}{216}$ (C) $\frac{288}{125}$ (D) $\frac{280}{125}$

[Main 2022 29th June 2nd Shift]

17. Let 3, 6, 9, 12, ... up to 78 terms and 5, 9, 13, 17, ... up to 59 terms be two series. Then, the sum of the terms common to both the series is equal to ______.

[Main 2022 29th June 2nd Shift]

18. Let $a_1 = b_1 = 1, a_n = a_{n-1} + 2$ and $b_n = a_n + b_{n-1}$ for every natural number $n \geq 2$. Then $\sum_{n=1}^{15} a_n \cdot b_n$ is equal to ______.

[Main 2022 25th July 1st Shift]

19. The sum $\sum_{n=1}^{21} \frac{3}{(4n-1)(4n+3)}$ is equal to

(A) $\frac{7}{87}$ (B) $\frac{7}{29}$ (C) $\frac{14}{87}$ (D) $\frac{21}{29}$

[Main 2022 25th July2nd Shift]

20. Consider two G.P.s $2, 2^2, 2^3, \ldots$ and $4, 4^2, 4^3, \ldots$ of 60 and n terms respectively. If the geometric mean of all the $60+n$ terms is $(2)^{\frac{225}{8}}$, then $\sum_{k=1}^{n} k(n-k)$ is equal to:

(A) 560 (B) 1540 (C) 1330 (D) 2600

[Main 2022 26th July 1st Shift]

21. The series of positive multiples of 3 is divided into sets:

$\{3\}, \{6, 9, 12\}, \{15, 18, 21, 24, 27\}, \ldots$

Then the sum of the elements in the 11th set is equal to ______.

[Main 2022 26th July 1st Shift]

22. If $\sum_{k=1}^{10} \frac{k}{k^4 + k^2 + 1} = \frac{m}{n}$, where m and n are co-prime, then $m + n$ is equal to______.

[Main 2022 26th July 2nd Shift]

23. Different A.P.s are constructed with the first term 100, the last term 199, and integral common differences. The sum of the common differences of all such, A.P.s having at least 3 terms and at most 33 terms is ______.

[Main 2022 26th July 2nd Shift]

24. Suppose $a_1, a_2, \ldots, a_n, \ldots$ be an arithmetic progression of natural numbers. If the ratio of the sum of the first five terms of the sum of first nine terms of the progression is 5 : 17 and $110 < a_{15} < 120$, then the sum of the first ten terms of the progression is equal to

(A) 290 (B) 380 (C) 460 (D) 510

[Main 2022 27th July 1st Shift]

25. Let $f(x) = 2x^2 - x - 1$ and $S = \{n \in Z : |f(n)| \le 800\}$. Then, the value of $\sum_{n \in S} f(n)$ is equal to ______.

[Main 2022 27th July 1st Shift]

26. Let the sum of an infinite G.P., whose first term is a and the common ratio is r, be 5. Let the sum of its first five terms be $\frac{98}{25}$. Then the sum of the first 21 terms is a_n and the common difference is $10ar^2$, is equal to

(A) $21a_{11}$ (B) $22a_{11}$ (C) $15a_{16}$ (D) $14a_{16}$

[Main 2022 27th July2nd Shift]

27. $$\frac{2^3 - 1^3}{1 \times 7} + \frac{4^3 - 3^3 + 2^3 - 1^3}{2 \times 11} + \frac{6^3 - 5^3 + 4^3 - 3^3 + 2^3 - 1^3}{3 \times 15} + \ldots + \frac{30^3 - 29^3 + 28^3 - 27^3 + \ldots + 2^3 - 1^3}{15 \times 63}$$ is equal to ______.

[Main 2022 27th July2nd Shift]

28. Consider the sequence $a_1, a_2, a_3, \ldots$ such that $a_1 = 1, a_2 = 2$ and $a_{n+2} = \frac{2}{a_{n+1}} + a_n$ for $n = 1, 2, 3, \ldots$.

If $$\left(\frac{a_1 + \frac{1}{a_2}}{a_3}\right) \cdot \left(\frac{a_2 + \frac{1}{a_3}}{a_4}\right) \cdot \left(\frac{a_3 + \frac{1}{a_4}}{a_5}\right) \ldots \left(\frac{a_{30} + \frac{1}{a_{31}}}{a_{32}}\right) = 2^{\alpha}\left({}^{61}C_{31}\right),$$

then α is equal to:

(A) −30 (B) −31 (C) −60 (D) −61

[Main 2022 28th July 1st Shift]

29. For $p, q \in R$, consider the real valued function $f(x) = (x - p)^2 - q, x \in R$ and $q > 0$. Let a_1, a_2, a_3 and a_4 be in an arithmetic progression with mean p and positive common difference. If $|f(a_i)| = 500$ for all $i = 1, 2, 3, 4$ then the absolute difference between the roots of $f(x) = 0$ is?

[Main 2022 28th July 1st Shift]

30. Let $x_1, x_2, x_3, \ldots, x_{20}$ be in geometric progression with $x_1 = 3$ and the common ratio $\frac{1}{2}$. A new data is constructed replacing each x_i by $(x_i - i)^2$. If $\bar{x}$ is the mean of new data, then the greatest integer less than or equal to $\bar{x}$ is ______.

[Main 2022 28th July 1st Shift]

31. If $\frac{6}{3^{12}} + \frac{10}{3^{11}} + \frac{20}{3^{10}} + \frac{40}{3^9} + \ldots + \frac{10240}{3} = 2^n \cdot m$, where m is odd, then $m \cdot n$ is equal to ______.

[Main 2022 28th July 2nd Shift]

32. If $$\frac{1}{(20-a)(40-a)} + \frac{1}{(40-a)(60-a)} + \ldots + \frac{1}{(180-a)(200-a)} = \frac{1}{256},$$ then the maximum value of a is

(A) 198 (B) 202 (C) 212 (D) 218

[Main 2022 29th July 1st Shift]

33. Let $a_1, a_2, a_3, \ldots$ be an A.P. If $\sum_{r=1}^{\infty} \frac{a_r}{2^r} = 4$, then $4a_2$ is equal to ______?

[Main 2022 29th July 1st Shift]

34. If $$\frac{1}{2 \times 3 \times 4} + \frac{1}{3 \times 4 \times 5} + \frac{1}{4 \times 5 \times 6} + \ldots + \frac{1}{100 \times 101 \times 102} = \frac{k}{101},$$ then $34k$ is equal to ______

[Main 2022 29th July 1st Shift]

35. Let $\{a_n\}_{n=0}^{\infty}$ be a sequence such that $a_0 = a_1 = 0$ and $a_{n+2} = 3a_{n+1} - 2a_n + 1, \forall n \ge 0$. Then $a_{25}a_{23} - 2a_{25}a_{22} - 2a_{23}a_{24} + 4a_{22}a_{24}$ is equal to

(A) 483 (B) 528 (C) 575 (D) 624

[Main 2022 29th July 2nd Shift]

36. $\sum_{r=1}^{20}(r^2+1)(r!)$ is equal to:

(A) $22!-21!$ (B) $22!-2(21!)$
(C) $21!-2(20!)$ (D) $21!-20!$

[Main 2022 29th July 2nd Shift]

37. For three positive integers $p, q, r, x^{pq^2} = y^{qr} = z^{p^2r}$ and $r = pq+1$ such that $3, 3\log_y x, 3\log_2 y, 7\log_x z$ are in A.P. with common difference $\frac{1}{2}$. Then $r-p-q$ is equal to

(A) 2 (B) 6 (C) 12 (D) −6

[Main 2023 24th Jan 1st Shift]

38. The 4th term of a G.P. is 500 and its common ratio is $\frac{1}{m}$, $m \in N$. Let S_n denote the sum of the first n terms of this G.P. If $S_6 > S_5 + 1$ and $S_7 < S_6 + \frac{1}{2}$, then the number of possible values of m is ______.

[Main 2023 24th Jan 1st Shift]

39. If $\frac{1^3+2^3+3^3+\ldots\text{up o } n \text{ terms}}{1\cdot3+2\cdot5+3\cdot7+\ldots\text{up to } n \text{ terms}} = \frac{9}{5}$, then the value of n is?

[Main 2023 24th Jan 2nd Shift]

40. Let A_1, A_2, A_3 be the three A.P.s with the same common difference d and having their first terms as $A, A+1, A2$, respectively. Let a, b, c be the 7th, 9th, 17th terms of A_1, A_2, A_3, respectively such that $\begin{vmatrix} a & 7 & 1 \\ 2b & 17 & 1 \\ c & 17 & 1 \end{vmatrix} + 70 = 0$.

If $a = 29$, then the sum of first 20 term of an A.P. whose first term is $c-a-b$ and common difference is $\frac{d}{12}$, is equal to ______.

[Main 2023 25th Jan 1st Shift]

41. For the two positive numbers a, b, if a, b and $\frac{1}{18}$ are in a geometric progression, while $\frac{1}{a}, 10$ and $\frac{1}{b}$ are in an arithmetic progression, then, $16a+12b$ is equal to ______.

[Main 2023 25th Jan 2nd Shift]

42. Let $a_1, a_2, a_3, \ldots$ be a G.P. of increasing positive numbers. If the product of fourth and sixth terms is 9 and the sum of fifth and seventh terms is 24, then $a_1 a_9 + a_2 a_4 a_9 + a_5 + a_7$ is equal to ______?

[Main 2023 29th Jan 1st Shift]

43. Let $\{a_k\}$ and $\{b_k\}$, $k \in N$, be two G.P.s with common ratio r_1 and r_2 respectively such that $a_1 = b_1 = 4$ and $r_1 < r_2$. Let $c_k = a_k + b_k, k \in N$. If $c_2 = 5$ and $c_3 = \frac{13}{4}$ then $\sum_{k=1}^{\infty} c_k - (12a_6 + 8b_4)$ is equal to ______.

[Main 2023 29th Jan 2nd Shift]

44. Let $a_1 = b_1 = 1$ and $a_n = a_{n-1} + (n-1)$, $b_n = b_{n-1} + a_{n-1}, \forall n \geq 2$.

If $S = \sum_{n=1}^{10} \frac{b_n}{2^n}$ and $T = \sum_{n=1}^{8} \frac{n}{2^{n-1}}$, then $2^7(2S-T)$ is equal to ______.

[Main 2023 29th Jan 2nd Shift]

45. If $a_n = \frac{-2}{4n^2-16n+15}$, then $a_1 + a_2 + \ldots + a_{25}$ is equal to:

(A) $\frac{51}{144}$ (B) $\frac{49}{138}$ (C) $\frac{50}{141}$ (D) $\frac{52}{147}$

[Main 2023 30th Jan 1st Shift]

46. Let $\sum_{n=0}^{\infty} \frac{n^3((2n)!) + (2n-1)(n!)}{(n!)((2n)!)} = ae + \frac{b}{e} + c$, where $a, b, c \in Z$ and $e = \sum_{n=0}^{\infty} \frac{1}{n!}$.

Then $a^2 - b + c$ is equal to ______.

[Main 2023 30th Jan 1st Shift]

47. Let $a, b, c > 1, a^3, b^3$ and c^3 be in A.P., and $\log_a b, \log_c a$ and $\log_b c$ be in G.P. If the sum of first 20 terms of an A.P., whose first term is $\frac{a+4b+c}{3}$ and the common difference is $\frac{a-8b+c}{10}$ is −444, then abc is equal to

(A) 343 (B) 216 (C) $\frac{343}{8}$ (C) $\frac{125}{8}$

[Main 2023 30th Jan 2nd Shift]

48. The parabolas $ax^2+2bx+cy=0$ and $dx^2+2ex+fy=0$ intersect on the line $y=1$. If a, b, c, d, e, f are positive real numbers and a, b, c are in G.P., then

(A) d, e, f are in A.P.

(B) $\frac{d}{a}, \frac{e}{b}, \frac{f}{c}$ are in G.P.

(C) $\frac{d}{a}, \frac{e}{b}, \frac{f}{c}$ are in A.P.

(D) d, e, f are in G.P.

[Main 2023 30th Jan 2nd Shift]

49. The 8th common term of the series

$S_1 = 3+7+11+15+19+\ldots,$

$S_2 = 1+6+11+16+21+\ldots$

is _______. **[Main 2023 30th Jan 2nd Shift]**

50. Let $a_1, a_2, \ldots, a_n$ be in A.P. If $a_5 = 2a_7$ and $a_{11} = 18$, then

$$12\left(\frac{1}{\sqrt{a_{10}}+\sqrt{a_{11}}}+\frac{1}{\sqrt{a_{11}}+\sqrt{a_{12}}}+\ldots+\frac{1}{\sqrt{a_{17}}+\sqrt{a_{18}}}\right)$$

is equal to _______.

[Main 2023 31st Jan 1st Shift]

51. Let $a_1, a_2, a_3, \ldots$ be an A.P. If $a_7 = 3$; the product $a_1 a_4$ is minimum and the sum of its first n terms is zero, then $n! - 4a_{n(n+2)}$ is equal to:

(A) 24 (B) $\frac{33}{4}$ (C) $\frac{381}{4}$ (D) 9

[Main 2023 31st Jan 2nd Shift]

52. The sum

$1^2 - 2.3^3 + 3.5^2 - 4.7^2 + 5.9^2 - \ldots + 15.29^2$

is _______.

[Main 2023 31st Jan 2nd Shift]

53. The sum to 10 terms of the series

$$\frac{1}{1+1^2+1^4}+\frac{2}{1+2^2+2^4}+\frac{3}{1+3^2+3^4}+\ldots \text{ is}$$

(A) $\frac{59}{111}$ (B) $\frac{55}{111}$ (C) $\frac{56}{111}$ (D) $\frac{58}{111}$

[Main 2023 1st Feb 1st Shift]

54. Let $a_1 = 8, a_2, a_3, \ldots a_n$ be an A.P. If the sum of its first four terms is 50 and the sum of its last four terms is 170, then the product of its middle two terms is _______.

[Main 2023 1st Feb 1st Shift]

55. The number of 3-digit numbers, that are divisible by either 2 or 3 but not divisible by 7 is _______.

[Main 2023 1st Feb 1st Shift]

56. The sum of the common terms of the following three arithmetic progressions.

$3, 7, 11, 15, \ldots, 399,$

$2, 5, 8, 11, \ldots, 359$ and

$2, 7, 12, 17, \ldots, 197$, is equal to_______.

[Main 2023 1st Feb 2nd Shift]

57. The sum of the first 20 terms of the series $5+11+19+29+41+\ldots$ is

(A) 3520 (B) 2450 (C) 3250 (D) 3420

[Main 2023 6th April 1st Shift]

58. Let $a_1, a_2, a_3, \ldots, a_n$ be n positive consecutive terms of an arithmetic progression. If $d > 0$ is its common difference, then

$$\lim_{n\to\infty}\sqrt{\frac{d}{n}}\left(\frac{1}{\sqrt{a_1}+\sqrt{a_2}}+\frac{1}{\sqrt{a_2}+\sqrt{a_3}}+\ldots+\frac{1}{\sqrt{a_{n-1}}+\sqrt{a_n}}\right)$$

is

(A) $\frac{1}{\sqrt{d}}$ (B) $\sqrt{d}$ (C) 1 (D) 2

[Main 2023 6th April 1st Shift]

59. If G.C.D. $(m, n) = 1$ and

$$1^2 - 2^2 + 3^2 - 4^2 + \ldots + (2021)^2 - (2022)^2 + (2023)^2 = 1012m^2n$$

then $m^2 - n^2$ is equal to

(A) 240 (B) 200 (C) 220 (D) 180

[Main 2023 6th April 2nd Shift]

60. If

$$(20)^{19} + 2(21)(20)^{18} + 3(21)^2(20)^{17} + \ldots + 20(21)^{19} = k(20)^{19},$$

then k is equal to _______.

[Main 2023 6th April 2nd Shift]

61. Let $S_K = \dfrac{1+2+\ldots+K}{K}$ and $\sum_{j=1}^{n} S_j^2 = \dfrac{n}{A}\left(Bn^2+Cn+D\right)$ where $A, B, C, D \in N$ and A has least value then

(A) $A+C+D$ is not divisible by D
(B) $A+B=5(D-C)$
(C) $A+B+C+D$ is divisible by 5
(D) $A+B$ is divisible by D

[Main 2023 8th April 1st Shift]

62. Let a_n be the nth term of the series $5+8+14+23+35+50+\ldots$ and $S_n = \sum_{k=1}^{n} a_k$. Then $S_{30}-a_{40}$ is equal to

(A) 11310 (B) 11260
(C) 11290 (D) 11280

[Main 2023 8th April 2nd Shift]

63. Let $0<z<y<x$ be three real numbers such that $\dfrac{1}{x}, \dfrac{1}{y}, \dfrac{1}{z}$ are in an arithmetic progression and $x, \sqrt{2y}, z$ are in a geometric progression. If $xy+yx+zx = \dfrac{3}{\sqrt{2}}xyz$, then $3(x+y+z)^2$ is equal to

[Main 2023 8th April 2nd Shift]

64. Let the first term a and the common ratio r of a geometric progression be positive integers. If the sum of squares of its first three terms is 33033, then the sum of these three terms is equal to

(A) 241 (B) 231 (C) 210 (D) 220

[Main 2023 10th April 1st Shift]

65. The sum of all those terms, of the arithmetic progression 3, 8, 13, ..., 373, which are not divisible by 3, is equal to ______.

[Main 2023 10th April 1st Shift]

66. If $S_n = 4+11+21+34+50+\ldots$ to n terms, then $\dfrac{1}{60}(S_{29}-S_9)$ is equal to

(A) 223 (B) 226 (C) 220 (D) 227

[Main 2023 10th April 2nd Shift]

67. Suppose $a_1, a_2, 2, a_3, a_4$ are in an arithmetico–geometric progression. If the common ratio of the corresponding geometric progression is 2 and the sum of all 5 terms of the arithmetico–geometric progression is $\dfrac{49}{2}$, then a_4 is equal to ______.

[Main 2023 10th April 2nd Shift]

68. Let $x_1, x_2, \ldots, x_{100}$ be in an arithmetic progression, with $x_1=2$ and their mean equal to 200. If $y_i = i(x_i - i), 1 \le i \le 100$, then the mean of $y_1, y_2, \ldots, y_{100}$ is

(A) 10100 (B) 10101.50
(C) 10049.50 (D) 10051,50

[Main 2023 11th April 1st Shift]

69. Let $S = 109+\dfrac{108}{5}+\dfrac{107}{5^2}+\ldots+\dfrac{2}{5^{107}}+\dfrac{1}{5^{108}}$. Then the value of $\left(16S-(25)^{-54}\right)$ is equal to

[Main 2023 11th April 1st Shift]

70. Let a, b, c and d be positive real numbers such that $a+b+c+d=11$. If the maximum value of $a^5b^3c^2d$ is 3750β, then the value of β is

(A) 90 (B) 110 (C) 55 (D) 108

[Main 2023 11th April 2nd Shift]

71. For $k \in N$, if the sum of the series $1+\dfrac{4}{k}+\dfrac{8}{k^2}+\dfrac{13}{k^3}+\dfrac{19}{k^4}+\ldots$ is 10, then the value of k is

[Main 2023 11th April 2nd Shift]

72. Let $<a_n>$ be a sequence such that $a_1+a_2+\ldots+a_n = \dfrac{n^2+3n}{(n+1)(n+2)}$.

If $28\sum_{k=1}^{10}\dfrac{1}{a_k} = p_1 p_2 p_3 \cdots p_m$,

where $p_1, p_2, \ldots, p_m$ are the first m prime numbers, then m is equal to

(A) 5 (B) 8 (C) 6 (D) 7

[Main 2023 12th April 1st Shift]

73. Let $s_1, s_2, s_3, \ldots, s_{10}$ respectively be the sum of 12 terms of 10 A.P.s whose first terms are 1, 2, 3, ..., 10 and the common differences are 1, 3, 5, ..., 19 respectively. Then $\sum_{i=1}^{10} s_i$ is equal to

(A) 7220 (B) 7360 (C) 7260 (D) 7380

[Main 2023 13th April 1st Shift]

74. The sum to 20 terms of the series $2\cdot 2^2 - 3^2 + 2\cdot 4^2 - 5^2 + 2\cdot 6^2 - \ldots$ is equal to ______.

[Main 2023 13th April 1st Shift]

75. Let $a_1, a_2, a_3, \ldots$ be a G.P. of increasing positive numbers. Let the sum of its 6th and 8th terms be 2 and the product of its 3rd and 5th terms be $\frac{1}{9}$. Then $6(a_2 + a_4)(a_4 + a_6)$ is equal to

(A) 3 (B) $3\sqrt{3}$ (C) 2 (D) $2\sqrt{2}$

[Main 2023 13th April 2nd Shift]

76. Let $[\alpha]$ denotes the greatest integer $\le \alpha$. Then $[\sqrt{1}] + [\sqrt{2}] + [\sqrt{3}] + \ldots + [\sqrt{120}]$ is equal to ______.

[Main 2023 13th April 2nd Shift]

77. Let $f(x) = \sum_{k=1}^{10} k \cdot x^k$, $x \in R$, if

$2f(2) + f'(2) = 119(2)^n + 1$ then n is equal to ______.

[Main 2023 13th April 2nd Shift]

78. Let A_1 and A_2 be two arithmetic means and G_1, G_2 and G_3 be three geometric means of two distinct positive numbers. Then $G_1^4 + G_2^4 + G_3^4 + G_1^2 + G_3^2$ is equal to

(A) $(A_1 + A_2)^2 G_1 G_3$

(B) $2(A_1 + A_2) G_1 G_3$

(C) $(A_1 + A_2) G_1^2 G_3^2$

(D) $2(A_1 + A_2) G_1^2 G_3^2$

[Main 2023 15th April 1st Shift]

79. If the sum of the series

$$\left(\frac{1}{2} - \frac{1}{3}\right) + \left(\frac{1}{2^2} - \frac{1}{2\cdot 3} + \frac{1}{3^2}\right) + \left(\frac{1}{2^3} - \frac{1}{2^2\cdot 3} + \frac{1}{2\cdot 3^2} - \frac{1}{3^3}\right) + \left(\frac{1}{2^4} - \frac{1}{2^3\cdot 3} + \frac{1}{2^2\cdot 3^2} - \frac{1}{2\cdot 3^3} + \frac{1}{3^4}\right) + \ldots$$ is $\frac{\alpha}{\beta}$,

where α and β are co-prime, then $\alpha + 3\beta$ is equal to ______.

[Main 2023 15th April 1st Shift]

80. Let $3, a, b, c$ be in A.P. and $3, a-1, b+1, c+9$ be in G.P. Then, the arithmetic mean of a, b and c is:

(A) –4 (B) –1 (C) 13 (D) 11

[Main 2024 1st Feb 1st Shift]

81. Let $3, 7, 11, 15, \ldots, 403$ and $2, 5, 8, 11, \ldots, 404$ be two arithmetic progressions. Then the sum of the common terms in them, is equal to ______. **[Main 2024 1st Feb 1st Shift]**

82. Let S_n denote the sum of the first n terms of an arithmetic progression. If $S_{10} = 390$ and the ratio of the tenth and the fifth terms is $15:7$, then $S_{15} - S_5$ is equal to:

(A) 800 (B) 890 (C) 790 (D) 690

[Main 2024 1st Feb 2nd Shift]

83. If three successive terms of a G.P. with common ratio $r(r>1)$ are the length of sides of a triangle and $[r]$ denotes the greatest integer less than or equal to r, then $3[r] + [-r]$ is equal to ______.

[Main 2024 1st Feb 2nd Shift]

84. The number of common terms in the progression 4, 9, 14, 19, …, up to to the 25th term and 3, 6, 9, 12, up to the 37th term is

(A) 9 (B) 5 (C) 7 (D) 8

[Main 2024 27th Jan 1st Shift]

85. If

$$8 = 3 + \frac{1}{4}(3+p) + \frac{1}{4^2}(3+2p) + \frac{1}{4^3}(3+3p) + \ldots\infty,$$

then the value of p is ______

[Main 2024 27th Jan 1stShift]

86. The 20th term from the end of the progression $20, 19\frac{1}{4}, 18\frac{1}{2}, 17\frac{3}{4}, \ldots, -129\frac{1}{4}$ is:

(A) –118 (B) –110 (C) –115 (D) –100

[Main 2024 27th Jan 2nd Shift]

87. If in a G.P. of 64 terms, the sum of all the terms is 7 times the sum of the odd terms of the G.P., then the common ratio of the G.P. is equal to

(A) 7 (B) 4 (C) 5 (D) 6

[Main 2024 29th Jan 1st Shift]

88. In an A.P., the sixth term is $a_6 = 2$. If the $a_1 a_4 a_5$ is the greatest, then the common difference of the A.P., is equal to

(A) $\frac{3}{2}$ (B) $\frac{8}{5}$ (C) $\frac{2}{3}$ (D) $\frac{5}{8}$

[Main 2024 29th Jan 1stShift]

89. If $\log_e a, \log_e b, \log_e c$ are in an A.P. and $\log_e a - \log_e 2b, \log_e 2b - \log_e 3c,$ $\log_e 3c - \log_e a$ are also in an A.P., then $a:b:c$ is equal to

(A) $9:6:4$ (B) $16:4:1$
(C) $25:10:4$ (D) $6:3:2$

[Main 2024 29th Jan 2nd Shift]

90. If each term of a geometric progression $a_1, a_2, a_3, \ldots$ with $a_1 = \frac{1}{8}$ and $a_2 \neq a_1$, is the arithmetic mean of the next two terms and $S_n = a_1 + a_2 + \ldots + a_n$, then $S_{2o} - S_{18}$ is equal to

(A) 2^{15} (B) -2^{18} (C) 2^{18} (D) -2^{15}

[Main 2024 29th Jan 2nd Shift]

91. Let S_a denote the sum of first n terms an arithmetic progression. If $S_{20} = 790$ and $S_{10} = 145$, Then $S_{15} - S_5$ is

(A) 395 (B) 390 (C) 405 (D) 410

[Main 2024 30th Jan 1st Shift]

92. Let $\alpha = 1^2 + 4^2 + 8^2 + 13^2 + 19^2 + 26^2 + \ldots$ up to 10 terms and $\beta = \sum_{n=1}^{10} n^4$.

If $4\alpha - \beta = 55k + 40$, then k is equal to

[Main 2024 30th Jan 1st Shift]

93. Let a and b be two distinct positive real numbers. Let 11th term of a G.P., whose first term is a and third term is b, is equal to p^{th} term of another G.P., whose first term is a and fifth term is b. Then p is equal to

(A) 20 (B) 25 (C) 21 (D) 24

[Main 2024 30th Jan 2nd Shift]

94. Let S_n be the sum of n-terms of an arithmetic progression 3, 7, 11, ... If $40 < \left(\frac{6}{n(n+1)}\sum_{k=1}^{n} S_k\right) < 42$, then n equals ________.

[Main 2024 30th Jan 2nd Shift]

95. The sum of the series

$$\frac{1}{1-3\cdot1^2+1^4} + \frac{2}{1-3\cdot2^2+2^4} + \frac{3}{1-3\cdot3^2+3^4} + \ldots$$

up to 10 terms is

(A) $\frac{45}{109}$ (B) $-\frac{45}{109}$ (C) $\frac{55}{109}$ (D) $-\frac{55}{109}$

[Main 2024 31st Jan 1st Shift]

96. Let 2nd, 8th and 44th, terms of a non-constant A.P. be respectively the 1st, 2nd and 3rd terms of G.P. If the first term of A.P. is 1 then the sum of first 20 terms is equal to

(A) 980 (B) 960 (C) 990 (D) 970

[Main 2024 31st Jan 2nd Shift]

2.12 Answer Key

2.12.1 Know Your Concepts

1. $\frac{1}{2}$ **2.** (11) **3.** (2, 5, 8 or 26, 5, –16) **4.** (18, 6, 2)
5. (6, 3, 2 or 2, 3, 6) **7.** (9) **8.** (20) **9.** (–5050) **10.** (3, 6)
11. (15, 12, 10) **12.** (6) **13.** (1, 4, 16, 64 or 64, 16, 4, 1) **14.** (3)
18. (3, 6, 12 or 12, 6, 3) **19.** (3) **20.** $\frac{1}{12}n(n+1)^2(n+2)$

2.12.2 JEE Main: Sample Questions

Single Answer Type

1. (A)	**2.** (A)	**3.** (B)	**4.** (B)	**5.** (B)	**6.** (A)	**7.** (D)	**8.** (B)
9. (C)	**10.** (B)	**11.** (C)	**12.** (D)	**13.** (B)	**14.** (C)	**15.** (B)	**16.** (B)
17. (D)	**18.** (C)	**19.** (A)	**20.** (C)	**21.** (D)	**22.** (D)	**23.** (B)	**24.** (A)
25. (B)	**26.** (C)	**27.** (C)	**28.** (B)	**29.** (B)	**30.** (B)	**31.** (A)	**32.** (B)

33. (D) **34.** (C) **35.** (B) **36.** (D) **37.** (A) **38.** (A) **39.** (A) **40.** (C)
41. (B) **42.** (B) **43.** (A) **44.** (C) **45.** (A) **46.** (D) **47.** (B) **48.** (C)
49. (D) **50.** (B) **51.** (B) **52.** (B) **53.** (B) **54.** (A) **55.** (A) **56.** (B)
57. (A) **58.** (D) **59.** (C) **60.** (B)

Integer Answer Type

1. (3) **2.** (2) **3.** (13) **4.** (7) **5.** (2) **6.** (400) **7.** (13) **8.** (201)
9. (9) **10.** (6) **11.** (5) **12.** (4) **13.** (5) **14.** (7) **15.** (8) **16.** (2)
17. (5) **18.** (8) **19.** (23) **20.** (6)

2.12.3 JEE Advanced: Sample Questions

Single Answer Type

1. (B) **2.** (C) **3.** (B) **4.** (A) **5.** (D) **6.** (B) **7.** (C) **8.** (C)
9. (D) **10.** (A) **11.** (B) **12.** (A) **13.** (A) **14.** (D) **15.** (C) **16.** (B)
17. (D) **18.** (D) **19.** (A) **20.** (A) **21.** (D) **22.** (C) **23.** (D) **24.** (D)
25. (C)

Multiple Answer Type

1. (B, D) **2.** (A, B, C) **3.** (B, C) **4.** (A, C, D)
5. (A, B, C, D) **6.** (A, B, C, D) **7.** (A, B, C, D) **8.** (B, D)
9. (A, B, C) **10.** (A, C, D) **11.** (C, D) **12.** (A, C, D)
13. (A, C) **14.** (B, C) **15.** (B, C) **16.** (A, C, D)
17. (A, B, D) **18.** (B, C, D) **19.** (A, B, C, D) **20.** (A, C, D)
21. (A, B, C, D) **22.** (A,B,C,D) **23.** (B, C) **24.** (A,C)
25. (A, B, D)

Comprehension Based Questions

1. (C) **2.** (B) **3.** (C) **4.** (B) **5.** (B) **6.** (A) **7.** (B) **8.** (A)
9. (D) **10.** (C) **11.** (A) **12.** (B) **13.** (C) **14.** (D) **15.** (A)

Match Matrix

1. (A – R; B – P; C – Q; D – P) **2.** (A – R; B – P; C – S; D – Q)
3. (A – R; B – Q; C – Q) **4.** (A – P; B – T; C – R; D – Q)
5. (A – Q, R, S; B – R; C – P, R, S; D – R, S)

Subjective Answer Type: Hard Nut

1. ($p=-2, q=-32$) **2.** (Prove it yourself.) **3.** (4) **4.** (7) **5.** $\frac{1}{2}\left(\frac{1}{2}-\frac{1}{2^{n+1}(n+1)}\right)$

6. $(1-\frac{1}{4^n(n+1)})$ **7.** (0) **8.** (Composite) **9.** (Prove it yourself.)

10. (Prove it yourself.) **11.** $(100.(101)!)$ **12.** (1) **13.** $(\frac{15}{16})$ **14.** $(\frac{2(n+1)!-(n+2)}{2(n+2)})$

15. (12) **16.** $(\frac{1}{2^n})$ **17.** $(\frac{256}{75})$

18. $n=18,19,20,21,22,23$ and removed numbers can be (1,6,11),(4,6,8),(3,6,9)

19. $(\frac{4n^2+4n}{2n^4+2n+1}, 2)$ **20.** $(\frac{3}{2})$ **21.** (16) **22.** $(\frac{2}{3})$ **23.** (4) **24.** (10) **25.** (4°)

26. (10) **27.** (5) **28.** (6) **29.** (13) **30.** (13) **31.** (9) **32.** (2) **33.** (3)
35. (4)

2.12.4 JEE Main: Archive

1. (B)	**2.** (D)	**3.** (98)	**4.** (B)	**5.** (1100)	**6.** (C)	**7.** (40)	**8.** (C)
9. (276)	**10.** (C)	**11.** (D)	**12.** (C)	**13.** (C)	**14.** (41651)	**15.** (B)	**16.** (C)
17. (2223)	**18.** (27560)	**19.** (B)	**20.** (C)	**21.** (6993)	**22.** (166)	**23.** (53)	**24.** (B)
25. (10620)	**26.** (A)	**27.** (120)	**28.** (C)	**29.** (50)	**30.** (142)	**31.** (12)	**32.** (C)
33. (16)	**34.** (286)	**35.** (B)	**36.** (B)	**37.** (A)	**38.** (12)	**39.** (5)	**40.** (495)
41. (3)	**42.** (60)	**43.** (9)	**44.** (461)	**45.** (C)	**46.** (26)	**47.** (B)	**48.** (C)
49. (151)	**50.** (8)	**51.** (A)	**52.** (6952)	**53.** (B)	**54.** (754)	**55.** (514)	**56.** (321)
57. (B)	**58.** (C)	**59.** (A)	**60.** (400)	**61.** (D)	**62.** (C)	**63.** (150)	**64.** (B)
65. (9525)	**66.** (A)	**67.** (16)	**68.** (C)	**69.** (2175)	**70.** (A)	**71.** (B)	**72.** (C)
73. (C)	**74.** (1310)	**75.** (A)	**76.** (825)	**77.** (10)	**78.** (A)	**79.** (7)	**80.** (D)
81. (6699)	**82.** (C)	**83.** (1)	**84.** (C)	**85.** (9)	**86.** (C)	**87.** (D)	**88.** (B)
89. (A)	**90.** (D)	**91.** (A)	**92.** (353)	**93.** (C)	**94.** (9)	**95.** (D)	**96.** (D)

2.13 Solutions

2.13.1 Know Your Concepts

1. $(\frac{1}{2})$

Let a, a_1 and d, d_1 be respectively first term and c.d. of two series in A.P.

$$\frac{\frac{n}{2}\left[2a+(n-1)d\right]}{\frac{n}{2}\left[2a_1+(n-1)d_1\right]}=\frac{3n-13}{5n+21}$$

$$\Rightarrow \frac{2a+(n-1)d}{2a_1+(n-1)d_1}=\frac{3n-13}{5n+21}$$

Put $n = 47$

$$\frac{a+23d}{a_1+23d_1}=\frac{141-13}{235+21}$$

$$\Rightarrow \frac{a+23d}{a_1+23d_1}=\frac{128}{256}=\frac{1}{2}$$

2. **(11)**

Let $A_1, A_2, \ldots, A_n$ be n A.M.s inserted between 20 and 80

$\therefore 20, A_1, A_2, \ldots, A_n, 80$ are in A.P.

$80 = (n+2)^{\text{th}}$ term

$$= 20+(n+2-1)d \Rightarrow d=\frac{60}{n+1}$$

$$A_1 = 20+\frac{60}{n+1}=\frac{20n+20+60}{n+1}=\frac{20n+80}{n+1}$$

$$A_n = 20+\frac{60n}{n+1}=\frac{80n+20}{n+1}$$

Now, $\frac{20n+80}{80n+20}=\frac{1}{3}\Rightarrow 60n+240=80n+20$

$\Rightarrow 20n=220 \Rightarrow n=11$

3. $(2, 5, 8 \text{ or } 26, 5, -16)$

$\alpha-\beta+\alpha+\alpha+\beta=15 \Rightarrow \alpha=5$

$\alpha-\beta+1, \alpha+4, \alpha+\beta+19$ are in G.P.

$\Rightarrow 6-\beta, 9, 24+\beta$ are in G.P.

$\Rightarrow 81=(6-\beta)(24+\beta)$

$\Rightarrow 81=144-18\beta-\beta^2$

$\Rightarrow \beta^2+18\beta-63=0$

$\Rightarrow \beta^2+21\beta-3\beta-63=0$

$\Rightarrow \beta=3, -21$

When $\beta=3$, the numbers are 2, 5, 8

when $\beta=-21$ numbers are 26, 5, -16

4. $(18, 6, 2 \text{ or } 2, 6, 18)$

$\frac{a}{r}\cdot a\cdot ar=216 \Rightarrow a^3=6^3 \Rightarrow a=6$

Again, $\frac{a}{r}\cdot a+a\cdot ar+\frac{a}{r}\cdot ar=156$

$$\Rightarrow 36\left(\frac{1}{r}+r+1\right)=156$$

$$\Rightarrow 6\frac{(1+r^2+r)}{r}=26$$

$\Rightarrow 3r^2+3r+3=13r$

$\Rightarrow 3r^2-10r+3=0$

$\Rightarrow 3r^2-9r-r+3=0$

$\Rightarrow 3r(r-3)-1(r-3)=0 \Rightarrow r=3, \frac{1}{3}$

When $r=\frac{1}{3}$, then 18, 6, 2

When $r-3$ then 2, 6, 18

5. (6, 3, 2 or 2, 3, 6)

Let the three numbers in A.P. be $\alpha-\beta, \alpha, \alpha+\beta$

Since, $\alpha-\beta+\alpha+\alpha+\beta=1 \Rightarrow \alpha=\frac{1}{3}$

Again, $\frac{1}{\frac{1}{3}-\beta}+3+\frac{1}{\frac{1}{3}+\beta}=11$

$\Rightarrow \frac{\frac{2}{3}}{\frac{1-9\beta^2}{9}}=8 \Rightarrow 6=8\left(1-9\beta^2\right)$

$\Rightarrow 1-9\beta^2=\frac{6}{8}$

$\Rightarrow 9\beta^2=1-\frac{6}{8}=\frac{2}{8}$

$\beta^2=\frac{1}{36} \Rightarrow \beta=\pm\frac{1}{6}$

Hence, numbers are 6, 3, 2 or 2, 3, 6

6. Since, a^2, b^2, c^2 are in A.P.

$\Rightarrow b^2-a^2=c^2-b^2$

$\Rightarrow \frac{(b+a)(b-a)}{(a+b)(b+c)(c+a)}=\frac{(c+b)(c-b)}{(a+b)(b+c)(c+a)}$

$\Rightarrow \frac{b-a}{(b+c)(c+a)}=\frac{c-b}{(a+b)(c+a)}$

$\Rightarrow \frac{(b+c)-(c+a)}{(b+c)(c+a)}=\frac{(c+a)-(a+b)}{(a+b)(c+a)}$

$\Rightarrow \frac{1}{c+a}-\frac{1}{b+c}=\frac{1}{a+b}-\frac{1}{c+a}$

$\Rightarrow \frac{1}{b+c}, \frac{1}{c+a}, \frac{1}{a+b}$ are in A.P.

$\Rightarrow (b+c), (c+a), (a+b)$ are in H.P.

7. **(9)**

Let the number of sides of the polygon be n. The sum of the interior angles of the polygon $=(n-2)180°$

Now, $a=120°$

$\therefore \frac{n}{2}\left[2\cdot 120°+(n-1)5°\right]=(n-2)180°$

$\Rightarrow \frac{5n}{2}\left[48+(n-1)\right]=(n-2)180°$

$\Rightarrow n\left[48+(n-1)\right]=72(n-2)$

$\Rightarrow n^2-25n+144=0$

$\Rightarrow (n-9)(n-16)=0 \Rightarrow n=9$ or 16

For $n=16$, the largest angle $=120°+15\times 5°$ $=195°$

This is not possible as an interior angle of a polygon can not be greater than 180°

Hence, $n=9$

8. **(20)**

Since, $\sqrt{a-x}, \sqrt{x}, \sqrt{a+x}$ are in A.P.

$\Rightarrow 2\sqrt{x}=\sqrt{a-x}+\sqrt{a+x}$

$\Rightarrow 4x=a-x+a+x+2\sqrt{a^2-x^2}$

$\Rightarrow (2x-a)^2=a^2-x^2$

$\Rightarrow 4x^2-4ax+a^2=a^2-x^2$

$\Rightarrow 5x^2=4ax$

$\Rightarrow x=\frac{4a}{5}$

$\therefore \sqrt{a-x}=\sqrt{a-\frac{4a}{5}}=\sqrt{\frac{a}{5}}$

$\sqrt{x}=2\sqrt{\frac{a}{5}}, \sqrt{a+x}=3\sqrt{\frac{a}{5}}$

As $\sqrt{\frac{a}{5}}$ is an integer $\Rightarrow a=5n^2$

$\Rightarrow a=20$

9. **(–5050)**

Coefficient of x^{99}

$=-(1+2+3+\ldots+100)$

$=-5050$

10. **(6,3)**

Let a and b be any two numbers

$4=\frac{2ab}{a+b}$

$A=\frac{a+b}{2}, G=\sqrt{ab}$

$2A+G^2=27 \Rightarrow a+b+ab=27$

$\Rightarrow \frac{2ab}{4}+ab=27$

$\Rightarrow 6ab=27\times 4$

$\Rightarrow ab=9\times 2=18$

$\therefore a+b=\frac{18}{2}=9$

Now, $(a-b)^2=81-72$

$\Rightarrow a-b=3 \Rightarrow\ 2a=12\ \ \Rightarrow a=6\ \Rightarrow b=3$

11. **(15,12,10)**

12. **(n = 6)**

$a+ar^{n-1}=66$

$ar \cdot ar^{n-2} = 128 \Rightarrow a^2 r^{n-1} = 128$

$\frac{a(r^n - 1)}{(r-1)} = 126$

$\therefore a + \frac{128}{a} = 66$

$\Rightarrow a^2 - 66a + 128 = 0$

$\Rightarrow a = 2,\ 64$

$\therefore r^{n-1} = 32 = 2^5$

$\Rightarrow 2(r \cdot 32 - 1) = 126(r-1)$

$\Rightarrow 32r - 1 = 63r - 63$

$\Rightarrow 31r = 62 \Rightarrow r = 2$ Hence, $n = 6$

13. 1, 4, 16, 64 or 64, 16, 4, 1

$\frac{a}{r^3} + \frac{a}{r} + ar + ar^3 = 85$ (1)

$\frac{a}{r^3} \cdot \frac{a}{r} \cdot ar \cdot ar^3 = 4096$

$\Rightarrow a^4 = 8^4 \Rightarrow a = 8$

From (1),

$8\left(\frac{1}{r^3} + \frac{1}{r} + r + r^3\right) = 85$

$\Rightarrow 8(1 + r^2 + r^4 + r^6) = 85r^3$

$\Rightarrow 8r^6 + 8r^4 - 85r^3 + 8r^2 + 8 = 0$

$r = \frac{1}{2}, 2$

Hence numbers are

1, 4, 16, 64 or 64, 16, 4, 1

14. **(3)**

$2\log_3(2^x - 5) = \log_3 2 + \log_2\left(2^x - \frac{7}{2}\right)$

$\Rightarrow (2^x - 5)^2 = 2\left(2^x - \frac{7}{2}\right)$

Put $2^x = y$

$(y-5)^2 = 2y - 7$

$\Rightarrow y^2 - 10y + 25 = 2y - 7$

$\Rightarrow y^2 - 12y + 32 = 0$

$\Rightarrow y^2 - 8y - 4y + 32 = 0$

$\Rightarrow y = 4, 8$

when $y = 4,\ 2^x = 4 \Rightarrow x = 2$

$y = 8,\ 2^x = 8 \Rightarrow x = 3$

But $x = 2$ is not possible

$\therefore x = 3$

15. $b = \frac{a+c}{2},\ c^2 = bd$ and $d = \frac{2ce}{c+e}$

$\therefore c^2 = \frac{a+c}{2} \cdot \frac{2ce}{c+e}$

$\Rightarrow c(c+e) = e(a+c)$

$\Rightarrow c^2 = ae \Rightarrow a, c, e$ are in G.P.

16. Total number of terms in the nth group

$1 + (n-1) \cdot 2 = 2n - 1$

Now, let t_n denote the first term of the nth group

$S = 1 + 2 + 5 + 10 + 17 + \ldots + t_n$

$S = 1 + 2 + 5 + 10 + \ldots + t_{n-1} + t_n$

Subtracting

$0 = 1 + 1 + 3 + 5 + 7 + \ldots$ to n terms - t_n

$\Rightarrow t_n = 1 + (1 + 3 + 5 + 7 + \ldots \text{to} (n-1) \text{terms})$

$= 1 + \frac{n-1}{2}[2 + (n-1-1) \cdot 2]$

$= 1 + (n-1)(1 + n - 2)$

$= 1 + (n-1)^2 = n^2 - 2n + 2$

Now, the sum of the numbers in the nth group

$= \frac{2n-1}{2}\left[2(n^2 - 2n + 2) + (2n - 1 - 1)\right]$

$= \frac{2n-1}{2}\left[2n^2 - 4n + 4 + 2n - 2\right]$

$= (2n-1)(n^2 - n + 1) = (n-1)^3 + n^3$

17. $a_1 - a_2 = a_3 - a_4 = a_{2n-1} - a_{2n} = -d$

$\therefore a_1^2 - a_2^2 + a_3^2 - a_4^2 + \ldots + a_{2n-1}^2 - a_{2n}^2$

$= -d\left[a_1 + a_2 + a_3 + a_4 + \ldots + a_{2n-1} + a_{2n}\right]$

$= -d \cdot \frac{2n}{2}[a_1 + a_{2n}] = -dn[a_1 + a_{2n}]$

Now, $a_{2n} = a_1 + (2n-1)d$

$\Rightarrow -d = \frac{a_1 - a_{2n}}{2n-1}$

Hence, $\frac{n}{(2n-1)}(a_1^2 - a_{2n}^2)$

18. 3, 6, 12 or 12, 6, 3

$\frac{a}{r} + a + ar = 21 \Rightarrow \frac{a(r^2 + r + 1)}{r} = 21$ (1)

and, $\frac{a^2}{r^2} + a^2 + a^2r^2 = 189$

$\Rightarrow \frac{a^2}{r^2}(r^4 + r^2 + 1) = 189$ (2)

Squaring (1) and dividing by (2)

$$\Rightarrow \frac{\left(r^2+r+1\right)^2}{r^4+r^2+1} = \frac{441}{189}$$

$$\Rightarrow \frac{r^2+r+1}{r^2-r+1} = \frac{441}{189} = \frac{7}{3}$$

$$\Rightarrow 3r^2+3r+3 = 7r^2-7r+7$$

$$\Rightarrow 4r^2-10r+4=0 \Rightarrow 2r^2-5r+2=0$$

$$\Rightarrow 2r^2-4r-r+2=0$$

$$\Rightarrow 2r(r-2)-1(r-2)=0 \Rightarrow r=\frac{1}{2}, 2$$

Hence the numbers are 3, 6, 12 or 12, 6, 3

19. **(3)**

a, x, y, z, b are in A.P.

$\Rightarrow b = a+4d$ where d is c.d.

$$\Rightarrow d = \frac{b-a}{4}$$

$$x = \frac{a+b-a}{4}, y = a+\frac{2(b-a)}{4}, z = a+\frac{3(b-a)}{4}$$

$$x+y+z = \frac{6a+6b}{4} \Rightarrow 6(a+b) = 60$$

$$\Rightarrow a+b=10$$

Again, $\frac{1}{b} = \frac{1}{a} + 4d_1$ where d_1 is c.d.

$$4d_1 = \frac{a-b}{ab} \Rightarrow d_1 = \frac{a-b}{4ab}$$

$$\frac{1}{x} = \frac{1}{a} + \frac{a-b}{4ab}, \frac{1}{y} = \frac{1}{a} + \frac{2(a-b)}{4ab},$$

$$\frac{1}{z} = \frac{1}{a} + \frac{3(a-b)}{4ab}$$

$$\therefore \frac{5}{3} = \frac{3}{a} + 6\left(\frac{a-b}{4ab}\right)$$

$$\Rightarrow \frac{5}{3} = \frac{6(a+b)}{4ab} \Rightarrow ab = 9 \Rightarrow \sqrt{ab} = 3$$

$\therefore$ G.M. of a and $b = 3$

20. $\left(\frac{1}{12}n(n+1)^2(n+2)\right)$

$$t_n = 1^2+2^2+3^2+\ldots+n^2$$

$$= \frac{n(n+1)(2n+1)}{6} = \frac{n(2n^2+3n+1)}{6}$$

$$\Rightarrow t_n = \frac{1}{3}n^3 + \frac{1}{2}n^2 + \frac{1}{6}n$$

$$\therefore \sum t_n = \frac{1}{3}\sum n^3 + \frac{1}{2}\sum n^2 + \frac{1}{6}\sum n$$

$$= \frac{1}{12}n(n+1)^2(n+2)$$

2.13.2 JEE Main: Sample Questions

Single Answer Type

1. **(A)**

$$\frac{11}{2}[2a+10d] = \frac{19}{2}[2a+18d]$$

$$\Rightarrow 11 \times 2(a+5d) = 19 \times 2(a+9d)$$

$$\Rightarrow 11a + 55d = 19a + 171d$$

$$\Rightarrow 8a + 116d = 0$$

$$\Rightarrow 2a + 29d = 0 \Rightarrow S_{30} = 0$$

2. **(A)**

Let roots be $\frac{a}{r}$, a, ar

$$\frac{3}{4} + a + ar = 5 \quad (1)$$

$$\frac{a^2}{r} + a^2 + a^2r = k \quad (2)$$

$$a^3 = 27 \Rightarrow a = 3 \quad (3)$$

Using $a = 3$ in (1) and (2) we get

$k = 15$

3. **(B)**

$$E = \left(a^4+4+\frac{1}{a^4}\right)\left(b^2+3+\frac{1}{b^2}\right)\left((c+1)^2+1\right)$$

$$E_{\min} = (6)(5)(1) = 30$$

4. **(B)**

Let the three numbers in G.P. be a, ar, ar^2

$$\therefore a(1+r+r^2) = 21 \quad (1)$$

$$\text{and } a^2(1+r^2+r^4) = 189 \quad (2)$$

$$\Rightarrow \frac{a^2(1+r^2+r^4)}{a^2(1+r+r^2)^2} = \frac{189}{(21)^2}$$

$$\Rightarrow \frac{r^2-r+1}{r^2+r+1} = \frac{3}{7} \Rightarrow 4r^2 - 10r + 4 = 0$$

$$\Rightarrow 2r^2 - 4r - r + 2 = 0$$

$$\Rightarrow 2r(r-2) - 1(r-2) = 0$$

$\therefore r = 2$ or $r = \frac{1}{2}$ (but $r = 2$ is rejected)

5. **(B)**

$5050 - [1 + 2 + 3 + \ldots + 50]$

$= 50\,50 - 25 \cdot 51 = 5050 - 1275 = 3775.$

6. **(A)**

Let first term and common difference of A.P. be a and d respectively.

Given, $a_3 + a_9 = 8$

$$\Rightarrow (a+2d) + (a+8d) = 8 \Rightarrow a + 5d = 4 \quad (1)$$

Also, $S_7 = 14$

$\Rightarrow \frac{7}{2}(2a+6d) = 14 \Rightarrow a + 3d = 2$ (2)

$\therefore$ From (1) and (2), we get

$d = 1$ and $a = -1$. Now verify alternatives.

7. **(D)**

$a_1, a_2, \ldots, a_{10}$ are in A.P.

$h_1, h_2, \ldots h_{10}$ are in H.P.

$\therefore \frac{1}{h_1}, \frac{1}{h_2}, \ldots, \frac{1}{h_{10}}$ are in A.P.

$\therefore a_{10} = a_1 + 9d \quad 3 = 2 + 9d$

$\therefore d = \frac{1}{9}$

and $\frac{1}{h_{10}} = \frac{1}{h_1} + 9D; \frac{1}{3} = \frac{1}{2} + 9D; -\frac{1}{6} = 9D;$

$\therefore D = -\frac{1}{54}$

$\therefore a_4 = a_1 + 3d = 2 + \frac{1}{3} = \frac{7}{3}$

$\frac{1}{h_7} = \frac{1}{h_1} + 6D \Rightarrow \frac{1}{h_7} = \frac{1}{2} + 6\times\left(-\frac{1}{54}\right)$

$\Rightarrow \frac{1}{h_7} = \frac{1}{2} - \frac{1}{9} \Rightarrow \frac{1}{h_7} = \frac{7}{18}$

$\therefore h_7 = \frac{18}{7};$

$\therefore a_4 h_7 = \frac{7}{3} \times \frac{18}{7} = 6$

8. **(B)**

$\frac{\alpha+\beta}{2} = 7 \Rightarrow \alpha + \beta = 14$

$\frac{\frac{1}{\alpha}+\frac{1}{\beta}}{2} = \frac{7}{2} \Rightarrow \frac{\alpha+\beta}{\alpha\beta} = 7$

$\alpha\beta = 2$. Hence, QE: $x^2 - 14x + 2 = 0$

9. **(C)**

$\frac{a}{1-r} = 7$ (1)

and $\frac{ar}{1-r^2} = 3$ (2)

Solving $r = \frac{3}{4}$ and $a = \frac{7}{4} \Rightarrow a + r = \frac{5}{2}$

10. **(B)**

Let the roots be $a - d, a, a + d$

$\therefore 3a = \frac{144}{64} = \frac{9}{4} \Rightarrow a = \frac{3}{4}$

Also $a(a^2 - d^2) = \frac{15}{64}; \frac{3}{4}\left(\frac{9}{16} - d^2\right)$

$= \frac{15}{64}; \frac{9}{16} - d^2 = \frac{5}{16}$

$d^2 = \frac{1}{4} \Rightarrow d = \frac{1}{2}$ or $-\frac{1}{2}$

if $a = \frac{3}{4}$ and $d = \frac{1}{2}$; if $a = \frac{3}{4}$ and $d = -\frac{1}{2}$

the numbers are $\frac{1}{4}, \frac{3}{4}, \frac{5}{4}; \frac{5}{4}, \frac{3}{4}, \frac{1}{4}$

$\therefore$ Difference between largest and smallest root is equal to $\frac{5}{4} - \frac{1}{4} = 1$

11. **(C)**

Let $6 - a = x, 6 - b = y, 6 - c = z$

$\Rightarrow x + y + z = 15$

$\therefore \frac{a}{6-a} + \frac{b}{6-b} + \frac{c}{6-c} = 6\left(\frac{1}{x} + \frac{1}{y} + \frac{1}{z}\right) - 3$

Here x, y, z are real and positive

So, $\frac{x+y+z}{3} \geq \frac{3}{\frac{1}{x}+\frac{1}{y}+\frac{1}{z}}$

From equations (1) and (2), we get

$6\left(\frac{1}{x} + \frac{1}{y} + \frac{1}{z}\right) - 3 \geq \frac{3}{5}.$

12. **(D)**

$\underbrace{A_1H_9G_8}\ \underbrace{A_2H_8G_6}\ \underbrace{A_3H_7G_4}\ \underbrace{A_4H_6G_2}$

Now $A_1H_9 = A_2H_8 = A_3H_7 = ab = 16$

$\therefore$ Expression $= 16^4 \times G_2G_4G_6G_8$

Now, $G_2G_8 = G_4G_6 = ab = 16$

$\therefore 16^4 \times 16^2 = 4^{12}$.

13. **(B)**

Let $S = \begin{array}{c|c} \alpha+\beta=1 & \gamma+\delta=4 \\ \alpha\beta=p & \gamma\delta=q \end{array} + \ldots$ (1)

So, $\frac{S}{10} = 0 + \frac{1}{10^2} + \frac{2}{10^3} + \ldots$ (2)

$\therefore$ Subtracting (2) from (1), we get

$$\frac{9S}{10}=\frac{5}{4}+\frac{1}{10^2}+\frac{1}{10^3}+\ldots=\frac{\frac{1}{10}}{1-\frac{1}{10}}=\frac{1}{9}$$

$$\Rightarrow S=\frac{1}{9}\left(\frac{10}{9}\right)=\frac{10}{81}$$

14. (C)

Given $b^2 = ac$

$D = b^2 - 4ac = b^2 - 4b^2 = -3b^2 < 0$

also $a > 0 \Rightarrow$ (C)

15. (B)

$$S(n) = 1+\frac{1}{2}+\frac{1}{3}+\ldots+\frac{1}{n} \qquad (1)$$

$$\sum_{k=1}^{n}(2k+1)S(k)$$

$= 3S(1) + 5S(2) + 7S(3) + \ldots + (2n + 1)\ S(n)$

$$= 3 + 5\left(1+\frac{1}{2}\right)+7\left(1+\frac{1}{2}+\frac{1}{3}\right)+\ldots+ (2n+1)\left(1+\frac{1}{2}+\frac{1}{3}+\ldots+\frac{1}{n}\right)$$

$$= 1(3 + 5 + 7 + \ldots (2n + 1)) + \frac{1}{2}(5 + 7 + \ldots (2n + 1)) + \frac{1}{3}(7 + 9 + \ldots + (2n + 1)) + \ldots + \frac{1}{n}(2n + 1)$$

$$= \left((n+1)^2-1\right)+\frac{1}{2}\left((n+1)^2-2^2\right)+ \frac{1}{3}\left((n+1)^2-3^2\right)+\ldots+\frac{1}{n}\left((n+1)^2-n^2\right)$$

$$= (n + 1)^2\left(1+\frac{1}{2}+\frac{1}{3}+\ldots+\frac{1}{n}\right) - (1 + 2 + 3 + \ldots n)$$

$$= (n + 1)^2\ S(n) - \frac{n(n+1)}{2}$$

16. (B)

$$S_N = 50^2;\ S_D = \frac{2\cdot 50\cdot 51}{2} = 50\cdot 51 \Rightarrow \frac{N}{D} = \frac{50}{51}$$

17. (D)

$\alpha+\beta=1$	$\gamma+\delta=4$
$\alpha\beta=p$	$\gamma\delta=q$

$\alpha, \beta, \gamma, \delta$ are in A.P.

Let $\alpha = a - 3d, \beta = a - d, \gamma = a + d, \delta = a + 3d$

$$\left.\begin{aligned}\alpha+\beta=1 \Rightarrow 2a-4d=1\\ \gamma+\delta=4\Rightarrow 2a+4d=4\end{aligned}\right]\ a=\frac{5}{4},\ d=\frac{3}{8}$$

Hence, $p=\alpha\beta=\left(\frac{1}{8}\right)\left(\frac{13}{8}\right)=\left(\frac{19}{8}\right)$

Also, $q=\gamma\delta\left(\frac{13}{8}\right)\left(\frac{19}{8}\right)=\frac{247}{64}$

18. (C)

G.P.: $\frac{a}{r}, a, ar, \ldots$

$a = 3$ and $r=\frac{1}{3}$ or 3 (rejected)

G.P. is $9, 3, 1, \frac{1}{3}, \ldots \Rightarrow$ C is not correct

19. (A)

$$S=1+\left(1+\frac{1}{5}\right)\left(\frac{1}{2}\right)+\left(1+\frac{1}{5}+\frac{1}{5^2}\right)\frac{1}{2^2}+\ldots(1)$$

$$\therefore\ \frac{S}{2}=\left(\frac{1}{2}\right)+\left(1+\frac{1}{5}\right)\frac{1}{2^2}+\ldots \qquad (2)$$

(Subtract)

$$\Rightarrow \frac{S}{2}=1+\frac{1}{5}\cdot\frac{1}{2}+\frac{1}{5^2\cdot 2^2}+\ldots$$

$$\Rightarrow \frac{S}{2}=\frac{1}{1-\frac{1}{10}}\Rightarrow\frac{S}{2}=\frac{10}{9}\Rightarrow S=\frac{20}{9}$$

20. (C)

$$S_n=\sum_{r=1}^{n}\frac{2r+1}{r^2(r+1)^2}=\sum_{r=1}^{n}\left[\frac{1}{r^2}-\frac{1}{(r+1)^2}\right]$$

$$\therefore\ S_{10}=\left(\frac{1}{1}-\frac{1}{2^2}\right)+\left(\frac{1}{2^2}-\frac{1}{3^2}\right)+\ldots+\left(\frac{1}{10^2}-\frac{1}{11^2}\right)$$

$$S_{10}=1-\frac{1}{121}=\frac{120}{121}$$

21. (D)

$$S=3\left(\frac{1}{2}+\frac{1}{6}+\frac{1}{12}+\frac{1}{20}+\ldots\right)$$

$$S=3\left(\left(1-\frac{1}{2}\right)+\left(\frac{1}{2}-\frac{1}{3}\right)+\left(\frac{1}{3}-\frac{1}{4}\right)+\left(\frac{1}{4}-\frac{1}{5}\right)+\ldots\right)$$

$\therefore\ S = 3$

22. (D)

$$(2011)^{\log_{10}x}=11^{\log_{10}(1+3+5+\ldots+4021)}$$

Now $\underbrace{1+3+5+\ldots+4021}_{\text{sum of the first 2011 odd natural number}}=(2011)^2$

$\therefore \dfrac{1}{ab}=11^{\log_{10}(2011)^2}$

$=11^{2\log_{10}2011}=(121)^{\log_{10}2011}$

Hence $x = 121$

23. **(B)**
Let the 1st term be '*T*' and the common ratio be '*r*'

$\dfrac{a(1-r^{101})}{1-r}=125$

$\sum_{r=1}^{101}\dfrac{1}{a_i}=\left(\dfrac{1}{a_1}+\dfrac{1}{a_2}+\ldots+\dfrac{1}{a_{101}}\right)$

$=\dfrac{1}{a}\left(\dfrac{1-r^{101}}{1-r}\right)\cdot\dfrac{1}{r^{100}}\;\dfrac{125}{(ar^{50})^2}=\dfrac{125}{625}=\dfrac{1}{5}$

24. **(A)**
Let x and y be the 1st term and common ratio respectively of a G.P.

$\therefore b=xy^{p-1},\ a=xy^{p+q-1},\ c=xy^{p-q-1}$

Thus, $ac=x^2y^{2p-2}=(xy^{p-1})^2=b^2$

$\therefore$ Discriminant $=4(b^2-ac)=0$

Hence, $af(x)=0$ at $x=-\dfrac{b}{a}$

and $af(x)>0$ at $x\neq-\dfrac{b}{a}$

That is, $af(x)\geq 0$

25. **(B)**

$\left(5+\sqrt{2}\right)x^2-\left(4+\sqrt{5}\right)x+8+2\sqrt{5}=0<^{\alpha}_{\beta}$

H.M. of roots is $\dfrac{2\alpha\beta}{\alpha+\beta}$

$\alpha\beta=\dfrac{8+2\sqrt{5}}{5+\sqrt{2}};\ \alpha+\beta=\dfrac{4+\sqrt{5}}{5+\sqrt{2}}$

$\therefore$ H.M. of $\alpha,\beta=\dfrac{2\times 2(4\times\sqrt{5})}{4+\sqrt{5}}=4$

26. **(C)**

$b^2=ac\Rightarrow 2\log_2 b=\log_2 a+\log_2 c$

$\log_2 a, \log_2 b, \log_2 c$ are in A.P.

$\dfrac{1}{\log_a 2},\dfrac{1}{\log_b 2},\dfrac{1}{\log_c 2}$ are in A.P.

$\therefore \log_a 2, \log_b 2, \log_c 2$ are in HP.

27. **(C)**

$1+\cos 60°+\cos^2 60°+\cos^3 60°+\ldots$

$=1+\dfrac{1}{2}+\dfrac{1}{4}+\dfrac{1}{8}+\ldots=\dfrac{1}{1-\dfrac{1}{2}}=2.$

28. **(B)**
A.M. $\geq$ G.M. (for positive numbers)

29. **(B)**
Expression =

$\dfrac{\left[(\sqrt{a_2}-\sqrt{a_1})+(\sqrt{a_3}-\sqrt{a_2})+\ldots+(\sqrt{a_{50}}-\sqrt{a_{49}})\right]}{d}$

where d = common difference of A.P.

$=\dfrac{\sqrt{a_{50}}-\sqrt{a_1}}{d}=\dfrac{\sqrt{144}-\sqrt{4}}{\left(\dfrac{140}{49}\right)}=\dfrac{49}{14}.$

30. **(B)**
Given $100(a+99d)=50(a+49d)$

$\Rightarrow a+149d=0$ (1)

and $T_{150}=a+149d=0$ (Using (1))

31. **(A)**
Here, $a\Rightarrow$ first term $=-4$ and
$d\Rightarrow$ common difference $=1$

$\therefore S_n=\dfrac{n(n-9)}{2}$

Also, $V_n=\left(\dfrac{n^2-5n-12}{2}\right)$

$\therefore \sum_{n=1}^{10}V_n=\dfrac{1}{2}\sum_{n=1}^{10}(n^2-5n-12)=-5$

Also, $(V_n)_{\text{min.}}=-9$ when $n=2$ or $n=3$

32. **(B)**
$a;\ a+9;\ a+18;\ \ldots$
Now, a_1, a_3 and a_8 in G.P., that is, a, $a+18$ and $a+63$ in G.P.

$\therefore (a+18)^2=a(a+63)=a^2+63a$

$\Rightarrow a=12;\ d=9$

So, $a_7=a+6d=12+54=66$

33. **(D)**

$\because$ first term $a_1=2$ {given}

$\therefore a_n=\dfrac{a_{n-1}^2}{a_{n-2}}\Rightarrow\dfrac{a_n}{a_{n-1}}=\dfrac{a_{n-1}}{a_{n-2}}$

The ratio of consecutive terms is constant

$\therefore$ sequence is G.P.

Let $a_2=x$

then common ratio $=\dfrac{x}{2}$

$\therefore a_5 = 2\left(\frac{x}{2}\right)^4 \Rightarrow \boxed{a_5 = \frac{x^4}{8}}$

$\because a_5 \to$ +ve integer

$\therefore x$ must be an even and integer.

$\because$ Given $\Rightarrow a_5 \le 512$

$\frac{x^4}{8} \le 512 \Rightarrow x^4 \le 4096 \Rightarrow x \le 8$

$\{\because x \in \text{even}\}$

$\therefore x = 2, 4, 6, 8$

So $a_5 = 2, 32, 162, 512$

34. **(C)**

Given $\frac{ar(r^{10}-1)}{r-1} = 18$ (1)

Also $\frac{\frac{1}{ar}\left(1-\frac{1}{r^{10}}\right)}{1-\frac{1}{r}} = 6 \Rightarrow \frac{1}{ar^{11}}\cdot\frac{(r^{10}-1)}{r-1} = 6$

$\frac{1}{a^2r^{11}}\cdot\frac{ar(r^{10}-1)}{r-1} = 6$ (2)

From (1) and (2)

$\frac{1}{a^2r^{11}} \times 18 = 6 \Rightarrow a^2r^{11} = 3$

Now $P = a^{10}r^{55} = (a^2r^{11})^5 = 3^5 = 243.$

35. **(B)**

$417 = 17 + (n-1)4$

$\Rightarrow 400 = 4(n-1) \Rightarrow n = 101$

Similarly $466 = 16 + (m-1)5$

$\Rightarrow 450 = 5(m-1) \Rightarrow m = 91$

Let T_n is common to both for some n for which m is an integer.

$17 + (n-1)4 = 16\ (m-1)5$

$1 + 4n - 4 = 5m - 5$

$5m = 4n + 2$

Hence $n = 2, 7, 12, \ldots, 97 \to 20$

36. **(D)**

$D = q^2 - 4pr$ also $q = \frac{2pr}{p+r}$

$= \left(\frac{2pr}{p+r}\right)^2 - 4pr = -4pr\left[1 - \frac{pr}{(p+r)^2}\right]$

$\Rightarrow -4pr\left[\frac{p^2+r^2+pr}{(p+r)^2}\right]$

$\therefore D < 0$ roots are imaginary

37. **(A)**

$\frac{1}{d}\left[\frac{a_2-a_1}{a_1a_2} + \frac{a_3-a_2}{a_2a_3} + \ldots + \frac{a_{n+1}-a_n}{a_n.a_{n+1}}\right]$

$= \frac{1}{d}\left[\frac{1}{a_1} - \frac{1}{a_2} + \frac{1}{a_2} - \frac{1}{a_3} + \ldots + \frac{1}{a_n} - \frac{1}{a_{n+1}}\right]$

$= \frac{1}{d}\left[\frac{1}{a_1} - \frac{1}{a_{n+1}}\right]$

$= \frac{1}{d}\left[\frac{a_{n+1}-a_1}{(a_1)(a_{n+1})}\right] = \frac{1}{d}\left[\frac{a_1+nd-a}{(a_1)(a_{n+1})}\right]$

$= \frac{n}{a_1[a_1+nd]} = \frac{1}{a_1\left[\frac{a_1}{n}+d\right]}$

$\Rightarrow$ as $n \to \infty$ then $S = \frac{1}{a_1d}$

38. **(A)**

$a_6 = a_1 + 5d = 2$

Now, $P = a_1\,a_4\,a_5 = a_1\,(a_1 + 3d)\,(a_1 + 4d)$

$= (2 - 5d)\,(2 - 2d)\,(2 - d)$

$= 2(4 - 16d + 17d^2 - 5d^3)$

Consider $f(d) = -5d^3 + 17d^2 - 16d + 4$

$\therefore f'(d) = -15d^2 + 34d - 16,$

for minima, $f'(d) = 0$

$\Rightarrow d = \frac{2}{3}, \frac{8}{5}$

$f''(d) = -30d + 34$

$f''\left(\frac{2}{3}\right) = -20 + 34 > 0 \Rightarrow d = \frac{2}{3}$ gives minima

39. **(A)**

We have

$A = \left(\frac{1}{2}\times\frac{1}{3}\times 1\right) + \left(\frac{1}{2}\times\frac{1}{9}\times 2\right) + \left(\frac{1}{2}\times\frac{1}{27}\times 3\right) + \ldots\ \infty$

$= \frac{1}{2}\underbrace{\left(\frac{1}{3} + \frac{2}{9} + \frac{3}{27} + \ldots\infty\right)}_{S} = \frac{S}{2}$

Let $S = \frac{1}{3} + \frac{2}{3^2} + \frac{3}{3^3} + \ldots\infty$, then

$\frac{S}{3} = 0 + \frac{1}{3^2} + \frac{2}{3^3} + \frac{3}{3^4} + \ldots\infty$

(Subtracting)

$\Rightarrow \frac{2S}{3} = \frac{1}{3} + \frac{1}{3^2} + \frac{1}{3^3} + \ldots\infty$

$$=\frac{\frac{1}{3}}{1-\frac{1}{3}}=\frac{\frac{1}{3}}{\frac{2}{3}}=\frac{1}{2}\Rightarrow S=\frac{3}{4}$$

$$\therefore A=\frac{S}{2}=\frac{3}{8}\Rightarrow 8A=3$$

40. **(C)**

41. **(B)**

The above series in an infinite G.P., whose first term $=\frac{1}{1+x}$ and common ratio $=\frac{-(1-x)}{(1+x)}$

$$\therefore S_\infty=\frac{\frac{1}{1+x}}{1+\left(\frac{1-x}{1+x}\right)}=\frac{1}{2}$$

42. **(B)**

We have, $t_n = S_n - S_{n-1}$

$= n(3n + 5) - (n - 1)(3n + 2)$

$= 6n + 2 = 32$ (Given)

$\Rightarrow n = 5$

43. **(A)**

$a_1, a_2, a_3, \dots$ with $d = 2$

$$\frac{S_{3n}}{S_n}=\frac{\frac{3n}{2}[2a_1+(3n-1)2]}{\frac{n}{2}[2a_1+(n-1)2]}=\frac{3[(a_1-1)+3n]}{(a_1-1)+n}$$

This is independent of n if $a_1 = 1$

$\therefore$ Arithmetic progression is

$1 + 3 + 5 + 7 + \dots$

$S_{10} = 10^2 = 100$

44. **(C)**

$$S=\frac{a}{1-r}=20\Rightarrow\frac{a^2}{1-r^2}=80\Rightarrow r=\frac{2}{3}$$

$$\therefore a=\frac{20}{3}=\frac{p}{q}\Rightarrow(p+q)=23.$$

45. **(A)**

For arithmetic progression, if $S_n = an^2 + bn$

$S_{n-1} = a(n - 1)^2 + b(n - 1)$

$\Rightarrow T_n = a(2n - 1) + b$

$\therefore T_{n-1} = a(2n - 3) + b$

$\Rightarrow d = 2a =$ common difference of arithmetic progression.

$S_n = 6n\sec^2\theta + n(n - 1)(\sin^2\theta(4 + \tan^2\theta) + \cos^2\theta(4 + \cot^2\theta))$

Common difference, $d = 2 \times$ coefficient fn^2

$= 2 \times (\sin^2\theta(3 + \sec^2\theta) + \cos^2\theta(3 + \mathrm{cosec}^2\theta))$

$$=2\left(\underbrace{3+\tan^2\theta+\cot^2\theta}_{\geq 2}\right)$$

$$d_{\min}\left(\theta=\frac{\pi}{4}\right)=10.$$

46. **(D)**

$$\sum_{n=1}^{\infty}\frac{5^n-2^n}{10^n}=\sum_{n=1}^{\infty}\left(\frac{5^n}{10^n}-\frac{2^n}{10^n}\right)$$

$$=\sum_{n=1}^{\infty}\left(\frac{1}{2}\right)^n-\left(\frac{1}{5}\right)^n$$

the difference between two infinite geometric sequences.

$$=\frac{\frac{1}{2}}{1-\frac{1}{2}}-\frac{\frac{1}{5}}{1-\frac{1}{5}}=1-\frac{1}{4}=\frac{3}{4}$$

47. **(B)**

$72x^3 - 108x^2 + 46x - 5 = 0$

Let roots are $a - d, a, a + d$

$$\therefore 3a=\frac{108}{72}=\frac{3}{2}\Rightarrow a=\frac{1}{2}$$

Now $a(a^2 - d^2) = \frac{5}{72}$

$$\frac{1}{4}-d^2=\frac{5}{36}\Rightarrow d^2=\frac{1}{9}$$

$$\Rightarrow d=\frac{1}{3}\Rightarrow 2d=\frac{2}{3}$$

Hence, difference between largest and smallest root $2d=\frac{2}{3}$ which lies in the interval $\left(\frac{1}{2},\frac{1}{\sqrt{2}}\right)$.

48. **(C)**

$$\frac{\frac{1}{x}}{1-\frac{1}{2x}}=\frac{1}{64}\Rightarrow\frac{\frac{1}{x}}{\frac{2x-1}{2x}}=\frac{1}{64}$$

$$\frac{2}{2x-1}=\frac{1}{64}\Rightarrow 128=2x-1$$

$$\Rightarrow x=\frac{129}{2}$$

49. **(D)**

$S_n = \frac{n}{2}(a+\ell) \Rightarrow 400 = \frac{n}{2}(5+45) \Rightarrow n = 16$

Again, $15 = a + (n-1)d = 15 + 15d$

$\therefore d = \frac{40}{15} = \frac{8}{3} \Rightarrow \left(\frac{n}{d}\right) = 3\left(\frac{16}{8}\right) = 6$

50. **(B)**

Let $a_1 = a$

$S = a + ar + ar^2 + \ldots + ar^9$

$$S = \frac{a(1-r^{10})}{1-r} \quad (1)$$

$$T = T = \frac{1}{a} + \frac{1}{ar} + \frac{1}{ar^2} + \ldots \frac{1}{ar^9}$$

$$= \frac{1}{a}\frac{\left(\frac{1}{ar^{10}} - 1\right)}{\left(\frac{1}{r} - 1\right)} = \frac{1}{a}\frac{(1-r^{10})\cdot r}{r^{10}(1-r)}$$

$$T = \frac{1}{a}\frac{(1-r^{10})}{r^9(1-r)} \quad (2)$$

Also, $\prod_{n=1}^{10} a_n = a \cdot ar \cdot ar^2 \ldots ar^9 = a^{10} \times r^{45}$ (3)

Now from (1) and (2)

$$\frac{S}{T} = \frac{a(1-r^{10})}{1-r} \cdot \frac{ar^9(1-r)}{(1-r^{10})}$$

$$\therefore \frac{S}{T} = a^2 r^9$$

$$\therefore \left(\frac{S}{T}\right)^5 = a^{10} \cdot r^{45} = \prod_{n=1}^{10} a_n$$

51. **(B)**

$a + ar^2 + ar^4 = 455$

$ar + ar^3 + ar^5 = 1365$

$$\Rightarrow \frac{ar(1+r^2+r^4)}{a(1+r^2+r^4)} = \frac{1365}{455} \Rightarrow r = 3.$$

52. **(B)**

$$\frac{ar^2}{a} = 9 \Rightarrow r = 3$$

$$a + ar = \frac{4}{3} \Rightarrow a \cdot 4 = \frac{4}{3} \Rightarrow a = \frac{1}{3}$$

$$\therefore a_4 = ar^3 = \frac{1}{3} \cdot 3^3 = 3^2 = 9$$

53. **(B)**

Let $S = 1 + \frac{3}{4} + \frac{6}{4^2} + \frac{10}{4^3} + \frac{15}{4^4} + \ldots$

$$\frac{S}{4} = \frac{1}{4} + \frac{3}{4^2} + \frac{6}{4^3} + \frac{10}{4^4} + \ldots$$

$$\Rightarrow \frac{3S}{4} = 1 + \frac{2}{4} + \frac{3}{4^2} + \frac{4}{4^3} + \frac{5}{4^4} + \ldots$$

$$\Rightarrow \frac{3S}{16} = \frac{1}{4} + \frac{2}{4^2} + \frac{3}{4^3} + \frac{4}{4^4} + \ldots$$

$$\Rightarrow \frac{9S}{16} = 1 + \frac{1}{4} + \frac{1}{4^2} + \frac{1}{4^3} + \ldots$$

$$\Rightarrow \frac{9S}{16} = \frac{1}{1-\frac{1}{4}} \Rightarrow \frac{9S}{16} = \frac{4}{3} \Rightarrow S = \frac{64}{27}$$

54. **(A)**

A.M. of 1, 2, 3, …, $n = \frac{n+1}{2}$

If k, $k+1$ are removed then

$$\text{A.M.} = \frac{\frac{n(n+1)}{2} - (2k+1)}{n-2}$$

$$\Rightarrow 1 = \frac{n+1}{2} - \frac{n(n+1) - 2(2k+1)}{2(n-2)} \Rightarrow n - k = 1$$

55. **(A)**

$a_3 = a_2 + a_1$

$r^2 - r - 1 = 0,\ r = \frac{1+\sqrt{5}}{2}$

56. **(B)**

$$\frac{1}{a_2} - \frac{1}{a_1} = \frac{1}{a_3} - \frac{1}{a_2} = \ldots \frac{1}{a_n} - \frac{1}{a_{n-1}} = d$$

$$a_1a_2 = \frac{a_1 - a_2}{d},\ a_2a_3 = \frac{a_2 - a_3}{d}, \ldots$$

$$\Rightarrow a_1a_2 + a_2a_3 + \ldots + a_{n-1}a_n$$

$$= \frac{a_1 - a_n}{d} = (n-1)a_1a_n$$

57. **(A)**

$b = \log_2 3 \times \log_3 4 \times \log_4 5 \times \ldots \times \log_{15} 16$

$\Rightarrow b = \log_2 16 \Rightarrow b = 4$ and $a = 2^b \Rightarrow a = 16$

Hence $\log_2(\log_{16} a) = 0$

58. **(D)**

$3200 = 100r^{10} \Rightarrow r = \pm\sqrt{2}$

Consider $r = 2^{\frac{1}{2}}$

$\Rightarrow G_1 = 100r = 100\sqrt{2}$ (not rational)

$\Rightarrow G_2, G_4, G_6, G_8$ are rational terms.

59. **(C)**

$k^2 + k(a-1) = 153$

$\Rightarrow k^2 + k(a-1) - 153 = 0$

where a is the first term

Number of terms = Number of factors of 153 excluding 1 = 5

60. **(B)**

Integer Answer Type

1. **(3)**

$$\text{Expression} = \frac{1+e^{-2\alpha}+e^{-4\alpha}+\ldots\infty}{1-e^{-2\alpha}+e^{-4\alpha}+\ldots\infty} = \frac{1+e^{-2\alpha}}{1-e^{-2\alpha}}$$

$$= \left(\frac{e^{2\alpha}+1}{e^{2\alpha}-1}\right)$$

As, $e^{\alpha} = \sqrt{\cot\frac{\pi}{12}} \Rightarrow e^{2\alpha} = (2+\sqrt{3})$

$\therefore$ Expression $= \sqrt{3}$

2. **(2)**

$5m^2 - 16 + p^2 = 4pm$

$5m^2 - 4pm + p^2 - 16 = 0$

$\therefore D \geq 0 \Rightarrow p \in [-\sqrt{80}, \sqrt{80}]$

$\therefore$ Greatest integral value of 'p' is 8.

Now, $8m^2 - 32m + 48 = 0$ < $m = 12/5$; $m = 4$ (rejected)

$5m^2 - 16, 2pm, p^2$

$\Rightarrow \frac{64}{5}, \frac{192}{5}, \frac{320}{5}$

$\therefore$ Common difference $= \frac{128}{5}$

3. **(13)**

Given, a_1 a_2 a_3 a_4 (a_2, a_3, a_4: G.P.; a_1, a_2, a_3: A.P.)

Let $a_2 = a, a_3 = ar, a_4 = ar^2$

Also, given $a_1 = a_4 + 5$

$\therefore a_1 = ar^2 + 5$

Now, $a_1\ a_2\ a_3 \rightarrow$ A.P.

$\therefore d = a_2 - a_1 = a_3 - a_2$

$d = a - ar^2 - 5 = ar - a$

$\therefore a(r^2 + r - 2) = -5$ (1)

Also, $d + 4r = 0$

$\therefore ar - a + 4r = 0 \Rightarrow a(1-r) = 4r$ (2)

From (1) and (2)

$$\frac{r^2+r-2}{1-r} = \frac{-5}{4r}$$

$4r^3 + 4r^2 - 13r + 5 = 0$

$(2r-1)(2r^2 + 3r - 5) = 0$

$\therefore r = \frac{1}{2} \Rightarrow a = 4$

Hence, $a_1 = 6, a_2 = 4, a_3 = 2, a_4 = 1$

$$\sum_{i=1}^{4} a_1 = 13$$

Aliter: $a, \underbrace{a+d, a+2d, a-5}_{\text{G.P.}}$

$$r = \underset{(1)}{\frac{a+2d}{a+d}} = \underset{(2)}{\frac{a-5}{a+2d}} = \underset{(3)}{\frac{-d}{4}} \qquad \because d + 4r = 0$$

using ratio and proportion

$$r = \frac{2d+5}{-d} = \frac{-d}{4}$$

$8d + 20 = d^2$

$d^2 - 8d - 20 = 0$

$(d-10)(d+2) = 0$

$\Rightarrow d = 10$ or -2

$\Rightarrow r = -\frac{5}{2}$ or $\frac{1}{2}$

$\downarrow$

rejected

$\because$ numbers are natural

$\therefore$ If $d = -2$ $r = \frac{1}{2}$ $a = 6$

$\therefore$ digits are $a_1 = 6$

$a_2 = 4$

$a_3 = 2$

$a_4 = 1$

$\therefore \Sigma a_i = 13$

4. **(7)**

$2\gamma = \alpha + \delta$ and $\alpha + \beta + \gamma = 9 \Rightarrow \gamma = 3$

$\alpha\gamma\delta = 24 \Rightarrow \gamma = 3, \delta = 4, \alpha = 2$

$\because \frac{1}{\alpha}, \frac{1}{\beta}, \frac{1}{\gamma}, \frac{1}{\delta}$ are in A.P. $\Rightarrow \beta = \frac{12}{5}$

$$\left|\frac{P(x)}{Q(x)}\right| = |5(x-\beta)| = 7$$

5. (2)

$a_1 + a_3 + \ldots + a_{99} = 50$

$\Rightarrow a + (a + 2d) + (a + 4d) + \ldots + (a + 98d) = 50$

$\Rightarrow 50a + 2d(1 + 2 + \ldots + 49) = 50$

$$\Rightarrow 50a + \frac{2d(50)(49)}{2} = 50$$

$\Rightarrow a + 49d = 1$

$$\therefore a_{50} = 1; \left|\sum_{r=1}^{\theta}(-1)^{\frac{r(r+1)}{2}} \cdot a_{2r-1}\right|$$

$= -a_1 - a_3 + a_5 + a_7 - a_9 - a_{11} + \ldots + a_{93} + a_{95} - a_{97} - a_{99}$

(26 negative terms and 24 positive)

$\Rightarrow |-a_1 - a_{99}| = |-2a - 98d| \Rightarrow |-2| \Rightarrow 2$

6. (400)

We have $\text{A.M.} = \dfrac{100+n}{2}$,

$\text{G.M.} = 10\sqrt{n}$ and $\text{H.M.} = \dfrac{200n}{100+n}$

For A.M. to be an integer, n must be even.

For G.M. to be an integer, n must be a perfect square.

$\Rightarrow n = 4k^2$

$$\therefore \text{H.M.} = \frac{200k^2}{25+k^2}$$

Since numerator is multiple of 25, denominator must be a multiple of 25 as H.M. is an integer.

Also $100 < n \le 500$

$\Rightarrow 25 < k^2 \le 125$

$\Rightarrow k^2 = 50, 75, 100$ and 125

$\therefore$ Satisfying these values, we get $k^2 = 100$ as the only solutions.

Hence $n = 4k^2 = 4(100) = 400$

7. (13)

We have $x^3 + 4x^2 - ax + 1 = 0$ with roots r_1, r_2, r_3

$\therefore \Sigma r_1 = -4, \Sigma r_1 r_2 = -a$ and $r_1 r_2 r_3 = -1$

Now let

$$E = \left(r_1 + \frac{1}{r_1}\right)^2 + \left(r_2 + \frac{1}{r_2}\right)^2 + \left(r_3 + \frac{1}{r_3}\right)^2$$

$$= (\Sigma r_1)^2 - 2\Sigma r_1 r_2 + \left[\frac{(r_2r_3)^2 + (r_3r_1)^2 + (r_1r_2)^2}{1}\right] + 6 = 22 + 2a + \left[(\Sigma r_1 r_2)^2 - 2\left(r_1^2 r_2 r_3 + r_2^2 r_3 r_1 + r_3^2 r_2 r_1\right)\right]$$

$= 22 + 2a + [a^2 + 2(-4)]$

$= 22 - 8 + a^2 + 2a = 22 - 8 + a^2 + 2a$

$\therefore E = (a + 1)^2 + 13$

Hence minimum value of $E = 13$ (when $a = -1$)

8. (201)

We have,

$\sigma = 0.9[(10^2 - 1) + (10^3 - 1) + \ldots + (10^{101} - 1)$

$= 0.9[(10 - 1) + (10^2 - 1) + (10^3 - 1) + \ldots + (10^{101} - 1) - 9]$

$$= 0.9\left[\frac{10\left(10^{101}-1\right)}{10-1} - (101) - 9\right]$$

$$= 0.9\left[\frac{10\left(10^{101}-1\right)}{9} - 110\right]$$

$= 10^{101} - 1 - 99 = 10^{101} - 100$

$= (10^a - b)$ (Given). So $a = 101$ and $b = 100$

Hence, $(a + b) = 101 + 100 = 201$

9. (9)

Let d be the common difference of the A.P., then $a_{2r} = a_{2r-1} = d$

$$\Rightarrow \sum_{r=1}^{10^{99}} a_{2r} = \sum_{r=1}^{10^{99}} \left(a_{2r-1} + d\right)$$

$$\Rightarrow \sum_{r=1}^{10^{09}} a_{2r} = \left(\sum_{r=1}^{10^{99}} a_{2r-1}\right) + 10^{99}\,\mathrm{d}$$

$\Rightarrow 10^{100} = 10^{99} + 10^{99}d$

$$\Rightarrow d = \frac{10^{100} - 10^{99}}{10^{99}} = \frac{10^{99}(10-1)}{10^{99}} = 9$$

10. (6)

Given x, y, z, w are 4 numbers such that x, y, z are in A.P. and y, z, w are in G.P.

Let $x = \frac{2a}{r} - a$; $y = \frac{a}{r}$; $z = a$; $w = ar$

Now, $x + w = 16$ and $y + z = 8$

$\Rightarrow \frac{2a}{r} - a + ar = 16$ and $\frac{a}{r} + a = 8$

$\Rightarrow a\left(\frac{2}{r} - 1 + r\right) = 16$ and $a(1 + r) = 8r$

$\Rightarrow 8r\left(\frac{2}{r} - 1 + r\right) = 16(1 + r)$

$\Rightarrow 2 - r + r^2 = 2 + 2r$

$\therefore r^2 = 3r$

$r = 0$ (reject) or 3

$\therefore a = 6$

$\therefore x = -2$; $y = 2$; $z = 6$; $w = 18$

Hence, $x^2 - y^2 + 2z - w$

$= (-2)^2 - (2)^2 + 2(6) - 18$

$= 4 - 4 + 12 - 18 = -6 \Rightarrow |-6| = 6$

Aliter: Given x, y, z, w are 4 numbers such that x, y, z are in A.P. and y, z, w are in G.P.

Let $x = a - d$, $y = a$ and $z = a + d$

$\therefore w = \frac{(a+d)^2}{a}$

Now, given $x + w = 16$

$\Rightarrow (a - d) + \frac{(a+d)^2}{a} = 16$ (1)

and $y + z = 8 \Rightarrow 2a + d = 8$ (2)

$\therefore$ (1) and (2)

$\Rightarrow a^2 - ad + a^2 + d^2 + 2ad = 16a$

$2a^2 + d^2 + ad = 16a$

$2a^2 + (8 - 2a)^2 + a(8 - 2a) = 16a$

$\Rightarrow 2a^2 + 4(4 - a)^2 + 8a - 2a^2 = 16a$

$\Rightarrow (4 - a)^2 + 2a = 4a$

$\Rightarrow a^2 - 10a + 16 = 0$

$\Rightarrow (a - 8)(a - 2) = 2$

$\therefore a = 2$ or $a = 8$

If $\begin{bmatrix} a = 2 & \text{and} & \text{if } a = 8 \\ d = 4 & & d = -8 \end{bmatrix}$

(Reject as $z = a + d = 0$)

$\therefore$ Numbers are $-2, 2, 6, 18$

$\therefore x^2 - y^2 + 2z - w = (-2)^2 - (2)^2 + 2(6) - 18$

$= 4 - 4 + 12 - 18 = -6 \Rightarrow |-6| = 6$

11. (5)

$T_n = \frac{n(n+1)}{2}$

$\therefore T_i = \frac{i(i+1)}{2}$

Now,

$$\sum_{k=1}^{\infty}\left(\frac{3}{T_i}\right)^k = \left(\frac{3}{T_i}\right) + \left(\frac{3}{T_i}\right)^2 + \left(\frac{3}{T_i}\right)^3 + \ldots\infty$$

$$= \frac{\frac{3}{T_i}}{1 - \frac{3}{T_i}} = \frac{3}{T_i - 3} = \frac{3}{\frac{i(i+1)}{2} - 3} = \frac{6}{i^2 + i - 6}$$

$$\therefore \sum_{i=3}^{\infty}\left(\frac{6}{i^2 + i - 6}\right) = \sum_{i=3}^{\infty}\left(\frac{6}{(i+3)(i-2)}\right)$$

$$= \frac{1}{5}\sum_{i=3}^{\infty}\left(\frac{6}{i-2} - \frac{6}{i+3}\right) = \frac{6}{5}\sum_{i=3}^{\infty}\left(\frac{1}{i-2} - \frac{1}{i+3}\right)$$

$$= \frac{137}{50} \equiv \frac{p}{q} \Rightarrow (3q - p - 8) = 5$$

12. (4)

$$S = \frac{7}{2^2 5^2} + \frac{13}{5^2 8^2} + \frac{19}{8^2 11^2} + \ldots$$

$$3S = \frac{21}{2^2 5^2} + \frac{39}{5^2 8^2} + \frac{57}{8^2 11^2} + \ldots$$

$$3S = \sum_{r=1}^{10}\frac{(3r+2)^2 - (3r-1)^2}{(3r-1)^2(3r+2)^2}$$

$$3S = \sum_{r=1}^{10}\left(\frac{1}{(3r-1)} - \frac{1}{(3r+2)^2}\right)$$

$$3S = \frac{1}{2^2} - \frac{1}{2^{10}}$$

$$3S = \frac{2^8 - 1}{2^{10}}$$

$$S = \frac{85}{1024} = \frac{m}{n}$$

Hence, $(n - 12m) = 4$

13. (5)

$$3\left(x - \frac{1}{2}\right)^2 + 2\left(y - \frac{3}{2}\right)^2 = 0$$

$\therefore\ x = \frac{1}{2}$ and $y = \frac{3}{2}$

Hence, $\sum_{r=0}^{10}(2y+rx) = k = \frac{121}{2}$

14. **(7)**

Let the roots be $A - 3d, A - d, A + d, A + 3d$

Sum of roots = 0

$4A = 0 \Rightarrow A = 0$

Product of roots $\Rightarrow 9d^4 = 16k^2$

$$d^2 = \frac{4k}{3} \qquad (1)$$

Sum of roots taken two at a time, that is,

$\Sigma x_1x_2 = -10d^2 = -(12k + 5)$

$10\cdot\left(\frac{4k}{3}\right) = 12k + 5$

$40k = 36k + 15$

$k = \frac{15}{4} = \frac{a}{b}$

Hence, $(a - 2b) = 7$

15. **(8)**

Let $a = A - d$

$b = A$

$c = A + d$

$\therefore (2A - 2d)(4A + 4d)(A) = \lambda(A - d)A(A + d)$

Hence, $\lambda = 8$

16. **(2)**

Given $a_1 = g_1 = h_1 = 1$

$a_{10} = a_1 = 9d = 1 + 9d = 2$

$d = \frac{1}{9}$

$g_{10} = g_1 = r^9 = 2$

$r = 2^{1/9}$

Given $h_1, h_2, \ldots, h_{10}$ are in H.P.

$\therefore \frac{1}{a_1}, \frac{l}{a_1 + d'}, \ldots$ are in H.P.

$a_1 + 9d' = 5$

$d' = \frac{4}{9}$

$\therefore [a_{19}g_{10}h_{28} + a_{28}g_{19}h_{19}]$

$$= \left(1+18\times\frac{1}{9}\right) 2 \left(\frac{1}{1+27\times\frac{4}{9}}\right) + \left(1+27\times\frac{1}{9}\right) 2^2 \left(\frac{1}{1+18\times\frac{4}{9}}\right)$$

$$3\times2\times\frac{1}{13} + 4\times4\times\frac{1}{9} = \left[\frac{262}{117}\right] = 2$$

17. **(5)**

$x^3 - 6x^2 + kx - 8 = 0$ with roots α, β, γ

Arithmetic mean of roots = 2

Geometric mean of roots = 2

$\Rightarrow \alpha = \beta = \gamma = 2$

So, $k = 12$

18. **(8)**

$$\frac{\frac{n(n+1)}{2} - n}{n-1} \le \frac{71}{2} \le \frac{\frac{n(n+1)}{2} - 1}{n-1}$$

$$\frac{n^2 - n}{2(n-1)} \le \frac{71}{2} \le \frac{n^2 + n - 2}{2(n-1)}$$

$$\frac{n}{2} \le \frac{71}{2} \le \frac{n+2}{2}$$

Hence, n can be 69, 70 or 71

For $n = 69$, $\frac{\frac{69(70)}{2} - x}{68} = \frac{71}{2} \Rightarrow x = 1$

For $n = 70$, $\frac{\frac{70(71)}{2} - x}{69} = \frac{71}{2}$

$\Rightarrow$ we de not get $x \in N$

For $n = 71$, we get $x = 71$

Hence $\lambda = 1 + 71 = 72$; $\sqrt{72} = 8$

19. **(23)**

Using the formula for the sum of an infinite geometric series, the first two sentences state that

$\frac{a}{1-r} = 20$ and $\frac{a^2}{1-r^2} = 80$

(If the terms are squared, the common ratio is squared.) Because the second equation factors

$$\frac{a}{1-r}\cdot\frac{a}{1+r}=80 \Rightarrow \frac{a}{1+r}=4$$

Using the first equation and the new equation, we have

$$\frac{a}{1-r}=20 \text{ and } \frac{a}{1+r}=4$$

$a = 20 - 20r$ and $a = 4 + 4r$

$20 - 20r = 4 + 4r$

$$16 = 24r \Rightarrow r = \frac{2}{3}$$

Putting into $a = 4 + 4r$,

we deduce $a = 4 + \frac{8}{3} = \frac{20}{3} \equiv \frac{p}{q}$

So, the sum of the numerator and denominator of the first term (a) is

$20 + 3 = 23$

20. **(6)**

For the given A.P., we have

$2(2a + b) = (5a - b) + (a + 2b) \Rightarrow b = 2a$ (1)

Also for the given G.P.,

we have $(ab + 1)^2 = (a - 1)^2 (b + 1)^2$ (2)

$\therefore$ Putting $b = 2a$ from (1) in (2),

we get $a = 0, -2$ or $\frac{1}{4}$

But $a > 0$, so $a = \frac{1}{4}$ and $b = 2a = \frac{1}{2}$

Hence $(a^{-1} + b^{-1}) = 2 + 4 = 6$.

2.13.4 JEE Advanced: Sample Questions

Single Answer Type

1. **(B)**

$T_r = r(r + 1)\,[(r + 1)^2 - r^2]$

2. **(C)**

$$\frac{\alpha+1}{\alpha}+\frac{\alpha}{\alpha-1}=-\frac{b}{a} \quad (1)$$

$$\left(\frac{\alpha}{\alpha-1}\right)\cdot\left(\frac{\alpha+1}{\alpha}\right)=\frac{c}{a} \quad (2)$$

$$\Rightarrow \frac{\alpha+1}{\alpha-1}=\frac{c}{a} \quad (3)$$

$$\alpha=\frac{c+a}{c-a}$$

$$\Rightarrow \alpha-1=\frac{2a}{c-a} \Rightarrow \frac{\alpha}{\alpha-1}=\frac{c}{2a}$$

$$\Rightarrow a\left(\frac{c+a}{2a}\right)^2+b\left(\frac{c+a}{2a}\right)+c=0$$

$$\Rightarrow (a + b + c)^2 = b^2 - 4ac$$

3. **(B)**

$g_2g_4g_6 \cdots g_{200} = \alpha$ (1)

$g_1g_3g_5 \cdots g_{199} = \beta$ (2)

Divide (1) by (2)

$$\left(\frac{g_2}{g_1}\right)\left(\frac{g_4}{g_3}\right)\cdots\left(\frac{g_{200}}{g_{199}}\right)=\frac{\alpha}{\beta}$$

$$\Rightarrow r^{100}=\frac{\alpha}{\beta} \Rightarrow r=\left(\frac{\alpha}{\beta}\right)^{\frac{1}{100}}$$

4. **(A)**

degree $n = 1 + 3 + 5 + \ldots + 21 = 121$

5. **(D)**

$$\frac{16x+\frac{4}{x}}{2} \ge \left(16x\cdot\frac{4}{x}\right)^{1/2} \Rightarrow 16x+\frac{4}{x} \ge 16$$

$\therefore$ Minimum value of $16x + \frac{4}{x}$ is 16.

6. **(B)**

$$S=\frac{1}{1}-\frac{1}{2}+\frac{1}{3}-\frac{1}{4}+\ldots-\frac{1}{2014}+\frac{1}{2015}$$

$$S=\sum_{r=1}^{2015}\frac{1}{r}-2\sum_{r=1}^{1007}\frac{1}{2r}=\sum_{r=1}^{2015}\frac{1}{r}-\sum_{r=1}^{1007}\frac{1}{r}$$

$$\Rightarrow S=\sum_{r=1008}^{2015}\frac{1}{r}$$

7. **(C)**

$$\sum_{n=2}^{\infty}\frac{1}{t_n}=\sum\frac{2}{n(n-1)(n+1)}$$

$$=\sum\left(\frac{(n+1)-(n-1)}{n(n-1)(n+1)}\right)$$

$$=\sum\left(\frac{1}{n(n-1)}-\frac{1}{(n)(n+1)}\right)$$

$$= \left(\frac{1}{2.1} - \frac{1}{2.3} + \frac{1}{3.2} - \frac{1}{3.4} + \ldots\right) = \frac{1}{2}$$

8. (C)

$\sqrt{ac} \geq b$ (G.M. ≥ H.M.)

and $\dfrac{a^{101} + c^{101}}{2} \geq \left(a^{101} \cdot c^{101}\right)^{1/2} = (\sqrt{ac})^{101}$

$\Rightarrow a^{101} + c^{101} \geq 2b^{101}$

⇒ Product of roots of equation is (–)ve

⇒ Real and distinct roots.

9. (D)

$$\frac{x + y + \frac{z}{2} + \frac{z}{2}}{4} \geq \left(x, y \cdot \frac{z}{2} \cdot \frac{z}{2}\right)^{1/4}$$

$$\Rightarrow 1 \geq \left(\frac{x^2 z^2}{4}\right)^{1/4} \quad \Rightarrow xyz^2 \leq 4$$

10. (A)

$\operatorname{cosec} x = \dfrac{1}{1 - \sin x}$; $1 - \sin x = \sin x$

$\sin x = \dfrac{1}{2}$; $x = n\pi + (-1)^n \dfrac{\pi}{6}, n \in I$

11. (B)

$$S_1 = \frac{n}{2}(2.1 + (n-1)1) = \frac{n}{2}(n+1)$$

$$S_2 = \frac{n}{2}(2.1 + (n-2)) = \frac{n^2}{2}$$

$$S_3 = \frac{n}{2}(2.1 + (n-3)) = \frac{n(n-1)}{2}$$

$$S_1 + S_3 = \frac{a^2 + n + n^2 - n}{2} = \frac{2n^2}{2} = 2S_2$$

12. (A)

$$\left(x_1^2 - x_2^2\right) + \left(x_3^2 - x_4^2\right) + \ldots$$

$$= -d(x_1 + x_2 + x_3 + \ldots + x_{2n})$$

$$= -d.n.(x_1 + x_{2n}),\ d = \frac{x_{2n} - X_1}{2n-1}$$

$$= -\frac{(x_{2n} - x_1)(x_{2n} + x_1)}{2n-1} \cdot n = \frac{n\left(x_1^2 - x_{2n}^2\right)}{2n-1}$$

13. (A)

For maximum value $\dfrac{2p}{3} = \dfrac{3q}{5} = \dfrac{4r}{7}$

$\Rightarrow p : q : r = \dfrac{3}{2} : \dfrac{5}{3} = 18 : 20 : 21$

14. (D)

Let T_{15} be the 15th term from the end

$T_{15} = a_n - 14d$

$= 100 - \dfrac{42}{5} = \dfrac{458}{5}$

15. (C)

$a(a_1 r) = a_1 + a_1 r + a_1 r^2$

$\Rightarrow r^2 + (1 - a)r + 1 = 0$

$D \geq 0 \Rightarrow (1 - a)^2 \geq 4; a \in (-\infty, -1] \cup [3, \infty)$

If $a = 3$ then $r = 1$

All numbers will be the same if $a = -1$ then $r = -1$

Two numbers will be the same

16. (B)

If $\log_e b_1, \log_e b_2, \ldots, \log_e b_{101} \to$ A.P.;

$D = \log_e 2$

$\Rightarrow b_1, b_2, b_3, \ldots, b_{101} \to$ G.P.; $r = 2$

$\therefore b_1, 2b_1, 2^2 b_1, \ldots, 2^{100} b_1 \ldots$ G.P.

$a_1, a_2, a_3, \ldots, a_{101} \ldots$ A.P.

Given, $a_1 = b_1$ and $a_{51} = b_{51}$

$\Rightarrow a_1 + 50D = 2^{50} b_1$

$\therefore (a_1 + 50D) = 250 a_1$ (As $b_1 = a_1$)

Now, $t = b_1(2^{51} - 1)$; $s = \dfrac{51}{2}(2a_1 + 50D)$

$\therefore t = a_1 \cdot 2^{51} - a_1 \Rightarrow t < a_1 2^{51}$ (1)

Also, $s = \dfrac{51}{2}(a_1 + a_1 + 50D)$

$\Rightarrow s = \dfrac{51}{2}\left(a_1 + 2^{50} a_1\right)$

$\therefore s = \dfrac{51 a_1}{2} + \dfrac{51}{2} \cdot 2^{50} a_1$

$\Rightarrow s > a_1 \cdot 2^{51}$ (2)

So, clearly, $s > t$ [from equation (1) and (2)]

Also $a_{101} = a_1 + 100D$; $b_{101} = b_1 \cdot 2^{100}$

$\therefore a_{101} = a_1 + 100\left(\dfrac{2^{50} a_1 - a_1}{50}\right)$; $b_{101} = 2^{100} a_1$

(3)

$\Rightarrow a_{101} = a_1 + 2^{51}a_1 - 2a_1 \Rightarrow a_{101} = 2^{51}a_1 - a_1$

$\Rightarrow a_{101} < 2^{51}a_1$ (4)

Clearly $b_{101} > a_{101}$

[from equations (3) and (4)]

17. (D)

Roots of the equation $px^2 + qx + r = 0$ with the given condition p, q, r, are in H.P. and p, r are different having the same sign.

If p, q, r are in H.P. then,

$$q = \frac{2pr}{p+r}$$

So, for equation $px^2 + qx + r = 0$

Discriminant (D) is $\Rightarrow q^2 - 4pr$ (1)

Put the value of q in equation (1)

$$\Rightarrow \left(\frac{2pr}{p+r}\right)^2 - 4pr$$

$$\Rightarrow \frac{4p^2r^2}{(p+r)^2} - 4pr$$

$$\Rightarrow 4pr\left[\frac{pr}{(p+r)^2} - 1\right]$$

$$\Rightarrow 4pr\left[\frac{pr(p+r)^2}{(p+r)^2}\right]$$

$$\Rightarrow \frac{4pr}{(p+r)^2}\left[pr - p^2 - r^2 - 2pr\right]$$

$$\Rightarrow \frac{4pr}{(p+r)^2}\left[-(p^2 + r^2 + pr)\right]$$

$$\Rightarrow -\frac{4pr}{(p+r)^2}\left[p^2 + r^2 + pr\right]$$

We have p and r of same sign.

So, 'pr' is +ve

So, $D < 0$

So, imaginary roots.

18. (D)

$\Rightarrow x^3 - ax^2 + bx - 16\sqrt{6} = 0$

$\{\alpha, \beta, \gamma$ are positive real roots$\}$

Here, we use (A.M $\geq$ G.M)

So, we have 3 numbers $\frac{4}{\alpha\beta}, \frac{3}{\beta\gamma}$ and $\frac{2}{\gamma\alpha}$

So; $$\frac{\frac{4}{\alpha\beta} + \frac{3}{\beta\gamma} + \frac{2}{\alpha\gamma}}{3} \geq \left(\frac{4}{\alpha\beta} \times \frac{3}{\beta\gamma} \times \frac{2}{\alpha\gamma}\right)^{1/3}$$

$$\Rightarrow \frac{\frac{4}{\alpha\beta} + \frac{3}{\beta\gamma} + \frac{2}{\alpha\gamma}}{3} \geq \left(\frac{24}{\alpha^2\beta^2\gamma^2}\right)^{1/3}$$

$\Rightarrow$ So, in $x^3 - x^2 + bx - 16\sqrt{6} = 0$

We have product of roots $(\alpha\beta\gamma)$

$$= \frac{-d}{a} = \left(\frac{-16\sqrt{6}}{1}\right) = 16\sqrt{6}$$

$$\Rightarrow \frac{\frac{4}{\alpha\beta} + \frac{3}{\beta\gamma} + \frac{2}{\alpha\gamma}}{3} \geq \left(\frac{24}{\left(16\sqrt{6}\right)^2}\right)^{1/3}$$

$$\frac{4}{\alpha\beta} + \frac{3}{\beta\gamma} + \frac{2}{\alpha\gamma} \geq 3\left(\frac{24}{16 \times 16 \times 6}\right)^{1/3}$$

$$\frac{4}{\alpha\beta} + \frac{3}{\beta\gamma} + \frac{2}{\alpha\gamma} \geq \frac{3}{4}$$

So minimum value is $\frac{3}{4}$

19. (A)

$\Rightarrow \sqrt{1 + 2(0.4) + 3(0.4)^2 + 4(0.4)^3 + \ldots\infty}$

$\Rightarrow$ we know that

$1 + 2(0.4) + 3(0.4)^2 + 4(0.4)^3 + \ldots\infty$ is form an A.G.P.

So, use the formula

$$1 + 2x + 3x^2 + 4x^3 + \ldots\infty = \frac{1}{(1-x)^2}$$

So, here $x = 0.4$

So, $$\sqrt{\frac{1}{(1-x)^2}} \Rightarrow \frac{1}{1-x} \Rightarrow \frac{1}{1-0.4} = \frac{1}{0.6} = \frac{5}{3}$$

20. (A)

$$\sum_{k=1}^{n}(K)^3 \Rightarrow 1^3 + 2^3 + 3^3 + \ldots + n^3$$

This is the sum of first 'n' natural numbers

So, $$\sum_{k=1}^{n}(K)^3 = \left(\frac{n(n+1)}{2}\right)^2$$

Let $n = 2; \left(\frac{2\times 3}{2}\right)^2 = 9$

$n = 3; \left(\frac{3\times 4}{2}\right)^2 = 36$

$n = 4; \left(\frac{4\times 5}{2}\right)^2 = 100$

$n = 5; \left(\frac{5\times 6}{2}\right)^2 = 225$

$n = 6; \left(\frac{6\times 7}{2}\right)^2 = 441$

So, it will not take '25'

21. **(D)**

Minimum value of $\frac{x^4+2x^2+101}{x^2+1}; x \in R$

So, $\frac{x^4+2x^2+1+100}{(x^2+1)}$

$\therefore \frac{(x^2+1)+100}{(x^2+1)}$

$\Rightarrow (x^2+1)+\frac{100}{(x^2+1)}$

Now we have to apply A.M. ≥ G.M. for (x^2+1) and $\frac{100}{(x^2+1)}$

So,

$$\frac{(x^2+1)+\frac{100}{(x^2+1)}}{2} \geq \left((x^2+1)\times\frac{100}{(x^2+1)}\right)^{1/2}$$

$$\Rightarrow \frac{(x^2+1)+\frac{100}{(x^2+1)}}{2} \geq (100)^{1/2}$$

$$\Rightarrow (x^2+1)+\frac{100}{(x^2+1)} \geq 2\times 10$$

$$\Rightarrow \frac{(x^2+1)+\frac{100}{(x^2+1)}}{2} \geq (100)^{1/2}$$

$$(x^2+1)+\frac{100}{(x^2+1)} \geq 20$$

22. **(C)**

$a = 159°$

$l = 177°$

Sum of the interior angles of a polygon $= 180°(n-2)$

$$S = \frac{n}{2}(a+l) = \frac{n}{2}(159°+177°)$$

$$180°(n-2) = \frac{n}{2}(336°)$$

$$360(n-2) = n(336)$$

$$360n - 720 = 336n$$

$$\therefore 24n = 720$$

$$n = 30$$

23. **(D)**

$$\sum_{K=1}^{\infty}\sum_{n=1}^{\infty}\frac{K}{2^{n+K}}$$

$$\Rightarrow \sum_{n=1}^{\infty}\frac{K}{2^{n+K}} \Rightarrow \frac{K}{2^K}\underbrace{\sum_{n=1}^{\infty}\frac{K}{2^{n+K}}}_{\text{Infinite (G.P.)}} \Rightarrow \frac{K}{2^K}\frac{\frac{1}{2}}{1=\frac{1}{2}} = \frac{K}{2^K}$$

$$\text{Now;} \Rightarrow \sum_{K=1}^{\infty}\frac{K}{2^K} \Rightarrow \underbrace{\frac{1}{2}+\frac{2}{2^2}+\frac{3}{2^3}+\ldots\infty}_{\text{A.G.P.}}$$

$$S = \frac{1}{2}+\frac{2}{2^2}+\frac{3}{2^3}+\frac{4}{2^4}+\ldots\infty$$

$$\frac{1}{2}S = \frac{1}{2^2}+\frac{2}{2^3}+\frac{3}{2^4}+\ldots$$

$$\frac{1}{2}S = \underbrace{\frac{1}{2}+\frac{1}{2^2}+\frac{1}{2^3}+\frac{1}{2^4}+\ldots}_{\text{Infinite G.P.}}$$

$$\frac{1}{2}S = \frac{\frac{1}{2}}{1-\frac{1}{2}} = 1 \Rightarrow S = 2$$

24. **(D)**

The given G.P. has first term unity

$\therefore a_1 = 1$

Hence the G.P. will become $1, r, r^2, \ldots$

As given $4a_2 + 5a_3$ is least

$\therefore 4r + 5r^2$ is least

We can say $f(r) = 4r + 5r^2$

Now the given function will be least if

$$\frac{(df(r))}{dr}=0$$

$$\therefore \frac{(df(r))}{dr}=\frac{d(4r+5r^2)}{dr}=0$$

$$\Rightarrow 4+10r=0$$

$$r=\frac{-2}{5}$$

$$a=1$$

$$r=\frac{-2}{5}$$

Sum of infinite G.P. is $\left(\frac{a}{1-r}\right)$

$$\therefore \frac{1}{1-\left(\frac{-2}{5}\right)}\Rightarrow\frac{1}{\left(\frac{7}{5}\right)}\Rightarrow\frac{5}{7}$$

25. **(C)**

Let roots be $\frac{a}{b}$, a, ar

then $\frac{a}{b}\times a\times ar=a^3=8$

$a=2$

Now $\frac{a}{r}+a+ar=7$ (1)

$\frac{a^2}{r}+a^2+a^2r=C$ (2)

$(a=2)$

From equation (1)

$\frac{2}{r}+2+2r=7$ (3)

From equation (2)

$\frac{4}{r}+4+4r=C$ (4)

$\therefore$ Multiply equation (3) and subtract from equation (4)

From (3) $\frac{4}{r}+4r+4r=14$

From (4) $-\frac{4}{r}+4r+4r=C$

$0=14-C$

$C=14$

Multiple Answer Type

1. **(B, D)**

If a, b, c are p^{th}, q^{th} and r^{th} terms of an A.P. then $\frac{b-c}{a-b}$ is always a rational number.

Hence, (B), (D)

2. **(A, B, C)**

Since H.M. of a and $c=b$

A.M. of a and $c=\frac{a+c}{2}$

$$\therefore \frac{a+c}{2}>b\Rightarrow a+c>2b$$

Also, $\frac{a^2+c^2}{2}>ac\Rightarrow a^2+c^2>2ac\left[ac>b^2\right]$

$$\Rightarrow a^2+c^2>2b^2$$

Hence, (A), (B), (C)

3. **(B, C)**

$$\underbrace{111\ldots1}_{91}\text{ times}=10^{90}+10^{89}+\ldots+10^2+10+1$$

$$=\frac{10^{91}-1}{10-1}=\frac{\left(10^7\right)^{13}-1}{10^7-1}\times\frac{10^7-1}{10-1}$$

Hence (B) and (C)

4. **(A, C, D)**

Let K be the c.d.,

$b=a+K, c=a+2K, d=a+3K$

Now, $ad-bc=a(a+3K)-(a+K)(a+2K)$

$=a^2+3aK-a^2-3aK-2K^2$

$=-2K^2<0$

$\Rightarrow ad<bc$

Again, $a+d=a+a+3K=2a+3K$

$b+c=2a+3K$

$\Rightarrow a+d=b+c$

Now, $(a+1)(d+1)-(b+1)(c+1)$

$=(a+1)(a+3K+1)-(a+K+1)(a+2K+1)$

$=(a+1)^2+3(a+1)K-(a+1)^2-(a+1)2K-(a+1)K-2K^2$

$=-2K^2<0$

$\Rightarrow (a+1)(d+1)<(b+1)(c+1)$

5. **(A, B, C, D)**

Let $b_2=b_1r, b_3=b_1r^2$

Since, $b_3>4b_2-3b_1$

$\Rightarrow b_1r^2>4b_1r-3b_1$

$\Rightarrow r^2 - 4r + 3 > 0$

$\Rightarrow r < 1$ or $r > 3$

Hence (A), (B), (C), (D)

6. (A, B, C, D)

(A) If $\frac{a}{b+c-a}, \frac{b}{c+a-b}, \frac{c}{a+b-c}$ are in H.P.

Then, $\frac{b+c-a}{a}, \frac{a+c-b}{b}, \frac{a+b-c}{c}$ are in A.P.

Adding 2 in each term

$\frac{b+c+a}{a}, \frac{a+c+b}{b}, \frac{a+b+c}{c}$ are in A.P.

$\Rightarrow \frac{1}{a}, \frac{1}{b}, \frac{1}{c}$ are in A.P.

$\Rightarrow a, b, c$ are in H.P.

(B) $\frac{2}{b} = \frac{1}{b-a} + \frac{1}{b-c}$

$\Rightarrow 2(b-a)(b-c) = b(2b-a-c)$

$\Rightarrow 2(b^2 - bc - ab + ac) = 2b^2 - b(a+c)$

$= -2b(a+c) + 2ac = -b(a+c)$

$\Rightarrow b = \frac{2ac}{a+c} \Rightarrow a, b, c$ are in H.P.

(C) If $a - \frac{b}{2}, \frac{b}{2}, c - \frac{b}{2}$ are in G.P.

$\Rightarrow \frac{b^2}{4} = \left(a - \frac{b}{2}\right)\left(c - \frac{b}{2}\right)$

$\Rightarrow \frac{b^2}{4} = ac - \frac{ab}{2} - \frac{bc}{2} + \frac{b^2}{4}$

$\Rightarrow b = \frac{2ac}{a+c}$

$\Rightarrow a, b, c$ are in H.P.

(D) If $\frac{a}{b+c}, \frac{b}{c+a}, \frac{c}{a+b}$ are in H.P.

$\Rightarrow \frac{b+c}{a}, \frac{c+a}{b}, \frac{a+b}{c}$ are in A.P.

$\Rightarrow \frac{b+c}{a} + 1, \frac{c+a}{b} + 1, \frac{a+b}{c} + 1$ are in A.P.

$\Rightarrow \frac{1}{a}, \frac{1}{b}, \frac{1}{c}$ are in A.P.

$\Rightarrow a, b, c$ are in H.P.

7. (A, B, C, D)

Since, $a(n) = 1 + \left(\frac{1}{2} + \frac{1}{3}\right) + \left(\frac{1}{4} + \frac{1}{5} + \frac{1}{6} + \frac{1}{7}\right) + \ldots + \left(\frac{1}{2^{n-1}} + \ldots + \frac{1}{2^n - 1}\right)$

$< 1 + 2\left(\frac{1}{2}\right) + 4\left(\frac{1}{4}\right) + \ldots$

$\Rightarrow a(n) < n$

Thus, [(A), (B), (C), (D)]

8. (B, D)

Consider the A.P. since a is equidistant from the first term α and last term β of the A.P., therefore, α, a, β are in A.P. $\Rightarrow a$ is the A.M. of α and β

$\therefore a = \frac{\alpha + \beta}{2}$

Similarly, b and c are the G.M. and H.M. of α and β respectively, then

$b = \sqrt{\alpha\beta}$ and $c = \frac{2\alpha\beta}{\alpha + \beta}$

$\therefore (\text{G.M.})^2 = (\text{A.M.})(\text{H.M.})$

$\Rightarrow b^2 = ac$

and A.M. $\geq$ G.M. $\geq$ H.M.

$\Rightarrow a \geq b \geq c$ [(B), (D)]

9. (A, B, C)

Since,

$$\frac{(b+c)+(c+a)+(a+b)}{3} \geq \frac{3}{\frac{1}{b+c} + \frac{1}{c+a} + \frac{1}{a+b}}$$

$$\Rightarrow 2(a+b+c)\left(\frac{1}{b+c} + \frac{1}{c+a} + \frac{1}{a+b}\right) \geq 9$$

Also,

$$\frac{a^2+1}{b+c} + \frac{b^2+1}{c+a} + \frac{c^2+1}{a+b} \geq 2\left(\frac{a}{b+c} + \frac{b}{c+a} + \frac{c}{a+b}\right)$$

$$= 2\left[(a+b+c)\left(\frac{1}{b+c} + \frac{1}{c+a} + \frac{1}{a+b} - 3\right)\right]$$

$$\Rightarrow \frac{a^2+1}{b+c} + \frac{b^2+1}{c+a} + \frac{c^2+1}{a+b} \geq 9 - 6 = 3$$

Hence (A), (B), (C)

10. (A, C, D)

Let $b = a + K, c = a + 2K, d = a + 3K$

Clearly, (A), (C), (D)

11. (C, D)

Since, $2b = a + c$ and $b^2 = \frac{2a^2c^2}{a^2 + c^2}$

$\Rightarrow \left(\frac{a+c}{2}\right)^2 = \frac{2a^2c^2}{a^2+c^2}$

$\Rightarrow 4\left(b^2 + \frac{1}{2}ac\right)(a-c)^2 = 0$

$\Rightarrow a - c = 0$ or $b^2 = -\frac{1}{2}ac$

if $a = c \Rightarrow a = b = c$

If $b^2 = -\frac{1}{2}ac \Rightarrow -\frac{a}{2}, b, c$ are in G.P.

12. (A, C, D)

Since, $n^2 < n^2 + 1 < n^2 + 2 < \ldots < n^2 + n$

$\Rightarrow \frac{1}{n^2+1} + \frac{2}{n^2+2} + \ldots + \frac{n}{n^2+n} <$

$\frac{1}{n^2} + \frac{2}{n^2} + \ldots + \frac{n}{n^2}$

$= \frac{1}{2} + \frac{1}{2n}$

and,

$\frac{1}{n^2+1} + \frac{2}{n^2+2} + \ldots + \frac{n}{n^2+n} >$

$\frac{1}{n^2+n} + \frac{2}{n^2+n} + \ldots + \frac{n}{n^2+n}$

$= \frac{1}{2}$. Hence, (A), (C), (D)

13. (A, C)

Since,

$s(n) = 1 + \left(\frac{1}{2} + \frac{1}{3}\right) + \left(\frac{1}{4} + \ldots + \frac{1}{7}\right)$

$+ \left(\frac{1}{8} + \ldots + \frac{1}{15}\right) + \ldots$

$+ \left(\frac{1}{2^{n-1}} + \frac{1}{2^{n-1}+1} + \ldots + \frac{1}{2^n - 1}\right)$

$< 1 + \frac{2}{2} + \frac{4}{4} + \ldots + \frac{2^{n-1}}{2^{n-1}} = n$

$\Rightarrow s(100) < 100$

Also, $s(200) > 100$

14. (B, C)

Let the roots are $a - d, a, a + d$

$\therefore a - d + a + a + d = 6 \Rightarrow a = 2$

Again the product of roots taken 2 at a time

$a(a - d) + a(a + d) + (a + d)(a - d) = -4$

$d^2 = 36 \Rightarrow d = 6$ or -6.

Hence the A.P.'s are

$-4, 2, 8, \ldots; S_{10} = 5[-8 + 9(6)]$

$= (46)(5) = 230 \Rightarrow$ (C)

or $8, 2, -4, \ldots; S_{10} = 5[16 + 9(-6)]$

$= (5)(-38) = -190 \Rightarrow$ (B)

15. (B, C)

Here $A_1 = \frac{3}{2}, A_2 = \frac{7}{4}, A_3 = \frac{15}{8}, A_4 = \frac{31}{16}$

Here, $A_{10} = \frac{2047}{1024}$ and $\sum_{n=1}^{10} A_n > 19$.

Verify alternative.

16. (A, C, D)

Consider number $\frac{a}{2}, \frac{a}{2}, \frac{b}{3}, \frac{b}{3}, \frac{b}{3}$

Apply A.M. ≥ G.M. $\frac{\frac{a}{2}+\frac{a}{2}+\frac{b}{3}+\frac{b}{3}+\frac{b}{3}}{5} \geq \left(\frac{a^2 \cdot b^3}{2^2 3^3}\right)^{\frac{1}{5}}$

$a + b \geq 5\left(\frac{4}{9(4)(27)}\right)^{\frac{1}{5}}$. Hence, $(a + b) \geq \frac{5}{3}$

17. (A, B, D)

$b = \frac{2ac}{a+c}$ and $b^2 = -2ac$

$b = \frac{-b}{a+c} \Rightarrow a + b + c = 0$

Now, check the options.

18. (B, C, D)

$X^2 + Y^2 = Z^2$

least pythogorean triplet (3, 4, 5)

$3^2 + 4^2 = 5^2$

9 + 16 = 25
(1 + 3 + 5) + (1 + 3 + 5 + 7)
= (1 + 3 + 5 + 7 + 9)
$a = 5, b = 7, c = 9$
$a + b + c$
least = 21

19. (A, B, C, D)

$a_1 + (a_1 + d), (a_1 + 2d), \ldots$

$b_1 + (b_1 + d_1), (b_1 + 2d_1), \ldots$

hence $a_{100} = a_1 + 99d$

$b_{100} = b_1 = 99d_1$

add

$a_{100} + b_{100} = 100 + 99(d + d_1)$

hence $d + d_1 = 0 \Rightarrow d = -d_1 \Rightarrow$ (A)
(B) and (C) are obviously true.

$$\sum_{r=1}^{100}(a_r + b_r) = \frac{100}{2}\,[(a_1 + b_1) + (a_{100} + b_{100})]$$

$$= \frac{100 \times 200}{2} = 10^4 \Rightarrow \text{(D)}$$

(using $S_n = \dfrac{n}{2}\,(a + d)$)

20. (A, C, D)

Roots are $\frac{a}{r}$, a, ar : where $a > 0$, $r > 1$. Now

$$\frac{a}{r} + a + ar = -p \qquad (1)$$

$$a \cdot \frac{a}{r} + a \cdot ar + ar \cdot \frac{a}{r} = q \qquad (2)$$

$$\frac{a}{r} \cdot a \cdot ar = 1 \qquad (3)$$

$a^3 = 1 \Rightarrow a = 1 \Rightarrow$ (C) is correct
From (1) putting $a = 1$ we get

$$\frac{1}{r} + 1 + r = -p \qquad (4)$$

$$\left(\sqrt{r} - \frac{1}{\sqrt{r}}\right)^2 + 3 = -p \Rightarrow -p > 3$$

$\Rightarrow p < -3 \Rightarrow$ (B) is not correct
from (2) putting $a = 1$ we get

$$\frac{1}{r} + r + 1 = q \qquad (5)$$

from (4) and (5) we have $-p = q$
$\Rightarrow p + q = 0 \Rightarrow$ (A) is correct
Now as, $r > 1$

$$\frac{a}{r} = \frac{1}{r} < 1$$

and $ar = r > 1 \Rightarrow$ (D) is correct

21. (A, B, C, D)

$S_n = n^2p \Rightarrow S_p = p^3 \Rightarrow$ (C)
$t_n = S_n - S_{n-1} = p[n^2 - (n-1)^2] = (2n - 1)p$
$t_1 = a_1 = p \Rightarrow$ (A)
$t_p = a_p = 2p^2 - p \Rightarrow$ (D)
common difference $a_2 - a_1 = 3p - p = 2p$
$\Rightarrow$ (B)

22. (A,B,C,D)

Let the series be 21, $21r$, $21r^2$, …

Sum = $\dfrac{21}{1-r}$ is a positive integer

also $21r$ is a positive integer

$S = \dfrac{(21)(21)}{21 - 21r}$ as $21\,r \in N$

hence $21 - 21r$ must be an integer
also $21r < 21$
Hence $21 - 21r$ may be equal to 1, 3, 7 or 9 (think!)
That is, must be a divisor of (21) (21)
hence $21 - 21r = 1$ or 3 or 7 or 9
$21r = 20$, 18, 14 or 12.

23. (B, C)

(A) As $\displaystyle\sum_{i=1}^{n} \frac{1}{H_i} = n\frac{1}{H} \Rightarrow$ (A) is incorrect.

(B) Obviously true.

(C) Continuous sum of n arithmetic mean between a and b is equals to n(single arithmic mean between a and b)

(D) We have $g_1 = 1, g_2 = r, g_3 = r^2$, where r is common ratio of G.P.
Now, $4g_2 + 5g_3 = 4r + 5r^2$

$$= 5\left(r^2 + \frac{4r}{5}\right) = 5\left(\left(r + \frac{2}{5}\right)^2 - \frac{4}{25}\right),$$

which is clearly minimum at $r = \dfrac{-2}{5}$

24. (A,C)

Given $a = 10$, $a_{a_2} = 100$; $a_{a_3} = ?$

Let d be the common difference.

$\therefore a_{a_2} = a + (a_2 - 1)d$

$= 10 + (a_2 - 1)d$

$100 = 10 + (10 + d - 1)d$

$90 = (9 + d)d$

$\therefore d^2 + 9d - 90 = 0$

$(d + 15)(d - 6) = 0$

$\therefore d = 6$

Now $a_3 = a + 2d = 10 + 12 = 22$

Now, $a_{a_3} = a_{22}$

$\therefore a_{22} = a + 21d$

$= 10 + (21)(6) = 10 + 126 = 136.$

25. (A, B, D)

If a, b, c, d are in H.P.

$\Rightarrow \frac{1}{a}, \frac{1}{b}, \frac{1}{c}, \frac{1}{d}$ are in A.P.

(A) Let $\sqrt{\frac{a+d}{b+c}} = p, \sqrt{\frac{ad}{bc}} = q$

$p + \frac{1}{p} = q + \frac{1}{q} \Rightarrow (p-q) = \frac{1}{q} - \frac{1}{p}$

$p - q = \frac{p-q}{pq}, p - q = 0 \text{ or } pq = 1$

$\Rightarrow p = q \text{ or } pq = 1$

$\Rightarrow \sqrt{\frac{a+d}{b+c}} = \sqrt{\frac{ad}{bc}}$

or $\left(\sqrt{\frac{a+d}{b+c}}\right)\left(\sqrt{\frac{ad}{bc}}\right) = 1$

$\Rightarrow \frac{a+d}{b+c} = \frac{ad}{bc}$ or $\frac{a+d}{b+c} = \frac{bc}{ad}$

$\Rightarrow \left(\frac{1}{d} + \frac{1}{a} = \frac{1}{b} + \frac{1}{c}\right)$

(B) $abc + bcd$

$= abd + acd \left(\frac{1}{d} + \frac{1}{a}\right) = \left(\frac{1}{c} + \frac{1}{b}\right)$

(C) $\frac{1}{a} + \frac{1}{c} = \frac{1}{b} + \frac{1}{d}$

(D) If $\frac{1}{a}, \frac{1}{b}, \frac{1}{c}, \frac{1}{d}$ are in A.P.

$\Rightarrow \frac{1}{a} \times \frac{1}{d} < \frac{1}{b} \times \frac{1}{c}$

$\frac{1}{a}+1, \frac{1}{b}+1, \frac{1}{c}+1, \frac{1}{d}+1$ are also in A.P.

$\Rightarrow \left(\frac{1}{a}+1\right)\left(\frac{1}{d}+1\right) < \left(\frac{1}{b}+1\right)\left(\frac{1}{c}+1\right)$

Comprehension Based Questions

1. **(C)**
2. **(B)**
3. **(C)**

Let the four integers be $a-d, a, a+d$ and $a+2d$ where a and d are integers and $d > 0$

Since, $a + 2d = (a-d)^2 + a^2 + (a+d)^2$

$\Rightarrow 2d^2 - 2d + 3a^2 - a = 0$

$\therefore d = \frac{1}{2}\left(1 \pm \sqrt{1 + 2a - 6a^2}\right)$

As d is a positive integer

$1 + 2a - 6a^2 > 0$

$\Rightarrow \frac{1-\sqrt{7}}{6} < a < \frac{1+\sqrt{7}}{6}$

Since, a is integer $\Rightarrow a = 0$

$\therefore d = 1$

$\therefore$ The four numbers are $-1, 0, 1, 2$

4. **(B)**
5. **(B)**
6. **(A)**

(iv) $G_1 \cdot G_2 \cdot \ldots G_n = \left(\sqrt{1 \times 1024}\right)^n = 2^{5n}$

$\Rightarrow 2^{5n} = 2^{45} \Rightarrow n = 9$

(v) $\frac{m(-2+1027)}{2} = 1025 \times 171$

$\Rightarrow m = 342$

(vi) Since, $n = 9$

$\therefore r = (1024)^{\frac{1}{9+1}} = 2$

$\therefore G_1 = 2, r = 2$

7. **(B)**
8. **(A)**

9. (D)

Given $P(x) = 3x^2 - x + \dfrac{25}{12}$

$$= 3\left[x^2 - \frac{1}{3}x + \frac{25}{36}\right] = 3\left(\left(x - \frac{1}{6}\right)^2 + \frac{2}{3}\right)$$

$\Rightarrow P(x)|_{\min} = 2$, occurs at $x = \dfrac{1}{6}$.

Given, $\dfrac{a}{1-r} = 2$ (1)

and $a = r^2$ (2)

$\therefore$ Solving (1) and (2), we get

$$\frac{r^2}{1-r} = 2 \Rightarrow r^2 = 2 - 2r \Rightarrow r^2 + 2r = 2$$

$\therefore r = (\sqrt{3} - 1)$ [As $r \in (0, 1)$]

$\Rightarrow a = r^2 = 4 - 2\sqrt{3}$

Let $f(x) = x^2 + (k + 1)x + a + 2\sqrt{3}$

Now, $f(x) = 0 \Rightarrow x^2 + (k + 1)x + 4 = 0$.

Since 2 lies between the roots, hence $f(2) < 0$

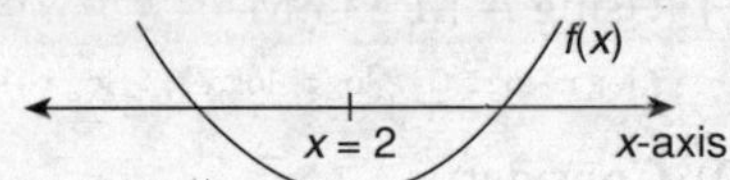

$\Rightarrow 4 + 2k + 2 + 4 < 0$

$\Rightarrow 2k + 10 < 0$

$\Rightarrow k < -5$

So, $k_{\text{max.}} = -6$

Let $y = \dfrac{P(x)}{P'(x)} = \dfrac{3x^2 - x + \frac{25}{12}}{6x - 1}$

$\Rightarrow 3x^2 - x + \dfrac{25}{12} = 6xy - y$

$\Rightarrow 3x^2 - x(1 + 6y) + \dfrac{25}{12} + y = 0$

As, $x \in R$ so $D \geq 0$

$\Rightarrow (6y + 1)^2 - 12\left(\dfrac{25 + 12y}{12}\right) \geq 0$

$\Rightarrow 36y^2 + 12y + 1 - 25 - 12y \geq 0$

$\Rightarrow 36y^2 - 24 \geq 0$

$\Rightarrow 3y^2 - 2 \geq 0$

Range of $y = \left(-\infty, -\sqrt{\dfrac{2}{3}}\right] \cup \left[\sqrt{\dfrac{2}{3}}, \infty\right)$

$-\sqrt{2/3}$ $\sqrt{2/3}$

$= (-\infty, p] \cup [q, \infty)$

So, $p = -\sqrt{\dfrac{2}{3}}$ and $q = \sqrt{\dfrac{2}{3}}$

Hence, $(p + q) = 0$

10. (C)

11. (A)

12. (B)

α, β, γ are in H.P., hence

(i) $\dfrac{1}{\alpha}, \dfrac{1}{\beta}, \dfrac{1}{\gamma}$ are in A.P.

$$\frac{1}{\alpha} + \frac{1}{\gamma} = \frac{2}{\beta}$$

$$\Rightarrow \beta = \frac{2\alpha\gamma}{\alpha + \gamma}; \frac{1}{\alpha} + \frac{1}{\gamma} + \frac{1}{\beta} = \frac{3}{\beta}$$

$$= \frac{\Sigma\alpha\beta}{\alpha\beta\gamma} = \frac{q}{q} = 1$$

$\beta = 3$ which is a root.

$\Rightarrow 27 + 9p + 3q - q = 0$

$\Rightarrow 9p + 2q + 27 = 0$

$\Rightarrow 9p + 2q = -27$

(ii) $2\left(\dfrac{1}{\alpha^2} + \dfrac{1}{\beta^2} + \dfrac{1}{\gamma^2}\right) \geq 2\left(\dfrac{1}{\alpha\beta} + \dfrac{1}{\beta\gamma} + \dfrac{1}{\gamma\alpha}\right)$

Using ($a^2 + b^2 + c^2 \geq ab + bc + ca$ with $a = \dfrac{1}{\alpha}$, $b = \dfrac{1}{\beta}$ and $c = \dfrac{1}{\gamma^2}$)

$$\Rightarrow 3\left(\frac{1}{\alpha^2} + \frac{1}{\beta^2} + \frac{1}{\gamma^2}\right) \geq \left(\frac{1}{\alpha} + \frac{1}{\beta} + \frac{1}{\gamma}\right)^2 = 1$$

(add $\dfrac{1}{\alpha^2} + \dfrac{1}{\beta^2} + \dfrac{1}{\gamma^2}$ both sides)

$$\frac{1}{\alpha^2} + \frac{1}{\beta^2} + \frac{1}{\gamma^2} \geq \frac{1}{3}$$

(iii) for equality, $\alpha = \beta = \gamma = 3$

$\Rightarrow \Sigma\alpha\beta = 27$ and $\alpha\beta\gamma = 27$

$$\frac{1}{\alpha\beta} + \frac{1}{\beta\gamma} + \frac{1}{\gamma\alpha} \leq \frac{1}{\alpha^2} + \frac{1}{\beta^2} + \frac{1}{\gamma^2}$$

$(ab + bc + ca \le a^2 + b^2 + c^2)$

$$3\left(\frac{1}{\alpha\beta}+\frac{1}{\beta\gamma}+\frac{1}{\gamma\alpha}\right) \le \left(\frac{1}{\alpha}+\frac{1}{\beta}+\frac{1}{\gamma}\right)^2 = 1$$

$$\left(\text{adding } 2\left(\frac{1}{\alpha\beta}+\frac{1}{\beta\gamma}+\frac{1}{\gamma\alpha}\right) \text{to both sides}\right)$$

$$\left(\frac{\alpha+\beta+\gamma}{\alpha\beta\gamma} \le \frac{1}{3}\right)$$

$$\Rightarrow -\frac{p}{q} \,{}'' \frac{1}{3} \Rightarrow \frac{p}{q} \ge -\frac{1}{3}.$$

13. (C)

14. (D)

15. (A)

1, $\log_y x$, $\log_z y$, $-15 \log_x z$ in A.P.

$\text{loy}_y x = 1 + d$

$\log_z y = 1 + 2d$

$-15\log_x z = 1 + 3d$

$\therefore -15\log_y x \log_z y \log_x z$

$= (1 + d)(1 + 2d)(1 + 3d)$

$\therefore -15 = (1 + d)(1 + 2d)(1 + 3d)$

$-15 = (1 + 3d + 2d^2)(1 + 3d)$

(i) $6d^3 + 11d^2 + 6d + 16 = 0$

(ii) Put $d = -2$

$6(-8) + 11(4) + 6(-2) + 16 = 0$

$\Rightarrow d = -2$ is a root

$6d^2(d + 2) - d(d + 2) + 8(d + 2)$

$$(d+2)\underbrace{\left(6d^2-d+8\right)}_{\text{no real root}}=0$$

$\Rightarrow d = -2$

$\log_y x = 1 - 2 = -1$

$x = y^{-1} \Rightarrow xy = 1$

$\log_z y = -3 \Rightarrow z^{-3} = y;\ y^3 = 1$

$-15 \log_x z = -5 \Rightarrow \log_x z = \frac{1}{3}$

$\Rightarrow z = x^{\frac{1}{3}} \Rightarrow z^3 = x \Rightarrow xz^{-3} = 1$

$\Rightarrow xz^{-3} + xy + yz^3 = 3$

(iii) Sequence is 1, –1, –3, –5, –7, …

$$S_n = \frac{25}{2}[2+(25-1)(-2)] = 25\ [1 - 24]$$

$= -25 \times 23 = -575$

absolute value = 575

Match Matrix

1. (A – R; B – P; C – Q; D – P)

(A) $2\log_{100} a + \log_a 10^4$

$$= \log_{10} a + \frac{4}{\log_{10} a} \ge \sqrt{\log_{10} a \frac{4}{\log_{10} a}}$$

$$\Rightarrow \log_{10} a + \frac{4}{\log_{10} a} \ge 4$$

$\Rightarrow$ least value = 4

(B) $$\frac{ab^2+ac^2+bc^2+ba^2+ca^2+cb^2}{6} \ge \left(a^6b^6c^6\right)^{\frac{1}{6}}$$

$$\Rightarrow a\left(b^2+c^2\right)+b\left(c^2+a^2\right)+ c\left(a^2+b^2\right) \ge 6abc$$

$\Rightarrow \lambda = 6$

(C) Using A.M $\ge$ G.M

$$x^{\log y-\log z} + y^{\log z-\log x} + z^{\log x-\log y} \ge 3$$

(D) Consider

$$\frac{a}{2},\frac{a}{2},\frac{b}{3},\frac{b}{3},\frac{b}{3},\frac{c}{4},\frac{c}{4},\frac{c}{4},\frac{c}{4}$$

$$\frac{\frac{a}{2}+\frac{a}{2}+\frac{b}{3}+\frac{b}{3}+\frac{b}{3}+\frac{c}{4}+\frac{c}{4}+\frac{c}{4}+\frac{c}{4}}{9} \ge \left(\frac{a^2}{4}\frac{b^3}{27}\frac{c^4}{256}\right)^{\frac{1}{9}}$$

$$\Rightarrow \frac{18}{9} \ge \left(\frac{a^2b^3c^4}{4\times27\times256}\right)^{\frac{1}{9}}$$

$$\Rightarrow \frac{a^2b^3c^4}{4\times27\times256} \le 2^9$$

$a^2b^3c^4 \le 2^9 2^2 2^8 3^3$

$a^2b^3c^4 \le 2^{19} 3^3$

Maximum value $= 2^{19} \cdot 3^3$

$\therefore \lambda_1\lambda_2 = 6$

2. (A – R; B – P; C – S; D – Q)

(A) $F(n+1)=\frac{2F(n)+1}{2}=F(n)+\frac{1}{2}$

$\therefore F(1),F(2),F(3),\ldots$ are in A.P. with

c.d. $\cdot\frac{1}{2}$

$\therefore F(101)=52$

(B) Since,

$a_1+2d+a_1+4d+a_1+10d+$
$a_1+16d+a_1+18d=10$

$\Rightarrow 5(a_1+10d)=10 \Rightarrow a_1+10d=2$

Now,

$$\sum_{i=1}^{21} a_i=\frac{21}{2}[2a_1+20d]=21(a_1+10d)=42$$

(C) $s=1+5+13+29+\ldots+t_{10}$

$s=1+5+13+\ldots+t_9+t_{10}$

Subtracting

$t_{10}=1+4+8+16+\ldots$ up to 10 terms

$=1+(4+8+16+\ldots$up to 9 terms$)$

$=2045$

(D) Sum of all two-digit numbers

$=\frac{90}{2}(10+99)=(45)(109)$

Sum of all two-digit numbers divisible by 2

$=\frac{45}{2}(10+98)=(45)(54)$

Sum of all two-digit numbers divisible by 3

$=\frac{30}{2}(12+99)=15(111)$

Sum of all two-digit numbers divisible by 6

$=\frac{15}{2}(12+96)=15(54)$

Therefore, the required sum

$=1620$

3. (A – R; B – Q; C – Q)

(A) $a_1, a_2, a_3, \ldots a_n, \ldots$

$a_1=\frac{5}{2}$; $a_{10}=16$

$\frac{5}{2}+9d=16 \Rightarrow 9d=\frac{27}{2} \Rightarrow d=\frac{3}{2}$

given $a_1+a_2+\ldots+a_n=110$

$\frac{n}{2}[2a+(n-1)d]=110$

$\Rightarrow \frac{n}{2}\left[5+(n-1)\frac{3}{2}\right]=110$

$\Rightarrow n[10+3n-3]=440$

$\Rightarrow n(3n+7)=440 \Rightarrow 3n^2+7n-440=0$

$\Rightarrow 3n^2+40n-33n-440=0$

$\Rightarrow n(3n+40)-11(3n+40)=0$

$\Rightarrow n=11$

(B) $a, a+d, \ldots, a+8d$

$\therefore 9a+36d=7(180°)$

or $a+4d=140°$

Also, $a+8d<180°$

$\therefore 4d<40°, d<10°$

That is, the common difference is less than 10° Ans $\Rightarrow$ (Q)

(C) $a_1, a_2, a_3, \ldots, a_n, \ldots$ is increasing G.P.

Let $a_1=a$; $a_2=ar$, etc $r>1$

$a_6=4a_4$; $ar^5=4ar^3$

$\Rightarrow r^2=4 \Rightarrow r=2$

($\because$ it is an increasing G.P.)

$\because a_9-a_7=192$

$\therefore ar^8-ar^6=192$ that is, $256a-64a=192$

$\Rightarrow 192a=192 \Rightarrow a=1$

$$\sum_{i=4}^{n} a_i=1016$$

$\Rightarrow (a_1+a_2+a_3+\ldots+a_n)-$
$(a_1+a_2+a_3)=1016$

$\Rightarrow \frac{1(2^n-1)}{2-1}=1016+1+2+4=1023$

$\therefore 2^n=1024 \Rightarrow n=10 \Rightarrow$ (Q)

4. (A – P; B – T; C – R; D – Q)

(A) $S_{2n}=S_n$

$=\frac{2n}{2}(2a+(2n-1)d)-\frac{n}{2}(2a+(n-1)d)$

$=\frac{n}{2}[2a+(3n-1)d]$

$=\frac{1}{3}\frac{3n}{2}[2a+(3n-1)d]=\frac{1}{3}S_{3n}]$

(B) The sum equals $\frac{n(n+1)(n+2)}{6} = 220$

which is true for $n = 10$

(C) Let first terms and common ratio be a and r respectively.

Now,

$$\sum_{n=1}^{100} g_{2n} = ar + ar^3 + ar^5 + \ldots + ar^{199} = \frac{10}{3} \quad (1)$$

Also,

$$\sum_{n=1}^{100} g_{2n-1} = a + a^2 + ar^4 + \ldots + ar^{198} = \frac{5}{9} \quad (2)$$

Now,

$$\frac{\text{Equation (1)}}{\text{Equation (2)}} \Rightarrow \frac{r\left(a+ar^2+\ldots+ar^{198}\right)}{\left(a+ar^2+\ldots+ar^{198}\right)}$$

$$= \frac{\frac{10}{3}}{\frac{5}{9}} \Rightarrow r = 6.$$

(D) Let a and r be first term and common ratio of G.P.

Also, total number of terms $= 2n$

so, G.P. is $a, ar, ar^2, ar^3, ar^4, \ldots, ar^{2n-1}$

$$\therefore S_n = \frac{a\left(r^{2n}-1\right)}{(r-1)} = \text{sum of all terms}$$

Also, sum of terms occupy the odd places

$$= \frac{a\left(r^{2n}-1\right)}{\left(r^2-1\right)}$$

Now, given that

$$\frac{a\left(r^{2n}-1\right)}{(r-1)} = \frac{5a\left(r^{2n}-1\right)}{(r-1)(r+1)}$$

$\Rightarrow r + 1 = 5$

$\therefore r = 4$

5. (A – Q, R, S; B – R; C – P, R, S; D – R, S)

(A) ≥ 4

(B) Given $\log_b\left(a^{\log_2 b}\right) = \log_a\left(b^{\log_4 a}\right)$

or $\log_2 b \cdot \log_b a = \log_4 a \cdot \log_a b$

$2\log_2 b \cdot \log_b a = \log_2 a \cdot \log_a b$

$2\log_2 a = \log_2 b \Rightarrow a^2 = b$

Now $p - (b-9)^2 = a^2 = b$

$p - (b^2 + 81 - 18b) = b$

$p = b^2 + 81 - 17b$

$$= \left(b - \frac{17}{2}\right)^2 + 81 - \frac{289}{4}$$

$$= \left(b - \frac{17}{2}\right)^2 + \frac{35}{4}$$

Hence $p_{\min} = \frac{35}{4} = 8.75$

$\Rightarrow$ minimum integral value $= 9$

(C) $\frac{(x-6)(x-3)}{x+2} \geq 0$

$$\frac{x-5}{x+1} - 3 \leq 0$$

$$\frac{x-5-3x-3}{x+1} \leq 0$$

$$\frac{-2x-8}{x+1} \leq 0$$

$$\frac{x+4}{x+1} \geq 0$$

$\Rightarrow$ common solution is $(-1, 3] \cup [6, \infty)$

(D) $a = 3^{1+x} + 3^{1-x} + 9^x + 5^x$

$3(3^x + 3^{-x}) + (9^x + 9^{-x}) \geq 6 + 3 \geq 9$

Subjective Answer Type: Hard Nut

1. $(p = -2, q = -32)$

$\alpha + \beta = 1, \alpha\beta = p$

and, $\gamma + \delta = 4$, $\gamma\delta = q$

Now, $\frac{\alpha}{\beta} = \frac{\gamma}{\delta} \Rightarrow \frac{\alpha+\beta}{\gamma+\delta} = \frac{\beta}{\delta}$

$$\Rightarrow \frac{(\alpha+\beta)^2}{(\gamma+\delta)^2} = \frac{\alpha\beta}{\gamma\delta} \Rightarrow \frac{p}{q} = 16$$

$\Rightarrow p = -2,\ q = -32$

2. Let three distinct real numbers be

$$\frac{\alpha}{r}, \alpha, \alpha r$$

Now, $\frac{\alpha^2}{r^2} + \alpha^2 + \alpha^2 r^2 = S^2$

$$\Rightarrow \alpha^2\left(\frac{1+r^2+r^4}{r^2}\right)=S^2 \quad (1)$$

Again, $\dfrac{\alpha\left(1+r+r^2\right)}{r}=aS$

$$\Rightarrow \frac{\alpha^2\left(1+r+r^2\right)^2}{r^2}=a^2s^2 \quad (2)$$

Dividing (1) by (2)

$$\frac{1-r+r^2}{1+r+r^2}=\frac{1}{a^2}$$

$$\Rightarrow \left(a^2-1\right)r^2-\left(a^2+1\right)r+\left(a^2-1\right)=0$$

$$\Rightarrow r^2+\frac{1+a^2}{1-a^2}r+1=0$$

$$\Rightarrow \left(\frac{1+a^2}{1-a^2}\right)^2-4\geq 0$$

$$\Rightarrow 3a^4-10a^2+3\leq 0$$

$$\Rightarrow \left(3a^2-1\right)\left(a^2-3\right)\leq 0$$

$$\Rightarrow \frac{1}{3}\leq a^2\leq 3 \Rightarrow a^2\in\left[\frac{1}{3},1\right)\cup(1,3]$$

3. **(4)**

$$\frac{b}{a}=\frac{c}{b}=\text{integer}$$

$$b^2=ac\Rightarrow c=\frac{b^2}{a}$$

Also, $\dfrac{a+b+c}{3}=b+2$

$$\Rightarrow a+b+c=3b+6$$

$$\Rightarrow a-2b+\frac{b^2}{a}=6$$

$$\Rightarrow 1-\frac{2b}{a}+\frac{b^2}{a^2}=\frac{6}{a}$$

$$\Rightarrow \left(\frac{b}{a}-1\right)^2=\frac{6}{a} \qquad [\frac{b}{a}\text{ is integer}]$$

$$\Rightarrow a=6 \text{ only}$$

$$\therefore \frac{a^2+a-14}{a+1}=\frac{42-14}{7}=\frac{28}{7}=4$$

4. **(7)**

Total number of runs scored $=\sum_{K=1}^{n}k2^{n-K+1}$

$$=2^{n+1}\sum_{K=1}^{n}\frac{k}{2^K}$$

$$=2^{n+1}\left[1\cdot\frac{1}{2}+2\cdot\frac{1}{2^2}+3\cdot\frac{1}{2^3}+\ldots+n\frac{1}{2^n}\right] \quad (1)$$

Let

$$S=1\cdot\frac{1}{2}+2\cdot\frac{1}{2^2}+3\cdot\frac{1}{2^3}+\ldots+(n-1)\left(\frac{1}{2}\right)^{n-1}+n\left(\frac{1}{2}\right)^n$$

$$\Rightarrow \frac{1}{2}S=1\cdot\left(\frac{1}{2}\right)^2+2\cdot\left(\frac{1}{2}\right)^3+\ldots+(n-1)\left(\frac{1}{2}\right)^n+n\left(\frac{1}{2}\right)^{n+1}$$

Subtracting, we get

$$\frac{1}{2}S=\frac{1}{2}+\left(\frac{1}{2}\right)^2+\left(\frac{1}{2}\right)^3+\ldots+\left(\frac{1}{2}\right)^n-n\left(\frac{1}{2}\right)^{n+1}$$

$$\Rightarrow \frac{1}{2}S=\frac{\frac{1}{2}\left[1-\frac{1}{2^n}\right]}{1-\frac{1}{2}}-n\left(\frac{1}{2}\right)^{n+1}$$

$$\Rightarrow \frac{1}{2}S=1-\frac{1}{2^n}-n\left(\frac{1}{2}\right)^{n+1}$$

$$\Rightarrow S=2\left(1-\frac{1}{2^n}\right)-n\left(\frac{1}{2}\right)^n$$

Hence, from (1), the total number of runs scored

$$=2^{n+1}\left(\frac{2^{n+1}-n-2}{2^n}\right)$$

$$=2\left[2^{n+1}-n-2\right]$$

From question,

$$\frac{n+1}{4}\left[2^{n+1}-n-2\right]=2\left[2^{n+1}-n-2\right]$$

$$\Rightarrow n+1=8\Rightarrow n=7$$

5. $\frac{1}{2}\left(\frac{1}{2}-\frac{1}{2^{n+1}(n+1)}\right)$

$\frac{3}{2^3}+\frac{4}{2^4\cdot 3}+\frac{5}{2^6\cdot 3}+\frac{6}{2^7\cdot 5}+\ldots$ to n terms

$$T_r=\frac{(r+2)}{2^{r+1}r(r+1)}$$

$$=\frac{1}{2^{r+1}}\left[\frac{1}{r}-\frac{1}{2(r+1)}\right]$$

$$=\frac{1}{2}\left[\frac{1}{2^r\cdot r}-\frac{1}{2^{r+1}(r+1)}\right]$$

Put $1,2,3,\ldots,n$

$$T_1=\frac{1}{2}\left[\frac{1}{2^1\cdot 1}-\frac{1}{2^2\cdot 2}\right]$$

$$T_2=\frac{1}{2}\left[\frac{1}{2^2\cdot 2}-\frac{1}{2^3\cdot 3}\right]$$

$$T_n=\frac{1}{2}\left[\frac{1}{2^n\cdot n}-\frac{1}{2^{n+1}(n+1)}\right]$$

Adding

$$T_1+T_2+\ldots+T_n=\frac{1}{2}\left[\frac{1}{2}-\frac{1}{2^{n+1}(n+1)}\right]$$

6. $(1-\frac{1}{4^n(n+1)})$

$$T_r=\frac{3r+4}{4^r\cdot r(r+1)}$$

$$=\frac{4(r+1)-r}{4^r r(r+1)}$$

$$=\frac{1}{4^{r-1}r}-\frac{1}{4^r(r+1)}$$

Put $r=1,2,3,\ldots,n$

$$T_1=\frac{1}{4^0\cdot 1}-\frac{1}{4^1(2)}$$

$$T_2=\frac{1}{4^1(2)}-\frac{1}{4^2(3)}$$

$$T_n=\frac{1}{4^{n-1}(n)}-\frac{1}{4^n(n+1)}$$

Adding

$$T_1+T_2+\ldots T_n=1-\frac{1}{4^n(n+1)}$$

7. **(0)**

$a_K=2a_{K-1}-a_{K-2}\Rightarrow a_1,a_2,\ldots,a_{11}$ are in A.P.

Now, $\frac{a_1^2+a_2^2+\ldots+a_{11}^2}{11}=90$

$$\Rightarrow\frac{(15)^2+(15+d)^2+\ldots+(15+10d)^2}{11}=90$$

$$\Rightarrow\frac{(15)^2\times 11+2\times 15d(1+2+\ldots+10)+d^2(1^2+2^2+\ldots+10^2)}{11}=90$$

$$\Rightarrow 35d^2+150d+135=0$$

$$\Rightarrow 7d^2+30d+27=0$$

$$\Rightarrow 7d^2+21d+9d+27=0$$

$$\Rightarrow(d+3)(7d+9)=0$$

$$\Rightarrow d=-3,\frac{-9}{7}$$

Given, $a_2<\frac{27}{2}\Rightarrow d=-3$

Therefore, $\frac{a_1+a_2+\ldots+a_{11}}{11}$

$$=\frac{\frac{11}{2}[30-10\times 3]}{11}=0$$

8. **(Composite)**

$$\underbrace{1111\ldots11}_{91\text{ terms}}$$

$$=10^{90}+10^{89}+\ldots+10^2+10+1$$

$$=1\frac{[10^{91}-1]}{(10-1)}$$

$$=\frac{(10^{13})^7-1}{(10^{13}-1)}\times\frac{(10^{13}-1)}{(10-1)}$$

$$=\left\{1+(10)^{13}+\ldots+(10^{13})^6\right\}\left\{1+10+\ldots+10^{12}\right\}$$

$\Rightarrow$ It is a composite number.

9.

Let us consider

$1, 2, 2^2, \ldots, 2^{n-1}$

Now, $\left(1\cdot 2\cdot 2^2 \ldots 2^{n-1}\right)^{\frac{1}{n}}$

$$\Rightarrow \frac{1\left(2^n-1\right)}{(2-1)n} > \left(2^{1+2+\ldots+(n-1)}\right)^{\frac{1}{n}}$$

$$\Rightarrow \frac{2^n-1}{n} > \left(2^{\frac{(n-1)n}{2}}\right)^{\frac{1}{n}}$$

$$\Rightarrow \frac{2^n-1}{n} > \sqrt{2^{n-1}}$$

$$\Rightarrow 2^n > 1 + n\left(2^{n-1}\right)$$

10. A.M. of a and $c = \frac{a+c}{2}$

G.M. of a and $c = \sqrt{ac}$

H.M. of a and $c = b$

Also, A.M. of b and $d = \frac{b+d}{2}$

G.M of b and $d = \sqrt{bd}$

and, H.M of b and $d = c$

(i) A.M. > H.M.

$\Rightarrow \frac{a+c}{2} > b$ and $\frac{b+d}{2} > c$

$\Rightarrow a+c+b+d > 2b+2c$

$\Rightarrow a+d > b+c$

(ii) G.M. > H.M.

$\Rightarrow \sqrt{ac} > b$ and $\sqrt{bd} > c$

$\Rightarrow acbd > b^2c^2 \Rightarrow ad > bc$

11. (100.(101)!)

$a_K = \left\{K(K+1)-(K-1)\right\}K!$

$= K(K+1)! - (K-1)K!$

Put $K = 1, 2, \ldots, 100$

$a_1 = 1\cdot 2! - 0.1!$

$a_2 = 2\cdot 3! - 1.2!$

$a_{100} = 100(101)! - 99(100)!$

$a_1 + a_2 + \ldots + a_{100} = 100(101)!$

12. **(1)**

Let

$$S = \sum_{r=1}^{1999} \sqrt{1 + \frac{1}{r^2} + \frac{1}{(r+1)^2}}$$

$$= \sum_{r=1}^{1999} \sqrt{\frac{r^4 + 2r^3 + 3r^2 + 2r + 1}{r^2(r+1)^2}}$$

$$= \sum_{r=1}^{1999} \frac{r^2+r+1}{r(r+1)} = \sum_{r=1}^{1999}\left(1 + \frac{1}{r(r+1)}\right)$$

$$= 1999 + \sum_{r=1}^{1999}\left(\frac{1}{r} - \frac{1}{r+1}\right)$$

$$= 1999 + \left(1 - \frac{1}{2000}\right) = 1999 + \frac{1999}{2000}$$

$\therefore p = 1999$ and $q = 2000$

$\therefore |p-q| = 1$

13. $(\frac{15}{16})$

$$\sum_{r=1}^{\infty} \frac{1}{(2r-1)^4} = \frac{1}{1^4} + \frac{1}{3^4} + \frac{1}{5^4} + \ldots \text{to } \infty$$

$$= \left(\frac{1}{1^4} + \frac{1}{2^4} + \frac{1}{3^4} + \ldots\right) - \left(\frac{1}{2^4} + \frac{1}{4^4} + \frac{1}{6^4} + \ldots\right)$$

$$= 1 - \frac{1}{2^4}\left(\frac{1}{1^4} + \frac{1}{2^4} + \frac{1}{3^4} + \ldots \text{to} \infty\right)$$

$$= 1 - \frac{1}{16}(1) = \frac{15}{16}$$

14. $\frac{2(n+1)! - (n+2)}{2(n+2)}$

$$T_r = \frac{\left(r^2+r-1\right)\cdot r!}{\left(r^2+3r+2\right)}$$

$$= \frac{\left[(r+1)^2 - (r+2)\right]r!}{(r+1)(r+2)}$$

$$T_r = \frac{(r+1)!}{(r+2)} - \frac{r!}{(r+1)}$$

$$\Rightarrow T_r = v_r - v_{r-1} \text{ where } v_r = \frac{(r+1)!}{(r+2)}$$

Put $r = 1, 2, 3, \ldots, n$

$T_1 = v_1 - v_0$
$T_2 = v_2 - v_1$
$T_3 = v_3 - v_2$

$T_n = v_n - v_{n-1}$

Adding

$S_n = v_n - v_0$

$$\frac{(n+1)!}{(n+2)} - \frac{1}{2} = \frac{2(n+1)! - (n+2)}{2(n+2)}$$

15. $$\frac{\left(a+\frac{1}{a}\right)^2 + \left(b+\frac{1}{b}\right)^2}{2} \geq \left[\left(a+\frac{1}{a}\right)^2\left(b+\frac{1}{b}\right)^2\right]^{\frac{1}{2}}$$

$$= \left(a+\frac{1}{a}\right)\left(b+\frac{1}{b}\right) \quad (1)$$

Since, $\left(a+\frac{1}{a}\right)\left(b+\frac{1}{b}\right) = ab + \frac{a}{b} + \frac{b}{a} + \frac{1}{ab}$

$$= ab + \frac{1}{ab} + \frac{(a+b)^2 - 2ab}{ab}$$

$$= ab + \frac{16}{ab} - 2 \geq 6[\text{A.M.} \geq \text{G.M.}]$$

Therefore, $\left(a+\frac{1}{a}\right)\left(b+\frac{1}{b}\right) \geq 6$

Hence from (1)

$$\left(a+\frac{1}{a}\right)^2 + \left(b+\frac{1}{b}\right)^2 \geq 12$$

Hence the minimum value of the given expression is 12.

16. $(\frac{1}{2^n})$

Let $S = \sum_{r=0}^{n}\left(\frac{n-3r+1}{n-r+1}\right)\frac{{}^nC_r}{2^r}$

$$= \sum_{r=0}^{n}\left(1 - \frac{2r}{(n-r+1)}\right)\frac{{}^nC_r}{2^r}$$

$$\sum_{r=0}^{n}\frac{{}^nC_r}{2^r} - \sum_{r=0}^{n}\frac{2r}{(n-r+1)} \cdot \frac{n!}{r!(n-r)!2^r}$$

$$\sum_{r=0}^{n}\frac{{}^nC_r}{2^r} - \sum_{r=0}^{n}\frac{{}^nC_{r-1}}{2^{r-1}} = \frac{1}{2^n}$$

17. $(\frac{256}{75})$

Area of ΔABC

= Area of ΔBCP + Area of ΔCAP + Area of ΔABP

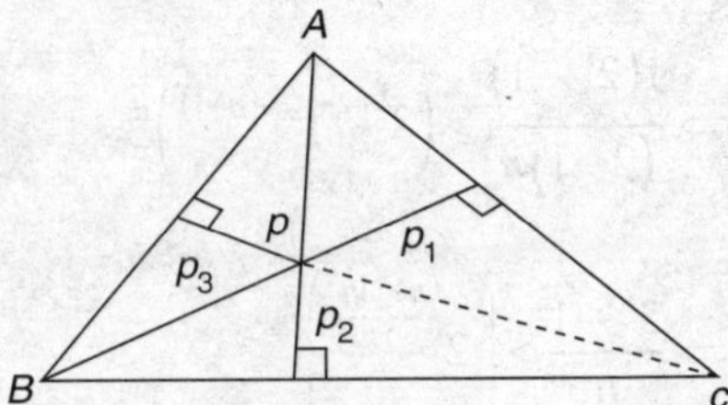

$$\Rightarrow 12 = 3p_1 + \frac{5}{2}p_2 + \frac{5}{2}p_3$$

Applying A.M. ≥ G.M.

$$\Rightarrow \frac{3p_1 + \frac{5}{2}p_2 + \frac{5}{2}p_3}{3} \geq \left[(3p_1)\left(\frac{5}{2}p_2\right)\left(\frac{5}{2}p_3\right)\right]^{\frac{1}{3}}$$

$$\Rightarrow \frac{12}{3} \geq \left(\frac{75}{4}p_1p_2p_3\right)^{\frac{1}{3}} \Rightarrow p_1p_2p_3 \leq \frac{256}{75}$$

$$\Rightarrow \text{ Maximum value of } p_1p_2p_3 = \frac{256}{75}$$

18. **[n = 18, 19, 20, 21, 22, 23 and removed numbers can be (1, 6, 11), (4, 6, 8), (3, 6, 9)]**

Three numbers in A.P. with the least and the greatest sum are (1, 2, 3) and $(n-2, n-1, n)$, respectively. The sum of removed numbers

$$= \frac{43}{4} \times (n-3)$$

$$\Rightarrow \frac{n(n+1)}{2} - 3(n-1) \leq \frac{43}{4}(n-3) \leq \frac{n(n+1)}{2} - 6$$

$$\Rightarrow \frac{(n-3)(n-2)}{2} \leq \frac{43}{4}(n-3) \leq \frac{(n-3)(n+4)}{2}$$

$$\Rightarrow 18 \leq n \leq 23$$

So possible numbers are $18, 19, 20, 21, 22, 23$

As $\frac{43}{4}(n-3)$ is a natural number

$\Rightarrow (n-3)$ should be divisible by 4

so, the only possibilities are 19 and 23

Also, $\frac{n(n+1)}{2}-\frac{43}{4}(n-3)$ is the sum of the removed numbers

$\Rightarrow$ This should be divisible by 3

$\Rightarrow$ 19 is only possibility

So, sum of removed number $=190-172=18$

$\Rightarrow$ middle term of removed A.P. $=\frac{18}{3}=6$

So, removed numbers can be

$(1,6,11),(4,6,8),(3,6,9)$.

19. $(\frac{4n^2+4n}{2n^4+2n+1}, 2)$

$$T_r=\frac{8r}{4r^4+1}=\frac{8r}{(2r^2+2r+1)(2r^2-2r+1)}$$

$$=2\left[\frac{1}{2r^2-2r+1}-\frac{1}{2r^2+2r+1}\right]$$

$$S_n=\sum_{r=1}^{n}T_r=2\left[1-\frac{1}{2n^2+2n+1}\right]$$

$$=\frac{4n^2+4n}{2n^2+2n+1}$$

$$\therefore \lim_{n\to\infty} S_n=\lim_{n\to\infty}\frac{4+\frac{4}{n}}{2+\frac{2}{n}+\frac{1}{n^2}}=\frac{4}{2}=2$$

20. $(\frac{3}{2})$

Since, a,b,c are in G.P.

$\therefore \frac{b}{a}=\frac{c}{b}=r\Rightarrow a=\frac{b}{r}, c=br$

Also, $\log_c a, \log_b c, \log_a b$ are in A.P.

$\therefore d=\log_b c-\log_c a=\log_a b-\log_b c$

$$\Rightarrow d=\frac{\log_{10} c}{\log_{10} b}-\frac{\log_{10} a}{\log_{10} c}=\frac{\log_{10} b}{\log_{10} a}-\frac{\log_{10} c}{\log_{10} b}$$

$$\Rightarrow d=\frac{(\log_{10} c)^2-\log_{10} a\log_{10} b}{\log_{10} b\log_{10} c}$$

$$=\frac{(\log_{10} b)^2-\log_{10} a\log_{10} c}{\log_{10} a\log_{10} b}$$

$$\Rightarrow \frac{(\log_{10} c)^2-\log_{10} a\log_{10} b-(\log_{10} b)^2+\log_{10} a\log_{10} c}{\log_{10} b\log_{10} c-\log_{10} a\log_{10} b}$$

$$\Rightarrow d=\frac{(\log_{10} c-\log_{10} b)(\log_{10} c+\log_{10} b+\log_{10} a)}{\log_{10} b(\log_{10} c-\log_{10} a)}$$

$$\Rightarrow d=\frac{\log\frac{c}{b}\cdot\log abc}{\log\frac{c}{a}\log b}$$

$$=\frac{\log r}{\log r^2}\frac{\log b^3}{\log b}=\frac{3}{2}$$

21. **(16)**

Since, $b-a=c-b=d-c=D$, where D is the c.d.

$\therefore b=a+D, c=a+2D, d=a+3D$

Since,

$2(a-b)+K(b-c)^2+(c-a)^3$

$=2(a-d)+(b-d)^2+(c-d)^3$

$\Rightarrow -2D+KD^2+(2D)^3=-6D+4D^2-D^3$

$\Rightarrow 9D^2+(K-4)D+4=0$

Since D is real

$\Rightarrow (K-4)^2-4(4)(9)\geq 0$

$\Rightarrow K^2-8K-128\geq 0\Rightarrow (K-16)(K+8)\geq 0$

$\Rightarrow K\leq -8$ or $K\geq 16$

Hence the smallest positive value of $K=16$

22. $(\frac{2}{3})$

Let a be the first term and r be the c.r. of an infinite G.P.

$$\therefore S_\infty=\frac{a}{1-r} \quad (1)$$

Now, $f(x)=x^3+3x-9$

$\Rightarrow f'(x)=3x^2+3>0$ for all $x\in[-5,3]$

$\therefore$ greatest value will occur at $x=3$

$f(3)=27+9-9=27$

From equation (1),

$$\frac{a}{1-r}=27 \quad (2)$$

Again, $f'(0)=3$

$\therefore a-ar=3\Rightarrow a(1-r)=3$

$\therefore 1-r=\frac{3}{a}$

From (2)

$\frac{\frac{a}{3}}{\frac{3}{a}}=27 \Rightarrow a^2=81$

$\Rightarrow a=9$

$\therefore 1-r=\frac{1}{3}$

$\Rightarrow r=\frac{2}{3}$

23. (4)

Since, $P(r)=9r^5+7r^2-9=0$

$\Rightarrow 9\left(1-r^5\right)=7r^2$

Also, $S=r^4+2r^9+3r^{14}+4r^{19}+\ldots$ to ∞

$\Rightarrow r^5S=r^9+2r^{14}+3r^{19}+\ldots$ to ∞

$S\left(1-r^5\right)=r^4+r^9+r^{14}+r^{19}+\ldots$ to ∞

$=\frac{r^4}{1-r^5}$

$\Rightarrow S=\frac{r^4}{\left(1-r^5\right)^2}$

$\Rightarrow S=\frac{r^4}{\frac{49}{81}r^4}=\frac{81}{49}=\frac{a}{b}$

$\Rightarrow (a+b)=130$

$\therefore$ Sum of digits $=4$

24. (10)

Let the roots be $a^2,(a+1)^2$, where $a\in N$

Therefore, the equation is

$\left(x-a^2\right)\left(x-(a+1)^2\right)=0$

From question

$\sqrt{a^2(a+1)^2}=1+\left|(a+1)^2-a^2\right|$

$\Rightarrow a(a+1)=1+(2a+1)$ [As $a\in N$]

$\Rightarrow a^2-a-2=0$

$\Rightarrow a=2$ or -1 $[a=-1,\text{rejected}]$

Hence, the roots are 4 and 9

Let $f(x)=x^2-cx+1$

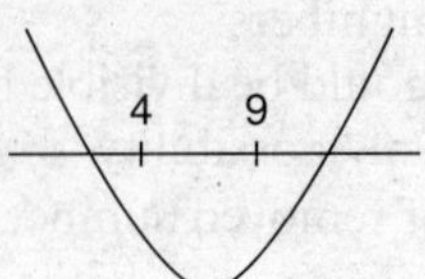

Clearly, $f(4)<0 \Rightarrow 17-4c<0 \Rightarrow c>\frac{17}{4}$ (1)

and, $f(9)<0 \Rightarrow 82-9c<0 \Rightarrow c>\frac{82}{9}$ (2)

From (1) and (2) $c>\frac{82}{9} \Rightarrow$ minimum integral value of $c=10$

25. (4°)

Since,

$S=\frac{1}{\sin 8°}+\frac{1}{\sin 16°}+\frac{1}{\sin 32°}+\ldots+\frac{1}{\sin 4096°}+\frac{1}{\sin 8192°}$

$t_1=\frac{1}{\sin 8°}=\frac{\sin(8°-4°)}{\sin 4°\sin 8°}$

$=\cot 4°-\cot 8°$

$\therefore S=(\cot 4°-\cot 8°)+(\cot 8°-\cot 16°)+(\cot 16°-\cot 32°)+\ldots+(\cot 2048°-\cot 4096°)+(\cot 4096°-\cot 8192°)=(\cot 4°-\cot 8192°)$

Since, $\cot 8192°=\cot(45\times 180°+92°)$

$=\cot 92°$

$\therefore S=\cot 4°-\cot 92°$

$=\frac{1-\tan^2 2°}{2\tan 2°}+\tan 2°=\frac{1+\tan^2 2°}{2\tan 2°}$

$=\frac{\sec^2 2°}{2\tan 2°}=\frac{1}{\sin 4°}$

$\Rightarrow \alpha=4°$

26. (10)

$S=\frac{7}{2^2 5^2}+\frac{13}{5^2 8^2}+\frac{19}{8^2 11^2}+\ldots$

$\Rightarrow 3S=\frac{21}{2^2 5^2}+\frac{39}{5^2 8^2}+\frac{57}{8^2 11^2}+\ldots$

$$\Rightarrow 3S=\sum_{r=1}^{10}\frac{(3r+2)^2-(3r-1)^2}{(3r-1)^2(3r+2)^2}$$

$$\Rightarrow 3S=\sum_{r=1}^{10}\left(\frac{1}{(3r-1)^2}-\frac{1}{(3r+2)^2}\right)$$

$$\Rightarrow 3S=\frac{1}{2^2}-\frac{1}{2^{10}}=\frac{2^8-1}{2^{10}}$$

$$\Rightarrow S=\frac{85}{1024}=\frac{85}{2^{10}}=\frac{\alpha}{\beta^n}$$

Hence, $n=10$.

27. **(5)**

$$T_n=\frac{n(n+1)}{2}\Rightarrow T_i=\frac{i(i+1)}{2}$$

Now,

$$\sum_{K=1}^{\infty}\left(\frac{3}{T_i}\right)^K=\left(\frac{3}{T_i}\right)+\left(\frac{3}{T_i}\right)^2+\left(\frac{3}{T_i}\right)^3+\ldots \text{ to } \infty$$

$$=\frac{\frac{3}{T_i}}{1-\frac{3}{T_i}}=\frac{3}{T_i-3}=\frac{3}{\frac{i(i+1)}{2}-3}$$

$$=\frac{6}{i^2+i-6}$$

$$\therefore \sum_{i=3}^{\infty}\left(\frac{6}{i^2+i-6}\right)=\sum_{i=3}^{\infty}\frac{6}{(i+3)(i-2)}$$

$$=\frac{6}{5}\sum_{i=3}^{\infty}\left(\frac{1}{i-2}-\frac{1}{i+3}\right)$$

$$=\frac{137}{50}=\frac{p}{q}\Rightarrow(3q-p-8)=5$$

28. **(6)**

Since, x, y, z are in A.P. and y, z, w are in G.P.

Let, $x=\frac{2a}{r}-a, y=\frac{a}{r}, z=a, w=ar$

Now, $x+w=16$ and $y+z=8$

$$\Rightarrow \frac{2a}{r}-a+ar=16 \text{ and } \frac{a}{r}+a=8$$

$$\Rightarrow a\left(\frac{2}{r}-1+r\right)=16 \text{ and } a\left(\frac{1}{r}+1\right)=8$$

$$\Rightarrow a=\frac{16}{\frac{2}{r}-1+r} \text{ and } a=\frac{8}{\frac{1}{r}+1}$$

$$\Rightarrow 8r\left(\frac{2}{r}-1+r\right)=16(1+r)$$

$$\Rightarrow r^2=3r\Rightarrow r=3 \ (r=0 \text{ not possible})$$

$\therefore a=6$

$\therefore x=-2, y=2, z=6, w=18$

Hence, $x^2-y^2+2z-w=-6$

Hence, $\left|x^2-y^2+2z-w\right|=6$

29. **(13)**

Since, $\sum\alpha=-4, \sum\alpha\beta=-a$ and $\alpha\beta\gamma=-1$

$$\text{Now, }\left(\alpha+\frac{1}{\alpha}\right)^2+\left(\beta+\frac{1}{\beta}\right)^2+\left(\gamma+\frac{1}{\gamma}\right)^2$$

$$=\sum\alpha^2-\sum\frac{1}{\alpha^2}+6$$

$$=\left(\sum\alpha\right)^2-2\sum\alpha\beta+\left[\frac{(\alpha\beta)^2+(\beta\gamma)^2+(\alpha\gamma)^2}{1}\right]+6$$

$$=22+2a+\left[\left(\sum\alpha\beta\right)^2-2\left(\alpha^2\beta\gamma+\alpha\beta^2\gamma+\alpha\beta\gamma^2\right)\right]$$

$$=22+2a+\left[a^2+2(-4)\right]$$

$$=a^2+2a+1+13$$

$$=(a+1)^2+13$$

Hence minimum value = 13 when $a=-1$

30. **(13)**

Let, $a_2=a, a_3=ar, a_4=ar^2$

It is given, $a_1=a_4+5$

$$\Rightarrow a_1=ar^2+5$$

As, a_1, a_2, a_3 are in A.P.

$$\therefore d=a_2-a_1=a_3-a_2$$

$$\Rightarrow d=a-ar^2-5=ar-a$$

$$\Rightarrow a\left(r^2+r-2\right)=-5 \qquad (1)$$

Also, $d + 4r = 0$

$\Rightarrow ar - a + 4r = 0 \Rightarrow a(1-r) = 4r$ (2)

from (1) and (2)

$$\frac{r^2 + r - 2}{1-r} = \frac{-5}{4r}$$

$\Rightarrow (2r-1)(2r^2 + 3r - 5) = 0$

$\therefore r = \frac{1}{2} \Rightarrow a = 4$

Hence, $a_1 = 6, a_2 = 4, a_3 = 2, a = 4$

$\therefore \sum_{i=1}^{4} a_i = 13$

31. **(9)**

$s = \frac{a+b+c+d}{3} \Rightarrow s - a = \frac{b+c+d-2a}{3}$

Similarly, $(s-b) = \frac{a+c+d-2b}{3}$

$(s-c) = \frac{a+b+d-2c}{3}$

$(s-d) = \frac{a+b+c-2d}{3}$

Using A.M > G.M.

We get

$abcd > 81(s-a)(s-b)(s-c)(s-d)$

$\Rightarrow \frac{abcd}{(s-a)(s-b)(s-c)(s-d)} > 9^2$

$\Rightarrow \alpha = 9$

32. **(2)**

Using A.M. ≥ G.M.

$a = 4, b = 6, c = 4$

Hence H.C.F. = 2

33. **(3)**

Since,

$a_1 = 0, |a_2| = |a_1 + 1|, |a_3|$

$= |a_2 + 1|, \ldots, |a_n| = |a_{n-1} + 1|$

Also let, $a_{n+1} = |a_n + 1|$

Squaring,

$a_1^2 = 0$

$a_2^2 = a_1^2 + 2a_1 + 1$

$a_3^2 = a_2^2 + 2a_2 + 1$

$a_4^2 = a_3^2 + 2a_3 + 1$

$a_{n+1}^2 = a_n^2 + 2a_n + 1$

Adding,

$a_1^2 + a_2^2 + a_3^2 + \ldots + a_{n+1}^2$

$= (a_1^2 + a_2^2 + a_3^2 + \ldots + a_n^2) +$

$2(a_1 + a_2 + a_3 + \ldots + a_n) + r$

$\Rightarrow a_{n+1}^2 = 2(a_1 + a_2 + \ldots + a_n) + n$

$\Rightarrow 2(a_1 + a_2 + \ldots + a_n) = a_{n+1}^2 - n$

$\Rightarrow 2(a_1 + a_2 + \ldots + a_n) \geq -n$

$\Rightarrow \frac{a_1 + a_2 + \ldots + a_n}{n} \geq -\frac{1}{2}$

$\Rightarrow \lambda = 1, \mu = 2 \Rightarrow \lambda + \mu = 3$

34. Since, a, b, c are in H.P.

$\frac{2}{b} = \frac{1}{a} + \frac{1}{c}$

Let

$$\frac{a+b}{2a-b} + \frac{c+b}{2c-b} = K \Rightarrow \frac{\frac{1}{b}+\frac{1}{a}}{\frac{2}{b}-\frac{1}{a}} + \frac{\frac{1}{b}+\frac{1}{c}}{\frac{2}{b}-\frac{1}{c}} = K$$

$$\left(\frac{3}{2a} + \frac{1}{2c}\right)c + \left(\frac{3}{2c} + \frac{1}{2a}\right)a = K$$

$\Rightarrow \frac{3c}{a} + \frac{3a}{c} + 2 = 2K$

$\Rightarrow 3\left(\frac{a}{c}\right)^2 - 2(K-1)\frac{a}{c} + 3 = 0$

Since, $\frac{a}{c}$ is real

$4(K-1)^2 - 36 \geq 0 \Rightarrow (16-1)^2 \geq 9$

$\Rightarrow K \geq 4$, But $K = 4$ gives $a = c = b$

Therefore $K > 4$

35. (4)

Since,

$(1+a)(1+b)(1+c)(1+d)$

$=1+(a+b+c+d)+(ab+ac+\ldots+cd)$

$+(abc+abd+\ldots+bcd)+abcd \quad (1)$

Using A.M. $\geq$ G.M.

We get

$(1+a)(1+b)(1+c)(1+d)\geq(1+e)^4$

$\Rightarrow \lambda=4$

2.13.5 JEE Main: Archive

1. (B)

$$\sum_{i=1}^{n} a_i=\frac{n}{2}\{2a_1+(n-1)\}=192$$

$$\Rightarrow 2a_1+(n-1)=\frac{384}{n} \quad (1)$$

$$\sum_{i=1}^{n/2} a_{2i}=\frac{n}{4}\left[2a_1+2+\left(\frac{n}{2}-1\right)2\right]=120$$

$$2a_1+n=\frac{480}{n} \quad (2)$$

From equations (2) and (1)

$$1=\frac{480}{n}-\frac{384}{n}$$

$n=480-384=96$

2. (D)

$$\frac{1}{2\cdot 3^{10}}+\frac{1}{2^2\cdot 3^9}+\ldots+\frac{1}{2^{10}\cdot 3}=\frac{K}{2^{10}\cdot 3^{10}}$$

$K=2^9+2^8\cdot 3+2^7\cdot 3^2+\ldots+3^9$

$$=\frac{2^9\left(\left(\frac{3}{2}\right)^{10}-1\right)}{\frac{3}{2}-1}=3^{10}-2^{10}$$

Now, $3^{10}-2^{10}=(3^5-2^5)(3^5+2^5)$

$=(211)(275)$

$=(35\times 6+1)(45\times 6+5)$

$6\lambda+5$

The remainder is 5

3. (98)

$$\frac{1}{3},\frac{5}{9},\frac{19}{27},\frac{65}{81},\ldots$$

$$\left(1-\frac{2}{3}\right)+\left(1-\frac{4}{9}\right)+\left(1-\frac{8}{27}\right)+\left(1-\frac{16}{81}\right)\ldots 100$$

terms

$$100-\left[\frac{2}{3}+\left(\frac{2}{3}\right)^2+\ldots\right]$$

$$100-\frac{\frac{2}{3}\left(1-\left(\frac{2}{3}\right)^{100}\right)}{1-\frac{2}{3}}$$

$$100-2\left(1-\left(\frac{2}{3}\right)^{100}\right)$$

$$S=98+2\left(\frac{2}{3}\right)^{100}$$

$\Rightarrow [S]=98$

4. (B)

$S=1\cdot 3^0+2\cdot 3^1+3\cdot 3^2+\ldots+10\cdot 3^9$

$3S=1\cdot 3^1+2\cdot 3^2\ldots+9\times 3^9+10\times 3^{10}$

$-2S=(1\cdot 3^0+3^1+3^2\ldots+3^9)-10\cdot 3^{10}$

$$S=5\times 3^{10}-\left(\frac{3^{10}-1}{4}\right)$$

$$S=\frac{20\cdot 3^{10}-3^{10}+1}{4}=\frac{19\cdot 3^{10}+1}{4}$$

5. (1100)

$$A=\sum_{i=1}^{10}\sum_{j=1}^{10}\min\{i,j\}$$

$$B=\sum_{i=1}^{10}\sum_{j=1}^{10}\max(i,j)$$

$$A=\sum_{j=1}^{10}\min(i,1)+\min(j,2)+\ldots\min(i,10)$$

$$=\underbrace{(1+1+1+\ldots+1)}_{19\text{ times}}+\underbrace{(2+2+2\ldots+2)}_{17\text{ times}}+$$

$$\underbrace{(3+3+3\ldots+3)}_{15\text{ times}}+\ldots(1)\ 1\text{ times}$$

$$B=\sum_{j=1}^{10}\max(i,1)+\max(j,2)+\ldots\max(i,10)$$

$$=\underbrace{(10+10+\ldots+10)}_{19\text{ times}}+$$

$$\underbrace{(9+9+\ldots+9)}_{17\text{ times}}+\ldots+11\text{ times}$$

$$A+b=20(1+2+3+\ldots+10)$$

$$=20\times\frac{10\times11}{2}=10\times110=1100$$

6. (C)

$$A=\left(\frac{1}{2}+\frac{1}{4^2}+\frac{1}{2^3}+\frac{1}{4^4}+\ldots\infty\right)$$

$$A=\left(\frac{1}{2}+\frac{1}{2^3}+\ldots\infty\right)+\left(\frac{1}{4^2}+\frac{1}{4^4}+\ldots\infty\right)$$

$$A=\left(\frac{\frac{1}{2}}{1-\frac{1}{4}}+\frac{\frac{1}{16}}{1-\frac{1}{16}}\right)$$

$$\Rightarrow A=\frac{1}{2}\times\frac{4}{3}+\frac{1}{16}\times\frac{16}{15}\Rightarrow A=\frac{11}{15}$$

$$B=-\frac{1}{2}+\frac{1}{4^2}+\frac{-1}{2^3}+\frac{1}{4^4}+\ldots\infty$$

$$B=\left(-\frac{1}{2}+\frac{-1}{2^3}+\ldots\infty\right)+\left(\frac{1}{4^2}+\frac{1}{4^4}+\ldots\infty\right)$$

$$B=\frac{-\frac{1}{2}}{1-\frac{1}{4}}+\frac{\frac{1}{16}}{1-\frac{1}{16}}$$

$$\Rightarrow B=-\frac{1}{2}\times\frac{4}{3}+\frac{1}{16}\times\frac{16}{15}$$

$$B=-\frac{9}{15}$$

$$\frac{A}{B}=\frac{11}{15}\times\frac{15}{(-9)}$$

$$\frac{A}{B}=-\frac{11}{9}$$

7. (40)

$a_1>0,\ a_2,a_3,a_4,a_5\to$ G.P.

$$3a_2+a_3=2a_4$$

$$3ar+ar^2=2ar^3$$

$$3+r=2r^2$$

$$2r^2-r-3=0$$

$$r=-1\text{ and }r=\frac{3}{2}$$

$$a_2+a_4=2a_3+1$$

$$ar+ar^3=2ar^2+1$$

$$a\left(r+r^3-2r^2\right)=1$$

$$a\left(\frac{3}{2}+\frac{27}{8}-\frac{18}{4}\right)=1$$

$$a=\frac{8}{3}$$

When $r=-1,\ a=-\frac{1}{4}$ (rejected, $a_1>0$)

$r=\frac{2}{3},\ a=\frac{8}{3}$ (selected)

Now

$$a_2+a_4+2a_5$$

$$=\frac{8}{3}\times\frac{3}{2}+\frac{8}{3}\times\frac{27}{8}+2\times\frac{8}{3}\times\frac{81}{16}$$

$$=4+9+27=40$$

8. (C)

$$X=1+a+a^2=\ldots$$

$$x=\frac{1}{1-a}\Rightarrow a=1-\frac{1}{x}$$

$$y=\frac{1}{1-b}\Rightarrow b=1-\frac{1}{y}$$

$$z=\frac{1}{1-c}\Rightarrow c=1-\frac{1}{z}$$

a, b, c are in A.P.

$\Rightarrow 1-\frac{1}{x},1-\frac{1}{y},1-\frac{1}{z}$ are in A.P.

$\Rightarrow -\frac{1}{x},-\frac{1}{y},-\frac{1}{z}$ are in A.P.

$\Rightarrow \frac{1}{x},\frac{1}{y},\frac{1}{z}$ are in A.P.

9. (276)

$$\frac{1}{5}+\frac{2}{65}+\frac{3}{325}+\frac{4}{1025}+\frac{5}{2501}+\ldots$$

$$T_n=\frac{n}{4n^4+1}$$

$$=\frac{n}{\left(2n^2+1\right)^2-(2n)^2}$$

$$=\frac{n}{\left(2n^2+2n+1\right)\left(2n^2-2n+1\right)}$$

$$= \frac{1}{4}\left[\frac{1}{2n^2-2n+1} - \frac{1}{2n^2+2n+1}\right]$$

So,

$$\sum_{n=1}^{10} T_n = \frac{1}{4}\left[\frac{1}{1} - \frac{1}{5} + \frac{1}{5} - \frac{1}{13} + \ldots \frac{1}{200+20+1}\right]$$

$$= \frac{1}{4}\left[1 - \frac{1}{221}\right] = \frac{1}{4} \times \frac{220}{221} - \frac{55}{221} = \frac{m}{n}$$

$$m + n = 55 + 221 = 276$$

10. (C)

$$S = 2 + \frac{6}{7} + \frac{12}{7^2} + \frac{20}{7^3} + \frac{30}{7^4} + \ldots$$

Considering infinite sequence,

$$S = 2 + \frac{6}{7} + \frac{12}{7^2} + \frac{20}{7^3} + \frac{30}{7^4} + \ldots$$

$$\frac{S}{7} = \frac{2}{7} + \frac{6}{7^2} + \frac{12}{7^3} + \frac{20}{7^4} + \ldots$$

$$\Rightarrow \frac{6S}{7} = 2 + \frac{4}{7} + \frac{6}{7^2} + \frac{8}{7^3} + \frac{10}{7^4} + \ldots$$

$$\Rightarrow \frac{6S}{7^2} = \frac{2}{7} + \frac{4}{7^2} + \frac{6}{7^3} + \frac{8}{7^4} + \ldots$$

$$\frac{6S}{7}\left(1 - \frac{1}{7}\right) = 2 + \frac{2}{7} + \frac{2}{7^2} + \frac{2}{7^3} + \ldots$$

$$\Rightarrow \frac{6^2 S}{7^2} = \frac{2}{1 - \frac{1}{7}} = \frac{2}{6} \times 7$$

$$\Rightarrow S = \frac{2 \times 7^3}{6^3} \Rightarrow 4S = \frac{7^3}{3^3} = \left(\frac{7}{3}\right)^3$$

11. (D)

$a_1, a_2, a_3, \ldots$ A.P.; $a_1 = 2$; $a_{10} = 3$; $d_1 = \frac{1}{9}$

$b_1, b_2, b_3 \ldots$ A.P.; $b_1 = \frac{1}{2}$; $b_{10} = \frac{1}{3}$; $d_2 = \frac{-1}{54}$

[Using $a_1 b_1 = 1 = a_{10} b_{10}$; d_1 and d_2 are common difference respectively]

$$a_4 b_4 = (2 + 3d_1)\left(\frac{1}{2} + 3d_2\right) = \left(2 + \frac{1}{3}\right)\left(\frac{1}{2} - \frac{1}{18}\right)$$

12. (C)

$$A_1 \cdot A_3 \cdot A_5 \cdot A_7 = \frac{1}{1296}$$

$$(A_4)^4 = \frac{1}{1296}$$

$$A_4 = \frac{1}{6} \qquad (1)$$

$$A_2 + A_4 = \frac{7}{36}$$

$$A_2 = \frac{1}{36}$$

$$A_2 = \frac{1}{36} \qquad (2)$$

$$A_6 = 1$$

$$A_8 = 6$$

$$A_{10} = 36$$

$$A_6 + A_8 + A_{10} = 43$$

13. (C)

$$d = \frac{100 - a}{n+1}$$

$$A_1 = a + d$$

$$A_n = 100 - d$$

$$\Rightarrow \frac{A_1}{A_n} = \frac{1}{7} \Rightarrow \frac{a+d}{100-d} = \frac{1}{7}$$

$$\Rightarrow 7a + 8d = 100$$

$$\Rightarrow 7a + 8\left(\frac{100-a}{n+1}\right) = 100 \qquad (1)$$

$$\because a + n = 33 \qquad (2)$$

Now, by equations (1) and (2)

$$7n^2 - 132n - 667 = 0$$

$\boxed{n = 23}$ and $n = \frac{-29}{7}$ reject.

14. (41651)

$$S_n = \frac{n^2}{1 - \frac{1}{(n+1)^2}} - \frac{n(n+1)^2}{(n+2)}$$

$$S_n = \frac{n(n^2 + 2n + 1)}{(n+2)}$$

$$S_n = \frac{n[n(n+2)+1]}{(n+2)}$$

$$S_n = n\left[n + \frac{1}{n+2}\right]$$

$$S_n = n^2 + \frac{n+2-2}{(n+2)}$$

$$S_n = n^2 + 1 - \frac{2}{(n+2)}$$

Now $\frac{1}{26}+\sum_{n=1}^{50}\left[\left(n^2-n\right)-2\left(\frac{1}{n+2}-\frac{1}{n+1}\right)\right]$

$=\frac{1}{26}+\left[\frac{50\times 51\times 101}{6}-\frac{50\times 51}{2}-2\left(\frac{1}{52}-\frac{1}{2}\right)\right]$

$= 41651$

15. (B)

$a_2 = 1, a_3 = 3, a_4 = 6$

$a_n = \frac{n(n-1)}{2}$

$S=\sum_{n=2}^{\infty}\frac{n(n-1)}{2(7^n)}$

$S=\frac{1}{7^2}+\frac{3}{7^3}+\frac{6}{7^4}+\frac{10}{7^5}+\frac{10}{7^6}+\ldots$

$6\frac{S}{7}=\frac{1}{7^2}+\frac{2}{7^3}+\frac{3}{7^4}+\frac{4}{7^5}+\ldots$

$6\frac{S}{7^2}=\frac{1}{7^3}+\frac{2}{7^4}+\frac{3}{7^5}+\ldots$

$6\frac{S}{7}\cdot\frac{6}{7}=\frac{1}{7^2}+\frac{1}{7^3}+\ldots=\frac{\frac{1}{7^2}}{1-\frac{1}{7}}$

$6\times 6\frac{S}{7^2}=\frac{1}{7\times 6}$

$S=\frac{7}{6^3}=\frac{7}{216}$

16. (C)

$S=1+\frac{5}{6}+\frac{12}{6^2}+\frac{22}{6^3}+\frac{35}{6^4}+\ldots$

$\frac{S}{6}=\frac{1}{6}+\frac{5}{6^2}+\frac{12}{6^3}+\frac{22}{6^4}+\ldots$

On subtraction

$\frac{5}{6}S=1+\frac{4}{6}+\frac{7}{6^2}+\frac{10}{6^3}+\frac{13}{6^4}+\ldots$

$\frac{5}{36}S=1+\frac{4}{6^2}+\frac{7}{6^3}+\frac{10}{6^4}+\frac{13}{6^5}+\frac{13}{6^5}+\ldots$

On subtraction

$\frac{25}{36}S=1+\frac{3}{6}+\frac{3}{6^2}+\frac{3}{6^3}+\ldots=\frac{8}{5}$

$S=\frac{288}{125}$

17. (2223)

For series of common terms

$a = 9, d = 12, n = 19$

$S_{19}=\frac{19}{2}\left[2(9)+18(12)\right]=2223$

18. (27560)

$a_1 = b_1 = 1$

$a_2 = a_1 + 2 = 3$

$a_3 = a_2 + 2 = 5$

$a_4 = a_2 + 2 = 7$

$\Rightarrow a_n = 2n-1$

$b_2 = a_1 + b_1 = 4$

$b_3 = a_3 + b_2 = 9$

$b_4 = a_4 + b_3 = 16$

$b_n = n^2$

$\sum_{n=1}^{15} a_n b_n$

$\sum_{n=1}^{15}(2n-1)n^2$

$\sum_{n=1}^{15}\left(2n^3-n^2\right)$

$=2\frac{n^2(n+1)^2}{4}-\frac{n(n+1)(2n+1)}{6}$

Put $n = 15$

$=\frac{2\times 225\times 16\times 16}{4}-\frac{15\times 16\times 31}{6}$

$= 27560$

19. (B)

20. (C)

$\left(\left(2^1 2^2\ldots 2^{60}\right)\left(4^1\cdot 4^2\ldots 4^n\right)\right)^{\frac{1}{60+n}}=2^{\frac{225}{8}}$

$\left(2^{30\times 61}4^{\frac{n(n+1)}{2}}\right)^{\frac{1}{60+n}}=2^{\frac{225}{8}}$

$2^{1830+n^2+n}=2^{\frac{(225)(60+n)}{8}}$

$= 8n^2 - 217n + 1140 = 0$

$n = 20, \frac{57}{8}$

$$\sum_{k=1}^{n} nk-k^2=\frac{n^2(n+1)}{2}-\frac{n(n+1)(2n+1)}{6}$$
$$=1330$$

21. (6993)

$$S_{11}=3[101+102+\ldots+121]$$
$$=\frac{3}{2}(222)\times 21=6993$$

22. (166)

$$\sum_{k=1}^{10}\frac{k}{k^4+k^2+1}=\frac{m}{n}$$
$$\Rightarrow \frac{1}{2}\sum_{k=1}^{10}\frac{(k^2+k+1)-(k^2-k+1)}{(k^2+k+1)(k^2-k+1)}$$
$$\Rightarrow \frac{1}{2}\left(\sum_{k=1}^{10}\left(\frac{1}{(k^2-k+1)}-\frac{1}{k^2+k+1}\right)\right)$$
$$\Rightarrow \frac{55}{111}=\frac{m}{n}$$
$$\therefore m+n=166$$

23. (53)

1st term = 100 = a

Last term = 199 = ℓ

If 3 terms

$a, a+d, a+2d$

$a_n=\ell=a+(n-1)d$

$$d_1=\frac{\ell-a}{n-1}$$

$n\to n$umber of terms

$$n=3,\ d_1=\frac{199-100}{2}=\frac{99}{2}\notin I$$
$$n=4,\ d_2=\frac{99}{3}=33\in I$$
$$n=10,\ d_3=\frac{99}{9}=11\in I$$
$$n=12,\ d_4=\frac{99}{11}=9\in I$$

24. (B)

$$\frac{S_5}{S_9}=\frac{5}{17}\Rightarrow \frac{\frac{5}{2}(2a+4d)}{\frac{9}{2}(2a+8d)}=\frac{5}{17}$$
$$\Rightarrow d=4a$$

$a_{15}=a+14d=57a$

Now, $110<a_{15}<120$

$\Rightarrow 110<57a<120$

$\Rightarrow a=2 \therefore d=8$

$$S_{10}=\frac{10}{2}(2\times 2+9\times 8)=380$$

25. (10620)

$$f(x)=2x^2-x-1$$
$$|f(x)|\le 800$$
$$2n^2-n-801\le 0$$
$$n^2-\frac{1}{2}n-\frac{801}{2}\le 0$$
$$\left(n-\frac{1}{4}\right)^2-\frac{801}{2}-\frac{1}{16}\le 0$$
$$\left(n-\frac{1}{4}\right)^2-\frac{6409}{16}\le 0$$
$$\left(n-\frac{1}{4}-\frac{\sqrt{6409}}{4}\right)\left(n-\frac{1}{4}+\frac{\sqrt{6409}}{16}\right)\le 0$$
$$\frac{1-\sqrt{6409}}{4}\le n\le \frac{1+\sqrt{6409}}{4}$$
$$n=\{-19,-18-17,\ldots 0,1,2,\ldots,20\}$$
$$\sum_{n\in S} f(x)=\sum(2x^2-x-1)$$
$$=2\left[19^2+18^2+\ldots+1^2+1^2+2^2+\ldots+19^2+20^2\right]$$
$$=4\left[1^2+2^2+\ldots+19^2\right]+2\left[20^2\right]-20-40$$
$$=\frac{4\times 19\times 20\times(2\times 19+1)}{6}+2\times 400-60$$
$$=\frac{4\times 19\times 20\times 39}{6}+800-60-9880+800-60$$
$$=10620$$

26. (A)

$$S_{21}=\frac{21}{2}\left[20ar+20\cdot 10ar^2\right]$$
$$=21\left[10aar+100ar^2\right]$$
$$=21\cdot a_{11}$$

27. (120)

$$T_n=\frac{2\sum_{r=1}^{n}(2r)^3-\left(\sum_{r=1}^{2n}r^3\right)}{n(4n+3)}$$

$\Rightarrow T_n = n$

So, $\sum_{n=1}^{15} T_n = 120$

28. **(C)**

$a_{n+2}a_{n+1} - a_{n+1} \cdot a_n = 2$

Series will satisfy

$a_1a_2, a_2a_3, a_3a_4, a_4a_5$

$1\times2, 2\times2, 2\times3, 2\times4$

$$\frac{a_n + \frac{1}{a_{n+1}}}{a_{n+2}} = \frac{a_{n+2} - \frac{1}{a_{n+1}}}{a_{n+2}}$$

$$= 1 - \frac{1}{a_{n+1}a_{n+2}}$$

$$= 1 - \frac{1}{2(r+1)}$$

$$= \frac{2r+1}{2(r+1)}$$

Now proof is given by

$$= \prod_{r=1}^{30} \frac{(2r+1)}{2(r+1)}$$

$$= \frac{(1\times3\times5\times\ldots\times61)}{2^{30}\times(2\times3\times\ldots\times31)}$$

$$\Rightarrow \frac{(1\times3\times5\times\ldots\times61)}{\lfloor 31\times2^{30}} \times \frac{2^{30}\times\lfloor 30}{2^{30}\times\lfloor 30}$$

$$= \frac{\lfloor 61}{2^{60}\lfloor 31\times\lfloor 30}$$

$\alpha = -60$

29. **(50)**

$f(x) = 0 \Rightarrow (x-p)^2 - q = 0$

Roots are $p+\sqrt{q}, p-\sqrt{q}$ absolute difference between roots $2\sqrt{q}$.

Now, $|f(a_i)| = 500$

Let a_1, a_2, a_3, a_4 are $a_1 a+d, a+2d, a+3d$

$|f(a_4)| = 500$

$|(a_1-p)^2 - q| = 500$

$\Rightarrow (a_1-p)^2 - q = 500$

$$\Rightarrow \frac{9}{4}d^2 - q = 500 \qquad (1)$$

and $|f(a_1)|^2 = |f(a_2)|^2$

$$\left((a_1-p)^2-q\right)^2 = \left((a_2-p)^2-q\right)^2$$

$$\Rightarrow \left((a_1-p)^2-(a_2-p)^2\right)\left((a_1-p)^2-q+(a_2-p)^2-q\right) = 0$$

$$\Rightarrow \frac{9}{4}d^2 - q + \frac{d^2}{4} - q = 0$$

$$2q = \frac{10d^2}{4} \Rightarrow q = \frac{5d^2}{4}$$

$$\Rightarrow d^2 = \frac{4q}{5}$$

From equation (1)

$$\frac{9}{4}\cdot\frac{4\cdot q}{5} - q = 500$$

$$\frac{4q}{5} = 500 = 50$$

30. **(142)**

$$\sum x_0^1 = \frac{3\left(1-\left(\frac{1}{2}\right)\right)^{20}}{1\frac{-1}{2}} = 6\left(1-\frac{1}{2^{20}}\right)$$

$$= \sum_{i=1}^{20}(x_{i-i})^2$$

$$= \sum_{i=1}^{20}(x_i)^2 + (i)^2 - 2x_i i$$

Now

$$= \sum_{i=1}^{20}(x_i)^2 = \frac{9\left(1-\left(\frac{1}{4}\right)\right)^{20}}{1-\frac{1}{4}} = 12\left(1-\frac{1}{2^{40}}\right)$$

$$\sum_{i=1}^{20} i^2 = \frac{1}{6}\times20\times21\times41 = 2870$$

$$\sum_{i=1}^{20} x_i u = s = 3 + 2.3\frac{1}{2} + 3.3\frac{1}{2^2} + 4.3\frac{1}{2^3} + \ldots \text{A.G.P.}$$

$= 6\left(2 - \dfrac{22}{2^{20}}\right)$

$$\bar{x} = \frac{12 - \dfrac{12}{2^{40}} + 2870 - 12\left(2 - \dfrac{22}{2^{20}}\right)}{20}$$

$$\bar{x} = \frac{2858}{20} + \left(-\frac{12}{2^{40}} + \frac{22}{2^{20}}\right) \times \frac{1}{20}$$

$[\bar{x}] = 142$

31. (12)

$$\frac{6}{3^{12}} + 10\left(\frac{1}{3^{11}} + \frac{2}{3^{10}} + \frac{2^2}{3^9} + \frac{2^3}{3^8} + \ldots + \frac{2^{10}}{3}\right)$$

$$\frac{6}{3^{12}} + \frac{10}{3^{11}}\left(\frac{6^{11}-1}{6-1}\right)$$

$= 2^{12} \cdot 1;\ m \cdot n = 12$

32. (C)

By splitting

$$\frac{1}{20}\left[\left(\frac{1}{20-a} - \frac{1}{40-a}\right) + \left(\frac{1}{40-a} - \frac{1}{60-a}\right) + \ldots + \left(\frac{1}{180-a} - \frac{1}{200-a}\right)\right]$$

$$\Rightarrow \frac{1}{20}\left(\frac{1}{20-a} - \frac{1}{200-a}\right) = \frac{1}{256}$$

$(20-a)(200-a) = 256 \times 9$

$a^2 - 220a + 1696 = 0$

$a = 8, 212$

Hence maximum value of a is 212.

33. (16)

$$S = \frac{a_1}{2} + \frac{a_2}{2^2} + \frac{a_3}{2^3} + \ldots$$

$$\frac{S}{2} = \frac{a_1}{2^2} + \frac{a_2}{2^3} + \ldots$$

$$\frac{S}{2} = \frac{a_1}{2} + d\left(\frac{1}{2^2} + \frac{1}{2^3} + \ldots\right)$$

$$\frac{S}{2} = \frac{a_1}{2} + d\left(\frac{\frac{1}{4}}{1 - \frac{1}{2}}\right)$$

$\therefore S = a_1 + d = a_2 = 4$

Or $4a_2 = 16$

34. (286)

$$\frac{1}{2\times3\times4} + \frac{1}{3\times4\times5} + \ldots + \frac{1}{100\times101\times102} = \frac{k}{101}$$

$$\frac{4-2}{2\times3\times4} + \frac{5-3}{3\times4\times5} + \ldots + \frac{102-100}{100\times101\times102} = \frac{2k}{101}$$

$$\frac{1}{2\times3} - \frac{1}{3\times4} + \frac{1}{3\times4} - \frac{1}{4\times5} + \ldots + \frac{1}{100\times101} - \frac{1}{101\times102} = \frac{2k}{101}$$

$$\frac{1}{2\times3} - \frac{1}{101\times102} = \frac{2k}{101}$$

$$\therefore 2k = \frac{101}{6} - \frac{1}{102}$$

$\therefore 34k = 286$

35. (B)

$a_0 = 0, a_1 = 0$

$a_{n+2} = 3a_{n+1} - 2a_{n+1} : n \geq 0$

$a_{n+2} - a_{n+1} = 2(a_{n+1} - a_n) + 1$

$n - 0 \quad a_2 \quad a_1 - 2(a_1 - a_0) + 1$

$n = 1 \quad a_3 - a_2 = 2(a_2 - a_1) + 1$

$n = 2 \quad a_4 - a_3 = 2(a_3 - a_2) + 1$

$n = n \quad a_{n+2} - a_{n+1} = 2(a_{n+1} - a_n) + 1$

$(a_{n+2} - a_1) - 2(a_{n+1} - a_0) - (n+1) = 0$

$a_{n+2} = 2a_{n+1} + (n+1)$

$n \to n-2$

$a_n - 2a_{n-1} = n - 1$

Now $a_{25}a_{23} - 2a_{25}a_{22} - 2a_{23}a_{24} + 4a_{22}a_{24}$

36. (B)

$$\sum_{x=1}^{20}(r^2+1)r!$$

$$\sum_{x=1}^{20}\left((r+1)^2 - 2r\right)r!$$

$$\sum_{x=1}^{20}\left((r+1)(r+1)! - r.r!\right) - \sum_{x=1}^{20} r.r!$$

$$\sum_{x=1}^{20}\left((r+1)(r+1)! - r.r!\right) - \sum_{x=1}^{20}\left((r+1)! - r!\right)$$

$= (21.\underline{|21} - 1) - (\underline{|21} - 1)$

$= 20.21! = 22! - 2.21!$

37. (A)

$pq^2 = \log_x \lambda$

$qr = \log_y \lambda$

$p^2 r = \log_z \lambda$

$\log_y x = \dfrac{qr}{pq^2} = \dfrac{r}{pq}$ (1)

$\log_x z = \dfrac{pq^2}{p^2 r} = \dfrac{q^2}{pr}$ (2)

$\log_z y = \dfrac{p^2 r}{qr} = \dfrac{p^2}{q}$ (3)

$3, \dfrac{3r}{pq}, \dfrac{3p^2}{q}, \dfrac{7q^2}{pr}$ in A.P.

$\dfrac{3r}{pq} - 3 = \dfrac{1}{2}$

$r = \dfrac{7}{6}pq$ (4)

$r = pq + 1$

$pq = 6$ (5)

$r = 7$ (6)

$\dfrac{3p^2}{q} = 4$

38. (12)

$T_4 = 500$

where a = first term,

r = common ratio $= \dfrac{1}{m}, m \in N$

$ar^3 = 500$

$\dfrac{a}{m^3} = 500$

$S_n - S_{n-1} = ar^{n-1}$

$S_6 > S_5 + 1$ and $S_7 - S_6 < \dfrac{1}{2}$

$S_6 - S_5 > 1$ $\qquad \dfrac{a}{m^6} < \dfrac{1}{2}$

$ar^5 > 1$ $\qquad m^3 > 10^3$

$\dfrac{500}{m^2} > 1$ $\qquad m > 10$ (2)

$m^2 < 500$ (1)

From (1) and (2)

$m = 11, 12, 13, \ldots, 22$

So number of possible values of m is 12

39. (5)

$1^3 + 2^3 + 3^3 + \ldots + n^3 = \left(\dfrac{n(n+1)}{2}\right)^2$

$1\times3 + 2\times5 + 3\times7 + \ldots + n$ terms

$= \sum_{r=1}^{n} r(2r+1) = \sum_{r=1}^{n}(2r^2 + r)$

$= \dfrac{2\cdot n(n+1)(2n+1)}{6} + \dfrac{n(n+1)}{2}$

$= \dfrac{n(n+1)}{6}(2(2n+1)+3)$

$= \dfrac{n(n+1)}{2} \times \dfrac{(4n+5)}{3}$

$= \dfrac{\dfrac{n^2(n+1)^2}{4}}{\dfrac{n(n+1)}{2} \times \dfrac{(4n+5)}{3}} = \dfrac{9}{5}$

$\Rightarrow \dfrac{5n(n+1)}{2} = \dfrac{9(4n+5)}{3}$

$\Rightarrow 15n(n+1) = 18(4n+5)$

$\Rightarrow 15n^2 + 15n = 72n + 90$

$\Rightarrow 15n^2 - 57n - 90 = 0 \Rightarrow 5n^2 - 19n - 30 = 0$

$\Rightarrow (n-5)(5n+6) = 0$

$\Rightarrow n = \dfrac{-6}{5}$ or 5

$\Rightarrow n = 5$

40. (495)

$$\begin{vmatrix} A+6d & 7 & 1 \\ 2(A+1+8d) & 17 & 1 \\ A+2+16d & 17 & 1 \end{vmatrix} + 70 = 0$$

$\Rightarrow A = -7$ and $d = 6$

$\therefore c - a - b = 20$

$S_{20} = 495$

41. (3)

$a, b, \dfrac{1}{18} \to$ G.P.

$\dfrac{a}{18} = b^2$ (1)

$\dfrac{1}{a}, 10, \dfrac{1}{b} \to$ A.P.

$\dfrac{1}{a} + \dfrac{1}{b} = 20$

$\Rightarrow a+b=20ab$, from equation (1); we get

$\Rightarrow 18b^2+b=360b^3$

$\Rightarrow 360b^2-18b-1=0 \qquad \{\because b\neq 0\}$

$\Rightarrow b=\dfrac{18\pm\sqrt{324+1440}}{720}$

$\Rightarrow b=\dfrac{18+\sqrt{1764}}{720} \qquad [\because b>0]$

$\Rightarrow b=\dfrac{1}{12}$

$\Rightarrow a=18\times\dfrac{1}{144}=\dfrac{1}{8}$

Now, $16a+12b=16\times\dfrac{1}{8}+12\times\dfrac{1}{12}=3$

42. (60)

$a_4\cdot a_6=9\Rightarrow (a_5)^2=9\Rightarrow a_5=3$

and

$a_5+a_7=24\Rightarrow a_5+a_5r^2=24$

$\Rightarrow (1+r^2)=8\Rightarrow r=\sqrt{7}$

$\Rightarrow a=\dfrac{3}{49}$

$\Rightarrow a_1a_9+a_2a_4a_9+a_5+a_7=9+27+3+21=60$

43. (9)

Given that

$c_k=a_k+b_k$ and $a_1=b_1=4$

also $a_2=4r_1 \quad a_3=4r_1^2$

$b_2=4r_2 \quad b_3=4r_2^2$

Now $c_2=a_2+b_2=5$ and $c_3=a_3+b_3=\dfrac{13}{4}$

$\Rightarrow r_1+r_2=\dfrac{5}{4}$ and $r_1^2+r_2^2=\dfrac{13}{16}$

Hence $r_1r_2=\dfrac{3}{8}$ which gives $r_1=\dfrac{1}{2}$ and $r_2=\dfrac{3}{4}$

$\sum_{k=1}^{\infty}c_k-(12a_6+8b_4)$

$=\dfrac{4}{1-r_1}+\dfrac{4}{1-r_2}-\left(\dfrac{48}{32}+\dfrac{27}{2}\right)$

$=24-15=9$

44. (461)

As, $S=\dfrac{S_1}{2}+\dfrac{S_2}{2^2}+\ldots+\dfrac{S_9}{2^9}+\dfrac{S_{10}}{2^{10}}$

$\Rightarrow \dfrac{S}{2}=\dfrac{b_1}{2^2}+\dfrac{b_2}{2^3}+\ldots+\dfrac{b_9}{2^{10}}+\dfrac{b_{10}}{2^{11}}$

Subtracting

$\Rightarrow \dfrac{S}{2}=\dfrac{b_1}{2}+\left(\dfrac{a_1}{2^2}+\dfrac{a_2}{2^3}\ldots+\dfrac{a_9}{2^{10}}\right)-\dfrac{b_{10}}{2^{11}}$

$\Rightarrow S=b_1-\dfrac{b_{10}}{2^{11}}+\left(\dfrac{a_1}{2^2}+\dfrac{a_2}{2^3}+\ldots+\dfrac{a_9}{2^{10}}\right)$

Subtracting

$\Rightarrow \dfrac{S}{2}=\dfrac{b_1}{2}-\dfrac{b_{10}}{2^{11}}+\left(\dfrac{a_1}{2}-\dfrac{a_9}{2^{10}}\right)+\left(\dfrac{1}{2^2}+\dfrac{2}{2^3}+\ldots+\dfrac{8}{2^9}\right)$

$\Rightarrow \dfrac{S}{2}=\dfrac{a_1+b_1}{2}-\dfrac{(b_{10}+2a_9)}{2^{11}}+\dfrac{T}{4}$

$\Rightarrow 2S=2(a_1+b_1)-\dfrac{(b_{10}+2a_9)}{2^9}+T$

$\Rightarrow 2^7(2S-T)=2^8(a_1+b_1)-\dfrac{(b_{10}+2a_9)}{4}$

Given $a_n-a_{n-1}=n-1$

$\therefore a_2-a_1=1$

$a_3-a_2=2$

$\vdots$

$a_9-a_8=8$

$a_9-a_1=1+2+\ldots+8=36$

$\Rightarrow a_9=37(a_1=1)$

Also, $b_n-b_{n-1}=a_{n-1}$

$\therefore b_{10}-b_1=a_1+a_2+\ldots+a_9$

$=1+2+4+7+11+16+22+29+37$

$\Rightarrow b_{10}=130 \qquad$ (As $b_1=1$)

$2^7(2S-T)=461$

45. (C)

If $a_n=\dfrac{-2}{4n^2-16n+15}$ then $a_1+a_2+\ldots a_{25}$

$\Rightarrow \sum_{n=1}^{25}a_n=\sum\dfrac{-2}{4n^2-16n+15}$

$=\sum\dfrac{-2}{4n^2-6n-10n+15}$

$=\sum\dfrac{-2}{2n(2n-3)-5(2n-3)}$

$\sum\dfrac{-2}{(2n-3)(2n-5)}$

$$= \sum \frac{1}{2n-3} - \frac{1}{2n-5}$$

$$= \frac{1}{47} - \frac{1}{(-3)}$$

$$= \frac{1}{47} - \frac{1}{(-3)} = \frac{50}{141}$$

46. (26)

$$\sum_{n=0}^{\infty} \frac{n^3((2n)!) + (2n-1)(n!)}{(n!)((2n)!)}$$

$$= \sum_{n=0}^{\infty} \frac{1}{(n-3)!} + \sum_{n=0}^{\infty} \frac{3}{(n-2)!}$$

$$+ \sum_{n=0}^{\infty} \frac{1}{(n-1)!} + \sum_{n=0}^{\infty} \frac{1}{(2n-1)!} - \sum_{n=0}^{\infty} \frac{1}{(2n)!}$$

$$= e + 3e + e + \frac{1}{2}\left(e - \frac{1}{e}\right) - \frac{1}{2}\left(e + \frac{1}{e}\right)$$

$$= 5e - \frac{1}{e}$$

$$a^2 - b + c = 26$$

47. (B)

As a^3, b^3, c^3 be in A.P. $\rightarrow \boxed{a^3 + c^3 = 2b^3}$ (1)

$\log_a b, \log_c a, \log_b c$ are in G.P.

$$\therefore \frac{\log b}{\log a} \cdot \frac{\log c}{\log b} = \left(\frac{\log a}{\log c}\right)^2$$

$$\therefore (\log a)^3 = (\log c)^3 \Rightarrow \boxed{a = c} \quad (2)$$

From (1) and (2)

$\boxed{a = b = c}$

$$T_1 = \frac{a + 4b + c}{3} = 2a;$$

$$d = \frac{a - 8b + c}{10} = \frac{-6a}{10} = \frac{-3}{5}a$$

$$\therefore S_{20} = \frac{20}{2}\left[4a + 19\left(-\frac{3}{5}a\right)\right]$$

$$= 10\left[\frac{20a - 57a}{5}\right]$$

$$= -74a$$

$$\therefore -74a = -444 \Rightarrow \boxed{a = 6}$$

$$\therefore abc = 6^3 = 216$$

48. (C)

$$ax^2 + 2bx + c = 0$$

$$\Rightarrow ax^2 + 2\sqrt{ac}x + c = 0 \quad (\because b^2 = ac)$$

$$\Rightarrow \left(x\sqrt{a} + \sqrt{c}\right)^2 = 0$$

$$x^2 - \frac{\sqrt{c}}{\sqrt{a}} \quad (1)$$

Now, $dx^2 + 2ex + f = 0$

$$\Rightarrow d\left(\frac{c}{a}\right) + 2e\left[-\frac{\sqrt{c}}{\sqrt{a}}\right] + f = 0$$

$$\Rightarrow \frac{dc}{a} + f = 2e\sqrt{\frac{c}{a}}$$

$$\Rightarrow \frac{d}{a} + \frac{f}{c} = 2e\sqrt{\frac{1}{ac}}$$

$$\Rightarrow \frac{d}{a} + \frac{f}{c} = \frac{2e}{b} \quad [\text{as } b = \sqrt{ae}]$$

$\therefore \frac{d}{a}, \frac{e}{b}, \frac{f}{c}$ are in A.P.

49. (151)

$$T_8 = 11 + (8-1) \times 20$$

$$= 11 + 140 = 151$$

50. (8)

$2a_7 = a_5$ (given)

$$2(a_1 + 6d) = a_1 + 4d$$

$$a_1 + 8d = 0 \quad (1)$$

$$a_1 + 10d = 18 \quad (2)$$

By (1) and (2) we get $a_1 = -72, d = 9$

$$a_{18} = a_1 + 17d = -72 + 153 = 81$$

$$a_{10} = a_1 + 9d = 9$$

$$12\left(\frac{\sqrt{a_{11}} - \sqrt{a_{10}}}{d} + \frac{\sqrt{a_{12}} - \sqrt{a_{11}}}{d} + \ldots + \frac{\sqrt{a_{18}} - \sqrt{a_{17}}}{d}\right)$$

$$12\left(\frac{\sqrt{a_{18}} - \sqrt{a_{10}}}{d}\right) = \frac{12(9-3)}{9} = \frac{12 \times 6}{6} = 8$$

51. (A)

$$a + 6d = 3 \quad (1)$$

$$Z - a(a + 3d)$$

$= (3-6d)(3-3d)$
$= 18d^2 - 27d + 9$
Differentiating with respect to d
$\Rightarrow 36d - 27 = 0$
$\Rightarrow d = \frac{3}{4}$, from (1) $a = \frac{-3}{2}$, $(Z = \text{minimum})$

Now, $S_n = \frac{n}{2}\left(-3+(n-1)\frac{3}{4}\right) = 0$

$\Rightarrow n = 5$
Now,
$n! - 4a_{n(n+2)} = 120 - 4(a_{35})$
$= 120 - 4(a + (35-1)d)$
$= 120 - 4\left(\frac{-3}{2} + 34\cdot\left(\frac{3}{4}\right)\right)$
$= 120 - 4\left(\frac{-6+102}{4}\right)$
$= 120 - 96 = 24$

52. **(6952)**
Separating odd placed and even placed terms we get

$$S = \left(1\cdot 1^2 + 3\cdot 5^2 + \ldots 15\cdot(29)^2\right) - \left(2\cdot 3^2 + 4\cdot 7^2 + \ldots + 14\cdot(27)^2\right)$$

$$S = \sum_{n=1}^{8}(2n-1)(4n-3)^2 - \sum_{n=1}^{7}(2n)(4n-1)^2$$

Applying the summation formula we get
$= 29856 - 22904 = 6952$

53. **(B)**

$$T_r = \frac{(r^2+r+1)-(r^2-r+1)}{2(r^4+r^2+1)}$$

$$\Rightarrow T_r = \frac{1}{2}\left[\frac{1}{r^2-r+1} - \frac{1}{r^2+r+1}\right]$$

$$T_1 = \frac{1}{2}\left[\frac{1}{1} - \frac{1}{3}\right]$$

$$T_2 = \frac{1}{2}\left[\frac{1}{3} - \frac{1}{7}\right]$$

$$T_3 = \frac{1}{2}\left[\frac{1}{7} - \frac{1}{13}\right]$$

$$T_{10} = \frac{1}{2}\left[\frac{1}{91} - \frac{1}{111}\right]$$

$$\Rightarrow \sum_{r=1}^{10} T_r = \frac{1}{2}\left[1 - \frac{1}{111}\right] = \frac{55}{111}$$

54. **(754)**
$a_1 + a_2 + a_3 + a_4 = 50$
$\Rightarrow 32 + 6d = 50$
$\Rightarrow d = 3$
and $a_{n-3} + a_{n-2} + a_{n-1} + a_n = 170$
$\Rightarrow 32 + (4n-10)\cdot 3 = 170$
$\Rightarrow n = 14$
$a_7 = 26$, $a_8 = 29$
$\Rightarrow a_7 \cdot a_8 = 754$

55. **(514)**
Divisible by 2 $\rightarrow$ 450
Divisible by 3 $\rightarrow$ 300
Divisible by 7 $\rightarrow$ 128
Divisible by 2 and 7 $\rightarrow$ 64
Divisible by 3 and 7 $\rightarrow$ 43
Divisible by 2 and 3 $\rightarrow$ 150
Divisible by 2, 3 and 7 $\rightarrow$ 21
$\therefore$ Total numbers
$= 450 + 300 - 150 - 64 - 43 + 21 = 514$

56. **(321)**
$3, 7, 11, 15, \ldots, 399 \quad d_1 = 4$
$2, 5, 8, 11, \ldots, 359 \quad d_2 = 3$
$2, 7, 12, 17, \ldots, 197 \quad d_3 = 5$
LCM $(d_1, d_2, d_3) = 60$
Common terms are 47, 107, 167
Sum = 321

57. **(B)**
Let the given series be S_n, then we can write as
$S_n = 5 + 11 + 19 + 29 + 41 + \ldots + T_n$
$S_n = 5 + 11 + 19 + 29 + 41 + \ldots + T_{n-1} + T_n$
Subtracting above equations, we get
$0 = 5 + 6 + 8 + 10 + \ldots - T_n$

$$\Rightarrow T_n = 5 + \left[\frac{(n-1)}{2}(12+(n-2)2)\right]$$

$\Rightarrow T_n = 5 + (n-1)(n+4)$

$\Rightarrow T_n = n^2 + 3n + 1$

So,

$$S_{29} = \sum_{n=1}^{20} T_n$$

$$\Rightarrow S_{20} = \sum_{n=1}^{20}\left(n^2 + 3n + 1\right)$$

$$\Rightarrow S_{20} = \frac{20\times 21\times 41}{6} + \frac{3\times 20\times 21}{2} + 20$$

$$\Rightarrow S_{20} = 3520$$

58. (C)

Given,

$a_1, a_2, a_3, \ldots, a_n$ are terms of an A.P.,

So common difference will be,

$d = a_2 - a_1 = a_3 - a_2 = \ldots = a_n - a_{n-1}$

Now solving,

$$\lim_{n\to\infty}\sqrt{\frac{d}{n}}\left(\frac{1}{\sqrt{a_1}+\sqrt{a_2}} + \frac{1}{\sqrt{a_2}+\sqrt{a_3}} + \ldots + \frac{1}{\sqrt{a_{n-1}}+\sqrt{a_n}}\right)$$

$$= \lim_{n\to\infty}\sqrt{\frac{d}{n}}\times\frac{1}{d}\left(\sqrt{a_n}-\sqrt{a_1}\right)$$

Now using the formula $a_n = a_1 + (n-1)d$ we get,

$$= \lim_{n\to\infty}\frac{1}{\sqrt{d}}\left(\frac{\sqrt{a_1+(n-1)d}-\sqrt{a_1}}{\sqrt{n}}\right)$$

$$= \lim_{n\to\infty}\frac{1}{\sqrt{d}}\left(\frac{\sqrt{n}\left(\sqrt{\frac{a_1}{n}+d-\frac{d}{n}}-\sqrt{\frac{a_1}{n}}\right)}{\sqrt{n}}\right)$$

$$= \frac{1}{\sqrt{d}}\left(\left(\sqrt{0+d-0}-\sqrt{0}\right)\right)$$

$$= \frac{1}{\sqrt{d}}\times\sqrt{d} = 1$$

59. (A)

Let

$$S = 1^1 - 2^2 + 3^2 - 4^2 + \ldots + (2021)^2 - (2022)^2 + (2023)^2$$

$$\Rightarrow S = (1-2)(1+2) + (3-4)(3+4) + \ldots + (2021-2022)(2021+2022) + (2023)^2$$

$$\Rightarrow S = -[3+7+11+15+\ldots..+4043] + (2023)^2$$

The number of terms in the bracket are $\frac{2022}{2} = 1011$

$$\Rightarrow S = -\frac{1011}{2}(6+1010\times 4) + (2023)^2$$

$$\Rightarrow S = -1011\times 2023 + (2023)^2$$

$$\Rightarrow S = 2023\times 1012$$

So,

$m = 17, n = 7$ and G.C.D.$(17, 7) = 1$

Hence, $m^2 - n^2 = 17^2 - 7^2 = 240$

Hence this is the correct option.

60. (400)

Let,

$$S = (20)^{19} + 2(21)(20)^{18} + 3(21)^2(20)^{17} + \ldots + 20(21)^{19}$$

$$\Rightarrow S = 20^{19} + 2\times(20)^{19}\times\left(\frac{21}{20}\right) + 3(20)^{19}\left(\frac{21}{20}\right)^2 + \ldots + 20(20)^{19}\times\left(\frac{21}{20}\right)^{19}$$

$$\Rightarrow S = 20^{19}\left(1 + 2\times\frac{21}{20} + 3\left(\frac{21}{20}\right)^2 + \ldots + 20\left(\frac{21}{20}\right)^{19}\right)$$

Now on comparing with $S = k\times(20)^{19}$ we get,

$$\Rightarrow k = 1 + 2\times\left(\frac{21}{20}\right) + 3\left(\frac{21}{20}\right)^2 + \ldots + 20\left(\frac{21}{20}\right)^{19} \quad (1)$$

$$\Rightarrow \frac{21}{20}k = \frac{21}{20} + 2\left(\frac{21}{20}\right)^2 + \ldots + 19\left(\frac{21}{20}\right)^{19} + 20\left(\frac{21}{20}\right)^{20} \quad (2)$$

Now on subtracting above equations we get,

$$\Rightarrow k - \frac{21}{20}k = 1 + \frac{21}{20} + \left(\frac{21}{20}\right)^2 + \ldots + \left(\frac{21}{20}\right)^{19} - 20\left(\frac{21}{20}\right)^{20}$$

$$\Rightarrow \frac{-k}{20}=\frac{\left(\frac{21}{20}\right)^{20}-1}{\frac{21}{20}-1}-20\left(\frac{21}{20}\right)^{20}$$

$$\Rightarrow -\frac{k}{20}=\left(\left(\frac{21}{20}\right)^{20}-1\right)\times 20-20\left(\frac{21}{20}\right)^{20}$$

$$\Rightarrow \frac{-k}{20}=-20$$

$$\Rightarrow k=400$$

61. (D)

Given that $S_K=\frac{1+2+\ldots+K}{K}$

We know that $1+2+3\ldots+n=\frac{n(n+1)}{2}$

$$\Rightarrow S_K=\frac{K(K+1)}{2K}=\frac{K+1}{2}$$

Now,

$$\sum_{j=1}^{n}(S_j)^2=\sum_{j=1}^{n}\frac{1}{4}\left(2^2+3^2+4^2+\ldots(n+1)^2\right)$$

$$=\sum_{j=1}^{n}\frac{1}{4}\left(1^2+2^2+3^2+4^2\ldots(n+1)^2-1^2\right)$$

$$\Rightarrow \frac{1}{4}\left(\frac{(n+1)(n+2)(2n+3)}{6}-1\right)$$

$$=\frac{1}{4}\left(\frac{(n^2+3n+2)(2n+3)-6}{6}\right)$$

$$=\frac{1}{4}\left(\frac{2n^3+6n^2+4n+3n^2+9n+6-6}{6}\right)$$

$$=\frac{1}{4}\left(\frac{2n^3+9n^2+13n}{6}\right)$$

$$=\frac{n}{24}\left(2n^2+9n+13\right)$$

On comparing this with

$\sum_{j=1}^{n}S_j^2=\frac{n}{A}\left(Bn^2+Cn+D\right)$ we get,

$A=24, B=2, C=9, D=13$

$$\Rightarrow \frac{A+B}{D}=\frac{26}{13}=2$$

That means, $A+B$ is divisible by D

Hence this is the correct option.

62. (C)

The given series is a_n, then we can write as

$a_n=5+8+14+23+\ldots+a_n$

$a_n=5+8+14+23+\ldots+a_{n-1}+a_n$

Subtracting above equations, we get

$0=5+3+6+9+\ldots+(a_n-a_{n-1})-a_n$

$$\Rightarrow a_n=5+\left[\frac{(n-1)}{2}\left(2\times 3+(n-1-1)3\right)\right]$$

$$\Rightarrow a_n=5+\frac{(n-1)}{2}(3n)$$

$$\Rightarrow a_n=\frac{1}{2}\left(3n^2-3n+10\right)$$

So,

$$\Rightarrow a_{40}=\frac{1}{2}\left(3\times 40^2-3\times 40+10\right)$$

$$\Rightarrow a_{40}=\frac{1}{2}(4800-120+10)$$

$$\Rightarrow a_{40}=2345$$

$$\Rightarrow S_{30}=\sum_{n=1}^{30}a_n$$

$$\Rightarrow S_{30}=\frac{1}{2}\left(3\sum_{n=1}^{30}(n^2)-3\sum_{n=1}^{30}(n)+\sum_{n=1}^{30}10\right)$$

$$=\frac{1}{2}\left(\frac{3(30)(31)(61)}{6}-\frac{3(30)(31)}{2}+10\times 30\right)$$

$$\Rightarrow S_{30}=13635$$

$$\Rightarrow S_{30}-a_{40}=13635-2345=11290$$

Therefore, the required value is 11290.

63. (150)

Given that $\frac{1}{x},\frac{1}{y},\frac{1}{z}$ are in A.P. and $x,\sqrt{2y},z$ are in G.P.

As given, $\frac{2}{y}=\frac{1}{x}+\frac{1}{y}$ (1)

Also, $2y^2=xz$ (2)

Also given that $xy+yz+zx=\frac{3}{\sqrt{2}}xyz$

$\Rightarrow \frac{1}{x}+\frac{1}{y}+\frac{1}{z}=\frac{3}{\sqrt{2}}$ (3)

From (1) and (3) we get $\frac{3}{y}=\frac{3}{\sqrt{2}}$

$y=\sqrt{2}$ (4)

Now from (2) $xy = 4$ (5)

Now using (2), (4) and (5)

$\Rightarrow x+z = 4\sqrt{2}$

Hence $3(x+y+z)^2 = 3(\sqrt{2}+4\sqrt{2})^2$

$= 150$

64. (B)

Given that,

$a^2 + (ar)^2 + (ar^2)^2 = 33033, (a, r \in N)$

$\Rightarrow a^2(1+r^2+r^4) = 11^2 \times 273$

On comparing both sides of the equation we get,

$a = 11$ and $1+r^2+r^4 = 273$

$\Rightarrow r^2 + r^4 = 272$

$\Rightarrow r^2(1+r^2) = 16\times 17$

$\Rightarrow r = 4$

Now $a + ar + ar^2 = 11 + 11\times 4 + 11\times 16$

$= 11 + 44 + 176 = 231$

65. (9525)

Given sequence is $3, 8, 13, \ldots, 373$

Here, $a = 3, d = 5$ and $a_n = 373$

Now,$a_n = a + (n-1)d$

$\Rightarrow 373 = 3 + (n-1)5$

$\Rightarrow n = 75$

We know that $S_n = \frac{n}{2}(a + a_n)$

$\Rightarrow S_{75} = \frac{75}{2}(3+373)$

$\Rightarrow S_{75} = 14100$

Now let us write the sequence of terms which are divisible by 3.

We get, $3, 18, 33, \ldots, 363$

$\Rightarrow 363 = 3(n-1)15$

$\Rightarrow n = 25$

Now let us find the sum of terms divisible by 3.

$\Rightarrow S_{\text{div by 3}} = \frac{25}{2}(3+363) = 4575$

Required sum = Sum of 75 terms – Sum of terms divisible by 3

$= 14100 - 4575 = 9525$

66. (A)

Given,

$S_n = 4 + 11 + 21 + 34 + \ldots + T_n$

$S_n = 4 + 11 + 21 + 34 + \ldots + T_{n-1} + T_n$

Subtracting above equations, we get

$0 = 4 + 7 + 10 + 13 + \ldots - T_n$

$\Rightarrow T_n = 4 + 7 + 10 + 13 + \ldots$

The above series is in A.P.

$\Rightarrow T_n = \frac{n}{2}(2\times 4 + (n-1)3)$

$\Rightarrow T_n = \frac{n}{2}(3n+5)$

So,

$S_n = \sum \frac{3n^2 + 5n}{2}$

$\Rightarrow S_n = \frac{1}{2}\left(\frac{3n(n+1)(2n+1)}{6} + \frac{5n(n+1)}{2}\right)$

$\Rightarrow S_n = \frac{n(n+1)}{2}\left(\frac{(2n+1)}{2} + \frac{5}{2}\right)$

$\Rightarrow S_{29} = \frac{29\times 30}{2}\left(\frac{59}{2} + \frac{5}{2}\right)$

$\Rightarrow S_{29} = 29\times 15\times 32 = 13920$

$\Rightarrow S_9 = \frac{9\times 10}{2}\left(\frac{19}{2} + \frac{5}{2}\right)$

$\Rightarrow S_9 = 9\times 5\times 12 = 540$

$\Rightarrow \frac{S_{29} - S_9}{60} = \frac{13920 - 540}{60} = 223$

67. (16)

Given, $a_1, a_2, 2, a_3, a_4$ are in an arithmetico–geometric progression, and the common ratio of the corresponding geometric progression is 2 and the sum of all 5 terms of the arithmetico–geometric progression is $\frac{49}{2}$

So, common ratio Here $r = 2$ and let common difference $= d$

And we know that,

$a_2 = (a_1 + d)\times 2$

$2 = (a_1 + 2d)\times 2^2$ (1)

$a_3 = (a_1 + 3d)\times 2^3$

$a_4 = (a_1 + 4d) \times 2^4$

Now adding all above equations we get,

$$31a_1 + 98d = \frac{49}{2} \quad (2)$$

Now from (1) and (2)

$$a_1 = 0, d = \frac{1}{4}$$

Hence, $a_4 = 16(0+1) = 16$

68. **(C)**

Given,

Mean of $x_1, x_2, \ldots, x_{100}$ is 200

So, the sum of observation will be,

$$\sum x_i = 100 \times 200$$

Now using the sum of A.P. formula in above equation as all terms are in a arithmetic progression we get,

$$\frac{100}{2}(x_1 + x_{100}) = 100 \times 200$$

$$\Rightarrow 50(2 + x_{100}) = 100 \times 200$$

$$\Rightarrow x_{100} = 398$$

$$\Rightarrow x_1 + 99d = 398$$

$$\Rightarrow d = 4$$

Now, $x_i = 2 + (i-1)4 = 4i - 2$

So, $y_i = i(x_i - i) = 3i^2 - 2i$

Now finding mean we get,

$$\bar{y} = \frac{1}{100}\sum y_i$$

$$\Rightarrow \bar{y} = \frac{1}{100}\sum(3i^2 - 2i)$$

$$\Rightarrow \bar{y} = \frac{1}{100}\left[3 \times \frac{100 \times 101 \times 201}{6} - 2\frac{100 \times 101}{2}\right]$$

$$\Rightarrow \bar{y} = \left[\frac{101 \times 201}{2} - 101\right]$$

$$\Rightarrow \bar{y} = 10049.5$$

69. **(2175)**

Given,

$$S = 109 + \frac{108}{5} + \frac{107}{5^2} + \ldots + \frac{2}{5^{107}} + \frac{1}{5^{108}} \quad (1)$$

$$\frac{S}{5} = \frac{109}{5} + \frac{108}{5^2} + \ldots + \ldots + \frac{2}{5^{108}} + \frac{1}{5^{109}} \quad (2)$$

Now subtracting above equations we get,

$$\frac{-4S}{5} = -109 + \left(\frac{1}{5} + \frac{1}{5^2} + \ldots + \frac{1}{5^{108}} + \frac{1}{5^{109}}\right)$$

$$\Rightarrow \frac{-4S}{5} = -109 + \frac{1}{5}\frac{\left(1 - \left(\frac{1}{5}\right)^{109}\right)}{1 - \frac{1}{5}}$$

$$\Rightarrow \frac{4S}{5} = -109 + \frac{1}{4}\left(1 - \frac{1}{5^{109}}\right)$$

$$\Rightarrow \frac{16S}{5} = -436 + \left(1 - \frac{1}{5^{109}}\right)$$

$$\Rightarrow 16S = -436 \times 5 + \left(5 - 5^{-108}\right)$$

$$\Rightarrow 16S - (25)^{-54} = 2175$$

70. **(A)**

Given

$$a + b + c + d = 11$$

We know that,

A.M. $\geq$ G.M.

$$\Rightarrow \frac{\frac{a}{5} + \frac{a}{5} + \frac{a}{5} + \frac{a}{5} + \frac{a}{5} + \frac{b}{3} + \frac{b}{3} + \frac{b}{3} + \frac{c}{2} + \frac{c}{2} + \frac{d}{1}}{11} \geq \sqrt[11]{\frac{a^5b^3c^2d}{5^5 3^3 2^2}}$$

$$\Rightarrow \frac{a+b+c+d}{11} \geq \sqrt[11]{\frac{a^5b^3c^2d}{5^5 3^3 2^2}}$$

$$\Rightarrow \frac{11}{11} \geq \sqrt[11]{\frac{a^5b^3c^2d}{5^5 3^3 2^2}}$$

$$\Rightarrow \sqrt[11]{\frac{a^5b^3c^2d}{5^5 3^3 2^2}} \leq 1$$

$$\Rightarrow a^5b^3c^2d \leq 5^5 3^2 2^2$$

$$\Rightarrow a^5b^3c^2d \leq 90 \times 3750$$

So, $\beta = 90$

71. **(B)**

Given,

$$10 = 1 + \frac{4}{k} + \frac{8}{k^2} + \frac{13}{k^3} + \frac{19}{k^4} + \ldots\infty \quad (1)$$

$$\frac{10}{k} = \frac{1}{k} + \frac{4}{k^2} + \frac{8}{k^3} + \frac{13}{k^4} + \ldots\infty \quad (2)$$

On subtracting above two equations we get,

$$10-\frac{10}{k}=1+\frac{3}{k}+\frac{4}{k^2}+\frac{5}{k^3}+\frac{6}{k^4}\ldots\infty$$

$$\Rightarrow 10\left(1-\frac{1}{k}\right)=1+\frac{3}{k}+\frac{4}{k^2}+\frac{5}{k^3}+\frac{6}{k^4}\ldots\infty \quad (3)$$

$$\Rightarrow 10\left(1-\frac{1}{k}\right)\left(\frac{1}{k}\right)=\frac{1}{k}+\frac{3}{k^2}+\frac{4}{k^3}+\frac{5}{k^4}+\frac{6}{k^5}\ldots\infty \quad (4)$$

Again subtracting both equations we get,

$$\Rightarrow 10\left(1-\frac{1}{k}\right)^2=1+\frac{2}{k}+\frac{1}{k^2}+\frac{1}{k^3}+\frac{1}{k^4}\ldots\infty$$

$$\Rightarrow 10\left(1-\frac{1}{k}\right)^2=\left(1+\frac{1}{k}+\frac{1}{k^2}+\frac{1}{k^3}+\frac{1}{k^4}\ldots\infty\right)+\frac{1}{k}$$

$$\Rightarrow 10\left(1-\frac{1}{k}\right)^2=\left(\frac{1}{1-\frac{1}{k}}\right)+\frac{1}{k}$$

$$\Rightarrow 10(k-1)^3=k\left(k^2+k-1\right)$$

$$\Rightarrow 9k^3-31k^2+31k-10=0$$

$$\Rightarrow k=2$$

72. **(C)**

Given,

$$S_n=\sum_{i=1}^{n}a_i=\frac{n^2+3n}{(n+1)(n+2)}$$

Now we know that,

$$a_n=S_n-S_{n-1}$$

$$\Rightarrow a_n=\frac{n^2+3n}{(n+1)(n+2)}-\frac{(n-1)^2\,3(n-1)}{n(n+1)}$$

$$\Rightarrow a_n=\frac{4}{n(n+1)(n+2)}$$

Now solving, $\sum_{k=1}^{10}\frac{1}{a_k}=\frac{1}{4}\sum_{k=1}^{10}k(k+1)(k+2)$

$$\Rightarrow \sum_{k=1}^{10}\frac{1}{a_k}=\frac{1}{16}\sum_{k=1}^{10}k(k+1)(k+2)(k+3)-(k-1)k(k+1)(k+2)$$

$$\Rightarrow \sum_{k=1}^{10}\frac{1}{a_k}=\frac{1}{16}\big[(1\times2\times3\times4-0)+(2\times3\times4\times5-1\times2\times3\times4)\ldots+(10\times11\times12\times13-9\times10\times11\times12)\big]$$

$$\Rightarrow \sum_{k=1}^{10}\frac{1}{a_k}=\frac{1}{16}[10\times11\times12\times13-0]$$

$$\Rightarrow \sum_{k=1}^{10}\frac{1}{a_k}=\frac{1}{2}(5\times11\times3\times13)$$

$$\Rightarrow 28\sum_{k=1}^{10}\frac{1}{a_k}=\frac{28\times5\times11\times3\times13}{2}$$

$$\Rightarrow 28\sum_{k=1}^{10}\frac{1}{a_k}=2\times3\times5\times7\times11\times13$$

Hence, there are six prime numbers in multiplications,

So, $m=6$

73. **(C)**

The first terms of the A.P. are 1, 2, 3, …10, so the general term of the first term will be i

And common differences are 1, 3, 5 ,…, so the general term of common difference is given by $2i-1$

Now the sum of the A.P. is given by,

$$S_i=\frac{12}{2}\left[2\times i+(12-1)(2i-1)\right]$$

$$\Rightarrow S_i=6\left[2\times i+11(2i-1)\right]$$

$$\Rightarrow S_i=144i-66$$

So, $\sum_{i=1}^{10}S_i=\sum_{i=1}^{10}144i-66\sum_{i=1}^{10}1$

$$\Rightarrow \sum_{i=1}^{10}S_i=144\left(\frac{10\times11}{2}\right)-66\times10$$

$$\Rightarrow \sum_{i=1}^{10}S_i=7260$$

74. **(1310)**

Given,

$2\times2^2-3^2+2\times4^2-5^2+2\times6^2-\ldots$ 20 terms,

$$\sum_{r=1}^{10}\left(2\times(2r)^2-(2r+1)^2\right)$$

$$=\sum_{r=1}^{10}\left(8r^2-4r^2-4r-1\right)$$

$$=\sum_{r=1}^{10}\left(4r^2-4r-1\right)$$

$$4\sum_{r=1}^{10}r^2-4\sum_{r=1}^{10}r-\sum_{r=1}^{10}1$$

$$=\frac{4\times10\times11\times21}{6}-4\frac{10\times11}{2}-10$$

$$=44\times35-220-10=1540-230=1310$$

75. **(A)**

Given,

$a_1, a_2, a_3, \ldots$ be a G.P. of increasing positive numbers,

And the sum of its 6th and 8th terms be 2

So, $a_6+a_8=2$

$$\Rightarrow ar^5+ar^7=2 \qquad (1)$$

And the product of its 3rd and 5th terms be $\frac{1}{9}$,

So, $a_3\cdot a_5=\frac{1}{9}\Rightarrow a^2\cdot r^2\cdot r^4=\frac{1}{9}$

$$\Rightarrow ar^3=\frac{1}{3}$$

Now from putting the value of $ar^3=\frac{1}{3}$ in equation (1) we get,

$$\frac{r^2}{3}+\frac{r^4}{3}=2$$

$$\Rightarrow r^4+r^2=6$$

$$\Rightarrow (r^2+3)(r^2-2)=0$$

$$\Rightarrow r^2=2$$

$$ar^3=\frac{1}{3}\Rightarrow ar\times2=\frac{1}{3}\Rightarrow ar=\frac{1}{6}$$

Now finding the value of

$6(a_2+a_4)(a_4+a_6)$, we get

$$=6(ar+ar^3)(ar^3+ar^5)$$

$$=6\left(\frac{1}{6}+\frac{1}{3}\right)\left(\frac{1}{3}+\frac{2}{3}\right)$$

$$=6\cdot\frac{1}{2}\cdot1=3$$

76. **(825)**

Given,

$$[\sqrt{1}]+[\sqrt{2}]+[\sqrt{3}]+\ldots+[\sqrt{120}]$$

Now we now that,

$[\sqrt{x}]=1$, when $x\in[1,4)$, $[\sqrt{x}]=2$, where $x\in[4,9)$ and similarly $[\sqrt{x}]=10$, when $x\in[100,120)$

Now using the above formula we get,

$$E=1+1+1+2+2+2+2+2+3+3+3+3+3+3+3+4+4+\ldots$$

$$\Rightarrow E=3\times1+5\times2+7\times3+\ldots+19\times9+10\times21$$

$$\Rightarrow E=\sum_{r=1}^{10}(2r+1)r$$

$$\Rightarrow E=2\sum_{r=1}^{10}(r^2+r)$$

$$\Rightarrow E=2\left[\frac{10\times11\times21}{6}\right]+\frac{10\times11}{2}$$

$$\Rightarrow E=770+55$$

$$\Rightarrow E=825$$

77. **(10)**

Given,

$$f(x)=\sum_{k=1}^{10}k\cdot x^k,\ x\in R$$

$$\Rightarrow f(x)=x+2x^2+3x^3+\ldots+10x^{10}$$

Now let $S=x+2x^2+3x^3+\ldots+10x^{10}$

So, $S\cdot x=x^2+2x^3+\ldots+9x^{10}+10x^{11}$

Now subtracting above equations we get,

$$S(1-x)=x+x^2+x^3+\ldots+x^{10}-10x^{11}$$

$$\Rightarrow S(1-x)=\frac{x(1-x^{10})}{1-x}-10x^{11}$$

$$\Rightarrow S=\frac{x(1-x^{10})}{(1-x)^2}-\frac{10x^{11}}{(1-x)}=f(x)$$

Now finding, $f(2)=2(1-2^{10})+10\times2^{11}$

$$\Rightarrow f(2)=2+18\times2^{10}$$

Now finding $f'(x)$ we get,

$$f'(x)=\frac{-10x^{11}}{(1-x)^2}-\frac{110x^{10}}{1-x}-\frac{10x^{10}}{(1-x)^2}+\frac{2x(1-x^{10})}{(1-x)^3}+\frac{1-x^{10}}{(1-x)^2}$$

$$\Rightarrow f'(2)=\frac{-10\times2^{11}}{(1-2)^2}-\frac{110\times2^{10}}{1-2}-\frac{10\times2^{10}}{(1-2)^2}+\frac{2\times2\times(1-2^{10})}{(1-2)^3}+\frac{1-2^{10}}{(1-2)^2}$$

$\Rightarrow f'(2) = 83\times 2^{10} - 3$

$2f(2) + f'(2) = 2(2 + 18\times 2^{10}) + 83\times 2^{10} - 3$

$= 119(2)^{10} + 1$

$\therefore n = 10$

78. **(A)**

Let the two numbers be a, b

$\Rightarrow a, A_1, A_2, b$ are in A.P.

$\Rightarrow b = a(4-1)d$

$\Rightarrow d = \dfrac{b-a}{3}$

$\Rightarrow A_1 = a + \dfrac{b-a}{3} = \dfrac{2a+b}{3}$

$\Rightarrow A_2 = a + \dfrac{b-a}{3}\times 2 = \dfrac{a+2b}{3}$

Similarly a, G_1, G_2, G_3, b are in G.P.

$\Rightarrow b = a(r)^{5-1}$

$\Rightarrow r = \left(\dfrac{b}{a}\right)^{\frac{1}{4}}$

$\Rightarrow G_1 = a\left(\dfrac{b}{a}\right)^{\frac{1}{4}}$

$\Rightarrow G_2 = a\left(\dfrac{b}{a}\right)^{\frac{2}{4}}$

$\Rightarrow G_3 = a\left(\dfrac{b}{a}\right)^{\frac{3}{4}}$

$= (G_1)^4 + (G_2)^4 + (G_3)^4 + (G_1)^2\times(G_3)^2$

$\Rightarrow a^4\times\dfrac{b}{a} + a^4\times\dfrac{b^2}{a^2} + a^4\times\dfrac{b^3}{a^3} + a^4\times\dfrac{b^2}{a^2}$

$= ba^3 + b^2a^2 + b^3a + a^2b^2$

$= ab(a^2 + b^2 + 2ab) = ab(a+b)^2$

$\Rightarrow (A_1 + A_2)^2 \times G_1G_3 = (a+b)^2\times ab$

79. **(7)**

Given,

Series,

$\left(\dfrac{1}{2} - \dfrac{1}{3}\right) + \left(\dfrac{1}{2^2} - \dfrac{1}{2\times 3} + \dfrac{1}{3^2}\right) +$

$\left(\dfrac{1}{2^3} - \dfrac{1}{2^2\times 3} + \dfrac{1}{2\times 3^2} - \dfrac{1}{3^3}\right)$

$+\left(\dfrac{1}{2^4} - \dfrac{1}{2^3\times 3} + \dfrac{1}{2^2\times 3^2} - \dfrac{1}{2\times 3^3} + \dfrac{1}{3^4}\right) + \ldots$

Now let $a = \dfrac{1}{2}, b = \dfrac{1}{3}$

So, the series will be:

$(a-b) + (a^2 - ab + b^2) + (a^3 - a^2b + ab^2 - b^3) + \ldots$

$= \dfrac{1}{a+b}\left((a^2 - b^2) + (a^3 + b^3) + (a^4 - b^4) + \ldots\right)$

$= \left(\dfrac{1}{a+b}\right)\left(a^2 + a^3 + a^4 + \ldots - (b^2 - b^3 + b^4 + \ldots)\right)$

$= \left(\dfrac{1}{a+b}\right)\left(\dfrac{a^2}{1-a} - \dfrac{b^2}{1+b}\right)$

Now putting the values of a and b we get,

$= \dfrac{1}{\frac{1}{2}+\frac{1}{3}}\left(\dfrac{\frac{1}{4}}{1-\frac{1}{2}} - \dfrac{\frac{1}{9}}{1+\frac{1}{3}}\right)$

$= \dfrac{6}{5}\left(\dfrac{1}{2} - \dfrac{1}{12}\right)$

$= \dfrac{6}{5}\left(\dfrac{5}{12}\right) = \dfrac{1}{2} = \dfrac{\alpha}{\beta}$

Hence, $\alpha + 3\beta = 1 + 6 - 7$

80. **(D)**

$3, a, b, c \to$ A.P. $\Rightarrow 3, 3+d, 3+2d, 3+3d$

$3, a-1, b+1, c+9 \to$ G.P.

$\Rightarrow 3, 2+d, 4+2d, 12+3d$

$a = 3 + d(2+d)^2 = 3(4+2d)$

$b = 3 + 2d$

$d = 4, -2$

$c = 3 + 3d$

If $d = 4$ G.P. $\Rightarrow$ 3, 6, 12, 24

$a = 7$

$b = 11$

$c = 15$

$\dfrac{a+b+c}{3} = 11$

81. **(6699)**

$3, 7, 11, 15, \ldots, 403$

$2, 5, 8, 11, \ldots, 404$

LCM $(4, 3) - 12$

$11, 23, 35, \ldots \text{let } (403)$

$403 = 11 + (n-1) \times 12$

$\frac{392}{12} = n-1$

$33 \times 66 = n$

$n = 33$

$\text{Sum } \frac{33}{2}(22 + 32 \times 12) = 6699$

82. **(C)**

$S_{10} = 390$

$\frac{10}{2}[2a + (10-1)d] = 390$

$\Rightarrow 2a + 9d = 78 \quad (1)$

$\frac{t_{10}}{t_5} = \frac{15}{7} \Rightarrow \frac{a+9d}{a+4d} = \frac{15}{7} \Rightarrow 8a = 3d \quad (2)$

From (1) and (2), $a = 3$ and $d = 8$

$S_{15} - S_5 = \frac{15}{2}(6 + 14 \times 8) - \frac{5}{2}(6 + 4 \times 8)$

$= \frac{15 \times 118 - 5 \times 38}{2} = 790$

83. **(1)**

$a, ar, ar^2 \rightarrow$ G.P.

Sum of any two sides > third side

$a + ar > ar^2, a + ar^2 > ar, ar + ar^2 > a$

$r^2 - r - 1 < 0$

$r \in \left(\frac{1-\sqrt{5}}{2}, \frac{1+\sqrt{5}}{2}\right) \quad (1)$

$r^2 - r + 1 > 0$

always true

$r^2 + r - 1 > 0$

$r \in \left(-\infty, -\frac{1-\sqrt{5}}{2}\right) \cup \left(\frac{-1+\sqrt{5}}{2}, \infty\right) \quad (2)$

Taking intersection of (1), (2)

$r \in \left(-\frac{1+\sqrt{5}}{2}, \frac{1+\sqrt{5}}{2}\right)$

As $r > 1$

$r \in \left(1, \frac{1+\sqrt{5}}{2}\right)$

$[r] = 1 [-r] = -2$

$3[r] + [-r] = 1$

84. **(C)**

4, 9, 14, 19, ..., up to the 25th term

$T_{25} = 4 + (25-1)5 = 4 + 120 = 124$

3, 6, 9, 12, ..., up to 37th term

$T_{37} = 3 + (37-1)3 = 3 + 108 = 111$

Common difference of Ist series $d_1 = 5$

Common difference of IInd series $d_2 = 3$

First common term = 9, and

their common difference = 15

(LCM of d_1 and d_2)

then common terms are

9, 24, 39, 54, 69, 84, 99

85. **(9)**

$8 = \frac{3}{1-\frac{1}{4}} + \frac{p \cdot \frac{1}{4}}{\left(1-\frac{1}{4}\right)^2}$

(Sum of infinite terms of A.G.P.)

$= \frac{a}{1-r} + \frac{dr}{(1-r)^2}$

$\Rightarrow \frac{4p}{9} = 4 \Rightarrow p = 9$

86. **(C)**

$20, 19\frac{1}{4}, 18\frac{1}{2}, 17\frac{3}{4}, \ldots, -129\frac{1}{4}$

This is an A.P. with a common difference

$d_1 = -1 + \frac{1}{4} = -\frac{3}{4}$

$-129\frac{1}{4}, \ldots 19\frac{1}{4}, 20$

This is also an A.P. $a = -129\frac{1}{4}$ and $d = \frac{3}{4}$

Required term

$= -129\frac{1}{4} + (20-1)\left(\frac{3}{4}\right)$

$= -129 - \frac{1}{4} + 15 - \frac{3}{4} = -115$

87. **(D)**

$a + ar + ar^2 + ar^3 + \ldots + ar^{63}$

$= 7(a + ar^2 + ar^4 \ldots + ar^{62})$

$\Rightarrow \frac{a(1-r^{64})}{1-r} = \frac{7a(1-r^{64})}{1-r^2}$

$r = 6$

88. (B)

$a_6 = 2 \Rightarrow a + 5d = 2$

$a_1a_4a_5 = a(a+3d)(a+4d)$

$= (2-5d)(2-2d)(2-d)$

$f(d) = 8 - 32d + 34d^2 - 20d + 30d^2 - 10d^3$

$f'(d) = -2(5d-8)(3d-2)$

(sign scheme: $-$, $+$, $-$ with critical points $\frac{2}{3}$, $\frac{8}{5}$)

$d = \frac{8}{5}$

89. (A)

$\log_e a, \log_e b, \log_e c$ are in A.P.

$\therefore b^2 = ac$ (i)

Also

$\log_e\left(\frac{a}{2b}\right), \log_e\left(\frac{2b}{3c}\right), \log_e\left(\frac{3c}{a}\right)$ are in A.P.

$\left(\frac{2b}{3c}\right)^2 = \frac{a}{2b} \times \frac{3c}{a}$

$\frac{b}{c} = \frac{3}{2}$

putting in equation (i) $b^2 = a \times \frac{2b}{3}$

$\frac{a}{b} = \frac{3}{2}$

$a:b:c = 9:6:4$

90. (D)

Let r' th term of the G.P. be ar^{n-1}. Given,

$2a_r = a_{r+1} + a_{r+2}$

$2ar^{n-1} = ar^n + ar^{n+1}$

$\frac{2}{r} = 1 + r$

$r^2 + r - 2 = 0$

Hence, we get, $r = -2 (as\ r \neq 1)$

So, $S_{20} - S_{18}$ = (Sum up to 20 terms) – (Sum up to 18 terms) = $T_{19} + T_{20}$

$T_{19} + T_{20} = ar^{18}(1+r)$

Putting the values $a = \frac{1}{8}$ and $r = -2$

we get $T_{19} + T_{20} = 2^{15}$

91. (A)

$S_{20} = \frac{20}{2}[2a + 19d] = 790$

$2a + 19d = 79$ (1)

$S_{10} = \frac{10}{2}[2a + 9d] = 145$

$2a + 9d = 29$ (2)

From (1) and (2) $a = -8, d = 5$

$S_{15} - S_5 = \frac{15}{2}[2a + 14d] - \frac{5}{2}[2a + 4d]$

$= \frac{15}{2}[-16 + 70] - \frac{5}{2}[-16 + 20]$

$= 405 - 10 = 395$

92. (353)

$\alpha = 1^2 + 4^2 + 8^2 \ldots$

$t_n = a^2 + bc + c$

$1 = a + b + c$

$4 = 4a + 2b + c$

$8 = 9a + 3b + c$

On solving we get,

$a = \frac{1}{2}, b = \frac{3}{2}, c = -1$

$\alpha = \sum_{n=1}^{10}\left(\frac{n^2}{2} + \frac{3n}{2} - 1\right)^2$

$4a = \sum_{n=1}^{10}(n^2 + 3n - 2)^2, \beta = \sum_{n=1}^{10} n^4$

$4\alpha - \beta = \sum_{n=1}^{10}(6n^3 + 5n^2 - 12n + 4)$

$= 55(353) + 40$

93. (C)

1st G.P. $\Rightarrow t_1 = a, t_3 = b = ar^2 \Rightarrow r^2 = \frac{b}{a}$

$t_{11} = ar^{10} = a(r^2)^5 = a \cdot \left(\frac{b}{a}\right)^5$

2nd G.P. $\Rightarrow T_1 = a, T_5 = ar^4 = b$

$\Rightarrow r^4 = \left(\frac{b}{a}\right) \Rightarrow r = \left(\frac{b}{a}\right)^{\frac{1}{4}}$

$T_p = ar^{p-1} = a\left(\frac{b}{a}\right)^{\frac{p-1}{4}}$

$$t_{11} = T_p \Rightarrow a\left(\frac{b}{a}\right)^5 = a\left(\frac{b}{a}\right)^{\frac{p-1}{4}}$$

$$\Rightarrow 5 = \frac{p-1}{4} \Rightarrow p = 21$$

94. **(9)**

$S_n = 3+7+11+\ldots n$ terms

$$= \frac{n}{2}(6(n-1)4) = 3n + 2n^2 - 2n$$

$$= 2n^2 + n$$

$$\sum_{k=1}^{n} S_k = 2\sum_{k=1}^{n} K^2 + \sum_{k=1}^{n} K$$

$$= 2\cdot\frac{n(n+1)(2n+1)}{6} + \frac{n(n+1)}{2}$$

$$= n(n+1)\left[\frac{2n+1}{3} + \frac{1}{2}\right]$$

$$= \frac{n(n+1)(4n+5)}{6}$$

$$\Rightarrow 40 < \frac{6}{n(n+1)}\sum_{k=1}^{n} S_k < 42$$

$$40 < 4n+5 < 42$$

$$35 < 4n < 37$$

$$n = 9$$

95. **(D)**

General term of the sequence,

$$T_r = \frac{r}{1-3r^2+r^4}$$

$$T_r = \frac{r}{r^4-2r^2+1-r^2}$$

$$T_r = \frac{r}{\left(r^2-1\right)^2 - r^2}$$

$$T_r = \frac{r}{\left(r^2-r-1\right)\left(r^2+r-1\right)}$$

$$T_r = \frac{\frac{1}{2}\left[\left(r^2+r-1\right)-\left(r^2-r-1\right)\right]}{\left(r^2-r-1\right)\left(r^2+r-1\right)}$$

$$= \frac{1}{2}\left[\frac{1}{r^2-r-1} - \frac{1}{r^2+r-1}\right]$$

Sum of 10 terms,

$$\sum_{r=1}^{10} T_r = \frac{1}{2}\left[\frac{1}{-1} - \frac{1}{109}\right] = \frac{-55}{109}$$

96. **(D)**

$1+d, 1+7d, 1+43d$ are in G.P.

$$(1+7d)^2 = (1+d)(1+43d)$$

$$1+49d^2+14d = 1+44d+43d^2$$

$$6d^2 - 30d = 0$$

$$d = 5$$

$$S_{20} = \frac{20}{2}\left[2\times1+(20-1)\times5\right]$$

$$= 10[2+95]$$

$$= 970$$

2.14 Summary

- Progression is divided into three parts A.P., G.P. and H.P.

Arithmetic Progression

The nth term of an A.P. is denoted by t_n and $t_n = a + (n-1)d$ where 'a' is the first term, 'd' is the common difference and 'n' denotes the number of terms. The last term of an A.P. is denoted by l and is defined as $l = t_n = a + (n-1)d$.

- Sum of n terms of an A.P. is denoted by S_n and is defined by $S_n = \frac{n}{2}[2a + (n-1)d]$.
- A.M. of any two given numbers a and b is equal to $\frac{a+b}{2}$ and we denote it by A.
- If n A.M.s $A_1, A_2, \ldots, A_n$ are inserted between a and b, then
  - $A_1 = \frac{na+b}{n+1}$
  - $A_2 = \frac{(n-1)a+2b}{n+1}$

  - $A_n = \frac{a+nb}{n+1}$
- The sum of n A.M. inserted between a and b is equal to $n\left(\frac{a+b}{2}\right)$.
- In an A.P. the sum of the terms equidistant from beginning and end is equal to the sum of first and last term.

Geometric Progression

The nth term of a G.P. = ar^{n-1}, where a is the first term and r is the c.r. of a G.P.

- The sum of n terms of a G.P. is $S_n = \dfrac{a(r^n - 1)}{r-1}$, if $r > 1$ and $\dfrac{a(1-r^n)}{1-r}$ if $r < 1$.
- The sum of infinite term of a G.P. = $\dfrac{a}{1-r}$, where $|r| < 1$.
- The G.M. between any two positive real number 'a' and 'b' $= \sqrt{ab}$.
- If $G_1, G_2, \ldots, G_n$ are n G.M.s inserted between any two given positive numbers a and b, then
  - $G_1 = a\left(\dfrac{b}{a}\right)^{\frac{1}{n+1}}$
  - $G_2 = a\left(\dfrac{b}{a}\right)^{\frac{2}{n+1}}$
  - $G_n = a\left(\dfrac{b}{a}\right)^{\frac{n}{n+1}}$

  The product of these n G.M.s $= \left(\sqrt{ab}\right)^n$.

Harmonic Progression

Harmonic progression is the reciprocal of the A.P.

- The H.M. between any two numbers a and $b = \dfrac{2}{\frac{1}{a}+\frac{1}{b}}$.
- The A.M. of n numbers $a_1, a_2, \ldots,$ $a_n = \dfrac{a_1 + a_2 + \ldots + a_n}{n}$.
- The G.M. of n numbers $a_1, a_2, \ldots,$ $a_n = (a_1 \cdot a_2 \ldots a_n)^{\frac{1}{n}}$.
- The H.M. of n numbers $a_1, a_2, \ldots,$ $a_n = \dfrac{n}{\frac{1}{a_1}+\frac{1}{a_2}+\ldots+\frac{1}{a_n}}$.
- If A, G and H respectively denote the A.M., G.M. and H.M. between any two numbers a and b, then
  - A, G, H are in G.P.
  - $A \geq G \geq H$.
- The sum of the first n natural numbers $= \dfrac{n(n+1)}{2}$.
- The sum of the squares of first n natural numbers $= \dfrac{n(n+1)(2n+1)}{6}$.
- The sum of the cubes of first n natural numbers $= \left(\dfrac{n(n+1)}{2}\right)^2$.
- For an A.G.P. series, we multiply by a common ratio and then subtract.
- To find the sum of series, we use the method of difference and the V_n method.

CHAPTER

3 Complex Numbers

OBJECTIVES

3.1 Introduction
3.2 Conjugate of a Complex Number
3.3 Modulus of a Complex Number
3.4 Argument or Amplitude of a Complex Number
3.5 Polar Form of a Complex Number
3.6 Euler's Formula
3.7 DeMoivre's theorem
3.8 Cube Roots of Unity
3.9 n^{th} Roots of Unity
3.10 Rotation
3.11 Equation of a Straight Line
3.12 Equation of Circle
3.13 Know Your Concepts
3.14 JEE Main: Sample Questions
3.15 JEE Advanced: Sample Questions
3.16 JEE Main: Archive
3.17 Answer Key
3.18 Solutions
3.19 Summary

3.1 Introduction

The introduction of complex numbers plays a very important role in the theory of numbers. The square root of any negative real number is called an *imaginary number*. The concept of imaginary numbers was first introduced by the Mathematician, Euler. He was the one who introduced i (read as 'Iota') to represent $\sqrt{-1}$. He also defined $i^2 = -1$. Complex numbers are used in electronics and electromagnetism. Complex numbers help us study the flow of fluid around objects, such as water around a pipe.

Clearly, complex numbers are amazing. They have many uses, more than we realise. They have a fascinating history.

3.1.1 Definition

A number of the form $a + ib$ where $a, b \in R$ and $i = \sqrt{-1}$ is called a *complex number*. It is denoted by $z = a + ib$. It is also defined as an ordered pair (a, b), that is, $z = (a, b)$. The complex numbers do not possess the property of order, that is, $(a + ib) >$ or $< (c + id)$ is not defined.

3.2 Conjugate of a Complex Number

Conjugate of a complex number $z = a + ib$ is defined as $\bar{z} = a - ib$, it is the image of z in real axis.

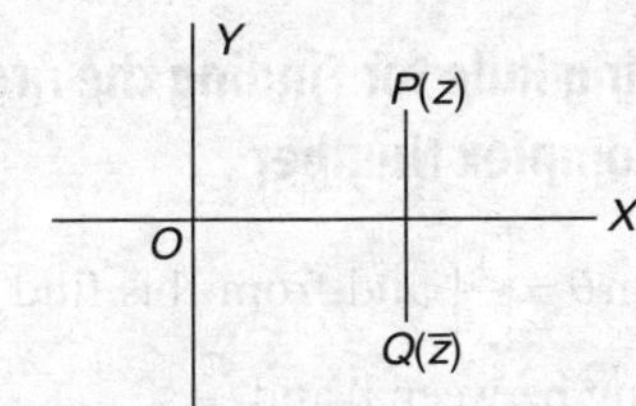

Thus, $z = a + ib$

$\bar{z} = a - ib$

$$\Rightarrow \quad z + \bar{z} = 2a \Rightarrow \text{Re}(z) = \frac{z + \bar{z}}{2}$$

$$z - \bar{z} = 2ib \Rightarrow \text{Im}(z) = \frac{z - \bar{z}}{2i}$$

3.3 Modulus of a Complex Number

Let $z = a + ib$, then the modulus of z is denoted by $|z|$ and is defined as $|z| = \sqrt{a^2 + b^2}$.

(i) $|z|^2 = z\bar{z}$

(ii) $|z_1 + z_2 + \ldots + z_n| = |\bar{z}_1 + \bar{z}_2 + \ldots + \bar{z}_n| = |\overline{z_1 + z_2 + \ldots + z_n}|$

3.4 Argument or Amplitude of a Complex Number

The argument or amplitude of a complex number $z = x + iy$ is the value of θ which satisfies the two equations,

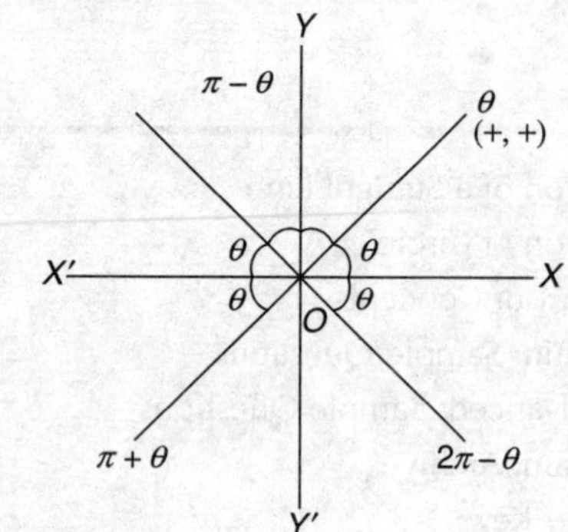

$$\cos\theta = \frac{x}{\sqrt{x^2+y^2}} \text{ and } \sin\theta = \frac{y}{\sqrt{x^2+y^2}}.$$

Argument of z is denoted by arg z or amp z. There will be an infinite number of values of θ satisfying the above equations and all these values will be the argument of z but usually we take only that value of θ for which $0 \le \theta < 2\pi$.

Note: Argument of 0 is not defined.

3.4.1 Working Rule for Finding the Argument of a Complex Number

(i) Take $\tan\theta = \left|\frac{y}{x}\right|$ and from this find the value of θ lying between 0 and $\frac{\pi}{2}$.

(ii) Then find in which quadrant the point z lies.

(iii) Argument of z will be $\theta, \pi - \theta, \pi + \theta$ or $2\pi - \theta$ according to whether the point z lies in the 1st, 2nd, 3rd or 4th quadrants.

3.4.2 Principal Value of the Argument

The value of argument θ satisfying the inequality $-\pi < \theta \le \pi$ is called the principal value of the argument. If the argument of a complex number is greater than π, then we subtract 2π from it to get its principal argument and if the argument is less than $-\pi$, we add 2π to get its principal argument.

$\pi - \theta$ II | θ I | $\theta - \pi$ III | $-\theta$ IV

In another method, we find the value of θ in the first quadrant. Then $\theta, \pi - \theta, \theta - \pi$ and $-\theta$ will be the principal argument in I, II, III and IVth quadrant respectively.

3.5 Polar Form of a Complex Number

Every non-zero complex number $z = x + iy$ can be put in the polar form $r(\cos\theta + i\sin\theta)$ where $r = |z|$ and $\theta = \arg z$.

Example

Let $z = -1 - i$

$$|z| = \sqrt{2}, \tan\theta = \left|\frac{y}{x}\right| = 1 \Rightarrow \theta = \frac{\pi}{4}$$

As the complex number lies in the IIIrd quadrant,

$$\text{the principal argument} = \frac{\pi}{4} - \pi = \frac{-3\pi}{4}.$$

$$\text{Polar form of } z = \sqrt{2}\left(\cos\left(-\frac{3\pi}{4}\right) + i\sin\left(-\frac{3\pi}{4}\right)\right).$$

Note:

(i) $\overline{z_1 z_2} = \bar{z}_1 \bar{z}_2$ **(ii)** $\overline{\left(\frac{Z_1}{Z_2}\right)} = \frac{\overline{Z_1}}{\overline{Z_2}}$

(iii) $\overline{Z^n} = \left(\bar{Z}\right)^n$ **(iv)** $|z_1 z_2| = |z_1||z_2|$

(v) $\left|\frac{Z_1}{Z_2}\right| = \frac{|Z_1|}{|Z_2|}$

(vi) $|z_1 + z_2| \le |z_1| + |z_2|$ (Equality holds if and only if origin, z_1 and z_2 are collinear and z_1 And z_2 are on the same side of the origin)

(vii) $|z_1 - z_2| \ge \big||z_1| - |z_2|\big|$

(viii) $\arg(z_1 z_2) = \arg z_1 + \arg z_2$

(ix) $\arg\left(\frac{z_1}{z_2}\right) = \arg z_1 - \arg z_2$

(x) $\arg(z^n) = n \arg z$

3.5.1 Remember This Expansion Formula

(i) $e^x = 1 + \frac{x}{1!} + \frac{x^2}{2!} + \frac{x^3}{3!} + \frac{x^4}{4!} + \ldots$ to ∞

(ii) $\cos x = 1 - \frac{x^2}{2!} + \frac{x^4}{4!} - \frac{x^6}{6!} + \ldots$ to ∞

(iii) $\sin x = x - \frac{x^3}{3!} + \frac{x^5}{5!} - \frac{x^7}{7!} + \ldots$ to ∞

3.6 Euler's Formula

As, $e^x = 1 + \frac{x}{1!} + \frac{x^2}{2!} + \frac{x^3}{3!} + \frac{x^4}{4!} + \frac{x^5}{5!} + \frac{x^6}{6!} + \ldots$ to ∞

Replace x by ix

$$e^{ix} = 1 + \frac{ix}{1!} + \frac{(ix)^2}{2!} + \frac{(ix)^3}{3!} + \frac{(ix)^4}{4!} + \frac{(ix)^5}{5!} + \frac{(ix)^6}{6!} + \ldots \text{ to } \infty$$

$$e^{ix} = \left(1 - \frac{x^2}{2!} + \frac{x^4}{4!} - \frac{x^6}{6!} + \ldots\right) + i\left(x - \frac{x^3}{3!} + \frac{x^5}{5!} - \ldots\right)$$

$$\Rightarrow \quad e^{ix} = \cos x + i \sin x$$

Similarly, $e^{-ix} = \cos x - i \sin x$

Thus, $e^{ix} = \cos x + i \sin x\}$

$e^{-ix} = \cos x - i \sin x\}$

This is known as Euler's formula.

3.7 Demoivre's Theorem

If n is an integer, positive or negative, the value of $(\cos\theta + i\sin\theta)^n = \cos\ n\theta + i\sin\ n\theta$.

If n is a rational number, then one of the values of $(\cos\theta + i\sin\theta)^n = \cos\ n\theta + i\sin\ n\theta$

Since $(\cos\theta_1 + i\sin\theta_1)(\cos\theta_2 + i\sin\theta_2)$

$$= (\cos\theta_1\cos\theta_2 - \sin\theta_1\sin\theta_2) + i(\sin\theta_1\cos\theta_2 + \cos\theta_1\sin\theta_2)$$

$$\Rightarrow \quad (\cos\theta_1 + i\sin\theta_1)(\cos\theta_2 + i\sin\theta_2) = \cos(\theta_1 + \theta_2) + i\sin(\theta_1 + \theta_2)$$

Generalising this, we get

$$(\cos\theta_1 + i\sin\theta_1)(\cos\theta_2 + i\sin\theta_2)\ldots(\cos\theta_n + i\sin\theta_n) = \cos(\theta_1 + \theta_2 + \ldots + \theta_n) + i\sin(\theta_1 + \theta_2 + \ldots\theta_n)$$

Put, $\theta_1 = \theta_2 = \ldots = \theta_n = \theta$, we get

$$(\cos\theta + i\sin\theta)(\cos\theta + i\sin\theta)\ldots \text{to } n \text{ terms}$$

$$= \cos(\theta + \theta \ldots \text{to } n \text{ terms}) + i\sin(\theta + \theta + \ldots \text{to } n \text{ terms})$$

$$\Rightarrow \quad (\cos\theta + i\sin\theta)^n = \cos\ n\theta + i\sin\ n\theta$$

3.8 Cube Roots of Unity

Let $x = \sqrt[3]{1} \Rightarrow x^3 = 1$

$$\Rightarrow \quad x^3 - 1 = 0 \Rightarrow (x-1)(x^2 + x + 1) = 0$$

If $x - 1 = 0 \Rightarrow x = 1$

If $x^2 + x + 1 = 0 \Rightarrow x = \frac{-1 \pm \sqrt{1-4}}{2}$

$$\Rightarrow \quad x = \frac{-1 \pm i\sqrt{3}}{2}$$

Thus, cube roots of unity are

$$x = 1, \frac{-1 + i\sqrt{3}}{2}, \frac{-1 - i\sqrt{3}}{2}$$

$\frac{-1 + i\sqrt{3}}{2}$ is denoted by ω

And $\frac{-1 - i\sqrt{3}}{2}$ is denoted by ω^2

Thus, the cube roots of unity are 1, ω, ω^2.

3.8.1 Properties

(i) $1 + \omega + \omega^2 = 0$

(ii) $\omega^3 = 1$

(iii) If n is a positive integer, then $\omega^n = \omega^r$ where r is the remainder when n is divided by 3.

(iv) If $n \in I$, then $1 + \omega^n + \omega^{2n} = 3$ or 0 according to whether n is a multiple of 3 or not

3.9 n^{th} Roots of Unity

Let $x = \sqrt[n]{1} \Rightarrow x^n = 1$

$$\Rightarrow \quad x^n = e^{i2r\pi}; r \in I$$

$$[e^{i2r\pi} = \cos 2r\pi + i\ \sin 2r\pi = 1]$$

$$\Rightarrow \quad x = e^{\frac{i2r\pi}{n}}$$

Where $r = 0, 1, 2 \ldots, (n-1)$

$$\Rightarrow \quad 1, e^{\frac{i\,2\pi}{n}}, e^{\frac{i\,4\pi}{n}}, \ldots, e^{\frac{i\,2(n-1)\pi}{n}}$$

are n roots of the equation, $x^n = 1$

Let $\alpha = e^{\frac{i\,2\pi}{n}}$

Hence the nth roots are $1, \alpha, \alpha^2, \ldots, \alpha^{n-1}$

(which are in GP)

Now, $1 + \alpha + \alpha^2 + \ldots + \alpha^{n-1} = 0$

$$\Rightarrow \quad 1 + e^{\frac{i\,2\pi}{n}} + e^{\frac{i\,4\pi}{n}} + \ldots + e^{\frac{i\,2(n-1)\pi}{n}} = 0$$

$$\Rightarrow 1 + \left(\cos\frac{2\pi}{n} + i\sin\frac{2\pi}{n}\right) + \left(\cos\frac{4\pi}{n} + i\sin\frac{4\pi}{n}\right) + \ldots + \left(\cos\frac{2(n-1)\pi}{n} + i\sin\frac{2(n-1)\pi}{n}\right) = 0$$

Comparing real and imaginary parts from both sides, we get

$$\cos\frac{2\pi}{n}+\cos\frac{4\pi}{n}+\ldots+\cos\frac{2(n-1)\pi}{n}=-1$$

$$\sin\frac{2\pi}{n}+\sin\frac{4\pi}{n}+\ldots+\sin\frac{2(n-1)\pi}{n}=0$$

Also, product of roots $=1\cdot\alpha\cdot\alpha^2\ldots\alpha^{n-1}$

$$=\alpha^{\frac{n(n-1)}{2}}=\left(\cos\frac{2\pi}{n}+i\sin\frac{2\pi}{n}\right)^{\frac{n(n-1)}{2}}$$

$$=\cos(n-1)\pi+i\sin(n-1)\pi$$

$=-1$ if n is even

$=1$ if n is odd

Note:

(i) The nth roots of unity denote the vertices of a regular polygon of n sides inscribed in a circle of radius unity whose centre is at the origin, one vertex being on the positive real axis.

(ii) Cube roots of unity denote the vertices of an equilateral triangle whose circumcentre is the origin and the circumradius is unity.

3.10 Rotation

Let $\qquad OP=|OP|\,e^{i\theta} \qquad (3.1)$

If we rotate OP in anti-clock direction at an angle ϕ and its new position is OQ, then

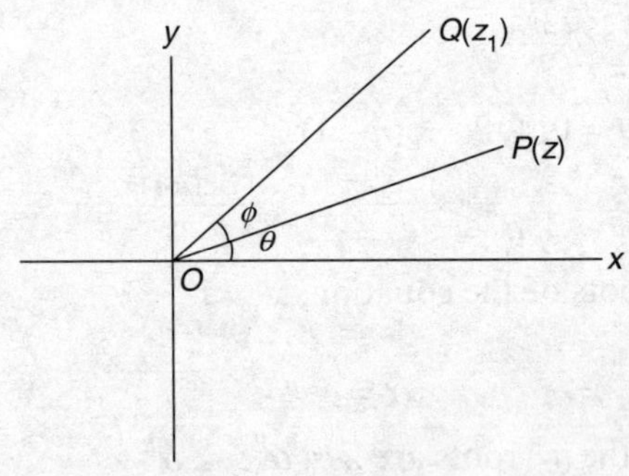

$$OQ=|OQ|\,e^{i(\theta+\phi)}$$

$\Rightarrow \qquad OQ=|OQ|e^{i\theta}\cdot e^{i\phi}$

$\Rightarrow \qquad OQ=|OQ|\dfrac{OP}{|OP|}e^{i\phi}$

$\Rightarrow \qquad OQ=\dfrac{|OQ|}{|OP|}OP\,e^{i\phi} \qquad (3.2)$

3.11 Equation of a Straight Line

Let z be any point on the straight line joining the points z_1 and z_2, then

$$\arg\left(\frac{z-z_1}{z-z_2}\right)=0 \text{ or } \pi$$

$\Rightarrow \qquad \dfrac{z-z_1}{z-z_2}$ is real

$\Rightarrow \qquad \left(\dfrac{z-z_1}{z-z_2}\right)=\left(\overline{\dfrac{z-z_1}{z-z_2}}\right)$

$\Rightarrow \qquad \dfrac{z-z_1}{z-z_2}-\left(\overline{\dfrac{z-z_1}{z-z_2}}\right)=0$

$\Rightarrow \qquad z(\bar{z}_1-\bar{z}_2)-\bar{z}(z_1-z_2)+(z_1\bar{z}_2-\bar{z}_1z_2)=0 \quad (3.3)$

We can write this as $\begin{vmatrix} z & \bar{z} & 1 \\ z_1 & \bar{z}_1 & 1 \\ z_2 & \bar{z}_2 & 1 \end{vmatrix}=0$

Multiplying (3.3) by i

$$iz(\bar{z}_1-\bar{z}_2)-i\,\bar{z}(z_1-z_2)+i(z_1\bar{z}_2-\bar{z}_1z_2)=0$$

Since $\bar{z}_1z_2$ is conjugate to $z_1\bar{z}_2$, the number $z_1\bar{z}_2-\bar{z}_1z_2$ is purely imaginary and hence $i(z_1\bar{z}_2-\bar{z}_1z_2)$ is real.

Let $i(z_1\bar{z}_2-\bar{z}_1z_2)=c$, where $c\in R$

We denote $-i(z_1-z_2)=b$

$\Rightarrow \qquad i(\bar{z}_1-\bar{z}_2)=\bar{b}$

$\Rightarrow \qquad \bar{b}z+b\bar{z}+c=0$ where $b\neq 0$ and $c\in R$

Which is the general equation of straight line.

3.11.1 Reflection Points for a Straight Line

Two points z_1 and z_2 are the reflection points for a given straight line if the line is the right bisector of the line segment joining the points z_1 and z_2. For any point z on the right bisector, we have $|z-z_1|=|z-z_2|$.

Let z_1 and z_2 be the reflection points for the straight line

$$\bar{b}z+b\bar{z}+c=0 \qquad (3.4)$$

such that for any point z on the line,

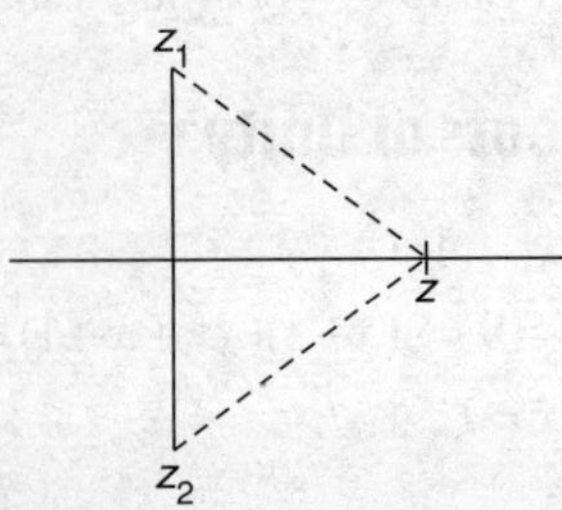

We have

$$|z - z_1| = |z - z_2|$$

$$\Rightarrow \quad (z - z_1)(\bar{z} - \bar{z}_1) = (z - z_2)(\bar{z} - \bar{z}_2)$$

$$\Rightarrow \quad z(\bar{z}_2 - \bar{z}_1) + \bar{z}(z_2 - z_1) + (z_1\bar{z}_1 - z_2\bar{z}_2) = 0 \quad (3.5)$$

As z is a point on the line (3.4), the equation (3.5) can be thought of as the equation of the line (3.4)

Comparing (3.4) and (3.5), we get

$$\frac{\bar{z}_2 - \bar{z}_1}{\bar{b}} = \frac{z_2 - z_1}{b} = \frac{z_1\bar{z}_1 - z_2\bar{z}_2}{c} = k \quad \text{(say)}$$

$$\Rightarrow \quad \bar{z}_2 - \bar{z}_1 = \bar{b}k,\ z_2 - z_1 = bk,\ z_1\bar{z}_1 - z_2\bar{z}_2 = ck$$

Now, $\bar{b}z_1 + b\bar{z}_2 = \frac{1}{k}[z_1(\bar{z}_2 - \bar{z}_1) + \bar{z}_2(z_2 - z_1)]$

$$= \frac{1}{k}(z_2\bar{z}_2 - z_1\bar{z}_1) = \frac{1}{k}(-ck) = -c$$

$\Rightarrow \quad \bar{b}z_1 + b\bar{z}_2 + c = 0$ is the required condition

3.12 Equation of Circle

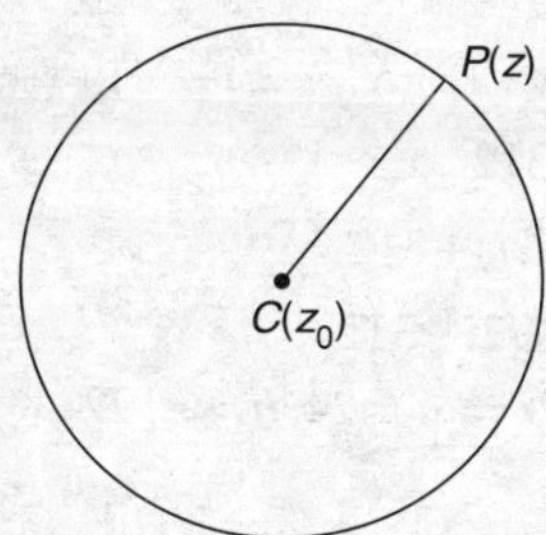

Let z_0 be the centre of the circle and $P(z)$ be any point on the circle.

Then, $|z - z_0| = r \Rightarrow |z - z_0|^2 = r^2$

$$\Rightarrow \quad (z - z_0)(\bar{z} - \bar{z}_0) = r^2$$

$$\Rightarrow \quad z\bar{z} - z_0\bar{z} - z\bar{z}_0 + z_0\bar{z}_0 = r^2$$

$$\Rightarrow \quad z\bar{z} - \bar{z}_0 z - z_0\bar{z} + |z_0|^2 - r^2 = 0$$

$$\Rightarrow \quad z\bar{z} + \bar{a}z + a\bar{z} + b = 0$$

Where, $a = -z_0, b = |z_0|^2 - r^2$ (a real number)

Clearly centre of circle is z_0, that is, $-a$

and radius $= \sqrt{|z_0|^2 - r^2} = \sqrt{|a|^2 - r^2} = \sqrt{a\bar{a} - r^2}$

Note:

(i) $\left(\frac{z_2 - z_4}{z_1 - z_4}\right)\left(\frac{z_1 - z_3}{z_2 - z_3}\right)$ is real and positive then z_1, z_2, z_3, z_4 are concyclic and they are in order (clockwise or anti-clockwise order).

(ii) $\left(\frac{z_2 - z_4}{z_1 - z_4}\right)\left(\frac{z_1 - z_3}{z_2 - z_3}\right)$ is real and negative then z_1, z_2, z_3, z_4 are concyclic but they are not in order.

(iii) Complex slope of the line $\bar{a}z + a\bar{z} + b = 0$ is given by $\omega = -\frac{a}{\bar{a}} = \frac{-\text{co-eff of } \bar{z}}{\text{co-eff of } z}$.

Important Points to Remember

(i) $|z - z_1|^2 + |z - z_2|^2 = |z_1 - z_2|^2$ represents a circle.

(ii) $|z - z_1|^2 + |z - z_2|^2 = k$ represents a circle if $k \geq \frac{1}{2}|z_1 - z_2|^2$.

(iii) If $\left|\frac{z - z_1}{z - z_2}\right| = k$ where $k \neq 1, 0$, then locus of z is a circle.

(iv) $|z - z_1| + |z - z_2| = k$ represents ellipse if $k > |z_1 - z_2|$.
If $k = |z_1 - z_2|$, then the locus is a straight line segment joining z_1 and z_2. If $k < |z_1 - z_2|$, no locus.

(v) $|z - z_1| - |z - z_2| = k$, represents hyperbola if $k < |z_1 - z_2|$.
If $k = |z_1 - z_2|$, then the locus is a straight line segment joining z_1 and z_2 but z does not lie between z_1 and z_2.

(vi) If $|z + z_1| = |z| + |z_1|$
$\Rightarrow$ Locus of z is a ray starting from origin and passing through z_1.

(vii) $|z - z_1| = |z - z_2|$, then the locus of z represents the perpendicular bisector of the line segment joining z_1 and z_2.

(viii) $\text{Arg}\left(\frac{z - z_1}{z - z_2}\right) = \theta$, then locus of z is circular arc (part of circle),
If $\theta < \frac{\pi}{2}$, then major arc of circle
$\theta > \frac{\pi}{2}$, then minor arc of circle
If $\theta = 0, \pi$, then straight line joining z_1 and z_2.

(ix) If $|z - z_0| = \frac{|\alpha\bar{z}_0 + \bar{\alpha}z_0 + b|}{2|\alpha|}$, then the locus of z is a parabola.

SOLVED EXAMPLES

Example 3.1 Write $\frac{5+5i}{3-4i}+\frac{20}{4+3i}$ in the form of $a+ib$

Solution

$$\frac{5+5i}{3-4i}+\frac{20}{4+3i}=\frac{5+5i}{3-4i}\times\frac{3+4i}{3+4i}+\frac{20}{4+3i}\times\frac{4-3i}{4-3i}$$
$$=\frac{15+20i+15i+20i^2}{9-16i^2}+\frac{20(4-3i)}{16-9i^2}$$
$$=\frac{-5+35i}{25}+\frac{80-60i}{25}$$
$$=\frac{75-25i}{25}=3-i$$

Example 3.2 Solve the following equation for x and y

$$\frac{(1+i)x-2i}{3+i}+\frac{(2+3i)y+i}{3-i}=i$$

Solution $\frac{x+i(x-2)}{3+i}+\frac{2y+i(3y+1)}{3-i}=i$

$\Rightarrow\quad \{x+i(x-2)\}(3-i)+\{2y+i(3y+1)\}(3+i)$
$=i(9-i^2)$

$\Rightarrow\quad 3x+(x-2)+i(3x-6-x)+6y-(3y+1)+$
$i(2y+9y+3)=10i$

$\Rightarrow\quad (4x-2+3y-1)+i(2x-6+11y+3)=10i$

Comparing real and imaginary part, we get

$$4x+3y=3$$
$$2x+11y=13$$

Solving we get

$$x=\frac{-3}{19},\ y=\frac{23}{19}$$

Example 3.3 Find the value of

$$a^6+a^4+a^2+1 \text{ if } a=\frac{1+i}{\sqrt{2}}.$$

Solution $a=\frac{1+i}{\sqrt{2}}\Rightarrow\sqrt{2}a=1+i$

$\Rightarrow\ 2a^2=1+i^2+2i$

$\Rightarrow\ 2a^2=2i\Rightarrow a^2=i$

$\Rightarrow\ a^4=i^2=-1$

$\Rightarrow\ a^4+1=0$

Now, $a^6+a^4+a^2+1=(a^6+a^2)+(a^4+1)$

$$=a^2(a^4+1)+1(a^4+1)$$
$$=(a^2+1)(a^4+1)=0$$

Example 3.4 If $f(x)=x^4-8x^3+4x^2+4x+39$ and $f(3+2i)=a+ib$, then evaluate $\frac{a}{b}$.

Solution As, $x=3+2i$

$\Rightarrow\quad (x-3)^2=4i^2$

$\Rightarrow\quad x^2-6x+9=-4\Rightarrow x^2-6x+13=0 \quad (1)$

Now,

$x^2-6x+13)\ x^4-8x^3+4x^2+4x+39\ (x^2-2x-21$

$+x^4-6x^3+13x^2$
$-\quad+\quad-$

$-2x^3-9x^2+4x+39$
$-2x^3+12x^2-26x$
$+\quad-\quad+$

$-21x^3+30x+39$
$-21x^3+126x-273$
$+\quad-\quad+$

$=-96x+312$

Thus, $x^4-8x^3+4x^2+4x+39$

$$=(x^2-6x+13)(x^2-2x-21)+(-96x+312)$$
$$=-96x+312$$
$$=-96(3+2i)+312$$
$$=-288-192i+312$$
$$=24-192i\Rightarrow a=24,\ b=-192$$

$\Rightarrow\quad \frac{a}{b}=-\frac{1}{8}$

Example 3.5 Find the square root of $7-30\sqrt{2}i$

Solution Let $\sqrt{7-30\sqrt{2}i}=x+iy \quad (1)$

Squaring both sides, we get

$$7-30\sqrt{2}i=x^2-y^2+2ixy$$

Comparing real and imaginary parts, we get

$$x^2-y^2=7 \quad (2)$$
$$2xy=-30\sqrt{2} \quad (3)$$

Now, $(x^2+y^2)^2=(x^2-y^2)^2+4x^2y^2$

$\Rightarrow\quad (x^2+y^2)^2=49+1800=1849$

$\Rightarrow\quad x^2+y^2=43 \quad (4)$

Adding (4) and (2)

$$2x^2=50\Rightarrow x=\pm5$$

Subtracting (2) from (4), $2y^2-36\Rightarrow y=\pm3\sqrt{2}$

From equation (3) as sign of xy is negative

$\Rightarrow$ x and y will be of opposite sign

$$x+iy=(5-3\sqrt{2}i)$$

or $$-5+3\sqrt{2}i=-(5-3\sqrt{2}i)$$

Thus, $$x+iy=\pm(5-3\sqrt{2}i)$$

Alternative method

$$\begin{aligned}&7-30\sqrt{2}i\\&=25-18-30\sqrt{2}i\\&=25-30\sqrt{2}i+18i^2\\&=(5-3\sqrt{2}i)^2\end{aligned}$$

Square root $=\pm(5-3\sqrt{2}i)$

Example 3.6 Find the roots of $\sqrt[4]{-81}$.

Solution $(-81)^{1/4}=(81i^2)^{1/4}=((\pm 9i)^2)^{1/4}$

$$=(\pm 9i)^{1/2}$$
$$=3\sqrt{i},3\sqrt{-i}$$

Now, $3\sqrt{i}=3\sqrt{\dfrac{1+2i+i^2}{2}}=3\left(\pm\dfrac{1}{\sqrt{2}}(1+i)\right)$

and, $$3\sqrt{-i}=3\left(\pm\frac{1}{\sqrt{2}}(1-i)\right)$$

Thus, $$\sqrt[4]{-81}=\pm\frac{3}{\sqrt{2}}(1\pm i)$$

Example 3.7 Find the principal argument of the complex number $\dfrac{2+i}{4i+(1+i)^2}$ Where $i=\sqrt{-1}$

Solution $$\frac{2+i}{4i+1+2i+i^2}=\frac{2+i}{6i}=\frac{i(2+i)}{-6}$$

$$=\frac{-1+2i}{-6}$$

$$=+\frac{1}{6}-\frac{1}{3}i$$

$$\tan\theta=\left|\frac{-\frac{1}{3}}{\frac{1}{6}}\right|=2\Rightarrow\theta=\tan^{-1}2$$

As the complex number lies in the IVth quadrant, the principal argument $=-\tan^{-1}2$.

Example 3.8 z_1 and z_2 are two complex numbers, such that $\dfrac{z_1-2z_2}{2-z_1\bar{z}_2}$ is unimodular, while z_2 is not unimodular. Find $|z_1|$.

Solution Since, $\dfrac{z_1-2z_2}{2-z_1\bar{z}_2}$ is unimodular

$$\Rightarrow\left|\frac{z_1-2z_2}{2-z_1\bar{z}_2}\right|=1$$

$$\Rightarrow |z_1-2z_2|^2=|2-z_1\bar{z}_2|^2$$

$$\Rightarrow (z_1-2z_2)(\bar{z}_1-2\bar{z}_2)=(2-z_1\bar{z}_2)(2-\bar{z}_1z_2)$$

$$\Rightarrow z_1\bar{z}_1-2z_1\bar{z}_2-2\bar{z}_1z_2+4z_2\bar{z}_2$$
$$=4-2\bar{z}_1z_2-2z_1\bar{z}_2+z_1\bar{z}_1z_2\bar{z}_2$$

$$\Rightarrow |z_1|^2+4|z_2|^2=4+|z_1|^2|z_2|^2$$

$$\Rightarrow (|z_1|^2-4)-|z_2|^2(|z_1|^2-4)=0$$

$$\Rightarrow (|z_1|^2-4)(1-|z_2|^2)=0$$

As $|z_2|\neq 1$; sincez_2 is not unimodular

$$\Rightarrow |z_1|^?=4\Rightarrow|z_1|=2$$

Example 3.9 If $|z_1|=1,|z_2|=2,|z_3|=3$ and

$$|9z_1z_2+4z_3z_1+z_2z_3|=6,$$

find the value of $|z_1+z_2+z_3|$.

Solution Since, $|z_1|=1\Rightarrow|z_1|^2=1\Rightarrow z_1\bar{z}_1=1$

$$|z_2|=2\Rightarrow z_2\bar{z}_2=4$$
$$|z_3|=3\Rightarrow z_3\bar{z}_3=9$$

Now, $|9z_1z_2+4z_3z_1+z_2z_3|=6$

$$\Rightarrow |z_3\bar{z}_3z_1z_2+z_2\bar{z}_2z_3z_1+z_1\bar{z}_1z_2z_3|=6$$

$$\Rightarrow |z_1z_2z_3(\bar{z}_1+\bar{z}_2+\bar{z}_3)|=6$$

$$\Rightarrow |z_1||z_2||z_3||\bar{z}_1+\bar{z}_2+\bar{z}_3|=6$$

$$\Rightarrow 6|z_1+z_2+z_3|=6$$

$$\Rightarrow |z_1+z_2+z_3|=1$$

Example 3.10 If the polynomial $7x^3+ax+b$ is divisible by x^2-x+1, find the value of $2a+b$.

Solution Let $f(x)=7x^3+ax+b$ (1)

$$x^2-x+1=0\Rightarrow x=\frac{1\pm\sqrt{1-4}}{2}=-\omega\text{ or }-\omega^2$$

As $f(x)$ is divisible by $x^2 - x + 1$

$\Rightarrow \quad f(-\omega) = 0$ and $f(-\omega^2) = 0$

$\Rightarrow \quad -7\omega^3 - a\omega + b = 0$ and $-7\omega^6 - a\omega^2 + b = 0$

$\Rightarrow \quad -7 - a\omega + b = 0$ and $-7 - a\omega^2 + b = 0$

Adding $-14 - a(\omega + \omega^2) + 2b = 0$

$\Rightarrow \quad -14 + a + 2b = 0 \Rightarrow a + 2b = 14$

Subtracting $-a(\omega - \omega^2) = 0 \Rightarrow a = 0 \Rightarrow b = 7$

Hence, $2a + b = 7$

Example 3.11 If α, β, γ are the roots of

$$x^3 - 3x^2 + 3x + 7 = 0,$$

find the value of $\dfrac{\alpha - 1}{\beta - 1} + \dfrac{\beta - 1}{\gamma - 1} + \dfrac{\gamma - 1}{\alpha - 1}$.

Solution $x^3 - 3x^2 + 3x + 7 = 0 \Rightarrow (x-1)^3 + 8 = 0$

$$\Rightarrow \left(\frac{x-1}{-2}\right)^3 = 1$$

$$\Rightarrow \frac{x-1}{-2} = 1, \omega, \omega^2$$

If $\dfrac{x-1}{-2} = 1 \Rightarrow x = 1 - 2 = -1$

$$\frac{x-1}{-2} = \omega \Rightarrow x = 1 - 2\omega; \frac{x-1}{-2} = \omega^2 \Rightarrow x = 1 - 2\omega^2$$

Let $\alpha = -1, \beta = 1 - 2\omega, \gamma = 1 - 2\omega^2$

Now,

$$\frac{\alpha - 1}{\beta - 1} + \frac{\beta - 1}{\gamma - 1} + \frac{\gamma - 1}{\alpha - 1}$$

$$= \frac{-1-1}{1-2\omega-1} + \frac{1-2\omega-1}{1-2\omega^2-1} + \frac{1-2\omega^2-1}{-1-1}$$

$$= \frac{1}{\omega} + \frac{1}{\omega} + \omega^2 = \omega^2 + \omega^2 + \omega^2 = 3\omega^2$$

Example 3.12 If $\left|z - \dfrac{6}{z}\right| = 4$, find the greatest integral value of $|z|$.

Solution $|z| = \left|\left(z - \dfrac{6}{z}\right) + \dfrac{6}{z}\right| \le \left|z - \dfrac{6}{z}\right| + \dfrac{6}{|z|}$

$\Rightarrow \quad |z| \le 4 + \dfrac{6}{|z|} \Rightarrow |z|^2 - 4|z| - 6 \le 0$

$\Rightarrow \quad (|z| - 2)^2 \le 10$

$\Rightarrow \quad -\sqrt{10} \le |z| - 2 \le \sqrt{10}$

$\Rightarrow \quad 2 - \sqrt{10} \le |z| \le 2 + \sqrt{10}$

$\Rightarrow \quad |z| \le 5.162$

$\Rightarrow \quad$ Greatest integral value of $|z| = 5$

Example 3.13 If z is any non-zero complex number, then show that $\left|\dfrac{z}{|z|} - 1\right| \le |\arg z|$.

Solution Let, $|z| = r$ and $\arg z = \theta$

$\Rightarrow \quad z = r(\cos\theta + i\sin\theta)$

$\Rightarrow \quad \dfrac{z}{|z|} = \cos\theta + i\sin\theta$

$\Rightarrow \quad \left|\dfrac{z}{|z|} - 1\right| = |(\cos\theta - 1) + i\sin\theta|$

$$= \sqrt{(\cos\theta - 1)^2 + \sin^2\theta} = \sqrt{2(1 - \cos\theta)}$$

$$= 2\left|\sin\frac{\theta}{2}\right| \le 2\left|\frac{\theta}{2}\right| \qquad [\text{As } |\sin\theta| \le \theta]$$

$\Rightarrow \quad \left|\dfrac{z}{|z|} - 1\right| \le |\theta| = |\arg z|$

Example 3.14 If $|z - 3| = 3$, then show that

$$\frac{z-6}{z} = i\tan(\arg z)$$

Solution

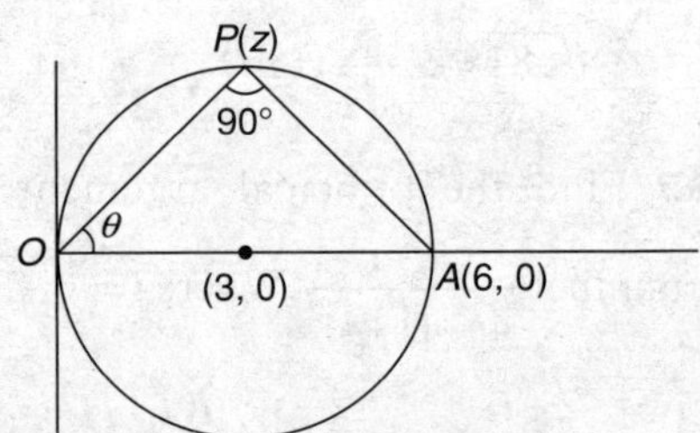

$|z - 3| = 3$, represents a circle whose centre is $(3, 0)$ and radius is 3.

Using rotation,

$$\frac{z-6}{z-0} = \frac{|z-6|}{|z-0|} e^{\frac{i\pi}{2}}$$

$$\Rightarrow \quad \frac{z-6}{z} = \frac{PA}{PO} i = i\tan\theta$$

$$= i\tan(\arg z)$$

Example 3.15 If $\arg\left(\frac{z_1}{z_2}\right)=\frac{\pi}{2}$, then find the value of $\left|\frac{z_1+z_2}{z_1-z_2}\right|$.

Solution $\arg\left(\frac{z_1}{z_2}\right)=\frac{\pi}{2}\Rightarrow\frac{z_1}{z_2}$ is purely imaginary

$$\Rightarrow \quad \frac{z_1}{z_2}=ik(k\in R)$$

$$\Rightarrow \quad \frac{z_1+z_2}{z_1-z_2}=\frac{ik+1}{ik-1}\Rightarrow\left|\frac{z_1+z_2}{z_1-z_2}\right|=1$$

Example 3.16 If

$\cos\alpha+\cos\beta+\cos\gamma=\sin\alpha+\sin\beta+\sin\gamma=0$,

prove that $\cos 3\alpha+\cos 3\beta+\cos 3\gamma=3\cos(\alpha+\beta+\gamma)$

Solution Let, $a=\cos\alpha+i\sin\alpha$

$$b=\cos\beta+i\sin\beta$$

$$c=\cos\gamma+i\sin\gamma$$

Therefore,

$$a+b+c$$

$$=(\cos\alpha+\cos\beta+\cos\gamma)+i(\sin\alpha+\sin\beta+\sin\gamma)$$

$$\Rightarrow \quad a+b+c=0$$

$$\Rightarrow \quad a^3+b^3+c^3=3abc$$

$$\Rightarrow \quad (\cos\alpha+i\sin\alpha)^3+(\cos\beta+i\sin\beta)^3+(\cos\gamma+i\sin\gamma)^3$$

$$=3(\cos\alpha+i\sin\alpha)(\cos\beta+i\sin\beta)(\cos\gamma+i\sin\gamma)$$

$$\Rightarrow \quad (\cos 3\alpha+\cos 3\beta+\cos 3\gamma)+i(\sin 3\alpha+\sin e\beta+\sin 3\gamma)$$

$$=3\left[\cos(\alpha+\beta+\gamma)+i\sin(\alpha+\beta+\gamma)\right]$$

Comparing the real parts, we get

$$\cos 3\alpha+\cos 3\beta+\cos 3\gamma=3\cos(\alpha+\beta+\gamma)$$

Example 3.17 If the vertices of a triangle are $3+4i$, $4+3i$, $2\sqrt{6}+i$, then find the distance between the orthocentre and the circumcentre of the triangle.

Solution Let $z_1=3+4i$; $z_2=4+3i$; $z_3=2\sqrt{6}+i$

Then, $|z_1|=|z_2|=|z_3|=5$

Hence, z_1, z_2, z_3 lies on a circle whose centre is at origin and radius = 5

Hence, circumcentre = (0, 0)

Now centroid $=\left(\frac{7+2\sqrt{6}}{3},\frac{8}{3}\right)$

Now, distance between circumcentre and centroid

$$=\sqrt{\left(\frac{7+2\sqrt{6}}{3}\right)^2+\frac{64}{9}}=\sqrt{\frac{49+24+28\sqrt{6}+64}{9}}$$

$$=\frac{1}{3}\sqrt{137+28\sqrt{6}}$$

Centroid divides the line joining orthocentre and circumcentre in the ratio 2 : 1.

$\Rightarrow$ Distance between orthocentre and circumcentre

$$=3\frac{1}{3}\sqrt{137+28\sqrt{6}}=\sqrt{137+28\sqrt{6}}$$

Example 3.18 If $\arg(z)<0$, then find $\arg(-z)-\arg(z)$

Solution Let $z=re^{-i\theta}\Rightarrow\arg(z)=-\theta$

$$\Rightarrow \quad -z=re^{i\pi}e^{-i\theta}$$

$$=re^{i(\pi-\theta)}$$

$$\Rightarrow \quad \arg(-z)=\pi-\theta$$

Hence, $\arg(-z)-\arg(z)=\pi-\theta+\theta=\pi$

Example 3.19 If α and β are roots of $x^2-2x+4=0$ prove that $\alpha^n+\beta^n=2^{n+1}\cos\frac{n\pi}{3}$.

Solution Since, $x^2-2x+4=0\Rightarrow x=\frac{2\pm\sqrt{4-16}}{2}$

$$\Rightarrow \quad x=\frac{2\pm 2i\sqrt{3}}{2}=1\pm i\sqrt{3}$$

Let, $\alpha=1+i\sqrt{3}$ and $\beta=1-i\sqrt{3}$

$$\Rightarrow \quad \alpha^n=\left(1+i\sqrt{3}\right)^n=2^n\left[\frac{1}{2}+i\frac{\sqrt{3}}{2}\right]^n$$

$$=2^n\left[\cos\frac{\pi}{3}+i\sin\frac{\pi}{3}\right]^n$$

$$\Rightarrow \quad \alpha^n=2^n\left[\cos\frac{n\pi}{3}+i\sin\frac{n\pi}{3}\right]$$

Similarly, $\beta^n=2^n\left[\cos\frac{n\pi}{3}-i\sin\frac{n\pi}{3}\right]$

$$\Rightarrow \quad \alpha^n+\beta^n=2^n\cdot 2\cos\frac{n\pi}{3}=2^{n+1}\cos\frac{n\pi}{3}$$

Example 3.20 For the complex numbers z_1 and z_2, prove that

$$|1-z_1\bar{z}_2|^2-|z_1-z_2|^2=\left(1-|z_1|^2\right)\left(1-|z_2|^2\right)$$

Solution $|1-z_1\bar{z}_2|^2-|z_1-z_2|^2$

$$= (1 - z_1\bar{z}_2)(1 - \bar{z}_1 z_2) - (z_1 - z_2)(\bar{z}_1 - \bar{z}_2)$$
$$= (1 - \bar{z}_1 z_2 - z_1\bar{z}_2 + z_1\bar{z}_1 z_2\bar{z}_2) - (z_1\bar{z}_1 - z_1\bar{z}_2 - \bar{z}_1 z_2 + z_2\bar{z}_2)$$
$$= 1 + |z_1|^2|z_2|^2 - |z_1|^2 - |z_2|^2$$
$$= 1(1 - |z_1|^2) - |z_2|^2(1 - |z_1|^2)$$
$$= (1 - |z_1|^2)(1 - |z_2|^2)$$

Example 3.21 If z_1, z_2 and z_3 represent the vertices A, B and C of the triangle ABC such that $|z_1| = |z_2| = |z_3| = 5$.

Find the value of $z_1 \sin 2A + z_2 \sin 2B + z_3 \sin 2C$.

Solution

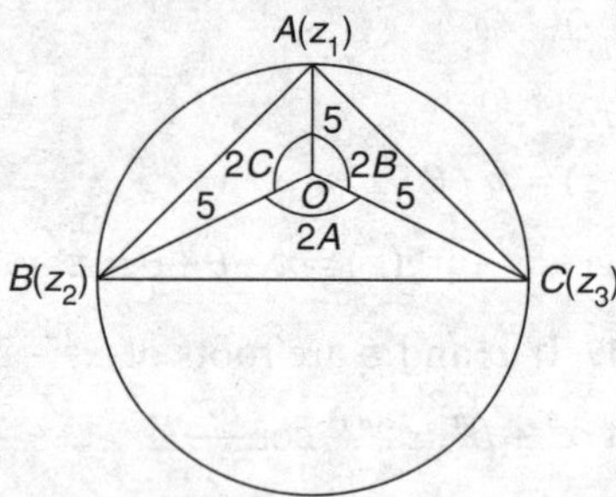

$$\frac{z_2}{z_1} = e^{i2C}$$
$$\frac{z_3}{z_1} = e^{-i2B}$$

Now, $z_1 \sin 2A + z_2 \sin 2B + z_3 \sin 2C$

$= z_1[\sin 2A + \sin 2Be^{i2C} + \sin 2Ce^{-i2B}]$

$= z_1[\sin 2A + \sin 2B \cos 2C + i \sin 2B \sin 2C + \sin 2C \cos 2B - i \sin 2C \sin 2B]$

$= z_1[\sin 2A + \sin(2B + 2C)]$

$= z_1[\sin 2A + \sin(2\pi - 2A)] = z_1(\sin 2A - \sin 2A) = 0$

Example 3.22 If $\text{Re}\left(\frac{1}{z}\right) \geq 4$, then find the greatest value of $|z_1 - z_2|$; where z_1 and z_2 satisfy the given inequality.

Solution Let $z = x + iy$

$$\text{Re}\left(\frac{1}{z}\right) \geq 4 \Rightarrow \text{Re}\left(\frac{\bar{z}}{|z|^2}\right) \geq \frac{1}{4}$$

$$\Rightarrow \quad \frac{x}{x^2 + y^2} \geq \frac{1}{4} \Rightarrow 4x \geq x^2 + y^2$$

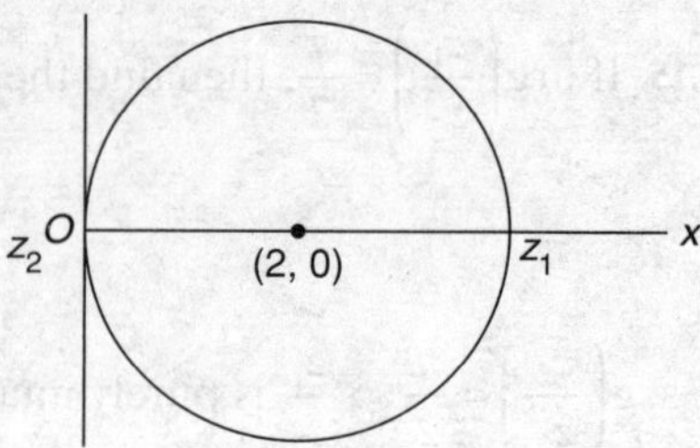

$\Rightarrow \quad x^2 - 4x + y^2 + 4 \leq 4$

$\Rightarrow \quad (x - 2)^2 + y^2 \leq 4$

Which represent a circle with centre (2, 0) and radius 2.

$$z_1 = 4 \text{ and } z_2 = 0$$

$\Rightarrow \quad \max|z_1 - z_2| = 4$

Example 3.23 If the area of the region represented by the inequality $\log_{\sqrt{2}}\left(\frac{|z|^2 - |z| + 8}{|z| + 1}\right) < 4$ is $k\pi$, then evaluate $\sqrt{k + 1}$.

Solution $\frac{|z|^2 - |z| + 8}{|z| + 1} < (\sqrt{2})^4$

$\Rightarrow \quad |z|^2 - |z| + 8 < 4(|z| + 1)$

$\Rightarrow \quad |z|^2 - 5|z| + 4 < 0$

$\Rightarrow \quad 1 < |z| < 4$

This gives the area, $16\pi - \pi = 15\pi$

$\Rightarrow \quad k = 15$

$\Rightarrow \quad \sqrt{k + 1} = 4$

Example 3.24 If $1, \alpha_1, \alpha_2, \ldots, \alpha_{n-1}$ are the nth roots of unity, find the value of

$$\frac{1}{3 - \alpha_1} + \frac{1}{3 - \alpha_2} + \ldots + \frac{1}{3 - \alpha_{n-1}}.$$

Solution Since $1, \alpha_1, \alpha_2, \ldots, \alpha_{n-1}$ are the n roots of the equation $x^n - 1 = 0$.

Hence, $x^n - 1$

$$= (x - 1)(x - \alpha_1)(x - \alpha_2)\ldots(x - \alpha_{n-1})$$

Taking log on both sides, we get

$$\log_e(x^n - 1) = \log_e(x - 1) + \log_e(x - \alpha_1) + \log_e(x - \alpha_2) + \ldots + \log_e(x - \alpha_{n-1})$$

Differentiating both sides with respect to x, we get

$$\frac{nx^{n-1}}{x^n - 1} = \frac{1}{x - 1} + \frac{1}{x - \alpha_1} + \frac{1}{x - \alpha_2} + \ldots + \frac{1}{x - \alpha_{n-1}}$$

Put $x = 3$, we get

$$\frac{1}{3-\alpha_1}+\frac{1}{3-\alpha_2}+\ldots+\frac{1}{3-\alpha_{n-1}}=\frac{n3^{n-1}}{3^n-1}-\frac{1}{2}$$

Example 3.25 The three points z_1, z_2, z_3 are connected by the relation $az_1+bz_2+cz_3=0$, where, a, b, c are real and $a + b + c = 0$. Prove that the three points are collinear.

Solution As $a+b+c=0 \Rightarrow c=-(a+b)$

Now $az_1+bz_2+cz_3=0$

$\Rightarrow \quad az_1+bz_2-(a+b)z_3=0$

$\Rightarrow \quad z_3=\dfrac{az_1+bz_2}{a+b}$

$\Rightarrow z_3$ divides the line joining z_1 and z_2 in the ratio $b:a$

Hence z_1, z_2, z_3 are collinear.

3.13 Know Your Concepts

1. Find the square root of the following:
   (i) $-5+12i$ (ii) $-8-15i$
2. Find the values of x and y so that $-3+ix^2y$ and x^2+y+4i may represent complex conjugate numbers.
3. Find the square root of
   $x^2+\dfrac{1}{x^2}-\dfrac{4}{i}\left(x-\dfrac{1}{x}\right)-6$, where $x \in R$.
4. Find the value of $\left[\dfrac{1+\sin\frac{\pi}{8}+i\cos\frac{\pi}{8}}{1+\sin\frac{\pi}{8}-i\cos\frac{\pi}{8}}\right]^8$.
5. If $3^{49}(x+iy)=\left(\dfrac{3}{2}+\dfrac{\sqrt{3}}{2}i\right)^{100}$
   and $x = ky$, then find the value of k.
6. If $x^2-x+1=0$, then find the value of $\displaystyle\sum_{n=1}^{5}\left(x^n+\frac{1}{x^n}\right)^2$.
7. If α is non-real and $\alpha=\sqrt[5]{1}$, then find the value of $2^{|1+\alpha+\alpha^2+\alpha^{-2}-\alpha^{-1}|}$.
8. If ω is a non-real cube root of unity, then find the value of $\dfrac{1+2\omega+3\omega^2}{2+3\omega+\omega^2}+\dfrac{2+3\omega+\omega^2}{3+\omega+2\omega^2}$.
9. If $z^2-z+1=0$, then find the value of
   $\left(z+\dfrac{1}{z}\right)^2+\left(z^2+\dfrac{1}{z^2}\right)^2+\left(z^3+\dfrac{1}{z^3}\right)^2+\ldots+\left(z^{24}+\dfrac{1}{z^{24}}\right)^2$.
10. If $\text{Im}\left(\dfrac{2z+1}{iz+1}\right)=-2$, then find the locus of the point representing z in the complex plane.
11. Find the radius of the circle represented by the complex number
    $z\bar{z}+(4-3i)z+(4+3i)\bar{z}+5=0$.
12. If $a=\cos\alpha+i\sin\alpha$, $b=\cos\beta+i\sin\beta$, $c=\cos\gamma+i\sin\gamma$ and $\dfrac{a}{b}+\dfrac{b}{c}+\dfrac{c}{a}=1$, then find the value of
    $\cos(\alpha-\beta)+\cos(\beta-\gamma)+\cos(\gamma-\alpha)$.
13. Find the value of the expression
    $2\left(1+\dfrac{1}{\omega}\right)\left(1+\dfrac{1}{\omega^2}\right)+3\left(2+\dfrac{1}{\omega}\right)\left(2+\dfrac{1}{\omega^2}\right)+4\left(3+\dfrac{1}{\omega}\right)\left(3+\dfrac{1}{\omega^2}\right)+\ldots+(n+1)\left(n+\dfrac{1}{\omega}\right)\left(n+\dfrac{1}{\omega^2}\right)$
14. If $x_r=\cos\dfrac{\pi}{2^r}+i\sin\dfrac{\pi}{2^r}$, prove that
    $x_1x_2x_3\ldots$ to $\infty=-1$.
15. If $iz^3+z^2-z+i=0$, then prove that $|z|=1$.
16. If ω, ω^2 are the imaginary cube roots of unity, then find the value of following:
    (i) $(3+3\omega+5\omega^2)^6-(2+6\omega+2\omega^2)^3$
    (ii) $(2-\omega)(2-\omega^2)(2-\omega^{10})(2-\omega^{11})$
17. If α and β are different complex numbers with $|\beta|=1$, then find the value of $\left|\dfrac{\beta-\alpha}{1-\bar{\alpha}\beta}\right|$.
18. Find the complex number z such that $z^2+|z|=0$.
19. If $\sqrt[3]{a+ib}=x+iy$, then prove that
    $\dfrac{a}{x}+\dfrac{b}{y}=4(x^2-y^2)$.

20. If $\frac{3}{2+\cos\theta+i\sin\theta}=a+ib$, then find the value of $|a^2+b^2-4a|$.

3.14 JEE Main: Sample Questions

Single Answer Type

1. Let $z \in C$ and satisfies
$\log_{0.5}\left(\frac{|z-1|+3}{2|z-1|-2}\right) \le 1$, then
(A) maximum value of $|z-1|$ is 4
(B) maximum value of $|z-1|$ is 8
(C) minimum value of $|z-1|$ is $\frac{3}{2}$
(D) minimum value of $|z-1|$ is $\frac{2}{3}$

2. The reflection of the complex number $\frac{2+i}{3-i}$ in the straight line $(1+i)z=(1-i)\bar{z}$, is
(A) $\frac{2+i}{3-i}$ (B) $\frac{-2+i}{-3+i}$
(C) $\frac{2-i}{3-i}$ (D) $\frac{2+i}{-3+i}$

3. Let z and ω be complex numbers such that $z\omega=|z|^2$ and $|z-\bar{z}|+|\omega+\bar{\omega}|=4$. If ω varies then the area enclosed by the locus of z, is
(A) 4 (B) 6 (C) 8 (D) 12

4. If $|Z-i| \le 2$ and $Z_1 = 5+3i$, then the maximum value of $|iZ+Z_1|$ is
(A) $7+\sqrt{13}$ (B) $7+\sqrt{12}$
(C) 7 (D) 9

5. If $z+z^3=0$, then which of the following must be true on the complex plane?
(A) $\text{Re}(z)<0$ (B) $\text{Re}(z)=0$
(C) $\text{Im}(z)=0$ (D) $z^4=1$

6. Let $i=\sqrt{-1}$. The product of the real part of the roots of $z^2-z=5-5i$ is
(A) –25 (B) –6 (C) –5 (D) 25

7. There is only one way to choose real numbers M and N such that when the polynomial $5x^4+4x^3+3x^2+Mx+N$ is divided by the polynomial x^2+1, the remainder is 0. If M and N assume these unique values, then $M-N$ is
(A) –6 (B) –2 (C) 6 (D) 2

8. In the quadratic equation $x^2+(p+iq)x+3i=0$, p and q are real. If the sum of the squares of the roots is 8 then
(A) $p=3, q=-1$ (B) $p=-3, q=-1$
(C) $p=\pm 3, q=\pm 1$ (D) $p=-3, q=1$

9. The complex number z satisfies $z+|z|=1+7i$; then the value of $|z|^2$ equals
(A) 625 (B) 169 (C) 49 (D) 25

10. If $z=(3+7i)(p+iq)$ where $p, q \in I-\{0\}$, is purely imaginary then the minimum value of $|z|^2$ is
(A) 0 (B) 58 (C) $\frac{3364}{3}$ (D) 3364

11. The number of complex numbers z satisfying $z^3=\bar{z}$ is
(A) 1 (B) 2 (C) 4 (D) 5

12. If z is a complex number satisfying the equations $|z-(1+i)|^2=2$ and $\omega=\frac{2}{z}$, then the locus traced by 'ω' in the complex plane is
(A) $x-y-1=0$ (B) $x+y-1=0$
(C) $x-y+1=0$ (D) $x+y+1=0$

13. If the expression $(1+ir)^3$ is of the form of $s(1+i)$ for some real 's' where 'r' is also real and $i=\sqrt{-1}$, then the value of 'r' can be
(A) $\cot\frac{\pi}{8}$ (B) $\sec\pi$
(C) $\tan\frac{\pi}{12}$ (D) $\tan\frac{5\pi}{12}$

14. The diagram shows several numbers in the complex plane. The circle is the unit circle centered at the origin. One of these numbers is the reciprocal of F, which is

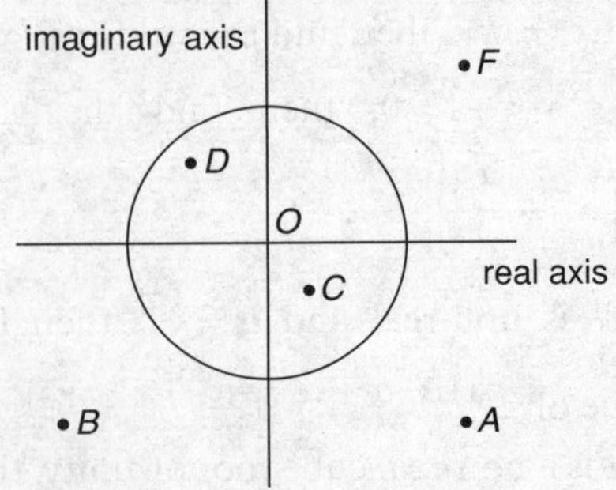

(A) A (B) B (C) C (D) D

15. If $z = x + iy$ and $\omega = \dfrac{1 - iz}{z - i}$ then $|\omega| = 1$ implies that, in the complex plane,
(A) z lies on the imaginary axis
(B) z lies on the real axis
(C) z lies on the unit circle
(D) z lies on the $|z - 1| = |z - 2|$

16. The complex number z satisfies $z + |z| = 2 + 8i$. The value of $|z|$ is
(A) 10 (B) 13 (C) 17 (D) 23

17. The locus of z, for $\arg z = -\dfrac{\pi}{3}$ is
(A) same as the locus of z for $\arg z = \dfrac{2\pi}{3}$
(B) same as the locus of z for $\arg z = \dfrac{\pi}{3}$
(C) the part of the straight line $\sqrt{3}x + y = 0$ with $(y < 0, x > 0)$
(D) the part of the straight line $\sqrt{3}x + y = 0$ with $(y > 0, x < 0)$

18. If P and Q are represented by the complex numbers z_1 and z_2 such that $\left|\dfrac{1}{z_1} + \dfrac{1}{z_2}\right| = \left|\dfrac{1}{z_1} - \dfrac{1}{z_2}\right|$, then the circumcentre of ΔOPQ (where O is the origin) is
(A) $\dfrac{z_1 - z_2}{2}$ (B) $\dfrac{z_1 + z_2}{2}$
(C) $\dfrac{z_1 + z_2}{3}$ (D) $z_1 + z_2$

19. All real numbers x which satisfy the inequality $|1 + 4i - 2^{-x}| \le 5$ where $i = \sqrt{-1}$, $x \in R$ are
(A) $[-2, \infty)$ (B) $(-\infty, 2]$
(C) $[0, \infty)$ (D) $[-2, 0]$

20. For $Z_1 = \sqrt[6]{\dfrac{1-i}{1+i\sqrt{3}}}$; $Z_2 = \sqrt[6]{\dfrac{1-i}{\sqrt{3}+i}}$; $Z_3 = \sqrt[6]{\dfrac{1+i}{\sqrt{3}-i}}$ which of the following holds good?
(A) $\sum |Z_1|^2 = \dfrac{3}{2}$
(B) $|Z_1|^4 + |Z_2|^4 = |Z_3|^{-8}$
(C) $\sum |Z_1|^3 + |Z_2|^3 + |Z_3|^{-6}$
(D) $|Z_1|^4 + |Z_2|^4 = |Z_3|^8$

21. The total number of real or purely imaginary solution of the equation, $z^3 + iz - 1 = 0$ is:
(A) zero (B) one (C) two (D) three

22. A point 'z' moves on the curve $|z - 4 - 3i| = 2$ in an argand plane. The maximum and minimum values of $|z|$ are
(A) 2, 1 (B) 6, 5 (C) 4, 3 (D) 7, 3

23. If z is a complex number satisfying the equation $|z + i| + |z - i| = 8$ on the complex plane, then the maximum value of $|z|$ is
(A) 2 (B) 4 (C) 6 (D) 8

24. Let z_r $(1 \le r \le 4)$ be complex numbers such that $|z_r| = \sqrt{r+1}$ and $|30z_1 + 20z_2 + 15z_3 + 12z_4| = k|z_1z_2z_3 + z_2z_3z_4 + z_3z_4z_1 + z_4z_1z_2|$
Then the value of k equals
(A) $|z_1z_2z_2|$ (B) $|z_2z_3z_4|$
(C) $|z_3z_4z_1|$ (D) $|z_4z_1z_2|$

25. Let Z be a complex number satisfying the equation $(Z^3 + 3)^2 = -16$ then $|Z|$ has the value equal to
(A) $5^{\frac{1}{2}}$ (B) $5^{\frac{1}{3}}$ (C) $5^{\frac{2}{3}}$ (D) 5

26. If z_1, z_2, z_3 are 3 distinct complex numbers such that $\dfrac{3}{|z_2 - z_3|} = \dfrac{4}{|z_3 - z_1|} = \dfrac{5}{|z_1 - z_2|}$, then the value of $\dfrac{9}{z_2 - z_3} + \dfrac{16}{z_3 - z_1} + \dfrac{25}{z_1 - z_2}$ equals
(A) 0 (B) 3 (C) 4 (D) 5

27. The area of the triangle whose vertices are the roots of $z^3 + iz^2 + 2i = 0$ is
(A) 2 (B) $\dfrac{3}{2}\sqrt{7}$ (C) $\dfrac{3}{4}\sqrt{7}$ (D) $\sqrt{7}$

28. Consider two complex number α and β as $\alpha = \left(\dfrac{a+bi}{a-bi}\right)^2 + \left(\dfrac{a+bi}{a+bi}\right)^2$, where $a, b \in R$ and $\beta = \dfrac{z-1}{z+1}$, where $|z| = 1$, then
(A) Both α and β are purely real
(B) Both α and β are purely imaginary
(C) α is purely real and β is purely imaginary
(D) β is purely real and α is purely imaginary

29. Let Z be a complex number satisfying the equation $z^2 - (3 + i)z + m + 2i = 0$, where $m \in R$. Suppose the equation has a real root. The additive inverse of non-real root, is

(A) $1 - i$ (B) $1 + i$
(C) $-1 - i$ (D) -2

30. The minimum value of $|z - 1 + 2i| + |4i - 3 - z|$ is

(A) $\sqrt{5}$ (B) 5 (C) $2\sqrt{13}$ (D) $\sqrt{15}$

31. Let $z = x + iy$ then the locus of the moving point $P(z)$ such that $\frac{1+\bar{z}}{z} \in R$, is (where $i^2 = -1$)

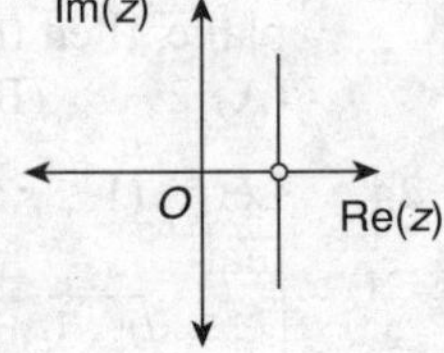

(A) union of lines with equations $x = 0$ and $y = \frac{-1}{2}$ but excluding origin
(B) union of lines with equations $x = 0$ and $y = \frac{1}{2}$ but excluding origin.
(C) union of lines with equations $x = \frac{-1}{2}$ and $y = 0$ but excluding origin.
(D) union of lines with equations $x = \frac{1}{2}$ and $y = 0$ but excluding origin.

32. The equation $z\bar{z} + (4 - 3i)z + (4 + 3i)\bar{z} + 5 = 0$ represents a circle whose radius is

(A) 5 (B) $\frac{5}{2}$ (C) $2\sqrt{5}$ (D) $\sqrt{5}$

33. The equation of the radical axis of the two circles represented by the equations, $|z - 2| = 3$ and $|z - 2 - 3i| = 4$ on the complex plane is

(A) $3y + 1 = 0$ (B) $3y - 1 = 0$
(C) $2y - 1 = 0$ (D) none

34. Given $z = f(x) + ig(x)$ where $f, g: (0, 1) \to (0, 1)$ are real valued functions, then, which of the following holds good?

(A) $z = \frac{1}{1-ix} + i\left(\frac{1}{1+ix}\right)$
(B) $z = \frac{1}{1+ix} + i\left(\frac{1}{1-ix}\right)$
(C) $z = \frac{1}{1+ix} + i\left(\frac{1}{1+ix}\right)$
(D) $z = \frac{1}{1-ix} + i\left(\frac{1}{1-ix}\right)$

35. $z_1 = \frac{a}{1-i}$; $z_2 = \frac{a}{2+i}$; $z_3 = a - bi$ for $a, b \in R$ if $z_1 - z_2 = 1$ then the centroid of the triangle formed by the points z_1, z_2, z_3 in the argand's plane is given by

(A) $\frac{1}{9}(1+7i)$ (B) $\frac{1}{3}(1+7i)$
(C) $\frac{1}{3}(1-3i)$ (D) $\frac{1}{9}(1-3i)$

36. Consider the equation $10z^2 - 3iz - k = 0$, where z is a complex variable and $i^2 = -1$. Which of the following statements is true?

(A) For all real positive numbers k, both roots are purely imaginary.
(B) For negative real numbers k, both roots are purely imaginary.
(C) For all purely imaginary numbers k, both roots are real and irrational.
(D) For all complex numbers k, neither root is real.

37. Number of complex numbers z such that $|z| = 1$ and $\left|\frac{z}{\bar{z}} + \frac{\bar{z}}{z}\right| = 1$ is

(A) 4 (B) 6
(C) 8 (D) more than 8

38. The number of ordered pairs(s) (a, b) of real numbers such that $(a + ib)^{2008} = a - ib$ holds good, is

(A) 2008 (B) 2009 (C) 2010 (D) 1

39. Let z be a complex number having the argument θ, $0 < \theta < \frac{\pi}{2}$ and satisfying the equality $|z - 3i| = 3$. Then $\cot\theta - \frac{6}{z}$ is equal to

(A) 1 (B) -1 (C) i (D) $-i$

40. If the complex number z satisfies the condition $|z| \geq 3$, then the least value of $\left|z + \frac{1}{z}\right|$ is equal to

(A) $\frac{5}{3}$ (B) $\frac{8}{3}$ (C) $\frac{11}{3}$ (D) $\frac{13}{3}$

41. The maximum and minimum values of $|z + 1|$ when $|z + 3| \leq 3$ are

(A) (5, 0) (B) (6, 0)
(C) (7, 1) (D) (5, 1)

42. If $z^3 + (3 + 2i)z + (-1 + ia) = 0$ has one real root, then the value of 'a' lies in the interval ($a \in R$)
(A) $(-2, -1)$ (B) $(-1, 0)$
(C) $(0, 1)$ (D) $(1, 2)$

43. If $|z| = 1$ and $|\omega - 1| = 1$ where $z, \omega \in C$, then the largest set of value of $|2z - 1|^2 + |2\omega - 1|^2$ equals
(A) $[1, 9]$ (B) $[2, 6]$
(C) $[2, 12]$ (D) $[2, 18]$

44. If z_1, z_2, z_3 are the vertices of the ΔABC on the complex plane which are also the roots of the equation, $z^3 - 3\alpha z^2 + 3\beta z + x = 0$, then the condition for the ΔABC to be equilateral triangle is
(A) $\alpha^2 = \beta$ (B) $\alpha = \beta^2$
(C) $\alpha^2 = 3\beta$ (D) $\alpha = 3\beta^2$

45. Let $ABCDEF$ be a regular hexagon with vertices in the anticlockwise sense. If z_1 and z_2 represent A and B respectively, the complex number representing C is
(A) $wz_1 + (1 + w)z_2$ (B) $w^2z_1 + (1 - w^2)z_2$
(C) $wz_1 + (w - 1)z_2$ (D) $wz_1 + (1 - w)z_2$

46. The points $z_1 = 3 + \sqrt{3}i$ and $z_2 = 2\sqrt{3} + 6i$ are given on a complex plane. The complex number lying on the bisector of the angle formed by the vectors z_1 and z_2 is:
(A) $z = \frac{(3+2\sqrt{3})}{2} + \frac{\sqrt{3}+2}{2}i$
(B) $z = 5 + 5i$
(C) $z = -1 - i$
(D) none

47. On the Argand plane point 'A' denotes a complex number z_1. A triangle OBQ is made directily similar to the triangle OAM, where $OM = 1$ as shown in the figure.

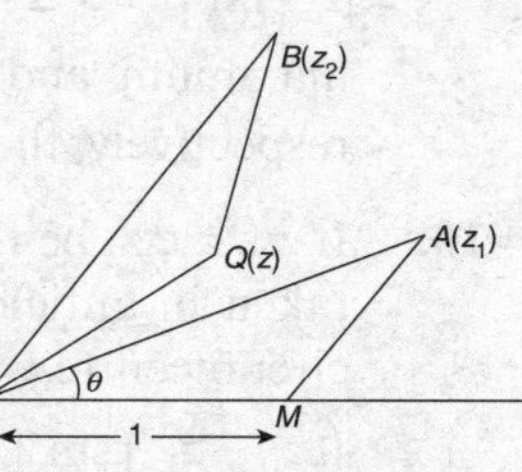

If the point B denotes the complex number z_2, then the complex number corresponding to the point 'Q' is
(A) z_1z_2 (B) $\frac{z_1}{z_2}$
(C) $\frac{z_2}{z_1}$ (D) $\frac{z_1 + z_2}{z_2}$

48. The number of solutions of $\sqrt{2}\,|z+1| = z - i$ is, (where $z = x + iy$)
(A) 1 (B) 2
(C) infinite (D) none of these

49. The equation
$z\left(\overline{z+1+i\sqrt{3}}\right) + \bar{z}\left(z+1+i\sqrt{3}\right) = 0$ represents
a circle with
(A) centre $\left(-\frac{1}{2}, \frac{\sqrt{3}}{2}\right)$ and radius 1
(B) centre $\left(\frac{1}{2}, \frac{-\sqrt{3}}{2}\right)$ and radius 1
(C) centre $\left(-\frac{1}{2}, \frac{\sqrt{3}}{2}\right)$ and radius 2
(D) centre $\left(\frac{1}{2}, \frac{-\sqrt{3}}{2}\right)$ and radius 2

50. If z lies on the circle centered at the origin, if the area of the triangle whose vertices are z, ωz and $z + \omega z$, where ω is the cube roots of unity, is $4\sqrt{3}$ sq. unit. Then the radius of the circle is
(A) 1 unit (B) 2 units
(C) 3 units (D) 4 units

51. The complex number z satisfying
$|z+2+i| + |z-2+i| = 4$,
$0 \le \arg(z+2+2i) \le \frac{\pi}{4}$
and $\frac{3\pi}{4} \le \arg(z-2+2i) \le \pi$ will lie on a line segment of length
(A) 4 (B) $\sqrt{2}$
(C) 2 (D) none of these

52. The complex number z and w satisfying $z^3 + \bar{w}^7 = 0$ and $z^5 \cdot w^{11} = 1$ are
(A) $w = \pm i$, $z = \pm 1$ (B) $w = \pm 1$, $z = i$
(C) $w = \pm i$, $z = \pm i$ (D) none of these

53. If $az^2 + bz + 1 = 0$, where $a, b \in c$ and $|a| = \frac{1}{2}$, have a root α such that $|\alpha| = 1$, then $|a\bar{b} - b|$ is equal to
(A) $\frac{1}{4}$ (B) $\frac{1}{2}$ (C) $\frac{3}{4}$ (D) $\frac{5}{4}$

54. If z is a complex number lying in the fourth quadrant of the Argand plane and $\left|\frac{kz}{k+1}+2i\right|>\sqrt{2}$ for all real value of $k(k\neq -1)$, then range of $\arg(z)$ is

(A) $\left(-\frac{\pi}{8},0\right)$ (B) $\left(-\frac{\pi}{6},0\right)$

(C) $\left(-\frac{\pi}{4},0\right)$ (D) none of these

55. If $A(z_1)$ and $B(z_2)$ be two complex numbers satisfying $\frac{A}{B}+\frac{B}{A}=1$, then the two points represented by A and B, the origin form the vertices of

(A) a right angled triangle
(B) an equilateral triangle
(C) an isosceles triangle which is not right angled
(D) none of these

56. If $z=\frac{\pi}{4}(1+i)^4\left(\frac{1-\sqrt{\pi}i}{\sqrt{\pi}+i}+\frac{\sqrt{\pi}-i}{1+\sqrt{\pi}i}\right)$, then $\left(\frac{\bar{z}}{\text{amp } z}\right)$ is equal to

(A) π (B) 3π (C) 1 (D) 4

57. Let z be a complex number having the argument θ, $0<\theta<\frac{\pi}{2}$ and satisfying the equality $|z-3i|=3$, then $\cot\theta-\frac{6}{z}$ is equal to

(A) $-i$ (B) i (C) -1 (D) 1

58. If the imaginary part of $(1-i)^n(1+i)^{-n}$ is negative (where $n\in N, n<100$), then the sum of all the possible values of n is

(A) 1025 (B) 1225 (C) 625 (D) 1238

59. If z_1 and z_2 are two complex numbers such that $|z_1|\leq 1, |z_2|\leq 1$ and $|z+iz_1|=|z-i\bar{z}_2|=2$, then z is equal to

(A) 1 or i (B) i or $-i$
(C) 1 or -1 (D) i or -1

Integer Answer Type

1. If $|z-i|\leq 2$ and $z_0=5+3i$, then find the sum of the maximum and minimum values of $\frac{1}{2}|iz+z_0|$.

2. If a, b, c, a_1, b_1, c_1 are non-zero complex numbers satisfying $\frac{a}{a_1}+\frac{b}{b_1}+\frac{c}{c_1}=1+i$ and $\frac{a_1}{a}+\frac{b_1}{b}+\frac{c_1}{c}=0$, then find the value of $\left|\frac{a^2}{a_1^2}+\frac{b^2}{b_1^2}+\frac{c^2}{c_1^2}\right|$.

3. If z_1, z_2 are two complex numbers such that $|z_1z_2|=\sqrt{2}$ and $\arg(z_1)-\arg(z_2)=\frac{\pi}{4}$, then find the value of $|\bar{z}_1z_2|^2$.

4. If $A=\{z:(1+2i)\bar{z}+(1-2i)z+2=0\}$ and $B=\{z:(3+2i)\bar{z}+(3-2i)z+3=0\}$, then find the value of $n(A\cap B)$.

5. If z_1, z_2 are complex numbers such that $|z_1|=\frac{5}{3}$, then find the value of $\left|\frac{75-27z_1\bar{z}_2}{5(z_1-z_2)}\right|$.

6. If $|z-2-i|=4$, then find the minimum value of $|3z-6+15i|$.

7. Find the sum of all the solutions of $z^2+|z|=\bar{z}^2$.

8. If $A=\left\{z:\left(\frac{z-\bar{z}}{2i}\right)^2\leq 2(z+\bar{z})\right\}$ and $B=\{z:|z|\leq\sqrt{5}\}$.

Find the number of points lying in $A\cap B$.

9. If z is a complex number such that $|z-1|+|z-3|\leq 4$. If M and m be the maximum and minimum values of $|z-5|$ respectively, then find the value of $M+m$.

10. If z_1, z_2, z_3 be the vertices of triangle ABC, taken in anticlockwise direction and z_0 its circumcentre, then find the value of $\left|\left(\frac{z_0-z_1}{z_0-z_2}\right)\frac{\sin 2A}{\sin 2B}+\frac{(z_0-z_3)}{(z_0-z_2)}\frac{\sin 2C}{\sin 2B}\right|$.

11. Find the number of distinct complex numbers z such that $|z|=1$ and z^3 is purely imaginary.

12. Let C be the set of complex numbers and $z\in C$. If $S=\{z:|z+\bar{z}|+2|z-\bar{z}|=4$ and $|z|$ is minimum$\}$ then the area of the polygon

formed by all the points in S taking as a vertices is $\frac{p}{q}$ where $p, q \in N$. Find the least value of $(p - q)$.

13. Let α^k where $k = 0, 1, 2, \ldots, 2013$ be 2014th root of unity. If z_1 and z_2 be any two complex numbers such that $|z_1| = |z_2| = \frac{1}{\sqrt{1007}}$, then find the value of $\sum_{k=0}^{2013} |z_1 + \alpha^k z_2|^2$.

14. Let $S = \{z | z\bar{z} - (3-4i)z - (3+4i)\bar{z} + 21 = 0\}$. If M and m be the maximum and minimum value of $\left(\frac{z-\bar{z}}{i(z+\bar{z})}\right)$ then find $\left(\frac{1}{M} + \frac{1}{m}\right)$.

15. Consider a complex number $z = \frac{2\omega + 1}{i\omega + 1}$ where $\omega = x + iy$ ($\omega \neq i$ and $x, y \in R$). If $\text{Im}(z) = -2$, then the locus of ω is a straight line. Find the y-intercept of the line.
[**Note:** $\text{Im}(z)$ denotes the imaginary part of z and $i^2 = -1$]

16. If ω is the imaginary cube root of unity, then find the number of pairs of integers (a, b) such that $|a\omega + b| = 1$.

17. A polynomial $f(z)$ when divided by $(z - w)$ leaves a remainder $2 + i\sqrt{3}$ and when divided by $(z - w^2)$ leaves a remainder $2 - i\sqrt{3}$. If the remainder obtained when $f(z)$ is divided by $z^2 + z + 1$ is $az + b$ (where w is a non-real cube root of unity and $a, b \in R^+$), then find the value of $(a + b)$.

18. $ABCD$ is a rhombus. Its diagonals AC and BD intersect at the point M and satisfying $BD = 2AC$. If the points D and M represent the complex numbers $1 + i$ and $2 - i$ respectively and if $(x_1 + iy_1)$ and $(x_2 + iy_2)$ are the complex numbers of the points A and C, then find the value of $\frac{x_1 x_2}{y_1 y_2}$.

19. Let $z = x + iy$, where $x, y \in R$ and $i = \sqrt{-1}$. Find the radius of circle inscribed in the triangle formed by locus of $P(z)$ satisfying $\frac{z^2 - 1}{z^2 + 1} \in R$ and the line $4x + 3y = 12$.

3.15 JEE Advanced: Sample Questions

Single Answer Type

1. A root of unity is a complex number that is a solution to the equation, $z^n = 1$ for some positive integer n. The number of non-real complex roots of unity that are also the roots of the equation $z^2 + az + b = 0$, for some integer a and b is
(A) 6 (B) 8 (C) 9 (D) 10

2. The complex number ω satisfying the equation $\omega^3 = 8i$ and lying in the second quadrant on the complex plane is
(A) $-\sqrt{3} + i$ (B) $-\frac{\sqrt{3}}{2} + \frac{1}{2}i$
(C) $-2\sqrt{3} + i$ (D) $-\sqrt{3} + 2i$

3. If $z^4 + 1 = \sqrt{3}i$
(A) z^3 is purely real
(B) z represents the vertices of a square of side $2^{1/4}$
(C) z^9 is purely imaginary
(D) z represents the vertices of a square of side $2^{3/4}$

4. If $z_1 = 2 + 3i$, $z_2 = 3 - 2i$ and $z_3 = -1 - 2\sqrt{3}i$, then which of the following is true?
(A) $\arg\left(\frac{z_3}{z_2}\right) = \arg\left(\frac{z_3 - z_1}{z_2 - z_1}\right)$
(B) $\arg\left(\frac{z_3}{z_2}\right) = \arg\left(\frac{z_2}{z_1}\right)$
(C) $\arg\left(\frac{z_3}{z_2}\right) = 2\arg\left(\frac{z_3 - z_1}{z_2 - z_1}\right)$
(D) $\arg\left(\frac{z_3}{z_2}\right) = \frac{1}{2}\arg\left(\frac{z_3 - z_1}{z_2 - z_1}\right)$

5. If m and n are the smallest positive integers satisfying the relation $\left(2\text{C is}\frac{\pi}{6}\right)^m = \left(4\text{C is}\frac{\pi}{4}\right)^n$, then $(m + n)$ has the value equal to
(A) 120 (B) 96 (C) 72 (D) 60

6. If z is a complex number satisfying the equation $Z^6 + Z^3 + 1 = 0$. If this equation has a root $re^{i\theta}$ with $90° < \theta < 180°$ then the value of 'θ' is
(A) 100° (B) 110° (C) 160° (D) 170°

7. Least positive argument of the 4th root of the complex number $2 - i\sqrt{12}$ is

(A) $\frac{\pi}{6}$ (B) $\frac{5\pi}{12}$ (C) $\frac{7\pi}{12}$ (D) $\frac{11\pi}{12}$

8. If A and B be two complex numbers satisfying $\frac{A}{B}+\frac{B}{A}=1$. Then the two points represented by A and B and the origin form the vertices of

(A) an equilateral triangle
(B) an isosceles triangle which is not equilateral
(C) an isosceles triangle which is not right angled
(D) a right angled triangle

9. If 1, $\alpha_1, \alpha_2, \ldots, \alpha_{2008}$ are (2009)th roots of unity, then the value of $\sum_{r=1}^{2008} r(\alpha_r + \alpha_{2009-r})$ equals

(A) 2009 (B) 2008
(C) 0 (D) –2009

10. If the six solutions of $x^6 = -64$ are written in the form $a + bi$, where a and b are real, then the product of those solutions with $a > 0$, is

(A) 4 (B) 8 (C) 16 (D) 64

11. Number of imaginary complex numbers satisfying the equation, $z^2 = \bar{z}\, 2^{1-|z|}$ is

(A) 0 (B) 1 (C) 2 (D) 3

12. If $z_n = \cos\frac{\pi}{(2n+1)(2n+3)} + i\sin\frac{\pi}{(2n+1)(2n+3)}$, then $\lim_{n\to\infty}(z_1 \cdot z_2 \cdot z_3 \cdot \ldots \cdot z_n) =$

(A) $\cos\frac{\pi}{3}+i\sin\frac{\pi}{3}$

(B) $\cos\frac{\pi}{6}+i\sin\frac{\pi}{6}$

(C) $\cos\frac{5\pi}{6}+i\sin\frac{5\pi}{6}$

(D) $\cos\frac{3\pi}{2}+i\sin\frac{3\pi}{2}$

13. The straight line $(1 + 2i)z + (2i - 1)\bar{z} = 10i$ on the complex plane has intercept on the imaginary axis equal to

(A) 5 (B) $\frac{5}{2}$ (C) $-\frac{5}{2}$ (D) –5

14. If $\cos\theta + i\sin\theta$ is a root of the equation $x^n + a_1x^{n-1} + a_2x^{n-2} + \ldots + a_{n-1}x + a_n = 0$ then the value of $\sum_{r=1}^{n} a_r \cos r\theta$ equals (where all coefficients are real)

(A) 0 (B) 1 (C) –1 (D) none

15. Let $A(z_1)$ and $B(z_2)$ represent two complex numbers on the complex plane. Suppose the complex slope of the lines joining A and B is defined as $\frac{z_1 - z_2}{\bar{z}_1 - \bar{z}_2}$. Then the lines ℓ_1 with complex slope ω_1 and ℓ_2 with complex slope ω_2 on the complex plane will be perpendicular to each other if

(A) $\omega_1 + \omega_2 = 0$ (B) $\omega_1 - \omega_2 = 0$
(C) $\omega_1\omega_2 = -1$ (D) $\omega_1\omega_2 = 1$

16. If the equation, $z^4 + a_1z^3 + a_2z^2 + a_3z + a_4 = 0$, where a_1, a_2, a_3, a_4 are real coefficients different from zero, has a purely imaginary root, then the expression $\frac{a_3}{a_1a_2} + \frac{a_1a_4}{a_2a_3}$ has the value equal to

(A) 0 (B) 1 (C) –2 (D) 2

17. Suppose A is a complex number and $n \in N$, such that $A^n = (A + 1)^n = 1$, then the least value of n is

(A) 3 (B) 6 (C) 9 (D) 12

18. The intercept made by the circle $z\bar{z} + \bar{\alpha}z + \alpha\bar{z} + r = 0$ on the real axis on complex plane, is

(A) $\sqrt{(\alpha+\bar{\alpha}) - r}$ (B) $\sqrt{(\alpha+\bar{\alpha})^2 - 2r}$

(C) $\sqrt{(\alpha+\bar{\alpha})^2 + r}$ (D) $\sqrt{(\alpha+\bar{\alpha})^2 - 4r}$

19. If Z_r; $r = 1, 2, 3, \ldots, 50$ are the roots of the equation $\sum_{r=0}^{50} (Z)^r = 0$, then the value of $\sum_{r=1}^{50} \frac{1}{Z_r - 1}$ is

(A) –85 (B) –25 (C) 25 (D) 75

20. All roots of the equation, $(1+z)^6+z^6=0$
    (A) lie on a unit circle with centre at the origin
    (B) lie on a unit circle with centre at $(-1, 0)$
    (C) lie on the vertices of a regular polygon with centre at the origin
    (D) are collinear

21. If z and w are two complex numbers simultaneously satisfying the equations, $z^3+w^5=0$ and $z^2\,\bar{w}^4=1$, then
    (A) z and w are both purely real
    (B) z is purely real and w is purely imaginary
    (C) w is purely real and z is purely imaginary
    (D) z and w are both imaginary

22. A function f is defined by $f(z)=(4+i)z^2=\alpha z+\gamma$ for all complex numbers z, where α and γ are complex numbers. If $f(1)$ and $f(i)$ are both real then the smallest possible value of $|\alpha|+|\gamma|$ equals
    (A) 1 (B) $\sqrt{2}$ (C) 2 (D) $2\sqrt{2}$

23. Given $f(z)$ = the real part of a complex number z. For example, $f(3-4i)=3$. If $a\in N$, $n\in N$ then the value of $\sum_{n=1}^{6a}\log_2\left|f\left(\left(1+i\sqrt{3}\right)^n\right)\right|$ has a value equal to
    (A) $18a^2+9a$ (B) $18a^2+7a$
    (C) $18a^2-3a$ (D) $18a^2-a$

24. It is given that complex numbers z_1 and z_2 satisfy $|z_1|=2$ and $|z_2|=3$. If the included angle of their corresponding vectors is 60° then $\left|\frac{z_1+z_2}{z_1-z_2}\right|$ can be expressed as $\frac{\sqrt{N}}{7}$ where N is a natural number, then N equals
    (A) 126 (B) 119 (C) 133 (D) 19

25. All complex numbers 'z' which satisfy the relation $|z-|z+1||=|z+|z-1||$ on the complex plane lie on the
    (A) line $y=0$ or an ellipse with foci $(-1, 0)$ and $(1, 0)$
    (B) radical axis of the circles $|z-1|=1$ and $|z+1|=1$
    (C) circle $x^2+y^2=1$
    (D) line $x=0$ or on a line segment joining $(-1, 0)$ to $(1, 0)$

Multiple Answer Type

1. If $z=\sqrt{20i-21}+\sqrt{21+20i}$, then the principle value of arg z can be
    (A) $\frac{\pi}{4}$ (B) $\frac{3\pi}{4}$
    (C) $\frac{-\pi}{4}$ (D) none of these

2. If $|z_1|=|z_2|$ and $\arg(z_1)+\arg(z_2)=\frac{\pi}{2}$, then
    (A) z_1z_2 is purely real
    (B) z_1z_2 is purely imaginary
    (C) $(z_1+z_2)^2$ is purely imaginary
    (D) $(\arg z_1^{-1})+\arg(z_2^{-1})=-\frac{\pi}{2}$

3. Roots of the equation $x^n-1=0, n\in I$
    (A) are collinear
    (B) lie on a circle
    (C) form a regular polygon of unit circumradius
    (D) are non-collinear

4. One vertex of the triangle of maximum area that can be inscribed in the curve $|z-2i|=2$, is $2+2i$, the remaining vertices is/are
    (A) $-1+i\left(2+\sqrt{3}\right)$ (B) $-1-i\left(2+\sqrt{3}\right)$
    (C) $-1+i\left(2-\sqrt{3}\right)$ (D) $-1-i\left(2-\sqrt{3}\right)$

5. Let z be a complex number and a be a real parameter such that $z^2+az+a^2=0$, then
    (A) the locus of z is a pair of straight lines
    (B) the locus of z is a circle
    (C) $\arg(z)=\pm\frac{2\pi}{3}$
    (D) $|z|=|a|$

6. If $\frac{\tan\alpha-i\left(\sin\frac{\alpha}{2}+\cos\frac{\alpha}{2}\right)}{1+2i\sin\frac{\alpha}{2}}$ is purely imaginary, then α is given by
    (A) $n\pi+\frac{\pi}{4}$ (B) $n\pi-\frac{\pi}{4}$
    (C) $2n\pi$ (D) $2n\pi+\frac{\pi}{4}$

7. If $z_1 = a+ib$ and $z_2 = c+id$ are complex numbers such that $|z_1| = |z_2| = 1$, and $\text{Re}(z_1\bar{z}_2) = 0$, then the pair of complex numbers $a+ic = w_1$ and $b+id = w_2$ satisfies

(A) $|w_1| = 1$ (B) $|w_2| = 1$

(C) $\text{Re}(w_1\bar{w}_2) = 0$ (D) none of these

8. If z_1, z_2 are two complex numbers satisfying the equation $\left|\frac{z_1+z_2}{z_1-z_2}\right| = 1$, then

(A) $z_1\bar{z}_2 + z_2\bar{z}_1 = 0$ (B) $\overline{\left(\frac{z_1}{z_2}\right)} = \frac{-z_1}{z_2}$

(C) $z_1\bar{z}_2 - z_2\bar{z}_1 = 0$ (D) none of these

9. If $(1+x)^n = c_0 + c_1x + c_2x^2 + \ldots + c_nx^n$, n is a positive integer, then

(A) $c_0 - c_2 + c_4 - \ldots = 2^{n/2}\cos\left(\frac{n\pi}{4}\right)$

(B) $c_1 - c_3 + c_5 - \ldots = 2^{n/2}\sin\left(\frac{n\pi}{4}\right)$

(C) $c_0 + c_4 + c_8 + \ldots = 2^{n+2} + 2^{\frac{(n-2)}{2}}\cos\left(\frac{n\pi}{4}\right)$

(D) none of these

10. A rectangle of maximum area is inscribed in the circle $|z-3-4i| = 1$. If one vertex of the rectangle is $4+4i$, then another adjacent vertex of their rectangle can be

(A) $2+4i$ (B) $3+3i$

(C) $3+5i$ (D) $3-3i$

11. If $z = x+iy$, and the equation $\left|\frac{2z-i}{z+1}\right| = m$ represents a circle, then m can be

(A $\frac{1}{2}$ (B) 1 (C) 2 (D) 3

12. If $\sqrt{5-12i} + \sqrt{-5-12i} = z$, then the principle value of arg$z$ can be

(A) $-\frac{\pi}{4}$ (B) $\frac{\pi}{4}$ (C) $\frac{3\pi}{4}$ (D) $-\frac{3\pi}{4}$

13. If $z^3 + (3+2i)z + (-1+ia) = 0$ has one real root, then the value of a lies in the interval $(a \in R)$

(A) (–2, 1) (B) (–1, 0)

(C) (0, 1) (D) (–2. 3)

14. If $1, z_1, z_2, z_3, \ldots, z_{n-1}$ are 'n' n^{th} roots of unity, then the value of $\frac{1}{3-z_1} + \frac{1}{3-z_2} + \frac{1}{3-z_3} + \ldots + \frac{1}{3-z_{n-1}}$ is equal to

(A) $\frac{\sum_{r=1}^{n-1} r3^{r-1}}{\sum_{r=1}^{n} 3^{r-1}}$ (B) $\frac{n3^{n-1}}{3^n-1} - \frac{1}{2}$

(C) $\frac{n3^n}{3^n-1} - \frac{1}{2}$ (D) none of these

15. If the lines $\alpha_1\bar{z} + \bar{\alpha}_1 z + \beta_1 = 0$ and $\alpha_2\bar{z} + \bar{\alpha}_2 z + \beta_2 = 0$, $\beta_1, \beta_2 \in R$ are mutually perpendicular, then

(A) $\alpha_1\bar{\alpha}_2 + \bar{\alpha}_1\alpha_2 = 0$ (B) $\alpha_1\bar{\alpha}_2 = \bar{\alpha}_1\alpha_2$

(C) $\left|\arg\left(\frac{z_1}{z_2}\right)\right| = \frac{\pi}{2}$ (D) $\arg\left(\frac{z_1}{z_2}\right) = 0$ or π

16. If $|2z+5| = |6z-9|$, and $|z|^2 = a\text{Re}(z) + b$, where a and b are real numbers, then

(A) $a = 4$ (B) $4(a+b) = 9$

(C) $2(a+b) = 9$ (D) $a+4b+3 = 0$

17. If $2 \le |z| \le 4$, minimum value of $\left|z+\frac{1}{z}\right|$ is m and maximum value of $\left|z+\frac{1}{z}\right|$ is M, then

(A) $m = \frac{3}{2}$ (B) $m = \frac{5}{2}$

(C) $M = \frac{15}{4}$ (D) $M = \frac{17}{4}$

18. a, b, c are three complex numbers representing the vertices of a triangle ABC, such that $\begin{vmatrix} a & b & c \\ b & c & a \\ c & a & b \end{vmatrix} = 0$, then which of the following **CAN BE** true:

(A) centroid of ΔABC is 0

(B) ΔABC is an equilateral triangle

(C) $|a| = |b| = |c|$

(D) $|a^3| + |b^3| + |c^3| = 3|abc|$

19. If the points $A(z - i)$, $B(z + i)$ and $C(iz)$ are the vertices of an equilateral triangle ABC, then

(A) $\text{Re}(z)$ is $\frac{\sqrt{3}}{2}$ (B) $\text{Im}(z)$ is $\frac{\sqrt{3}}{2}$

(C) $\text{Re}(z)$ is $-\frac{\sqrt{3}}{2}$

(D) $\text{Im}(z) + \text{Re}(z)$ is $\pm\sqrt{3}$

20. The complex number z satisfying the condition $|z + 9 - 3\sqrt{3}\, i| = 3\sqrt{3}$. Let A and B respectively be the points on the curve where the complex number z has the least and greatest positive argument, then which is/are true

(A) area of ΔAOB is $\frac{81\sqrt{3}}{4}$, (where O is the origin)

(B) the greatest value of $|z|$ is $9\sqrt{3}$

(C) least positive argument is $\frac{2\pi}{3}$

(D) ΔAOB is an equilateral

21. Complex numbers z_1 and z_2 satisfying $|z_1 - 12i| = \sqrt{3}$, $|z_2 - 2\sqrt{3} - 6i| = \sqrt{3}$ and z_3 is a complex number for which $|z_3 - z_1| + |z_3 - z_2|$ is minimum, then select the correct option(s) (consider $\arg z \in (-\pi, \pi]$)

(A) maximum value of $|z_3|$ is $\sqrt{111}$

(B) maximum value of $|z_3|$ is $\sqrt{\frac{369}{2}}$

(C) sum of minimum and maximum values of $\arg z_3$ is $\frac{\pi}{2} + \cos^{-1}\left(\frac{13}{17\sqrt{3}}\right)$

(D) sum of minimum and maximum values of $\arg z_3$ is $\frac{\pi}{2} + \tan^{-1}\left(\frac{13}{17\sqrt{3}}\right)$

22. Let $w = \frac{\sqrt{3}+i}{2}$ and $P = \{w^n : n = 1, 2, 3, \ldots\}$. Further, $H_1 = \left\{z \in C : \text{Re}\, z > \frac{1}{2}\right\}$ and $H_2 = \left\{z \in C : \text{Re}\, z < \frac{-1}{2}\right\}$, where C is the set of all complex numbers. If $z_1 \in P \cap H_1$, $z_2 \in P \cap H_2$ and O represents the origin, then $\angle z_1 O z_2 =$

(A) $\frac{\pi}{2}$ (B) $\frac{\pi}{6}$ (C) $\frac{2\pi}{3}$ (D) $\frac{5\pi}{6}$

23. Let $a, b \in R$ and $a^2 + b^2 \neq 0$. Suppose $S = \left\{z \in C : z = \frac{1}{a + ibt}, t \in R, t \neq 0\right\}$, where $i = \sqrt{-1}$. If $z = x + iy$ and $z \in S$, then (x, y) lies on

(A) the circle with radius $\frac{1}{2a}$ and centre $\left(\frac{1}{2a}, 0\right)$ for $a > 0, b \neq 0$

(B) the circle with radius $\frac{-1}{2a}$ and centre $\left(\frac{-1}{2a}, 0\right)$ for $a < 0, b \neq 0$

(C) the x-axis for $a \neq 0, b = 0$

(D) the y-axis for $a = 0, b \neq 0$

24. The locii of a point $P(z)$ in the complex plane satisfying $\left|z + \frac{1}{z}\right| = 2$ are two circles C_1 and C_2. These circles

(A) have centres on real axis

(B) cut each other orthogonally

(C) are congruent

(D) have exactly two common tangents

25. If $|z_1| = 1$, $|z_2 - 2| = 3$ and $|z_3 - 5| = 6$, then maximum value of $|2z_1 - 3z_2 - 4z_3|$ is not less than (where z_1, z_2, z_3 are complex numbers)

(A) 43 (B) 44 (C) 45 (D) 46

Comprehension Based Questions

Paragraph for Question No. 1 to 3

If the complex numbers z_1, z_2 and z_3 are such that $z_1z_2 = -80 - 320i$, $z_2z_3 = 60$ and $z_3z_1 = -96 + 24i$, then

1. $|z_1 + z_2 + z_3|^2$ is equal to

(A) 59 (B) 64 (C) 74 (D) 79

2. The possible value of $\arg(z_2)$ is

(A) $\frac{\pi}{3}$ (B) $\frac{\pi}{6}$ (C) $\frac{\pi}{2}$ (D) $\frac{\pi}{4}$

3. The possible value of z_1 is

(A) $4(5 - 3i)$ (B) $-10(1 + i)$

(C) $4(5 + 3i)$ (D) $4(4 + 5i)$

Paragraph for Question No. 4 to 6

Let two curves C_1 and C_2 be defined as

C_1 : Locus of z which satisfies

$|z^2 - 2iz| + |z(z-9+7i)| \le 18$ such that $|z|$ is maximum.

C_2 : Locus of z which satisfies

$\left||z+3\sqrt{2}| - |z-3\sqrt{2}|\right| = 2\sqrt{2}$

4. If z_0 lies on C_1 then maximum value of $|z_0 + 4 + 4i|$ is

(A) $5\sqrt{2}$ (B) $6\sqrt{2}$ (C) $8\sqrt{2}$ (D) $16\sqrt{2}$

5. If $z_1 \left(\text{Re}(z_1) > 0\right)$ lies on C_1 and C_2 both and z_2 is obtained by rotating z_1 about origin in anticlockwise direction through 135° and $z_3 = \bar{z}_2$, then real part of $(z_1 + z_2 + z_3)$ equals

(A) $-2-\sqrt{2}$ (B) $-\sqrt{2}+1$
(C) $-2+\sqrt{2}$ (D) $-\sqrt{2}-1$

6. The locus of z satisfying $|\arg(z-1)| = \tan^{-1}(4)$ meets the curve C_2 at A and B, then area of the triangle formed by A, B and C where the complex number corresponding to C is $e^{i2\pi}$, is

(A) 4 sq. units (B) $8\sqrt{2}$ sq. units
(C) 16 sq. units (D) $16\sqrt{2}$ sq. units

Paragraph for Question No. 7 to 9

For the complex number $w = \dfrac{4z-5i}{2z+1}$

7. The locus of z, when w is a real number other than 2, is

(A) a point circle

(B) a straight line with slope $-\dfrac{5}{2}$ and y-intercept $\dfrac{5}{4}$

(C) a straight line with slope $\dfrac{5}{2}$ and y-intercept $\dfrac{5}{4}$

(D) a straight line passing through the origin

8. The locus of z, when w is a purely imaginary number is

(A) a circle with centre $\left(\dfrac{1}{2}, \dfrac{-5}{4}\right)$ passing through origin

(B) a circle with centre $\left(\dfrac{-1}{4}, \dfrac{5}{8}\right)$ passing through origin

(C) a circle with centre $\left(\dfrac{1}{4}, \dfrac{-5}{8}\right)$ and radius $\dfrac{\sqrt{29}}{8}$

(D) any other circle

9. The locus of z, when $|w| = 1$ is

(A) a circle with centre $\left(-\dfrac{5}{8}, \dfrac{1}{4}\right)$ and radius $\dfrac{1}{2}$

(B) a circle with centre $\left(\dfrac{1}{4}, -\dfrac{5}{8}\right)$ and radius $\dfrac{1}{2}$

(C) a circle with centre $\left(\dfrac{5}{8}, -\dfrac{1}{4}\right)$ and radius $\dfrac{1}{2}$

(D) any other circle

Paragraph for Question No. 10 to 12

Let A, B, C be three sets of complex numbers as defined below.

$A = \{z : |z+1| \le 2 + \text{Re}(z)\}$, $B = \{z : |z-1| \ge 1\}$ and

$C = \left\{z : \left|\dfrac{z-1}{z+1}\right| \ge 1\right\}$

10. The number of point(s) having integral coordinates in the region $A \cap B \cap C$ is

(A) 4 (B) 5 (C) 6 (D) 10

11. The area of region bounded by $A \cap B \cap C$ is

(A) $2\sqrt{3}$ (B) $\sqrt{3}$
(C) $4\sqrt{3}$ (D) 2

12. The real part of the complex number in the region $A \cap B \cap C$ and having maximum amplitude is

(A) -1 (B) $\dfrac{-3}{2}$
(C) $\dfrac{1}{2}$ (D) -2

Paragraph for Question No. 13 to 15

Let $A(z_1)$ be the point of intersection of curves arg $(z-2+i) = \dfrac{3\pi}{4}$ and arg $(z+i\sqrt{3}) = \dfrac{\pi}{3}$. $B(z_2)$ be the point on the curve arg $(z+i\sqrt{3}) = \dfrac{\pi}{3}$ such that

$|z_2 - 5|$ is minimum and $C(z_3)$ be the centre of circle $|z - 5| = 3$. [**Note:** $i^2 = -1$]

13. The area of triangle ABC is equal to

(A) $4\sqrt{3}$ (B) $\frac{3\sqrt{3}}{2}$ (C) $2\sqrt{3}$ (D) 4

14. The equation of straight line passing through origin and perpendicular to line joining $A(z_1)$ and $B(z_2)$ on the complex plane is equal to

(A) $z = \lambda(2 + i\sqrt{3})$ (B) $z = \lambda(-\sqrt{3} + i)$

(C) $z = \lambda(1 + i\sqrt{3})$ (D) $z = \lambda(\sqrt{3} + i)$

15. If $|z - z_1| = 1$ and $w = \text{Re}(z + 2)$ then ω lies on

(A) real axis

(B) line not passing through the origin

(C) line segment joining (2, 0) and (4, 0)

(D) circle centred at (1, 0) and radius 2

Match Matrix

1.

Column I	Column II
(A) If a complex number z satisfying $\lvert z-3-4i\rvert = \alpha$ and z_a and z_b be the corresponding complex numbers for which $\lvert z+2+2i\rvert$ is minimum and maximum respectively, then the odd integral values of α satisfying $\arg\left\lvert\frac{z_a+2+2i}{z_b-3-4i}\right\rvert$ is	(P) 2
(B) The imaginary part of complex number z satisfying $\lvert z-1-2i\rvert \le 1$ and having the least positive argument, is	(Q) 3π
(C) Area bounded by $-8z^2 - 8\bar{z}^2 + 20z\bar{z} = 36$ (where z is a complex number) is	(R) 9
(D) Number of values of z satisfying $\lvert z-2i\rvert = 2$ and $z(1-i) - \bar{z}(1+i) = 4i$ simultaneously is	(S) $\frac{6}{5}$

2.

Column I	Column II
(A) If $P = \cos\frac{2\pi}{7} + i\sin\frac{2\pi}{7}$, $a = p + p^2 + p^4$ and $b = p^3 + p^5 + p^6$ then the value of $\lvert(a+b)+abi\rvert^2$ is	(P) 1
(B) If locus of point $P(z)$ which satisfies $\arg(3 + z + \bar{z} + \lvert z\rvert i) = \frac{\pi}{4}$ is a part of conic C, then length of latus rectum of 'C' is	(Q) 0
(C) If the equation $z^4 + a_1z^3 + a_2z^2 + a_3z + a_4 = 0$, where a_1, a_2, a_3, a_4 are real coefficients different from zero has a purely imaginary root, then the value of the expression $\frac{a_3}{a_1a_2} + \frac{a_1a_4}{a_2a_3}$ is	(R) 6
(D) If z_1, z_2, z_3 are 3 distinct complex numbers such that $\frac{3}{\lvert z_2-z_3\rvert} = \frac{4}{\lvert z_3-z_1\rvert} = \frac{5}{\lvert z_1-z_2\rvert}$, then the value of $\frac{9}{z_2-z_3} + \frac{16}{z_3-z_1} + \frac{25}{z_1-z_2}$ is	(S) 5

3. Match the entries in column-I with those in column-II.

Column I	Column II
(A) If $|z - (3 + 4i)| = 2$ then the maximum value of $|z|$ is	(P) 10
(B) If $|z - i| < 1$, then the value of $|z + 12 - 6i|$ is less than	(Q) 14
(C) If $|z_1| = 1$ and $|z_2| = 2$, then $|z_1 + z_2|^2 + |z_1 - z_2|^2$ is equal to	(R) 4
(D) If $z = 1 + i$, then $4(z^4 - 4z^3 + 7z^2 - 6z + 3)$ is equal to	(S) 7
	(T) 5

4. Match the statements in column I with those in Column II.
[**Note:** Here z takes values in the complex plane and Im z and Re z denote, respectively, the imaginary part and the real part of z.]

Column I	Column II
(A) The set of points z satisfying $|z - i|z|| = |z + i\,|z||$ is contained in or equal to	(P) an ellipse with eccentricity $\frac{4}{5}$
(B) The set of points z satisfying $|z + 4| + |z - 4| = 10$ is contained in or equal to	(Q) the set of points z satisfying Im $z = 0$
(C) If $|\omega| = 2$, then the set of points $z = \omega - \frac{1}{\omega}$ is contained in or equal to	(R) the set of points z satisfying $|\text{Im } z| \le 1$
(D) If $|\omega| = 1$, then the set of points $z = \omega + \frac{1}{\omega}$ is contained in or equal to	(S) the set of points z satisfying $|\text{Re } z| \le 2$
	(T) the set of points z satisfying $|z| \le 3$

5. Match the statements in column I with those in Column II.

Column I	Column II
(A) The set $\left\{ \text{Re}\left(\frac{2iz}{1-z^2}\right) : z \text{ is a complex number}, |z| = 1, z \ne \pm 1 \right\}$ is	(P) $(-\infty, -1) \cup (1, \infty)$
(B) The domain of the function $f(x) = \sin^{-1}\left(\frac{8(3)^{x-2}}{1-3^{2(x-1)}}\right)$, is	(Q) $(-\infty, 0) \cup (0, \infty)$
(C) If $f(\theta) = \begin{vmatrix} 1 & \tan\theta & 1 \\ -\tan\theta & 1 & \tan\theta \\ -1 & -\tan\theta & 1 \end{vmatrix}$, then the set $\left\{ f(\theta) : 0 \le \theta < \frac{\pi}{2} \right\}$	(R) $[2, \infty)$
(D) If $f(x) = x^{\frac{3}{2}}(3x - 10)$, $x \ge 0$ then $f(x)$ is increasing in	(S) $(-\infty, -1] \cup [1, \infty)$
	(T) $(-\infty, 0] \cup [2, \infty)$

Subjective Answer Type (Hard Nut)

1. If 1, ω, ω^2 are the three cube roots of unity, show that

$(a+b\omega+c\omega^2)^3+(a+b\omega^2+c\omega)^3$

$=(2a-b-c)(2b-c-a)(2c-a-b)=27abc$

if $a+b+c=0$.

2. Consider an equilateral triangle formed by the points

$A\left(\frac{2}{\sqrt{3}}e^{i\frac{\pi}{2}}\right), B\left(\frac{2}{\sqrt{3}}e^{-i\frac{\pi}{6}}\right), C\left(\frac{2}{\sqrt{3}}e^{-i\frac{5\pi}{6}}\right)$.

Let $P(z)$ be any point on its incircle. Find the value of $AP^2+BP^2+CP^2$.

3. Let the complex numbers z_1, z_2, z_3 be the vertices of an equilateral triangle. Let z_0 be the circumcentre of the triangle. Prove that $z_1^2+z_2^2+z_3^2=3z_0^2$.

4. If

$\cos(B-C)+\cos(C-A)+\cos(A-B)=-\frac{3}{2}$,

then find the value of $\sum\cos 2(A+B)$.

5. Let z_1, z_2 be two complex numbers with $|z_1|=|z_2|$. Prove that $\frac{(z_1+z_2)^2}{z_1z_2}$ is real. Hence deduce that

$\frac{(z_1+z_2)^2}{z_1z_2}+(\arg z_1-\arg z_2)^2\geq 4$.

6. A cubic equation $f(x)=0$ has one real root α and two complex roots $\beta\pm i\gamma$. Points A, B, C represent roots α, $\beta+i\gamma$ and $\beta-i\gamma$ respectively on the Argand diagram. Show that the roots of the derived equation $f'(x)=0$ are complex if A falls inside one of the two equilateral triangles described on the base BC.

7. If $\frac{1}{a+\omega}+\frac{1}{b+\omega}+\frac{1}{c+\omega}+\frac{1}{d+\omega}=\frac{1}{\omega}$ where $a, b, c, d\in R$ and ω is cube root of unity, find the value of $\sum\frac{1}{a^2-a+1}$.

8. Let z_1, z_2, z_3 be the complex numbers such that $|z_1-1|=|z_2-1|=|z_3-1|$ and $\arg\left(\frac{z_3-z_1}{z_2-z_1}\right)=\frac{\pi}{6}$, then find the value of $|z_2(z_2-1)-z_3(z_2+1)+(z_3+1)(z_3-1)|$.

9. Let z_1, z_2, z_3, z_4 represent the vertices of a square all of which lie on the sides of the triangle with vertices (0, 0), (2, 1), (3, 0). Given that z_1 and z_2 are purely real, find z_3 and z_4.

10. Let A, B, C be the set of complex numbers defined as $A=\{z:||z+2|-|z-2||=2\}$,

$B=\left\{z:\arg\left(\frac{z-1}{z}\right)=\frac{\pi}{2}\right\}$ and

$C=\{z:\arg(z-1)=\pi\}$, find $A\cap B\cap C$.

11. Find the maximum and minimum value of $|2z-1+i|$ if $|z-1|+|z+i|=4$.

12. If we draw a circle by taking the line segment OP as diameter where O is origin and P is any point. Points Q and R are taken on the circle, both points lie on the same side of OP such that $\angle POQ=\angle QOR=\theta$. If P, Q, R are z_1, z_2, z_3 such that $2\sqrt{3}z_2^2=(2+\sqrt{3})z_1z_3$, then find angle θ.

13. If $A_1, A_2, \ldots, A_n$ are the vertices of a regular polygon inscribed in a circle of radius unity, then find the value of $|A_1A_2|^2+|A_1A_3|^2+\ldots+|A_1A_n|^2$.

14. If $|z|\leq 1, |w|\leq 1$, show that

$|z-w|^2\leq(|z|-|w|^2)+(\arg z-\arg w)^2$.

15. Prove that all the roots of the equation

$z^n\cos\theta_0+z^{n-1}\cos\theta_1+\ldots+z\cos\theta_{n-1}+\cos\theta_n=2$,

where $\theta_0, \theta_1, \theta_2, \ldots, \theta_n\in R$

lie outside the circle $|z|=\frac{1}{2}$.

16. Let $A_1, A_2, \ldots, A_n$ be the vertices of an n-sided regular polygon such that $\frac{1}{A_1A_2}=\frac{1}{A_1A_3}+\frac{1}{A_1A_4}$. Find the value of n.

17. If the complex number z lies on the curve $|z-(-1+i)|=1$, then find the locus of the complex number $w=\frac{z+i}{1-i}$.

18. For any integer k, let $\alpha_k = \cos\frac{k\pi}{7} + i\sin\frac{k\pi}{7}$, then find the value of $\frac{\sum_{k=1}^{12}|\alpha_{k+1}-\alpha_k|}{\sum_{k=1}^{3}|\alpha_{4k-1}-\alpha_{4k-2}|}$.

19. If the complex numbers α and $\frac{1}{\bar{\alpha}}$ lie on the circle $(x-x_0)^2+(y-y_0)^2=r^2$ and $(x-x_0)^2+(y-y_0)^2=4r^2$, respectively. $z_0 = x_0 + iy_0$ satisfies the equation $2|z_0|^2 = r^2+2$, then find $|\alpha|$.

20. If $z_1^2+z_2^2 \pm 2z_1z_2\cos\theta = 0$, prove that the points represented by z_1, z_2 and origin form an isosceles triangle.

21. If $|z-2-3i|^2+|z-4-3i|^2=\lambda$ represents the equation of a circle, then find the value of λ.

22. Find the locus of z if $\frac{z^3+1}{z^3-1}$ is purely imaginary.

23. Find the locus of z if $\frac{z^2+1}{z^2-1}$ is real.

24. If t and c are two complex numbers such that $|t| \neq |c|, |t|=1$ and $z=\frac{at+b}{t-c}, z=x+iy$, find the locus of z where a, b are complex numbers.

25. If $|3z-2|+|3z+2|=4$, then find the locus of z.

26. If $\alpha_0, \alpha_1, \alpha_2, \ldots \alpha_{n-1}$ are the n, n^{th} roots of the unity, then find the value of $\sum_{i=0}^{n-1}\frac{\alpha_i}{2-\alpha_i}$.

27. Find the equation of circle inscribed in an equilateral triangle with vertices as complex numbers $1+i\sqrt{3}, 1-i\sqrt{3}$ and -2. Find its radius.

28. Find the parameter 'a' such that there is at least one complex number z satisfying $|z+\sqrt{2}| < a^2-3a+2$ and $|z+i\sqrt{2}| < a^2$.

29. If the curves $2|z|^2-4a\,\text{Re}(z)+c^2=0$ and $2|z|^2-4b\ \text{Im}(z)+c^2=0$ touch each other where $a,b,c \in R-\{0\}, |a| \geq \frac{|c|}{\sqrt{2}}$ and $|b| \geq \frac{|c|}{\sqrt{2}}$, then find the value of $\frac{c^2(a^2+b^2)}{a^2b^2}$.

30. If A and B are supplementary angles and n is an odd integer, then find the value of $\sum_{r=0}^{n} {}^{n}C_r \cos(nA+r(B-A))$.

3.16 JEE Main: Archive

1. Let $A=\{z \in C : 1 \leq |z-(1+i)| \leq 2\}$ and $B=\{z \in A : |z-(1-i)|=1\}$. Then, B:

(A) is an empty set
(B) contains exactly two elements
(C) contains exactly three elements
(D) is an infinite set

[Main 2022 24th June 1st Shift]

2. Let $S=\{z \in C \,|\, |z-3| \leq 1$ and

$$z(4+3i)+\bar{z}(4-3i) \leq 24\}.$$

If $\alpha+i\beta$ is the point in S which is closest to $4i$, then $25(\alpha+\beta)$ is equal to ______.

[Main 2022 24th June 2nd Shift]

3. Let a circle C in the complex plane pass through the points $z_1 = 3+4i$, $z_2 = 4+3i$ and $z_3 = 5i$. If $z(\neq z_1)$ is a point on C such that the line through z and z_1 is perpendicular to the line through z_2 and z_3, then $\arg(z)$ is equal to:

(A) $\tan^{-1}\left(\frac{2}{\sqrt{5}}\right)-\pi$ (B) $\tan^{-1}\left(\frac{24}{7}\right)-\pi$

(C) $\tan^{-1}(3)-\pi$ (D) $\tan^{-1}\left(\frac{3}{4}\right)-\pi$

[Main 2022 25th June 1st Shift]

4. Let z_1 and z_2 be two complex numbers such that $\bar{z}_1 = i\bar{z}_2$ and $\arg\left(\frac{z_1}{\bar{z}_2}\right) = \pi$. Then

(A) $\arg z_2 = \frac{\pi}{4}$ (B) $\arg z_2 = -\frac{3\pi}{4}$

(C) $\arg z_1 = \frac{\pi}{4}$ (D) $\arg z_1 = -\frac{3\pi}{4}$

[Main 2022 25th June 2nd Shift]

5. Let $A = \left\{ z \in C : \left| \frac{z+1}{z-1} < 1 \right| \right\}$ and $B = \left\{ z \in C : \arg\left(\frac{z-1}{z+1} \right) = \frac{2\pi}{3} \right\}$.

Then $A \cap B$ is:

(A) a portion of a circle centred at $\left(0, -\frac{1}{\sqrt{3}}\right)$ that lies in the second and third quadrants only

(B) a portion of a circle centred at $\left(0, -\frac{1}{\sqrt{3}}\right)$ that lies in the second quadrant only

(C) an empty set

(D) a portion of a circle of radius $\frac{2}{\sqrt{3}}$ that lies in the third quadrant only

[Main 2022 26th June 1st Shift]

6. If $z^2 + z + 1 = 0$, $z \in C$, then $\left| \sum_{n=1}^{15} \left(Z^n + (-1)^n \frac{1}{Z^n} \right)^2 \right|$ is equal to ______.

[Main 2022 26th June 2nd Shift]

7. The area of the polygon, whose vertices are the non-real roots of the equation $\bar{z} = iz^2$ is:

(A) $\frac{3\sqrt{3}}{4}$ (B) $\frac{3\sqrt{3}}{2}$ (C) $\frac{3}{2}$ (D) $\frac{3}{4}$

[Main 2022 27th June 1st Shift]

8. The number of points of intersection of $|z - (4+3i)| = 2$ and $|z| + |z-4| = 6$, $z \in C$ is:

(A) 0 (B) 1 (C) 2 (D) 3

[Main 2022 27th June 2nd Shift]

9. The number of elements in the set $\{z = a + ib \in C : a, b \in Z \text{ and } 1 < |z - 3 + 2i| < 4\}$ is ______. **[Main 2022 28th June 1st Shift]**

10. Sum of squares of modulus of all the complex numbers z satisfying $\bar{z} = iz^2 + z^2 - z$ is equal to

[Main 2022 28th June 2nd Shift]

11. Let α and β be the roots of the equation $x^2 + (2i-1) = 0$. Then, the value of $|\alpha^8 + \beta^8|$ is equal to:

(A) 50 (B) 250 (C) 1250 (D) 1500

[Main 2022 29th June 1st Shift]

12. Let $S = \{z \in C : |z-2| \le 1, z(1+i) + \bar{z}(1-i) \le 2\}$. Let $|z - 4i|$ attain minimum and maximum values, respectively, at $z_1 \in S$ and $z_2 \in S$. If $5\left(|z_1|^2 + |z_2|^2\right) = \alpha + \beta\sqrt{5}$, where α and β are integers, then the value of $\alpha + \beta$ is equal to ______. **[Main 2022 29th June 1st Shift]**

13. Let $\arg(z)$ represent the principal argument of the complex number z. The, $|z| = 3$ and $\arg(z-1) - \arg(z+1) = \frac{\pi}{4}$ intersect:

(A) exactly at one point

(B) exactly at two points

(C) nowhere

(D) at infinitely many points

[Main 2022 29th June 2nd Shift]

14. If $\alpha, \beta, \gamma, \delta$ are the roots of the equation $x^4 + x^3 + x^2 + x + 1 = 0$, then $\alpha^{2021} + \beta^{2021} + \gamma^{2021} + \delta^{2021}$ is equal to

(A) –4 (B) –1 (C) 1 (D) 4

[Main 2022 25th July 1st Shift]

15. For $n \in N$, let $S_n = \left\{ z \in C : |z - 3 + 2i| = \frac{n}{4} \right\}$ and $T_n = \left\{ z \in C : |z - 2 + 3i| = \frac{1}{n} \right\}$.

Then the number of elements in the set $\{n \in N : S_n \cap T_n = \phi\}$ is:

(A) 0 (B) 2

(C) 3 (D) infinite

[Main 2022 25th July 1st Shift]

16. For $z \in C$ if the minimum value of $\left(|z - 3\sqrt{2}| + |z - p\sqrt{2}i| \right)$ is $5\sqrt{2}$, then a value of p is ______.

(A) 3 (B) $\frac{7}{2}$ (C) 4 (D) $\frac{9}{2}$

[Main 2022 25th July 2nd Shift]

17. Let O be the origin and A be the point $z_1 = 1+2i$. If B is the point z_2, $\text{Re}(z_2) < 0$, such that OAB is a right angled isosceles triangle with OB as hypotenuse, then which of the following is NOT true?

(A) $\arg z_2 = \pi - \tan^{-1} 3$

(B) $\arg(z_1 - 2z_2) = -\tan^{-1}\frac{4}{3}$

(C) $|z_2| = \sqrt{10}$

(D) $|2z_1 - z_2| = 5$

[Main 2022 26th July 1st Shift]

18. If $z = x+iy$ satisfies $|z| - 2 = 0$ and $|z-i| - |z+5i| = 0$, then

(A) $x+2y-4=0$ (B) $x^2+y-4=0$

(C) $x+2y+4=0$ (D) $x^2-y+3=0$

[Main 2022 26th July 2nd Shift]

19. Let the minimum value v_0 of $v = |z|^2 + |z-3|^2 + |z-6i|^2$, $z \in C$ is attained at $z = z_0$. Then $\left|2z_0^2 - \bar{z}_0^3 + 3\right|^2 + v_0^2$ is equal to

(A) 1000 (B) 1024 (C) 1105 (D) 1196

[Main 2022 27th July 1st Shift]

20. Let $S = \left\{z \in C : z^2 + \bar{z} = 0\right\}$. Then $\sum_{z \in S}(\text{Re}(z) + \text{Im}(z))$ is equal to ______.

[Main 2022 27th July 1st Shift]

21. Let S be the set of all (α, β), $\pi < \alpha, \beta < 2\pi$, for which the complex number $\frac{1 - i\sin\alpha}{1 + 2i\sin\alpha}$ is purely imaginary and $\frac{1 + i\cos\beta}{1 - 2i\cos\beta}$ is purely real. Let $Z_{\alpha\beta} = \sin 2\alpha + i\cos 2\beta$, $(\alpha, \beta) \in S$.

Then $\sum_{(\alpha,\beta)\in S}\left(iZ_{\alpha\beta} + \frac{1}{i\bar{Z}_{\alpha\beta}}\right)$ is equal to:

(A) 3 (B) $3i$ (C) 1 (D) $2-i$

[Main 2022 27th July 2nd Shift]

22. Let $S_1 = \left\{z_1 \in C : |z_1 - 3| = \frac{1}{2}\right\}$ and $S_2 = \left\{z_2 \in C : \left|z_2 - |z_2+1|\right| = \left|z_2 + |z_2 - 1|\right|\right\}$.

Then, for $z_1 \in S_1$ and $z_2 \in S_2$, the least value of $|z_2 - z_1|$ is:

(A) 0 (B) $\frac{1}{2}$ (C) $\frac{3}{2}$ (D) $\frac{5}{2}$

[Main 2022 28th July 1st Shift]

23. Let $z = a+ib, b \neq 0$ be a complex number satisfying $z^2 = \bar{z} \cdot 2^{1-|z|}$. Then the least value of $n \in N$, such that $z^n = (z+1)^n$, is equal to ______.

[Main 2022 28th July 2nd Shift]

24. If $z = 2+3i$, then $z^5 + (\bar{z})^5$ is equal to:

(A) 244 (B) 224

(C) 245 (D) 265

[Main 2022 29th July 1st Shift]

25. If $z \neq 0$ be a complex number such that $\left|z - \frac{1}{z}\right| = 2$, then the maximum value of $|z|$ is:

(A) $\sqrt{2}$ (B) 1

(C) $\sqrt{2}-1$ (D) $\sqrt{2}+1$

[Main 2022 29th July 2nd Shift]

26. Let $S = \{z = x+iy : |z-1+i| \geq |z|, |z| < 2, |z+i| = |z-1|\}$.

Then the set of all values of x, for which $w = 2x+iy \in S$ for some $y \in R$, is

(A) $\left(-\sqrt{2}, \frac{1}{2\sqrt{2}}\right]$ (B) $\left(-\frac{1}{\sqrt{2}}, \frac{1}{4}\right]$

(C) $\left(-\sqrt{2}, \frac{1}{2}\right]$ (D) $\left(-\frac{1}{\sqrt{2}}, \frac{1}{2\sqrt{2}}\right]$

[Main 2022 29th July 2nd Shift]

27. Let $p, q \in R$ and $\left(1-\sqrt{3}i\right)^{200} = 2^{199}(p+iq)$, $i = \sqrt{-1}$. Then $p+q+q^2$ and $p-q+q^2$ are the roots of the equation.

(A) $x^2+4x-1=0$ (B) $x^2-4x+1=0$

(C) $x^2+4x+1=0$ (D) $x^2-4x-1=0$

[Main 2023 24th Jan 1st Shift]

28. The value of $\left(\frac{1+\sin\frac{2\pi}{9}+i\cos\frac{2\pi}{9}}{1+\sin\frac{2\pi}{9}-i\cos\frac{2\pi}{9}}\right)^3$ is

(A) $\frac{-1}{2}\left(1-i\sqrt{3}\right)$ (B) $\frac{1}{2}\left(1-i\sqrt{3}\right)$

(C) $-\frac{1}{2}\left(\sqrt{3}-i\right)$ (D) $\frac{1}{2}\left(\sqrt{3}+i\right)$

[Main 2023 24th Jan 2nd Shift]

29. Let $z_1 = 2+3i$ and $z_2 = 3+4i$. The set $S=\left\{z \in C : |z-z_1|^2 - |z-z_2|^2 = |z_1 - z_2|^2\right\}$ represents a

(A) straight line with sum of its intercepts on the coordinates axes equalling 14

(B) hyperbola with the length of the transverse axis 7

(C) straight line with the sum of its intercepts on the coordinate axes equalling −18

(D) hyperbola with eccentricity 2

[Main 2023 25th Jan 1st Shift]

30. Let z be a complex number such that $\left|\frac{z-2i}{z+i}\right| = 2, z \neq -1$. Then z lies on the circle of radius 2 and centre

(A) $(2, 0)$ (B) $(0, 0)$

(C) $(0, 2)$ (D) $(0, -2)$

[Main 2023 25th Jan 2nd Shift]

31. For two non-zero complex number z_1 and z_2, if $\text{Re}(z_1 z_2) = 0$ and $\text{Re}(z_1 + z_2) = 0$, then which of the following are possible?

(1) $\text{Im}(z_1) > 0$ and $\text{Im}(z_2) > 0$

(2) $\text{Im}(z_1) < 0$ and $\text{Im}(z_2) > 0$

(3) $\text{Im}(z_1) > 0$ and $\text{Im}(z_2) < 0$

(4) $\text{Im}(z_1) < 0$ and $\text{Im}(z_2) < 0$

Choose the correct answer from the options given below

(A) 2 and 4 (B) 2 and 3

(C) 1 and 2 (D) 1 and 3

[Main 2023 29th Jan 1st Shift]

32. Let

$$\alpha = 8-14i,\ A = \left\{z \in C : \frac{\alpha z - \bar{\alpha}\bar{z}}{z^2 - (\bar{z})^2 - 112i} = 1\right\}$$

and $B = \{z \in C : |z+3i| = 4\}$. Then $\sum_{z \in A \cap B} (\text{Re}\, z - \text{Im}\, z)$ is equal to ______.

[Main 2023 29th Jan 2nd Shift]

33. Let $z = 1+i$ and $z_1 = \dfrac{1+i\bar{z}}{\bar{z}(1-z)+\frac{1}{z}}$. Then $\frac{12}{\pi}\arg(z_1)$ is equal to ______.

[Main 2023 30th Jan 1st Shift]

34. For all $z \in C$ on the curve $C_1 : |z| = 4$, let the locus of the point $z + \frac{1}{z}$ be the curve C_2. Then

(A) the curves C_1 and C_2 intersect at 4 points

(B) the curves C_1 lies inside C_2

(C) the curves C_1 and C_2 intersect at 2 points

(D) the curves C lies inside C.

[Main 2023 31st Jan 1st Shift]

35. The complex number $z = \dfrac{i-1}{\cos\frac{\pi}{3} + i\sin\frac{\pi}{3}}$ is equal to:

(A) $\sqrt{2}\left(\cos\frac{5\pi}{12} + i\sin\frac{5\pi}{12}\right)$

(B) $\cos\frac{\pi}{12} - i\sin\frac{\pi}{12}$

(C) $\sqrt{2}\left(\cos\frac{\pi}{12} + i\sin\frac{\pi}{12}\right)$

(D) $\sqrt{2}i\left(\cos\frac{5\pi}{12} - i\sin\frac{5\pi}{12}\right)$

[Main 2023 31st Jan 2nd Shift]

36. If the centre and radius of the circle $\left|\frac{z-2}{z-3}\right| = 2$ are respectively (α, β) and γ, then $3(\alpha+\beta+\gamma)$ is equal to

(A) 11 (B) 9 (C) 10 (D) 12

[Main 2023 1st Feb 1st Shift]

37. Let a, b be two real numbers such that $ab < 0$. If the complex number $\frac{1+ai}{b+i}$ is of unit modulus and $a+ib$ lies on the circle $|z-1| = |2z|$, then a possible value of $\frac{1+[a]}{4b}$, where $[t]$ is greatest integer function, is:

(A) $-\frac{1}{2}$ (B) -1 (C) 1 (D) $\frac{1}{2}$

[Main 2023 1st Feb 2nd Shift]

38. Let $a \neq b$ be two non-zero real numbers. Then the number of elements in the set $X = \{z \in C : \text{Re}(az^2 + bz) = a \text{ and } \text{Re}(bz^2 + az) = b\}$ is equal to

(A) 0 (B) 1 (C) 3 (D) 2

[Main 2023 6th Apr 2nd Shift]

39. For $\alpha, \beta, z \in C$ and $\lambda > 1$, if $\sqrt{\lambda - 1}$ is the radius of the circle $|z-\alpha|^2 + |z-\beta|^2 = 2\lambda$, then $|\alpha - \beta|$ is equal to ______.

[Main 2023 6th Apr 2nd Shift]

40. If for $z = \alpha + i\beta, |z+2| = z + 4(1+i)$, then $\alpha + \beta$ and $\alpha\beta$ are the roots of the equation

(A) $x^2 + 3x - 4 = 0$
(B) $x^2 + 7x + 12 = 0$
(C) $x^2 + x - 12 = 0$
(D) $x^2 + 2x - 3 = 0$

[Main 2023 8th Apr 1st Shift]

41. Let $A = \left\{\theta \in (0, 2\pi) : \frac{1+2i\sin\theta}{1-i\sin\theta} \text{ is purely imaginary}\right\}$

Then the sum of the elements is in A is

(A) 4π (B) 3π (C) π (D) 2π

[Main 2023 8th Apr 2nd Shift]

42. Let the complex number $z = x + iy$ be such that $\frac{2z-3i}{2z+i}$ is purely imaginary. If $x + y^2 = 0$, then $y^4 + y^2 - y$ is equal to

(A) $\frac{2}{3}$ (B) $\frac{3}{2}$ (C) $\frac{3}{4}$ (D) $\frac{4}{3}$

[Main 2023 10th Apr 1st Shift]

43. Let

$$S = \left\{z = x + iy : \frac{2z-3i}{4z+2i} \text{ is a real number}\right\}.$$

Then which of the following is **NOT** correct?

(A) $y + x^2 + y^2 \neq -\frac{1}{4}$
(B) $(x, y) = \left(0, -\frac{1}{2}\right)$
(C) $x = 0$
(D) $y \in \left(-\infty, -\frac{1}{2}\right) \cup \left(-\frac{1}{2}, \infty\right)$

[Main 2023 10th Apr 2nd Shift]

44. Let w_1 be the point obtained by the rotation of $z_1 = 5 + 4i$ about the origin through a right angle in the anticlockwise direction, and w_2 be the point obtained by the rotation of $z_2 = 3 + 5i$ about the origin through a right angle in the clockwise direction. Then the principal argument $w_1 - w_2$ is equal to

(A) $\pi - \tan^{-1}\frac{8}{9}$ (B) $-\pi + \tan^{-1}\frac{33}{5}$

(C) $-\pi + \tan^{-1}\frac{8}{9}$ (D) $\pi - \tan^{-1}\frac{33}{5}$

[Main 2023 11th Apr 1st Shift]

45. For $a \in C$, let

$A = \{z \in C : \text{Re}(a + \bar{z}) > \text{Im}(\bar{a} + z)\}$ and

$B = \{z \in C : \text{Re}(a + \bar{z}) < \text{Im}(\bar{a} + z)\}$.

Then, among the two statements:

(S1): If $\text{Re}(a), \text{Im}(a) > 0$, then the set A contains all the real numbers

(S2): If $\text{Re}(a), \text{Im}(a) < 0$, then the set B contains all the real numbers,

(A) only (S2) is true (B) only (S1) is true
(C) both are true (D) both are false

[Main 2023 11th Apr 2nd Shift]

46. Let $S = \left\{z \in C - \{i, 2i\} : \frac{z^2 + 8iz - 15}{z^2 - 3iz - 2} \in R\right\}$.

$\alpha - \frac{13}{11}i \in S, \alpha \in R - \{0\}$, then $242\alpha^2$ is equal to ______.

[Main 2023 11th Apr 2nd Shift]

47. Let C be the circle in the complex plane with centre $z_0 = \frac{1}{2}(1+3i)$ and radius $r = 1$. Let $z_1 = 1 + i$ and the complex number z_2 be outside circle C such that $|z_1 - z_0||z_2 - z_0| = 1$. If z_0, z_1 and z_2 are collinear, then the smaller value of $|z_2|^2$ is equal to

(A) $\frac{5}{2}$ (B) $\frac{7}{2}$ (C) $\frac{13}{2}$ (D) $\frac{3}{2}$

[Main 2023 12th Apr 1st Shift]

48. Let $w = z\bar{z} + k_1 z + k_2 iz + \lambda(1+i)$, $k_1, k_2 \in R$. Let $\text{Re}(w) = 0$ be the circle C of radius 1 in the first quadrant touching the line $y = 1$ and the y-axis. If the curve $\text{Im}(w) = 0$ intersects C at A and B, then $30(AB)^2$ is equal to ______.

[Main 2023 13th Apr 1st Shift]

49. Let $S = \left\{z \in C : \bar{z} = i\left(z^2 + \text{Re}(\bar{z})\right)\right\}$.

Then $\sum_{z \in S} |z|^2$ is equal to

(A) $\frac{5}{2}$ (B) 4 (C) $\frac{7}{2}$ (D) 3

[Main 2023 13th Apr 2nd Shift]

50. If the set $\left\{\text{Re}\left(\frac{z - \bar{z} + z\bar{z}}{2 - 3z + 5\bar{z}}\right) : z \in C, \text{Re}\, z = 3\right\}$ is equal to the interval $(\alpha, \beta]$, then $24(\beta - \alpha)$ is equal to

(A) 36 (B 27 (C) 30 (D) 42

[Main 2023 15th Apr 1st Shift]

51. Let $S = \{z \in C : |z - 1| = 1$ and

$(\sqrt{2} - 1)(z + \bar{z}) - i(z - \bar{z}) = 2\sqrt{2}\}$.

Let $z_1, z_2 \in S$ be such that $|z_1| = \max_{z \in S}|z|$ and $|z_2| = \min_{z \in S}|z|$. Then $\left|\sqrt{2}z_1 - z_2\right|^2$ equals

(A) 1 (B) 4 (C) 3 (D) 2

[Main 2024 1st Feb 1st Shift]

52. Let $P = \{z \in C : |z + 2 - 3i| \le 1\}$ and $Q = \{z \in C : z(1+i) + \bar{z}(1-i) \le -8\}$.

Let in $P \cap Q$, $|z - 3 + 2i|$ be maximum and minimum at z_1 and z_2 respectively.

If $|z_1|^2 + 2|z|^2 = \alpha + \beta\sqrt{2}$, where α, β are integer, then $\alpha + \beta$ equals ______.

[Main 2024 1st Feb 1st Shift]

53. If z is a complex number such that $|z| \ge 1$, then the minimum value of $\left|z + \frac{1}{2}(3 + 4i)\right|$ is:

(A) $\frac{5}{2}$ (B) 2 (C) 3 (D) 0

[Main 2024 1st Feb 2nd Shift]

54. If $S = \{z \in C : |z - i| = |z + i| = |z - 1|\}$, then, $n(S)$ is:

(A) 1 (B) 0 (C) 3 (D) 2

[Main 2024 27th Jan 1st Shift]

55. If a satisfies the equation $x^2 + x + 1 = 0$ and $(1 + \alpha)^7 = A + B\alpha + C^2$, $A, B, C \ge 0$, then $5(3A - 2B - C)$ is equal to ______.

[Main 2024 27th Jan 1st Shift]

56. Let the complex numbers α and $\frac{1}{\alpha}$ lie on the circles $|z - z_0|^2 = 4$ and $|z - z_0|^2 = 16$ respectively, where $z_0 = 1 + i$. Then, the value of $100|\alpha|^2$ is______.

[Main 2024 27th Jan 2nd Shift]

57. If $z = \frac{1}{2} - 2i$, is such that $|z + 1| = \alpha z + \beta(1 + i)$, $i = \sqrt{-1}$ and $\alpha, \beta \in R$, then $\alpha + \beta$ is equal to _____.

(A) –4 (B) 3 (C) 2 (D) –1

[Main 2024 29th Jan 1st Shift]

58. Let α, β be the roots of the equation $x^2 - x + 2 = 0$ with $\text{Im}(\alpha) > \text{Im}(\beta)$. Then $\alpha^6 + \alpha^4 + \beta^4 - 5\alpha^2$ is to _____.

[Main 2024 29th Jan 1st Shift]

59. Let r and θ respectively be the modulus and amplitude of the complex number $z = 2 - i\left(2\tan\frac{5\pi}{8}\right)$, then (r, θ) is equal to

(A) $\left(2\sec\frac{3\pi}{8}, \frac{3\pi}{8}\right)$ (B) $\left(2\sec\frac{3\pi}{8}, \frac{5\pi}{8}\right)$

(C) $\left(2\sec\frac{5\pi}{8}, \frac{3\pi}{8}\right)$ (D) $\left(2\sec\frac{11\pi}{8}, \frac{11\pi}{8}\right)$

[Main 2024 29th Jan 2nd Shift]

60. Let α, β be the roots of the equation $x^2 - \sqrt{6}x + 3 = 0$ such that $\text{Im}(\alpha) > \text{Im}(\beta)$. Let a, b be integers not divisible by 3 and n be a natural number such that $\frac{\alpha^{99}}{\beta} + \alpha^{98} = 3^n(a+ib), i = \sqrt{-1}$.
Then $n+a+b$ is equal to ______.
[Main 2024 29th Jan 2nd Shift]

61. If $z = x+iy, xy \neq 0$, satisfies the equation $z^2 + i\bar{z} = 0$, then $|z^2|$ is equal to:
(A) 9 (B) 1 (C) 4 (D) $\frac{1}{4}$
[Main 2024 30th Jan 1st Shift]

62. If z is a complex number, then the number of common roots of the equation $z^{1985} + z^{100} + 1 = 0$ and $z^3 + 2z^2 + 2z + 1 = 0$, is equal to:
(A) 1 (B) 2 (C) 0 (D) 3
[Main 2024 30th Jan 2nd Shift]

63. If α denotes the number of solutions of $|1-i|^x = 2^x$ and $\beta = \left(\frac{|z|}{\arg(z)}\right)$, where $z = \frac{\pi}{4}(1+i)^4\left(\frac{1-\sqrt{\pi}i}{\sqrt{\pi}+i} + \frac{\sqrt{\pi}-i}{1+\sqrt{\pi}i}\right), i = \sqrt{-1}$, then the distance of the point (α, β) from the line $4x - 3y = 7$ is ______.
[Main 2024 31st Jan 1st Shift]

64. Let z_1 and z_2 be two complex number such that $z_1 + z_2 = 5$ and $z_1^3 + z_2^3 = 20 + 15i$. Then $|z_1^4 + z_2^4|$ equals:
(A) $30\sqrt{3}$ (B) 75
(C) $15\sqrt{15}$ (D) $25\sqrt{3}$
[Main 2024 31st Jan 2nd Shift]

3.17 Answer Key

3.17.1 Know Your Concepts

1. (i) $\pm(2+3i)$ (ii) $\pm\frac{(3-5i)}{\sqrt{2}}$ **2.** $x = \pm 1, y = -4$ **3.** $\pm\left(x - \frac{1}{x} + 2i\right)$ **4.** (–1)

5. $-\frac{1}{\sqrt{3}}$ **6.** (8) **7.** (4) **8.** (2ω) **9.** (48) **10.** $(x+2y=2)$ **11.** $(2\sqrt{5})$

12. (1) **13.** $\frac{n^2(n+1)^2 + 4n}{4}$ **16.** (i) 0 (ii) 49 **17.** (1) **18.** $(0, \pm i)$ **20.** (3)

3.17.2 JEE Main: Sample Questions

Single Answer Type

1. (B)	**2.** (D)	**3.** (C)	**4.** (C)	**5.** (B)	**6.** (B)	**7.** (C)	**8.** (C)
9. (A)	**10.** (D)	**11.** (D)	**12.** (A)	**13.** (B)	**14.** (C)	**15.** (B)	**16.** (C)
17. (C)	**18.** (B)	**19.** (A)	**20.** (B)	**21.** (A)	**22.** (D)	**23.** (B)	**24.** (D)
25. (B)	**26.** (A)	**27.** (A)	**28.** (C)	**29.** (C)	**30.** (C)	**31.** (C)	**32.** (C)
33. (B)	**34.** (B)	**35.** (A)	**36.** (B)	**37.** (C)	**38.** (C)	**39.** (C)	**40.** (B)
41. (A)	**42.** (B)	**43.** (D)	**44.** (A)	**45.** (B)	**46.** (B)	**47.** (C)	**48.** (D)
49. (B)	**50.** (D)	**51.** (C)	**52.** (C)	**53.** (C)	**54.** (C)	**55.** (B)	**56.** (D)
57. (B)	**58.** (B)	**59.** (C)					

Integer Answer Type

1. (5)	**2.** (2)	**3.** (2)	**4.** (1)	**5.** (9)	**6.** (6)	**7.** (0)	**8.** (9)
9. (6)	**10.** (1)	**11.** (6)	**12.** (7)	**13.** (4)	**14.** (2)	**15.** (1)	**16.** (6)
17. (5)	**18.** (4)	**19.** (1)					

3.17.3 JEE Advanced: Sample Questions

Single Answer Type

1. (B) **2.** (A) **3.** (D) **4.** (C) **5.** (C) **6.** (C) **7.** (B) **8.** (A)
9. (D) **10.** (A) **11.** (C) **12.** (B) **13.** (A) **14.** (C) **15.** (A) **16.** (B)
17. (B) **18.** (D) **19.** (B) **20.** (D) **21.** (A) **22.** (B) **23.** (D) **24.** (C)
25. (D)

Multiple Answer Type

1. (A, B, C) **2.** (B, C, D) **3.** (B, C, D) **4.** (A, C)
5. (A, C, D) **6.** (A, C, D) **7.** (A, B, C) **8.** (A, B)
9. (A, B, C) **10.** (B, C) **11.** (A, B, D) **12.** (A, B, C, D)
13. (A, B, D) **14.** (A, B) **15.** (A, C) **16.** (A, B, D)
17. (A, D) **18.** (A, B, C, D) **19.** (A, B, C, D) **20.** (A, B, C, D)
21. (A,C) **22.** (C, D) **23.** (A, C, D) **24.** (B, D)
25. (A, B, C, D)

Comprehension Based Questions

1. (C) **2.** (D) **3.** (B) **4.** (A) **5.** (C) **6.** (A) **7.** (C) **8.** (B)
9. (D) **10.** (B) **11.** (A) **12.** (B) **13.** (C) **14.** (B) **15.** (C)

Match Matrix

1. (A – R; B – S; C – Q; D – P) **2.** (A – S; B – R; C – P; D – Q)
3. (A – S; B – Q; C – P; D – R) **4.** (A – Q, R; B – P; C – P, S, T; D – Q, R, S, T)
5. (A – S; B – T; C – R; D – R)

Subjective Answer Type: Hard Nut

2. (5) **4.** (0) **7.** (1) **8.** (2) **9.** $z_3 = \frac{9}{4} + \frac{3}{4}i;\ z_4 = \frac{3}{2} + \frac{3}{4}i$ **10.** (4)
11. $2\sqrt{\frac{7}{2}}$ **12.** (15°) **13.** (2n) **16.** (7) **17.** (Circle) **18.** (4)
19. $\frac{1}{\sqrt{7}}$ **21.** (4) **22.** (Circle) **23.** (pair of lines except the point (±1, 0))
24. (circle) **25.** (line segment) **26.** $\frac{n}{2^n - 1}$ **27.** $|z| = 1$
28. $a \in (0,1] \cup [2, \infty)$ **29.** (2) **30.** (0)

3.17.4 JEE Main: Archive

1. (D) **2.** (80) **3.** (B) **4.** (C) **5.** (B) **6.** (2) **7.** (A) **8.** (C)
9. (40) **10.** (2) **11.** (A) **12.** (26) **13.** (C) **14.** (B) **15.** (D) **16.** (C)
17. (D) **18.** (C) **19.** (A) **20.** (0) **21.** (C) **22.** (C) **23.** (6) **24.** (A)
25. (D) **26.** (B) **27.** (B) **28.** (C) **29.** (A) **30.** (D) **31.** (B) **32.** (14)
33. (9) **34.** (A) **35.** (A) **36.** (D) **37.** (A) **38.** (A) **39.** (B) **40.** (B)
41. (A) **42.** (C) **43.** (B) **44.** (A) **45.** (D) **46.** (1680) **47.** (A) **48.** (24)
49. (B) **50.** (C) **51.** (D) **52.** (36) **53.** (D) **54.** (A) **55.** (5) **56.** (20)
57. (B) **58.** (13) **59.** (A) **60.** (49) **61.** (B) **62.** (B) **63.** (3) **64.** (B)

3.18 Solutions

3.18.1 Know Your Concepts

1. (i) $-5+12i = 4-9+12i$

$= 4+12i+9i^2 = (2+3i)^2$

$\therefore \sqrt{-5+12i} = \pm(2+3i)$

(ii) Let $\sqrt{-8-15i} = x+iy$ (1)

squaring both sides

$-8-15i = x^2-y^2+2ixy$

$\therefore x^2-y^2 = -8$ (2)

$2xy = -15$ (3)

Now, $\left(x^2+y^2\right)^2 = \left(x^2-y^2\right)^2+4x^2y^2$

$= 64+225$

$\Rightarrow x^2+y^2 = 17$ (4)

Adding (4) and (2)

$2x^2 = 9 \Rightarrow x = \pm\dfrac{3}{\sqrt{2}}$

Subtracting (2) from (4)

$2y^2 = 25 \Rightarrow y = \pm\dfrac{5}{\sqrt{2}}$

From (3) it is clear that x and y should be of opposite sign

$\therefore x+iy = \dfrac{3}{\sqrt{2}} - i\dfrac{5}{\sqrt{2}}$

or

$-\dfrac{3}{\sqrt{2}}+i\dfrac{5}{\sqrt{2}} = -\dfrac{(3-5i)}{\sqrt{2}}$

$\therefore x+iy = \pm\dfrac{(3-5i)}{\sqrt{2}}$

2. Since, $-3+ix^2y = x^2+y-4i$

$\therefore x^2+y = -3$ and $x^2y = -4$

$\therefore x^2 - \dfrac{4}{x^2} = -3$

$\Rightarrow x^4+3x^2-4 = 0$

$\Rightarrow \left(x^2+4\right)\left(x^2-1\right) = 0 \Rightarrow x = \pm 1$

when $x = \pm 1$, $y = -4$

3. $x^2+\dfrac{1}{x^2}+4i\left(x-\dfrac{1}{x}\right)-6$

$= \left(x-\dfrac{1}{x}\right)^2+4i\left(x-\dfrac{1}{x}\right)-4$

$= \left(x-\dfrac{1}{x}\right)^2+4i\left(x-\dfrac{1}{x}\right)+(2i)^2$

$= \left(x-\dfrac{1}{x}+2i\right)^2$

$\therefore$ square root $= \pm\left(x-\dfrac{1}{x}+2i\right)^2$

4. $\left[\dfrac{1+\sin\frac{\pi}{8}+i\cos\frac{\pi}{8}}{1+\sin\frac{\pi}{8}-i\cos\frac{\pi}{8}}\right]^8$

$$= \left[\frac{\left(\cos\frac{\pi}{16}+\sin\frac{\pi}{16}\right)^2+i\left(\cos^2\frac{\pi}{16}-\sin^2\frac{\pi}{16}\right)}{\left(\cos\frac{\pi}{16}+\sin\frac{\pi}{16}\right)^2-i\left(\cos^2\frac{\pi}{16}-\sin^2\frac{\pi}{16}\right)}\right]^8$$

$$= \left[\frac{\left(\cos\frac{\pi}{16}+\sin\frac{\pi}{16}\right)+i\left(\cos\frac{\pi}{16}-\sin\frac{\pi}{16}\right)}{\left(\cos\frac{\pi}{16}+\sin\frac{\pi}{16}\right)-i\left(\cos\frac{\pi}{16}-\sin\frac{\pi}{16}\right)}\right]^8$$

$$= \left[\frac{\cos\left(\frac{\pi}{4}-\frac{\pi}{16}\right)+i\sin\left(\frac{\pi}{4}-\frac{\pi}{16}\right)}{\cos\left(\frac{\pi}{4}-\frac{\pi}{16}\right)-i\sin\left(\frac{\pi}{4}-\frac{\pi}{16}\right)}\right]^8$$

$$= \left[\frac{e^{i\left(\frac{\pi}{4}-\frac{\pi}{16}\right)}}{e^{-i\left(\frac{\pi}{4}-\frac{\pi}{16}\right)}}\right]^8 = \left[e^{i\left(\frac{\pi}{2}-\frac{\pi}{8}\right)}\right]^8$$

$= e^{i3\pi} = -1$

5. $3^{49}(x+iy) = \left(\dfrac{3}{2}+\dfrac{\sqrt{3}}{2}i\right)^{100}$

$\left[i\sqrt{3}\left(\dfrac{1-i\sqrt{3}}{2}\right)\right]^{100}$

$= i^{100}3^{50}(-\omega)^{100}$

$\Rightarrow 3^{49}(x+iy) = 3^{50}\omega = 3^{50}\left(-\dfrac{1}{2}+\dfrac{i\sqrt{3}}{2}\right)$

$\Rightarrow x+iy = -\dfrac{3}{2}+\dfrac{i3\sqrt{3}}{2}$

$\Rightarrow x = -\dfrac{1}{\sqrt{3}}y \Rightarrow k = -\dfrac{1}{\sqrt{3}}$

6. Since, $x^2 - x + 1 = 0$

$$\Rightarrow x = \frac{1 \pm i\sqrt{3}}{2} = -\omega, -\omega^2$$

$$\therefore \sum_{n=1}^{5}\left(x^{2n} + \frac{1}{x^{2n}} + 2\right)$$

$$= \left(x^2 + \frac{1}{x^2} + 2\right) + \left(x^4 + \frac{1}{x^4} + 2\right) + \left(x^6 + \frac{1}{x^6} + 2\right) + \left(x^8 + \frac{1}{x^8} + 2\right) + \left(x^{10} + \frac{1}{x^{10}} + 2\right)$$

$$= \left(\omega^2 + \omega^4 + \omega^6 + \omega^8 + \omega^{10}\right) + \left(\frac{1}{\omega^2} + \frac{1}{\omega^4} + \frac{1}{\omega^6} + \frac{1}{\omega^8} + \frac{1}{\omega^{10}}\right) + 10$$

$$= -1 - 1 + 10 = 8$$

7. Since, $\alpha^5 = 1$

$$\therefore \left|1 + \alpha + \alpha^2 + \alpha^{-2} - \alpha^{-1}\right|$$

$$= \left|1 + \alpha + \alpha^2 + \alpha^3 - \alpha^4\right|$$

$$= \left|1 + \alpha + \alpha^2 + \alpha^3 + \alpha^4 - 2\alpha^4\right|$$

$$= \left|\frac{1 - \alpha^5}{1 - \alpha} - 2\alpha^4\right| = 2|\alpha|^4 = 2 \times 1 = 2$$

8. $$\frac{1 + 2\omega + 3\omega^2}{2 + 3\omega + \omega^2} + \frac{2 + 3\omega + \omega^2}{3 + \omega + 2\omega^2}$$

$$= \frac{1 + 2\omega + 3\omega^2}{2\omega^3 + 3\omega^4 + \omega^2} + \frac{2 + 3\omega + \omega^2}{3\omega^3 + \omega^4 + 2\omega^2}$$

$$= \frac{1}{\omega^2}\left(\frac{1 + 2\omega + 3\omega^2}{1 + 2\omega + 3\omega^2}\right) + \frac{\left(2 + 3\omega + \omega^2\right)}{\omega^2\left(2 + 3\omega + \omega^2\right)}$$

$$= \omega + \omega = 2\omega$$

9. Since, $z^2 - z + 1 = 0$

$$\Rightarrow z = \frac{1 \pm \sqrt{3}i}{2} \Rightarrow z = -\omega, -\omega^2$$

For, $z = -\omega$

$$z + \frac{1}{z} = -\omega - \frac{1}{\omega} = -\left(\omega + \omega^2\right) = 1$$

$$z^2 + \frac{1}{z^2} = \omega^2 + \frac{1}{\omega^2} = \omega^2 + \omega = -1$$

$$z^3 + \frac{1}{z^3} = -\omega^3 - \frac{1}{\omega^3} = -2$$

$$\Rightarrow \left(z + \frac{1}{z}\right)^2 + \left(z^2 + \frac{1}{z^2}\right)^2 + \left(z^3 + \frac{1}{z^3}\right)^2 + \ldots + \left(z^{24} + \frac{1}{z^{24}}\right)^2$$

$$= 48$$

10. Let $z = x + iy$

$$\therefore \text{Im}\left(\frac{2z + 1}{iz + 1}\right) = -2$$

$$\Rightarrow \text{Im}\,\frac{2(x + iy) + 1}{i(x + iy) + 1} = -2$$

$$\Rightarrow \text{Im}\,\frac{(2x + 1) + i(2y)}{(1 - y) + ix} = -2$$

$$\Rightarrow \text{Im}\,\frac{(2x + 1) + i(2y)}{(1 - y) + ix} \times \frac{(1 - y) - ix}{(1 - y) - ix} = -2$$

$\Rightarrow x + 2y = 2$, which is a straight line.

11. Put $z = x + iy$

$$x^2 + y^2 + (4 - 3i)(x + iy) + (4 + 3i)(x - iy) + 5 = 0$$

$$\Rightarrow x^2 + y^2 + 8x + 6y + 5 = 0$$

$\Rightarrow$ Radius $= 2\sqrt{5}$

12. Since, $\frac{a}{b} + \frac{b}{c} + \frac{c}{a} = 1$

$$\Rightarrow \frac{\cos\alpha + i\sin\alpha}{\cos\beta + i\sin\beta} + \frac{\cos\beta + i\sin\beta}{\cos\gamma + i\sin\gamma} + \frac{\cos\gamma + i\sin\gamma}{\cos\alpha + i\sin\alpha} = 1$$

$$\Rightarrow \{\cos(\alpha - \beta) + \cos(\beta - \gamma) + \cos(\gamma - \alpha)\} + i\{\sin(\alpha - \beta) + \sin(\beta - \gamma) + \sin(\gamma - \alpha)\} = 1$$

Equating real parts,

$$\cos(\alpha - \beta) + \cos(\beta - \gamma) + \cos(\gamma - \alpha) = 1$$

13. $$t_n = (n + 1)\left(n + \frac{1}{\omega}\right)\left(n + \frac{1}{\omega^2}\right)$$

$$= n^3 + n^2\left(\frac{1}{\omega^2} + \frac{1}{\omega} + 1\right) + n\left(1 + \frac{1}{\omega^2} + \frac{1}{\omega}\right) + 1$$

$$= n^3 + n^2\left(1 + \omega + \omega^2\right) + n\left(1 + \omega + \omega^2\right) + 1$$

$$= n^3 + 1$$

$$\therefore S_n = \sum_{n=1}^{n}\left(n^3 + 1\right) = \frac{n^2(n + 1)^2 + 4n}{4}$$

14. $$x_r = \cos\frac{\pi}{2^r} + i\sin\frac{\pi}{2^r}$$

$x_1 = \cos\frac{\pi}{2} + i\sin\frac{\pi}{2}$

$x_2 = \cos\frac{\pi}{2^2} + i\sin\frac{\pi}{2^2}$

$x_3 = \cos\frac{\pi}{2^3} + i\sin\frac{\pi}{2^3}$

$\therefore x_1x_2x_3\ldots$ to ∞

$= \cos\frac{\pi}{2}\left(1+\frac{1}{2}+\frac{1}{2^2}+\ldots \text{ to } \infty\right)$

$+ i\sin\frac{\pi}{2}\left(1+\frac{1}{2}+\frac{1}{2^2}+\ldots \text{ to } \infty\right)$

$= \cos\pi + i\sin\pi = -1$

15. $iz^3 + z^2 - z + i = 0$

$\Rightarrow iz^2(z-i) - 1(z-i) = 0$

$\Rightarrow$ either $iz^2 - 1 = 0$ or $z - i = 0$

when, $z - i = 0 \Rightarrow |z| = 1$

when, $iz^2 - 1 = 0 \Rightarrow z^2 = -i \Rightarrow |z|^2 = 1$

$\Rightarrow |z| = 1$

$\therefore$ In any case, $|z| = 1$.

16. (i) $\left(3+3\omega+5\omega^2\right)^6 - \left(2+6\omega+2\omega^2\right)^3$

$= \left(3\left(1+\omega+\omega^2\right)+2\omega^2\right)^6 -$

$\left(2\left(1+\omega+\omega^2\right)+4\omega\right)^3$

$= 64\omega^6 - 64\omega^3 = 0$

(ii) $(2-\omega)\left(2-\omega^2\right)\left(2-\omega^{10}\right)\left(2-\omega^{11}\right)$

$= (2-\omega)\left(2-\omega^2\right)(2-\omega)\left(2-\omega^2\right)$

$= (2-\omega)^2\left(2-\omega^2\right)^2 = \left[(2-\omega)\left(2-\omega^2\right)\right]^2$

$= \left[4-2\omega^2-2\omega+\omega^3\right]^2$

$= \left[5-2(-1)\right]^2 = 49$

17. Since, $|\beta| = 1$

$\Rightarrow \beta\bar{\beta} = 1 \Rightarrow \beta = \frac{1}{\bar{\beta}}$

Now, $\left|\frac{\beta-\alpha}{1-\bar{\alpha}\beta}\right| = \left|\frac{\frac{1}{\bar{\beta}}-\alpha}{1-\bar{\alpha}\beta}\right| = \left|\frac{1}{\bar{\beta}}\right|\left|\frac{1-\alpha\bar{\beta}}{1-\bar{\alpha}\beta}\right|$

$= \frac{1}{|\beta|}\left|\frac{\overline{1-\alpha\bar{\beta}}}{1-\bar{\alpha}\beta}\right|$

$= \frac{1}{|\beta|}\left|\frac{1-\bar{\alpha}\beta}{1-\bar{\alpha}\beta}\right| = 1$

18. $z^2 + |z| = 0$

Put $z = x + iy$

$x^2 - y^2 + 2ixy + \sqrt{x^2+y^2} = 0$

and $2xy = 0$

If $x = 0 \Rightarrow -y^2 + \sqrt{y^2} = 0$

$\Rightarrow -|y|^2 + |y| = 0$

$\Rightarrow |y| = 0$ or 1

$\Rightarrow y = 0$ or $y = \pm 1$

if $y = 0$, then

$x^2 + \sqrt{x^2} = 0 \Rightarrow |x|^2 + |x| = 0$

$\Rightarrow x = 0$

Hence the complex number is $0, \pm i$.

19. $a + ib = (x+iy)^3$

$\Rightarrow a + ib = x^3 + 3ix^2y - 3xy^2 - iy^3$

$\Rightarrow a = x\left(x^2-3y^2\right)$

$b = y\left(3x^2-y^2\right)$

$\Rightarrow \frac{a}{x}+\frac{b}{y} = 4\left(x^2-y^2\right)$

20. $\frac{3((2+\cos\theta)-i\sin\theta)}{(2+\cos\theta)^2+\sin^2\theta} = a+ib$

$\Rightarrow \frac{3(2+\cos\theta)-i(3\sin\theta)}{5+4\cos\theta} = a+ib$

$\Rightarrow a = \frac{3(2+\cos\theta)}{5+4\cos\theta}, b = \frac{-3\sin\theta}{5+4\cos\theta}$

Now,

$$a^2+b^2-4a = \frac{9(2+\cos\theta)^2+9\sin^2\theta - 12(2+\cos\theta)(5+4\cos\theta)}{\left(5+4\cos^2\theta\right)}$$

$$= \frac{36+36\cos\theta+9 - 12\left(10+8\cos\theta+5\cos\theta+4\sin^2\theta\right)}{(5+4\cos\theta)^2}$$

$$=\frac{45+36\cos\theta-120-12\times13\cos\theta-48\cos^2\theta}{(5+4\cos\theta)^2}$$

$$=-75-120\cos\theta-48\cos^2\theta$$

$$=-\frac{3(25+40\cos\theta+16\cos^2\theta)}{(5+4\cos\theta)^2}$$

$$=\frac{-3(5+4\cos\theta)^2}{(5+4\cos\theta)^2}$$

$\therefore |a^2+b^2-4a|=3$

3.18.2 JEE Main: Sample Questions

Single Answer Type

1. **(B)**
$3|z-1|-2>0$ (1)
For valid logarithm definition
and $\frac{|z-1|+3}{3|z-1|-2}\geq\left(\frac{1}{2}\right)$ (2)
$\therefore$ From (1) and (2), $\frac{2}{3}<|z-1|\leq 8$.

2. **(D)**
$\frac{2+i}{3-i}\equiv P\left(\frac{1}{2},\frac{1}{2}\right)$
Also, $(1+i)z=(1-i)\bar{z}\Rightarrow x+y=0$
So, reflection of $P\left(\frac{1}{2},\frac{1}{2}\right)$ in the line
$x+y=0$, is $Q\left(\frac{-1}{2},\frac{-1}{2}\right)=\frac{2+i}{-3+i}$

3. **(C)**
$z\omega=|z|^2\Rightarrow|z|\,|\omega|=|z|^2\Rightarrow|z|=|\omega|$
Let $z=re^{i\theta}\Rightarrow\omega=re^{-i\theta}\Rightarrow\omega=\bar{z}$
$\therefore$ Given equation becomes
$|z-\bar{z}\,|+|z+\bar{z}\,|=4$

4. **(C)**
$|iZ+Z_1|=|i(Z-i)+(Z_1-1)|$
$\leq|i(Z-i)|+|Z_1-1|$
$\leq|i|\,|Z-i|+|4+3i|$
$\leq 2+\sqrt{4^2+9}=7$

5. **(B)**
$z(1+z^2)=0\Rightarrow z=0$ or $z^2=i^2$
$\Rightarrow z=0$ or $z=\pm i\Rightarrow \text{Re}(z)=0$

6. **(B)**
roots are $3-i$ and $-2+i\Rightarrow-6$

7. **(C)**
Let $P(x)=5x^4+4x^3+3x^2+Mx+N$
let $Q(x)=x^2+1$
if the quotient is Q
then $P(x)=Q(x^2+1)$
if $x=i$ then $P(i)=0$
if $x=-1$ then $P(-i)=0$
Hence $5-4i-3+Mi+N=0$.
Hence $N+Mi=-2+4i$
$\therefore N=-2;\ M=4$
$\therefore M-N=6$

8. **(C)**
$\alpha+\beta=-(p+iq);\ \alpha\beta+3i$
Given: $\alpha^2+\beta^2=8$
$(\alpha+\beta)^2-2\alpha\beta=8$
$(p+iq)^2-6i=8$
$p^2-q^2+i(2pq-6)=8$
$\Rightarrow p^2-q^2=8$ and $pq=3$
$\Rightarrow p=3$ and $q=1$ or $p=-3$ and $q=-1$

9. **(A)**
$z=x+iy$
$\therefore x+iy+\sqrt{x^2+y^2}=1+7i$
$x+\sqrt{x^2+y^2}=1$ (1)
and $y=7$ (2)
$\therefore x+\sqrt{x^2+49}=1$
$x^2+49=1+x^2-2x$
$2x=-48\quad x=-24$
$\therefore |z|^2=x^2+y^2=625$

10. **(D)**
$z=(3p-7q)+i(3q+7p)$
for purely imaginary $3p=7q$
$\Rightarrow p=7$ or $q=3$ (for least value)
$|z|=|3+7i|\,|p+iq|$
$\Rightarrow|z|^2=58(p^2+q^2)$
$=58[7^2+9]=58^2$

11. **(D)**
$z=0;\ z=\pm1;\ z=\pm i;$
$z^3=\bar{z}\Rightarrow|z|^3=|\bar{z}|=|z|$
note that $z^n=|\bar{z}|$ has $n+2$ solutions
hence $|z|=0$ or $|z|^2=1$
again $z^4=z\,\bar{z}=|z|^2=1$

$\Rightarrow z^4 = 1$

$\Rightarrow$ total number of roots are 5

Note:

(a) The equation $z^n = \bar{z}$ will have $(n + 2)$ solutions.

(b) All non-zero solutions of $z^3 = \bar{z}$ are (1, 0), (–1, 0), (0, 1), (0, –1), that is, a square.

12. (A)

We have $|z - (1 - i)|^2 = 2$

$\Rightarrow (x - 1)^2 + (y - 1)^2 = 2$ (Put $z = x + iy$)

$\Rightarrow x^2 + y^2 = 2(x + y)$ (1)

Let $\omega = h + ik = \dfrac{2}{z} = \dfrac{2}{x+iy} = \dfrac{2(x-iy)}{x^2+y^2}$, so

$h = \dfrac{2x}{x^2+y^2}$, $k = \dfrac{-2y}{x^2+y^2}$

$\Rightarrow h - k = \dfrac{2(x+y)}{x^2+y^2} = 1$ [(from equation (1)]

$\therefore$ Locus of the point $\omega(h, k)$ will be $x - y = 1$.

13. (B)

We have $(1 + ri)^3 = s(1 + i)$

$1 + 3ri + 3r^2i^2 + r^3i^3 = s(1 + i)$

$1 - 3r^2 + i(3r - r^3) = s + si$

$\Rightarrow 1 - 3r^2 = s = 3r - r^3$

Hence $1 - 3r^2 = 3r - r^3$

$\Rightarrow r^3 - 3r^2 - 3r + 1 = 0$

$\Rightarrow (r^3 + 1) - 3r(r + 1) = 0$

$\Rightarrow (r + 1)(r^2 + 1 - r - 3r) = 0$

$\therefore r = -1$ or $r^2 - 4r + 1 = 0$

$\Rightarrow r = \dfrac{4 \pm \sqrt{16-4}}{2} = \dfrac{4 \pm 2\sqrt{3}}{2}$

$\Rightarrow r = 2 + \sqrt{3}$ or $2 - \sqrt{3}$

14. (C)

Let F as $a + bi$, $a, b \in R$

Where we see from the diagram that $a, b > 0$ and $a^2 + b^2 > 1$

(as F lies outside the unit circle)

Since $\alpha = \dfrac{1}{a+bi} = \dfrac{a-bi}{a^2+b^2}$

$= \dfrac{a}{a^2+b^2} - \dfrac{b}{a^2+b^2}i$,

(real part +ve and imaginary part –ve and both less than unity) we see that the reciprocal of F is in the IV quadrant, since the real part is positive and the imaginary part is negative. Also, the magnitude of the reciprocal is

$\dfrac{1}{a^2+b^2}\sqrt{a^2+(-b)^2} = \dfrac{1}{\sqrt{a^2+b^2}} < 1$

Thus the only possibility is point C.

15. (B)

$w = \dfrac{-i(z+i)}{z-i}$; $|w| = \left|\dfrac{z+i}{z-i}\right| = 1 \Rightarrow |z+i| = |z-i|$

$\Rightarrow$ z lies on the perpendicular bisector of the segment joining (0, 1) and (0, –1) which is the x-axis

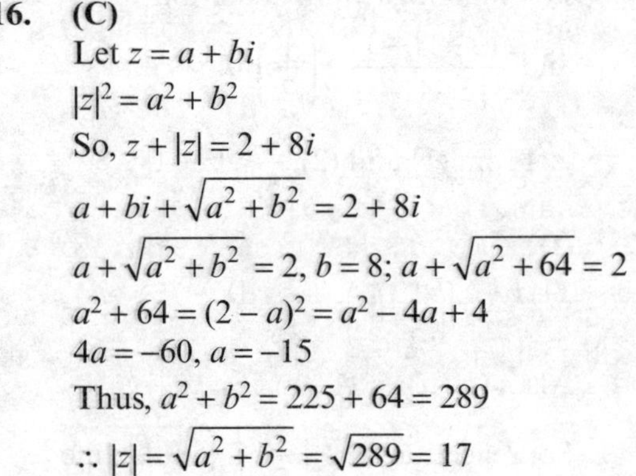

$\Rightarrow z$ lies on the x-axis

$\Rightarrow$ Im(z) is real

16. (C)

Let $z = a + bi$

$|z|^2 = a^2 + b^2$

So, $z + |z| = 2 + 8i$

$a + bi + \sqrt{a^2+b^2} = 2 + 8i$

$a + \sqrt{a^2+b^2} = 2$, $b = 8$; $a + \sqrt{a^2+64} = 2$

$a^2 + 64 = (2 - a)^2 = a^2 - 4a + 4$

$4a = -60$, $a = -15$

Thus, $a^2 + b^2 = 225 + 64 = 289$

$\therefore |z| = \sqrt{a^2+b^2} = \sqrt{289} = 17$

17. (C)

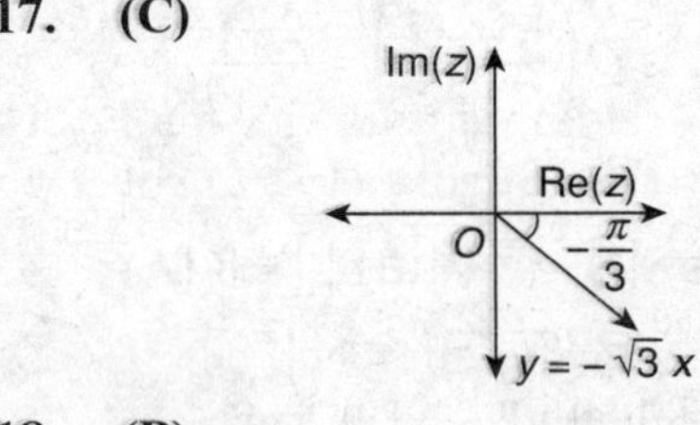

18. (B)

We have $\left|\dfrac{1}{z_2} + \dfrac{1}{z_1}\right| = \left|\dfrac{1}{z_2} - \dfrac{1}{z_1}\right|$

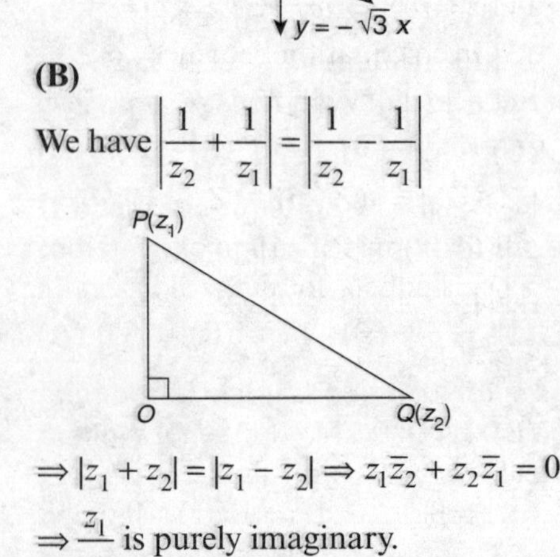

$\Rightarrow |z_1 + z_2| = |z_1 - z_2| \Rightarrow z_1\bar{z}_2 + z_2\bar{z}_1 = 0$

$\Rightarrow \dfrac{z_1}{z_2}$ is purely imaginary.

Hence ΔPQR is right angled at O.

$\therefore$ Circumcentre of ΔPOQ is the mid point of PQ, that is, $\frac{1}{2}(z_1 + z_2)$

19. **(A)**

$(1 - 2^{-x})^2 + 16 \le 25$; $(1 - 2^{-x}) - 3^2 \le 0$;

$(4 - 2^{-x})(-2 - 2^{-x}) \le 0$; $(2^{-x} - 4)(2^{-x} + 2) \ge 0$

20. **(B)**

$$|z_1| = \left|\frac{1-i}{1+i\sqrt{3}}\right|^{1/6} = \left|\frac{\sqrt{2}}{2}\right|^{1/6} = 2^{-\frac{1}{12}}$$

|||ly $|z_2| = 2^{-\frac{1}{12}}$; $|z_3| = 2^{-\frac{1}{12}}$ hence the result.

21. **(A)**

Let x be the real solution.

$\Rightarrow x^3 - 1 + xi = 0$

$\Rightarrow x^3 - 1 = 0$ and $x = 0$ which is not possible.

Note that the equation has no purely imaginary root as well.

22. **(D)**

$|(x-4) + i(y-3)| = 2$

circle with centre (4, 3) and radius 2

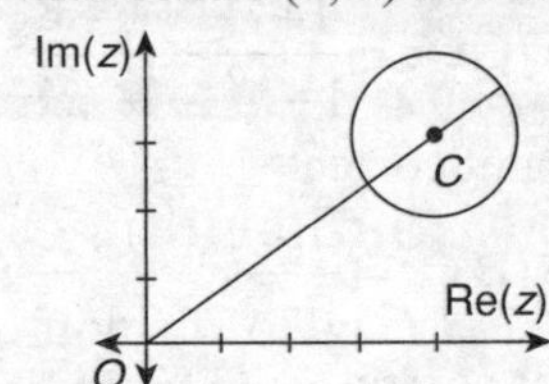

Hence $OC = 5$

$|z|_{max} = 5 + 2 = 7$ $\quad |z|_{min} = 5 - 2 = 3$

23. **(B)**

If $|z + i| + |z - i| = 8$

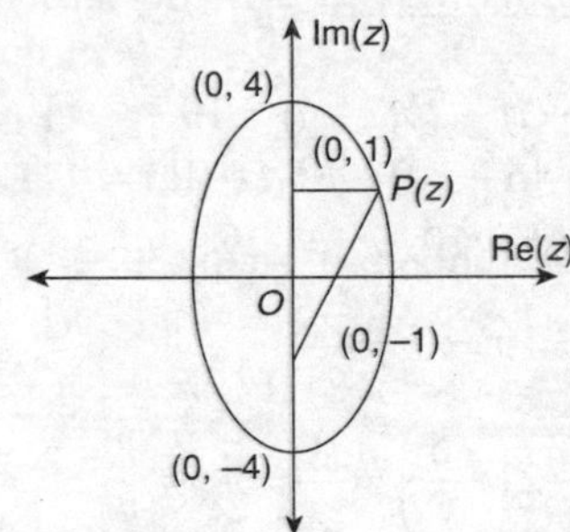

$PF_1 + PF_2 = 8$

$\therefore |z|_{max} = 4$

24. **(D)**

We have $\left|\frac{z_1}{2} + \frac{z_2}{3} + \frac{z_3}{4} + \frac{z_4}{5}\right|$

$$= \frac{k}{60}|z_1z_2z_3z_4|\left|\frac{1}{z_1} + \frac{1}{z_2} + \frac{1}{z_3} + \frac{1}{z_4}\right|$$

Now, $z_1\bar{z}_1 = 2$, $z_2\bar{z}_2 = 3$, $z_3\bar{z}_3 = 4$ and $z_4\bar{z}_4 = 5$

$$\text{So, } k = \frac{60}{|z_1z_2z_3z_4|} = \frac{60}{\sqrt{2}\sqrt{3}\sqrt{4}\sqrt{5}}$$

$$= \sqrt{30} = |z_4z_1z_2|$$

Note for objective take

$z_1 = \sqrt{2}$; $z_2 = \sqrt{3}$; $z_3 = 2$; $z_4 = \sqrt{5}$

Aliter: Given $|z_1| = \sqrt{2}$; $|z_2| = \sqrt{3}$; $|z_3| = \sqrt{4}$; $|z_4| = \sqrt{5}$

Hence given that $\left|\frac{z_1}{2} + \frac{z_2}{3} + \frac{z_3}{4} + \frac{z_4}{5}\right|$

$$= \frac{k}{60}|z_1z_2z_3z_4|\left|\frac{1}{z_1} + \frac{1}{z_2} + \frac{1}{z_3} + \frac{1}{z_4}\right|$$

$$\therefore \left|\frac{z_1}{2} + \frac{z_2}{3} + + \frac{z_3}{4} + \frac{z_4}{5}\right|$$

$$= \frac{k}{60}\sqrt{120}\left|\frac{1}{z_1} + \frac{1}{z_2} + \frac{1}{z_3} + \frac{1}{z_4}\right|$$

$$= \frac{k}{\sqrt{30}}\left|\frac{1}{z_1} + \frac{1}{z_2} + \frac{1}{z_3} + \frac{1}{z_4}\right|$$

Hence $k = \sqrt{30} = |z_1z_2z_3z_4| \Rightarrow$ (D)

using $z_1\bar{z}_1 = 2$; $\frac{\pi}{6} = 3$; $z_3\bar{z}_3 = 4$; $z_4\bar{z}_4 = 5$

25. **(B)**

$(Z^3 + 3)^2 = 16i^2$ $\quad Z^3 + 3 = 4i$ or $-4i$

$Z^3 = -3 + 4i$ or $-3 - 4i$

$|Z|^3 = |-3 + 4i| = 5$ $\quad |Z|^3 = 5 \Rightarrow |Z| = 5^{1/3}$

26. **(A)**

$$\text{We have } \frac{3}{|z_2 - z_3|} = \frac{4}{|z_3 - z_1|} = \frac{5}{|z_1 - z_2|} = k^2 \quad \text{(let)}$$

$$\Rightarrow \frac{9}{|z_2 - z_3|^2} = \frac{16}{|z_3 - z_1|^2} = \frac{25}{|z_1 - z_2|^2} = k^2$$

$$\text{Now } \frac{9}{|z_2 - z_3|^2} = k^2$$

$$\Rightarrow \frac{9}{z_2 - z_3} = k^2(\bar{z}_2 - \bar{z}_3) \quad (1)$$

[As $|z|^2 = z\bar{z}$]

$$\text{|||ly } \frac{16}{|z_3 - z_1|^2} = k^2$$

$$\Rightarrow \frac{16}{z_3 - z_1} = k^2(\bar{z}_3 - \bar{z}_1) \quad (2)$$

|||ly $\dfrac{25}{|z_1 - z_2|^2} = k^2$

$\Rightarrow \dfrac{25}{z_1 - z_2} = k^2(\bar{z}_1 - \bar{z}_2)$ (3)

$\therefore$ On adding (1), (2) and (3), we get

$\dfrac{9}{z_2 - z_3} + \dfrac{16}{z_3 - z_1} + \dfrac{25}{z_1 - z_2}$

$= k^2(\bar{z}_2 - \bar{z}_3 + \bar{z}_3 - \bar{z}_1 + \bar{z}_1 - \bar{z}_2) = 0$

27. **(A)**

$(z - 1)(z^2 + 2iz - 2) = 0$

$\Rightarrow z = -1, (1 - i), -(1 + i)$

$$\text{Area} = \frac{1}{2}\begin{vmatrix} 0 & 1 & 1 \\ 1 & -1 & 1 \\ -1 & -1 & 1 \end{vmatrix} = |-2| = 2$$

(D) β is purely real and α is purely imaginary.

28. **(C)**

Note that $\alpha = \bar{\alpha} \Rightarrow \alpha$ is real

and $\beta + \bar{\beta} = \dfrac{z-1}{z+1} + \dfrac{\bar{z}-1}{\bar{z}+1}$

$= \dfrac{(z-1)(\bar{z}+1)(z+1)(\bar{z}-1)}{(z+1)(\bar{z}+1)} = \dfrac{2z\bar{z}-2}{D'} = 0$

as $z\,\bar{z} = |z|^2 = 1$ (given)

29. **(C)**

Let α be the real root

$\alpha^2 - (3 + i)\alpha + m + 2i = 0$

$(\alpha^2 - 3\alpha + m) + i(2 - \alpha) = 0$

$\therefore \alpha = 2$ (real root) $\quad \therefore 4 - 6 + m = 0 \Rightarrow m = 2$

Product of the roots $= 2(1 + i)$ with one root as 2. Non-real root $= 1 + i$, additive inverse is $-1 - i$.

30. **(C)**

The expression is the sum of the distance of z from the two points $1 - 2i$ and $-3 + 4i$. The minimum value is the distance between these two points $= \sqrt{4^2 + 6^2} = 2\sqrt{13}$.

31. **(C)**

Given $\dfrac{1+\bar{z}}{z}$ is real $\Rightarrow \dfrac{1+\bar{z}}{z} = \dfrac{1+z}{\bar{z}}$

$\Rightarrow \bar{z} + \bar{z}^2 = z + z^2$

$\Rightarrow (\bar{z} - z) + (\bar{z} - z)(\bar{z} + z) = 0$

$\Rightarrow (\bar{z} - z)(1 + \bar{z} + z) = 0,$

so either $\bar{z} = z (z \neq 0)$ or $z + \bar{z} + 1 = 0$

$\Rightarrow y = 0$ or $x = \dfrac{-1}{2}$ but excluding origin.

32. **(C)**

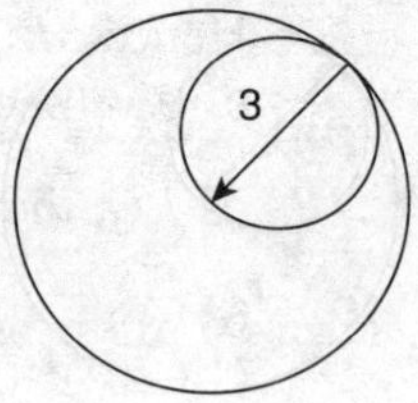

The $z = x + iy$

$\Rightarrow x^2 + y^2 + (4 - 3i)(x + iy) + (4 + 3i)(x - iy) + 5 = 0$

$\Rightarrow x^2 + y^2 + 8x + 6y + 5 = 0$, a circle

Radius $= \sqrt{4^2 + 3^2 - 5} = 2\sqrt{5}$

33. **(B)**

Square both the sides, use $z\,\bar{z} = |z|^2$ and subtract.

34. **(B)**

Choice (A) on simplification gives,

$z = \dfrac{1+x}{1+x^2} + i\dfrac{1+x}{1+x^2}$

for $x = 0.5$; $f(0.5) > 1$ which is out of range

$\Rightarrow$ (A) is not correct

Choice (B): $z = \dfrac{1-x}{1+x^2} + i\dfrac{1-x}{1+x^2}$

$f(x)$ and $g(x) \in (0, 1)$ if $x \in (0, 1)$

$\Rightarrow$ is correct

Choice (C): $z = \dfrac{1+x}{1+x^2} + \dfrac{1-x}{1+x^2}i$

$\Rightarrow$ C is not correct

Choice (D): $z = \dfrac{1-x}{1+x^2} + \dfrac{1+x}{1+x^2}i$

$\Rightarrow$ is not correct

35. **(A)**

$z_1 = \dfrac{a(1+i)}{2}$; $z_2 = \dfrac{b(2-i)}{5}$

$\dfrac{a(1+i)}{2} - \dfrac{b(2-i)}{5} = 1$

$5a(1 + i) - 2b(2 - i) = 10$

$(5a - 4b - 10) + i(5a + 2b) = 0$

$5a - 4b = 10$; $5a = -2b$

$-6b = 10 \Rightarrow b = \dfrac{-5}{3}$

$5a = -2\left(\dfrac{-5}{3}\right) = \dfrac{10}{3} \Rightarrow a = \dfrac{2}{3}$

36. **(B)**

Use the quadratic formula to obtain

$z = \dfrac{3i \pm \sqrt{-9 + 40k}}{20}$

which has discriminant $D - -9 + 40k$.

If $k = 1$, then $D = 31$, so (A) is false.

If k is a negative real number, then D is a negative real number, so (B) is true.
If $k = i$, then $D = -9 + 40i = 16 + 40i + 25i^2$

$= (4 + 5i)^2$, and the roots are $\frac{1}{5}+\frac{2}{5}i$ and $-\frac{1}{5}-\frac{1}{10}i$, so (C) is false.

If $k = 0$ (which is a complex number), then the roots are 0 and $\frac{3}{10}i$, so (D) is false.

37. **(C)**
Let $z = \cos x + i \sin x$, $x \in [0, 2\pi)$. Then

$$1 = \left|\frac{z}{\bar{z}}+\frac{\bar{z}}{z}\right| = \frac{|z^2+\bar{z}^2|}{|z|^2}$$
$$= |\cos 2x + i \sin 2x + \cos 2x - i \sin 2x|$$
$$= 2\,|\cos 2x|$$

hence $\cos 2x = \frac{1}{2}$ or $\cos 2x = -\frac{1}{2}$

If $\cos 2x = \frac{1}{2}$, then

$$x_1 = \frac{\pi}{6}, x_2 = \frac{5\pi}{6}, x_3 = \frac{7\pi}{6}, x_4 = \frac{11\pi}{6}$$

If $\cos 2x = -\frac{1}{2}$, then

$$x_5 = \frac{\pi}{3}, x_6 = \frac{2\pi}{3}, x_7 = \frac{4\pi}{3}, x_8 = \frac{5\pi}{3}$$

Hence there are eight solutions
$z_k = \cos x_k + i \sin x_k$, $k = 1, 2, \ldots, 8$

38. **(C)**
Let $z = a + ib \Rightarrow \bar{z} = a - ib$
hence we have $z^{2008} = \bar{z}$
$\therefore |z|^{2008} = |\bar{z}| = |z|$
$|z|\,[|z|^{2007} - 1] = 0$
$|z| = 0$ or $|z| = 1$; if $|z| = 0$
$\Rightarrow z = 0 \Rightarrow (0, 0)$
if $|z| = 1$ $z^{2009} = z\,\bar{z} = |z|^2 = 1$
$\Rightarrow$ 2009 values of z $\Rightarrow$ Total 2010

39. **(C)**
$z = r(\cos \theta + i \sin \theta)$
now $r = OA \sin \theta = 6 \sin \theta$

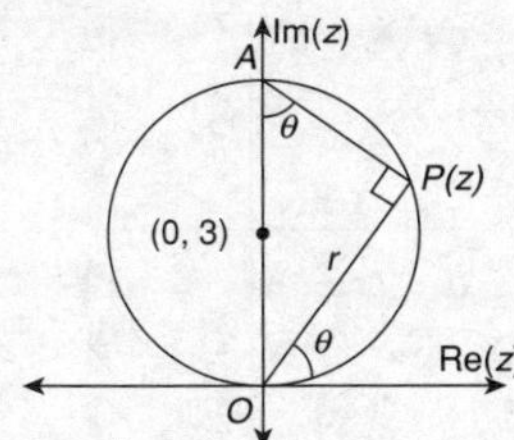

$z = 6 \sin \theta (\cos \theta + i \sin \theta)$

$$\frac{6}{z} = \frac{1}{\sin\theta(\cos\theta + i\sin\theta)}$$
$$= \frac{\cos\theta - i\sin\theta}{\sin\theta} = -i + \cot\theta$$
$$\Rightarrow \cot\theta - \frac{6}{z} = i$$

40. **(B)**

$$\left|z+\frac{1}{z}\right| \ge |z| - \frac{1}{|z|}$$
$$\left|z+\frac{1}{z}\right|_{\text{least}} \ge 3 - \frac{1}{3} \ge \frac{8}{3}$$

41. **(A)**
$|z + 3| \le 3$ denotes set of points on or inside a circle with centre $(-3, 0)$ and radius 3. $|z + 1|$ denotes the distance of P from A of $|z + 1|_{\min} = 0$ and $|z + 1|_{\max} = AD$

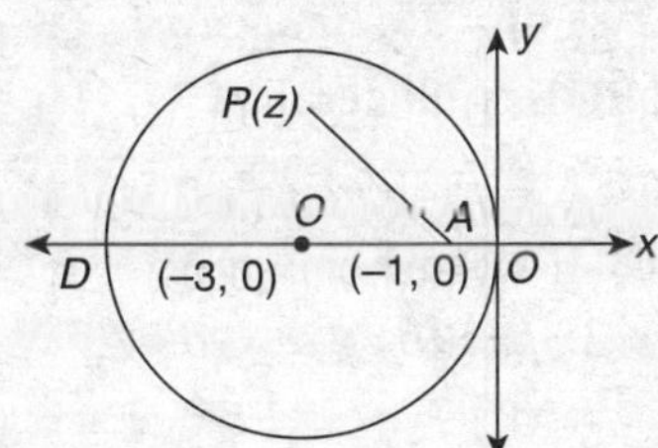

42. **(B)**
Let $z = \alpha$ be a real root
$\alpha^3 + (3 + 2i)\alpha + (-1 + ia) = 0$
$(\alpha^3 + 3\alpha - 1) + i(a + 2\alpha) = 0$
$\therefore \alpha^3 + 3\alpha - 1 = 0$ and $\alpha = -\frac{a}{2}$
$\therefore -\frac{a^3}{8} - \frac{3a}{2} - 1 = 0$
$a^3 + 12a + 8 = 0$
Let $f(a) = a^3 + 12a + 8$
$\therefore f(-1) < 0$ and $f(0) > 0$.
Hence $a \in (-1, 0)$.

43. **(D)**
$E = |2z - 1|^2 + |2\omega - 1|^2$

$$E = \left[\left|z-\frac{1}{2}\right|^2 + \left|\omega-\frac{1}{2}\right|^2\right]$$

$\left|z-\frac{1}{2}\right|$ distance of z tan $\left(\frac{1}{2}, 0\right)$ and $\left|\omega-\frac{1}{2}\right|$ distance of ω from $\left(\frac{1}{2}, 0\right)$.

$$E_{\text{max.}} = 4\left[\left(-1, -\frac{1}{2}\right)^2 + \left(2-\frac{1}{2}\right)^2\right]$$

$$= 4\left[\frac{9}{4}+\frac{9}{4}\right] = 18$$

$$E_{\text{min}} = 4\left[\left(1-\frac{1}{2}\right)^2 + \left(2-\frac{1}{2}\right)^2\right]$$

$$= 4\left[\frac{1}{4}+\frac{1}{4}\right] = 2$$

rejected $E = [2, 18]$

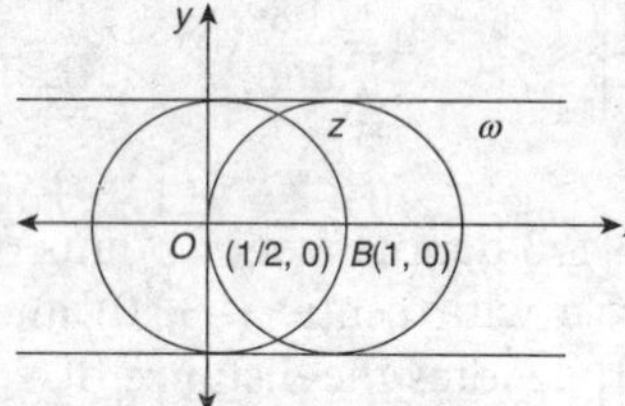

44. (A)

$z_1 + z_2 + z_3 = 3\alpha; \Sigma z_1 z_2 = 3\beta$

If ABC is equilateral

$z_1^2 + z_2^2 + z_3^2 = z_1 z_2 + z_2 z_3 + z_3 z_1$

$(z_1 + z_2 + z_3)^2 = 3\Sigma z_1 z_2$

$9\alpha^2 = 3.3\beta = 9\beta \Rightarrow \alpha^2 = \beta$

45. (B)

Rotation about B gives

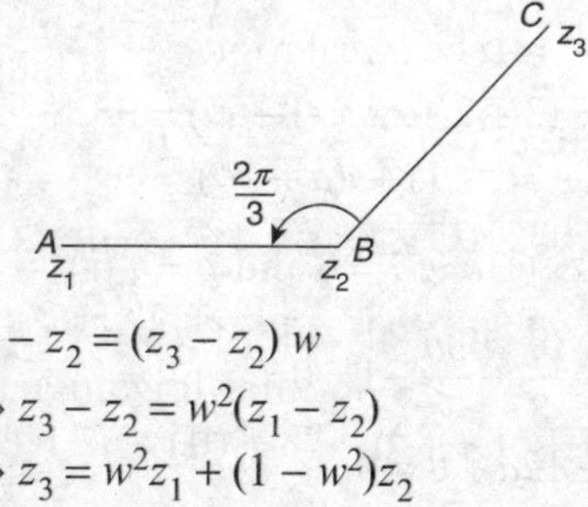

$z_1 - z_2 = (z_3 - z_2)\, w$

$\Rightarrow z_3 - z_2 = w^2(z_1 - z_2)$

$\Rightarrow z_3 = w^2 z_1 + (1 - w^2) z_2$

46. (B)

Note that $z_1 = 3 + \sqrt{3}\, i$ lies on the line $y = \frac{1}{\sqrt{3}}$ x and $z_2 = 2\sqrt{3} + 6i$ lies on the line $y = \sqrt{3}x$. Hence $z = 5 + 5i$ will only lie on the bisector of z_1 and z_2, that is, $y = x$

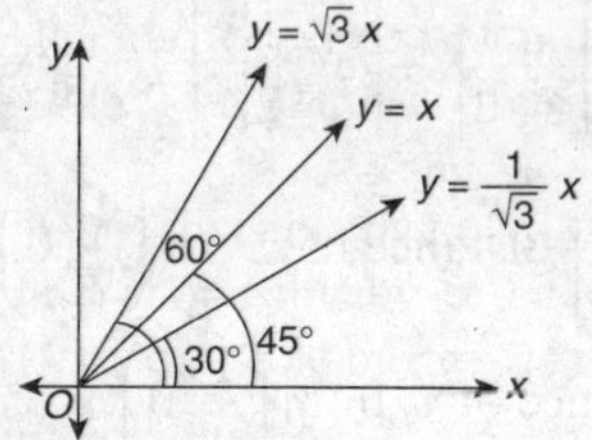

47. (C)

$$\frac{z_2}{|z_2|} = \frac{z}{|z|}e^{i\theta} \qquad (1)$$

$$\frac{z_1}{|z_1|} = 1e^{i\theta} \qquad (2)$$

Substitute the value of $e^{i\theta}$ from (2) in (1)

$$\frac{z}{|z|} = \frac{z_2}{|z_2|}\cdot\frac{z_1}{|z_1|} \Rightarrow \frac{z}{|z|} = \frac{\frac{z_2}{z_1}}{\left|\frac{z_2}{z_1}\right|};\ z = \frac{z_2}{z_1}$$

48. (D)

$\sqrt{2}|x + iy + 1| = x + iy - i$

$\Rightarrow \sqrt{2}\sqrt{(x+1)^2 + y^2} = x + i(y-1)$

$\Rightarrow \sqrt{2}\sqrt{(x+1)^2 + y^2} = x \qquad (1)$

$y - 1 = 0 \Rightarrow y = 1 \qquad (2)$

From (1)

$\sqrt{2}\sqrt{(x+1)^2 + 1^2} = x \Rightarrow x$ is positive as LHS is positive

$\Rightarrow x^2 + 4x + 4 = 0 \Rightarrow x = -2$

but x is positive, therefore we get a contradiction. Hence no solution.

49. (B)

The equation can be written as

$$z\left(\overline{z + 1 + i\sqrt{3}}\right) + z\bar{z} + \left(1 + i\sqrt{3}\right)\bar{z} = 0$$

$$\Rightarrow z\bar{z} + \frac{1 - i\sqrt{3}}{2}z + \frac{1 + i\sqrt{3}}{2}\bar{z} = 0$$

which is a circle with centre $\left(\frac{1}{2}, \frac{\sqrt{3}}{2}\right)$ and

$$\text{radius} = \sqrt{\left(\frac{1 - i\sqrt{3}}{2}\right)\left(\frac{1 + i\sqrt{3}}{2}\right)} = 1$$

Hence (B).

50. (D)

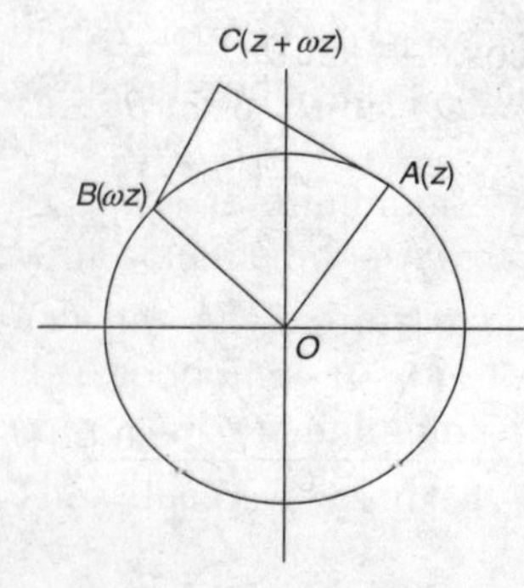

$\angle AOB = \dfrac{2\pi}{3}$

$\Delta OAB = \Delta ABC$

$\therefore$ area of ΔABC = area of ΔOAB

$\Rightarrow \dfrac{1}{2}|z||\omega z|\sin\dfrac{2\pi}{3} = 4\sqrt{3}$

$\Rightarrow |z| = 4$ = radius of the circle [(D)]

51. **(C)**

From the figure it is clear that the required length of the line segment is $AB = 2$

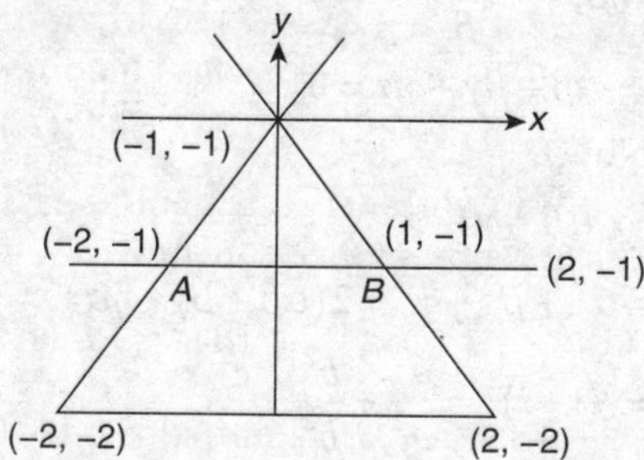

52. **(C)**

$z^3 + \bar{\omega}^7 = 0 \Rightarrow z^3 = -\bar{\omega}^7 \Rightarrow z^{15} = -\bar{\omega}^{35}$

$z^5\omega^{11} = 1 \Rightarrow z^{15} = \dfrac{1}{\omega^{33}}$

$\Rightarrow -\bar{\omega}^{35} = \dfrac{1}{\omega^{33}} \Rightarrow \bar{\omega}^{35}\omega^{33} = -1$

$\Rightarrow \left|\bar{\omega}^{35}\omega^{33}\right| = 1 \Rightarrow |\omega| = 1$

Again, $\bar{\omega}^{35}\omega^{33} = -1$

$\Rightarrow \left(\bar{\omega}^{33}\omega^{33}\right)\bar{\omega}^2 = 1$

$\Rightarrow \left(|\omega|^2\right)^{33}\bar{\omega}^2 = 1$

$\Rightarrow w = \pm i \Rightarrow z = \pm i$

Hence (C).

53. **(C)**

Given

$a\alpha^2 + b\alpha + 1 = 0 \qquad (1)$

Also $\bar{a}\bar{\alpha}^2 + \bar{b}\bar{\alpha} + 1 = 0$

$\Rightarrow \dfrac{\bar{a}}{\alpha^2} + \dfrac{\bar{b}}{\alpha} + 1 = 0 \qquad [\text{As } |\alpha| = 1]$

$\Rightarrow \alpha^2 + \bar{b}\alpha + \bar{a} = 0 \qquad (2)$

From (1) and (2)

$\dfrac{\alpha^2}{\bar{a}b - \bar{b}} = \dfrac{\alpha}{1-|a|^2} = \dfrac{1}{a\bar{b} - b}$

$\Rightarrow \dfrac{\bar{a}b - \bar{b}}{1-|a|^2} = \dfrac{1-|a|^2}{a\bar{b} - b}$

$\Rightarrow |a\bar{b} - b| = 1 - |a|^2 = \dfrac{3}{4}$

Hence (C).

54. **(C)**

$\dfrac{Kz}{K+1}$ represents any point lying on the line joining origin and z. So, $\dfrac{Kz}{K+1}$ should lie outside the circle $|z+2| > \sqrt{2}$

So, z should lie in the shaded region.

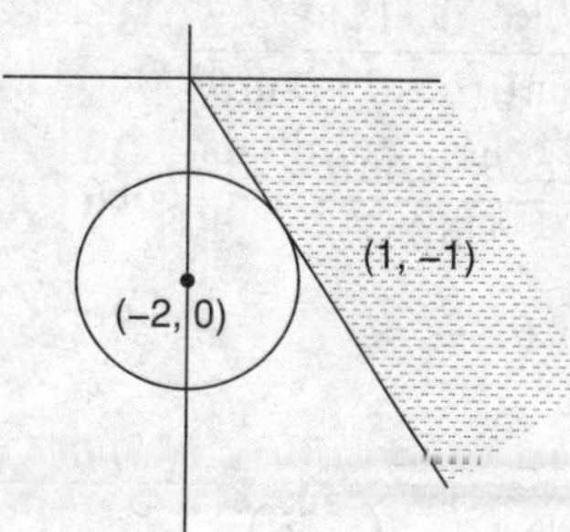

$\Rightarrow -\dfrac{\pi}{4} < \arg z < 0$

Hence (C).

55. **(B)**

$\dfrac{A}{B} + \dfrac{B}{A} = 1 \Rightarrow A^2 + B^2 = AB$

$\Rightarrow z_1^2 + z_2^2 + 0^2 = z_1 z_2 + z_1 \cdot 0 + z_2 \cdot 0$

$\Rightarrow$ points represented by A, B and origin form the vertices of an equilateral triangle. [(B)]

56. **(D)**

$z = \dfrac{\pi}{4}(1+i)^4\left(\dfrac{(1+\pi)+(\pi+1)}{\sqrt{\pi}+\pi i + i - \sqrt{\pi}}\right)$

$\Rightarrow z = \dfrac{\pi}{2}(1+i)^4\dfrac{1}{i}$

$\therefore |z| = \dfrac{\pi}{2}\cdot 4 = 2\pi;\ \text{amp}(z) = \dfrac{\pi}{2}$

$\therefore \dfrac{|z|}{\text{amp}(z)} = \dfrac{2\pi}{\frac{\pi}{2}} = 4$

57. **(B)**

$z = r(\cos\theta + i\sin\theta)$

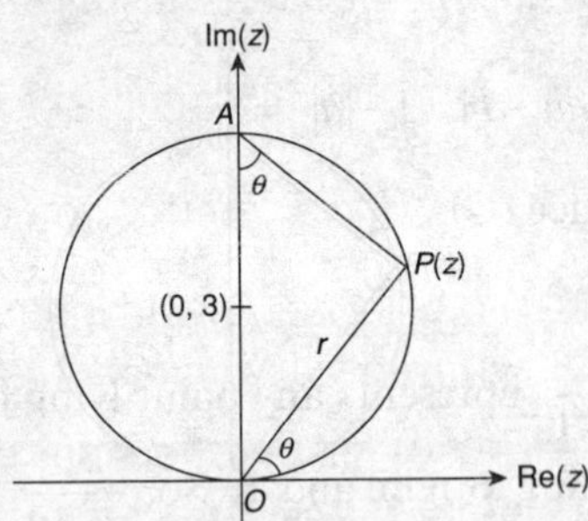

$r = OA\sin\theta - 6\sin\theta$

$z = 6\sin\theta(\cos\theta + i\sin\theta)$

$$\frac{6}{z} = \frac{1}{\sin\theta(\cos\theta + i\sin\theta)}$$

$$= \frac{\cos\theta - i\sin\theta}{\sin\theta} = -i + \cot\theta$$

$\Rightarrow \cot\theta - \frac{6}{z} = i$ [Ans. (B)]

58. **(B)**

$$z = \left(\frac{1-i}{1+i}\right)^n = \left(-\frac{2i}{2}\right)^n = (-i)^n$$

$\text{Im}(z) < 0$ for $n = 4K + 1, K \in I$

Sum $= 1 + 5 + 9 + \ldots + 97 = 1225$

59. **(C)**

z lies on the perpendicular bisector of two conjugate complex numbers $-i\omega$ and $i\bar{\omega}$

$\Rightarrow z$ is purely real

$\therefore z = K, K \in R$

$\Rightarrow |z + i\omega| \le |z| + |i\omega| \le 2$

$\Rightarrow |z| = 1$

$\Rightarrow z = 1$ or -1 Ans (C)

Integer Answer Type

1. **(5)**

$|iz + z_0| = |i(z-i) - 1 + 5 + 3i|$

$= |i(z-i) + (4+3i)|$

$\le |i||z-i| + |4+3i|$

$\le 1 \cdot 2 + 5 = 7$

$\therefore$ maximum value of $\frac{1}{2}|iz + z_0| = \frac{7}{2}$

Also, $|iz + z_0| = |i(z-i) + (4+3i)|$

$\ge \big||i||z-i| - |4+3i|\big|$

$= |2-5| = 3$

$\therefore$ minimum value of $\frac{1}{2}|iz + z_0| = \frac{3}{2}$

Hence sum of maximum and minimum = 5

2. **(2)**

Let, $\frac{a}{a_1} = \alpha, \frac{b}{b_1} = \beta, \frac{c}{c_1} = \gamma$

$\therefore \alpha + \beta + \gamma = 1 + i$

and, $\frac{1}{\alpha} + \frac{1}{\beta} + \frac{1}{\gamma} = 0$

$\Rightarrow \alpha\beta + \beta\gamma + \gamma\alpha = 0$

Now,

$(\alpha + \beta + \gamma)^2$

$= \alpha^2 + \beta^2 + \gamma^2 + 2(\alpha\beta + \beta\gamma + \gamma\alpha)$

$$\Rightarrow (1+i)^2 = \frac{a^2}{a_1^2} + \frac{b^2}{b_1^2} + \frac{c^2}{c_1^2}$$

$$\Rightarrow \left|\frac{a^2}{a_1^2} + \frac{b^2}{b_1^2} + \frac{c^2}{c_1^2}\right| = 2$$

3. **(2)**

Since, $|z_1 z_2| = \sqrt{2} \Rightarrow |\bar{z}_1 z_2| = \sqrt{2}$

$\arg(z_1) - \arg(z_2) = \frac{\pi}{4}$

$\Rightarrow -\arg(\bar{z}_1) - \arg(z_2) = \frac{\pi}{4}$

$\Rightarrow \arg(\bar{z}_1 z_2) = -\frac{\pi}{4}$

$\therefore \bar{z}_1 z_2 = \sqrt{2}e^{-\frac{i\pi}{4}} = 1 - i$

$\therefore |\bar{z}_1 z_2|^2 = 2$

4. **(1)**

Since, sets A and B represents two non parallel lines. Hence the number of intersection points = 1

$\therefore n(A \cap B) = 1$

5. **(9)**

Since, $|z_1| = \frac{5}{3} \Rightarrow |z_1|^2 = \frac{25}{9}$

$$\text{Now,} \left|\frac{75 - 27z_1\bar{z}_2}{5(z_1 - z_2)}\right| = \left|\frac{27\left(\frac{25}{9} - z_1\bar{z}_2\right)}{5(z_1 - z_2)}\right|$$

$$=\left|\frac{27(z_1\bar{z}_1-z_1\bar{z}_2)}{5(z_1-z_2)}\right|=\left|\frac{27z_1(\bar{z}_1-\bar{z}_2)}{5(z_1-z_2)}\right|$$

$$=\frac{27}{5}|z_1|\frac{|\bar{z}_1-\bar{z}_2|}{|z_1-z_2|}$$

$$=\frac{27}{5}\times\frac{5}{3}=9$$

6. (6)

Since, $|z-2-i|=4$

$\Rightarrow |3z-6-3i|=12$

$\Rightarrow |3z-6+15i-18i|=12$

Let $3z-6+15i=\alpha$

$\Rightarrow |\alpha-18i|=12$

$\Rightarrow$ minimum value = 18 – 12 = 6

7. (0)

$z^2+|z|=\bar{z}^2$

$\Rightarrow \bar{z}^2+|z|=z^2$

Adding, $2|z|=0 \Rightarrow z=0$

8. (9)

Put $z=x+iy$

$$\left(\frac{z-\bar{z}}{zi}\right)^2 \le 2(z+\bar{z})$$

$\Rightarrow y^2 \le 4x$

Also, $|z|\le\sqrt{5} \Rightarrow x^2+y^2\le 5$

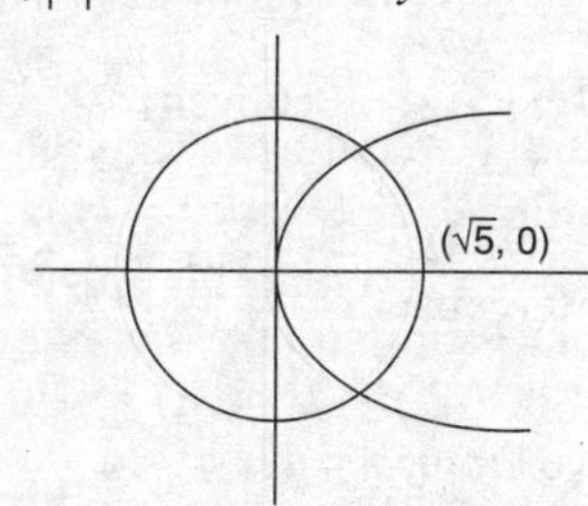

Hence, the number of points lying in $(A\cap B)$ is

$(0,0),(1,0),(2,0),(1,1),(1,-1),(1,2),$
$(1,-2),(2,1),(2,-1)$

9. (6)

$M=PA'=5$

$m=PA=1$

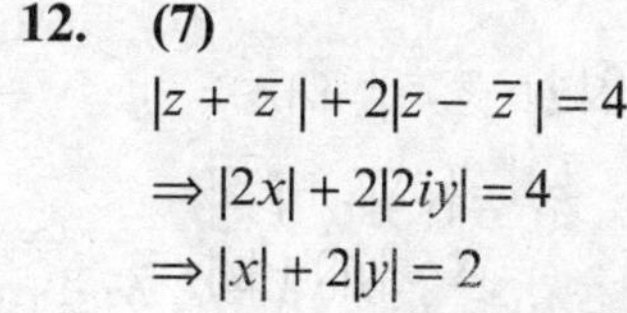

10. (1)

Using rotation formula

$z_0-z_2=(z_0-z_1)e^{i2C}$

and $z_0-z_3=(z_0-z_2)e^{i2A}$

$$\therefore \left|\frac{z_0-z_1}{z_0-z_2}\frac{\sin 2A}{\sin 2B}+\frac{z_0-z_3}{z_0-z_2}\frac{\sin 2C}{\sin 2B}\right|$$

$$=\left|\frac{(\cos 2C-i\sin 2C)\sin 2A+(\cos 2B+i\sin 2A)\sin 2C}{\sin 2B}\right|$$

$$=\left|\frac{\sin 2(A+C)}{\sin 2B}\right|-1$$

11. (6)

Let $z=e^{i\theta}\Rightarrow z^3=\cos 3\theta+i\sin 3\theta$

$\because z^3$ is purely imaginary

$\Rightarrow \cos 3\theta \Rightarrow \theta=\frac{2n\pi}{3}\pm\frac{\pi}{6}$

$\therefore$ Possible value of θ= 30°, 150°, 90°, 270°, 210°, 330°.

12. (7)

$|z+\bar{z}|+2|z-\bar{z}|=4$

$\Rightarrow |2x|+2|2iy|=4$

$\Rightarrow |x|+2|y|=2$

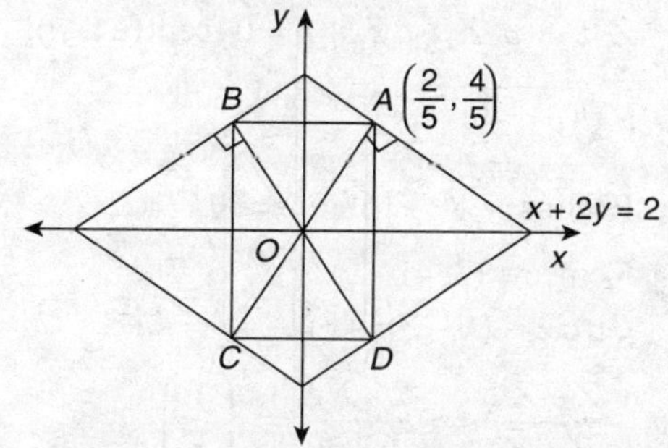

[**Note:** $|z|$ is minimum when it is perpendicular.]

Required area = Area of rectangle

$$= \frac{4}{5} \times \frac{8}{5} = \frac{32}{25} \text{ sq. units}$$

13. **(4)**

$z^{2014} \Rightarrow 1$

$z = e^{\left(\frac{2k\pi}{2014}\right)i} = \alpha^k$ (As given in the questions)

Let $z_1 = re^{i\theta_1}$ and $z_2 = re^{i\theta_2}$

where $r = \frac{1}{\sqrt{1007}}$

$\therefore |z_1 + \alpha^k z_2|^2 = (z_1 + \alpha^k z_2)\left(\bar{Z}_1 + \overline{\alpha^k}\,\bar{z}_2\right)$

$$= r^2\left(e^{i\theta_1} + e^{\left(\frac{2k\pi}{2014}+\theta_2\right)}\right)\left(e^{-i\theta_1} + e^{-i\left(\frac{2k\pi}{2014}+\theta_2\right)}\right)$$

$$= r^2\left[2 + 2\cdot\cos\left(\theta_1 - \theta_2 - \frac{k\pi}{2007}\right)\right]$$

Let $\theta_1 - \theta_2 = \theta$ and $\beta = \frac{\pi}{2007}$

$$\sum_{k=0}^{2013}\left|Z_1 + \alpha^k z_2\right|$$

$$= 2r^2\left[2014 + \underbrace{\begin{array}{c}\cos\theta + \cos(\theta-\alpha) + \cos(\theta - 2\alpha)\\ + \ldots + \cos(\theta - 2013\alpha)\end{array}}_{0}\right]$$

$$= 2 \times \frac{1}{1007} \times 2014 = 4$$

14. **(2)**

$S = \{z | z\,\bar{z} - (3-4i)z - (3+4i)\,\bar{z} + 21 = 0\}$

$Z\,\bar{z} + a\,\bar{z} + a\bar{z} + b = 0$ (centre $(-a)$,

$r\ \sqrt{|a|^2 - b}$

Radius $= \sqrt{9+16-21} = \sqrt{4} = 2$

Centre $\equiv (3, 4)$

Circle is $(x-3)^2 + (y-4)^2 = 4$

$$\frac{Z - \bar{Z}}{i(Z+\bar{Z})} \Rightarrow \frac{2\times 2iy}{i(2x)} \Rightarrow \left(\frac{y}{x}\right)$$

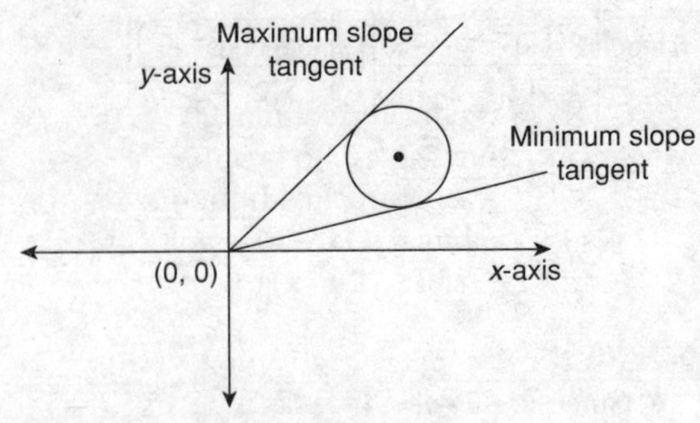

Equation of tangent

$(y-4) = m(x-3) \pm 2\sqrt{1+m^2}$

Satisfies (0, 0)

$3m - 4 = \pm 2\sqrt{1+m^2}$

$\Rightarrow 9m^2 + 16 - 24m = 2(1+m^2)$

$\Rightarrow 5m^2 - 24m + 12 = 0$

$$\frac{1}{M} + \frac{1}{m} = \left(\frac{M+m}{Mm}\right) \Rightarrow \frac{24}{5} \times \frac{5}{12} = 2$$

15. **(1)**

Put $\omega = x + iy$ in $z = \frac{2\omega+1}{i\omega+1}$

$$\Rightarrow z = \frac{[(2x+1)+i2y][(1-y)-ix]}{[(1-y)+ix][(1-y)-ix]}$$

$$= \frac{(2x+1)(1-y) + 2xy + i(2y(1-y) - x(2x+1))}{(1-y)^2 + x^2}$$

As $I_m(z) = -2$ (Given)

$\Rightarrow 2y - 2y^2 - 2x^2 - x = -2(x^2 + y^2 - 2y + 1)$

$\Rightarrow 2y - x = 4y - 2 \Rightarrow x + 2y - 2 = 0$

$\Rightarrow$ y-intercept = 1

16. **(6)**

We have $|a\omega + b| = 1$

$\Rightarrow |a\omega + b|^2 = 1$

$\Rightarrow (a\omega + b)(a\bar{\omega} + b) = 1$

$\Rightarrow a^2 + ab(\omega + \bar{\omega}) + b^2 = 1$

$\Rightarrow a^2 - ab + b^2 = 1$

$\Rightarrow (a-b)^2 + ab = 1$ (1)

(As $1 + \omega + \omega^2 = 0$)

When $(a-b)^2 = 0$ and $ab = 1$ then (1, 1); (–1, –1)

When $(a-b)^2 = 1$ and $ab = 0$ then (0, 1); (1, 0); (0, –1); (–1, 0)

Hence (0, 1); (1, 0); (0, $N-1$); (−1, 0); (1, 1); (−1, −1), that is, 6 ordered pairs.

17. (5)

We have $f(w) = 2+i\sqrt{3}$, $f(w^2) = 2-i\sqrt{3}$

Let $f(z) = (z^2 + z + 1)\, g(z) + (az + b)$

$\therefore aw + b = 2+i\sqrt{3}$ (1)

and $aw^2 + b = 2-i\sqrt{3}$ (2)

$\therefore$ On solving (1) and (2), we get $a = 2, b = 3 \Rightarrow a + b = 5$

18. (4)

$m_{BD} = -2$

$m_{AC} = \dfrac{1}{2}$

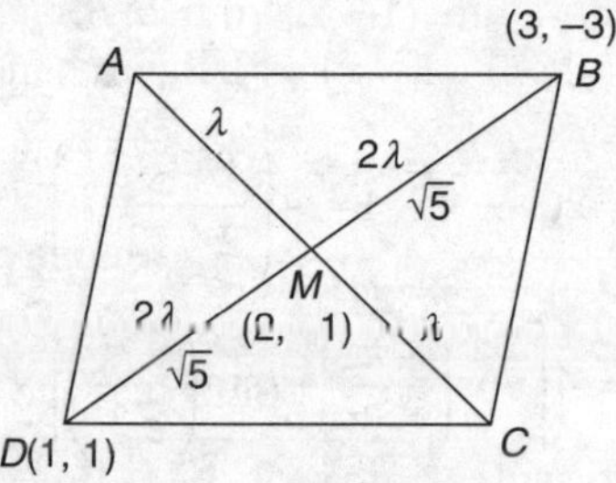

Parametric equation of BD is

$$\frac{x-2}{\cos\theta} = \frac{y+1}{\sin\theta} = \pm\frac{\sqrt{5}}{2}$$

where $\cos\theta = \dfrac{2}{\sqrt{5}}$; $\sin\theta = \dfrac{1}{\sqrt{5}}$

$\therefore x = 2 \pm 1$ and $y = -1 \pm \dfrac{1}{2}$

So, $\left(3, \dfrac{-1}{2}\right)$ or $\left(1, \dfrac{-3}{2}\right) \Rightarrow \dfrac{x_1x_2}{y_1y_2} = \dfrac{3\cdot 4}{3} = 4$

19. (1)

As $\dfrac{z^2-1}{z^2+1} \in R \Rightarrow z^2 = \text{Real} \Rightarrow xy = 0$

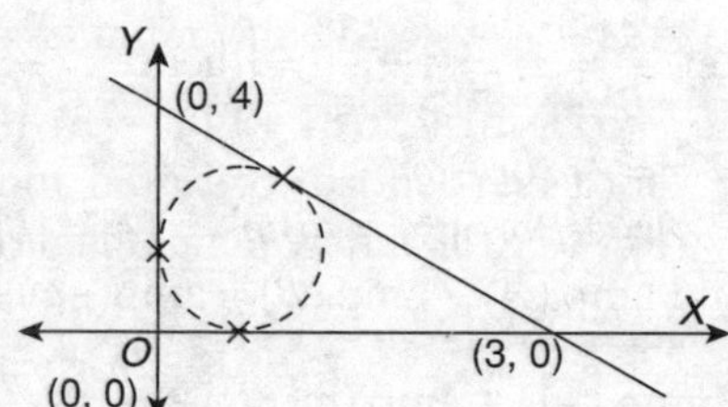

So, clearly locus of $P(z)$ is pair of straight line $xy = 0$.

$$\text{Clearly, } r = \frac{\Delta}{s} = \frac{\frac{1}{2}(3)(4)}{\frac{3+4+5}{2}} = \frac{12}{12} = 1$$

3.18.3 JEE Advanced: Sample Questions

Single Answer Type

1. (B)

Let α is a non-real complex root of unity that is also a root of the equation $z^2 + az + b = 0$, then $\bar{\alpha}$ will also be its root. ($|\alpha| = 1$)

Hence $\alpha + \bar{\alpha} = -a$

$\therefore |a| = |\alpha + \bar{\alpha}| \le |\alpha| + |\bar{\alpha}| = 2$

and $b = \alpha\bar{\alpha} = 1$

Hence we must check those equations for which

$-2 \le a \le 2$ and $b = 1$; that is,

$z^2 + 2z + 1 = 0$; $z^2 + z + 1 = 0$; $z^2 + 1 = 0$

$z^2 - 2z + 1 = 0$; $z^2 - z + 1 = 0$

Hence roots are $\pm 1, \pm i$; $\dfrac{-1\pm\sqrt{-3}}{2}$, $\dfrac{1\pm\sqrt{-3}}{2}$

That is, 8.

2. (A)

$$\omega = 2\cdot i^{1/3} = 2\left(\cos\frac{\pi}{2} + i\sin\frac{\pi}{2}\right)^{\frac{1}{3}}$$

$$= 2\left[\cos\frac{2n\pi + \frac{\pi}{2}}{3} + i\sin\frac{2n\pi + \frac{\pi}{2}}{3}\right]$$

Put $n = 1$

$$= 2\left[\cos\frac{5\pi}{6} + i\sin\frac{5\pi}{6}\right] = -\sqrt{3} + i$$

3. (D)

$$z^4 = -1 + \sqrt{3}i = 2\left(\cos\frac{2\pi}{3} + i\sin\frac{2\pi}{3}\right)$$

$z^4 = 2w \Rightarrow A, C$ are not possible

$$\text{Root are } z_1 = 2^{1/4}\left(\cos\frac{\pi}{6} + i\sin\frac{\pi}{6}\right)$$

a

$2^{1/4}$ $2^{1/4}$

$$z_2 = 2^{1/4}\left(\cos\frac{2\pi}{3} + i\sin\frac{2\pi}{3}\right)\ldots$$

$$\therefore\ a = \sqrt{2^{1/2} + 2^{1/2}}$$

$$= (2^{3/2})^{1/2} = 2^{3/4}$$

4. (C)

Note that $|z_1| = |z_2| = |z_3| = \sqrt{13}$

Hence z_1, z_2, z_3 lies on a circle with centre (0, 0) and $r = \sqrt{13}$ as shown.

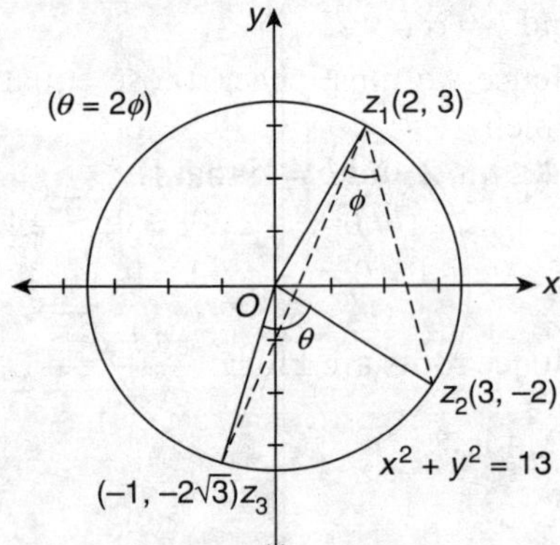

$$\text{Now Arg. } \frac{Z_2}{Z_3} = 2\text{ Arg. } \frac{z_2 - z_1}{z_3 - z_1}$$

$$\therefore\ \text{Arg. } \frac{Z_3}{Z_2} = 2\text{ Arg. } \frac{z_3 - z_1}{z_2 - z_1}$$

5. (C)

$$2^{m-2n}\cdot\left[\cos\frac{m\pi}{6} + i\sin\frac{m\pi}{6}\right]$$

$$= \left[\cos\frac{n\pi}{4} + i\sin\frac{n\pi}{4}\right]$$

For equality $m = 2n$

$$\frac{n\pi}{4} = \frac{m\pi}{6} + 2k\pi\ k \in I$$

Put $m = 2n$

$$\frac{n\pi}{4} = \frac{n\pi}{3} + 2k\pi;$$

$$-\left(\frac{n\pi}{12}\right) = 2k\pi \text{ [ignore (–)ve sign]}$$

$n = 24k$; $m = 48k$; for m, n to be smallest $m + n = 72$.

6. (C)

Let $Z^3 = t$

hence equation becomes $t^2 + t + 1 = 0$

$\Rightarrow t = \omega$ or ω^2

$$Z^3 = \cos\frac{2\pi}{3} + i\sin\frac{2\pi}{3} = e^{\left(2m\pi + \frac{2\pi}{3}\right)i}$$

$$Z = e^{\frac{\left(2m\pi + \frac{2\pi}{3}\right)}{e^3}i}$$

Put $m = 1$ to get $\theta = \dfrac{8\pi}{9} \in (90°, 180°) = 160°$

7. (B)

$$z^4 = 2(1 - \sqrt{3}i) = 4\left(\frac{1}{2} - \frac{\sqrt{3}}{2}i\right)$$

$$= 4\left[\cos\left(-\frac{\pi}{3}\right) + i\sin\left(-\frac{\pi}{3}\right)\right]$$

$$z = \sqrt{2}\left[\cos\frac{2m\pi - \left(\frac{\pi}{3}\right)}{4} + i\sin\frac{2m\pi - \left(\frac{\pi}{3}\right)}{4}\right]$$

$$m = 1,\ z = \sqrt{2}\left[\cos\left(\frac{5\pi}{12}\right) + i\sin\left(\frac{5\pi}{12}\right)\right]$$

8. (A)

$A^2 - AB + B^2 = 0$

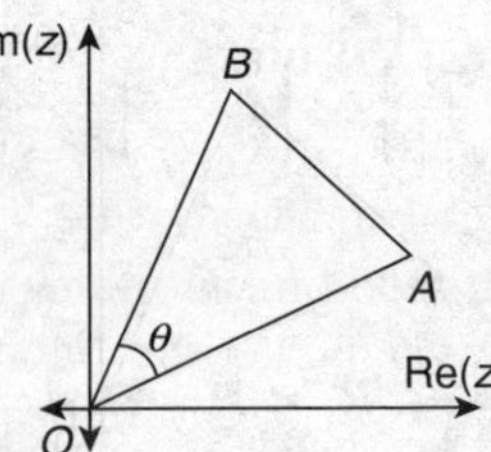

Let $\dfrac{A}{B} = z; \Rightarrow z + \dfrac{1}{z} = 1$ hence $z^2 - z + 1 = 0$

$\Rightarrow z = -\omega$ or $-\omega^2$

$\Rightarrow A = -B\omega$ or $A = -B\omega^2 \Rightarrow |A| = |B|$

and amp.(A) – amp.(B) = amp$(-\omega)$

$= \text{amp }(-1) + \text{amp.}(\omega) = \pi + \dfrac{2\pi}{3}$

9. **(D)**
Let $S = 1(\alpha_1 + \alpha_{2008}) + 2(\alpha_2 + \alpha_{2007}) + 3(\alpha_3 + \alpha_{2006}) + \ldots + 2008(\alpha_{2008} + \alpha_1)$ (1)
Also $S = 2008(\alpha_{2008} + \alpha_1) + 2007(\alpha_2 + \alpha_{2007}) + \ldots + 2(\alpha_2 + \alpha_{2007}) + 1(\alpha_1 + \alpha_{2008})$ (2)
(Writing in reverse order)
$\therefore$ On adding (1) and (2), we get
$2S = 2009[2(\alpha_1 + \alpha_2 + \alpha_3 + \ldots + \alpha_{2008})]$
$2S = 2009[2\,(\underbrace{1+\alpha_1+\alpha_2+\alpha_3+\ldots\alpha_{2008}}_{\text{zero}}-1)]$

Hence $S = -2009$
Note: (α_1 and α_{2008}), (α_2 and α_{2007}), (α_3 and α_{2006}), ..., (α_{1004} and α_{1005}) are conjugates of each other.

10. **(A)**
Use De Moivre's theorem get the product of roots with +ve real part
$z = 2(-1)^{1/6}$

$$-2\left[\cos\frac{(2m+1)\pi}{6} + i\sin\frac{(2m+1)\pi}{6}\right]$$

put $m = 0$ or $m = 5$ for positive real part to get
$z_1 z_2 = 4e^{i\pi/6} \cdot e^{(11\pi/6)i} = 4e^{2\pi i} = 4$

11. **(C)**
$z^2 = \bar{z} \cdot 2^{1-|z|}$
$z^3 = |z|^2\, 2^{1-|z|}$ (1)
$\Rightarrow |z| = 2^{1-|z|}$
Hence z^3 is purely +ve real (as $z \neq 0$)
$\Rightarrow z$ is +ve real.
Hence $z = re^{i\frac{2k\pi}{3}}$ $k = 0, 1, 2$
we therefore need to solve
$r = 2^{1-r} \Rightarrow 2r = \dfrac{2}{r} \Rightarrow r = 1$
$\therefore z = e^{i\frac{2k\pi}{3}}$
hence $z = 1, \omega, \omega^2$
but 1 is not imaginary
hence $z = w$ or w^2

12. **(B)**

$$z_n = e^{\frac{i\pi}{2}\left(\frac{1}{2n+1} - \frac{1}{2n+3}\right)}$$

$$\prod_{n=1}^{\infty} z_n = e^{\frac{i\pi}{2}\left[\left(\frac{1}{3}-\frac{1}{5}\right)+\left(\frac{1}{5}-\frac{1}{7}\right)+\ldots+\left(\frac{1}{2n+1}-\frac{1}{2n+3}\right)\right]}$$

$$= e^{\frac{i\pi}{2}\left[\frac{1}{3} - \frac{1}{2n+3}\right]} = e^{\frac{i\pi}{6}}$$

as $n \to \infty$

13. **(A)**
Put $z = iy$ $(1 + 2i)iy - (2i - 1)iy = 10i$
$2y + 0y = 10 \Rightarrow y = 5$
Note : For x-intercept put $z = x + 0i \Rightarrow x = \dfrac{5}{2}$

14. **(C)**
Divide the equation by x^n and put $x = \cos\theta + i\sin\theta$.
Equate the real and imaginary parts.

15. **(A)**
ℓ_1 is perpendicular to ℓ_2
$\Rightarrow \dfrac{Z_1 - Z_2}{Z_3 - Z_4}$ is purely imaginary

$$\frac{z_1 - z_2}{z_3 - z_4} + \frac{\bar{z}_1 - \bar{z}_2}{\bar{z}_3 - \bar{z}_4} = 0$$

$$\frac{z_1 - z_2}{\bar{z}_1 - \bar{z}_2} + \frac{z_3 - z_4}{\bar{z}_3 - \bar{z}_4} = 0$$

$\Rightarrow \omega_1 + \omega_2 = 0$
Note: If ℓ_1 parallel to ℓ_2 then

$$\frac{z_1 - z_2}{z_3 - z_4} + \frac{\bar{z}_1 - \bar{z}_2}{\bar{z}_3 - \bar{z}_4} \Rightarrow \omega_1 = \omega_2$$

16. **(B)**
Let xi be the root where $x \neq 0$ and $x \in R$ (as if $x = 0$ satisfies then $a_4 = 0$ which contradicts)
$x^4 - a_1x^3 i - a_2x^2 + a_3xi + a_4 = 0$
$x^4 - a_2x^2 + a_4 = 0$ (1)
and $a_1x^3 - a_3x = 0$ (2)
From equation (2); $a_1x^2 - a_3 = 0$
$\Rightarrow x^2 = \dfrac{a_3}{a_1}$ (as $x \neq 0$)

Putting the value of x^2 in equation (1)

$\dfrac{a_3^2}{a_1^2} - \dfrac{a_2a_3}{a_1} + a_4 = 0$ or $a_3^2 + a_4a_1^2 = a_1a_2a_3$

or $\dfrac{a_3}{a_1a_2} + \dfrac{a_1a_4}{a_2a_3} = 1$ (dividing by $a_1a_2a_3$)

17. **(B)**

Let $A = x + iy$; $|A| = 1 \Rightarrow x^2 + y^2 = 1$ and $|A + 1| = 1 \Rightarrow (x + 1)^2 + y^2 = 1$

$\Rightarrow x = -\frac{1}{2}$ and $y = \pm\frac{\sqrt{3}}{2} \Rightarrow (A) = \omega$ or ω^2

$\Rightarrow (\omega)^n = (1 + \omega)^n = (-\omega^2)^n$

$\Rightarrow n$ must be even and divisible by 3

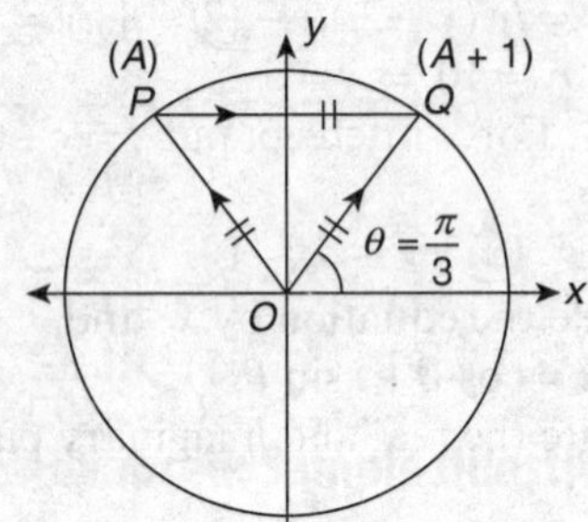

Alternatively: $A = \frac{-1+i\sqrt{3}}{2}$; $A+1 = \frac{1+i\sqrt{3}}{2}$

and n arg $A = n$ arg $(A - 1)$ = arg 1 = $2n\pi$

$\Rightarrow$ arg A = arg $(A + 1) = 2\pi$

Now let $\vec{OQ} = A + 1$ and $\vec{OP} = A$

$\Rightarrow \vec{OP}$ and $\vec{OQ}$ vectors must be turned a minimum number of times to coincide with positive x-axis

$\Rightarrow 6$

18. **(D)**

Points where the circle cuts the x-axis $z = \bar{z}$. Hence substituting $z = \bar{z}$ in the equation of circle

$z^2 + \alpha\bar{Z} + \bar{\alpha}_Z + r = 0$

$z^2 + (\alpha + \bar{\alpha})z + r = 0$

$AB = |z_1 - z_2| = \sqrt{(z_1 + z_2)^2 - 4z_1z_2}$

$= \sqrt{(\alpha+\bar{\alpha})^2 - 4r}$

19. **(B)**

$$E = \frac{1}{z_1 - 1} + \frac{1}{z_2 - 1} + \cdots + \frac{1}{z_{50} - 1}$$

where $z_1, z_2, \ldots, z_{50}$ are the roots of the equation $z^{51} - 1 = 0$ other than 1.

$$= -25 + \left(\frac{1}{2} + \frac{1}{z_1 - 1}\right) + \left(\frac{1}{2} + \frac{1}{z_2 - 1}\right) + \ldots + \left(\frac{1}{2} + \frac{1}{z_{50} - 1}\right)$$

Note that (1st + last) and (2nd + 2nd last) will vanish using $z_r = z^r$ and $z^{51} = 1$

20. **(D)**

$$z = -\frac{1}{2}\left(1 + i\cot\frac{2r+1}{12}\pi\right), \; r = 1, 2, 3, 4, 5$$

21. **(A)**

$z^3 = -\omega^5 \Rightarrow |z|^3 = |\omega|^5 \Rightarrow |z|^6 = |\omega|^{10}$ (1)

and $z^2 = \frac{1}{\bar{\omega}^4} \Rightarrow |z|^2 = \frac{1}{|\omega|^4}$

$\Rightarrow |z|^6 = \frac{1}{|\omega|^{12}}$ (2)

From (1) and (2)

$|\omega| = 1$ and $|z| = 1 \Rightarrow Z\bar{Z} = \omega\bar{\omega} = 1$

Again $z^6 = \omega^{10}$ (3)

and $z^6 \cdot \bar{\omega}^{12} = 1$; $z^6 = \frac{1}{\bar{\omega}^{12}} = \omega^{10}$ (from 3)

$\Rightarrow (\omega\bar{\omega})^{10}(\bar{\omega})^2 = 1 \Rightarrow (\bar{\omega})^2 = 1$

$\Rightarrow \bar{\omega} = 1$ or $-1 \Rightarrow \omega = 1$ or -1

if $\omega = 1$ then $z^3 + 1 = 0$ and $z^2 = 1 \Rightarrow z = -1$

if $\omega = -1$ then $z^3 - 1 = 0$ and $z^2 = 1 \Rightarrow z = 1$

Hence $z = 1$ and $\omega = -1$ or $z = -1$ and $\omega = 1$

22. **(B)**

Let $\alpha = a + ib$ and $\gamma = c + id$

where $a, b, c, d \in R$. We have to minimise

$\sqrt{a^2 + b^2} + \sqrt{c^2 + d^2}$

now $f(z) = (4 + i)z^2 + z(a + ib) + (c + id)$

$f(1) = 4 + i + a + ib + c + id$ is real

or $(4 + a + c) + i(1 + b + d)$ is real

Hence $b + d + 1 = 0$ (1)

$f(i) = -(4 + i) + i(a + ib) + (c + id)$ is real

$f(i) = -4 - b + c + i(a + d - 1)$ is real

$a + d = 1$ (2)

From (1) and (2) $a - b = 2$ (3)

Hence there is no restriction on 'c'.

Let $c = 0$.

Hence $|\alpha| + |\gamma| = \sqrt{a^2 + b^2} + \sqrt{d^2}$

$= \sqrt{4+2ab}+|d| \geq \sqrt{4+2ab} \geq \sqrt{2}$

With equality if $d = 0$; $a = 1$ and $b = -1$

23. **(D)**

$$(1+i\sqrt{3})^n = \left[2\left(\cos\frac{\pi}{3}+i\sin\frac{\pi}{3}\right)\right]^n$$

$$= 2^n\left(\cos\frac{n\pi}{3}+i\sin\frac{n\pi}{3}\right)$$

$f\left((1+i\sqrt{3})^n\right)$ = real part of $z = 2^n \cos\dfrac{n\pi}{3}$

$$\therefore \sum_{n=1}^{6a}\log_2\left|2^n\cos\frac{n\pi}{3}\right|$$

$$= \sum_{n=1}^{6a}\left(n+\log_2\left|\cos\frac{n\pi}{3}\right|\right)$$

$$= 6a(6a+1) + \underbrace{(-1-1+0-1-1+0)}_{a \text{ such term}}$$

$= 3a(6a + 1) - 4a = 18a^2 - a$

24. **(C)**

Using the cosine rule

$$|z_1+z_2| = \sqrt{|z_1|^2+|z_2|^2-2|z_1||z_2|\cos 120°}$$

$$= \sqrt{4+9+2\cdot 3} = \sqrt{19}$$

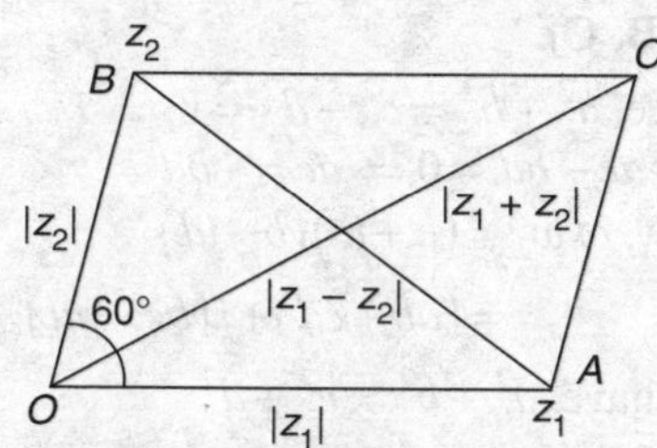

and $|z_1 - z_2| = \sqrt{|z_1|^2+|z_2|^2-2|z_1||z_1|\cos 60°}$

$$= \sqrt{4+9-6} = \sqrt{7}$$

$$\therefore \left|\frac{z_1+z_2}{z_1+z_2}\right| = \sqrt{\frac{19}{7}} = \sqrt{\frac{133}{7}} \Rightarrow N = 133$$

25. **(D)**

Given $|z - |z + 1||^2 = |z + |z - 1||^2$

$\therefore (z - |z + 1|)(\bar{z}\ |z + 1|)$

$= (z + |z - 1|)(\bar{z} + |z - 1|)$

$z\bar{z} - z|z + 1| - \bar{z}|z + 1| + |z + 1|^2$

$= z\bar{z} + z|z - 1| + \bar{z}|z - 1| + |z - 1|^2$

$|z + 1|^2 - |z - 1|^2 = (z + \bar{z})[|z - 1| + |z + 1|]$

$(z + 1)(\bar{z} + 1) - (z - 1)(\bar{z} - 1)$

$= (z + \bar{z})[|z - 1| + z + 1]$

$(z\bar{z} + z + \bar{z} + 1) - (z\bar{z} - z - \bar{z} + 1)$

$= (z + \bar{z})[|z - 1| + |z + 1|]$

$2(z + \bar{z}) = (z + \bar{z})[|z + 1| + |z - 1|]$

$(z + \bar{z})[|z + 1| + |z - 1| - 2] = 0$

A (−1, 0) — z — B (1, 0)

$\Rightarrow$ either $z + \bar{z} = 0 \Rightarrow z$ is purely imaginary

$\Rightarrow z$ lies on y axis $\Rightarrow x = 0$

or $|z + 1| + |z - 1| = 2$

$\Rightarrow z$ lie on the line segment joining (−1, 0) and (1, 0).

Multiple Answer Type

1. **(A, B, C)**

$\sqrt{20i-21} = \sqrt{-25+4+20i} = \pm(2+5i)$

and $\sqrt{21+20i} = \sqrt{25-4+20i} = \pm(5+2i)$

So, $z = 2 + 5i + 5 + 2i = 7 + 7i$

$\Rightarrow$ Principal value of $\arg z = \dfrac{\pi}{4}$

$z = 2+5i-5-2i = -3+3i$

$\Rightarrow$ Principal value of $\arg z = \dfrac{3\pi}{4}$

$z = -2-5i-5-2i = -7-7i$

$\Rightarrow$ Principal value of $\arg z = -\dfrac{3\pi}{4}$

$z = -2-5i+5+2i = 3-3i$

$\Rightarrow$ Principal value of $\arg z = -\dfrac{\pi}{4}$

Hence, (A), (B), (C)

2. **(B, C, D)**

Let $|z_1| = |z_2| = r$

$\Rightarrow z_1 = r(\cos\theta + i\sin\theta)$

and $z_2 - r\left(\cos\left(\dfrac{\pi}{2}-\theta\right)+i\sin\left(\dfrac{\pi}{2}-\theta\right)\right)$

$\Rightarrow z_1z_2 = r^2 i$, which is purely imaginary

$z_1 + z_2 = r\left[(\cos\theta+\sin\theta)+i(\cos\theta+\sin\theta)\right]$

$\Rightarrow (z_1+z_2)^2 = r^2\left[2i(\cos\theta+\sin\theta)^2\right]$

which is purely imaginary

Also, $\arg\left(z_1^{-1}\right)+\arg\left(z_2^{-1}\right)=-\dfrac{\pi}{2}$

Hence (B), (C), (D)

3. **(B, C, D)**

Since, roots are $1, \alpha, \alpha^2, \ldots, \alpha^{n-1}$, where $\alpha = \cos\dfrac{2\pi}{n} + i\sin\dfrac{2\pi}{n}$.

The distance of the complex numbers represented by these roots from origin is 1, that is, all these points lie on a circle.

$\Rightarrow$ They are non-collinear

$\Rightarrow$ They form a regular polygon of unit circum-radius.

Hence (B), (C), (D)

4. **(A, C)**

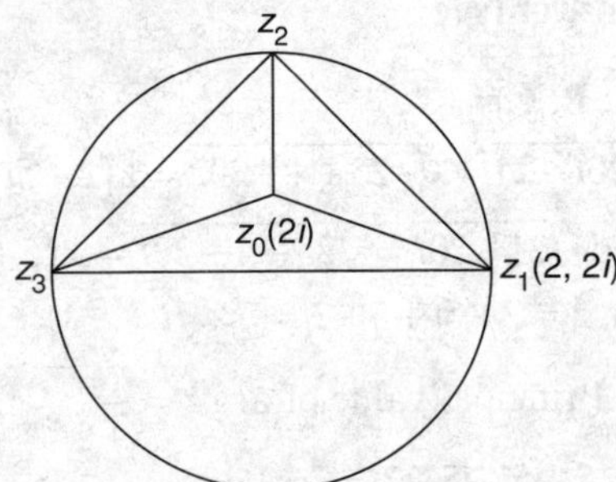

Clearly the inscribed triangle is equilateral

$\Rightarrow \dfrac{z_2 - z_0}{z_1 - z_0} = e^{\frac{i2\pi}{3}}, \dfrac{z_3 - z_0}{z_1 - z_0} = e^{-\frac{i2\pi}{3}}$

$\Rightarrow z_2 = -1 + i\left(2+\sqrt{3}\right)$ and $z_3 = -1 + i\left(2-\sqrt{3}\right)$

Hence (A), (C)

5. **(A, C, D)**

$z^2 + az + a^2 = 0$

$\Rightarrow z = a\omega, a\omega^2$

(where 'ω' is non-real root of cube unity)

$\Rightarrow$ locus of z is a pair of straight lines and

$\arg(z) = \arg(a) + \arg(\omega)$

or $\arg(a) + \arg\left(\omega^2\right)$

$\Rightarrow \arg(z) = \pm\dfrac{2\pi}{3}$

Also, $|z| = |a||\omega|$ or $|a|\left|\omega^2\right|$

$\Rightarrow |z| = |a|$ hence, (A), (C), (D)

6. **(A, C, D)**

$$\operatorname{Re}\left(\frac{\tan\alpha - i\left(\sin\frac{\alpha}{2}+\cos\frac{\alpha}{2}\right)}{1+2i\sin\frac{\alpha}{2}}\right) = 0$$

$$\Rightarrow \operatorname{Re}\frac{\left[\left(\tan\alpha - i\left(\sin\frac{\alpha}{2}+\cos\frac{\alpha}{2}\right)\right)\left(1-2i\sin\frac{\alpha}{2}\right)\right]}{1-4i^2\sin^2\frac{\alpha}{2}} = 0$$

$\Rightarrow \tan\alpha = 2\sin\dfrac{\alpha}{2}\left(\sin\dfrac{\alpha}{2}+\cos\dfrac{\alpha}{2}\right)$

$\Rightarrow \tan\alpha = 2\sin^2\dfrac{\alpha}{2} + \sin\alpha$

$\Rightarrow \dfrac{\sin\alpha}{\cos\alpha} = \sin\alpha + 1 - \cos\alpha$

$\Rightarrow \sin\alpha = \sin\alpha\cos\alpha + \cos\alpha - \cos^2\alpha$

$\Rightarrow \sin\alpha(1-\cos\alpha) = \cos\alpha(1-\cos\alpha)$

$\Rightarrow \sin\alpha = \cos\alpha, \cos\alpha = 1$

$\Rightarrow \alpha = n\pi + \dfrac{\pi}{4}, \alpha = 2n\pi$

Hence, (A), (C), (D)

7. **(A, B, C)**

Since, $a^2 + b^2 = c^2 + d^2 = 1$

and $ac + bd = 0 \Rightarrow ac = -bd$

Now, $\omega_1\bar{\omega}_2 = (a+ic)(b-id)$

$= (ab+cd) + i(bc-ad)$

We have, $a^2 + b^2 = c^2 + d^2$

$\Rightarrow a^2 - c^2 = d^2 - b^2$

$\Rightarrow a^2 - c^2 + 2iac = d^2 - b^2 - 2ibd$

$\Rightarrow (a+ic)^2 = (d-ib)^2$

$\Rightarrow a + ic = d - ib$ or $-d + ib$

$\Rightarrow a = d$ and $c = -b$ or $a = -d, c = b$

$\Rightarrow c^2 + d^2 = b^2 + d^2, a^2 + c^2 = a^2 + b^2$

$\Rightarrow a^2 + c^2 = 1, b^2 + d^2 = 1$

$\Rightarrow |\omega_1| = |\omega_2| = 1$ Also, $ab + cd = -cd + cd = 0$

$\Rightarrow \operatorname{Re}(\omega_1\bar{\omega}_2) = 0$

Hence, (A), (B), (C)

8. (A, B)

$|z_1+z_2|^2=|z_1-z_2|^2$

$\Rightarrow (z_1+z_2)(\bar{z}_1+\bar{z}_2)=(z_1-z_2)(\bar{z}_1-\bar{z}_2)$

Hence, (A), (B)

9. (A, B, C)

$(1+x)^n=C_0+C_1x+C_2x^2+C_3x^3+C_4x^4+\ldots$

Put $x=i$

$(1+i)^n=(C_0-C_2+C_4+\ldots)+$

$i(C_1-C_3+C_5+\ldots)$

$=2^{\frac{n}{2}}\left(\cos\frac{n\pi}{4}+i\sin\frac{n\pi}{4}\right)$

$=(C_0-C_2+C_4-\ldots)+i(C_1-C_3+C_5+\ldots)$

$\therefore C_0-C_2+C_4-\ldots=2^{\frac{n}{2}}\cos\frac{n\pi}{4}$

$C_1-C_3+C_5\ldots=2^{\frac{n}{2}}\cos\frac{n\pi}{4}$

Also, we know that

$C_0+C_2+C_4+\ldots=2^{n-1}$

$\therefore C_0+C_4+C_8+\ldots=2^{n-2}+2^{\frac{n-2}{2}}\cos\frac{n\pi}{4}$

10. (B, C)

Clearly, the inscribed rectangle is a square. Let the adjacent vertex be z. Then

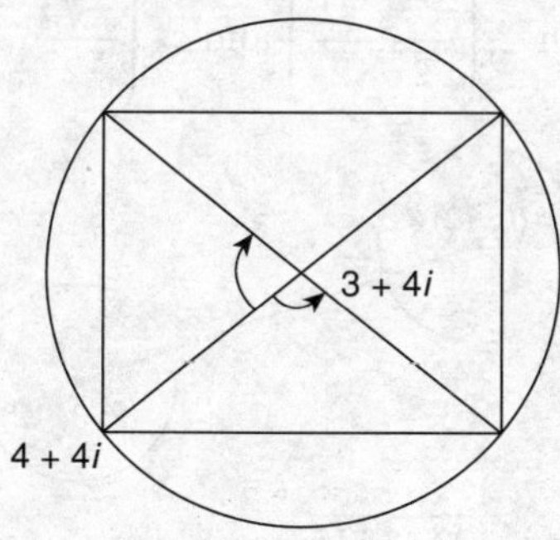

$\frac{(3+4i)-(z)}{(3+4i)-(4+4i)}=e^{\pm\frac{i\pi}{2}}$

$\Rightarrow 3+4i-z=\pm i(-1)$

$\Rightarrow z=3+4i\pm i=3+5i$ or $3+3i$

Hence (B), (C)

11. (A, B, D)

$\left|\frac{2z-i}{z+1}\right|=m$

$\Rightarrow \left|z-\frac{i}{2}\right|=\frac{m}{2}|z+1|$

Which shows that the given equation will represent a circle, if $\frac{m}{2}\neq 0, 1$, that is, $m\neq 0, 2$

Hence, (A), (B), (D)

12. (A, B, C, D)

$\sqrt{5-12i}=\sqrt{(3-2i)^2}=\pm(3-2i)$

$\sqrt{-5-12i}=\sqrt{(2-3i)^2}=\pm(2-3i)$

$\therefore z=-1-i,-5+5i,5-5i,1+i$

Therefore, the principal values of $\arg z$ are

$-\frac{3\pi}{4},\frac{3\pi}{4},-\frac{\pi}{4},\frac{\pi}{4}$

Hence (A), (B), (C), (D)

13. (A, B, D)

Let $z=\alpha$ be a real root, then

$\alpha^3+(3+2i)\alpha+(-1+ia)=0$

$\Rightarrow (\alpha^3+3\alpha-1)+i(2\alpha+a)=0$

$\Rightarrow \alpha^3+3\alpha-1=0$ and $\alpha=\frac{-a}{2}$

$\Rightarrow -\frac{a^3}{8}-\frac{3a}{2}-1=0$

$\Rightarrow a^3+12a+8=0$

Let $f(a)=a^3+12a+8$

$\Rightarrow f'(a)=3a^2+12>0$

$\Rightarrow f(a)$ cuts the x-axis only once.

Now, $f(-1)<0, f(0)>0$

$f(-2)<0, f(1)>0, f(3)>0$

$\Rightarrow$ (A), (B), (D)

14. (A, B)

$x^n-1=(x-1)(x-z_1)(x-z_2)\ldots(x-z_{n-1})$

$\Rightarrow \log(x^n-1)=\log(x-1)+\log(x-z_1)+$

$\log(x-z_2)+\ldots+\log(x-z_{n-1})$

Differentiating both sides

$\frac{nx^{n-1}}{x^n-1}=\frac{1}{x-1}+\frac{1}{x-z_1}+\frac{1}{x-z_2}+\ldots+\frac{1}{x-z_{n-1}}$

Put $x=3$

$\frac{1}{3-z_1}+\frac{1}{3-z_2}+\ldots+\frac{1}{3-z_{n-1}}=\frac{n3^{n-1}}{3^n-1}-\frac{1}{2}$

$$= \frac{\sum_{r=1}^{n-1} r3^{r-1}}{\sum_{r-1}^{n} 3^{r-1}}$$

Hence (A) (B)

15. (A, C)

Since, $\frac{\alpha_1}{\alpha_2}$ is purely imaginary $\Rightarrow \left|\arg\frac{z_1}{z_2}\right| = \frac{\pi}{2}$

Also, $\frac{\alpha_1}{\alpha_2} + \frac{\bar{\alpha}_1}{\alpha_2} = 0 \Rightarrow \alpha_1\bar{\alpha}_2 + \bar{\alpha}_1\alpha_2 = 0$

Hence (A), (C)

16. (A, B, D)

Since, $|2z+5|^2 = |6z-9|^2$

$\Rightarrow (2z+5)(2\bar{z}+5) = (6z-9)(6\bar{z}-9)$

$\Rightarrow 4|z|^2 + 10z + 10\bar{z} + 25 = 36|z|^2 - 54z - 54\bar{z} + 81$

$\Rightarrow 32|z|^2 = 64(z+\bar{z}) + 56$

$\Rightarrow 32|z|^2 = 64(2\,\text{Re}(z)) - 56$

$\Rightarrow |z|^2 = 4\,\text{Re}(z) - \frac{7}{4}$

$\therefore a = 4, b = -\frac{7}{4}$

Hence (A), (B), (D)

17. (A, D)

$\left|z + \frac{1}{z}\right| \le |z| + \frac{1}{|z|} \le 4 + \frac{1}{4} = \frac{17}{4}$

$\left|z + \frac{1}{2}\right| \ge \left||z| - \frac{1}{|z|}\right| \ge 2 - \frac{1}{2} = \frac{3}{2}$

Hence (A), (D)

18. (A, B, C, D)

Given determinant is equivalent to

$(a + b + c)(a^2 + b^2 + c^2 - ab - bc - ca) = 0$

$\Rightarrow$ either Δ is equilateral or centroid of triangle is zero or both.

19. (A, B, C, D)

Use $z_1^2 + z_2^2 + z_3^2 = z_1z_2 + z_2z_3 + z_3z_1$

$\Rightarrow z^2 = \frac{3}{2}i$

$\Rightarrow z = \frac{\sqrt{3}}{2}(1+i)$ and $z = -\frac{\sqrt{3}}{2}(1+i)$

20. (A, B, C, D)

$C \equiv (-9, 3\sqrt{3})$

$AB = 9$

$OB = 9$

$OA = 9$

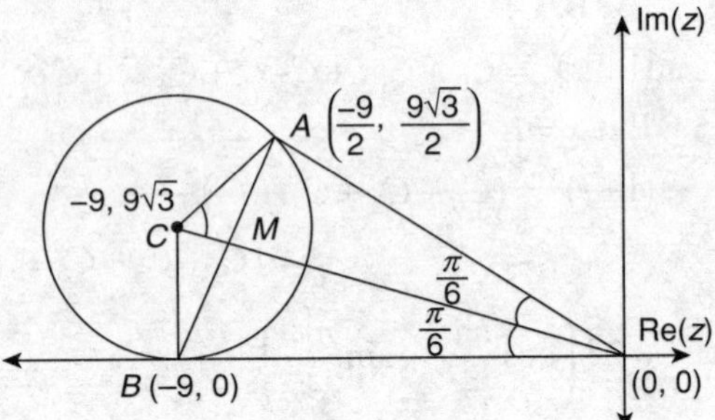

$\therefore \Delta AOB$ is equilateral

$\therefore$ area $= \frac{\sqrt{3}}{4}(a)^2 = \frac{81\sqrt{3}}{4}$

$\because$ greatest value of $|z|$ = distance of OC + radius.

$= 6\sqrt{3} + 3\sqrt{3} = 9\sqrt{3}$

21. (A,C)

$\because |z_3 - z_1| + |z_3 - z_2|$ is minimum

$\Rightarrow z_3$ lies on line segment z_1z_2

and $|z_1 - z_2| = 2\sqrt{3}$

then $z_1\left(\frac{\sqrt{3}}{2}, \frac{21}{2}\right)$ and $z_2\left(\frac{3\sqrt{3}}{2}, \frac{15}{2}\right)$

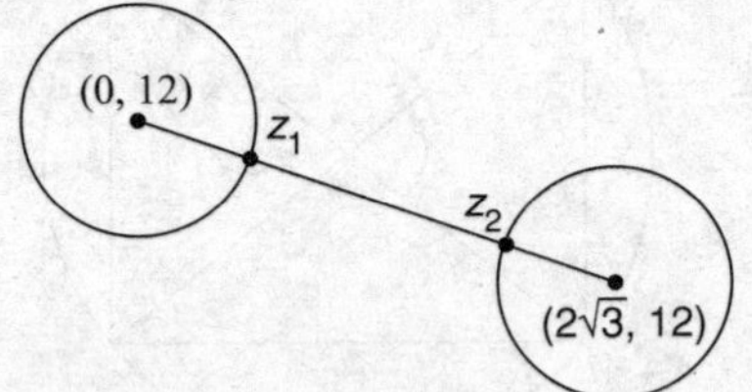

$|z_2| = \sqrt{\frac{27}{4} + \frac{225}{4}} = \sqrt{63}$ and

$|z_1| = \sqrt{\frac{441+3}{4}} = \sqrt{111} \Rightarrow |z_3|_{max} = \sqrt{111}$

Minimum $\arg(z_3) = \arg(z_2) = \tan^{-1}\left(\frac{5}{\sqrt{3}}\right)$

Maximum $\arg(z_3) = \arg(z_1) = \tan^{-1}\left(\frac{21}{\sqrt{3}}\right)$

Sum of arguments $= \pi + \tan^{-1}\left(\frac{26}{(1-35)\sqrt{3}}\right)$

$= \pi - \tan^{-1}\left(\frac{13}{17\sqrt{3}}\right) = \frac{\pi}{2} + \cot^{-1}\left(\frac{13}{17\sqrt{3}}\right)$

22. (C, D)

$w = \frac{\sqrt{3}+i}{2} = \cos\frac{\pi}{6} + i\sin\frac{\pi}{6}$

$P = w^n = \cos\frac{n\pi}{6} + i\sin\frac{n\pi}{6}, n = 0, 1, 2, \ldots, 11.$

$H_1 = \left\{z \in C : \text{Re}(z) > \frac{1}{2}\right\}$

$z_1 \in P \cap H_1$

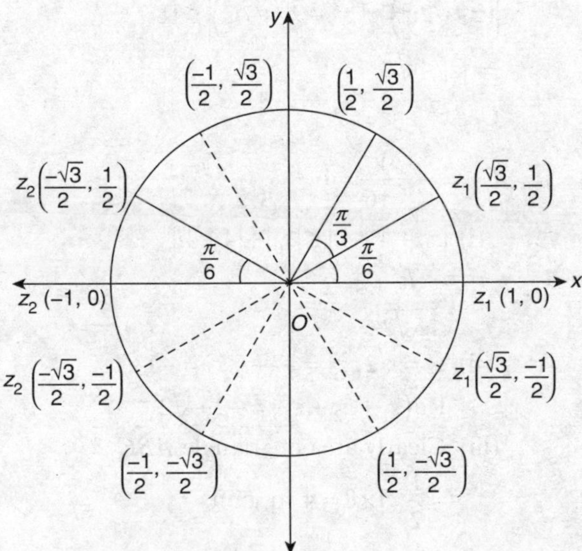

$\therefore$ possible value of z_1 are

$\left\{\left(\frac{\sqrt{3}}{2}, \frac{1}{2}\right), (0, 1), \left(\frac{\sqrt{3}}{2}, \frac{-1}{2}\right)\right\}$

$H_2 = \left\{z \in C : \text{Re}(z) < \frac{-1}{2}\right\}$

$z_2 \in p \cap H_2$

$\therefore$ possible values of z_2 are

$\left\{\left(\frac{-\sqrt{3}}{2}, \frac{1}{2}\right), (-1, 0), \left(\frac{-\sqrt{3}}{2}, \frac{-1}{2}\right)\right\}$

Now, possible angle z_1Oz_2 is $= \frac{2\pi}{3}, \frac{5\pi}{6}, \pi$

23. (A, C, D)

$x + iy = \frac{1}{a+ibt}$

$\Rightarrow x + iy = \frac{a-ibt}{a^2+b^2t^2}$

Let $a \neq 0$ and $b \neq 0$

$\therefore x = \frac{a}{a^2+b^2t^2}$ (1)

$y = \frac{-bt}{a^2+b^2t^2}$ (2)

$\Rightarrow \frac{y}{x} = \frac{-bt}{a} \Rightarrow t = \frac{-ay}{bx}$

So, put in (1)

$\Rightarrow x\left\{a^2 + b^2, \frac{a^2y^2}{b^2x^2}\right\} = a$

$\Rightarrow a^2(x^2 + y^2) = ax$

$\therefore x^2 + y^2 - \frac{1}{a}x = 0$

Gives $\left(x - \frac{1}{2a}\right)^2 + y^2 = \frac{1}{4a^2}$

$\Rightarrow$ Option (A) is correct

For $a \neq 0, b = 0$

$x + iy = \frac{1}{a}$

$\Rightarrow x = \frac{1}{a}, y = 0 \Rightarrow z$ lies on x-axis

$\Rightarrow$ option (C) is correct

For $a = 0, b \neq 0$

$\Rightarrow x + iy = \frac{1}{ibt}$

$y = \frac{-1}{bt}, x = 0 \Rightarrow z$ lies on the y-axis.

$\Rightarrow$ option (D) is correct

24. (B, D)

We have $\left|z + \frac{1}{z}\right| = 2 \Rightarrow \left|z + \frac{1}{z}\right|^2 = 4$

$\Rightarrow \left(z + \frac{1}{z}\right)\left(\bar{z} + \frac{1}{\bar{z}}\right) = 4$

$\Rightarrow z\bar{z} + \frac{\bar{z}}{z} + \frac{z}{\bar{z}} + \frac{1}{z\bar{z}} = 4$

$\Rightarrow z\bar{z}+\frac{\bar{z}^2+z^2}{z\bar{z}}+\frac{1}{z\bar{z}}=4$

$\Rightarrow (z\bar{z})^2-4(z\bar{z})+z^2+\bar{z}^2+1=0$

$\Rightarrow (z\bar{z})^2-2z\bar{z}+1-2z\bar{z}+z^2+\bar{z}^2=0$

$\Rightarrow (z\bar{z}-1)^2+(z-\bar{z})^2=0$

$\Rightarrow (z\bar{z}-1)^2-i^2(z-\bar{z})^2=0$

$\Rightarrow (z\bar{z}-1+i(z-\bar{z})(z\bar{z}-1-i(z-\bar{z})=0$

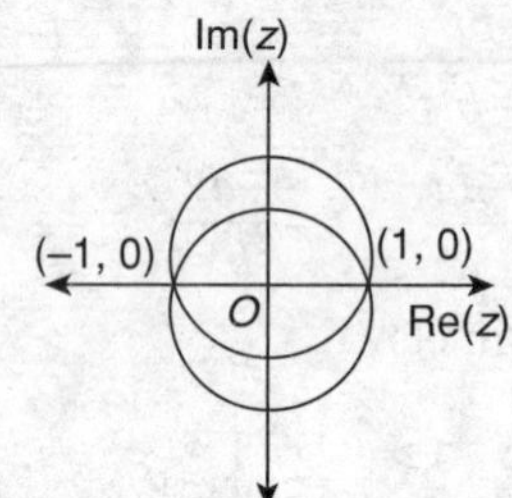

Each of the above factors represents a circle with centre and radius (in ordered pair) as $(i, \sqrt{i\bar{i}+1})$ and $(-i, \sqrt{i\bar{i}+1})$, that is, ((0, 1), $\sqrt{2}$) and ((0, –1), $\sqrt{2}$) respectively. The union of the circle with equations $x^2+y^2-2y-1=0$ and $x^2+y^2+2y-1=0$

Note: These circles cut each other orthogonally.

25. **(A, B, C, D)**

$|2z_1-3z_2-4z_3|$

$=|2z_1-3(z_2-2)-4(z_3-5)-26|$

$\le 2|z_1|+3|z_2-2|+4|z_3-5|+|26|$

$\le 2+9+24+26\le 61$

Comprehension Based Questions

1. **(C)**

2. **(D)**

3. **(B)**

Since, $(z_1z_2)(z_2z_3)(z_3z_1)$

$=(-80-320i)(60)(-96+24i)$

$\Rightarrow (z_1z_2z_3)^2=-80(1+4i)(60)(24)(-4+i)$

Now,

$z_1^2=\frac{(z_1z_2z_3)^2}{(z_2z_3)^2}=\frac{-80(1+4i)\times 24(-4+i)}{60}$

$=32(8+15i)$

Similarly, $z_2^2=200i$, $z_3^2=-18i$

$(z_1+z_2+z_3)^2$

$=z_1^2+z_2^2+z_3^2+2(z_1z_2+z_2z_3+z_3z_1)$

$=24+70i$

$\therefore |z_1+z_2+z_3|^2=\sqrt{(24)^2+(70)^2}=74$

$\text{Arg}(z_2)=\frac{\pi}{4}$

and possible value of z_1 is $4(5+3i)$

4. **(A)**

5. **(C)**

6. **(A)**

Since

$\left|(z^2-2iz)-(z^2-9z+7iz)\right|\le 18$

$\Rightarrow |z|\le\sqrt{2}$

$C_1: x^2+y^2=2$

$C_2: \frac{x^2}{2}-\frac{y^2}{16}=1$

(i) $|z_0+4+4i|\le|z_0|+|4+4i|$

$\le\sqrt{2}+4\sqrt{2}$

$\le 5\sqrt{2}$

(ii) $z_1=\sqrt{2}$, $z_2=-i+i$, $z_3=-1-i$

$\text{Re}(z_1+z_2+z_3)=-2+\sqrt{2}$

(iii) Clearly area of triangle ABC

$=\frac{1}{2}\times 1\times 8=4$ sq. units

7. **(C)**

8. **(B)**

9. **(D)**

(i) $w=\frac{4z-5i}{2z+1}$ $(w\ne z)$

if w is real then $w=\bar{W}$

$\bar{W}=\underbrace{\frac{4\bar{z}+5i}{2\bar{z}+1}}_{\bar{w}}=\underbrace{\frac{4z-5i}{2z+17}}_{w}$

$\overline{\quad 2 \quad}$

$(4\bar{z}+5i)(2z+1)=(4z-5i)(2\bar{z}+1)$

$8z\bar{z}+4\bar{z}+10zi+5i$

$=8z\bar{z}+4\bar{z}-10\bar{z}i-5i$

$$4(z - \bar{z}) - 10i(z + \bar{z}) - 10i = 0$$
$$8iy - 20ix - 10i = 0$$
$$4y - 10x - 5 = 0 \Rightarrow 10x - 4y + 5 = 0$$

10. (B)

11. (A)

12. (B)

For A, $|z + 1| \le 2 + \text{Re}(z)$

$\Rightarrow (x + 1)^2 + y^2 \le 4 + 4x + x^2$

$\Rightarrow y^2 \le 3 + 2x$

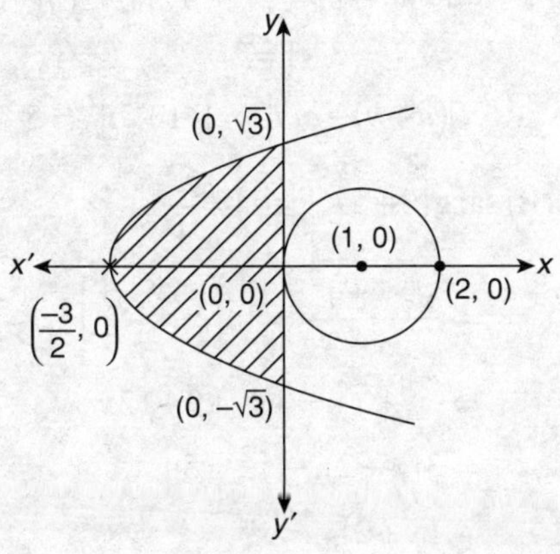

$\Rightarrow y^2 \le 2\left(x + \dfrac{3}{2}\right)$ (1)

For B, $|z - 1| \ge 1$

$\Rightarrow (x - 1)^2 + y^2 \ge 1$ (2)

For C, $|z - 1|^2 \ge |z + 1|^2$

$\Rightarrow (z - 1)(\bar{z} - 1) \ge (z + 1)(\bar{z} + 1)$

$\Rightarrow (z\bar{z} - \bar{z} - z + 1) \ge (z\bar{z} + \bar{z} + z + 1)$

$\Rightarrow z + \bar{z} \le 0$

That is, $x \le 0$ (3)

(i) $(-1, 1)$, $(-1, 1)$, $(-1, -1)$, $(0, 0)$, $(0, 1)$, $(0, -1)$ but $z = -1$ is not in the domain in set C.

$\therefore$ Total number of point(s) having integral coordinates in the region $A \cap B \cap C$ is 5.

(ii) Required area $= 2\displaystyle\int_{\frac{-3}{2}}^{0} \sqrt{2\left(x + \frac{3}{2}\right)}\,dx$

$= 2\sqrt{3}$ (square units)

(iii) Clearly $z = \dfrac{-3}{2} + i0$ is the complex number in the region $A \cap B \cap C$ and having maximum amplitude.

$\therefore \text{Re}(z) = \dfrac{-3}{2}$

13. (C)

Clearly, $A(z_1)$ is the point of intersection of $\arg(z - 2 + i) = \dfrac{3\pi}{4}$ and $\arg(z + i\sqrt{3}) = \dfrac{\pi}{3}$

$\Rightarrow z_1 = 1$

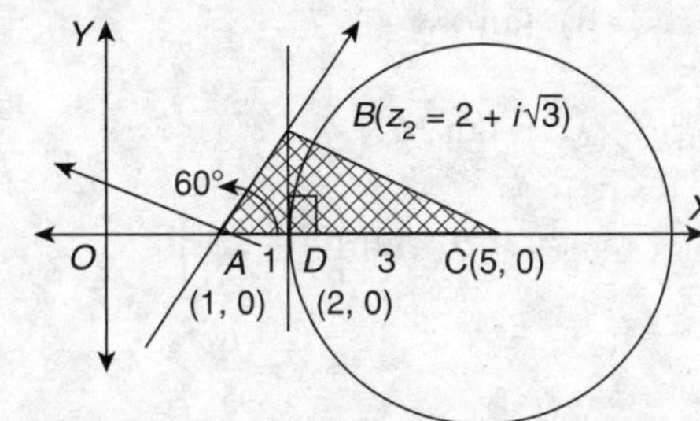

Also, $B(z_2)$ is the point on $\arg(z + i\sqrt{3}) = \dfrac{\pi}{3}$

such that $|z_2 - 5|$ = minimum,

so $z_2 = 2 + i\sqrt{3}$.

Ans $C(z_3)$ be the centre of the circle

$|z - 5| = 3 \Rightarrow z_3 = 5$.

Hence, area $(\Delta ABC) = \dfrac{1}{2}(AC)(AD)$

$= \dfrac{1}{2}(4)(\sqrt{3}) = 2(\sqrt{3})$ (square units)

14. (B)

Clearly, required equation of straight line is

$z = \lambda i(z_2 - z_1) = \lambda i(2 + i\sqrt{3} - 1)$

$= \lambda i(1 + i\sqrt{3}) = \lambda(-\sqrt{3} + i)$

(where λ is a real parameter.)

15. (C)

We have $|z - 1| = 1$

$\Rightarrow z = 1 + e^{i\theta}$, $\theta \in [0, 2\pi)$

$\Rightarrow z = 1 + e^{i\theta} = 1 + \cos\theta + i\sin\theta$

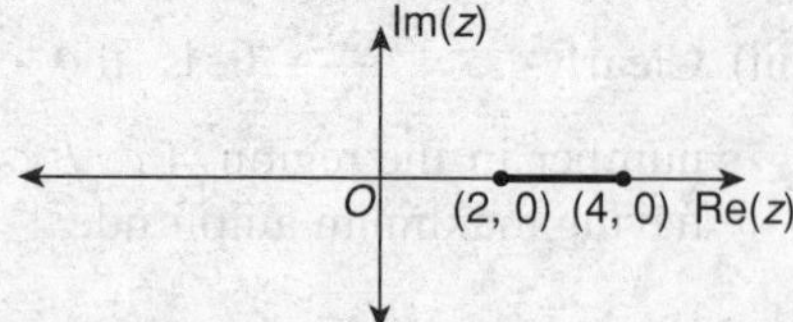

$\therefore \omega = \text{Re}(z+2) = 3 + \cos\theta =$ purely real

Clearly, locus of ω is the line segment joining (2, 0) and (4, 0).

Match Matrix

1. **(A – R; B – S; C – Q; D – P)**

(B) $\tan\phi = \frac{1}{2}$

$\therefore \tan 2\phi = \frac{4}{3}$

$\tan\theta = \tan\left(\frac{\pi}{2} - 2\phi\right)$

$= \frac{3}{4} \Rightarrow y = \frac{6}{5}$

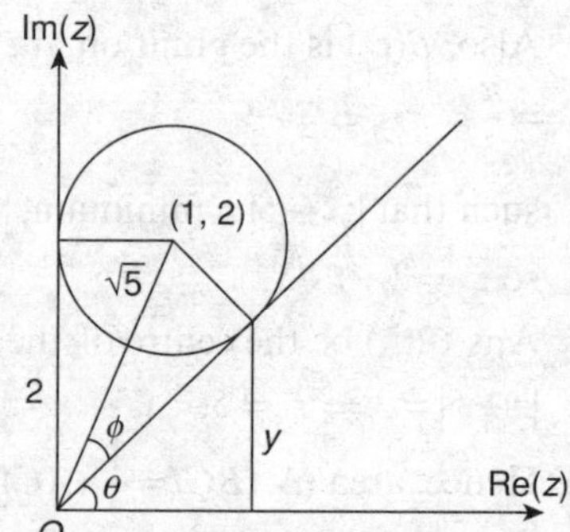

(C) Put $z = x + iy$

$-8(x+iy)^2 - 8(x-iy)^2 + 20(x^2+y^2) = 36$

$\Rightarrow -8\left(2\left(x^2-y^2\right)\right) + 20\left(x^2+y^2\right) = 36$

$\Rightarrow 4x^2 + 36y^2 = 36$

$\Rightarrow \frac{x^2}{9} + \frac{y^2}{1} = 1$

$\Rightarrow a = 3, b = 1$

$\therefore$ Area $= \pi ab = 3\pi$

(D) $|z - 2i| = 2$ is a circle with centre $2i$

$z(1-i) - \bar{z}(1+i) = 4i$ is a line

This line passes through centre of circle

$\therefore$ Total number of points of intersection = 2

2. **(A – S; B – R; C – P; D – Q)**

(A) $P = \cos\frac{2\pi}{7} + i\sin\frac{2\pi}{7}$, which is one of the 7th roots of unity

Therefore, seventh roots of unity are $1, p, p^2, p^3, p^4, p^5$ and p^6

Also, $p^7 = 1$

Now,

$a + b = p + p^2 + p^3 + p^4 + p^5 + p^6 = -1$

and $ab = \left(p + p^2 + p^4\right)\left(p^3 + p^5 + p^6\right)$

$= 3 - 1 = 2$

$\therefore \left|(a+b) + abi\right|^2 = |-1+2i|^2 = 5$

(B) $\arg\left(3 + 2x + |z|i\right) = \frac{\pi}{4}$

$\Rightarrow \frac{\sqrt{x^2+y^2}}{2x+3} = 1 \quad \left[x > -\frac{3}{2}\right]$

$\Rightarrow x^2 + y^2 = 4x^2 + 9 + 12x$

$\Rightarrow 3(x+2)^2 - y^2 = 3$

$\Rightarrow \frac{(x+2)^2}{1} - \frac{y^2}{3} = 1$

$\therefore$ Length of latus rectum $= \frac{2b^2}{a} = 6$

(C) Let αi be the root where $\alpha \neq 0$ and $\alpha \in R$

$\therefore \alpha^4 - a_1\alpha^3 i - a_2\alpha^2 + a_3\alpha i + a_4 = 0$

$\Rightarrow \alpha^4 - a_2\alpha^2 + a_4 = 0 \quad (1)$

and $a_1\alpha^3 - a_3\alpha = 0 \quad (2)$

$\Rightarrow a_1\alpha^2 - a_3 = 0$

$\Rightarrow \alpha^2 = \frac{a_3}{a_1}$

Putting in (1)

$\frac{a_3^2}{a_1^2} - a_2\frac{a_3}{a_1} + a_4 = 0$

$\Rightarrow a_3^2 + a_4 a_1^2 = a_1 a_2 a_3$

$\Rightarrow \frac{a_3}{a_1 a_2} + \frac{a_1 a_4}{a_2 a_3} = 1$

(D) Since, $\frac{3}{|z_2 - z_3|} = \frac{4}{|z_3 - z_1|} = \frac{5}{|z_1 - z_2|} = K$

$\Rightarrow \frac{9}{|z_2 - z_3|^2} = \frac{16}{|z_3 - z_1|^2}$

$$= \frac{25}{|z_1 - z_2|^2} = K^2$$

$$\frac{9}{z_2 - z_3} = K^2\left(\bar{z}_2 - \bar{z}_3\right)$$

Similarly, $\dfrac{16}{z_3 - z_1} = K^2\left(\bar{z}_3 - \bar{z}_1\right)$

and, $\dfrac{25}{z_1 - z_2} = K^2\left(\bar{z}_1 - \bar{z}_2\right)$

$$\Rightarrow \frac{9}{z_2 - z_3} = \frac{16}{z_3 - z_1} + \frac{25}{z_1 - z_2} = 0$$

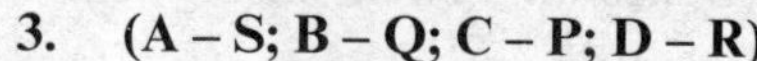

3. **(A – S; B – Q; C – P; D – R)**

(A) $|z-(3+4i)| = 2$ represents a circle whose centre is (3, 4) and radius is equal to 2.

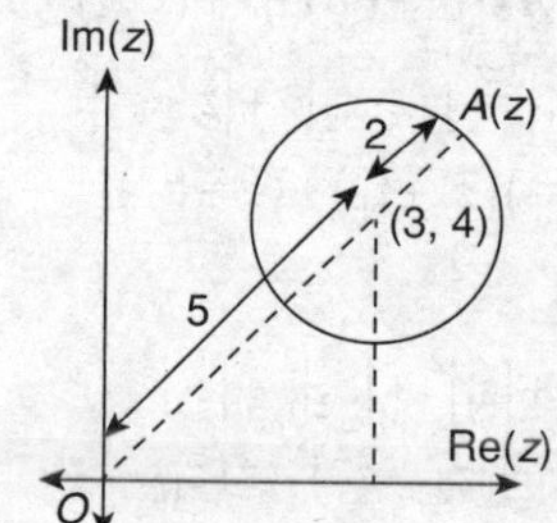

From the figure it is clear that maximum value of $|z| = 7$.

(B) $|z - 12 - 6i| = |(z - i)| + (12 - 5i) \le |z - i| + |12 - 5i| < 1 + 13 = 14$

(C) $|z_1 + z_2|^2 + |z_1 - z_2|^2 = 2(|z_1|^2 + |z_2|^2)$
$= 2(1 + 4) = 10$

(D) If $z = 1 + i$, then $(z - 1)^4 = i^4$
$\Rightarrow z^4 - 4z^3 + 6z^2 - 4z + 1 = 1$
$\Rightarrow (z^4 - 4z^3 + 7z^2 - 6z + 3) - z^2 + 2z - 2 = 1$
$\Rightarrow z^4 - 4z^3 + 7z^2 - 6z + 3 = z^2 - 2z + 3$
$= (z - 1)^2 + 2 = i^2 + 2 = 1$
$\therefore 4(z^4 - 4z^3 + 7z^2 - 6z + 3) = 4.$

4. **(A – Q, R; B – P; C – P, S, T; D – Q, R, S, T)**

(A) $|z - i|z|| = |z + i|z||$

$\Rightarrow (z - i|z|)(\bar{z} + i\,|\,z\,|) = (z + i\,|\,z\,|)(\bar{z} - i\,|\,z)$

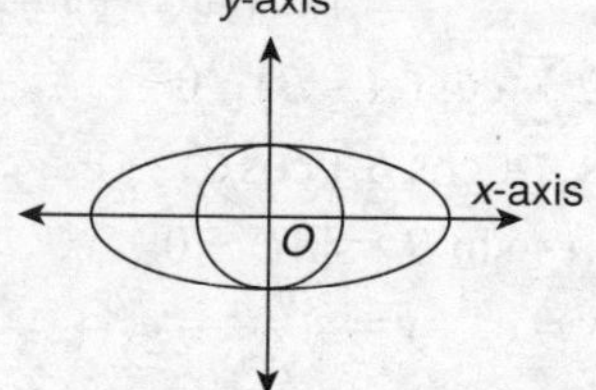

$\Rightarrow 2i(\bar{z}\,|\,z\,| - z\,|\,z\,|) = 0$

$\therefore \text{Im}(z) = 0$

(B) $|z + 4| + |z - 4| = 10$
$\Rightarrow 2ae = 8 \Rightarrow 2a = 10$

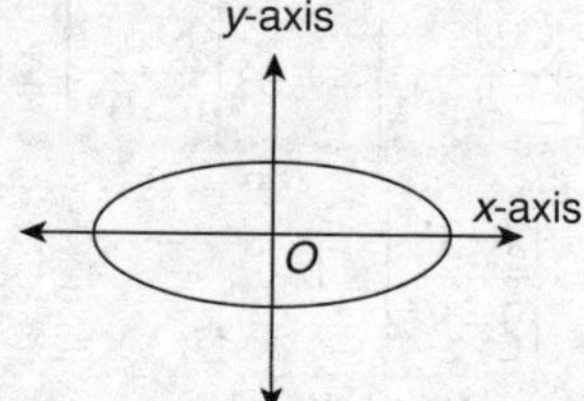

$$e = \frac{4}{5};\ a = 5;\ b^2 = 25\left(1 - \frac{16}{25}\right)$$

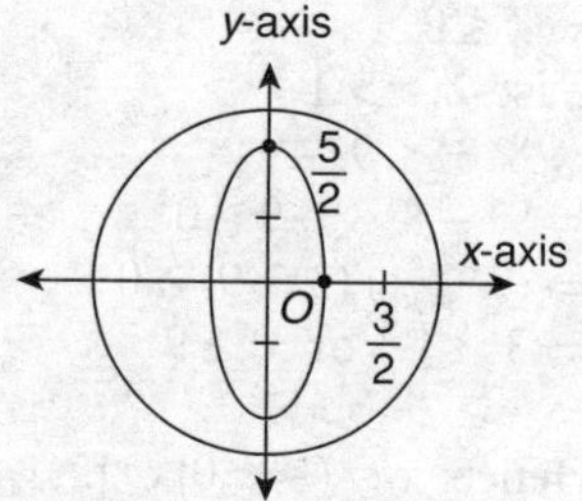

$b^2 = 9$

(C) $z = \omega - \dfrac{1}{\omega} = \omega - \dfrac{\bar{\omega}}{\omega\bar{\omega}} = \omega - \dfrac{\bar{\omega}}{4}$

$$z = x + iy - \frac{x - iy}{4};\ a + ib = \frac{3x}{4} + i\frac{5y}{4}$$

$$\Rightarrow \left(\frac{4a}{3}\right)^3 + \left(\frac{4b}{5}\right)^2 = 4;\ \frac{x^2}{9} + \frac{y^2}{25} = \frac{1}{4}$$

$$e = \sqrt{1 - \frac{9}{25}} = \frac{4}{5}$$

(D) $z = \omega + \dfrac{\omega\bar{\omega}}{\omega} = 2\,\text{Re}(\omega)$

$\therefore \text{Im}(z) = 0$

5. **(A – S; B – T; C – R; D – R)**

(A) $|z| = 1 \Rightarrow z \cdot \bar{z} = 1$, Let $z = x + iy$

$$\frac{2iz}{1 - z^2} = \frac{2iz\bar{z}}{\bar{z} - z\bar{z}^2} = \frac{2i}{\bar{z} - z}$$

$$= \frac{2i}{(x - iy) - (x + iy)} = -\frac{1}{y}$$

$$\text{Re}\left(\frac{2iz}{1 - z^2}\right) = -\frac{1}{y}$$

But $-1 \le y < 0$ or $0 < y \le 1$

$\Rightarrow -\infty < -\dfrac{1}{y} \le -1$ or $1 \le \dfrac{-1}{y} < \infty$

$\Rightarrow \dfrac{-1}{y} \in (-\infty, -1] \cup [1, \infty)$

(B) $\left|\dfrac{8\cdot 3^{x-2}}{1-3^{2(x-1)}}\right| \le 1 \Rightarrow \left|\dfrac{8\cdot 3^{x}}{9-9^{x}}\right| \le 1$

$\Rightarrow \left|\dfrac{8\cdot 3^{x}}{9-9^{x}}\right| \le 1$

Case-I: $x < 1$:

$\Rightarrow 8\cdot 3^x \le 9 - 9^x \Rightarrow 9^x + 8\cdot 3^x - 9 \le 0$

$\Rightarrow (3^x - 1)(3^x + 9) \le 0$

$\Rightarrow -9 \le 3^x \le 1$

$\Rightarrow x \le 0$ (1)

Case-2: $x > 1$

$\Rightarrow 8\cdot 3^x \le 9^x - 9$

$\Rightarrow 9^x - 8\cdot 3^x - 9 \ge 0$

$\Rightarrow (3^x + 1)(3^x - 9) \ge 0$

$\Rightarrow 3^x \le -1$ or $3^x \ge 9$

$\Rightarrow x \ge 2$ (2)

Hence, $x \in (-\infty, 0] \cup [2, \infty)$

(C) $f(\theta) = 2\sec^2\theta,\ 0 \le \theta < \dfrac{\pi}{2}$

(expanding along C_1)

$\Rightarrow f(\theta) \in [2, \infty)$

(D) $f(x) = x^{3/2}(3x - 10),\ x \ge 0$

$\therefore f'(x) = \dfrac{3}{2}\sqrt{x}\,(5x - 10)$

$f'(x) \ge 0 \Rightarrow x \ge 2$

$\Rightarrow x \in [2, \infty)$

Subjective Answer Type: Hard Nut

1. Since,

$a^3 + b^3 = (a+b)(a^2 - ab + b^2)$

$= (a+b)\{(a^2 + (\omega+\omega^2)ab + \omega^3 b^2)\}$

$= (a+b)(a+\omega b)(a+\omega^2 b)$

Using this we get the required result

2. Let $z_1 = \dfrac{2}{\sqrt{3}} e^{\frac{i\pi}{2}}$

$z_2 = \dfrac{2}{\sqrt{3}} e^{-\frac{\pi}{6}}$

$z_3 = \dfrac{2}{\sqrt{3}} e^{-i\frac{5\pi}{6}}$

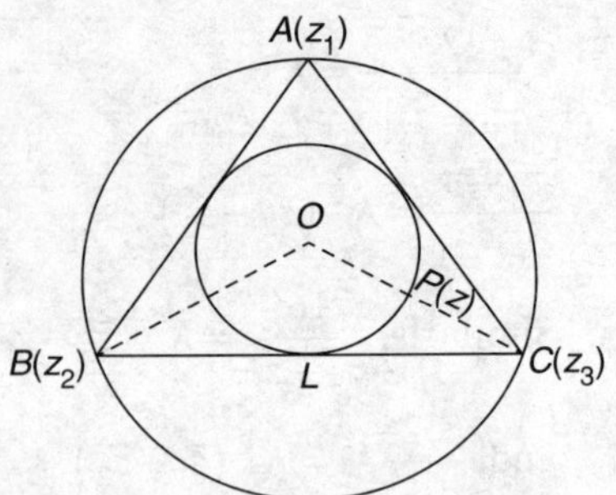

Clearly $|z_1| = |z_2| = |z_3| = \dfrac{2}{\sqrt{3}}$

$\Rightarrow z_1, z_2, z_3$ lies on a circle having centre at origin and radius $\dfrac{2}{\sqrt{3}}$

Now, $(AP)^2 + (BP)^2 + (CP)^2$

$= |z - z_1|^2 + |z - z_2|^2 + |z - z_3|^2$

$= (z - z_1)(\bar{z} - \bar{z}_1) + (z - z_2)(\bar{z} - \bar{z}_2) + (z - z_3)(\bar{z} - \bar{z}_3)$

$= 3|z|^2 + |z_1|^2 + |z_2|^2 + |z_3|^2$

$= -\bar{z}(z_1 + z_2 + z_3) - z(\bar{z}_1 + \bar{z}_2 + \bar{z}_3)$

$= 3\left(\dfrac{1}{3}\right) + 3\left(\dfrac{4}{3}\right) - 0 - 0$ [As $z_1 + z_2 + z_3 = 0$]

$= 5$

3. We know that if z_1, z_2, z_3 are the vertices of an equilateral triangle, then

$z_1^2 + z_2^2 + z_3^2 = z_1 z_2 + z_2 z_3 + z_3 z_1$ (1)

Since, for an equilateral triangle Circumcentre and centroid coincide

$\Rightarrow z_0 = \dfrac{z_1 + z_2 + z_3}{3}$

$\Rightarrow 9z_0^2 = z_1^2 + z_2^2 + z_3^2 + 2(z_1 z_2 + z_2 z_3 + z_3 z_1)$

$\Rightarrow 9z_0^2 = 3(z_1^2 + z_2^2 + z_3^2)$

$\Rightarrow z_1^2 + z_2^2 + z_3^2 = 3z_0^2$

4. Since,

$(\cos A + \cos B + \cos C)^2 + (\sin A + \sin B + \sin C)^2$

$= 3 + 2\sum \cos(A - B) = 0$

$\Rightarrow \cos A + \cos B + \cos C$

$= \sin A + \sin B + \sin C = 0$

Let, $x = e^{iA}, y = e^{iB}, z = e^{iC}$

$x+y+z=0$ and $\frac{1}{x}+\frac{1}{y}+\frac{1}{z}=0$

$\Rightarrow xy+yz+zx=0$

$\Rightarrow x^2y^2+y^2z^2+z^2x^2+2xyz(x+y+z)=0$

$\Rightarrow e^{i2(A+B)}+e^{i2(B+C)}+e^{i2(C+A)}=0$

$\Rightarrow \sum \cos 2(A+B)=0$

5. Let, $\arg z_1-\arg z_2=\alpha \Rightarrow \frac{z_1}{z_2}=e^{i\alpha}$

$$\Rightarrow \frac{z_1+z_2}{z_2}=1+e^{i\alpha}=1+\cos\alpha+i\sin\alpha$$

$$=2\cos\frac{\alpha}{2}\left(\cos\frac{\alpha}{2}+i\sin\frac{\alpha}{2}\right)$$

$$=2\cos\frac{\alpha}{2}e^{\frac{i\alpha}{2}}$$

$$\Rightarrow \frac{(z_1+z_2)^2}{z_2^2}=4\cos^2\frac{\alpha}{2}e^{i\alpha}=4\cos^2\frac{\alpha}{2}\frac{z_1}{z_2}$$

$$\Rightarrow \frac{(z_1+z_2)^2}{z_1z_2}=4\cos^2\frac{\alpha}{2} \text{ which is real}$$

$$\frac{(z_1+z_2)^2}{z_1z_2}=4\left(1-\sin^2\frac{\alpha}{2}\right)$$

$$\Rightarrow 4-\frac{(z_1+z_2)^2}{z_1z_2}=4\sin^2\frac{\alpha}{2}$$

$$=\left(\frac{\sin\frac{\alpha}{2}}{\frac{\alpha}{2}}\right)^2\alpha^2\le\alpha^2$$

$$\Rightarrow 4-\frac{(z_1+z_2)^2}{z_1z_2}\le(\arg z_1-\arg z_2)^2$$

$$\Rightarrow \frac{(z_1+z_2)^2}{z_1z_2}+(\arg z_1-\arg z_2)^2\ge 4$$

6. Let $f(x)=k(x-\alpha)(x-\beta-i\gamma)(x-\beta+i\gamma)$

$$=K(x-\alpha)\left[(x-\beta)^2+\gamma^2\right]$$

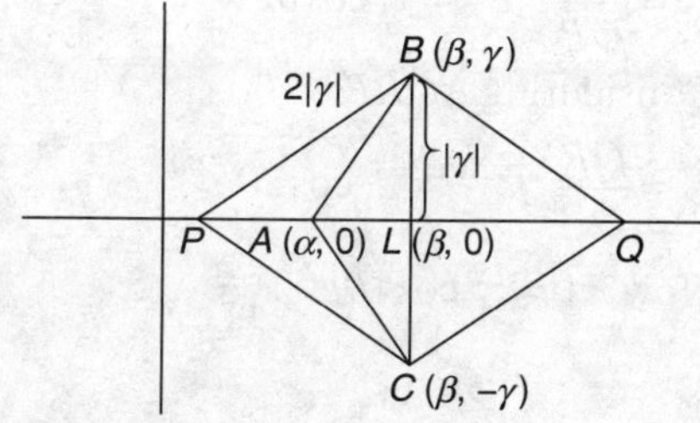

$$f'(x)=K\left[\left\{(x-\beta)^2+\gamma^2\right\}+(x-\alpha)2(x-\beta)\right]$$

Discriminant of equation $f'(x)=0$ is given by

$$D=4\left[(\alpha+2\beta)^2-3\left(\beta^2+\gamma^2+2\alpha\beta\right)\right]$$

$$=4\left(\alpha^2+\beta^2-3\gamma^2-2\alpha\beta\right) \quad (1)$$

$PL=\sqrt{3}|\gamma|$

If A lies inside the equilateral triangles having BC as base, then $|\beta-\alpha|<\sqrt{3}|\gamma|$

$\Rightarrow \beta^2+\alpha^2-3\gamma^2-2\alpha\beta<0$

$\therefore$ from (1), $D<0$ and hence roots of equation $f'(x)=0$ are imaginary if A lies between P and Q on the line segment PQ.

7. Since,

$$\frac{1}{a+\omega}+\frac{1}{b+\omega}+\frac{1}{c+\omega}+\frac{1}{d+\omega}=\frac{1}{\omega} \quad (1)$$

$$\Rightarrow \frac{1}{a+\omega^2}+\frac{1}{b+\omega^2}+\frac{1}{c+\omega^2}+\frac{1}{d+\omega^2}=\frac{1}{\omega^2} \quad (2)$$

$$\Rightarrow \sum\left(\frac{1}{a+\omega}-\frac{1}{a+\omega^2}\right)=\frac{1}{\omega}-\frac{1}{\omega^2}$$

$$=(\omega^2-\omega)\sum\frac{1}{(a+\omega)(a+\omega^2)}=(\omega^2-\omega)$$

$$\Rightarrow \sum\frac{1}{a^2-a+1}=1$$

8. Clearly z_1, z_2 and z_3 lie on a circle centred at (1, 0)

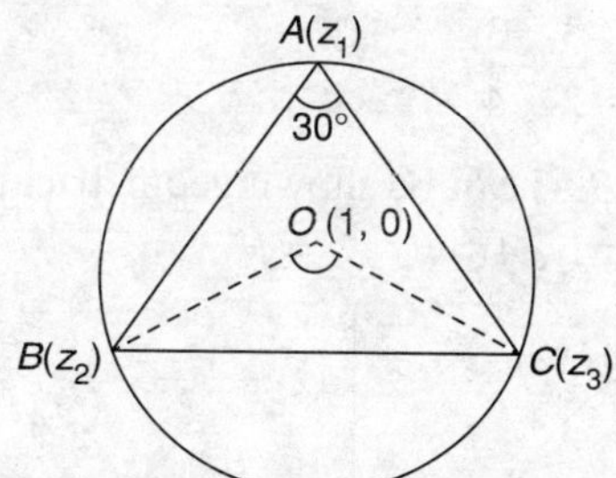

Also, $\angle BAC=30°$

$\Rightarrow \angle BOC=60° \Rightarrow \Delta BOC$ is equilateral

$\Rightarrow z_2^2+z_3^2+1=z_2+z_3+z_2z_3$

$\Rightarrow z_2(z_2-1)-z_3(z_2+1)+(z_3+1)(z_3-1)+2$
$=0$

$\Rightarrow |z_2(z_2-1)-z_3(z_2+1)+(z_3+1)(z_3-1)|=2$

9. Let $|z_1| = a$

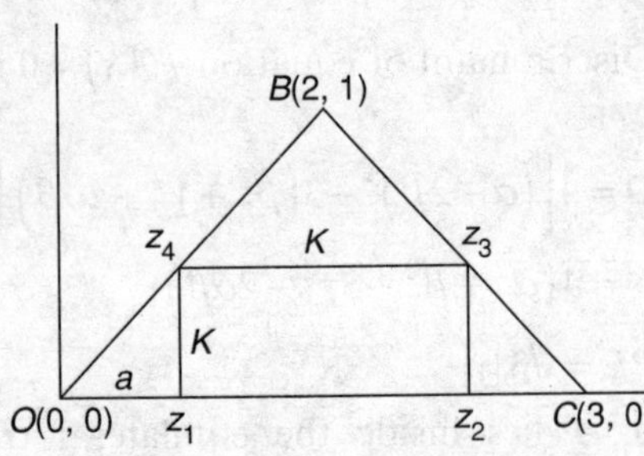

$|z_1 - z_2| = |z_1 - z_4| = K$

Now, $(z_4 - z_1) = i(z_2 - z_1)$

$\Rightarrow z_4 = a + iK$

Therefore, $z_3 = a + iK + K$

$= (a+K) + iK$

As z_0, z_4, z_B are collinear

$$\begin{vmatrix} 0 & 0 & 1 \\ a & K & 1 \\ 2 & 1 & 1 \end{vmatrix} = 0 \Rightarrow a - 2K = 0 \qquad (1)$$

Also, z_B, z_3, z_C are collinear

$$\begin{vmatrix} 2 & 1 & 1 \\ a+K & K & 1 \\ 3 & 0 & 1 \end{vmatrix} = 0 \Rightarrow 2K + a - 3 \qquad (2)$$

$\therefore 4K - 3 = 0 \Rightarrow K = \frac{3}{4}$

$\Rightarrow a = \frac{3}{2}$

$z_3 = \left(\frac{3}{2} + \frac{3}{4}\right) + i\frac{3}{4} = \frac{9}{4} + \frac{3}{4}i$

$z_4 = \frac{3}{2} + \frac{3}{4}i$

10. A, B, C can be shown geometrically clearly, $A \cap B \cap C = \phi$

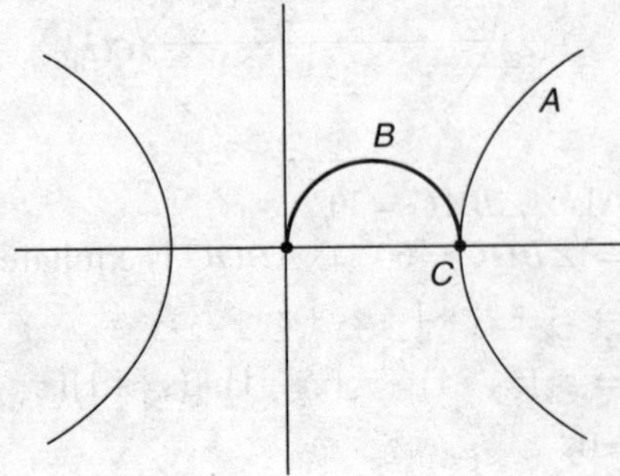

11. The equation $|z-1| + |z+i| = 4$ represents an ellipse with foci (1, 0) and (0, −1) and length of the major axis = 4, centre of the ellipse is $\left(\frac{1}{2}, -\frac{1}{2}\right)$, that is, $\frac{1-i}{2}$

$\left|z - \frac{(1-i)}{2}\right|$ represents the distance of any point on the ellipse from its centre. Its maximum value is the length of semi-major axis and its minimum value is the length of semi-minor axis.

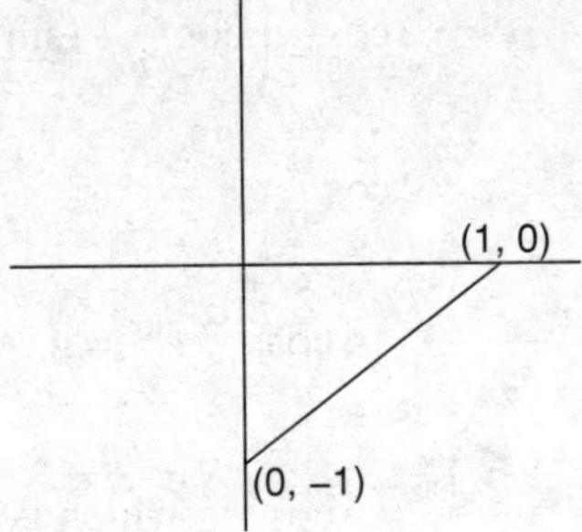

Here, $2a = 4$, $2ae = \sqrt{2}$

$\Rightarrow e = \frac{1}{2\sqrt{2}}$

From, $b^2 = a^2(1 - e^2)$

we get $b = \sqrt{\frac{7}{2}}$

$\left(\frac{1}{2}|2z - 1 + i|\right)_{max} = 2$

and $\left(\frac{1}{2}|2z - 1 + i|\right)_{min} = \sqrt{\frac{7}{2}}$

$\Rightarrow$ max value of $|2z - 1 + i| = 4$

and mini value of $|2z - 1 + i| = 2\sqrt{\frac{7}{2}}$

12. Considering ΔOQP, using rotation formula,

$z_2 = \frac{OQ}{OP} z_1 e^{i\theta} = z_1 \cos\theta e^{i\theta}$

Considering ΔORP

$z_3 = \frac{OR}{OP} z_1 e^{2i\theta} = z_1 \cos 2\theta e^{2i\theta}$

Now, $z_2^2 = z_1^2 \cos^2\theta e^{2i\theta}$

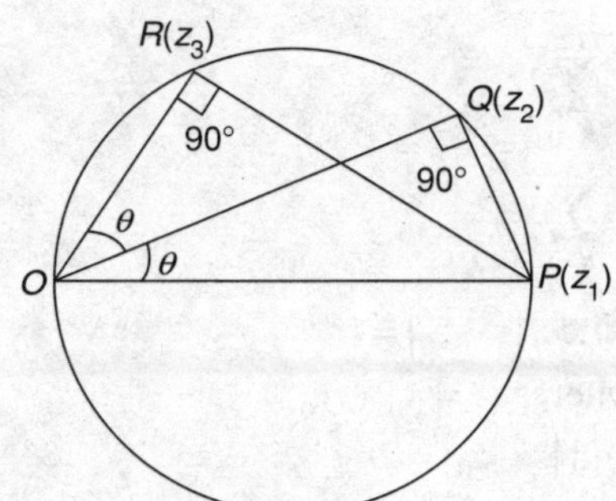

$\Rightarrow z_2^2 = \dfrac{z_1^2 z_3 \cos^2\theta}{z_1 \cos 2\theta}$

$\Rightarrow z_2^2 \cos 2\theta = z_1 z_3 \cos^2\theta$

$\Rightarrow 2\sqrt{3} z_2^2 \cos 2\theta = 2\sqrt{3} z_1 z_3 \cos^2\theta$ (1)

It is given that

$2\sqrt{3} z_2^2 = (2+\sqrt{3}) z_1 z_3$

Using this in equation (1)

$(2+\sqrt{3}) z_1 z_3 \cos 2\theta = 2\sqrt{3} z_1 z_3 \cos^2\theta$

$\Rightarrow (2+\sqrt{3})(2\cos^2\theta - 1) = 2\sqrt{3}\cos^2\theta$

$\Rightarrow \cos^2\theta = \dfrac{2+\sqrt{3}}{4}$

$= \left(\dfrac{\sqrt{3}+1}{2\sqrt{2}}\right)^2$

(where θ is acute and hence $\cos\theta > 0$)

$\Rightarrow \theta = 15°$

13. Since,

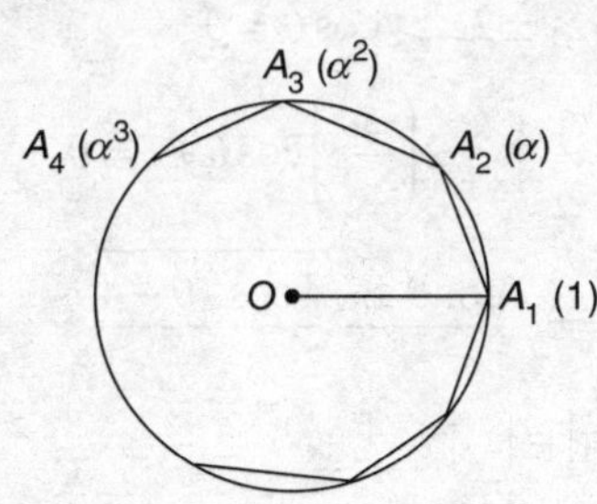

$|A, A_{r+1}|^2$

$= |\alpha^r - 1|^2$

$= |1 - \alpha^r|^2$

$= \left|1 - \dfrac{\cos 2\pi r}{n} - i \sin\dfrac{2\pi r}{n}\right|^2$

$= \left(1 - \cos\dfrac{2\pi r}{n}\right)^2 + \sin^2\dfrac{2\pi r}{n}$

$= 2 - 2\cos\dfrac{2\pi r}{n}$

$\therefore \sum_{r=1}^{n-1} |A_1 A_{r+1}|^2 = 2(n-1) - 2(-1) = 2n$

14. Let $z = r_1(\cos\theta_1 + i\sin\theta_1)$

and, $\omega = r_2(\cos\theta_2 + i\sin\theta_2)$

$\therefore |z| = r_1, |\omega| = r_2$

$\arg z = \theta_1$ and $\arg\omega = \theta_2$

Now, $|z-\omega|^2$

$= |(r_1\cos\theta_1 - r_2\cos\theta_2) + i(r_1\sin\theta_1 - r_2\sin\theta_2)|^2$

$= (r_1\cos\theta_1 - r_2\cos\theta_2)^2 + (r_1\sin\theta_1 - r_2\sin\theta_2)^2$

$= r_2^2 + r_2^2 - 2r_1r_2\cos(\theta_1 - \theta_2)$

$= r_1^2 + r_2^2 - 2r_1r_2\left(1 - 2\sin^2\dfrac{\theta_1-\theta_2}{2}\right)$

$= (r_1 - r_2)^2 + 4r_1r_2\sin^2\dfrac{\theta_1-\theta_2}{2}$

$\leq (r_1 - r_2)^2 + 4.1.1.\left(\dfrac{\theta_1-\theta_2}{2}\right)^2$

$\Rightarrow |z-\omega|^2 \leq (|z| - |\omega|)^2 + (\arg z - \arg\omega)^2$

15. $|2| = |z^n\cos\theta_0 + z^{n-1}\cos\theta_1 + \ldots +$
$z\cos\theta_{n-1} + \cos\theta_n|$

$\leq 1 + |z| + |z|^2 + \ldots + |z|^n$

$\leq 1 + |z| + |z|^2 + \ldots \text{to } \infty$

$= \dfrac{1}{1-|z|}$

$1 - |z| \leq \dfrac{1}{2} \Rightarrow 1 - \dfrac{1}{2} \leq |z| \Rightarrow |z| \geq \dfrac{1}{2}$

16. Since,

$\dfrac{1}{A_1A_2} = \dfrac{1}{A_1A_3} + \dfrac{1}{A_1A_4}$

$\Rightarrow \dfrac{1}{\sin\frac{\pi}{n}} = \dfrac{1}{\sin\frac{2\pi}{n}} + \dfrac{1}{\sin\frac{3\pi}{n}}$

$\Rightarrow \sin\dfrac{2\pi}{n}\sin\dfrac{3\pi}{n} = \sin\dfrac{\pi}{n}\left(\sin\dfrac{2\pi}{n} + \sin\dfrac{3\pi}{n}\right)$

$\Rightarrow 2\sin\frac{\pi}{n}\cos\frac{\pi}{n}\sin\frac{3\pi}{n}$

$= \sin\frac{\pi}{n}\left(\sin\frac{2\pi}{n}+\sin\frac{3\pi}{n}\right)$

$\Rightarrow \sin\frac{4\pi}{n}+\sin\frac{2\pi}{n} = \sin\frac{2\pi}{n}+\sin\frac{3\pi}{n}$

$\Rightarrow \frac{3\pi}{n} = m\pi + (-1)^m\frac{4\pi}{n}$

If $m = 0$, we get

$\frac{3\pi}{n} = \frac{4\pi}{n} \Rightarrow 3 = 4$ (not possible)

$m = 1$

$\frac{3\pi}{n} = \pi - \frac{4\pi}{n} \Rightarrow \frac{7\pi}{n} = \pi$

$\Rightarrow n = 7$

17. Since, $|z+1-i| = 1$

Also, $\omega = \frac{z+i}{1-i}$

$\Rightarrow (1-i)\omega - i = z$

$\Rightarrow |(1-i)\omega - i + 1 - i| = |z+1-i|$

$\Rightarrow |1-i|\left|\omega + \frac{1-2i}{1-i}\right| = 1$

$\Rightarrow \left|\omega + \frac{(1-2i)(1+i)}{(1+i)(1-i)}\right| = \frac{1}{\sqrt{2}}$

$\Rightarrow \left|\omega + \frac{3-i}{2}\right| = \frac{1}{\sqrt{2}}$

$\Rightarrow \left|\omega - \left(-\frac{3}{2}+\frac{1}{2}i\right)\right| = \frac{1}{\sqrt{2}}$

$\Rightarrow$ Locus of ω is a circle centred at $\left(-\frac{3}{2},\frac{1}{2}\right)$ and radius $\frac{1}{\sqrt{2}}$

18. Since,

$$\frac{\sum_{K=1}^{12}\left|e^{\frac{i(K+1)\pi}{7}} - e^{\frac{iK}{7}}\right|}{\sum_{K=1}^{3}\left|e^{\frac{i(4K-1)\pi}{7}} - e^{\frac{i(4K-2)\pi}{7}}\right|}$$

$$= \frac{\sum_{K=1}^{12}\left|e^{\frac{iK\pi}{7}}\right|\left|e^{\frac{i\pi}{7}}-1\right|}{\sum_{K=1}^{3}\left|e^{\frac{i(4K-2)\pi}{7}}\right|\left|e^{\frac{i\pi}{7}}-1\right|}$$

$$= \frac{\sum_{K=1}^{12}1}{\sum_{K=1}^{3}1} = \frac{12}{3} = 4$$

19. Since, $|z-z_0| = r$ (1)

where, $z_0 = x_0 + iy_0$

and $|z-z_0| = 2r$ (2)

$\Rightarrow |\alpha - z_0|^2 = r^2$ and $\left|\frac{1}{\bar{\alpha}} - z_0\right|^2 = 4r^2$

$\Rightarrow |1-\bar{\alpha}z_0|^2 = 4r^2|\alpha|^2$

Subtracting

$|1-\bar{\alpha}z_0|^2 - |\alpha - z_0|^2 = 4r^2|\alpha|^2 - r^2$

$\Rightarrow 1 + |\alpha z_0|^2 - \bar{\alpha}z_0 - \alpha\bar{z}_0 - |\alpha|^2 - |z_0|^2 + \bar{\alpha}z_0 + \alpha\bar{z}_0$

$= 4r^2|\alpha|^2 - r^2$

$\Rightarrow 1 + |\alpha|^2|z_0|^2 - |\alpha|^2 - |z_0|^2 = 4r^2|\alpha|^2 - r^2$

$\Rightarrow (1-|\alpha|^2)\left(1-\frac{r^2+2}{2}\right) = 4r^2|\alpha|^2 - r^2$

$\Rightarrow (1-|\alpha|^2)\left(-\frac{r^2}{2}\right) = 4r^2|\alpha|^2 - r^2$

$\Rightarrow |\alpha|^2 - 1 = 8|\alpha|^2 - 2$

$\Rightarrow |\alpha|^2 = \frac{1}{7} \Rightarrow |\alpha| = \frac{1}{\sqrt{7}}$

20. $z_1^2 + z_2^2 \pm 2z_1z_2\cos\theta = 0$

$\Rightarrow \left(\frac{z_1}{z_2}\right)^2 \pm 2\left(\frac{z_1}{z_2}\right)\cos\theta + 1 = 0$

$\frac{z_1}{z_2} = \pm\frac{2\cos\theta \pm \sqrt{4\cos^2\theta - 4}}{2}$

$\Rightarrow |z_1| = |z_2|$

Hence the points represented by z_1, z_2 and origin form an isosceles triangle.

21. We know that

$|z-z_1|^2 + |z-z_2|^2 = |z_1 - z_2|^2$ represents a circle

Therefore,

$|z-(2+3i)|^2 + |z-(4+3i)|^2 = |2+3i-4-3i|^2$

represents a circle

$\therefore \lambda = 4$

22. As $\frac{z^3+1}{z^3-1}$ is purely imaginary

$\Rightarrow \frac{z^3+1}{z^3-1}+\frac{\bar{z}^3+1}{\bar{z}^3-1}=0 \qquad [z^3 \neq 1]$

$\Rightarrow z^3\bar{z}^3-z^3+\bar{z}^3-1+z^3\bar{z}^3+z^3-\bar{z}^3-1=0$

$\Rightarrow 2|z|^6-2=0 \Rightarrow |z|=1$

As, $z^3 \neq 1 \Rightarrow z \neq 1, \omega, \omega^2$

$\Rightarrow$ The locus of z is a circle of unit radius centred at origin (except $1, \omega, \omega^2$).

23. $\frac{z^2+1}{z^2-1}=\frac{\bar{z}^2+1}{\bar{z}^2-1}$

$\Rightarrow z^2\bar{z}^2-z^2+\bar{z}^2-1=z^2\bar{z}^2+z^2-\bar{z}^2-1$

$\Rightarrow 2(z^2-\bar{z}^2)=0$

$\Rightarrow z+\bar{z}=0$ or $z-\bar{z}=0$

$\Rightarrow$ either z is purely real, that is, z lies on the x-axis

or z is purely imaginary, that is, z lies on the y-axis

Hence, the locus is a pair of straight lines.

$xy=0$ except the point $(\pm 1, 0)$

24. $z=\frac{at+b}{t-c} \Rightarrow t=\frac{b+cz}{z-a}$

$\Rightarrow \left|\frac{b+cz}{z-a}\right|=|t|=1 \Rightarrow \frac{\left|z+\frac{b}{c}\right|}{|z-a|}=\frac{1}{|c|}$

$\Rightarrow$ Locus of z is a circle.

25. $|3z-2|+|3z+2|=4$

$\Rightarrow \left|z-\frac{2}{3}\right|+\left|z+\frac{2}{3}\right|=\frac{4}{3}$

$\Rightarrow P(z)$ is a point on lines segment AB,

where $A=\left(\frac{2}{3},0\right), B=\left(-\frac{2}{3},0\right)$

26. Since,

$$\sum_{i=0}^{n-1}\frac{\alpha i}{2-\alpha_i}=-\sum_{i=0}^{n-1}\frac{(2-\alpha_i)-2}{2-\alpha_i}$$

$$=2\sum_{i=0}^{n-1}\frac{1}{2-\alpha_i}-\sum_{i=0}^{n-1}-1$$

Now

$x^n-1=(x-\alpha_0)(x-\alpha_1)(x-\alpha_2)\ldots(x-\alpha_{n-1})$

Taking log on both sides

$$\log(x^n-1)=\log(x-\alpha_0)+\log(x-\alpha_1)+\log(x-\alpha_2)+\ldots+\log(x-\alpha_{n-1})$$

Differentiating with respect to x

$$\frac{nx^{n-1}}{x^n-1}=\frac{1}{x-\alpha_0}+\frac{1}{x-\alpha_1}+\frac{1}{x-\alpha_2}+\ldots+\frac{1}{x-\alpha_{n-1}}$$

Put $x=2$

$$\sum_{i=0}^{n-1}\frac{1}{2-\alpha_i}=\frac{n2^{n-1}}{2^n-1}$$

Hence, $\sum_{i=0}^{n-1}\frac{\alpha_i}{2-\alpha_i}=\frac{n}{2^n-1}$

27. $D=\frac{(z_2+z_3)}{2}$, centroid

$G=\frac{(z_1+z_2+z_3)}{3}$

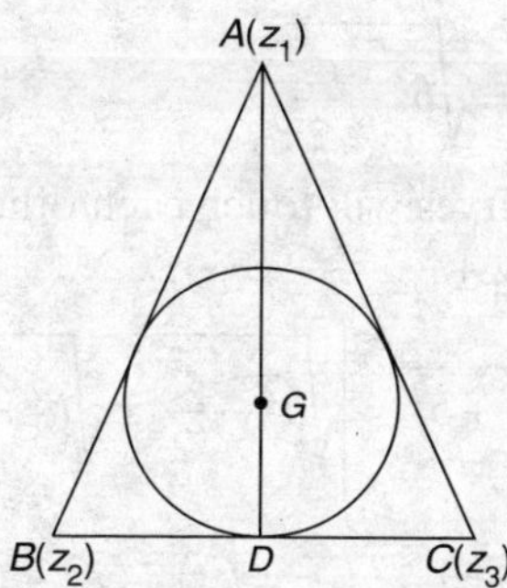

Therefore, the equation of the required circle is

$$\left|z-\frac{z_1+z_2+z_3}{3}\right|=\left|\frac{z_1+z_2+z_3}{3}-\frac{z_2+z_3}{2}\right|$$

$$=\left|\frac{(z_1-z_2)+(z_1-z_3)}{6}\right|$$

Here, $z_1=i+i\sqrt{3}$, $z_2=1-i\sqrt{3}$, $z_3=-2$

$\Rightarrow$ Equation of the circle is

$|z-0|=\left|\frac{1+i\sqrt{3}}{2}\right| \Rightarrow |z|=1$

$\Rightarrow$ Radius $=1$

28. $|z+\sqrt{2}|=a^2-3a+2$

and, $|z+i\sqrt{2}|<a^2$

$\Rightarrow a^2-3a+2=\left|z+\sqrt{2}\right|$

$=\left|\left(z+i\sqrt{2}\right)+\left(\sqrt{2}-i\sqrt{2}\right)\right|$

$\Rightarrow a^2-3a+2\le\left|z+i\sqrt{2}\right|+\left|\sqrt{2}\,(1-i)\right|$

$\Rightarrow a^2-3a+2<a^2+2$

$\Rightarrow a>0$ (1)

Also, $a^2-3a+2\ge 0\Rightarrow(a-1)(a-2)\ge 0$

$\Rightarrow a\le 1$ or $a\ge 2$ (2)

$a\in(0,1]\cup[2,\infty)$

29. Since, $2|z|^2-4a\,\text{Re}(z)+c^2=0$

$\Rightarrow |z|^2-2a\,\text{Re}(z)+\frac{c^2}{2}=0$

$\Rightarrow z\bar{z}-a(z+\bar{z})+a^2=a^2-\frac{c^2}{2}$

$\Rightarrow |z-a|=\sqrt{a^2-\frac{c^2}{2}}$

Similarly, for the second curve

$|z-ib|=\sqrt{b^2-\frac{c^2}{2}}$

The curves will touch each other if

$c_1c_2=r_1\pm r_2$

$\Rightarrow \sqrt{a^2+b^2}=\left|\sqrt{a^2-\frac{c^2}{2}}\pm\sqrt{b^2-\frac{c^2}{2}}\right|$

$\Rightarrow 2a^2b^2=c^2\left(a^2+b^2\right)\Rightarrow c^2=\frac{2a^2b^2}{a^2+b^2}$

Therefore, $\frac{c^2\left(a^2+b^2\right)}{a^2b^2}=2$

30. $\sum_{r=0}^{n}{}^nC_r\cos\left(nA+r(B-A)\right)$

$=$ R.P. of $\sum_{r=0}^{n}{}^nC_r e^{i(n-r)A}e^{irB}$

$=$ R.P.$\left[e^{iA}+e^{iB}\right]^n$

$=$ R.P.$\left[(\cos A+\cos B)+i(\sin A+\sin B)\right]^n$

$A+B=\pi$

$=$ R.P.$[0+2i\sin A]^n=0$

3.18.4 JEE Main: Archive

1. **(D)**

$A=\left\{z\in C:1\le\left|z-(1+i)\right|\le 2\right\}$

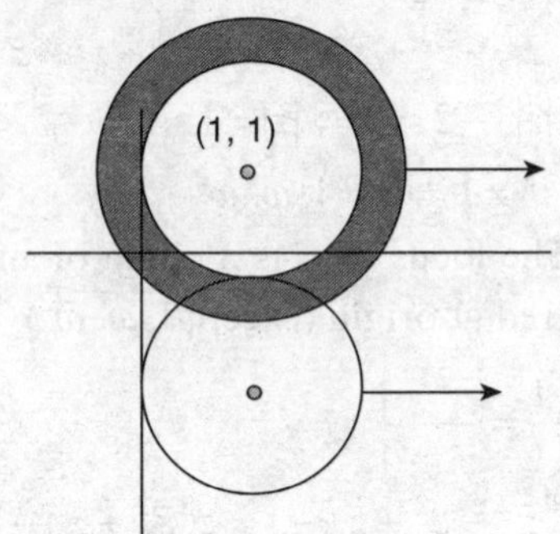

$B=\left\{z\in A:\left|z-(1-i)\right|=1\right\}$

$A\cap B$ is an infinite set.

2. **(80)**

$|z-3|\le 1$

represents a circle of radius 1 and centred at (3, 0)

$z(4+3i)+\bar{z}(4-3i)\le 24$

$(x+iy)(4+3i)+(x-iy)(4-3i)\le 24$

$4x+3xi+4iy-3y+4x-3ix-4iy-3y\le 24$

$8x-6y\le 24$

$4x-3y\le 12$

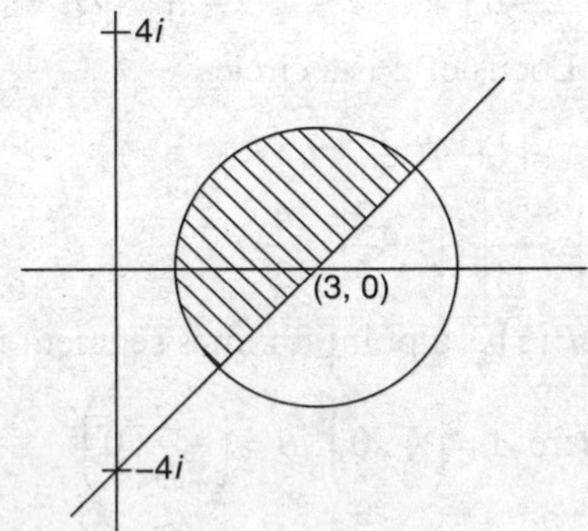

minimum of $(0,4)$ from circle

$=\sqrt{3^2+4^2}-1=4$

will lie along line joining (0, 4) and (3, 0)

$\therefore$ equation line

$\frac{x}{3}+\frac{y}{4}=1\Rightarrow 4x+3y=12$ (1)

equation circle $(x-3)^2+y^2=1$ (2)

$$\left(\frac{12-3y}{4}-3\right)^2+y^2=1$$

$$\left(\frac{-3y}{4}\right)^2+y^2=1$$

$$\frac{25y^2}{16}=1 \Rightarrow y=\pm\frac{4}{5}$$

for minimum distance $y=\frac{4}{5}$

$$\therefore x=\frac{12}{5}$$

$$\therefore 25(\alpha+\beta)=25\left(\frac{4}{5}+\frac{12}{5}\right)$$

$$=16\times 5=80$$

3. **(B)**

Slope of $BC=\frac{3-5}{4-0}=-\frac{1}{2}$

Slope of $AP=2$

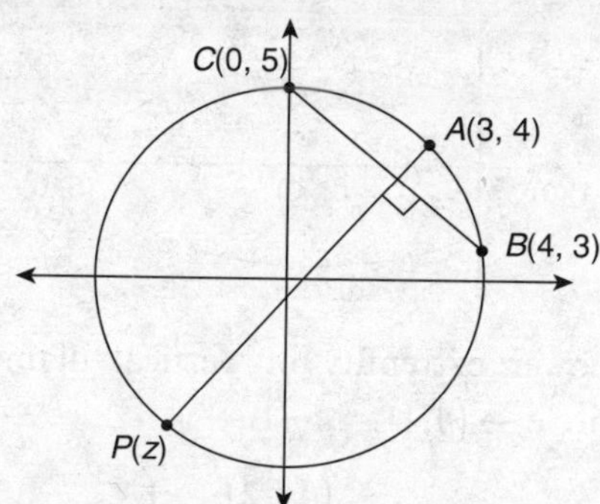

equation of AP: $y-4=2(x-3)$

$\Rightarrow y=2(x-1)$

P lies on circle $x^2+y^2=25$

$\Rightarrow x^2+(2(x-1))^2=25$

$\Rightarrow x=-\frac{7}{5}$ and $y=-\frac{24}{5}$

$\Rightarrow \arg(z)=\tan^{-1}\left(\frac{24}{7}\right)-\pi$

4. **(C)**

$\bar{z}_1=i\bar{z}_2$

$z_1=-iz_2$

$$\arg\left(\frac{z_1}{\bar{z}_2}\right)=\pi$$

$$\arg\left(-i\frac{z_2}{\bar{z}_2}\right)=\pi$$

$$\arg(z_2)=\theta$$

$$-\frac{\pi}{2}+\theta+\theta=\pi$$

$$2\theta=\frac{3\pi}{2}$$

$$\arg(z_2)=\theta=\frac{3\pi}{4}, \arg z_1=\frac{\pi}{4}$$

5. **(B)**

Set A

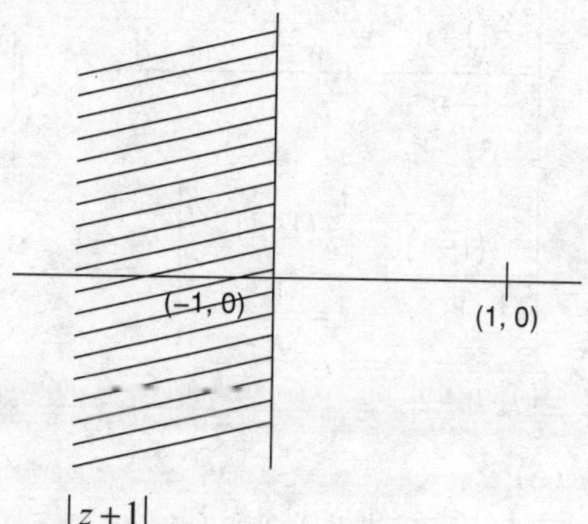

$$\Rightarrow\left|\frac{z+1}{z-1}\right|<1$$

$$\Rightarrow|z+1|<|z-1|$$

$$\Rightarrow(x+1)^2+y^2<(x-1)^2+y^2$$

$$\Rightarrow x<0$$

Set B

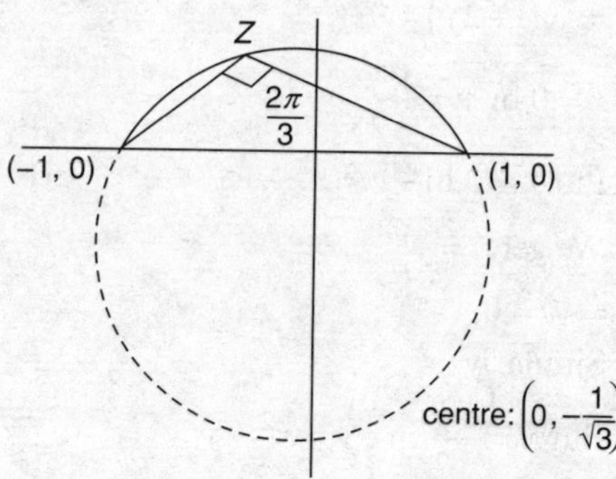

$$\Rightarrow\arg\left(\frac{z-1}{z+1}\right)=\frac{2\pi}{3}$$

$$\Rightarrow\tan^{-1}\left(\frac{y}{x-1}\right)-\tan^{-1}\left(\frac{y}{x+1}\right)=\frac{2\pi}{3}$$

$$\Rightarrow x^2+y^2+\frac{2y}{\sqrt{3}}-1=0$$

$A \cap B$

$\Rightarrow \text{Centre}\left(0, -\frac{1}{\sqrt{3}}\right)$

6. **(2)**

$z^2 + z + 1 = 0 \Rightarrow z = w, w^2$

$\left|\sum_{n=1}^{15}\left(z^n + (-1)\frac{1}{z^n}\right)\right| = \left|\sum_{n=1}^{15}\left(z^{2n} + \frac{1}{z^{2n}} + 2(-1)^n\right)\right|$

$= \left|\sum_{n=1}^{15} w^{2n} + \frac{1}{w^{2n}} + 2(-1)^n\right|$

$= \left|\frac{w^2(1-w^{30})}{1-w^2} + \frac{\frac{1}{w^2}\left(1-\frac{1}{w^{30}}\right)}{1-\frac{1}{w^2}} + 2(-1)\right|$

$= \left|\frac{w^2(1-1)}{1-w^2} + \frac{\frac{1}{w^2}(1-1)}{1-\frac{1}{w^2}} - 2\right|$

$= |0+0-2| = 2$

7. **(A)**

$\Rightarrow$ Let $z = x + iy$, $x, y \in R$

Now, $\bar{z} = iz^2$

then $x - iy = i\left(x^2 - y^2 + 2xyi\right)$

$x - iy = i\left(x^2 - y^2\right) - 2xy$

$\Rightarrow x = -2xy$ and $-y = x^2 - y^2$

$\Rightarrow x(1+2y) = 0$

$x = 0$ or $y = -\frac{1}{2}$

Put $x = 0$ in $-y = x^2 - y^2$

We get $y = y^2$

$\Rightarrow y = 0, 1$

Similarly

Put $y = -\frac{1}{2}$ in $-y = x^2 - y^2$

$\Rightarrow \frac{1}{2} = x^2 - \frac{1}{4}$

$\Rightarrow x^2 = \frac{3}{4}$

$x = \pm\frac{\sqrt{3}}{2}$

$z = \left(0, i, \frac{\sqrt{3}}{2} - \frac{1}{2}i, -\frac{\sqrt{3}}{2} - \frac{1}{2}i\right)$

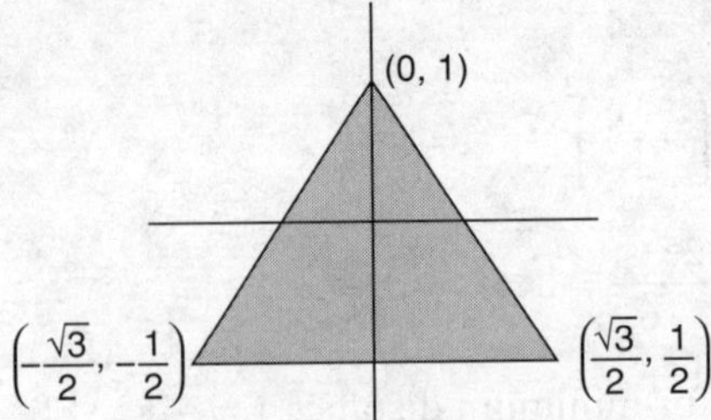

$\text{Area} = \frac{1}{2}\left(\sqrt{3}\right)\left(\frac{3}{2}\right)$

$= \frac{3\sqrt{3}}{4}$

8. **(C)**

$C: (x-4)^2 + (y-3)^2 = 4$

$E: \frac{(x-2)^2}{9} + \frac{y^2}{5} = 1$

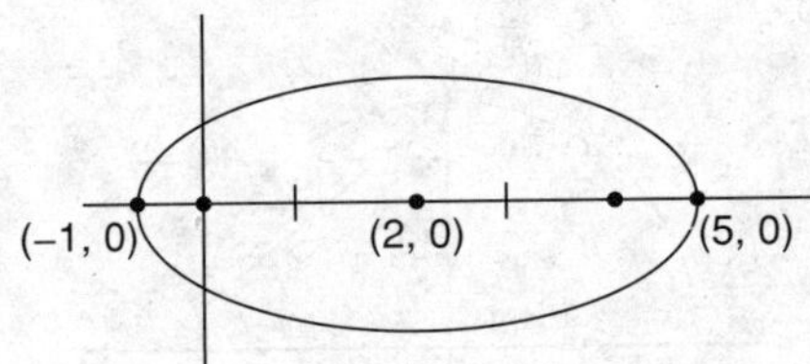

Lower extremity of vertical diameter of circle $\rightarrow (4, 1)$

Put in ellipse $\Rightarrow \frac{(4-2)^2}{9} + \frac{1}{5} - 1$

$= \frac{4}{9} + \frac{1}{5} - 1$

$= \frac{29}{45} - 1 < 0$

Two solutions

9. **(40)**

$1 < |Z - 3 + 2i| < 4$

$1 < (a-3)^2 + (b+2)^2 < 16$

$(0, \pm 2), (\pm 2, 0), (\pm 1, \pm 2), (\pm 2, \pm 1)$

$(\pm 2, \pm 3), (3\pm, \pm 2), (\pm 1, \pm 1), (2\pm, \pm 2)$

$(\pm 3, 0), (0, \pm 3), (\pm 3, \pm 1), (\pm 1, \pm 3)$

Total 40 points

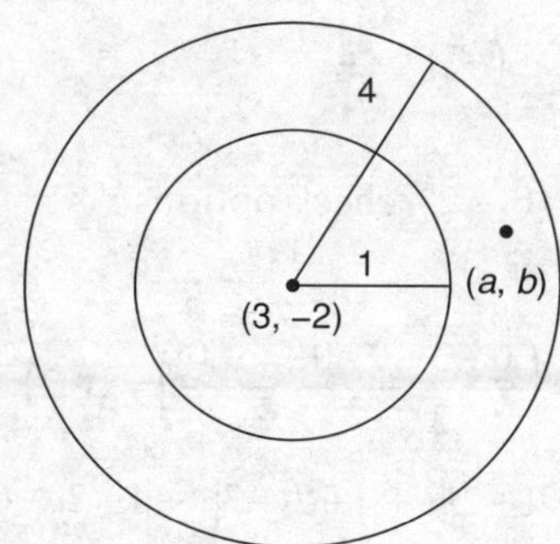

10. **(2)**

$z+\bar{z}=iz^2+z^2$

Consider $z=x+iy$

$2x=(i+1)(x^2-y^2+2xyi)$

$\Rightarrow 2x=x^2-y^2-2xy$ and $x^2-y^2+2xy=0$

$\Rightarrow 2x=-4xy$

$\Rightarrow x=0$ or $y=-\dfrac{1}{2}$

Case 1: $x=0 \Rightarrow y=0$ here $z=0$

Case 2: $y=-\dfrac{1}{2}$

$\Rightarrow 4x^2-4x-1=0$

$(2x-1)^2=2$

$2x-1=\pm\sqrt{2}$

$x=\dfrac{1\pm\sqrt{2}}{2}$

Here $z=\dfrac{1+\sqrt{2}}{2}-\dfrac{i}{2}$ or $z=\dfrac{1-\sqrt{2}}{2}-\dfrac{i}{2}$

Sum of squares of modulus of z

$=0+\dfrac{(1+\sqrt{2})^2+1}{4}+\dfrac{(1-\sqrt{2})^2+1}{4}=\dfrac{8}{4}=2$

11. **(A)**

$x^2=1-2i \Rightarrow \alpha^2=1-2i, \beta^2=1-2i$

Hence $\alpha^8=\beta^8$

$|\alpha^8+\beta^8|=|2\alpha^8|=2|\alpha^2|^4$

$=2\cdot(\sqrt{5})^4=50$

12. **(26)**

$|z-2|\le 1$

$(x-2)^2+y^2\le 1$ (1)

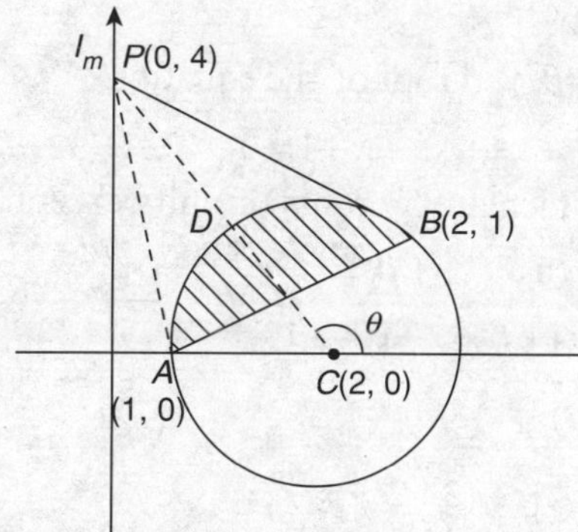

and $z(1+i)+\bar{z}(1-i)\le 2$

Put $z=x+iy$

$\therefore x-y\le 1$ (2)

$PA=\sqrt{17}, PB=\sqrt{13}$

Maximum is PA and minimum is PD

Let $D(2+\cos\theta, 0+\sin\theta)$

$\therefore m_{cp}=\tan\theta=-2$

$\cos\theta=-\dfrac{1}{\sqrt{5}}, \sin\theta=\dfrac{2}{\sqrt{5}}$

$\therefore D\left(2-\dfrac{1}{\sqrt{5}}, \dfrac{2}{\sqrt{5}}\right)$

$\Rightarrow z_1=\left(2-\dfrac{1}{\sqrt{5}}\right)+\dfrac{2i}{\sqrt{5}}$

$|z_1|=\dfrac{25-4\sqrt{5}}{5}$ and $z_2=1$

$\therefore |z_2|^2=1$

$\therefore 5\left(|z_1|^2+|z_2|^2\right)=30-4\sqrt{5}$

$\therefore \alpha=30$

$\beta=-4$

$\therefore \alpha+\beta=26$

13. **(C)**

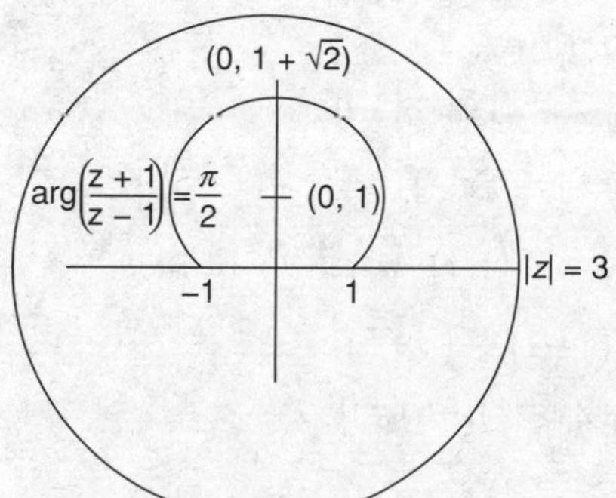

14. **(B)**

$\alpha, \beta, \gamma, \delta$ root of the equation

$x^4 + x^3 + x^2 + x + 1 = 0$

Which are 5th roots of unity except 1.

then $\alpha^{2021} + \beta^{2021} + \gamma^{2021} + \delta^{2021}$

$= \alpha + \beta + \gamma + \delta = -1$

15. **(D)**

$S_n : |z - (3 - 2i)| = \frac{n}{4}$

is a circle center $C_1(3, -2)$ and radius $\frac{n}{4}$

$T_n : |z - (2 - 3i)| = \frac{1}{n}$

is a circle center $C_2(2, -3)$ and radius $\frac{1}{n}$

Here $S_n \cap T_n = \phi$

Both circles do not intersect each other

Case-1: $C_1C_2 > \frac{n}{4} + \frac{1}{n}$

$\sqrt{2} > \frac{n}{4} + \frac{1}{n}$

then $n = 1, 2, 3, 4$

Case-2: $C_1C_2 < \left|\frac{n}{4} - \frac{1}{n}\right|$

$\Rightarrow \sqrt{2} < \left|\frac{n^2 - 4}{4n}\right|$

$\Rightarrow n$ has infinite solutions for $n \in N$

16. **(C)**

17. **(D)**

$AB = AO \cdot z^{-i\pi/2} = -2 + i$

So $OB = (-2 + i) + (1 + 2i)$

$z_2 = -1 + 3i$

$\therefore |2z_1 - z_2| = \sqrt{10}$

18. **(C)**

$|z - i| - |z + 5i| = 0$

$\Rightarrow |x + (y - 1)i| = |x + (y + 5)i|$

$x^2 + (y - 1)^2 = x^2 + (y + 5)^2$

$(y - 1)^2 - (y + 5)^2 = 0$

$(2y + 4)(-6) = 0$

$y = -2$

$\therefore x^2 + (-2)^2 = 4$

$x = 0$

$Z \equiv (0, -2)$, check options

19. **(A)**

$z_0 = \left(\frac{0 + 3 + 0}{3}, \frac{0 + 6 + 0}{3}\right) = (1, 2)$

$v_0 = |1 + 2i|^2 + |1 + 2i - 3|^2 + |1 + 2i - 6i|^2 = 30$

Then $\left|2z_0^2 - \bar{z}_0^3 + 3\right|^2 + v_0^2$

$= \left|2(1 + 2i)^2 - (1 - 2i)^3 + 3\right|^2 + 900$

$= |2(1 - 4 + 4i) - (1 - 4 - 4i)(1 - 2i) + 3|^2 + 900$

$= |8 + 6i|^2 + 900 = 100 + 900 = 1000$

20. **(0)**

$S = \{z \in C : z^2 + \bar{z} = 0\}$

Let $z = x + iy$

$z^2 = x^2 - y^2 + 2ixy$

$\bar{z} = x - iy$

$z^2 + \bar{z} = x^2 - y^2 + x + i(2xy - y) = 0$

$\Rightarrow x^2 + x - y^2 = 0$ and $2xy - y = 0$

$y = 0$ or $x = \frac{1}{2}$

If $y = 0$; $x = 0, -1$

If $x = \frac{1}{2}$; $y = \frac{\sqrt{3}}{2}, \frac{-\sqrt{3}}{2}$

$\sum_{z \in S} (\text{Re}(z) + \text{Im}(z)$

$= \left(0 - 1 + \frac{1}{2} + \frac{1}{2}\right) + 0 + 0 + \frac{\sqrt{3}}{2} - \frac{\sqrt{3}}{2}\right)$

21. **(C)**

$\pi < \alpha, \beta < 2\pi$

$\frac{1 - i\sin\alpha}{1 + i(2\sin\alpha)}$ = Purely imaginary

$\Rightarrow \frac{1 - i\sin\alpha}{1 + i(2\sin\alpha)}$ = Purely imaginary

$\Rightarrow \frac{(1 - i\sin a)(1 - i(2\sin\alpha))}{1 + 4\sin^2\alpha} =$

Purely imaginary

$\Rightarrow \dfrac{1-2\sin^2\alpha}{1+4\sin^2\alpha}=0$

$\Rightarrow \sin^2\alpha=\dfrac{1}{2}$

$\Rightarrow \alpha=\left\{\dfrac{5\pi}{4},\dfrac{7\pi}{4}\right\}$

and $\dfrac{1+i\cos\beta}{1+i(-2\cos\beta)}$ = Purely real

$\Rightarrow \dfrac{(1+i\cos\beta)(1+2i\cos\beta)}{1+4\cos^2\beta}$ = Purely real

$\Rightarrow 3\cos\beta=0$

$\Rightarrow \boxed{\beta=\dfrac{3\pi}{2}}$

$\Rightarrow Z_{\alpha\beta}=\sin\dfrac{5\pi}{2}+i\cos 3\pi=1-i$

or $Z_{\alpha\beta}=\sin\dfrac{7\pi}{2}+i\cos 3\pi=-1-i$

Required value

$=\left[i(1-i)+\dfrac{1}{i(1+i)}\right]+\left[i(-1-i)+\dfrac{1}{i(-1+i)}\right]$

$=i(-2i)+\dfrac{1}{i}\dfrac{2i}{(-2)}\Rightarrow 2-1=1$

22. (C)

$|z_2+|z_2-1||^2=|z_2-|z_2+1||^2$

$\Rightarrow |z_2+|z_2-1||(\bar{z}_2+|z_2-1|)$

$=(z_2-|z_2+1|)(\bar{z}_2-(z_2+1))$

$\Rightarrow z_2|\bar{z}_2+12_2-1|-(\bar{z}_2-|z_2+1|)+$
$\bar{z}_2(|z_2-1|+|z_2+1|)$

$=|z_2+1|^2=|z_2-1|^2$

$\Rightarrow [z_2+\bar{z}_2](|z_2-1|)+(z_2+1|)=2(z_2+\bar{z}_2)$

$\Rightarrow (z_2+\bar{z}_2)(|z_2-1|+|z_2+1|-2)=0$

$\therefore z_2+\bar{z}_2=0$ or $|z_2-1|+|z_2+1|-2=0$

$\therefore z_2$ lie on imaginary axis. Or on real axis with in $[-1,1]$

Also, $|z_1-3|=\dfrac{1}{2}$ lie on circle having centre 3 and radius $\dfrac{1}{2}$.

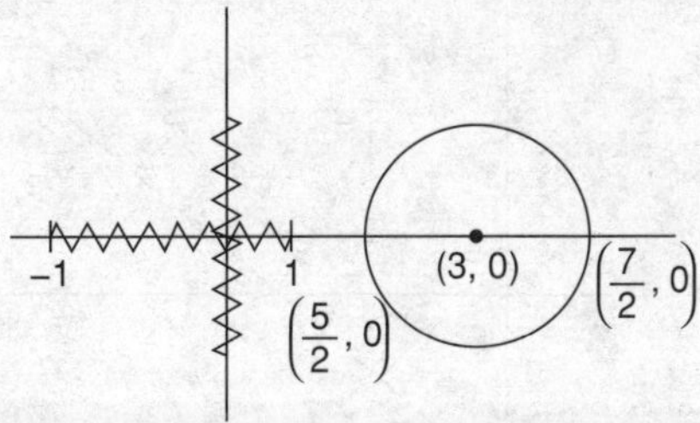

Clearly $|z_1-z_2|\min=\dfrac{5}{2}-1=\dfrac{3}{2}$

23. (6)

$|z^2|=|\bar{z}|\cdot 2^{1-|z|}\Rightarrow |z|=1$

$z^2=\bar{z}\Rightarrow z^3=1 \qquad \therefore z=\omega$ or ω^2

$\omega^n=(1+\omega)^n=(-\omega^2)^n$

Least natural value of n is 6.

24. (A)

$z^5+(\bar{z})^5=(2+3i)^5+(2-3i)^5$

$=2\left({}^5C_0 2^5+{}^5C_2 2^3(3i)^2+{}^5C_4 2^1(3i)^4\right)$

$=2(32+10\times 8(-9)+5\times 2\times 81)=244$

25. (D)

$\left|z-\dfrac{1}{z}\right|=2$

$\left||z|-\dfrac{1}{|z|}\right|\le\left|z-\dfrac{1}{z}\right|\le|z|+\dfrac{1}{|z|}$

Let $|z|=r$

$\left|r-\dfrac{1}{r}\right|\le 2\le r+\dfrac{1}{r}$

$\left|r-\dfrac{1}{r}\right|\le 2$ and $r+\dfrac{1}{r}\ge 2$ always true

$r-\dfrac{1}{r}\ge -2$ and $r-\dfrac{1}{r}\le 2$

$r^2-1\le 2r$

$r^2-2r\le 1$

$(r-1)^2\le 2$

$r-1\le\sqrt{2}$

$\therefore |z|_{\max}=1+\sqrt{2}$

26. (B)

$|z-1+i|\ge|z|;|z|<2;|z+i|=|z-1|$

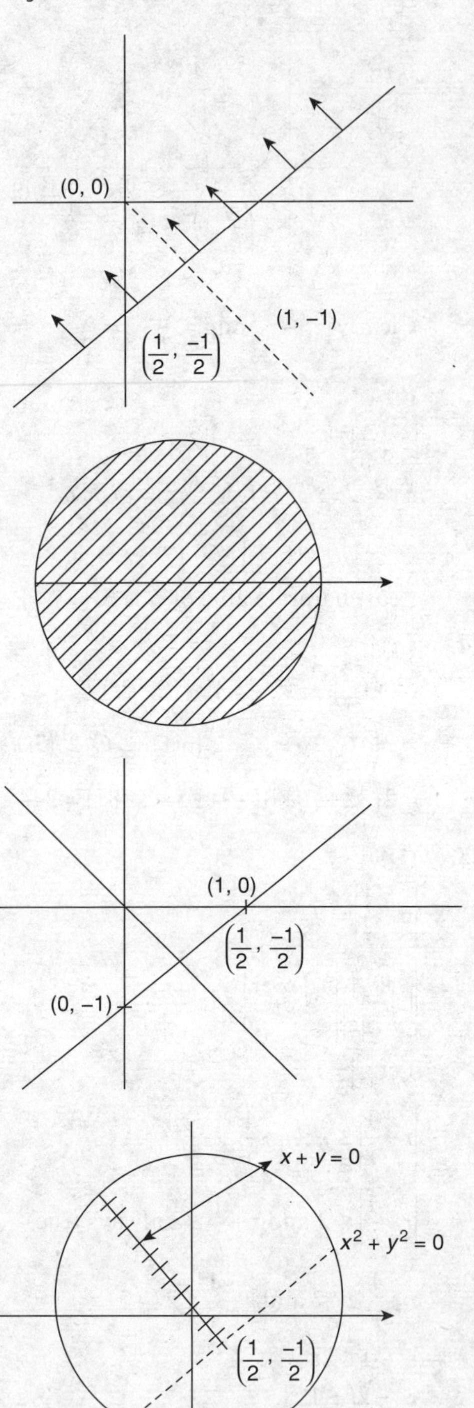

Hence

$w = 2x + iy \in S$

$2x \le \frac{1}{2} \qquad \therefore x \le \frac{1}{4}$

Now,

$(2x)^2 + (2x)^2 < 4$

$x^2 < \frac{1}{2} \Rightarrow x \in \left(-\frac{1}{\sqrt{2}}, \frac{1}{\sqrt{2}}\right)$

$\therefore x \in \left(-\frac{1}{\sqrt{2}}, \frac{1}{4}\right]$

27. (B)

$\left(1-\sqrt{3}i\right)^{200} = 2^{199}(p+iq)$

$2^{200}\left(\cos\frac{\pi}{3} - i\sin\frac{\pi}{3}\right)^{200} = 2^{199}(p+iq)$

$2\left(-\frac{1}{2} - i\frac{\sqrt{3}}{2}\right) = p + iq$

$p = -1, q = -\sqrt{3}$

$\alpha = p + q + q^2 = 2 - \sqrt{3}$

$\beta = p - q + q^2 = 2 + \sqrt{3}$

$\alpha + \beta = 4$

Equation $x^2 - 4x + 1 = 0$

28. (C)

Let $\sin\frac{2\pi}{9} + i\cos\frac{2\pi}{9} = z$

$\left(\frac{1+z}{1+\bar{z}}\right)^3 = \left(\frac{1+z}{1+\frac{1}{z}}\right)^3 = z^3$

$\Rightarrow \left(i\left(\cos\frac{2\pi}{9} - i\sin\frac{2\pi}{9}\right)\right)^3$

$\Rightarrow i\left(\cos\frac{2\pi}{3} - i\sin\frac{2\pi}{3}\right) = -i\left(\frac{-1}{2} - i\frac{\sqrt{3}}{2}\right)$

$\Rightarrow -\frac{1}{2}\left(\sqrt{3} - i\right)$

29. (A)

$\left((x-2)^2 + (y-3)^2\right) - \left((x-3)^2 - (y-4)^2\right)$

$= 1 + 1$

$\Rightarrow x + y = 7$

30. (D)

$(z - 2i)(\bar{z} + 2i) = 4(z + i)(\bar{z} - i)$

$z\bar{z} + 4 + 2i(z - \bar{z}) = 4\left(z\bar{z} + 1 + i(\bar{z} - z)\right)$

$$3z\bar{z}-6i(z-\bar{z})=0$$
$$x^2+y^2-2i(2iy)=0$$
$$x^2+y^2+4y=0$$

31. **(B)**

$z_1=x_1+iy_1$

$z_2=x_2+iy_2$

$\text{Re}(z_1z_2)=x_1x_2-y_1y_2=0$

$\text{Re}(z_1+z_2)=x_1+x_2=0$

x_1 and x_2 are of opposite sign

y_1 and y_2 are of opposite sign

32. **(14)**

$\alpha=8-14i$

$z=x+iy$

$az=(8x+14y)+i(-14x+8y)$

$z+\bar{z}=2x$

$z-\bar{z}=2iy$

Set ***A*:** $\dfrac{2i(-14x+8y)}{i(4xy-112)}=1$

$(x-4)(y+7)=0$

$x=4$ or $y=-7$

Set ***B*:** $x^2+(y+3)^2=16$

Set ***B*:** $x^2+(y+3)^2=16$

When $x=4$ $\quad y=-3$

When $y=-7$ $\quad x=0$

$\therefore A\cap B=\{4-3i, 0-7i\}$

So,

$$\sum_{z\in A\cap B}(\text{Re}\,z-\text{Im}\,z)=4-(-3)+(0-(-7))=14$$

33. **(9)**

$z=1+i$

$$z_1=\frac{1+i\bar{z}}{\bar{z}(1-z)+\frac{1}{z}}$$

$$z_1=\frac{1+i(1-i)}{(1-i)(1-1-i)+\frac{1}{1+i}}$$

$$=\frac{1+i-i^2}{(1-i)(-i)+\frac{1-i}{2}}$$

$$=\frac{2+i}{\frac{-3i-1}{2}}=\frac{4+2i}{-3i-1}$$

$$=\frac{-(4+2i)(3i-1)}{(3i)^2-(1)^2}$$

$\text{Arg}(z_1)=\dfrac{3\pi}{4}$

$\therefore \dfrac{12}{\pi}\arg(z_1)=\dfrac{12}{\pi}\times\dfrac{3\pi}{4}=9$

34. **(A)**

Let $w=z+\dfrac{1}{z}=4e^{i\theta}+\dfrac{1}{4}e^{-i\theta}$

$\Rightarrow w=\dfrac{17}{4}\cos\theta+i\dfrac{15}{4}\sin\theta$

So locus of w is ellipse $\dfrac{x^2}{\left(\frac{17}{4}\right)^2}+\dfrac{y^2}{\left(\frac{15}{4}\right)^2}=1$

Locus of z is circle $x^2+y^2=16$

35. **(A)**

$$Z=\frac{i-1}{\cos\frac{\pi}{3}+i\sin\frac{\pi}{3}}=\frac{i-1}{\frac{1}{2}+\frac{\sqrt{3}}{2}i}$$

$$=\frac{i-1}{\frac{1}{2}+\frac{\sqrt{3}}{2}i}\times\frac{\frac{1}{2}-\sqrt{\frac{3}{2}}i}{\frac{1}{2}-\sqrt{\frac{3}{2i}}}=\frac{\sqrt{3}-1}{2}+\frac{\sqrt{3}+1}{2}i$$

Apply polar form

$r\cos\theta=\dfrac{\sqrt{3}-1}{2}$

$r\sin\theta=\dfrac{\sqrt{3}+1}{2}$

Now, $\tan\theta=\dfrac{\sqrt{3}+1}{\sqrt{3}-1}$

$\theta=\dfrac{5\pi}{12}$

36. **(D)**

$$\sqrt{(x-2)^2+y^2}=2\sqrt{(x-3)^2+y^2}$$
$$=x^2+y^2-4x+4=4x^2+4y^2-24x+36$$
$$=3x^2+3y^2-20x+32=0$$

$= x^2 + y^2 - \frac{20}{3}x + \frac{32}{3} = 0$

$= (\alpha, \beta) = \left(\frac{10}{3}, 0\right)$

$\gamma = \sqrt{\frac{100}{9} - \frac{32}{3}} = \sqrt{\frac{4}{9}} = \frac{2}{3}$

$3(\alpha, \beta, \gamma) = 3\left(\frac{10}{3} + \frac{2}{3}\right)$

$= 12$

37. **(A)**

$ab < 0 \left|\frac{1+ai}{b+i}\right| = 1$

$|1+ai| = |b+i|$

$a^2 + 1 = b^2 + 1 \Rightarrow a = \pm b \Rightarrow b = -a$ as $ab < 0$

(a, b) lies on $|z-1| = |2z|$

$|a+ib-1| = 2|a+ib|$

$(a-1)^2 + b^2 = 4(a^2 + b^2)$

$(a-1)^2 = a^2 = 4(2a^2)$

$1 - 2a = 6a^2 \Rightarrow 6a^2 + 2a - 1 = 0$

$a = \frac{-2 \pm \sqrt{28}}{12} = \frac{-1 \pm \sqrt{7}}{6}$

$a = \frac{\sqrt{7}-1}{6}$ and $b = \frac{1-\sqrt{7}}{6}$

$[a] = 0$

$\therefore \frac{1+[a]}{4b} = \frac{6}{4(1-\sqrt{7})} = -\left(\frac{1+\sqrt{7}}{4}\right)$

or $[a] = 0$

Similarly it is not matching with $a = \frac{-1-\sqrt{7}}{6}$

No answer is matching

38. **(A)**

Given,

$X = \{z \in C : \text{Re}(az^2 + bz) = a$

and $\text{Re}(bz^2 + az) = b\}$

Now, let $z = x + iy$

So, $\text{Re}(az^2 + bz) = a$

$\Rightarrow \text{Re}\left(a(x+iy)^2 + b(x+iy)\right) = a$

$\Rightarrow \text{Re}\left(a(x^2 - y^2 + 2ixy) + b(x+iy)\right) = a$

$\Rightarrow a(x^2 - y^2) + bx = a$ (1)

And, $\text{Re}(bz^2 + az) = b$

$\Rightarrow \text{Re}\left(b(x+iy)^2 + a(x+iy)\right) = b$

$\Rightarrow \text{Re}\left(b(x^2 - y^2 + 2ixy) + a(x+iy)\right) = b$

$\Rightarrow b(x^2 - y^2 + 2ixy) + a(x+iy) = b$ (2)

From (1) and (2), (1) – (2) we get,

$(x^2 - y^2)(a-b) - x(a-b) = a - b$

$\Rightarrow x^2 - y^2 - x = 1$ (3)

From (1) and (2), (1) + (2)

$\left((x^2 - y^2) + x - 1\right)(a+b) = 0$

(**Note:** here $a + b \neq 0$ is considered but it is not clear from the question)

$\Rightarrow x^2 - y^2 + x - 1 = 0$ (4)

Now from (3) and (4) we get,

$x = 0, y^2 = -1$ (No solution)

39. **(B)**

Given

$\alpha, \beta, z \in C$ and $\lambda > 1$,

And $\sqrt{\lambda - 1}$ is the radius of the circle

$|z^2 - \alpha|^2 + |z - \beta|^2 = 2\lambda$

Now we know that equation of the circle is given by,

$|z - z_1|^2 + |z - z_2|^2 = |z_1 - z_2|^2$,

where radius is given by $r = \frac{|z_1 - z_2|}{2}$,

On comparing with $|z - \alpha|^2 + |z - \beta|^2 = 2\lambda$, we get

Radius $\left|\frac{z_1 - z_2}{2}\right| = \frac{|\alpha - \beta|}{2}$ and $|\alpha - \beta|^2 = 2\lambda$

$\Rightarrow \frac{|\alpha - \beta|}{2} = \sqrt{\lambda - 1}$ (given)

$\Rightarrow |\alpha - \beta| = 2\sqrt{\lambda - 1}$

$\Rightarrow |\alpha - \beta|^2 = 4(\lambda - 1)$

$\Rightarrow 2\lambda = 4(\lambda - 1)$

$\Rightarrow \lambda = 2$

Hence, $|\alpha-\beta|^2 = 4 \Rightarrow |\alpha-\beta| = 2$

40. (B)

Given,

$z = \alpha + i\beta$ and $|z+2| = z + 4(1+i)$

Now putting the value of $z = \alpha + i\beta$ in $|z+2| = z+4(1+i)$ we get,

$|z+2| = z+4(1+i)$

$\Rightarrow \sqrt{(\alpha+2)^2+\beta^2} = (\alpha+4)+i(\beta+4)$

Now on comparing real and imaginary part, we get

$\sqrt{(\alpha+2)^2+\beta^2} = \alpha+4$ (1)

and $\beta+4=0 \Rightarrow \beta=-4$ (2)

Now solving,

$\sqrt{(\alpha+2)^2+16} = \alpha+4$

$\Rightarrow \alpha^2+4\alpha+20 = \alpha^2+8\alpha+16$

$\Rightarrow \alpha = 1$

So, $\alpha+\beta=-3$, $\alpha\beta=-4$

We know that quadratic equation is given by,

$x^2-(\text{sum of roots})x+(\text{product of roots})=0$

So, equation with roots -3 and -4 will be,

$x^2+7x+12=0$

41. (A)

Let

$$z = \frac{1+2i\sin\theta}{1-i\sin\theta}$$

$$\Rightarrow z = \frac{1+2i\sin\theta}{1-i\sin\theta}\times\frac{1+i\sin\theta}{1+i\sin\theta}$$

$$\Rightarrow z = \frac{1-2\sin^2\theta+3i\sin\theta}{1+\sin^2\theta}$$

$$\Rightarrow z = \left(\frac{1-2\sin^2\theta}{1+\sin^2\theta}\right)+i\left(\frac{3\sin\theta}{1+\sin^2\theta}\right)$$

Since, z is a purely imaginary number, so the real part must be zero, hence

$$\frac{1-2\sin^2\theta}{1+\sin^2\theta}=0$$

$\Rightarrow 1-2\sin^2\theta = 0$

$\Rightarrow \cos 2\theta = 0$

$$\Rightarrow \theta = \frac{\pi}{4}, \frac{3\pi}{4}, \frac{5\pi}{4}, \frac{7\pi}{4}, \text{ for } \theta\in(0, 2\pi)$$

$$\therefore \text{Sum of all values} = \frac{16\pi}{4} = 4\pi$$

42. (C)

Given,

The complex number $z = x+iy$ be such that $\frac{2z-3i}{2z+i}$ is purely imaginary,

Now putting the value of $z=x+iy$ in $\frac{2z-3i}{2z+i}$ we get,

$$\frac{2(x+iy)-3i}{2(x+iy)+i}$$

$$=\frac{(2x+i(2y-3))(2x-i(2y+1))}{(2x+i(2y+1))(2x-i(2y+1))}$$

Now taking real part as,

$$\frac{4x^2+(2y-3)(2y+1)}{4x^2+(2y+1)^2}=0$$

$\Rightarrow 4x^2+4y^2-4y-3=0$

$$\Rightarrow x^2+y^2-y-\frac{3}{4}=0$$

Now using,

$x+y^2=0 \Rightarrow x=-y^2$ we get,

$$x^2+y^2-y-\frac{3}{4}=0$$

$$\Rightarrow y^4+y^2-y-\frac{3}{4}=0$$

$$\Rightarrow y^4+y^2-y=\frac{3}{4}$$

43. (B)

We have,

$$Z=\frac{2z-3i}{4z+2i}$$

$$\Rightarrow X = \frac{2x+i(2y-3)}{4x+i(4y+2)}$$

$$\Rightarrow Z = \frac{2x+i(2y-3)}{4x+i(4y+2)}\times\frac{4x-i(4y+2)}{4x-i(4y+2)}$$

$$\Rightarrow Z=\frac{8x^2+(4y+2)(2y-3)+i\left[4x(2y-3)-(8xy+4x)\right]}{16x^2+(4y+2)^2}$$

Since, Z is purely real, so

$$\frac{4x(2y-3)-(8xy+4x)}{16x^2+(4y+2)^2}=0$$

$$\Rightarrow -12x+8xy-(8xy+4x)=0$$

$$\Rightarrow x=0$$

And,

$$4x+i(4y+2)\neq 0$$

$$\Rightarrow i(4y+2)\neq 0$$

$$\Rightarrow y\neq -\frac{1}{2}$$

Hence this is the required option.

44. (A)

Given

two complex numbers w_1 and w_2

$w_1=3+5i$ and $w_2=5+4i$ are both rotated by 90° with respect to the origin, anticlockwise and clockwise respectively,

So by concept of rotation we get,

$w_3=iw_1=i(3+5i)=-5+3i$

And $w_4=iw_2=-i(5+4i)=4-5i$

Now principal argument of $w_3-w_4=-9+8i$ will be $\pi-\tan^{-1}\frac{8}{9}$ {as complex number is in second quadrant so principal argument is given by $\pi-\tan^{-1}\left|\frac{y}{x}\right|$}

45. (D)

Let $a=x_1+iy_1, \bar{a}=x_1-iy_1$

And $z=x_2+iy_2, \bar{z}=x_2-iy_2$

Now,

$a+\bar{z}=(x_1+x_2)+i(y_1-y_2)$

and $\bar{a}+z=(x_1+x_2)+i(y_2-y_1)$

Now according to the question,

$\text{Re}(a+\bar{z})=x_1+x_2$ and

$\text{Im}(\bar{a}+z)=y_2-y_1$

$$A=\{z:x_1+x_2>y_2-y_1\}$$
$$=\{z:x_1+y_1+x_2>y_2\}$$

$$B=\{z:x_1+x_2<y_2-y_1\}$$
$$=\{z:x_1+y_1+x_2<y_2\}$$

If $y_2=0$ and $x_1, y_1>0$ then

$$A=\{z:x_2>-(x_1+y_1)\}$$

A covers a part of negative real axis and therefore, does not contain whole real axis

Similarly if $y_2=0$, and $x_1, y_1<0$, then

$$B=\{z:x_2<-(x_1+y_1)\}$$

$\therefore B$ covers part of positive real axis and therefore does not cover whole real axis.

Hence, both are false.

46. (1680)

Put $z=x+iy$ in the given expression and equate the imaginary part to zero.

$$\Rightarrow \frac{z^2+8iz-15}{z^2-3iz-2}=\frac{(x+iy)^2+8i(x+iy)-15}{(x+iy)^2-3i(x+iy)-2}$$

$$\Rightarrow \frac{x^2-y^2-8y-15+i(2xy+8x)}{x^2-y^2+3y-2+i(2xy-3x)}$$

$$=\frac{x^2-y^2-8y-15+i(2xy+8x)}{x^2-y^2+3y-2+i(2xy-3x)}\times\frac{(x^2-y^2+3y-2-i(2xy-3x))}{(x^2-y^2+3y-2-i(2xy-3x))}$$

$$=\frac{\begin{array}{l}(x^2-y^2-8y-15)(x^2-y^2+3y-2)\\+(2xy+8x)(2xy-3x)\\+i(2xy+8x)(x^2-y^2+3y-2)\\-i(2xy-3x)(x^2-y^2-8y-15)\end{array}}{(x^2-y^2+3y-2)^2-(2xy-3x)^2}$$

But $\text{Im}\left(\frac{z^2+8iz-15}{z^2-3iz-2}\right)=0$,

$$\Rightarrow -(x^2-y^2-8y-15)(2xy-3x)+(2xy+8x)(x^2-y^2+3y-2)=0$$

$$\Rightarrow (x^2-y^2)(2xy+8x-2xy+3x)+(8y+15)(2xy-3x)+(2xy+8x)(3y-2)=0$$

$$\Rightarrow 11x^3-11xy^2+16xy^2-24xy+30xy-45x+6xy^2-4xy+24xy-16x-0$$

$\Rightarrow 11x^3 + 11xy^2 + 26xy - 61x = 0$

$\Rightarrow 11x^2 + 11y^2 + 26y - 61 = 0$

$\because \alpha \neq 0$

So, put $y = -\frac{13}{11}, x = \alpha$

$11\alpha^2 + 11 \cdot \frac{13^2}{11^2} - 26 \cdot \frac{13}{11} - 61 = 0$

$\Rightarrow 121\alpha^2 = 840$

$\Rightarrow 242\alpha^2 = 1680$

Hence this is the required answer.

47. (A)

Given,

$z_0 = \frac{1+3i}{2}, z_1 = (1+i)$

So,

$|z_1 - z_0| = \sqrt{\left(1-\frac{1}{2}\right)^2 + \left(1-\frac{3}{2}\right)^2}$

$= \sqrt{\frac{1}{4} + \frac{1}{4}} = \frac{1}{\sqrt{2}}$

And z_2 be outside circle, $|z_1 - z_0||z_2 - z_0| = 1$

$\Rightarrow \frac{1}{\sqrt{2}}|z_2 - z_0| = 1$

$\Rightarrow |z_2 - z_0| = \sqrt{2}$

Now given, z_0, z_1 and z_2 are collinear,

So, by concept of rotation we get,

$\frac{z_2 - z_0}{z_1 - z_0} = \frac{|z_2 - z_0|}{|z_1 - z_0|} e^{i0}$

$\Rightarrow \frac{z_2 - z_0}{z_1 - z_0} = \frac{|z_2 - z_0|}{|z_1 - z_0|}(\pm 1)$

$\Rightarrow \frac{z_2 - z_0}{z_1 - z_0} = \pm 2$

$\Rightarrow z_2 = z_0 \pm 2(z_1 - z_0)$

So, $z_2 = 2z_1 - z_0 = \frac{3}{2} + \frac{1}{2}i \Rightarrow |z_2|^2 = \frac{5}{2}$

Or $z_2 = 3z_0 - 2z_1 = \frac{-1}{2} + \frac{5}{2}i \Rightarrow |z_2|^2 = \frac{13}{2}$

Hence, the smaller value will be, $|z_2|^2 = \frac{5}{2}$

48. (24)

Given,

$w = z\bar{z} + k_1 z + k_2 iz + \lambda(1+i), k_1, k_2 \in R.$

And $\text{Re}(w) = 0$,

So, taking $z = x + iy$ and $\bar{z} = x - iy$

We get,

$w = x^2 + y^2 + k_1(x+iy) + k_2 i(x+iy) + \lambda(1+i)$

$\Rightarrow \text{Re}(w) = 0 \Rightarrow x^2 + y^2 + k_1 x - k_2 y + \lambda = 0$

And, $\text{Im}(w) = 0 \Rightarrow k_1 y + k_2 x + \lambda = 0$

Now given $\text{Re}(w) = 0$ circle is touching $y = 1$ and y-axis, with radius 1, so we have the equation of a circle

$x^2 + y^2 - 2x - 4y + 4 = 0$

So, on comparing with $\text{Re}(w) = 0$ we get,

$k_1 = -2, k_2 = 4, \lambda = 4$

So, $\text{Im}(w) = 0 \Rightarrow -2y + 4x + 4 = 0$

And given $-2y + 4x + 4 = 0$ intersects the circle,

So, solving $x^2 + y^2 - 2x - 4y + 4 = 0$ and $-2y + 4x + 4 = 0$ we get,

Intersecting point as $A(0, 2)$ and $B\left(\frac{2}{5}, \frac{14}{5}\right)$

Hence, by distance formula we get,

$30(AB)^2 = 30\left[\left(0-\frac{2}{5}\right)^2 + \left(2-\frac{14}{5}\right)^2\right] = 24$

49. (B)

Given,

$\bar{z} = i(z^2 + \text{Re}(\bar{z}))$

Now, let $z = x + iy$, so $\bar{z} = x - iy$

Now putting the value in, $\bar{z} = i(z^2 + \text{Re}(\bar{z}))$ we get,

$\bar{z} = i(z^2 + \text{Re}(\bar{z}))$

$\Rightarrow x - iy = i(x^2 - y^2 + 2ixy + x)$

$\Rightarrow x - iy = i(x^2 - y^2 + x) - 2xy$

Now comparing real part we get,

$\Rightarrow x = -2xy \Rightarrow x(2y+1) = 0$

$\Rightarrow x = 0 y = -\frac{1}{2}$ (1)

And imaginary part we get,

$= y = x^2 - y^2 + x$ (2)

Now taking Case (I) when $x = 0$ in equation (2) we get,

$\Rightarrow -y = -y^2$

$\Rightarrow y^2 - y = 0 \Rightarrow y = 0, 1$

So, $z = 0, i$

Now taking Case (II) when $y = \frac{-1}{2}$ in equation (2) we get

$\Rightarrow \frac{1}{2} = x^2 - \frac{1}{4} + x$

$\Rightarrow x^2 + x - \frac{3}{4} = 0$

$\Rightarrow 4x + 4x - 3 = 0$

$\Rightarrow (2x-1)(2x+3) = 0$

$\Rightarrow x = \frac{1}{2}, \frac{-3}{2}$

So, $z = \frac{1}{2} - \frac{1}{2}i, \frac{-3}{2} - \frac{1}{2}i$

Now finding, $\sum |z|^2 = 0 + 1 + \frac{1}{2} + \frac{5}{2} = 4$

50. **(C)**

Given that the set

$\left\{\text{Re}\left(\frac{z - \bar{z} + z\bar{z}}{2 - 3z + 5\bar{z}}\right) : z \in C, \text{Re}\, z = 3\right\}$ is equal to the interval $(\alpha, \beta]$.

Let $z = x + iy$

$= \text{Re}\left(\frac{x + iy - (x - iy) + x^2 + y^2}{2 - 3(x + iy) + 5(x - iy)}\right)$

$= \text{Re}\left(\frac{x^2 + y^2 + i(2y)}{2 + 2x - 8iy}\right)$

$= \text{Re}\left(\frac{(x^2 + y^2 + 2yi)(2(1+x) + 8iy)}{(2(1+x))^2 + (8y)^2}\right)$

$= \frac{2(x^2 + y^2)(1+x) - 16y^2}{4(1+x)^2 + (8y)^2}$

Now using, $\text{Re}(z) = 3 \Rightarrow x = 3$ we get,

$= \frac{8(9 + y^2) - 16y^2}{64 + 64y^2}$

$\Rightarrow f(y) = \frac{1}{8}\frac{(9 - y^2)}{1 + y^2}$

Let $t = \frac{1}{8}\frac{(9 - y^2)}{1 + y^2}$

$\Rightarrow 8t + 8ty^2 = 9 - y^2$

$\Rightarrow y^2(8t + 1) + 8t - 9 = 0$

Now $D \geq 0$

$\Rightarrow 0^2 - 4(8t + 1)(8t - 9) \geq 0$

$\Rightarrow \left(t - \left(\frac{-1}{8}\right)\right)\left(t - \frac{9}{8}\right) \leq 0$

$\Rightarrow t \in \left(\frac{-1}{8}, \frac{9}{8}\right]$ as $t \neq \frac{-1}{8}$

Range of $f(y) = (-0.125, 1.125)$

$\Rightarrow \alpha = -0.125$

$\Rightarrow \beta = 1.125$

$\Rightarrow \beta - \alpha = 1.25$

$\Rightarrow 24(\beta - \alpha) = 30$

Hence this is the required solution.

51. **(D)**

Let $Z = x + iy$

Then $(x-1)^2 + y^2 = 1$ (1)

and $(\sqrt{2} - 1)(2x) - i(2iy) = 2\sqrt{2}$

$\Rightarrow (\sqrt{2} - 1)x + y = \sqrt{2}$ (2)

Solving (1) and (2)
we get

Either $x = 1$ or $x = \frac{1}{2 - \sqrt{2}}$ (3)

On solving (3) and (2)
we get

For $x = 1 \Rightarrow y = 1 \Rightarrow Z_2 = 1 + i$ and for

$x = \frac{1}{2 - \sqrt{2}} \Rightarrow y = \sqrt{2} - \frac{1}{\sqrt{2}}$

$\Rightarrow Z_1 = \left(1 + \frac{1}{\sqrt{2}}\right) + \frac{i}{\sqrt{2}}$

Now

$\left|\sqrt{2}z_1 - z_2\right|^2$

$$=\left|\left(\frac{1}{\sqrt{2}}+1\right)\sqrt{2}+i-(1+i)\right|^2$$

$$=\left(\sqrt{2}\right)^2$$

$$=2$$

52. **(36)**

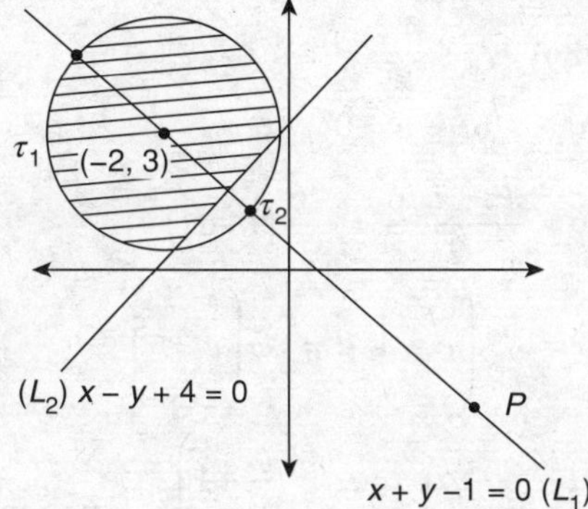

Clearly for the shaded region z_1 is the intersection of the circle and the line passing through $P(L_1)$ and z_2 is intersection of line L_1 and L_2.

Circle: $(x+2)^2+(y-3)^2=1$

$L_1: x+y-1=0$

$L_2: x-y+4=0$

On solving circle and L_1 we get

$$z_1:\left(-2-\frac{1}{\sqrt{2}},3+\frac{1}{\sqrt{2}}\right)$$

On solving L_1 andz_2 is intersection of line L_1 and L_2

we get $z_2:\left(\frac{-3}{2},\frac{5}{2}\right)$

$$|z_1|^2+2|z_2|^2=14+5\sqrt{2}+17$$

$$=31+5\sqrt{2}$$

$\alpha=31$

So $\beta=5$

$\alpha+\beta=36$

53. **(D)**

$|z|\geq 1$

Min. value of $\left|z+\frac{3}{2}+2i\right|$ is actually zero.

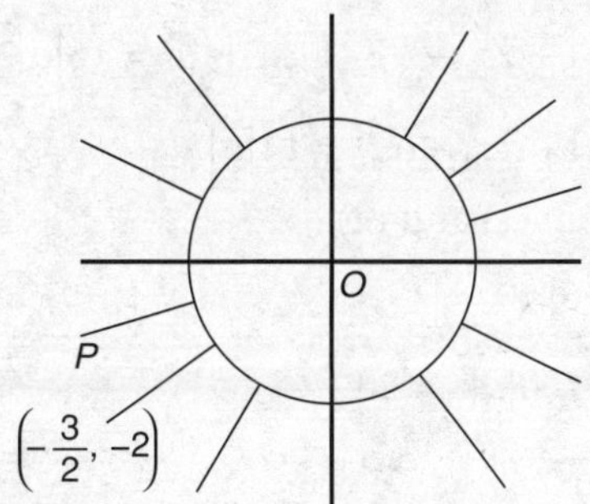

54. **(A)**

$|z-i|=|z+i|=|z-1|$

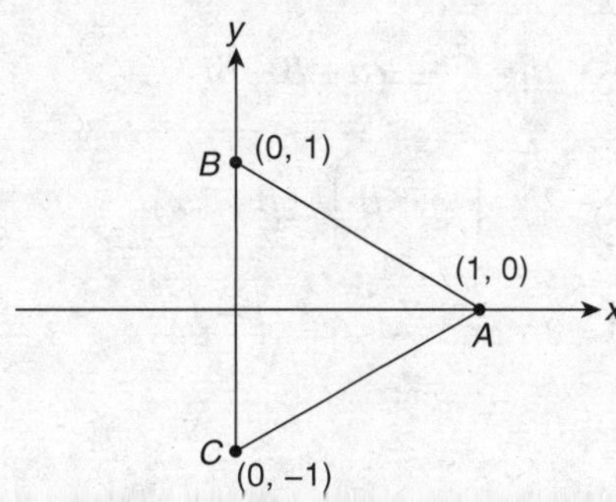

ABC is a triangle. Hence its circumcentre will be the only point whose distance from A, B, C will be same.

So $n(S)=1$

55. **(5)**

$x^2+x+1=0\Rightarrow x=\omega,\ \omega^2=\alpha$

Let $\alpha=\omega$

Now $(1+\alpha)^7=-\omega^{14}=-\omega^2=1+\omega$

$A=1, B=1, C=0$

$\therefore 5(3A-2B-C)=5(3-2-0)=5$

56. **(20)**

$|z-z_0|^2=4$

$\Rightarrow(\alpha-z_0)(\bar{\alpha}-\bar{z}_0)=4$

$\Rightarrow\alpha\bar{\alpha}-\alpha\bar{z}_0-z_0\bar{\alpha}+|z_0|^2=4$

$\Rightarrow|\alpha|^2-\alpha\bar{z}_0-z_0\bar{\alpha}=2$ (1)

$|z-z_0|^2=16$

$$\Rightarrow\left(\frac{1}{\bar{\alpha}}-z_0\right)\left(\frac{1}{\alpha}-\bar{z}_0\right)=16$$

$\Rightarrow(1-\bar{\alpha}z_0)(1-\alpha\bar{z}_0)=16|\alpha|^2$

$\Rightarrow 1-\bar{\alpha}z_0-\alpha\bar{z}_0+|\alpha|^2|z_0|^2=16|\alpha|^2$

$\Rightarrow 1-\bar{\alpha}z_0-\alpha\bar{z}_0=14|\alpha|^2$ (2)

from (1) and (2)

$\Rightarrow 5|\alpha|^2=1$

$\Rightarrow 100|\alpha|^2=20$

57. **(B)**

$z=\frac{1}{2}-2i$

$|z+1|=\alpha z+\beta(1+i)$

$\left|\frac{3}{2}-2i\right|=\frac{\alpha}{2}-2\alpha i+\beta+\beta i$

$\left|\frac{3}{2}-2i\right|=\left(\frac{\alpha}{2}+\beta\right)+(\beta-2\alpha)i$

$\beta=2\alpha$ and $\frac{\alpha}{2}+\beta=\sqrt{\frac{9}{4}+4}$

$\alpha+\beta=3$

58. **(13)**

$\alpha^6+\alpha^4+\beta^4-5\alpha^2$

$=\alpha^4(\alpha-2)+\alpha^4-5\alpha^2+(\beta-2)^2$

$=\alpha^5-\alpha^4-5\alpha^2+\beta^2-4\beta+4$

$=\alpha^3(\alpha-2)-\alpha^4-5\alpha^2+\beta-2-4\beta+4$

$=-2\alpha^3-5\alpha^2-3\beta+2$

$=-2\alpha(\alpha-2)-5\alpha^2-3\beta+2$

$=-7\alpha^2+4\alpha-3\beta+2$

$=-7(\alpha-2)+4\alpha-3\beta+2$

$=-3\alpha-3\beta+16=-3(1)+16=13$

59. **(A)**

$z=2-i\left(2\tan\frac{5\pi}{8}\right)=x+iy$ (let)

$r=\sqrt{x^2+y^2}$ and $\theta=\tan^{-1}\frac{y}{x}$

$r=\sqrt{(2)^2+\left(2\tan\frac{5\pi}{8}\right)^2}$

$=\left|2\sec\frac{5\pi}{8}\right|=\left|2\sec\left(\pi-\frac{3\pi}{8}\right)\right|$

$=2\sec\frac{3\pi}{8}$

and $\theta=\tan^{-1}\left(\frac{-2\tan\frac{5\pi}{8}}{2}\right)$

$=\tan^{-1}\left(\tan\left(\pi-\frac{5\pi}{8}\right)\right)$

$=\frac{3\pi}{8}$

60. **(49)**

$x^2-\sqrt{6}x+6=0^{\alpha}_{\Delta\beta}$

$x=\frac{\sqrt{6}\pm i\sqrt{6}}{2}=\frac{\sqrt{6}}{2}(1+\pm i)$

$\alpha=\sqrt{3}\left(e^{i\frac{\pi}{4}}\right),\beta=\sqrt{3}\left(e^{-i\frac{\pi}{4}}\right)$

$\therefore\frac{\alpha^{99}}{\beta}+\alpha^{98}=\alpha^{98}\left(\frac{\alpha}{\beta}+1\right)$

$=\frac{\alpha^{98}(\alpha+\beta)}{\beta}=3^{49}\left(e^{i99\frac{\pi}{4}}\right)\times\sqrt{2}$

$=3^{49}(-1+i)$

$=3^n(a+ib)$

$\therefore n=49, a=-1, b=1$

$\therefore n+a+b=49-1+1=49$

61. **(B)**

$z^2=-i\bar{z}$

$|z^2|=|iz|$

$|z^2|=|z|$

$|z^2|-|z|=0$

$|z|(|z|-1)=0$

$|z|=0$ (not acceptable)

$\therefore|z|=1$

$\therefore|z|^2=1$

62. **(B)**

$z^{1985}+z^{100}+1=0$ and $z^3+2z^2+2z+1=0$

$(z+1)(z^2-z+1)+2z(z+1)=0$

$(z+1)(z^2+z+1)=0$

$\Rightarrow z=-1, z=\omega,\omega^2$

Now putting $z=-1$ does not satisfy

Now put $z = w$

$\Rightarrow w^{1985} + w^{100} + 1$

$\Rightarrow w^2 + w + 1 = 0$

Also, $z = w^2$

$\Rightarrow w^{3970} + w^{200} + 1$

$\Rightarrow w + w^2 + 1 = 0$

Two common roots.

63. **(3)**

$(\sqrt{2})^x = 2^x \Rightarrow x = 0 \Rightarrow \alpha = 1$

$$z = \frac{\pi}{4}(1+i)^4\left[\frac{\sqrt{\pi} - \pi i - i - \sqrt{\pi}}{\pi + 1} + \frac{\sqrt{\pi} - i - \pi i - \sqrt{\pi}}{1+\pi}\right]$$

$$= -\frac{\pi i}{2}\left(1 + 4i + 6i^2 + 4i^3 + 1\right)$$

$$= 2\pi i$$

$$\beta = \frac{2\pi}{\frac{\pi}{2}} = 4$$

Distance from (1, 4) to $4x - 3y = 7$

Will be $\frac{15}{5} = 3$

64. **(B)**

$z_1 + z_2 = 5$

$z_1^3 + z_2^3 = 20 + 15i$

$z_1^3 + z_2^3 = (z_1 + z_2)^3 - 3z_1z_2(z_1 + z_2)$

$z_1^3 + z_2^3 = 125 - 3z_1 \cdot z_2(5)$

$\Rightarrow 20 + 15i = 125 - 15z_1z_2$

$\Rightarrow 3z_1z_2 = 25 - 4 - 3i$

$\Rightarrow 3z_1z_2 = 21 - 3i$

$\Rightarrow z_1 \cdot z_2 = 7 - i$

$\Rightarrow (z_1 + z_2)^2 = 25$

$\Rightarrow z_1^2 + z_2^2 = 25 - 2(7 - i)$

$\Rightarrow 11 + 2i$

$(z_1^2 + z_2^2)^2 = 121 - 4 + 44i$

$\Rightarrow z_1^4 + z_2^4 = 117 + 44i - 2(49 - 1 - 14i)$

$\Rightarrow z_1^4 + z_2^4 = 117 + 44i - 2(49 - 1 - 14i)$

$\Rightarrow \left|z_1^4 + z_2^4\right| = 75$

3.19 Summary

- A number of the form $a + ib$ where $a, b \in R$ and $i = \sqrt{-1}$ is called a compex number and is denoted by z, that is, $z = a + ib$.
- If $z = a + ib$, then its conjugate is denoted by $\bar{z}$ and defined as $\bar{z} = a - ib$
  - $\text{Re}(z) = \frac{z + \bar{z}}{2}$
  - $\text{Im}(z) = \frac{z - \bar{z}}{2i}$
- If $z = a + ib$, then modulus of z is denoted by $|z|$ and is defined as $|z| = \sqrt{a^2 + b^2}$
  - $|z|^2 = z\,\bar{z}$
  - $|\overline{z_1 + z_2 + \ldots + z_n}| = |\bar{z}_1 + \bar{z}_2 + \ldots + \bar{z}_n|$ $= |\overline{z_1 + z_2 + \cdots + z_n}|$
- The principal value of the argument lies in the interval $(-\pi, \pi]$.
- θ, $\pi - \theta$, $\theta - \pi$ and $-\theta$ are the principal arguments in I, II, III and IVth quadrants.
- Euler's formula is $e^{ix} = \cos x + i \sin x$, $e^{-ix} = \cos x - i \sin x$.
- De Moivre's theorem states that if n is an integer positive or negative the value of $(\cos\theta + i\sin\theta)^n = \cos n\theta + i \sin n\theta$
- Cube roots of unity are 1, ω, ω^2 where $\omega = \frac{-1 + i\sqrt{3}}{2}$ and $\omega^2 = \frac{-1 - i\sqrt{3}}{2}$
  - $1 + \omega + \omega^2 = 0$
  - $\omega^3 = 1$
  - $1 + \omega^n + \omega^{2n} = 3$ or 0; according to whether n is a multiple of 3 or not.
- n^{th} roots of unity are 1, α, α^2, ..., α^{n-1}, where $\alpha = e^{i\frac{2\pi}{n}}$.
- n^{th} roots of unity denote the vertices of a regular polygon of n sides inscribed in a circle of radius unity.
- Concept of rotation has been discussed in theory.

- Equation of straight line passing through z_1 and z_2 is given by $\begin{vmatrix} z & \bar{z} & 1 \\ z_1 & \bar{z}_1 & 1 \\ z_2 & \bar{z}_2 & 1 \end{vmatrix} = 0$.
- Equation of circle having the centre at z_0 and radius r is given by
  $z\bar{z} - \bar{z}_0 z - z_0\bar{z} + |z_0|^2 - r^2 = 0$.
- Geometrical properties of complex numbers are illustrated point wise.

CHAPTER 4

Permutation and Combination

OBJECTIVES

4.1 Introduction
4.2 Permutation
4.3 Fundamental Principle of Counting
4.4 Circular Permutation
4.5 Sum of Numbers Formed Using the Digits
4.6 Tests of Divisibility
4.7 Rank of Words
4.8 Combination
4.9 De-arrangement
4.10 Multinomial Theorem
4.11 Non-negative Integral Solution Method
4.12 Principle of Inclusion and Exclusion
4.13 Formation of Geometrical Figures
4.14 Know Your Concepts
4.15 JEE Main: Sample Questions
4.16 JEE Advanced: Sample Questions
4.17 JEE Main: Archive
4.18 Answer Key
4.19 Solutions
4.20 Summary

4.1 Introduction

This chapter is concerned with various methods of counting and choosing. We can use 'Permutation and Combination' in any kind of problem where many possibilities of different events exist. These concepts are now used to solve many problems that we come across in modern subjects like meteorology, astronomy, botany, zoology and software industries. The process of counting without actually counting each possibility such as, addition and multiplication rule, Bijection rule, inclusion–exclusion principle. Here, we are going to discuss some of the above methods in particular.

4.2 Permutation

Each of the different arrangements that can be made by taking some or all of a number of given things or objects at a time is called a *permutation*. In permutation, the order of appearance of things is taken into account.

Example

Suppose we want to find the number of ways in which 2 letters from the set of three letters A, B, C can be arranged. The complete list of possible permutations would be AB, AC, BA, BC, CA and CB. Thus, six arrangements can be made with three distinct letters taking two at a time.

Notation

The number of permutations of n different things taken r at a time $(1 \le r \le n)$ is denoted by the symbol ${}^nP_r$.

4.2.1 To Find the Number of Permutations of *n* Different Things Taken *r* at a Time

The number of permutations of r different things out of n different things is equal to number of ways of filling up r vacant places with n different things and is denoted by ${}^nP_r$.

Let the n different things be $a_1, a_2, a_3 \ldots, a_n$

1st	2nd	3rd	----------------	$(r-1)^{th}$	r^{th}
n ways	$(n-1)$ ways	$(n-2)$ ways			$n-(r-1)$ ways

Number of ways of filling up 1st place $= n$
Number of ways of filling up 2nd place $= (n-1)$
Number of ways of filling up 3rd place $= (n-2)$

Number of ways of filling up the rth place $= n-(r-1)$
$= (n-r+1)$

By the multiplication rule of counting first, second, third ... r^{th} place can be together filled up by

$$n(n-1)(n-2)\ldots(n-r+1) \text{ ways}$$

Hence,

$$n_{p_r} = \frac{[n(n-1)(n-2)\ldots(n-r+1)][(n-r)\ldots 3\cdot 2\cdot 1]}{[(n-r)\ldots 3\cdot 2\cdot 1]}$$

$$\Rightarrow \qquad n_{p_r} = \frac{n!}{(n-r)!}$$

Number of permutations of n different things taken all at a time.

1st	2nd	3rd	-----------------	n^{th}

Number of ways of filling up the first place = n
Number of ways of filling up the second place = $(n-1)$

Number of ways of filling up the nth place = 1

By multiplication rule,

The number of ways of filling up first, second, and n^{th} places together = $n(n-1)(n-2)\ldots 2\cdot 1 = n!$
Thus, $n_{p_n} = n!$

4.2.2 Factorial 0!

Since, the number of permutations of n different things taken all at a time = $n!$ (4.1)

Also, $$n_{p_n} = \frac{n!}{0!} \qquad (4.2)$$

Hence, $n! = \frac{n!}{0!}$, which is valid only when $0! = 1$
Hence, $0! = 1$

4.2.3 Find the Number of Permutations of *n* Different Things Taken *r* at a Time when Repetition is Allowed

Number of permutations of n different things taken r at a time when repetition is allowed = Number of ways in which r blank places can be filled up by n different things when repetition is allowed.

Therefore, Number of ways in which first place can be filled up = n

Number of ways in which second place can be filled up = n

Number of ways in which the rth place can be filled up = n

Hence by multiplication rule of fundamental principle of counting, number of ways in which first, second..., r^{th} places can be together filled up by $n \times n \times n \times \ldots$ to r factors = n^r.

4.3 Fundamental Principle of Counting

1. **Multiplication rule:** If a work A can be done in m ways and another work B can be done in n ways, then the work A and B can be done in $(m \times n)$ ways.

   If three operations can be separately performed in m, n and p ways respectively, then the three operations together can be performed in $(m \times n \times p)$ ways. Similar result holds for any finite number of operations.

2. **Addition rule:** If a work A can be done in m ways and work B (independent of A) be done in n ways, then either of the two works (A or B) can be done in $(m + n)$ ways. This can be extended to any finite number of mutually exclusive operations.

   Let $A_1, A_2, \ldots A_k$ be any k finite sets, where $k \geq 1$. If the given sets are pair wise disjoint, that is, $A_i \cap A_j = \phi$ for $i, j \in \{1, 2, \ldots, k\}, i \neq j$,

   Then

$$\left|\bigcup_{i=1}^{k} A_i\right| = |A_1 \cup A_2 \cup \ldots \cup A_k| = \sum_{i=1}^{k} |A_i|$$

Prove that

$$n_{p_r} = r \cdot {}^{n-1}p_{r-1} + {}^{n-1}p_r$$

Proof

$$r \cdot {}^{n-1}p_{r-1} + {}^{n-1}p_r$$

$$= r\frac{(n-1)!}{(n-1-r+1)!} + \frac{(n-1)!}{(n-1-r)!}$$

$$= r\frac{(n-1)!}{(n-r)!} + \frac{(n-1)!}{(n-1-r)!}$$

$$= \frac{(n-1)!}{(n-r-1)!}\left[\frac{r}{n-r} + 1\right]$$

$$= \frac{(n-1)!}{(n-r-1)!} \cdot \frac{n}{(n-r)} = \frac{n!}{(n-r)!} = n_{p_r}$$

Note:

(i) The number of permutations of n things taken all at a time when p of them are identical of one type, q of them are identical of second type, r of them are identical of third type and rest are all different $= \dfrac{n!}{p!q!r!}$

(ii) If n different things are to be arranged out of which r particular things are to be together, then we count these r particular things as one thing and the remaining $(n - r)$ things as separate things.

The number of things $= (n - r + 1)$.

Now arrange these $(n - r + 1)$ different things and then arrange the r particular things among themselves.

Required number $= (n - r + 1)! \cdot r!$

(iii) The number of permutations of n different things, taken r at a time, when a particular thing is to be always included in each arrangement is $r \cdot {}^{n-1}P_{r-1}$.

4.4 Circular Permutation

Let us consider 4 persons A, B, C and D, they can be arranged in a straight line in 4! Ways, that is, 24 ways but when we have to arrange them around a circle then we have the following 6 cases.

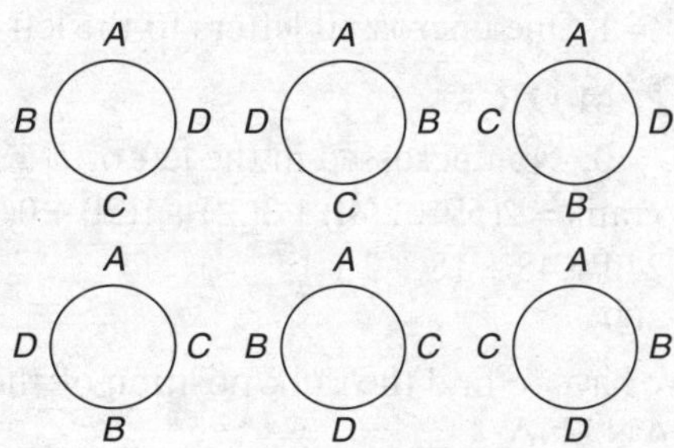

Hence, we can say that 'n' articles can be arranged in a straight line in n! ways. And 'n' articles can be arranged in a circle in $(n - 1)!$ ways (if we consider clockwise and anti-clockwise arrangements are different).

Consider 4 beads and we make a necklace from these 4 beads. As we observe that we have to arrange these four beads A, B, C, D in a circle.

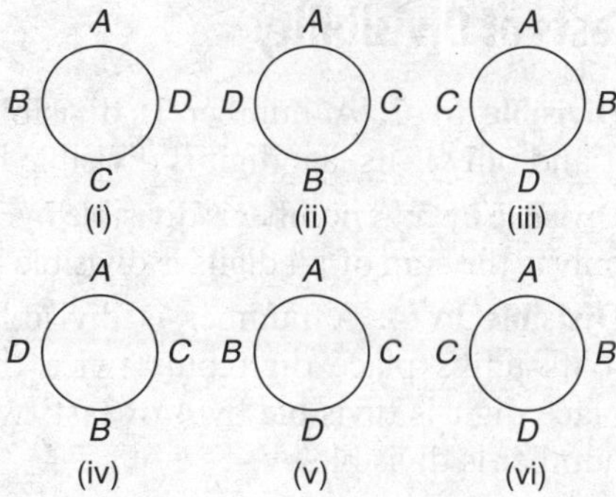

If we see carefully, we observe that figure **(i) and (iv)** represents same necklace. Similarly **(ii) and (v)**; and **(iii) and (vi)** represents the same necklace.

Therefore, total number of necklaces $= \dfrac{6}{2} = 3$.

When clockwise and anti-clockwise arrangements are not different, that is, when observation can be made from both sides, then number of circular arrangements of n different things is $\dfrac{(n-1)!}{2}$.

4.5 Sum of Numbers Formed by Using the Digits

I. Sum of numbers formed by using the digits $D_1, D_2, \ldots, D_r$ all at a time when none of the digits is zero.

$$= \left(\sum_{k=1}^{r} D_k\right)(r-1!)\underbrace{(111\ldots1)}_{r \text{ times}}$$

II. When one of the digits is zero, then

$$\left(\sum_{k=1}^{r} D_k\right)\left[(r-1!)\underbrace{(111\ldots1)}_{r \text{ times}} - (r-2)!\underbrace{(111\ldots1)}_{(r-1)\text{times}}\right]$$

III. Sum of the numbers formed by using the digits $D_1, D_2, D_3 \ldots, D_r$ when repetition of digits is allowed and none of the digits is zero.

$$= \left(\sum_{k=1}^{r} D_k\right) r^{r-1}\underbrace{(111\ldots1)}_{r \text{ times}}$$

IV. When one of the digits is zero.

$$= \left(\sum_{k=1}^{r} D_k\right)\left[r^{r-1}\underbrace{(111\ldots1)}_{r \text{ times}} - (r-1)^{r-2}\underbrace{(111\ldots1)}_{(r-1)\text{times}}\right]$$

4.6 Tests of Divisibility

I. Divisible by 2: A number is divisible by 2 if and only if its last digit is divisible by 2.

II. Divisible by 3: A number is divisible by 3 if and only if the sum of all digits is divisible by 3.

III. Divisible by 4: A number is divisible by 4 if its unit's place digit plus twice the tens place digit is divisible by 4 or last two-digit number is divisible by 4.

IV. Divisible by 5: A number is divisible by 5 if and only if its last digit is divisible by 5.

V. Divisible by 6: A number is divisible by 6 if and only if its unit place digit is even and the sum of its digit is divisible by 3.

VI. Divisible by 7: A number is divisible by 7 if and only if $1 \times$ unit digit $+ 3 \times$ tens digit $+ 2 \times$ hundred digit $- 1 \times$ thousand digit $-3 \times$ ten thousand digit $-2 \times$ hundred thousand digits is divisible by 7. If there are more digits present, the sequence of multipliers 1, 3, 2, −1, −3, −2 is repeated as often necessary.

VII. Divisible by 8: A number is divisible by 8 if and only if its unit place digit $+ 2 \times$ ten's place digit $+ 4 \times$ hundred's place digit is divisible by 8 or last three-digit number is divisible by 8.

VIII. Divisible by 9: A number is divisible by 9 if and only if the sum of its digits is divisible by 9.

IX. Divisible by 11: A number is divisible by 11 if and only if the difference between the sum of the digits in odd places (starting from right) and the sum of the digits in even places (starting from right) is multiple of 11.

4.7 Rank of Words

When all the letters of a word are arranged to form different words and the words are arranged as per dictionary, the position occupied by that word is the rank of that word. Let us assume that we have three letters A, B and C. We arrange them as per dictionary as

ABC – 1st	BCA – 4th
ACB – 2nd	CAB – 5th
BAC – 3rd	CBA – 6th

So, if I ask you what the rank of word CAB, the answer is 5th.

4.7.1 Method to Find the Rank/position of the Word as in Dictionary Format

I. We have to find the rank of the word MOTHER

3	4	6	2	1	5
M	O	T	H	E	R
2	2	3	1	0	0
5!	4!	3!	2!	1!	0!

$2 \times 5! + 2 \times 4! + 3 \times 3! + 1 \times 2! + 0 \times 1! + 0 \times 0!$

$= 240 + 48 + 12 + 2$

$= 308$

Add 1 in this we get the rank

Hence rank is $308 + 1 = 309$

Alternate

E H M O R T (6 letters)

So rank $= x_1(5!) + x_2(4!) + x_3(3!) + x_4(2!) + x_5(1!) + 0!$

Now, $x_1 = 2$, As two uncrossed letters on the left of M

EH~~M~~ORT

$x_2 = 2$, Two uncrossed letters to the left of o

EH~~M~~~~O~~RT

$x_3 = 3$, Three uncrossed letters to the left of T

EH~~M~~~~O~~R~~T~~

$x_4 = 1$, One uncrossed letters to the left of H

E~~H~~~~M~~~~O~~R~~T~~

$x_5 = 0$, No uncrossed in the left of E

So rank $= 2(5!) + 2(4!) + 3(3!) + 1(2!) + 0(1!) + 0!$

$= 240 + 48 + 18 + 2 + 1$

$= 309$

II. We have to find the rank/position of the word BANANA

2	1	3	1	3	1
B	A	N	A	N	A
$\frac{3}{3!2!}$	$\frac{0}{3!2!}$	$\frac{2}{2!2!}$	$\frac{0}{2!}$	1	0
5!	4!	3!	2!	1!	0!

$\frac{3}{3!2!} \times 5! + 0 + \frac{2}{2!2!} \times 3! + 0 + 1 \times 1! + 0$

$= 30 + 0 + 3 + 1$

Hence the rank is $34 + 1 = 35$

Highest Power of a Number in *N*!

Highest power of a prime number p that divides $n!$ exactly can be calculated by

$$\left[\frac{n}{p}\right]+\left[\frac{n}{p^2}\right]+\left[\frac{n}{p^3}\right]+\ldots$$

Where [ ] denotes greatest integer.

4.8 Combination

4.8.1 Definition

Combination is defined as the selection which does not follow the rule of order. Suppose there are three persons *A*, *B*, *C* and two chairs, the total number of ways to select two persons from these three persons are *AB*, *BC*, *CA*. We denote this by $^3C_2$.

The number of combinations of n different things taken r at a time is denoted by ^nC_r and is defined as $^nC_r = \frac{n!}{r!(n-r)!}$.

4.8.2 Properties of Combination

I. $^nC_r = {}^nC_{n-r}$

$$^nC_r = \frac{n!}{r!(n-r)!} = \frac{n!}{(n-r)!(n-(n-r))!}$$

$$= {}^nC_{n-r}$$

Thus $^nC_r = {}^nC_{n-r}$

II. $^nC_x = {}^nC_y \Rightarrow x = y$ or $x + y = n$

III. $^nC_r + {}^nC_{r-1} = {}^{n+1}C_r$ (Pascal rule)

Proof:

$$^nC_r + {}^nC_{r-1} = \frac{n!}{r!(n-r)!} + \frac{n!}{(r-1)!(n-(r-1))!}$$

$$= \frac{n!}{r!(n-r)!} + \frac{n!}{(r-1)!(n-r+1)!}$$

$$= \frac{n!}{(r-1)!(n-r)!}\left[\frac{1}{r} + \frac{1}{n-r+1}\right]$$

$$= \frac{n!}{(r-1)!(n-r)!}\left[\frac{n-r+1+r}{r(n-r+1)}\right]$$

$$= \frac{n!(n+1)}{(r-1)!r(n-r)!(n-r+1)}$$

$$= \frac{(n+1)!}{r!(n+1-r)!} = {}^{n+1}C_r$$

Thus, $^nC_r + {}^nC_{r-1} = {}^{n+1}C_r$

IV. $^nC_r = \frac{n}{r}\,{}^{n-1}C_{r-1}$

Proof:

$$^nC_r = \frac{n!}{r!(n-r)!} = \frac{(n-1)!n}{(r-1)!r\cdot(n-r)!}$$

$$= \frac{n}{r}\frac{(n-1)!}{(r-1)!((n-1)-(r-1))!} = \frac{n}{r}\,{}^{n-1}C_{r-1}$$

Thus, $^nC_r = \frac{n}{r}\,{}^{n-1}C_{r-1}$

V. $^nC_r = {}^{n+1}C_{r+1}\cdot\left(\frac{r+1}{n+1}\right)$

$$= \frac{(r+1)(r+2)}{(n+1)(n+2)}\,{}^{n+2}C_{r+2}$$

VI. $^nC_r\,{}^rC_t = {}^nC_t\,{}^{n-t}C_{r-t}\ (n \ge r \ge t)$

VII. $\frac{^nC_r}{^nC_{r-1}} = \frac{(n-r+1)}{r}$

VIII. $^nC_0 + {}^nC_1 + {}^nC_2 + \ldots + {}^nC_n = 2^n$

IX. $^nC_0 + {}^nC_2 + {}^nC_4 + \ldots = {}^nC_1 + {}^nC_3 + {}^nC_5 + \ldots = 2^{n-1}$

X. $^nC_n + {}^{n+1}C_n + {}^{n+2}C_n + \ldots + {}^{2n+1}C_n = {}^{2n+2}C_{n+1}$

XI. The greatest value of ^nC_r is $^nC_{n/2}$ if n is even and $^nC_{\frac{n-1}{2}}$ or $^nC_{\frac{n+1}{2}}$ if n is odd.

4.8.3 Restricted Selection/Arrangement

I. The number of ways in which r objects can be selected from n different objects if k particular objects are

(i) always included $= {}^{n-k}C_{r-k}$

(ii) never included $= {}^{n-k}C_r$

II. The number of arrangements of n distinct objects taken r at a time so that k particular objects are

(i) always included $= {}^{n-k}C_{r-k}\cdot r!$

(ii) never included $= {}^{n-k}C_r\cdot r!$

4.8.4 All Possible Selections

I. Selection from distinct objects: The total number of selections of n different things taken one or more at a time is $2^n - 1$.

Proof: Consider the 1st thing there are two ways, it may be included or excluded. Similarly, for 2nd thing there are two ways either include or exclude and so on for the next article.

Hence from the fundamental principle of multiplication, the total number of ways is $2 \times 2 \times 2 \times \ldots \times 2(n \text{ times}) = 2^n$, but the total number of ways in which nothing is selected = $1 \times 1 \times 1 \ldots \times 1$ (n times) = 1 way.

Hence, the total number of ways of selecting one or more at a time out of n different things $= 2^n - 1$.

In other ways, the total number of ways of selecting one or more thing from n different things $= {}^nC_1 + {}^nC_2 + \ldots + {}^nC_n = 2^n - 1$

II. Selection from identical objects:

(i) The number of ways of selecting r objects out of n identical objects is 1.

(ii) The total number of ways of selecting zero or more objects from n identical objects is $n + 1$.

(iii) The total number of selections of at least one object out of $a_1 + a_2 + \ldots + a_n$ objects, where a_1 are alike (of one kind), a_2 are alike of second kind and so on … a_n are alike of the nth kind is

$$[(a_1+1)(a_2+1)(a_3+1)\ldots(a_n+1)] - 1$$

III. Selection when both identical and distinct objects are present: The number of ways of selecting at least one thing out of $a_1 + a_2 + a_3 + \ldots + a_n + k$ things, where a_1 are alike of first kind, a_2 are alike of 2nd kind and so on and k are distinct is equal to $[(a_1+1)(a_2+1)\ldots(a_n+1)]2^k - 1$.

IV. Total number of divisors of a given natural number N: Every natural number N can always be put in the form

$N = p_1^{\alpha_1} p_2^{\alpha_2} p_3^{\alpha_3} \ldots p_k^{\alpha_k}$, Where $p_1, p_2, p_3 \ldots p_k$ are distinct prime and $\alpha_1, \alpha_2, \alpha_3, \ldots, \alpha_k \in W$

(i) The total number of divisors of N including 1 and N

$= (\alpha_1+1)(\alpha_2+1)\ldots(\alpha_k+1)$

(ii) The total number of divisors of N excluding 1 and N (which is called proper divisors)

$= (\alpha_1+1)(\alpha_2+1)\ldots(\alpha_{k+1}) - 2$

(iii) Sum of all divisors $= \left(p_1^0 + p_1^1 + \ldots + p_1^{\alpha_1}\right)$

$\left(p_2^0 + p_2^1 + \ldots + p_2^{\alpha_2}\right)$

$\left(p_k^0 + p_k^1 + \ldots + p_k^{\alpha_k}\right)$

$$= \left(\frac{1 - p_1^{\alpha_1+1}}{1 - p_1}\right)\left(\frac{1 - p_2^{\alpha_2+1}}{1 - p_2}\right)\ldots\left(\frac{1 - p_k^{\alpha_k+1}}{1 - p_k}\right)$$

(iv) Sum of proper divisors = sum of all divisors – $(N + 1)$

(v) The number of even divisors of N is possible only if $p_1 = 2$; otherwise there is no even divisor.

Therefore, required number of even divisors

$= \alpha_1(\alpha_2+1)(\alpha_3+1)\ldots(\alpha_k+1)$

(vi) The number of odd divisors of N

Case I: If $p_1 = 2$, the number of odd divisors

$= (\alpha_2+1)(\alpha_3+1)\ldots(\alpha_k+1)$

Case II: If $p_1 \neq 2$ the number of odd divisors

$= (\alpha_1+1)(\alpha_2+1)\ldots(\alpha_k+1)$

4.8.5 Division and Distribution of Objects (with Fixed Number of Objects in Each Group)

I. Into groups of unequal size (different number of objects in each group)

(i) The number of ways in which $(m + n)$ distinct objects can be divided into two unequal groups containing m and n objects is ${}^{m+n}C_m \cdot {}^nC_n = \dfrac{(m+n)!}{m!n!}$.

(ii) The number of ways to distribute $(m + n)$ distinct objects into two unequal groups containing m and n objects

= (Number of ways to divide) × (Number of groups)

$$= \frac{(m+n)!}{m!n!} \times 2!$$

Generalising this, the number of ways in which $(x_1 + x_2 + \ldots + x_n)$ distinct objects can be divided into n unequal groups containing $x_1, x_2, x_3, \ldots, x_n$ objects is

$$\frac{(x_1+x_2+x_3+\ldots+x_n)!}{x_1!x_2!x_3!\ldots x_n!}$$

The number of ways to distribute $(x_1+x_2+\ldots+x_n)$ distinct objects among n persons in the groups containing $x_1, x_2, \ldots, x_n$ objects

= (Number of ways to divide) × (Number of groups)

$$= \frac{(x_1+x_2+x_3+\ldots+x_n)!}{x_1!x_2!\ldots x_n!}\times n!$$

II. Into groups of equal size (each group containing same number of objects)

(i) The number of ways in which $2n$ distinct objects can be divided equally into two groups of equal size $= \frac{2n!}{(n!)^2 2!}$.

(ii) Number of ways in which $2n$ different objects can be distributed equally among 2 groups

= (number of ways of dividing into groups) × (number of groups)!

$$= \frac{2n!}{(n!)^2 2!}\times 2! = \frac{2n!}{(n!)^2}$$

Generalising this the number of ways in which (mn) distinct objects can be divided equally into n groups (unmarked) $= \frac{(mn)!}{(m!)^n n!}$ and the number of ways in which (mn) distinct objects can be distributed equally among n groups =

$$\frac{(mn)!}{(m!)^n n!}\times n! = \frac{(mn)!}{(m!)^n}.$$

4.9 De-arrangement

Suppose three friends A, B, C stay in a hostel on a particular night; none of them slept in their own room. In how many ways this is possible?

Let the room of A be 'a', that of B be 'b' and that of C be 'c', then

Own room	Person	Case I	Case II
a	A	b	c
b	B	c	a
c	C	a	b

⇒ 3 friends can be de-arranged in two ways we can write this as

$$D(3) = 3!\left(\frac{1}{0!}-\frac{1}{1!}+\frac{1}{2!}-\frac{1}{3!}\right)$$

$$= 6\left(1-1+\frac{1}{2}-\frac{1}{6}\right)$$

$$= 6\left(\frac{2}{6}\right) = 2$$

Now let's generalise it

Let $S = \{1, 2, 3, \ldots, n\}$, then a function f from S to S known as de-arrangement if f is a bijective function and $f(i) \neq i$ for any $i \in S$.

In other words, re-arrangement of objects such that no object goes to its original place is called derangement.

If n things are arranged in a row, the number of ways in which they can be deranged so that none of them occupies its original place is $n!\left(1-\frac{1}{1!}+\frac{1}{2!}-\frac{1}{3!}+\ldots+(-1)^n\frac{1}{n!}\right)$.

Note:

(i) De-arrangement of 1 object is not possible.

(ii) Partial de-arrangement: Consider n articles only r articles are de-arranged remaining $(n - r)$ articles are at their own place. Then this can be done in $\left(^nC_r\right)(D(r))$

4.10 Multinomial Theorem

Let $x_1, x_2, \ldots x_m$ be integers, then there are a number of solutions to the equation

$$x_1 + x_2 + \ldots + x_m = n \quad (4.3)$$

Subject to the condition

$$a_1 \le x_1 \le b_1, a_2 \le x_2 \le b_2, \ldots, a_m \le x_m \le b_m$$

is equal to the coefficient of x^n in

$$\left(x^{a_1}+x^{a_1+1}+\ldots+x^{b_1}\right)\left(x^{a_2}+x^{a_2+1}+\ldots+x^{b_2}\right)$$
$$\left(x^{a_3}+x^{a_3+1}+\ldots..+x^{b_3}\right)\ldots\left(x^{a_m}+x^{a_m+1}+\ldots+x^{b_m}\right) \quad (4.4)$$

Thus, the sum of m integers in (4.3) is equal to the number of times x^n occurs in (4.4).

In another way, the number of selections of r things out of n things of which p are of one type, q of second type, s of third type and so on.

= coefficient of x^r in

$$\left[(1+x+x^2+\ldots+x^p)(1+x+x^2+\ldots+x^q)\right.$$
$$\left.(1+x+x^2+\ldots+x^s)\ldots\right]$$

Note: Restricted selection

Number of selections of r things out of n things of which p are alike of one kind; q are alike of second kind; r are alike of third kind such that $p+q+r=n$ and each things is taken at least once = coefficient of x^r in $(x^1+x^2+\ldots+x^p)$ $(x^1+x^2+\ldots+x^q)$ $(x^1+x^2+\ldots+x^r)$

= coefficient of x^r in $(1+x+x^2+\ldots+x^{p-1})$ $(1+x+x^2+\ldots+x^{q-1})(1+x+x^2+\ldots+x^{r-1})$

Note:

(i) The number of ways in which r identical things can be distributed among n persons when each person can get zero or more things.

= Coefficient of x^r in $(1+x+x^2+\ldots+x^r)^n$

= Coefficient of x^r in $\left(\dfrac{1-x^{r+1}}{1-x}\right)^n$

= Coefficient of x^r in $(1-x)^{-n} = {}^{n+r-1}C_r$

(ii) $(1-x)^{-n} = 1 + {}^{n}C_1 x + {}^{n+1}C_2 x^2 + {}^{n+2}C_3 x^3 + \ldots +$ to ∞

Coefficient of x^r in $(1-x)^{-n} = {}^{n+r-1}C_r$

4.11 Non-negative Integral Solution Method

I. The number of non-negative integral solution of the equation

$x_1+x_2+\ldots+x_n=r$

$\Rightarrow$ Number of ways of distributing r identical things among n persons where each person can get zero or more things $= {}^{n+r-1}C_r$

II. The number of non-negative integral solution of $x_1+x_2+x_3+x_4=16$

Where $x_1, x_2, x_3 x_4 \geq 3$

Let

$y_1 = x_1-3,\ y_2 = x_2-3,\ y_3 = x_3-3,\ y_4 - x_4 - 3$

Then,

$y_1+y_2+y_3+y_4 = x_1+x_2+x_3+x_4-12$

$=16-12=4$

Thus, $y_1+y_2+y_3+y_4=4$ where

$y_1, y_2, y_3, y_4 \geq 0$

Hence the required number of integral solutions is

${}^{4+4-1}C_4 = {}^7C_4 = {}^7C_3$

III. The number of non-negative integral solutions of the equation

$x_1+2x_2+3x_3+4x_4+\ldots+nx_n=m$

= Coefficient of x^m in

$$\left[(1+x+x^2+x^3+\ldots)(1+x^2+x^4+x^6+\ldots)\right.$$
$$\left.(1+x^3+x^6+x^9+\ldots)(1+x^n+x^{2n}+x^{3n}+\ldots)\right]$$

= Coefficient of x^m in

$$\left[(1-x)^{-1}(1-x^2)^{-1}(1-x^3)^{-1}\ldots(1-x^n)^{-1}\right]$$

IV. Number of positive integral solution of equation $x_1+2x_2+3x_3+\ldots+nx_n=m$

= Coefficient of x^m in

$$\left[(x+x^2+x^3+\ldots)(x^2+x^4+x^6+\ldots)\right.$$
$$\left.(x^3+x^6+x^9+\ldots)\ldots(x^n+x^{2n}+x^{3n}+\ldots)\right]$$

= Coefficient of $x\, m-(1+2+3+\ldots+n)$in

$$\left[(1+x+x^2+\ldots)(1+x^2+x^4+\ldots)\right.$$
$$\left.(1+x^3+x^6+\ldots)(1+x^n+x^{2n}+\ldots)\right]$$

Permutation with Repetition

The number of permutations of r things out of n things of which p are of one kind, q is of second kind and so on

$= r!\cdot$ coefficient of x^r in

$$\left[\left(1+\frac{x}{1!}+\frac{x^2}{2!}+\ldots+\frac{x^p}{p!}\right)\left(1+\frac{x}{1!}+\frac{x^2}{2!}+\ldots+\frac{x^q}{q!}\right)\ldots\right]$$

4.12 Principle of Inclusion and Exclusion

If A and B are finite sets, then from the Venn diagram it is clear that

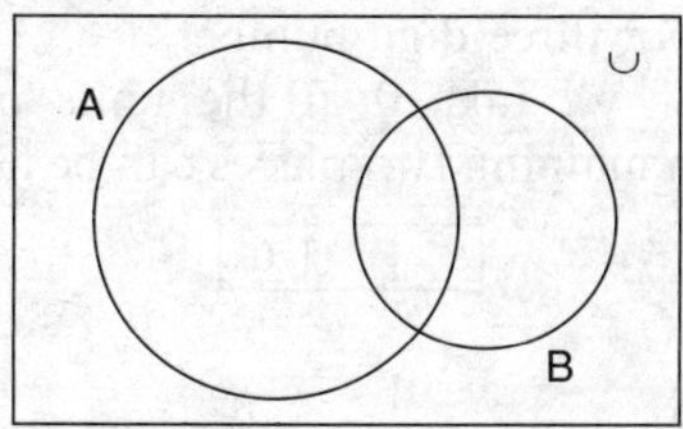

$$n(A \cup B) = n(A) + n(B) - n(A \cap B)$$

and $n(A' \cap B') = n(\cup) - n(A \cup B)$

If A, B and C are three finite sets then from the Venn diagram it is clear that

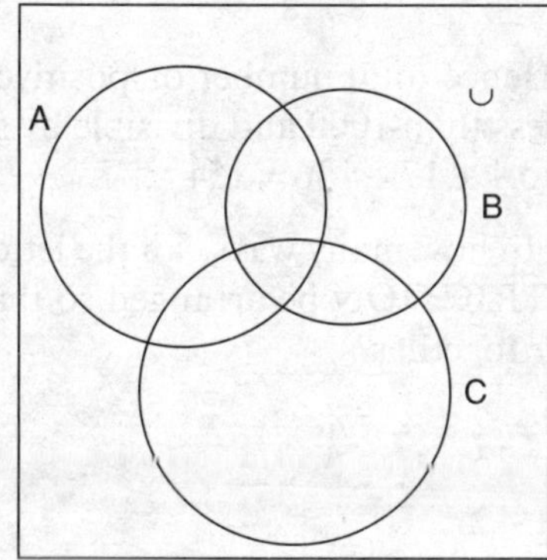

$$n(A \cup B \cup C) = n(A) + n(B) + n(C) - n(A \cap B) - n(B \cap C) - n(C \cap A) + n(A \cap B \cap C)$$

and $n(A' \cap B' \cap C') = n(\cup) - n(A \cup B \cup C)$

4.13 Formation on Geometrical Figures

1. The total number of straight lines from given 'n' points $= n_{C_2}$
2. The total number of straight lines from given 'n' points out of which exactly 'k' points are collinear $= n_{C_2} - k_{C_2} + 1$.
3. The total number of triangles formed from given 'n' points if no three of them are collinear $= n_{C_3}$.
4. Total number of polygons (of k sides) from given 'n' points if no three of them are collinear $= n_{C_k}$.
5. Total number of diagonals of a polygon of 'n' sides $= n_{C_2} - n = \dfrac{n(n-3)}{2}$
6. Total number of squares in a chess board of different size of squares = $1^2 + 2^2 + \ldots + 8^2 = 204$
7. Number of rectangles in a chess board = $9_{C_2} \times 9_{C_2}$
8. In a chess board of $n \times n$
   Number of squares $= 1^2 + 2^2 + \ldots + n^2$
   Number of rectangles $= n+1_{C_2} \times n+1_{C_2}$
9. Number of shortest paths: Let there be h number of horizontal steps and v number of vertical steps, then number of ways to reach from one corner to another corner is $h+v_{C_h} = h+v_{C_v}$.

SOLVED EXAMPLES

Example 4.1 How many 3-digit numbers can be formed by using the digits 0, 1, 2, 3, 4, 5 such that

(i) Repetition of digits is not allowed.
(ii) Repetition of digits is allowed.

Solution

(i) Let the 3-digit number be XYZ. The position (X) can be filed by 1, 2, 3, 4, 5 but not 0. So, it can be filled in 5 ways. Position (Y) can be filled in 5 ways again. Since 0 can be placed in this position, position Z can be filled in 4 ways. Hence by the fundamental principle of counting, total number of ways is $5 \times 5 \times 4$ = 100 ways.

(ii) Let the 3-digit number be XYZ position (X) can be filled in 5 ways (excluding 0) position (Y) can be filled in 6 ways position (Z) can be filled in 6 ways. Hence by the fundamental principle of counting, total number of ways is $5 \times 6 \times 6 = 180$.

Example 4.2 How many numbers between 10 and 10,000 can be formed by using the digits 1, 2, 3, 4, 5, if

(i) Repetition of digits is not allowed.
(ii) Repetition of digits is allowed.

Solution

(i) Number of two digit numbers $= 5 \times 4 = 20$
Number of three-digit numbers $= 5 \times 4 \times 3 = 60$
Number of four-digit numbers $= 5 \times 4 \times 3 \times 2 = 120$
Hence, the total number of numbers formed $= 20 + 60 + 120 = 200$

(ii) Number of two-digit numbers $= 5 \times 5 = 25$

Number of three-digit numbers $= 5 \times 5 \times 5 = 125$

Number of four-digit numbers $= 5 \times 5 \times 5 \times 5 = 625$

Hence the total number of numbers formed $= 25 + 125 + 625 = 775$

Example 4.3 A 5-digit number divisible by 3 is to be formed using the numbers 0, 1, 2, 3, 4 and 5 without repetition. Find the total number of ways in which this can be done.

Solution A number will be divisible by 3, if sum of the digits in number is divisible by 3.

Since the sum of the digit is $0 + 1 + 2 + 3 + 4 + 5 = 15$ which is divisible by 3.

Therefore, the digit that can be left out, while the sum still is multiple of 3, is either 0 or 3.

If we leave out 0, then the possible numbers $= 5_{P5} = 5! = 120$

If we leave out 3, then the possible numbers $5_{P5} - 4_{P4} = 5! - 4! = 96$

Hence, the required total numbers $= 120 + 96 = 216$

Example 4.4 Find the total number of positive integers less than 1000 and divisible by 5 (repetition of digit is not allowed).

Solution Since the digits are 0, 1, 2, 3, 4, 5, 6, 7, 8, 9, the number of positive integers less than 1000 will be of one digit, two digits and 3-digits.

Case I: Total number of positive integers of one digit divisible by 5 is only one number $5 = 1$.

Case II: For two-digit numbers

	0
9	1

If we take '0' at unit place, then tens place can be filled up by 9 ways.

Hence total number of positive integer $9 \times 1 = 9$

If we take 5 at the units place then ten's place has 8 options.

	5
8	1

Hence total number of positive integers $= 8 \times 1 = 8$

Thus, total number of two digit positive integer $= 9 + 8 = 17$

Case III: For three-digit numbers

If we take 0 at the units place then remaining two places can be filled up by

		0
		1

9_{P2} ways $= \dfrac{9!}{7!} = 8 \times 9 = 72$

If we take 5 at the units place then hundred place has 8 options and ten's place has 8 options $= 8 \times 8 \times 1 = 64$

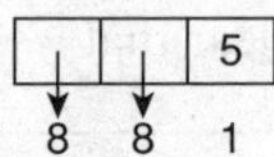

Hence total number of positive integers less than 1000 and divisible by 5 is equal to $1 + 17 + 136 = 154$

Example 4.5 In how many ways can the letters of the word RESTRICTION be arranged so that vowels never occur together?

Solution R – 2

T – 2

I – 2

Total number of arrangements $= \dfrac{11!}{2!2!2!} = \dfrac{11!}{(2!)^3}$

If the four vowels (2I's, 1 E, 1 O) are regarded as one group, then the remaining 7 letters and the one group can be arranged in $\dfrac{8!}{2!2!}$ ways while the vowels can be arranged among themselves in $\dfrac{4!}{2!}$ ways.

Therefore, the total number of ways in which the vowels occur together is $\dfrac{8!}{2!2!} \times \dfrac{4!}{2!}$.

Hence, the required number ways so that the vowels never occur together

$$= \frac{11!}{(2!)^3} - \frac{8! \times 4!}{(2!)^3}$$

$$= \frac{8!}{8}(9 \times 10 \times 11 - 24)$$

$$= \frac{8!}{8}(990 - 24) = \frac{8!}{8}(966)$$

$$= 966 \times 7!$$

$$= 966 \times 720 \times 7$$

$$= 4868640$$

Example 4.6 How many five-digit numbers divisible by 4 can be formed from the digits 0, 1, 2, 3, 4, 5, 6 (repetition is not allowed)?

Solution A number is divisible by 4 if its last two digits are divisible by 4.

Case I: When '0' is included among the last two digits, then last two digits may be (04) (20) (40) (60)

$\Rightarrow$ last two places can be filled in 4 ways and the first three places by 5_{p_3} ways

$\Rightarrow$ Number of numbers in this case = $5_{p_3} \times 4 = 240$

Case II: When 0 is not included among the last two digits.

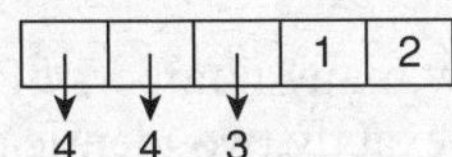

The last two digits may be

(1, 2) (1, 6), (2, 4), (3, 2), (3, 6), (5, 2), (5, 6), (6, 4), that is 8 ways

Number of numbers in this case

$= 4 \times 4 \times 3 \times 8 = 48 \times 8 = 384$

Hence the required solution is

$240 + 384 = 624$

Example 4.7 Find the number of 3 letter words that can be formed from the word 'series'.

Solution In the word 'series', the letters are not distinct

s – 2
e – 2
r – 1
i – 1

Case I: When all the three letters are distinct $= 4_{C_3} \times 3! = 4! = 24$

Case II: When two letters alike and one letter different

$$2_{C_1} \times 3_{C_1} \times \frac{3!}{2!} = 2 \times 3 \times 3 = 18$$

Hence the total number of 3 letter words

$= 24 + 18 = 42$

Example 4.8 How many 5 letters word can be formed using the letter of word MANAGEMENT such that any two alike letters are there then they are always together.

Solution In the word 'Management'

M – 2
A – 2
N – 2
E – 2
G – 1
T – 1

Case I: When all the 5 letters are distinct

$6_{C_5} \times 5! = 6! = 720$

Case II: If two letters alike and 3 letters different

$$4_{C_1} \times 5_{C_3} \times 4! = 4 \times \frac{5!}{3!2!} \times 4!$$
$$= 4 \times 4 \times 5 \times 3 \times 4$$
$$= 80 \times 12 = 960$$

Case III: When two letters alike of one type, two letters alike of second type and one different.

$$4_{C_2} \times 4_{C_1} \times 3! = \frac{4!}{2!2!} \times 4 \times 3!$$
$$= 3 \times 4 \times 4 \times 3$$
$$= 144$$

Hence the required 5 letters word

$= 720 + 960 + 144 = 1824$

Example 4.9 In how many ways can a mixed double game in badminton be arranged from 5 husbands and wives if no husband and wife play in the same game?

Solution Every game involves two males and two females.

Two males can be chosen in 5_{C_2} ways. The two females must now be chosen from 3 females barring the wife whose husband already selected which can be done in 3_{C_2} ways.

But the two females can now be paired with the male in 2! ways.

Hence the required number of ways

$$5_{C_2} \times 3_{C_2} \times 2! = 60$$

Example 4.10 Find the number of seven-digit integers with sum of the digits equal to 10 and formed by using the digits 1, 2 and 3 only.

Solution

Case I: We take 3 one time, 2 one time and 1 five times

In this case the number of seven-digit numbers is $\frac{7!}{5!} = 7 \times 6 = 42$

Case II: Take three 2 and four 1

In this case the number of 7-digit number $= \frac{7!}{3!4!} = 35$

Hence the required seven-digit number $= 42 + 35 = 77$

Example 4.11 Find the number of integral points that lie exactly in the interior of the triangle with vertices (0, 0), (0, 21), (21, 0)

Solution Since equation of AB is $x + y = 21$ (1)

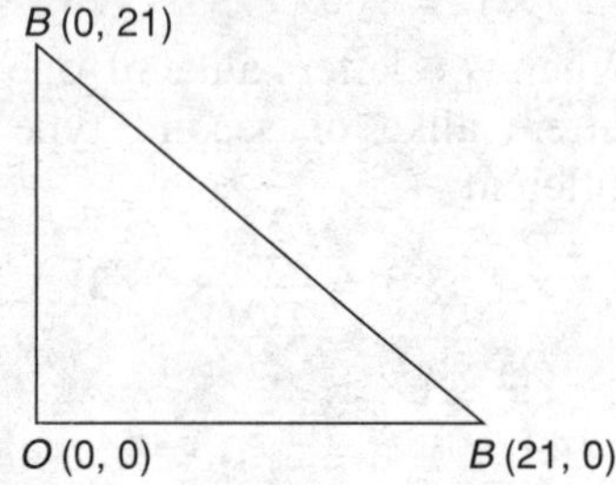

We will find the point inside this

(1, 1), (1, 2), …, (1, 19)

(2, 1), (2, 2), …, (2, 18)

(18, 1), (18, 2)

(19, 1)

Hence the number of integral points

$= 19 + 18 + 17 + \ldots + 1$

$= \frac{19}{2}(1+19) = 19 \times 10 = 190$

Example 4.12 Find the number of 6-digit numbers which can be formed using the digits 1, 2, 3, 4, 5, 6 such that sum of the digits at even places is divisible by 3 (repetition of digits is not allowed)

Solution Sum of the digits at even places can be 6, 9, 12, 15. Corresponding to each way in table digits at even places can be arranged in 3! Ways and the remaining 3 digits can be arranged in 3! Ways. So, the total number of such numbers is

Sum	Digits
6	1, 2, 3 }
9	1, 2, 3 1, 3, 5 2, 3, 4 }
12	1, 5, 6 2, 4, 6 3, 4, 5 }
15	4, 5, 6 }

$= 8 \times 3! \times 3! = 288$

Example 4.13 How many numbers of five digits can be made having exactly two identical digits?

Solution

Case I: Two identical digits are 0, 0

Number of ways to select 3 more digits = $^9C_3$

Number of arrangements of these 5 digits $= \frac{5!}{2!} - 4! = 36$

[When 0 comes in first place number of numbers = 4!]

$\Rightarrow$ Number of such numbers = $^9C_3 \times 36 = 3024$

Case II: Two identical digits are any of (1, 1), (2, 2), …, (9, 9)

For 1, 1, 0 selection of two more = $^8C_2$

Arrangement of these 5 digits $= \frac{5!}{2!} - \frac{4!}{2!} = 48$

Therefore, number of such numbers

$= {}^8C_2 \times 48$

For 1, 1 selection of 3 more (non-zero)

$= {}^8C_3 \times \frac{5!}{2!}$

$= {}^8C_3 \times 60$

So total number of cases II

$= 9 \times \left({}^8C_2 \times 48 + {}^8C_3 \times 60\right) = 42336$

Therefore, the required numbers

$= 3024 + 42336 = 45360$

Example 4.14 Find the rank/position of the word TITANIC.

Solution

5	3	5	1	4	3	2
T	I	T	A	N	I	C
$\frac{5}{2!2!}$	$\frac{2}{2!}$	4	0	2	1	0
6!	5!	4!	3!	2!	1!	0!

$$\frac{5}{2!2!}\times 6!+\frac{2}{2!}\times 5!+4\times 4!+0\times 3!+2\times 2!+1\times 1!+0$$
$$=900+120+96+4+1$$
$$=1221+1=1222$$

Example 4.15 Suvrojyoti writes letter to his 5 students but while putting letters in the addressed envelopes he did not put a single letter in its correct envelope. In how many ways he can do so?

Solution

$$D(5)=5!\left[\frac{1}{0!}-\frac{1}{1!}+\frac{1}{2!}-\frac{1}{3!}+\frac{1}{4!}-\frac{1}{5!}\right]$$
$$=5!\left(1-1+\frac{1}{2}-\frac{1}{6}+\frac{1}{24}-\frac{1}{120}\right)$$
$$=120\frac{(60-20+5-1)}{120}$$
$$=65-21=44$$

Example 4.16 Find the number of positive integers from 1 to 1000 which are divisible by at least 2, 3 or 5.

Solution Let A be the set of positive integers from 1 to 1000 which are divisible by k. We have to find $n(A_2\cup A_3\cup A_5)$.

$$n(A_2)=\left[\frac{1000}{2}\right]=500$$
$$n(A_3)=\left[\frac{1000}{3}\right]=333$$
$$n(A_5)=\left[\frac{1000}{5}\right]=200$$
$$n(A_2\cap A_3)=166,\ n(A_3\cap A_5)=66$$
$$n(A_2\cap A_5)=100,\ n(A_2\cap A_3\cap A_5)=33$$

Thus,

$$n(A_2\cup A_3\cup A_5)=n(A_2)+n(A_3)+n(A_5)-n(A_2\cap A_3)-n(A_3\cap A_5)-n(A_2\cap A_5)+n(A_2\cap A_3\cap A_5)=734$$

Example 4.17 If all permutations of the letters of the word AGAIN are arranged as in dictionary then find the 50th word.

Solution AAGIN (Alphabetical order)

Starting with A and arranging the other four letters = 4! Ways = 24

Then starting with G and arranging the other four letters $=\frac{4!}{2!}=12$

Start with I and arranges the other 4 letters $=\frac{4!}{2!}=12$

Thus, we have so constructed 48 words.

The 49th word is NAAGI and hence 50th word is NAAIG

Example 4.18 $a\vec{i}+b\vec{j}+c\vec{k}$ where $a, b, c\in\{1, 2, 3, 4, 5\}$ such that $2^a+3^b+5^c$ is divisible by 4.

Solution $4\lambda=2^a+3^b+5^c=2^a+(4-1)^b+(4+1)^c$

$=2^a+4k+(-1)^b+(1)^c$

(i) $a=1$, $b=$ even, c is any number

(ii) $a\neq 1$, $b=$ odd, c is any number

$=1\times 2\times 5+4\times 3\times 5=10+60=70$

Example 4.19 Find the maximum number of points of intersection of 7 straight lines and 5 circles when 3 straight lines are parallel and 2 circles are concentric.

Solution Points of intersection of 7 straight lines

$$={}^7C_2-{}^3C_2=18$$

Two concentric circles can intersect these 7 lines at a maximum of (7 × 4) points = 28 points.

Third circle can intersect the given system at a maximum of 14 + 2 + 2 = 18 points.

Fourth circle can intersect the system at a maximum of 14 + 2 + 2 + 2 = 20 points.

Fifth circle can intersect the system at a maximum of 14 + 2 + 2 + 2 + 2 = 22 points.

Hence the maximum number of points of intersection

= 18 + 28 + 18 + 20 + 22 = 106 points

Example 4.20 Find the number of integral solutions of the equation $x_1x_2x_3x_4 = 210$

Solution $210 = 2 \times 3 \times 5 \times 7$

All the factors are prime. Clearly any number can be assigned to x_i's in 4 ways. Hence total ways of assigning the factors to x_i's are 4^4. If two x_i's are negative, we have 4_{C_2} solutions and if all x_i's are negative we have 4_{C_4} solutions.

Hence, the total number of solutions

$$= 4^4\left(1 + 4_{C_2} + 4_{C_4}\right) = 8.4^4 = 2048$$

Example 4.21 Find the possible number of sets $\{m, n, p\}$ such that $1 \le m \le 100$, $1 \le n \le 50$, $1 \le p \le 25$ and $2^m + 2^n + 2^p$ is divisible by 3.

Solution Consider 2^k, $k \in N$

If k is odd $2^k + 1$ is divisible by 3 and if K *is* even, $2^k + 2$ is divisible by 3.

If 2^k is divided by 3, then the remainder is 1 if k is even.

2^k is divided by 3, then the remainder is 2 if k is odd.

So, $2^m + 2^n + 2^p$ is divisible by 3 either all m, n, p are odd or all m, n, p are even.

Hence possible numbers of ordered pairs are

$$50 \times 25 \times 13 + 50 \times 25 \times 12 = 31250$$

Example 4.22 In how many ways can we select 4 letters from the letters of the word MISSISSIPI?

Solution MIIIISSSSPP

Number of ways of selecting 4 alike letters $= 2_{C_1} = 2$

Number of ways of selecting 3 alike and one different letter $= 2_{C_1} \times 3_{C_1} = 6$

Number of ways of selecting 2 alike and 2 alike $= 3_{C_2} = 3$

Number of ways of selecting 2 alike and 2 different $= 3_{C_1} \times 3_{C_2} = 9$

Number of ways of selecting 4 different letters $= 4_{C_4} = 1$

Hence the total number of ways of selection $= 2 + 6 + 3 + 9 + 1 = 21$

Example 4.23 Find the number of distinct rational numbers x such that $0 < x < 1$ and $x = \dfrac{p}{q}$ where $p, q \in \{1, 2, 3, 4, 5, 6\}$.

Solution As $0 < x < 1$, we have $p < q$.

p	q
1	2, 3, 4, 5, 6
2	3, 4, 5, 6
3	4, 5, 6
4	4, 6
5	6

Clearly the number of rational numbers $= 5 + 4 + 3 + 2 + 1 = 15$

When p and q have a common factor, we get some rational numbers which are not different from those already counted. Here there are four such numbers $\dfrac{2}{4}, \dfrac{2}{6}, \dfrac{3}{6}, \dfrac{4}{6}$.

Hence the required number of rational numbers $= 15 - 4 = 11$

Example 4.24 Find the sum of all 4-digit numbers formed by using 2, 3, 7, 8, when

(A) repetition of digits is not allowed

(B) repetition of digits is allowed

Solution

(A) $(2 + 3 + 7 + 8)\,(4 - 1)!\,(1111)$

$= 20 \times 6 \times 1111 = 133320$

(B) $(2 + 3 + 7 + 8) \times 4^{4-1}(1111) = 1422080$

Example 4.25 Find the number of proper divisors of 38808. Also find the sum of all these divisors.

Solution $38808 = 2^3 \times 3^2 \times 7^2 \times 11$

Hence, the total number of proper divisors

$= (3 + 1)\,(2 + 1)\,(2 + 1)\,(1 + 1) - 2$

$= 72 - 2 = 70$

Sum of all these proper divisors

$= (2^0 + 2^1 + 2^2 + 2^3)\,(3^0 + 3^1 + 3^2)\,(7^0 + 7^1 + 7^2)\,(11^0 + 11^1) - 1 - 38808$

$= (15)\,(13)\,(57)\,(12) - 38809$

$= 133380 - 38809 = 94571$

Example 4.26 Find the number of divisors of the number $N = 2^3\,3^5\,5^7\,7^9\,9^{11}$ which are perfect squares.

Solution Here, $N = 2^3\,3^5\,5^7\,7^9\,9^{11}$

$= 2^3\,3^{27}\,5^7\,7^9$

For perfect square of N, each prime factor must occur even number of times.

2 can be taken in 2 ways (2^0 or 2^2)

3 $(3^0, 3^2, 3^4, 3^6, 3^8, 3^{10}, 3^{12}, 3^{14}, 3^{16}, 3^{18}, 3^{20}, 3^{22}, 3^{24}, 3^{26}) = 14$ ways

5 $(5^0, 5^2, 5^4, 5^6)$ 4 ways

7 $(7^0, 7^2, 7^4, 7^6, 7^8)$ 5 ways

$\Rightarrow$ Total divisors which are perfect squares.

$$= 2 \times 14 \times 4 \times 5 = 560$$

Example 4.27 Find the number of ways in which 22 different books can be given to 5 students, so that two students get 5 books each and all the remaining students get 4 books each.

Solution First, we divide 22 books into 5 groups of the size 5, 5, 4, 4, 4. The number of ways is $\dfrac{22!}{(5!)^2(4!)^3 3!2!}$. Now the number of ways of distribution of these groups among 5 students is 5! Thus, the total number of ways of distribution is

$$\frac{22!}{3!2!(5!)^2(4!)^3} \times 5! = \frac{22!}{3!2!5!(4!)^3}.$$

Example 4.28 Find the number of ways in which 16 constables can be assigned to patrol 8 villages, 2 for each.

Solution The number of ways to divide 16 constables into 8 batches is $\dfrac{16!}{8!(2!)^8}$

Hence the number of ways of assigning 8 villages

$$= \frac{16!}{8!(2!)^8} \times 8! = \frac{16!}{(2!)^8}$$

Example 4.29 Find the total number of seven digit numbers $x_1\, x_2\, x_3\, x_4\, x_5\, x_6\, x_7$ having the property that $x_1 \le x_2 < x_3 < x_4 \le x_5 < x_6 < x_7$.

Solution Since the digits are

0, 1, 2, 3, 4, 5, 6, 7, 8, 9

Case I: When $x_1 < x_2 < x_3 < x_4 < x_5 < x_6 < x_7$ then, the total number of ways of selection $= 9_{C_7}$

Case II: When $x_1 = x_2 < x_3 < x_4 < x_5 < x_6 < x_7$ then, the total number of ways of selection $= 9_{C_6}$

Case III: When $x_1 < x_2 < x_3 < x_4 = x_5 < x_6 < x_7$, then, the total number of ways of selection $= 9_{C_6}$

Case IV: When $x_1 = x_2 < x_3 < x_4 = x_5 < x_6 < x_7$, then, the total number of ways of selection $= 9_{C_5}$

Thus, the required seven-digit number

$$= 9_{C_7} + 9_{C_6} + 9_{C_6} + 9_{C_5}$$

$$= 10_{C_7} + 10_{C_6} = 11_{C_7}$$

Example 4.30 In how many ways can an examiner assign 30 marks to 8 questions, giving not less than 2 marks to any question?

Solution From question

$$x_1 + x_2 + \ldots + x_8 = 30$$

where $2 \le x_i \le 16 (i = 1, 2, \ldots, 8)$

Using multinomial theorem, the required number of ways

$= \text{Coefficient of } x^{30} \text{ in } (x^2 + x^3 + \ldots + x^{16})^8$

$= \text{Coefficient of } x^{30} \text{ in } x^{16}(1 + x + \ldots + x^{14})^8$

$= \text{Coefficient of } x^{14} \text{ in } \left(\dfrac{1 - x^{15}}{1 - x}\right)^8$

$= \text{Coefficient of } x^{14} \text{ in } (1 - x^{15})^8 (1 - x)^{-8}$

$= \text{Coefficient of } x^{14} \text{ in } (1 - x)^{-8}$

$= 8 + 14 - 1_{C_{14}}$

$= 21_{C_{14}} = 21_{C_7}$

Example 4.31 Find the number of ways in which 14 identical toys can be distributed among 3 boys so that each one gets at least one toy and no two boys get an equal number of toys.

Solution Let the boys get a, $a + b$ and $a + b + c$ toys respectively.

Then, $a + (a + b) + (a + b + c) = 14$, $a \ge 1$, $b \ge 1$, $c \ge 1$

$\Rightarrow \quad 3a + 2b + c = 14, a \ge 1, b \ge 1, c \ge 1$

$= \text{Coefficient of } x^{14} \text{ in } (x^3 + x^6 + x^9 + \ldots)$
$(x^2 + x^4 + x^6 + \ldots)$
$(x + x^2 + x^3 + \ldots)$

= Coefficient of x^8 in

$(1+x^3+x^6+\ldots)(1+x^2+x^4+\ldots)(1+x+x^2+\ldots)$

= Coefficient of x^8 in

$(1+x^3+x^6+x^2+x^5+x^8+x^4+x^7)$

$$(1+x+x^2+\ldots)=10$$

Three distinct numbers can be assigned to three boys in 3! ways.

So, total number of ways = 10 × 3! = 60

Example 4.32 In how many ways can 2 small squares be selected from an (8 × 8) chess board so that they are not from the same row and from the same column.

Solution Total number of small squares = 64

Number of ways to select 2 small squares from 64 squares = 64_{C_2}

Hence the number of ways to select two small squares which are not from the same row and column

$$=64_{C_2}-(8\times 8_{C_2}+8\times 8_{C_2})$$

[Selecting from same row and column]

$$=\frac{63\times 64}{2}-16\times\frac{7\times 8}{2}$$

$$=63\times 32-16\times 28$$

$$=2016-48=1568$$

Example 4.33 A student is allowed to select at most n books from a collection of $(2n + 1)$ books. If the total number of ways in which he can select a book is 63, find the value of n.

Solution Since it is given that

$$2n+1_{C_1}+2n+1_{C_2}+\ldots+2n+1_{C_n}=63 \quad (1)$$

But we know that

$$2n+1_{C_0}+2n+1_{C_1}+\ldots+2n+1_{C_n}+2n+1_{C_{n+1}}+\ldots+2n+1_{C_{2n}}+2n+1_{C_{2n+1}}=2^{2n+1}$$

$$\Rightarrow \quad 2+(2n+1_{C_1}+\ldots+2n+1_{C_n})+(2n+1_{C_{2n}}+\ldots+2n+1_{C_{n+1}})=2^{2n+1}$$

$$\Rightarrow \quad 2+2(2n+1_{C_1}+\ldots+2n+1_{C_n})=2^{2n+1}$$

$$\Rightarrow \quad 2+2\times 63=2^{2n+1}$$

$$\Rightarrow \quad 64=2^{2n}\Rightarrow 2^{2n}=2^6\Rightarrow 2n=6\Rightarrow n=3$$

Example 4.34 There are n straight lines in a plane, no two of which are parallel and no three of them are concurrent. Find the number of new lines that can be formed by joining the intersection points of the straight lines.

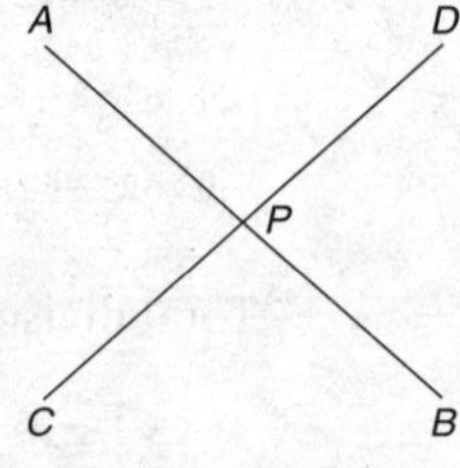

Solution Every two lines have one intersection point. Thus the total number of intersection points is n_{C_2}.

Consider two lines AB and CD and their intersection point P.

There are n straight lines. Thus, each line intersects with $(n - 1)$ lines and therefore, has $(n - 1)$ points on it.

Thus, apart from the point P, there must be $(n - 2)$ points on AB and $(n - 2)$ points on CD.

Together there are $2(n - 2) + 1 = (2n - 3)$ points lying on lines AB and CD and hence the number of points not lying on the line AB and CD is $n_{C_2}-(2n-3)$

Now if P is connected to any point which does not lie on AB or CD, we get a new line. There are n_{C_2} such P's, hence the number of new lines that can be drawn is

$$\frac{1}{2}\times n_{C_2}\left(n_{C_2}-(2n-3)\right)$$

$$=\frac{1}{8}n(n-1)(n-2)(n-3)$$

[Every new line has been counted twice]

Example 4.35 If $100! = k\,(2!)^x\,(5!)^y\,(7!)^z$, then find the maximum possible value of $x+y+z$.

Solution Since, $(2!)^x = 2^x$

$$(5!)^y=2^{3y}3^y5^y$$

$$(7!)^z=2^{4z}3^{2z}5^z7^z$$

$$\text{But } 100!=2^{97}\times 3^{48}\times 5^{24}\times 7^{16}$$

Maximum value of $z = 16$

Maximum value of $y + z = 24$

Maximum value of $x + 3y + 4z = 97$

Hence for maximum value of $x + y + z$, $y = 0$, $z = 0$ and $x = 97$. Then $x + y + z = 97$

Example 4.36 Find the number of integral solutions of the equation

$$x+y+z+t=29, x>0, y>1, z>2 \text{ and } t\geq 0 .$$

Solution It is given that

$x+y+z+t=29$, where

$$x\geq 1, y\geq 2, z\geq 3, t\geq 0$$

Let $\alpha=x-1, \beta=y-2, \gamma=z-3, \delta=t-0;$

$\Rightarrow \quad x=\alpha+1, y=\beta+2, z=\gamma+3, t=\delta$

Clearly $\alpha, \beta, \gamma, \delta\geq 0$

Therefore, we get

$$\alpha+1+\beta+2+\gamma+3+\delta=29;$$

$\Rightarrow \quad \alpha+\beta+\gamma+\delta=23$

Hence the non-negative integral solution

$$= {}^{4+23-1}C_4 = {}^{26}C_4$$

4.14 Know Your Concepts

1. How many 3 letter words can be formed from the letters of the word CALCUTTA?
2. In a plane there are two families of lines $y=x+r$, $y=-x+r$, where $r\in\{0,1,2,3,4\}$. Find the total number of squares of diagonals of length 2 formed by the lines.
3. Find the number of words which can be formed out of the letters *a*, *b*, *c*, *d*, *e*, *f* taken 3 together, each word containing one vowel at least.
4. How many words can be formed out of the letter of the ARTICLE so that the vowels occupy the even places?
5. How many numbers divisible by 5 and lying between 3000 and 4000 can be formed from the digits 3, 4, 5, 6, 7, 8, no digit being repeated in any number?
6. If ${}^{15}C_{3r} : {}^{15}C_{r+1} = 11:3$ find r
7. The sides *AB*, *BC*, *CA* of a triangle *ABC* have 3, 4 and 5 interior points respectively on them. Find the total number of triangles that can be constructed by using these interior points as vertices.
8. Everybody in a room shakes hands with everybody else. The total number of handshakes is 153. Find the total number of persons in the room.
9. A mathematics paper has 12 questions divided into 3 sections *A*, *B*, *C*. Each section has 4 questions. In how many ways can you answer 5 questions selecting at least one question from each part?
10. Between two stations *A* and *B* there are twelve intermediate stations. Find the number of ways in which a train can stop at four of these stations so that no two of these stations are consecutive.
11. Three integers are selected from the integers 1, 2, ..., 1000. In how many ways can these integers be selected such that their sum is divisible by 4.
12. Find the number of zeroes at the end of 300!
13. If the 11 letters *A*, *B*, ..., *K* denote an arbitrary permutation of the integers (1, 2, ..., 11), then prove that $(A-1)(B-2)(C-3)\ldots(K-11)$ will be always even.
14. A regular polygon of 10 sides is constructed. In how many ways can 3 vertices be selected so that no two vertices are consecutive?
15. Find the number of all the 6-digit numbers $a_1a_2a_3a_4a_5a_6$ (all distinct digits) such that $a_1>a_2>a_3>a_4<a_5<a_6$.
16. In how many ways can two natural numbers be selected which are less than or equal to 100 and differ at most by 10.
17. A boat is to be manned by eight men of whom 2 can only row on the bow side and 3 can only row on the stroke side. Find the number of ways in which the crew can be arranged.
18. Find the number of positive integral solutions of the equation $x_1x_2x_3x_4x_5=420$.
19. A man has 3 friends. In how many ways can he invite one friend every day for dinner on 6 successive nights so that no friend is invited more than 3 times?
20. How many 4-digit numbers are there which contain no more than 2 different digits?
21. In how many ways can 8 persons be seated at a round table
    (a) If two of them (say *A* and *B*) must not sit in adjacent seats.
    (b) If 4 of the persons are men and 4 ladies and if no two men are to be in adjacent seats.

(c) If 8 persons constitute 4 married couples and if no husband and wife as well as no two men are to be in adjacent seats.

22. During a draw of lottery, tickets bearing numbers 1, 2, 3, ..., 40, 6 tickets are drawn out and then arranged in the descending order of their numbers. In how many ways, it is possible to have 4th ticket bearing number 25.

23. (a) How many divisors are there of the number $x = 21600$. Find also the sum of these divisors.

(b) In how many ways the number 7056 can be resolved as a product of 2 factors.

(c) Find the number of ways in which the number 300300 can be split into 2 factors which are relatively prime.

4.15 JEE Main: Sample Questions

Single Answer Type

1. The number of 4-digit numbers in which repetition of digits is not allowed, even digits appear at even places, odd digits appear at odd places and the number is divisible by 4 is equal to

(A) 160 (B) 320 (C) 480 (D) 520

2. A swimming team has eight members, only two of which are boys. The coach wants to take a delegation from the team to a special swimming camp. If the delegation must have either five or six members, and must include at least one boy, then the number of ways to select the delegation, is

(A) 66 (B) 77 (C) 88 (D) 99

3. Number of ways in which 3 men and their wives can be arranged in a line such that none of the 3 men stand in a position that is ahead of his wife, is

(A) $3!\cdot 3!$ (B) $2!\cdot 3!\cdot 3!$

(C) $3!$ (D) $\dfrac{6!}{2!2!2!}$

4. Number of 5-digit numbers which are divisible by 5 and each number containing the digit 5, digits being all different is equal to $k(4!)$, the value of k is

(A) 84 (B) 168 (C) 188 (D) 208

5. The number of 10-digit numbers such that the product of any two consecutive digits in the number is a prime number, is

(A) 1024 (B) 2048 (C) 512 (D) 64

6. Six fair coins are tossed simultaneously. Number of ways in which at most three of the coins turn up as heads in a definite order, is

(A) 32 (B) 41 (C) 42 (D) 6

7. Number of zeroes at the end of $\prod_{n=5}^{30}(n)^{n+1}$ is

(A) 111 (B) 147 (C) 127 (D) 137

8. Number of six-digit numbers divisible by 5 that can be formed using the digits 4, 2, 5, 0, 6, 7 each digit to be used exactly once, is

(A) 308 (B) 240 (C) 216 (D) 256

9. There are 9 children comprising 3 boys and 6 girls. Let m denote the number of ways when the 3 boys are in succession and n denote the corresponding figure when the two-end positions are occupied by boys. If $m = kn$ then k is equal to

(A) 3 (B) 2 (C) 1 (D) $\left(\dfrac{1}{3}\right)$

10. In a bag there is a minimum of six old Indian coins of every denomination (Athanni, Chavanni, Duanni, Ekanni). Number of ways in which one can take 6 coins from the bag is

(A) 120 (B) 90 (C) 84 (D) 60

11. We have 19 identical gems available with us which are needed to be distributed among A, B and C such that A always gets an even number of gems. The number of ways this can be done is

(A) 72 (B) 92 (C) 100 (D) 110

12. Which of the following statements holds good?

(A) The number of ways 1, 2 and 4 inspectors can be sent to A, B and C cities respectively is 630.

(B) The sum of all the 3-digit numbers which can be formed using the digits 1, 5, 7 and 8 without repetition is 13986.

(C) The number of ways in which the letters of word 'TUILERIES' can be rearranged without changing the relative order of vowels and consonants is 720.

(D) If the letters of the word 'AKALA' are arranged as in dictionary then the word which appears on 8th position is AKAAL.

13. The number of ways of selection of 4 letters of the word 'PERFECT' is
(A) 15 (B) 21 (C) 25 (D) 31

14. 3 Indian and 3 American men and their wives are to be seated round a circular table. Let m denote the number of ways when the Indian couples are together and n denote the number of ways when all the six couples are together. If $m = kn$ then k equals
(A) 36 (B) 42 (C) 45 (D) 48

15. There are 7 persons which include a group of 3 friends F_1, F_2, F_3. The number of ways they can be seated at a round table if at least two out of F_1, F_2, F_3 are seated next to each other is
(A) 144 (B) 576 (C) 360 (D) 432

16. If the letters of the word LAUGH are arranged by a lexicographer, then the number of words that appear before the word LAUGH is
(A) 76 (B) 77 (C) 78 (D) 79

17. The total number of 5-digit numbers having all different digits and divisible by 4 that can be formed using the digits {1, 3, 2, 6, 8, 9}, is equal to
(A) 192 (B) 32 (C) 1152 (D) 384

18. The number of 7-digit telephone numbers that can be created if the first three digits must be 426, 427 or 428 and the right most digits is divisible by 3, is k thousand then k is equal to
(A) 12 (B) 9 (C) 6 (D) 3

19. How many six-digit numbers are there in which no digit is repeated, even digits appear at even places, odd digits appear at odd places and the number is divisible by 4?
(A) 1440 (B) 2160 (C) 2700 (D) 3600

20. Numbers from 1 to 1000 which are divisible by 60 but not by 24 is
(A) 16 (B) 41 (C) 8 (D) 12

21. The number of 4-digit positive integers with four different digits, where the leading digit is not zero, and the integer is a multiple of 5 and 5 is the largest digit, is
(A) 24 (B) 48 (C) 60 (D) 84

22. The number of permutations of 1, 2, 3, 4, 5, 6, 7, 8 and 9 taken all at a time are such that the digit 1 appearing somewhere to the left of 2, 3 appearing to the left of 4 and 5 somewhere to the left of 6, is $k \times 7!$ Then the value of k is
(A) 6 (B) 7 (C) 8 (D) 9

23. The number of ways in which all the letters of the word 'HUSSEY' be arranged so that the two S are never together is
(A) 120 (B) 240 (C) 360 (D) 480

24. The number of ways in which all the letters of the word 'COCONUT' be arranged such that at least one 'C' comes at an odd place, is
(A) 360 (B) 720 (C) 1080 (D) 1260

25. There are 12 monkeys sitting in a line. Let M denote the number of ways of selecting 3 monkeys when no two of them are consecutive and N is the corresponding figure when exactly two are consecutive, then
(A) $3M = 4N$ (B) $3N = 4M$
(C) $M = N$ (D) $4M = 5N$

26. Let $A = \{1, 2, 3, 4\}$ and $B = \{1, 2, 3, 4\}$. If $f: A \to B$ is an one-one function and $f(x) \neq x$ for all $x \in A$, then the number of such possible functions, is
(A) 6 (B) 9 (C) 24 (D) 44

27. Consider all possible permutations of the letters of the word W = CORPORATION. Then which of the following is/are correct?
(A) Number of permutations of the word W if Os are separated is $42 \times 8!$
(B) Number of permutations of the word W if vowels are together is $10 \times 7!$
(C) Number of permutations of the word W if consonants appear in alphabetical order is 9240.
(D) Number of permutations containing the word CAPTION is 30.

28. The number of six-digit numbers between 700000 and 800000 such that the first digit is equal to the sum of the other five digits is
(A) 1512 (B) 330 (C) 290 (D) 130

29. Let N be the number of points (x, y, z) in space such that $x+y+z=12$, where $x, y, z \in N$. The number of divisors of N is equal to
(A) 2 (B) 4 (C) 9 (D) 16

30. If the number of fourteen-digit numbers which can be formed by using the digits 1, 2, 3, 4, 7, 7, 7, …, 7 (digit 7 is used 10 times) such that there are at least two identical digits between distinct digits (distinct digits are 1, 2, 3 and 4) is $k \cdot 5!$ then k is
(A) 3 (B) 7 (C) 14 (D) 5

31. If the number of ways in which a selection of 100 balls can be made out of 100 identical red balls, 100 identical blue balls and 100 identical white balls is *abcd* (where *abcd* is a four-digit number), then $(a + c - b - d)$ is equal to
(A) 6 (B) 7 (C) 8 (D) 9

32. All possible 120 permutations of WDSMC are arranged in dictionary order, as if each were an ordinary five-letter word. The last letter of the 86th word in the list, is
(A) W (B) D (C) M (D) C

33. 5 different objects are to be distributed among 3 persons such that no two persons get the same number of objects. The number of ways this can be done, is
(A) 60 (B) 90 (C) 120 (D) 150

34. The number of ways in which 7 different red roses, 4 different blue roses and 5 alike green roses can form a garland so that flowers of same colour should be together, is (do not distinguish clockwise and anticlockwise circular permutation)
(A) 7! 4! 5! 3! (B) 7! 4!
(C) 7! 4! 5! 2! (D) 7! 4! 2!

35. The number of words that can be formed using all the letters of the word GARGEE if no two alike letters are together, is
(A) 84 (B) 62 (C) 240 (D) 280

36. In how many ways can six boys and five girls stand in a row if all the girls are to stand together but the boys cannot all stand together?
(A) 172,800 (B) 432,000
(C) 86,400 (D) 423,000

37. The total number of ways of selecting two numbers from the set {1, 2, 3, 4, …, 90} so that their sum is divisible by 3, is
(A) 885 (B) 1335 (C) 1770 (D) 3670

38. The total number of words that can be formed using all letters of the word 'BRIJESH' that neither begins with 'I' nor ends with 'B' is
(A) 4920 (B) 4800 (C) 3600 (D) 3720

39. The number of ways in which 15 indistinguishable oranges can be distributed in 3 different boxes so that every box has at most 8 oranges, is
(A) 52 (B) 108 (C) 76 (D) 28

40. The number of ways in which 19 indistinguishable books can be packaged into 3 numbered boxes if no box may be empty, is
(A) 153 (B) 171 (C) 243 (D) 306

41. There are 2 women participating in a chess tournament. Every participant played 2 games with the other participants. The number of games that the men played between themselves exceeded by 66 as compared to the number of games that the men played with the women. If the number of participants is n, then the value of $(n - 6)$ is
(A) 5 (B) 6 (C) 7 (D) 8

42. The number of ways in which 5A's and 6B's can be arranged in a row which reads the same backwards and forwards is N, then the value of $\frac{N}{2}$ is
(A) 4 (B) 5 (C) 7 (D) 9

43. If N is the number of ways in which a person can walk up a stairway which has 7 steps if he can take 1 or 2 steps up the stairs at a time, then the value of $\frac{N}{3}$ is
(A) 7 (B) 9 (C) 11 (D) 13

44. Two boys and six girls are officers of a math club. When the photographer takes a picture for the school yearbook, she asks the eight officers of the club and one faculty sponsor to sit in a row with the faculty sponsor in the middle and the two boys not next to one another. The number of different seating arrangements, are

(A) 31680 (B) 30960
(C) 20240 (D) 33840

45. A three-digit number *abc* is such that
(i) its hundreds digit is equal to the sum of the other two digits, and
(ii) $b(c + 1) = 52 - 4a$
The number of such numbers are
(A) 2 (B) 3 (C) 4 (D) 5

46. What is the greatest power of 5 which can divide 80! exactly?
(A) 16 (B) 20 (C) 19 (D) 22

47. The number of triangles with positive area whose vertices are points in the *xy*-plane with integer coordinates (x, y) satisfying $1 \le x \le 4$ and $1 \le y \le 4$, is
(A) 500 (B) 532 (C) 516 (D) 538

48. The number of ways in which 5 boys and 5 girls can be seated alternatively on a round table if a particular boy and a particular girl are never adjacent to each other in any arrangement, is
(A) 288 (B) 552 (C) 1584 (D) 1728

49. At an election, a voter may vote for any number of candidates not exceeding half the number of candidates standing for election. If the number of candidates contesting is eight, the number of ways in which a voter may vote assuming he votes for at least one, is
(A) 161 (B) 162 (C) 163 (D) 164

50. The number of ordered quadruples (a_1, a_2, a_3, a_4) of positive odd integers that satisfy $a_1 + a_2 + a_3 + a_4 = 32$ is equal to
(A) 286 (B) 4495 (C) 680 (D) 4040

51. Consider a determinant $D = \begin{vmatrix} a & b \\ c & d \end{vmatrix}$ where $a, b, c, d \in \{0, 1\}$. If n denotes the total number of determinants whose entries are 0 or 1 only and m denotes the number of determinants with a non-zero value, then the ratio $\frac{m}{n}$ equals

(A) $\frac{3}{16}$ (B) $\frac{4}{16}$ (C) $\frac{6}{16}$ (D) $\frac{8}{16}$

52. A man wants to buy '*m*' mangoes in '*n*' different varieties, mangoes of the same variety being identical and they are available in abundance. The number of different ways he can plan his purchases, if he has to buy at least two mangoes of the same variety is
(A) ${}^{m-2n}C_{n-1}$
(B) ${}^{m+n-3}C_{m-1}$
(C) ${}^{m+n-1}C_m - {}^{n}C_m$
(D) $\frac{1}{m!}\left[{}^{m+n-1}P_{n-1} - {}^{n}P_m\right]$

53. Sixteen players $S_1, S_2, S_3, \ldots, S_{16}$ play in a tournament. The number of ways in which they can be grouped into eight pairs so that S_1 and S_2 are in different groups, is equal to

(A) $\frac{(14)!}{2^6 \cdot 6!}$ (B) $\frac{(15)!}{2^7 \cdot 7!}$ (C) $\frac{(14)!}{2^7 \cdot 6!}$ (D) $\frac{(14)!}{2^6 \cdot 7!}$

54. There are five different peaches and three different apples. The number of ways in which they can be divided into two packs of four fruits if each pack must contain at least one apple, is
(A) 95 (B) 65 (C) 60 (D) 30

55. The number of ways in which 7 people can occupy 6 seats, 3 seats on each side in a first-class railway compartment if two specified persons are to be always included and occupy adjacent seats on the same side is $(5!) \cdot k$, then k has the value equal to
(A) 7 (B) 8 (C) 9 (D) 10

56. In an eleven-storey building, 9 people enter a lift cabin. It is known that they will leave the lift in groups of 2, 3 and 4 at different residential storeys. The number of ways in which they can get down is

(A) $\frac{(12)!}{2!(3!)^2 \cdot 4!}$ (B) $\frac{10!}{4}$

(C) $\frac{10!}{4!}$ (D) $\frac{(12)!}{(3!)^2 \cdot 4!}$

57. The number of ways in which two Americans, two British, one Chinese, one Dutch and one Egyptian can sit at a round table so that persons of the same nationality are separated is
(A) 48 (B) 240 (C) 336 (D) none

58. There are 7 persons which include a group of 3 friends F_1, F_2, F_3. The number of ways they can be seated at a round table if no two out of F_1, F_2, F_3 are seated next to each other is
(A) 144 (B) 576 (C) 36 (D) 216

Integer Answer Type

1. All the seven-digit numbers containing each of the digits 1, 2, 3, 4, 5, 6, 7 exactly once and not divisible by 5, are arranged in the increasing order. If N is the 2018th number, then find the last digit of N.

2. If N is the number of all four-digit numbers having non-zero digits and which are divisible by 4 but not by 8, then find the last digit of N.

3. If the number of common divisors of 10800 and 9000 is a, then find the value of $\sqrt{a}$.

4. A five-digit number which is divisible by 5 and each number containing the digit 5, digits being all different is equal to $K(4!)$, then find the value of $\frac{K}{24}$.

5. Find the value of $\lim_{n\to\infty}\sum_{K=1}^{n}\frac{\left(K^2+K-1\right)}{(K+1)!}$.

6. The number of permutations of 1, 2, 3, 4, 5, 6, 7, 8 and 9 taken all at a time are such that the digit 1 appearing somewhere to the left of 2, 3 appearing to the left of 4 and 5 somewhere to the left of 6 is $K\times 7!$, then find the value of K.

7. Find the numbers from 1 to 1000 which are divisible by 60 but not by 24.

8. Let N be the number of ways in which a person can walk up a stairway which has 7 steps. If he can take 1 or 2 steps up to the stairs at a time, then find the value of $\frac{N}{3}$.

9. A three-digit number abc is such that
(i) its hundreds digit is equal to the sum of the other two digits, and
(ii) $b(c+1)=52-4a$, then find the such numbers.

10. Let $f(n)=\sum_{r=0}^{n}\sum_{K=r}^{n}\left({}^{K}C_r\right)$, then find the total number of divisors of $f(9)$.

11. If the number of arrangements of 4 alike apples, 5 alike mangoes, 1 banana and 1 orange in which all the apples are together or all the mangoes are together is K, then find the sum of digits in K.

12. Let S be the set of all five-digit positive integers formed from the digits in the set {1, 2, 3, 4, 5, 6, 7, 8, 9} without repetition. Find the number of these integers in which the digits 3, 5, 7 appear in order (not necessarily consecutively) from left to right.

13. Let A_i where $i = 1, \ldots, 12$ be the vertices of a regular dodecagon and G its centre. Let 's' be the number of straight lines that can be formed with these 13 points, 't' be number of triangles that can be formed and 'd' be number of diagonals in a dodecagon. Find the value of $(s + t + d)$.

14. Let T_n denote the number of triangles that can be formed using the vertices of a regular polygon of side n. If $T_{n+1} - T_n = 10$ then find the number of different selections of n-digit numbers that can be formed using the digits, one 1, two 2's, three 3's and four 4's.

15. Seven different lectures are to be delivered in 7 periods of a class on a particular day. Out of 7 lecturers, A, B and C are three of them. Find the number of ways in which a routine for the day can be made such that 'A' delivers his lecture before 'B' and 'B' before 'C'.

16. Let n be the number of ways in which 5 boys and 5 girls can stand in a queue in such a way that all the girls stand consecutively in the queue. Let m be the number of ways in which 5 boys and 5 girls can stand in a queue in such a way that exactly four girls stand consecutively in the queue. Find the value of $\frac{m}{n}$.

17. Find the number of 4-digit numbers with distinct digits chosen from the set {0, 1, 2, 3, 4, 5} in which no two adjacent digits are even.

18. Let N denote the number of odd integers between 550 and 800 using the digits 4, 5, 6, 7, 8 and 9. Find the sum of the digits in N.

19. Find the number of quadrilaterals which can be constructed by joining the vertices of a convex polygon of 20 sides, if none of the sides of the polygon is also the side of the quadrilateral.

20. Consider a determinant of order 3 all whose entries are either 0 or 1. Five of these entries are 1 and four of them are '0'. Also, $a_{ij} = a_{ji}\ \forall\ 1 \le i, j \le 3$. Find the number of such determinants.

4.16 JEE Advanced: Sample Questions

Single Answer Type

1. In a certain test, there are n questions. In this test 2^{n-i} students gave wrong answers to at least i questions, where $i = 1, 2, \ldots, n$. If the total number of wrong answers given is 2047, then n is equal to

(A) 10 (B) 11 (C) 12 (D) 13

2. The number of natural numbers which are less than 2.10^8 and which can be written by means of the digits 1 and 2 is

(A) 772 (B) 870 (C) 900 (D) 766

3. If the sides of a triangle are determined by throwing a triplet of dice then the number of different triangles with all distinct sides is

(A) 7 (B) 2
(C) 3 (D) none of these

4. How many four digits number can be formed using digit {0, 1, 2, 3} (repetition is allowed) which are divisible by 11

(A) 34 (B) 31
(C) 32 (D) none of these

5. The number of ways in which 9 identical balls can be placed in three identical boxes is

(A) 55 (B) $\dfrac{9!}{(3!)^4}$ (C) $\dfrac{9!}{(3!)^3}$ (D) 12

6. The sum of the factors of 7! which are odd and are of the form $3t + 1$ where t is a whole number, is

(A) 10 (B) 8 (C) 9 (D) 15

7. The number of permutations of the letters of the word HINDUSTAN such that neither the pattern 'HIN' nor 'DUS' nor 'TAN' appears, are

(A) 166674 (B) 169194
(C) 166680 (D) 181434

8. The number of 3-digit odd numbers divisible by 3 using the digits 3, 4, 5, 6 when repetition is allowed is

(A) 12 (B) 10
(C) 11 (D) none of these

9. A seven-digit number divisible by 9 is to be formed by using 7 out of numbers 1, 2, 3, 4, 5, 6, 7, 8, 9. The number of ways in which this can be done is

(A) 9! (B) 2(7!) (C) 4(7!) (D) 4(9!)

10. Let $E = \left[\dfrac{1}{3} + \dfrac{1}{50}\right] + \left[\dfrac{1}{3} + \dfrac{2}{50}\right] + \ldots$ up to 50 terms, then, the exponent of 2 in $(E)!$ is

(A) 13 (B) 12 (C) 15 (D) 14

11. The number of points in the cartesian plane with integral coordinates satisfying the inequalities $|x| \le k, |y| \le k, |x - y| \le k$; is

(A) $(k+1)^3 - k^3$ (B) $(k+2)^3 - (k+1)^3$
(C) $(k^2 + 1)$ (D) none of these

12. Line L_1 contains l_1 points and line L_2 contains l_2 points. If the points on L_1 are joined to the points on L_2, then the number of points of intersection of new lines, is

(A) ${}^{l_1}C_2 \times {}^{l_2}C_2$ (B) $4.{}^{l_1}C_2 \times {}^{l_2}C_2$
(C) $2.{}^{l_1}C_2 \times {}^{l_2}C_2$ (D) none of these

13. The number of different matrices that can be formed with elements 0, 1, 2 or 3 each matrix having 4 elements, is

(A) 3×2^4 (B) 2×4^4 (C) 3×4^4 (D) 4^4

14. The number of ways in which the sum of the upper faces of four distinct dice can be six is

(A) 4 (B) 6 (C) 7 (D) 10

15. The number of ways in which we can choose 3 squares on a chess board such that one of the squares has its two sides common to two other squares is

(A) 290 (B) 292 (C) 296 (D) 298

16. The maximum number of points into which 4 circles and 4 straight lines intersect, is
(A) 26 (B) 56 (C) 50 (D) 72

17. The number of numbers greater than a million that can be formed with the digits 2, 3, 0, 3, 4, 2 and 3 is
(A) 340 (B) 360
(C) 370 (D) none of these

18. A gentleman invites 13 guests to a dinner and places 8 of them at one table and the remaining 5 at the other, the tables being round. The number of ways he can arrange the guests is
(A) 9! (B) $\frac{11!}{40}$ (C) $\frac{12!}{40}$ (D) $\frac{13!}{40}$

19. How many different nine-digit numbers can be formed from the number 223355888 by arranging its digits so that the odd digits occupy odd positions?
(A) 60 (B) 36 (C) 300 (D) 180

20. The number of ways in which 5X's can be placed in the squares of the figure so that no horizontal row remains empty is

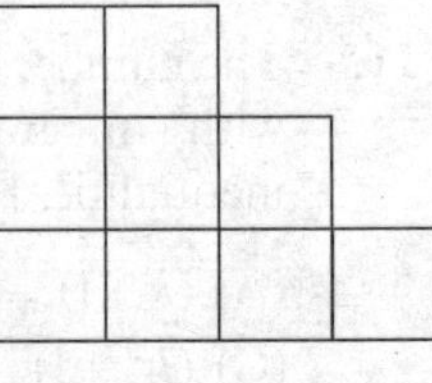

(A) 97 (B) 98 (C) 100 (D) 126

21. The number of n-digit numbers which consist of the digits 1 and 2 only if each digit is to be used at least once is equal to 510, then n is equal to
(A) 6 (B) 7 (C) 8 (D) 9

22. Sum of all the six-digit numbers formed by using the digits 2, 3, 4, 5, 6 and 7 (without repetition) is 360 N where N equals
(A) 111111 (B) 333333
(C) 666666 (D) 999999

23. The vowels in the word FACETIOUS are in alphabetic order. Consider all the possible words made from arrangements of all the letters in this word. Number of words in which vowels are in alphabetic order, is
(A) $C(9, 5)\cdot 4!$ (B) $C(9, 4)$
(C) $C(9, 4)\cdot 5!$ (D) $4!\cdot 5!$

24. Let P_n denote the number of ways in which three people can be selected out of n people sitting in a row, if no two of them are consecutive. If, $P_{n+1} - P_n = 15$ then the value of n is
(A) 6 (B) 8 (C) 10 (D) 12

25. A man has 3 friends. If N is the number of ways he can invite one friend every day for dinner on 6 successive nights so that no friend is invited more than 3 times, then the value of $\frac{N}{170}$ is
(A) 2 (B) 3 (C) 4 (D) 5

26. If N is the number of ways in which 3 distinct numbers can be selected from the set $\{3^1, 3^2, 3^3, \ldots, 3^{10}\}$ so that they form a G.P., then the value of $\frac{N}{5}$ is
(A) 3 (B) 4 (C) 5 (D) 6

27. Let $f(n) = \sum_{r=0}^{n}\sum_{k=r}^{n}\binom{k}{r}$. The total number of divisors of $f(9)$ is equal to
(A) 4 (B) 6 (C) 8 (D) 10

Multiple Answer Type

1. Let $\vec{a} = \vec{i} + \vec{j} + \vec{k}$ and $\vec{r}$ be a variable vector such that $\vec{r}.\vec{i}, \vec{r}.\vec{j}$ and $\vec{r}\cdot\vec{k}$ are positive integers. If $\vec{r}\cdot\vec{a} \le 12$. Then the number of values of $\vec{r}$ is
(A) $12_{C_9} - 1$ (B) 12_{C_3}
(C) 12_{C_9} (D) none of these

2. If $x \in N$ and
$$(x-1)_{C_4} - (x-1)_{C_3} - \frac{5}{4}x - 2_{P_2} < 0,$$
then x is equal to
(A) 4, 5, 6, 7, 8, 9, 10 (B) $3 < x < 11$
(C) $3 \le x \le 10$ (D) $3 < x \le 10$

3. A class has 30 students. The following prizes are to be awarded to the students of this class. First and second in Mathematics, first and second in Physics, first in Chemistry, first in Biology. If N denotes the number of ways in which this can be done, then N is divisible by

(A) 400
(B) 600
(C) 8100
(D) divisible by 4 distinct prime numbers

4. A letter lock consists of three rings marked with 15 different letters. If N denotes the number of ways in which it is possible to make successful attempt to open the lock then,
(A) N is divisible by 482
(B) N is the product of 3 distinct prime numbers
(C) N is the product of 4 distinct prime numbers
(D) none of these

5. If m and n are positive integers more than or equal to 2. $M > n$, then $(mn)!$ is divisible by
(A) $(m!)^n$ (B) $(n!)^m$
(C) $(m+n)!$ (D) $(m-n)!$

6. The number of ways in which we can choose 2 distinct integers from 1 to 100 such that the difference between them is at most 10 is
(A) $100_{C_2} - 90_{C_2}$ (B) $100_{C_{98}} - 90_{C_{88}}$
(C) $100_{C_2} - 90_{C_{88}}$ (D) none of these

7. There are n married couples at a party. Each person shakes hand with every person other than her or his spouse. The total number of handshakes must be
(A) $2n_{C_2} - n$ (B) $2n_{C_2} - (n-1)$
(C) $2n(n-1)$ (D) $2n_{C_2}$

8. For the equation $x+y+z+w=19$, the number of positive integral solution is equal to
(A) the number of ways in which 15 identical things can be distributed among 4 persons.
(B) the number of ways in which 19 identical things can be distributed among 4 persons.
(C) coefficient of x^{19} in $(x^0+x^1+x^2+\ldots+x^{19})^4$
(D) coefficient of x^{19} in $(x+x^2+x^3+\ldots+x^{19})^4$

9. There are n lines in a plane, no two of which are parallel and no three of which are concurrent. Let the plane be divided into u_n parts, then
(A) $u_4 = 1$ (B) $u_3 = 11$
(C) $u_2 = 3$ (D) $u_n = u_{n-1} + 4$

10. Let x be the number of 5-digit numbers, the sum of whose digits is even and y be the number of 5 digits numbers, the sum of whose digit is odd, then
(A) $x = y$ (B) $x+y = 90000$
(C) $x = 45000$ (D) $x < y$

11. If $n_{C_\alpha} = n_{C_\beta}$, then it may be true that
(A) $\alpha = \beta$ (B) $\alpha+\beta = 2n$
(C) $\alpha+\beta = n$ (D) all of these

12. The number of 5-letter words formed with the letters of the word CALCULUS is divisible by
(A) 2 (B) 3 (C) 5 (D) 7

13. If ${}^{100}C_{50}$ can be prime factorised as $2^\alpha 3^\beta 5^\gamma 7^\delta$ where $\alpha, \beta, \gamma, \delta$, are non-negative integers, then the correct relation is/are
(A) $\alpha < \beta$ (B) $\gamma < \delta$
(C) $\alpha+\delta = \beta+\gamma-1$ (D) $\gamma+\delta = 0$

14. Words are formed by using all the letters of the word AAIOCEAL. Which of the following statement(s) is/are correct?
(A) The number of words which contain the word IOCL is $\frac{5!}{3!}$.
(B) The number of words in which the word IOCL is not at last four places is $\frac{4!}{3!}$.
(C) The number of words in which first three letters are not A is ${}^5C_3 \cdot 5!$.
(D) The number of words in which the letters of the word IOCL are separated is $5\frac{4!}{3!}$.

15. The number of ways in which 10 candidates $A_1, A_2, \ldots, A_{10}$ can be ranked, so that A_1 is always above A_2 is
(A) $\frac{10!}{2}$ (B) $8! \times {}^{10}C_2$
(C) ${}^{10}P_2$ (D) ${}^{10}C_2$

16. All the five-digit numbers in which each successive digit exceeds its predecessor are arranged in the increasing order. The 105th number does not contain the digit

(A) 1 (B) 2 (C) 3 (D) 4

17. A multiple-choice question in an examination consists of 5 alternatives in which one or more than one may be correct. The number of different possible answers that can be given by an examinee is also equal to

(A) The number of all possible forecasts of 5 football matches in which at least one forecast is incorrect if each forecast is either a win or loss

(B) The value of $(p + q)$ where p and q satisfy the equations $C(p, p - 2) = 10$ and $C(q, 15) = C(q, 11)$

(C) The number of ways in which a normal dice tossed twice can fall in a definite order given that both throws are different.

(D) The number of different signals that can be made with 5 different flags by raising any number of flags at a time.

18. Let P_n denote the number of permutations of n distinct things taken all at a time and $x_n = {}^{n+5}C_4 - \left(\frac{143}{96}\right)\left(\frac{P_{n+5}}{P_{n+3}}\right)$ (where $n \in B$).

The possible value of n for which x_n is negative, can be

(A) 1 (B) 2 (C) 3 (D) 4

19. The number of linear arrangements of 4 alike of one kind and 5 alike of another kind is not equal to

(A) the number of 5-digit number whose digits are in increasing order

(B) the number of linear arrangements of 8 identical apples and 4 identical oranges if no two oranges are together

(C) the number of ways of distribution of 9 distinct toys between two children when one is having exactly one toy more than the other

(D) the number of ways in which 8 students can be divided into two teams, that need not necessarily be of equal size

20. The square of any odd number is of the form

(A) $8n + 1$ (B) $6n + 1$

(C) $4n + 1$ (D) $2n + 1$

[where n is a non-negative integer]

21. There are 10 seats in the first row of a theatre of which 4 are to be occupied. The number of ways of arranging 4 persons so that no two persons sit side by side is

(A) ${}^7C_4$ (B) $4 \cdot {}^7C_3$

(C) ${}^7C_3 \times 4!$ (D) 840

22. The number of ways in which three numbers in an A.P. can be selected from 1, 2, 3, …, n is

(A) $\left(\frac{n-1}{2}\right)^2$ if n is even

(B) $\frac{n(n-1)}{4}$ if n is even

(C) $\frac{(n-1)^2}{4}$ if n is odd

(D) $\frac{(n-1)^2}{2}$ if n is odd

23. The number of ways in which 6 different toys can be distributed equally among 3 children is also equal to

(A) the number of ways in which all the letters of the word KUMKUM can be arranged in a definite order

(B) the number of different signals which can be obtained from 10 different flags taken two at a time, keeping one above the other

(C) the number of ways in which five different marbles can be placed in three boxes

(D) the number of ways in which all the letters of the word CALCULUS can be arranged without changing the order and position of non-alike letters.

24. Identify the correct statement(s).

5 letters are to be kept in 3 envelopes. Each envelope can hold any number of letters. The letters are to be kept in envelopes so that no envelope remains empty. Then the number of ways letters can be kept in envelopes if

(A) letters are identical but envelopes are different is 36
(B) letters are different but envelopes are identical is 25
(C) letters as well as envelopes are identical is 6
(D) letters as well as envelopes are identical but envelopes are kept in row is 6.

Comprehension Based Questions

Paragraph for Question No. 1 to 3
Consider the letters of the word MATHEMATICS.

1. The possible number of words taking all letters at a time such that at least one repeating letter is at an odd position in each word is

(A) $\frac{11!}{2!2!2!}$ (B) $\frac{11!}{2!2!2!}-\frac{9!}{2!}$
(C) $\frac{11!}{2!2!2!}-\frac{8!}{2!}$ (D) $\frac{9!}{2!2!2!}$

2. The possible number of words taking all letters at a time such that in each word both M's are together and both T's are together but both A's are not together is

(A) $\frac{9!}{2!2!2!}$ (B) $\frac{6!4!}{2!2!}$
(C) $\frac{11!}{2!2!2!}-\frac{10!}{2!2!}$ (D) $7!\,{}^{8}C_2$

3. The possible number of words in which no two vowels are together is

(A) $7!\,{}^{8}C_4\frac{4!}{2!}$ (B) $\frac{7!}{2!}\,{}^{8}C_4\frac{4!}{2!}$
(C) $\frac{7!}{2!2!}\,{}^{8}C_4\frac{4!}{2!}$ (D) $\frac{7!}{2!2!2!}\,{}^{8}C_4\frac{4!}{2!}$

Paragraph for Question No. 4 to 6
Let $S=\{1, 2, 3, \ldots, n\}$ and $A\subseteq S$. Let $n(A)$ denote a cardinal number and $\min(A)$ represent the least number among the elements of set A.

4. The greatest value of $\min(A)$, where $A\subseteq S$ and $n(A)=r, 1\le r\le n$; is

(A) r (B) $(n-r)$
(C) $(n-r+1)$ (D) $(r+1)$

5. The number of subsets A of S for which $n(A)=r$ and $\min(A)=K$, is

(A) ${}^{n-K}C_{r-1}$ (B) ${}^{n}C_{r-1}$
(C) ${}^{n-K+1}C_{r-1}$ (D) ${}^{n-K-1}C_{r-1}$

6. The value of $\sum_{n(A)=r}(\min(A)) = K$ is

(A) $n\cdot{}^{n-K}C_{r-1}$
(B) $(n+1)^{n-K}C_{r-1}-r^{n-K+1}C_r$
(C) $K\cdot{}^{n-K}C_{r-1}+n^{n-K+1}C_r$
(D) ${}^{n}C_r$

Paragraph for Question No. 7 to 9
Consider the word 'CARCASSONNE'. Words are formed using all the letters of this word.

7. If the number of words which contain the word 'CAR' and vowels are separated is $k(5!)$ then k is equal to
(A) 157 (B) 315 (C) 630 (D) 710

8. If the number of arrangements of the letters of the given word if position of each vowel is fixed is N then number of ways in which N can be resolved as a product of 2 divisors which are co-prime is
(A) 32 (B) 16 (C) 8 (D) 4

9. The number of arrangements using only 5 letters of the given word is
(A) 8220 (B) 8120
(C) 13020 (D) 15040

Paragraph for Question No. 10 to 12
Consider the word W = SOLICITATION consisting of 6 vowels; namely, three I's, one A and two O's and 6 consonants namely two T's and C, L, N, S one each. Words are formed using only the letters from the word W.

10. If the number of ways in which all the letters of the word W can be arranged so that vowels appear alphabetically is $K(7!)$, then the value of K is
(A) 99 (B) 33 (C) 66 (D) 72

11. The number of 4-lettered words that can be made if each word contains exactly 2 consonants and exactly 2 vowels, is
(A) 1008 (B) 504 (C) 1056 (D) 168

12. If the number of words that can be made by using all the letters of the word W, if O's as well as I's are separated is $N(8!)$, then the value of N is
(A) 158 (B) 228 (C) 260 (D) none

Paragraph for Question No. 13 to 15
Different words are formed by arranging the letters of the word SUCCESS.

13. The numbers of words in which C's are together but S's are separated, is
(A) 120 (B) 96 (C) 24 (D) 420

14. The number of words in which no two C's and no two S's are together is
(A) 120 (B) 96 (C) 24 (D) 180

15. The number of words in which the consonants appear in alphabetic order is
(A) 42 (B) 40 (C) 420 (D) 280

Match Matrix

1. Only one matching option is possible

Column I	Column II
(A) If r, s, t are different prime numbers and a, b, c are the positive integers such that LCM of a, b, c is $r^2s^4t^2$ and HCF of a, b, c is rs^2t, then the number of ordered triplets (a, b, c) is	(P) 826
(B) If a, b, c are different prime numbers and p, q, r are positive integers of the form $a^xb^yc^z$ such that the LCM of q, r, is $a^2b^5c^3$ and HCF of p, r is $a^2b^2c^3$, then the number of possible ordered triplets (p, q, r) is (where x, y, z are natural numbers and less than equal to 5)	(Q) 792
(C) The number of odd numbers between 3000 to 63000 that have different digits is	(R) 432

(Continued)

Column I	Column II
(D) The number of quadrilaterals constructed by joining the vertices of a convex polygon of 20 sides, if none of the sides of the polygon is also the side of the quadrilateral.	(S) 2275

2. Only one matching option is possible

Column I	Column II
(A) The number of ways in which the letters of the word AAKRITI can be arranged if no two alike letters are together, is	(P) 65
(B) The number of permutations of the digits 1, 2, 3, 4 and 5 taken all at a time so that the sum of the digits at the first two places is smaller than the sum of the digits at the last two place, is	(Q) 8
(C) If the number of solutions of equation $x+2y+z=13$ is ${}^nC_r$ where x and z are even natural numbers and y is a positive number, then the least value of $(n+r)$ is	(R) 48
(D) If LCM of $(a, b)=5^2\times3^4$, LCM of $(b, c)=5^3\times3^4$ and LCM of $(c, a)=5^3\times3^4$, then the possible ordered triplet of (a, b, c) is	(S) 660

3. Consider the letters of the word MATHEMATICS. Set of repeating letters $=\{M, A, T\}$ and set of non-repeating letters $=\{H, E, I, C, S\}$.

List I	List -II
(A) The number of words taking all letters of the given word such that at least one repeating letter is at an odd position is	(P) $28 \cdot 7!$
(B) The number of words formed taking all letters of the given word in which no two vowels are together is	(Q) $\frac{11!}{(2!)^3}$
(C) The number of words formed taking all letters of the given word such that in each word both M's are together and both T's are together but both A's are not together is	(R) $210 \cdot 7!$
(D) The number of words formed taking all letters of the given word such that the relative order of vowels and consonants does not change is	(S) $\frac{4!\,7!}{(2!)^3}$

4.

Column I	Column II
(A) Consider a box containing n balls numbered from 1 to n. If 455 is the number of ways to get three balls from the box such that no two balls are consecutively numbered, then the value of n, is	(P) 12
(B) Let S be the set of points (a, b) in the coordinate plane, where each of a and b may be –1, 0, or 1. The number of distinct lines that pass through at least two numbers of S, is	(Q) 17
(C) All non-empty subsets of {2, 4, 5, 7} are selected. The number of different sums which the elements of these subsets can have, is	(R) 18
	(S) 20

5.

Column I	Column II
(A) In a birthday party there are n couples. Everyone's shaking hands with everyone. If the number of hand shakes is 28, then the value of n is equal to	(P) 8
(B) If the coefficients of $(4r - 4)^{th}$ term and $(r + 14)^{th}$ term in the expansion of $(1 + x)^{28}$ are equal, then the value of r, is(are)	(Q) 6
(C) If the number of circular permutations of n persons if two specified persons are always together is 48 then the value of n, is	(R) 4
(D) A convex polygon has 20 diagonals, then the number of sides in the polygon in divisible by	(S) 3
	(T) 1

Subjective Answer Type (Hard Nut)

1. A shop sells 6 different flavours of ice cream. In how many ways can a customer choose 4 ice cream cones if

(i) they are all of different flavours

(ii) they are not necessarily of different flavours

(iii) they contain only 3 different flavours
(iv) they contain only 2 or 3 different flavours

2. Find the number of integers between 1 and 10000 with at least one 8 and at least one 9 as digits.

3. Find the sum of all numbers greater than 10000 formed by using the digits 0, 1, 2, 4, 5 with no digit being repeated in any number.

4. How many 15-letter arrangements of 5A's, 5B's and 5C's have no A's in the first 5 letters, no B's in the next 5 letters, and no C's in the last 5 letters.

5. Find the number of elements sets of positive integers $\{a, b, c\}$ such that $abc = 2310$.

6. An 8 oared boat is to be manned by a crew chosen from 14 men of which 4 can only steer but cannot row and the rest can row but cannot steer. Of those who can row, 2 can row on the bow side. In how many ways can the crew be arranged?

7. Find the number of three digits numbers from 100 to 999 inclusive which have any one digit that is the average of the other two.

8. 12 persons are to be seated at a square table, three on each side. 2 persons wish to sit on the north side and two wish to sit on the east side. One other person insists on occupying the middle seat (which may be on any side). Find the number of ways in which they can be seated.

9. There are 5 white, 4 yellow, 3 green, 2 blue and 1 red ball. The balls are all identical except for colour. These are to be arranged in a line in 5 places. Find the number of distinct arrangements.

10. 10 identical balls are distributed in 5 different boxes kept in a row and labelled A, B, C D and E. Find the number of ways in which the ball can be distributed in the boxes if no two adjacent boxes remain empty.

11. In how many ways 4 letters can be selected from the letters of the word INEFFECTIVE?

12. There are 12 points in a plane, 5 of which are concyclic and out of remaining 7 points, no three are collinear and none concyclic with the previous 5 points. Find the number of circles passing through at least 3 points out of 12 given points.

13. In how many ways can 7 departments be divided among 3 ministers such that every minister gets at least one and at most 4 departments to control?

14. In a box there are 10 balls, 4 red, 3 black, 2 white and 1 yellow. In how many ways can a child select 4 balls out of these 10 balls? (Assume that the balls of the same colour are identical.)

15. There are three papers of 100 marks each in an examination. Find the number of ways in which a student can get 150 marks such that he gets at least 60% in two papers.

16. In a shooting competition a man can score 5, 4, 3, 2 or 0 points for each shot. Find the number of different ways in which he can score 30 in seven shots.

17. Find the number of ways to choose an ordered pair (a, b) of numbers from the set $\{1, 2, \ldots, 10\}$ such that $|a-b| \leq 5$.

18. A, B and C are the set of all the positive divisors of 10^{60}, 20^{50} and 30^{40} respectively. Find $n(A \cup B \cup C)$.

19. There are 5 boxes of 5 different colours. Also, there are 5 balls of colours same as those of the boxes. In how many ways can we place 5 balls in 5 boxes such that
   (i) all balls are placed in the boxes of colours not same as those of the ball
   (ii) at least 2 balls are placed in boxes of the same colour

20. Find the number of positive integral solutions of $x+y+z+w=20$ under the following conditions:
   (i) zero values of x, y, z, w is included
   (ii) zero values are excluded
   (iii) no variable may exceed 10, zero values excluded

(iv) each variable is an odd number

(v) $0 < x < y < z < w$

21. Find the number of integers which lie between 1 and 10^6 and which have the sum of the digits equal to 12.

22. Find the number of triangles whose angular points are at the angular points of a given polynomial of n sides, but none of whose sides are the sides of the polygon.

23. In how many ways can a pack of 52 cards be

(i) distributed equally among four players in order

(ii) divided into 4 groups of 13 cards each

(iii) divided into four sets of 20, 15, 10, 7 cards

(iv) divided into four sets, three of them having 15 cards each and the fourth having 7 cards

24. Find the number of ways of distributing 10 identical balls in 3 boxes so that no box contains more than four balls and less than 2 balls.

25. In how many ways can 6 letters be placed in 6 envelopes such that

(i) no letter is placed in its corresponding envelope

(ii) at least 4 letters are placed in correct envelopes

(iii) at most 3 letters are placed in wrong envelopes

4.17 JEE Main: Archive

1. In an examination, there are 5 multiple choice questions with 3 choices, out of which exactly one is correct. There are 3 marks for each correct answer, –2 marks for each wrong answer and 0 marks if the question is not attempted. Then, the number of ways a student appearing in the examination gets 5 marks is ______.

[Main 2022 24th June 1st Shift]

2. The number of 7-digit numbers which are multiples of 11 and are formed using all the digits 1,2, 3, 4,5, 7 and 9 is ______.

[Main 2022 24th June 2nd Shift]

3. The sum of all the elements of the set $\{\alpha \in \{1, 2, \ldots, 100\} : \text{HCF}(\alpha, 24) = 1\}$ is ______.

[Main 2022 24th June 2nd Shift]

4. The number of 3-digit odd numbers, whose sum of digits is a multiple of 7, is ______.

[Main 2022 25th June 1st Shift]

5. The total number of three-digit numbers, with one digit repeated exactly two times, is

[Main 2022 25th June 2nd Shift]

6. There are ten boys $B_1, B_2, \ldots, B_{10}$ and five girls $G_1, G_2, \ldots, G_5$ in a class. Then the number of ways of forming a group consisting of three boys and three girls, if both B_1 and B_2 together should not be the members of a group, is ______.

[Main 2022 26th June 1st Shift]

7. The total number of 3-digit numbers, whose greatest common divisor with 36 is 2, is ______.

[Main 2022 26th June 2nd Shift]

8. The number of ways, 16 identical cubes, of which 11 are blue and rest are red, can be placed in a row so that between any two red cubes there should be at least 2 blue cubes, is ______.

[Main 2022 27th June 1st Shift]

9. The total number of 5-digit numbers, formed by using the digits 1, 2, 3, 5, 6, 7 without repetition, which are multiple of 6, is

(A) 36 (B) 48 (C) 60 (D) 72

[Main 2022 28th June 1st Shift]

10. Let $A = (1, a_1, a_2, \ldots, a_{18}, 77)$ be a set of integers with $1 < a_1 < a_2 < \ldots < a_{18} < 77$. Let the set $A + A = \{x + y : x, y \in A\}$ contain exactly 39 elements. Then, the value of $a_1 + a_2 + \ldots + a_{18}$ is equal to ______.

[Main 2022 28th June 1st Shift]

11. The number of ways to distribute 30 identical candies among four children C_1, C_2, C_3 and C_4 so that C_2 receives at least 4 and at most 7 candies, C_3 receives at least 2 and at most 6 candies, is equal to

(A) 205 (B) 615 (C) 510 (D) 430

[Main 2022 28th June 2nd Shift]

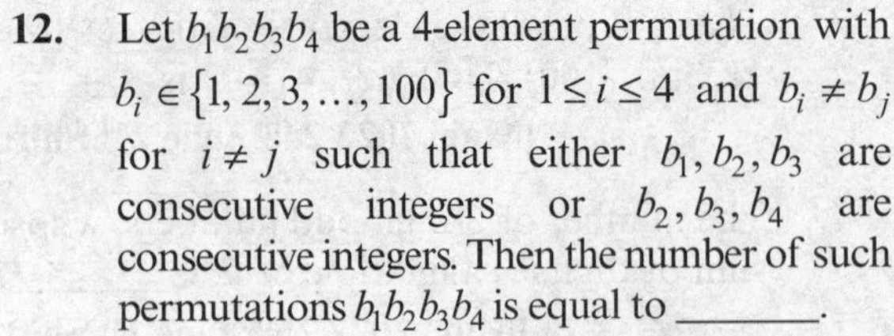

12. Let $b_1b_2b_3b_4$ be a 4-element permutation with $b_i \in \{1, 2, 3, \ldots, 100\}$ for $1 \le i \le 4$ and $b_i \ne b_j$ for $i \ne j$ such that either b_1, b_2, b_3 are consecutive integers or b_2, b_3, b_4 are consecutive integers. Then the number of such permutations $b_1b_2b_3b_4$ is equal to _______.

[Main 2022 29th June 1st Shift]

13. The total number of four-digit numbers such that each of the first three digits is divisible by the last digit, is equal to _______.

[Main 2022 29th June 2nd Shift]

14. The letters of the word 'MANKIND' are written in all possible orders and arranged in serial order as in an English dictionary. Then the serial number of the word 'MANKIND' is _______.

[Main 2022 25th July 1st Shift]

15. The number of 5-digit natural numbers, such that the product of their digits is 36, is _______.

[Main 2022 26th July 1st Shift]

16. Numbers are to be formed between 1000 and 3000, which are divisible by 4, using the digits 1, 2, 3, 4, 5 and 6 without repetition of digits. Then the total number of such numbers is _______.

[Main 2022 26th July 2nd Shift]

17. Let S be the sample space of all five-digit numbers. If p is the probability that a randomly selected number from S, is a multiple of 7 but not divisible by 5, then $9p$ is equal to

(A) 1.0146 (B) 1.2085
(C) 1.0285 (D) 1.1521

[Main 2022 27th July 1st Shift]

18. Let S be the set of all passwords which are six to eight characters long, where each character is either a letter from $\{A, B, C, D, E\}$ or a number from $\{1, 2, 3, 4, 5\}$ with the repetition of characters allowed. If the number of passwords in S whose at least one character is a number from $\{1, 2, 3, 4, 5\}$ is $\alpha \times 5^6$, then α is equal to _______.

[Main 2022 28th July 1st Shift]

19. A class contains b boys and g girls. If the number of ways of selecting 3 boys and 2 girls from the class is 168, then $b + 3g$ is equal to **[Main 2022 28th July 2nd Shift]**

20. Let $S = \{4, 6, 9\}$ and $T = \{9, 10, 11, \ldots, 1000\}$. If $A = \{a_1 + a_2 + \ldots + a_k : k \in N,$

$a_1, a_2, a_3, \ldots, a_k \in S\}$,

then the sum of all the elements in the set $T - A$ is equal to _______.

[Main 2022 29th July 1st Shift]

21. The number of natural numbers lying between 1012 and 23421 that can be formed using the digits 2, 3, 4, 5, 6 (repetition of digits is not allowed) and divisible by 55 is _______.

[Main 2022 29th July 2nd Shift]

22. The number of 9-digit numbers that can be formed using all the digits of the number 123412341 so that the even digits occupy only even places, is _______.

[Main 2023 24th Jan 1st Shift]

23. A boy needs to select five courses from 12 available courses, out of which 5 courses are language courses. If he can choose at most two language courses, then the number of ways he can choose five courses is

[Main 2023 24th Jan 1st Shift]

24. The number of integers, greater than 7000, that can be formed using the digits 3, 5, 6, 7, 8 without repetition is

(A) 120 (B) 168 (C) 220 (D) 48

[Main 2023 24th Jan 2nd Shift]

25. Let $S = \{1, 2, 3, 5, 7, 10, 11\}$. The number of non-empty subsets of S that have the sum of all elements as multiple of 3, is _______.

[Main 2023 25th Jan 1st Shift]

26. The number of numbers, strictly between 5,000 and 10,000 can be formed using the digits 1, 3, 5, 7, 9 without repetition, is

(A) 6 (B) 12 (C) 120 (D) 72

[Main 2023 25th Jan 2nd Shift]

27. Suppose Anil's mother wants to give 5 whole fruits to Anil from a basket of 7 red apples, 5 white apples and 8 oranges. If in the selected 5 fruits, at least 2 oranges, at least one red apple and at least one white apple must be given, then the number of ways

Anil's mother can offer 5 fruits to Anil is _______. **[Main 2023 25th Jan 2nd Shift]**

28. If all the six-digit numbers $x_1x_2x_3x_4x_5x_6$ with $0 < x_1 < x_2 < x_3 < x_4 < x_5 < x_6$ are arranged in the increasing order, then the sum of the digits in the 72nd number is _______.

[Main 2023 29th Jan 1st Shift]

29. Five-digit numbers are formed using the digits 1, 2, 3, 5, 7 with repetitions and are written in descending order with serial numbers. For example, the number 77777 has a serial number 1. Then the serial number of 35337 is _______.

[Main 2023 29th Jan 1st Shift]

30. The number of 3-digit numbers, that are divisible by either 3 or 4 but not divisible by 48, is

(A) 472 (B) 432 (C) 507 (D) 400

[Main 2023 29th Jan 2nd Shift]

31. The letters of the word OUGHT are written in all possible ways and these words are arranged as in a dictionary, in a series. Then the serial number of the word TOUGH is:

(A) 89 (B) 84 (C) 86 (D) 79

[Main 2023 29th Jan 2nd Shift]

32. The total number of 4-digit numbers whose greatest common divisor with 54 is 2, is _______. **[Main 2023 29th Jan 2nd Shift]**

33. The number of 4-digit numbers (repetition of digits is allowed) which are made using the digits 1, 2, 3 and 5, and are divisible by 15, is equal to _______.

[Main 2023 30th Jan 1st Shift]

34. The number of ways of selecting two numbers a and b, $a \in \{2, 4, 6, \ldots, 100\}$ and $b \in \{1, 3, 5, \ldots, 99\}$ such that 2 is the remainder when $a+b$ is divided by 23 is

(A) 186 (B) 54 (C) 108 (D) 268

[Main 2023 30th Jan 2nd Shift]

35. The number of seven-digit odd numbers that can be formed using all the seven digits 1, 2, 2, 2, 3, 3, 5 is _______.

[Main 2023 30th Jan 2nd Shift]

36. Number of 4-digit numbers that are less than or equal to 2800 and either divisible by 3 or by 11, is equal to _______.

[Main 2023 31st Jan 1st Shift]

37. Let 5-digit numbers be constructed using the digits 0, 2, 3, 4, 7, 9 with repetition allowed, and are arranged in ascending order with serial numbers. Then the serial number of the number 42923 is _______.

[Main 2023 31st Jan 1st Shift]

38. If ${}^{2n+1}P_{n-1} : {}^{2n-1}P_n = 11:21$, then n^2+n+15 is equal to:

[Main 2023 31st Jan 2nd Shift]

39. The number of words, with or without meaning, that can be formed using all the letters of the word ASSASSINATION so that the vowels occur together, is _______.

[Main 2023 01st Feb 1st Shift]

40. The number of integral solutions to the equation $x+y+z=21$, where $x \geq 1$, $y \geq 3$, $z \geq 4$, is equal to _______.

[Main 2023 01st Feb 2nd Shift]

41. The total number of six-digit numbers formed using the digits 4, 5, 9 only and divisible by 6, is _______.

[Main 2023 01st Feb 2nd Shift]

42. The number of ways of giving 20 distinct oranges to 3 children such that each child gets at least one orange is

[Main 2023 6th April 1st Shift]

43. All the letters of the word PUBLIC are written in all possible orders and these words are written as in a dictionary with serial numbers. Then the serial number of the word PUBLIC is

(A) 576 (B) 578 (C) 580 (D) 582

[Main 2023 6th April 2nd Shift]

44. The number of 4-letter words, with or without meaning, each consisting of 2 vowels and 2 consonants, which can be formed from the letters of the word UNIVERSE without repetition is

[Main 2023 6th April 2nd Shift]

45. The number of arrangements of the letters of the word 'INDEPENDENCE' in which all the vowels always occur together is

(A) 16800 (B) 33600
(C) 18000 (D) 14800

[Main 2023 8th April 1st Shift]

46. The number of ways in which 5 girls and 7 boys can be seated at a round table so that no two girls sit together is

(A) 720 (B) $126(5!)^2$
(C) $7(360)^2$ (D) $7(720)^2$

[Main 2023 8th April 1st Shift]

47. If the number of words, with or without meaning, which can be made using all the letters of the word MATHEMATICS in which C and S do not come together, is (6!) k then k is equal to

(A) 2835 (B) 5670 (C) 1890 (D) 945

[Main 2023 8th April 2nd Shift]

48. The number of permutations of the digits 1, 2, 3, …, 7, without repetition, which neither contain the string 153 nor the string 2467, is ______. **[Main 2023 10th April 1st Shift]**

49. Some couples participated in a mixed doubles badminton tournament. If the number of matches played, so that no couple played in a match, is 840, then the total numbers of persons who participated in the tournament is

[Main 2023 10th April 1st Shift]

50. Eight persons are to be transported from city A to city B in three cars of different makes. If each car can accommodate at most three persons, then the number of ways in which they can be transported, is

(A) 1120 (B) 3360
(C) 1680 (D) 560

[Main 2023 10th April 2nd Shift]

51. The sum of all the four-digit numbers that can be formed using all the digits 2, 1, 2, 3 is equal to ______.

[Main 2023 10th April 2nd Shift]

52. The number of triplets (x, y, z) where x, y, z are distinct non negative integers satisfying $x + y + z = 15$, is

(A) 80 (B) 136 (C) 114 (D) 92

[Main 2023 11th April 1st Shift]

53. In an examination, 5 students have been allotted their seats as per their roll numbers. The number of ways in which none of the students sits in their allotted seat, is

[Main 2023 11th April 1st Shift]

54. If the letters of the word MATHS are permuted and all possible words so formed are arranged as in a dictionary with serial numbers, then the serial number of the word THAMS is

(A) 103 (B) 102 (C) 101 (D) 104

[Main 2023 11th April 2nd Shift]

55. The number of five-digit numbers, greater than 40000 and divisible by 5, which can be formed using the digits 0, 1, 3, 5, 7 and 9 without repetition, is equal to

(A) 132 (B) 120 (C) 72 (D) 96

[Main 2023 12th April 1st Shift]

56. Let the digits a, b, c be in A.P. Nine-digit numbers are to be formed using each of these three digits thrice such that three consecutive digits are in A.P. at least once. How many such numbers can be formed?

[Main 2023 12th April 1st Shift]

57. The number of seven-digit positive integers formed using the digits 1, 2, 3 and 4 only, and sum of the digits equal to 12 is

[Main 2023 13th April 1st Shift]

58. All words, with or without meaning, are made using all the letters of the word MONDAY. These words are written as in a dictionary with serial numbers. The serial number of the word MONDAY is

(A) 327 (B) 328 (C) 324 (D) 326

[Main 2023 13th April 2nd Shift]

59. The total number of 3-digit numbers that are divisible by 6 and can be formed by using the digits 1, 2, 3, 4, 5 with repetition, is ______. **[Main 2023 13th April 2nd Shift]**

60. The total number of three-digit numbers, divisible by 3, which can be formed using the digits 1, 3, 5, 8, if repetition of digits is allowed, is

(A) 21 (B) 20 (C) 22 (D) 18

[Main 2023 15th April 1st Shift]

61. A person forgets his 4-digit ATM pin code. But he remembers that in the code all the digits are different, the greatest digit is 7 and the sum of the first two digits is equal to the sum of the last two digits. Then the maximum number of trials necessary to obtain the correct code is

[Main 2023 15th April 1st Shift]

62. If the number of ways five different employees can sit into four indistinguishable offices where any office may have any number of persons including zero, then is equal to:

(A) 47 (B) 53 (C) 51 (D) 43

[Main 2024 1st Feb 1st Shift]

63. The lines $L_1, L_2, \ldots, L_{20}$ are distinct. For $n = 1, 2, 3, \ldots, 10$ all the lines L_{2n-1} are parallel to each other and all the lines L_{2n} pass through a given point P. The maximum number of point of intersection of pairs of lines from the set $\{L_1, L_2, \ldots, L_{20}\}$ is equal to_______.

[Main 2024 1st Feb 2nd Shift]

64. Let $\alpha = \frac{(4!)!}{(4!)^{3!}}$ and $\beta = \frac{(5!)!}{(5!)^{4!}}$. Then

(A) $\alpha \in N$ and $\beta \notin N$

(B) $\alpha \notin N$ and $\beta \in N$

(C) $\alpha \in N$ and $\beta \in N$

(D) $\alpha \notin N$ and $\beta \notin N$

[Main 2024 27th Jan 2nd Shift]

65. All the letters of the word 'GTWENTY' are written in all possible ways with or without meaning and these words are written as in a dictionary. The serial number of the word 'GTWENTY' is

[Main 2024 29th Jan 1st Shift]

66. The number of ways of arranging 8 identical books into 4 identical shelves where any number of shelves may remain empty is equal to

(A) 18 (B) 16 (C) 12 (D) 15

[Main 2024 29th Jan 2nd Shift]

67. In a mathematics examination paper, there are 20 questions of equal marks and the question paper is divided into three sections: A, B and C. A student is required to attempt a total of 15 questions taking at least 4 questions from each section. If section A has 8 questions, section B has 6 questions and section C has 6 questions, then the total number of ways a student can select 15 questions is_______.

[Main 2024 30th Jan 2nd Shift]

68. The total number of words (with or without meaning) that can be formed out of the letters of the word 'DISTRIBUTION' taken four at a time, is equal to _______

[Main 2024 31st Jan 1st Shift]

69. The number of ways in which 21 identical apples can be distributed among three children such that each child gets at least 2 apples, is

(A) 406 (B) 130 (C) 142 (D) 136

[Main 2024 31st Jan 2nd Shift]

4.18 Answer Key

4.18.1 Know Your Concepts

1. (96) **2.** (9) **3.** (96) **4.** (144) **5.** (12) **6.** (3) **7.** (205) **8.** (18)
9. (624) **10.** $({}^9C_4)$ **11.** (41541750) **12.** (74) **13.** (C) **14.** (50) **15.** $({}^{10}C_6 \times {}^5C_3)$
16. $({}^{100}C_2 - {}^{90}C_2)$ **17.** (5760) **18.** (1875) **19.** (510) **20.** (576)
21. (a) 5.6! (b) 3!.4! (c) 12 **22.** $(24C_2 \times 15C_3)$ **23.** (a) 72, 78120 (b) 23 (c) 32

4.18.2 JEE Main: Sample Questions

Single Answer Type

1. (A)	**2.** (B)	**3.** (D)	**4.** (B)	**5.** (B)	**6.** (C)	**7.** (D)	**8.** (C)
9. (C)	**10.** (C)	**11.** (D)	**12.** (B)	**13.** (C)	**14.** (B)	**15.** (B)	**16.** (A)
17. (A)	**18.** (A)	**19.** (A)	**20.** (C)	**21.** (D)	**22.** (D)	**23.** (B)	**24.** (C)
25. (A)	**26.** (B)	**27.** (A)	**28.** (B)	**29.** (B)	**30.** (C)	**31.** (C)	**32.** (B)
33. (B)	**34.** (B)	**35.** (A)	**36.** (B)	**37.** (B)	**38.** (D)	**39.** (A)	**40.** (A)
41. (C)	**42.** (B)	**43.** (A)	**44.** (A)	**45.** (C)	**46.** (C)	**47.** (C)	**48.** (D)
49. (B)	**50.** (C)	**51.** (C)	**52.** (C)	**53.** (A)	**54.** (D)	**55.** (B)	**56.** (B)
57. (C)	**58.** (A)						

Integer Answer Type

1. (1) **2.** (9) **3.** (6) **4.** (7) **5.** (2) **6.** (9) **7.** (8) **8.** (7)
9. (4) **10.** (8) **11.** (9) **12.** (300) **13.** (400) **14.** (22) **15.** (840) **16.** (5)
17. (150) **18.** (6) **19.** (2275) **20.** (12)

4.18.3 JEE Advanced: Sample Questions

Single Answer Type

1. (B) **2.** (D) **3.** (A) **4.** (A) **5.** (D) **6.** (B) **7.** (B) **8.** (C)
9. (C) **10.** (C) **11.** (A) **12.** (C) **13.** (D) **14.** (D) **15.** (B) **16.** (C)
17. (B) **18.** (D) **19.** (C) **20.** (B) **21.** (D) **22.** (D) **23.** (A) **24.** (B)
25. (B) **26.** (B) **27.** (C)

Multiple Answer Type

1. (B, C) **2.** (A, B, D) **3.** (A, B, C, D) **4.** (A, B)
5. (A, B, C, D) **6.** (A, B, C) **7.** (A, C) **8.** (A, D)
9. (A, B) **10.** (A, B, C) **11.** (A, C) **12.** (A, B, C)
13. (A, C, D) **14.** (A, C) **15.** (A, B) **16.** (A, C, D)
17. (A, B) **18.** (A, B, C) **19.** (C, D) **20.** (A, C, D)
21. (B, C, D) **22.** (B, C) **23.** (A, B, D) **24.** (B, D)

Comprehension Based Questions

1. (A) **2.** (D) **3.** (C) **4.** (C) **5.** (A) **6.** (B) **7.** (B) **8.** (C)
9. (A) **10.** (C) **11.** (A) **12.** (B) **13.** (C) **14.** (B) **15.** (A)

Match Matrix

1. (A – R; B – Q; C – P; D – S) **2. (A – S; B – R; C – Q; D – P)**
3. (A – Q; B – R; C – P; D – S) **4. (A – Q; B – S; C – P)**
5. (A – R; B – Q, R; C – Q; D – P, R, T)

Subjective Answer Type: Hard Nut

1. (i) 15 (ii) 126 (iii) 60 (iv) 105 **2.** (974) **3.** (3119976) **4.** (2252) **5.** (40)
6. $4(4!)2\ {}^8C_4 \times {}^6C_2$ **7.** (121) **8.** (2! 3! 8!) **9.** (2111) **10.** (711)
11. (1422) **12.** (211) **13.** (1680) **14.** (20) **15.** $({}^3C_2 \times {}^{32}C_2)$
16. (420) **17.** (80) **18.** (73001) **19.** (i) 44 (ii) 31
20. (i) 1771 (ii) 969 (iii) 633 (iv) 165 (v) 23 **21.** (6062) **22.** $\frac{n}{6}(n-4)(n-5)$
23. (i) $\frac{52!}{(13!)^4}$ (ii) $\frac{52!}{4!(13!)^4}$ (iii) $\frac{52!}{20!15!10!7!}$ (iv) $\frac{52!}{(15!)^3\,3!7!}$ **24.** (6) **25.** (i) 265 (ii) 16 (iii) 56

4.18.4 JEE Main: Archive

1. (40) **2.** (576) **3.** (1633) **4.** (63) **5.** (243) **6.** (1120) **7.** (150) **8.** (56)
9. (D) **10.** (702) **11.** (D) **12.** (18915) **13.** (1086) **14.** (1492) **15.** (180)
16. (30) **17.** (C) **18.** (7073) **19.** (17) **20.** (11) **21.** (6) **22.** (60) **23.** (546)
24. (B) **25.** (43) **26.** (D) **27.** (6860) **28.** (32) **29.** (1436) **30.** (B) **31.** (A)
32. (3000) **33.** (21) **34.** (C) **35.** (240) **36.** (710) **37.** (2997) **38.** (45) **39.** (50400)
40. (105) **41.** (81) **42.** (171) **43.** (D) **44.** (432) **45.** (A) **46.** (B) **47.** (B)
48. (4898) **49.** (16) **50.** (C) **51.** (26664) **52.** (C) **53.** (44) **54.** (A)
55. (B) **56.** (1260) **57.** (413) **58.** (A) **59.** (16) **60.** (C) **61.** (72) **62.** (A)
63. (101) **64.** (C) **65.** (553) **66.** (D) **67.** (11376) **68.** (3734) **69.** (D)

4.19 Solutions

4.19.1 Know Your Concepts

1. CALCUTTA has letters CC, AA, TT, U, L
Case I: When all the three letters are different
$${}^{5}P_3 = \frac{5!}{(5-3)!} = 3\times4\times5 = 60$$
Case II: When two are alike and one is different
$${}^{3}C_1 \times {}^{4}C_1 \times \frac{3!}{2!} = 3\times4\times3 = 36$$
Hence, total number of 3 letter words
$= 60 + 36 = 96$

2. Draw all five lines $y = x + r$ and other five lines $y = -x + r$. We can observe that the required square is $3^2 = 9$.

3. Total number of 3 letter words $= {}^{6}P_3 = 120$
The number of words containing no vowels
$= {}^{4}P_3 = 24$
$\therefore$ Number of required words $= 120 - 24 = 96$

4. In the word ARTICLE,
vowels are A, I, E
There are 3 even places
So, required words in which vowels occupy even places $= {}^{3}P_3 \times 4! = 144$

5. The number lying between 3000 and 4000 and divisible by 5
$\Rightarrow$ unit place will be filled up by only one digit (5) and thousand by (3)
Hence remaining two places can be filled up by ${}^{4}P_2$
Hence, $1\times {}^{4}P_2 \times 1 = 12$

6. $$\frac{{}^{15}C_{3r}}{{}^{15}C_{r+1}} = \frac{11}{3} \Rightarrow \frac{(r+1)!(14-r)!}{(3r)!(15-3r)!} = \frac{11}{3} \quad (1)$$
As ${}^{15}C_{3r}$ is meaningful if
$r = 0, 1, 2, 3, 4, 5$
Also, ${}^{15}C_{3r}$ should be multiple of 11
For $r = 2, 3$ the numerator is multiple of 11
For, $r = 3$
$$\frac{(r+1)!(14-r)!}{(3r)!(15-3r)!} = \frac{11}{3}$$

7. Total number of triangles by taking interior points as vertices

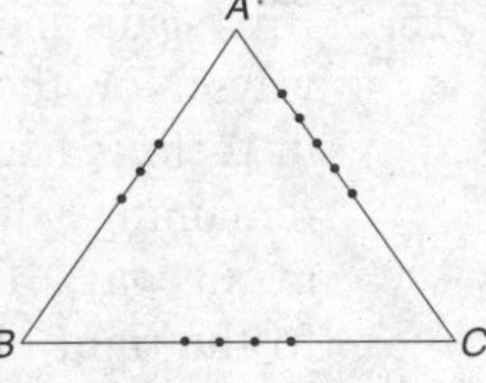

$= {}^{12}C_3 - {}^{3}C_3 - {}^{4}C_3 - {}^{5}C_3 = 205$

8. Let total number of persons $= n$.
$\therefore {}^{n}C_2 = 153$
$\Rightarrow n^2 - n - 306 = 0 \Rightarrow (n-18)(n+17) = 0$
$\therefore n = 18$

9. Required number of ways
= Total number of ways – number of ways in which he does not select any question from any one section
$= {}^{12}C_5 - 3\times {}^{8}C_5 = \frac{12!}{5!7!} - 3\times\frac{8!}{5!3!} = 624$

10. Total number of stations is 12 out of which train stopped at 4 stations and does not stop at 8 stations. Consider these 8 stations $a_1, a_2, \ldots, a_8$, this will create 9 places, so we just have to select 4 out of 9 and that can be done in ${}^{9}C_4$ ways.

11. We know a number is divisible by 4 if the last two digit are divisible by 4 using this.

12. $$E_5 300! = \left[\frac{300}{5}\right] + \left[\frac{300}{25}\right] + \left[\frac{300}{125}\right]$$
$= 60 + 12 + 2 = 74$
Hence, the number of zeros = 74

13. Given set of numbers is $\{1, 2, \ldots, 11\}$ in which 5 are even and six are odd. Hence in the given product, it is not possible to arrange but only to subtract the even number from odd number.
There must be at least one factor involving subtraction of an odd number from another odd number. Hence, product is always even.

14. The required number of selections is given as the number of selections without restriction. (The number of selection when 3 vertices are consecutive) – (The number of selections when 2 vertices are consecutive).
Therefore, the required number of selections:
${}^{10}C_3 - 10 - 10\times 6C_1 = 120 - 70 = 50$

15. First, 6 distinct digits can be selected in ${}^{10}C_6$ ways. Now the position of the smallest digit in them is fixed, that is, position 4. Of the remaining 5 digits, two digits can be selected in ${}^{5}C_2$ ways. These two digits can be placed to the right of the 4th position in one way only. The remaining three digits to the left of 4th position are in the required order automatically.

Hence ${}^{10}C_6 \times {}^{5}C_2 = {}^{10}C_6 \times {}^{5}C_3$

16. Let the chosen natural number be x_1 and x_2

Let there be a natural number 'a' before x_1, b between x_1 and x_2 and c after x_2.

$\therefore a+b+c=98$ where $a \ge b, b \ge 10, c \ge 0$

Now if we consider the choices where difference is at least 11, then the number of solutions is ${}^{90}C_2$

$\therefore$ Number of ways in which b is less than 10 is ${}^{100}C_2 - {}^{90}C_2$.

17. First, we have to select 2 men for bow side and 3 for stroke side

$\therefore$ The number of selections of the crew for two sides $= {}^{5}C_2 \times {}^{3}C_3$

For each selection there are 4 persons each on both sides who can be arranged in $4! \times 4!$ ways

$\therefore$ Required number of arrangements

$= {}^{5}C_2 \times {}^{3}C_3 \times 4! \times 4! = 5760$

18. $x_1x_2x_3x_4x_5 = 2^2 \times 3 \times 5 \times 7$

Here, 2^2 can be assigned to a_1, a_2, a_3, a_4, a_5

$\Rightarrow a_1+a_2+a_3+a_4+a_5 = 2 = {}^{2+5-1}C_{5-1}$

$= {}^{6}C_4 = 15$

Also, 3′ can be assigned to b_1, b_2, b_3, b_4, b_5

$\Rightarrow b_1+b_2+b_3+b_4+b_5 = 1 = {}^{1+5-1}C_{5-1} = {}^{5}C_4 = 5$

Similarly, for 5 and 7

$\therefore$ Total number of positive integral solutions $= 15 \times 5 \times 5 \times 5 = 1875$

19. We have the following possibilities

$(1, 2, 3)$ or (330) or (222)

[grouping of 6 days of week]

Hence, the total number of ways is

$$\frac{6!}{1!2!3!}3! + \frac{6!}{3!3!2!}3! + \frac{6!}{(2!2!2!)3!}3!$$

$= 360 + 60 + 90 = 510$

20. **Case I:** If 0 is one of the digits and is used only once, then the possible numbers are in the form $(aaa0, aa0a, a0aa)$ and a can take 9 values, hence number of numbers in this case is $3 \times 9 = 27$

Case II: If 0 is one of the digits and is used twice, then the possible numbers are in the form $(aa00, a00a, a0a0)$ and a can take 9 values

Hence number of numbers in this case is $3 \times 9 = 27$

Case III: If 0 is one of the digits and is used thrice, then the possible numbers are $(a000)$ and a can take 9 values.

Hence number of numbers in this case is 9.

Case IV: If only one digit is used (other than 0) then number of numbers = 9

Case V: If two digits a and b are used, then the possible numbers are (*aaab, aaba, abaa, baaa*).The number of numbers in this case is

${}^{9}C_2 \times 2! \times 4 = 36 \times 2 \times 4 = 288$

Case VI: If two digits a and b are used then possible numbers are

$(aabb, abba, abab, bbaa, baab, baba)$

Number of numbers in this case is ${}^{9}C_2 \times 6 = 216$

So, the total number of numbers is

$27+27+9+9+288+216 = 576$.

21. (a) 5.6! (b) 3!.4! (c) 12

22. $24_{C_2} \times 15_{C_3}$

23. (a) $21600 = 2^5 3^3 5^2$

Total number of divisors $= (5+1)(3+1)(2+1) = 72$

Sum of all divisors

$= (3^0+3^1+3^2+3^3)(5^0+5^1+5^2)$

$(2^0+2^1+2^2+2^3+2^4+2^5)$

$$= \frac{(3^4-1)}{(3-1)}\left(\frac{5^3-1}{5-1}\right)\left(\frac{2^6-1}{2-1}\right)$$

$= 78120$

(b) $7056 = 2^4 3^2 7^2$

Number of divisors $= (4+1)(2+1)(2+1)$

$= 5 \times 3 \times 3 = 45$

$\therefore$ Number of ways of resolving into 2 factors

$$= \frac{45+1}{2} = 23$$

(c) $300300 = 2^2 \times 3^1 \times 5^2 \times 7^1 \times 11^1 \times 13^1$

We have to make two factors x and y of 300300, which are relatively prime

That is. $xy = 300300$ but $\text{GCD}(x, y) = 1$

$\Rightarrow 2^2 3^1 5^2 7^1 11^1 13^1$ should have as single entities

So, number of unordered pair divisors

$= \{(1+1)(1+1)(1+1)(1+1)(1+1)(1+1)\}$

$= 2^5 = 32$

$\therefore$ Number of ways of splitting into two factors = 32

4.19.2 JEE Main: Sample Questions

Single Answer Type

1. **(A)**

1st, 3rd place must be filled with odd numbers and 2nd, 4th place must be filled with even numbers and last two digits must be divisible by 4; hence, the possibilities for the 3rd and 4th places are

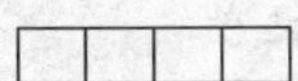

12, 32, 52,72, 92
16, 36, 56, 76, 96
Total 10 possibilities
Hence total number of ways are $4 \cdot 4 \cdot 10 = 160$

2. **(B)**

Method-1

Case-1: Delegation consists of 5 members

(a) 1 Boy + 4 Girls $\quad {}^2C_1 \times {}^6C_4 = 30$

(b) 2 Boys + 3 Girls $\quad {}^2C_2 \times {}^6C_3 = 20$

50

Case-2: Delegation consists of 6 members

(a) 1 Boy + 5 Girls $\quad {}^2C_1 \times {}^6C_5 = 12$

(b) 2 Boys + 4 Girls $\quad {}^2C_2 \times {}^6C_4 = 15$

27 ways

Method-2

Case-1: Delegation consists of 5 members.
Total number of ways of selecting 5 member team – number of ways when no boy is selected
$= {}^8C_5 - {}^6C_5 = 50$

Case-2: Delegation consists of 6 members
$\therefore$ Total – no. boys selected
$= {}^8C_6 - {}^6C_6 = 27$
Total = 50 + 27 = 77

3. **(D)**

$$\text{Let } \left.\begin{matrix} H_1 & W_1 \\ H_2 & W_2 \\ H_3 & W_3 \end{matrix}\right]$$

Treat $H_1 W_1$ as alike objects say XX, $H_2 W_2$ as YY and $H_3 W_3$ as ZZ

Now arrangements of $X\,X\,Y\,Y\,Z\,Z$, that is, $\frac{6!}{2!2!2!}$ is this required number of ways (think!)

Aliter: ${}^6C_2$ (for H_1 W_1) and ${}^4C_2$ (for H_2 W_2) ...

Hence total ways $({}^6C_2 \cdot 1)({}^4C_2 \cdot 1) = \frac{6!}{2!2!2!}$

4. **(B)**

0, 1, 2, 3, 4, 5, 6, 7, 8, 9

Case-1: 5 is always included and zero is excluded; select 4 non-zero digits in ${}^8C_4$ ways

${}^8C_4 \cdot 4!$ (say 1 2 3 4 5) (1)

				5

Case-2: Zero included ${}^8C_3$ (say 0 5 1 2 3)

${}^8C_3 \cdot [4! + (4! - 3!)]$

				0

$= {}^8C_3 \times 42$ (2)

				5

Add (1) and (2) to get

Total = $70 \times 24 + 56 \times 42$

$= 7 \times 8\,[10 \times 3 + 42] = 7 \times 8 \times 6\,(5 + 7)$

$= 6 \times 7 \times 8 \times 12 = 4!(168) \Rightarrow k = 168$

Aliter:

Case-1: When 5 is the last digit

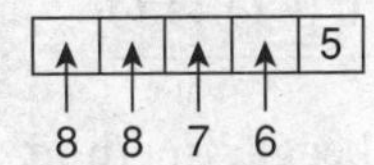

$= 8 \times 8 \times 7 \times 6 = 4! \times 112$

Case-2: When 0 is the last digit. Select the place to fill 5 in ${}^4C_1$ ways then select 3 digits out of the remaining 8 digits excluding 0 and 5 in ${}^8C_3$ ways and arrange them in 3! ways

				0

$\therefore$ Total number of ways $= 4C_1 \times 1 \times 8C_3 \times 3! = 4! \times 56$

$\therefore$ Total of both the cases $= 4!(112 + 56)$
$= 4! \times 168 \Rightarrow k = 168$

5. **(B)**
As per the given condition, digit 1 should occur at alternate places of the number and at the remaining 5 places one of 2, 3, 5 or 7 should appear. Now when the number starts with 1, number of numbers $= 4^5$ and when the number starts with one of 2, 3, 5 or 7, number of numbers $= 4^5$. So, total number $= 2 \times 4^5 = 2048$

6. **(C)**
${}^6C_0 + {}^6C_1 + {}^6C_2 + {}^6C_3 = 42$

7. **(D)**
One pair of 2 and 5 makes 10 (that is, $2 \times 5 = 10$) which gives one zero at the end.
Now $5^6, 10^{11}, 15^{16}, 20^{21}, 25^{26}, 30^{31}$ contains 6, 11, 16, 21, 52, 31 fives.
Therefore, total number of fives
$= 6 + 11 + 16 + 21 + (2 \times 26) + 31 = 137$

8. **(C)**
Total $= 6! - 5! = 600$
divisible by 5

(a) $5! = 120$ [| | | | | 0]
+
(b) $5! - 4! = 96$ [| | | | | 5]

$a + b = 216$

9. **(C)**
$m = 3! \cdot 7!$ $\boxed{B_1 B_2 B_3}$ $G_1 G_2 \ldots G_6$
$m = {}^3C_2\, 2!\, 7! = 3!\, 7!$. Hence $k = 1$.

10. **(C)**

Athanni (A) Chavanni (B) Duanni (C) Ekanni (D) (beggar)
0 0 0 0 0 0 Ø Ø Ø (False)

Using beggar method number of ways $= {}^9C_3 = 84$

11. **(D)**
$A + B + C = 19$
Case-I: If $A = 0$ then number of ways $= {}^{20}C_1$
Case-II: If $A = 2$ then number of ways $= {}^{18}C_1$
Case-III: If $A = 3$ then number of ways $= {}^{16}C_1$
⋮ ⋮ ⋮ ⋮
Case X: If $A = 18$ then number of ways $= {}^2C_1$
$\therefore$ Total no. of ways $= {}^{20}C_1 + {}^{18}C_1 + {}^{16}C_1 + \ldots + {}^2C_1 = 110$

12. **(B)**

(A) $\dfrac{7!}{1!2!4!} = 105$

(B) $1 \times 6\ (1 + 10 + 100) + 5 \times 6\ (1 + 10 + 100) + 7 \times 6\ (1 + 10 + 100) + 8 \times 6\ (1 + 10 + 100)$

(C) vowels will rotate in their own places and consonants will also rotate in their own places
$\dfrac{5!}{2!2!} \times 4! = 720$
Ans. $720 - 1 = 719$

(D) word which appears in the 8th position is AKALA

13. **(C)**
7 letters
$P \rightarrow 1$
$E \rightarrow 2$
$R \rightarrow 1$
$F \rightarrow 1$
$C \rightarrow 1$
$T \rightarrow 1$
2 alike + 2 diff. $= 1 \times {}^5C_2 = 10$
All 4 different $= {}^6C_4 = 15 \Rightarrow 25$

14. **(B)**

$m = 8!\, 2^3$ $\boxed{\begin{matrix} H_1\, W_1 \\ M_1F_1 \end{matrix}}$ $\boxed{\begin{matrix} H_2\, W_2 \\ M_2F_2 \end{matrix}}$ $\boxed{\begin{matrix} H_3\, W_3 \\ M_3F_3 \end{matrix}}$

$n = 5!\, 2^6$

$\dfrac{r}{R} = \dfrac{4}{65}$ $\boxed{M_1F_1}$ $\boxed{M_2F_2}$ $\boxed{M_3F_3}$

$k = 42$

15. **(B)**

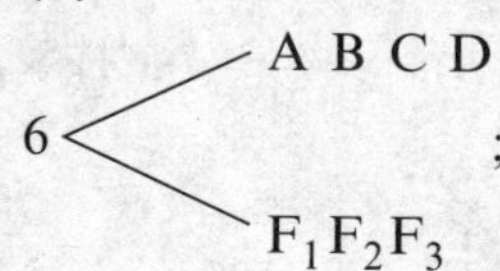

$3! \times {}^4C_3 \times 3! = 6 \times 4 \times 6$
$= 36 \times 4 = 144$

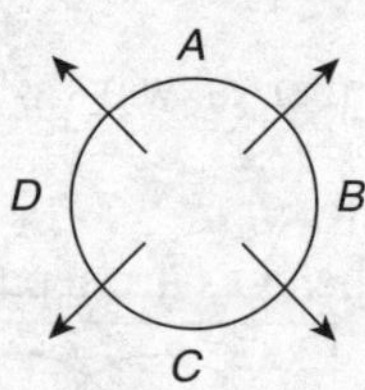

$\therefore$ Required ways = Total −144 = 720 −144
= 576

16. **(A)**
Answer is 76.

17. **(A)**
A number is divisible by four, if the last two digits are divisible by four. In this case the last two digits can be 12, 16, 28, 32, 36, 68, 92 or 96.
Total number of such numbers $= 8({}^4C_3 \cdot 3!)$
= 192.

18. **(A)**
(3) (10^3) (4) $\underbrace{x\,x\,x}_{\text{3 ways}}$ $\underbrace{x\,x\,x}_{\text{each place in 10 ways}}$ $\underset{\uparrow}{x}$ filled in 4 ways (0, 3, 6, 9)

$\Rightarrow (12)\ 10^3 \Rightarrow k = 12.$

19. **(A)**
Number of 6-digit numbers
$= 4 \times 3 \times 4 \times 3 \times 10 = 1440$

N = | ▨ | • | ▨ | • | ▨ | • |
O E O E O E

1 2
1 6
3 2
3 6
5 2
5 6 } 10 ways
7 2
7 6
9 2
9 6

20. **(C)**
Let $n(A)$ = Number of divisible by 60
= (60, 120, …, 960) = 16
$n(B)$ = Number of divisible by 24 = (24, 48m, …, 984) = 41
$n(A \cap B)$ = Number divisible by both
= 120 + 240 + … + 960 = 8
Hence, $n(A \cap \bar{B}) = n(A) - n(A \cap B) =$
$16 - 8 = 8.$

21. **(D)**

$N =$ | a | b | c | d |

(i) $d = 0$
$\Rightarrow {}^4C_2 \times 3! = 36$

(ii) $N =$ | a | b | c | d |
$d = 5$
'0' is taken
'0' is not taken $= {}^4C_3 \times 3! = 24$
$= {}^4C_2 \times (3! - 2!) = 24$
$\Rightarrow$ Total ways = 36 + 24 + 24 = 84

22. **(D)**
Number of digits are 9.
Select 2 places for the digit 1 and 2 in ${}^9C_2$ ways
From the remaining 7 places, select any two places for 3 and 4 in ${}^7C_2$ ways and from the remaining 5 places, select any two for 5 and 6 in ${}^5C_2$ ways.
Now, the remaining 3 digits can be filled in 3! ways
$\therefore$ Total ways $= {}^9C_2 \cdot {}^7C_2 \cdot {}^5C_2 \cdot 3!$

$$= \frac{9!}{2!\,7!} \cdot \frac{7!}{2!\,5!} \cdot \frac{5!}{2!\,3!} \cdot 3! = \frac{9!}{8} = \frac{9 \cdot 8 \cdot 7!}{8} = 9 \cdot 7!$$

23. **(B)**
| H | U | E | Y |
Number of ways $= {}^5C_2 \times 4! = 10 \times 24 = 240$

Aliter: Number of ways $= \dfrac{6!}{2!} - 5!$
= 360 −120 = 240

24. **(C)**
C = 02; O = 02; N = 1; U = 1; T = 1

$\therefore$ Number of ways $= \dfrac{7!}{2!2!} - \left({}^3C_2 \times \dfrac{5!}{2!}\right)$

= 1080

25. **(A)**
$M = {}^{10}C_3 = 120$ (using gap method)
Also, ${}^{12}C_3 = (10 + N + {}^{10}C_3) \Rightarrow N = 90$

26. **(B)**
Derangement
So, $D(4) = 9$

27. **(A)**
Do it yourself.

28. **(B)**
$u + v + w + x + y = 7$; where $u, v, w, x, y, \geq 0$
Using beggar's method, ${}^{11}C_4 = 330$

29. **(B)**
$N = {}^{11}C_2 = 11 \times 5$
$\therefore$ Number of divisors $= (1 + 1)(1 + 1) = 4$

30. **(C)**
$x_1 \;\; 1 \;\; x_2 \;\; 2 \;\; x_3 \;\; 3 \;\; x_4 \;\; 4 \;\; x_5$
where x_i denotes no. of 7's
where $x_2, x_3, x_4 \geq 2$
and $x_1, x_5 \geq 0$
Using beggar method, we get
${}^8C_4 \times 4! = 14 \times 5!$

31. **(C)**
x = number of red balls, y = number of blue balls, z = number of white balls
$x + y + z = 100$
$\therefore$ Number of ways $= {}^{102}C_2 = 5151$ (using beggar's method)
$\Rightarrow a = 5; b = 1; c = 5; d = 1$
$\Rightarrow (a + c - b - d) = 5 + 5 - 1 - 1 = 8$

32. **(B)**
85th SMCDW; 86th SMCWD

33. **(B)**

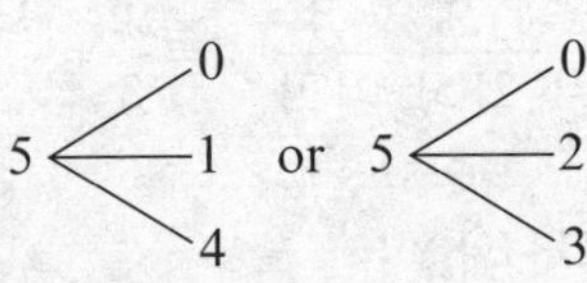

$= \frac{5!}{0! \cdot 1! \cdot 4!} \times 3! + \frac{5!}{0! \cdot 2! \cdot 3!} \cdot 3! = 30 + 60 =$
90 ways

34. **(B)**
Make 3 bundles each of the same colour,
Number of ways to arrange them = 2!

Total ways $= \frac{2! \times 7! \times 4! \times 1}{2} = 7!\,4!$
(as clockwise and anticlockwise are same)

35. **(A)**
Total $- n(A \cup B) = \frac{6!}{2!2!} - (n(A) + n(B) - n(A \cap B))$

$= \frac{6!}{2!2!} - \left(\frac{5!}{2!} + \frac{5!}{2!} - 4!\right) = 180 - 96 = 84$

Set A represents number of ways when G's are together
Set B represents number of ways when E's are together

Aliter: GG EE AR
Number of words when G's are separated

$= \frac{4!}{2!} \cdot {}^5C_2 = 120$

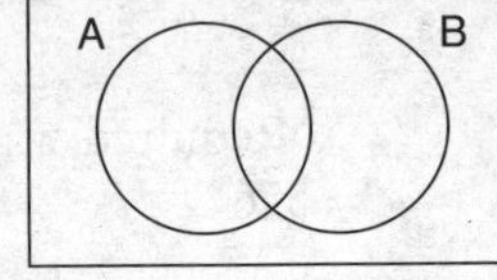

Number of words when G's are separated but E's are together $= 3! \times {}^4C_2 = 36$
$\therefore$ Number of ways when no two alike letters are together $= 120 - 36 = 84$

36. **(B)**
Number of ways when all girls together $= 7! \cdot 5!$
Number of ways when all boys and all girls are together $= 6! \cdot 5! \cdot 2!$
$\therefore$ required number of ways $= 7! \cdot 5! - 6! \cdot 5! \cdot 2!$
$= 6! \cdot 5!\,(7 - 2) = 432{,}000$

37. **(B)**
Given numbers can be rearranged as
1 4 7 .... 88 $\rightarrow$ 30
2 5 8 .... 89 $\rightarrow$ 30
3 6 9 .... 90 $\rightarrow$ 30
That means we must take two numbers from last row or one number each from first and second row.
Total ways $= {}^{30}C_2 + {}^{30}C_1 \cdot {}^{30}C_1 = 435 + 900 = 1335$

38. **(D)**
Total words, without any restriction = 7!
Total words beginning with I = 6!
Total words ending with B = 6!
Total words with I and ending with B = 5!
Thus, total number of required words $= 7! - (6! + 6! - 5!)$
$= 7! - 2(6!) + 5!$

39. **(A)**

$x + y + z = 15$; $x \le 8$; $y \le 8$, $z \le 8$

Total solution = ${}^{17}C_2 = 136$

suppose x or y or z is given 9 oranges

$x + y + z = 6$

${}^3C_1 \cdot {}^8C_2 = 84$ invalid

effective ways = 136 – 84 = 52

40. **(A)**

Using beggar ${}^{18}C_2 = \dfrac{18 \cdot 17}{2} = 153$

41. **(C)**

There are 2 women, and let number of men be n

According to question

$2 \times {}^nC_2 = 66 + 2 \times {}^nC_1 \times {}^2C_1$

$\Rightarrow 2 \times \dfrac{n(n-1)}{2} = 2[33 + 2n]$ or

$\dfrac{n(n-1)}{1 \times 2} = 33 + n(2)$

or $n^2 - 5n - 66 = 0$ or $(n - 11)(n + 6) = 0$

$\therefore n = 11$

$\Rightarrow$ Total participants = 2 + 11 = 13

42. **(B)**

A AAAA | B BBBBB

Since the word reads the same backwards and forwards, the middle letter must be A.

M

××××× ↓ ×××××

so that even number of A's and B's are available for arrangement about the middle position M in the above figure.

Take AABBB on one side of M (6th place) and then their image about M in a unique way

$\therefore$ Number of ways, $N = \dfrac{5!}{2!\, 3!} = 10$

43. **(A)**

If x denotes the number of times he can take unit step and y denotes the number of times he can take 2 steps, then $x + 2y = 7$.

We must have $x = 1, 3, 5$.

If $x = 1$, the steps will be 1 2 2 2

$\Rightarrow$ Number of ways $= \dfrac{4!}{3!} = 4$

If $x = 3$, the steps will be 1 1 1 2 2

$\Rightarrow$ Number of ways $= \dfrac{5!}{2!.3!} = 10$

If $x = 5$, the steps will be 1 1 1 1 1 2

$\Rightarrow$ Number of ways $= {}^6C_1 = 6$

If $x = 7$, the steps will be 1 1 1 1 1 1 1

$\Rightarrow {}^7C_0 = 1$

Hence total number of ways = $N = 21$

$\Rightarrow \dfrac{N}{3} = 7$

44. **(A)**

×××× $\boxed{F \cdot S}$ ××××

$\therefore$ Required number of ways = Total – no. of ways in which the two boys are together

$= 8! - \underset{\text{two adjacent positions for boys}}{6} \cdot \underset{\text{no.of ways in which the boys can sit}}{2} \cdot \underset{\text{other 6 girls can sit}}{6!}$

$= 8 \cdot 7 \cdot 6! - 12 \cdot 6!$

$= 6!\,[56 - 12] = 44 \cdot 6! = 31680$

45. **(C)**

$N = abc$

$a = b + c$

$b(c + 1) = 52 - 4a$

$b(c + 1) = 52 - 4(b + c)$

$bc + b = 52 - 4b - 4c$

$bc + 5b + 4c + 20 = 52 + 20$ (add 20 on both sides)

$(b + 4)(c + 5) = 72$

Hence $\left\{\begin{matrix} b = 8 & c = 1 \\ b = 5 & c = 3 \\ b = 4 & c = 4 \\ b = 2 & c = 7 \end{matrix}\right\} \Rightarrow$ 4 numbers

46. **(C)**

All the 16 multiples of 5 from 1 to 80 will have a factor of 5, 25, 50 and 75 will have a second factor of 5. So the factor 5 occurs 16 + 3 = 19 times in 80!, and 19 is the greatest power of 5 that can divide 80! Exactly.

47. **(C)**

${}^{16}C_3 - 2[2 \times {}^3C_3 + {}^4C_3] - 8 \times {}^4C_3$

$= {}^{16}C_3 - 12 - 32$

1 2 3 4

$\therefore 560 - 44 = 516$

48. **(D)**

$B_1 | G_1$ always separated.

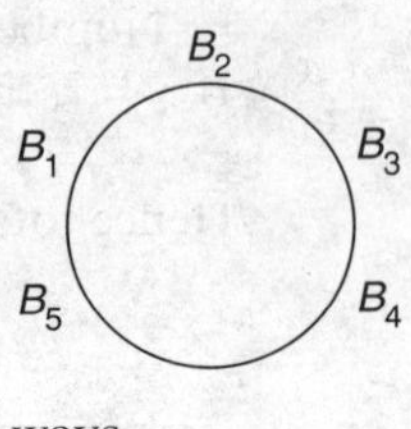

5 boys can be seated in 4! now for G_1 cannot sit between B_1 and B_2 or B_1 and B_5 and therefore can sit only in 3 ways and the remaining 4 girls in 4! ways.

Total ways

(4!) (3) (4!) = 576 × 3 = 1728.

49. **(B)**

$^8C_1 + {^8C_2} + {^8C_3} + {^8C_4} = 162$

50. **(C)**

Each a_i can be replaced by $2b_i - 1$, where b_i is a positive integer.

$$\text{Now } 32 = \sum_{i=1}^{4}(2b_i - 1) = \left(2\sum_{i=1}^{4} b_i\right) - 4$$

$$\Rightarrow 18 = \sum_{i=1}^{4} b_i \qquad (1)$$

Now assign 1 to each b_i, so $14 = \sum_{i=1}^{4} b'_i$

∴ By using the beggar method, the number of ways

$$= \frac{17!}{14!3!} = \frac{17\times16\times15}{6} = 17\times40 = 680$$

51. **(C)**

$D = ad - bc$; $n = 2^4 = 16$; D can be non-zero if

Case-I: $ad = 1$ and $bc = 0$, that is,

$b = 1; c = 0$

$b = 0; c = 1 \Rightarrow$ 3 cases

$b = 0; c = 0$

Case-II: $bc = 1$ and $ad = 0$, that is,

$a = 1; d = 0$

$a = 0; d = 1 \Rightarrow$ 3 cases

$a = 0; d = 0$

Hence total determinant with non-zero value = 6.

$$\therefore \frac{m}{n} = \frac{6}{16}$$

52. **(C)**

Treat each of V_1 V_2, ... V_n varieties as n beggars.

Total ways when m mangoes can be taken without any restriction = $^{m+n-1}C_m$

Now $^{m+n-1}C_m$ = All of same variety + (m – 1) of same variety and 1 diff. + (m – 2) of same variety and 2 diff. + ... + all 'm' of diff. variety.

At least two mangoes of the same variety (say x)

= Total – All m of different variety

$= {^{m+n-1}C_m} - {^nC_m}$

53. **(A)**

Required number of ways = (number of ways in which 16 players can be divided in 8 couples) – (number of ways when S_1 and S_2 are in the same group)

$$= \frac{16!}{2^8\cdot 8!} - \frac{(14!)}{2^7\cdot 7!} = \frac{16\cdot15\cdot14!}{2\cdot2^7\cdot8\cdot7!} - \frac{(14!)}{2^7\cdot7!}$$

$$= \frac{(14!)}{2^7\cdot7!}\left[\frac{16\cdot15}{16} - 1\right] = \frac{(14)\cdot(14)!}{14\cdot(6!)\cdot2^6} = \frac{(14)!}{2^6\cdot6!}$$

54. **(D)**

Peaches 5 p_1, p_2, p_3, p_4, p_5

Apples 3 a_1, a_2, a_3

Hence number of ways = $^3C_1 \times {^5C_3} = 30$

55. **(B)**

Including the two specified people, 4 others can be selected in $^8C_4$ ways.

The two adjacent seats can be taken in 4 ways and the two specified people can be arranged in 2! ways, remaining 4 people can be arranged in 4! ways. Therefore,

$^5C_4 \cdot 4 \cdot 2!\, 4! = 5!\, 8 = 8 \cdot 5!$

56. **(B)**

9 people can be divided into groups of 2, 3 and 4 in $\frac{9!}{2!3!4!}$ ways. Now each group can be distributed in $^{10}C_3 \times 3!$ ways

$$\Rightarrow \frac{9!}{2!3!4!}\, 10\cdot9\cdot8 = \frac{10!}{4}$$

Note that in an eleven-storeyed building there will be 10 floors.

57. **(C)**

Total = 6!

$n(A)$ = when A_1A_2 together = 5! 2!

$n(B)$ = when B_1B_2 together = 5! 2!

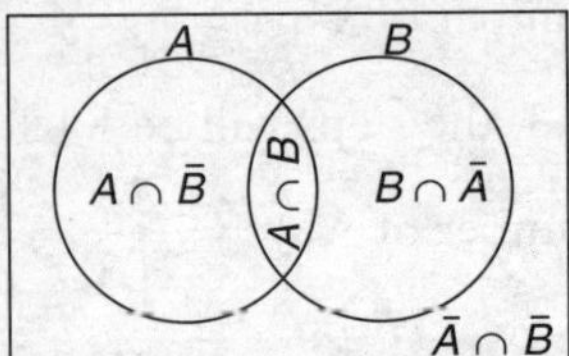

So $n(A \cap B) = 4!\,2!\,2! = 96$

Hence $n(\bar{A} \cap \bar{B}) = 6![2 \cdot 5!\,2! - 96]$

$= 720 - 384 = 336$

58. **(A)**

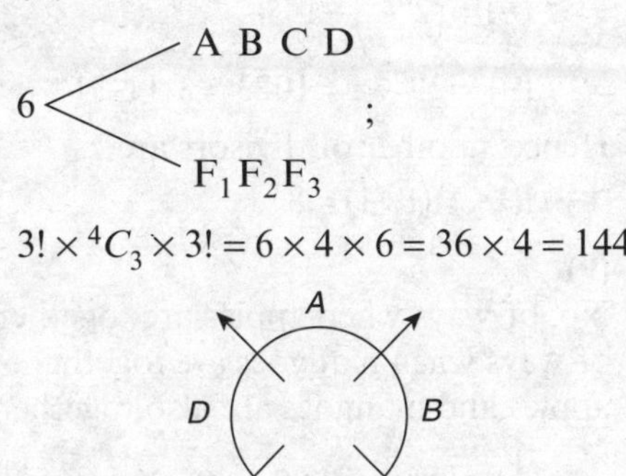

Integer Answer Type

1. **(1)**

We can arrange as follows

1 ---------------------------- $= 5 \times 5! = 600$

2 ---------------------------- $= 5 \times 5! = 600$

3 ---------------------------- $= 5 \times 5! = 600$

41 ---------------------------- $= 4 \times 4! = 96$

42 ---------------------------- $= 4 \times 4! = 96$

431 ---------------------------- $= 3 \times 3! = 18$

4321 ---------------------------- $= 2 \times 2! = 4$

43251 ---------------------------- $= 2!$

$\therefore$ 2018th number is 4325671

$\therefore$ Last digit = 1

2. **(9)**

Out of all groups of 4 consecutive even numbers, there is exactly one of the form $8k+4$ which is divisible by 4 but not by 8.

Number of four digit even numbers

$= 9 \times 9 \times 9 \times 4$

$\therefore N = 9 \times 9 \times 9$

Last digit = 9

3. **(6)**

$10800 = 2^4 3^3 5^2$

$9000 = 2^3 3^2 5^3$

$\therefore$ Number of common divisors $= 4 \times 3 \times 3$

$= 36 = a$

$\Rightarrow \sqrt{a} = 6$

4. **(7)**

Case I: When 5 is the last digit

$= 8 \times 8 \times 7 \times 6 = 4! \times 112$

Case II: When 0 is the last digit, then 5 can be filled by ${}^4C_1$ ways then select 3 digits out of remaining 8 digits excluding 0 and 5 in ${}^8C_3$ ways and arrange them in 3! ways

$\therefore$ Total number of ways $= {}^4C_1 \times 1 \times {}^8C_3 \times 3!$

$= 4!\ x\ 56$

$\Rightarrow$ Total number of ways $= 4!(112 + 56)$

$\Rightarrow k = 168$

$\Rightarrow \dfrac{k}{24} = 7$

5. **(2)**

$$\sum_{K=1}^{n} \frac{K^2 + K - 1}{(K+1)!} = \sum_{K=1}^{n} \left[\frac{1}{(K-1)!} - \frac{1}{(K+1)!}\right]$$

$$= \left(\frac{1}{0!} - \frac{1}{2!}\right) + \left(\frac{1}{1!} - \frac{1}{3!}\right) + \left(\frac{1}{3!} - \frac{1}{4!}\right) + \ldots + \left(\frac{1}{(n-1)!} - \frac{1}{(n+1)!}\right)$$

$$\Rightarrow \frac{1}{0!} + \frac{1}{1!} - \left(\frac{1}{n!} + \frac{1}{(n+1)!}\right)$$

$$\therefore \lim_{n \to \infty} \sum_{K=1}^{n} \frac{(K^2 + K - 1)}{(K+1)!} = 2$$

6. **(9)**

Total number of digits = 9

Select 2 places for the digit 1 and 2 in ${}^9C_2$ ways

From the remaining 7 places, select any two places for 3 and 4 in ${}^7C_2$ ways and from the remaining 5 places for 5 and 6 in ${}^5C_2$ ways, remaining 3 digits can be filled in 3! ways

Total ways

$= {}^9C_2 \cdot {}^7C_2 \cdot {}^5C_2 \cdot 3! = 9 \cdot 7! \Rightarrow K = 9$

7. **(8)**

Let $n(A)$ = Number divisible by 60

$(60, 120, \ldots, 960) = 16$

$n(B)$ = Number divisible by 24

$=(24, 48, \ldots, 984)=41$

$n(A\cap\bar{B})$ = Number divisible by both

$=120+240+\ldots+960=8$

Hence, $n(A\cap\bar{B})=n(A)-n(A\cap B)$

$=16-8=8$

8. **(7)**

Let x denote the number of times he can take one step and y denote the number of times he can take 2 steps.

Then, $x+2y=7$, we must have $x=1, 3, 5$

If $x=1$, the steps will be $1, 2, 2, 2=\frac{4!}{3!}=4$

If $x=3$, then steps will be

$1, 1, 1, 2, 2=\frac{5!}{2!3!}=10$

If $x=5$, then steps will be $1, 1, 1, 1, 1, 2=\frac{6!}{5!}=6$

If $x=7$, then steps will be

$1, 1, 1, 1, 1, 1, 1={}^7C_0=1$

Hence, total number of ways $=N=21$

$\Rightarrow \frac{N}{3}=7$

9. **(4)**

Let $N=abc$

$\Rightarrow a=b+c$

$b(c+1)=52-4a \Rightarrow b(c+1)=52-4(b+c)$

$\Rightarrow bc+b=52-4b-4c$

$\Rightarrow bc+5b+4c+20=52+20$

$\Rightarrow (b+a)(c+5)=72$

$$\text{Hence, }\begin{cases} b=8, & c=1 \\ b=5, & c=3 \\ b=4, & c=4 \\ b=2, & c=7 \end{cases} \Rightarrow 4 \text{ Numbers}$$

10. **(8)**

$$\sum_{K=r}^{n} {}^KC_r={}^rC_r+{}^{r+1}C_r+{}^{r+2}C_r+\ldots+{}^nC_r$$

$=1+{}^{r+1}C_1+{}^{r+2}C_2+{}^{r+3}C_3+\ldots+{}^nC_{n-r}$

$={}^{r+1}C_0+{}^{r+1}C_1+{}^{r+2}C_2+\ldots+{}^nC_{n-r}$

$={}^{n+1}C_{n-r}={}^{n+1}C_{r+1}$

$$\therefore f(n)=\sum_{r=0}^{n} {}^{n+1}C_{r+1}$$

$={}^{n+1}C_1+{}^{n+1}C_2+{}^{n+1}C_3+\ldots+{}^{n+1}C_{n+1}$

$\Rightarrow f(n)=(2^{n+1})-1$

$\Rightarrow f(9)=2^{10}-1=1023=3\cdot 11\cdot 31$

Hence, number of divisors are

$(1+1)(1+1)(1+1)=8$

11. **(9)**

No. of ways when apples are together + no. of ways when mangoes are together – when apples and mangoes are also together.

$\frac{8!}{5!}+\frac{7!}{4!}-4! \Rightarrow k=522 \Rightarrow 9$

12. **(300)**

$\underbrace{{}^5C_3}_{\text{Select 3 places out of five}}$ (for 3, 5, 7) $\cdot \underbrace{{}^6C_2}_{\text{Select two more digit out of the remaining 6}} \cdot 2!$

$=10\cdot 15\cdot 2=300\cdot$ | | | | | |

13. **(400)**

$s={}^{12}C_2=66$

$t={}^{13}C_3-6=\frac{13\cdot 12\cdot 11}{6}-6=286-6=280$

$d={}^{12}C_2-12=66-12=54$

Hence, $(s+t+d)=66+280+54=120+280=400$

14. **(22)**

$T_n={}^nC_3$

$\therefore T_{n+1}={}^{n+1}C_3$

Also ${}^{n+1}C_3-{}^nC_3=10$

$\therefore {}^nC_2=10$

$\therefore n=5$

Now we have 1, 2, 2, 3, 3, 3, 4, 4, 4, 4 digits to form a five-digit number.

Category	Selection
4 alike and 1 different	$1\times {}^3C_1=3$
3 alike and 2 others alike	${}^2C_1\times {}^2C_1=4$
3 alike and 2 different	${}^2C_1\times {}^3C_2=6$
2 alike; 2 other alike and 1 different	${}^3C_2\times {}^2C_1=6$
2 alike and 3 different	${}^3C_1\times 1=3$

Total = 22

15. (840)

Select 3 places for A, B, C in

${}^7C_3$ ways ×××××××

and arrange them only in 1 way

Remaining 4 places can be arranged in 4! ways

$\therefore$ Total ways $= {}^7C_3 \cdot 4! = 35 \cdot 24 = 840$

16. (5)

$n = 5!\, 6!$

$m = 5!\, {}^6C_2 \cdot {}^5C_4 \cdot {}^2C_1 \cdot 4!$ (Gap method)

$\therefore \dfrac{m}{n} = 5$

17. (150)

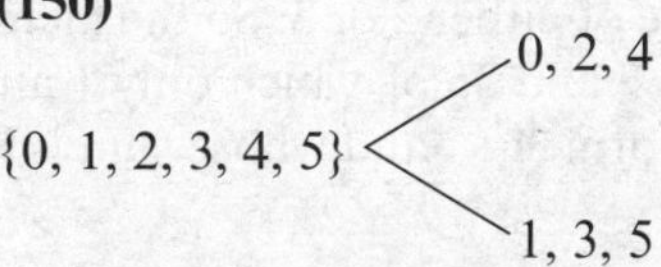

$N = \boxed{abcd}$ no two consecutive digits are even

Not 3 even + 1 odd is not possible (think!)

Case-1: 1 even + 3 odd digits

(a) $\underbrace{2\,1\,3\,5 \mid 4\,1\,3\,5 \mid 0\,1\,3\,5}$

$= 2 \cdot 4! + (4! - 3!) = 48 + 18 = 66$

Case-2: 2 even + 2 odd digits

(a) non-zero even + 2 odd

For example, 2, 4, 1, 3 or 2, 4, 1, 5 or 2, 4, 3, 5

2 non-consecutive places

$$\begin{bmatrix} X & X & X & X & \text{or} & X & X & X & X & \text{or} & X & X & X & X \\ 2 & & 4 & & & 2 & & 4 & & & 2 & & 4 & \end{bmatrix}$$

Can be taken in 3 ways.

Even number can be integer changed in 2! ways. Two odd also can be placed in 2! ways. 2 odds can be taken in 3 ways, that is, 13, 35, 51

Hence total for this case $3 \times 2!\, 2!\, 3 = 36$

(b) 0 and 2 as even + 2 odd or 0 and 4 as even + 2 odd. For example,

$\left.\begin{matrix} 0213, 0215, 0235 \\ 0413, 0, 415, 0435 \end{matrix}\right\}$ 6 possibilities.

consider the case 0 2 1 3 which can appear as

[2][ ][0][ ] or [2][ ][ ][0]

[ ][2][ ][0] or [ ][0][ ][2]

[digit 2 and 0 is fixed]

Number of numbers = 8

Since there are six such cases. Hence total for this case $= 8 \cdot 6 = 48$.

$\therefore$ Final answer $= 66 + 36 + 48 = 150$

18. (6)

Case 1: Consider the numbers between 550 – 599

4, 5, 6, 7, 8, 9

1st place in one way (that is, 5)

2nd place in 5 ways (that is, 5, 6, 7, 8, 9)

3rd place in 3 ways (that is. 5, 7, 9)

Number of numbers in this case $= 1 \cdot 5 \cdot 3 = 15$.

Case 2: 600 – 799 (both inclusive)

Number of numbers $= 2 \cdot 6 = 36$

Total $= 15 + 36 = 51$

$\therefore$ Number of digits in $N = 6$

19. (2275)

We have number of quadrilaterals

$= {}^{17}C_4 - {}^{15}C_2 = \dfrac{17 \cdot 16 \cdot 15 \cdot 14}{1 \cdot 2 \cdot 3 \cdot 4} - \dfrac{15 \cdot 14}{1 \cdot 2}$

$= (10)(17)(14) \quad (15)(7) - 2380 - 105 = 2275$

20. (12)

We have $a_{ij} = a_{ji} \Rightarrow a_{12} = a_{21} = a$ (say)

$$\begin{vmatrix} \times & a & b \\ a & \times & c \\ b & c & \times \end{vmatrix}$$

$a_{31} = a_{13} = b$ and $a_{23} = a_{32}$

1, 1, 1, 1, 1 and 0, 0, 0, 0

Case 1: If diagonal elements are (1, 1, 1)

Conjugate elements are, (0, 0), (0, 0), (1, 1) as shown.

$$\begin{vmatrix} 1 & 0 & 0 \\ 0 & 1 & 1 \\ 0 & 1 & 1 \end{vmatrix}$$

This can be done in $\dfrac{3!}{2!} = 3$ ways

Case 2: If diagonal has (1, 0, 0) then the conjugate elements are (1, 1), (1, 1), (0, 0)

Now conjugate elements can be arranged in

$\dfrac{3!}{2} = 3$ ways

$$\begin{vmatrix} 1 & \bullet & \bullet \\ \bullet & 0 & \bullet \\ \bullet & \bullet & 0 \end{vmatrix}$$

and diagonal elements can be filled in $\frac{3!}{2} = 3$ ways

Hence $3 \times 3 = 9$ ways

$\therefore$ Total determinant $= 3 + 9 = 12$

4.19.3 JEE Advanced: Sample Questions

Single Answer Type

1. **(B)**
The number of students answering exactly $i(1 \le i \le n-1)$ questions wrong is $2^{n-i} - 2^{n-i-1}$
The number of students answering all n questions wrong is 2°
Hence the total number of wrong answers is
$$\sum_{i=1}^{n-1} i\left(2^{n-i} - 2^{n-i-1}\right) + n\left(2^\circ\right) = 2047$$
$$\Rightarrow 2^{n-1} + 2^{n-2} + 2^{n-3} + \ldots + 2^\circ = 2047$$
$$\Rightarrow 2^n - 1 = 2047 \Rightarrow 2^n = 2048 \Rightarrow 2^n = 2^{11}$$
$$\Rightarrow n = 11$$

2. **(D)**
The required numbers are
1, 2, 11, 12, 21, 22, …, 122222222
Let us calculate how many numbers are there.
There are 2 one-digit such numbers. There are 2^2 two-digit such numbers and so on. There are 2^8 eight-digit such numbers. All the nine-digit numbers beginning with 1 and written by means of 1 and 2 are smaller than $2 \cdot 10^8$. Thus there are 2^8 such nine-digit numbers.
Hence the required number of numbers is
$$2 + 2^2 + 2^3 + \ldots + 2^8 + 2^8 = \frac{2\left(2^8 - 1\right)}{(2-1)} + 2^8$$
$$= 2^9 - 2 + 2^8 = 766$$

3. **(A)**
The different triangles possible are
$(4, 3, 2)(5, 4, 2)(5, 4, 3), (6, 5, 2),$
$(6, 5, 3), (6, 5, 4), (6, 4, 3)$
$\Rightarrow$ Total number of possible triangles is 7.

4. **(A)**
In this case, the number will be divisible by 11 if the sum of even place digits = sum of odd place digits
Total numbers = 34

5. **(D)**
Obviously total number of ways
$= 1 + 4 + 7 = 12$

6. **(B)**
$7! = 2^4 \times 3^2 \times 5 \times 7$
Since the factor should be odd as well as of the form $3t + 1$, the factor cannot be a multiple of either 2 or 3. So the factors may be 1, 5, 7 and 35 of which only 1 and 7 are of the form $3t + 1$ and whose sum is 8.

7. **(B)**
Total number of permutations $= \frac{9!}{2!}$
Number of those containing 'HIN' = 7!
Number of those containing 'DUS' $= \frac{7!}{2!}$
Number of those containing 'TAN' = 7!
Number of those containing 'HIN' and 'DUS' = 5!
Number of those containing 'HIN' and 'TAN' = 5!
Number of those containing 'TAN' and 'DUS' = 5!
Number of those containing 'HIN', 'DUS' and 'TAN' = 3!
Required number =
$$\frac{9!}{2!} - \left(7! + 7! + \frac{7!}{2}\right) + 3 \times 5! - 3! = 169194$$

8. **(C)**
Three-digit odd numbers using only 3 and only 5 are 2
using 3, 4 and 5 are 4
using 4, 5 and 6 are 2
using two 3 and one 6 are 2
using two 6 and one 3 are 1
so total three-digit numbers = 11

9. **(C)**
Sum of 7 digits = a multiple of 9
Sum of numbers 1, 2, 3, 4, 5, 6, 7, 8, 9 = 45
So, two left should also have sum of 9
The pair of left numbers is (1, 8), (2, 7),

(3, 6), (4, 5) with each pair left number of 7-digit number = 7!
so, with all 4 pairs = $4 \times 7!$

10. (C)

For, $1 \le x \le 33, \frac{1}{3} < \frac{1}{3} + \frac{x}{50} < 1$

Therefore, $\left[\frac{1}{3} + \frac{x}{50}\right] = 0$ for $1 \le x \le 33$

For, $34 \le x \le 50, 1 < \frac{1}{3} + \frac{x}{50} < \frac{4}{3}$

$\Rightarrow \left[\frac{1}{3} + \frac{x}{50}\right] = 1$ for $34 \le x \le 50$

Thus, $E = 17$

Exponent of 2 in $(17)!$

$= \left[\frac{17}{2}\right] + \left[\frac{17}{4}\right] + \left[\frac{17}{8}\right] + \left[\frac{17}{16}\right] = 15$

11. (A)

$|x| \le K \Rightarrow -K \le x \le K$

$|y| \le K \Rightarrow -K \le y \le K$

and

$|x - y| \le K \Rightarrow |y - x| \le K \Rightarrow x - K \le y \le x + K$

Clearly number of points having integral coordinates =

$(2K+1)^2 - 2[K + (K-1) + \ldots + 2 + 1]$

$= 3K^2 + 3K + 1$

12. (C)

Required number of points on intersection of new lines

$= 2 \times$ number of choosing 1 point on L_1 and 2 points on L_2

$= 2 \cdot {}^{\ell_1}C_1 \times {}^{\ell_2}C_2.$

13. (D)

The matrix will be of the order 4×1 or 1×4 or 2×2

For each order, the number of different matrices = the number of ways to fill four places by 0, 1, 2, 3

$= 4 \times 4 \times 4 \times 4$

14. (D)

The number of required ways will be equal to the number of solutions of

$x_1 + x_2 + x_3 + x_4 = 6$

where $x_1, x_2, x_3, x_4 \ge 1$

Let

$x_1 - 1 = y_1, x_2 - 1 = y_2, x_3 - 1 = y_3, x_4 - 1 = y_4$

Where $y_1, y_2, y_3, y_4 \ge 0$

$\Rightarrow y_1 + y_2 + y_3 + y_4 = 2$

Number of solutions $= {}^{2+4-1}C_{4-1} = {}^5C_3 = 10$

15. (B)

Case 1: Every square of $(2 \text{ by } 2) = 196$

Case 2: In every line it is possible to have 6 shaped figures.

So, total number of such figures =

$6 \times 8 \times 2 = 96$

Hence, total number of ways

$= 196 + 96 = 292$

16. (C)

4 lines intersect each other in ${}^4C_2 = 6$points

and 4 circles intersect in ${}^4P_2 = 12$ points

Each line cut 4 circles into 8 points

$\therefore$ 4 lines cut 4 circles into 32 points

$\therefore$ Required number = 6 + 12 + 32 = 50

17. (B)

Required number =

$\frac{7!}{2!3!} - \frac{6!}{2!3!} = 420 - 60 = 360$

18. (D)

Solve it yourself.

19. (C)

22 33 55 888

3, 3, 5, 5 Occupy the 5 odd places in the 9-digit number

$${}^5C_4 \cdot \underbrace{\frac{4!}{2!2!}}_{\text{Odd nos on even places}} \times \underbrace{\frac{5!}{2!3!}}_{\text{even nos on odd places}} = 5 \times 6 \times 10 = 300$$

20. (B)

${}^9C_5 -$ [only top remains empty + middle empty + bottom empty]

${}^9C_5 - [{}^7C_5 + {}^6C_5 + {}^5C_5] = 126 - (21 + 6 + 1)$

$= 98$

21. (D)

We have $2^n - 2 = 510$

or $2n = 512$

or $n = 9$

22. (D)
Do it yourself.

23. (A)
F A C E T I O U S; 9 letters; ${}^9C_5 \cdot 1 \cdot 4!$

24. (B)
Here $P_n = {}^{n-2}C_3$ and $P_{n+1} = {}^{n-1}C_3$
Hence, ${}^{n-1}C_3 - {}^{n-2}C_2 = 15$
or ${}^{n-2}C_3 + {}^{n-2}C_2 - {}^{n-2}C_3 = 15$
or ${}^{n-2}C_2 = 15$ or $n = 8$

25. (B)
(i) He can invite 2 friends three times each
Let us select first those 2 friends in ${}^3C_2$ ways
Now these two friends each three times can be invited on 6 days in $\frac{6!}{3!3!}$
Thus total number of ways 2 friends can be invited three times $= {}^3C_2 \frac{6!}{3!3!}$
(ii) Another possibility is that he invites all three friends two times each. Then
Number of ways $= \frac{6!}{2!2!2!}$
(iii) One more possibility is that he invites one friend three times, one two times, and one three times. Then
Number of ways $= \frac{6! \times 6}{3!2!}$
Hence, total number of ways
$= {}^3C_2 \times \frac{6!}{3!3!} + \frac{6! \times 6}{3!2!} + \frac{6!}{2!2!2!} = 510$

26. (B)
If three numbers are in G.P, then their exponents must be in A.P.
If a, b, c are the selected numbers in G.P., then the exponents of a and c both are either odd or both even, or otherwise exponent b will not be an integer.
Now two odd exponent (from 1, 2, 3, …, 10) can be selected in ${}^5C_2$ ways and two even exponents can be selected in ${}^5C_2$ ways. Hence, number of G.P.s are $2\,{}^5C_2 = 20$.

27. (C)
$$\sum_{k=r}^{n} {}^kC_r = {}^rC_r + {}^{r+1}C_r + {}^{r+2}C_r + \ldots + {}^nC_r$$
$= 1 + {}^{r+1}C_1 + {}^{r+2}C_2 + {}^{r+3}C_3 + \ldots + {}^nC_{n-r}$
$= {}^{r+1}C_0 + {}^{r+1}C_1 + {}^{r+2}C_2 + \ldots + {}^nC_{n-r}$
$= {}^{n+1}C_{n-r}$
Now, ${}^{n+1}C_{n-r} = {}^{n+1}C_{n+r}$
$\therefore f(n) = \sum_{r=0}^{n} {}^{n+1}C_{r+1}$
$= {}^{n+1}C_1 + {}^{n+1}C_2 + {}^{n+1}C_3 + \ldots + {}^{n+1}C_{n+1}$
$= {}^{n+1}C_0 + {}^{n+1}C_1 + {}^{n+1}C_2 + \ldots + {}^{n+1}C_1 - 1$
$\Rightarrow f(n) = (2^{n+1}) - 1$
$\Rightarrow f(9) = 2^{10} - 1 = 1023 = 3 \cdot 11 \cdot 31$

Multiple Answer Type

1. (B, C)
Let $\vec{r} = x\hat{i} + y\hat{j} + z\hat{k}$, where x, y, z are positive integer (given)
Also, $\vec{r} \cdot \vec{a} \le 12 \Rightarrow x + y + z \le 12$
Therefore, the number of values of $\vec{r}$ = the number of positive integral solution of $x + y + z \le 12$
$= \sum_{n=3}^{12} {}^{n-1}C_2 = {}^2C_2 + {}^3C_2 + \ldots + {}^{11}C_2$
$= {}^3C_0 + {}^3C_1 + {}^4C_2 + \ldots + {}^{11}C_9 \;\left[{}^nC_r = {}^nC_{n-r}\right]$
$= {}^4C_1 + {}^4C_2 + \ldots + {}^{11}C_9$

$= {}^{12}C_9$

2. (A, B, D)

3. (A, B, C, D)
First and second prizes in mathematics (Physics) can be awarded in
$\left({}^{30}P_2\right)\left({}^{30}P_2\right) = (30)^2\,(29)^2$ ways
First prize in chemistry (Biology) can be awarded in $\left({}^{30}P_1\right)\left({}^{30}P_1\right) = (30)^2$ ways.
Hence,
$N = (30)^2\,(29)^2\,(30)^2 = 2^4 \cdot 3^4 \cdot 5^4\,(29)^2$
Since, $400 = 2^4 \cdot 5^2$; $600 = 2^3 \cdot 3 \cdot 5^2$,
$8100 = 2^2 3^4 5^2$
$\therefore N$ is divisible by 400, 600 and 8100

Also, N is divisibly by four distinct prime numbers 2, 3, 5, 29

4. **(A, B)**

The total number of attempts that can be made to open the lock is $(15)^3$ out of these, there is just one attempt in which the lock will open.

Then, $N=(15)^3-1=(15-1)(225+15+1)$

$=2\times 7\times 241$

Clearly N is divisible by 482 and N is a product of three distinct prime numbers.

5. **(A, B, C, D)**

$\dfrac{(mn)!}{(m!)^n}$ is the number of ways of distributing (mn) distinct objects among m persons equally

Hence, $\dfrac{(mn)!}{(m!)^n}$ is an integer

$\Rightarrow \dfrac{(m!)^n}{(mn)!}$, Similarly $\dfrac{(n!)^m}{(mn)!}$

Further $m+n<2m\le mn\Rightarrow \dfrac{(m+n)!}{(mn)!}$

and $m-n<m<mn\Rightarrow \dfrac{(m-n)!}{(mn)!}$

6. **(A, B, C)**

Let us choose the integer x_1 and x_2

Let integer a be before x_1, integer b between x_1 and x_2

and integer c after x_2.

$\therefore a+b+c=98$, where $a\ge 0, b\ge 10, c\ge 0$

Now, if we consider the choices where difference is at least 11, then the number of solutions is

${}^{88+3-1}C_{3-1}={}^{90}C_2$

$\therefore$ The number of ways in which b is less than 10 is

${}^{100}C_2-{}^{90}C_2$, which is equal to (A), (B), (C).

7. **(A, C)**

Total handshakes possible $={}^{2n}C_2$. This will include n handshakes in which a person shakes hand with her or his spouse.

$\Rightarrow$ Required number $={}^{2n}C_2-n=2n(2n-1)$.

8. **(A, D)**

$x+y+z+w=19$, $x, y, z, w\ge 1$ has solution as coefficient of x^{19} in $\left(x+x^2+\ldots+x^{19}\right)^4$

which is same as coefficient of x^{15} in $\left(1+x+\ldots+x^{15}\right)^4$

9. **(A, B)**

The nth line will rise to n additional parts, when u_{n-1} parts are already there

$\Rightarrow u_n=u_{n-1}+n\Rightarrow (D)$ is incorrect

Again, $u_0=1, u_1=2$

We easily get $u_2=4, u_3=7, u_4=11$

$\Rightarrow$ (A), (B) are correct and (C) is incorrect.

10. **(A, B, C)**

It is evident $x=y$ and since sum of digits is either even or odd

$\Rightarrow x+y=$ Total 5-digit numbers

$=9\times 10\times 10\times 10\times 10$

$\Rightarrow 2x=90{,}000\Rightarrow x=45{,}000$

11. **(A, C)**

Since, ${}^nC_r={}^nC_{n-r}$

$\Rightarrow$ (A) and (B)

12. **(A, B, C)**

$C\to 2; L\to 2; U\to 2; A\to 1; S\to 1$

When all are different $\to 120$

2 identical and 3 different $\to$

${}^3C_1\cdot {}^4C_3\cdot \dfrac{5!}{2!}=720$

2 identical, 2 identical and 1 different

$\to {}^3C_2\cdot {}^3C_1\cdot \dfrac{5!}{2!2!}=270$

$\therefore$ Total number of words $=1110$

13. **(A, C, D)**

${}^{100}C_{50}=\dfrac{100!}{50!50!}=2^\alpha 3^\beta 5^\gamma 7^\delta\ldots$

where $\alpha, \beta, \gamma, \delta, \ldots$ are non-negative integers

Exponent of 2 in $(100)!=97$

Exponent of 2 in $50!=47$

Exponent of 2 in ${}^{100}C_{50} = 3$

$\left[\because {}^{100}C_{50} = \frac{2^{97}}{2^{47} \cdot 2^{47}} = 2^3\right]$

Similarly, exponent of 3, 5 and 7 in ${}^{100}C_{50}$ are 4, 0 and 0 respectively.

$\therefore {}^{100}C_{50} = 2^3 \cdot 3^4 \cdot 5^0 \cdot 7^0 \ldots$

$\Rightarrow \alpha = 3, \beta = 4, \gamma = 0, \delta = 0$

14. **(A, C)**

(A) IOCL, A, A, A, E

Required number of words $= \frac{5!}{3!}$

(B) Required number of words
= Total – 'IOCL' is at last four places
$= \frac{8!}{3!} - \frac{4!}{3!} \Rightarrow$ incorrect

(C) There A can be filled at last five places in ${}^5C_3 \cdot 1$ ways
Now, remaining five places are filled by I, O, C, L, E in 5! ways
$\Rightarrow$ required number of words =
${}^5C_3 \cdot 1 \cdot 5!$

(D) $A / A / A / A / E / {}^5C_4 \cdot 4! \cdot \frac{4!}{3!}$ (gap method)

incorrect

15. **(A, B)**

$\frac{(10)!}{2}$

16. **(A, C, D)**

1 .... ${}^8C_4 = 70$

23 .... ${}^6C_3 = 20$

24 .... ${}^5C_3 = 10$

25 .... ${}^4C_3 = 4$

26789

Hence (A), (C), (D)

17. **(A, B)**

(A) Number of possible forecasts of 5 football matches when none is correct $= 2^5 - 1 = 31$

(B) $p = 5, q = 26$
$\therefore p + q = 31$

(C) On first throw there are 6 ways to throw a dice
On second throw there are 5 ways to throw a dice
$\therefore$ it can fall in 30 different ways

(D) ${}^5C_1 \cdot 1 + {}^5C_2 \cdot 2! + {}^5C_3 \cdot 3! + {}^5C_4 \cdot 4! + {}^5C_5 \cdot 5!$
$= 5 + 20 + 60 + 120 + 120 = 325$

18. **(A, B, C)**

$x_n = {}^{n+5}C_4 - \left(\frac{143}{96}\right)\left(\frac{P_{n+5}}{P_{n+3}}\right)$

given $x_n < 0$

hence ${}^{n+5}C_4 < \frac{143}{96} \times \frac{(n+5)!}{(n+3)!}$

$\frac{(n+5)!}{4!(n+1)!} < \frac{143}{96} \cdot \frac{(n+5)!}{(n+3)(n+2)(n+1)!}$

$\frac{1}{24} < \frac{143}{96} \cdot \frac{1}{(n+3)(n+2)}$

$\Rightarrow 4(n+3)(n+2) < 143 \Rightarrow 4(n^2 + 5n + 6) - 143 < 0$

$-\frac{17}{2}$ 0 $\frac{7}{2}$

$\Rightarrow 4n^2 + 20n - 119 < 0 \Rightarrow 4n^2 + 34n - 14n - 119 < 0$

$\Rightarrow 2n(2n + 17) - 7(2n + 17) < 0$

$\Rightarrow (2n + 17)(2n - 7) < 0$

$\therefore n \in \{1, 2, 3\}$.

19. **(C, D)**

Answer is ${}^9C_4 = 126$

(A) 126 (B) 126 (C) ${}^9C_4 \times 2! = 252$

(D) ${}^8C_1 + {}^8C_2 + {}^8C_3 + \frac{{}^8C_4}{2} = 8 + 28 + 56 + 35 = 127$

20. **(A, C, D)**

$(2n + 1)^2 = 4n^2 + 4n + 1 = 4n(n + 1) + 1 = 8m + 1$

($n(n + 1)$ is even).

21. **(B, C, D)**

4 to be occupied say s s s s

Remaining 6 are | x | x | x | x | x | x |. Now 4 can be selected in ${}^7C_4$ ways and can be arranged in ${}^7C_4 \times 4!$ ways.

22. **(B, C)**

If a, b, c are in A.P., then a and c are both odd or both even.

Case-I: n is even.

The number of ways of selection of two even numbers a and c is ${}^{\frac{n}{2}}C_2$. Number of ways of selection of two odd numbers is ${}^{\frac{n}{2}}C_2$. Hence the number of A.P.s is

$$2\,{}^{\frac{n}{2}}C_2 = 2\frac{\frac{n}{2}\left(\frac{n}{2}-1\right)}{2} = \frac{n(n-2)}{4}$$

Case-II: n is odd.

The number of ways of selection of two odd numbers a and c is ${}^{\frac{(n+1)}{2}}C_2$. The number of ways of selection of two even numbers a and c is ${}^{\frac{(n-1)}{2}}C_2$. Hence, the number of A.P.s is

$${}^{\frac{(n+1)}{2}}C_2 + {}^{\frac{(n-1)}{2}}C_2$$

$$= \frac{\left(\frac{n+1}{2}\right)\left(\frac{n+1}{2}-1\right)}{2} + \frac{\left(\frac{n-1}{2}\right)\left(\frac{n-1}{2}-1\right)}{2}$$

$$= \frac{1}{8}(n-1)\,((n+1)+(n-3)) = \frac{(n-1)^2}{4}$$

23. **(A, B, D)**

$$\left(\frac{6!}{2!2!2!3!}\right)\times 3! = \frac{6!}{2!\,2!\,2!} = 90$$

(A) $\frac{6!}{2!\,2!\,2!} \Rightarrow$ (A) is correct

(B) ${}^{10}C_2\cdot 2! \Rightarrow$ (B) is correct

(C) 5 → 1, 2, 2 ; 1, 1, 3

$$= \left(\frac{5!}{1!\,1!\,3!}\frac{1}{2!} + \frac{5!}{2!\,2!\,1!}\frac{1}{2!}\right)3!$$

$= 150 \Rightarrow$ (C) is incorrect.

(Using Division and Distribution)

(D) A and S remains in 2nd and 8th position

Hence, number of ways $= \frac{6!}{2!\,2!\,2!}$

24. **(B, D)**

(A) give 1 letter to each of the 3 envelopes and use the beggar method to get answer ${}^4C_2 = 6$

(B) make bundles of 2, 2, 1 and 3, 1, 1. Each bundle is to be distributed in only one way since envelops are identical

$$\frac{5!}{2!\,2!\,2!} + \frac{5!}{3!\,1!\,1!\,2!} = 25$$

(C) answer is 2

(D) using the beggar method, the answer is 6 (since envelopes are kept in a row, they are treated as different)

Comprehension Based Questions

1. (A)

2. (D)

3. (C)

(i) Since there are 5 even places and 3 pairs of repeated letters, therefore at least one of these must be at an odd place. Hence, the number of ways is $\frac{11!}{(2!\,2!\,2!)}$

(ii) Total number of arrangements $= 7! \times {}^8C_2$

(iii) Consonants can be placed in $\frac{7!}{2!\,2!}$ ways. Then there are 8 places and 4 vowels. Therefore, number of ways is $\frac{7!}{2!\,2!}\,{}^8C_4\,\frac{4!}{2!}$

4. (C)

5. (A)

6. (B)

(i) Consider

$A = \{n-r+1, n-r+2, \ldots, n\} \subset s$

Clearly $n(A) = r$ and $\min(A) = n-r+1$

$\therefore$ Greatest value of $\min(A) = n-r+1$

(ii) K is the least member of A and rest of the $(r-1)$ members of A are among the numbers $K+1, K+2, K3, \ldots, n$

$\therefore$ Number of ways $= {}^{n-K}C_{r-1}$

(iii) $\sum_{n(A)=r} (\min A) = \sum_{K=1}^{n-r+1} k \cdot {}^{n-K}C_{r-1}$

Now,

$k \cdot {}^{n-K}C_{r-1} = [(n+1)-(n-k+1)] \cdot {}^{n-K}C_{r-1}$

$= (n+1)\, {}^{n-K}C_{r-1} - (n-k+1)\, {}^{n-K}C_{r-1}$

Also, $(n-k+1) \cdot {}^{n-K}C_{r-1} = (n-k-r+1)!$

7. **(B)**

$C \rightarrow 2$ $\boxed{CAR}$ CSSNN

$A \rightarrow 2 = \frac{6!}{2!\,2!} \times {}^7C_3 \times 3!$

$S \rightarrow 2$

$N \rightarrow 2$

$O \rightarrow 1$

$E \rightarrow 1$

$R \rightarrow 1$

8. **(C)**

$N = \frac{7!}{2!\,2!\,2!} = 2 \times 3^2 \times 5 \times 7$

$\therefore$ number of ways in which N can be resolved as a product of two divisors which are coprime $= 2^{4-1} = 2^3 = 8$

9. **(A)**

Category	Selection	Arrangement	Req. no. of ways
(1) 2 Alike + 2 other alike + 1 different	${}^4C_2\,{}^5C_1$	$30 \times \frac{5!}{2!\,2!}$	$= 900$
(2) 2 Alike + 3 different	${}^4C_1\,{}^6C_3$	$80 \times \frac{5!}{2!}$	$= 4800$
(3) All 5 different	${}^7C_5$	$21 \times 5!$	$= 2520$
			8220

10. (C)

11. (A)

12. (B)

Vowels: I = 3; A = 1; O = 2

Consonants: S = 1; L = 1; C = 1; N = 1; T = 2

(i) Vowels appear in alphabetical order

$\therefore {}^{12}C_6 \times 1 \times \frac{6!}{2!}$

$\Rightarrow \frac{12!}{6!\,2!} = \frac{12 \times 11 \times 10!}{12 \times 5!} = 66 \cdot 7!$

$\Rightarrow K = 66.$

(ii) 2 consonants can be selected in ${}^2C_2 + {}^5C_2 = 1 + 10 = 11$

Number of arrangements of 2 consonants

$= 1 + 10 \times 2! = 21$

2 vowels can be selected in ${}^2C_1 + {}^3C_2 = 5$

and arrangement is $2 + 3 \times 2! = 8$

$\therefore$ Number of 4 letter words

${}^4C_2 \times 21 \times 8 = 6 \times 21 \times 8 = 126 \times 8 = 1008$

(iii) Number of words when I's are separated

$\frac{9!}{2!\,2!} \cdot {}^{10}C_3 \cdot 1$

and number of words when I's are separated and O's are together

$\frac{8!}{2!} \cdot {}^9C_3 \cdot 1$

$\therefore \text{Total} = \frac{9!}{2!\,2!} \cdot {}^{10}C_3 - \frac{8!}{2!} \cdot {}^9C_3 = 228 \times 8!$

$\Rightarrow N = 228$

13. **(C)**

Consider CC as single object U, $\boxed{CC}$, E can be arranged in 3! ways ⁎ U ⁎ CC ⁎ E ⁎.

Now the three S are to be placed in four available places.

Hence, required number of ways $= {}^4C_3 \times 3! = 24$

14. **(B)**

Let us first find the word when S's are separated

(C may or may not be together)

(i) Arrange the remaining letters $= \frac{4!}{2!} = 12$ ways

(ii) × U × C × C × E ×

There are five available places for three SSS.

Hence total number of ways no two S together

$= 12 \times {}^5C_3 = 120$

Hence both C as well as S are separated

$= 120 - 24 = 96$

15. **(A)**

Consonants in SUCCESS are S, C, C, S, S (represent them by V)

Number of words when c consonants are in alphabetical order $= \frac{7!}{5!} = 42$

Match Matrix

1. **(A – R; B – Q; C – P; D – S)**

(A) $a\ b\ c$

$\left.\begin{matrix} r & r & r^2 & 3 \text{ ways} \\ r & r^2 & r^2 & 3 \text{ ways}\end{matrix}\right\} \Rightarrow 6 \text{ ways}$

$\left.\begin{matrix} t & t & t^2 & 3 \text{ ways} \\ t & t^2 & t^2 & 3 \text{ ways}\end{matrix}\right\} \Rightarrow 6 \text{ ways}$

$\left.\begin{matrix} s^2 & s^2 & s^4 & 3 \text{ ways} \\ s^2 & s^3 & s^4 & 6 \text{ ways} \\ s^2 & s^4 & s^4 & 3 \text{ ways}\end{matrix}\right\} \Rightarrow 12 \text{ ways}$

Total ways $= 6 \times 6 \times 12 = 432$

(B) Number of possible ways $=$

$8 \times 11 \times 9 = 792$

(C) **Case I:** Number of odd numbers with all four digits different from 3000 to 6000

4th □□□□ 1st
↓ 3/4/5   1/3/5/7/9

If 4th place has the digit 4 fixed, then

$= (1)(8)(7)(5)$

3 or 5 fixed at 4th place, then

$= (2)(8)(7)(4)$

Sum of case I $= 56(5+8) = 728$

Case II: From 6000 to 6300

Then total numbers $= 98$

Hence, total numbers $= 826$

(D) Number of quadrilaterals $= {}^{17}C_4 - {}^{15}C_2$

$= 2275$

2. **(A – S; B – R; C – Q; D – P)**

(A) $A = 2$; $I = 2$; K, R, T different

$\therefore$ Total $= \frac{7!}{2!\,2!} = 1260$

If P denote when A's together and Q denote when I's together

$\therefore n(P) = \frac{6!}{2!} = 360 = n(Q)$

$n(P \cap Q) = 5! = 120$

$n(P \cup Q) = 360 + 360 - 120 = 600$

$\therefore$ Required number of ways

$= 1260 - 600 = 660$

(B) Required number $= \frac{(120) - 24}{2} = 48$

(C) Total solutions $= {}^{7-1}C_{3-1} = {}^6C_2 = 8$

(D) Let $a = 5^{\alpha_1} \cdot 3^{\beta_1}$, $b = 5^{\alpha_2} \cdot 3^{\beta_2}$ and $c = 5^{\alpha_3} \cdot 3^{\beta_3}$

a_1	a_2	a_3
2	0	3
2	1	3
2	2	3
0	2	3
1	2	3

$\alpha_1, \alpha_2, \alpha_3$ can be selected in 5 ways, where $\beta_1, \beta_2, \beta_3$ can be selected in 13 ways

$\therefore$ Total ways $= 5 \times 13 = 65$

3. **(A – Q; B – R; C – P; D – S)**

(P) Odd positions 1, 3, 5, 7, 9, 11 (6)

Even positions 2, 4, 6, 8, 10 (5)

Minimum one letter goes at odd positions

Hence, answer is total number of words formed without any conditions

$\frac{11!}{(2!)(2!)(2!)} = \frac{11!}{(2!)^3}$

(Q) 4 vowels (AA I E)

Rest = 7 (8 gaps formed); gap selection $= {}^8C_4$

arrangement of vowels $= \frac{6!}{2!\,2!}$

Arrangement of rest letters $= \frac{7!}{2!\,2!}$

$\Rightarrow {}^8C_4 \times \frac{4!}{2!} \times \frac{7!}{2!\,2!} \Rightarrow 210 \times 7!$

(R) (MM) (TT) H E I CS (7)
AA separate
8 gaps
${}^8C_2 \times (1!) \times 7!$ (for AA to be separated)
$\Rightarrow 28 \times 7!$

(S) 4 vowels A A I E
7 consonants M M T T H C S
$\frac{4!}{2!} \times \frac{7!}{2!\,2!}$
(arrangement of vowels and consonants at their respective positions)
$\Rightarrow \frac{4! \cdot 7!}{(2!)^3}$

4. **(A – Q; B – S; C – P)**

(A) 1 2 3 … n
(S) (S) (S) selected
(n – 3) not to be selected | × | × | × | … |×|
$\therefore$ total (n – 2) gaps
$\therefore n - {}^2C_3 = 455 = {}^{15}C_3$
$n - 2 = 15 \Rightarrow n = 17$

(B) Number of ordered pairs (a, b) where $a, a \in \{-1, 0, 1\}$ is 9.
Hence Total $= {}^9C_2 = 36$
But there are 8 lines consisting of 3 points
$= 8 \times {}^3C_2 = 24$
$\therefore$ Distinct lines = 36 – 16 = 20 ⇒ B

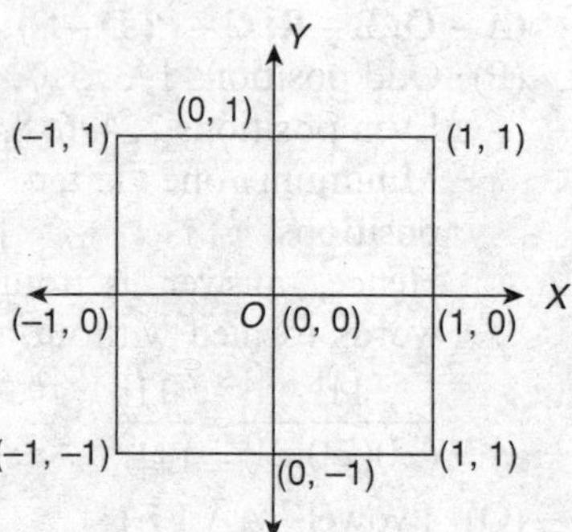

(C) Total number of subsets $= 2^4 - 1 = 16 - 1 = 15$ sum of elements of following pair of subsets are equal
{7}, {2, 5}
{4, 2, 5}, {7, 4}
{4, 5}, {7, 2}
$\therefore$ Different sums = 15 – 3 = 12

5. **(A – R; B – Q, R; C – Q; D – P, R, T)**

(A) n couples ⇒ $2n$ persons
${}^{2n}C_2 = 28 = {}^8C_2$
$\therefore 2n = 8 \Rightarrow n = 4$

(B) $t_{p+1} = {}^nC_p x^p$ in $(1 + x)^n$
Here $p + 1 = (4r - 4)$
$p = 4r - 5$
also, $p + 1 = r + 14$
$\Rightarrow p = r + 13$
$\therefore {}^{28}C_{4r-5} = {}^{28}C_{r+13}$
either $4r - 5 = r + 13 \Rightarrow r = 6$
or $4r - 5 + r + 13 = 28$
$5r = 20 \Rightarrow r = 4$

(C) n ⟨ AB, $(n - 2)$
Number of ways $= (n - 2)! \cdot 2! = 48 \Rightarrow n = 6$

(D) $d = \frac{n(n-3)}{2} \Rightarrow 20 = \frac{n(n-3)}{2} \Rightarrow n = 8$

Subjective Answer Type: Hard Nut

1. (i) 15 (ii) 126 (iii) 60 (iv) 105

2. 974

3. Required sum
= Sum of the numbers formed by using digits 0, 1, 2, 4, 5 – sum of the numbers formed by using digits 1, 2, 4, 5

$$= (0+1+2+4+5)(5-1)!\left(\frac{10^5-1}{10-1}\right)$$
$$-(1+2+4+5)(4-1)!\frac{(10^4-1)}{(10-1)}$$
$$= 288\left(\frac{10^5-1}{10-1}\right) - 76\left(\frac{10^4-1}{10-1}\right)$$
$$= 3119976$$

4. Total number of ways
$$= \left({}^5C_0\right)^3 + \left({}^5C_1\right)^3 + \left({}^5C_2\right)^3 + \left({}^5C_3\right)^3 + \left({}^5C_4\right)^3 + \left({}^5C_5\right)^3$$
$$= 1^3 + 5^3 + 10^3 + 10^3 + 5^3 + 1^3 = 2252$$

5. Since, $2310 = 2 \times 3 \times 5 \times 7 \times 11$
Clearly
Number of elements
sets of positive integers $\{a, b, c\}$
such that $abc = 2310$ equal to 40

6.

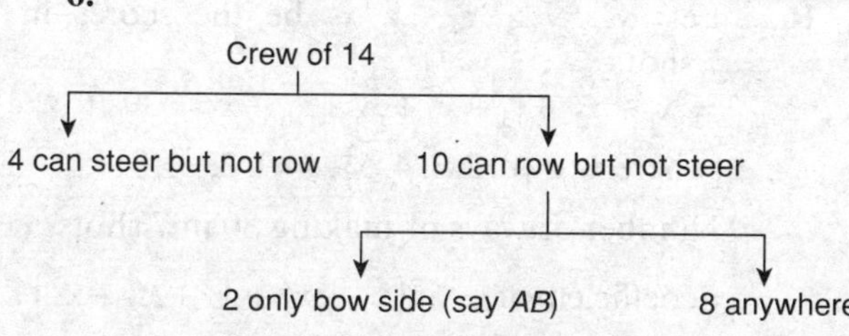

Hence total number of ways crew can be arranged $= 4(4!)^2\ {}^8C_4 \cdot {}^6C_2$.

7. Clearly 121

8. Let twelve persons be
$P_1, P_2, P_3, P_4, P_5, P_6, P_7, N_1, N_2, E_1, E_2, M$.
where N_1, N_2 are the persons who wish to sit on the north side of the table, E_1, E_2 on the east side and M who wants to sit in the middle seat (which may be on any side).

Case I: When M sits on south or west side of the table
$\Rightarrow$ Number of ways $= 2$
N_1 and N_2 can occupy their seats in $({}^3C_2 \times 2!)$ ways
Similarly, for E_1, E_2
Rest of persons can be arranged in 7! ways
$\Rightarrow$ Number of ways $=$
${}^3C_2 \times 2! \times {}^3C_2 \times 2! \times 2 \times 7!$ (1)

Case II: When M sits on either north or east side
Then, number of ways $= 2 \times 2! \times {}^3C_2 \times 2! \times 7!$ (2)
Adding (1) and (2)
$2! \cdot 3! \cdot 8!$

9. 2111

10. 771 ways

11. INEFFECTIVE contains 11 letters EEE, FF, II, C, T, N, V

Case I: All distinct
Number of words $= {}^7C_4 \times 4! = 840$

Case II: Two alike, two distinct
Number of words $= {}^3C_1 \times {}^6C_2 \times \dfrac{4!}{2!} = 540$

Case III: Two alike, two alike
Number of words $= {}^3C_2 \times \dfrac{4!}{2!\,2!} = 18$

Case IV: 3 alike, one distinct
Number of words $= 1 \times {}^6C_1 \times \dfrac{4!}{3!} = 24$

Hence total 4 letter words
$= 840 + 540 + 18 + 24 = 1422$

12. Select three points out of 12 in ${}^{12}C_3$ ways. This number includes the number of circles obtained from 3 points out of 5 concyclic points. Note that we get the same circle by selecting any three points out of 5 concyclic points but we count it ${}^5C_3$ times.
Required number $= {}^{12}C_3 - {}^5C_3 + 1 = 211$

13. Let 3 minister be M_1, M_2, M_3.
We can divide 7 departments among 3 ministers such that each minister gets at least 1 and at most 4 departments as follows.

S. No.	M_1	M_2	M_3
1	4	2	1
2	2	2	3
3	3	3	1

If we have a case of (2, 2, 3) then there is no need to make cases (3, 2, 2) or (2, 3, 2) because we will include them when we apply distribution formula to distribute ways of division among ministers.

Case I: Number of ways to divide and distribute departments
in numbers 4, 2,1

$$= \left(\frac{7!}{4!\,2!\,1!}\right) \times 3! = 630 \quad (1)$$

Case II: Number of ways to divide and distribute departments
in numbers 2, 2, 3

$$= \left[\frac{7!}{2!\,2!\,3!} \cdot \frac{1}{2!}\right] \times 3! = 630$$

Case III: Number of ways to divide and distribute departments in numbers 3, 3, 1

$= \left[\dfrac{7!}{(3!)^2(1!)}\cdot\dfrac{1}{2!}\right]\times 3! = 420$

Thus, total number of ways

$= 630 + 630 + 420 = 1680$

14. The 10 balls are R R R R B B B W W Y

Case I: All alike

Number of ways of selecting all alike balls

$= {}^1C_1 = 1$

Case II: 3 alike and 1 distinct

${}^2C_1 \times {}^3C_1 = 6$

Case III: 2 alike, 2 alike

${}^3C_2 = 3$

Case IV: 2 alike, 2 distinct

${}^3C_1 \times {}^3C_2 = 9$

Case V: All distinct

$= {}^4C_4 = 1$

Total number of ways to select 4 balls

$= 1 + 6 + 3 + 9 + 1 = 20$

15. Suppose that a student gets at least 60% marks in the first two papers, then he must get at most 30% marks in the third paper to make a total of 150 marks.

Let x_1, x_2, x_3 be marks obtained in 3 papers respectively

$\Rightarrow x_1 + x_2 + x_3 = 150$ (1)

where

$60 \le x_1 \le 100, 60 \le x_2 \le 100, 0 \le x_3 \le 30$

The required number of ways

= Number of integral solutions of (1)

= Coefficient of x^{150} in

$\left(x^{60} + x^{61} + \ldots + x^{100}\right)^2\left(1 + x + x^2 + x^{30}\right)$

= coefficient of x^{30} in

$\left(1 + x + x^2 + \ldots + x^{40}\right)^2\left(1 + x + x^2 + \ldots x^{30}\right)$

= coefficient of x^{30} in

$\left(\dfrac{1 - x^{41}}{1 - x}\right)^2\left(\dfrac{1 - x^{31}}{1 - x}\right)$

= coefficient of x^{30} in $(1 - x)^{-3}$

$= {}^{30+3-1}C_{3-1} = {}^{32}C_2$

But the two papers of at least 60% marks can be chosen out of 3 papers in ${}^3C_2$ ways

Hence, the required number of ways

$= {}^3C_2 \times {}^{32}C_2$

16. Let $x_1, x_2, x_3, x_4, \ldots, x_7$ be the scores in 7 shots

$\Rightarrow x_1 + x_2 + x_3 + x_4 + x_5 + x_6 + x_7 = 30$

[where $x_i \in \{0, 2, 3, 4, 5\}, i = 1, 2, \ldots, 7\}$

Number of ways of making 30 in 7 shots

=coefficient of x^{30} in $\left(x^0 + x^2 + x^3 + x^4 + x^5\right)^7$

= coefficient of x^{30} in

$\left\{\left(x^0 + x^2 + x^3\right) + x^4(x + 1)\right\}^7$

= coefficient of x^{30} in

$\left\{x^{28}(x + 1)^7 + {}^7C_1x^{24}(x + 1)^6\left(1 + x^2 + x^3\right) + {}^7C_2x^{20}(x + 1)^5\left(1 + x^2 + x^3\right) + \ldots\right\}$

$= {}^7C_2 + {}^7C_1\left({}^6C_3 + {}^6C_2 + {}^6C_0\right) + {}^7C_2\left({}^5C_1 + 2\right)$

$= 21 + 252 + 147 = 420$

17. Total ways (without condition) $= 10^2 = 100$

Let $b - a \ge 6$

$1 \le a < b \le 10 \Rightarrow 1 \le a < b - 5 \le 5 \Rightarrow {}^5C_2 = 10$

Similarly, for $a - b \ge 6$, we will get 10 ways

Hence required number $= 100 - 10 - 10 = 80$

18. Let $n(A)$ = number of positive divisors of $10^{60} = 2^{60} \times 5^{60}$ is $61 \times 61 = 61^2$

$n(B)$ = Number of positive divisors of $20^{50} = 2^{100} \times 5^{50}$ is 101×51

and $n(C)$ = Number of positive divisors of $30^{40} = 2^{40} \times 3^{40} \times 5^{40} = 41^3$

The set of common factors of A and B will be of the form $2^m \cdot 5^n$ where $0 \le m \le 60$ and $0 \le n \le 50$

So, $n(A \cap B) = 61 \times 51$

Similarly, the common factors of (B and C) and (A and C) is of the form $2^m \times 5^n$

$\therefore n(B \cap C) = 41^2$ and $n(A \cap C) = 41^2$

and $n(A\cap B\cap C)$ is also 41^2

Now, $n(A\cap B\cap C)$

$= n(A)+n(B)+n(C)-n(A\cap B)-$

$n(B\cap C)-n(A\cap C)+n(A\cap B\cap C)$

$= 61^2+101\times 51+41^3-61\times 51-41^2-41^2+41^2$

$= 61(61-51)+41^2(41-1)+101\times 51$

$= 610+1681\times 40+5151 = 73001$

19. (i) All the balls placed in the wrong boxes using derangement formulae

$= 5!\left[1-\frac{1}{1!}+\frac{1}{2!}-\frac{1}{3!}+\frac{1}{4!}-\frac{1}{5!}\right]$

$= 120\left[1-1+\frac{1}{2}-\frac{1}{6}+\frac{1}{24}-\frac{1}{120}\right]$

$= 44$

(ii) At least 2 balls are placed in the correct boxes

= Total number of ways to place balls in boxes – number of ways to place balls such that all balls are placed in wrong boxes – number of ways to place balls in boxes such that 1 ball is placed in the correct box.

$= 5!-44-{}^5C_1\times 4!\left[1-\frac{1}{1!}+\frac{1}{2!}-\frac{1}{3!}+\frac{1}{4!}\right]$

$= 120-44-5\times 9 = 31$

20. (i) $x+y+z+w = 20;\ x\ge 0,\ y\ge 0,$

$z\ge 0,\ w\ge 0$

$= {}^{20+4-1}C_{20}$ [using ${}^{n+r-1}C_n$]

$= {}^{23}C_3 = 1771$

(ii) ${}^{n-1}C_{r-1} = {}^{19}C_3 = 969$

(iii) If no variable exceeds 10, then sum of rest should be less than or equal to 10

Let $x\le 10$, then $y+z+w\ge 10$

and

$\max(y+z+w) = 20-\min(x) = 20-1 = 19$

$10\le y+z+w\le 19$

[where, $y\ge 1,\ z\ge 1,\ w\ge 1$]

$\Rightarrow 10\le (y_1+1)+(z_1+1)+(w_1+1)\le 19$

$\Rightarrow 7\le y_1+z_1+w_1\le 16,\ 0\le y_1\le 9,$

$0\le z_1\le 9,\ 0\le w_1\le 9$

Number of solutions of $y_1+z_1+w_1\le 16$

Can be obtained by adding a dummy variable $x_1\ (x_1\ge 0)$ such that

$x_1+y_1+z_1+w_1 = 16$

Number of solutions

$=$ coefficient of x^{16} in $\left(1-x^{10}\right)^3(1-x)^{-4}$

$= {}^{19}C_4-3\cdot{}^9C_3$

Similarly ${}^9C_3$

Total number of solutions =

${}^{19}C_3-4\cdot{}^9C_3$

$= 633$

(iv) Each variable is an odd number

$\therefore x = 2x_1+1,\ y = 2y_1+1,$

$z = 2z_1+1,\ w = 2w_1+1$

$x+y+z+w = 20$

$\Rightarrow x_1+y_1+z_1+w_1 = 8$

[where $x_1, y_1, z_1, w_1\ge 0$]

Number of solutions $= {}^{8+4-1}C_{4-1}$

$= {}^{11}C_3 = 165$

(v) Let $x = x_1$

$y = x+x_2 = x_1+x_2$

$z = y+x_3 = (x_1+x_2)+x_3$

$w = z+x_4 = (x_1+x_2+x_3)+4$

Now, $x+y+z+w = 20$

$\Rightarrow 4x_1+3x_2+2x_3+x_4 = 20$

$[x_1, x_2, x_3, x_4\ge 1]$

Let

$x_1 = y_1+1,\ x_2 = y_2+1,$

$x_3 = y_3+1,\ x_4 = y_4+1$

$\Rightarrow 4y_1+3y_2+2y_3+y_4 = 10$

[where $y_1, y_2, y_3, y_4\ge 0$]

$\therefore$ Total number of solutions = 23

21. Required number of ways

$=$ coefficient of x^{12} in $\left(x^0+x^1+x^2+\ldots+x^9\right)^6$

$= 6062$

22. An n-sided polygon has n angular points. Number of triangles formed from these n angular points $= {}^nC_3$. But it also includes the triangle with sides on the polygon.

Let us consider a side PQ. If each angular point of the remaining $(n - 2)$ points is joined with PQ, we get a triangle with one side PQ.

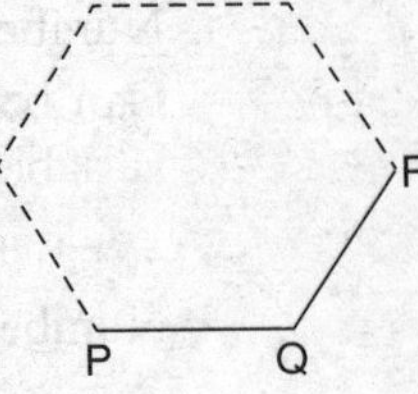

$\therefore$ Number of triangles with PQ as one side $= (n-2)$

In similar ways n sides like QR can be considered. Hence number of triangles $= n(n-2)$.

But some triangles have been counted twice. Number of such triangles $= n$.

$\therefore$ The number of triangles of which one side is the side of the triangle $= n(n-2)-n$

$= n(n-3)$

Hence number of required triangles

$= {}^nC_3 - n(n-3) = \frac{n}{6}(n-4)(n-5)$

23. Do it yourself.

(i) $\frac{52!}{(13!)^4}$ (ii) $\frac{52!}{4!(13!)^4}$

(iii) $\frac{52!}{20!15!10!7!}$ (iv) $\frac{52!}{(15!)^3 3!7!}$

24. Let x_1, x_2 and x_3 be the number of balls placed in boxes 1, 2 and 3 respectively.

Number of ways of distributing 10 balls in 3 boxes.

= Number of integral solutions of the equation

$x_1 + x_2 + x_3 = 10$ (1)

Since, $2 \le x_i \le 4$ for $i = 1, 2, 3$

So the number of ways of distributing balls in boxes

$=$ coefficient of x^{10} in $\left(x^2 + x^3 + x^4\right)^3$

$= 6$

25. (i) Number of ways to place 6 letters in 6 envelopes such that all are placed in wrong envelopes

$= 6!\left[1 - \frac{1}{1!} + \frac{1}{2!} - \frac{1}{3!} + \ldots + \frac{1}{6!}\right]$

$= 360 - 120 + 30 - 6 + 1 = 265$

(ii) Number of ways to place letters such that at least 4 letters are placed in correct envelopes

$= {}^6C_4 \times 1 + 1 = 16$

(iii) Number of ways to place 6 letters in 6 envelopes such that at most 3 letters are placed in wrong envelopes

$= 1 + \frac{6\times 5}{2} + \frac{6\times 5\times 4}{6}\left(\frac{3!}{2!} - \frac{3!}{3!}\right)$

$= 1 + 15 + 20\times 2 = 56$

4.19.4 JEE Main: Archive

1. **(40)**

$x_1 + x_2 + x_3 + x_4 + x_5 = 5$

Only one possibility 3, 3, 3, – 2, – 2

Number of ways is $= \frac{5!}{3!\,2!} \times 2 \times 2 = 40$

2. **(576)**

Digits are 1, 2, 3, 4, 5, 7, 9

Multiple of $11 \rightarrow$ Difference of sum at even and odd

Place is divisibly by 11.

Let number of the form *abcdefg*

$\therefore (a+c+e+g) - (b+d+f) = 11x$

$a+b+c+d+e+f = 31$

$\therefore$ either $a+c+e+g = 21$ or 10

$\therefore b+d+f = 10$ or 21

Case-1:

$a+c+e+g = 21$

$b+d+f = 10$

$(b, d, f) \in \{(1, 2, 7)(2, 3, 5), (1, 4, 5)\}$

$(a, c, e, g) \in \{(1, 4, 7, 9), (3, 4, 5, 9), (2, 3, 7, 9)\}$

$\therefore$ Total number in case-1 $= (3!\times 3)(4!) = 432$

Case-2:

$a+c+e+g = 10$

$b+d+f = 21$

$(a, b, e, g) \in \{(1, 2, 3, 4)\}$

(b, d, f) and $\{(5, 7, 9)\}$

$\therefore$ Total number in case -2 $= 3! \times 4! = 144$

$\therefore$ Total numbers = 144 + 432 = 576

3. **(1633)**

HCF $(\alpha, 24) = 1$

Now, $24 = 2^2 \cdot 3$

$\rightarrow \alpha$ is not the multiple of 2 or 3

Sum of values of α

$$= S(U) - \begin{cases} S(\text{Multiple of } 2) + S(\text{Multiple of } 3) \\ -S(\text{Multiple of } 6) \end{cases}$$

$$= (1+2+3+\ldots 100) - (2+4+6\ldots+100) - (3+6+\ldots 99) + (6+12+\ldots+96)$$

$$= \frac{100\times101}{2} - 50\times51 - \frac{33}{2}\times(3+99) + \frac{16}{2}(6+96)$$

$= 5050 - 2550 - 1683 + 816 = 1633$

4. **(63)**

$x\,y\,z \rightarrow$ odd number

$z = 1,3,5,7,9$

$x + y + z = 7, 14, 21$

[sum of digit multiple of 7]

$\underset{1 \text{ to } 9}{x} + \underset{0 \text{ to } 9}{y} = 6, 4, 2, 13, 11, 9, 7, 5, 20, 18, 16, 14, 12$

$x + y = 6 \Rightarrow (1, 5), (2, 4), (3, 3), (4, 2), (5, 1)$ $(6, 0)$

$\rightarrow$ TN = 6

$x + y = 4 \Rightarrow (1, 3), (2, 2), (3, 1), (4, 0)$

$\rightarrow$ TN = 4

$x + y = 2 \Rightarrow (1, 1), (2, 0)$

$\rightarrow$ TN = 2

$x + y = 13 \Rightarrow$

$(4, 9), (5, 8), (6, 7), (7, 6), (8, 5), (9, 4)$

$\rightarrow$ TN = 6

$x + y = 11 \Rightarrow (2, 9), (3, 8), (4, 7), (5, 6), (6, 5), (6, 5), (7, 4), (8, 3), (9, 2)$

$\rightarrow$ TN = 8

$x + y = 9 \Rightarrow (1, 8), (2, 7), (3, 8), (4, 5), (5, 4), (8, 1), (9, 0)$

$\rightarrow$ TN = 9

$x + y = 7 \Rightarrow (1, 8), (2, 5), (3, 4), (8, 1), (7, 0)$

$\rightarrow$ TN = 7

$x + y = 5 \Rightarrow (1, 4), (2, 3), (3, 2), (4, 1), (5, 0)$

$\rightarrow$ TN = 5

$x + y = 20$ Not possible

$x + y = 18 \Rightarrow (9, 9) \rightarrow$ T.N. = 1

$x + y = 16 \Rightarrow (7, 9), (8, 8), (9, 7)$

$\rightarrow$ TN = 3

$x + y = 14 \Rightarrow (5, 9), (6, 8), (7, 7), (8, 6), (9, 5)$

$\rightarrow$ TN = 5

$x + y = 12 \Rightarrow (3, 9), (4, 8), (5, 7), (6, 6), \ldots, (9, 3)$

$\rightarrow$ TN = 7

5. **(243)**

If 0 taken twice then ways = 9

If 0 taken once then ${}^9C_1 \times 2 = 18$

If 0 not taken then ${}^9C_1\, {}^8C_1 \cdot 3 = 216$

Total = 243

6. **(1120)**

$n(B) = 10$

$n(a) = 5$

The number of ways of forming a group of 3 girls of 3 boys

$= {}^{10}C_3 \times {}^5C_3$

$= \frac{10\times9\times8}{3\times2} \times \frac{5\times4}{2} = 1200$

The number of ways when two particular boys B_1 of B_2 be the member of group together

$= {}^8C_1 \times {}^5C_3 = 8\times10 = 80$

Number of ways when boys B_1 of B_2 hot in the same group together

$= 1200 \times 80 = 1120$

7. **(150)**

$36 = 2\times2\times3\times3$

Number should be an odd multiple of 2 and does not have factors 3 and 9

Odd multiples of 2 are

$102, 106, 110, 114\ldots998$ (225 no.)

No. of multiples of 3 are

$102, 114, 126\ldots990$ (75 no.)

Which also include multiples of 9

Hence,

Required = 225 – 75 = 150

8. **(56)**

$x_1 + x_2 + x_3 + x_4 + x_5 + x_6 = 11$

$x_1, x_6 \geq 0, \qquad x_2, x_3, x_4, x_5 \geq 2$

$x_2 = t_1 + 2$

$x_3 = t_3 + 2$

$x_4 = t_4 + 2$

$x_5 = t_5 + 2$ 16 cubes — 11 Blue, 5 Red

$x_1, t_2, t_3, t_4, t_5, x_6 \geq 0$

No. of solutions $= {}^{6+3-1}C_3 = {}^{8}C_3 = 56$

9. **(D)**
To make a number divisible by 3 we can use the digits 1, 2, 5, 6, 7 or 1, 2, 3, 5, 7.
Using 1, 2, 5, 6, 7, number of even numbers is
$= 4 \times 3 \times 2 \times 1 \times 2 = 48$
Using 1, 2, 3, 5, 7, number of even numbers is
$= 4 \times 3 \times 2 \times 1 \times 1 = 24$
Required answer is 72.

10. **(702)**
$a_1, a_2, a_3, \ldots, a_{18}, 77$
are in A.P., that is, $1, 5, 9, 13, \ldots, 77$.
Hence,
$a_1 + a_2 + a_3 + \ldots + a_{18} = 5 + 9 + 13 + \ldots + 18$ terms
$= 702$

11. **(D)**
$t_1 + t_2 + t_3 + t_4 = 30$

Coefficient of x^{30} in $\left(1 + x + x^2 + \ldots + x^{30}\right)^2$

$\left(x^4 + x^5 + x^6 + x^7\right)\left(x^2 + x^3 + x^4 + x^5 + x^6\right)$

$x^6\left(\dfrac{1 - x^{31}}{1 - x}\right)^2\left(1 + x + x^2 + x^3\right)\left(1 + x + x^2 + x^3 + x^4\right)$

$x^6\left(1 - x^{31}\right)^2\left(1 - x^4\right)\left(1 - x^5\right)(1 - x)^{-4}$

$x^6\left(1 - x^4 - x^5 + x^9\right)\left(1 + x^{62} - 2x^{31}(1 - x)^{-4}\right)$

$x^6\left(1 - x^4 - x^5 + x^9\right)(1 - x)^{-4}$

Coefficient of x^n in $(1 - x)^{-r}$ is ${}^{n+r-1}C_{r-1}$

$\Rightarrow {}^{27}C_3 - {}^{23}C_3 - {}^{22}C_3 + {}^{18}C_3$

$2925 - 1771 - 1540 + 816$

$= 430$

OR

$x_2 \in [4, 7],\ x_3 \in [2, 6]$

$\Rightarrow t_1 + t_2 + t_3 + t_4 = 24$

total ways =

${}^{24+4-1}C_{4-1} - {}^{20+4-1}C_{4-1} - {}^{19+4-1}C_{4-1} + {}^{15+4-1}C_{4-1}$

$= {}^{27}C_3 - {}^{23}C_3 - {}^{22}C_3 + {}^{18}C_3 = 430$

12. **(18915)**
$b_i \in \{1, 2, 3, \ldots 100\}$

Let A = set when b_1, b_2, b_3 are consecutive

$n(A) = \dfrac{97 + 97 + \ldots + 97}{98 \text{ times}} = 97 \times 98$

Similarly, when $b_2 b_3 b_4$ are consecutive

$n(A) = 97 \times 98$

$n(A \cap B) = \dfrac{97 + 97 + \ldots 97}{98 \text{ times}} = 97 \times 98$

Similarly, when $b_2 b_3 b_4$ are consecutive

$n(B) = 97 \times 98$

$n(A \cap B) = 97$

$n(A \cup B) = n(A) + n(B) - n(A \cap B)$

Number of permutations = 18915

13. **(1086)**
Let the number be $abcd$, where a, b, c are divisible by d.

	No. of such numbers
$d = 1,$	$9 \times 10 \times 10 = 900$
$d = 2$	$4 \times 5 \times 5 = 100$
$d = 3$	$3 \times 4 \times 4 = 48$
$d = 4$	$2 \times 3 \times 3 = 18$
$d = 5$	$1 \times 2 \times 2 = 4$
$d = 6, 7, 8, 9$	$4 \times 4 = 16$

$= 1086$

14. **(1492)**

M A N K I N D

$\left(\dfrac{4 \times 6!}{2!}\right) + (5! \times 0) + \left(\dfrac{4! \times 3}{2!}\right) + (3! \times 2) +$

$(2! \times 1) + (1! \times 1) + (0! \times 0) + 1 = 1492$

15. **(180)**

$$3\times\frac{5!}{2!\,2!}+\frac{5!}{3!\times 2!}+\frac{5!}{2!}+\frac{5!}{3!}=180$$

16. **(30)**

Here 1st digit is 1 or 2 only

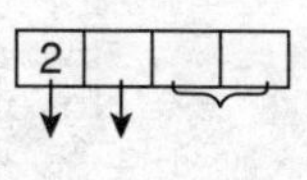

Case-I:

If first digit is 1

Then last two digits can be 24, 32, 36, 52, 56, 64

$1\times 3\times 6 = 18$ ways

Case-II:

If first digit is 2 then last digit can be 16, 36, 56, 64

$1\times 3\times 4 = 12$ ways

17. **(C)**

$n(S)$ = all 5-digit numbers = 9×10^4

A : no is multiple of 7 but not divisible by 5

Smallest 5 digit divisible by 7 is 10003

Largest 5 digit divisible by 7 is 99995

$\therefore 99995 = 10003 + (n-1)\,7 \therefore n = 12857$

Numbers divisible by 35

$99995 = 10010 + (P-1)\,35 \Rightarrow P = 2572$

$\therefore$ Numbers divisible by 7 but not by 35 are

$12857 - 2572 = 10285$

$\therefore P = \dfrac{10285}{90000}$ $\qquad \therefore 9P = 1.0285$

18. **(7073)**

Required no. = Total – no character from $\{1, 2, 3, 4, 5\}$

$= \left(10^6-5^6\right)+\left(10^7-5^7\right)+\left(10^8-5^8\right)$

$= 10^6\left(1+10+100\right)-5^6\left(1+5+25\right)$

$= 10^6\times 111-5^6\times 31$

$= 2^6\times 5^6\times 111-5^6\times 31$

$= 5^6\left(2^6\times 111-31\right)$

$= 5^6\times \underbrace{7073}_{\alpha}$

$\therefore \alpha = 7073$

19. **(17)**

${}^{b}C_3\times {}^{g}C_2 = 168$

$b(b-1)(b-2)(g)(g-1) = 8\times 7\times 6\times 3\times 2$

$b+3g = 17$

20. **(11)**

$S=\{4, 6, 9\}$ $\qquad T=\{9, 10, 11, \ldots 1000\}$

$A\{a_1+a_2+\ldots+a_k : K\in N\}$ and $a_i\in S$

Here by the definition of set 'A'

$A=\{a : a = 4x+6y+9z\}$

Except the element 11, every element of set T is of the form $4x+6y+9z$ for some $x, y, z\in W$

$\therefore T-A=\{11\}$

21. **(6)**

4-digit numbers

For divisibility by 55, the number should be div. by 5 and 11 both

Also, for divisibility by 11

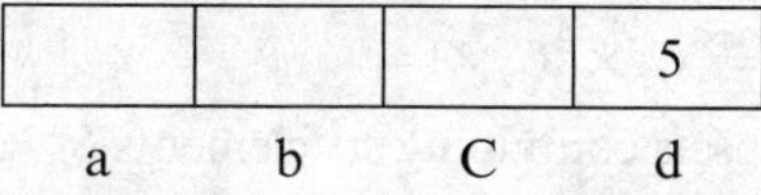

$a+c=b+5$

for $b=1$; $\quad a=2, c=4$

$a=4, c=2$

for $b=2$ $\quad a=3, c=4$

$a=4, c=3$

for $b=3$ $\quad a=6, c=2$

$a=2, c=6$

$\therefore$ 6 possible four digit numbers are divisible by 55

(II) 5-digit number is not possible

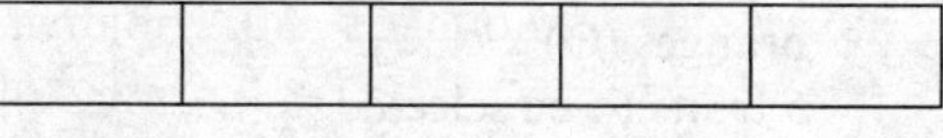

(Not possible)

22. **(60)**

Even digits occupy at even places

$$\frac{4!}{2!\,2!}\times\frac{5!}{2!\,3!}=\frac{24\times 120}{4\times 12}=60$$

23. **(546)**

For at most two language courses

$= {}^{5}C_2\times {}^{7}C_3+{}^{5}C_1\times {}^{7}C_4+{}^{7}C_5 = 546$

24. **(B)**

Four-digit numbers greater than 7000

$= 2\times 4\times 3\times 2 = 48$

Five-digit numbers = 5! = 120
Total number greater than 7000
= 120 + 48 = 168

25. **(43)**
Elements of the type $3k = 3$
Elements of the type $3k + 1 = 1, 7, 9$
Elements of the type $3k + 2 = 2, 5, 11$
Subsets containing one element $S_1 = 1$
Subsets containing two elements
$S_2 = {}^3C_1 \times {}^3C_1 = 9$
Subsets containing three elements
$S_3 = {}^3C_1 \times {}^3C_1 + 1 + 1 = 11$
Subsets containing four elements
$S_4 = {}^3C_3 + {}^3C_3 + {}^3C_2 \times {}^3C_2 = 11$
Subsets containing five elements
$S_5 = {}^3C_2 \times {}^3C_2 \times 1 = 9$
Subsets containing six elements $S_6 = 1$
Subsets containing seven elements $S_7 = 1$
$\Rightarrow$ sum = 43

26. **(D)**
Number between 5000 and 10000
Using digits 1, 3, 5, 7, 9

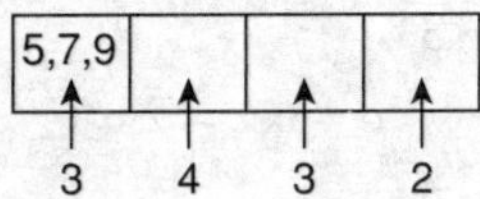

Total numbers $= 3 \times 4 \times 3 \times 2 = 72$

27. **(6860)**
7 Red apple (RA), 5 white apple (WA), oranges (O)
5 fruits to be selected
(**Note:** fruits taken different)
Possible selections: (2O, 1RA, 2WA) or (2O, 2RA, 1WA) or (3O, 1RA, 1WA)
$\Rightarrow {}^8C_2\,{}^7C_1\,{}^5C_2 + {}^8C_2\,{}^7C_2\,{}^5C_1 + {}^8C_3\,{}^7C_1\,{}^5C_1$
$\Rightarrow 1960 + 2940 + 1960$
$= 6860$

28. **(32)**

1	2				

$= {}^7C_4 = 35$

1	3				

$= {}^6C_4 = 15$

1	4				

$= {}^5C_4 = 5$

1	5				

$= {}^4C_4 = 1$

2	3				

$= {}^6C_4 = 15$
71 words

2 4 5 6 7 8 $\rightarrow$ 72nd word
$2 + 4 + 5 + 6 + 7 + 8 = 32$

29. **(1436)**
No. of 5-digit numbers starting with digit 1
$= 5 \times 5 \times 5 \times 5 = 625$
No. of 5-digit numbers starting with digit 2
$= 5 \times 5 \times 5 \times 5 = 625$
No. of 5-digit numbers starting with 31
$= 5 \times 5 \times 5 = 125$
No. of 5-digit numbers starting with 32
$= 5 \times 5 \times 5 = 125$
No. of 5-digit numbers starting with 33
$= 5 \times 5 \times 5 = 125$
No. of 5-digit numbers starting with 351
$= 5 \times 5 = 25$
No. of 5-digit numbers starting with 352
$= 5 \times 5 = 25$
No. of 5-digit numbers starting with 3531
$= 5$
No. of 5-digit numbers starting with 3532
$= 5$
Before 35337 will be 4 numbers,
So rank of 35337 will be 1690
So, in descending order serial number will be
$3125 - 1690 + 1 = 1436$

30. **(B)**
Total 3-digit numbers = 900
Divisible by 3 = 300 (Using $\frac{900}{3} = 300$)
Divisible by 4 = 225 (Using $\frac{900}{4} = 225$)
Divisible by 3 and 4 = 108, …
(Using $\frac{900}{12} = 75$)
Number divisible by either 3 or 4
= 300 + 2250 − 75 = 450
We have to remove divisible by 48,
144, 192, …, 18 terms
Required number of numbers
= 450 − 18 = 432

31. **(A)**

Lets arrange the letters of OUGHT in alphabetical order.

G, H, O, T, U

Words starting with

G ------------ $\to 4!$

H ------------ $\to 4!$

O------------ $\to 4!$

TG---------- $\to 3!$

TH --------- $\to 3!$

TOG ------- $\to 2!$

TOH ------- $\to 2!$

TOUGH $\to 2!$

Total = 89

32. **(3000)**

N should be divisible by 2 but not by 3

N = (Number divisible by 2) – (Numbers divisible by 6)

$$N = \frac{9000}{2} - \frac{9000}{6} = 4500 - 1500 = 3000$$

33. **(21)**

For number to be divisible by 15, the last digit should be 5 and the sum of the digits must be divisible by 3.

Possible combinations are

1	2	1	5

Numbers = 3

2	2	3	5

Numbers = 3

3	3	1	5

Numbers = 3

1	1	5	5

Numbers = 3

2	3	5	5

Numbers = 6

3	5	5	5

Numbers = 3

Total numbers = 21

34. **(C)**

$a \in \{2, 4, 6, 8, 10, \ldots, 100\}$

$b \in \{1, 3, 5, 7, 9, \ldots, 99\}$

Now, $a + b \in \{25, 71, 117, 163\}$

(i) $a + b = 25$, no. of ordered pairs (a, b) is 12

(ii) $a + b = 71$, no. of ordered pairs (a, b) is 35

(iii) $a + b = 117$, no. of ordered pairs (a, b) is 42

(iv) $a + b = 163$, no. of ordered pairs (a, b) is 19

$\therefore$ Total = 108 pairs

35. **(240)**

Digits are 1, 2, 2, 2, 3, 3, 5

If unit digit 5, then total numbers $= \dfrac{6!}{3!\,2!}$

If unit digit 3, then total numbers $= \dfrac{6!}{3!}$

If unit digit 1, then total numbers $= \dfrac{6!}{3!\,2!}$

$\therefore$ Total numbers $= 60 + 60 + 120 = 240$

36. **(710)**

$1000 - 2799$

Divisible by 3

$1002 + (n-1)3 = 2799$

$n = 600$

Divisible by 11

$$1 - 2799 \to \left[\frac{2799}{11}\right] = [254] = 254$$

$$1 - 999 = \left[\frac{999}{11}\right] = 90$$

$1000 - 2799 = 254 - 90 = 164$

Divisible by 33

$$1 - 2799 \to \left[\frac{2799}{33}\right] = 84$$

$$1 - 999 \to \left[\frac{999}{33}\right] = 30$$

$1000 - 2799 \to 54$

$\therefore n(3) + n(11) - n(33)$

$600 + 164 - 54 = 710$

37. **(2997)**

$2\underset{6}{+}\underset{6}{+}\underset{6}{+}\underset{6}{+}=1296$

$3\underset{6}{+}\underset{6}{+}\underset{6}{+}\underset{6}{+}=1296$

$40\underset{6}{+}\underset{6}{+}\underset{6}{+}\underset{6}{+}=216$

$420\underset{6}{+}\underset{6}{+}=36$

$4\,\underline{2}\,\underline{2}\underset{6}{+}\underset{6}{+}=36$

$423\underset{6}{+}\underset{6}{+}=36$

$424\underset{6}{+}\underset{6}{+}=36$

$427\underset{6}{+}\underset{6}{+}=36$

$429\,\underline{0}\underset{6}{+}=6$

$429\;2\;0=1$

$429\;2\;2=1$

$429\;2\;3=1$

$=2997$

38. **(45)**

$$\frac{(2n+1)!(n-1)!}{(n+2)!(2n-1)!}=\frac{11}{21}$$

$$\Rightarrow \frac{(2n+1)(2n)}{(n+2)(n+1)n}=\frac{11}{21}$$

$$\Rightarrow \frac{2n+1}{(n+1)(n+2)}=\frac{11}{42}$$

$\Rightarrow n=5$

$\Rightarrow n^2+n+15=25+5+15=45$

39. **(50400)**

Vowels: A, A, A, I, I, O

Consonants: S, S, S, S, N, N, T

$\because$ Total number of ways in which vowels come together

$$=\frac{|\underline{8}}{|\underline{4}\,|\underline{2}}\times\frac{|\underline{6}}{|\underline{3}\,|\underline{2}}=50400$$

40. **(105)**

$${}^{15}C_2=\frac{15\times 14}{2}=105$$

41. **(81)**

Taking single digit $\rightarrow$ 444444 $\frac{6!}{6!}=1$

Taking two digits $\rightarrow$

$(4,5)$ 444555 $\quad$ $(4,9)$ 444999

$\frac{5!}{3!\,2!}=10 \qquad \frac{5!}{3!\,2!}=10$

Taking three digits

$4,5,9,4,4,4=\frac{5!}{3!}=20$

$4,5,9,5,5,5\Rightarrow\frac{5!}{4!}=5$

$4,5,9,9,9,9\Rightarrow\frac{5!}{4!}=5$

$4,5,9,4,5,9\Rightarrow\frac{5!}{2!\,2!}=30$

Total = 81

42. **(171)**

Given,

We have to distribute 20 identical oranges to 3 children,

Let child A get O_A oranges, child B get O_B oranges and child C get O_C oranges,

So, $O_A+O_B+O_C=20$

Now this can be distributed using the multinomial theorem,

So, number of ways will be,

$=$ coefficient of x^{20} in $\left(x+x^2+\ldots+x^{18}\right)^3$

$=$ coefficient of x^{20} in $x^3\left(1+x+x^2+\ldots x^{17}\right)^3$

$=$ coefficient of x^{17} in $\left(\frac{1-x^{18}}{1-x}\right)^{-3}$

$=$ coefficient of x^{17} in $(1-x)^{-3}$

$={}^{19}C_2=171$

As we know that expansion of $(1-x)^{-3}$ is $1+{}^3C_1x+{}^4C_2x^2+{}^5C_3x^3+\ldots+{}^{19}C_{17}x^{17}\ldots\infty$

Note: This is bonus question as here 20 oranges are considered identical but in question it is considered distinct.

43. **(D)**

The given word is PUBLIC

Arranging the letters alphabetically, we get BCILPU

When the word starts with any of the letters B/C/I/L, the number of possibilities $=5!\times 4$ $=480$

When the word starts with PB, then the number of possibilities = 4! = 24
When the word starts with PC, then the number of possibilities = 4! = 24
When the word starts with PI, then the number of possibilities = 4! = 24
When the word starts with PL, then the number of possibilities = 4! = 24
When the word starts with PUBC, then the number of possibilities = 2! = 2
When the word starts with PUBI, then the number of possibilities = 2! = 2
When the word starts with PUBLC, then the number of possibilities = 1
When the word starts with PUBLIC, then the number of possibilities = 1
Rank = 480 + 24 × 4 + 2 × 2 + 1 × 2 = 582
Hence, rank of the work PUBLIC is 582.

44. **(432)**
Given,
We have to make four letter words from UNIVERSE each consisting of 2 vowels and 2 consonants. Now, in UNIVERSE, we have, E, E, I, U, (vowels) + N, R, S, V (consonants)
2 different vowels + 2 consonants can be arranged in $^3C_2 \cdot {^4C_2} \cdot 4! = 3 \times 6 \times 24 = 432$ ways
{$^3C_2$ is taken because of repetition of letter E}

45. **(A)**
There are 5 vowels in the given word which are 4E's and 1I.
Since they have to always occur together we take them as a single object E E E E I for the time being.
This single object together with 7 remaining objects will account for 8 objects.
There 8 objects in which there are 3N's and 2D's can be arranged in $\frac{8!}{3!\,2!}$ ways.
Corresponding to each of their arrangements the 5 vowels E, E, E, E and I which can be arranged in $\frac{5!}{4!}$ ways.
Hence, required number of arrangements
$= \frac{8!}{3!\,2!} \times \frac{5!}{4!} = 16800$
Hence this is the correct option.

46. **(B)**
Given,
7 boys and 5 girls are to be seated around a circular table such that no two girls are to be seated together. Now we know that n objects can be arranged in a circle in $(n-1)!$ ways.
Let us first arrange 7 boys in a circular arrangement in (7 – 1)! ways.
Now there will be 7 gaps.
So let us select any 5 gaps out of 7 gaps and arrange 5 girls in the chosen gaps. This can be done in $^7C_5 \times 5!$ ways.
Hence, required arrangements are
$6! \times {^7C_5} \times 5!$
$= 6 \times 5! \times \frac{7 \times 6}{2} \times 5!$
$= 126(5!)^2$.
Therefore, required arrangements are $126(5!)^2$

47. **(B)**
The word 'MATHEMATICS' consists of 11 letters including two M's, two T's and two A's,
Now, first we will arrange letters
_M_A_T _H_E_M_A_T _I_.
Here, we have 10 gaps to fill C and S.
The number of words with or without meaning can be formed from the word MATHEMATICS when C, S does not come together is
$\frac{9!}{2!\cdot 2!\cdot 2!} \times {^{10}C_2} \times 2!$
$= \frac{9 \times 8 \times 7 \times 6!}{2!\cdot 2!} \times \frac{10 \times 9}{2}$
= 5670
Hence this is the correct option.

48. **(4898)**
Given,
The number 1, 2, 3, 7
Now total numbers of 7-digit numbers will be 7! without repetition,
Now total numbers which contain string 154 will be 5!
Total numbers which contain string 2367 will be 4!
And number which contain both string 2367 and 154 will be 2

So, the total number which does not contain string 154 or 2367 will be 7! – (5! + 4! –2)
= 5040 – (120 + 24 –2)
= 5040 –142 = 4898
Hence this is the required answer.

49. **(16)**
Given,
Some couples participated in a mixed doubles badminton tournament. If the number of matches played, so that no couple played in a match, is 840,
Let the total number of persons be $2n$, so number of couples will be n
Now number of matches played, so that no couple played in a match is given by,
$\Rightarrow {}^nC_2\, {}^{n-2}C_2\, 2 = 840$
{as, if we choose a couple as $\{A, B\}$ and another couple as $\{C, D\}$ and taking the other two couples as $\{x, y\}$ and $\{w, z\}$, so the possible combination of the match where no couple played in a match will be, $\{A, x\} \rightarrow \{C, w\}$ and $\{C, x\} \rightarrow \{A, w\}$, hence there are two possible combinations after choosing}
$\Rightarrow n(n-1)(n-2)(n-3) = 5 \cdot 6 \cdot 7 \cdot 8$
So, $n = 8$
Hence, the number of persons will be, $2n = 16$

50. **(C)**
Let us find the different ways so that 8 persons can travel in 3 cars.

C_1	C_2	C_3
3	3	2
2	3	3
3	2	3

Hence, we have 3 ways.
Now the number of ways to distribute 8 persons such that they can travel in 3 cars with any car carrying maximum of 3 persons is $\left(\dfrac{8!}{3!\,3!\,2!}\right) \times 3$
= 1680
Therefore, the required answer is 1680

51. **(26664)**
Total 4-digit numbers using the digits 1, 2, 2, 3 is
$= \dfrac{4!}{2!} = 12$
Numbers are
----- 3
----- 3
----- 3
----- 2
----- 2
----- 2
----- 2
----- 2
----- 2
----- 1
----- 1
----- 1
Required sum is

$$= \underbrace{[3\times3+2\times6+1\times3]}_{\text{unit digit's sum}}$$
$$+\underbrace{[(3\times3+2\times6+1\times3)\times10]}_{\text{ten's place sum}}$$
$$+\underbrace{[(3\times3+2\times6+1\times3)\times100]}_{\text{hundred's place sum}}$$
$$+\underbrace{[(3\times3+2\times6+1\times3)\times1000]}_{\text{thousand's place sum}}$$
$$= [24]+[24\times10]+[24\times100]+[24\times1000]$$
$$= 24+240+2400+2400 = 26664$$

52. **(C)**
Given,
$x + y + z = 15$
Now we know that,
Non-negative integral solution of equation $a + b + c = n$ is given by ${}^{n+3-1}C_{3-1}$ where $a = b = c$ and $a = b \neq c$ are also possibilities
So by the above formula we get,
Total number of non-negative solutions will be, ${}^{15+3-1}C_{3-1} = {}^{17}C_2$
Now solving, if any of these 2 are equal.
So, the equation will become $x + 2y = 15$
Now finding possible cases we get,

$y = 0$ $x = 15$
$y = 1$ $x = 13$
$y = 2$ $x = 11$
$y = 3$ $x = 9$
$y = 4$ $x = 7$
$y = 5$ $x = 5 \rightarrow x = y = z = 5$
$y = 6$ $x = 3$
$y = 7$ $x = 1$

So, total possibilities where x, y and z are distinct will be,

$= {}^{17}C_2 - {}^{3}C_2 \times 8 + 2$

{**Note:** Adding 2 because the cases $x = y = z$ is subtracted three times}

$= 136 - 24 + 2 = 114$

53. **(44)**

5 boys with allotted roll numbers and seat numbers are seated in such a way that no one sits in the allotted seats. The number of ways of such seating arrangement is equal to the de-arrangement of 5 boys, that is,

$= 5!\left(1 - \frac{1}{1!} + \frac{1}{2!} - \frac{1}{3!} + \frac{1}{4!} - \frac{1}{5!}\right)$

$= 5! - 5! + 60 - 20 + 5 - 1 = 44$

54. **(A)**

The given word is MATHS

Arranging the letters alphabetically, we get AHMST

When the word starts with any of the letters A/H/M/S, the number of possibilities $= 4! \times 4 = 96$

Now when the word starts with TA, then the number of possibilities $= 3! = 6$

Now when the word starts with THAMS, then the number of possibilities = 1

Rank = 96 + 6 + 1 = 103

Hence, rank of the work THAMS is 103.

55. **(B)**

We need to find the number of 5-digit numbers greater than 40000 and divisible by 5.

Case I: Numbers start with 5.

$\underline{5}\ \underset{\downarrow}{\_}\ \underset{\downarrow}{\_}\ \underset{\downarrow}{\_}\ \underline{0} = 4 \times 3 \times 2 = 24$

Here if the number starts with 5 it should end with 0 to be divisible by 5.

Case II: Numbers start with 7 and the last digit can have 0 or 5.

$\underline{7}\ \underset{4}{\underset{\downarrow}{\_}}\ \underset{3}{\underset{\downarrow}{\_}}\ \underset{2}{\underset{\downarrow}{\_}}\ \underset{2}{\underset{\downarrow}{\_}} = 48$

Case III: Numbers start with 9 and the last digit should have 0 or 5.

$\underline{9}\ \underset{4}{\underset{\downarrow}{\_}}\ \underset{3}{\underset{\downarrow}{\_}}\ \underset{2}{\underset{\downarrow}{\_}}\ \underset{2}{\underset{\downarrow}{\_}} = 48$

Total numbers = 120.

Hence this is the correct option.

56. **(1260)**

Three numbers a, b, c are in A.P. and they are used for making a 9-digit number using each digit thrice such that at least three consecutive digits are in A.P., So, c, b, a will also be in A.P.

Now, we have 7 position to put (a, b, c) or (c, b, a) as $\underbrace{\_\ \_\ \_\ \_\ \_\ \_\ \_}_{\text{7 positions}}$

Number of such numbers is

$= {}^{2}C_1 \cdot {}^{7}C_1 \cdot \frac{6!}{2!\,2!\,2!}$

$= 1260$

57. **(413)**

The given numbers are 1, 2, 3, 4 and sum of digits should be 12.

Case 1: 4 3 1 1 1 1 1, Here the sum of the digits is 12.

The number of ways arranging them is $= \frac{7!}{5!}$

$= 42.$

Case 2: 4 2 2 1 1 1 1.

The number of ways arranging them is

$= \frac{7!}{2!\,4!} = 105$

Case 3: 3 2 2 2 1 1 1.

The number of ways arranging them is

$= \frac{7!}{3!\,3!} = 140$

Case 4: 3 3 2 1 1 1 1.

The number of ways arranging them is

$= \frac{7!}{2!\,4!} = 105$

Case 5: 2 2 2 2 2 1 1

The number of ways arranging them is

$= \frac{7!}{5!\,2!} = 21$

The total number of ways are = 42 + 105 + 140 + 105 + 21 = 413

Therefore, the required answer is 413.

58. **(A)**

The given word is MONDAY

Arranging the letters alphabetically, we get ADMNOY

When the word starts with any of the letters A/D, the number of possibilities

$= 5! \times 2 = 240$

When the word starts with MA, then the number of possibilities = 4! = 24
When the word starts with MD, then the number of possibilities = 4! = 24
When the word starts with MN, then the number of possibilities = 4! = 24
When the word starts with MOA, then the number of possibilities = 3! = 6
When the word starts with MOD, then the number of possibilities = 3! = 6
When the word starts with MONA, then the number of possibilities = 2! = 2
When the word starts with MONDAY, then the number of possibilities = 1
Rank = 240 + 24 × 3 + 6 × 2 + 2 + 1 = 327
Hence, rank of the word MONDAY is 327.

Shortcut Method

Another method is a shortcut, first number the letter and then check how many numbers are there on right hand side less that number and then start number in factorial in decreasing order and then multiply and add them and in last add one to find rank,

3	5	4	2	1	6
M	*O*	*N*	*D*	*A*	*Y*
2	3	2	1	0	0
5!	4!	3!	2!	1!	0!

$\therefore$ Rank = (2 × 5! + 3 × 4! + 2 × 3! + 1 × 2!)
= 240 + 72 + 12 + 2 + 1 = 327

59. **(16)**

A number is divisible by 6 when it is divisible by 2 and 3 both.
The three-digit number can be $\underline{a}\ \underline{b}\ \underline{2}$.
$a + b$ can be 4, 7, 10
If $a + b = 4$ then (a, b) can be (1, 3), (3, 1), (2, 2)
If $a + b = 7$ then (a, b) can be (2, 5), (5, 2), (3, 4), (4, 3)
If $a + b = 10$ then (a, b) can be (5, 5)
So, 8 such cases when 2 is at the units place
Similarly, there exist 8 such cases when 4 is at the units place.
Total = 16 cases.
Hence this is the required answer.

60. **(C)**

Given,
Three-digit number to be formed which are divisible by 3 using the digits 1, 3, 5, 8 and repetition is allowed,

Now taking case (1), where all the digits are same, we get
(1, 1, 1), (3, 3, 3), (5, 5, 5), (8, 8, 8) → 4 ways
Now taking case (2), where 2-digits are the same and one is distinct, we get
$(5, 5, 8) \to \frac{3!}{2!} = 3$ ways, (8, 8, 5) → 3 ways
Now taking case (3), where all are distinct, we get
(1, 3, 5) → 6 ways, (1, 8, 3) → 6 ways
So, total ways will be 4 + 3 + 3 + 6 + 6 = 22

61. **(72)**

Let the 4-digit pin be *abcd.*
Now given the sum of the first two is equal to the last two, so $a + b = c + d$
And the numbers available are 0, 1, 2, 3, 4, 5, 6, 7
Now, Let $a + b = c + d = \lambda$
Also the greatest digit used is 7
So, $a + b = c + d = \lambda \geq 7$
Now taking all cases starting from $\lambda = 7$ we get,
$\lambda = 7$, $(a, b) \to (0, 7)$ or $(7, 0)$, then $(c, d) \to (1, 6), (2, 5), (3, 4), (4, 3), (5, 2), (6, 1)$ and vice versa, so 2 × 12 = 24 numbers
$\lambda = 8$, $(a, b) \to (1, 7)$ or $(7, 1)$, then $(c, d) \to (2, 6), (3, 5), (6, 2), (5, 3)$ and vice versa so total 2 × 8 = 16 numbers
$\lambda = 9$, $(a, b) \to (2, 7)$ or $(7, 2)$then $(c, d) \to (3, 6), (4, 5), (5, 4), (6, 3)$ and vice versa, so total 2 × 8 = 16 numbers
$\lambda = 10$, $(a, b) \to (3, 7), (7, 3)$, so $(c, d) \to (6, 4), (4, 6) \to$ 8 numbers
$\lambda = 11$, $(a, b) \to (4, 7), (7, 4)$, so $(c, d) \to (6, 5), (5, 6) \to$ 8 numbers
$\lambda = 12$, 13 and 14 are not possible,
So, Total numbers = 24 + 16 + 16 + 8 + 8 = 72

62. **(A)**

Total ways to partition 5 into 4 parts are:
$5, 0, 0, 0 \Rightarrow 1$ way
$4, 1, 0, 0 \Rightarrow \frac{5!}{4!} = 5$ ways
$3, 2, 0, 0 \Rightarrow \frac{5!}{3!\,2!} = 10$ ways
$2, 2, 0, 1 \Rightarrow \frac{5!}{2!\,2!\,2!} = 15$ ways
$2, 1, 1, 1 \Rightarrow \frac{5!}{2!(1!)^3\,3!} = 10$ ways

$3,1,1,0 \Rightarrow \frac{5!}{3!\,2!} = 10$ ways

Total $\Rightarrow 1+5+10+15+10+10 = 51$ ways

63. (101)

$L_1, L_3, L_5, \ldots, L_{19}$ are Parallel

$L_2, L_4, L_6, \ldots, L_{20}$ are Concurrent

Total points of intersection =

${}^{20}C_2 - {}^{10}C_2 - {}^{10}C_2 + 1$

$= 101$

64. (C)

$$\alpha = \frac{(4!)}{(4!)^{3!}}, \beta = \frac{(5!)!}{(5!)^{4!}}$$

$$\alpha = \frac{(24)!}{(4!)^6}, \beta = \frac{(120)!}{(5!)^{24}}$$

Let 24 distinct objects be divided into 6 groups of 4 objects in each group.

No. of ways of formation of group

$$= \frac{24!}{(4!)^6 \cdot 6!} \in N$$

Similarly,

Let 120 distinct objects be divided into 24 groups of 5 objects in each group.

No. of ways of formation of groups

$$= \frac{(120)!}{(5!)^{24} \cdot 24!} \in N$$

65. (553)

Words starting with E = 360

Words starting with GE = 60

Words starting with GN = 60

Words starting with GTE = 24

Words starting with GTN = 24

Words starting with GTT = 24

GTWENTY = 1

Total = 553

66. (D)

3 shelves empty: $(8, 0, 0, 0) \rightarrow 1$ way

2 shelves empty: $\left.\begin{matrix}(7,1,0,0)\\(6,2,0,0)\\(5,3,0,0)\\(4,4,0,0)\end{matrix}\right] \rightarrow 4$ ways

1 shelf empty:

$\left.\begin{matrix}(6,1,1,0) & (3,3,2,0)\\(4,2,1,0) & (4,2,2,0)\end{matrix}\right] \rightarrow 5$ ways

0 shelves empty:

$\left.\begin{matrix}(1,2,3,2) & (5,1,1,1)\\(2,2,2,2) & \\(3,3,1,1) & \\(4,2,1,1) & \end{matrix}\right] \rightarrow 5$ ways

Total = 15 ways

67. (11376)

If 4 questions from each section are selected, the remaining 3 questions can be selected either in (1, 1, 1) or (3, 0, 0) or (2, 1, 0)

$\therefore$ Total ways =

$${}^8C_5 \cdot {}^6C_5 \cdot {}^6C_5 + {}^8C_6\,{}^6C_5 \cdot {}^6C_4 \times 2 + {}^8C_5 \cdot {}^6C_6 \cdot {}^6C_4 \times 2$$

$$+ {}^8C_4 \cdot {}^6C_6\,{}^6C_5 \times 2 + {}^8C_7 \cdot {}^6C_4 \cdot {}^6C_4$$

$$= 56 \cdot 6 \cdot 6 + 28 \cdot 6 \cdot 15 \cdot 2 + 56 \cdot 15 \cdot 2 + 70 \cdot 6 \cdot 2 + 8 \cdot 15 \cdot 15$$

$$= 2016 + 5040 + 1680 + 840 + 1800 = 11376$$

68. (3734)

We have III, TT, D, S, R, B, U, O, N

Number of words with selection (a, a, a, b)

$= {}^8C_1 \times \frac{4!}{3!} = 32$

Number of words with selection (a, a, b, b)

$= \frac{4!}{2!\,2!} = 6$

Number of words with selection (a, a, b, c)

$= {}^2C_1 \times {}^8C_2 \times \frac{4!}{2!} = 672$

Number of words with selection (a, b, c, d)

$= {}^9C_4 \times 4! = 3024$

$\therefore$ total = 3024 + 672 + 6 + 32

= 3734

69. (D)

After giving 2 apples to each child, 15 apples are left. Now, 15 apples can be distributed in ${}^{15+3-1}C_2 = {}^{17}C_2$ ways

$= \frac{17 \times 16}{2} = 136$

4.20 Summary

Permutation

- Permutation is an arrangement.
- Permutation follows the rule of order.
- The number of permutations of n different things taken r at a time is denoted by n_{P_r} and is defined as $n_{P_r} = \frac{n!}{(n-r)!}$
- The number of permutations of n different things taken all at a time = $n!$
- The number of permutations of n different things taken r at a time when repetition is allowed = n^r
- Multiplication rule and Addition rule have been explained in detail.
- $n_{P_r} = r\, n\text{-}1_{P_{r-1}} + n\text{-}1_{P_r}$
- The number of permutations of n things taken all at a time, if p things are identical of one type, q things are identical of second type, r things are identical of third type and rest are distinct $= \frac{n!}{p!\,q!\,r!}$.
- In a circular permutation if we consider clockwise and anti-clockwise arrangements different then n articles can be arranged in a circle in $(n-1)!$ ways.
- When clockwise and anti-clockwise arrangements are not different then number of arrangements $= \frac{1}{2}(n-1)!$

The detailed explanation, how to find sum of numbers formed by using the digit, test of divisibility, how to find rank of word has been mentioned in detail in theory.

Combination

- Combination is a selection.
- Combination does not follow the rule of order.
- The number of combinations of n different things taken r at a time is denoted by n_{c_r} and is defined as $n_{c_r} = \frac{n!}{r!(n-r)!}$
- Properties of combination
  - o $n_{c_r} = n_{c_{n-r}}$
  - o $n_{c_r} + n_{c_{r-1}} = n+1_{cr}$
  - o $n_{c_r} = \frac{n}{r}\, n\text{-}1_{c_{r-1}}$
  - o $\frac{n_{c_r}}{n_{c_{r-1}}} = \frac{(n-r+1)}{r}$
- The number of ways in which r objects can be selected from n distinct objects if k particular objects are

  i. Always included $= n-k_{c_{r-k}}$

  ii. Never included $= n-k_{c_r}$
- The total numbers of selections of at least one object out of $a_1 + a_2 + \ldots + a_n$ objects where a_1 are alike of first type, a_2 are alike of second type and so on and a_n are alike of nth type is $(a_1 + 1)(a_2 + 1) \ldots (a_n + 1) - 1$.
- Selection when both identical and distinct objects are present is also explained.
- The detailed explanation of divisors of a given natural number and how to find the sum of all divisors are given in theory.
- How to divide and distribute objects are explained in theory.
- Also in theory the concept of de-arrangement, multinational theorem, non-negative integral solution method and principle of inclusion and exclusion are explained in detail.
- Some geometrical concepts by using permutation and combination are explained in theory.

CHAPTER

5 Binomial Theorem

OBJECTIVES

5.1 Introduction
5.2 Properties of Binomial Theorem
5.3 Greatest Term
5.4 Properties of Binomial Coefficients
5.5 Know Your Concepts
5.6 JEE Main: Sample Questions
5.7 JEE Advanced Sample Questions
5.8 JEE Main: Archive
5.9 Answer Key
5.10 Solutions
5.11 Summary

5.1 Introduction

The *binomial theorem* is a powerful tool of expansion, which has application in Algebra, Probability, and so on. An expression such as $(a+b)$ is called a binomial. Expanding an expression such as $(a+b)^{15}$ is called a *binomial expansion.*

The binomial theorem is a formula for carrying out such an expansion without having to multiply it out. The basics of combinations, permutations and factorials are required to fully understand binomial coefficients. It starts from basic concepts, for example, binomial expansion, general term in the expansion, middle term, the greatest binomial coefficient to more fundamental concepts.

All concepts are supplemented with illustrations and assignments.

5.2 Properties of Binomial Theorem

An algebraic expression consisting of only two terms is called a binomial expression.

Binomial theorem for a positive integer

Since, $(a+b)^2 = a^2 + 2ab + b^2$

$$= 2c_0a^2 + 2c_1ab + 2c_2b^2$$

$$(a+b)^3 = 3c_0a^3 + 3c_1a^2b + 3c_2ab^2 + 3c_3b^3$$

$$(a+b)^4 = 4c_0a^4 + 4c_1a^3b + 4c_2a^2b^2 + 4c_3ab^3 + 4c_4b^4$$

$$(a+b)^n = nc_0a^n + nc_1a^{n-1}b + nc_2a^{n-2}b^2 + \ldots + nc_ra^{n-r}b^4 + \ldots + nc_nb^n$$

- $(r+1)$ th term is called a *general term* and it is denoted by $t_{r+1} = nc_ra^{n-r}b^r$
- The total number of terms in $(a+x)^n = (n+1)$.
- $(a+x)^n = \sum_{r=0}^{n} nc_ra^{n-r}x^r$
- $(a-x)^n = nc_0a^n - nc_1a^{n-1}x + nc_2a^{n-2}x^2 \ldots + (-1)^n nc_nx^n$
- $(1+x)^n = nc_0 + nc_1x + nc_2x^2 + \ldots + nc_nx^n$
- The kth term from the end in the expansion of $(x+y)^n = (n-k+2)^{\text{th}}$ terms from the beginning in $(x+y)^n$ or the kth term from the beginning in $(y+x)^n$.

5.2.1 Middle Term in the Expansion of $(a+x)^n$

In the expansion of $(a+x)^n$, the total number of terms $= (n+1)$.

If n is even, the total number of terms is odd. Suppose, $n = 6$, the total number of terms $= 7$.

For finding the middle term, add 1 and then divide it by 2, that is, $\frac{7+1}{2}$, that is, the 4th term will be the middle term. If $n = 5$, then total no. of terms $= 6$, then $\frac{6}{2}$, that is, the 3rd and 4th terms will be the middle term. In a binomial expansion, the middle term has greatest binomial coefficient.

5.3 Greatest Term

Let t_r and t_{r+1} be the rth and $(r + 1)^{th}$ terms in the expansion of $(a + x)^n$.

$$\text{Then } \frac{t_{r+1}}{t_r} = \frac{nc_r a^{n-r} x^r}{nc_{r-1} a^{n-r+1} x^{r-1}} = \left(\frac{n-r+1}{r}\right)\left(\frac{x}{a}\right)$$

Let, t_{r+1} be the numerically greatest term in the above expansion, then $t_{r+1} \geq t_r$.

$$\Rightarrow \left(\frac{n-r+1}{r}\right)\left|\frac{x}{a}\right| \geq 1$$

$$n - r + 1 \geq r\left|\frac{a}{x}\right|$$

$$(n+1) \geq r\left(1 + \left|\frac{a}{x}\right|\right)$$

$$\Rightarrow r \leq \frac{(n+1)}{1 + \left|\frac{a}{x}\right|}$$

5.4 Properties of Binomial Coefficients

(i) $^nc_r = {^nc_{n-r}} = \frac{n!}{r!(n-r)!}$

(ii) $^nc_{r-1} + {^nc_r} = {^{n+1}c_r}$

(iii) $^nc_r = \frac{n}{r}\,{^{n-1}c_{r-1}}$

(iv) $\frac{^nc_r}{^nc_{r-1}} = \frac{n-r+1}{r}$

5.4.1 Other Properties of the Binomial Coefficients

(i) $(1+x)^n = nc_0 + nc_1 x + nc_2 x^2 + \ldots + nc_n x^n$

Putting, $x = 1$, we get

$2^n = nc_0 + nc_1 + nc_2 + \ldots + nc_n$

or, $c_0 + c_1 + c_2 + \ldots + c_n = 2^n$

That is, the sum of binomial coefficient $= 2^n$

(ii) $(1+x)^n = c_0 + c_1 x + c_2 x^2 + \ldots + c_n x^n$

Put, $x = -1$, we get

$c_0 - c_1 + c_2 - c_3 + c_4 - c_5 + \ldots = 0$

(iii) $c_0 + c_1 + c_2 + c_3 + c_4 + c_5 + \ldots = 2^n$

$c_0 - c_1 + c_2 - c_3 + c_4 - c_5 + \ldots = 0$

Adding and subtracting respectively, we get

$c_0 + c_2 + c_4 + \ldots = 2^{n-1}$

$c_1 + c_3 + c_5 + \ldots = 2^{n-1}$

(iv) $(1 + x)^n = c_0 + c_1 x + c_2 x^2 + c_3 x^3 + \ldots + c_n x^n$

Differentiating with respect to x, we get

$n(1+x)^{n-1} = 1 \cdot c_1 + 2x \cdot c_2 + 3x^2 \cdot c_3 + \ldots + nx^{n-1} \cdot c_n$

Put $x = 1$, we get

$n \cdot 2^{n-1} = 1 \cdot c_1 + 2 \cdot c_2 + 3 \cdot c_3 + \ldots + n \cdot c_n$

Put $x = -1$, we get

$0 = 1 \cdot c_1 - 2 \cdot c_2 + 3 \cdot c_3 - \ldots$

(v) $(1+x)^n = c_0 + c_1 x + c_2 x^2 + \ldots + c_n x^n$

Put, $x = i$

$(1+i)^n = c_0 + c_1 i + c_2 i^2 + c_3 i^3 + c_4 i^4 + c_5 i^5 + \ldots$

$$\Rightarrow \left[\sqrt{2}\left(\frac{1}{\sqrt{2}} + i \cdot \frac{1}{\sqrt{2}}\right)\right]^n = (c_0 - c_2 + c_4 - \ldots) + i(c_1 - c_3 + c_5 - \ldots)$$

$$\Rightarrow 2^{n/2}\left(\cos\frac{n\pi}{4} + i\sin\frac{n\pi}{4}\right) = (C_0 - C_2 + C_4 - \ldots) + i(C_1 - C_3 + C_5 - \ldots)$$

Comparing real and imaginary parts from both sides, we get

$$c_0 - c_2 + c_4 - \ldots = 2^{n/2}\cos\frac{n\pi}{4}$$

$$c_1 - c_3 + c_5 - \ldots = 2^{n/2}\sin\frac{n\pi}{4}$$

(vi) $(1+x)^n = C_0 + C_1 x + C_2 x^2 + \text{--------} + C_n x^n$

$(x+1)^n = C_0 x^n + C_1 x^{n-1} + C_2 x^{n-2} + \text{--------}\ C_n$

Multiplying we get

$(1+x)^{2n} = (C_0 + C_1 x + C_2 x^2 + \text{--------} + C_n x^n)X$

$(C_0 x^n + C_1 x^{n-1} + C_2 x^{n-2} + \text{--------} + C_n)$

Comparing the coefficient of x^n from both sides, we get

$C_0^2 + C_1^2 + C_2^2 + \text{------} + C_n^2 = {^{2n}C_n}$

Comparing the coefficient of x^{n-1} from both sides, we get

$$c_0c_1 + c_1c_2 + \ldots + c_{n-1}c_n = {}^{2n}C_{n-1}$$

(vii) $(1+x)^n = c_0 + c_1x + c_2x^2 + \ldots + c_nx^n$

Integrating it between 0 and 1, we get

$$\int_0^1 (1+x)^n\,dx = c_0\int_0^1 dx + c_1\int_0^1 x\,dx + c_2\int_0^1 x^2dx + \ldots + c_n\int_0^1 x^n dx$$

$$\Rightarrow \left[\frac{(1+x)^{n+1}}{n+1}\right]_0^1 = c_0 + \frac{c_1}{2} + \frac{c_2}{3} + - - - - + \frac{c_n}{n+1}$$

$$\Rightarrow \frac{c_0}{1} + \frac{c_1}{2} + \frac{c_2}{3} + \ldots + \frac{c_n}{n+1} = \frac{2^{n+1}-1}{n+1}$$

Integrating between –1 and 0, we get

$$\int_{-1}^0 (1+x)^n dx$$

$$= c_0\int_{-1}^0 dx + c_1\int_{-1}^0 x\,dx + c_2\int_{-1}^0 x^2dx + \ldots + c_n\int_{-1}^0 x^n dx$$

$$\Rightarrow \left[\frac{(1+x)^{n+1}}{n+1}\right]_{-1}^0 = c_0[x]_{-1}^0 + c_1\left[\frac{x^2}{2}\right]_{-1}^0 + c_2\left[\frac{x^3}{3}\right]_{-1}^0 + \ldots + c_n\left[\frac{x^{n+1}}{n+1}\right]_{-1}^0$$

$$\Rightarrow \frac{1}{n+1} = c_0 - \frac{c_1}{2} + \frac{c_2}{3} - \frac{c_3}{4} + \ldots$$

SOLVED EXAMPLES

Example 5.1 Find the coefficient of x^{-7} in the expansion of $\left(2x - \frac{1}{3x^2}\right)^{11}$.

Solution $(r+1)^{\text{th}}$ term, that is,

$$t_{r+1} = {}^{11}c_r(2x)^{11-r}\left(-\frac{1}{3x^2}\right)^r$$

$$= {}^{11}c_r 2^{11-r}x^{11-r}\frac{(-1)^r}{3^r x^{2r}}$$

$$= {}^{11}c_r\frac{2^{11-r}}{3^r}(-1)^r x^{11-3r}$$

Put, $11 - 3r = -7$

$\Rightarrow 3r = 18$

$\Rightarrow r = 6$

Hence, coefficient of $x^{-7} = {}^{11}c_6\frac{2^5}{3^6}(-1)^6 = {}^{11}c_6\frac{2^5}{3^6}$

Example 5.2 If in the expansion of $\left(px + \frac{1}{x}\right)^n$, the 4th term is $\frac{5}{2}$, find n and p.

Solution From question, ${}^nc_3(px)^{n-3}\frac{1}{x^3} = \frac{5}{2}$

$$\Rightarrow {}^nc_3p^{n-3}x^{n-6} = \frac{5}{2}$$

Clearly the left-hand side should be independent of x, that is, $n - 6 = 0 \Rightarrow n = 6$

$$\Rightarrow {}^6c_3p^{6-3} = \frac{5}{2}$$

$$\Rightarrow \frac{6!}{3!\,3!}p^3 = \frac{5}{2}$$

$$\Rightarrow 20p^3 = \frac{5}{2}$$

$$\Rightarrow p^3 = \frac{1}{8}$$

$$\Rightarrow p = \frac{1}{2}$$

Hence, $p = \frac{1}{2}, n = 6$

Example 5.3 Find the 4th term from the end in the expansion of $\left(\frac{a^3}{2} - \frac{2}{a^2}\right)^7$.

Solution 4th term from the end = $(7 - 4 + 2)^{\text{th}}$ term from beginning, that is, 5th term from the beginning.

$$= {}^7c_4\left(\frac{a^3}{2}\right)^{7-4}\left(-\frac{2}{a^2}\right)^4$$

$$= {}^7c_4\frac{a^9}{2^3}\frac{2^4}{a^8} = 70a$$

Example 5.4 Find the total number of rational terms in the expansion of $\left(\sqrt[3]{4} + \frac{1}{\sqrt[4]{6}}\right)^{20}$.

Solution $t_{r+1} = {}^{20}c_r \left(4^{1/3}\right)^{20-r} \left(6^{-1/4}\right)^r$

$$= {}^{20}c_r\, 4^{(20-r)/3}\, 6^{-r/4}$$

$20 - r$ should be a multiple of 3. For this,

$$r = 2, 5, 8, 11, 14, 17, 20$$

Also, r should be a multiple of 4. For this,

$$r = 0, 4, 8, 12, 16, 20$$

Hence the total number of rational terms = 2

$\Rightarrow$ Total no. of irrational terms = 21 – 2 = 19

Example 5.5 Find the term independent of x in the expansion of $\left(\sqrt{\dfrac{x}{3}} + \dfrac{3}{2x^2}\right)^{10}$.

Solution $T_{r+1} = {}^{10}c_r \left(\dfrac{x}{3}\right)^{\frac{10-r}{2}} \left(\dfrac{3}{2x^2}\right)^r$

$$\Rightarrow T_{r+1} = {}^{10}c_r \frac{3^r}{3^{\frac{10-r}{2}}\, 2^r} x^{\frac{10-r}{2} - 2r}$$

$$= {}^{10}c_r \frac{3^r}{3^{\frac{10-r}{2}}\, 2^r} x^{\frac{10-5r}{2}}$$

Put $\dfrac{10-5r}{2} = 0$

$\Rightarrow r = 2$

Hence, the term independent of $x = {}^{10}c_2 \dfrac{3^2}{3^4 2^2}$

$$= {}^{10}c_2 \frac{1}{36}$$

Example 5.6 Find the value of k such that the term independent of x in $\left(\sqrt{x} + \dfrac{k}{x^2}\right)^{10}$ is 405.

Solution $t_{r+1} = {}^{10}c_r \left(\sqrt{x}\right)^{10-r} \left(\dfrac{k}{x^2}\right)^r$

$$= {}^{10}c_r k^r x^{\frac{10-r}{2}} x^{-2r}$$

$$= {}^{10}c_r k^r x^{\frac{10-5r}{2}}$$

Put $\dfrac{10-5r}{2} = 0$

$\Rightarrow r = 2$

$\Rightarrow$ term independent of $x = {}^{10}C_2 k^2$

From question, ${}^{10}C_2 k^2 = 405$

$$\Rightarrow \frac{10!}{2!8!} k^2 = 405$$

$$\Rightarrow 45k^2 = 405$$

$$\Rightarrow k^2 = \frac{405}{45} = 9$$

$$\Rightarrow k = \pm 3$$

Example 5.7 Find the coefficient of $\dfrac{1}{x}$ in $\left(1 + 3x^2 + x^4\right)\left(1 + \dfrac{1}{x}\right)^8$.

Solution $(r+1)^{\text{th}}$ term in $\left(1 + \dfrac{1}{x}\right)^8 = {}^8c_r \left(\dfrac{1}{x}\right)^r = {}^8c_r x^{-r}$

Now the $(r + 1)^{\text{th}}$ term will be

$${}^8c_r x^{-r} + 3\,{}^8c_r x^{2-r} + {}^8c_r x^{4-r}$$

For the first term, $r = 1$,
For the second term, $2 - r = -1 \Rightarrow r = 3$
For the 3rd term, $4 - r = -1 \Rightarrow r = 5$

Hence coefficient of $\dfrac{1}{x} = {}^8c_1 + 3\,{}^8c_3 + {}^8c_5 = 232$

Example 5.8 If a_r denotes the coefficient of x^r in $(1 - x)^{2n-1}$, find the value of $a_{r-1} + a_{2n-r}$

Solution $(r + 1)^{\text{th}}$ term in $(1 - x)^{2n-1} = {}^{2n-1}c_r (-x)^r$

$= {}^{2n-1}c_r (-1)^r x^r$

From the question, $a_r = {}^{2n-1}c_r (-1)^r$

$$a_{r-1} = {}^{2n-1}c_{r-1} (-1)^{r-1}$$

$$a_{2n-r} = {}^{2n-1}c_{2n-r} (-1)^{2n-r}$$

$$= {}^{2n-1}c_{r-1} (-1)^{-r}$$

$$a_{r-1} + a_{2n-r} = {}^{2n-1}c_{r-1}\left[(-1)^{r-1} + (-1)^{-r}\right] = 0$$

Example 5.9 Find the numerically greatest term in the expansion of $(4 - 3x)^7$, when $x = \dfrac{2}{3}$.

Solution Let $(r + 1)^{\text{th}}$ term be the numerically greatest term.

$$\Rightarrow |t_{r+1}| \geq |t_r|$$

$$\Rightarrow {}^7c_r\, 4^{7-r} |-3x|^r \geq {}^7c_{r-1}\, 4^{7-r+1} |-3x|^{r-1}$$

$$\Rightarrow {}^7c_r\, 4^{7-r}\, 3^r \left(\frac{2}{3}\right)^r \geq {}^7c_{r-1}\, 4^{8-r}\, 3^{r-1} \left(\frac{2}{3}\right)^{r-1}$$

$\Rightarrow \dfrac{7!}{r!(7-r)!}2^r \geq \dfrac{7!}{(r-1)!(8-r)!}4\,2^{r-1}$

$\Rightarrow \dfrac{2}{r} \geq \dfrac{1}{8-r}\cdot 4$

$\Rightarrow 2(8-r) \geq 4r$

$\Rightarrow 8 - r \geq 2r$

$\Rightarrow 3r \leq 8$

$\Rightarrow r \leq \dfrac{8}{3}$

$\Rightarrow$ greatest integral value of $r = 2$.

Hence the 3rd term will be numerically greatest term

$\Rightarrow |t_3| = {}^7c_2\, 4^{7-2}\, |3x|^2$

$= \dfrac{7!}{2!\,5!}\,4^5\,2^2$

$= \dfrac{6.7}{2}\times 4^5 \times 4$

$= 21 \times 4^6 = 86016$

Example 5.10 Find the greatest coefficient in the expansion of $(1 + x)^{2n}$.

Solution As we know, in a binomial expansion, the middle term has the greatest binomial coefficient.

Here, the total number of terms $= 2n + 1$

$\Rightarrow \dfrac{2n+1+1}{2}$, that is, $(n + 1)^{\text{th}}$ term will be the middle term

$\Rightarrow$ Coefficient of the $(n + 1)^{\text{th}}$ term in $(1 + x)^{2n} = {}^{2n}c_n$

Note: Multinomial Theorem

As

(i) $(a+x)^n = \sum_{r=0}^{n} {}^nc_r\, a^{n-r}\, x^r,\ n \in N$

$= \sum_{r=0}^{n} \dfrac{n!}{(n-r)!r!} a^{n-r} x^r$

$= \sum_{r+s=n} \dfrac{n!}{r!\,s!} a^s x^r$ where $s = n - r$

Total no. of terms $= {}^{n+2-1}c_{2-1}$

$= {}^{n+1}c_1 = (n+1)$

(ii) $(x+y+z)^n = \sum_{r+s+t=n} \dfrac{n!}{r!\,s!\,t!} x^r y^s z^t$

The given expansion has ${}^{n+3-1}c_{3-1}$ terms, that is, ${}^{n+2}c_2$ terms

(iii) $(x+y+z+w)^n$

$= \sum_{p+q+r+s=n} \dfrac{n!}{p!\,q!\,r!\,s!} x^p y^q z^r w^s$

The total no. of terms $= {}^{n+4-1}c_{4-1}$, that is, ${}^{n+3}c_3$ terms

Generalising this, we get,

$(x_1 + x_2 + \text{--------} + x_k)^n$

$= \sum_{r_1+r_2+\text{---}+r_k=n} \dfrac{n!\, x_1^{r_1} x_2^{r_2} \ldots x_k^{r_k}}{r_1!\, r_2! \ldots r_k!}$

The total no. of terms $= {}^{n+k-1}c_{k-1}$

The general term in the above expansion is

$\dfrac{n!}{r_1!\,r_2!\ldots r_k!} x_1^{r_1} x_2^{r_2} \ldots x_k^{r_k}$

Example 5.11 Find the coefficient of $x^3y^4z^2$ in the expansion of $(2x-3y+4z)^9$.

Solution The general term in the expansion of

$(2x - 3y + 4z)^9 = \dfrac{9!}{\alpha_1!\,\alpha_2!\,\alpha_3!}(2x)^{\alpha_1}(-3y)^{\alpha_2}(4z)^{\alpha_3}$

$= \dfrac{9!}{\alpha_1!\,\alpha_2!\,\alpha_3!} 2^{\alpha_1}(-3)^{\alpha_2} 4^{\alpha_3} x^{\alpha_1} y^{\alpha_2} z^{\alpha_3}$

Put $\alpha_1 = 3, \alpha_2 = 4, \alpha_3 = 2$

Coefficient of $x^3y^4z^2 = \dfrac{9!}{3!\,4!\,2!}\,2^3\,3^4\,4^2$

Example 5.12 Find the term independent of x in the expansion of $\left(\dfrac{x+1}{x^{2/3}-x^{1/3}+1} - \dfrac{x-1}{x-x^{1/2}}\right)^{10}$.

Solution The given expansion can be written as

$\left[\dfrac{(x^{1/3})^3+1^3}{x^{2/3}-x^{1/3}+1} - \dfrac{(\sqrt{x}+1)(\sqrt{x}-1)}{\sqrt{x}(\sqrt{x}-1)}\right]^{10}$

$= \left[(x^{1/3}+1) - \left(1+\dfrac{1}{\sqrt{x}}\right)\right]^{10}$

$= (x^{1/3} - x^{-1/2})^{10}$

Now, $t_{r+1} = {}^{10}c_r (x^{1/3})^{10-r}(-x^{-1/2})^r$

$= {}^{10}c_r x^{\frac{10-r}{3}} (-1)^r x^{-r/2}$

$= {}^{10}c_r (-1)^r x^{\frac{20-2r-3r}{6}}$

$= {}^{10}c_r (-1)^r x^{\frac{20-5r}{6}}$

Put, $\dfrac{20-5r}{6} = 0 \Rightarrow r = 4$

That is, the 5th term will be independent of x.

Now, $T_5 = {}^{10}c_4(-1)^4 = {}^{10}c_4$

Example 5.13 If $(2+\sqrt{3})^n = I + f$, where I and n are the positive integers and $0 < f < 1$, show that I is an odd integer and $(1-f)(I+f) = 1$.

Solution Since,

$$I + f = (2+\sqrt{3})^n = 2^n + {}^nc_1 2^{n-1}\sqrt{3} + {}^nc_2 2^{n-2}(\sqrt{3})^2 + \ldots \quad (1)$$

Let $f_1 = (2-\sqrt{3})^n$ where $0 < f_1 < 1$

Now, $f_1 = 2^n - {}^nc_1 2^{n-1}\sqrt{3} + {}^nc_2 2^{n-2}(\sqrt{3})^2 - \ldots \quad (2)$

Adding (1) and (2), we get

$$I + f + f_1 = 2\cdot[2^n + {}^nc_2 2^{n-2}(\sqrt{3})^2 + \ldots]$$

$\Rightarrow I + f + f_1 = \text{even integer} \quad (3)$

$\Rightarrow f + f_1 = \text{even integer} - I = \text{an integer} \quad (4)$

Now, $0 < f < 1$

$0 < f_1 < 1$

$0 < f + f_1 < 2$

$\Rightarrow f + f_1 = 1$

Hence from (3), I is an odd integer.

Now $(1-f)(I+f)$

$= f_1(I+f)$ [As, $f + f_1 = 1 \Rightarrow f_1 = 1 - f$]

$= (2-\sqrt{3})^n(2+\sqrt{3})^n = 1$

Example 5.14 Find the sum of the rational terms in the expansion of $(\sqrt{2} + 3^{1/5})^{10}$.

Solution $t_{r+1} = {}^{10}c_r 2^{\frac{10-r}{2}} 3^{r/5}$

Terms should be free from radicals. Hence, $10-r$ should be a multiple of 2.

$\Rightarrow r = 0, 2, 4, 6, 8, 10$

Also, r should be a multiple of 5

$\Rightarrow r = 0, 5, 10$

Hence, $r = 0, 10$

Sum of rational terms $= {}^{10}c_0\, 2^5\, 3^0 + {}^{10}c_{10}\, 2^0\, 3^2$

$= 32 + 9 = 41$

Example 5.15 If the number is 17^{256}, find the

(i) last digit **(ii)** last two digits **(iii)** last three digits

Solution $17^{256} = (17^2)^{128} = (289)^{128} = (290-1)^{128}$

$= {}^{128}c_0(290)^{128} - {}^{128}c_1(290)^{127} + \ldots - {}^{128}c_{127}(290) + 1$

(i) For last digit

$17^{256} = 290(k) + 1$

$\Rightarrow$ Last digit $= 1$

(ii) For last two digits

$= (290)^2[\,] - 128c_{127}(290) + 1$

$= 100m - {}^{128}c_1(290) + 1$

$= 100m - 128(300-10) + 1$

$= 100(m-384) + 1281$

$\Rightarrow$ Last two digits $= 00 + 81 = 81$

(iii) For last three digits

$= 1000m + {}^{128}c_2(290)^2 - {}^{128}c_1(290) + 1$

$= 1000m + 128 \times 127 \times 290 \times 145 - 128 \times 290 + 1$

$= 1000m + (128)(290)(127 \times 145 - 1) + 1$

$= 1000m + 683527680 + 1$

$= {}^{100}0\, m + 683427000 + 680 + 1$

$= 1000(m + 683527) + 681$

$\Rightarrow$ Last three digits $= 681$

Example 5.16 If the sum of the coefficients in the expansion of $(\alpha x^2 - 2x + 1)^{35}$ is equal to the sum of the coefficients in the expansion of $(x - \alpha y)^{35}$, find the value of α.

Solution Since, the sum of coefficients in $(\alpha x^2 - 2x + 1)^{35}$ = sum of coefficients in $(x - \alpha y)^{35}$

Put $x = y = 1$

$\Rightarrow (\alpha - 1)^{35} = (1-\alpha)^{35}$

$\Rightarrow 2(\alpha - 1)^{35} = 0$

$\Rightarrow \alpha = 1$

Example 5.17 If $(1 + x - 2x^2)^{20} = \sum_{r=0}^{40} a_r x^r$, then find the value of $a_1 + a_3 + a_5 + \ldots + a_{39}$.

Solution Since, $(1 + x - 2x^2)^{20}$

$= a_0 + a_1x + a_2x^2 + a_3x^3 + \ldots + a_{39}x^{39} + a_{40}x^{40}$

Put, $x = 1$, we get

$$a_0 + a_1 + a_2 + a_3 + \ldots + a_{39} + a_{40} = 0 \quad (1)$$

Again, put $x=-1$, we get

$$a_0-a_1+a_2-a_3+\ldots-a_{39}+a_{40}=2^{20} \quad (2)$$

Subtracting (2) from (1), we get

$$2[a_1+a_3+a_5+\ldots+a_{39}]=-2^{20}$$

$$\Rightarrow a_1+a_3+a_5+\ldots+a_{39}=-2^{19}$$

Example 5.18 If $(1+x)^n=c_0+c_1x+c_2x^2+\ldots+c_nx^n$, then find the value of $c_0+3c_1+5c_2+\ldots+(2n+1)c_n$.

Solution $c_0+3c_1+5c_2+\ldots+(2n+1)c_n$

$$=(c_0+c_1+c_2+\ldots+c_n)+2(1.c_1+2.c_2+\ldots+n.c_n)$$

$$=2^n+2n2^{n-1}=(n+1)2^n$$

Example 5.19 If $(1+x)^n=c_0+c_1x+c_2x^2+\ldots+c_nx^n$, then find the value of $3c_0+3^2\frac{c_1}{2}+3^3\frac{c_2}{3}+3^4\frac{c_3}{4}+\ldots+3^{n+1}\frac{c_n}{n+1}$

Solution $(1+x)^n=c_0+c_1x+c_2x^2+\ldots+c_nx^n$

Integrating within the limits 0 to 3, we get

$$\int_0^3(1+x)^n dx=c_0\int_0^3 dx+c_1\int_0^3 x dx+c_2\int_0^3 x^2dx+\ldots+c_n\int_0^3 x^n dx$$

$$\Rightarrow\left[\frac{(1+x)^{n+1}}{n+1}\right]_0^3=3\cdot c_0+3^2\frac{c_1}{2}+3^3\frac{c_2}{3}+3^4\frac{c_3}{4}+\ldots+3^{n+1}\frac{c_n}{n+1}$$

$$\Rightarrow 3c_0+3^2\frac{c_1}{2}+3^3\frac{c_2}{3}+3^4\frac{c_3}{4}+\ldots+3^{n+1}\frac{c_n}{n+1}=\frac{4^{n+1}-1}{n+1}$$

Example 5.20 If n is a positive integer, then find the value of $\sum_{k=1}^{n}k^3\left(\frac{{}^nc_k}{{}^nc_{k-1}}\right)^2$.

Solution $\sum_{k=1}^{n}k^3\left(\frac{{}^nc_k}{{}^nc_{k-1}}\right)^2=\sum_{k=1}^{n}k^3\left(\frac{n-k+1}{k}\right)^2$

$$=\sum_{k=1}^{n}k((n+1)-k)^2$$

$$=(n+1)^2\sum_{k=1}^{n}k-2(n+1)\sum_{k=1}^{n}k^2+\sum_{k=1}^{n}k^3$$

$$=(n+1)^2\frac{n(n+1)}{2}-2(n+1)\frac{n(n+1)(2n+1)}{6}+\frac{n^2(n+1)^2}{4}$$

$$=\frac{n(n+1)^2}{2}\left[(n+1)-2\frac{(2n+1)}{3}+\frac{n}{2}\right]$$

$$=\frac{n(n+1)^2}{2}\left[\frac{6n+6-8n-4+3n}{6}\right]$$

$$=\frac{1}{12}n(n+1)^2(n+2)$$

Example 5.21 Prove that

$$({}^nc_0+{}^nc_1)({}^nc_1+{}^nc_2)({}^nc_2+{}^nc_3)\ldots({}^nc_{n-1}+{}^nc_n)=\frac{(n+1)^n}{n!}{}^nc_1{}^nc_2{}^nc_3\ldots{}^nc_n.$$

Solution

$$({}^nc_0+{}^nc_1)({}^nc_1+{}^nc_2)({}^nc_2+{}^nc_3)\ldots({}^nc_{n-1}+{}^nc_n)$$

$$={}^nc_0\left(1+\frac{{}^nc_1}{{}^nc_0}\right)\cdot{}^nc_1\left(1+\frac{{}^nc_2}{{}^nc_1}\right)\cdot{}^nc_2\left(1+\frac{{}^nc_3}{{}^nc_2}\right)\ldots{}^nc_{n-1}\left(1+\frac{{}^nc_n}{{}^nc_{n-1}}\right)$$

$$=\left(1+\frac{n}{1}\right)\left(1+\frac{n-1}{2}\right)\left(1+\frac{n-2}{3}\right)\ldots\left(1+\frac{1}{n}\right){}^nc_0{}^nc_1{}^nc_2\ldots{}^nc_{n-1}$$

$$=\frac{(n+1)}{1}\frac{(n+1)}{2}\frac{(n+1)}{3}\ldots\frac{(n+1)}{n}{}^nc_1{}^nc_2{}^nc_3---{}^nc_{n-1}{}^nc_n$$

$$=\frac{(n+1)^n}{n!}{}^nc_1{}^nc_2{}^nc_3\ldots{}^nc_n$$

Example 5.22 If a, b, c are three consecutive coefficients in the expansion of $(1+x)^n$, then prove that $n=\frac{2ac+b(a+c)}{b^2-ac}$.

Solution Let three consecutive terms be r, $(r+1)$ and $(r+2)$, respectively.

$$t_r={}^nc_{r-1}=a$$

$$t_{r+1}={}^nc_r=b$$

$$t_{r+2}={}^nc_{r+1}=c$$

$$\frac{b}{a}=\frac{n-r+1}{r}$$

$\Rightarrow br = (n + 1)\, a - ra$

$$\Rightarrow r = \frac{(n+1)a}{a+b} \qquad (1)$$

Also,
$$\frac{c}{b} = \frac{\dfrac{n!}{(r+1)!(n-r-1)!}}{\dfrac{n!}{r!(n-r)!}}$$

$$\Rightarrow \frac{c}{b} = \frac{n-r}{r+1}$$

$\Rightarrow cr + c = bn - br$

$\Rightarrow (b + c)\, r = (bn - c)$

$$\Rightarrow r = \frac{bn-c}{b+c} \qquad (2)$$

From (1)

$$\frac{(n+1)a}{a+b} = \frac{bn-c}{b+c}$$

$\Rightarrow (na + a)\,(b + c) = bn\,(a + b) - c\,(a + b)$

$\Rightarrow n\,(ab + ac) + ab + ac = n\,(ab + b^2) - (ac + bc)$

$\Rightarrow n\,(b^2 - ac) = 2ac + ab + bc$

$$\Rightarrow n = \frac{2ac + b(a+c)}{(b^2 - ac)}$$

Example 5.23 Prove that

$$^{n}c_0\,{}^{n}c_0 - {}^{n+1}c_1\,{}^{n}c_1 + {}^{n+2}c_2\,{}^{n}c_2 - \ldots = (-1)^n$$

Solution $^{n}c_0\,{}^{n}c_0 - {}^{n+1}c_1\,{}^{n}c_1 + {}^{n+2}c_2\,{}^{n}c_2 - \ldots$

$= {}^{n}c_n\,{}^{n}c_0 - {}^{n+1}c_n\,{}^{n}c_1 + {}^{n+2}c_n\,{}^{n}c_2 - \ldots \;[{}^{n}c_r = {}^{n}c_{n-r}]$

= coefficient of x^n in

$$\left[(1+x)^n \times {}^{n}c_0 - (1+x)^{n+1} \times {}^{n}c_1 + (1+x)^{n+2} \times {}^{n}c_2 - \ldots\right]$$

= coefficient of x^n in $(1+x)^n\,[{}^{n}c_0 - {}^{n}c_1\,(1+x) + {}^{n}c_2\,(1+x)^2 - \ldots]$

= coefficient of x^n in $(1 + x)^n\,[1 - (1 + x)]^n$

= coefficient of x^n in $(-1)^n\, x^n(1 + x)^n$

$= (-1)^n$

Example 5.24 If $(1+x)^n = c_0 + c_1x + c_2x^2 + \ldots + c_nx^n$, then find the value of $3c_0 - 8c_1 + 13c_2 - 18c_3 + \ldots$ up to $(n + 1)$ terms.

Solution $3c_0 - 8c_1 + 13c_2 - 18c_3 + \ldots$ up to $(n + 1)$ terms

$= 3(c_0 - c_1 + c_2 - c_3 + \ldots) - 5\,(1 \times c_1 - 2 \times c_2 + 3 \times c_3 - \ldots)$

$= 3 \times 0 - 5 \times 0 = 0$

Example 5.25 Let $R = (5\sqrt{5} + 11)^{2n+1}$ and $f = R - [R]$ where [ ] denotes the greatest integer function. Prove that $Rf = 4^{2+1}$.

Solution Let I denote the integral part of R, then $R = I + f$

$$I + f = (5\sqrt{5} + 11)^{2n+1} \text{ where } 0 < f < 1$$

If $\quad f' = (5\sqrt{5} - 11)^{2n+1}$, then $0 < f' < 1$

Now, $\quad I + f - f' = (5\sqrt{5} + 11)^{2n+1} - (5\sqrt{5} - 11)^{2n+1}$

$= 2\left[{}^{2n}c_1 (5\sqrt{5})^{2n} \cdot 11 + {}^{2n+1}c_3 (5\sqrt{5})^{2n-2} \cdot (11)^3 + \ldots\right]$

= An even integer

$\Rightarrow f - f'$ must also be an integer

$\Rightarrow f - f' = 0 \; [0 < f < 1, 0 < f' < 1]$

$\Rightarrow f = f'$

$Rf = Rf'$

$= (5\sqrt{5} + 11)^{2n+1} (5\sqrt{5} - 11)^{2n+1}$

$= (125 - 121)^{2n+1}$

$= 4^{2n+1}$

Example 5.26 Find the larger of $99^{50} + 100^{50}$ and 101^{50}.

Solution Since, $(101)^{50} = (100 + 1)^{50}$

$$= 100^{50} + 50 \times 100^{49} + \frac{50 \times 49}{1 \times 2} 100^{48} + \ldots$$

And, $99^{50} = (100 - 1)^{50}$

$$= 100^{50} - 50 \times 100^{49} + \frac{50 \times 49}{1 \times 2} 100^{48} - \ldots$$

Subtracting, we get

$$(101)^{50} - 99^{50} = 2\left[50 \times 100^{49} + \frac{50 \times 49 \times 48}{1 \times 2 \times 3} 100^{47} + \ldots\right]$$

$$= 100^{50} + 2\frac{50 \times 49 \times 48}{1 \times 2 \times 3} 100^{47} + \ldots$$

$> 100^{50}$

$\Rightarrow (101)^{50} > 99^{50} + 100^{50}$

Example 5.27 Using the binomial theorem, prove that $3^{2n+2} - 8n - 9$ is divisible by 64, $\forall n \in N$.

Solution $3^{2n+2} - 8n - 9 = 9(1+8)^n - 8n - 9$

$= 9[1 + {}^{n}c_1 \times 8 + {}^{n}c_2 8^2 + \ldots + 8^n) - 8n - 9$

$= 64n + 9({}^{n}c_2 \times 8^2 + {}^{n}c_3 8^3 + \ldots + 8^n)$

$= 64k$, where k is an integer.

Hence, $3^{2n+2} - 8n - 9$ is divisible by 64.

Example 5.28 Prove that

$$\frac{{}^nc_1 \sin 2x + {}^nc_2 \sin 4x + {}^nc_3 \sin 6x + \ldots}{1 + {}^nc_1 \cos 2x + {}^nc_2 \cos 4x + {}^nc_3 \cos 6x + \ldots} = \tan nx, \forall n \in N.$$

Solution The given expression can be written as

$$\frac{\sum_{r=0}^{n} {}^nc_r \sin 2rx}{\sum_{r=0}^{n} {}^nc_r \cos 2rx} = \frac{\sum_{r=0}^{n} {}^nc_r \sin(2n-2r)x}{\sum_{r=0}^{n} {}^nc_r \cos(2n-2r)x}$$

$$= \frac{\sum_{r=0}^{n} {}^nc_r \sin 2rx + \sum_{r=0}^{n} {}^nc_r \sin(2n-2r)x}{\sum_{r=0}^{n} {}^nc_r \cos 2rx + \sum_{r=0}^{n} {}^nc_r \cos(2n-2r)x}$$

We know, $\left[\frac{a}{b} = \frac{c}{d} = \frac{a+c}{b+d}\right]$

$$= \frac{\sum_{r=0}^{n} {}^nc_r \left[\sin 2rx + \sin(2n-2r)x\right]}{\sum_{r=0}^{n} {}^nc_r \left[\cos 2rx + \cos(2n-2r)x\right]}$$

$$= \frac{\sum_{r=0}^{n} {}^nc_r \left[2\sin\frac{2rx + 2nx - 2rx}{2} \cos\frac{2rx - 2nx + 2rx}{2}\right]}{\sum_{r=0}^{n} {}^nc_r \left[2\cos\frac{2rx + 2nx - 2rx}{2} \cos\frac{2rx - 2nx + 2rx}{2}\right]}$$

$$= \frac{\sin nx \sum_{r=0}^{n} {}^nc_r \cos(2r-n)x}{\cos nx \sum_{r=0}^{n} {}^nc_r \cos(2r-n)x} = \tan nx$$

Example 5.29 Find the value of $\sum_{i=0}^{n}\sum_{J=1}^{n} {}^nc_J\, {}^Jc_i$.

Solution $\sum_{i=0}^{n}\sum_{J=1}^{n} {}^nc_J\, {}^Jc_i = {}^nc_1\ (1c_0 + 1c_1) + {}^nc_2\ (2c_0 + 2c_1 + 2c_2) + {}^nc_3\ (3c_0 + 3c_1 + 3c_2 + 3c_3) + \ldots + {}^nc_n\ ({}^nc_0 + {}^nc_1 + \ldots + {}^nc_n)$

$$= {}^nc_1 \cdot 2^1 + {}^nc_2 \cdot 2^2 + {}^nc_3 \cdot 2^3 + \ldots + {}^nc_n \cdot 2^n$$
$$= (1 + 2)^n - 1 = 3^n - 1$$

Example 5.30 Find the coefficient of x^n in the expansion of $\left(1 + \frac{x}{1!} + \frac{x^2}{2!} + \frac{x^3}{3!} + \ldots + \frac{x^n}{n!}\right)^2$.

Solution Since coefficient of x^n in

$$\left(1 + \frac{x}{1!} + \frac{x^2}{2!} + \frac{x^3}{3!} + \ldots + \frac{x^n}{n!}\right)^2$$

= coefficient of x^n in

$$\left(1 + \frac{x}{1!} + \frac{x^2}{2!} + \frac{x^3}{3!} + \ldots + \frac{x^n}{n!} + \frac{x^{n+1}}{(n+1)!} + \ldots + \infty\right)^2$$

Coefficient of x^n in $(e^x)^2$ = coefficient of x^n in e^{2x}

Coefficient of x^n in

$$1 + (2x) + \frac{(2x)^2}{2!} + \frac{(2x)^3}{3!} + \ldots + \frac{(2x)^n}{n!} + \ldots + \infty = \frac{2^n}{n!}$$

5.5 Know Your Concepts

1. (i) Find the 4th term in the expansion of $(x - 2y)^{12}$.

(ii) If the third term in the expansion of $\left(\frac{1}{x} + x^{\log_{10} x}\right)^5$ is 1000 where $x > 1$, then find the value of x.

(iii) In the binomial expansion of $(a - b)^n$, $n \geq 5$, the sum of 5th and 6th term is zero, then find the value of $\frac{a}{b}$.

(iv) If the 17th and 18th terms of the expansion $(2 + a)^{50}$ are equal, find the value of a.

2. (i) Find the coefficient of $\frac{1}{y^2}$ in $\left(y + \frac{c^3}{y^2}\right)^{10}$.

(ii) Find the term independent of x in $(1+x)^m\left(1+\frac{1}{x}\right)^n$.

(iii) If in the expansion of $(1+x)^m(1-x)^n$, the coefficient of x and x^2 are 3 and –6 respectively, find the value of m.

(iv) The sum of the coefficients of the first three terms of the expansion $\left(x - \frac{3}{x^2}\right)^m$, $x \neq 0$, m being a natural number is 559. Find the term of the expansion containing x^3.

3. (i) Find the middle term in the expansion of $\left(\frac{x}{3} + 9y\right)^{10}$.

(ii) Find the middle term in the expansion of $\left(2x^2 - \frac{1}{x}\right)^7$.

(iii) Prove that the middle term in the expansion of $\left(x+\frac{1}{x}\right)^{2n}$ is $\frac{1\times3\times5...(2n-1)}{n!}2^n$.

(iv) Prove that the coefficient of the middle term in $(1+x)^{2n}$ is equal to the sum of the coefficient of the two middle terms in $(1+x)^{2n-1}$.

(v) Find the coefficient of x^{65} in the expansion of $(1+x)^{131}(x^2-x+1)^{130}$.

4. Find the total number of irrational terms in the expansion of $(5^{1/6}+2^{1/8})^{100}$.

5. Find the value of

$$\frac{1}{81^n}-{}^{2n}c_1\frac{10}{81^n}+{}^{2n}c_2\frac{10^2}{81^n}-{}^{2n}c_3\frac{10^3}{81^n}+...+\frac{10^{2n}}{81^n}.$$

6. Find the value of the second largest term in the expansion of $(4+5x)^{20}$ when $x=\frac{1}{3}$.

7. If in the expansion of $(2x+5)^{10}$, the numerically greatest term is equal to the middle term, then find the value of x.

8. Find the coefficient of the term independent of x in the expansion of $(1+x+2x^3)\left(\frac{3}{2}x^2-\frac{1}{3x}\right)^9$.

9. Find the remainder, when 5^{99} is divided by 13.

10. Prove that $2^{4n}-15n-1$ is divisible by 225.

11. If $x=(8+3\sqrt{7})^n$, where n is natural number, prove that the integral part of x is an odd integer and also find the value of $x-x^2+x[x]$, where [ ] denotes the greatest integer function.

12. Find the greater of the number of 100^{100} and (300)!

13. Find the values of

$$\frac{1}{(12)!}+\frac{1}{10!\,2!}+\frac{1}{8!\,4!}+...+\frac{1}{12!}.$$

14. If ${}^{n-1}c_r=(k^2-3)\cdot{}^{n}c_{r+1}$, then find the interval in which k lies.

15. If n is a positive integer, show that $3^{2n}-1+24n-32n^2$ is divisible by 512 if $n>2$.

16. Find the remainder when $3^{2002}+7^{2002}+2002$ is divided by 29.

17. If P_n denotes the product of all the coefficients of $(1+x)^n$ and $9!\,P_{n+1}=10^9P_n$, then find the value of n.

18. Find the value of ${}^nc_1+2^2\,{}^nc_2+2^4\,{}^nc_3+2^6\,{}^nc_4+...$

19. If $(1+x)^n=c_0+c_1x+c_2x^2+...+c_nx^n$, find the value of $c_0+(c_0+c_1)+(c_0+c_1+c_2)+...+(c_0+c_1+c_2+...+c_n)$.

20. If ${}^9c_n+3\cdot{}^9c_{n+1}+3\cdot{}^9c_{n+2}+{}^9c_{n+3}>{}^{12}c_{n+2}$, then find the greatest possible value of n.

5.6 JEE Main: Sample Questions

Single Answer Type

1. The coefficient of t^4 in $(1-t)^6(1+t^2)^5$ is equal to
(A) 60 (B) 80 (C) 100 (D) 125

2. If the constant term of the binomial expansion $\left(2x-\frac{1}{x}\right)^n$ is -160, then n is equal to
(A) 4 (B) 6 (C) 8 (D) 10

3. If the coefficients of x^7 and x^8 in the expansion of $\left[2+\frac{x}{3}\right]^n$ are equal, then the value of n is:
(A) 15 (B) 45 (C) 55 (D) 56

4. The number of rational terms in the expansion of $\left(2^{\frac{1}{7}}+3^{\frac{1}{4}}\right)^{198}$ is equal to
(A) 7 (B) 8 (C) 9 (D) 10

5. What is the sum of the coefficients of the polynomial after expansion of $(25-60x+22x^2+18x^3-6x^8)^{2013}$?
(A) 1 (B) −1 (C) 5262 (D) 0

6. Let the term independent of x in the expansion of $\left(x^2-\frac{1}{x}\right)^9$ have the value p, and q be the sum of the coefficients of its middle terms, then $(p-q)$ equals
(A) 0 (B) $2\times{}^9C_4$ (C) ${}^9C_5$ (D) ${}^9C_3$

7. Which of the following statement(s) is/are correct?

(A) The coefficient of x^{10} in $(1 + x^2 - x^3)^8$ is 476.
(B) The remainder when $(9^7 + 7^9)$ is divided by 32 is 0.
(C) The coefficient of the term independent of 'x' in the expansion of $\left(2x^2 - \frac{1}{3x}\right)^{11}$ is $\frac{-{}^{11}C_7 \cdot 2^4}{3^7}$.
(D) The only value of 'r' for which the coefficient of $(2r + 3)^{\text{rd}}$ term is equal to the coefficient of $(r + 5)^{\text{th}}$ term in the expansion of $(1 + x)^{15}$ is '2'.

8. The greatest value of the terms independent of x in the expansion of $\left(x\sin\theta + \frac{\cos\theta}{x}\right)^{10}$ is

(A) ${}^{10}C_5$ (B) 2^5 (C) $2^5 \times {}^{10}C_5$ (D) $\frac{{}^{10}C_5}{2^5}$

9. The sum of the binomial coefficient of $\left[2x + \frac{1}{x}\right]^n$ is equal to 256. The constant term in the expansion is

(A) 1120 (B) 2110 (C) 1210 (D) 1225

10. The coefficient of x^{49} in the expansion of $(x - 1)\left(x - \frac{1}{2}\right)\left(x - \frac{1}{2^2}\right) \ldots \left(x - \frac{1}{2^{49}}\right)$ is equal to

(A) $-2\left(1 - \frac{1}{2^{50}}\right)$ (B) + ve coefficient of x

(C) $2\left(1 - \frac{1}{2^{49}}\right)$ (D) $-2\left(1 - \frac{1}{2^{49}}\right)$

11. $(2n + 1)(2n + 3)(2n + 5) \ldots (4n - 1)$ is equal to

(A) $\frac{(4n)!}{2^n \cdot (2n)!\,(2n)!}$ (B) $\frac{(4n)!\,n!}{2^n \cdot (2n)!\,(2n)!}$

(C) $\frac{(4n)!\,n!}{(2n)!\,(2n)!}$ (D) $\frac{(4n)!\,n!}{2^n\,!\,(2n)!}$

12. If $n \in N$ and n is even, then

$$\frac{1}{1.(n-1)!} + \frac{1}{3!(n-3)!} + \frac{1}{5!(n-5)!} + \ldots + \frac{1}{(n-1)!\;1!} =$$

(A) 2^n (B) $\frac{2^{n-1}}{n!}$ (C) $2^n n!$ (D) $\frac{2^{n+1}}{n!}$

13. In the expansion of $\left(3^{\frac{x}{4}} + 3^{\frac{5x}{4}}\right)^n$ the sum of the binomial coefficients is 64 and the term with the greatest binomial coefficient exceeds the third term by $(n - 1)$, then the value of x must be

(A) 1 (B) 2 (C) 0 (D) –1

14. If the second term of the expansion $\left[a^{\frac{1}{13}} + \frac{a}{\sqrt{a^{-1}}}\right]^n$ is $14a^{\frac{5}{2}}$ then the value of $\frac{{}^nC_3}{{}^nC_2}$ is

(A) 4 (B) 3 (C) 12 (D) 6

15. The sum of the coefficients of all the even powers of x in the expansion of $(2x^2 - 3x + 1)^{11}$ is

(A) 2×6^{10} (B) 3×6^{10} (C) 6^{11} (D) 6^{10}

16. If $(1 + x - 3x^2)^{2145} = a_0 - a_1x + a_2x^2 + \ldots$ then $a_0 - a_1 + a_2 - a_3 + \ldots$ ends with

(A) 1 (B) 3 (C) 7 (D) 9

17. The number of rational terms in the expansion of $(\sqrt{2} + \sqrt[4]{3})^{100}$ is

(A) 25 (B) 26 (C) 27 (D) 28

18. In the expansion of $\left(\sqrt{\frac{q}{p}} + \sqrt[10]{\frac{p^7}{q^3}}\right)$, there is a term similar to pq, then that term is equal to

(A) $210pq$ (B) $252pq$ (C) $120pq$ (D) $45pq$

19. The number of values of 'r' satisfying the equation, ${}^{39}C_{3r-1} - {}^{39}C_{r^2} = {}^{39}C_{r^2-1} - {}^{39}C_{3r}$ is

(A) 1 (B) 2 (C) 3 (D) 4

20. The greatest term in the binomial expansion of $(a + 2x)^9$ when $a = 1$ and $x = \frac{1}{3}$ is

(A) 3rd and 4th (B) 4th and 5th
(C) only 4th (D) only 5th

21. Given $(1 - 2x + 5x^2 - 10x^3)(1 + x)^n = 1 + a_1x + a_2x^2 + \ldots$ and that $a_1^2 = 2a_2$ then the value of n is

(A) 6 (B) 2 (C) 5 (D) 3

22. The largest real value for x such that $\sum_{k=0}^{4}\left(\frac{5^{4-k}}{(4-k)!}\right)\left(\frac{x^k}{k!}\right)=\frac{8}{3}$ is

(A) $2\sqrt{2}-5$ (B) $2\sqrt{2}+5$
(C) $-2\sqrt{2}-5$ (D) $-2\sqrt{2}+5$

23. The coefficient of term that contains $x^{10}y^3z^2$ in the expansion of the binomial $(x^2-2y+z)^{10}$ is

(A) $-\frac{8\cdot(10)!}{2!\cdot3!\cdot5!}$ (B) $\frac{8\cdot(10)!}{2!\cdot3!\cdot5!}$
(C) $\frac{(10)!}{2!\cdot3!\cdot5!}$ (D) $-\frac{(10)!}{2!\cdot3!\cdot5!}$

24. In the expansion of $\left(3-\sqrt{\frac{17}{4}+3\sqrt{2}}\right)^{15}$, the 11th term is a

(A) positive integer
(B) positive irrational number
(C) negative integer
(D) negative irrational number

25. If the coefficient x^2 and x^3 in the expansion of $(1+8x+bx^2)(1-3x)^9$ in the power of x are equal, then b is

(A) $\frac{54}{7}$ (B) $\frac{27}{7}$
(C) $\frac{108}{7}$ (D) 1

26. If $2a+2b+3c=\frac{1}{5}$ and a, b, c > 0 then maximum value of term independent of x in the expansion of $\left(abx^{\frac{1}{2}}+cx^{\frac{-1}{3}}\right)^{25}$ is

(A) ${}^{25}C_{10}$ (B) ${}^{25}C_{15}(35)^{25}$
(C) ${}^{25}C_{15}\left(\frac{1}{35}\right)^{35}$ (D) ${}^{25}C_{10}\left(\frac{1}{25}\right)^{35}$

27. In the expansion of $(1+x+x^2+\ldots+x^{27})(1+x+x^2+\ldots+x^{14})^2$, the coefficient of x^{28} is

(A) 195 (B) 224 (C) 378 (D) 405

28. The ratio of coefficient of x^2 to coefficient x^{10} in the expansion of $\left(x^5+4.3^{-\log_{\sqrt{3}}\sqrt{x^3}}\right)^{10}$ is

(A) 4 : 7 (B) 10 : 3 (C) 3 : 10 (D) 7 : 4

29. If last term in the expansion of $\left(2^{\frac{1}{3}}-\frac{1}{\sqrt{2}}\right)^n$ is $\left(\frac{1}{3^{\frac{5}{3}}}\right)^{\log_3 8}$, then the 5th term from the beginning is

(A) 210 (B) 420 (C) 105 (D) 315

30. The number of terms with rational coefficients in the expansion of $\left(\sqrt[3]{5x}+\sqrt[2]{3y}+z\right)^6$ is

(A) 7 (B) 6 (C) 8 (D) 9

31. Coefficient of x^5 in the expansion of $\left(x+\frac{1}{x}\right)^{10}\left(x-\frac{1}{x}\right)^{9}$ is

(A) ${}^9C_3$ (B) ${}^9C_4$ (C) $-{}^9C_3$ (D) $-{}^9C_4$

32. The number of rational terms in the expansion of $\left(2^{\frac{1}{3}}+5^{\frac{1}{9}}\right)^{6551}$ is equal to

(A) 729 (B) 730 (C) 735 (D) 736

33. If $(1-x^2+x^3)(1+x)^n=a_0+a_1x+a_2x^2+\ldots$, $a_i's\in R$, then the value of $a_1^2-a_1-2a_2$ is

(A) –2 (B) 2 (C) 4 (D) –4

34. Term independent of x in $\left(x^3-\frac{4}{x^2}\right)^{15}$ is

(A) $-{}^{15}C_6 2^{18}$ (B) $-{}^{15}C_9 4^{10}$
(C) $-{}^{15}C_6 2^{16}$ (D) $-{}^{15}C_9 4^{8}$

35. The sum of the series

$$\frac{{}^{101}C_1}{{}^{101}C_0}+\frac{2\times{}^{101}C_2}{{}^{101}C_1}+\frac{3\times{}^{101}C_3}{{}^{101}C_2}+\ldots+\frac{101\times{}^{101}C_{101}}{{}^{101}C_{100}}$$

equals

(A) 5050 (B) 4950 (C) 5151 (D) 5051

36. The value of

$$\binom{50}{1}^2+2\binom{50}{2}^2+3\binom{50}{3}^2+\ldots+50\binom{50}{50}^2$$

is equal to $\left(\text{where}\binom{n}{r}={}^nC_r\right)$

(A) $\binom{100}{50}$ (B) $50\binom{100}{50}$

(C) $100\binom{100}{50}$ (D) $\frac{2^{49}}{49!}(1\times3\times5\ldots99)$

37. We have $3m$ balls, out of which m are identical and the rest are all different. The number of ways of selecting m objects from the group is

(A) $\frac{^{2m}C_m}{2}$ (B) $\frac{1}{2}\,^{2m}C_{m-1}+2^{2m-1}$

(C) $\frac{1}{2}\,^{2m}C_m+2^{2m+1}$ (D) $\frac{1}{2}\,^{2m}C_m+2^{2m-1}$

38. If (201)! is divided by 24^k then the largest value of k is
(A) 98 (B) 65 (C) 49 (D) 66

39. When $(2x + 3y)^7$ is expanded and the like terms are combined, then the coefficient of the term x^5y^2, is
(A) 288 (B) 972 (C) 3024 (D) 6048

40. Let C_r denote the combinational coefficient in the expansion of the binomial $(1 + x)^{14}$, then the value of the sum $\sum_{r=1}^{15} r\cdot C_{r-1}$ is
(A) 2^{18} (B) 2^{17} (C) 2^{16} (D) 2^{15}

41. The value of $\sum_{r=1}^{10} r^2\cdot\left(\frac{^{26}C_r}{^{26}C_{r-1}}\right)$ is equal to
(A) 900 (B) 1100 (C) 1300 (D) 1800

42. If $6^{83} + 8^{83}$ is divided by 49, then the remainder is
(A) 35 (B) 5 (C) 1 (D) 0

43. Last three digits of the number $N = 7^{100} - 3^{100}$ are
(A) 100 (B) 300 (C) 500 (D) 000

44. Coefficient of α^t in the expansion of, $(\alpha+p)^{m-1}+(\alpha+p)^{m-2}(\alpha+q)+(\alpha+p)^{m-3}(\alpha+q)^2+\ldots(\alpha+q)^{m-1}$ where $\alpha\neq-q$ and $p\neq q$ is

(A) $\frac{^mC_t\cdot(p^t-q^t)}{p-q}$ (B) $\frac{^mC_t(p^{m-t}-q^{m-t})}{p-q}$

(C) $\frac{^mC_t(p^t+q^t)}{p-q}$ (D) $\frac{^mC_t(p^{m-t}+q^{m-t})}{p-q}$

45. Last two digits of $(21)^{100}$, is
(A) 00 (B) 01 (C) 11 (D) 21

46. The remainder, if $1+2+2^2+2^3+\ldots+2^{1999}$ is divided by 5 is
(A) 0 (B) 1 (C) 2 (D) 3

47. If $\sum_{r=1}^{10} r(r-1)\,^{10}C_r = k\cdot 2^9$, then k is equal to
(A) 10 (B) 45 (C) 90 (D) 100

48. The sum $\frac{\binom{11}{0}}{1}+\frac{\binom{11}{1}}{2}+\frac{\binom{11}{2}}{3}+\ldots+\frac{\binom{11}{11}}{12}$ equals $\left(\text{where}\binom{n}{r}\text{denotes }^nC_r\right)$

(A) $\frac{2^{11}}{12}$ (B) $\frac{2^{12}}{12}$

(C) $\frac{2^{11}-1}{12}$ (D) $\frac{2^{12}-1}{12}$

49. Let $\binom{n}{k}$ represent the combination of 'n' things taken 'k' at a time, then the value of sum $\binom{99}{97}+\binom{98}{96}+\binom{97}{95}+\ldots+\binom{3}{1}+\binom{2}{0}$ equals

(A) $\binom{99}{97}$ (B) $\binom{100}{98}$ (C) $\binom{99}{98}$ (D) $\binom{100}{97}$

50. The value of $\sum_{r=1}^{10} {}^{15}C_r\,{}^{35}C_{(10-r)}$ is

(A) $^{10}C_{10}$ (B) $^{50}C_{10} - {}^{35}C_{10}$

(C) $^{40}C_{10}$ (D) $(^{40}C_{10}-1)$

51. The total number of ways in which a king can distribute 100 identical diamonds among 3 princes Ram, Shyam and Balram such that each gets at least one and Ram, Shyam and Balram get at most 75, 50 and 25 diamonds respectively
(A) $^{99}C_{97} - {}^{49}C_{47} - {}^{74}C_{72}$
(B) $^{99}C_{97} - {}^{50}C_{48} - {}^{75}C_{73}$
(C) $^{99}C_2 - {}^{50}C_2 - {}^{75}C_2$
(D) $^{99}C_2 - {}^{49}C_2 - {}^{74}C_2$

52. Greatest term in the expansion of $(7-4x)^{10}$ is (when $x = 1$)

(A) T_4 and T_5 (B) T_4 only
(C) T_5 only (D) T_5 and T_6

53. $({}^{10}C_1)^2 + ({}^{10}C_3)^2 + ({}^{10}C_5)^2 + \ldots + ({}^{10}C_9)^2$ is equal to

(A) $\dfrac{{}^{20}C_{10} + {}^{10}C_5}{2}$ (B) $\dfrac{{}^{20}C_{10} - {}^{10}C_5}{2}$

(C) ${}^{20}C_{10}$ (D) ${}^{10}C_5$

54. Sum of the series

$$\frac{1\times {}^{10}C_1}{2^9} - \frac{2\times {}^{10}C_2 \times 3^1}{2^8} + \frac{3\times {}^{10}C_3 3^2}{2^7}$$

$- \ldots - 10 \cdot {}^{10}C_{10} 3^9$ is equal to $\dfrac{k}{2^8}$, then k is equal to

(A) 5^8 (B) -5^9 (C) -5^{10} (D) 5^{11}

55. If $r, k, p \in W$, then $\sum\limits_{r+k+p=10} {}^{30}C_r \cdot {}^{20}C_k \cdot {}^{10}C_p$ is equal to

(A) $\begin{pmatrix}60\\50\end{pmatrix}$ (B) $\begin{pmatrix}60\\30\end{pmatrix}$

(C) $\begin{pmatrix}60\\20\end{pmatrix}$ (D) $\begin{pmatrix}30\\10\end{pmatrix}\begin{pmatrix}30\\20\end{pmatrix}$

56. Value of $\sum\limits_{r=0}^{15} ({}^{15}C_r {}^{40}C_{15} {}^{20}C_r - {}^{35}C_{15} {}^{15}C_r {}^{25}C_r)$ is

(A) 0 (B) ${}^{40}C_{15} - {}^{35}C_{15}$

(C) ${}^{35}C_{15} - {}^{40}C_{15}$ (D) ${}^{40}C_{15}$

57. The value of expression ${}^{n+1}C_2 + 3({}^2C_2 + {}^3C_2 + {}^4C_2 + \ldots + {}^nC_2)$ is

(A) $\sum n^3$ (B) $\sum n^2$

(C) $\sum n$ (D) $\dfrac{n^2(n+1)}{2}$

58. $\sum\limits_{r=2}^{n} \dfrac{1}{{}^rC_2}$ is equal to

(A) $2\left(\dfrac{n-1}{n}\right)$ (B) $\dfrac{n-1}{n}$

(C) $\dfrac{n}{n-1}$ (D) $\dfrac{2n}{n-1}$

Integer Answer Type

1. If $\left(1-2x+5x^2-10x^3\right)$ $(1+x)^n = 1 + a_1x + a_2x^2 + \ldots$ and $a_1^2 = 2a_2$, then find the value of n.

2. If the greatest value of the terms independent of x in the expansion of $\left(x\sin\theta + \dfrac{\cos\theta}{x}\right)^{10}$ is $\dfrac{{}^{10}C_K}{2^K}$, where $0 \le K \le 10$, then find the value of K.

3. Find the number of terms with rational coefficients in the expansion of $\left(\sqrt[3]{5x} + \sqrt[2]{3y} + z\right)^6$.

4. If in the expansion of $(1+x)^{14}$, the value of $\sum\limits_{r=1}^{15} r\cdot C_{r-1}$ is a^b then find the value of $\left[\dfrac{b}{a}\right]$, where [ ] denotes GIF.

5. Find the value of $\sum\limits_{K=0}^{21} K \cdot {}^{21}C_K \left(\dfrac{1}{3}\right)^K \left(\dfrac{2}{3}\right)^{21-K}$.

6. If the coefficient of x^8 in $(x-1)(x-2)(x-3)\ldots(x-10)$ is N, then find the sum of digits of N.

7. If coefficient of x^{14} in $\left(1+2x+3x^2+\ldots+16x^5\right)^2$ is K, then find unit digit in K.

8. If $\sum\limits_{r=0}^{n} \dfrac{(r+5)}{{}^nC_r} = \sum\limits_{r=0}^{n} \dfrac{n}{{}^nC_r}$ and if the value of n is K, then find the sum of the digits in K.

9. If the term independent of x in the expansion of $\left(1+2x+bx^2\right)\left(x+\dfrac{1}{x}\right)^6$ is equal to 80, then find the value of b.

10. Find the coefficient of x^9 in the expansion of $(1+x)\left(1+x^2\right)\left(1+x^3\right)\ldots\left(1+x^{100}\right)$.

11. Find the coefficient of t^8 in the expansion of $(1+2t^2-t^3)^9$.

12. If the sum $S = {}^{20}C_2 + 2\times{}^{20}C_3 + 3\times{}^{20}C_4 + \ldots + 19\times{}^{20}C_{20}$ can be expressed as $1 + p\times 2^q$, then find the least value of $(p + q)$.

13. Find the index n of the binomial $\left(\frac{x}{5}+\frac{2}{5}\right)^n$ if the 9th term of the expansion has numerically the greatest coefficient ($n \in N$).

14. Find the coefficient of x^3 in $(1 + x)^{25}(1 - x)^5$.

15. When $(3x + 5)^{100}$ is expanded, find the largest power of 2 dividing the coefficient of x^{39}.

16. Let $f(x) = 1 - x + x^2 - x^3 + \ldots + x^{16} - x^{17} = a_0 + a_1(1+x) + a_2(1+x)^2 + \ldots + a_{17}(1+x)^{17}$. Find the value of a_2.

17. If a is the remainder when 5^{40} is divided by 11 and b is the remainder when 2^{2003} is divided by 17, then the value of $b - a$ is ________.

18. If the number of distinct terms in the expansion of $\left(x+\frac{1}{x}+y+\frac{1}{y}\right)^{14}$ is $\alpha\beta\gamma$, where $\alpha\beta\gamma$ is a three-digit number, then $(\alpha+\beta+\gamma)$ is ________.

5.7 JEE Advanced: Sample Questions

Single Answer Type

1. Total number of rational terms in $\left(\frac{2^{2/3}+2^{4/3}+2^2}{2^{1/3}}\right)^{20}$ is
(A) 6 (B) 7 (C) 8 (D) 9

2. If $\sum_{r=1}^{n}(r^2+3r+3)(r+1)! = (998)(998)!$, then n is equal to
(A) 994 (B) 996 (C) 998 (D) 1000

3. The term independent of a in the expansion of $\left(1+\sqrt{a}+\frac{1}{\sqrt{a}-1}\right)^{-30}$ is
(A) 18th (B) 19th
(C) 20th (D) none of these

4. If A is a general homogenous polynomial of degree 10 in 6 variables and B is a general polynomial of the degree 6 in 10 variables, then
(A) A has more terms than B
(B) B has more terms than A
(C) both have the same number of terms
(D) none of these

5. If $(1+3x)^n = \sum_{r=0}^{n} a_r x^r (1+x)^{n-r}$, then the value of a_r is
(A) $2^r\,{}^nc_0$ (B) $2^r\,{}^nc_r$ (C) ${}^nc_r$ (D) $3^r\,{}^nc_r$

6. If $x_1, x_1, x_2, x_3, \ldots, x_{n-1}$ are n roots of the equation $x^n + \alpha x + \beta = 0$, then $(x_1-x_2)(x_1-x_3)(x_1-x_4)\ldots(x_1-x_{n-1})$ is equal to
(A) $n(n-1)x_1^{n-2}$ (B) ${}^nc_2 x_1^{n-2}$
(C) $nx_1^{n-1}+\alpha$ (D) none of these

7. The value of
${}^nc_0\times{}^{2n}c_n - {}^nc_1\times{}^{2n-2}c_2 + {}^nc_2\times{}^{2n-4}c_4 - \ldots$ is
(A) 2^{2n} (B) 2^n (C) 2^{2n-1} (D) 2^{n+1}

8. The coefficient of x^{10} in the expansion of $\left(x+\frac{x^2}{2!}+\frac{x^3}{3!}+\ldots+\frac{x^{10}}{10!}\right)^3$ is equal to
(A) $\frac{3^{10}+3\cdot 2^{10}+2}{10!}$ (B) $\frac{3^{10}-2^{10}+3}{10!}$
(C) $\frac{3^{10}-2^{10}}{10!}$ (D) $\frac{3^{10}-3\cdot 2^{10}+3}{10!}$

9. The value of
$(k+1)^n c_{k+1} - 2k\,{}^nc_k + 3(k-1)^n c_{k-1} - \ldots + (-1)^k (k+1)^n c_1$
is equal to
(A) $n^n c_k$ (B) $n^{n-1}c_k$
(C) $n^{n-2}c_k$ (D) $n^{n-3}c_k$

10. If $f(x) = 1 - x + x^2 - x^3 + \ldots - x^{15} + x^{16} - x^{17}$, then the coefficient of x^2 in $f(x - 1)$ is
(A) 816 (B) 822
(C) 826 (D) none of these

11. The constant term in the expansion of $\left(1+\frac{1}{x}+x+x^2\right)^4$ is
(A) 35 (B) 30 (C) 32 (D) 31

12. Last three digits of $(5! + 1)^{41}$ are
(A) 281 (B) 381 (C) 741 (D) 881

13. $\lim_{n\to\infty} \dfrac{\sum\sum_{1\le i<J\le n}(i+1)(J+1)}{n^4}$ is equal to

(A) $\frac{1}{4}$ (B) $\frac{1}{6}$

(C) $\frac{1}{8}$ (D) none of these

14. The greatest integer which divides the number $101^{100} - 1$ is

(A) 100 (B) 1000

(C) 10000 (D) 100000

15. The sum of the coefficients of all the integral powers of x in the expansion of $(1+2\sqrt{x})^{40}$, is

(A) $3^{40}+1$ (B) $3^{40}-1$

(C) $\frac{1}{2}(3^{40}-1)$ (D) $\frac{1}{2}(3^{40}+1)$

16. The coefficient of $x^3y^4z^5$ in the expansion of $(xy+yz+zx)^6$ is

(A) 60 (B) 70

(C) 50 (D) none of these

17. Let $f(x)=a_0+a_1x+a_2x^2+\ldots+a_{2n}x^{2n}$ and $g(x)=b_0+b_1x+b_2x^2+\ldots+b_{n-1}x^{n-1}+x^n+x^{n+1}+\ldots+x^{2n}$

If $f(x)=g(x+1)$, then a_n in terms of n is equal to

(A) ${}^{2n+1}c_{n+1}$ (B) ${}^{2n-1}c_{n-1}$

(C) ${}^{2n-1}c_{n+1}$ (D) ${}^{n+1}c_{n-1}$

18. The term independent of 'x' in the expansion of $\left(9x-\frac{1}{3\sqrt{x}}\right)^{18}$, $x>0$, is α times the corresponding binomial coefficient, then α is equal to

(A) 1 (B) 2 (C) 3 (D) 4

19. Sum of the infinite series

$$1+\frac{2}{3}\cdot\frac{1}{2}+\frac{2}{3}\cdot\frac{5}{6}\cdot\frac{1}{2^2}+\frac{2}{3}\cdot\frac{5}{6}\cdot\frac{8}{9}\cdot\frac{1}{2^3}+\ldots \text{ to } \infty \text{ is}$$

(A) $2^{\frac{1}{3}}$ (B) $4^{\frac{1}{3}}$

(C) $8^{\frac{1}{3}}$ (D) none of these

20. If $C_0, C_1, C_2, \ldots, C_n$ denote the binomial coefficients in the expansion of $(1+x)^n$, then $\sum_{r=0}^{n}(-1)^r\,{}^nC_r\dfrac{(1+r\log_e 10)}{(1+\log_e 10^n)^r}$ is equal to

(A) 0 (B) 1 (C) 2 (D) none of these

21. The integer just greater than $(\sqrt{3}+1)^{2m}$ contains

(A) 2^{m+2} as a factor (B) 2^{m+1} as a factor

(C) 2^{m+3} as a factor (D) none of these

22. Let $f(x)=a_0+a_1x+a_2x^2+\ldots+a_{2n}x^{2n}$ and $g(x)=b_0+b_1x+b_2x^2+\ldots+b_{n-1}x^{n-1}+x^n+x^{n+1}+\ldots+x^{2n}$.

If $f(x)=g(x+1)$, then a_n in terms of n is equal to

(A) ${}^{n+1}C_{n+1}$ (B) ${}^{2n-1}C_{n-1}$

(C) ${}^{2n-1}C_{n+1}$ (D) ${}^{2n+1}C_{n+1}$

23. The sum of $\sum_{r=0}^{n}(-1)^r\dfrac{{}^nC_r}{{}^{r+2}C_r}$ is

(A) $\frac{2}{n+1}$ (B) $\frac{1}{n+2}$ (C) $\frac{2}{n-2}$ (D) $\frac{2}{n+2}$

24. If $x>2$ then $\sum_{r=1}^{n}\dfrac{1}{\log_{4^r}x}$ is equal to

(A) $\frac{n(n+1)}{n^4}\log_x 2$ (B) $n(n+1)\log_x 2$

(C) $\frac{n(n+1)}{2}\log_x 2$ (D) $n(n+1)\log_x 4$

25. The value of 'r' for which x^{-17} is contained by the rth term in the expansion of $(x^4-x^{-3})^{15}$, is

(A) 10 (B) 11 (C) 12 (D) 13

Multiple Answer Type

1. If $a=\sum_{r=0}^{33}{}^{33}C_r$, $b=\sum_{r=0}^{33}{}^{66}C_{2r}$, $c=\sum_{r=0}^{33}{}^{99}C_{3r}$, then

(A) $3c=2(ab+1)$ (B) $a^3=3c+2$

(C) $3c=2(ab-1)$ (D) $a^2=2b$

2. Which of the following is/are correct?

(A) ${}^{n-1}C_r+{}^{n-1}C_{r-1}={}^nC_r$

(B) ${}^{n-1}P_r+r\cdot{}^{n\,1}P_{r-1}={}^nP_r$

(C) ${}^nC_r = \frac{n}{r}\,{}^{n-1}C_{r-1}$

(D) $\frac{n-r+1}{r}\cdot{}^nC_{r-1} = {}^nC_r$

3. In the expansion of $\left(5^{\frac{1}{6}} + 2^{\frac{1}{8}}\right)^{100}$, which of the following statements hold(s) good?

(A) The number of irrational terms is 97.
(B) The number of rational terms is 5.
(C) The binomial coefficient of the term in which exponent of '5' is 15 is ${}^{100}C_{90}$.
(D) The term in which the exponent of 2 and 5 will be equal is the 58th term from the beginning.

4. Sum of the series $({}^{100}C_1)^2 + 2({}^{100}C_2)^2 + 3({}^{100}C_3)^2 + \ldots + 100({}^{100}C_{100})^2$ equals

(A) $\frac{2^{99}[1\cdot3\cdot5\ldots(199)]}{(99)!}$ (B) $100\cdot{}^{200}C_{100}$

(C) $50\cdot{}^{200}C_{100}$ (D) $100\ {}^{199}C_{99}$

5. In the expansion of $\left(x^3 + 3.2^{-\log_{\sqrt{2}}\sqrt{x^3}}\right)^{11}$

(A) there appears a term with the power x^2
(B) there does not appear a term with the power x^2
(C) there appears a term with the power x^{-3}
(D) the ratio of the coefficient of x^3 to that of x^{-3} is $\frac{1}{3}$

6. If it is known that the third term of the binomial expansion $\left(x + x^{\log_{10}x}\right)^5$ is 10^6 then x is equal to

(A) 10 (B) $10^{-5/2}$ (C) 100 (D) 5

7. Let $(1 + x^2)^2 (1 + x)^n = A_0 + A_1x + A_2x^2 + \ldots$ If A_0, A_1, A_2 are in AP, then the value of n is

(A) 2 (B) 3 (C) 5 (D) 7

8. If $\left(9+\sqrt{80}\right)^n = I + f$ where I, n are integers and $0 < f < 1$, then

(A) I is an odd integer
(B) I is an even integer
(C) $(I + f)(1 - f) = 1$
(D) $1 - f = (9 - \sqrt{80})^n$

9. Which of the following statement(s) is/are correct?

(A) $1+\frac{2}{2}+\frac{3}{2^2}+\frac{4}{2^3}+\ldots+\infty = 4.$

(B) Integral part of $\left(9+4\sqrt{5}\right)^n$, $n \in N$ is even.

(C) $({}^nC_0 + {}^nC_1 + {}^nC_2 + \ldots + {}^nC_n)^2 = {}^{2n}C_0 + {}^{2n}C_1 + {}^{2n}C_2 + \ldots + {}^{2n}C_{2n}.$

(D) $\frac{1}{(3+2x)^2}$ can be expanded as infinite series in ascending powers of x only if $|x| < \frac{2}{3}$.

10. $\sum_{r+5=0}^{10} (-1)^{r+5}.\ {}^{10}C_{r+5}.2^{5-r}$ is greater than

(A) −2 (B) −1 (C) 0 (D) 1

11. Let $(1+3x+2x^2)^6 = \sum_{r=0}^{12} a_r x^r$.

Identify which of the following statement(s) is/are correct.

(A) The coefficient of x^{12} is 2^{12}.
(B) The coefficient of x^{11} is 9.2^6.
(C) $\frac{\sum_{r=0}^{6} a_{2r}}{a_{12}} = \frac{3^6}{2}$.
(D) $\frac{\sum_{r=1}^{6} a_{2r-1}}{a_{12}} = 2.3^6$.

12. If the coefficient of x^{2r} is greater than half the coefficient of x^{2r+1} in the expansion of $(1 + x)^{15}$, then the possible value(s) of 'r' is/are

(A) 5 (B) 6 (C) 7 (D) 8

13. The value(s) of integer 'n' for which $3^{512} - 1$ is divisible by 2^n is/are

(A) 3 (B) 7 (C) 11 (D) 13

14. Consider $E = \left(\sqrt[8]{x} + \sqrt[5]{y}\right)^z = I + f, 0 \le f < 1.$

(A) If $x = 5$, $y = 2$, $z = 100$, then number of irrational terms in expansion of E is 98.
(B) If $x = 5$, $y = 2$, $z = 100$, then number of rational terms in expansion of E is 4.
(C) If $x = 16$, $y = 1$, $z = 6$, then $I = 197$.
(D) If $x = 16$, $y = 1$ and $z = 6$, then $f = (\sqrt{2} - 1)^6$.

15. Let $N = \sum_{r=0}^{6} {}^6C_r 2^r$. Then which of following statements holds good?

(A) The number of ways to distribute 6 different things to Ravi, Raheem, Sohan so that each child receives none, one, more than one thing is N.

(B) The sum of coefficients in expansion of $(2x + 1)^6$ is N.

(C) The sum of binomial coefficients in expansion of $(2x + 1)^6$ is N.

(D) The number of ways in which two disjoint subsets A and B can be formed from set {1, 2, 3, 4, 5, 6} is N.

16. If $\sum_{k=1999}^{2015} {}^kC_x = {}^aC_{x+1} - {}^bC_{x+1}$, then $|a - b|$ cannot be

(A) 16 (B) 17 (C) 18 (D) 19

17. If $(1 + x + x^2 + x^3)^{12} = a_0 + a_1x + a_2x^2 + \ldots + a_{36}x^{36}\ \forall\ x \in R$, then

(A) $a_0 + a_1 + a_2 + \ldots + a_{36}$ is divisible by 2^{18}

(B) $a_0 + 2a_1 + 3a_{36} = 28$

(C) $a_r = a_{36-r}$ for each $r \in \{0, 1, 2, \ldots, 36\}$

(D) $a_0 + a_2 + a_4 + \ldots + a_{36} = a_1 + a_3 + a_5 + \ldots + a_{35}$

18. The coefficient of x^{324} in the expansion of $(x + 1)(x - 2)^2(x + 3)^3(x - 4)^4 \ldots (x - 24)^{24}(x + 25)^{25}$ is

(A) 325

(B) $\left(\sum_{k=0}^{8} {}^8C_k + 4.\sum_{s=0}^{4} {}^4C_s + \sum_{r=0}^{2} {}^2C_r + {}^5C_0\right)$

(C) $1^2 - 2^2 + 3^2 - 4^2 + \ldots - 24^2 + 25^2$

(D) $1^1 + 2^2 + 3^3 + 4^4 + \ldots + 25^{25}$

19. If $ac > b^2$, then the sum of the coefficients in the expansion of $(a\alpha^2x^2 + 2b\alpha x + c)^n$, $(a, b, c, \alpha \in R, n \in N)$ is

(A) positive, if $a > 0$

(B) positive if $c > 0$

(C) negative, if $a < 0$, n is odd

(D) positive if $c < 0$, n is even

20. Number of values of r satisfying the equation ${}^{69}c_{3r-1} - {}^{69}c_{r^2} = {}^{69}c_{r^2-1} - {}^{69}c_{3r}$ is

(A) 1 (B) 2 (C) 3 (D) 7

21. The value(s) of integer n for which $3^{512} - 1$ is divisible by 2^n is/are

(A) 3 (B) 7 (C) 11 (D) 13

22. If

$$\frac{{}^9C_0}{8} - \frac{{}^9C_1}{9} + \frac{{}^9C_2}{10} - \frac{{}^9C_3}{11} + \ldots + \frac{{}^9C_8}{16} - \frac{{}^9C_9}{17} = \frac{1}{n},$$

then n is divisible by

(A) 7 (B) 11 (C) 13 (D) 17

23. If in the expansion of $(x + a)^n$, $n \in N$, the sum of odd numbered terms be α and the sum of even numbered terms be β, then

(A) $\alpha^2 + \beta^2 = (x+a)^{2n} + (x-a)^{2n}$

(B) $4\alpha\beta = (x+a)^{2n} - (x-a)^{2n}$

(C) $\alpha^2 - \beta^2 = (x^2 - a^2)^n$

(D) $2(\alpha^2 + \beta^2) = (x+a)^{2n} + (x-a)^{2n}$

24. If

$$\frac{1}{\sqrt{4x+1}}\left[\left(\frac{1+\sqrt{4x+1}}{2}\right)^n - \left(\frac{1-\sqrt{4x+1}}{2}\right)^n\right],$$

then the possible values of 'n' can be $(n \in N)$

(A) 10 (B) 11 (C) 12 (D) 13

25. Let $n \in N, n > 3$ then $2^{4n} - 2^n(7n+1)$ is divisible by

(A) 125 (B) 196 (C) 16 (D) 49

Comprehension Based Questions

Paragraph for Question No. 1 to 3

If $(1 + x + x^2)^n = \sum_{r=0}^{2n} a_r x^r$, where $n \in N$.

1. Which of the following is true?

(A) $a_r = a_{n-r}$ (B) $a_{2r} = a_{n-r}$

(C) $a_r = a_{2n-r}$ (D) $a_r = a_{n-r+1}$

2. Value of $2(a_0 + a_1 + \ldots + a_{n-1}) + a_n$ is

(A) 2^{2n-1} (B) 3^n (C) $\frac{3^n}{2}$ (D) $\frac{3^n - 1}{2}$

3. If $a_0^2 - a_1^2 + a_2^2 - a_3^2 + \ldots + a_{2n}^2 = Ka_n$, then K is equal to

(A) 0 (B) $\frac{1}{2}$ (C) 1 (D) 2

Paragraph for Question No. 4 to 6

Let the 2nd, 3rd and 4th terms in the expansion of $(x+a)^n$ where $n \in N$ and $a, x \in R$ be 240, 720 and 1080 respectively.

4. The value of $(x-a)^n$ is equal to
   (A) –32 (B) –1 (C) 32 (D) 64

5. The smallest term in the expansion of $(x+a)^n$ is equal to
   (A) 16 (B) 32 (C) 81 (D) 160

6. The sum of odd numbered terms in the expansion of $(x+a)^n$ is equal to
   (A) 1486 (B) 1562 (C) 1664 (D) 2376

Paragraph for Question No. 7 to 9

Let $S_k = 3^k \binom{100}{0}\binom{100}{k} - 3^{k-1}\binom{100}{1}\binom{99}{k-1} + 3^{k-2}\binom{100}{2}\binom{98}{k-2} - \ldots + (-1)^k \binom{100}{k}\binom{100-k}{0}$

where $\binom{n}{r}$ represents the coefficient of x^r in the binomial expansion of $(1+x)^n \; \forall \; n \in N$ and $V_k = \left(\frac{1}{2}\right)^k S_k = M(100, k)$.

7. The value of $\sum_{r=0}^{100} S_k S_{100-k}$ equals
   (A) $\binom{200}{100}$ (B) $2^{100}\binom{200}{100}$
   (C) $2^{200}\binom{200}{100}$ (D) $2^{99}\binom{200}{99}$

8. The value of the $M(100, 49) + M(100, 50)$ is equal to
   (A) $M(99, 50)$ (B) $M(100, 51)$
   (C) $M(101, 49)$ (D) $M(101, 51)$

9. In an examination, there are 100 true–false type questions. The number of ways in which one student can make his choice to attempt or not to attempt the questions is given by
   (A) $\sum_{k=0}^{100} S_k$ (B) $\sum_{k=0}^{100} V_k$
   (C) $\sum_{k=1}^{100} S_k$ (D) $\sum_{k=1}^{100} V_k$

Paragraph for Question No. 10 to 12

A path of length n is a sequence of points (x_1, y_1), (x_2, y_2), …, (x_n, y_n) with integer coordinates such that for all i between 1 and $n-1$ both inclusive, **either** $x_{i+1} = x_i + 1 = y_i$ (in which case we say the ith step is rightward), **or** $x_{i+1} = x_i$ and $y_{i+1} = y_i + 1$ (in which case we say that the ith step is upward). This path is said to start at (x_1, y_1) and end at (x_n, y_n). Let $P(a, b)$, for a and b non-negative integers, denotes the number of paths that start at (0, 0) and end at (a, b).

10. The value of $\sum_{i=0}^{10} P(i, 10-i)$, is
    (A) 1024 (B) 512 (C) 256 (D) 128

11. Number of ordered pairs (i, j) where $i \neq j$ for which $P(i, 100-i) = P(j, 100-j)$, is
    (A) 50 (B) 99 (C) 100 (D) 101

12. The sum $P(43, 4) + \sum_{j=1}^{5} P(49-j, 3)$ is equal to
    (A) $P(4, 48)$ (B) $P(3, 49)$
    (C) $P(4, 47)$ (D) $P(5, 47)$

Match Matrix

1. **Only one matching option is possible**

Column I	Column II
(A) If the sum $S = {}^{20}C_2 + 2\cdot{}^{20}C_3 + 3\cdot{}^{20}C_4 + \ldots + 19.\,{}^{20}C_{20}$ can be expressed as $1 + p\cdot 2^q$, then the least value of $(p+q)$ is	(P) 0
(B) The remainder when $\left(15^{23} + 23^{23}\right)$ is divided by 19, is	(Q) 106
(C) If $\sum_{r=1}^{10} r(r-1)\,{}^{10}C_r = K\cdot 2^9$, then K is	(R) 29
(D) The total number of terms that are dependent on x in the expansion of $\left(x^2 - 2 + \frac{1}{x^2}\right)^{53}$ is	(S) 45

2. Only one matching option is possible

Column I	Column II
(A) $(20)^{13}+(13)^{20}$ is divided by 9, then the remainder is	(P) 8
(B) $\sum_{r=0}^{4} {}^4C_r (r-2)^2$ is divisible by	(Q) 198
(C) Let $n \in N$ and $n<(\sqrt{2}+1)^6$, then the integer just greater than n is	(R) 0
(D) If the ninth term in the expansion of $\left\{3^{\log_3\sqrt{25^{x-1}+7}}+3^{-\frac{1}{8}\log_3(5^{x-1}+1)}\right\}^{10}$ is equal to 180, then x is	(S) 1

3.

Column I	Column II
(A) The number of solution(s) of the equation $[x]=\|2x-1\|$ [**Note:** $[x]$ denotes greatest integer less than or equal to x.]	(P) 1
(B) If the sum of the coefficients in the expansion of $\left(\frac{1}{3}-15x+15x^3\right)^{2014}(17x-17x^5+3)^{2015}$ is n, then the number of dissimilar terms in the expansion of $(1+x)^n$ is greater than	(Q) 2
(C) If the numerically greatest term in the expansion of $(1-3x)^8$ for $x=\frac{1}{2}$ is T_r then r is greater than	(R) 3
	(S) 4

4. Match List I with List II and select the correct answer using the code given below the list.

List I	List II
(P) If the binomial coefficients of the rth, $(r+1)^{th}$ and $(r+2)^{th}$ term in the expansion of $(1+x)^{14}$ are in A.P., then r is equal to	(1) 5
(Q) The sum of the coefficients in the polynomial expansion of $[(1+x+x^2+x^3+x^4 \ldots+(x)^{n-1})(1-x)]^m$, $(n, m \in N)$, is	(2) 4
(R) Sum of the series $\sum_{r=1}^{n}(-1)^r\cdot{}^nC_r\left[1+i^r+i^{2r}+\ldots+i^{(n-1)r}\right]$, where $(n=4k, k \in I)$ is equal to (where $i=\sqrt{-1}$)	(3) 0
(S) If $3^{37}=80\lambda+k$, where $\lambda \in N$, then k is	(4) 3

Subjective Answer Type (Hard Nut)

1. Prove that no three consecutive binomial coefficients can be in (i) GP and (ii) HP.

2. If $(1+x)^n=\sum_{r=0}^{n} a_r x^r$, $br=1+\frac{a_r}{a_{r-1}}$ and $\Pi b_r=\frac{(101)^{100}}{100\,!}$, then find the value of $\frac{n}{20}$

3. If $\frac{2}{1!13!}+\frac{2}{3!11!}+\frac{2}{5!9!}+\frac{1}{7!7!}=\frac{2^m}{n!}$, then find the value of m and n.

4. If $1.0!+3.1!+7.2!+13.3!+21.4!+\ldots$ up to $(n+1)$ terms $=4000.(4000!)$, then find the value of n.

5. If $\sin^2\frac{\pi}{7}\sin\frac{2\pi}{7}\sin\frac{4\pi}{7}=\sum_{r=0}^{n} c_r\cos\frac{r\pi}{7}$, where $c_0, c_1, -\ldots, c_n$ are constants and $c_n \neq 0$, find the value of n where $n<5$.

6. If $x + y = 1$, then find the value of $\sum_{r=0}^{n} r\, n_{c_r}$ $x^r y^{n-r}$.

7. If $(1 + x)^n = \sum_{r=0}^{n} {}^n c_r x^r$ and $\sum_{r=0}^{n} \frac{1}{{}^n c_r} = a$, then find the value of $\sum\sum_{0 \le i < j \le n} \left(\frac{i}{{}^n c_i} + \frac{j}{{}^n c_j} \right)$.

8. Find the sum of the infinite series $a_1 + a_2 + a_3 + \ldots$, where $a_n = (\log 3)^n \sum_{k=1}^{n} \frac{(2k+1)}{k!(n-k)!}$

9. Prove that $\sum_{r=1}^{k} (-3)^{r-1}\ {}^{3n}c_{2r-1} = 0$ where $k = \frac{3n}{2}$ and n is an even positive integer.

10. If $\sum_{r=0}^{n} (r^2 + r + 1) r! = (2000)(2000)!$, then find the coefficient of x^{n+1} in the expansion of $\left(x + \frac{1}{x} \right)^{2000}$.

11. If $(1 - x^3)^n = \sum_{r=0}^{n} a_r x^r (1 - x)^{3n-2r}$, find a_r where $n \in N$.

12. If

$$\frac{{}^n c_r + 4 \cdot {}^n c_{r+1} + 6 \cdot {}^n c_{r+2} + 4 \cdot {}^n c_{r+3} + {}^n c_{r+4}}{{}^n c_r + 3 \cdot {}^n c_{r+1} + 3 \cdot {}^n c_{r+2} + {}^n c_{r+3}} = \frac{n + \lambda}{r + \lambda},$$

find the value of λ.

13. Sum the series $\frac{1^2}{3}\,{}^n c_1 + \frac{1^2 + 2^2}{5}\,{}^n c_2 + \frac{1^2 + 2^2 + 3^2}{7}\,{}^n c_3 + \ldots + \frac{1^2 + 2^2 + \ldots + n^2}{(2n+1)}\,{}^n c_n$.

14. Sum the series $\sum_{r=0}^{n} (r+1)\,{}^{2n} c_r$.

15. If $(1 + x + x^2)^n = \sum_{r=0}^{2n} a_r x^r$, find the value of $a_0 + 3a_1 + 5a_2 + \ldots + (4n + 1)a_{2n}$

16. Prove that

$${}^n c_1 \sin x \cos(n-1)x + {}^n c_2 \sin 2x \cos(n-2)x + {}^n c_3 \sin 3x \cos(n-3)x + \ldots + {}^n c_n \sin nx = 2^{n-1} \sin nx$$

17. Prove that

$$\sum_{r=0}^{n} {}^n c_r a^r b^{n-r} \cos(rB - (n-r)A) = c^n,$$

where a, b, c, A, B, C are the elements of triangle ABC.

18. Prove that $\sum_{r=0}^{n} (-1)^r \left(\frac{{}^n c_r}{(r+3)_{c_r}} \right) = \frac{3!}{2(n+3)}$.

19. If $\sum_{r=0}^{n} \left(\frac{r+2}{r+1} \right) {}^n c_r = \frac{2^8 - 1}{6}$, then find the value of n.

20. Let $1 + \sum_{r=1}^{10} (3^r \cdot {}^{10} c_r + r \cdot {}^{10} c_r) = 2^{10} (\alpha \cdot 4^5 + \beta)$ where $\alpha, \beta \in N$ and $f(x) = x^2 - 2x - k^2 + 1$. If α, β lies between the roots of $f(x) = 0$, then find the smallest positive integral value of k.

5.8 JEE Main: Archive

1. The remainder when 3^{2022} is divided by 5 is
(A) 1 (B) 2 (C) 3 (D) 4

[Main 2022 24th June 1st Shift]

2. The remainder on dividing $1 + 3 + 3^2 + 3^3 + \ldots + 3^{2021}$ by 50 is ______.

[Main 2022 24th June 2nd Shift]

3. Let C_r denote the binomial coefficient of x^r in the expansion of $(1+x)^{10}$. If $\alpha, \beta \in R$. $C_1 + 3 \cdot 2C_2 + 5 \cdot 3C_3 + \ldots$ up to 10 terms $=$

$$\frac{\alpha \times 2^{11}}{2^\beta - 1} \left(C_0 + \frac{C_1}{2} + \frac{C_2}{3} + \ldots \text{up to 10 terms} \right)$$

then the value of $\alpha + \beta$ is equal to ________.

[Main 2022 25th June 1st Shift]

4. The coefficient of x^{101} in the expression $(5+x)^{500}+x(5+x)^{499}+x^2(5+x)^{498}+\ldots x^{500}, x>0$, is

(A) ${}^{501}C_{101}(5)^{399}$ (B) ${}^{501}C_{101}(5)^{400}$

(C) ${}^{501}C_{100}(5)^{400}$ (D) ${}^{500}C_{101}(5)^{399}$

[Main 2022 25th June 2nd Shift]

5. If the sum of the coefficients of all the positive even powers of x in the binomial expansion of $\left(2x^3+\frac{3}{x}\right)^{10}$ is $5^{10}-\beta\cdot 3^9$, then β is equal to ________.

[Main 2022 25th June 2nd Shift]

6. The remainder when $(2021)^{2023}$ is divided by 7 is:

(A) 1 (B) 2 (C) 5 (D) 6

[Main 2022 26th June 1st Shift]

7. If

$$\left({}^{40}C_0\right)+\left({}^{41}C_1\right)+\left({}^{42}C_2\right)+\ldots+\left({}^{60}C_{20}\right)$$

$=\frac{m}{n}\,{}^{60}C_{20}, m$

and n are coprime, then $m+n$ is equal to ________.

[Main 2022 26th June 2nd Shift]

8. If the coefficient of x^{10} in the binomial expansion of $\left(\frac{\sqrt{x}}{5^{\frac{1}{4}}}+\frac{\sqrt{5}}{x^{\frac{1}{3}}}\right)^{60}$ is $5^k\ell$, where $\ell, k\in N$ and ℓ is co-prime to 5, then k is equal to ________.

[Main 2022 27th June 1st Shift]

9. If the sum of the coefficients of all the positive powers of x, in the binomial expansion of $\left(x^n+\frac{2}{x^5}\right)^7$ is 939, then the sum of all the possible integral values of n is:

[Main 2022 27th June 2nd Shift]

10. If $\sum_{k-1}^{31}\left({}^{31}C_k\right)\left({}^{31}C_{k-1}\right)-\sum_{k=1}^{30}\left({}^{30}C_k\right)\left({}^{30}C_{k-1}\right)$ $=\frac{\alpha(60!)}{(30!)(31!)}$, where $\alpha\in R$, then the value of 16α is equal to

(A) 1411 (B) 1320 (C) 1615 (D) 1855

[Main 2022 28th June 1st Shift]

11. The number of positive integers k such that the constant term in the binomial expansion of $\left(2x^3+\frac{3}{x^k}\right)^{12}$, $x\neq 0$ is $2^8\cdot\ell$, where ℓ is an odd integer, is ________.

[Main 2022 28th June 1st Shift]

12. The term independent of x in the expression of $\left(1-x^2+3x^3\right)\left(\frac{5}{2}x^3-\frac{1}{5x^2}\right)^{11}$, $x\neq 0$ is

(A) $\frac{7}{40}$ (B) $\frac{33}{200}$ (C) $\frac{39}{200}$ (D) $\frac{11}{50}$

[Main 2022 28th June 2nd Shift]

13. If the constant term in the expansion of $\left(3x^3-2x^2+\frac{5}{x^5}\right)^{10}$ is $2^k, \ell$, where ℓ is an odd integer, then the value of k is equal to:

(A) 6 (B) 7 (C) 8 (D) 9

[Main 2022 29th June 1st Shift]

14. Let $n\geq 5$ be an integer. If $9^n-8n-1=64\alpha$ and $6^n-5n-1=25\beta$, then $\alpha-\beta$ is equal to

(A) $1+{}^nC_2(8-5)+{}^nC_3\left(8^2-5^2\right)+\ldots+{}^nC_n\left(8^{n-1}-5^{n-1}\right)$

(B) $1+{}^nC_3(8-5)+{}^nC_4\left(8^2-5^2\right)+\ldots+{}^nC_n\left(8^{n-2}-5^{n-2}\right)$

(C) ${}^nC_3(8-5)+{}^nC_4\left(8^2-5^2\right)+\ldots+{}^nC_n\left(8^{n-2}-5^{n-2}\right)$

(D) ${}^nC_4(8-5)+{}^nC_5\left(8^2-5^2\right)+\ldots+{}^nC_n\left(8^{n-3}-5^{n-3}\right)$

[Main 2022 29th June 2nd Shift]

15. Let the coefficients of x^{-1} and x^{-3} in the expression of $\left(2x^{\frac{1}{5}}-\frac{1}{x^{\frac{1}{5}}}\right)^{15}$, $x>0$, be m and n respectively. If r is a positive integer such $mn^2 = {}^{15}C_r \cdot 2^r$, then the value of r is equal to ______.

[Main 2022 29th June 2nd Shift]

16. If the maximum value of the term independent of t in the expansion of $\left(t^2x^{\frac{1}{5}}+\frac{(1-x)^{\frac{1}{10}}}{t}\right)^{15}$, $x\geq 0$, is K, then $8K$ is equal to ______.

[Main 2022 25th July 1st Shift]

17. The remainder when $(11)^{1011}+(1011)^{11}$ is divided by 9 is

(A) 1 (B) 4 (C) 6 (D) 8

[Main 2022 25th July 2nd Shift]

18. If the coefficients of x and x^2 in the expansion of $(1+x)^p(1-x)^q$, $p, q \leq 15$, are -3 and -5 respectively, then the coefficient of x^3 is equal to ______.

[Main 2022 26th July 1st Shift]

19. $\sum\limits_{\substack{i,j=0 \\ i\neq j}}^{n} {}^nC_i \, {}^nC_j$ is equal to

(A) $2^{2n} - {}^{2n}C_n$ (B) $2^{2n-1} - {}^{2n-1}C_{n-1}$

(C) $2^{2n} - \frac{1}{2}\,{}^{2n}C_n$ (D) $2^{n-1} + {}^{2n-1}C_n$

[Main 2022 26th July 2nd Shift]

20. The remainder when $(2021)^{2022}+(2022)^{2021}$ is divided by 7 is

(A) 0 (B) 1 (C) 2 (D) 6

[Main 2022 27th July 1st Shift]

21. Let the 9th term in the binomial expansion of $(3+6x)^n$, in the increasing powers of $6x$, be the greatest for $x=\frac{3}{2}$, the least value of n is n_0. If k is the ratio of the coefficient of x^6 to the coefficient of x^3, then $k+n_0$ is equal to:

[Main 2022 27th July 2nd Shift]

22. The remainder when $7^{2022}+3^{2022}$ is divided by 5 is

(A) 0 (B) 2 (C) 3 (D) 4

[Main 2022 28th July 1st Shift]

23. Let the coefficients of the middle terms in the expansion of $\left(\frac{1}{\sqrt{6}}+\beta x\right)^4$, $(1-3\beta x)^2$ and $\left(1-\frac{\beta}{2}x\right)^6$, $\beta>0$, respectively form the first three terms of an A.P. If d is the common difference of this A.P., then $50-\frac{2d}{\beta^2}$ is equal to ______.

[Main 2022 28th July 2nd Shift]

24. If

$$1+\left(2+{}^{49}C_1+{}^{49}C_2+\ldots+{}^{49}C_{49}\right)\left({}^{50}C_2+{}^{50}C_4+\ldots+{}^{50}C_{50}\right)$$

is equal to $2^n \cdot m$, where m is odd, then $n+m$ is equal to ______.

[Main 2022 28th July 2nd Shift]

25. Let the ratio of the fifth term from the beginning to the fifth term form the end in the binomial expansion of $\left(\sqrt[4]{2}+\frac{1}{\sqrt[4]{3}}\right)^n$, in the increasing powers of $\frac{1}{\sqrt[4]{3}}$ be $\sqrt[4]{6}:1$. If the sixth term from the beginning is $\frac{\alpha}{\sqrt[4]{3}}$, then α is equal to ______.

[Main 2022 29th July 1st Shift]

26. If $\sum\limits_{k=1}^{10} K^2\left({}^{10}C_k\right)^2 = 22000L$, then L is equal to ______.

[Main 2022 29th July 2nd Shift]

27. The value $\sum\limits_{r=0}^{22} {}^{22}C_r \, {}^{23}C_r$ is

(A) ${}^{45}C_{23}$ (B) ${}^{44}C_{23}$ (C) ${}^{45}C_{24}$ (D) ${}^{44}C_{22}$

[Main 2023 24th Jan 1st Shift]

28. Suppose $\sum\limits_{r=0}^{2023} r^2 \, {}^{2023}C_r = 2023\times\alpha\times 2^{2022}$. Then the value of α is ______.

[Main 2023 24th Jan 1st Shift]

29. If

$$\left({}^{30}C_1\right)^2+2\left({}^{30}C_2\right)^2+3\left({}^{30}C_3\right)^2+\ldots+30\left({}^{30}C_{30}\right)^2=\frac{\alpha 60!}{(30!)^2},$$

then α is equal to

(A) 30 (B) 60 (C) 15 (D) 10

[Main 2023 24th Jan 2nd Shift]

30. Let the sum of the coefficients of the first three terms in the expansion of $\left(x-\frac{3}{x^2}\right)^n, x\neq 0, n\in N$, be 376. Then the coefficient of x^4 is ________.

[Main 2023 24th Jan 2nd Shift]

31. If a_r is the coefficient of x^{10-r} in the binomial expansion of $(1+x)^{10}$, then $\sum_{r=1}^{10} r^3\left(\frac{a_r}{a_{r-1}}\right)^2$ is equal to

(A) 4895 (B) 1210 (C) 5445 (D) 3025

[Main 2023 25th Jan 1st Shift]

32. The constant term in the expansion of $\left(2x+\frac{1}{x^7}+3x^2\right)^5$ is ________.

[Main 2023 25th Jan 1st Shift]

33. $\sum_{k=0}^{6} {}^{51-k}C_3$ is equal to

(A) ${}^{51}C_4-{}^{45}C_4$ (B) ${}^{51}C_3-{}^{45}C_3$

(C) ${}^{52}C_4-{}^{45}C_4$ (D) ${}^{52}C_3-{}^{45}C_3$

[Main 2023 25th Jan 2nd Shift]

34. The remainder when $(2023)^{2023}$ is divided by 35 is ________.

[Main 2023 25th Jan 2nd Shift]

35. If the coefficient of x^9 in $\left(\alpha x^3+\frac{1}{\beta x}\right)^{11}$ and the coefficient of x^{-9} in $\left(\alpha x-\frac{1}{\beta x^3}\right)^{11}$ are equal, then $(\alpha\beta)^2$ is equal to ________.

[Main 2023 29th Jan 1st Shift]

36. Let the coefficients of three consecutive terms in the binomial expansion of $(1+2x)^n$ be in the ratio 2 : 5 : 8. Then the coefficient of the term, which is in the middle of these three terms, is

[Main 2023 29th Jan 1st Shift]

37. Let K be the sum of the coefficients of the odd powers of x in the expansion of $(1+x)^{99}$. Let a be the middle term in the expansion of $\left(2+\frac{1}{\sqrt{2}}\right)^{200}$.

If $\frac{{}^{200}C_{99}K}{a}=\frac{2^\ell m}{n}$, where m and n are odd numbers, then the ordered pair (ℓ, n) is equal to:

(A) (50, 51) (B) (51, 99)

(C) (50, 101) (D) (51, 101)

[Main 2023 29th Jan 2nd Shift]

38. If the coefficient of x^{15} in the expansion of $\left(ax^3+\frac{1}{bx^{\frac{1}{3}}}\right)^{15}$ is equal to the coefficient of x^{-15} in the expansion of $\left(ax^{\frac{1}{3}}-\frac{1}{bx^3}\right)^{15}$, where a and b are positive real numbers, then for each such ordered pair (a, b):

(A) $a = b$ (B) $ab = 1$

(C) $a = 3b$ (D) $ab = 3$

[Main 2023 30th Jan 1st Shift]

39. The coefficient of x^{301} in

$$(1+x)^{500}+x(1+x)^{499}+x^2(1+x)^{498}+\ldots+x^{500}$$ is:

(A) ${}^{501}C_{302}$ (B) ${}^{500}C_{301}$

(C) ${}^{500}C_{300}$ (D) ${}^{501}C_{200}$

[Main 2023 30th Jan 1st Shift]

40. Let $x=\left(8\sqrt{3}+13\right)^{13}$ and $y=\left(7\sqrt{2}+9\right)^9$.

If $[t]$ denotes the greatest integer $\leq t$, then

(A) $[x]+[y]$ is even

(B) $[x]$ is odd but $[y]$ is even

(C) $[x]$ is even but $[y]$ is odd

(D) $[x]$ and $[y]$ are both odd

[Main 2023 30th Jan 2nd Shift]

41. The 50th root of a number x is 12 and the 50th root of another number y is 18. Then the remainder obtained on dividing $(x+y)$ by 25 is ________.

[Main 2023 30th Jan 2nd Shift]

42. Let $\alpha > 0$, be the smallest number such that the expansion of $\left(x^{\frac{2}{3}}+\frac{2}{x^3}\right)^{30}$ has a term $\beta x^{-\alpha}, \beta \in N$. Then α is equal to ________.

[Main 2023 31st Jan 1st Shift]

43. The remainder on dividing 5^{99} by 11 is ________.

[Main 2023 31st Jan 1st Shift]

44. The coefficient of x^{-6}, in the expansion of $\left(\frac{4x}{5}, \frac{5}{2x^2}\right)^9$, is ________.

[Main 2023 31st Jan 2nd Shift]

45. If the constant term in the binomial expansion of $\left(\frac{x^{\frac{5}{2}}}{2}-\frac{4}{x}\right)^9$ is -84 and the coefficient of $x^{-3\ell}$ is $2^{\alpha}\beta$, where $\beta < 0$ is an odd number, Then $|\alpha\ell-\beta|$ is equal to ________.

[Main 2023 31st Jan 2nd Shift]

46. The value of

$$\frac{1}{1!\,50!}+\frac{1}{3!\,48!}+\frac{1}{5!\,46!}+\ldots+\frac{1}{49!\,2!}+\frac{1}{51!\,1!}$$ is

(A) $\frac{2^{50}}{50!}$ (B) $\frac{2^{50}}{51!}$ (C) $\frac{2^{51}}{51!}$ (D) $\frac{2^{51}}{50!}$

[Main 2023 01st Feb 1st Shift]

47. The remainder when $19^{200}+23^{200}$ is divided by 49, is ________.

[Main 2023 01st Feb 1st Shift]

48. Sum of the squares of all possible values of $x = 4$. If the term without x in the expansion of $\left(x^{\frac{2}{3}}+\frac{a}{x^3}\right)^{22}$ is 7315, then $|\alpha|$ is equal to ________.

[Main 2023 01st Feb 2nd Shift]

49. Let the sixth term in the binomial expansion of $\left(\sqrt{2^{\log_2\left(10-3^x\right)}}+\sqrt[5]{2^{(x-2)\log_2 3}}\right)^m$, in the increasing powers of $2^{(x-2)\log_2 3}$, be 21. If the binomial coefficients of the second, third and fourth terms in the expansion are respectively the first, third and fifth terms of an A.P., then the sum of the squares of all possible values of x is

[Main 2023 01st Feb 2nd Shift]

50. If the ratio of the fifth term from the beginning to the fifth term from the end in the expansion of $\left(\sqrt[4]{2}+\frac{1}{\sqrt[4]{3}}\right)^n$ is $\sqrt{6}:1$, then the third term from the beginning is

(A) $30\sqrt{2}$ (B) $30\sqrt{3}$ (C) $60\sqrt{2}$ (D) $60\sqrt{3}$

[Main 2023 6th April 1st Shift]

51. If ${}^{2n}C_3 : {}^{n}C_3 = 10:1$, then the ratio $\left(n^2+3n\right):\left(n^2-3n+4\right)$ is

(A) 35 : 16 (B) 27 : 11
(C) 65 : 37 (D) 2 : 1

[Main 2023 6th April 1st Shift]

52. The coefficient of x^{18} in the expansion of $\left(x^4-\frac{1}{x^3}\right)^{15}$ is ________.

[Main 2023 6th April 1st Shift]

53. If the coefficient of x^7 in $\left(ax^2+\frac{1}{2bx}\right)^{11}$ and x^{-7} in $\left(ax-\frac{1}{3bx^2}\right)^{11}$ are equal, then

(A) $729ab = 32$ (B) $32ab = 729$
(C) $64ab = 243$ (D) $243ab = 64$

[Main 2023 6th April 2nd Shift]

54. Among the statements:

$(S1): 2023^{2022}-1999^{2022}$ is divisible by 8.

$(S2): 13(13)^n-11n-13$ is divisible by 144 for infinitely many $n \in N$.

(A) only (S2) is correct
(B) only (S1) is correct
(C) both (S1) and (S2) are correct
(D) both (S1) and (S2) are incorrect

[Main 2023 6th April 2nd Shift]

55. If the coefficient of three consecutive terms in the expansion of $(1+x)^n$ are in the ratio $1:5:20$ then the coefficient of the fourth term is
(A) 2436 (B) 5481 (C) 1827 (D) 3654
[Main 2023 8th April 1st Shift]

56. Let $[t]$ denote the greatest integer $\le t$. If the constant term in the expansion of $\left(3x^2-\frac{1}{2x^5}\right)^7$ is α then $|\alpha|$ is equal to ______.
[Main 2023 8th April 1st Shift]

57. If a_α is the greatest term in the sequence $a_n=\frac{n^3}{n^4+147}$, $n=1,2,3,\ldots$, then α is equal to ______. **[Main 2023 8th April 1st Shift]**

58. The largest natural number n such that $3n$ divides 66! is ______.
[Main 2023 8th April 1st Shift]

59. The absolute difference of the coefficients of x^{10} and x^7 in the expansion of $\left(2x^2+\frac{1}{2x}\right)^{11}$ is equal to
(A) 13^3-13 (B) 11^3-11
(C) 10^3-10 (D) 12^3-12
[Main 2023 8th April 2nd Shift]

60. $25^{190}-19^{190}-8^{190}+2^{190}$ is divisible by
(A) neither 14 nor 34 (B) 14 but not by 34
(C) 34 but not by 14 (D) both 14 and 34
[Main 2023 8th April 2nd Shift]

61. If the coefficient of x^7 in $\left(ax-\frac{1}{bx^2}\right)^{13}$ and the coefficient of x^{-5} in $\left(ax+\frac{1}{bx^2}\right)^{13}$ are equal, then a^4b^4 is equal to
(A) 11 (B) 44 (C) 22 (D) 33
[Main 2023 10th April 1st Shift]

62. The coefficient of x^7 in $\left(1-x+2x^3\right)^{10}$ is ______. **[Main 2023 10th April 1st Shift]**

63. Let the number $(22)^{2022}+(2022)^{22}$ leave the remainder α when divided by 3 and β when divided by 7. Then $\left(\alpha^2+\beta^2\right)$ is equal to
(A) 20 (B) 13 (C) 5 (D) 10
[Main 2023 10th April 2nd Shift]

64. If the coefficients of x and x^2 in $(1+x)^p(1-x)^q$ are 4 and -5 respectively, then $2p+3q$ is equal to
(A) 60 (B) 69 (C) 66 (D) 63
[Main 2023 10th April 2nd Shift]

65. The number of integral terms in the expansion of $\left(3^{\frac{1}{2}}+5^{\frac{1}{4}}\right)^{680}$ is equal to
[Main 2023 11th April 1st Shift]

66. The mean of the coefficients of $x, x^2, \ldots, x^7$ in the binomial expression of $(2+x)^9$ is ______. **[Main 2023 11th April 1st Shift]**

67. If the 1011th term from the end in the binomial expansion of $\left(\frac{4x}{5}-\frac{5}{2x}\right)^{2022}$ is 1024 times 1011th term from the beginning, then $32|x|$ is equal to
(A) 15 (B) 10 (C) 12 (D) 8
[Main 2023 11th April 2nd Shift]

68. The sum of the coefficients of three consecutive terms in the binomial expansion of $(1+x)^{n+2}$, which are in the ratio $1:3:5$, is equal to
(A) 92 (B) 63 (C) 41 (D) 25
[Main 2023 11th April 2nd Shift]

69. If $\frac{1}{n+1}\,{}^nC_n+\frac{1}{n}\,{}^nC_{n-1}+\ldots+\frac{1}{2}\,{}^nC_1+{}^nC_0=\frac{1023}{10}$ then n is equal to
(A) 9 (B) 8 (C) 7 (D) 6
[Main 2023 12th April 1st Shift]

70. The sum of the coefficients of the first 50 terms in the binomial expansion of $(1-x)^{100}$, is equal to
(A) ${}^{101}C_{50}$ (B) ${}^{99}C_{49}$
(C) $-{}^{101}C_{50}$ (D) $-{}^{99}C_{49}$
[Main 2023 12th April 1st Shift]

71. Fractional part of the number $\frac{4^{2022}}{15}$ is equal to
(A) $\frac{8}{15}$ (B) $\frac{4}{15}$ (C) $\frac{14}{15}$ (D) $\frac{1}{15}$
[Main 2023 13th April 1st Shift]

72. Let α be the constant term in the binomial expansion of $\left(\sqrt{x}-\frac{6}{x^{\frac{3}{2}}}\right)^n$, $n \le 15$. If the sum of the coefficients of the remaining terms in the expansion is 649 and the coefficient x^{-n} is $\lambda\alpha$, then λ is equal to ______. **[Main 2023 13th April 1st Shift]**

73. Let for
$x \in R, S_0(x) = x, S_k = C_k x + k\int_0^x S_{k-1}(t)\,dt$
where
$C_0 = 1, C_k = 1 - \int_0^1 S_{k-1}(x)\,dx, k = 1, 2, 3, \ldots$
Then $S_2(3) + 6C_3$ is equal to ______.
[Main 2023 13th April 1st Shift]

74. The coefficient of x^5 in the expansion of $\left(2x^3 - \frac{1}{3x^2}\right)^5$ is

(A) $\frac{80}{9}$ (B) 9(C) 8 (D) $\frac{26}{3}$

[Main 2023 13th April 2nd Shift]

75. The remainder, when 7^{103} is divided by 17, is ______. **[Main 2023 13th April 2nd Shift]**

76. Let $\left(a+bx+cx^2\right)^{10} = \sum_i^{20} p_i x^i$, $a, b, c \in N$.
If $p_1 = 20$ and $p_2 = 210$, then $2(a+b+c)$ is equal to

(A) 6 (B) 15 (C) 12 (D) 8

[Main 2023 15th April 1st Shift]

77. If the coefficient of x^{30} in the expansion of $\left(1+\frac{1}{x}\right)^6\left(1+x^2\right)^7\left(1-x^3\right)^8$; $x \ne 0$ is α, then $|\alpha|$ equals ______.

[Main 2024 1st Feb 1st Shift]

78. Let m and n be the coefficients of seventh and thirteenth terms respectively in the expansion of $\left(\frac{1}{3}x^{\frac{1}{3}} + \frac{1}{2x^{\frac{2}{3}}}\right)^{18}$. Then $\left(\frac{n}{m}\right)^{\frac{1}{3}}$ is:

(A) $\frac{4}{9}$ (B) $\frac{1}{9}$ (C) $\frac{1}{4}$ (D) $\frac{9}{4}$

[Main 2024 1st Feb 2nd Shift]

79. ${}^{n-1}C_r = \left(k^2-8\right){}^nC_{r+1}$ if and only if:

(A) $2\sqrt{2} < k \le 3$ (B) $2\sqrt{3} < k \le 3\sqrt{2}$
(C) $2\sqrt{3} < k < 3\sqrt{3}$ (D) $2\sqrt{2} < k < 2\sqrt{3}$

[Main 2024 27th Jan 1st Shift]

80. If A denotes the sum of all the coefficients in the expansion of $\left(1-3x+10x^2\right)^n$ and B denotes the sum of all the coefficients in the expansion of $\left(1+x^2\right)^n$, then

(A) $A = B^3$ (B) $3A = B$
(C) $B = A^3$ (D) $A = 3B$

[Main 2024 27th Jan 1st Shift]

81. The coefficient of x^{2012} in the expansion of $(1-x)^{2008}\left(1+x+x^2\right)^{2007}$ is equal to ______.

[Main 2024 27th Jan 2nd Shift]

82. If $\frac{{}^{11}C_1}{2} + \frac{{}^{11}C_2}{3} + \ldots + \frac{{}^{11}C_9}{10} = \frac{n}{m}$ with $\gcd(n, m) = 1$, then $n+m$ is equal to

[Main 2024 29th Jan 1st Shift]

83. Remainder when $64^{32^{32}}$ is divided by 9 is equal to ______. **[Main 2024 29th Jan 2nd Shift]**

84. The number of integral terms in the expansion of $\left\{7^{\left(\frac{1}{2}\right)} + 11^{\left(\frac{1}{6}\right)}\right\}^{824}$ is equal to ______. **[Main 2024 30th Jan 1st Shift]**

85. Suppose $2-p, p, 2-\alpha, \alpha$ are the coefficients of four consecutive terms in the expansion of $(1+x)^n$. Then the value of $p^2 - \alpha^2 + 6\alpha + 2p$ equals

(A) 4 (B) 10
(C) 8 (D) data inconsistent

[Main 2024 30th Jan 2nd Shift]

86. Let $\alpha = \sum_{k=0}^{n}\left(\frac{\left({}^nC_k\right)^2}{k+1}\right)$ and

$\beta = \sum_{k=0}^{n-1}\left(\frac{{}^nC_k\,{}^nC_{k+1}}{k+2}\right)$

If $5\alpha = 6\beta$, then n equals

[Main 2024 30th Jan 2nd Shift]

87. In the expansion of
$(1+x)\left(1-x^2\right)\left(1+\frac{3}{x}+\frac{3}{x^2}+\frac{1}{x^3}\right)^5$, $x \ne 0$,
the sum of the coefficient of x^3 and x^{-13} is equal to ______. **[Main 2024 31st Jan 1st Shift]**

88. If for some m, n; ${}^6C_m + 2\left({}^6C_{m+1}\right) + {}^6C_{m+2} > {}^8C_3$ and ${}^{n-1}P_3 : {}^nP_4 = 1:8$, then ${}^nP_{m+1} + {}^{n+1}C_m$ is equal to

(A) 380 (B) 376 (C) 384 (D) 372

[Main 2024 31st Jan 2nd Shift]

89. Let the coefficient of x^r in the expansion of $(x+3)^{n-1} + (x+3)^{n-2}(x+2) + (x+3)^{n-3}(x+2)^2 + \ldots + (x+2)^{n-1}$ be α_r. If $\sum_{r=0}^{n} \alpha_r = \beta^n - \gamma^n, \beta, \gamma \in N$, then the value of $\beta^2 + \gamma^2$ equals ________.

[Main 2024 31st Jan 2nd Shift]

5.9 Answer Key

5.9.1 Know Your Concepts

1. (i) $-1760x^9y^3$ (ii) 100 (iii) $\frac{n-4}{5}$ (iv) 1 **2.** (i) 210 C^{12} (ii) ${}^{m+n}C_n$ (iii) 12 (iv) $-5940x^3$

3. (i) $61236x^5y^5$ (ii) $t_4 = -560x^5, t_5 = -280x^2$ (v) 0 **4.** (97) **5.** (1) **6.** (6th term)

7. $\left(-3, \frac{-25}{12}\right) \cup \left(\frac{25}{12}, 3\right)$ **8.** $\frac{17}{54}$ **9.** (8) **11.** (1) **12.** (300!) **13.** $\frac{2^{11}}{12!}$

14. $\left[-2, -\sqrt{3}\right) \cup \left(\sqrt{3}, 2\right]$ **16.** (1) **17.** (9) **18.** $\frac{5^n - 1}{4}$ **19.** $(n+2)2^{n-1}$ **20.** (3)

5.9.2 JEE Main: Sample Questions

Single Answer Type

1. (C) **2.** (B) **3.** (C) **4.** (A) **5.** (B) **6.** (D) **7.** (A) **8.** (D)
9. (A) **10.** (A) **11.** (B) **12.** (B) **13.** (C) **14.** (A) **15.** (B) **16.** (B)
17. (B) **18.** (B) **19.** (B) **20.** (B) **21.** (A) **22.** (A) **23.** (A) **24.** (A)
25. (A) **26.** (C) **27.** (B) **28.** (B) **29.** (A) **30.** (A) **31.** (C) **32.** (B)
33. (B) **34.** (A) **35.** (C) **36.** (D) **37.** (D) **38.** (B) **39.** (D) **40.** (B)
41. (B) **42.** (A) **43.** (D) **44.** (B) **45.** (B) **46.** (A) **47.** (B) **48.** (D)
49. (D) **50.** (B) **51.** (D) **52.** (C) **53.** (A) **54.** (C) **55.** (A) **56.** (A)
57. (D) **58.** (A)

Integer Answer Type

1. (6) **2.** (5) **3.** (7) **4.** (8) **5.** (7) **6.** (6) **7.** (0) **8.** (1)
9. (4) **10.** (8) **11.** (2520) **12.** (29) **13.** (12) **14.** (1040) **15.** (0006) **16.** (816)
17. (7) **18.** (9)

5.9.3 JEE Advanced: Sample Questions

Single Answer Type

1. (B) **2.** (C) **3.** (D) **4.** (B) **5.** (B) **6.** (B) **7.** (B) **8.** (D)
9. (D) **10.** (A) **11.** (D) **12.** (A) **13.** (C) **14.** (C) **15.** (D) **16.** (A)
17. (A) **18.** (A) **19.** (B) **20.** (A) **21.** (B) **22.** (D) **23.** (D) **24.** (B)
25. (C)

Multiple Answer Type

1. (B, C, D) **2.** (A, B, C, D) **3.** (A, C) **4.** (A, C, D)
5. (B, C, D) **6.** (A, B) **7.** (A, B) **8.** (A, C, D)
9. (A, C) **10.** (A, B, C) **11.** (B, C) **12.** (A, B, C)
13. (A, B, C) **14.** (A, C) **15.** (A, B, D) **16.** (A, C, D)
17. (A, B, C, D) **18.** (A, B, C) **19.** (A, B, C, D) **20.** (C, D)
21. (A, B, C) **22.** (B, C, D) **23.** (B, C, D) **24.** (B, C)
25. (B, C, D)

Comprehension Based Questions

1. (C) **2.** (D) **3.** (C) **4.** (B) **5.** (B) **6.** (B) **7.** (B) **8.** (D)
9. (A) **10.** (A) **11.** (C) **12.** (A)

Match Matrix

1. (A – R; B – P; C – S; D – Q) **2.** (A – R; B – P; C – Q; D – S)
3. (A – Q; B – P, Q, R; C – P, Q, R, S) **4.** (A – P; B – R; C – R; D – S)

Subjective Answer Type (Hard Nut)

2. (5) **3.** ($m = 13$ and $n = 14$) **4.** (3999) **5.** (2) **6.** (nx) **7.** $\frac{n^2a}{2}$

8. (18 log 3+6) **10.** (1) **11.** ${}^nC_r3^r$ **12.** (4) **13.** $\frac{n(n+3)}{3}2^{n-3}$

14. $(n+1)2^{2n-1}+\frac{1}{2}{}^{2n}c_n$ **15.** $(2n+1)3n$ **19.** (5) **20.** (5)

5.9.4 JEE Main: Archive

1. (D) **2.** (4) **3.** (286) **4.** (A) **5.** (83) **6.** (C) **7.** (102) **8.** (5)
9. (57) **10.** (A) **11.** (2) **12.** (B) **13.** (D) **14.** (C) **15.** (5) **16.** (6006)
17. (D) **18.** (23) **19.** (A) **20.** (A) **21.** (24) **22.** (C) **23.** (57) **24.** (99)
25. (84) **26.** (221) **27.** (A) **28.** (1012) **29.** (C) **30.** (405) **31.** (B) **32.** (1080)
33. (C) **34.** (7) **35.** (A) **36.** (1120) **37.** (C) **38.** (D) **39.** (D) **40.** (A)
41. (23) **42.** (2) **43.** (9) **44.** (5040) **45.** (98) **46.** (B) **47.** (29) **48.** (1)
49. (D) **50.** (D) **51.** (D) **52.** (5005) **53.** (A) **54.** (B) **55.** (D) **56.** (1275)
57. (5) **58.** (31) **59.** (D) **60.** (C) **61.** (C) **62.** (960) **63.** (C) **64.** (D)
65. (171) **66.** (2736) **67.** (B) **68.** (B) **69.** (A) **70.** (D) **71.** (D) **72.** (36)
73. (18) **74.** (A) **75.** (12) **76.** (C) **77.** (678) **78.** (D) **79.** (A) **80.** (A)
81. (0) **82.** (2041) **83.** (1) **84.** (138) **85.** (D) **86.** (10) **87.** (118) **88.** (D)
89. (25)

5.10 Solutions

5.10.1 Know Your Concepts

1. (i) 4th term $= {}^{12}C_3x^{12-3}(-2y)^3$

$= -1760x^9y^3$

(ii) $T_3 = {}^5C_2\left(\frac{1}{x}\right)^{5-2}\left(x^{\log_{10}x}\right)^2 = 10^3$

$\Rightarrow 10\frac{1}{x^3}\left(x^{\log_{10}x}\right)^2 = 10^3$

$\Rightarrow \frac{1}{x^3}\left(x^{\log_{10}x}\right)^2 = 10^2$

$\Rightarrow x^{-3+2\log_{10}x} = 10^2$

$\Rightarrow (-3+2\log_{10}x)\log_{10}x = 2\log_{10}10$

$\Rightarrow$ Put $\log_{10}x = y$

$\Rightarrow 2y^2-3y-2=0$

$\Rightarrow 2y^2-4y+y-2=0$

$\Rightarrow 2y(y-2)+1(y-2)=0$

$\Rightarrow y = 2, -\frac{1}{2}$

$\log_{10}x = 2 \Rightarrow x = 100$

$\log_{10}x = -\frac{1}{2} \Rightarrow x = \frac{1}{\sqrt{10}}$

But $x > 1 \Rightarrow x = 100$

(iii) Since, $T_5 + T_6 = 0$

$\Rightarrow {}^nC_4a^{n-4}b^4 - {}^nC_5a^{n-5}b^5 = 0$

$\Rightarrow {}^nC_4a^{n-4}b^4 = {}^nC_5a^{n-5}b^5$

$\Rightarrow \frac{a}{b} = \frac{{}^nC_5}{{}^nC_4} = \frac{n-4}{5}$

(iv) $t_{17} = t_{18}$

$\Rightarrow {}^{50}C_{16}2^{50-16}a^{16} = {}^{50}C_{17}2^{50-17}a^{17}$

$\Rightarrow a = \frac{{}^{50}C_{16}}{{}^{50}C_{17}}\frac{2^{34}}{2^{33}}$

$\Rightarrow a = \frac{\frac{50!}{16!\,34!}}{\frac{50!}{17!\,33!}}\times 2 = \frac{17}{34}\times 2 = 1$

2. (i) $t_{r+1} = {}^{10}C_r y^{10-r}\left(\frac{C_3}{y^2}\right)r$

$= {}^{10}C_r y^{10-3r} C^{3r}$

Put $10 - 3r = -2$

$\Rightarrow r = 4$

$\therefore$ Coeff. of $\frac{1}{y^2} = {}^{10}C_4 C^{12} = 210\ C^{12}$

(ii) $(1+x)^m\left(1+\frac{1}{x}\right)^n = \frac{(1+x)^{m+n}}{x^n}$

Term independent of x = coeff. of x^n in $(1+x)^{m+n}$

$= {}^{m+n}C_n$

(iii) $(1+x)^m(1-x)^n$

$= \left(1 + mx + \frac{m(m-1)}{2}x^2 + \ldots\right) \times \left(1 - nx + \frac{n(n-1)}{2}x^2 \ldots\right)$

Coefficient of $x = m - n = 3$ (1)

Coefficient of

$x^2 = \frac{m(m-1)}{2} - mn + \frac{n(n-1)}{2} = -6$

$\Rightarrow m^2 - m - 2mn + n^2 - n = -12$

$\Rightarrow (m-n)^2 - (m+n) = -12$

$\Rightarrow 9 - (m+n) = -12 \Rightarrow m + n = 21$ (2)

Adding (1) and (2), $m = 12$

(iv) Sum of coefficients of first three terms

$= {}^mC_0 - {}^mC_1(3) + {}^mC_2(3)^2 = 559$

$\Rightarrow 1 - 3m + \frac{9m(m-1)}{2} = 559$

$\Rightarrow 3m\left[\frac{3m-3}{2} - 1\right] = 558$

$\Rightarrow m(3m-5) = 186 \times 2$

$\Rightarrow 3m^2 - 5m - 372 = 0$

$\Rightarrow 3m^2 - 36m + 31m - 372 = 0$

$\Rightarrow 3m(m-12) + 31(m-12) = 0$

$\Rightarrow m = 12$

$\therefore$ Expansion is $\left(x - \frac{3}{x^2}\right)^{12}$

$t_{r+1} = {}^{12}C_r x^{12-r}\left(-\frac{3}{x^2}\right)^r$

$= {}^{12}C_r(-3)^r x^{12-3r}$

Put $12 - 3r = 3 \Rightarrow r = 3$

$\therefore$ term containing

$x^3 = {}^{12}C_3(-3)^3 = -5940x^3$

3. (i) $\left(\frac{x}{3} + 9y\right)^{10}$

Total number of terms = 11

$\therefore \frac{11+1}{2} = 6^{\text{th}}$ term will be the middle term

$\therefore$ Middle term $= {}^{10}C_5\left(\frac{x}{3}\right)^{10-5}(9y)^5$

$= {}^{10}C_5 \frac{x^5}{3^5} 9^5 y^5 = 61236x^5y^5$

(ii) $\left(2x^2 - \frac{1}{x}\right)^7$

Total number of terms = 8

$\therefore$ 4th and 5th term will be the middle term

4th term $= {}^7C_3(2x^2)^{7-3}\left(-\frac{1}{x}\right)^3$

$= {}^7C_3 2^4 x^8\left(-\frac{1}{x^3}\right) = -560x^5$

5th term $= {}^7C_4(2x^2)^{7-4}\left(-\frac{1}{x}\right)^4$

$= {}^7C_4 2^3 x^6 \frac{1}{x^4} = 280x^2$

(iii) $\left(x + \frac{1}{x}\right)^{2n}$

Total number of terms $= 2n + 1$

$\therefore \frac{2n+1+1}{2} \Rightarrow (n+1)^{\text{th}}$ term will be middle term

$\therefore (n+1)^{\text{th}}$ term $= {}^{2n}C_n = \frac{2n!}{n!\,n!}$

$= \frac{1 \times 3 \times 5 \ldots (2n-1)(2 \times 4 \times 6 \ldots 2n)}{n!\,n!}$

$$=\frac{(1\times3\times5\ldots(2n-1)2^n n!)}{n!\,n!}$$

$$=\frac{1\times3\times5\times\ldots(2n-1)2^n}{n!}$$

(iv) $(1+x)^{2n}$ total number of terms $= 2n+1$

$\Rightarrow (n+1)^{\text{th}}$ term will be the middle term

Coefficient of $(n+1)^{\text{th}}$ term $= {}^{2n}C_n$

Again, $(1+x)^{2n-1} \Rightarrow$ Total number of terms $= 2n$

$\Rightarrow n^{\text{th}}$ and $(n+1)^{\text{th}}$ term will be the middle term.

$\therefore$ Sum of coefficients of middle term

$= {}^{2n-1}C_{n-1} + {}^{2n-1}C_n = {}^{2n}C_n$

(v) $(1+x)^{131}(x^2-x+1)^{130}$

$= ((x+1)(x^2-x+1))^{130}\ (x+1)$

$= (x^3+1)^{130}\ (x+1)$

Now in $(1+x^3)^{130}$

$t_{r+1} = 130C_r x^{3r}$

Clearly coefficient of $x^{65} = 0$

4. $\left(5^{\frac{1}{6}} + 2^{\frac{1}{8}}\right)^{100}$

Total number of terms = 101

Now, $t_{r+1} = {}^{100}C_r \left(5^{\frac{1}{6}}\right)^{100-r}\left(2^{\frac{1}{8}}\right)^r$

$= {}^{100}C_r 5^{\frac{100-r}{6}} 2^{\frac{r}{8}}$

We find the total number of rational terms

$\Rightarrow 100-r$ is divisible by 6 and r is divisible by 8

Only four terms are rational

$\Rightarrow$ Total number of irrational terms = 97

5. $\frac{1}{81^n}(1-10\times{}^{2n}C_1+10^2\times{}^{2n}C_2$

$-10^3\ {}^{2n}C_3+\ldots+10^{2n})$

$=\frac{1}{81^n}(1-10)^{2n}=1$

6. For the largest term $\frac{n-r+1}{r}\left|\frac{y}{x}\right| \geq 1$

$\Rightarrow \frac{21-r}{r}\left(\frac{5x}{4}\right) \geq 1 \Rightarrow \frac{(21-r)^5}{12r} \geq 1$

$\Rightarrow 105-5r \geq 12r$

$\Rightarrow 17r \leq 105$

$\Rightarrow r \leq 6.1$

Largest term $= t_7$

$\therefore$ Second largest term = 6th term

7. Since, 6th term is middle term, which is numerically the greatest term

$\therefore \left|\frac{t_6}{t_7}\right| > 1, \left|\frac{t_6}{t_5}\right| > 1$

$\therefore x \in \left(-3, \frac{-25}{12}\right) \cup \left(\frac{25}{12}, 3\right)$

8. $t_{r+1} = {}^9C_r\left(\frac{3}{2}x^2\right)^{9-r}\left(-\frac{1}{3x}\right)^r$

$= {}^9C_r \frac{3^{9-r}}{2^{9-r}} x^{18-2r} \frac{(-1)^r}{3^r x^r}$

$= {}^9C_r \frac{3^{9-2r}}{2^{9-r}}(-1)^r x^{18-3r}$

Put $18-3r=0 \Rightarrow r=6$

$17-3r=0 \Rightarrow r=\frac{17}{3}$

$21-3r=0 \Rightarrow r=7$

$\therefore$ Coeff. of independent term

$= {}^9C_6\frac{3^{-3}}{2^3} + 2\times{}^9C_7\frac{3^{-5}}{2^2}(-1)$

$= {}^9C_6\frac{1}{6^3} - {}^9C_7\frac{2}{6^2 3^3}$

$= \frac{9!}{6!\,3!}\times\frac{1}{6^3} - {}^9C_7\frac{2}{6^2 3^3}$

$= \frac{7\times8\times9}{6}\times\frac{1}{6^3} - \frac{8\times9}{2}\times\frac{2}{6^2 3^3}$

$= \frac{14}{6^2} - \frac{8}{6^2 3}$

$= \frac{1}{6^2}\left[14-\frac{8}{3}\right] = \frac{1}{36}\times\frac{42-8}{3}$

$= \frac{1}{36}\times\frac{34}{3} = \frac{17}{54}$

9. Since $5^{99} = 5(5^2)^{49} = 5(26-1)^{49}$

$= 5\left[{}^{49}C_0(26)^{49} - 49C_1 26^{48} + {}^{49}C_2(26)^{47} + \ldots + {}^{49}C_{48}(26) - {}^{49}C_{49}\right]$

$= 5(26K-1) = 130K - 5$

$\therefore \frac{5^{99}}{13} = \frac{130K-13+8}{13} = (10K-1) + \frac{8}{13}$

Hence, remainder = 8

10. $2^{4n} - 15n - 1 = (1+15)^n - 15n - 1$

$= \left[1 + 15n + {}^nC_2(15)^2 + \ldots + {}^nC_n(15)^n\right] - 15n - 1$

$= (15)^2\left[{}^nC_2 + {}^nC_3(15) + \ldots + {}^nC_n(15)^{n-1}\right]$

$\Rightarrow$ Divisible by 225

11. Let

$x = I + f = (8+3\sqrt{7})^n$ where I is integral part and $0 < f < 1$

$= 8^n + {}^nC_1 8^{n-1}(3\sqrt{7}) + {}^nC_2 8^{n-2}(3\sqrt{7})^2 \ldots$

Let $f_1 = (8-3\sqrt{7})^n$ where $0 < f_1 < 1$

$= 8^n - {}^nC_1 8^{n-1}(3\sqrt{7}) + {}^nC_2 8^{n-2}(3\sqrt{7})^2 \ldots$

$\therefore I + f + f_1 = 2\left[8^n + {}^nC_2 8^{n-2}(3\sqrt{7})^2 + \ldots\right]$

$\Rightarrow I + f + f_1 =$ even integer

$\Rightarrow f + f_1 =$ integer

As, $0 < f < 1$ and $0 < f_1 < 1$

$\Rightarrow f + f_1 = 1$

$\therefore I =$ even integer $-1 =$ odd integer

$\Rightarrow$ Integral part of x is an odd integer

Now, $x - x^2 + x[x]$

$= x - x(x - [x])$

$= x - xf = x(1-f) = xf_1$

$= (8+3\sqrt{7})^n(8-3\sqrt{7})^n = 1$

12. We know that if $n > 6$, then

$\left(\frac{n}{3}\right)^n < n!$

Put $n = 300$

$(100)^{300} < 300!$

But $(100)^{100} < (100)^{300}$

$\therefore (100)^{100} < 300!$

Hence, the greater number is 300!

13. $\frac{1}{12!} + \frac{1}{10!\,2!} + \frac{1}{8!\,4!} + \ldots + \frac{1}{12!}$

$= \frac{1}{12!}\left[\frac{12!}{12!} + \frac{12!}{10!\,2!} + \frac{12!}{8!\,4!} + \ldots + \frac{12!}{12!}\right]$

$= \frac{1}{12!}\left[{}^{12}C_0 + {}^{12}C_2 + {}^{12}C_4 + \ldots + {}^{12}C_{12}\right]$

We know that

$(1+x)^{12} = {}^{12}C_0 + {}^{12}C_1 x + {}^{12}C_2 x^2 + \ldots + {}^{12}C_{12}x^{12}$

Put $x = 1$

$2^{12} = {}^{12}C_0 + {}^{12}C_1 + {}^{12}C_2 + {}^{12}C_3 + {}^{12}C_4 + \ldots + {}^{12}C_{12}$

Put $x = -1$

$0 = {}^{12}C_0 - {}^{12}C_1 + {}^{12}C_2 - {}^{12}C_3 + \ldots + {}^{12}C_{12}$

Adding

$2^{12} = 2\left[{}^{12}C_0 + {}^{12}C_4 + \ldots + {}^{12}C_{12}\right]$

$\Rightarrow {}^{12}C_0 + {}^{12}C_4 + \ldots + {}^{12}C_{12} = 2^{11}$

Hence $\frac{2^{11}}{12!}$

14. ${}^{n-1}C_r = (K^2-3)\,{}^nC_{r+1}$

$\Rightarrow \frac{(n-1)!}{r!(n-r-1)!} = (K^2-3)\frac{n!}{(r+1)(n-r-1)!}$

$\Rightarrow 1 = (K^2-3)\frac{n}{(r+1)}$

$\Rightarrow (K^2-3) = \frac{r+1}{n}$ (1)

Since, $0 \le r \le n-1$

$\Rightarrow 1 \le r+1 \le n$

$\Rightarrow \frac{1}{n} \le \frac{r+1}{n} \le 1$

$\Rightarrow 0 < \frac{r+1}{n} \le 1$

Hence, from (1)

$0 < K^2 - 3 \le 1$

$\Rightarrow 3 < K^2 \le 4$

When, $K^2 \le 4, K \in [-2, 2]$

$K^2 > 3 \Rightarrow K < -\sqrt{3}$ or $K > \sqrt{3}$

Hence, $K \in \left[-2, -\sqrt{3}\right) \cup \left(\sqrt{3}, 2\right]$

15. $3^{2n} - 1 + 24n - 32n^2$

$= (1+8)^n - 1 + 24n - 32n^2$

$= \left(1 + 8n + \frac{n(n-1)}{2!}8^2 + {}^nC_3 8^3 + \ldots\right) - 1 + 24n - 32n^2$

$= \left(8n + 32n^2 - 32n + {}^nC_3 8^3 + {}^nC_4 8^4 + \ldots + 24n - 32n^2\right)$

$= 8^3\left({}^nC_3 + {}^nC_4 8 + \ldots\right)$

$\Rightarrow$ Divisible by 512

16. $3^{2002} + 7^{2002} + 2002$

$= 9^{1001} + 49^{1001} + 2002$

We know that $x^n + y^n$ is always divisible by $x + y$ if n is odd

$2002 = 29 \times 69 + 1$

$\Rightarrow$ Remainder = 1

17. $P_n = {}^nC_0 \times {}^nC_1 \times {}^nC_2 \ldots {}^nC_n$

$\frac{P_{n+1}}{P_n} = \frac{{}^{n+1}C_1}{{}^nC_1}\frac{{}^{n+1}C_2}{{}^nC_2}\ldots\frac{{}^{n+1}C_n}{{}^nC_n}.{}^{n+1}C_{n+1}$

$= \frac{n+1}{n}\frac{n+1}{n-1}\frac{n+1}{n-2}\ldots\frac{n+1}{1}$

$= \frac{(n+1)^n}{n!} = \frac{10^9}{9!} \Rightarrow n = 9$

18. $\left(1+x^2\right)^n = {}^nC_0 + {}^nC_1x^2 + {}^nC_2x^4 + \ldots$

Put $x = 2$

$5^n = {}^nC_0 + {}^nC_1 2^2 + {}^nC_2 2^4 + {}^nC_3 2^6 + \ldots$

$\Rightarrow 5^n - 1 = 2^2\left[{}^nC_1 + 2^2\ {}^nC_2 + 2^4\ {}^nC_3 + \ldots\right]$

$\Rightarrow {}^nC_1 + 2^2\ {}^nC_2 + 2^4\ {}^nC_3 + \ldots = \frac{5^n - 1}{4}$

19. Let

$S = C_0 + (C_0 + C_1) + (C_0 + C_1 + C_2) + \ldots + (C_0 + C_1 + C_2 + \ldots + C_n)$ (1)

$\Rightarrow S = C_n + (C_n + C_{n-1}) + (C_n + C_{n-1} + C_{n-2}) + \ldots(C_n + C_{n-1} + C_{n-2} + \ldots + C_0)$ (2)

$\Rightarrow S = (n+1)C_n + nC_{n-1} + (n-1)C_{n-2} + \ldots + C_0$

$\Rightarrow S = (C_0 + C_1 + C_2 + \ldots + C_n) + (C_1 + 2C_2 + 3C_3 + \ldots + nC_n)$

$\Rightarrow S = 2^n + n2^{n-1} = (n+2)2^{n-1}$

20. $n = 3$

5.10.2 JEE Main: Sample Questions

Single Answer Type

1. (C)

The coefficient of t^4 in $(1 - t)^6 (1 + t^2)^5$

$= {}^6C_0\,{}^5C_2 + {}^6C_2 \cdot {}^5C_1 + {}^6C_4 \cdot {}^5C_0$

$= (1)(10) + (15)(5) + (15)(1) = 10 + 75 + 15 = 100$

2. (B)

$T_{r+1} = {}^nC_r(2x)^{n-r} \cdot \left(-\frac{1}{x}\right)^r = {}^nC_r \cdot 2^{n-r} \cdot x^{n-2r}(-1)^r.$

For constant term $2r = n$ (1)

$\therefore$ term $= {}^{2r}C_r \cdot 2^r(-1)^r = -160$

Hence r must be odd.

Hence check $r = 3$

$-{}^6C_3 \cdot 2^3 = -160$

$\therefore r = 3 \Rightarrow n = 6$

3. (C)

$2^n\left(1 + \frac{x}{6}\right)^n \Rightarrow T_{r+1} = 2^n.{}^nC_r\left(\frac{x}{6}\right)^r$

$\Rightarrow 2^n.{}^nC_7 \cdot \frac{1}{6^7} = 2^n.{}^nC_8 \cdot \frac{1}{6^8}$

$\Rightarrow 6\,{}^nC_7 = {}^nC_8 \Rightarrow n - 7 = 48 \Rightarrow n = 55$

4. (A)

$T_{r+1} = {}^{198}C_r(2)^{\frac{198-r}{7}}(3)^{\frac{r}{4}}$

$r = 0, 1, 2, \ldots, 198$

$\therefore$ For r = 16, 44, 72, 100, 128, 156, 184, terms in the expansion are rational.

5. (B)

Put $x = 1 \Rightarrow$ sum $= 25 - 60 + 22 + 18 - 6 = -1$

6. (D)

T_{r+1} in $\left(x^2 - \frac{1}{x}\right)^9$ is ${}^9C_r x^{2(9-r)}\left(-\frac{1}{x}\right)^r$

$= {}^9C_r \cdot x^{18-3r} \cdot (-1)^r$
for term independent of x, $18 - 3r = 0 \Rightarrow r = 6$
$\therefore$ 7th term is independent of x and equals ${}^9C_6 = {}^9C_3 = 84$.
Also there are 10 terms, hence 5th and 6th are the two middle terms
$T_5 = {}^9C_4 \cdot x^6$; $T_6 = -{}^9C_5 \cdot x^3$
$\therefore q$ = coefficient of 5th + coefficient of 6th term
$= {}^9C_4 - {}^9C_5 = 0$
Hence $p = 84$; $q = 0$
$\therefore p - q = {}^9C_3$

7. (A)
(A) $[(1 + x^2) - x^3]^8$
$= {}^8C_0(1 + x^2)^8 + {}^8C_1(1 + x^2)^7 (-x^3) + {}^8C_2(1+x^2)^6(-x^3)^2 + {}^8C_3(1+x^2)^5(-x^3)^3$
$\therefore$ Coefficient of $x^{10} = {}^8C_0\,{}^8C_5 + {}^8C_2\,{}^6C_2$
(B) Consider $(1+8)^7 - (1-8)^9 = 128 + 8^2 k$ remainder is 0
(C) not possible
(D) $r = 2$ or $r = 3$

8. (D)

$$\left(x\sin\theta + \frac{\cos\theta}{x}\right)^{10}$$

$$T_{r+1} = {}^{10}C_r(x\sin\theta)^{10-r} \times \left(\frac{\cos\theta}{x}\right)^r$$

$= {}^{10}C_r (\sin\theta)^{10-r} (\cos\theta)^r, x^{10-2r}$
Hence for independent of x, $r = 5$

$$T_6 = {}^{10}C_5 (\sin\theta)^5 \times (\cos\theta)^5 \Rightarrow \frac{{}^{10}C_5(\sin 2\theta)^5}{2^5}$$

9. (A)
$C_0 + C_1 + C_2 + \ldots + C_n = 256$
$\Rightarrow 2^n = 2^8 \Rightarrow n = 8$

10. (A)
Coefficient of x^{49} in this series is

$$-\left[1+\frac{1}{2}+\frac{1}{2^2}+\ldots+\frac{1}{2^{49}}\right] = -\left[\frac{1-\frac{1}{2^{50}}}{1-\frac{1}{2}}\right]$$

$$= -2\left[1-\frac{1}{2^{50}}\right]$$

11. (B)
$E = (2n + 1)(2n + 3)(2n + 5) \ldots (4n - 1)$
Multiply numerator and denominator by $(2n + 2)(2n + 4) \ldots (4n)$ and also by $(2n)!$ and $n!$.

$$E = \frac{(2n)!(2n+1)(2n+2)(2n+3)\ldots(4n-1)\cdot 4n}{(2n)!(2n+2)(2n+4)\ldots(2n+2n)}$$

$$= \frac{(4n)! \times (n)!}{(2n)!\, 2^n[(n+1)(n+2)\ldots(2n)]n!}$$

$$= \frac{(n!)\cdot(4n)!}{2^n \cdot ((2n)!)^2}$$

12. (B)

$${}^nC_r = \frac{n!}{r!(n-r)!}; \frac{{}^nC_r}{n!} = \frac{1}{r!(n-r)!};$$

put $r = 1, 3, 5, \ldots$ and add

$$\Rightarrow \text{LHS} = \frac{1}{n!}[{}^nC_1 + {}^nC_3 + {}^nC_5 + \ldots] = \frac{2^{n-1}}{n!}$$

13. (C)
$C_0 + C_1 + \ldots + C_n$
$\Rightarrow 2^n = 2^6 \Rightarrow n = 6$

14. (A)
$T_2 = {}^nC_1(a^{1/13})^{n-1} \cdot a\sqrt{a} = 14a^{5/2}$

or $n \cdot a^{\frac{n-1}{13}} = 14a$ or $n \cdot a^{\frac{n-14}{13}} = 14$

Hence $\dfrac{n-14}{13} = 0 \Rightarrow n = 14$

Now $\dfrac{{}^{14}C_3}{{}^{14}C_2} = \dfrac{14!}{3!\,11!}\cdot\dfrac{2!\,2!}{14!} = \dfrac{12}{3} = 4$

15. (B)
Let $(2x^2 - 3x + 1)^{11} = a_0 + a_1x + a_2x^2 + \ldots + a_{22}x^{22}$; to find $a_0 + a_2 + a_4 + \ldots a_{22}$?
$S_E + S_0 = P(1) = 0$ (1)
where $P(x) = (2x^2 - 3x + 1)^{11}$
$S_E - S_0 = P(-1) = 6^{11}$ (2)
$\Rightarrow 2S_E = 6^{11} \Rightarrow S_E = 3.6^{10}$

16. (B)
Put $x = -1$; $(-3)^{2145} = a_0 - a_1 + a_2 - a_3 + \ldots$; $-(3)^{2145}$
$= -(3^4)^{536} \times 3 \Rightarrow$ ends in 3

17. (B)

$$T_{r+1} = {}^{100}C_r\, 2^{\frac{100-r}{2}} \cdot 3^{r/4}$$

$\Rightarrow r$ must be even and divisible by 4
$\Rightarrow r = 0, 4, 8, \ldots, 100$

18. **(B)**
$5n - 8r = 10$ and $12r - n = 10$
$\Rightarrow r = 5 \Rightarrow n = 10$
$\Rightarrow$ coefficient of $pq = {}^{10}C_5 = 252$

19. **(B)**
$r = 3$ or 5; $r = 0$ is not possible.

20. **(B)**

$$\frac{T_{r+1}}{T_r} = \frac{{}^9C_r\, a^{9-r}.(2x)^r}{{}^9C_{r-1}\, a^{9-r+1}(2x)^{r-1}}$$

$$= \frac{9-r+1}{r}\frac{2x}{a} = \frac{9-r+1}{r}\frac{2}{3} = \frac{10-r}{r}\frac{2}{3}$$

$\therefore T_{r+1} > T_r$ if $\frac{10-r}{r}\frac{2}{3} > 1$, that is, $20 > 5r$, that is, $r < 4$
Hence for values of 'r' up to 3, $T_{r+1} < T_r$.
When $r = 4$, $T_{r+1} = T_r$, that is, $T_4 = T_5$ and both of them are the greatest terms.

21. **(A)**
$(1 - 2x + 5x^2 + 10x^3)\,[C_0 + C_1x + C_2x^2 + \ldots]$
$= 1 + a_1x + a_2x^2 + \ldots$

$$a_1 = n - 2 \text{ and } a_2 = \frac{n(n-1)}{2} - 2n + 5$$

put $a_1^2 = 2a_2$
$(n - 2)^2 = n(n - 1) - 4n + 10$
$n^2 - 4n + 4 = n^2 - 5n + 10 \Rightarrow n = 6$

22. **(A)**

$$\sum_{k=0}^{4}\left(\frac{5^{4-k}}{(4-k)!}\right)\left(\frac{x^k}{k!}\right) = \sum_{k=0}^{4}\left(\frac{5^{4-k}}{(4-k)!}\right)\left(\frac{x^k}{k!}\right)\frac{4!}{4!}$$

$$= \sum_{k=0}^{4}\frac{{}^4C_k \times 5^{4-k}\times x^k}{4!} = \frac{(5+x)^4}{4!}$$

$$\text{Now } \frac{(5+x)^4}{4!} = \frac{8}{3} \Rightarrow (5+x)^4 = 64$$

$x + 5 = 2^{3/2} = 2\sqrt{2} \Rightarrow x = 2\sqrt{2} - 5$

23. **(A)**
Consider $(x^2 + (z - 2y))^{10}$ and non-expand binomially.

24. **(A)**

$$\frac{17}{4} + 3\sqrt{2} = \left(\frac{3}{2} + \sqrt{2}\right)^2$$

$$\Rightarrow \left(3 - \sqrt{\frac{17}{4} + 3\sqrt{2}}\right)^{15} = \left(\frac{3}{2} - \sqrt{2}\right)^{15}$$

$\therefore\ T_{11} = {}^{15}C_{10}\left(\frac{3}{2}\right)^5\left(-\sqrt{2}\right)^{10}$ is a positive integer.

25. **(A)**

$$(1 + 8x + bx^2)(1 - 3x)^9$$

$$(1 + 8x + bx^2)\sum_{r=0}^{9} {}^9C_r(-3x)^r$$

$\therefore$ Coefficient of $x^2 = {}^9C_2 \times 9 - 8 \times {}^9C_1 \times 3 + b$
$\therefore$ Coefficient of $x^3 = -\,{}^9C_3 \times 27 + 8 \times {}^9C_2 \times 9 - b^9 \times C_1 \times 3$

Both are equal $\Rightarrow b = \frac{54}{7}$

26. **(C)**

$$T_{r+1} = {}^{25}C_r(ab)^{(h-r)}.x^{\left(\frac{h-r}{2}\right)}.c^r.x^{\left(\frac{-r}{3}\right)}$$

$h = 25 \rightarrow r = 15$
$= {}^{25}C_{15}\, a^{10}.b^{10}.c^{15}$
$10a + 10b + 15c = 1$

$$\frac{(a+a+\ldots+a)+(b+b+\ldots+b)+(c+c+\ldots+c)}{35}$$

$$\geq (a^{10}b^{10}c^{15})^{\frac{1}{35}} \Rightarrow \left(\frac{1}{35}\right)^{35} \geq a^{10}b^{10}c^{15}$$

27. **(B)**
Concept: Coefficient of x^r in $(1 - x)^{-n}$, $n \in N$ is ${}^{n+r-1}C_r$

$$\text{Now given product is} = \frac{1-x^{28}}{1-x}.\left(\frac{1-x^{15}}{1-x}\right)^2$$

$$= \frac{(1-x^{28})(1-x^{15})^2}{(1-x)^3}$$

$$= \frac{(1-x^{28})(1-2x^{15})}{(1-x)^3}$$

$= (1 - 2x^{15} - x^{28})(1 - x)^{-3}$
Hence coefficient of $x^{28}(1 - 2x^5 - x^{28})(1 - x)^{-3} \times$ Coefficient of x^{13} in $(1 - x)^{-3} - 1$
$= {}^{30}C_2 - 2 \times {}^{15}C_2 - 1 = 435 - 210 - 1 = 224$

28. (B)

$\left(x^5 + 4\times 3^{-\log_{\sqrt{3}} \sqrt{x^3}}\right)^{10} = (x^5 + 4\times x^{-3})^{10}$

$T_{r+1} = {}^{10}C_r (x^5)^{10-r} (4x^{-3})^r$

$= {}^{10}C_r 4^r x^{50-5r-3r} = {}^{10}C_r 4^r x^{50-8r}$

$50 - 8r = 2 \Rightarrow r = 6$

and $50 - 8r = 10 \Rightarrow r = 5$

Required ratio

$\therefore \dfrac{{}^{10}C_6 4^6}{{}^{10}C_5 4^5} = \dfrac{10}{3}$

29. (A)

Last term $T_{n+1} = {}^nC_n \left(-\dfrac{1}{\sqrt{2}}\right)^n = \dfrac{(-1)^n}{2^{n/2}}$

Now $\left(\dfrac{1}{3^{5/3}}\right)^{\log_3 8} = 3^{\log_3 8^{-5/3}} = 2^{-5}$

$\therefore \dfrac{(-1)^n}{2^{n/2}} = 2^{-5} \Rightarrow n = 10$

$\therefore T_5 = {}^{10}C_4 (2^{1/3})^6 (-1)^4 (2^{-1/2})^4 = 210$

30. (A)

Any term of the expansion is of the form

$= 5^{\frac{p}{3}} \times 3^{\frac{q}{2}} \times x^p y^q z^r$ where $p, q, r \in I$ and

$0 < p, q, r \le 6$ such that $p + q + r = 6$

for rational coefficient

$\therefore p = 0, 3, 6; q = 0, 2, 4, 6$

The possibilities are

$(p, q, r) = (0, 0, 6), (0, 2, 4), (0, 4, 2), (0, 6, 0),$
$(3, 0, 3)$ $(3, 2, 1)$ $(6, 0, 0)$

$\therefore$ Seven possibilities.

31. (C)

$\left(x + \dfrac{1}{x}\right)\left(x^2 - \dfrac{1}{x^2}\right)^9$

$\Rightarrow \left(x + \dfrac{1}{x}\right)\sum {}^9C_r (-1)^r x^{18-4r}$

Coefficient of x^5 is given by $r = 3$.

32. (B)

$T_{r+1} = {}^{6561}C_r (2^{1/3})^{6561-r} (5^{1/9})^r$

$= {}^{6561}C_r 2^{2187-r/3} . 5^{r/9}$

$\dfrac{r}{3}$ and $\dfrac{r}{9}$ must be integers.

$\therefore 0 \le r \le 6561 \Rightarrow 0 \le \dfrac{r}{9} \le 729$

$\therefore \dfrac{r}{9} = 0, 1, 2 \ldots 729$

$\therefore$ Number of terms = 730

33. (B)

$(1 - x^2 + x^3)$ $({}^nC_0 + {}^nC_1 x + {}^nC_2 x^2 + {}^nC_3 x^3 + \ldots)$

$= a_0 + a_1 x + \ldots$

$a_0 = 1, a_2 = {}^nC_2 - {}^nC_0 = \dfrac{n^2 - n - 2}{2} = \dfrac{a_1^2 - a_1 - 2}{2}$

$a_1 = n \Rightarrow a_1^2 - a_1 - 2a_2 = 2$

34. (A)

$T_{r+1} = {}^{15}C_r (-4)^r x^{45-5r}$

$\therefore 45 - 5r = 0; r = 9$

$\therefore$ Constant term $= -4^9 . {}^{15}C_9$

35. (C)

Consider the general term

$T_r = \dfrac{r \cdot {}^{101}C_r}{{}^{101}C_{r-1}}$ which simplifies to $(102 - r)$

Hence sum $= \sum_{r=1}^{101} (102 - r)$

$= 101 + 100 + 99 + \ldots + 2 + 1$

$= 5151$

36. (D)

Let $S = 0\,({}^{50}C_0)^2 + 1({}^{50}C_1)^2 + 2({}^{50}C_2)^2 + \ldots + 50({}^{50}C_{50})^2$ (1)

Also $S = 50({}^{50}C_{50})^2 + 49({}^{50}C_{49})^2 + \ldots + 0({}^{50}C_0)^2$ (2)

(writing in reverse order)

$\therefore$ On adding (1) and (2), we get

$2S = 50[({}^{50}C_0)^2 + ({}^{50}C_1)^2 + \ldots + ({}^{50}C_{50})^2]$

$\Rightarrow 2S = 50 \times {}^{100}C_{50}$

[As $({}^nC_0)^2 + ({}^nC_1)^2 + ({}^nC_2)^2 + \ldots + ({}^nC_n)^2 = {}^{2n}C_n$]

$\Rightarrow 2S = 50 \times \dfrac{100!}{50!\,50!}$

$\Rightarrow 2S = \dfrac{(50)50!\,2^{50}(1\times 3\times 5\ldots 99)}{50!\,50!}$

$\Rightarrow S = \dfrac{2^{49}}{49!}(1\times 3\times 5 \ldots 99)$

37. **(D)**

Number of ways $S = {}^{2m}C_m + {}^{2m}C_{m-1} + {}^{2m}C_{m-2} + \ldots + {}^{2m}C_0$

$S = {}^{2m}C_m + {}^{2m}C_{m+1} + {}^{2m}C_{m+2} + \ldots + {}^{2m}C_{2m}$

$\Rightarrow 2S = {}^{2m}C_m + 2^{2m}$

$\therefore S = \dfrac{{}^{2m}C_m}{2} + 2^{2m-1}$

38. **(B)**

In (201)! number of 2s = 100 + 50 + 25 + 12 + 6 + 3 + 1 = 197

$\therefore$ Number of 8's $= \left[\dfrac{197}{3}\right] = 65$

Number of 3's = 67 + 22 + 7 + 2 = 98

$\therefore$ Number of 24's = 65

39. **(D)**

$T_{r+1} = {}^7C_r(2x)^{7-r}(3y)^r$

Put $r = 2 = {}^7C_2\, 2^5 \times x^5 \times 3^2 \times y^2$

Coefficient of $x^5y^2 = {}^7C_2 \times 2^5 \times 3^2$

$= 21 \times 32 \times 9 = 6048$

40. **(B)**

$S = 1 \times C_0 + 2 \times C_1 + 3 \times C_2 + \ldots + 15 \times C_{14}$

Re-writing the sum as reverse order

$S = 15 \times C_0 + 14C_1 + 13C_2 + \ldots + C_{14} \times 1$

$2S = 16[C_0 + C_1 + C_2 + \ldots + C_{14}] = 16 \times 2^{14}$

$S = 2^3 \times 2^{14} = 2^{17}$

41. **(B)**

$$\sum_{r=1}^{10} r^2 \times \left(\frac{{}^{26}C_r}{{}^{26}C_{r-1}}\right) = \sum_{r=1}^{10} r(27-r) = 1100$$

42. **(A)**

$(1 + 7)^{83} + (7 - 1)^{83} = (1 + 7)^{83} - (1 - 7)^{83}$

$= 2[{}^{83}C_1 \times 7 + {}^{83}C_3 \times 7^3 + \ldots + {}^{83}C_{83} \times 7^{83}]$

$= (2 \times 7 \times 83) + 49I$ where I is an integer

$\therefore 14 \times 83 = 1162 \Rightarrow \dfrac{1162}{49} = 23\dfrac{35}{49}$

$\therefore$ Remainder is 35.

43. **(D)**

Consider $(5 + 2)^{100} - (5 - 2)^{100}$

$= 2[{}^{100}C_1\ 5^{99} \times 2 + {}^{100}C_3\ 5^{97} \times 2^3 + \ldots + {}^{100}C_{99} 5 \times 2^{99}]$

$= 2[1000 \times 5^{98} + {}^{100}C_3\ 5^{94} + \ldots + 1000 \times 2^{98}]$

$\Rightarrow$ minimum 000 as last three digits.

44. **(B)**

$E = (\alpha + p)^{m-1}$

$$\left[1 + \frac{\alpha+q}{\alpha+p} + \left(\frac{\alpha+q}{\alpha+p}\right)^2 + \ldots + \left(\frac{\alpha+q}{\alpha+p}\right)^{m-1}\right]$$

$$= (\alpha+p)^{m-1}\left[\frac{\left(\frac{\alpha+q}{\alpha+p}\right)^m - 1}{\left(\frac{\alpha+q}{\alpha+p}\right) - 1}\right]$$

$$= \frac{(\alpha+p)^m[(\alpha+p)^m - (\alpha+q)^m]}{(\alpha+p)^m(p-q)}$$

Hence coefficient of α^t in $\dfrac{(\alpha+p)^m - (\alpha+q)^m}{p-q}$

$$= \frac{(p+\alpha)^m - (q+\alpha)^m}{p-q} = \frac{{}^mC_t(p^{m-t} - q^{m-t})}{p-q}$$

45. **(B)**

$(21)^{100} = (1 + 20)^{100} = 1 + {}^{100}C_1 \times 20 + {}^{100}C_2 \times 20^2 + \ldots + {}^{100}C_{100} \times 20^{100}$

Hence last two digits are 01.

46. **(A)**

$\dfrac{1(2^{2000} - 1)}{1} = 2^{2000} - 1 = (2^2)^{1000} - 1 = 4^{1000} - 1$

$(5 - 1)^{1000} - 1 = (1 - 5)^{1000} - 1$

$1 - 1000C_1 \times 5 + 1000C_2 \times 5^2 + \ldots + 1000C_{1000} \times 5^{1000} - 1$

Which is divisible by 5.

47. **(B)**

We have $\displaystyle\sum_{r=1}^{10} {}^{10}C_r \cdot r(r-1)$

$$= \sum_{r=1}^{10} \frac{10!}{r!\,(10-r)!} r(r-1)$$

$$= 90 \sum_{r=1}^{10} \frac{(8!)\, r(r-1)}{r(r-1)(r-2)!\,(10-r)!}$$

$$= 90 \sum_{r=1}^{10} \frac{8!}{(r-2)!\,(10-r)!}$$

$$= 90 \sum_{r=2}^{10} \frac{8!}{(r-2)!\,(10-r)!}$$

$= 90[{}^8C_0 + {}^8C_1 + {}^8C_2 + \ldots + {}^8C_8]$

$= 90 \times 2^8 = 45 \times 2^9 \Rightarrow k = 45$

48. (D)

Consider the expansion of $(1 + x)^{11} = {}^{11}C_0 + {}^{11}C_1x + {}^{11}C_2x^2 + \ldots + {}^{11}C_{11}\, x^{11}$. Integrating both sides between 0 and 1, we get

$$\frac{\binom{11}{0}}{1}+\frac{\binom{11}{1}}{2}+\frac{\binom{11}{2}}{3}+\ldots+\frac{\binom{11}{11}}{12}=\frac{2^{12}-1}{12}$$

49. (D)

We have $S = {}^2C_0 + {}^3C_1 + {}^4C_2 + {}^5C_3 + \ldots + {}^{99}C_{97}$

(writing in reverse order)

$$S = \underbrace{{}^3C_0 + {}^3C_1} + {}^4C_2 + {}^5C_3 + \ldots + {}^{97}C_{95} + {}^{98}C_{96} + {}^{99}C_{97}$$

$$S = \underbrace{{}^4C_1 + {}^4C_2} + {}^5C_3 + \ldots + {}^{97}C_{95} + {}^{98}C_{96} + {}^{99}C_{97}$$

(Using ${}^nC_r + {}^nC_{r-1} = {}^{n+1}C_r$)

$S = {}^5C_2 + {}^5C_3 + \ldots + {}^{97}C_{95} + {}^{98}C_{96} + {}^{99}C_{97}$

Similarly continuing in the same manner, we get,

$$S = {}^{99}C_{96} + {}^{99}C_{97} = {}^{100}C_{97} = \frac{100\cdot 97\cdot 98}{3\cdot 2}$$

$= 161700$

50. (B)

$$\sum_{r=1}^{10} {}^{15}C_r \times {}^{35}C_{(10-r)}$$

$$= {}^{15}C_1 \times {}^{35}C_9 + {}^{15}C_2 \times {}^{35}C_8 + \ldots + {}^{15}C_{10} \times {}^{35}C_0$$

$$= {}^{50}C_{10} - {}^{35}C_{10}$$

51. (D)

Number of ways = coeff. of x^{100} in $(x^1 + x^2 + x^3 + \ldots x^{75})\,(x^1 + x^2 + x^3 + \ldots x^{50})\,(x^1 + x^2 + x^3 + \ldots x^{25})$

$\Rightarrow (x^3)\,(x^{75} - 1)\,(x^{50} - 1)\,(x^{25} - 1)\,(x - 1)^{-3}$.

52. (C)

$$7^{10}\left[1-\frac{4}{7}x\right]^{10};$$

$$\text{Here } \frac{(n+1)\,|\,x\,|}{1+|\,x\,|} = \frac{(10+1)\left|-\frac{4}{7}\times 1\right|}{1+\left|-\frac{4}{7}\times 1\right|} = 4$$

T_4 and T_5 are numerically greatest but T_4 is negative so greatest term is T_5.

53. (A)

We know that

$({}^{10}C_0)^2 + ({}^{10}C_1)^2 + \ldots + ({}^{10}C_{10})^2 = {}^{20}C_{10}$ (1)

and $({}^{10}C_0)^2 - ({}^{10}C_1)^2 + \ldots + ({}^{10}C_{10})^2 = {}^{20}C_5\,(-1)^5$ (2)

(1) – (2)

$$\Rightarrow 2[({}^{10}C_1)^2 + ({}^{10}C_3)^2 + \ldots + ({}^{10}C_9)^2]$$

$$= {}^{20}C_{10} + {}^{10}C_5$$

$$\Rightarrow ({}^{10}C_1)^2 + ({}^{10}C_3)^2 + \ldots + ({}^{10}C_9)^2$$

$$= \frac{{}^{20}C_{10} + {}^{10}C_5}{2}$$

54. (C)

$$\text{Sum} = \sum_{r=1}^{10} r \times {}^{10}C_r \left(\frac{1}{2}\right)^{10-r} (-3)^{r-1}$$

$$= \sum_{r=1}^{10} 10 \times {}^9C_{r-1}\left(\frac{1}{2}\right)^{9-(r-1)} \times (-3)^{r-1} = 10\left[\frac{1}{2}-3\right]^9$$

$$= 10\sum_{r=1}^{10} {}^9C_{r-1}\left(\frac{1}{2}\right)^{9-(r-1)} (-3)^{r-1} = 10\left[\frac{-5}{2}\right]^9$$

$$= \frac{-10\times 5^9}{2^9} = \frac{-5^{10}}{2^8} = \frac{k}{2^8} \Rightarrow k = -5^{10}$$

55. (A)

Required sum of coefficient of x^{10} in the expansion of $(1 + x)^{30}\,(1 + x)^{20}\,(1 + x)^{10}$ that is, ${}^{60}C_{10}$.

56. (A)

$${}^{40}C_{15}\sum_{r=0}^{15} {}^{15}C_r\,{}^{20}C_r - {}^{35}C_{15}\sum_{r=0}^{15} {}^{15}C_r\,{}^{25}C_r$$

$$= {}^{40}C_{15}\,{}^{35}C_{15} - {}^{35}C_{15}\,{}^{40}C_{25} = 0$$

57. (D)

$${}^{n+1}C_2 + 3({}^3C_1 + {}^3C_2 + {}^4C_2 + \ldots + {}^nC_2)$$

(Using ${}^nC_r + {}^nC_{r-1} = {}^{n+1}C_r$)

$$= {}^{n+1}C_2 + 3({}^{n+1}C_3) = \frac{n^2(n+1)}{2}$$

58. (A)

$$\sum_{r=2}^{n} \frac{1}{{}^rC_2} = \sum_{r=2}^{n} \frac{2}{r(r-1)} = 2\sum_{r=2}^{n}\left(\frac{1}{r-1}-\frac{1}{r}\right)$$

$$= 2\left(1-\frac{1}{n}\right) = \frac{2(n-1)}{n}$$

Integer Answer Type

1. **(6)**

$\left(1-2x+5x^2-10x^3\right)\left({}^nC_0+{}^nC_1x+{}^nC_2x^2+\ldots\right)$

$=1+a_1x+a_2x^2+\ldots$

$\Rightarrow a_1=n-2$ and $a_2=\dfrac{n(n-1)}{2}-2n+5$

Since, $a_1^2=2a_2$

$\Rightarrow (n-2)^2=n(n-1)-4n+10$

$\Rightarrow n=6$

2. **(5)**

$t_{r+1}={}^{10}C_r\,(x\sin\theta)^{10-r}\left(\dfrac{\cos\theta}{x}\right)^r$

$={}^{10}C_r\,(\sin\theta)^{10-r}(\cos\theta)^r\,x^{10-2r}$

Put $10-2r=0\Rightarrow r=5$

$\Rightarrow$ Term independent of

$x={}^{10}C_5\,(\sin\theta\cos\theta)^5$

$=\dfrac{{}^{10}C_5\,(\sin2\theta)^5}{2^5}$

$\Rightarrow$ Greatest value of the term independent of $x=\dfrac{{}^{10}C_5}{2^5}\Rightarrow K=5$

3. **(7)**

Any term of the expansion is of the form

$5^{\frac{P}{3}}3^{\frac{q}{2}}x^py^qz^r$

where $p,q,r\in I$ and $0<p,q,r\le6$ such that $p+q+r=6$

For rational coefficient,

$p=0,3,6;\ q=0,2,4,6$. The possibilities are

$(p,q,r)=(0,0,6),(0,2,4),(0,4,2),(0,6,0),$
$(3,0,3)(3,2,1),(6,0,0)$

$\Rightarrow$ 7 possibilities

4. **(8)**

$(1+x)^{14}=C_0+C_1x+C_2x^2+\ldots+C_{14}x^{14}$

$\Rightarrow x(1+x)^{14}=C_0x+C_1x^2+C_2x^3+\ldots+C_{14}x^{15}$

Differentiating with respect to x

$(1+x)^{14}+x\times14(1+x)^{13}$

$=C_0+2C_1x+3C_2x^2+\ldots+15\times C_{14}x^{14}$

Put $x=1$

$1\times C_0+2\times C_1+3\times C_2+\ldots+15\times C_{14}$

$=2^{14}+14\times2^{13}$

$\Rightarrow\displaystyle\sum_{r=1}^{15}r\times C_{r-1}=2^{13}\times2^4=2^{17}$

$\Rightarrow a=2,b=17$

$\therefore\left[\dfrac{b}{a}\right]=8$

5. **(7)**

Consider

$K\times{}^{21}C_K\left(\dfrac{1}{3}\right)^K\left(\dfrac{2}{3}\right)^{21-K}$

$=K\dfrac{21\times(20)!}{K(K-1)!(21-K)!}\dfrac{1}{3}\times\dfrac{1}{3^{K-1}}\left(\dfrac{2}{3}\right)^{21-K}$

$=7\times{}^{20}C_{K-1}\dfrac{1}{3^{K-1}}\left(\dfrac{2}{3}\right)^{21-K}$

Let $K-1=p$ If $K=0,p=-1$

$K=21,p=20$

$=7\times{}^{20}C_p\dfrac{1}{3^p}\times\left(\dfrac{2}{3}\right)^{20-p}$

Hence, sum $=7\times\displaystyle\sum_{p=-1}^{20}{}^{20}C_p\left(\dfrac{1}{3}\right)^p\left(\dfrac{2}{3}\right)^{20-p}$

$=7\times\displaystyle\sum_{p=0}^{20}{}^{20}C_p\left(\dfrac{1}{3}\right)^p\left(\dfrac{2}{3}\right)^{20-p}=7\left(\dfrac{2}{3}+\dfrac{1}{3}\right)^{20}=7$

6. **(6)**

$N=\dfrac{(1+2+\ldots+10)^2-\left(1^2+2^2+\ldots+10^2\right)}{2}$

7. **(0)**

$\left(1+2x+\ldots+16x^{15}\right)\left(1+2x+\ldots+16x^{15}\right)$

Coefficient of

$x^{14}=(1\times15)+(2\times14)+(3\times13)+\ldots+(15\times1)$

$=\displaystyle\sum_{n=1}^{15}n(16-n)=\sum_{n=1}^{16}\left(16n-n^2\right)$

$=16\dfrac{n(n+1)}{2}-\dfrac{n(n+1)(2n+1)}{6}$

where $n=15=680$

8. **(1)**

Let, $S=\frac{5}{^nC_0}+\frac{6}{^nC_1}+\frac{7}{^nC_2}+\ldots+\frac{(n+5)}{^nC_n}$ (1)

$\Rightarrow S=\frac{n+5}{^nC_0}+\ldots+\frac{5}{^nC_n}$ (2)

Adding

$$2S=(n+10)\left[\frac{1}{^nC_0}+\frac{1}{^nC_1}+\frac{1}{^nC_2}+\ldots+\frac{1}{^nC_n}\right]$$

$$\Rightarrow S=\frac{(n+10)}{2}\left[\frac{1}{^nC_0}+\frac{1}{^nC_1}+\ldots+\frac{1}{^nC_n}\right]$$

$$=n\left[\frac{1}{^nC_0}+\frac{1}{^nC_1}+\ldots+\frac{1}{^nC_n}\right]$$

$\Rightarrow n+10=2n\Rightarrow n=10\Rightarrow K=10$

$\therefore$ Sum of digits = 1

9. **(4)**

$$\left(x+\frac{1}{x}\right)^6+2x\left(x+\frac{1}{x}\right)^6+bx^2\left(x+\frac{1}{x}\right)^6$$

term independent of x

$={}^6C_3+b\left({}^6C_4\right)=80$

$\Rightarrow 15b+20=80$

$\Rightarrow b=4$

10. **(8)**

There are 8 products, which generate x^9, so coefficient is 8.

11. **(2520)**

$T_{r+1}={}^9C_r\,t^{2r}(2-t)^r$

Now coefficient of t^8 appears in $r=3$, $r=4$

$r=3\Rightarrow {}^9C_3\,t^6(2-t)^3={}^9C_3\times t^6\times {}^3C_p\,2^{3-p}\,(-t)^p$

Put $p=2\Rightarrow {}^9C_3\times {}^3C_2\times 2^1\times t^8=84\times 6\times t^8$

Similarly when $r=4$.

${}^9C_4\,t^8(2-t)^4={}^9C_4\times t^8\times {}^4C_p\,2^{4-p}\,(-t)^p$

Put $p=0\Rightarrow {}^9C_4\times {}^4C_0\times 2^4\times t^8=126\times 16\times t^8$

Hence, the coefficient of t^8 in the expansion of $(1+2t^2-t^3)^9$ is $84\times 6+126\times 16=504+2016=2520$

12. **(29)**

Let $S={}^{20}C_2+2\times{}^{20}C_3+3\times{}^{20}C_4+\ldots+19\times{}^{20}C_{20}$ (1)

add ${}^{20}C_0+{}^{20}C_1+\ldots+{}^{20}C_{20}\,(=2^{20})$ on both sides of equation (1)

$S+2^{20}={}^{20}C_2+2\times{}^{20}C_3+3\times{}^{20}C_4+\ldots+19\times{}^{20}C_{20}$
$+{}^{20}C_0+{}^{20}C_1+{}^{20}C_2+{}^{20}C_3+{}^{20}C_4+\ldots+{}^{20}C_{20}$

$S+2^{20}=1+{}^{20}C_1+2\times{}^{20}C_2+3\times{}^{20}C_3+4\times{}^{20}C_4+\ldots+20\times{}^{20}C_{20}$ (2)

Now

$(1+x)^n=C_0+C_1x+C_2x^2+\ldots+C_nx^n$

$n(1+x)^{n-1}=C_1+2\times C_2x+3\times C_3x^2+\ldots+n\times C_nx^{n-1}$

Put $x=1$ and $n=20$

$20\times 2^{19}={}^{20}C_1+2\,{}^{20}C_2+3\,{}^{20}C_3+\ldots+20\,{}^{20}C_{20}$

Hence $S+2^{20}=1+20\times 2^{19}$

$S=1+20\times 2^{19}-2^{20}=1+10\times 2^{20}-2^{20}$

$=1+9\times 2^{20}$

$=1+p\times 2^9$; So, $p=9$, $q=20$

Hence, $(p+q)_{\text{least}}=9+20=29$

13. **(12)**

Note:

(i) Here the 9th term has numerically the greatest coefficient which does not mean it is the greatest binomial coefficient.

(ii) For the greatest coefficient, put $x=1$ and then calculate numerically greatest term.

Here, clearly

$$\frac{n+1}{1+\left|\frac{a}{x}\right|}=\frac{n+1}{1+\left|\frac{\frac{1}{5}}{\frac{2}{5}}\right|}=\frac{2(n+1)}{3}$$

As T_9 is numerically greatest term, so

$8<\frac{2(n+1)}{3}<9$

$\Rightarrow 12<n+1<13.5$

$\Rightarrow 11<n<12.5\Rightarrow n=12$

14. **(1040)**

Coefficient x^3 in $(1-x^2)^5\times(1+x)^{20}$

Coefficient x^3 in $(1-5x^2)(1+x)^{20}$

= coefficient x^3 in $(1+x)^{20}$ – 5 coefficient x in $(1+x)^{20}$

$={}^{20}C_3-5\times 20=1140-100=1040$

15. **(0006)**

Coefficient of x^{39} in $(3x+5)^{100}$ is

$T_{r+1}={}^{100}C_r(3r)^{100-r}\times 5^r$

put $r=61$

$T_{62} = {}^{100}C_{61}\, 3^{39}\, x^{39}\, 5^{61}$

Coefficient of x^{39} is ${}^{100}C_{39}\, \underbrace{3^{39} \times 5^{61}}$

Now $3^{39} \times 5^{61}$ has no 2's

Consider, ${}^{100}C_{39} = \dfrac{100!}{(39)!\,(61)!}$

$= \dfrac{2^{97} I_1}{(2^{35} I_2)(2^{56} I_3)}$

when I_1, I_2 and I_3 are odd integers.
Number of 2's in $N^r = 97$
Number of 2's in $D^r = 91$
Hence largest power of 2 is 6.

16. **(816)**
We have,
$f(x) = a_0 + a_1(1 + x) + a_2(1 + x)^2 + \ldots$
$f'(x) = a_1 + 2a_2\,(1 + x) + 3a_3(1 + x)^2 + \ldots$
$f''(x) = 2a_2 + 6a_3\,(1 + x) + \ldots$
$f''(-1) = 2a_2$(Note)
Now $f(x) = 1 - x + x^2 - x^3 + \ldots + x^{16} - x^{17}$
$f'(x) = -1 + 2x - 3x^2 + \ldots + 16x^{15} - 17x^{16}$
$f''(x)]_{x=-1} = 1\times 2 + 2\times 3 + 3\times 4 + \ldots + 16\times 17$

$= \sum_{r=1}^{16} r(r+1) = \sum r^2 + \sum r$

$= \dfrac{16\times 17\times 33}{6} + \dfrac{16\times 17}{2} = \dfrac{16\times 17}{2}\,[11 + 1]$

$f''(-1) = 16\times 17\times 6 \Rightarrow 2a_2 = 16\times 17\times 6$
Hence, $a_2 = 16\times 51 = 816$

17. **(7)**
$5^{40} = (11 - 6)^{40} = 11k + 6^{40} = 11k + (33 + 3)^{20}$
$= 11k + 11\lambda + (11 - 2)^{10} = 11k + 11\lambda + 11\mu + 2^{10}$
$= 11t + 1 \Rightarrow a = 1$
Also, $2^{2003} = 8(17 - 1)^{500} = 8(7\lambda + 1)$
$b = 8$

18. **(9)**

$\left(x + \dfrac{1}{x} + y + \dfrac{1}{y}\right)^{14} \Rightarrow \left((x+y) + \left(\dfrac{x+y}{xy}\right)\right)^{14}$

$\Rightarrow \dfrac{(x+y)^{14}(1+xy)^{14}}{(xy)^{14}}$

Both factors in the numerator have 15 independent terms.
$\Rightarrow 15 \times 15 = 225$ terms $\Rightarrow \alpha + \beta + \gamma = 9$

5.10.3 JEE Advanced: Sample Questions

Single Answer Type

1. **(B)**
The given expression

$= \left(\dfrac{1}{2^{1/3}}\right)^{20} \times \left(2^{2/3} + 2^{4/3} + 2^2\right)^{20}$

$= \dfrac{1}{\left(2^{1/3}\right)^{20}}\left(4^{\frac{1}{3}} + 4 + 4^{\frac{2}{3}}\right)^{20}$

$= \dfrac{1}{\left(2^{1/3}\right)^{20}}\left(2^{1/3} + 4^{1/3}\right)^{40}$

$\Rightarrow (r+1)^{\text{th}}$ term $= 2^{-20/3}\; {}^{40}C_r\, 2^{\frac{40-r}{3}} \cdot 4^{r/3}$

$= {}^{40}C_r\, 2^{\frac{20+r}{3}}$

Hence for rational terms,
$r = 1, 4, 7, 10, 13, 16, 19$
Total number of rational terms = 7

2. **(C)** Do it yourself.

3. **(D)**

$\left(1 + \sqrt{a} + \dfrac{1}{\sqrt{a} - 1}\right)^{-30}$

$= \left(\dfrac{\sqrt{a} - 1}{a}\right)^{30} = \dfrac{1}{a^{30}}\left(1 - \sqrt{a}\right)^{30}$

$= \dfrac{1}{a^{30}}\left[{}^{30}C_0 - {}^{30}C_1\sqrt{a} + \ldots + {}^{30}C_{30}\left(\sqrt{a}\right)^{30}\right]$

$\Rightarrow$ No term is independent of a

4. **(B)**
The number of terms in a general homogenous polynomial is evaluated by ${}^{n+r-1}C_{r-1}$, where n = degree and r = number of variables.
Hence the number of terms in a general polynomial of 10th degree and variable $6 = {}^{10+6-1}C_{6-1} = {}^{15}C_5 = 3003$
and number of terms in a polynomial of 6th degree in 10 variables is
${}^{6+10-1}C_{10-1} = {}^{15}C_9 = 5005$
Hence B has more terms than A.

5. **(B)**
It is given that

$$(1+3x)^n = \sum_{r=0}^{n} a_r x^r (1+x)^{n-r}$$

$$\Rightarrow \left(\frac{1+3x}{1+x}\right)^n = \sum_{r=0}^{n} a_r \left(\frac{x}{1+x}\right)^r$$

$$\Rightarrow \left(1+\frac{2x}{1+x}\right)^r = \sum_{r=0}^{n} a_r \left(\frac{x}{1+x}\right)^r$$

$$\Rightarrow (1+2A)^n = \sum_{r=0}^{n} a_r A^r \text{ where } A = \frac{x}{1+x}$$

Comparing coefficient of A^r, we get

$a_r = 2^r\ {}^nC_r$

6. **(B)**

$x^n + \alpha x + \beta = (x-x_1)^2 g(x)$

Differentiating two times

$$n(n-1)x^{n-2} = 4(x-x_1)g'(x) + (x-x_1)^2 g''(x) + 2g(x)$$

Put $x = x_1$

$g(x_1) = {}^nC_2 x_1^{n-2}$

7. **(B)**
Consider the expansion of

$$(2x+x^2)^n = \left[(x+1)^2 - 1\right]^n$$

$$\Rightarrow x^n(2+x)^n = (x+1)^{2n} - {}^nC_1(x+1)^{2n-2} + \ldots + (-1)^n$$

Equating coefficient of x^n, we get

$2^n = {}^nC_0 \cdot {}^{2n}C_n - {}^nC_1\ {}^{2n-2}C_n + {}^nC_2\ {}^{2n-4}C_n \ldots$

8. **(D)**

Coefficient of x^{10} in $\left(x+\frac{x^2}{2!}+\frac{x^3}{3!}+\ldots+\frac{x^{10}}{10!}\right)^3$

= coefficient of x^{10} in

$$\left(x+\frac{x^2}{2!}+\frac{x^3}{3!}+\ldots+\frac{x^{10}}{10!}+\ldots \text{ to } \infty\right)^3$$

= coefficient of x^{10} in $(e^x-1)^3$

$$= \frac{3^{10}}{10!} - {}^3C_1 \frac{2^{10}}{10!} + \frac{{}^3C_2}{10!} = \frac{3^{10} - 3\times 2^{10} + 3}{10!}$$

9. **(D)**

$$(1+x)^n = {}^nC_0 + {}^nC_1 x + {}^nC_2 x^2 + \ldots + {}^nC_K x^K + \ldots$$

Differentiating

$$n(1+x)^{n-1} = {}^nC_1 + 2.{}^nC_2 x + \ldots + K\ {}^nC_K x^{K-1} + (K+1)\ {}^nC_{K+1} x^K + \ldots$$

and $(1+x)^{-1} = 1 - x + x^2 - x^3 + x^4 + \ldots$ (1)

Differentiating

$-(1+x)^{-2} = -1 + 2x - 3x^2 + 4x^3 + \ldots$ (2)

Multiplying (1) and (2) and comparing coefficient of x^K from both sides, we get the required value of the given expression as $n \cdot {}^{(n-3)}C_K$.

10. **(A)**

$$f(x) = 1 - x + x^2 - x^3 + \ldots - x^{15} + x^{16} - x^{17}$$

$$= \frac{1-x^{18}}{1+x}$$

$$\Rightarrow f(x-1) = \frac{1-(x-1)^{18}}{x}$$

Required coefficient of x^2 = Coefficient of x^3 in $1-(x-1)^{18}$

$= {}^{18}C_3 = 816$

11. **(D)**

$$\left(1+\frac{1}{x}+x+x^2\right)^4 = \left(\frac{(1-x^4)}{x(1-x)}\right)^4 = \frac{1}{x^4}(1-x^4)^4(1-x)^{-4}$$

Constant term in the expansion of $\left(1+\frac{1}{x}+x+x^2\right)^4$ = Coefficient of x^4 in the expansion of $(1-x^4)^4(1-x)^{-4}$

$= {}^7C_4 - {}^4C_1 = 31$

12. **(A)**

$$(5!+1)^{4!} = (121)^{24} = (11)^{48}$$

$$= (10+1)^{48} = {}^{48}C_0 10^{48} + {}^{48}C_1 10^{47} + \ldots + {}^{48}C_{46}(10)^2 + {}^{48}C_{47}\times 10 + {}^{48}C_{48}$$

$= 1000(m) + 112800 + 480 + 1 \qquad (m \in I)$
$= 1000(m) + 113281$
$\Rightarrow$ Last three digits $= 281$

13. **(C)** Do it yourself.

14. **(C)**
Since,
$$(1+x)^n = 1 + nx + \frac{n(n-1)}{2}x^2 + \ldots + x^n$$
or, $(1+x)^n - 1 = nx + \frac{n(n-1)}{2}x^2 + \ldots + x^n$
If $x = n$
$$(1+n)^n - 1 = n^2 + \frac{n(n-1)}{2}n^2 + \ldots + n^n$$
$$\Rightarrow (1+n)^n - 1 = n^2\left[1 + \frac{n(n-1)}{2} + \ldots + n^{n-2}\right]$$
Put $n = 100$
$$(1+100)^n - 1$$
$$= (100)^2\left[1 + \frac{100(100-1)}{2} + \ldots + (100)^{98}\right]$$
Clearly $(101)^{100} - 1$ is divisibly by 10000

15. **(D)**
The coefficient of integral powers of x are
${}^{40}C_0,\ {}^{40}C_2 \times 2^2,\ {}^{40}C_4 \times 2^4, \ldots,\ {}^{40}C_{40} \times 2^{40}$
Since
$$(1+2)^{40} = {}^{40}C_0 + {}^{40}C_1 \times 2 + {}^{40}C_2 \times 2^2 + \ldots + {}^{40}C_{40} \times 2^{40} \qquad (1)$$
$$(1-2)^{40} = {}^{40}C_0 - {}^{40}C_1 \times 2 + {}^{40}C_2 \times 2^2 - \ldots + {}^{40}C_{40} \times 2^{40} \qquad (2)$$
Adding (1) and (2)
$$3^{40} + 1 = 2\left({}^{40}C_0 + {}^{40}C_2 \times 2^2 + \ldots + {}^{40}C_{40} \times 2^{40}\right)$$
$$\Rightarrow {}^{40}C_0 + {}^{40}C_2 \times 2^2 + \ldots + {}^{40}C_{40} \times 2^{40} = \frac{3^{40}+1}{2}$$

16. **(A)**
Since,
$$(xy + yz + zx)^6 = \sum_{r+s+t=6} \frac{6!}{r!s!t!}(xy)^r (yz)^s (zx)^t$$
$$= \sum_{r+s+t=6} \frac{6!}{r!\,s!\,t!} x^{r+t} y^{r+s} z^{s+t}$$
If the general term in the above expansion contains
$x^3y^4z^5$, then
$r + t = 3,\ r + s = 4$ and $s + t = 5$
Also, $r + s + t = 6$
$\Rightarrow r = 1,\ s = 3,\ t = 2$
$$\text{Coefficient of } x^3y^4z^5 = \frac{6!}{1!\,3!\,2!} = 60$$

17. **(A)**
Since, $f(x) = g(x+1)$
$$\Rightarrow \sum_{r=0}^{2n} a_r x^r = \sum_{r=0}^{n-1} b_r (x+1)^r + \sum_{r=n}^{2n} (x+1)^r$$
$$\Rightarrow \sum_{r=0}^{2n} a_r x^r = \sum_{r=0}^{n-1} b_r (x+1)^r + (x+1)^n + (x+1)^{n+1} + \ldots + (x+1)^{2n}$$
$$\Rightarrow \sum_{r=0}^{2n} a_r x^r$$
$$= \sum_{r=0}^{n-1} b_r (x+1)^r + (x+1)^n \left\{\frac{(x+1)^{n+1} - 1}{x+1-1}\right\}$$
$$= \sum_{r=0}^{n-1} b_r (x+1)^r + \frac{(x+1)^{2n+1} - (x+1)^n}{x}$$
Comparing coefficient of x^n from both sides
$$a_n = {}^{2n+1}C_{n+1}$$

18. **(A)**
Let $(r+1)^{\text{th}}$ term be independent of x
$$t_{r+1} = {}^{18}C_r (9x)^{18-r} \left(-\frac{1}{3\sqrt{x}}\right)^r$$
$$= {}^{18}C_r 9^{18-r} \frac{(-1)^r}{3^r} \times \frac{1}{x^{\frac{r}{2}}} \times x^{18-r}$$
$$\Rightarrow {}^{18}C_r (-1)^r \frac{9^{18-r}}{3^r} x^{\frac{36-3r}{2}}$$
Put $\frac{36-3r}{2} = 0 \Rightarrow r = 12$
Clearly $\alpha = 1$

19. **(B)**

Let

$$S=1+\frac{2}{3}\times\frac{1}{2}+\frac{2}{3}\times\frac{5}{6}\times\frac{1}{2^2}+\frac{2}{3}\times\frac{5}{6}\times\frac{8}{9}\times\frac{1}{2^3}+\ldots\text{to }\infty$$

We know that

$$(1+x)^n=1+nx+\frac{n(n-1)}{2!}x^2+\frac{n(n-1)(n-2)}{3!}x^3+\ldots\text{to }\infty$$

Comparing these two, we get

$$nx=\frac{2}{3}\times\frac{1}{2} \qquad (1)$$

and $$\frac{n(n-1)}{2\times1}x^2=\frac{2}{3}\times\frac{5}{6}\times\frac{1}{2^2} \qquad (2)$$

$$\Rightarrow \frac{\frac{n(n-1)}{2}x^2}{n^2x^2}=\frac{\frac{2}{3}\times\frac{5}{6}\times\frac{1}{2^2}}{\frac{2}{3}\times\frac{1}{2}\times\frac{2}{3}\times\frac{1}{2}}$$

$$\Rightarrow \frac{n-1}{2n}=\frac{5}{4}\Rightarrow 5n=2n-2$$

$$\Rightarrow n=-\frac{2}{3}$$

From (1),

$$-\frac{2}{3}x=\frac{2}{3}\times\frac{1}{2}\Rightarrow x=-\frac{1}{2}$$

$$\therefore \text{ Sum of given series}=\left(1-\frac{1}{2}\right)^{-\frac{2}{3}}=4^{\frac{1}{3}}$$

20. **(A)**

$$\sum_{r=0}^{n}(-1)^r\,{}^nC_r\frac{1+r\log_e 10}{(1+\log_e 10^n)^r}$$

$$=\sum_{r=0}^{n}(-1)^r\,{}^nC_r\frac{1+rx}{(1+nx)^r} \qquad [\text{Put } \log_e 10=x]$$

$$=\sum_{r=0}^{n}(-1)^r\,{}^nC_r\frac{1}{(1+nx)^r}+\sum_{r=0}^{n}(-1)^r\frac{n}{r}\,{}^{n-1}C_{r-1}\frac{rx}{(1+nx)^r}$$

$$\sum_{r=0}^{n}(-1)^r\,{}^nC_r\frac{1}{(1+nx)^r}-\frac{nx}{1+nx}\sum_{r=1}^{n}(-1)^{r-1}\,{}^{n-1}C_{r-1}\left(\frac{1}{1+nx}\right)^{r-1}$$

$$=\left(1-\frac{1}{1+nx}\right)^n-\frac{nx}{1+nx}\left(1-\frac{1}{1+nx}\right)^{n-1}$$

$$=\left(\frac{nx}{1+nx}\right)^n-\left(\frac{nx}{1+nx}\right)^n=0$$

21. **(B)**

Let, $\left(\sqrt{3}+1\right)^{2m}=I+F$, where $I\in N$ and $0<F<1$

Let $G=\left(\sqrt{3}-1\right)^{2m}$. Then

$$I+F+G=\left(\sqrt{3}+1\right)^{2m}+\left(\sqrt{3}-1\right)^{2m}$$

$$=2^m\left(2+\sqrt{3}\right)^m+2^m\left(2-\sqrt{3}\right)^m$$

$$=2^{m+1}\times\text{an integer} \qquad (1)$$

$\Rightarrow I+F+G=$ an even integer

$\Rightarrow F+G=$ an integer

$\Rightarrow F+G=1\,[0<F<1, 0<G<1]$

$\therefore$ From (1)

$I+1=2^{m+1}\times$ an integer

$\Rightarrow 2^{m+1}$ is factor of the integer just greater than $\left(\sqrt{3}+1\right)^{2m}$

22. **(D)** Do it yourself.

23. **(D)**

$$\sum_{r=0}^{n}(-1)^r\frac{{}^nC_r}{{}^{r+2}C_r}=\sum_{r=0}^{n}(-1)^r\frac{n!}{(n-r)!\,r!}\frac{2!\,r!}{(r+2)!}$$

$$=2\sum_{r=0}^{n}(-1)^r\frac{n!}{(n-r)!(r+2)!}$$

$$=\frac{2}{(n+1)(n+2)}\sum_{r=0}^{n}(-1)^r\frac{(n+2)!}{\{(n+2)-(r+2)\}!(r+2)!}$$

$$=\frac{2}{(n+1)(n+2)}\sum_{r=0}^{n}(-1)^{r+2}\,{}^{n+2}C_{r+2}$$

$$=\frac{2}{(n+1)(n+2)}\sum_{r=0}^{n}(-1)^{r+2}\,{}^{n+2}C_{r+2}$$

$$=\frac{2}{(n+1)(n+2)}\sum_{s=0}^{n+2}(-1)^s\,{}^{n+2}C_s$$

$$=\frac{2}{(n+1)(n+2)}\left[\left(\sum_{s=0}^{n+2}(-1)^s\,{}^{n+2}C_s\right)-\left({}^{n+2}C_0-{}^{n+2}C_1\right)\right]$$

$$=\frac{2}{n+2}$$

24. (B)

$$\sum_{r=1}^{n}\frac{1}{\log_{4^r} x}=\sum_{r=1}^{n}\log_x 4^r$$

$= \log_x 4 + \log_x 4^2 + \ldots + \log_x 4^n$

$= \log_x 4^{1+2+3+\ldots+n}$

$= \log_x 4^{\frac{n(n+1)}{2}} = \frac{n(n+1)}{2}\log_x 4 = n(n+1)\log_x 2$

25. (C)

$T_r = {}^{15}C_{r-1}(x^4)^{15-r+1}\left(-\frac{1}{x^3}\right)^{r-1}$

$= {}^{15}C_{r-1}(-1)^{r-1}x^{67-7r}$

$67 - 7r = -17 \Rightarrow 7r = 84;\ r = 12$

Multiple Answer Type

1. (B, C, D)

${}^nC_0 + {}^nC_1 + {}^nC_2 \ldots {}^nC_n = 2^n \Rightarrow a = 2^{33}$

${}^nC_0 + {}^nC_2 + {}^nC_4 \ldots = 2^{n-1} \Rightarrow b = 2^{65}$

$(1 + x)^n = {}^nC_0 + {}^nC_1x + {}^nC_2x^2 + {}^nC_3x^3 + \ldots + {}^nC_nx^n$

Put $x = 1$, ω and ω^2 and add them, we get

$c = \frac{2^{99}-2}{3}$. Now check the answer.

2. (A, B, C, D)

It is obvious.

3. (A, C)

$T_{r+1} = {}^{100}C_r 5^{\frac{100-r}{6}} 2^{\frac{r}{8}}$

For $\frac{r}{8} \in Q$

$r = 0, 8, 16, 24, \ldots 96$

For $\frac{100-r}{6}$ to be rational $r = 16, 40, 64, 88$

4 terms are rational $\frac{100-r}{6} = 15$

(C) $r = 10$

4. (A, C, D)

$S = 0\times({}^{100}C_0)^2 + 1\times({}^{100}C_1)^2 + 2\times({}^{100}C_2)^2 + 3\times({}^{100}C_3)^2 + \ldots + 100\times({}^{100}C_{100})^2$

$S = 100\times({}^{100}C_0)^2 + 99\times({}^{100}C_1)^2 + \ldots + 0\times({}^{100}C_{100})^2$

$2S = 100\,[({}^{100}C_0)^2 + ({}^{100}C_1)^2 + ({}^{100}C_2)^2 + \ldots + ({}^{100}C_{100})^2]$

$2S = 100\times{}^{200}C_{100} = 100\,\frac{(200)!}{100!\cdot 100!}$

$\Rightarrow 2S = \frac{100\cdot 200\cdot(199)!}{100\cdot(99)!\cdot(100)!} \Rightarrow S = 100\times{}^{199}C_{99}$

5. (B, C, D)

Write T_{r+1} and verify it.

6. (A, B)

$T_3 = 10^6$ (Given) $\Rightarrow x = 10$

7. (A, B)

$(1 + x^2)^2\,(1 + x)^n = (1 + 2x^2 + x^4)\,({}^nC_0 + {}^nC_1x + {}^nC_2x^2 + \ldots)$

$= {}^nC_0 + {}^nC_1x + ({}^nC_2 + 2\times{}^nC_0)x^2 + \ldots$

Hence $A_0 = 1$; $A_1 = {}^nC_1$

$A_2 = {}^nC_2 + 2$ which are in A.P.

8. (A, C, D)

$(9+4\sqrt{5})^n = I + f$

Let $(9-4\sqrt{5})^n = f'$

$I + f + f' = (9+4\sqrt{5})^n + (9-4\sqrt{5})^n$ = even integer

$0 < f + f' < 2 \Rightarrow f + f' = 1$

$\therefore$ I is an odd integer.

9. (A, C)

(A) $\left(1-\frac{1}{2}\right)^{-2} = \left(\frac{1}{2}\right)^{-2} = \frac{1}{2^{-2}} = 4$

$\Rightarrow$ (A) is correct

(B) it should be odd $\Rightarrow$ (B) is incorrect

(C) LHS $= (2^n)^2$; RHS $= 2^{2n} \Rightarrow$ (C) is correct

(D) $|x| < \frac{3}{2} \Rightarrow$ (D) is incorrect

10. (A, B, C)

$r + 5 = t;\ \sum_{t=0}^{10}(-1)^t\,{}^{10}C_t \times 2^{10-t}$

$= 2^{10}\sum_{t=0}^{10}(-1)^t \times {}^{10}C_t \times \frac{1}{2^t} = 2^{10}\left(1-\frac{1}{2}\right)^{10} = 1$

11. (B, C)

$(1+3x+2x^2)^6 = a_0 + a_1x + a_2x^2 + \ldots + a_{12}x^{12}$

$\therefore$ Coefficient of $x^{12} = 2^6$

Coefficient of $x^{11} = {}^6C_5.2^5.3 = 9.2^6$

Now, $a_0 + a_1 + a_2 + \ldots + a_{12} = 6^6$

$a_0 - a_1 + a_2 - \ldots + a_{12} = 0$

$\therefore a_0 + a_2 + \ldots a_{12} = \frac{6^6}{2}$

and $a_1 + a_3 + \ldots + a_{11} = \frac{6^6}{2}$

Now, $\frac{\sum_{r=0}^{6} a_{2r}}{a_{12}} = \frac{6^6}{2.2^6} = \frac{3^6}{2}$

and $\frac{\sum_{r=1}^{6} a_{2r-1}}{a_{12}} = \frac{6^6}{2.2^6} = \frac{3^6}{2}$

12. (A, B, C)

${}^{15}C_{2r} > \frac{1}{2}\,{}^{15}C_{2r+1}$

$\Rightarrow \frac{15!}{(2r)!\,(15-2r)!} > \frac{1}{2} \times \frac{15!}{(2r+1)!\,(14-2r)!}$

$\Rightarrow 4r + 2 > 15 - 2r \Rightarrow r > \frac{13}{6} \Rightarrow r > 2$

$\therefore r = 5, 6, 7$ but will not hold when $r = 8$.

13. (A, B, C)

$3^{512} - 1 = (8+1)^{256} - 1$

$= (1 + {}^{256}C_1 8 + {}^{256}C_2 8^2 \ldots + {}^{256}C_{256} 8^{256}) - 1$

$= {}^{256}C_1 8[1 + \ldots]$

$3^{512} - 1 = 2^{11} \times (\text{some integer})$

So $3^{512} - 1$ is divisible by $2^3, 2^7, 2^{11}$

14. (A, C)

$N = \left(\sqrt[8]{5} + \sqrt[5]{2}\right)^{100}$

$T_{r+1} = {}^{100}C_r (5)^{\frac{100-r}{8}} (2)^{\frac{r}{5}}$

For rational terms $r = 5\lambda$

$\frac{100 - 5\lambda}{8}$ must be integer, $\lambda = 8k + 4$

$\Rightarrow r = 40k + 20 \le 100 \Rightarrow k \le 2$

Number of rational terms = 3

Number of irrational terms = 98

$N = (\sqrt{2} + 1)^6$

$\therefore I + f = {}^6C_0(\sqrt{2})^6 + {}^6C_1(\sqrt{2})^5 + \ldots + {}^6C_6$

Also,

$f' = (\sqrt{2}-1)^6 = {}^6C_0(\sqrt{2})^6 - {}^6C_1(\sqrt{2})^5 + \ldots + {}^6C_6$

$\Rightarrow I + f + f' = 2\left\{{}^6C_0(\sqrt{2})^6 + {}^6C_2(\sqrt{2})^4 + {}^6C_4(\sqrt{2})^2 + {}^6C_6\right\}$

$\Rightarrow I + 1 = 2\{8 + 60 + 30 + 1\} \Rightarrow I = 197$

$\therefore f = 1 - (\sqrt{2} - 1)^6$

15. (A, B, D)

$N = (2 + 1)^6 = 3^6$

Now check options.

16. (A, C, D)

$\sum_{k=1999}^{2015} {}^kC_x = {}^{1999}C_x + {}^{2000}C_x + \ldots {}^{2015}C_x$

$= {}^{1999}C_{x+1} + {}^{1999}C_x + {}^{2000}C_x + \ldots {}^{2015}C_x - {}^{1999}C_{x+1}$

$= {}^{2016}C_{x+1} - {}^{1999}C_{x+1}$

$\Rightarrow a = 2016, b = 1999$

17. (A, B, C, D)

Given $(1 + x + x^2 + x^3)^{12}$

$= a_0 + a_1x + a_2x^2 + \ldots + a_{36}x^{36}$ (1)

replace x by $\frac{1}{x}$, we get

$(1 + x + x^2 + x^3)^{12} = a_0x^{36} + a_1x^{35} + a_2x^{34} + \ldots + a_{36}$ (2)

$\therefore$ Comparing (1) and (2), we get

$a_r = a_{36-r}$ for each $r \in \{0, 1, 2, \ldots, 36\}$

For option (A) put $x = 1$ in equation (1)

For option (D) put $x = -1$ in equation (1)

Also, from equation (1), $a_0 = a_{36} = 1$ and $a_1 = 12$

$\therefore$ option (B) $= 256 + 4 \times 16 + 4 + 1 = 325$

18. (A, B, C)

Coefficient of x^{324} is

$= 1^2 - 2^2 + 3^2 - 4^2 + \ldots 24^2 + 25^2 = 325$

$\because \sum_{k=0}^{8} {}^8C_k = 2^8 = 256, \sum_{s=0}^{4} {}^4C_s = 2^4 = 16 \sum_{r=0}^{2} {}^2C_r = 4$

19. (A, B, C, D)

In the expansion of $\left(a\alpha^2x^2 + 2b\alpha x + c\right)^n$

The sum of the coefficients $= \left(a\alpha^2 + 2b\alpha + c\right)^n$

Let, $f(\alpha) = a\alpha^2 + 2b\alpha + c$

Discriminant $= 4(b^2 - ac) < 0$

Hence, $f(\alpha) < 0$ or $f(\alpha) > 0$ for all $a \in R$

If $a > 0$, then $f(\alpha) > 0 \Rightarrow (a\alpha^2 + 2b\alpha + c)^n > 0$

If

$c > 0$, that is, $f(0) > 0 \Rightarrow f(\alpha) > 0$

$\Rightarrow (a\alpha^2 + 2b\alpha + c)^n > 0$

If $a < 0$, then $f(\alpha) < 0 \Rightarrow (a\alpha^2 + 2b\alpha + c)^n < 0$ if n is odd

If $c < 0$, that is, $f(0) < 0 \Rightarrow f(\alpha) < 0$

$\Rightarrow (a\alpha^2 + 2b\alpha + c)^n > 0$ if n is even

20. (C, D)

$^{69}C_{3r-1} - {}^{69}C_{r^2} = {}^{69}C_{r^2-1} - {}^{69}C_{3r}$

$\Rightarrow {}^{69}C_{3r-1} + {}^{69}C_{3r} = {}^{69}C_{r^2-1} + {}^{69}C_{r^2}$

$\Rightarrow {}^{70}C_{3r} = {}^{70}C_{r^2}$

$\Rightarrow$ Either $3r = r^2$ or $r^2 + 3r = 70$

$\Rightarrow r = 0, 3$ or $r = 7, -10$

But the given equation is not defined for $r = 0, -10$

Hence, $r = 3$ or 7

21. (A, B, C)

$3^{512} - 1 = (8+1)^{256} - 1$

$= (1 + {}^{256}C_1 \times 8 + {}^{256}C_2 \times 8^2 + \ldots + {}^{256}C_{256}8^{256}) - 1$

$= {}^{256}C_1 \times 8[1 + \ldots]$

$\Rightarrow 3^{512} - 1 = 2^{11} \times$ (some integer)

So, $3^{512} - 1$ is divisible by $2^3, 2^7, 2^{11}$

22. (B, C, D)

Let, $S = \sum_{r=0}^{9} \frac{{}^9C_r(-1)^r}{r+8}$

Since, $x^7(1-x)^9 = \sum_{r=0}^{9} {}^9C_r(-1)^r x^{r+7}$

$\therefore \int_0^1 x^7(1-x)^9\,dx = \sum_{r=0}^{9} \frac{{}^9C_r(-1)^r}{r+8} = \frac{7!9!}{17!}$

$= \frac{1}{5 \times 11 \times 13 \times 16 \times 17}$

23. (B, C, D)

$\alpha = {}^nC_0x^n + {}^nC_2x^{n-2}a^2 + {}^nC_4x^{n-4}a^4 + \ldots$

$\beta = {}^nC_1x^{n-1}a + {}^nC_3x^{n-3}a^3 + {}^nC_5x^{n-5}a^5 + \ldots$

$\alpha + \beta = (x+a)^n$ and $\alpha - \beta = (x-a)^n$

$(x+a)^{2n} - (x-a)^{2n} = (\alpha+\beta)^2 - (\alpha-\beta)^2 = 4\alpha\beta$

$(x+a)^{2n} + (x-a)^{2n} = 2(\alpha^2 + \beta^2)$

$(x+a)^n(x-a)^n = \alpha^2 - \beta^2$

24. (B, C)

$\frac{1}{\sqrt{4x+1}}\left[\left(\frac{1+\sqrt{4x+1}}{2}\right)^n - \left(\frac{1-\sqrt{4x+1}}{2}\right)^n\right]$

$= \frac{1}{2^{n-1}}\left[{}^nC_1 + {}^nC_3(4x+1) + {}^nC_5(4x+1)^2 + \ldots + {}^nC_{11}(4x+1)^5\right]$

$\Rightarrow n = 11$ or 12

25. (B, C, D)

$2^{4n} - 2^n(7n+1) = 2^n\left[(1+7)^n - 7n - 1\right]$

$= 2^n\left[{}^nC_27^2 + {}^nC_37^3 + \ldots\right]$

Comprehension Based Questions

1. (C)

2. (D)

3. (C)

Combined solutions (1 to 3)

Since, $(1 + x + x^2)^n = \sum_{r=0}^{2n} a_r x^r$

Replace x by $\frac{1}{x}$

$\Rightarrow \sum_{r=0}^{2n} a_r\left(\frac{1}{x}\right)^r = \frac{1}{x^{2n}}(x^2 + x + 1)^n$

$\Rightarrow \sum_{r=0}^{2n} a_r x^{2n-r} = \sum_{r=0}^{2n} a_r x^n$

(i) Equating coefficient of x^{2n-r} from both sides

$a_r = a_{2n-r}$, for $0 \le r \le 2n$

(ii) Putting $x = 1$

$a_0 + a_1 + a_2 + \ldots + a_{2n} = 3^n$

As $a_r = a_{2n-r}$

$$\therefore 2(a_0 + a_1 + \ldots + a_{n-1}) + a_n = \frac{3^n - 1}{2}$$

(iii) Replace x by $-\frac{1}{x}$

$$\left(1 - \frac{1}{x} + \frac{1}{x^2}\right)^n = \sum_{r=0}^{2n} (-1)^r \frac{a_r}{x^r}$$

Since, $a_0^2 - a_1^2 + a_2^2 - a_3^2 + \ldots + a_{2n}^2$

= Coefficient of the constant term in

$$(a_0 + a_1 x + \ldots + a_{2n} x^{2n})$$

$$\left(a_0 - \frac{a_1}{x} + \frac{a_2}{x^2} - \frac{a_3}{x^3} + \ldots + \frac{a_{2n}}{x^{2n}}\right)$$

= Coefficient of constant term in

$$(1 + x + x^2)^n \left(1 - \frac{1}{x} + \frac{1}{x^2}\right)^n$$

$= a_n$ Hence $K = 1$

4. **(B)**

5. **(B)**

6. **(B)**

Combined solutions (4 to 6)

2nd term $= {}^nC_1 x^{n-1} a = 240$ (1)

3rd term $= {}^nC_2 x^{n-2} a^2 = 720$ (2)

4th term $= {}^nC_3 x^{n-3} a^3 = 1080$ (3)

Multiplying (1) and (3) and dividing by square of (2), we get

$n = 5$

Hence, $5x^4 a = 240$ and $10x^3 a^2 = 720$

$$\Rightarrow \frac{(5x^4 a)^2}{10x^3 a^2} = \frac{(240)^2}{720} \Rightarrow x^5 = 32 \Rightarrow x = 2$$

$\therefore a = 3$

Hence, $x = 2, a = 3, n = 5$

$(x - a)^n = -1$

Also, $(2+3)^5 = 2^5 + {}^5C_1 2^4 \cdot 3 + \ldots$

Smallest term = 32

Sum of odd number of terms is 1562

7. **(B)**

8. **(D)**

9. (A)

Combined solutions (7 to 9)

We have

$$S_k = 3^k \binom{100}{0}\binom{100}{k} - 3^{k-1}\binom{100}{1}\binom{99}{k-1} + 3^{k-2}\binom{100}{2}\binom{98}{k-2} - \ldots + (-1)^k \binom{100}{k}\binom{100-k}{0}$$

$$= \sum_{r=0}^{k} (-1)^r \times 3^{k-r} \binom{100}{r} \times \binom{100-r}{k-r}$$

$$= \sum_{r=0}^{k} (-1)^r \times 3^{k-r} \times {}^{100}C_r \times {}^{100-r}C_{k-r}$$

$$= \sum_{r=0}^{k} (-1)^r \times 3^{k-r} \times \frac{100!}{r!(100-r)!} \times \frac{(100-r)!}{(k-r)!(100-k)!}$$

$$= \sum_{r=0}^{k} (-1)^r \times 3^{k-r} \times \frac{k!}{r!(k-r)!} \times \frac{100!}{k!(100-k)!}$$

$$= \sum_{r=0}^{k} (-1)^r \times 3^{k-r} \times {}^kC_r \times {}^{100}C_k$$

$$= 3^k \times {}^{100}C_k \times \sum_{r=0}^{k} (-1)^r \times {}^kC_r \times \left(\frac{1}{3}\right)^r$$

$$= 3^k \times {}^{100}C_k \left(1 - \frac{1}{3}\right)^k = 3^k \times {}^{100}C_k \times \left(\frac{2}{3}\right)^k = {}^{100}C_k \times 2^k$$

$$\therefore V_k = M(100, k) = S_k \cdot \left(\frac{1}{2}\right)^k = {}^{100}C_k$$

(i) We have

$$\sum_{k=0}^{100} S_k S_{100-k}$$

$$= \sum_{k=0}^{n} ({}^{100}C_k 2^k)({}^{100}C_{100-k} 2^{100-k})$$

$$\sum_{k=0}^{100} {}^{100}C_k \, {}^{100}C_{100-k} 2^{100} = {}^{2n}C_n 2^n$$

$$\left(\text{Using } \sum_{k=0}^{100} {}^{100}C_k \, {}^{100}C_{100-k} = {}^{200}C_{100}\right)$$

(ii) $M(100, k) = {}^{100}C_k$

$\therefore M(100, 49) + M(100, 50)$

$= {}^{100}C_{49} + {}^{100}C_{50} = {}^{101}C_{50} = {}^{101}C_{101-50}$

$= {}^{101}C_{51}$

$= M(101, 51)$
(Using ${}^{100}C_r + {}^{100}C_{r-1} = {}^{101}C_r$ and ${}^{100}C_r = {}^{100}C_{100-r}$ respectively)

(iii) The student has three choices, either to attempt true or false or not to attempt the question. Hence, the total number of ways of answering patterns $3 \cdot 3 \cdot 3 \ldots$ (n times) $= 3^n$ which is given by $\sum_{k=0}^{100} {}^{100}C_k \times 2^k = \sum_{k=0}^{100} S_k$.

10. **(A)**

$S = \sum_{i=0}^{10} P(i, 10-i) = P(0, 10) + P(1, 9) + \ldots +$
$P(10, 0) = {}^{10}C_0 + {}^{10}C_1 + \ldots + {}^{10}C_{10} = 2^{10}$
$= 1024$

11. **(C)**
$P(i, 100-i) = {}^{100}C_i$
and $P(j, 100-j) = {}^{100}C_j$
Given ${}^{100}C_i = {}^{100}C_j \Rightarrow i + j = 100$
Number of non-negative integral solutions for this equation = 101 (including $i = j = 50$)
Hence, the required number of ordered pairs = 100

12. **(A)**
$P(i, n-i) = {}^nP_i = P(n-i, i)$

Now, $P(43, 4) + \sum_{j=1}^{5} P(49-j, 3)$
$= ({}^{47}C_4 + {}^{47}C_3) + {}^{48}C_3 + {}^{49}C_3 + {}^{50}C_3 + {}^{51}C_3$
(using ${}^nC_r + {}^nC_{r-1} = {}^{n+1}C_r$, repeatedly)
$= ({}^{48}C_4 + {}^{48}C_3) + {}^{49}C_3 + {}^{50}C_3 + {}^{51}C_3$
$= ({}^{49}C_4 + {}^{49}C_3) + {}^{50}C_3 + {}^{51}C_3 = ({}^{51}C_4 + {}^{50}C_3)$
$= {}^{51}C_4 + {}^{51}C_3 = {}^{52}C_4 = P(48, 4) = P(4, 48)$

Match Matrix

1. **(A – R; B – P; C – S; D – Q)**

(A) $S = {}^{20}C_2 + 2\times{}^{20}C_3 + 3\times{}^{20}C_4 + \ldots + 19\times{}^{20}C_{20}$ (1)

Add, ${}^{20}C_0 + {}^{20}C_1 + \ldots + {}^{20}C_{20} = 2^{20}$ on both sides of (1)

$S + 2^{20} = 1 + {}^{20}C_1 + 2\times{}^{20}C_2 + 3\times{}^{20}C_3 + \ldots + 20\times{}^{20}C_{20}$ (2)

Since,
$(1+x)^n = C_0 + C_1x + C_2x^2 + \ldots + C_nx^n$
$\Rightarrow n(1+x)^{n-1} = C_1 + 2\times C_2x + 3\times C_3x^2 + \ldots + n\times C_nx^{n-1}$

Put $x = 1$ and $n = 20$
$20\times 2^{19} = {}^{20}C_1 + 2\times{}^{20}C_2 + 3\times{}^{20}C_3 + \ldots + 20\times{}^{20}C_{20}$
Hence, $S + 2^{20} = 1 + 20\times 2^{19}$
$\Rightarrow S = 1 + 9\times 2^{20} \Rightarrow p = 9, q = 20$
$\therefore p + q = 29$

(B) Let $E = (19-4)^{23} + (19+4)^{23}$
$= 2\left[19^{23} + {}^{23}C_2\times 19^{21}\times 4^2 + \ldots + {}^{23}C_{22}\times 19\times 4^{22}\right]$
$= 2\times 19\left[19^{22} + {}^{23}C_2\times 19^{20}\times 4^2 + \ldots + {}^{23}C_{22}\times 4^{22}\right]$
$\Rightarrow E$ is divisible by 19 $\Rightarrow$ Remainder $= 0$

(C) $\sum_{r=1}^{10} r(r-1)\,{}^{10}C_r = \sum_{r=1}^{10} \frac{10!}{r!(10-r)!} r(r-1)$
$= 90\sum_{r=2}^{10} \frac{8!}{(r-2)!(10-r)!}$
$= 90\left[{}^8C_0 + {}^8C_1 + {}^8C_2 + \ldots + {}^8C_8\right]$
$= 90\times 2^8 = 45\times 2^9 \Rightarrow K = 45$

(D) $\frac{(1-x^2)^{106}}{x^{106}}$

T_{54} is independent of x
$\therefore$ Total number of terms dependent on $x = 107 - 1 = 106$

2. **(A – R; B – P; C – Q; D – S)**

(A) $(18+2)^{13} + (9+4)^{20} = 9K + 2^{13} + 4^{20}$
$= 9K + 2(9-1)^4 + 2(9-1)^{13}$
$= 9\mu + 2 - 2 = 9\mu \Rightarrow$ remainder $= 0$

(B) $\sum_{r=0}^{4} {}^4C_r(r-2)^2 = 16 \Rightarrow$ divisible by 8

(C) $\left(\sqrt{2}+1\right)^6+\left(\sqrt{2}-1\right)^6=198$

(D) 1

3. (A – Q; B – P, Q, R; C – P, Q, R, S)

(A) LHS $\in I \Rightarrow x=\frac{1}{2}, 1$

(B) Put $x=1$
$n=3$
Number of dissimilar terms is 4

(C) $\frac{T_{r+1}}{T_r}=\frac{3}{2}\gtrless 1$
$27 \gtrless 5r$
for $r=5$ $T_6>T_5$
for $r=6$ $T_7<T_6$
$\therefore$ T_6 is numerically greater.

4. (A – P; B – R; C – R; D – S)

(P) We are given that $2\,{}^{14}C_r={}^{14}C_{r-1}+{}^{14}C_{r+1}$

$\Rightarrow 2(r+1)(15-r)=r(r+1)+(15-r)(14-r)$

or

$30+28r-2r^2=r^2+r+210-29r+r^2$

or $r^2-14r^2+45=0$

$\Rightarrow (r-5)(r-9)=0$

(Q) $[(1+x+x^2+x^3+x^4\ldots+(x)^{n-1})(1-x)]^m$

$=\left[\left\{\frac{1-(x)^n}{1-x}\right\}(1-x)\right]^m=(1-(x)^n)^m$

$={}^mC_0+{}^mC_1(-x)^n+\ldots.+(-1)^m\,{}^mC_m(x)^{mn}$

So, the sum of all the coefficients is 0.

(R)

$\sum_{r=1}^{n}(-1)^r\times{}^nC_r\left(\frac{1-i^{nr}}{1-i^r}\right)$

$=\sum_{r=1}^{n}(-1)^r\times{}^nC_r\left(\frac{1-i^{4kr}}{1-i^r}\right)=0$

(S) $3^{37}=3^{4\times9}\times3=3(81)^9=3(80+1)^9$

$=3({}^9C_0 80^9+{}^9C_1 80^8+\ldots+{}^9C_9)$

Hence, the remainder is 3.

Subjective Answer Type (Hard Nut)

1. (i) Let the rth, $(r+1)^{\text{th}}$ and $(r+2)^{\text{th}}$ coefficeint of $(1+x)^n$ be in G.P.

$\Rightarrow C_{r-1}, C_r, C_{r+1}$ are in G.P.

$\Rightarrow \frac{C_r}{C_{r-1}}=\frac{C_{r+1}}{C_r}$

$\Rightarrow \frac{n-r+1}{r}=\frac{n-r}{r+1}$

$\Rightarrow (n-r+1)(r+1)=r(n-r)$

$\Rightarrow nr+n-r^2-r+r+1=nr-r^2$

$\Rightarrow n+1=0 \Rightarrow n=-1$

Which is not possible, since n is a positive integer.

(ii) Let C_{r-1}, C_r, C_{r+1} are in H.P.

$\Rightarrow C_r=\frac{2}{\frac{1}{C_{r-1}}+\frac{1}{C_{r+1}}}$

$\Rightarrow 2=\frac{C_r}{C_{r-1}}+\frac{C_r}{C_{r+1}}$

$\Rightarrow 2=\frac{n-r+1}{r}+\frac{r+1}{n-r}$

$\Rightarrow 2r(n-r)=(n-r+1)(n-r)+r(r+1)$

$\Rightarrow 2nr-2r^2=n^2-nr-nr+r^2+n-r+r^2+r$

$\Rightarrow n^2-4nr+4r^2+n=0$

$\Rightarrow (n-2r)^2+n=0$

This is not possible, as $(n-2r)^2\geq 0$ and n is a positive integer.

2. (5)

3. $\frac{2}{1!13!}+\frac{2}{3!11!}+\frac{2}{5!9!}+\frac{1}{7!7!}$

$=\frac{1}{14!}\left(2\frac{14!}{1!13!}+2\frac{14!}{3!11!}+2\frac{14!}{5!9!}+\frac{14!}{7!7!}\right)$

$=\frac{1}{14!}\left(2\times{}^{14}C_1+2\times{}^{14}C_3+2\times{}^{14}C_5+{}^{14}C_7\right)$

$=\frac{1}{14!}\left({}^{14}C_1+{}^{14}C_3+{}^{14}C_5+{}^{14}C_7+{}^{14}C_9+{}^{14}C_{11}+{}^{14}C_{13}\right)$

$$= \frac{2^{14-1}}{14!} = \frac{2^{13}}{14!} \Rightarrow m = 3 \text{ and } n = 14$$

4. $t_r = \left(r^2 + r + 1\right)r!, r = 0, 1, 2, \ldots, n$

$$\Rightarrow t_r = \left[(r+1)^2 - r\right]r!$$
$$= (r+1)(r+1)! - r \cdot r!$$
$$= (r+1)(r+1)! - (r+1-1)r!$$
$$= (r+2-1)(r+1)! - (r+1)r! + r!$$
$$= (r+2)(r+1)! - (r+1)! - (r+1)! + r!$$
$$= \left[(r+2)! - (r+1)!\right] - \left[(r+1)! - r!\right]$$

Put $r = 0, 1, 2, \ldots, n$

$$t_0 = (2! - 1!) - (1! - 0!)$$
$$t_1 = (3! - 2!) - (2! - 1!)$$
$$t_2 = (4! - 3!) - (3! - 2!)$$

$$t_n = ((n+2)! - (n+1)!) - ((n+1)! - n!)$$

Adding

$$t_0 + t_1 + t_2 + \ldots + t_n$$
$$= (n+2)! - 1! - (n+1)! + 1$$
$$= (n+1)!\,(n+2-1)$$
$$= (n+1)(n+1)! = 4000(4000)!$$
$$\Rightarrow n+1 = 4000 \Rightarrow n = 3999$$

5. Since, $\sin^2\frac{\pi}{7}\sin\frac{2\pi}{7}\sin\frac{4\pi}{7}$

$$= \frac{\left(1-\cos\frac{2\pi}{7}\right)}{2} \cdot \frac{1}{2}\left(\cos\frac{2\pi}{7} - \cos\frac{6\pi}{7}\right)$$
$$= \frac{1}{4}\left(1-\cos\frac{2\pi}{7}\right)\left(\cos\frac{2\pi}{7} + \cos\frac{\pi}{7}\right)$$
$$= \frac{1}{4}\left(\cos\frac{2\pi}{7} + \cos\frac{\pi}{7} - \cos^2\frac{2\pi}{7} - \cos\frac{2\pi}{7}\cos\frac{\pi}{7}\right)$$
$$= \frac{1}{4}\left(\cos\frac{2\pi}{7} + \cos\frac{\pi}{7} - \frac{1}{2} - \frac{1}{2}\cos\frac{4\pi}{7} - \frac{1}{2}\cos\frac{3\pi}{7} - \frac{1}{2}\cos\frac{\pi}{7}\right)$$
$$= \frac{1}{4}\left(\cos\frac{2\pi}{7} + \frac{1}{2}\cos\frac{\pi}{7} - \frac{1}{2}\right)$$

$\Rightarrow$ The value of $n = 2$

6. $\sum_{r=0}^{n} r \times {}^nC_r x^r y^{n-r}$

$$= \sum_{r=0}^{n} r \times \frac{n}{r}\,{}^{n-1}C_{r-1} x^{r-1} \times x\left(\frac{y}{x}\right)^{(n-1)-(r-1)}$$
$$= nx(y+x)^{n-1} = nx(1)^{n-1} = nx$$

7. Let, $S = \sum_{0\le i<j\le n}\sum\left(\frac{i}{{}^nC_i} + \frac{j}{{}^nC_j}\right)$

$$= n\sum_{0\le i<j\le n}\sum\left(\frac{1}{{}^nC_i} + \frac{1}{{}^nC_j}\right) - S$$
$$\Rightarrow S = \frac{n}{2}\sum_{0\le i<j\le n}\sum\left(\frac{1}{{}^nC_i} + \frac{1}{{}^nC_j}\right)$$
$$\Rightarrow S = \frac{n}{2}\left[\left\{\left(\frac{1}{{}^nC_0} + \frac{1}{{}^nC_1}\right) + \left(\frac{1}{{}^nC_0} + \frac{1}{{}^nC_2}\right) + \ldots + \left(\frac{1}{{}^nC_0} + \frac{1}{{}^nC_n}\right)\right\}\right]$$
$$+ \left\{\left(\frac{1}{{}^nC_1} + \frac{1}{{}^nC_2}\right) + \ldots + \left(\frac{1}{{}^nC_1} + \frac{1}{{}^nC_n}\right)\right\}$$
$$+ \ldots + \left(\frac{1}{{}^nC_{n-1}} + \frac{1}{{}^nC_n}\right)\Big]\Big\}$$
$$\Rightarrow S = \frac{n}{2} \times n\left(\frac{1}{{}^nC_0} + \frac{1}{{}^nC_1} + \ldots + \frac{1}{{}^nC_n}\right)$$
$$= \frac{n^2a}{2}$$

8. $a_n = \frac{(\log 3)^n}{n!}\sum_{K=1}^{n}(2K+1)\,{}^nC_K$

$$= \frac{(\log 3)^n}{n!}\left[2 \times n2^{n-1} + 2^n - 1\right]$$
$$= \frac{(2\log 3)^n}{(n-1)!} + \frac{(2\log 3)^n}{n!} - \frac{(\log 3)^n}{n!}$$

$$\therefore \sum_{n=1}^{\infty} a_n = 2\log_3 \sum_{n=1}^{\infty} \frac{(2\log 3)^{n-1}}{(n-1)!} + \sum_{n=1}^{\infty} \frac{(2\log 3)^n}{n!} - \sum_{n=1}^{\infty} \frac{(\log 3)^n}{n!}$$

$$= 2\log 3 \times e^{2\log 3} + \left(e^{2\log 3} - 1\right) - \left(e^{\log 3} - 1\right)$$

$$= 18\log 3 + 8 - 2 = 18\log 3 + 6$$

9. Do it yourself.

10. $t_r = \left(r^2 + r + 1\right) r!$

$$= (r+1)(r+1)r! - r \cdot r!$$

$$= (r+1)(r+1)! - r \cdot r!$$

$$= (r+2-1)(r+1)! - (r+1-1)r!$$

$$= \left[(r+2)! - (r+1)!\right] - \left[(r+1)! - r!\right]$$

Put $r = 0, 1, 2, 3, \ldots, n$ respectively

$$t_0 = [2! - 1!] - [1! - 0!]$$

$$t_1 = [3! - 2!] - [2! - 1!]$$

$$t_2 = [4! - 3!] - [3! - 2!]$$

$$t_n = \left[(n+2)! - (n+1)!\right] - \left[(n+1)! - n!\right]$$

Adding

$$\sum_{r=0}^{n} t_r = \left[(n+2)! - 1!\right] - \left[(n+1)! - 0!\right]$$

$$= (n+2)! - (n+1)!$$

$$= (n+1)!\,(n+2-1)$$

$$= (n+1)(n+1)!$$

$$\Rightarrow (n+1)(n+1)! = 2000(2000!)$$

$$\Rightarrow n+1 = 2000$$

Now,

$$t_{r+1} = {}^{2000}C_r x^{2000-r} \frac{1}{x^r}$$

$$= {}^{2000}C_r x^{2000-2r}$$

Put $2000 - 2r = 2000 \Rightarrow r = 0$

$\therefore$ Coefficient of $x^{2000} = {}^{2000}C_0 = 1$

11. $\left(1 - x^3\right)^n = \sum_{r=0}^{n} a_r x^r (1-x)^{3n-2r}$

$$\Rightarrow (1-x)^n \left(1 + x + x^2\right)^n = \sum_{r=0}^{n} a_r x^r \frac{(1-x)^{3n}}{(1-x)^{2r}}$$

$$\Rightarrow \left[\frac{1 + x + x^2}{(1-x)^2}\right]^n = \sum_{r=0}^{n} a_r \frac{x^r}{(1-x)^{2r}}$$

$$\Rightarrow \left[\frac{(1-x)^2 + 3x}{(1-x)^2}\right]^n = \sum_{r=0}^{n} a_r \left(\frac{x}{(1-x)^2}\right)^r$$

$$\Rightarrow \left[1 + 3\left(\frac{x}{(1-x)^2}\right)\right]^n = \sum_{r=0}^{n} a_r \left[\frac{x}{(1-x)^2}\right]^r$$

Put $\dfrac{x}{(1-x)^2} = \alpha$

$$\Rightarrow (1 + 3\alpha)^n = \sum_{r=0}^{n} a_r \alpha^r$$

Comparing coefficient of α^r

$${}^nC_r 3^r = a_r \Rightarrow a_r = {}^nC_r 3^r$$

12. **(4)** Do it yourself.

13. $t_r = \dfrac{1^2 + 2^2 + \ldots + r^2}{(2r+1)}\, {}^nC_r$

$$= \frac{r(r+1)(2r+1)}{6(2r+1)}\, {}^nC_r$$

$$= \frac{1}{6} r(r+1)\, {}^nC_r$$

$$\therefore \sum_{r=1}^{n} t_r = \frac{1}{6} \sum_{r=1}^{n} (r+1) r\, {}^nC_r$$

$$= \frac{1}{6} \sum_{r=1}^{n} (r+1) n \times {}^{n-1}C_{r-1}$$

$$= \frac{1}{6} n \sum_{r=1}^{n} \left[(r-1) + 2\right] {}^{n-1}C_{r-1}$$

$$= \frac{1}{6} n \sum_{r=1}^{n} \left[(r-1)\, {}^{n-1}C_{r-1} + 2 \times {}^{n-1}C_{r-1}\right]$$

$$= \frac{1}{6} n \sum_{r=1}^{n} \left[(n-1)\, {}^{n-2}C_{r-2} + 2 \times {}^{n-1}C_{r-1}\right]$$

$$= \frac{1}{6}n(n-1)2^{n-2} + \frac{n}{3}2^{n-1}$$
$$= \frac{1}{6}n(n+3)2^{n-2}$$

14. $\sum_{r=0}^{n}(r+1)\ {}^{2n}C_r$

$$= \sum_{r=0}^{n} r \times {}^{2n}C_r + \sum_{r=0}^{n} {}^{2n}C_r$$
$$= \sum_{r=0}^{n} r \times \frac{2n}{r}\ {}^{2n-1}C_{r-1} + \sum_{r=0}^{n} {}^{2n}C_r$$
$$= n\sum_{r=0}^{n} 2 \times {}^{2n-1}C_{r-1} + \sum_{r=0}^{n} {}^{2n}C_r$$
$$= n\Big[\left({}^{2n-1}C_0 + {}^{2n-1}C_0\right) + \left({}^{2n-1}C_1 + {}^{2n-1}C_1\right) + \ldots + \left({}^{2n-1}C_{n-2} + {}^{2n-1}C_{n-2}\right) + \left({}^{2n-1}C_{n-1} + {}^{2n-1}C_{n-1}\right)\Big]$$
$$+ \frac{1}{2}\Big[\left({}^{2n}C_0 + {}^{2n}C_0\right) + \left({}^{2n}C_1 + {}^{2n}C_1\right) + \ldots + \left({}^{2n}C_n + {}^{2n}C_n\right)\Big]$$
$$= n\left[{}^{2n-1}C_0 + {}^{2n-1}C_1 + \ldots + {}^{2n-1}C_{2n-1}\right] + \frac{1}{2}\left[{}^{2n}C_0 + {}^{2n}C_1 + \ldots + {}^{2n}C_{2n}\right] + \frac{1}{2}\ {}^{2n}C_n$$
$$= n \times 2^{2n-1} + \frac{1}{2}2^{2n} + \frac{1}{2}\ {}^{2n}C_n.$$

15. $\left(1+x+x^2\right)^n$ (1)

$= a_0 + a_1x + a_2x^2 + \ldots + a_{2n}x^{2n}$

Put $x = 1$

$a_0 + a_1 + a_2 + \ldots + a_{2n} = 3^n$ (2)

Diff. (1)

$n\left(1+x+x^2\right)^{n-1} \times (1+2x)$

$= a_1 + 2a_2x + 3a_3x^2 + \ldots + 2n\ a_{2n}x^{2n-1}$

Put $x = 1$

$3n \cdot 3^{n-1} = a_1 + 2a_2 + 3a_3 + \ldots + 2n\ a_{2n}$

$\Rightarrow a_1 + 2a_2 + 3a_3 + \ldots + 2n\ a_{2n} = n3^n$ (3)

Now,

$a_0 + 3a_1 + 5a_2 + \ldots + (4n+1)a_{2n}$

$= (a_0 + a_1 + a_2 + \ldots + a_{2n}) + 2(a_1 + 2a_2 + 3a_3 + \ldots + 2n\ a_{2n})$

$= 3^n + 2n3^n = (2n+1)3^n$

16. $t_r = {}^nC_r \sin rx \cos(n-r)x$

$$\Rightarrow \sum_{r=1}^{n} t_r = \sum_{r=1}^{n} {}^nC_r \sin rx \cos(n-r)x$$
$$\Rightarrow \sum_{r=0}^{n} t_r = \sum_{r=0}^{n} {}^nC_r \sin rx \cos(n-r)x$$

Let $S = \sum_{r=0}^{n} {}^nC_r \sin rx \cos(n-r)x$

$$\Rightarrow S = \sum_{r=0}^{n} {}^nC_r \sin(n-r)x \cos rx$$
$$\therefore 2S = \sum_{r=0}^{n} {}^nC_r \sin x = 2^n \sin x$$
$$\Rightarrow S = 2^{n-1}\sin x$$

17. $\sum_{r=0}^{n} {}^nC_r a^r b^{n-r} \cos(rB-(n-r)A)$

$$= \text{R.P.}\sum_{r=0}^{n} {}^nC_r a^r b^{n-r} e^{i(rB-(n-r)A)}$$
$$= \text{R.P.}\sum_{r=0}^{n} {}^nC_r a^r b^{n-r} e^{irB} e^{-(n-r)iA}$$
$$= \text{R.P.}\sum_{r=0}^{n} {}^nC_r \left(ae^{iB}\right)^r \left(be^{-iA}\right)^{n-r}$$
$$= \text{R.P.}\left(ae^{iB} + be^{-iA}\right)^n$$
$$= \text{R.P.}\left[a(\cos B + i\sin B) + b(\cos A - i\sin A)\right]^n$$
$$= \text{R.P.}\left[(a\cos B + b\cos A) + i(a\sin B - b\sin A)\right]^n$$
$$= c^n$$

18. $\sum_{r=0}^{n}(-1)^r \frac{{}^nC_r}{{}^{r+3}C_r} = \sum_{r=0}^{n}(-1)^r \frac{n!\,3!}{(n-r)!\,(r+3)!}$

$$= 3!\sum_{r=0}^{n}(-1)^r \frac{n!}{(n-r)!(r+3)!}$$
$$= \frac{3!}{(n+1)(n+2)(n+3)}\sum_{r=0}^{n}(-1)^r \frac{(n+3)!}{\{(n+3)-(r+3)\}!\,(r+3)!}$$

$$= \frac{3!}{(n+1)(n+2)(n+3)} \sum_{r=0}^{n} (-1)^r \ {}^{n+3}C_{r+3}$$

$$= \frac{3!}{(n+1)(n+2)(n+3)} \left[\left({}^{n+3}C_3 - {}^{n+3}C_4 + {}^{n+3}C_5 \ldots\right)\right]$$

$$= \frac{3!}{(n+1)(n+2)(n+3)} \left[\frac{(n+1)(n+2)}{2}\right]$$

$$= \frac{3!}{2(n+3)}$$

19. $\sum_{r=0}^{n} \left(\frac{r+2}{r+1}\right) {}^nC_r = \sum_{r=0}^{n} \left(1 + \frac{1}{r+1}\right) {}^nC_r$

$$= \sum_{r=0}^{n} {}^nC_r + \sum_{r=0}^{n} \frac{{}^nC_r}{r+1}$$

$$= \sum_{r=0}^{n} {}^nC_r + \frac{1}{n+1} \sum_{r=0}^{n} {}^{n+1}C_{r+1}$$

$$= 2^n + \frac{1}{(n+1)}\left(2^{n+1} - 1\right)$$

$$= \frac{2^n (n+3) - 1}{n+1}$$

Since, $\sum_{r=0}^{n} \left(\frac{r+2}{r+1}\right) {}^nC_r = \frac{2^8 - 1}{6}$

$$\Rightarrow \frac{2^n (n+3) - 1}{(n+1)} = \frac{2^8 - 1}{6}$$

$\Rightarrow n = 5$

20. $1 + \sum_{r=1}^{10} \left(3^r \times {}^{10}C_r + r \times {}^{10}C_r\right)$

$$= 1 + \sum_{r=1}^{10} 3^r \times {}^{10}C_r + 10 \sum_{r=1}^{10} {}^9C_{r-1}$$

$$= 1 + 4^{10} - 1 + 10 \times 2^9 = 2^{10}\left(4^5 + 5\right)$$

$\Rightarrow \alpha = 1, \beta = 5$

Again, since, $f(1) < 0$ and $f(5) < 0$

$\Rightarrow 16 - K^2 < 0 \Rightarrow K \in (-\infty, -4) \cup (4, \infty)$

Hence, the smallest integral value of $K = 5$

5.10.4 JEE Main: Archive

1. (D)

$3^{2022} = 9^{1011} = (10-1)^{1011} = 10m - 1 = 10m - 5 + 4$

$= 5(2m-1) + 4$ (m is integer)

2. (4)

$$\frac{1 \cdot \left(3^{2022} - 1\right)}{2} = \frac{9^{1011} - 1}{2}$$

$$= \frac{(10-1)^{1011} - 1}{2}$$

$$= \frac{100\lambda + 10110 - 1 - 1}{2}$$

$$= 50\lambda + \frac{10108}{2}$$

$= 50\lambda + 5054$

$= 50\lambda + 50 \times 101 + 4$

$\text{Rem}(50) = 4$

3. (286)

$(1+x)^{10} = C_0 + C_1 x + C_2 x^2 + \ldots + C_{10} x^{10}$

Differentiating

$10(1+x)^9 = C_1 + 2C_2 x + 3C_3 x^2 + \ldots + 10C_{10} x^9$

replace $x \to x^2$

$10\left(1+x^2\right)^9 = C_1 + 2C_2 x^2 + 3C_3 x^4 + \ldots + 10C_{10} x^{18}$

$10 \cdot x\left(1+x^2\right)^9 = C_1 x + 2C_2 x^3 + 3C_3 x^5 + \ldots + 10C_{10} x^{19}$

Differentiating

$$10\left(\left(1+x^2\right)^9 \times 1 + x \times 9\left(1+x^2\right)^8 2x\right)$$

$= C_1 x + 2C_2 \times 3x^3 + 3 \times 5 \times C_3 x^4 + \ldots + 10 \times 19 C_{10} x^{18}$

Putting $x = 1$

$10\left(2^9 + 18 \times 2^8\right)$

$= C_1 + 3 \times 2 \times C_2 + 5 \times 3 \times C_3 + \ldots + 19 \times 10 \times C_{10}$

$C_1 + 3 \times 2 \times C_2 + \ldots + 19 \times 10 \times C_{10}$

$= 10 \times 2^9 \times 10 = 100 \times 2^9$

$$C_0 + \frac{C_1}{2} + \frac{C_2}{3} + \ldots + \underset{\substack{\uparrow \\ 10^{\text{th}} \text{ term}}}{\frac{C_9}{11}} + \underset{\substack{\uparrow \\ 11^{\text{th}} \text{ term}}}{\frac{C_{10}}{11}} = \frac{2^{11} - 1}{11}$$

$C_0 + \frac{C_1}{2} + \frac{C_2}{3} + \ldots + \frac{C_9}{11} = \frac{2^{11}-2}{11}$

Now, $100 \cdot 2^9 = \frac{\alpha \cdot 2^{11}}{2^\beta - 1}\left(\frac{2^{11}-2}{11}\right)$

Equation is of the form $y = k\left(2^x - 1\right)$.

It has infinite solutions even if we take $x, y \in N$.

4. **(A)**

$(5+x)^{500} + x(5+x)^{499} + x^2(5+x)^{498} + \ldots + x^{500}$

$= \frac{(5+x)^{501} - x^{501}}{(5+x)-x} = \frac{(5+x)^{501} - x^{501}}{5}$

$\Rightarrow$ Coefficient x^{101} in the given expression

$= \frac{{}^{501}C_{101}5^{400}}{5} = {}^{501}C_{101}5^{399}$

5. **(83)**

$T_{r+1} = {}^{10}C_r\left(2x^3\right)^{10-r}\left(\frac{3}{x}\right)^r$

$= {}^{10}C_r 20^{10-r} 3^r x^{30-4r}$

Put $r = 0, 1, 2, \ldots, 7$ and we get $\beta = 83$

6. **(C)**

$(2021)^{2023} = (7\lambda - 2)^{2023}$

$= {}^{2023}C_0(7A)^{2023} - \ldots {}^{2023}C_{2023}2^{2023}$

$= 7t - 2^{2023}$

$\therefore 7t - 2^{2023}$

$\therefore -2^{2023} = -2 \times 2^{2022}$

$= -2 \times \left(2^3\right)^{674}$

$= -2(1+7\mu)^{674}$

$= -(7\alpha + 2)$

$\Rightarrow$ remainder $= -2$ or $+5$

7. **(102)**

${}^{40}C_0 + {}^{41}C_1 + {}^{42}C_2 + \ldots {}^{59}C_{19} + {}^{60}C_{20}$

$\left(\frac{1}{41}+1\right){}^{41}C_1 + {}^{42}C_2 + \ldots$

$\left[\frac{42}{41}\left(\frac{2}{42}\right)+1\right]{}^{42}C_2 + {}^{43}C_3 + \ldots$

$\left(\frac{2}{41}+1\right){}^{42}C_2 + {}^{43}C_3 + \ldots$

$\left(\frac{43}{41} \times \frac{3}{43} + 1\right){}^{43}C_3 + {}^{44}C_4 + \ldots$

$= \frac{3+41}{4} \cdot {}^{43}C_3 + \ldots$

Similarly:

$\frac{20+41}{41}$

$\Rightarrow m = 61; n = 41$

$m + n = 102$

8. **(5)**

$\left(\frac{\sqrt{x}}{5^{\frac{1}{4}}} + \frac{\sqrt{5}}{x^{\frac{1}{3}}}\right)^{60}$

$T_{r+1} = {}^{60}C_r\left(\frac{x^{1/2}}{5^{1/4}}\right)^{60-r}\left(\frac{5^{1/2}}{x^{1/3}}\right)^r$

$= {}^{60}C_r 5^{\frac{3r-60}{4}} \times x^{\frac{180-5r}{6}}$

$= \frac{180-5r}{6} = 10 \Rightarrow r = 24$

Coeff. of $x^{10} = {}^{60}C_{24}5^3 = \frac{\lfloor 60}{\lfloor 24 \lfloor 36}5^3$

Powers of 5 in $= {}^{60}C_{24} \times 5^3 = \frac{5^{14}}{5^4 \times 5^8} \times 5^3 = 5^5$

9. **(57)**

Coefficients and their cumulative sums are:

Coefficient	Cumulative sum
$x^{7n} \to {}^7C_0$	1
$x^{6n-5} \to 2 \times {}^7C_1$	$1+14$
$x^{5n-10} \to 2^2 \times {}^7C_2$	$1+14+84$
$x^{4n-15} \to 2^3 \times {}^7C_3$	$1+14+84+280$
$x^{3n-20} \to 2^4 \times {}^7C_4$	$1+14+84+280+560 = 939$
$x^{2n-25} \to 2^5 \times {}^7C_5$	

$3n - 20 \geq 0 \cap 2n - 25 < 0 \cap n \in I$

$\therefore 7 \leq n \leq 12$

Sum $= 7 + 8 + 9 + 10 + 11 + 12 = 57$

10. (A)

$\sum_{R=1}^{31} {}^{31}C_R \times {}^{31}C_{R-1}$

$= {}^{31}C_1 \times {}^{31}C_0 + {}^{31}C_2 \times {}^{31}C_1 + \ldots + {}^{31}C_{31} \times {}^{31}C_{30}$

$= {}^{31}C_0 \times {}^{31}C_{30} + {}^{31}C_1 \times {}^{31}C_{29} + \ldots + {}^{31}C_{30} \times {}^{31}C_0$

$= {}^{62}C_{30}.$

Similarly

$\sum_{R=1}^{30} \left({}^{30}C_R \times {}^{30}C_{R-1}\right) = {}^{60}C_{29}$

$${}^{62}C_{30} - {}^{60}C_{29} = \frac{62!}{30!\,32!} - \frac{60!}{29!\,31!}$$

$$= \frac{60!}{29!\,31!}\left\{\frac{62\times 61}{30\times 32} - 1\right\}$$

$$= \frac{60!}{30!\,31!}\left(\frac{2822}{32}\right)$$

$$\therefore 16\alpha = 16 \times \frac{2822}{32} = 1411$$

11. (2)

$$\left(2x^3 + \frac{3}{x^k}\right)^{12}$$

$$t_{r+1} = {}^{12}C_r \left(2x^3\right)^r \left(\frac{3}{x^k}\right)^{12-r}$$

$x^{3r-(12-r)k} \rightarrow$ constant

$\therefore 3r - 12k + rk = 0$

$$\Rightarrow k = \frac{3r}{12-r}$$

$\therefore$ Possible values of r are $3, 6, 8, 9, 10$ and corresponding values of k are $1, 3, 6, 9, 15$

Now, ${}^{12}C_r = 220, 924, 495, 220, 66$

$\therefore$ Possible values of k for which we will get 2^8 are 3, 6

12. (B)

$$\left(1 - x^2 + 3x^3\right)\left(\frac{5}{2}x^3 - \frac{1}{5x^2}\right)^{11}$$

General term of $\left(\frac{5}{2}x^3 - \frac{1}{5x^2}\right)^{11}$ is

$${}^{11}C_r \left(\frac{5}{2}x^3\right)^{11-r} \left(-\frac{1}{5x^2}\right)^r$$

General term is ${}^{11}C_r \left(\frac{5}{2}\right)^{11-r} \left(-\frac{1}{5}\right)^r x^{33-5r}$

Now, the term independent of x

$1 \times$ coefficient of x^0 in $\left(\frac{5}{2}x^3 - \frac{1}{5x^2}\right)^{11}$

$-1 \times$ coefficient of x^{-2} in $\left(\frac{5}{2}x^3 - \frac{1}{5x^2}\right)^{11} +$

$3 \times$ coefficient of x^{-3} in $\left(\frac{5}{2}x^3 - \frac{1}{5x^2}\right)^{11}$

for coefficient of x^0, $33 - 5r = 0$ not possible

for coefficient of x^{-2}, $33 - 5r = -2$

$35 = 5r \Rightarrow r = 7$

for coefficient of x^{-3}, $33 - 5r = -3$

$36 = 5r$ not possible

So, the term independent of x is

$$(-1)\,{}^{11}C_7 \left(\frac{5}{2}\right)^4 \left(-\frac{1}{5}\right)^7 = \frac{33}{200}$$

13. (D)

General term

$$T_{r+1} = \frac{\lfloor 10}{\lfloor r_1 \lfloor r_2 \lfloor r_3} (3)^{r_1} (-2)^{r_2} (5)^{r_3} (x)^{3r_1 + 2r_2 - 5r_3}$$

$3r_1 + 2r_2 - 5r_3 = 0$ (1)

$r_1 + r_2 + r_3 = 10$ (2)

from equations (1) and (2)

$r_1 + 2(10 - r_3) - 5r_3 = 0$

$r_1 + 20 = 7r_3$

$(r_1, r_2, r_3) = (1, 6, 3)$

constant term $= \frac{\lfloor 10}{\lfloor 1 \lfloor 6 \lfloor 3} (3)^1 (-2)^6 (5)^3$

$= 2^9 \times 3^2 \times 5^4 \times 7^1$

$\ell = 9$

14. (C)

$$\alpha = \frac{(1+8)^n - 8n - 1}{64} = {}^nC_2 + {}^nC_3 8 + {}^nC_4 8^2 + \ldots$$

$\beta = {}^nC_2 + {}^nC_3 5 + {}^nC_4 5^2 + \ldots$

option (C) will be the answer.

15. (5)

$$T_{r+1} = (-1)^r \times {}^{15}C_r \times 2^{15-r} x^{\frac{15-2r}{5}}$$

$m = {}^{15}C_{10} 2^5$

$n = -1$

So $mn^2 = {}^{15}C_5 2^5$

16. **(6006)**

$$\left(t^2 x^{\frac{1}{5}} + \frac{(1-x)^{\frac{1}{10}}}{t}\right)^{15}$$

$$T_{r+1} = {}^{15}C_r\left(t^2 x^{\frac{1}{5}}\right)^{15-r} \cdot \frac{(1-x)^{\frac{r}{10}}}{t^r}$$

For the term independent of t,

$30 - 2r - r = 0$

$\Rightarrow r = 10$

So, maximum value of ${}^{15}C_{10}x(1-x)$ will be at $x = \frac{1}{2}$

That is, 6006.

17. **(D)** Do it yourself.

18. **(23)**

Since coefficient of x is -3

$\Rightarrow {}^{p}C_1 - {}^{q}C_1 = -3$

$\Rightarrow p - q = -3$ (1)

Comparing coefficients of x^2

$-{}^{p}C_1\,{}^{q}C_1 + {}^{p}C_2 + {}^{q}C_2 = -5$

$-pq + \frac{p(p-1)}{2} + \frac{q(q-1)}{2} = -5$ (2)

Solving (1) and (2)

$p = 8, q = 11$

Coefficient of x^3 is

$-{}^{q}C_3 + {}^{p}C_3 + {}^{p}C_1\,{}^{q}C_2 - {}^{p}C_2\,{}^{q}C_1$

$= -{}^{11}C_3 + {}^{8}C_3 + {}^{8}C_1\,{}^{11}C_2 - {}^{8}C_2\,{}^{11}C_1$

$= 23$

19. **(A)**

$$\sum_{\substack{i, j=0 \\ i \neq j}}^{n} {}^{n}C_i\,{}^{n}C_j$$

$$= \sum_{i=0}^{n} {}^{n}C_1 \times \sum_{j=0}^{n} {}^{n}C_j - \sum_{i=j=0}^{n} \left({}^{n}C_i\right)^2$$

$$= (2^n)(2^n) - {}^{2n}C_n$$

$$= 2^{2n} - {}^{2n}C_n$$

20. **(A)**

$(2021)^{2022} + (2022)^{2021}$

$= (2023-2)^{2022} + (2023-1)^{2021}$

$= 7n_1 + 2^{2022} + 7n_2 - 1$

$= 7(n_1 + n_2) + 8^{674} - 1$

$= 7(n_1 + n_2) + (7-1)^{674} - 1$

$= 7(n_1 + n_2) + 7n_3 + 1 - 1$

$= 7(n_1 + n_2 + n_3)$

$\therefore$ Given number is divisible by 7 hence the remainder is zero.

21. **(24)**

$(3+6x)^n = {}^{n}C_0 3^n + {}^{n}C_1 3^{n-1}(6x)^1 + \ldots$

$T_{r+1}\ {}^{n}C_r 3^{n-r} \times (6x) r = {}^{n}C_r 3^{n-r} \times 6^r \times x^r$

$= {}^{n}C_r 3^{n-r} \times 3^r \times 2^r \times \left(\frac{3}{2}\right)^r = {}^{n}C_r 3^n \times 3^r$

[for $x = \frac{3}{2}$]

T_9 is greatest of $x = \frac{3}{2}$

So, $T_9 > T_{10}$ and $T_9 > T_8$

(concept of numerically greatest term)

Here, $\frac{T_9}{T_{10}} > 1$ and $\frac{T_9}{T_8} > 1$

$\Rightarrow \frac{{}^{n}C_8 3^n \times 3^8}{{}^{n}C_9 3^n \times 3^9} > 1$ and $\frac{{}^{n}C_8 3^n \times 3^8}{{}^{n}C_7 3^n \times 3^7} > 1$

and $\frac{{}^{n}C_8}{{}^{n}C_7} > \frac{1}{3}$

and $\frac{n-7}{8} > \frac{1}{3}$

$\Rightarrow \frac{29}{3} < n < 11 \Rightarrow n = 10 = n_0$

So, in $(3+6x)^n$ for $n = n_0 = 10$

That is, in $(3+6x)^{10}$, here

$T_{r+1} = {}^{10}C_r 3^{10-r} 6^r x^r$

$T_7 = {}^{10}C_6 3^4 \times 6^6 \times x^6 = 210 \times 3^{10} \times 2^6 x^6$

$T_4 = {}^{10}C_3 3^7 6^3 x^3 = 120 \times 3^{10} \times 2^3 x^3$

Ratio of coefficient of x^6 and coefficient of $x^3 = k$

$\therefore k = \frac{210 \times 3 \times {}^{10}2^6}{120 \times 3^{10} \times 2^3} = \frac{7}{4} \times 2^3 = 14$

So, $k + n_0 = 14 + 10 = 24$

22. **(C)**

$7^{2022}+3^{2022}$

$=(49)^{1011}+(9)^{1011}$

$=(50-1)^{1011}+(10-1)^{1011}$

$=5\lambda-1+5K-1$

$=5m-2$

Remainder $=5-2=3$

23. **(57)**

${}^4C_2\times\frac{\beta^2}{6},-6\beta,-{}^6C_3\times\frac{\beta^3}{8}$ are in A.P.

$\beta^2-\frac{5}{2}\beta^3=-12\beta$

$\beta=\frac{12}{5}$ or $\beta=-2$ $\therefore\beta=\frac{12}{5}$

$d=-\frac{72}{5}-\frac{144}{25}=-\frac{504}{25}$

$\therefore 50-\frac{2d}{\beta^2}=57$

24. **(99)**

$1+\left(1+2^{49}\right)\left(2^{49}-1\right)=2^{98}$

$m=1,n=98$

$m+n=99$

25. **(84)**

$$\frac{T_5}{T_{n-3}}=\frac{{}^nC_4\left(2^{1/4}\right)^{n-4}\left(3^{-1/4}\right)^4}{{}^nC_{n-4}\left(2^{1/4}\right)\left(3^{-1/4}\right)^{n-4}}=\frac{\sqrt[4]{6}}{1}$$

$\Rightarrow 2^{\frac{n-8}{4}}3^{\frac{n-8}{4}}=6^{\frac{1}{4}}$

$\Rightarrow 6^{n-8}=6$

$\Rightarrow n-8=1\Rightarrow n=9$

$T_6={}^9C_5\left(2^{1/4}\right)^4\left(3^{-1/4}\right)^5=\frac{84}{\sqrt[4]{3}}$

$\therefore\alpha=84$

26. **(221)**

$$\sum_{K=1}^{10}K^2\left({}^{10}C_K\right)^2$$

$$\sum_{K=1}^{10}\left(K\times{}^{10}C_K\right)^2=\sum_{K=1}^{10}\left(10\times{}^9C_{K-1}\right)^2$$

$$=100\sum_{K=1}^{10}{}^9C_{K-1}\times{}^9C_{10-K}$$

$$=100\left({}^{18}C_9\right)=100\left(\frac{18!}{9!9!}\right)$$

$\Rightarrow 4862000=22000L$

Hence $L=221$

27. **(A)**

$$\sum_{r=0}^{22}{}^{22}C_r\times{}^{23}C_r=\sum_{r=0}^{22}{}^{22}C_r\times{}^{23}C_{23-r}$$

$={}^{45}C_{23}$

28. **(1012)**

$$\sum_{r=0}^{n}r^2\ {}^nC_r=n(n+1)\times2^{n-2}$$

Then $\sum_{r=0}^{2023}r^2\ {}^{2023}C_r=2023\times2024\times2^{2021}$

$=2023\times\alpha\times2^{2022}$ So,

$\Rightarrow\alpha=1012$

29. **(C)**

$$S=0\times\left({}^{30}C_0\right)^2+1\times\left({}^{30}C_1\right)^2+2\times\left({}^{30}C_2\right)^2+\ldots+30\times\left({}^{30}C_{30}\right)^2$$

$$S=30\times\left({}^{30}C_0\right)^2+29\times\left({}^{30}C_1\right)^2+28\times\left({}^{30}C_2\right)^2+\ldots+0\times\left({}^{30}C_0\right)^2$$

$$2S=30\times\left({}^{30}C_0^2+{}^{30}C_1^2+\ldots+{}^{30}C_{30}^2\right)$$

$$S=15\times{}^{60}C_{30}=15\times\frac{60!}{(30!)^2}$$

$$\frac{15\times10!}{(30!)^2}=\frac{\alpha\times60!}{(30!)^2}$$

$\Rightarrow\alpha=15$

30. **(405)**

Given binomial $\left(x-\frac{3}{x^2}\right)^n$, $x\neq0,n\in N$,

Sum of coefficients of the first three terms

${}^nC_0-{}^nC_1\times3+{}^nC_2\times3^2=376$

$\Rightarrow 3n^2-5n-250=0$

$\Rightarrow (n-10)(3n+25)=0$
$\Rightarrow n=10$

Now, the general term ${}^{10}C_r x^{10-r}\left(\frac{-3}{x^2}\right)^r$

$= {}^{10}C_r x^{10-r}(-3)^r \times x^{-2r}$

$= {}^{10}C_r (-3)^r \times x^{10-3r}$

Coefficient of $x^4 \Rightarrow 10-3r=4$
$\Rightarrow r=2$
${}^{10}C_2(-3)^2 = 405$

31. (B)

$a_r = {}^{10}C_{10-r} = {}^{10}C_r$

$\Rightarrow \sum_{r=1}^{10} r^3\left(\frac{{}^{10}C_r}{{}^{10}C_{r-1}}\right)^2$

$= \sum_{r=1}^{10} r^3\left(\frac{11-r}{r}\right)^2 = \sum_{r=1}^{10} r(11-r)^2$

$= \sum_{r=1}^{10}\left(121r + r^3 - 22r^2\right) = 1210$

32. (1080)

General term is $\sum \frac{5!(2x)^{n_1}\left(x^{-7}\right)^{n_2}\left(3x^2\right)^{n_3}}{n_1!\,n_2!\,n_3!}$

For constant term,
$n_1 + 2n_3 = 7n_2$
and $n_1 + 2n_3 = 7n_2$
Only possibility $n_1 = 1, n_2 = 1, n_3 = 3$
$\Rightarrow$ constant term = 1080

33. (C)

$\sum_{k=0}^{6} {}^{51-k}C_3$

$= {}^{51}C_3 + {}^{50}C_3 + {}^{49}C_3 + \ldots + {}^{45}C_3$

$= {}^{45}C_3 + {}^{46}C_3 + \ldots + {}^{51}C_3$

$= {}^{45}C_4 + {}^{45}C_3 + {}^{46}C_3 + \ldots + {}^{51}C_3 - {}^{45}C_4$

$\left({}^{n}C_r + {}^{n}C_{r-1} = {}^{n+1}C_r\right)$

$= {}^{52}C_4 - {}^{45}C_4$

34. (7)

$(2023)^{2023} = (2030-7)^{2023}$

$= (35K-7)^{2023}$

$= {}^{2023}C_0(35K)^{2023}(-7)^0 + {}^{2023}C_1(35K)^{2022}(-7) + \ldots + {}^{2023}C_{2023}(-7)^{2023}$

$= 35N - 7^{2023}$

Now, $-7^{2023} = -7\times 7^{2022} = -7\left(7^2\right)^{1011}$

$= -7(50-1)^{1011}$

$= -7\left({}^{1011}C_0 50^{1011} - {}^{1011}C_1(50)^{1010} + \ldots {}^{1011}C_{1011}\right)$

$= -7(5\lambda - 1)$

$= -35\lambda + 7$

$\therefore$ When $(2023)^{2023}$ is divided by 35, the remainder is 7.

35. (A)

Coefficient of x^9 in $\left(\alpha x^3 + \frac{1}{\beta x}\right) = {}^{11}C_6 \cdot \frac{\alpha^5}{\beta^6}$

$\because$ Both are equal

$\therefore \frac{11}{C_6}\cdot\frac{\alpha^5}{\beta^6} = -\frac{11}{C_5}\cdot\frac{\alpha^6}{\beta^5}$

$\Rightarrow \frac{1}{\beta} = -\alpha$

$\Rightarrow \alpha\beta = -1$

$\Rightarrow (\alpha\beta)^2 = 1$

36. (1120)

$t_{r+1} = {}^{n}C_r(2x)^r$

$\Rightarrow \frac{{}^{n}C_{r-1}(2)^{r-1}}{{}^{n}C_r(2)^r} = \frac{2}{5}$

$\Rightarrow \frac{\frac{n!}{(r-1)!\,(n-r+1)!}}{\frac{n!\,2}{r!\,(n-r)!}} = \frac{2}{5}$

$\Rightarrow \frac{r}{n-r+1} = \frac{4}{5} \Rightarrow 5r = 4n - 4r + 4$

$\Rightarrow 9r = 4(n+1)$ (1)

$\Rightarrow \frac{{}^{n}C_r(2)^r}{{}^{n}C_{r+1}(2)^{r+1}} = \frac{5}{8}$

$$\Rightarrow \frac{\frac{n!}{r!(n-r)!}}{\frac{n!}{(r+1)!(n-r-1)!}} = \frac{5}{4} \Rightarrow \frac{r+1}{n-r} = \frac{5}{4}$$

$\Rightarrow 4r+4 = 5n-5r \Rightarrow 5n-4 = 9r$ (2)

From (1) and (2)

$\Rightarrow 4n+4 = 5n-4 = 8$

$(1) \Rightarrow r = 4$

37. **(C)**

In the expansion of

$(1+x)^{99} = C_0 + C_1x + C_2x^2 + \ldots + C_{99}x^{99}$

$K = C_1 + C_3 + \ldots + C_{99} = 2^{98}$

$a \Rightarrow$ Middle in the expansion of $\left(2+\frac{1}{\sqrt{2}}\right)^{200}$

$$T_{\frac{200}{2}+1} = {}^{200}C_{100}(2)^{100}\left(\frac{1}{\sqrt{2}}\right)^{100}$$

$$= {}^{200}C_{100} \times 2^{50}$$

So, $\frac{{}^{200}C_{99} \times 2^{98}}{{}^{200}C_{100} \times 2^{50}} = \frac{100}{101} \times 2^{48}$

So, $\frac{25}{101} \times 2^{50} = \frac{m}{n}2$

$\therefore m, n$ are odd, so (ℓ, n) become $(50, 101)$.

38. **(D)**

Coefficient of x^{15} in $\left(ax^3 + \frac{1}{bx^{1/3}}\right)^{15}$

$$T_{r+1} = {}^{15}C_r\left(ax^3\right)^{15-r}\left(\frac{1}{bx^{1/3}}\right)^r$$

$45 - 3r - \frac{r}{3} = 15$

$30 - \frac{10r}{3}$

$r = 9$

Coefficient of $x^{15} = {}^{15}C_9 a^6 b^{-9}$

Coefficient of x^{-15} in $\left(ax^{\frac{1}{3}} - \frac{1}{bx^3}\right)^{15}$

$$T_{r+1} = {}^{15}C_r\left(ax^{\frac{1}{3}}\right)^{15-r}\left(-\frac{1}{bx^3}\right)^r$$

$5 - \frac{r}{3} - 3r = -15$

$\frac{10r}{3} = 20$

$r = 6$

Coefficient of $= {}^{15}C_6 a^9 \times b^{-6}$

$\Rightarrow \frac{a^9}{b^6} = \frac{a^6}{b^9}$

39. **(D)**

$(1+x)^{500} + x(1+x)^{499} + x^2(1+x)^{498} + \ldots + x^{500}$

$$= (1+x)^{500}\left\{\frac{1-\left(\frac{x}{1+x}\right)^{501}}{1-\frac{x}{1+x}}\right\}$$

$$= (1+x)^{500}\frac{\left((1+x)^{501} - x^{501}\right)}{(1+x)^{501}}(1+x)$$

$$= (1+x)^{501} - x^{501}$$

Coefficient of x^{301} in $(1+x)^{501} - x^{501}$ is given by

${}^{501}C_{301} = {}^{501}C_{200}$

40. **(A)**

$x = 8\sqrt{3}+13 = {}^{13}C_0\left(8\sqrt{3}\right)^{13} + {}^{13}C_1\left(8\sqrt{3}\right)(13)^1 + \ldots$

$x' = \left(8\sqrt{3}-13\right)^{13} = {}^{13}C_0\left(8\sqrt{3}\right)^{13} - {}^{13}C_1\left(8\sqrt{3}\right)^{12}(13)^1 + \ldots$

$x - x' = 2\left[{}^{13}C_1\left(8\sqrt{3}\right)^{12}(13)^1 + {}^{13}C_3\left(8\sqrt{3}\right)^{10}(13)^3 \ldots\right]$

therefore, $x - x'$ is an even integer, hence $[x]$ is even.

Now,

$y = \left(7\sqrt{2}+9\right)^9 = {}^9C_0\left(7\sqrt{2}\right)^9 + C_1\left(7\sqrt{2}\right)^8(9)^1 + {}^9C_2\left(7\sqrt{2}\right)^7(9)^2 \ldots$

$y' = \left(7\sqrt{2}-9\right)^9 = {}^9C_0\left(7\sqrt{2}\right)^9 - {}^9C_1\left(7\sqrt{2}\right)^8(9)^1 + {}^9C_2\left(7\sqrt{2}\right)^7(9)^2 \ldots$

$$y - y' = 2\left[{}^9C_1\left(7\sqrt{2}\right)^8(9)^1 + {}^9C_3\left(7\sqrt{2}\right)^6(9)^3 + \ldots\right]$$

$y - y' =$ Even integer, hence $[y]$ is even.

41. **(23)**

$$x + y = 12^{50} = (150-6)^{25} + (325-1)^{25}$$

$$= 25K - \left(6^{25}+1\right) = 25K - \left((5+1)^{25}+1\right)$$

$= 25K_1 - 2$ Remainder $= 23$

42. **(2)**

$$T_{r+1} = {}^{30}C_r\left(x^{\frac{2}{3}}\right)^{30-r}\left(\frac{2}{x^3}\right)^r$$

$$= {}^{30}C_r \times 2^r \times x^{\frac{60-11r}{3}}$$

$$\frac{60-11r}{3} < 0 \Rightarrow 11r > 60 \Rightarrow r > \frac{60}{11} \Rightarrow r = 6$$

$$T_7 = {}^{30}C_6 \times 2^6 x^{-2}$$

We have also observed $\beta = {}^{30}C_6(2)^6$ is a natural number.

$\therefore \alpha = 2$

43. **(9)**

$$5^{99} = 5^4 \times 5^{95}$$

$$= 625\left[5^5\right]^{19}$$

$$= 625[3125]^{19}$$

$$= 625[3124+1]^{19}$$

$$= 625[11k \times 19 + 1]$$

$$= 625 \times 11k \times 19 + 625$$

$$= 11k_1 + 616 + 9$$

$$= 11(k_2) + 9$$

44. **(5040)**

$$\left(\frac{4x}{5} + \frac{5}{2x^2}\right)^9$$

Now, $T_{r+1} = {}^9C_r \times \left(\frac{4x}{5}\right)^{9-r}\left(\frac{5}{2x^2}\right)^r$

$$= {}^9C_r \times \left(\frac{4}{5}\right)^{9-r}\left(\frac{5}{2}\right)^r \times x^{9-3r}$$

Coefficient of x^{-6}, that is, $9 - 3r = -6 \Rightarrow r = 5$

So, coefficient of

$$x^{-6} = {}^9C_5\left(\frac{4}{5}\right)^4 \times \left(\frac{5}{2}\right)^5 = 5040$$

45. **(98)**

In, $\left(\frac{x^{\frac{5}{2}}}{2} - \frac{4}{x^{\ell}}\right)^9$

$$T_{r+1} = {}^9C_r \frac{\left(x^{\frac{5}{2}}\right)^{9-r}}{2^{9-r}}\left(\frac{-4}{x^{\ell}}\right)^r$$

$$= (-1)^r \frac{{}^9C_r}{2^{9-r}} 4^r x^{\frac{45}{2} - \frac{5r}{2} - \ell r}$$

$$= 45 - 5r - 2\ell r = 0$$

$$r = \frac{45}{5+2\ell} \qquad (1)$$

Now, according to the question,

$$(-1)^r \frac{{}^9C_r}{2^{9-r}} 4^r = -84$$

$$= (-1)^r\, {}^9C_r 2^{3r-9} = 21 \times 4$$

Only natural value of r possible if $3r - 9 = 0$

$r = 3$ and ${}^9C_3 = 84$

$\therefore \ell = 5$ from equation (1)

Now, coefficient of $x^{-31} = x^{\frac{45}{2} - \frac{5r}{2} - \ell r}$ at $\ell = 5$, gives

$r = 5$

$$\therefore {}^9C_5(-1)\frac{4^5}{2^4} = 2^{\alpha} \times \beta$$

$$= -63 \times 2^7$$

$\Rightarrow \alpha = 7, \beta = -63$

$\therefore$ Value of $|\alpha\ell - \beta| = 98$

46. **(B)**

$$\sum_{r=1}^{26} \frac{1}{(2r-1)!\,(51-(2r-1))!} = \sum_{r=1}^{26} {}^{51}C_{(2r-1)} \frac{1}{51!}$$

$$= \frac{1}{51!}\left\{{}^{51}C_1 + {}^{51}C_3 + \ldots + {}^{51}C_{51}\right\} = \frac{1}{51!}\left(2^{50}\right)$$

47. (29)

$(21+2)^{200}+(21-2)^{200}$

$\Rightarrow 2\left[{}^{100}C_0 21^{200}+200C_2 21^{198}\times 2^2+\ldots+{}^{200}C_{198}21^2 2^{198}+2^{200}\right]$

$\Rightarrow 2\left[49I_1+2^{200}\right]=49I_1+2^{201}$

Now,

$2^{201}=(8)^{67}=(1+7)^{67}=49I_2+{}^{67}C_0{}^{67}C_1\times 7=$

$49I_2+470=49I_2+49\times 9+29$

$\therefore$ Remainder is 29.

48. (1)

$T_{r+1}={}^{22}C_r\times\left(x^{\frac{2}{3}}\right)^{22-r}\times(\alpha)^r, x^{-3r}$

$={}^{22}C_r\times x^{\frac{44}{3}-\frac{2r}{3}-3r}(\alpha)^r$

$\frac{44}{3}=\frac{11r}{3}$

$r=4$

${}^{22}C_4\times\alpha^4=7315$

$\frac{22\times 21\times 20\times 19}{24}\times\alpha^4=7315$

$\alpha=1$

49. (D)

$T_6={}^mC_5\left(10-3^x\right)^{\frac{m-5}{2}}\times\left(3^{x-2}\right)=21 \qquad (1)$

${}^mC_1, {}^mC_2, {}^mC_3$ are in A.P.

$2\times{}^mC_2={}^mC_1+{}^mC_3$

Solving for m, we get

$m=2(\text{rejected}), 7$

Put in equation (1)

$21\times\left(10-3^x\right)\frac{3^x}{9}=21$

$3^x=3^0, 3^2$

$x=0, 2$

50. (D)

Given expansion is $\left(\sqrt[4]{2}+\frac{1}{\sqrt[4]{3}}\right)^n$

We know that general term of an expansion $(x+a)^n$ is $T_{r+1}={}^nC_r x^{n-r}a^r$

$\left(5^{\text{th}}\text{ term from beginning}\right)$

We know that the rth term from the end in a binomial expansion is T_{n-r+2} from the beginning.

$T_5'=T_{n-4+1}={}^nC_4\left(\frac{1}{\sqrt[4]{3}}\right)^{n-4}\left(\sqrt[4]{2}\right)^4$

It is also given that the ratio of the fifth term from the beginning to the fifth term from the end is $\sqrt{6}:1$.

Now, $\frac{T_5}{T_5'}=\frac{\sqrt{6}}{1}$

$\Rightarrow 2^{\frac{n-8}{4}}\times 3^{\frac{n-8}{4}}=\sqrt{6}$

$\therefore\left(\sqrt{6}\right)^{\frac{n-8}{2}}=\left(\sqrt{6}\right)$

$\therefore n=10$

$\therefore T_3={}^{10}C_2\left(\sqrt[4]{2}\right)^8\left(\frac{1}{\sqrt[4]{3}}\right)^2=\frac{45\times 4}{\sqrt{3}}$

$=60\sqrt{3}$

Hence, this is the correct option.

51. (D)

We have been given that ${}^{2n}C_3:{}^nC_3=10$

We know that ${}^nC_r=\frac{n!}{r!(n-r)!}$

$\Rightarrow\frac{\frac{(2n)!}{3!(2n-3)!}}{\frac{n!}{3!(n-3)!}}=10$

$\Rightarrow\frac{(2n)!}{3!(2n-3)!}\times\frac{3!(n-3)!}{n!}=10$

$\Rightarrow\frac{(2n)(2n-1)(2n-2)(2n-3)!}{(2n-3)!}\times\frac{(n-3)!}{n(n-1)(n-2)(n-3)!}=10$

$\Rightarrow\frac{(2n)(2n-1)(2n-2)}{n(n-1)(n-2)}=10$

$\Rightarrow \dfrac{(4)(2n-1)}{(n-2)} = 10$

$\Rightarrow 8n - 4 = 10n - 20$

$\Rightarrow n = 8$

Hence, the value of $\dfrac{n^2+3n}{n^2-3n+4}$ is

$= \dfrac{8^2+3(8)}{8^2-3(8)+4} = \dfrac{88}{44} = 2$

Therefore, the value of $\dfrac{n^2+3n}{n^2-3n+4}$ is $2:1$

52. (5005)

Given,

Binomial expression $\left(x^4 - \dfrac{1}{x^3}\right)^{15}$

Now, general term in given by,

$T_{r+1} = {}^{15}C_r \left(x^4\right)^{15-r} \left(-\dfrac{1}{x^3}\right)^r$

$\Rightarrow T_{r+1} = {}^{15}C_r x^{60-7r} (-1)^r$

For coefficient of x^{18}, equate $60 - 7r = 18$

$\Rightarrow r = 6$

Coefficient of $x^{18} = (-1)^6 \cdot {}^{15}C_6 = 5005$

53. (A)

The coefficient of x^7 in $\left(ax^2 + \dfrac{1}{2bx}\right)^{11}$ and $\left(ax - \dfrac{1}{3bx^2}\right)^{11}$ are equal, then

$T_{r+1} = {}^{11}C_r \left(ax^2\right)^{11-r} \left(\dfrac{1}{2bx}\right)^r$

Now, solving $22 - 3r = 7 \Rightarrow r = 5$

Coefficient of x^{-7} in $\left(ax - \dfrac{1}{3bx^2}\right)^{11}$

$T_{r+1} = {}^{11}C_r (ax)^{11-r} \left(-\dfrac{1}{3bx^2}\right)^r$

$= {}^{11}C_r a^{11-r} \left(-\dfrac{1}{3b}\right)^r x^{11-3r}$

Now, solving $11 - 3r = -7 \Rightarrow r = 6$

Now equating coefficient of x^7 and x^{-7} we get,

$\therefore {}^{11}C_5 (a)^6 \left(\dfrac{1}{2b}\right)^5 = {}^{11}C_6 a^5 \left(-\dfrac{1}{3b}\right)^6$

$\Rightarrow 3^6 ab = 32$

$\Rightarrow 729ab = 32$

Hence this is the correct option.

54. (B)

Given,

(S1): $(2023)^{2022} - (1999)^{2022}$ is divisible by 8

Now we know that $(x - y)$ divides $\left(x^n - y^n\right) \forall n \in N$

So, $(2023 - 1999)$ divides $(2023)^{2022} - (1999)^{2022}$

$\Rightarrow$ 24 divides $(2023)^{2022} - (1999)^{2002}$

$\Rightarrow$ 8 will divide $(2023)^{2022} - (1999)^{2002}$

As 8 divides 24.

Hence, $(S1)$ is correct.

Now solving,

$(S2): 13(13)^n - 11n - 13$ is divisible by 144 for $n \in N$

So using binomial theorem in $(1+12)^n$ we get,

$13(1+12)^n - 11n - 13$

$= 13\left({}^nC_0 + {}^nC_1 12 + {}^nC_2 12^2 + \ldots + {}^nC_n 12^n\right)$

$-11n - 13$

$= 12 \times 13n - 11n + 12^2 \lambda$

$= 145n + 144\lambda$ which is not divisible by 144

Hence, (S2) is incorrect

55. (D)

Given,

${}^nC_{r-1} : {}^nC_r : {}^nC_{r+1} = 1 : 5 : 20$

Now using the formula $\dfrac{{}^nC_r}{{}^nC_{r-1}} = \dfrac{n-r+1}{r}$ we get,

$\dfrac{{}^nC_r}{{}^nC_{r-1}} = \dfrac{n-r+1}{r} = \dfrac{5}{1}$

$\Rightarrow n - 6r + 1 = 0$ (1)

and $\dfrac{{}^nC_{r+1}}{{}^nC_r} = \dfrac{n-r}{r+1} = \dfrac{20}{5}$

$\Rightarrow n-5r-4=0$ (2)

Solving the above equations, we get,

$n=29$ and $r=5$

So, the coefficient of fourth term will be

$${}^{n}C_3={}^{29}C_3=\frac{29\times28\times27}{3\times2\times1}=3654$$

56. (1275)

The given expansion is $\left(3x^2-\frac{1}{2x^5}\right)^7$.

The general term in the binomial expansion of $(x+a)^n$ is given by $T_{r+1}={}^{n}C_r x^{n-r}a^r$.

$$\Rightarrow T_{r+1}={}^{7}C_r\left(3x^2\right)^{7-r}\left(\frac{-1}{2x^5}\right)^r$$

$$\Rightarrow {}^{7}C_r(-1)^r\left(\frac{3^{7-r}}{2^r}\right)x^{14-2r-5r}$$

Now, for the term independent of x

$\Rightarrow 14-2r-5r=0$

$\Rightarrow r=2$

Coefficient of x^0 is ${}^{7}C_2(-1)^2\times\left(\frac{3^{7-2}}{2^2}\right)$

$$=\frac{7\times6}{2}\times\frac{3^5}{2^2}$$

$$\Rightarrow \alpha=\frac{5103}{4}$$

$\Rightarrow |\alpha|=1275$

Hence this is the required answer.

57. (5)

Let,

$$y=\frac{x^3}{x^4+147}=f(x)$$

We know that, for increasing function

$$\frac{dy}{dx}>0$$

So, differentiating the given function

$$f(x)=\frac{x^3}{x^4+147}$$

we get,

$$\frac{dy}{dx}=\frac{-x^2\left(x^4-441\right)}{\left(x^4+147\right)^2}$$

$$\Rightarrow \frac{-x^2\left(x^4-441\right)}{\left(x^4+147\right)^2}>0$$

$$\Rightarrow \frac{x^2\left(x^4-441\right)}{\left(x^4+147\right)}<0$$

$\Rightarrow x^4<441$

Now for maxima/minima $\frac{dy}{dx}=0$

$\Rightarrow x^4=441$

$\Rightarrow x=\alpha, 4<\alpha<5$

$\Rightarrow$ Maximum value of $f(x)$ is at $x=4$ or $x=5$

$$f(4)=\frac{64}{403},\ f(5)=\frac{125}{772}$$

$\because f(5)>f(4)$

$\Rightarrow \alpha=5$

58. (31)

We know that the maximum value of n for p^n where p is prime divides the number $a!$ is given by,

$$\left[\frac{a}{p}\right]+\left[\frac{a}{p^2}\right]+\left[\frac{a}{p^3}\right]+\left[\frac{a}{p^4}\right]+\ldots$$

Since 3 is a prime number, we have

$$=\left[\frac{66}{3}\right]+\left[\frac{66}{3^2}\right]+\left[\frac{66}{3^3}\right]+\left[\frac{66}{3^4}\right]+\ldots$$

$=22+7+2+0=31$

So, maximum value of n is 31.

59. (D)

The given expansion is $\left(2x^2+\frac{1}{2x}\right)^{11}$

The general term in the binomial expansion of $(x+a)^n$ is given by $T_{r+1}={}^{n}C_r x^{n-r}a^r$.

$$\Rightarrow T_{r+1}={}^{11}C_r\left(2x^2\right)^{11-r}\left(\frac{1}{2x}\right)^r$$

$$\Rightarrow {}^{11}C_r\left(2^{11-r-r}\right)x^{22-2r-r}$$

Now for coefficient of $x^7 \Rightarrow 22-3r=7$

$\Rightarrow r=4$

Coefficient of x^7 is ${}^{11}C_5\left(2^{11-5\times2}\right)$

Now for coefficient of $x^{10} \Rightarrow 22-3r=10$

$\Rightarrow r=4$

Coefficient of x^{10} is ${}^{11}C_4\left(2^{11-2\times4}\right)$

We need the absolute difference, we get it by

$\left|{}^{11}C_5\left(2^{11-5\times2}\right)-{}^{11}C_4\left(2^{11-4\times2}\right)\right|$

$=|924-2640|=1716$

$=12^3-12$

Hence the absolute difference is $=12^3-12$.

60. (C)

Given,

$25^{190}-19^{190}-8^{190}+2^{190}$

Now $\left(25^{190}-19^{190}\right), 8^{190}, 2^{190}$ are divisible by 2, as 25^{190} will give remainder as 1 and 19^{190} will give remainder as 1 when divided by 2, so, combined 25^{190} will give a remainder 0.

Now, $\left(25^{190}-8^{190}\right)$ and $\left(-19^{190}+2^{190}\right)$ are divisible by 17, a^n b^n is divisible by a b if n is even. So, the given number is divisible by 34 as it is divisible by 2 and 17 both,

Now, $25^{190}=(7\times3+4)^{190}$, dividing by 7 we get,

$(7\times3+4)^{190}=4^{190}=16^{95}=2^{95}=4\cdot8^{31}$

Now when $4\cdot(7+1)^{31}$ is divided by 7, it will give the remainder as $4\cdot1=4$

And $19^{190}=(7\times3-2)^{190}$, when divided by 7 we get,

$(7\times3-2)^{190}=(-2)^{190}=2^{190}$

$=2\times8^{63}=2\times(7+1)^{63}$

So, when $2\times(7+1)^{63}$ is divided by 7, it will give the remainder as 2.

Similarly, $8^{190}=(7+1)^{63}$ will give 1 remainder

And $2^{190}=2(7+1)^{63}$ will give 1 remainder

So, combining all, we get

$25^{190}-19^{190}-8^{190}+2^{190}=4-2-1+1=5$

$\Rightarrow$ Not divisible by 7, so it will not be divisible by 14

61. (C)

Given,

The coefficient of x^7 in the expansion of $\left(ax^2-\frac{1}{bx}\right)^{13}$ is equal to the coefficient of x^{-5} in $\left(ax+\frac{1}{bx^2}\right)^{13}$.

We know that, the general term T_{r+1} in the expansion $(a+b)^n$ is

$T_{r+1}={}^nC_r a^{n-r}b^r$

Applying to $\left(ax-\frac{1}{bx^2}\right)^{13}$, we get

$T_{r+1}={}^{13}C_r(ax)^{13-r}\left(-\frac{1}{bx^2}\right)^r$

$\Rightarrow T_{r+1}=(-1)^r\times{}^{13}C_r(a)^{13-r}(x)^{13-3r}(b)^{-r}$

Therefore, $13-3r=7\Rightarrow r=2$ for coefficient of x^7.

Thus, $T_3={}^{13}C_2\left(\frac{a^{11}}{b^2}\right)$

Similarly, applying to $\left(ax+\frac{1}{bx^2}\right)^{13}$, we get

$T_{r+1}={}^{13}C_r(ax)^{13-r}\left(\frac{1}{bx^2}\right)^r$

$\Rightarrow T_{r+1}={}^{13}C_r(a)^{13-r}(x)^{13-3r}(b)^{-r}$

Therefore, $13-3r=-5$ for coefficient of x^{-5}

$\Rightarrow r=6$

So, $T_7={}^{13}C_6(a)^7(b)^{-6}$

Hence, applying the given condition we get

${}^{13}C_2\left(\frac{a^{11}}{b^2}\right)={}^{13}C_6(a)^7(b)^{-6}$

$\Rightarrow a^4b^4=\frac{{}^{13}C_6}{{}^{13}C_2}$

$\Rightarrow a^4b^4=\frac{13!}{7!6!}\times\frac{2!\cdot11!}{13!}$

$\Rightarrow a^4b^4=\frac{11\times10\times9\times8}{6\times5\times4\times3}$

$\Rightarrow a^4b^4=22$

62. (960)

We need to find the coefficient of x^7 in $\left(1-x+2x^3\right)^{10}$.

We know that for $(x+y+z)^n$,

$$T_n=\frac{n!}{a!\,b!\,c!}(x)^a(y)^b(z)^c \text{ such that}$$

$$a+b+c=n$$

Now for $\left(1-x+2x^3\right)^{10}$

$$T_n=\frac{10!}{a!\,b!\,c!}(1)^a(-x)^b\left(2x^3\right)^c$$

$$=\frac{10!}{a!\,b!\,c!}(-1)^b\,2^c\,x^{b+3c}.\text{ with } a+b+c=10$$

Here we need coefficient of x^7

Hence the combinations would be

a	b	c
3	7	0
5	4	1
7	1	2

which satisfies both $a+b+c=10$ and $b+3c=7$.

Hence, the coefficient of x^7 will be,

$$=\frac{10!}{2!\,1!\,7!}(2)^2(-1)^1$$

$$+\frac{10!}{1!\,4!\,5!}(2)^1(-1)^4+\frac{10!}{0!\,7!\,3!}(2)^0(-1)^7$$

$$=960$$

63. (C)

Given that α be the remainder when $(22)^{2022}+(2022)^{22}$ is divided by 3 and β be the remainder when the same is divided by 7.

$$\Rightarrow (22)^{2022}+(2022)^{22}=(21+1)^{2022}+(2022)^{22}.$$

Here $(2022)^{22}$ is divisible by 3 as 2022 is divisible by 3

So, on expanding $(21+1)^{2022}$, we get

$$\Rightarrow (21+1)^{2022}={}^{2022}C_0(21)^{2022}+{}^{2022}C_1(21)^{2021}+\ldots+{}^{2022}C_{2022}(1)^{2022}$$

$$\Rightarrow 3\left(3^{2021}\times 7^{2022}+{}^{2022}C_1\times 3^{2020}\times 7^{2021}+\ldots\right)+1$$

$$=3k_1+1$$

In this case the remainder is 1

Hence, $\alpha=1$

Now

$$\Rightarrow (22)^{2022}+(2022)^{22}=(21+1)^{2022}+(2023-1)^{22}$$

Take $(2023-1)^{22}$

$$\Rightarrow (2023-1)^{22}={}^{22}C_0(2023)^{22}-{}^{22}C_1(2023)^{21}+\ldots+{}^{22}C_{22}(-1)^{22}$$

$$=7\left({}^{22}C_0(7)^{21}(289)^{22}-{}^{22}C_1(7)^{20}(289)^{21}+\ldots\right)+1$$

$$=7k_2+1$$

$$\Rightarrow (21+1)^{2022}+(2023-1)^{22}=7k_1+1+7k_2+1$$

$$=7\mu+2$$

$$\Rightarrow \beta=2$$

Hence, $\alpha^2+\beta^2=1^2+2^2=5$

Therefore, the required answer is 5.

64. (D)

Given that the coefficient of x and x^2 in $(1+x)^p(1-x)^q$ are 4 and -5.

$$\Rightarrow (1+x)^p(1-x)^q=\left(1+px+\frac{p(p-1)}{2!}x^2+\ldots\right)\left(1-qx+\frac{q(q-1)}{2!}x^2-\ldots\right)$$

Now coefficient of x from the above expansion will be $p-q$ which is equal to 4.

$$\Rightarrow p-q=4$$

Similarly, the coefficient of x^2 is -5.

$$\Rightarrow \frac{p(p-1)}{2}+\frac{q(q-1)}{2}-pq=-5$$

$$\Rightarrow \frac{p^2-2pq+q^2}{2}-\frac{(p+q)}{2}=-5$$

$$\Rightarrow \frac{(p-q)^2}{2}-\frac{(p+q)}{2}=-5$$

$$\Rightarrow \frac{16}{2}+5=\frac{(p+q)}{2}$$

$$\Rightarrow p+q=26 \text{ and } p-q=4$$

On solving the above equations, we get,
$p=15$ and $q=11$
$\Rightarrow 2p+3q=2(15)+3(11)=63$
Therefore, the required answer is 63.

65. (171)

Given expansion is $\left(3^{\frac{1}{2}}+5^{\frac{1}{4}}\right)^{680}$

The general term in the expansion $(x+a)^n$ is $T_{r+1}={}^nC_r\cdot x^{n-r}\cdot a^r$

$$\Rightarrow T_{r+1}={}^{680}C_r\left(3^{\frac{1}{2}}\right)^{680-r}\left(5^{\frac{1}{4}}\right)^r, 0\le r\le 680$$

$$={}^{680}C_r\,(3)^{\frac{680-r}{2}}(5)^{\frac{r}{4}}$$

For $(3)^{\frac{680-r}{2}}(5)^{\frac{r}{4}}$ to be rational, r should be a multiple of 4.
$\Rightarrow r=0,4,8,12,\ldots,680$
$\Rightarrow a_n=a+(n-1)d$
$\Rightarrow 680-0+(n-1)4$
$\Rightarrow n=171$
That means r can take 171 values
Hence, the required answer is 171.

66. (2736)
We know that,
Binomial expansion of $(2+x)^9$ is given by,
$(2+x)^9={}^9C_0x^02^9+{}^9C_1x^12^8+\ldots{}^9C_9x^92^0$
Now put $x=1$ we get,
$(2+1)^9={}^9C_02^9+{}^9C_12^8++{}^9C_22^7+\ldots{}^9C_92^0$
$\Rightarrow 3^9={}^9C_02^9+{}^9C_12^8++{}^9C_22^7+\ldots{}^9C_92^0$
So, sum of coefficients of $x, x^2, x^3,\ldots, x^7$ will be
$\Rightarrow {}^9C_12^8+{}^9C_22^7+\ldots{}^9C_72^2$
$=3^9-{}^9C_02^9-{}^9C_82^1-{}^9C_92^0$
$\Rightarrow {}^9C_12^8+{}^9C_22^7+\ldots{}^9C_72^2=3^9-2^9-18-1$
$\Rightarrow {}^9C_12^8+{}^9C_22^7+\ldots{}^9C_72^2=19683-512-19$
$\Rightarrow {}^9C_12^8+{}^9C_22^7+\ldots{}^9C_72^2=19152$

Now, the mean is given by $\frac{19152}{7}=2736$

67. (B)
Given,
The 1011th term from the end in the binomial expansion of $\left(\frac{4x}{5}-\frac{5}{2x}\right)^{2022}$ is 1024 times 1011th term from the beginning,
So, according to question we have,

$$\frac{T_{1013}}{1011}=1024$$

$$\Rightarrow \frac{{}^{2022}C_{1012}\left(\frac{4x}{5}\right)^{1010}\cdot\left(\frac{-5}{2x}\right)^{1012}}{{}^{2022}C_{1010}\left(\frac{4x}{5}\right)^{1012}\left(\frac{-5}{2x}\right)^{1010}}=1024$$

$$\Rightarrow \frac{\left(\frac{-5}{2x}\right)^2}{\left(\frac{4x}{5}\right)^2}=1024$$

$$\Rightarrow \frac{5^4}{2^6x^4}=2^{10}$$

$$\Rightarrow \left(\frac{5}{x}\right)^4=2^{16}$$

$$\Rightarrow |x|=\frac{5}{16}$$

$$\Rightarrow 32|x|=10$$

68. (B)
Ratio of three consecutive terms in the binomial expansion of $(1+x)^{n+2}$ is

${}^{n+2}C_{r-1}:{}^{n+2}C_r:{}^{n+2}C_{r+1}::1:3:5$
So,

$$\Rightarrow \frac{{}^{n+2}C_r}{{}^{n+2}C_{r-1}}=\frac{3}{1}$$

$$\Rightarrow \frac{n-r+3}{r}=3$$

$\Rightarrow n+3=4r$ (1)
And,

$$\frac{{}^{n+2}C_{r+1}}{{}^{n+2}C_r}=\frac{5}{3}$$

$$\Rightarrow \frac{n+2-r}{r+1}=\frac{5}{3}$$

$\Rightarrow 3n+6-3r=5r+5$

$\Rightarrow 3n+1=8r$ (2)

So, $3(4r-3)+1=8r$

$\Rightarrow 12r-9+1=8r$

$\Rightarrow 12r-9+1=8r$

$\Rightarrow 4r=8$

$\Rightarrow r=2$

So, $n=5$

Hence, we have

$(1+x)^{n+2}=(1+x)^7$

Sum of consecutive terms is

${}^7C_1+{}^7C_2+{}^7C_3$

$=7+21+35=63$

69. **(A)**

Given that

$$\frac{1}{n+1}\,{}^nC_n+\frac{1}{n}\,{}^nC_{n-1}+\ldots+\frac{1}{2}\,{}^nC_1+{}^nC_0=\frac{1023}{10}$$

$$\Rightarrow\sum_{r=0}^{n}\frac{1}{r+1}\,{}^nC_r=\frac{1023}{10}\left(\because\ {}^{n+1}C_{r+1}=\frac{n+1}{r+1}\,{}^nC_r\right)$$

$$\Rightarrow\sum_{r=0}^{n}\frac{1}{n+1}\,{}^{n+1}C_{r+1}=\frac{1023}{10}$$

$$\Rightarrow\frac{1}{n+1}\left[{}^{n+1}C_1+{}^{n+1}C_2+\ldots+{}^{n+1}C_{n+1}\right]=\frac{1023}{10}$$

We know that

${}^nC_0+{}^nC_1+{}^nC_2+\ldots+{}^nC_n=2^n$

$$\Rightarrow\frac{2^{n+1}-1}{n+1}=\frac{1023}{10}=\frac{2^{10}-1}{10}$$

$\Rightarrow n+1=10$

$\Rightarrow n=9$

Hence this is the correct option.

70. **(D)**

We know that

$$(1-x)^{100}={}^{100}C_0-{}^{100}C_1\times x+{}^{100}C_2\times x^2-\ldots-{}^{100}C_{49}\times x^{49}+\ldots+{}^{100}C_{100}\times x^{100}$$

Now let, the sum of the first fifty coefficients be S

$\Rightarrow S={}^{100}C_0-{}^{100}C_1+{}^{100}C_2-\ldots-{}^{100}C_{49}$

Let us substitute $x=1$ in the expansion

$$\Rightarrow(1-1)^{100}={}^{100}C_0-{}^{100}C_1\times1+{}^{100}C_2\times1^2-\ldots-{}^{100}C_{49}\times1^{49}+\ldots+{}^{100}C_{100}\times1^{100}$$

$$\Rightarrow 0={}^{100}C_0-{}^{100}C_1+{}^{100}C_2-\ldots-{}^{100}C_{49}+{}^{100}C_{50}-{}^{100}C_{51}+{}^{100}C_{52}\ldots+{}^{100}C_{100}$$

There are a total of 101 terms and the middle term is $T_{51}={}^{100}C_{50}$.

As ${}^nC_r={}^nC_{n-r}$

$$\Rightarrow 0={}^{100}C_0-{}^{100}C_1+{}^{100}C_2-\ldots-{}^{100}C_{49}+{}^{100}C_{50}-{}^{100}C_{49}+{}^{100}C_{48}\ldots+{}^{100}C_0$$

$$\Rightarrow 0={}^{100}C_{50}+{}^{100}C_0-{}^{100}C_1+{}^{100}C_2-\ldots-{}^{100}C_{49}-{}^{100}C_{49}+{}^{100}C_{48}\ldots+{}^{100}C_0$$

$$\Rightarrow 0={}^{100}C_{50}+2\left({}^{100}C_0-{}^{100}C_1+{}^{100}C_2-\ldots-{}^{100}C_{49}\right)$$

$$\Rightarrow-\frac{{}^{100}C_{50}}{2}=\left({}^{100}C_0-{}^{100}C_1+{}^{100}C_2-\ldots-{}^{100}C_{49}\right)$$

$$\Rightarrow {}^{100}C_0-{}^{100}C_1+{}^{100}C_2-\ldots-{}^{100}C_{49}=-\frac{{}^{100}C_{50}}{2}$$

Therefore, the required answer is

$$-\frac{{}^{100}C_{50}}{2}=-\frac{1}{2}\times\frac{100}{50}\,{}^{99}C_{49}=-\,{}^{99}C_{49}.$$

71. **(D)**

Given that $\dfrac{4^{2022}}{15}$

We know that

$(a+x)^n={}^nC_0a^nx^0+{}^nC_1a^{n-1}x^1+\ldots+{}^nC_na^0x^n$

$$\Rightarrow\frac{4^{2022}}{15}=\frac{16^{1011}}{15}=\frac{(15+1)^{1011}}{15}$$

$$=\frac{{}^{1011}C_0(15)^{1011}+{}^{1011}C_1(15)^{1010}+\ldots+{}^{1011}C_{1011}(15)^0}{15}$$

$$\Rightarrow\frac{{}^{1011}C_0(15)^{1011}+{}^{1011}C_1(15)^{1010}+\ldots+{}^{1011}C_{1010}(15)^1}{15}+\frac{1}{15}$$

$\therefore$ Fractional part of $\dfrac{4^{2022}}{15}=\dfrac{1}{15}$

Hence this is the required option.

72. **(36)**

Given,

$$\left(\sqrt{x}-\frac{6}{x^{\frac{3}{2}}}\right)^n, n\le 15$$

Now, the rth term is given by

$$T_{r+1}={}^nC_r\left(x^{\frac{1}{2}}\right)^{n-r}\left(-6x^{\frac{-3}{2}}\right)^r$$

$$\Rightarrow T_{r+1}=(-6)^r\times{}^nC_r\times x^{\frac{n-r}{2}}\times x^{\frac{-3r}{2}}$$

$$\Rightarrow T_{r+1}=(-6)^r\times{}^nC_r\times x^{\frac{n-4r}{2}}$$

Now we have to taken $n=4r$ for the constant term so n is divisible by 4.

So, $\alpha={}^nC_{\frac{n}{4}}(-6)^{\frac{n}{4}}$

And the sum of all coefficients will be

$$\left(\sqrt{1}-\frac{6}{1^{\frac{3}{2}}}\right)^n=(-5)^n$$

Also given, $(-5)^n-{}^nC_{\frac{n}{4}}(-6)^{\frac{n}{4}}=649$

Now by hit and trial we get, $n=4$

Hence, $\alpha={}^4C_1(-6)^{\frac{4}{4}}=-24$

Now for coefficient of x^{-4}

$$\frac{n-4r}{2}=-4$$

$$\Rightarrow n=4r-8\Rightarrow \boxed{r=3}$$

So, $\lambda(\alpha)=(-6)^3\times{}^4C_3$

$$\Rightarrow \lambda(-24)=(-6)^3\times{}^4C_3\Rightarrow \lambda=36$$

73. **(18)**

$$S_0(x)=x,\ S_k(x)=C_kx+k\int_0^x S_{k-1}(t)dt$$

Now for $S_0(x)=x,\ C_0=1$

Solving, $C_k=1-\int_0^1 S_{k-1}(x)dx$ for $k=1$

we get $C_1=1-\int_0^1 xdx=\frac{1}{2}$

Now, putting $k=1$ in

$$S_k(x)=C_kx+k\int_0^x S_{k-1}(t)dt$$

we get,

$$\Rightarrow S_1(x)=\frac{x}{2}+1\times\int_0^x tdt=\frac{x}{2}+\frac{x^2}{2}$$

For $k=2$ we get,

$$C_2=1-\int_0^1\left(\frac{x}{2}+\frac{x^2}{2}\right)dx=\frac{7}{12}$$

And $S_2(x)=\frac{7x}{12}+\frac{x^2}{2}+\frac{x^3}{3}$

Now taking $k=3$ we get,

$$C_3=1-\int_0^1\left(\frac{7x}{12}+\frac{x^2}{2}+\frac{x^3}{3}\right)dx=\frac{11}{24}$$

Finding the value of $S_2(3)+6\times C_3$ we get,

$$S_2(3)+6\times C_3=\frac{7\times 3}{12}+\frac{3^2}{2}+\frac{3^3}{3}+6\times\frac{11}{24}$$

$$\Rightarrow S_2(3)+6\times C_3=\frac{61}{4}+\frac{11}{4}=18$$

74. **(A)**

General term of the binomial expansion of $\left(2x^3-\frac{1}{3x^2}\right)^5$ is

$$T_{r+1}={}^5C_r\left(2x^3\right)^{5-r}\left(-\frac{1}{3x^2}\right)^r$$

$$\Rightarrow T_{r+1}={}^5C_r(2)^{5-r}\left(-\frac{1}{3}\right)^r x^{15-5r}$$

For coefficient of x^5, we must have

$15-5r=5\Rightarrow r=2$

So, the required coefficient is

$$={}^5C_2(2)^3\left(\frac{1}{9}\right)=\frac{80}{9}$$

75. **(12)**

Using the binomial expansion in

$7^{103}-\left(7^2\right)^{51}\times 7=(51-2)^{51}\times 7$ we get,

$$(51-2)^{51}\times 7=7\times\left({}^{51}C_0 51^{51}\times 2^0-{}^{51}C_1 51^{50}\times 2^1+\ldots-{}^{51}C_{51}2^{51}\right)$$

Now 51 is divisible by 17, so when the above equation is divided by 17 we get, $7\times\left(-2^{51}\right)$ as remainder.

Again using the binomial expansion in

$7\times(-2^{51}) = -56(2^4)^{12} = -56(17-1)^{12}$ we get,

$-56(17-1)^{12} = -56\left({}^{12}C_0 17^{12}\times 1^0 - {}^{12}C_1 17^{11}\times 1^1 + \ldots + {}^{12}C_{12}1^{12}\right)$

Dividing the above equation by 17 we get,

$(-56\times 1)\div 17 = -5,$

Now, changing the negative remainder to positive we get, $-5+17=12,$

Hence, when 7^{103} is divided by 17, it gives 12 as remainder.

76. **(C)**

Given that

$$\left(a+bx+cx^2\right)^{10} = \sum_{i=10}^{20} p_i x^i,\ a, b, c \in N$$

General term: $\frac{10!}{r_1!\,r_2!\,r_3!}(a)^{r_1}(bx)^{r_2}\left(cx^2\right)^{r_3}$

For coefficient of $x: r_2 + 2r_3 = 1$

Let $\begin{matrix} r_1 & r_2 & r_3 \\ 9 & 1 & 0 \end{matrix}$

$\therefore$ Coefficient of $x = \frac{10!}{9!}a^9 b^1 = 20$

$\Rightarrow a^9\times b = 2$ (1)

Coefficient of

$\Rightarrow x^2: \frac{10!}{8!\,2!\,0!}a^8\times b^2 + \frac{10!}{9!\,0!\,1!}\times a^9\times c = 210$

$\Rightarrow 45a^8\times b^2 + 10\times a^9\times c = 210$

$\Rightarrow 9a^8b^2 + 2a^9\times c = 42$ (2)

As $a, b, c \in N$, so by hit and trial in equations (1) and (2), we get $a=1, b=2, c=3$

Hence, $2(a+b+c) = 2(3+2+1) = 12$

77. **(678)**

The coefficient of x^{30} in $\frac{(x+1)^6\left(1+x^2\right)^7\left(1-x^3\right)^8}{x^6}$, coefficient of x^{36} in $(1+x)^6\left(1+x^2\right)^7\left(1-x^3\right)^8$

General term

${}^6C_{r_1}\ {}^7C_{r_2}\ {}^8C_{r_3}(-1)^{r_3} X^{r_1+2r_2+3r_3}$

$r_1 + 2r_2 + 3r_3 = 36$

Case-I:

r_1	r_2	r_3
0	6	8
2	5	8
4	4	8
6	3	8

$r_1 + 2r_2 = 12$ (Taking $r_3 = 8$)

Case-II:

r_1	r_2	r_3
1	7	7
3	6	7
5	5	7

$r_1 + 2r_2 = 15$ (Taking $r_3 = 7$)

Case-III:

r_1	r_2	r_3
4	7	6
6	6	6

$r_1 + 2r_2 = 18$ (Taking $r_3 = 6$)

Coeff. $= 7 + (15\times 21) + (15\times 35) + (35) - (6\times 8) - (20\times 7\times 8) - (6\times 21\times 8) + (15\times 28) + (7\times 28) = -678 = \alpha$

$|\alpha| = 678$

78. **(D)**

$$\left(\frac{x^{\frac{1}{3}}}{3} + \frac{x^{\frac{-2}{3}}}{18}\right)^{18}$$

$$t_7 = {}^{18}C_6\left(\frac{x^{\frac{1}{3}}}{3}\right)^{12}\left(\frac{x^{-\frac{2}{3}}}{2}\right)^6 = {}^{18}C_6\frac{1}{(3)^{12}}\times\frac{1}{2^6}$$

$$t_{13} = {}^{18}C_{12}\left(\frac{x^{\frac{1}{3}}}{3}\right)^6\left(\frac{x^{-\frac{2}{3}}}{2}\right)^{12}$$

$$= {}^{18}C_{12}\frac{1}{(3)^6}\times\frac{1}{2^{12}}\times x^{-6}$$

$m = {}^{18}C_6\times 3^{-12}\times 2^{-6} : n = {}^{18}C_{12}\times 2^{-12}\times 3^{-6}$

$$\left(\frac{n}{m}\right)^{\frac{1}{3}} = \left(\frac{2^{-12}\times 3^{-6}}{3^{-12}\times 2^{-6}}\right)^{\frac{1}{3}} = \left(\frac{3}{2}\right)^2 = \frac{9}{4}$$

79. **(A)**

${}^{n-1}C_r = \left(k^2-8\right){}^nC_{r+1}$

$\underbrace{r+1\ge 0, r\ge 0}_{r\ge 0}$

$\frac{{}^nC_r}{{}^nC_{r+1}} = k^2 - 8\ \frac{r+1}{n} = k^2 - 8$

$\Rightarrow k^2 - 8 > 0$

$\left(k-2\sqrt{2}\right)\left(k+2\sqrt{2}\right) > 0$

$k \in \left(-\infty, -2\sqrt{2}\right) \cup \left(2\sqrt{2}, \infty\right)$ (1)

$\therefore n \geq r+1, \dfrac{r+1}{n} \leq 1$

$\Rightarrow k^2 - 8 \leq 1$

$k^2 - 9 \leq 0$

$-3 \leq k \leq 3$ (2)

From equations (1) and (2) we get

$k \in \left[-3, -2\sqrt{2}\right) \cup \left(2\sqrt{2}, 3\right]$

80. **(A)**

Sum of coefficients in the expansion of

$\left(1-3x+10x^2\right)^n = A$

then $A = (1-3+10)^n = 8^n$ (put $x = 1$)

and sum of coefficients in the expansion of

$\left(1+x^2\right)^n = B$

then $B = (1+1)^n = 2^n$ $A = B^3$

81. **(0)**

$(1-x)(1-x)^{2007}\left(1+x+x^2\right)^{2007}$

$(1-x)\left(1-x^3\right)^{2007}$

$(1-x)\left({}^{2007}C_0 - {}^{2007}C_1\left(x^3\right)+\ldots\right)$

General term

$(1-x)\left((-1)^{r\,2007}C_r x^{3r}\right)$

$(-1)^r\ {}^{2007}C_r x^{3r} - (-1)^r\ {}^{2007}C_r x^{3r+1}$

$3r = 2012$

$r \neq \dfrac{2012}{3}$

$3r+1 = 2012$ $3r = 2011$

$r \neq \dfrac{2011}{3}$

Hence, there is no term containing x^{2012}

So, the coefficient of $x^{2012} = 0$.

82. **(2041)**

$\sum_{r=1}^{9} \dfrac{{}^{11}C_r}{r+1}$

$= \dfrac{1}{12}\sum_{r=1}^{9} {}^{12}C_{r+1}$

$= \dfrac{1}{12}\left[2^{12} - 26\right] = \dfrac{2035}{6}$

$\therefore m+n = 2041$

83. **(1)**

Let $32^{32} = t$

$64^{32^{32}} = 64^t = 8^{2t} = (9-1)^{2t}$

$= 9k+1$

Hence, remainder = 1

84. **(138)**

General term in expansion of

$\left((7)^{\frac{1}{2}} + (11)^{\frac{1}{6}}\right)^{824}$ is

$t_{r+1} = {}^{824}C_r (7)^{\frac{824-r}{2}} (11)^{\frac{r}{6}}$

Hence $r = 0, 6, 12, \ldots, 822$

85. **(D)**

No solution. Refer to answer key.

86. **(10)**

$\alpha = \sum_{k=0}^{n} \dfrac{{}^nC_k \times {}^nC_k}{k+1} \times \dfrac{n+1}{n+1}$

$= \dfrac{1}{n+1}\sum_{k=0}^{n} {}^{n+1}C_{k+1} \times {}^nC_{n-k}$

$\alpha = \dfrac{1}{n+1} \times {}^{2n+1}C_{n+1}$

$\beta = \sum_{k=0}^{n-1} {}^nC_k \times \dfrac{{}^nC_{k+1}}{k+2}\dfrac{n+1}{n+1}$

$\dfrac{1}{n+1}\sum_{k=0}^{n-1} C_{n-k} \times {}^{n+1}C_{k+2}$

$= \dfrac{1}{n+1} \times {}^{2n+1}C_{n+2}$

$\dfrac{\beta}{\alpha} = \dfrac{2n+1}{2n+1}C_{n+2}$

$\dfrac{\beta}{\alpha} = \dfrac{2n+1-(n+2)+1}{n+2} = \dfrac{5}{6}$ $n = 10$

87. **(118)**

$(1+x)\left(1-x^2\right)\left(1+\dfrac{3}{x}+\dfrac{3}{x^2}+\dfrac{1}{x^3}\right)^5$

$= (1+x)\left(1-x^2\right)\left(\left(1+\dfrac{1}{x}\right)^3\right)^5$

$$=\frac{(1+x)^2(1-x)(1+x)^{15}}{x^{15}}$$

$$=\frac{(1+x)^{17}-x(1+x)^{17}}{x^{15}}$$

$=$ coeff. (x^3) in the expansion $\approx$ coeff. (x^{18}) in

$(1+x)^{17}-x(1+x)^{17}$

$=0-1$

$=-1$

coeff. (x^{-13}) in the expansion $\approx$ coeff (x^2) in

$(1+x)^{17}-x(1+x)^{17}$

$=\binom{17}{2}-\binom{17}{1}$

$=17\times 8-17$

$=17\times 7$

$=119$

Hence Answer = 119 – 1 = 118

88. **(D)**

${}^6C_m+2\left({}^6C_{m+1}\right)+{}^6C_{m+2}>{}^8C_3$

${}^7C_{m+1}+{}^7C_{m+2}>{}^8C_3$

${}^8C_{m+2}>{}^8C_3$

$\therefore m=2$

And ${}^{n-1}P_3:{}^nP_4=1:8$

$$\frac{(n-1)(n-2)(n-3)}{n(n-1)(n-2)(n-3)}=\frac{1}{8}$$

$\therefore n=8$

$\therefore {}^nP_{m+1}+{}^{n+1}C_m={}^8P_3+{}^9C_2$

$$=8\times 7\times 6+\frac{9\times 8}{2}$$

$=372$

89. **(25)**

$(x+3)^{n-1}+(x+3)^{n-2}(x+2)+(x+3)^{n-3}$

$(x+2)^2+\ldots+(x+2)^{n-1}$

$$\sum\alpha_r=4^{n-1}+4^{n-2}\times 3+4^{n-3}\times 3^2\ldots+3^{n-1}$$

$$=4^{n-1}\left[1+\frac{3}{4}+\left(\frac{3}{4}\right)^2\ldots+\left(\frac{3}{4}\right)^{n-1}\right]$$

$$=4^{n-1}\times\frac{1-\left(\frac{3}{4}\right)^n}{1-\frac{3}{4}}$$

$$=4^n-3^n=\beta^n-\gamma^n$$

$\beta=4,\gamma=3$

$\beta^2+\gamma^2=16+9=25$

5.11 Summary

- If n is a positive integer, then

  $(a+b)^n = nc_0a^n+nc_1a^{n-1}b+nc_2a^{n-2}b^2+\ldots$
  $+nc_ra^{n-r}b^4+\ldots+nc_nb^n$

- The general term is denoted by t_{r+1} and is defined as $t_{r+1}=nc_ra^{n-r}b^r$
- Total number of terms in $(a+b)^n=(n+1)$.
- If the total number of terms is odd, then there will be only one middle term and if the total number of terms is even, there will be two middle terms.
- In a binomial expansion, the middle term has the greatest binomial coefficient.
- If t_{r+1} is the numerically greatest term, then $t_{r+1}\geq t_r$.
- Properties of binomial coefficient
  - $c_0+c_1+c_2+\ldots+c_n=2^n$
  - $c_0-c_1+c_2-c_3+c_4-c_5+\ldots=0$
  - $c_0+c_2+c_4+\ldots=c_1+c_3+c_5+\ldots=2^{n-1}$
  - $1c_1+2c_2+3c_3+\ldots+nc_n=n2^{n-1}$
  - $1c_1-2c_2+3c_3-\ldots=0$
  - $c_0-c_2+c_4-\ldots=2^{n/2}\cos\frac{n\pi}{4}$
  - $c_1-c_3+c_5-\ldots=2^{n/2}\sin\frac{n\pi}{4}$
  - $C_0^2+C_1^2+C_2^2+\text{------}+C_n^2={}^{2n}C_n$
  - $c_0c_1+c_1c_2+\ldots+c_{n-1}c_n={}^{2n}C_{n-1}$
  - $\frac{c_0}{1}+\frac{c_1}{2}+\frac{c_2}{3}+\ldots+\frac{c_n}{n+1}=\frac{2^{n+1}-1}{n+1}$
  - $c_0-\frac{c_1}{2}+\frac{c_2}{3}-\frac{c_3}{4}+\ldots=\frac{1}{n+1}$

CHAPTER 6

Determinants and Matrices

OBJECTIVES

6.1 Determinants
6.2 Matrices
6.3 Know Your Concepts
6.4 JEE Main: Sample Questions
6.5 JEE Advanced: Sample Questions
6.6 JEE Main: Archive
6.7 Answer Key
6.8 Solutions
6.9 Summary

6.1 Determinants

6.1.1 Order

Let us consider the system of homogeneous linear equations in two variables x and y as

$$a_1x+b_1y=0$$

and $$a_2x+b_2y=0$$

we get,

$$\frac{y}{x}=\frac{-a_1}{b_1}=\frac{-a_2}{b_2}$$

$$\Rightarrow \frac{a_1}{b_1}=\frac{a_2}{b_2}$$

$$\Rightarrow a_1b_2-a_2b_1=0$$

We can represent $a_1b_2-a_2b_1$ as $\begin{vmatrix} a_1 & b_1 \\ a_2 & b_2 \end{vmatrix}$ which is called the determinant of order two.

Similarly, let us consider a system of three homogeneous linear equations in three variables x, y, z.

$$a_1x+b_1y+c_1z=0 \quad (6.1)$$

$$a_2x+b_2y+c_2z=0 \quad (6.2)$$

$$a_3x+b_3y+c_3z=0 \quad (6.3)$$

Solving (6.2) and (6.3) by cross multiplication, we get

$$\frac{x}{b_2c_3-b_3c_2}=\frac{y}{c_2a_3-c_3a_2}=\frac{z}{a_2b_3-a_3b_2}=K$$

Putting the value of x, y, z in (6.1) we get

$$a_1(b_2c_3-b_3c_2)+b_1(c_2a_3-c_3a_2)+c_1(a_2b_3-a_3b_2)=0$$

We can write this as

$$a_1\begin{vmatrix} b_2 & c_2 \\ b_3 & c_3 \end{vmatrix}+b_1\begin{vmatrix} c_2 & a_2 \\ c_3 & a_3 \end{vmatrix}+c_1\begin{vmatrix} a_2 & b_2 \\ a_3 & b_3 \end{vmatrix}=0$$

This is written as

$$\begin{vmatrix} a_1 & b_1 & c_1 \\ a_2 & b_2 & c_2 \\ a_3 & b_3 & c_3 \end{vmatrix}=0$$

The expression

$\begin{vmatrix} a_1 & b_1 & c_1 \\ a_2 & b_2 & c_2 \\ a_3 & b_3 & c_3 \end{vmatrix}$ is called determinant of order 3.

Remark:

Sign system for order 2, order 3, order 4 are respectively

$$\begin{vmatrix} + & - \\ - & + \end{vmatrix}, \begin{vmatrix} + & - & + \\ - & + & - \\ + & - & + \end{vmatrix}, \begin{vmatrix} + & - & + & - \\ - & + & - & + \\ + & - & + & - \\ - & + & - & + \end{vmatrix}$$

6.1.2 Applications of Determinants

- If Δ denotes the area of a triangle with vertices $A(x_1, y_1), B(x_2, y_2), c(x_3, y_3)$, then

$$\Delta = \left|\frac{1}{2}\begin{vmatrix} x_1 & y_1 & 1 \\ x_2 & y_2 & 1 \\ x_3 & y_3 & 1 \end{vmatrix}\right|$$

- Cross product of vectors

$\bar{a} = a_1\hat{i} + a_2\hat{j} + a_3\hat{k},\ \bar{b} = b_1\hat{i} + b_2\hat{j} + b_3\hat{k}$

$$\text{Then, } \bar{a}\times\bar{b} = \begin{vmatrix} \hat{i} & \hat{j} & \hat{k} \\ a_1 & a_2 & a_3 \\ b_1 & b_2 & b_3 \end{vmatrix}$$

6.1.3 Properties of Determinants

(i) The value of the determinant remains unaltered if the rows and columns are interchanged.

Example: $\begin{vmatrix} a_1 & b_1 & c_1 \\ a_2 & b_2 & c_2 \\ a_3 & b_3 & c_3 \end{vmatrix} = \begin{vmatrix} a_1 & a_2 & a_3 \\ b_1 & b_2 & b_3 \\ c_1 & c_2 & c_3 \end{vmatrix}$

(ii) If any two rows or two columns of a determinant are interchanged, the determinant retains its absolute value but changes its sign.

(iii) If any two rows or columns of determinant are identical, its value will be zero.

(iv) If the elements of any row (or column) of a determinant are each multiplied by the same factor K, then the value of determinant is multiplied by K.

(v) $\begin{vmatrix} a_1+x & b_1 & c_1 \\ a_2+y & b_2 & c_2 \\ a_3+z & b_3 & c_3 \end{vmatrix} = \begin{vmatrix} a_1 & b_1 & c_1 \\ a_2 & b_2 & c_2 \\ a_3 & b_3 & c_3 \end{vmatrix} + \begin{vmatrix} x & b_1 & c_1 \\ y & b_2 & c_2 \\ z & b_3 & c_3 \end{vmatrix}$

That is, if each element in a row (or column) of a determinant is written as the sum of two or more terms, then the determinant can be written as the sum of two or more determinants.

NOTE:

I. Let, $\Delta = \begin{vmatrix} a_1 & b_1 & c_1 \\ a_2 & b_2 & c_2 \\ a_3 & b_3 & c_3 \end{vmatrix}$

Let A_1, B_1, C_1 be cofactors of a_1, b_1, c_1; A_2, B_2, C_2 be cofactors of a_2, b_2, c_2 and A_3, B_3, C_3 be cofactors of a_3, b_3, c_3 respectively.

If Δ denotes the value of the determinant, then

$$\Delta = a_1A_1 + b_1B_1 + c_1C_1 = a_2A_2 + b_2B_2 + c_2C_2$$
$$= a_3A_3 + b_3B_3 + c_3C_3$$

$$a_1A_2 + b_1B_2 + c_1C_2 = 0$$

II. Let $\Delta \neq 0$ be the value of the determinant

$$\begin{vmatrix} a_1 & b_1 & c_1 \\ a_2 & b_2 & c_2 \\ a_3 & b_3 & c_3 \end{vmatrix}.$$

If A_1, B_1, C_1, are respectively the cofactors of the elements a_1, b_1 and c_1 of the determinant,

$$\text{Then } \begin{vmatrix} A_1 & B_1 & C_1 \\ A_2 & B_2 & C_2 \\ A_3 & B_3 & C_3 \end{vmatrix} = \Delta^2$$

6.1.4 Multiplication of Two Determinants of the Same Order

Let the two determinants of the third order be

$$\Delta_1 = \begin{vmatrix} a_1 & b_1 & c_1 \\ a_2 & b_2 & c_2 \\ a_3 & b_3 & c_3 \end{vmatrix} \text{ and } \Delta_2 = \begin{vmatrix} \alpha_1 & \beta_1 & \gamma_1 \\ \alpha_2 & \beta_2 & \gamma_2 \\ \alpha_3 & \beta_3 & \gamma_3 \end{vmatrix}$$

Let $\Delta = \Delta_1 \cdot \Delta_2$

Hence,

$$\Delta = \begin{pmatrix} a_1\alpha_1 + b_1\beta_1 + c_1\gamma_1 & a_1\alpha_2 + b_1\beta_2 + c_1\gamma_2 & a_1\alpha_3 + b_1\beta_3 + c_1\gamma_3 \\ a_2\alpha_1 + b_2\beta_1 + c_2\gamma_1 & a_2\alpha_2 + b_2\beta_2 + c_2\gamma_2 & a_2\alpha_3 + b_2\beta_3 + c_2\gamma_3 \\ a_3\alpha_1 + b_3\beta_1 + c_3\gamma_1 & a_3\alpha_2 + b_3\beta_2 + c_3\gamma_2 & a_3\alpha_3 + b_3\beta_3 + c_3\gamma_3 \end{pmatrix}$$

We have multiplied **ROW by ROW.**

We can take multiplication row by column, or column by row, or column by column.

Express a determinant as the product of two determinants.

$$\begin{vmatrix} a_1\alpha_1 + b_1\beta_1 & a_1\alpha_2 + b_1\beta_2 \\ a_2\alpha_1 + b_2\beta_2 & a_2\alpha_2 + b_2\beta_2 \end{vmatrix} = \begin{vmatrix} a_1 & b_1 \\ a_2 & b_2 \end{vmatrix}\begin{vmatrix} \alpha_1 & \beta_1 \\ \alpha_2 & \beta_2 \end{vmatrix}$$

Example:

$$\begin{vmatrix} a^2-2ax+x^2 & b^2-2bx+x^2 & c^2-2cx+x^2 \\ a^2-2ay+y^2 & b^2-2by+y^2 & c^2-2cy+y^2 \\ a^2-2az+z^2 & b^2-2bz+z^2 & c^2-2cz+z^2 \end{vmatrix}$$

$$=\begin{vmatrix} 1 & x & x^2 \\ 1 & y & y^2 \\ 1 & z & z^2 \end{vmatrix}\begin{vmatrix} a^2 & b^2 & c^2 \\ -2a & -2b & -2c \\ 1 & 1 & 1 \end{vmatrix}$$

6.1.5 Differentiation of Determinant

Let $\Delta(x)=\begin{vmatrix} f(x) & g(x) & h(x) \\ f_1(x) & g_1(x) & h_1(x) \\ f_2(x) & g_2(x) & h_2(x) \end{vmatrix}$

Then,

$$\Delta'(x)=\begin{vmatrix} f'(x) & g'(x) & h'(x) \\ f_1(x) & g_1(x) & h_1(x) \\ f_2(x) & g_2(x) & h_2(x) \end{vmatrix}+\begin{vmatrix} f(x) & g(x) & h(x) \\ f_1'(x) & g_1'(x) & h_1'(x) \\ f_2(x) & g_2(x) & h_2(x) \end{vmatrix}+\begin{vmatrix} f(x) & g(x) & h(x) \\ f_1(x) & g_1(x) & h_1(x) \\ f_2'(x) & g_2'(x) & h_2'(x) \end{vmatrix}$$

6.1.6 Integration of Determinant

We know the determinant is a numerical value, so it can always be integrated by expanding but the integration of the determinant can be done without expansion, if it has only one variable row/column.

Let $\Delta(x)=\begin{vmatrix} f(x) & g(x) & h(x) \\ a & b & c \\ l & m & n \end{vmatrix}$

So, $\int_a^b \Delta(x)dx=\begin{vmatrix} \int_a^b f(x)dx & \int_a^b g(x)dx) & \int_a^b h(x)dx \\ a & b & c \\ l & m & n \end{vmatrix}$

Note: If in a determinant, the elements of more than one row or column are function of x, then the integration can be done only after evaluation.

Use of Σ in Determinant

If $\Delta(r)=\begin{vmatrix} f(r) & g(r) & h(r) \\ a & b & c \\ a_1 & b_1 & c_1 \end{vmatrix}$

Then, $\sum_{r=1}^{n}\Delta(r)=\begin{vmatrix} \sum_{r=1}^{n} f(r) & \sum_{r=1}^{n} g(r) & \sum_{r=1}^{n} h(r) \\ a & b & c \\ a_1 & b_1 & c_1 \end{vmatrix}$

Note: If in a determinant, the elements of more than one row or column are function of r, then Σ can be done only after evaluation.

6.1.7 Special Determinants

(i) Symmetric Determinant: The elements situated at equal distances from the diagonal are equal both in magnitude and sign.

$$\begin{vmatrix} a & h & g \\ h & b & f \\ g & f & c \end{vmatrix}=abc+2fgh-af^2-bg^2-ch^2$$

(ii) Skew Symmetric Determinant: If in a determinant all the diagonal elements are zero and the elements situated at equal distances from the diagonal are equal in magnitude but opposite in sign. The value of a skew symmetric determinant of odd order is zero.

$$\begin{vmatrix} 0 & b & -c \\ -b & 0 & a \\ c & -a & 0 \end{vmatrix}=0$$

(iii) Circulant Determinant: The elements of the rows (or columns) are in cyclic arrangement.

$$\begin{vmatrix} a & b & c \\ b & c & a \\ c & a & b \end{vmatrix}=-(a^3+b^3+c^3-3abc)$$

6.1.8 Solution of System of Linear Equations

Let us consider the system of linear equations

$$a_1x+b_1y+c_1z=d_1$$
$$a_2x+b_2y+c_2z=d_2$$

$$a_3x+b_3y+c_3z=d_3$$

Let
$$\Delta=\begin{vmatrix}a_1 & b_1 & c_1\\ a_2 & b_2 & c_2\\ a_3 & b_3 & c_3\end{vmatrix}$$

$$\Delta_1=\begin{vmatrix}d_1 & b_1 & c_1\\ d_2 & b_2 & c_2\\ d_3 & b_3 & c_3\end{vmatrix}$$

$$\Delta_2=\begin{vmatrix}a_1 & d_1 & c_1\\ a_2 & d_2 & c_2\\ a_3 & d_3 & c_3\end{vmatrix}$$

and
$$\Delta_3=\begin{vmatrix}a_1 & b_1 & d_1\\ a_2 & b_2 & d_2\\ a_3 & b_3 & d_3\end{vmatrix}$$

So, we can write

$$x=\frac{\Delta_1}{\Delta},\ y=\frac{\Delta_2}{\Delta},\ z=\frac{\Delta_3}{\Delta}$$

This is known as *Cramer's rule.*

Case I: If $\Delta\neq 0$, then the system of equations will have unique finite solutions and so equations are consistent.

Case II: If $\Delta=0$ and at least one of $\Delta_1, \Delta_2, \Delta_3$ is non-zero, then the system is inconsistent, that is, it has no solution.

Case III: If $\Delta=0$ and $\Delta_1=\Delta_2=\Delta_3=0$, then the system will have an infinite number of solutions, that is, the equations are consistent.

6.1.9 System of Homogeneous Linear Equations

Let us consider a system of homogenous linear equations in three variables x, y and z.

$$a_1x+b_1y+c_1z=0$$
$$a_2x+b_2y+c_2z=0$$

and,
$$a_3x+b_3y+c_3z=0$$

Here,
$$\Delta=\begin{vmatrix}a_1 & b_1 & c_1\\ a_2 & b_2 & c_2\\ a_3 & b_3 & c_3\end{vmatrix}$$

If $\Delta\neq 0$, then $x=\frac{\Delta_1}{\Delta}=y=\frac{\Delta_2}{\Delta}=z=\frac{\Delta_3}{\Delta}=0$

So $x=y=z=0$ is a solution of system of equations, the solution is called a trivial solution.

Non-trivial solutions exist iff, $\Delta=\Delta_1=\Delta_2=\Delta_3=0$, then all values of x, y, z satisfy the system of equations and thus the system has infinitely many solutions.

SOLVED EXAMPLES

Example 6.1 Evaluate $\begin{vmatrix}1 & \omega & \omega^2\\ \omega & \omega^2 & 1\\ \omega^2 & 1 & \omega\end{vmatrix}$

where ω is cube root of unity.

Solution Let, $\Delta=\begin{vmatrix}1 & \omega & \omega^2\\ \omega & \omega^2 & 1\\ \omega^2 & 1 & \omega\end{vmatrix}$

Applying $R_1\to R_1+R_2+R_3$

$$\Delta=\begin{vmatrix}1+\omega+\omega^2 & \omega+\omega^2+1 & \omega^2+1+\omega\\ \omega & \omega^2 & 1\\ \omega^2 & 1 & \omega\end{vmatrix}$$

$$\Delta=\begin{vmatrix}0 & 0 & 0\\ \omega & \omega^2 & 1\\ \omega^2 & 1 & \omega\end{vmatrix}=0$$

Example 6.2 If $\Delta=\begin{vmatrix}1 & \sin\theta & 1\\ -\sin\theta & 1 & \sin\theta\\ -1 & -\sin\theta & 1\end{vmatrix}$ such that $\Delta\in[\alpha, \beta]$ then find the value of $\alpha+\beta$.

Solution Since $\Delta=\begin{vmatrix}1 & \sin\theta & 1\\ -\sin\theta & 1 & \sin\theta\\ -1 & -\sin\theta & 1\end{vmatrix}$

$$=1\begin{vmatrix}1 & \sin\theta\\ -\sin\theta & 1\end{vmatrix}-\sin\theta\begin{vmatrix}-\sin\theta & \sin\theta\\ -1 & 1\end{vmatrix}+1\begin{vmatrix}-\sin\theta & 1\\ -1 & -\sin\theta\end{vmatrix}$$

$$=\left(1+\sin^2\theta\right)-\sin\theta(-\sin\theta+\sin\theta)+\left(\sin^2\theta+1\right)$$

$$=2\left(1+\sin^2\theta\right)$$ As $1\le 1+\sin^2\theta\le 2$

$$\Rightarrow 2\le 2\left(1+\sin^2\theta\right)\le 4$$

$$\Rightarrow \Delta\in[2, 4]$$

$$\Rightarrow \alpha+\beta-6$$

Example 6.3 If x, y, z are positive real numbers not equal to 1, find the value of $\begin{vmatrix} 1 & \log_x y & \log_x z \\ \log_y x & 1 & \log_y z \\ \log_z x & \log_z y & 1 \end{vmatrix}$

Solution Let $\Delta = \begin{vmatrix} 1 & \log_x y & \log_x z \\ \log_y x & 1 & \log_y z \\ \log_z x & \log_z y & 1 \end{vmatrix}$

$$\Rightarrow \Delta = \begin{vmatrix} 1 & \frac{\log y}{\log x} & \frac{\log z}{\log x} \\ \frac{\log x}{\log y} & 1 & \frac{\log z}{\log y} \\ \frac{\log x}{\log z} & \frac{\log y}{\log z} & 1 \end{vmatrix}$$

$$= \frac{1}{\log x \log y \log z} \begin{vmatrix} \log x & \log y & \log z \\ \log x & \log y & \log z \\ \log x & \log y & \log z \end{vmatrix} = 0$$

Example 6.4 Find the value of the determinant $\begin{vmatrix} 13 & 16 & 19 \\ 14 & 17 & 20 \\ 15 & 18 & 21 \end{vmatrix}$.

Solution Let, $\Delta = \begin{vmatrix} 13 & 16 & 19 \\ 14 & 17 & 20 \\ 15 & 18 & 21 \end{vmatrix}$

Operating, $R_2 \to R_2 - R_1$ and $R_3 \to R_3 - R_2$

$$\Delta = \begin{vmatrix} 13 & 16 & 19 \\ 1 & 1 & 1 \\ 1 & 1 & 1 \end{vmatrix} = 0$$

Example 6.5 Evaluate $\begin{vmatrix} 2^2 & 3^2 & 4^2 \\ 3^2 & 4^2 & 5^2 \\ 4^2 & 5^2 & 6^2 \end{vmatrix}$

Solution Let, $\Delta = \begin{vmatrix} 2^2 & 3^2 & 4^2 \\ 3^2 & 4^2 & 5^2 \\ 4^2 & 5^2 & 6^2 \end{vmatrix}$

Operating, $R_2 - R_1$ and $R_3 - R_2$

$$\Delta = \begin{vmatrix} 2^2 & 3^2 & 4^2 \\ 5 & 7 & 9 \\ 7 & 9 & 11 \end{vmatrix}$$

Now, $c_2 - c_1$ and $c_3 - c_2$

$$\Delta = \begin{vmatrix} 4 & 5 & 7 \\ 5 & 2 & 2 \\ 7 & 2 & 2 \end{vmatrix}$$

Now applying $c_3 - c_2$

$$\Delta = \begin{vmatrix} 4 & 5 & 2 \\ 5 & 2 & 0 \\ 7 & 2 & 0 \end{vmatrix} = 2(10-14) = -8$$

Example 6.6 Solve for x,

$$\begin{vmatrix} x-2 & 2x-3 & 3x-4 \\ x-4 & 2x-9 & 3x-16 \\ x-8 & 2x-27 & 3x-64 \end{vmatrix} = 0$$

Solution Let $\Delta = \begin{vmatrix} x-2 & 2x-3 & 3x-4 \\ x-4 & 2x-9 & 3x-16 \\ x-8 & 2x-27 & 3x-64 \end{vmatrix} = 0$

Applying $R_2 - R_1$ and $R_3 - R_1$, we get

$$\Delta - \begin{vmatrix} x-2 & 2x-3 & 3x-4 \\ -2 & -6 & 12 \\ -6 & -24 & -60 \end{vmatrix} = 0$$

$$\Delta = \begin{vmatrix} x-2 & 2x-3 & 3x-4 \\ 1 & 3 & 6 \\ 1 & 4 & 10 \end{vmatrix} = 0$$

[taking –2 and –6 common from R_2 and R_3]

$$\Rightarrow 6(x-2) - 4(2x-3) + (3x-4) = 0$$

$$\Rightarrow x = 4$$

Example 6.7 If a, b, c denote the pth, q^{th} and r^{th} term of a H.P. respectively, then find the value of

$$\begin{vmatrix} bc & ca & ab \\ p & q & r \\ 1 & 1 & 1 \end{vmatrix}.$$

Solution Let A be the first term and D be the c.d. of an A.P.

Now, $\quad \frac{1}{a} = A + (p-1)D$

$$\frac{1}{b} = A + (q-1)D$$

$$\frac{1}{c} = A + (r-1)D$$

Let $\Delta=\begin{vmatrix} bc & ca & ab \\ p & q & r \\ 1 & 1 & 1 \end{vmatrix}=abc\begin{vmatrix} \frac{1}{a} & \frac{1}{b} & \frac{1}{c} \\ p & q & r \\ 1 & 1 & 1 \end{vmatrix}$

$$=abc\begin{vmatrix} A+(p-1)D & A+(q-1)D & A+(r-1)D \\ p & q & r \\ 1 & 1 & 1 \end{vmatrix}$$

Operating $R_1-(A-D)R_3-DR_2$, we get

$$\Delta=abc\begin{vmatrix} 0 & 0 & 0 \\ p & q & r \\ 1 & 1 & 1 \end{vmatrix}=0$$

Example 6.8 If the lines

$$ax+y+1=0,\ x+by+1=0 \text{ and } x+y+c=0$$

(where a,b,c are distinct and different from 1) are concurrent, then find the value of $\frac{1}{1-a}+\frac{1}{1-b}+\frac{1}{1-c}$.

Solution As the given lines

$$ax+y+1=0$$
$$x+by+1=0$$

and $x+y+c=0$ are concurrent,

therefore, $\begin{vmatrix} a & 1 & 1 \\ 1 & b & 1 \\ 1 & 1 & c \end{vmatrix}=0$

Applying c_2-c_1 and c_3-c_1, we get

$$\begin{vmatrix} a & 1-a & 1-a \\ 1 & b-1 & 0 \\ 1 & 0 & c-1 \end{vmatrix}=0$$

$$\Rightarrow a(b-1)(c-1)-(1-a)(c-1)+(1-a)(1-b)=0$$

Dividing by $(a-1)(b-1)(c-1)$, we get

$$\Rightarrow \frac{a}{a-1}+\frac{1}{b-1}+\frac{1}{c-1}=0$$

$$\Rightarrow \frac{a}{a-1}-1+\frac{1}{b-1}+\frac{1}{c-1}=-1$$

$$\Rightarrow \frac{1}{a-1}+\frac{1}{b-1}+\frac{1}{c-1}=-1$$

$$\Rightarrow \frac{1}{1-a}+\frac{1}{1-b}+\frac{1}{1-c}=1$$

Example 6.9 If $\begin{vmatrix} 1 & x & x^2 \\ x & x^2 & 1 \\ x^2 & 1 & x \end{vmatrix}=3$, then find the value of $\begin{vmatrix} x^3-1 & 0 & x-x^4 \\ 0 & x-x^4 & x^3-1 \\ x-x^4 & x^3-1 & 0 \end{vmatrix}$.

Solution Let $\Delta=\begin{vmatrix} 1 & x & x^2 \\ x & x^2 & 1 \\ x^2 & 1 & x \end{vmatrix}$

Cofactor of first row $=x^3-1,\ 0,\ x-x^4$

Cofactor of second row $=0,\ x-x^4,\ x^3-1$

Cofactor of third row $=x-x^4,\ x^3-1,\ 0$

Let Δ_1 be a determinant formed by the cofactors of Δ.

$$\Rightarrow \Delta_1=\begin{vmatrix} x^3-1 & 0 & x-x^4 \\ 0 & x-x^4 & x^3-1 \\ x-x^4 & x^3-1 & 0 \end{vmatrix}$$

We know $\Delta_1=\Delta^2$

$\Rightarrow \Delta_1=9$

Example 6.10 If α,β,γ are such that $\alpha+\beta+\gamma=0$, then find the value of $\begin{vmatrix} 1 & \cos\gamma & \cos\beta \\ \cos\gamma & 1 & \cos\alpha \\ \cos\beta & \cos\alpha & 1 \end{vmatrix}$.

Solution Let $\Delta=\begin{vmatrix} 1 & \cos\gamma & \cos\beta \\ \cos\gamma & 1 & \cos\alpha \\ \cos\beta & \cos\alpha & 1 \end{vmatrix}$

Applying $c_2-c_1\cos\gamma$ and $c_3-c_1\cos\beta$

$$\Rightarrow \Delta=\begin{vmatrix} 1 & 0 & 0 \\ \cos\gamma & 1-\cos^2\gamma & \cos\alpha-\cos\beta\cos\gamma \\ \cos\beta & \cos\alpha-\cos\beta\cos\gamma & 1-\cos^2\beta \end{vmatrix}$$

$$\Rightarrow \Delta=\begin{vmatrix} 1 & 0 & 0 \\ \cos\gamma & \sin^2\gamma & -\sin\beta\sin\gamma \\ \cos\beta & -\sin\beta\sin\gamma & \sin^2\beta \end{vmatrix}$$

$$=\sin^2\beta\sin^2\gamma-\sin^2\beta\sin^2\gamma=0$$

Example 6.11 If $\begin{vmatrix} b^2+c^2 & a^2 & a^2 \\ b^2 & a^2+c^2 & b^2 \\ c^2 & c^2 & a^2+b^2 \end{vmatrix} = \lambda a^2b^2c^2$, find the value of λ.

Solution Let $\Delta = \begin{vmatrix} b^2+c^2 & a^2 & a^2 \\ b^2 & a^2+c^2 & b^2 \\ c^2 & c^2 & a^2+b^2 \end{vmatrix}$

Applying $R_1 - R_2 - R_3$, we get

$$\Delta = \begin{vmatrix} 0 & -2c^2 & -2b^2 \\ b^2 & a^2+c^2 & b^2 \\ c^2 & c^2 & a^2+b^2 \end{vmatrix}$$

$$\Delta = 2\begin{vmatrix} 0 & -c^2 & -b^2 \\ b^2 & a^2+c^2 & b^2 \\ c^2 & c^2 & a^2+b^2 \end{vmatrix}.$$

Applying $R_2 + R_1$ and $R_3 + R_1$

$$\Delta = 2\begin{vmatrix} 0 & -c^2 & -b^2 \\ b^2 & a^2 & 0 \\ c^2 & 0 & a^2 \end{vmatrix}$$

$$= 2\left[c^2\left(a^2b^2 - 0\right) - b^2\left(0 - a^2c^2\right)\right]$$

$$= 4a^2b^2c^2$$

$\Rightarrow \lambda = 4$

Example 6.12 If x, y, z are integers in A.P. lying between 1 and 9 and $x51$, $y41$ and $z31$ are three-digit numbers then find the value of

$$\begin{vmatrix} 5 & 4 & 3 \\ x51 & y41 & z31 \\ x & y & z \end{vmatrix}.$$

Solution Let $\Delta = \begin{vmatrix} 5 & 4 & 3 \\ x51 & y41 & z31 \\ x & y & z \end{vmatrix}$

$$\Rightarrow \Delta = \begin{vmatrix} 5 & 4 & 3 \\ 100x+50+1 & 100y+40+1 & 100z+30+1 \\ x & y & z \end{vmatrix}$$

Applying $R_2 - 100R_3 - 10R_1$

$$\Rightarrow \Delta = \begin{vmatrix} 5 & 4 & 3 \\ 1 & 1 & 1 \\ x & y & z \end{vmatrix}$$

$$= x - 2y + z$$

$$= 0 \qquad [\text{As } 2y = x + z]$$

Example 6.13 If the determinant

$$\begin{vmatrix} \cos 2x & \sin^2 x & \cos 4x \\ \sin^2 x & \cos 2x & \cos^2 x \\ \cos 4x & \cos^2 x & \cos 2x \end{vmatrix}$$

is expanded in terms of $\sin x$, then find constant term in the expansion.

Solution We know, to find the constant term, we put $x = 0$

$$= \begin{vmatrix} 1 & 0 & 1 \\ 0 & 1 & 1 \\ 1 & 1 & 1 \end{vmatrix}$$

$$= 1(1-1) + 1(0-1) = -1$$

Example 6.14 Find the value of determinant

$$\begin{vmatrix} \cos^2\theta & \cos\theta\sin\theta & -\sin\theta \\ \cos\theta\sin\theta & \sin^2\theta & \cos\theta \\ \sin\theta & -\cos\theta & 0 \end{vmatrix}$$

Solution Let $\Delta = \begin{vmatrix} \cos^2\theta & \cos\theta\sin\theta & -\sin\theta \\ \cos\theta\sin\theta & \sin^2\theta & \cos\theta \\ \sin\theta & -\cos\theta & 0 \end{vmatrix}$

Applying $c_1 - \sin\theta c_3$ and $c_2 + \cos\theta c_3$

$$\Rightarrow \Delta = \begin{vmatrix} 1 & 0 & -\sin\theta \\ 0 & 1 & \cos\theta \\ \sin\theta & -\cos\theta & 0 \end{vmatrix}$$

$$= 1\left(\cos^2\theta\right) - \sin\theta\left(-\sin\theta\right) = \cos^2\theta + \sin^2\theta = 1.$$

Example 6.15 Find the maximum value of

$$\begin{vmatrix} 1+\sin^2 x & \cos^2 x & 4\sin 2x \\ \sin^2 x & 1+\cos^2 x & 4\sin 2x \\ \sin^2 x & \cos^2 x & 1+4\sin 2x \end{vmatrix}.$$

Solution $R_1 - R_3, R_2 - R_3$

$$= \begin{vmatrix} 1 & 0 & -1 \\ 0 & 1 & -1 \\ \sin^2 x & \cos^2 x & 1+4\sin 2x \end{vmatrix}$$

$$= 1\left(1+4\sin 2x+\cos^2 x\right)-1\left(0-\sin^2 x\right)$$

$$= 2+4\sin 2x$$

$\Rightarrow$ Maximum value = 6

Example 6.16 If $\Delta(x) = \begin{vmatrix} a & b & c \\ 6 & 4 & 3 \\ x & x^2 & x^3 \end{vmatrix}$, then find the value of $\int_0^1 \Delta(x)\,d(x)$.

Solution $\int_0^1 \Delta(x)\,d(x) = \begin{vmatrix} a & b & c \\ 6 & 4 & 3 \\ \frac{1}{2} & \frac{1}{3} & \frac{1}{4} \end{vmatrix}$

$$= \frac{1}{12}\begin{vmatrix} a & b & c \\ 6 & 4 & 3 \\ 6 & 4 & 3 \end{vmatrix} = 0$$

Example 6.17 If $(x_1 - x_2)^2 + (y_1 - y_2)^2 = a^2$

$(x_2 - x_3)^2 + (y_2 - y_3)^2 = b^2$ and

$(x_3 - x_1)^2 + (y_3 - y_1)^2 = c^2$ and

A (x_3, y_3)
c
b
B (x_1, y_1) a
C (x_2, y_2)

$$K\begin{vmatrix} x_1 & y_1 & 1 \\ x_2 & y_2 & 1 \\ x_3 & y_3 & 1 \end{vmatrix}^2$$

$$= (a+b+c)(b+c-a)(c+a-b)(a+b-c)$$

Then find the value of K.

Solution Consider the triangle with vertices $B(x_1, y_1), C(x_2, y_2)$ and $A(x_3, y_3)$ and $AB = c, BC = a, CA = b$.

$$\text{Area of } \Delta ABC = \frac{1}{2}\begin{vmatrix} x_1 & y_1 & 1 \\ x_2 & y_2 & 1 \\ x_3 & y_3 & 1 \end{vmatrix}$$

Also, area of $\Delta ABC = \sqrt{s(s-a)(s-b)(s-c)}$

$$\Rightarrow 4\begin{vmatrix} x_1 & y_1 & 1 \\ x_2 & y_2 & 1 \\ x_3 & y_3 & 1 \end{vmatrix}^2$$

$$= (a+b+c)(b+c-a)(c+a-b)(a+b-c)$$

Hence, $K = 4$.

Example 6.18 Find the number of positive integral solution of the equation

$$\begin{vmatrix} x^3+1 & x^2y & x^2z \\ xy^2 & y^3+1 & y^2z \\ xz^2 & yz^2 & z^3+1 \end{vmatrix} = 30.$$

Solution Let $\Delta = \begin{vmatrix} x^3+1 & x^2y & x^2z \\ xy^2 & y^3+1 & y^2z \\ xz^2 & yz^2 & z^3+1 \end{vmatrix}$

$$= \begin{vmatrix} x^3 & x^2y & x^2z \\ xy^2 & y^3+1 & y^2z \\ xz^2 & yz^2 & z^3+1 \end{vmatrix} + \begin{vmatrix} 1 & x^2y & x^2z \\ 0 & y^3+1 & y^2z \\ 0 & yz^2 & z^3+1 \end{vmatrix}$$

$$= x\begin{vmatrix} x^2 & x^2y & x^2z \\ y^2 & y^3+1 & y^2z \\ z^2 & yz^2 & z^3+1 \end{vmatrix} + \left(y^3z^3+y^3+z^3+1\right) - y^3z^3$$

Now $c_2 - yc_1$ and $z_3 - zc_1$

$$\Rightarrow x\begin{vmatrix} x^2 & 0 & 0 \\ y^2 & 1 & 0 \\ z^2 & 0 & 1 \end{vmatrix} + y^3 + z^3 + 1$$

$$= x^3 + y^3 + z^3 + 1$$

As, $\Delta = 30$

$\Rightarrow x^3 + y^3 + z^3 = 29$

We can write

$29 = 3^3 + 1^3 + 1^3$,

Hence the solutions are $(3, 1, 1), (1, 3, 1), (1, 1, 3)$

Hence, the number of positive integral solutions = 3.

Example 6.19 If $a, b, c, d > 0, x \in R$ and

$$\left(a^2+b^2+c^2\right)x^2 - 2(ab+bc+cd)x + b^2 + c^2 + d^2 \le 0$$

then find the value of $\begin{vmatrix} 33 & 14 & \log a \\ 65 & 27 & \log b \\ 97 & 40 & \log c \end{vmatrix}$.

Solution Since $(a^2 + b^2 + c^2)x^2 - 2(ab + bc + cd)x + b^2 + c^2 + d^2 \le 0$

$\Rightarrow (ax-b)^2 + (bx-c)^2 + (cx-d)^2 \le 0$

$\Rightarrow (ax-b)^2 + (bx-c)^2 + (cx-d)^2 = 0$

$\Rightarrow \frac{b}{a} = \frac{c}{b} = \frac{d}{c} = x$

$\Rightarrow b^2 = ac$

$\Rightarrow 2\log b = \log a + \log c$

Now Let, $\Delta = \begin{vmatrix} 33 & 14 & \log a \\ 65 & 27 & \log b \\ 97 & 40 & \log c \end{vmatrix}$

$R_1 \to R_1 + R_3$

$$\Delta = \begin{vmatrix} 130 & 54 & \log a + \log c \\ 65 & 27 & \log b \\ 97 & 40 & \log c \end{vmatrix}$$

Now, $R_1 \to R_1 - 2R_2$

$$\Delta = \begin{vmatrix} 0 & 0 & 0 \\ 65 & 27 & \log b \\ 97 & 40 & \log c \end{vmatrix} = 0$$

Example 6.20 If

$$\Delta(x) = \begin{vmatrix} \sin 2x & e^x \sin x + x\cos x & \sin x + x^2 \cos x \\ \cos x + \sin x & e^x + x & 1 + x^2 \\ e^x \cos x & e^{2x} & e^x \end{vmatrix}$$

then find the value of $\Delta'\left(\frac{\pi}{4}\right)$.

Solution We can write

$$\Delta(x) = \begin{vmatrix} \sin x & \cos x & 1 \\ 1 & 1 & 1 \\ e^x & 0 & 0 \end{vmatrix} \times \begin{vmatrix} \cos x & \sin x & 0 \\ e^x & x & 0 \\ 1 & x^2 & 0 \end{vmatrix} = 0$$

$\Rightarrow \Delta(x)$ is independent of x

$\Rightarrow \Delta'(x) = 0$ for all x

$\Rightarrow \Delta'\left(\frac{\pi}{4}\right) = 0$

Example 6.21 If $f(x) = \begin{vmatrix} x^3 & \cos^2 x & 2^{x^4} \\ \tan^5 x & 1 & \sec 2x \\ \sin^3 x & x^4 & 5 \end{vmatrix}$

then find the value of $\int_{-\frac{\pi}{2}}^{\frac{\pi}{2}} f(x)dx$.

Solution As, $f(x) = \begin{vmatrix} x^3 & \cos^2 x & 2^{x^4} \\ \tan^5 x & 1 & \sec 2x \\ \sin^3 x & x^4 & 5 \end{vmatrix}$

$$\Rightarrow f(-x) = \begin{vmatrix} -x^3 & \cos^2 x & 2^{x^4} \\ -\tan^5 x & 1 & \sec 2x \\ -\sin^3 x & x^4 & 5 \end{vmatrix} = -f(x)$$

$\Rightarrow f(x)$ is an odd function, hence $\int_{-\frac{\pi}{2}}^{\frac{\pi}{2}} f(x)dx = 0$

Example 6.22 If $\Delta_r = \begin{vmatrix} 2^{r-1} & x & 2^n - 1 \\ 2.3^{r-1} & y & 3^n - 1 \\ 4.5^{r-1} & z & 5^n - 1 \end{vmatrix}$, then find the value of $\sum_{r=1}^{n} \Delta_r$.

Solution $\sum_{r=1}^{n} \Delta_r = \begin{vmatrix} \sum_{r=1}^{n} 2^{r-1} & x & 2^n - 1 \\ 2\sum_{r=1}^{n} 3^{r-1} & y & 3^n - 1 \\ 4\sum_{r=1}^{n} 5^{r-1} & z & 5^n - 1 \end{vmatrix}$

$$\sum_{r=1}^{n} \Delta_r = \begin{vmatrix} 2^n - 1 & x & 2^n - 1 \\ 3^n - 1 & y & 3^n - 1 \\ 5^n - 1 & z & 5^n - 1 \end{vmatrix} = 0$$

Example 6.23 Find the value of λ for which the equations $x + y - 3 = 0, (1+\lambda)x + (2+\lambda)y - 8 = 0$ and $x - (1+\lambda)y + (2+\lambda) = 0$ are consistent.

Solution As the given system of equation of two variables are consistent

$\Rightarrow \Delta = 0$

$$\Rightarrow \begin{vmatrix} 1 & 1 & -3 \\ 1+\lambda & 2+\lambda & -8 \\ 1 & -1-\lambda & 2+\lambda \end{vmatrix} = 0$$

Applying $c_2 - c_1$ and $c_3 + 3c_1$, we get

$$\begin{vmatrix} 1 & 0 & 0 \\ 1+\lambda & 1 & -8+3+3\lambda \\ 1 & -2-\lambda & \lambda+5 \end{vmatrix} = 0$$

$\Rightarrow (\lambda+5)+(2+\lambda)(-5+3\lambda)=0$

$\Rightarrow 3\lambda^2+2\lambda-5=0$

$\Rightarrow (\lambda-1)(3\lambda+5)=0$

$\Rightarrow \lambda = 1, \dfrac{-5}{3}$

Example 6.24 Find the real value of r for which the system of equations $2rx-2y+3z=0$, $x+ry+2z=0$, $2x+rz=0$ has non-trivial solution.

Solution As the given system of equations has a non-trivial solution, hence

$$\begin{vmatrix} 2r & -2 & 3 \\ 1 & r & 2 \\ 2 & 0 & r \end{vmatrix} = 0$$

$\Rightarrow r^3 - 2r - 4 = 0$

$\Rightarrow (r-2)(r^2+2r+2)=0$

$\Rightarrow$ Hence, $r = 2$

Example 6.25 For what values of p and q does the system of equations $2x+py+6z=8$, $x+2y+qz=5$ and $x+y+3z=4$ have

(i) a unique solution

(ii) no solution

(iii) infinitely many solutions

Solution Since, the given system of equation is $2x+py+6z=8$, $x+2y+qz=5$ and $x+y+3z=4$

$$\text{Let, } \Delta = \begin{vmatrix} 2 & p & 6 \\ 1 & 2 & q \\ 1 & 1 & 3 \end{vmatrix}$$

Applying $c_2 - c_1$ and $c_3 - 3c_1$

$$\Delta = \begin{vmatrix} 2 & p-2 & 0 \\ 1 & 1 & q-3 \\ 1 & 0 & 0 \end{vmatrix}$$

$$= (p-2)(q-3)$$

Case I: If $p \neq 2, q \neq 3$, then $\Delta \neq 0$.

$\Rightarrow$ The system of equations has a unique solution.

Case II: When $p = 2$,

$$\Delta_x = \begin{vmatrix} 8 & 2 & 6 \\ 5 & 2 & q \\ 4 & 1 & 3 \end{vmatrix} = 0$$

$$\Delta_y = \begin{vmatrix} 2 & 8 & 6 \\ 1 & 5 & q \\ 1 & 4 & 3 \end{vmatrix} = 0$$

$$\Delta_z = \begin{vmatrix} 2 & 2 & 8 \\ 2 & 2 & 5 \\ 1 & 1 & 4 \end{vmatrix} = 0$$

When $p = 2$, $\Delta_x = \Delta_y = \Delta_z = 0$, so the system of equations will have an infinite number of solutions for $p = 2$ and for any real value of q.

When $q = 3$, $\Delta = 0$

$$\Delta_x = \begin{vmatrix} 8 & p & 6 \\ 5 & 2 & 3 \\ 4 & 1 & 3 \end{vmatrix} = -3\,(p-2)$$

If $p \neq 2$, $\Delta_x \neq 0$, so the system of equations will have no solutions.

Thus,

(i) If $p \neq 2, q \neq 3$, unique solution

(ii) If $p \neq 2, q = 3$, no solution

(iii) $p = 2$ and $q \in R$, infinite solutions

6.2 Matrices

A matrix is a rectangular array of symbols (which could be real or complex numbers) arranged in rows and columns. If a matrix has m rows and n columns, then the order of the matrix is $m \times n$. Matrix A of order $(m \times n)$ is written as

$$A = \begin{bmatrix} a_{11} & a_{12} & \cdots\cdots & a_{1n} \\ a_{21} & a_{22} & \cdots\cdots & a_{2n} \\ \cdots & \cdots\cdots\cdots & & \cdots \\ \cdots & \cdots\cdots\cdots & & \cdots \\ a_{m1} & a_{m2} & \cdots\cdots & a_{mn} \end{bmatrix}$$

6.2.1 Different Types of Matrices

1. **Horizontal Matrix:** A matrix is called a horizontal matrix, if there are less number of rows than columns.

$$\begin{bmatrix} x_1 & y_1 & z_1 \\ x_2 & y_2 & z_2 \end{bmatrix}$$

2. **Vertical Matrix:** A matrix is called a vertical if there are more number of rows than columns.

$$\begin{bmatrix} x_1 & y_1 \\ x_2 & y_2 \\ x_3 & y_3 \end{bmatrix}$$

3. **Row Matrix:** A matrix having just one row is called a row matrix.

$\begin{bmatrix} a_{11} & a_{12} & \cdots & a_{1n} \end{bmatrix}$ is a $(1 \times n)$ row matrix.

4. **Column Matrix:** A matrix having just one column is called a column matrix.

$\begin{bmatrix} a_{11} \\ a_{21} \\ \cdots \\ \cdots \\ a_{m1} \end{bmatrix}$ having one column is a $(m \times 1)$ column matrix.

5. **Square Matrix:** If in a matrix, the numbers of rows are equal to number of columns, it is called a square matrix.

$$\begin{bmatrix} x_1 & x_2 & x_3 \\ y_1 & y_2 & y_3 \\ z_1 & z_2 & z_3 \end{bmatrix}$$

The elements x_1, y_2, z_3 are called diagonal elements and the line passing through them is known as the *principal* or *leading diagonal.* The other diagonal is known as the *off diagonal*.

6. **Diagonal Matrix:** A square matrix A is known as a diagonal matrix if all the elements except the principal diagonal elements are zero.

$$A = \begin{bmatrix} x_1 & 0 & 0 \\ 0 & y_2 & 0 \\ 0 & 0 & z_3 \end{bmatrix}$$

7. **Scalar Matrix:** If in a diagonal matrix, all the elements of the principal diagonal are equal, then it is called scalar matrix.

$$\begin{bmatrix} a & 0 & 0 \\ 0 & a & 0 \\ 0 & 0 & a \end{bmatrix}$$

8. **Unit Matrix or Identity Matrix:** If in a diagonal matrix, all the elements of the principal diagonal are 1, then it is called a unit matrix.

Example: $I_3 = \begin{bmatrix} 1 & 0 & 0 \\ 0 & 1 & 0 \\ 0 & 0 & 1 \end{bmatrix}$

9. **Upper Triangular Matrix:** A square matrix, in which all the elements below the principal diagonal are zero, is called an upper triangular matrix.

$$\begin{bmatrix} 1 & 4 & 3 \\ 0 & 5 & 7 \\ 0 & 0 & 8 \end{bmatrix}$$

10. **Lower Triangular Matrix:** A square matrix in which all the elements above the principal diagonal are zero, is called a lower triangular matrix.

$$\begin{bmatrix} 1 & 0 & 0 \\ 4 & 5 & 0 \\ 6 & 7 & 9 \end{bmatrix}$$

11. **Comparable Matrices:** Two matrices $A = [a_{ij}]_{m \times n}$ and $B = [b_{ij}]_{p \times q}$ are said to be comparable, if $m = p$ and $n = q$.
12. **Non-singular Matrix:** A square matrix A is said to be non-singular if $|A| \neq 0$.
13. **Singular Matrix:** A square matrix A is said to be singular if $|A| = 0$.
14. **Idempotent Matrix:** A square matrix A is called an idempotent matrix if it satisfies the relation $A^2 = A$.

    Note: $A^2 = A \ \forall \ n \geq 2, n \in N$
15. **Periodic Matrix:** A square matrix A is said to be periodic if $A^{K+1} = A$, where K is a positive integer. If K is the least positive integer for which $A^{K+1} = A$, then K is called the period of A. If $K = 1$, we get $A^2 = A$, which is called an idempotent matrix.
16. **Nilpotent Matrix:** A square matrix A is called a nilpotent matrix of order K if it satisfies the relation $A^K = 0$ and $A^{K-1} \neq 0$, where K is a positive integer.

17. **Involutary Matrix:** A square matrix A is called an involutary matrix, if it satisfies the relation $A^2 = I$, where I is a unit matrix.

Note: For an involutory Matrix, $A = A^{-1}$.

18. **Symmetric Matrix:** A square matrix A is said to be symmetric if $A' = A$.

19. **Skew-Symmetric Matrix:** Any square matrix A is said to be skew-symmetric if $A' = -A$.

Note: The trace of a skew-symmetric matrix is always zero.

20. **Orthogonal Matrix:** A square matrix A is said to be orthogonal, iff $AA' = I$, where I is an identity matrix.

Note:

(i) If $AA' = I$, then $A^{-1} = A$.

(ii) A and B are orthogonal, then AB is also orthogonal.

(iii) If A is orthogonal, then A^{-1} and A' are also orthogonal.

21. **Conjugate Matrix:** If a matrix A has complex numbers as its elements, then the matrix obtained from A by replacing each element of A by its conjugate is called the conjugate of matrix A and is denoted by $\bar{A}$.

22. **Hermitian Matrix:** A square matrix A is said to be a Hermitian matrix if all the principal diagonal elements are purely real.

23. **Skew-Hermitian Matrix:** A square matrix is said to be a skew Hermitian matrix if all the diagonal elements are purely imaginary or zero.

6.2.2 Matrix: Key Properties and Operations

1. **Trace of a Matrix:** The sum of all the principal diagonal elements of a square matrix A is called the trace of the matrix and is denoted by $t_r(A)$.

If $A = \begin{bmatrix} x_1 & y_1 & z_1 \\ x_2 & y_2 & z_2 \\ x_3 & y_3 & z_3 \end{bmatrix}$, then $t_r(A) = x_1 + y_2 + z_3$.

Note:

(i) $t_r(\lambda A) = \lambda t_r(A)$

(ii) $t_r(A+B) = t_r(A) + t_r(B)$

(iii) $t_r(AB) = t_r(BA)$

2. **Equal Matrices:** Any two matrices A and B are said to be equal, if

(i) they are of the same order, that is, they have same number of rows and columns,

(ii) the elements in the corresponding positions of the two matrices are equal.

3. **Addition of Matrices:** Any two matrices can be added or subtracted if they have same number of rows and columns.

4. **Scalar Multiplication:** Let A be a matrix and $K \neq 0$ be any number called a scalar, then the matrix obtained by multiplying every element of A by K is called the scalar multiple of A by K and is denoted by KA.

5. **Multiplication of Matrices:** Any two matrices A and B are said to be conformable for the product AB if the number of columns in A is equal to the number of rows in B.

Example: Let $A = \begin{bmatrix} 1 & 3 & 2 \\ 0 & 1 & 2 \\ 2 & 3 & 1 \end{bmatrix}$ and $B = \begin{bmatrix} 1 & 2 \\ 2 & 1 \\ 2 & 3 \end{bmatrix}$

$$AB = \begin{bmatrix} 1.1+3.2+2.2 & 1.2+3.1+2.3 \\ 0.1+1.2+2.2 & 0.2+1.1+2.3 \\ 2.1+3.2+1.2 & 2.2+3.1+1.3 \end{bmatrix}$$

$$= \begin{bmatrix} 11 & 11 \\ 6 & 7 \\ 10 & 10 \end{bmatrix}$$

6.2.3 Transpose of a Matrix

Let A be a given matrix, then the matrix obtained by interchanging the rows and columns of A is called the transpose of A. The transpose of the matrix A is denoted by A' or A^T.

Note: Let A' and B' denote the transposes of matrix A and B respectively, then

(i) $(A')' = A$ **(ii)** $(A \pm B)' = A' \pm B'$

(iii) $(KA)' = KA'$ **(iv)** $(AB)' = B'A'$

6.2.4 Adjoint of a Square Matrix

Let A be a square matrix and B be a matrix formed by the cofactors of the corresponding elements of $|A|$, then the transpose of B is called the Adjoint of A and is denoted by $\text{adj}(A)$.

Example: Let $A = \begin{bmatrix} 1 & 2 & 3 \\ -1 & 0 & 1 \\ 4 & 3 & 2 \end{bmatrix}$

Cofactor of the elements of the first row $= -3, 6, -3$.

Cofactor of the elements of the second row $= 5, -10, 5$.

Cofactor of the elements of the third row $= 2, -4, 2$.

Let B be a matrix formed by the cofactors of the corresponding elements of $|A|$.

then $B = \begin{bmatrix} -3 & 6 & -3 \\ 5 & -10 & 5 \\ 2 & -4 & 2 \end{bmatrix}$

$$\text{adj}(A) = B' = \begin{bmatrix} -3 & 5 & 2 \\ 6 & -10 & 4 \\ -3 & 5 & 2 \end{bmatrix}$$

Properties of the Adjoint of a Matrix

(i) If A is a square matrix of order n, then $A(\text{adj } A) = |A| I_n = (\text{adj } A)A$ where I_n is a square matrix of order n.

(ii) If A is a square matrix of order n, then $\text{adj}(A') = (\text{adj } A)'$.

(iii) If A and B are two square matrices of the same order, then $\text{adj}(AB) = (\text{adj } B)(\text{adj } A)$

(iv) $\text{adj}(\text{adj})A = |A|^{n-2} A$, where A is a non-singular matrix.

(v) Adjoint of a diagonal matrix is a diagonal matrix.

(vi) $|\text{adj}(\text{adj } A)| = |A|^{(n-1)^2}$, where A is a non-singular matrix.

(vii) If A is an n rowed square non-singular matrix, then $|\text{adj } A| = |A|^{n-1}$.

(viii) $\text{adj}(KA) = K^{n-1}(\text{adj } A)$, K is a scalar.

6.2.5 Inverse of a Matrix

A non-singular matrix of order n is said to be invertible if there exists a matrix B such that $AB = I_n = BA$, then B is called inverse of A and denoted by A^{-1}. The inverse of A is denoted by A^{-1} and is determined by the formula,

$$A^{-1} = \frac{1}{|A|}(\text{adj } A)$$

Properties of the Inverse of a Matrix

(i) A square matrix is invertible if and only if it is non-singular.

(ii) $\left(A^{-1}\right)^{-1} = A$

(iii) The inverse of the transpose of a matrix is the transpose of its inverse, that is, $(A')^{-1} = \left(A^{-1}\right)'$.

(iv) If A and B are two invertible matrices of the same order then AB is also invertible and moreover $(AB)^{-1} = B^{-1}A^{-1}$.

(v) If A is a non-singular matrix such that A is symmetric, then A^{-1} is also symmetric.

(vi) If A is a non-singular matrix, then $\left|A^{-1}\right| = |A|^{-1}$.

6.2.6 Solution of System of Linear Equations by Matrix Method

Consider a system of linear equations

$$a_{11}x_1 + a_{12}x_2 + \ldots + a_{1n}x_n = b_1$$
$$a_{21}x_1 + a_{22}x_2 + \ldots + a_{2n}x_n = b_2$$
$$\rule{5cm}{0.4pt}$$
$$\rule{5cm}{0.4pt}$$
$$a_{n1}x_1 + a_{n2}x_2 + \ldots + a_{nn}x_n = b_n$$

We can express these equations as a single matrix equation.

$$\begin{bmatrix} a_{11} & a_{12}\cdots\cdots & \cdots\cdots a_{1n} \\ a_{21} & a_{22}\cdots\cdots & \cdots\cdots a_{2n} \\ \cdots\cdots & \cdots\cdots\cdots & \cdots\cdots\cdots\cdots \\ \cdots\cdots & \cdots\cdots\cdots & \cdots\cdots\cdots\cdots \\ a_{n1} & a_{n2}\cdots\cdots & \cdots\cdots a_{nn} \end{bmatrix} \begin{bmatrix} x_1 \\ x_2 \\ \cdots\cdots \\ \cdots\cdots \\ x_n \end{bmatrix} = \begin{bmatrix} b_1 \\ b_2 \\ \cdots\cdots \\ \cdots\cdots \\ b_n \end{bmatrix}$$

Let $|A| \neq 0$, so that A^{-1} exists uniquely.

Multiplying both sides of $Ax = B$ by A^{-1}

$\Rightarrow A^{-1}(Ax) = A^{-1}B$

$\Rightarrow \left(A^{-1}A\right)x = A^{-1}B$

$\Rightarrow Ix = A^{-1}B$

$\Rightarrow x = A^{-1}B$

Hence,

$x = A^{-1}B$ is the unique solution of $Ax = B, |A| \neq 0$.

When the System of Linear Equations is Homogeneous

Consider the system of linear homogeneous equation

$$a_{11}x_1 + a_{12}x_2 + \ldots + a_{1n}x_n = 0$$
$$a_{21}x_1 + a_{22}x_2 + \ldots + a_{2n}x_n = 0$$
$$\text{--------------------------------}$$
$$\text{--------------------------------}$$
$$a_{m1}x_1 + a_{m2}x_2 + \ldots + a_{mn}x_n = 0$$

We can write this in matrix form as

$$\begin{bmatrix} a_{11} & a_{12}\ldots\ldots & \ldots\ldots a_{1n} \\ a_{21} & a_{22}\ldots\ldots & \ldots\ldots a_{2n} \\ \ldots\ldots & \ldots\ldots\ldots & \ldots\ldots\ldots\ldots \\ \ldots\ldots & \ldots\ldots\ldots & \ldots\ldots\ldots\ldots \\ a_{m1} & a_{m2}\ldots\ldots & \ldots\ldots a_{mn} \end{bmatrix} \begin{bmatrix} x_1 \\ x_2 \\ \ldots\ldots \\ \ldots\ldots \\ x_n \end{bmatrix} = \begin{bmatrix} 0 \\ 0 \\ \ldots\ldots \\ \ldots\ldots \\ 0 \end{bmatrix}$$

(i) If $|A| \neq 0$, the system of equations has only a trivial solution and that will be the only solution.

(ii) If $|A| = 0$, the system of equations has a non-trivial solution and it has infinite solutions.

6.2.7 The Reflection Matrix

1. **The reflection in the x-axis:** Let $P_1(x_1, y_1)$ be the image of the point $P(x, y)$ after reflection in the x-axis, then $x_1 = x$ and $y_1 = -y$

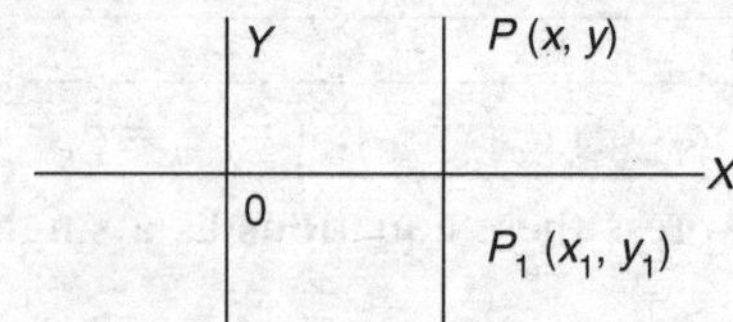

We can write this as

$$x_1 = 1 \cdot x + 0 \cdot y$$
$$y_1 = 0 \cdot x + (-1) \cdot y$$

In matrix form, we can write this as

$$\begin{bmatrix} x_1 \\ y_1 \end{bmatrix} = \begin{bmatrix} 1 & 0 \\ 0 & -1 \end{bmatrix} \begin{bmatrix} x \\ y \end{bmatrix}$$

Thus, the matrix $\begin{bmatrix} 1 & 0 \\ 0 & -1 \end{bmatrix}$ is the reflection of $P(x, y)$ in the x-axis.

2. **Reflection in the y-axis:** Let $P_1(x_1, y_1)$ be the image of the point $P(x, y)$ with respect to the y-axis, then $x_1 = -x$ and $y_1 = y$

We can write this as

$$x_1 = (-1) \cdot x + 0 \cdot y$$

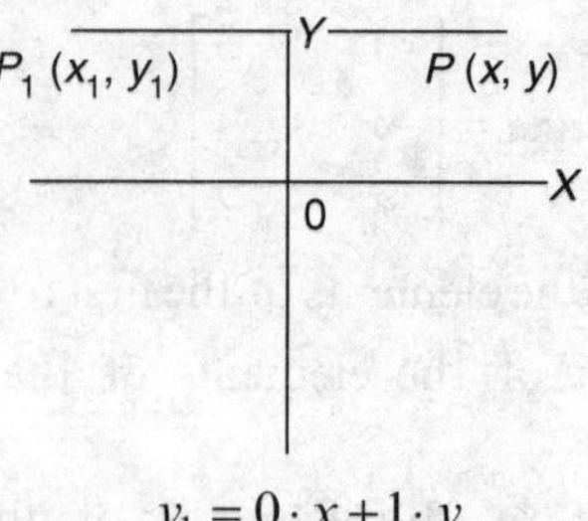

$$y_1 = 0 \cdot x + 1 \cdot y$$

These systems of equations in the matrix form can be written as

$$\begin{bmatrix} x_1 \\ y_1 \end{bmatrix} = \begin{bmatrix} -1 & 0 \\ 0 & 1 \end{bmatrix} \begin{bmatrix} x \\ y \end{bmatrix}$$

The matrix $\begin{bmatrix} -1 & 0 \\ 0 & 1 \end{bmatrix}$ is the reflection of a point $P(x, y)$ in the y-axis.

3. **Reflection through the origin:** Let $P_1(x_1, y_1)$ be the image of the point $P(x, y)$ after reflection through the origin, then

$$x_1 = -x, \ y_1 = -y$$

Y
P (x, y)
X1
X
0
P1 (x1, y1)
Y1

We can write this as

$$x_1 = (-1) \cdot x + 0 \cdot y$$
$$y_1 = 0 \cdot x + (-1) \cdot y$$

These systems of equations in the matrix form can be written as

$$\begin{bmatrix} x_1 \\ y_1 \end{bmatrix} = \begin{bmatrix} -1 & 0 \\ 0 & -1 \end{bmatrix} \begin{bmatrix} x \\ y \end{bmatrix}$$

The matrix $\begin{bmatrix} -1 & 0 \\ 0 & -1 \end{bmatrix}$ is the reflection of point $P(x, y)$ with respect to the origin.

4. **Reflection in the line $y = x$:** Let $P_1(x_1, y_1)$ be the image of the point $P(x, y)$ after reflection in the line $y = x$, then $x_1 = y$, $y_1 = x$

We can write this as

$$\begin{bmatrix} x_1 \\ y_1 \end{bmatrix} = \begin{bmatrix} 0 & 1 \\ 1 & 0 \end{bmatrix} \begin{bmatrix} x \\ y \end{bmatrix}$$

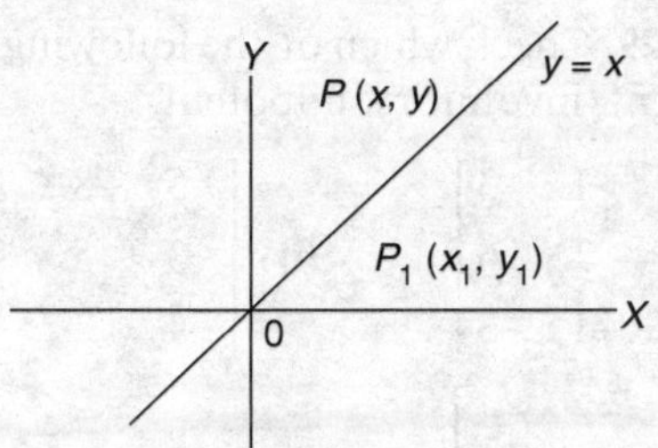

Thus, the matrix $\begin{bmatrix} 0 & 1 \\ 1 & 0 \end{bmatrix}$ is the reflection of point $P(x, y)$ in the line $y = x$.

5. **Reflection in the line $y = x \tan\theta$:** Let $P_1(x_1, y_1)$ be the image of the point $P(x, y)$ with respect to the line $y = x \tan\theta$, then we can write

$$x_1 = x\cos 2\theta + y\sin 2\theta$$
$$y_1 = x\sin 2\theta + y(-\cos 2\theta)$$

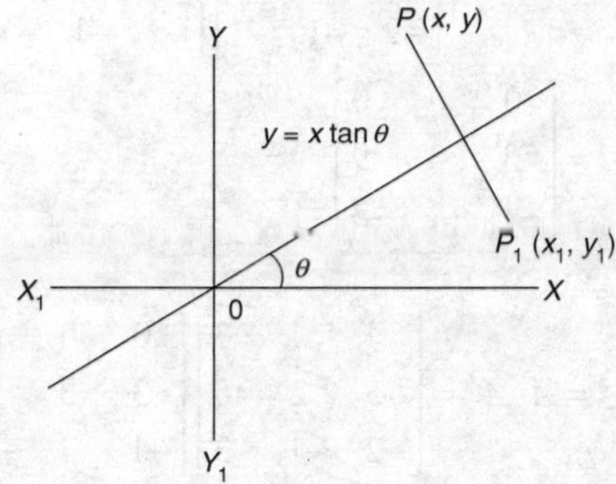

In matrix form we can write this as

$$\begin{bmatrix} x_1 \\ y_1 \end{bmatrix} = \begin{bmatrix} \cos 2\theta & \sin 2\theta \\ \sin 2\theta & -\cos 2\theta \end{bmatrix} \begin{bmatrix} x \\ y \end{bmatrix}$$

Thus, the matrix $\begin{bmatrix} \cos 2\theta & \sin 2\theta \\ \sin 2\theta & -\cos 2\theta \end{bmatrix}$ is called reflection of point $P(x, y)$ in the line $y = x \tan\theta$.

SOLVED EXAMPLES

Example 6.26 Let the three matrices be

$$A = \begin{bmatrix} 2 & 1 \\ 4 & 1 \end{bmatrix}, B = \begin{bmatrix} 3 & 4 \\ 2 & 3 \end{bmatrix} \text{ and } C = \begin{bmatrix} 3 & -4 \\ -2 & 3 \end{bmatrix},$$

then find the value of

$$t_r(A) + t_r\left(\frac{ABC}{2}\right) + t_r\left(\frac{A(BC)^2}{4}\right) + t_r\left(\frac{A(BC)^3}{8}\right) + \ldots + \infty.$$

Solution Since, $BC = \begin{bmatrix} 3 & 4 \\ 2 & 3 \end{bmatrix}\begin{bmatrix} 3 & -4 \\ -2 & 3 \end{bmatrix}$

$$= \begin{bmatrix} 9-8 & -12+12 \\ 6-6 & -8+9 \end{bmatrix} = \begin{bmatrix} 1 & 0 \\ 0 & 1 \end{bmatrix} = I$$

Therefore,

$$t_r(A) + t_r\left(\frac{ABC}{2}\right) + t_r\left(\frac{A(BC)^2}{4}\right) + t_r\left(\frac{A(BC)^3}{8}\right) + \ldots \text{to } \infty$$

$$= t_r(A) + t_r\left(\frac{A}{2}\right) + t_r\left(\frac{A}{2^2}\right) + \ldots \text{to } \infty$$

$$= t_r(A) + \frac{1}{2}t_r(A) + \frac{1}{2^2}t_r(A) + \ldots \text{to } \infty$$

$$= \frac{t_r(A)}{1-\frac{1}{2}} = 2t_r(A) = 2(2+1) = 6$$

Example 6.27 If

$$A = \begin{bmatrix} 0 & c & -b \\ -c & 0 & a \\ b & -a & 0 \end{bmatrix} \text{ and } B = \begin{bmatrix} a^2 & ab & ac \\ ab & b^2 & bc \\ ac & bc & c^2 \end{bmatrix}$$

then find AB.

Solution Since,

$$A = \begin{bmatrix} 0 & c & -b \\ -c & 0 & a \\ b & -a & 0 \end{bmatrix} \text{ and } B = \begin{bmatrix} a^2 & ab & ac \\ ab & b^2 & bc \\ ac & bc & c^2 \end{bmatrix}$$

Therefore,

$$AB = \begin{bmatrix} 0 & c & -b \\ -c & 0 & a \\ b & -a & 0 \end{bmatrix}\begin{bmatrix} a^2 & ab & ac \\ ab & b^2 & bc \\ ac & bc & c^2 \end{bmatrix}$$

$$= \begin{bmatrix} 0+abc-abc & 0+b^2c-b^2c & 0+bc^2-bc^2 \\ -a^2c+0+a^2c & -abc+0+abc & -ac^2+0+ac^2 \\ a^2b-a^2b+0 & ab^2-ab^2+0 & abc-abc+0 \end{bmatrix}$$

$$= \begin{bmatrix} 0 & 0 & 0 \\ 0 & 0 & 0 \\ 0 & 0 & 0 \end{bmatrix} = 0$$

Example 6.28 If the matrix $A = \begin{bmatrix} 0 & 2\beta & \gamma \\ \alpha & \beta & -\gamma \\ \alpha & -\beta & \gamma \end{bmatrix}$ is orthogonal, then find the values of α, β, and γ.

Solution Since, $A = \begin{bmatrix} 0 & 2\beta & \gamma \\ \alpha & \beta & -\gamma \\ \alpha & -\beta & \gamma \end{bmatrix}$

$$\Rightarrow A' = \begin{bmatrix} 0 & \alpha & \alpha \\ 2\beta & \beta & -\beta \\ \gamma & -\gamma & \gamma \end{bmatrix}$$

Since, A is an orthogonal matrix,

$\Rightarrow AA' = I$

Now,

$$AA' = \begin{bmatrix} 0 & 2\beta & \gamma \\ \alpha & \beta & -\gamma \\ \alpha & -\beta & \gamma \end{bmatrix}\begin{bmatrix} 0 & \alpha & \alpha \\ 2\beta & \beta & -\beta \\ \gamma & -\gamma & \gamma \end{bmatrix} = \begin{bmatrix} 1 & 0 & 0 \\ 0 & 1 & 0 \\ 0 & 0 & 1 \end{bmatrix}$$

$$\Rightarrow \begin{bmatrix} 4\beta^2+\gamma^2 & 2\beta^2-\gamma^2 & -2\beta^2+\gamma^2 \\ 2\beta^2-\gamma^2 & \alpha^2+\beta^2+\gamma^2 & \alpha^2-\beta^2-\gamma^2 \\ -2\beta^2+\gamma^2 & \alpha^2-\beta^2-\gamma^2 & \alpha^2+\beta^2+\gamma^2 \end{bmatrix}$$

$$= \begin{bmatrix} 1 & 0 & 0 \\ 0 & 1 & 0 \\ 0 & 0 & 1 \end{bmatrix}$$

Comparing the corresponding elements, we have

$$4\beta^2 + \gamma^2 = 1; 2\beta^2 - \gamma^2 = 0$$

Adding $6\beta^2 = 1$

$$\Rightarrow \beta = \pm\frac{1}{\sqrt{6}}$$

$$\Rightarrow \gamma = \pm\frac{1}{\sqrt{3}}$$

Also, $\alpha^2 + \beta^2 + \gamma^2 = 1$

$$\Rightarrow \alpha^2 + \frac{1}{6} + \frac{1}{3} = 1$$

$$\Rightarrow \alpha = \pm\frac{1}{\sqrt{2}}$$

$\Rightarrow$ Thus, $\alpha = \pm\frac{1}{\sqrt{2}}, \beta = \pm\frac{1}{\sqrt{6}}, \gamma = \pm\frac{1}{\sqrt{3}}$

Example 6.29 Check which of the following matrix is idempotent, involutory, nilpotent.

(i) $\begin{bmatrix} 1 & 1 & 3 \\ 5 & 2 & 6 \\ -2 & -1 & -3 \end{bmatrix}$ **(ii)** $\begin{bmatrix} -5 & -8 & 0 \\ 3 & 5 & 0 \\ 1 & 2 & -1 \end{bmatrix}$

(iii) $\begin{bmatrix} 2 & -2 & -4 \\ -1 & 3 & 4 \\ 1 & -2 & -3 \end{bmatrix}$

Solution

(i) Let $A = \begin{bmatrix} 1 & 1 & 3 \\ 5 & 2 & 6 \\ -2 & -1 & -3 \end{bmatrix}$

$$A^2 = A \cdot A = \begin{bmatrix} 1 & 1 & 3 \\ 5 & 2 & 6 \\ -2 & -1 & -3 \end{bmatrix}\begin{bmatrix} 1 & 1 & 3 \\ 5 & 2 & 6 \\ -2 & -1 & -3 \end{bmatrix}$$

$$= \begin{bmatrix} 0 & 0 & 0 \\ 3 & 3 & 9 \\ -1 & -1 & -3 \end{bmatrix}$$

Again,

$$A^3 = A \cdot A^2 = \begin{bmatrix} 1 & 1 & 3 \\ 5 & 2 & 6 \\ -2 & -1 & -3 \end{bmatrix}\begin{bmatrix} 0 & 0 & 0 \\ 3 & 3 & 9 \\ -1 & -1 & -3 \end{bmatrix}$$

$$= \begin{bmatrix} 0 & 0 & 0 \\ 0 & 0 & 0 \\ 0 & 0 & 0 \end{bmatrix} = 0$$

Thus, 3 is the least positive integer such that $A^3 = 0$, So A is a nilpotent matrix of order 3.

(ii) Let $A = \begin{bmatrix} -5 & -8 & 0 \\ 3 & 5 & 0 \\ 1 & 2 & -1 \end{bmatrix}$

$$A^2 = A \cdot A = \begin{bmatrix} -5 & -8 & 0 \\ 3 & 5 & 0 \\ 1 & 2 & -1 \end{bmatrix}\begin{bmatrix} -5 & -8 & 0 \\ 3 & 5 & 0 \\ 1 & 2 & -1 \end{bmatrix}$$

$$= \begin{bmatrix} 1 & 0 & 0 \\ 0 & 1 & 0 \\ 0 & 0 & 1 \end{bmatrix} = I$$

$\Rightarrow$ The matrix A is involutory.

(iii) Let $A=\begin{bmatrix}2 & -2 & -4\\ -1 & 3 & 4\\ 1 & -2 & -3\end{bmatrix}$, then

$$A^2=A\cdot A=\begin{bmatrix}2 & -2 & -4\\ -1 & 3 & 4\\ 1 & -2 & -3\end{bmatrix}\begin{bmatrix}2 & -2 & -4\\ -1 & 3 & 4\\ 1 & -2 & -3\end{bmatrix}$$

$$=\begin{bmatrix}2 & -2 & -4\\ -1 & 3 & 4\\ 1 & -2 & -3\end{bmatrix}=A$$

Hence A is an idempotent matrix.

Example 6.30 Find the inverse of $\begin{bmatrix}1 & 2 & 3\\ 2 & 3 & 4\\ 3 & 4 & 6\end{bmatrix}$.

Solution Let, $A=\begin{bmatrix}1 & 2 & 3\\ 2 & 3 & 4\\ 3 & 4 & 6\end{bmatrix}$

Let B be a matrix formed by the cofactor of the corresponding element of $|A|$.

Then $B=\begin{bmatrix}2 & 0 & -1\\ 0 & -3 & 2\\ -1 & 2 & -1\end{bmatrix}$

$$\text{adj}(A)=B'=\begin{bmatrix}2 & 0 & -1\\ 0 & -3 & 2\\ -1 & 2 & -1\end{bmatrix}$$

Also, $|A|=\begin{bmatrix}1 & 2 & 3\\ 2 & 3 & 4\\ 3 & 4 & 6\end{bmatrix}=1(2)-2(0)+3(-1)=-1$

$\Rightarrow |A|\neq 0$

Now, $A^{-1}=\dfrac{\text{adj}(A)}{|A|}=\begin{bmatrix}-2 & 0 & 1\\ 0 & 3 & -2\\ 1 & -2 & 1\end{bmatrix}$

Example 6.31 If the matrix $A=\begin{bmatrix}1 & 2 & 2\\ 2 & 1 & -2\\ a & 2 & b\end{bmatrix}$ satisfying $AA'=9I_3$, then find the value of $|a+b|$.

Solution Since

$$AA'=\begin{bmatrix}1 & 2 & 2\\ 2 & 1 & -2\\ a & 2 & b\end{bmatrix}\begin{bmatrix}1 & 2 & a\\ 2 & 1 & 2\\ 2 & -2 & b\end{bmatrix}$$

$$=\begin{bmatrix}9 & 0 & a+2b+4\\ 0 & 9 & 2a-2b+2\\ a+4+2b & 2a-2b+2 & a^2+b^2+4\end{bmatrix}=\begin{bmatrix}9 & 0 & 0\\ 0 & 9 & 0\\ 0 & 0 & 9\end{bmatrix}$$

Comparing the corresponding elements, we get

$$a+2b+4=0 \quad (1)$$

$$2a-2b+2=0 \quad (2)$$

$$a^2+b^2+4=9 \quad (3)$$

From (1) and (2), we get

$$a=-2 \text{ and } b=-1$$

$\Rightarrow |a+b|=3.$

Example 6.32 Find the real values of x for which the matrix $\begin{bmatrix}x+1 & 3 & 5\\ 1 & x+3 & 5\\ 1 & 3 & x+5\end{bmatrix}$ is non-singular.

Solution Let $A=\begin{bmatrix}x+1 & 3 & 5\\ 1 & x+3 & 5\\ 1 & 3 & x+5\end{bmatrix}$

$$\Rightarrow |A|=\begin{vmatrix}x+1 & 3 & 5\\ 1 & x+3 & 5\\ 1 & 3 & x+5\end{vmatrix}$$

Applying $C_1\to C_1+C_2+C_3$

$$|A|=\begin{vmatrix}x+9 & 3 & 5\\ x+9 & x+3 & 5\\ x+9 & 3 & x+5\end{vmatrix}$$

Now applying R_2-R_1 and R_3-R_1

$$|A|=\begin{vmatrix}x+9 & 3 & 5\\ 0 & x & 0\\ 0 & 0 & x\end{vmatrix}=x^2(x+9)$$

As A is a non-singular matrix,

$\Rightarrow |A|\neq 0$

$\Rightarrow x^2(x+9)\neq 0$

$\Rightarrow x\neq 0, -9$

Hence, $x\in R-\{0,-9\}$.

Example 6.33 Express the matrix $A=\begin{bmatrix}1 & 2 & 4\\ -2 & 5 & 3\\ -1 & 6 & 3\end{bmatrix}$ as the sum of a symmetric and a skew symmetric matrix.

Solution Since, $A=\begin{bmatrix}1 & 2 & 4\\ -2 & 5 & 3\\ -1 & 6 & 3\end{bmatrix}$

$$\Rightarrow A'=\begin{bmatrix}1 & -2 & -1\\ 2 & 5 & 6\\ 4 & 3 & 3\end{bmatrix}$$

Now, $\qquad A+A'=\begin{bmatrix}2 & 0 & 3\\ 0 & 10 & 9\\ 3 & 9 & 6\end{bmatrix} \qquad (1)$

And, $\qquad A-A'=\begin{bmatrix}0 & 4 & 5\\ -4 & 0 & -3\\ -5 & 3 & 0\end{bmatrix} \qquad (2)$

Adding (1) and (2), we get

$$2A=\begin{bmatrix}2 & 0 & 3\\ 0 & 10 & 9\\ 3 & 9 & 6\end{bmatrix}+\begin{bmatrix}0 & 4 & 5\\ -4 & 0 & -3\\ -5 & 3 & 0\end{bmatrix}$$

$$\Rightarrow A=\begin{bmatrix}1 & 0 & \frac{3}{2}\\ 0 & 5 & \frac{9}{2}\\ \frac{3}{2} & \frac{9}{2} & 3\end{bmatrix}+\begin{bmatrix}0 & 2 & \frac{5}{2}\\ -2 & 0 & -\frac{3}{2}\\ -\frac{5}{2} & \frac{3}{2} & 0\end{bmatrix}$$

$\Rightarrow A=$ Symmetric matrix + skew Symmetric matrix

Example 6.34 If $A=\begin{bmatrix}0 & -\tan\frac{\alpha}{2}\\ \tan\frac{\alpha}{2} & 0\end{bmatrix}$

and I is a (2×2) unit matrix, then prove that

$$(I-A)\begin{bmatrix}\cos\alpha & -\sin\alpha\\ \sin\alpha & \cos\alpha\end{bmatrix}=I+A$$

Solution Since, $(I-A)\begin{bmatrix}\cos\alpha & -\sin\alpha\\ \sin\alpha & \cos\alpha\end{bmatrix}$

$$=\begin{bmatrix}1 & \tan\frac{\alpha}{2}\\ -\tan\frac{\alpha}{2} & 1\end{bmatrix}\begin{bmatrix}\cos\alpha & -\sin\alpha\\ \sin\alpha & \cos\alpha\end{bmatrix}$$

$$=\begin{bmatrix}1 & \tan\frac{\alpha}{2}\\ -\tan\frac{\alpha}{2} & 1\end{bmatrix}\begin{bmatrix}\dfrac{1-\tan^2\frac{\alpha}{2}}{1+\tan^2\frac{\alpha}{2}} & \dfrac{-2\tan\frac{\alpha}{2}}{1+\tan^2\frac{\alpha}{2}}\\ \dfrac{2\tan\frac{\alpha}{2}}{1+\tan^2\frac{\alpha}{2}} & \dfrac{1-\tan^2\frac{\alpha}{2}}{1+\tan^2\frac{\alpha}{2}}\end{bmatrix}$$

$$=\begin{bmatrix}\dfrac{1-\tan^2\frac{\alpha}{2}}{1+\tan^2\frac{\alpha}{2}}+\dfrac{2\tan^2\frac{\alpha}{2}}{1+\tan^2\frac{\alpha}{2}} & \dfrac{-2\tan\frac{\alpha}{2}}{1+\tan^2\frac{\alpha}{2}}+\dfrac{\tan\frac{\alpha}{2}\left(1-\tan^2\frac{\alpha}{2}\right)}{1+\tan^2\frac{\alpha}{2}}\\ \dfrac{-\tan\frac{\alpha}{2}\left(1-\tan^2\frac{\alpha}{2}\right)}{1+\tan^2\frac{\alpha}{2}}+\dfrac{2\tan\frac{\alpha}{2}}{1+\tan^2\frac{\alpha}{2}} & \dfrac{2\tan^2\frac{\alpha}{2}}{1+\tan^2\frac{\alpha}{2}}+\dfrac{1-\tan^2\frac{\alpha}{2}}{1+\tan^2\frac{\alpha}{2}}\end{bmatrix}$$

$$=\begin{bmatrix}\dfrac{1+\tan^2\frac{\alpha}{2}}{1+\tan^2\frac{\alpha}{2}} & \dfrac{-\tan\frac{\alpha}{2}\left(1+\tan^2\frac{\alpha}{2}\right)}{1+\tan^2\frac{\alpha}{2}}\\ \dfrac{\tan\frac{\alpha}{2}\left(1+\tan^2\frac{\alpha}{2}\right)}{1+\tan^2\frac{\alpha}{2}} & \dfrac{1+\tan^2\frac{\alpha}{2}}{1+\tan^2\frac{\alpha}{2}}\end{bmatrix}$$

$$=\begin{bmatrix}1 & -\tan\frac{\alpha}{2}\\ \tan\frac{\alpha}{2} & 1\end{bmatrix}=I+A$$

Example 6.35 If $A=\begin{bmatrix}a & b & c\\ b & c & a\\ c & a & b\end{bmatrix}$, where a, b, c are real positive numbers. If $abc=1$ and $A^TA=I$, then find the value of $a^3+b^3+c^3$.

Solution Since, $A^T A = I$

$$\Rightarrow |A^T A| = |I|$$

$$\Rightarrow |A^2| = 1 \qquad (1)$$

$$\text{Now, } |A| = \begin{vmatrix} a & b & c \\ b & c & a \\ c & a & b \end{vmatrix} = a^3 + b^3 + c^3 - 3abc$$

$$\Rightarrow \left(a^3 + b^3 + c^3 - 3abc\right)^2 = 1$$

Since a, b, c are positive real numbers, using A.M. ≥ G.M.

$$\text{We get, } \frac{a^3 + b^3 + c^3}{3} \geq \left(a^3 b^3 c^3\right)^{1/3}$$

$$\Rightarrow a^3 + b^3 + c^3 \geq 3abc$$

$$\Rightarrow a^3 + b^3 + c^3 - 3abc = 1$$

$$\Rightarrow a^3 + b^3 + c^3 = 4$$

Example 6.36 Using the matrix method solve the following system of equations

$x + 2y + z = 7$, $x + 3z = 11$, $2x - 3y = 1$

Solution The given system of equations can be written as

$$\begin{bmatrix} 1 & 2 & 1 \\ 1 & 0 & 3 \\ 2 & -3 & 0 \end{bmatrix}\begin{bmatrix} x \\ y \\ z \end{bmatrix} = \begin{bmatrix} 7 \\ 11 \\ 1 \end{bmatrix}$$

$$\Rightarrow AX = B$$

$$\text{Where } A = \begin{bmatrix} 1 & 2 & 1 \\ 1 & 0 & 3 \\ 2 & -3 & 0 \end{bmatrix}, X = \begin{bmatrix} x \\ y \\ z \end{bmatrix}, B = \begin{bmatrix} 7 \\ 11 \\ 1 \end{bmatrix}$$

$$\text{Now, } |A| = \begin{vmatrix} 1 & 2 & 1 \\ 1 & 0 & 3 \\ 2 & -3 & 0 \end{vmatrix} = 18$$

⇒ The given system of equations has a unique solution given by $X = A^{-1}B$.

Let C be a matrix formed by the cofactors of the corresponding elements of $|A|$.

$$C = \begin{bmatrix} 9 & 6 & -3 \\ -3 & -2 & 7 \\ 6 & -2 & -2 \end{bmatrix}$$

$$\text{adj}(A) = C^T = \begin{bmatrix} 9 & -3 & 6 \\ 6 & -2 & -2 \\ -3 & 7 & -2 \end{bmatrix}$$

$$\text{Now, } A^{-1} = \frac{\text{adj}(A)}{|A|} = \frac{1}{18}\begin{bmatrix} 9 & -3 & 6 \\ 6 & -2 & -2 \\ -3 & 7 & -2 \end{bmatrix}$$

$$\text{Now, } X = A^{-1}B = \frac{1}{18}\begin{bmatrix} 9 & -3 & 6 \\ 6 & -2 & -2 \\ -3 & 7 & -2 \end{bmatrix}\begin{bmatrix} 7 \\ 11 \\ 1 \end{bmatrix}$$

$$= \frac{1}{18}\begin{bmatrix} 63 - 33 + 6 \\ 42 - 22 - 2 \\ -21 + 77 - 2 \end{bmatrix}$$

$$\Rightarrow \begin{bmatrix} x \\ y \\ z \end{bmatrix} = \frac{1}{18}\begin{bmatrix} 36 \\ 18 \\ 54 \end{bmatrix} = \begin{bmatrix} 2 \\ 1 \\ 3 \end{bmatrix}$$

$$\Rightarrow x = 2, y = 1, z = 3.$$

Example 6.37 If $A = \begin{bmatrix} x & 3 & 2 \\ 1 & y & 4 \\ 2 & 2 & z \end{bmatrix}$ and if $xyz = 60$ and $8x + 4y + 3z = 20$, then prove that $A(\text{adj } A) = 68I_3$, where I_3 is a unit matrix.

Solution We know that

$$A(\text{adj } A) = |A| I$$

$$\text{Now, } |A| = \begin{vmatrix} x & 3 & 2 \\ 1 & y & 4 \\ 2 & 2 & z \end{vmatrix}$$

$$= x(yz - 8) - 3(z - 8) + 2(2 - 2y)$$

$$= xyz - (8x + 4y + 3z) + 28$$

$$= 60 - 20 + 28 = 68.$$

$$\text{Hence } A(\text{adj } A) = 68\begin{vmatrix} 1 & 0 & 0 \\ 0 & 1 & 0 \\ 0 & 0 & 1 \end{vmatrix} = 68I_3$$

Example 6.38 If $\begin{bmatrix} a & b \\ c & 1-a \end{bmatrix}$ is an idempotent matrix and $f(x) = x - x^2$ and $bc = \frac{1}{4}$, then find the value of $f(a)$.

Solution Let, $A=\begin{bmatrix} a & b \\ c & 1-a \end{bmatrix}$.

As A is an indempotent matrix, hence $A^2 = A$.

$$\Rightarrow \begin{bmatrix} a & b \\ c & 1-a \end{bmatrix}\begin{bmatrix} a & b \\ c & 1-a \end{bmatrix}=\begin{bmatrix} a & b \\ c & 1-a \end{bmatrix}$$

$$\Rightarrow \begin{bmatrix} a^2+bc & ab+b-ab \\ ac+c-ac & bc+(1-a)^2 \end{bmatrix}=\begin{bmatrix} a & b \\ c & 1-a \end{bmatrix}$$

$$\Rightarrow \begin{bmatrix} a^2+bc & b \\ c & bc+(1-a)^2 \end{bmatrix}=\begin{bmatrix} a & b \\ c & 1-a \end{bmatrix}$$

$$\Rightarrow a^2+bc=a$$

$$\Rightarrow a-a^2=bc=\frac{1}{4}$$

$$\Rightarrow f(a)=\frac{1}{4}$$

Example 6.39 Let $a_r = r\left(7_{C_r}\right)$, $b_r=(7-r)\left(7_{C_r}\right)$ and $A_r=\begin{bmatrix} a_r & 0 \\ 0 & b_r \end{bmatrix}$. If $A=\sum_{r=0}^{7}\begin{bmatrix} a_r & 0 \\ 0 & b_r \end{bmatrix}=\begin{bmatrix} a & 0 \\ 0 & b \end{bmatrix}$, then find the value of $a+b$.

Solution Since $\sum_{r=0}^{7}A_r=\sum_{r=0}^{7}\begin{bmatrix} a_r & 0 \\ 0 & b_r \end{bmatrix}=\begin{bmatrix} a & 0 \\ 0 & b \end{bmatrix}$

$$\Rightarrow a=\sum_{r=0}^{7}a_r=\sum_{r=0}^{7}r\left(7_{C_r}\right)$$

and, $b=\sum_{r=0}^{7}b_r=\sum_{r=0}^{7}(7-r)\left(7_{C_r}\right)$

$$\Rightarrow a+b=7\sum_{r=0}^{7}7_{C_r}=7\left(2^7\right)=7\times128=896$$

Example 6.40 If in a ΔABC, $\begin{vmatrix} 1 & a & b \\ 1 & c & a \\ 1 & b & c \end{vmatrix}=0$, then find the value of $64\left(\sin^2 A+\sin^2 B+\sin^2 C\right)$.

Solution Since, $\begin{vmatrix} 1 & a & b \\ 1 & c & a \\ 1 & b & c \end{vmatrix}=0$

$$\Rightarrow 1\left(c^2-ab\right)-a(c-a)+b(b-c)=0$$

$$\Rightarrow c^2-ab-ac+a^2+b^2-bc=0$$

$$\Rightarrow a^2+b^2+c^2-ab-bc-ca=0$$

$$\Rightarrow \frac{1}{2}\left\{(a-b)^2+(b-c)^2+(c-a)^2\right\}=0$$

Which is possible only when $a=b=c$

$\Rightarrow$ Triangle is equilateral, hence $A=B=C=\frac{\pi}{3}$

Therefore, $64\left(\sin^2 A+\sin^2 B+\sin^2 C\right)$

$$=64\left(\frac{3}{4}+\frac{3}{4}+\frac{3}{4}\right)=16\times9=144$$

Example 6.41 If the matrix $A=\frac{1}{\sqrt{2}}\begin{bmatrix} 1 & i \\ -i & a \end{bmatrix}$, where $i=\sqrt{-1}$ is unitary, then find the value of a.

Solution Since, $A=\frac{1}{\sqrt{2}}\begin{bmatrix} 1 & i \\ -i & a \end{bmatrix}$

$$\bar{A}=\frac{1}{\sqrt{2}}\begin{bmatrix} 1 & -i \\ i & a \end{bmatrix}$$

$$\Rightarrow \left(\bar{A}\right)^T=\frac{1}{\sqrt{2}}\begin{bmatrix} 1 & i \\ -i & a \end{bmatrix}=A^\theta$$

$$\Rightarrow AA^\theta=I$$

$$\Rightarrow \frac{1}{\sqrt{2}}\begin{bmatrix} 1 & i \\ -i & a \end{bmatrix}.\frac{1}{\sqrt{2}}\begin{bmatrix} 1 & i \\ -i & a \end{bmatrix}=\begin{bmatrix} 1 & 0 \\ 0 & 1 \end{bmatrix}$$

$$\Rightarrow \frac{1}{2}\begin{bmatrix} 2 & i(1+a) \\ -i(1+a) & 1+a^2 \end{bmatrix}=\begin{bmatrix} 1 & 0 \\ 0 & 1 \end{bmatrix}$$

$$\Rightarrow 1+a=0$$

$$\Rightarrow a=-1$$

Example 6.42 If $P=\begin{bmatrix} \frac{\sqrt{3}}{2} & \frac{1}{2} \\ -\frac{1}{2} & \frac{\sqrt{3}}{2} \end{bmatrix}$, $A=\begin{bmatrix} 1 & 1 \\ 0 & 1 \end{bmatrix}$ and $Q=PAP^T$, then prove that $P^T\left(Q^{2005}\right)P=\begin{bmatrix} 1 & 2005 \\ 0 & 1 \end{bmatrix}$

Solution Since, $Q=PAP^T$

$$\Rightarrow P^TQ=P^TPAP^T\left[PP^T=I\right]$$

$$=AP^T$$

$$\Rightarrow P^T\left(Q^{2005}\right)P=AP^T\left(Q^{2004}\right)P$$

$$=A^2P^T\left(Q^{2003}\right)P$$

$= A^3 P^T \left(Q^{2002}\right) P$

= ____________

= ____________

$= A^{2004} P^T (PA)\left[Q = PAP^T \text{ and } QP = PA\right]$

$= A^{2005}$

$= \begin{bmatrix} 1 & 2005 \\ 0 & 1 \end{bmatrix}$

6.3 Know Your Concepts

1. Find the value of the determinant
$$\begin{vmatrix} 265 & 240 & 219 \\ 240 & 225 & 198 \\ 219 & 198 & 181 \end{vmatrix}.$$

2. Find the value of determinant
$$\begin{vmatrix} \sqrt{13}+\sqrt{3} & 2\sqrt{5} & \sqrt{5} \\ \sqrt{15}+\sqrt{26} & 5 & \sqrt{10} \\ 3+\sqrt{65} & \sqrt{15} & 5 \end{vmatrix}.$$

3. If $\begin{vmatrix} a+b+2c & a & b \\ c & b+c+2a & b \\ c & a & c+a+2b \end{vmatrix}$ $= K\left(a+b+c\right)^3$, find the value of K.

4. If $D_1 = \begin{vmatrix} 1 & 1 & 1 \\ x^2 & y^2 & z^2 \\ x & y & z \end{vmatrix}$ and $D_2 = \begin{vmatrix} 1 & 1 & 1 \\ yz & xz & xy \\ x & y & z \end{vmatrix}$ then find the value of $\dfrac{D_1}{D_2}$.

5. If $\begin{vmatrix} x & 3 & 7 \\ 2 & x & 2 \\ 7 & 6 & x \end{vmatrix} = 0$, then find the value of x.

6. If $\begin{vmatrix} x^3+4x & x+3 & x-2 \\ x-2 & 5x & x-1 \\ x-3 & x+2 & 4x \end{vmatrix} = ax^5 + bx^4 + cx^3 + dx^2 + ex + f$, then find the value of f.

7. If the value of a third order determinant is 11, then find the value of the determinant formed by its cofactors.

8. Find the maximum value of the determinant
$$\begin{vmatrix} 1 & 3\cos\theta & 1 \\ \sin\theta & 1 & 3\cos\theta \\ 1 & \sin\theta & 1 \end{vmatrix}.$$

9. If α, β, γ are the roots of the equation $x^3 + px + q = 0$, then find the value of the determinant
$$\begin{vmatrix} \alpha & \beta & \gamma \\ \beta & \gamma & \alpha \\ \gamma & \alpha & \beta \end{vmatrix}.$$

10. Find the solution of the system of equations
$x + 2y + 3z = 4$
$2x + 3y + 4z = 5$
$3x + 4y + 5z = 6$

11. If the system of equations
$2x + 5y + 8z = 0$
$x + 4y + 7z = 0$
$6x + 4y - \lambda z = 0$
has a non-trivial solution, then find the value of λ.

12. If $\begin{vmatrix} a^2 & b^2 & c^2 \\ (a+1)^2 & (b+1)^2 & (c+1)^2 \\ (a-1)^2 & (b-1)^2 & (c-1)^2 \end{vmatrix}$ $= K\begin{vmatrix} a^2 & b^2 & c^2 \\ a & b & c \\ 1 & 1 & 1 \end{vmatrix}$, then find the value of K.

13. If $a_1, a_2, a_3, \ldots, a_n, \ldots$ are in G.P., then find the value of the determinant
$$\begin{vmatrix} \log a_n & \log a_{n+1} & \log a_{n+2} \\ \log a_{n+3} & \log a_{n+4} & \log a_{n+5} \\ \log a_{n+6} & \log a_{n+7} & \log a_{n+8} \end{vmatrix}.$$

14. Prove that
$$\begin{vmatrix} a-b-c & 2a & 2a \\ 2b & b-c-a & 2b \\ 2c & 2c & c-a-b \end{vmatrix} = \left(a+b+c\right)^3.$$

15. If l, m, n are real numbers such that $l^2 + m^2 + n^2 = 0$, then find the value of determinant
$$\begin{vmatrix} 1+l^2 & lm & ln \\ lm & 1+m^2 & mn \\ ln & mn & 1+n^2 \end{vmatrix}.$$

16. If $\begin{vmatrix} xc_1 & xc_2 & xc_3 \\ yc_1 & yc_2 & yc_3 \\ zc_1 & zc_2 & zc_3 \end{vmatrix}$ $= \frac{\alpha}{\beta} xyz(x-y)(y-z)(z-x)$, then find the value of $\alpha\beta$.

17. If $A = \begin{vmatrix} a & b & 2c \\ d & e & 2f \\ l & m & 2n \end{vmatrix}$ and $B = \begin{vmatrix} f & 2d & e \\ 2n & 4l & 2m \\ c & 2a & b \end{vmatrix}$, then find the value of $\frac{B}{A}$.

18. If A, B and C are the angles of a non-right angled triangle ABC, then find the value of

$$\begin{vmatrix} \tan A & 1 & 1 \\ 1 & \tan B & 1 \\ 1 & 1 & \tan C \end{vmatrix}$$

19. Let $f(x) =$

$$\begin{vmatrix} \cos(x+x^2) & \sin(x+x^2) & -\cos(x+x^2) \\ \sin(x-x^2) & \cos(x-x^2) & \sin(x-x^2) \\ \sin 2x & 0 & \sin(2x^2) \end{vmatrix}$$

then find the value of $f'(0)$.

20. If $pqr \neq 0$ and the system of equations

$(p+a)x+by+cz=0$

$ax+(q+b)y+cz=0$

$ax+by+(r+c)z=0$

has a non-trivial solution,then find the value of $\frac{a}{p}+\frac{b}{q}+\frac{c}{r}$.

21. Let $A = \begin{bmatrix} x^2 & 6 & 8 \\ 3 & y^2 & 9 \\ 4 & 5 & z^2 \end{bmatrix}$ and

$B = \begin{bmatrix} 2x & 3 & 5 \\ 2 & 2y & 6 \\ 1 & 4 & 2z-3 \end{bmatrix}$

If $t_r(A) = t_r(B)$, then find the value of $x + y + z$.

22. If A is a square matrix of order n such that $|\text{adj}(\text{adj } A)| = |A|^9$, then find the value of n.

23. If A is an idempotent matrix and $A^n + B = I$, then find the value of AB.

24. If $A = \begin{bmatrix} 1 & 2 \\ 1 & 1 \end{bmatrix}$, then find $(I+A)(I-A)^{-1}$.

25. Find x so that $\begin{bmatrix} 1 & x & 1 \end{bmatrix} \begin{bmatrix} 1 & 3 & 2 \\ 0 & 5 & 1 \\ 0 & 3 & 2 \end{bmatrix} \begin{bmatrix} 1 \\ 1 \\ x \end{bmatrix} = 0$.

26. If $\begin{bmatrix} 1 & 2 & a \\ 0 & 1 & 4 \\ 0 & 0 & 1 \end{bmatrix}^n = \begin{bmatrix} 1 & 18 & 2007 \\ 0 & 1 & 36 \\ 0 & 0 & 1 \end{bmatrix}$, then evalute $(n + a)$.

27. If $a = \cos\theta + i \sin\theta$, $b = \cos 2\theta - i \sin 2\theta$, $c = \cos 3\theta + i \sin 3\theta$ and if $A = \begin{bmatrix} a & b & c \\ b & c & a \\ c & a & b \end{bmatrix}$, then find the value of θ if A is non-invertible.

28. If $\alpha, \beta, \gamma \in \left(0, \frac{\pi}{2}\right)$ and

$$\begin{bmatrix} 0 & 2\sin\beta & \tan\gamma \\ \cos\alpha & \sin\beta & -\tan\gamma \\ \cos\alpha & -\sin\beta & \tan\gamma \end{bmatrix}$$

is orthogonal, then find the value of $\alpha + \gamma$.

29. Suppose $a, b, c \in R$ and $abc = 1$ if

$A = \begin{bmatrix} 2a & b & c \\ b & 2c & a \\ c & a & 2b \end{bmatrix}$ is such that $AA' = 4I$ and $|A| > 0$ then find the value of $(a^3+b^3+c^3)$.

30. If a, b and c are all different from zero such that $\frac{1}{a}+\frac{1}{b}+\frac{1}{c} = 0$, then prove that the matrix

$A = \begin{bmatrix} 1+a & 1 & 1 \\ 1 & 1+b & 1 \\ 1 & 1 & 1+c \end{bmatrix}$ is non-singular.

6.4 JEE Main: Sample Questions

Single Answer Type

1. Let A is a matrix of order 3 defined as, $A = [a_{ij}]_{3\times 3}$ where $a_{ij} = \lim\limits_{x\to 0} \dfrac{\sin(ix)}{\tan(jx)}$, $\forall\ 1 \le i, j \le 3$. Then A^2 is
(A) $4A$ (B) $3A$ (C) $2A$ (D) A

2. For α, β, γ, let $A = \begin{bmatrix} \alpha^2 & 6 & 8 \\ 3 & \beta^2 & 9 \\ 4 & 5 & \gamma^2 \end{bmatrix}$ and $B = \begin{bmatrix} 2\alpha & 3 & 5 \\ 2 & 2\beta & 6 \\ 1 & 4 & 2\gamma - 3 \end{bmatrix}$. If $T_r(A) = T_r(B)$ then the value of $\left(\dfrac{1}{\alpha} + \dfrac{1}{\beta} + \dfrac{1}{\gamma}\right)$ is
(A) 1 (B) 2 (C) 3 (D) 4
(where $T_r(A)$ denotes trace of matrix A)

3. If the system of equations
$(\tan\alpha)\,x + (\cot\alpha)\,y + (8\cos 2\alpha)\,z = 0$
$(\cot\alpha)\,x + (8\cos 2\alpha)\,y + (\tan\alpha)\,z = 0$
$(8\cos 2\alpha)\,x + (\tan\alpha)\,y + (\cot\alpha)\,z = 0$
Has a non-trivial solution, then $\sin(4\alpha)$ is equal to
(A) $\dfrac{-\sqrt{3}}{2}$ (B) -1 (C) $\dfrac{-1}{2}$ (D) $\dfrac{1}{2}$

4. The number of positive integral values of p for which the system of equations $x + 2y + z = px$, $-3x + py = 2z$ and $2x - 2y + z = -pz$ has non-trivial solutions, is
(A) 0 (B) 1 (C) 2 (D) 3

5. If M is a square matrix of order 2, then $-\text{tr.}(M^2) + (\text{tr.}(M))^2$ is equal to
(A) $\dfrac{\det.(M)}{2}$ (B) $2\det.(M)$
(C) $3\det.(M)$ (D) $4\det.(M)$

6. The system of equations
$px + (p + 1)\,y + (p - 1)\,z = 0$
$(p + 1)\,x + py + (p + 2)\,z = 0$
$(p - 1)\,x + (p + 2)\,y + pz = 0$
has non-trivial solutions for
(A) exactly two real values of p
(B) no real value of p
(C) exactly one real value of p
(D) infinitely many real values of p

7. If the system of linear equations $(a - 1)\,x + z = \alpha$, $x + (b - 1)\,y = \beta$ and $y + (c - 1)\,z = \gamma$ where $a, b, c \in I$ does not have a unique solution, then the maximum possible value of $|a + b + c|$ is:
(A) 0 (B) 1 (C) 3 (D) 4

8. For $\theta = \dfrac{3\pi}{5}$, let $B = [b_{ij}]$ be a square matrix of order 2 such that $b_{ij} = \begin{cases} \cos\theta, & i = j \\ \cos\left(\dfrac{j\pi}{2} + \theta\right), & i > j \\ \sin\left(\dfrac{j\pi}{2} - \theta\right), & i < j \end{cases}$
The trace of matrix B^5 is equal to
(A) -2 (B) -1 (C) 1 (D) 2

9. Let M denote the matrix $\begin{pmatrix} 0 & i \\ i & 0 \end{pmatrix}$, where $i^2 = -1$, and let I denote the identity matrix $\begin{pmatrix} 1 & 0 \\ 0 & 1 \end{pmatrix}$. Then the matrix $I + M + M^2 + M^3 + M^4 + \ldots M^{2010}$ is equal to
(A) $\begin{pmatrix} 0 & 0 \\ 0 & 0 \end{pmatrix}$ (B) $\begin{pmatrix} 0 & i \\ i & 0 \end{pmatrix}$
(C) $\begin{pmatrix} 1 & i \\ i & 1 \end{pmatrix}$ (D) $\begin{pmatrix} -1 & 0 \\ 0 & -1 \end{pmatrix}$

10. Let $A = \begin{bmatrix} 2 & 2 & 1 \\ 2 & 5 & 2 \\ 1 & 2 & 2 \end{bmatrix}$ and $B = \begin{bmatrix} -x & -y & z \\ 0 & y & 2z \\ x & -y & z \end{bmatrix}$ where $x, y, z \in R$. If $B^T AB = \begin{bmatrix} 8 & 0 & 0 \\ 0 & 27 & 0 \\ 0 & 0 & 42 \end{bmatrix}$ then the number of ordered triplets (x, y, z) is
(A) 2 (B) 6 (C) 8 (D) 9

11. Let $P = \begin{bmatrix} 1 & 0 & 0 \\ 4 & 1 & 0 \\ 16 & 4 & 1 \end{bmatrix}$ and I be the identity matrix of order 3. If $Q = [q_{ij}]$ is a matrix such that $P^{50} - Q = I$, then $\dfrac{q_{31} + q_{32}}{q_{21}}$ equals

(A) 52 (B) 103 (C) 201 (D) 205

12. The value of a for which the system of equations $a^3x + (a + 1)^3y + (a + 2)^3z = 0$, $ax + (a + 1)y + (a + 2)z = 0$ and $x + y + z = 0$ has a non-zero solution is

(A) 1 (B) 0 (C) –1 (D) 2

13. Let $\Delta_0 = \begin{vmatrix} a_{11} & a_{12} & a_{13} \\ a_{21} & a_{22} & a_{23} \\ a_{31} & a_{32} & a_{33} \end{vmatrix}$ and let Δ_1 denote the determinant formed by the cofactors of elements of Δ_0 and Δ_2 denote the determinant formed by the cofactor of Δ_1, similarly Δ_n denotes the determinant formed by the cofactors of Δ_{n-1} then the determinant value of Δ_n is

(A) Δ_0^{2n} (B) $\Delta_0^{2^n}$ (C) $\Delta_0^{n^2}$ (D) Δ_0^2

14. Three distinct points $P(3u^2, 2u^3)$; $Q(3v^2, 2v^3)$ and $R(3w^2, 2w^3)$ are collinear. Then

(A) $uv + vw + wu = 0$
(B) $uv + vw + wu = 3$
(C) $uv + vw + wu = 2$
(D) $uv + vw + wu = 1$

15. Let a determinant be given by $A = \begin{vmatrix} a & b & c \\ p & q & r \\ x & y & z \end{vmatrix}$ and suppose that det. $A = 6$.

If $B = \begin{vmatrix} p+x & q+y & r+z \\ a+x & b+y & c+z \\ a+p & b+q & c+r \end{vmatrix}$ then

(A) det. $B = 6$ (B) det. $B = -6$
(C) det. $B = 12$ (D) det. $B = -12$

16. The values of θ, λ for which the following equations $(\sin\theta)x - (\cos\theta)y + (\lambda + 1)z = 0$, $(\cos\theta)x + (\sin\theta)y - \lambda z = 0$ and $\lambda x + (\lambda + 1)y + (\cos\theta)z = 0$ have a non trivial solution, is

(A) $\theta = n\pi$, $\lambda \in R - \{0\}$, $n \in I$
(B) $\theta = 2n\pi$, λ is any rational number, $n \in I$
(C) $\theta = (2n + 1)\pi$, $\lambda \in R^+$, $n \in I$
(D) $\theta = (2n + 1)\dfrac{\pi}{2}$, $\lambda \in R$, $n \in I$

17. The system of equations
$x - y\cos\theta + z\cos 2\theta = 0$
$-x\cos\theta + y - z\cos\theta = 0$
$x\cos 2\theta - y\cos\theta + z = 0$
has a non-trivial solution for θ equals

(A) $n\pi$ only, $n \in I$
(B) $n\pi + \dfrac{\pi}{4}$ only, $n \in I$
(C) $(2n - 1)\dfrac{\pi}{2}$ only, $n \in I$
(D) all values of θ

18. The following system of equations $3x - 7y + 5z = 3$, $3x + y + 5z = 7$ and $2x + 3y + 5z = 5$ is

(A) consistent with a trivial solution
(B) consistent with a unique non-trivial solution
(C) consistent with infinite solutions
(D) inconsistent with no solution

19. If the system of equations $x - 2y + z = a$, $2x + y - 2z = b$ and $x + 3y - 3z = c$ has at least one solution, then the relationship between a, b and c is

(A) $a + b + c = 0$ (B) $a - b + c = 0$
(C) $-a + b + c = 0$ (D) $a + b - c = 0$

20. Three-digit numbers $x17$, $3y6$ and $12z$ where x, y, z are integers from 0 to 9, are divisible by a fixed constant k. Then the determinant $\begin{vmatrix} x & 3 & 1 \\ 7 & 6 & z \\ 1 & y & 2 \end{vmatrix}$ must be divisible by

(A) k (B) k^2 (C) k^3 (D) k^2

21. Let $N = \begin{vmatrix} 28 & 25 & 38 \\ 42 & 38 & 65 \\ 56 & 47 & 83 \end{vmatrix}$, then the number of ways in which N can be resolved as a product of two divisors which are relatively prime is

(A) 4 (B) 8 (C) 9 (D) 16

22. If the system of linear equations
$x + 2ay + az = 0$
$x + 3by + bz = 0$
$x + 4cy + cz = 0$

has a non-zero solution, then a, b, c (taken in that order)

(A) are in G.P.
(B) are in H.P.
(C) satisfy $a + 2b + 3c = 0$
(D) are in A.P.

23. Let α, β, γ be the real roots of the equation $x^3 + ax^2 + bx + c = 0$ ($a, b, c \in R$ and $a \neq 0$). If the system of equations (in u, v and w) given by

$\alpha u + \beta v + \gamma w = 0$
$\beta u + \gamma v + \alpha w = 0$
$\gamma u + \alpha v + \beta w = 0$

has non-trivial solutions, then a^2 equals

(A) b (B) $2b$ (C) $3b$ (D) $4b$

24. Let $\Delta_1 = \begin{vmatrix} ap^2 & 2ap & 1 \\ aq^2 & 2aq & 1 \\ ar^2 & 2ar & 1 \end{vmatrix}$ and

$\Delta_2 = \begin{vmatrix} apq & a(p+q) & 1 \\ aqr & a(q+r) & 1 \\ arp & a(r+p) & 1 \end{vmatrix}$ then

(A) $\Delta_1 = \Delta_2$ (B) $\Delta_2 = 2\Delta_1$
(C) $\Delta_1 = 2\Delta_2$ (D) $\Delta_1 + 2\Delta_2 = 0$

25. Let

$F(x) = \begin{vmatrix} 1 & 1+\sin x & 1+\sin x+\cos x \\ 2 & 3+2\sin x & 4+3\sin x+2\cos x \\ 3 & 6+3\sin x & 10+6\sin x+3\cos x \end{vmatrix}$,

$x \in R$ then $F'\left(\frac{\pi}{2}\right)$ is equal to

(A) –1 (B) 0 (C) 1 (D) 2

26. Let $\Delta_1 = \begin{vmatrix} a_{11} & a_{12} & a_{13} \\ a_{21} & a_{22} & a_{23} \\ a_{31} & a_{32} & a_{33} \end{vmatrix}$, $\Delta_1 \neq 0$

$\Delta_2 = \begin{vmatrix} b_{11} & b_{12} & b_{13} \\ b_{21} & b_{22} & b_{23} \\ b_{31} & b_{32} & b_{33} \end{vmatrix}$ where b_{ij} is the cofactor

of a_{ij} $\forall$ $i, j = 1, 2, 3$ and $\Delta_3 = \begin{vmatrix} c_{11} & c_{12} & c_{13} \\ c_{21} & c_{22} & c_{23} \\ c_{31} & c_{32} & c_{33} \end{vmatrix}$

where c_{ij} is the cofactor of b_{ij} $\forall$ $i, j = 1, 2, 3$, then which one of the following is always correct?

(A) $\Delta_1, \Delta_2, \Delta_3$ are in A.P.
(B) $\Delta_1, \Delta_2, \Delta_3$ are in G.P.
(C) $\Delta_1^2 = \frac{\Delta_3}{\Delta_2}$
(D) $\Delta_1 = \frac{\Delta_2}{\sqrt{\Delta_3}}$

27. If the product of n matrices

$\begin{bmatrix} 1 & 1 \\ 0 & 1 \end{bmatrix}\begin{bmatrix} 1 & 2 \\ 0 & 1 \end{bmatrix}\begin{bmatrix} 1 & 3 \\ 0 & 1 \end{bmatrix}\cdots\begin{bmatrix} 1 & n \\ 0 & 1 \end{bmatrix}$

is equal to the matrix $\begin{bmatrix} 1 & 378 \\ 0 & 1 \end{bmatrix}$ then the value of n is equal to

(A) 26 (B) 27
(C) 377 (D) 378

28. If $A = \begin{bmatrix} 1 & 2 & x \\ 0 & 1 & 0 \\ 0 & 0 & 1 \end{bmatrix}$ and $B = \begin{bmatrix} 1 & -2 & y \\ 0 & 1 & 0 \\ 0 & 0 & 1 \end{bmatrix}$ and $AB = I_3$, then $x + y$ equals

(A) 0 (B) –1 (C) 2 (D) –2

29. If $A = \begin{bmatrix} 3 & 4 \\ 1 & -6 \end{bmatrix}$ and $B = \begin{bmatrix} -2 & 5 \\ 6 & 1 \end{bmatrix}$ then X such that $A + 2X = B$ equals

(A) $\begin{bmatrix} 2 & 3 \\ -1 & 0 \end{bmatrix}$ (B) $\begin{bmatrix} 3 & 5 \\ -1 & 0 \end{bmatrix}$
(C) $\begin{bmatrix} 5 & 2 \\ -1 & 0 \end{bmatrix}$ (D) none of these

30. If $A = \begin{bmatrix} a & b \\ c & d \end{bmatrix}$ satisfies the equations $x^2 - (a + d)\, x + k = 0$, then

(A) $k = bc$
(B) $k = ad$
(C) $k = ad - bc$
(D) $k = a^2 + b^2 + c^2 + d^2$

31. If $\begin{bmatrix} x & -5 & -1 \end{bmatrix}\begin{bmatrix} 1 & 0 & 2 \\ 0 & 2 & 1 \\ 2 & 0 & 3 \end{bmatrix}\begin{bmatrix} x \\ 4 \\ 1 \end{bmatrix} = 0$, then x equals

(A) $\pm 2\sqrt{3}$ (B) $\pm 4\sqrt{3}$
(C) $\pm 3\sqrt{2}$ (D) $\pm 4\sqrt{2}$

32. Let $A+2B=\begin{bmatrix}1 & 2 & 0\\ 6 & -3 & 3\\ -5 & 3 & 1\end{bmatrix}$ and

$2A-B=\begin{bmatrix}2 & -1 & 5\\ 2 & -1 & 6\\ 0 & 1 & 2\end{bmatrix}$

then $Tr(A)-Tr(B)$ has the value equal to

(A) 0 (B) 1 (C) 2 (D) 3

33. The number of solutions of the matrix equation $X^2=I$ other then I, is

(A) 0 (B) 1

(C) 2 (D) more than 2

(where I is the 2×2 unit matrix)

34. Given $A=\begin{bmatrix}1 & 3\\ 2 & 2\end{bmatrix}$; $I=\begin{bmatrix}1 & 0\\ 0 & 1\end{bmatrix}$. If $A-\lambda I$ is a singular matrix then

(A) $\lambda\in\phi$

(B) $\lambda^2-3\lambda-4=0$

(C) $\lambda^2+3\lambda+4=0$

(D) $\lambda^2-3\lambda-6=0$

35. Let $A=\begin{bmatrix}1 & \sin\theta & 1\\ -\sin\theta & 1 & \sin\theta\\ -1 & -\sin\theta & 1\end{bmatrix}$,

where $0\le\theta<2\pi$, then

(A) Det. $(A)=0$

(B) Det. $(A)\in(0,\infty)$

(C) Det. $(A)\in[2,4]$

(D) Det. $(A)\in[2,\infty)$

36. Number of real values of x for which the matrix $A=\begin{bmatrix}3-x & 2 & 2\\ 2 & 4-x & 1\\ -2 & -4 & -1-x\end{bmatrix}$ is singular, is

(A) 1 (B) 2

(C) 3 (D) infinite

37. $A=\begin{bmatrix}1 & \tan x\\ -\tan x & 1\end{bmatrix}$ then let us define a function $f(x)=$ det. (A^TA^{-1}) then which of the following can not be the value of $\underbrace{f(f(f(f\ldots f(x))))}_{n\text{ times}}$ is $(n\ge 2)$

(A) $f^n(x)$ (B) 1

(C) $f^{n-1}(x)$ (D) $nf(x)$

38. A is a 2×2 matrix such that $A\begin{bmatrix}1\\ -1\end{bmatrix}=\begin{bmatrix}-1\\ 2\end{bmatrix}$ and $A^2\begin{bmatrix}1\\ -1\end{bmatrix}=\begin{bmatrix}1\\ 0\end{bmatrix}$. The sum of the elements of A, is

(A) –1 (B) 0 (C) 2 (D) 5

39. In a square matrix A of order 3 the elements, a_{ii}'s are the sum of the roots of the equation $x^2-(a+b)x+ab=0$; $a_{i,\,i+1}$'s are the product of the roots, $a_{i,\,i-1}$'s are all unity and the rest of the elements are all zero. The value of the det. (A) is equal to

(A) 0 (B) $(a+b)^3$

(C) a^3-b^3 (D) $(a^2+b^2)(a+b)$

40. Let D_k is the $k\times k$ matrix with 0's in the main diagonal, unity as the element of 1st row and $(f(k))^{\text{th}}$ column and k for all other entries. If $f(x)=x-\{x\}$ where $\{x\}$ denotes the fractional part function then the value of det. (D_2) + det. (D_3) equals

(A) 32 (B) 34 (C) 36 (D) 44

41. For a matrix $A=\begin{bmatrix}1 & 2r-1\\ 0 & 1\end{bmatrix}$, the value of $\prod_{r=1}^{50}\begin{bmatrix}1 & 2r-1\\ 0 & 1\end{bmatrix}$ is equal to

(A) $\begin{bmatrix}1 & 100\\ 0 & 1\end{bmatrix}$ (B) $\begin{bmatrix}1 & 4950\\ 0 & 1\end{bmatrix}$

(C) $\begin{bmatrix}1 & 5050\\ 0 & 1\end{bmatrix}$ (D) $\begin{bmatrix}1 & 2500\\ 0 & 1\end{bmatrix}$

42. A is an involutory matrix given by $A=\begin{bmatrix}0 & 1 & -1\\ 4 & -3 & 4\\ 3 & -3 & 4\end{bmatrix}$ then the inverse of $\frac{A}{2}$ will be

(A) $2A$ (B) $\frac{A^{-1}}{2}$ (C) $\frac{A}{2}$ (D) A^2

43. Let three matrices be $A=\begin{bmatrix}2 & 1\\ 4 & 1\end{bmatrix}$; $B=\begin{bmatrix}3 & 4\\ 2 & 3\end{bmatrix}$ and $C=\begin{bmatrix}3 & -4\\ -2 & 3\end{bmatrix}$. $t_r(A)+t_r\left(\frac{ABC}{2}\right)+t_r\left(\frac{A(BC)^2}{4}\right)+t_r\left(\frac{A(BC)^3}{8}\right)+\ldots+\infty=$

(A) 6 (B) 9 (C) 12 (D) 44

44. If A is an idempotent matrix satisfying, $(I - 0.4A)^{-1} = I - \alpha A$ where I is the unit matrix of the same order as that of A then the value of α is equal to

(A) $\frac{2}{5}$ (B) $\frac{2}{3}$ (C) $\frac{-2}{3}$ (D) $\frac{1}{2}$

45. Consider a matrix $A = \begin{bmatrix} 3 & 1 \\ -6 & -2 \end{bmatrix}$, then $(I + A)^{99}$ equals (where I is a unit matrix of order 2)

(A) $I + 2^{98} A$ (B) $I + 2^{99} A$
(C) $I + (2^{99} + 1) A$ (D) $I + (2^{99} - 1) A$

46. If A is a non-null diagonal matrix of order 3 such that $A^2 = A$, then number of possible matrices A, is

(A) 2 (B) 3 (C) 8 (D) 7

47. Let $A = \begin{bmatrix} 2 & 0 & 7 \\ 0 & 1 & 0 \\ 1 & -2 & 1 \end{bmatrix}$ and

$B = \begin{bmatrix} -\alpha & 14\alpha & 7\alpha \\ 0 & 1 & 0 \\ \alpha & -4\alpha & -2\alpha \end{bmatrix}$. If $AB = I$, where I is an identity matrix of order 3 then trace B has value equal to

(A) 0 (B) $\frac{2}{5}$ (C) $\frac{1}{5}$ (D) 5

48. Let the matrix A and B be defined as $A = \begin{bmatrix} 3 & 2 \\ 2 & \alpha \end{bmatrix}$ and $B = \begin{bmatrix} 3 & 1 \\ 7 & 3 \end{bmatrix}$.

If det. $(2A^9 B^{-1}) = -2$, then the number of distinct possible real values of α equals

(A) 0 (B) 1 (C) 2 (D) 3

49. If A and B are non singular matrices of same order then Adj. (AB) is

(A) Adj. (A) (Adj. B)
(B) (Adj. B) (Adj. A)
(C) Adj. A + Adj. B
(D) none of these

50. Let $A = \begin{bmatrix} x+\lambda & x & x \\ x & x+\lambda & x \\ x & x & x+\lambda \end{bmatrix}$,

then A^{-1} exists if

(A) $x \neq 0$ (B) $\lambda \neq 0$
(C) $3x + \lambda \neq 0, \lambda \neq 0$ (D) $x \neq 0, \lambda \neq 0$

51. Which of the following statements is incorrect for a square matrix A. ($|A| \neq 0$)

(A) If A is a diagonal matrix, A^{-1} will also be a diagonal matrix.
(B) If A is a symmetric matrix, A^{-1} will also be a symmetric matrix.
(C) If $A^{-1} = A \Rightarrow A$ is an idempotent matrix.
(D) If $A^{-1} = A \Rightarrow A$ is an involutory matrix.

52. Identify the correct statement.

(A) If a system of n simultaneous linear equations has a unique solution, then the coefficient matrix is singular.
(B) If a system of n simultaneous linear equations has a unique solution, then the coefficient matrix is non-singular.
(C) If A^{-1} exists, (adj. A)$^{-1}$ may or may not exist.
(D) If $F(x) = \begin{bmatrix} \cos x & -\sin x & 0 \\ \sin x & \cos x & 0 \\ 0 & 0 & 0 \end{bmatrix}$,

then $F(x) \cdot F(y) = F(x - y)$.

53. If $A = \begin{bmatrix} 3 & -3 & 4 \\ 2 & -3 & 4 \\ 0 & -1 & 1 \end{bmatrix}$, then $A^{-1} =$

(A) A (B) A^2 (C) A^3 (D) A^4

54. If $A = \begin{bmatrix} 0 & 2b & c \\ a & b & -c \\ a & -b & c \end{bmatrix}$ is orthogonal, then $|abc|$ is equal to

(A) $\frac{1}{2}$ (B) $\frac{1}{3}$ (C) $\frac{1}{6}$ (D) 1

55. Let $A = [a_{ij}]_{3\times3}$ be such that $a_{ij} = \begin{cases} 3, \text{ when } i = j \\ 0, \text{ otherwise} \end{cases}$

then $\left\{ \frac{\text{det.}(\text{adj.}(\text{adj. } A))}{5} \right\}$ equals

[**Note:** $\{k\}$ denotes fractional part of k.]

(A) $\frac{2}{3}$ (B) $\frac{1}{5}$ (C) $\frac{2}{5}$ (D) $\frac{1}{3}$

56. Let A, B, C, D be (not necessarily square) real matrices such that $A^T = BCD$, $B^T = CDA$, $C^T = DAB$ and $D^T = ABC$. For the matrix $S = ABCD$, consider the two statements,

I: $S^3 = S$ II: $S^2 = S^4$

(A) II is true but not I
(B) I is true but not II
(C) both I and II are true
(D) both I and II are false

57. If A is a non-singular matrix satisfying $A = AB - BA$, then which one of the following holds true?
(A) det. $B = 0$
(B) $B = 0$
(C) det. $A = 1$
(D) det. $(B + I) = $ det. $(B - I)$

58. Let $A = \begin{pmatrix} 0 & \sin\alpha & \sin\alpha\cdot\sin\beta \\ -\sin\alpha & 0 & \cos\alpha\cdot\cos\beta \\ -\sin\alpha\cdot\sin\beta & -\cos\alpha\cdot\cos\beta & 0 \end{pmatrix}$, then
(A) $|A|$ is independent of α and β
(B) A^{-1} depends only on α
(C) A^{-1} depends only on β
(D) none of these

Integer Answer Type

1. If t is real and $\lambda = \dfrac{t^2 - 3t + 4}{t^2 + 3t + 4}$, then find the number of solutions of the system of equations $3x - y + 4z = 3$, $x + 2y - 3z = -2$, $6x + 5y + \lambda z = -3$.

2. Let $\Delta = \begin{vmatrix} \operatorname{cosec}\alpha & 1 & 0 \\ 1 & 2\operatorname{cosec}\alpha & 1 \\ 0 & 1 & 2\operatorname{cosec}\alpha \end{vmatrix}$, where $0 < \alpha < \dfrac{\pi}{2}$, then find the least value of Δ.

3. Find the least positive integral value of x for which
$$\begin{vmatrix} x & 2 & 2 \\ 3 & x & 2 \\ 3 & 3 & x \end{vmatrix} + \begin{vmatrix} 1-x & 2 & 4 \\ 2 & 4-x & 8 \\ 4 & 8 & 16-x \end{vmatrix} > 33$$

4. Let $\Delta = \begin{vmatrix} \alpha & \alpha+\beta & \alpha+\beta+\gamma \\ 3\alpha & 4\alpha+3\beta & 5\alpha+4\beta+3\gamma \\ 6\alpha & 9\alpha+6\beta & 11\alpha+9\beta+6\gamma \end{vmatrix}$.
If $\alpha = 1, \beta = w$ and $y = w^2$, then find the value of $[\Delta + 1.9]$, where $[\cdot]$ denotes GIF.

5. If $\Delta = \begin{vmatrix} [e] & [\pi] & [\pi^2 - 6] \\ [\pi] & [\pi^2 - 6] & [e] \\ [\pi^2 - 6] & [e] & [\pi] \end{vmatrix}$ $[\,]$ denotes GIF, then find the value of $\Delta + 12$.

6. Find the number of distinct values of a 2×2 determinant whose entries are from the set $\{-1, 0, 1\}$.

7. Find the coefficient of x in $f(x)$ if
$$f(x) = \begin{vmatrix} (1+x)^a & (1+2x)^b & 1 \\ 1 & (1+x)^a & (1+2x)^b \\ (1+2x)^b & 1 & (1+x)^a \end{vmatrix},$$
where a and b are positive integers.

8. If α, β, γ are the roots of the equation $x^3 + px + q = 0$, then find the value of the determinant $\begin{vmatrix} \alpha & \beta & \gamma \\ \beta & \gamma & \alpha \\ \gamma & \alpha & \beta \end{vmatrix}$.

9. Let $\Delta(x) = \begin{vmatrix} (x-2) & (x-1)^2 & x^3 \\ (x-1) & x^2 & (x+1)^3 \\ x & (x+1)^2 & (x+2)^3 \end{vmatrix}$, then find the coefficient of $(-x)$ in $\Delta(x)$.

10. If A and B are square matrices of the same order such that $|A| = |B| = 1$ and A(adj A + adj B) = B then find the value of $|A + B|$.

11. Let $P = \left\{(a, b) : A^3 = A, \text{ where } A = \begin{bmatrix} \frac{1}{2} & \frac{1}{2} \\ a & b \end{bmatrix}\right\}$, then find $n(P)$.

12. If A is 2×2 matrix such that $A\begin{bmatrix} 1 \\ -1 \end{bmatrix} = \begin{bmatrix} -1 \\ 2 \end{bmatrix}$ and $A^2\begin{bmatrix} 1 \\ -1 \end{bmatrix} = \begin{bmatrix} 1 \\ 0 \end{bmatrix}$, then find the sum of all elements of A.

13. Let $\begin{bmatrix} a_{n+1} \\ b_{n+1} \end{bmatrix} = \begin{bmatrix} \sqrt{3}+1 & 1-\sqrt{3} \\ \sqrt{3}-1 & 1+\sqrt{3} \end{bmatrix} \begin{bmatrix} a_n \\ b_n \end{bmatrix}, n \in N$.

If $a_1 = b_1 = 1$ and $a_{22} = 2^{(a)^b}$, then find the value of $(a+b)$.

14. Consider a square matrix A of order 3 such that $|A| = 3$. If $\text{adj}(\text{adj } A) = \lambda A$, then find the value of λ.

15. Let A, B, C be square matrices of the same order which satisfy $|A - B| \neq 0$, $A^4 = B^4$, $C^4A = C^3B$ and $B^3A = A^3B$, then find the numerical value of $A^3 + B^3 + C^3$.

16. If A is a non-singular periodic matrix with period 3 and $A^6 + B = I$, then find the value of $|A|$.

17. If all elements of a (2×2) matrix A are real and distinct such that $|A + A^T| = 0$, then find the least positive integral value of $|AA^T|$.

18. If A is a (3×1) column matrix such that $A^T BA = 0$, then find the value of $|B|$.

19. If $A = \begin{bmatrix} 1 & 1 \\ 0 & 1 \end{bmatrix}$ and the value of $|A + A^2 + A^3 + A^4 + A^5|$ is equal to $\sqrt{a^b}$, then find the value of b.

6.5 JEE Advanced: Sample Questions

Single Answer Type

1. If $\alpha \in \left(0, \frac{\pi}{2}\right)$, then the least value of the determinant

$$\begin{vmatrix} \operatorname{cosec}\alpha & 1 & 0 \\ 1 & 2\operatorname{cosec}\alpha & 1 \\ 0 & 1 & 2\operatorname{cosec}\alpha \end{vmatrix}$$ is

(A) 0 (B) –1 (C) 1 (D) 2

2. If $\begin{vmatrix} x^n & x^{n+2} & x^{n+3} \\ y^n & y^{n+2} & y^{n+3} \\ z^n & z^{n+2} & z^{n+3} \end{vmatrix}$

$= (x-y)(y-z)(z-x)\left(\frac{1}{x}+\frac{1}{y}+\frac{1}{z}\right)$,

then n is equal to

(A) 1 (B) –1 (C) 2 (D) –2

3. Let $f(x) = \begin{vmatrix} x^2+1 & x-1 & x+2 \\ 2x & x-3 & x+4 \\ 3x-1 & 4x & 8x-2 \end{vmatrix}$

$-px^4 - qx^3 - rx^2 - sx - t$ and $f(x) = 0 \; \forall \; x$, then the equation

$(p+4q-3t)x^2 + (q+2t-2p)x + p - 5q + t = 0$

has one root equal to

(A) $\frac{p}{t}$ (B) $\frac{q}{t}$ (C) $\frac{p}{q}$ (D) $\frac{q}{p}$

4. If $(1 + x + x^2)^n = a_0 + a_1x + a_2x^2 + \ldots + a_{2n}x^n$,

and $p = \begin{vmatrix} a_{n-3} & a_{n-1} & a_{n+1} \\ a_{n-6} & a_{n-3} & a_{n+3} \\ a_{n-14} & a_{n-7} & a_{n+7} \end{vmatrix}$, then

(A) $p = 0$ (B) $p > 0$
(C) $p < 0$ (D) none of these

5. If $\begin{vmatrix} \log_x y & \log_y z & \log_z x \\ \log_y z & \log_z x & \log_x y \\ \log_z x & \log_x y & \log_y z \end{vmatrix} = 0$ where $x, y, z \in R$ and $x, y, z > 1$, then

(A) x, y, z are in A.P.
(B) x, y, z are in G.P.
(C) x, y, z are in H.P.
(D) x, y, z are equal

6. If a, b, c are in G.P. and $(x - \alpha)$ is a factor of $ax^2 + 2bx + c$, then value of the determinant

$\begin{vmatrix} a & b & a\alpha+b \\ b & c & b\alpha+c \\ a\alpha+b & b\alpha+c & 0 \end{vmatrix}$ is equal to

(A) 1 (B) –1 (C) 2 (D) 0

7. If a, b, c are in A.P. with a common difference $d(\neq 0)$ and x, y, z are in G.P. with a common ratio $r(\neq 1)$, then the area of the triangle with vertices (a, x), (b, y) and (c, z) is independent of

(A) a (B) d (C) x (D) r

8. If $D = \begin{vmatrix} 2 & 1 & [\sin^2\theta] \\ [\sin^2\theta] & \cos\theta & i \\ i & 1 & \sin\theta \end{vmatrix}$

where [ ] denotes greatest integer function and $i=\sqrt{-1}$, then arg (D) belongs to

(A) $\left(-\tan^{-1}2, -\pi+\tan^{-1}2\right)$

(B) $\left(\frac{-3\pi}{4}, \frac{-\pi}{2}\right)\cup\left(-\tan^{-1}2, -\pi+\tan^{-1}2\right)$

(C) $\left[\frac{-3\pi}{4}, \frac{-\pi}{2}\right]$

(D) none of these

9. If α, β, γ are the roots of the equation $x^2(px+q) = r(x+1)$, then the value of the determinant $\begin{vmatrix} 1+\alpha & 1 & 1 \\ 1 & 1+\beta & 1 \\ 1 & 1 & 1+\gamma \end{vmatrix}$ is

(A) $\alpha\beta\gamma$ (B) $1+\frac{1}{\alpha}+\frac{1}{\beta}+\frac{1}{\gamma}$

(C) 0 (D) none of these

10. In a quadrilateral *ABCD*, let

$\Delta = \begin{vmatrix} \cos A & \sin A & \cos(A+D) \\ \cos B & \sin B & \cos(B+D) \\ \cos C & \sin C & \cos(C+D) \end{vmatrix}$, then Δ is

(A) independent of *A* and *B* only
(B) independent of *B* and *C* only
(C) independent of *A*, *B* and *C* only
(D) independent of *A*, *B*, *C* and *D* all

11. Given is a system of equations in *x*, *y*, *z*: $x+y+z=6$, $x+2y+3z=10$ and $x+2y+az=b$. If this system has an infinite number of solutions, then

(A) $a=3, b=10$ (B) $a=3, b\neq 10$

(C) $a\neq 3, b=10$ (D) $a\neq 3, b\neq 10$

12. If $f(x) = \begin{vmatrix} x^3 & \sin x & \tan x \\ \cos x & x^8 & 1 \\ e^{x^2} & \sec x & 2 \end{vmatrix}$,

then the value of $\int_{-\pi/2}^{\pi/2} f(x)\,dx$ is given by

(A) 1 (B) 2

(C) 3 (D) none of these

13. If $A = a^2+b^2+c^2$, $B = ab+bc+ca$, then $(a^3+b^3+c^3-3abc)^2$ is equal to

(A) $\begin{vmatrix} B & A & B \\ B & B & A \\ A & B & B \end{vmatrix}$ (B) $\begin{vmatrix} A & B & B \\ B & B & A \\ B & A & B \end{vmatrix}$

(C) $\begin{vmatrix} B & B & A \\ B & A & B \\ A & B & B \end{vmatrix}$ (D) none of these

14. If the maximum and minimum values of the determinant

$\begin{vmatrix} 1+\sin^2 x & \cos^2 x & \sin 2x \\ \sin^2 x & 1+\cos^2 x & \sin 2x \\ \sin^2 x & \cos^2 x & 1+\sin 2x \end{vmatrix}$

are α and β, then

(A) $\alpha+\beta^{99}=5$ (B) $\alpha^3-\beta^{17}=25$

(C) $(\alpha^{2n}-\beta^{2n})$ is always an even integer for $n\in N$

(D) a triangle can be constructed having its sides as α, β and $\alpha-\beta$

15. If $\begin{vmatrix} x+3 & x+4 & x+\lambda \\ x+4 & x+5 & x+\mu \\ x+5 & x+6 & x+\delta \end{vmatrix} = 0$, where λ, μ, δ are in A.P., is

(A) an identity in *x*

(B) an equation whose all roots are real

(C) an equation with only one real root

(D) none of these

16. If $\begin{vmatrix} \sum_{K=0}^{n-2} 1 & n(n-1) & n^2 \\ \sum_{K=1}^{n} 1 & (n+1)(n-1) & n(n+1) \\ \sum_{K=1}^{n-1} 1 & n^2+1 & n^2 \end{vmatrix} = 72$,

then *n* is equal to

(A) 6 (B) 9

(C) 8 (D) none of these

17. Let x_1, y_1, z_1; x_2, y_2, z_2; x_3, y_3, z_3 be three 3-digit even numbers and $\Delta = \begin{vmatrix} x_1 & y_1 & z_1 \\ x_2 & y_2 & z_2 \\ x_3 & y_3 & z_3 \end{vmatrix}$, then Δ is

(A) divisible by 2 but not necessarily by 4
(B) divisible by 4 but not necessarily by 8
(C) divisible by 8
(D) none of these

18. If $f(x) = \begin{vmatrix} 1 & 2(x-1) & 3(x-1)(x-2) \\ x-1 & (x-1)(x-2) & (x-1)(x-2)(x-3) \\ x & x(x-1) & x(x-1)(x-2) \end{vmatrix}$

then the value of $f(49)$ is

(A) $49x$ (B) $-49x$ (C) 0 (D) 1

19. The system of equations $2x + y - 5 = 0, x - 2y + 1 = 0, 2x - 14y - a = 0$ is consistent, then a is equal to

(A) 1 (B) 2
(C) 5 (D) none of these

20. The arbitrary constant on which the value of the determinant

$$\begin{vmatrix} 1 & \alpha & \alpha^2 \\ \cos(p-d)a & \cos pa & \cos(p-d)a \\ \sin(p-d)a & \sin pa & \sin(p-d)a \end{vmatrix}$$

does not depend, is

(A) α (B) p (D) d (D) a

21. If x, y, z are different from zero and $\Delta = \begin{vmatrix} a & b-y & c-z \\ a-x & b & c-z \\ a-x & b-y & c \end{vmatrix}$, then the value of $\frac{a}{x} + \frac{b}{y} + \frac{c}{z}$ is

(A) -1 (B) 0 (C) 1 (D) 2

22. The system of equations $x + y + z = 0$ $2x + 3y + z = 0$ and $x + 2y = 0$ has

(A) a unique solution; $x = 0, y = 0, z = 0$
(B) infinite solutions
(C) no solution
(D) finite number of non-zero solutions

23. If $\begin{bmatrix} 1 & x & 1 \end{bmatrix} \begin{bmatrix} 1 & 3 & 2 \\ 0 & 5 & 1 \\ 0 & 3 & 2 \end{bmatrix} \begin{bmatrix} x \\ 1 \\ -2 \end{bmatrix} = 0$,

then x is equal to

(A) $-\frac{1}{2}$ (B) $\frac{1}{2}$ (C) 1 (D) -1

24. If A is a skew-symmetric matrix, then the trace of A is

(A) 1 (B) -1
(C) 0 (D) none of these

25. The matrix $A = \begin{bmatrix} -5 & -8 & 0 \\ 3 & 5 & 0 \\ 1 & 2 & -1 \end{bmatrix}$ is

(A) an idempotent matrix
(B) an involutory matrix
(C) a nilpotent matrix
(D) none of these

26. Let $A = \begin{bmatrix} x & 3 & 2 \\ 1 & y & 4 \\ 2 & 2 & z \end{bmatrix}$, If $xyz = 60$ and $8x + 4y + 3z = 20$, then $A(\text{adj } A)$ is equal to

(A) $\begin{bmatrix} 64 & 0 & 0 \\ 0 & 64 & 0 \\ 0 & 0 & 64 \end{bmatrix}$ (B) $\begin{bmatrix} 88 & 0 & 0 \\ 0 & 88 & 0 \\ 0 & 0 & 88 \end{bmatrix}$

(C) $\begin{bmatrix} 68 & 0 & 0 \\ 0 & 68 & 0 \\ 0 & 0 & 68 \end{bmatrix}$ (D) $\begin{bmatrix} 34 & 0 & 0 \\ 0 & 34 & 0 \\ 0 & 0 & 34 \end{bmatrix}$

27. The value of a for which the system of equations $a^3x + (a+1)^3y + (a+2)^3z = 0$, $ax + (a+1)y + (a+2)z = 0$ and $x + y + z = 0$ has a non-zero solution is

(A) 1 (B) 0
(C) -1 (D) none of these

28. If $A = \begin{bmatrix} a & 0 & 0 \\ 0 & a & 0 \\ 0 & 0 & a \end{bmatrix}$, then the value of $|\text{adj } A|$ is

(A) a^3 (B) a^6 (C) a^9 (D) a^{27}

29. If A is an idempotent matrix and $A^n + B = I$, then AB is equal to

(A) I (B) 0
(C) nI (D) none of these

30. If ω is the complex cube root of unity, then $\begin{bmatrix} \omega^2 & -\omega^2 \\ \omega & \omega \end{bmatrix}\begin{bmatrix} \omega & \omega^2 \\ -\omega & \omega^2 \end{bmatrix}^{-1}$ is equal to

(A) $\begin{bmatrix} \omega & \omega^2 \\ \omega & \omega \end{bmatrix}$ (B) $\begin{bmatrix} 1 & 0 \\ 0 & 1 \end{bmatrix}$

(C) $\begin{bmatrix} \frac{1}{2} & 0 \\ 0 & \frac{1}{2} \end{bmatrix}$ (D) none of these

31. If $\begin{bmatrix} x & x+1 \\ x-1 & x \end{bmatrix}$ is an orthogonal matrix, then the trace of matrix A is

(A) 2 (B) –2
(C) 3 (D) none of these

32. If A and B are symmetric matrices of the same order and $X = AB + BA$ and $Y = AB - BA$, then $(XY)^T$ is equal to

(A) XY (B) YX
(C) $-YX$ (D) none of these

33. Let

$$a = \lim_{x\to 1}\left(\frac{x}{\ln x} - \frac{1}{x\ln x}\right),\ b = \lim_{x\to 0}\left(\frac{x^3-16x}{4x+x^2}\right),$$

$$c = \lim_{x\to 0}\left(\frac{\ln(1+\sin x)}{x}\right) \text{ and}$$

$$d = \lim_{x\to -1}\left(\frac{(x+1)^3}{3(\sin(x+1)-(x+1))}\right), \text{ then}$$

$\begin{bmatrix} a & b \\ c & d \end{bmatrix}$ is

(A) idempotent (B) involutory
(C) non-singular (D) nilpotent

34. There are two possible values of A in the solution of the matrix equation

$$\begin{bmatrix} 2A+1 & -5 \\ -4 & A \end{bmatrix}^{-1}\begin{bmatrix} A-5 & B \\ 2A-2 & C \end{bmatrix} = \begin{bmatrix} 14 & D \\ E & F \end{bmatrix}$$

where A, B, C, D, E, F are real numbers. The absolute value of the difference of these two solutions is

(A) $\frac{8}{3}$ (B) $\frac{11}{3}$ (C) $\frac{1}{3}$ (D) $\frac{19}{3}$

35. If A is a skew symmetric matrix of order 2 and B, C are matrices $\begin{bmatrix} 1 & 4 \\ 2 & 9 \end{bmatrix}, \begin{bmatrix} 9 & -4 \\ -2 & 1 \end{bmatrix}$ respectively, then $A^3(BC) + A^5(B^2C^2) + A^7(B^3C^3) + \ldots + A^{2n+1}(B^nC^n)$, is

(A) a symmetric matrix
(B) a skew-symmetric matrix
(C) an identity matrix
(D) none of these

36. If $A = \begin{bmatrix} 0 & 1 & 2 \\ 1 & 2 & 3 \\ 3 & a & 1 \end{bmatrix}$ and $A^{-1} = \begin{bmatrix} \frac{1}{2} & -\frac{1}{2} & \frac{1}{2} \\ -4 & 3 & b \\ \frac{5}{2} & -\frac{3}{2} & \frac{1}{2} \end{bmatrix}$ then

(A) $a = 1, b = -1$ (B) $a = 2, b = -\frac{1}{2}$

(C) $a = -1, b = 1$ (D) $a = \frac{1}{2}, b = \frac{1}{2}$

37. If A is a symmetric as well as a skew-symmetric matrix, then A is

(A) null matrix
(B) triangular matrix
(C) diagonal matrix
(D) none of these

38. If [illegible] I, $A + B$ are idempotent matrices, then AB is equal to

(A) BA (B) $-BA$
(C) I (D) 0

39. Which of the following is an orthogonal matrix?

(A) $\begin{bmatrix} \frac{6}{7} & -\frac{2}{7} & \frac{3}{7} \\ \frac{2}{7} & \frac{2}{7} & -\frac{3}{7} \\ -\frac{6}{7} & \frac{2}{7} & \frac{3}{7} \end{bmatrix}$ (B) $\begin{bmatrix} -\frac{6}{7} & -\frac{2}{7} & -\frac{3}{7} \\ \frac{2}{7} & \frac{3}{7} & \frac{6}{7} \\ -\frac{3}{7} & \frac{6}{7} & \frac{2}{7} \end{bmatrix}$

(C) $\begin{bmatrix} \frac{6}{7} & \frac{2}{7} & -\frac{3}{7} \\ \frac{2}{7} & \frac{3}{7} & \frac{6}{7} \\ \frac{3}{7} & -\frac{6}{7} & \frac{2}{7} \end{bmatrix}$ (D) $\begin{bmatrix} \frac{6}{7} & \frac{2}{7} & \frac{3}{7} \\ \frac{2}{7} & -\frac{3}{7} & \frac{6}{7} \\ \frac{3}{7} & \frac{6}{7} & -\frac{2}{7} \end{bmatrix}$

40. If for a real number α, the system of linear equations $\begin{bmatrix} 1 & \alpha & \alpha^2 \\ \alpha^2 & 1 & \alpha \\ \alpha^2 & \alpha & 1 \end{bmatrix}\begin{bmatrix} x \\ y \\ z \end{bmatrix} = \begin{bmatrix} 1 \\ -1 \\ 1 \end{bmatrix}$ has infinitely many solutions, then $1+\alpha+\alpha^2$ is

(A) 2 (B) 3 (C) 1 (D) 4

41. If $P = \begin{bmatrix} 1 & \alpha & 3 \\ 1 & 3 & 3 \\ 2 & 4 & 4 \end{bmatrix}$ is the adjoint of $a(3\times3)$ matrix A and $|A| = 4$, then α is equal to

(A) 0 (B) 4 (C) 5 (D) 11

Multiple Answer Type

1. If

$$\Delta_r = \begin{vmatrix} 2^{r-1} & \dfrac{1}{r(r+1)} & \sin r\theta \\ x & y & z \\ 2^n - 1 & \dfrac{n}{n+1} & \dfrac{\sin\left(\dfrac{n+1}{2}\right)\theta \sin\dfrac{n}{2}\theta}{\sin\dfrac{\theta}{2}} \end{vmatrix}$$

then $\sum_{r=1}^{n} \Delta_r$ is equal to

(A) 0
(B) independent of n
(C) independent of θ
(D) independent of x, y, z

2. Let $\Delta = \begin{vmatrix} x^2 & (y+z)^2 & yz \\ y^2 & (z+x)^2 & zx \\ z^2 & (x+y)^2 & xy \end{vmatrix}$, which of the following can be true?

(A) $\Delta = 0$
(B) Δ is divisible by $x+y+z$
(C) Δ is divisible by $x^2+y^2+z^2$
(D) Δ is divisible by both $(x-y)(x+y+z)$

3. Let $f(x) = a+bx+cx^2$ and ω and ω^2 are the cube roots of unity, other than 1 and $\Delta = \begin{vmatrix} a & b & c \\ b & c & a \\ c & a & b \end{vmatrix}$, then factors of Δ are

(A) $f(1)$ (B) $f(\omega)$
(C) $f(\omega^2)$ (D) none of these

4. The digits A, B, C are such that the three-digit numbers $A88$, $6B8$, $86C$ are divisible by 72, then the determinant $\begin{vmatrix} A & 6 & 8 \\ 8 & B & 6 \\ 8 & 8 & C \end{vmatrix}$ is divisible by

(A) 72 (B) 12 (C) 274 (D) 216

5. If a, b, c are non-zero real numbers such that $\begin{vmatrix} bc & ca & ab \\ ca & ab & bc \\ ab & bc & ca \end{vmatrix} = 0$, then

(A) $\frac{1}{a}+\frac{1}{b\omega}+\frac{1}{c\omega^2}=0$
(B) $\frac{1}{a}+\frac{1}{b\omega^2}+\frac{1}{c\omega}=0$
(C) $\frac{1}{a\omega}+\frac{1}{b\omega^2}+\frac{1}{c}=0$
(D) $\frac{1}{a}+\frac{1}{b}+\frac{1}{c}=0$

6. If A is a square matrix, then
(A) AA' is symmetric
(B) AA' is skew symmetric
(C) $A'A$ is symmetric
(D) $A'A$ is skew symmetric

7. If $A^2 - 3A + 2I = 0$, then A is equal to

(A) I (B) $2I$
(C) $\begin{bmatrix} 3 & -2 \\ 1 & 0 \end{bmatrix}$ (D) $\begin{bmatrix} 3 & 1 \\ -2 & 0 \end{bmatrix}$

8. Let $A_r = \begin{bmatrix} c_{r-1}^2 & 0 \\ 0 & c_r^2 \end{bmatrix}$ for $r \geq 1$ and $C_r = {}^nC_r$.

If $A_1 + A_2 + A_3 + \ldots + A_n = \begin{bmatrix} K_1 & 0 \\ 0 & K_2 \end{bmatrix}$, then

(A) $K_1 + K_2 = 2$ (B) $K_2 = {}^{2n}C_{n+1}$
(C) $K_1 = {}^{2n}C_n - 1$ (D) $K_1 = K_2$

9. Let $A = \begin{bmatrix} 3 & -3 & 4 \\ 2 & -3 & 4 \\ 0 & -1 & 1 \end{bmatrix}$, then

(A) $\text{adj}(\text{adj } A) = A$ (B) $|\text{adj}(\text{adj } A)| = 1$
(C) $|\text{adj}(A)| = 1$ (D) none of these

10. If the adjoint of a (3×3) matrix P is $\begin{bmatrix}1 & 4 & 4\\ 2 & 1 & 7\\ 1 & 1 & 3\end{bmatrix}$, then the possible value(s) of the determinant of P is(are)

(A) –1 (B) 1 (C) –2 (D) 2

11. If a square matrix A satisfies $A^2 + A + I = 0$, then

(A) $A^{-1} = -\left(A^4 + A^9\right)$ (B) $A^7 + A^8 = -I$

(C) $A^{-1} = A^5$ (D) $A^6 + A^9 = -I$

12. Consider a (3×3) matrix A whose elements $a_{ij} = \tan^{-1}\left(\tan(i+j)\right), 1 \le i, j \le 3$, then

(A) $A^T = A$

(B) $\text{Tr}(A) = 12 - 3\pi$

(C) $\text{Tr}(A) = 12 - 4\pi$

(D) $(\text{adj } A)^3$ is a symmetric matrix

13. If $A = \begin{bmatrix}1 & 1\\ -1 & 1\end{bmatrix}$, then

(A) $A^{10} = \begin{bmatrix}0 & 32\\ -32 & 0\end{bmatrix}$

(B) $A^{12} = \begin{bmatrix}-64 & 0\\ 0 & -64\end{bmatrix}$

(C) $A^{14} = \begin{bmatrix}0 & 128\\ -128 & 0\end{bmatrix}$

(D) $A^{16} = \begin{bmatrix}256 & 0\\ 0 & 256\end{bmatrix}$

14. Suppose a square matrix A satisfies $A^2 - 5A + 7I = 0$. If $A^5 = aA + bI$, then

(A) $a = 151$ (B) $b = -365$

(C) $a = 149$ (D) $b = -385$

15. If a, b, c are all different from zero such that $\frac{1}{a} + \frac{1}{b} + \frac{1}{c} = 0$, then matrix

$$A = \begin{bmatrix}1+a & 1 & 1\\ 1 & 1+b & 1\\ 1 & 1 & 1+c\end{bmatrix} \text{ is}$$

(A) skew symmetric (B) non-singular

(C) A^{-1} does not exist (D) symmetric

16. Let adj. $A = \begin{bmatrix}1 & -2 & 3\\ 4 & 0 & 1\\ -1 & 2 & 5\end{bmatrix}$, then which of the following statements is correct?

(A) $\text{tr}(A^{-1}) = 0.75$ (B) $|A| = -8$

(C) $|A| = 8$ (D) $\text{tr}(A^{-1}) = -0.75$

17. Let A, B, C be $n \times n$ order real matrices and product is pairwise commutative, also $ABC = O_n$, if $\lambda = \det.(A^3 + B^3 + C^3).\det.(A + B + C)$ then λ can be

(A) 0 (B) 2 (C) –2 (D) 4

18. If $A(x) = \begin{bmatrix}(2x+1) & (x-3) & x\\ x^2 & (x+5) & 2\\ 3x & (x-1) & (x+1)\end{bmatrix}$, then

(A) $|A(0)| = 7$

(B) $|A(1)| = 10$

(C) $A(0)A(1) = \begin{bmatrix}0 & 20 & 5\\ -11 & 30 & 14\\ 2 & -6 & -2\end{bmatrix}$

(D) $A(0)A(1) = \begin{bmatrix}0 & -20 & -5\\ 11 & 30 & 14\\ 2 & -6 & 0\end{bmatrix}$

19. If $A = \begin{bmatrix}1 & 1 & 3\\ 5 & 2 & 6\\ -2 & -1 & -3\end{bmatrix}$, where $A^x = O$ (where O is a null matrix and $x \in N$ less than 20), then which of the following is(are) false?

(A) The greatest value of x is 18.

(B) The sum of the values of x is 187.

(C) The difference between the largest and smallest value of x is 15.

(D) The number of values of x is 7.

20. If A and B are $n \times n$ matrices with real entries and $|A|$ denotes determinant of A, then

(A) $A^2 \cdot B^3 = O \Rightarrow$ neither A nor B is invertible

(B) $\| AB|AB| = |AB|^{n+1}$

(C) If $|A| = 2$, A cannot be an involutory matrix

(D) $B(AB)^{-1} A = I$, where I is the identity matrix

21. If A and B are square matrices of order 2 such that $A + \text{adj.}(B^T) = \begin{bmatrix} 2 & 1 \\ 1 & 2 \end{bmatrix}$ and $A^T - \text{adj.}(B) = \begin{bmatrix} 0 & 1 \\ 1 & 0 \end{bmatrix}$, then

(A) B is a symmetric matrix
(B) $A^n = A \ \forall \ n \in N$
(C) $|A + A^2 + A^3 + A^4 + A^5| = 0$
(D) $|B + B^2 + B^3 + B^4 + B^5| = 0$

22. Let $M = \begin{bmatrix} 1 & \lambda^3 & 0 \\ 0 & 1 & \mu^3 \\ \gamma^3 & 0 & 1 \end{bmatrix}$ where $\lambda, \mu, \gamma \in R$, then

(A) the sum of the roots of the equation $|M - xI| = 0$ is 3
(B) the sum of the non-real roots of the equation $|M - xI| = 0$ is $2 - \lambda\mu\gamma$
(C) the product of the roots of the equation $|M - xI| = 0$ is $1 + (\lambda\mu\gamma)^3$
(D) the product of the real roots of the equation $|M - xI| = 0$ is $1 + \lambda\mu\gamma$

23. If A is a symmetric and B is a skew symmetric matrix, then which of the following is/are incorrect?

(A) ABA^T is a symmetric matrix.
(B) $AB^T + BA^T$ is a symmetric matrix.
(C) $(A + B)(A - B)$ is skew symmetric.
(D) $(A + I)(B - I)$ is skew symmetric.

24. Let $A = a_{ij}$ be a 3×3 matrix where

$$a_{ij} = \begin{cases} (i^j - j^i + ij)x & ; \text{ if } i < j, x \in R \\ 1 & ; \text{ if } i > j \\ 0 & ; \ i = j \end{cases},$$

then choose the correct option.

(A) minimum value of det. $A = -\dfrac{1}{20}$
(B) $\displaystyle\int_{-2}^{2} (\text{det. } A)dx = \frac{80}{3}$
(C) minimum value of det. $(A) = \dfrac{1}{20}$
(D) $\displaystyle\int_{-2}^{2} (\text{det. } A)dx = \frac{61}{3}$

Comprehension Based Questions

Paragraph for Question No. 1 to 3

Consider the determinant $\Delta = \begin{vmatrix} p & q & r \\ x & y & z \\ l & m & n \end{vmatrix}$

M_{ij} denotes the minor of an element in the ith row and j^{th} column.

C_{ij} denotes the cofactor of an element in the ith row and j^{th} column.

1. The value of $p \cdot C_{21} + q \cdot C_{22} + r \cdot C_{23}$ is
(A) $-\Delta$ (B) Δ (C) 0 (D) Δ^2

2. The value of $x \cdot C_{21} + y \cdot C_{22} + z \cdot C_{23}$ is
(A) 0 (B) $-\Delta$ (C) Δ (D) Δ^2

3. The value of $q \cdot M_{12} - y \cdot M_{22} + m \cdot M_{32}$ is
(A) 0 (B) $-\Delta$ (C) Δ (D) Δ^2

Paragraph for Question No. 4 to 6

A determinant is called cyclic, if it follows the arrangement symmetrically with a, b, c. That is, $\begin{vmatrix} 1 & 1 & 1 \\ a & b & c \\ a^2 & b^2 & c^2 \end{vmatrix}$

If we increase the degree of any row in this determinant symmetrically, its value will be increased by over expression which is also cyclic and increases the degree of the value of the determinant, then

4. The value of $\begin{vmatrix} 1 & 1 & 1 \\ a^2 & b^2 & c^2 \\ bc & ca & ab \end{vmatrix}$ is equal to

(A) $(a-b)(b-c)(c-a)$
(B) $(a-b)(b-c)(c-a)(a+b+c)$
(C) $(a-b)(b-c)(c-a)(ab+bc+ca)$
(D) $(a-b)(b-c)(c-a)abc$

5. The value of $\begin{vmatrix} a & b & c \\ a^2 & b^2 & c^2 \\ bc & ca & ab \end{vmatrix}$ is equal to

(A) $(a-b)(b-c)(c-a)$
(B) $(a-b)(b-c)(c-a)(a+b+c)$

(C) $(a-b)(b-c)(c-a)(ab+bc+ca)$
(D) $(a-b)(b-c)(c-a)abc$

6. The value of $\begin{vmatrix} a & b & c \\ a^2 & b^2 & c^2 \\ a^3 & b^3 & c^3 \end{vmatrix}$ is equal to

(A) $(a-b)(b-c)(c-a)$
(B) $(a-b)(b-c)(c-a)(a+b+c)$
(C) $(a-b)(b-c)(c-a)(ab+bc+ca)$
(D) $(a-b)(b-c)(c-a)abc$

Paragraph for Question No. 7 to 9

Let A be the set of all (3×3) symmetric matrices all of whose entries are either 0 or 1. Five of these entries are 1 and four of them are 0.

7. The number of matrices of A is
(A) 3 (B) 6 (C) 9 (D) 12

8. The number of matrices in A for which the system of linear equation $A\begin{bmatrix} x \\ y \\ z \end{bmatrix}=\begin{bmatrix} 1 \\ 0 \\ 0 \end{bmatrix}$ has a unique solution is
(A) less than 4
(B) at least 4 but less than 7
(C) at least 7 but less than 10
(D) at least 10

9. The number of matrices in A for which the system of linear equation $A\begin{bmatrix} x \\ y \\ z \end{bmatrix}=\begin{bmatrix} 1 \\ 0 \\ 0 \end{bmatrix}$ is inconsistent, is
(A) 0 (B) more than 2
(C) 2 (D) 1

Paragraph for Question No. 10 to 12

If A and B are two square matrices of order (3×3) which satisfy $AB = A$ and $BA = B$, then

10. Which of the following is true?
(A) If matrix A is singular then matrix B is non-singular.
(B) If matrix A is non-singular then matrix B is singular.
(C) If matrix A is singular then matrix B is also singular
(D) Cannot say anything.

11. $(A+B)^7$ is equal to
(A) $7(A+B)$ (B) $7I_{3\times3}$
(C) $64(A+B)$ (D) $128I$

12. $(A+I)^5$ (where I is identity matrix) is equal to
(A) $I+60A$ (B) $I+16A$
(C) $I+31A$ (D) none of these

Match Matrix

1. Let $f(x)=\begin{vmatrix} \sec^2 x & 1 & 1 \\ \cos^2 x & \cos^2 x & \operatorname{cosec}^2 x \\ 1 & \cos^2 x & \cot^2 x \end{vmatrix}$

Column I	Column II
(A) $\int_0^{\pi/4} f(x)dx-\frac{1}{4}$ is equal to	(P) π
(B) Maximum value of $f(x)$ is	(Q) 0
(C) Minimum value of $f(x)$ is	(R) $\frac{3\pi}{32}$
(D) Period of $f(x)$ is	(S) 1

2. If $a_K=\int_0^{\pi/2}(K+\sin\theta)^2\cos\theta d\theta$ and the value of the determinant $\begin{vmatrix} a_K & K & K^2 \\ a_{K+1} & K+1 & (K+1)^2 \\ a_{K+2} & K+2 & (K+2)^2 \end{vmatrix}$ is $\frac{p}{q}$

Column I	Column II
(A) Value of p is	(P) 13
(B) Value of q is	(Q) 1
(C) Value of $\lvert p-q\rvert$ is	(R) 2
(D) Value of $\sqrt{p^2+q^2}$ is	(S) 3

3. Match the following.

Column I	Column II
(A) If A, B, C are square matrices of order $2\times 2, 3\times 3$ and 4×4 respectively such that $\lvert A\rvert = 2, \lvert B\rvert = 3$ and $\lvert C\rvert = 4$, then $\lvert ABC\rvert$ is equal to	(P) 7
(B) If A is an idempotent matrix and I is an identity matrix of the same order, such that $(A+I)^n = I+127$, then value of n is	(Q) 8
(C) If A is a matrix such that $a_{ij} = (i+j)(i-j)$, then A is singular if order of matrix is	(R) 9
(D) If A is a matrix such that $(I-A)^{-1} = I+A+A^2+\ldots+A^7$, then $A^n = 0$ if n is	(S) 24

4. Let A be the matrix of order 3 such that $A = \left[a_{ij}\right]_{3\times 3}$, where

$$a_{ij} = \begin{cases} 0 & \text{for } i = j \\ 1 & \text{for } i \neq j \end{cases}, \text{ then}$$

Column I	Column II
(A) $\lvert A^{-1}+I\rvert$ is equal to	(P) 0
(B) $\lvert A^3 - A^2 + 2A\rvert$ is equal to	(Q) 4
(C) $\lvert A^4 - 4A - 7I\rvert$ is equal to	(R) 8
(D) $\lvert A^4 - 8A^{-1} - A^2 - 6I\rvert$ is equal to	(S) 64
	(T) 128

Subjective Answer Type (Hard Nut)

1. If a, b, c are the sides of a triangle of area Δ and

$$\begin{vmatrix} (a-b)^2 & a^2 & b^2 \\ 1 & 1 & 1 \\ c^2 & a^2 & (c-a)^2 \end{vmatrix} + \begin{vmatrix} (b-c)^2 & b^2 & c^2 \\ 1 & 1 & 1 \\ a^2 & b^2 & (a-b)^2 \end{vmatrix} + \begin{vmatrix} (c-a)^2 & c^2 & a^2 \\ 1 & 1 & 1 \\ b^2 & c^2 & (b-c)^2 \end{vmatrix}$$

$= (\lambda\Delta)^2$, then find the value of λ.

2. If $x > 0, y > 0, z > 0$ and $\left(\dfrac{4x^2}{1+4x^2} - y\right)^2 + \left(\dfrac{4y^2}{1+4y^2} - z\right)^2 + \left(\dfrac{4z^2}{1+4z^2} - x\right)^2 = 0$,

then find the value of $\begin{vmatrix} x & y & z \\ y & z & x \\ z & x & y \end{vmatrix}$.

3. If $\begin{vmatrix} a^2+\lambda^2 & ab+c\lambda & ca-b\lambda \\ ab-c\lambda & b^2+\lambda^2 & bc+a\lambda \\ ca+b\lambda & bc-a\lambda & c^2+\lambda^2 \end{vmatrix} \begin{vmatrix} \lambda & c & -b \\ -c & \lambda & a \\ b & -a & \lambda \end{vmatrix}$

$= (1+a^2+b^2+c^2)^3$, find the value of λ.

4. If $-a^2x + aby + acz = 0$, $abx - b^2y + bcz = 0$ and $acx + bcy - c^2z = 0$ have a non-trivial solution, then find the value of $\begin{vmatrix} 0 & c & b \\ c & 0 & a \\ b & a & 0 \end{vmatrix}$.

5. If A, B, C are angles of a triangle, then find the value of $\begin{vmatrix} e^{2iA} & e^{-ic} & e^{-iB} \\ e^{-ic} & e^{2iB} & e^{-iA} \\ e^{-iB} & e^{-iA} & e^{2ic} \end{vmatrix}$.

6. If $a \neq p, b \neq q, c \neq r$ and $\begin{vmatrix} p & b & c \\ a & q & c \\ a & b & r \end{vmatrix} = 0$,

then find the value of $\dfrac{p}{p-a} + \dfrac{q}{q-b} + \dfrac{r}{r-c}$.

7. If 3^n is a factor of $\begin{vmatrix} 1 & 1 & 1 \\ n_{c_1} & n+3_{c_1} & n+6_{c_1} \\ n_{c_2} & n+3_{c_2} & n+6_{c_2} \end{vmatrix}$,

then find the maximum value of n.

8. If $f(\theta)=\begin{vmatrix} \cos^2\theta & \cos\theta\sin\theta & -\sin\theta \\ \cos\theta\sin\theta & \sin^2\theta & \cos\theta \\ \sin\theta & -\cos\theta & 0 \end{vmatrix}$,

then find the value of $f\left(\frac{\pi}{6}\right)$.

9. If $f(x)=\begin{vmatrix} 2\cos^2 x & \sin 2x & -\sin x \\ \sin 2x & 2\sin^2 x & \cos x \\ \sin x & -\cos x & 0 \end{vmatrix}$,

then find the value of $\int_0^{\pi/2}\left(f(x)+f'(x)\right)dx$.

10. If $f(x)=$

$$\begin{vmatrix} \sec x & \cos x & \sec^2 x+\cot x \operatorname{cosec} x+\cos x \\ \cos^2 x & \cos^2 x & \operatorname{cosec}^2 x+\cos^4 x \\ 1 & \cos^2 x & \cos^2 x \end{vmatrix},$$

then find the value of $\int_0^{\pi/2} f(x)dx$.

11. If $f(x)=$

$$\begin{vmatrix} x^2-4x+6 & 2x^2+4x+10 & 3x^2-2x+16 \\ x-2 & 2x+2 & 3x-1 \\ 1 & 2 & 3 \end{vmatrix}$$

and $\left\{\int_{-2}^{2} x^2\left[f(x)\right]dx\right\}=\frac{\alpha}{\beta}$, where [ ] denotes GIF and { } denotes the fractional part,then find the value of $\alpha+\beta$.

12. If $\begin{vmatrix} 3^2+K & 4^2 & 3^2+3+K \\ 4^2+K & 5^2 & 4^2+4+K \\ 5^2+K & 6^2 & 5^2+5+K \end{vmatrix}=0,$

then find the value of $\sqrt{2^K\sqrt{2^K\sqrt{2^K\ldots\infty}}}$.

13. If $a\neq 0, b\neq 0, c\neq 0$ and

$$\begin{vmatrix} 1+a & 1 & 1 \\ 1+b & 1+2b & 1 \\ 1+c & 1+c & 1+3c \end{vmatrix}=0,$$

then find the value of $\left|\frac{1}{a}+\frac{1}{b}+\frac{1}{c}\right|$.

14. Find the coefficient of x in the expansion of

$$\begin{vmatrix} (1+x)^{22} & (1+x)^{44} & (1+x)^{66} \\ (1+x)^{33} & (1+x)^{66} & (1+x)^{99} \\ (1+x)^{44} & (1+x)^{88} & (1+x)^{144} \end{vmatrix}.$$

15. If $\left(1+x+x^2\right)^n=a_0+a_1x+a_2x^2+\ldots+a_{2n}x^{2n}$,

then find the value of $\begin{vmatrix} a_{n-3} & a_{n-1} & a_{n+1} \\ a_{n-6} & a_{n-3} & a_{n+3} \\ a_{n-14} & a_{n-7} & a_{n+7} \end{vmatrix}$.

16. If $\Delta(x)=\begin{vmatrix} x & 1+x^2 & x^3 \\ \log\left(1+x^2\right) & e^x & \sin x \\ \cos x & \tan x & \sin^2 x \end{vmatrix}$,

then prove that $\Delta(x)$ is divisible by x.

17. If $A+B+C=\pi$, then find the value of the determinant $\begin{vmatrix} \sin^2 A & \cot A & 1 \\ \sin^2 B & \cot B & 1 \\ \sin^2 C & \cot C & 1 \end{vmatrix}$.

18. There are three points (a, x), (b, y) and (c, z) such that the straight lines joining any two of them are not equally inclined to the coordinate axis where $a, b, c, x, y, z \in R$.

If $\begin{vmatrix} x+a & y+b & z+c \\ y+b & z+c & x+a \\ z+c & x+a & y+b \end{vmatrix}=0$ and $a+c=-b$,

then prove that $x, \frac{-y}{2}, z$ are in A.P.

19. If α is a root of $x^4 = 1$ with a negative principal argument,then find the principal argument of $\Delta(\alpha)$ where

$$\Delta(\alpha)=\begin{vmatrix} 1 & 1 & 1 \\ \alpha^n & \alpha^{n+1} & \alpha^{n+3} \\ \frac{1}{\alpha^{n+1}} & \frac{1}{\alpha^n} & 0 \end{vmatrix}.$$

20. The set of natural numbers N is partitioned into arrays of rows and columns in the form of matrices as $M_1 = (1), M_2 = \begin{pmatrix} 2 & 3 \\ 4 & 5 \end{pmatrix}$,

$M_3 = \begin{pmatrix} 6 & 7 & 8 \\ 9 & 10 & 11 \\ 12 & 13 & 14 \end{pmatrix}, \ldots, M_n = ()$

and so on. Find the sum of the elements of the diagonal in M_n.

21. If A be an $(n \times n)$ orthogonal matrix and X be any matrix such that $X = (A + I)^{-1} (A - I)$, then prove that X is a skew symmetric matrix, whenever n is an odd integer.

22. Let A be any (3×2) matrix. Show that AA' is a singular matrix.

23. Three matrices are given as

$A = \begin{bmatrix} \alpha^2 & 4 & 6 \\ 9 & \beta^4 & 7 \\ 1 & 2 & 2\gamma^2 \end{bmatrix}$,

$B = \begin{bmatrix} 2\beta^2 & -1 & 0 \\ 2 & \gamma^2 - 2\gamma & 1 \\ 1 & 9 & 2\alpha - 1 \end{bmatrix}$,

$C = \begin{bmatrix} \gamma & 2 & 1 \\ 1 & \alpha & 1 \\ 2 & 0 & \beta \end{bmatrix}$.

If $t_r(A) = t_r(B) - 2$ and $\alpha, \beta, \gamma \in R$ then find det.(C).

24. A singular matrix $A = \left[a_{ij}\right]_{2 \times 2}$ always commutes with $B = \begin{bmatrix} 1 & 1 \\ 2 & 1 \end{bmatrix}$ such that $\left(\dfrac{a_{11}}{a_{12}}\right)^2 = K$, then find the value of K.

25. Let $A = \begin{bmatrix} 1 & 2 \\ 3 & 4 \end{bmatrix}$ and $B = \begin{bmatrix} a & b \\ c & d \end{bmatrix}$ be two matrices such that they are commutative and $c \neq 3b$, then find the value of $\left|\dfrac{a-d}{3b-c}\right|$.

26. If $A = \begin{bmatrix} u & 2 & -1 \\ -1 & 1 & 2 \\ 2 & -1 & 1 \end{bmatrix}$ where $\left(u \neq \dfrac{-11}{3}\right)$ and det(adj (adj A)) $= (23)^4$, then find the value of u.

27. Let A be a (4×4) matrix such that sum of elements in each row is 1. If the sum of all the elements in A^{10} is n, then find the total number of solutions of the equation $\sin^{-1} x + \cos^{-1} x^2 = \dfrac{2\pi}{n}$.

28. P is an orthogonal matrix and $A^{4n} = A^4$ and $A^{5n} = A$ $(n \in I)$ and $Q = PAP^T$ such that $X = P^T Q^{2005} P$ is equal to A^K, then find the value of K.

29. Let $A = \begin{bmatrix} 3x^2 \\ 1 \\ 6x \end{bmatrix}$, $B = \begin{bmatrix} a & b & c \end{bmatrix}$ and

$C = \begin{bmatrix} (x+2)^2 & 5x^2 & 2x \\ 5x^2 & 2x & (x+2)^2 \\ 2x & (x+2)^2 & 5x^2 \end{bmatrix}$

be three given matrices where a, b, c and $x \in R$. If $t_r(AB) = t_r(C)$ for all $x \in R$ and if $f(x) = ax^2 + bx + c$, then find the value of $f(1)$.

30. If $A = \begin{bmatrix} -1 & 3 & 5 \\ 1 & -3 & -5 \\ -1 & 3 & 5 \end{bmatrix}$ and X be the solution set of the equation $A^x = A$, where $X \in N - \{1\}$. Evaluate $\Pi\left(\dfrac{x^3+1}{x^3-1}\right)$ or all $x \in X$.

31. If $A = \begin{bmatrix} 3 & 2 & 2 \\ 2 & 4 & 1 \\ -2 & -4 & -1 \end{bmatrix}$ and X, Y are two non-zero column vectors such that $AX = \lambda X$; $AY = \mu Y$, $\lambda \neq \mu$, find the angle between X and Y.

32. If $A = \begin{bmatrix} 1 & x & 1 \\ x & 2 & y \\ 1 & y & 3 \end{bmatrix}$ and $B = \begin{bmatrix} 3 & -3 & z \\ -3 & 2 & -3 \\ z & -3 & 1 \end{bmatrix}$ and if the matrix AB is symmetric, then find the value of x, y, z.

33. If $A=\begin{bmatrix}3 & a & -1\\ 2 & 5 & c\\ b & 8 & 2\end{bmatrix}$ is symmetric and

$B=\begin{bmatrix}d & 3 & a\\ b-a & c & -2b-c\\ -2 & 6 & -1\end{bmatrix}$ is skew symmetric,

then find whether AB is symmetric, skew symmetric or neither of them. Justify your answer.

34. If $A=\begin{bmatrix}1 & 1 & 1\\ 2 & 4 & 1\\ 2 & 3 & 1\end{bmatrix}, B=\begin{bmatrix}2 & 3\\ 3 & 4\end{bmatrix}$, find P such that $BPA=\begin{bmatrix}1 & 0 & 1\\ 0 & 1 & 0\end{bmatrix}$.

35. Find the values of a and b for which the system

$\begin{bmatrix}3 & -2 & 1\\ 5 & -8 & 9\\ 2 & 1 & a\end{bmatrix}\begin{bmatrix}x\\ y\\ z\end{bmatrix}=\begin{bmatrix}b\\ 3\\ -1\end{bmatrix}$ has

a unique solution,
no solution,
infinitely many solutions.

36. If $A=\begin{bmatrix}a & b & c\\ b & c & a\\ c & a & b\end{bmatrix}$ where a, b, c are real positive numbers, $abc = 1$ and $A^TA = I$, then find the value of $a^3 + b^3 + c^3$.

37. If $\begin{bmatrix}a & b\\ c & 1-a\end{bmatrix}$ is an idempotent matrix, find the value of $f(a)$, where $f(x) = x - x^2$, when $bc=\frac{1}{4}$, otherwise evaluate a.

6.6 JEE Main: Archive

Determinants

1. The number of values of α for which the system of equations:

$x+y+z=\alpha$

$\alpha x+2\alpha y+3z=-1$

$x+3\alpha y+5z=4$

is inconsistent, is

(A) 0 (B) 1 (C) 2 (D) 3

[Main 2022 24th June 1st Shift]

2. Let the system of linear equations

$x+y+\alpha z=2$

$3x+y+z=4$

$x+2z=1$

have a unique solution (x^*, y^*, z^*). If $(\alpha, x^*), (y^*, \alpha)$ and $(x^*, -y^*)$ are collinear points, then the sum of absolute values of all possible values of α is:

(A) 4 (B) 3 (C) 2 (D) 1

[Main 2022 24th June 2nd Shift]

3. The system of equations

$-kx+3y-14z=25$

$-15x+4y-kz=3$

$-4x+y+3z=4$

is consistent for all k in the set

(A) R (B) $R-\{-11,13\}$

(C) $R-\{13\}$ (D) $R-\{-11,11\}$

[Main 2022 25th June 2nd Shift]

4. The ordered pair (a, b), for which the system of linear equations

$3x-2y+z=b$

$5x-8y+9z=3$

$2x+y+az=-1$

has no solution, is:

(A) $\left(3, \frac{1}{3}\right)$ (B) $\left(-3, \frac{1}{3}\right)$

(C) $\left(-3, -\frac{1}{3}\right)$ (D) $\left(3, -\frac{1}{3}\right)$

[Main 2022 26th June 1st Shift]

5. If the system of equations

$\alpha x+y+z=5, x+2y+3z=4, x+3y+5z=\beta$, has infinitely many solutions, then the ordered pair (α, β) is equal to

(A) $(1, -3)$ (B) $(-1, 3)$

(C) $(1, 3)$ (D) $(-1, -3)$

[Main 2022 26th June 2nd Shift]

6. Let the system of linear equations

$x+2y+z=2$,

$\alpha x+3y-z=\alpha, -\alpha x+y+2z=-\alpha$

be inconsistent. Then α is equal to:

(A) $\frac{5}{2}$ (B) $-\frac{5}{2}$ (C) $\frac{7}{2}$ (D) $-\frac{7}{2}$

[Main 2022 27th June 1st Shift]

7. Let for some real numbers α and β, $a=\alpha-i\beta$. If the system of equations $4ix+(1+i)y=0$ and

$8\left(\cos\frac{2\pi}{3}+i\sin\frac{2\pi}{3}\right)x+\bar{a}y=0$

has more than one solution then $\frac{\alpha}{\beta}$ is equal to:

(A) $-2+\sqrt{3}$ (B) $2-\sqrt{3}$
(C) $2+\sqrt{3}$ (D) $-2-\sqrt{3}$

[Main 2022 27th June 2nd Shift]

8. If the system of linear equations
$2x+3y-z=-2$
$x+y+z=4$
$x-y+|\lambda|z=4\lambda-4$
where $\lambda\in R$, has no solution, then

(A) $\lambda=7$ (B) $\lambda=-7$
(C) $\lambda=8$ (D) $\lambda^2=1$

[Main 2022 28th June 1st Shift]

9. If the system of linear equations
$2x-3y=\lambda+5$,
$\alpha x+5y=\beta+1$, where $\alpha,\beta,\gamma\in R$ has infinitely many solutions, then the value of $|9\alpha+3\beta+5\gamma|$ is equal to

[Main 2022 28th June 2nd Shift]

10. If the system of linear equations
$2x+y-z=7$
$x-3y+2z=1$
$x+4y+\delta z=k$, where $\delta, k\in R$ to

(A) –3 (B) 3 (C) 6 (D) 9

[Main 2022 29th June 1st Shift]

11. The number of $\theta\in(0,4\pi)$ for which the system of linear equations
$3(\sin 3\theta)x-y+z=2$
$3(\cos 2\theta)x+4y+3z=3$
$6x+7y+7z=9$
has no solution is:

(A) 6 (B) 7 (C) 8 (D) 9

[Main 2022 25th July 1st Shift]

12. The number of real values λ, such that the system of linear equations
$2x-3y+5z=9$
$x+3y-z=-18$
$3x-y+\left(\lambda^2-|\lambda|\right)z=16$
has no solution, is:

(A) 0 (B) 1 (C) 2 (D) 4

[Main 2022 25th July 2nd Shift]

13. If the system of linear equations
$8x+y+4z=-2$
$x+y+z=0$
$\lambda x-3y=\mu$
has infinitely many solutions, then the distance of the point $\left(\lambda,\mu,-\frac{1}{2}\right)$ from the plane $8x+y+4z+2=0$ is:

(A) $3\sqrt{5}$ (B) 4 (C) $\frac{26}{9}$ (D) $\frac{10}{3}$

[Main 2022 26th July 1st Shift]

14. Let p and $p+2$ be prime numbers and let

$$\Delta=\begin{vmatrix} P! & (P+1)! & (P+2)! \\ (P+1)! & (P+2)! & (P+3)! \\ (P+2)! & (P+3)! & (P+4)! \end{vmatrix}$$

Then the sum of the maximum values of α and β, such that p^α and $(p+2)^\beta$ divide Δ, is ______.

[Main 2022 29th July 1st Shift]

15. If the system of equations
$x+y+z=6$
$2x+5y+\alpha z=\beta$
$x+2y+3z=14$
has infinitely many solutions, then $\alpha+\beta$ is equal to:

(A) 8 (B) 36 (C) 44 (D) 48

[Main 2022 29th July 2nd Shift]

16. If the system of equations
$x+2y+3z=3$
$4x+3y-4z=4$
$8x+4y-\lambda z=9+\mu$
has infinitely many solutions, then the ordered pair (λ,μ) is equal to

(A) $\left(\frac{72}{5}, \frac{21}{5}\right)$ (B) $\left(-\frac{72}{5}, \frac{-21}{5}\right)$

(C) $\left(\frac{72}{5}, \frac{-21}{5}\right)$ (D) $\left(\frac{-72}{5}, \frac{21}{5}\right)$

[Main 2023 24th Jan 2nd Shift]

17. Let S_1 and S_2 be respectively the sets of all $a \in R - \{0\}$ for which the system of linear equations

$ax + 2ay - 3az = 1$

$(2a+1)x + (2a+3)y + (a+1)z = 2$

$(3a+5)x + (a+5)y + (a+2)z = 3$

has a unique solution and infinitely many solutions. Then

(A) $n(S_1) = 2$ and S_2 is an infinite set

(B) S_1 is an infinite set and $n(S_2) = 2$

(C) $S_1 = \Phi$ and $S_2 = R - \{0\}$

(D) $S_1 = R - \{0\}$ and $S_2 = \Phi$

[Main 2023 25th Jan 1st Shift]

18. Consider the following system of questions

$\alpha x + 2y + z = 1$

$2\alpha x + 3y + z = 1$

$3x + \alpha y + 2z = \beta$

For some $\alpha, \beta \in R$, which of the following is NOT correct?

(A) It has no solution if $\alpha = -1$ and $\beta \neq 2$.

(B) It has no solution for $\alpha = -1$ and for all $\beta \in R$.

(C) It has no solution for $\alpha = 3$ and for all $\beta \neq 2$.

(D) It has a solution for all $\alpha \neq -1$ and $\beta = 2$.

[Main 2023 29th Jan 1st Shift]

19. Let the system of linear equations

$x + y + kz = 2$

$2x + 3y - z = 1$

$3x + 4y + 2z = k$

have infinitely many solutions. Then the system

$(k+1)x + (2k-1)y = 7$

$(2k+1)x + (k+5)y = 10$ has:

(A) infinitely many solutions

(B) unique solution satisfying $x - y = 1$

(C) no solution

(D) unique solution satisfying $x + y = 1$

[Main 2023 30th Jan 1st Shift]

20. For $\alpha, \beta \in R$, suppose the system of linear equations

$x - y + z = 5$

$2x + 2y + \alpha z = 8$

$3x - y + 4z = \beta$

has infinitely many solutions. Then α and β are the roots of

(A) $x^2 - 10x + 16 = 0$

(B) $x^2 + 18x + 56 = 0$

(C) $x^2 - 18x + 56 = 0$

(D) $x^2 + 14x + 24 = 0$

[Main 2023 30th Jan 2nd Shift]

21. For the system of linear equations

$x + y + z = 6$

$\alpha x + \beta y + 7z = 3$

$x + 2y + 3z = 14,$

which of the following is NOT true?

(A) If $\alpha = \beta = 7$, then the system has no solution.

(B) If $\alpha = \beta$ and $\alpha \neq 7$, then the system has a unique solution.

(C) There is a unique point (α, β) on the line $x + 2y + 18 = 0$ for which the system has infinitely many solutions.

(D) For every point $(\alpha, \beta) \neq (7, 7)$ on the line $x - 2y + 7 = 0$, the system has infinitely many solutions.

[Main 2023 31st Jan 1st Shift]

22. Let S denote the set of all real values of λ such that the system of equations

$\lambda x + y + z = 1$

$x + \lambda y + z = 1$

$x + y + \lambda z = 1$

is inconsistent, then $\sum_{\lambda \in S}\left(|\lambda|^2 + |\lambda|\right)$ is equal to

(A) 2 (B) 12 (C) 4 (D) 6

[Main 2023 01st Feb 1st Shift]

23. For the system of linear equations $ax+y+z=1,\quad x+ay+z=1, x+y+az=\beta$, which one of the following statements is NOT correct?

(A) It has infinitely many solutions if $\alpha=2$ and $\beta=-1$.

(B) It has no solution if $\alpha=-2$ and $\beta=1$.

(C) $x+y+z=\frac{3}{4}$ if $\alpha=2$ and $\beta=1$.

(D) It has infinitely many solutions if $\alpha=1$ and $\beta=1$.

[Main 2023 01st Feb 2nd Shift]

24. If the system of equations

$x+y+az=b$

$2x+5y+2z=6$

$x+2y+3z=3$

has infinitely many solutions, then $2a+3b$ is equal to

(A) 25 (B) 20 (C) 23 (D) 28

[Main 2023 06th Apr 1st Shift]

25. For the system of equations

$x+y+z=6$

$x+2y+\alpha z=10$

$x+3y+5z=\beta$,

which one of the following is NOT true?

(A) The system has no solution for $\alpha=3, \beta=24$.

(B) The system has a unique solution for $\alpha=-3, \beta=14$.

(C) The system has infinitely many solutions for $\alpha=3, \beta=14$.

(D) The system has a unique solution for $\alpha=3, \beta\neq 14$.

[Main 2023 06th Apr 2nd Shift]

26. Let S be the set of all values of $\theta\in[-\pi,\pi]$ for which the system of linear equations

$x+y+\sqrt{3}z=0$

$-x+(\tan\theta)y+\sqrt{7}z=0$

$x+y+(\tan\theta)z=0$

has non-trivial solution. Then $\frac{12}{\pi}\sum_{0\in S}\theta$ is equal to

(A) 20 (B) 40 (C) 30 (D) 10

[Main 2023 08th Apr 2nd Shift]

27. For the system of linear equations

$2x-y+3z=5$

$3x+2y-z=7$

$4x+5y+\alpha z=\beta$,

which of the following is NOT correct?

(A) The system has infinitely many solutions for $\alpha=-5$ and $\beta=9$.

(B) The system has infinitely many solutions for $\alpha=-6$ and $\beta=9$.

(C) The system in inconsistent for $\alpha=-5$ and $\beta=8$.

(D) The system has a unique solution for $\alpha\neq-5$ and $\beta=8$.

[Main 2023 10th Apr 1st Shift]

28. Let S be the set of values of λ, for which the system of equations $6\lambda x-3y+3z=4\lambda^2$, $2x+6\lambda y+4z=1$ and $3x+2y+3\lambda z=\lambda$ has no solution. Then $12\sum_{\lambda\in S}|\lambda|$ is equal to ______.

[Main 2023 10th Apr 2nd Shift]

29. Let A be a 2×2 matrix with real entries such that $A'=\alpha A+1$, where $\alpha\in R-\{-1,1\}$. If $\det(A^2-A)=4$, the sum of all possible values of α is equal to

(A) 0 (B) $\frac{3}{2}$ (C) 2 (D) $\frac{5}{2}$

[Main 2023 11th Apr 1st Shift]

30. If the system of linear equations

$7x+11y+\alpha z=13$

$5x+4y+7z=\beta$

$175x+194y+57z=361$

has infinitely many solutions, then $\alpha+\beta+2$ is equal to

(A) 4 (B) 3 (C) 5 (D) 6

[Main 2023 11th Apr 2nd Shift]

31. If $\begin{vmatrix} x+1 & x & x \\ x & x+\lambda & x \\ x & x & x+\lambda^2 \end{vmatrix}=\frac{9}{8}(103x+81)$,

then $\lambda, \frac{\lambda}{3}$ are the roots of the equation

(A) $4x^2+24x-27=0$

(B) $4x^2-24x-27=0$

(C) $4x^2 + 24x + 27 = 0$
(D) $4x^2 - 24x + 27 = 0$

[Main 2023 11th Apr 2nd Shift]

32. Let $D_k = \begin{vmatrix} 1 & 2k & 2k-1 \\ n & n^2+n+2 & n^2 \\ n & n^2+n & n^2+n+2 \end{vmatrix}$.

If $\sum_{k=1}^{n} D_k = 96$, then n is equal to _______.

[Main 2023 12th Apr 1st Shift]

33. For the system of linear equations
$2x + 4y + 2az = b$
$x + 2y + 3z = 4$
$2x + 5y + 2z = 8$
which of the following is NOT correct?
(A) It has a unique solution if $a = b = 6$.
(B) It has infinitely many solutions if $a = 3, b = 6$.
(C) It has infinitely many solutions if $a = 3, b = 8$
(D) It has a unique solution if $a = b = 8$.

[Main 2023 13th Apr 1st Shift]

34. If the system of equations
$2x + y - z = 5$
$2x - 5y + \lambda z = \mu$
$x + 2y - 5z = 7$
has infinitely many solutions, then $(\lambda + \mu)^2 + (\lambda - \mu)^2$ is equal to
(A) 904 (B) 916 (C) 912 (D) 920

[Main 2023 13th Apr 2nd Shift]

35. Let the system of linear equations
$-x + 2y - 9z = 7$
$-x + 3y + 7z = 9$
$-2x + y + 5z = 8$
$-3x + y + 13z = \lambda$
have a unique solution $x = \alpha, y = \beta, z = \gamma$. Then the distance of the point (α, β, γ) from the plane is
(A) 11 (B) 7 (C) 9 (D) 13

[Main 2023 15th Apr 1st Shift]

36. If the system of equations
$2x + 3y - z = 5$
$x + \alpha y + 3z = -4$
$3x - y + \beta z = 7$
has infinitely many solutions, then $13\alpha\beta$ is equal to
(A) 1110 (B) 1120 (C) 1210 (D) 1220

[Main 2024 1st Feb 1st Shift]

37. Let the system of equations
$x + 2y + 3z = 5, 2x + 3y + z = 9,$
$4x + 3y + \lambda z = \mu$
have infinite number of solutions. Then $\lambda + 2\mu$ is equal to
(A) 28 (B) 17 (C) 22 (D) 15

[Main 2024 1st Feb 2nd Shift]

38. Let $A = \begin{bmatrix} 2 & 0 & 1 \\ 1 & 1 & 0 \\ 1 & 0 & 1 \end{bmatrix}, B = [B_1, B_2, B_3]$, where B_1, B_2, B_3 are column matrices, and $A_1 = \begin{bmatrix} 1 \\ 0 \\ 0 \end{bmatrix}, AB_2 = \begin{bmatrix} 2 \\ 3 \\ 0 \end{bmatrix}, AB_3 = \begin{bmatrix} 3 \\ 2 \\ 1 \end{bmatrix}$

If $\alpha = |B|$ and β is the sum of all the diagonal elements of B, then $\alpha^3 + \beta^3$ is equal to _______.

[Main 2024 27th Jan 1st Shift]

39. The values of α, for which
$\begin{vmatrix} 1 & \frac{3}{2} & \alpha + \frac{3}{2} \\ 1 & \frac{1}{3} & \alpha + \frac{1}{3} \\ 2\alpha + 3 & 3\alpha + 1 & 0 \end{vmatrix} = 0$, lie in the interval
(A) $(-2, 1)$ (B) $(-3, 0)$
(C) $\left(-\frac{3}{2}, \frac{3}{2}\right)$ (D) $(0, 3)$

[Main 2024 27th Jan 2nd Shift]

40. Let for any three distinct consecutive terms, a, b, c of an A.P., the lines $ax + by + c = 0$ be concurrent at the point P and $Q(\alpha, \beta)$ be a point such that the system of equations
$x + y + z = 6$
$2x + 5y + \alpha z = \beta$ and
$x + 2y + 3z = 4$,
have infinitely many solutions. Then $(PQ)^2$ is equal to _______.

[Main 2024 29th Jan 2nd Shift]

41. Consider the system of linear equations $x+y+z=4\mu$, $x+2y+2\lambda z=10\mu$,

$x+3y+4\lambda^2 z=\mu^2+15$,

where $\lambda, \mu \in R$. Which one of the following statements is NOT correct?

(A) The system has unique solution if $\lambda \neq \frac{1}{2}$ and $\mu \neq 1, 15$.

(B) The system is inconsistent if $\lambda = \frac{1}{2}$ and $mu \neq 1$.

(C) The system has an infinite number of solutions if $\lambda = \frac{1}{2}$ and $\mu = 15$.

(D) The system is consistent if $\lambda \neq \frac{1}{2}$.

[Main 2024 30th Jan 1st Shift]

42. If

$$f(x)=\begin{vmatrix} 2\cos^4 x & 2\sin^4 x & 3+\sin^2 2x \\ 3+2\cos^4 x & 2\sin^4 x & \sin^2 2x \\ 2\cos^4 x & 3+2\sin^4 x & \sin^2 2x \end{vmatrix}$$

then $\frac{1}{5}f'(0)$ is equal to ______.

(A) 0 (B) 1 (C) 2 (D) 6

[Main 2024 30th Jan 1st Shift]

43. Consider the system of linear equations $x+y+z=5$, $x+2y+\lambda^2 z=9$, $x+3y+\lambda z=\mu$, where $\lambda, \mu \in R$. Then, which of the following statements is NOT correct?

(A) The system has infinite number of solutions if $\lambda = 1$ and $\mu = 13$.

(B) The system is inconsistent if $\lambda = 1$ and $\mu \neq 13$.

(C) The system is consistent if $\lambda \neq 1$ and $\mu = 13$.

(D) The system has an unique solution if $\lambda \neq 1$ and $\mu \neq 13$.

[Main 2024 30th Jan 2nd Shift]

44. If the system of linear equations

$x-2y+z=-4$

$2x+\alpha y+3z=5$

$3x-y+\beta z=3$

has infinitely many solutions, then $12\alpha+13\beta$ is equal to

(A) 60 (B) 64 (C) 54 (D) 58

[Main 2024 31st Jan 1st Shift]

45. If $f(x)=\begin{vmatrix} x^3 & 2x^2+1 & 1+3x \\ 3x^2+2 & 2x & x^3+6 \\ x^3-x & 4 & x^2-2 \end{vmatrix}$ for all $x \in R$, then $2f(0)+f'(0)$ is equal to

(A) 48 (B) 24 (C) 42 (D) 18

[Main 2024 31st Jan 1st Shift]

46. Let A be a 3×3 real matrix such that

$$A\begin{pmatrix}1\\0\\1\end{pmatrix}=2\begin{pmatrix}1\\0\\1\end{pmatrix}, A\begin{pmatrix}-1\\0\\1\end{pmatrix}=4\begin{pmatrix}-1\\0\\1\end{pmatrix}, A\begin{pmatrix}0\\1\\0\end{pmatrix}=2\begin{pmatrix}0\\1\\0\end{pmatrix}.$$

Then, the system $(A-3I)\begin{pmatrix}x\\y\\z\end{pmatrix}=\begin{pmatrix}1\\2\\3\end{pmatrix}$ has

(A) an unique solution

(B) exactly two solutions

(C) no solution

(D) infinitely many solutions

[Main 2024 31st Jan 2nd Shift]

Matrices

1. Let $S=\left\{\sqrt{n}: 1 \leq n \leq 50 \text{ and } n \text{ is odd}\right\}$.

Let $a \in S$ and $A=\begin{bmatrix} 1 & 0 & a \\ -1 & 1 & 0 \\ -a & 0 & 1 \end{bmatrix}$.

If $\sum_{a\in S} \det(\text{adj } A) = 100\lambda$, then λ is equal to

(A) 218 (B) 221 (C) 663 (D) 1717

[Main 2022 24th June 1st Shift]

2. Let $S=\left\{\begin{pmatrix}-1 & a \\ 0 & b\end{pmatrix}; a, b \in \{1, 2, 3, \ldots, 100\}\right\}$ and let $T_n=\left\{A \in S : A^{n(n+1)} = I\right\}$. Then the number of elements in $\bigcap_{n=1}^{100} T_n$ is ______.

[Main 2022 24th June 2nd Shift]

3. Let A be a 3×3 real matrix such that

$$A\begin{pmatrix}1\\1\\0\end{pmatrix}=\begin{pmatrix}1\\1\\0\end{pmatrix}; A\begin{pmatrix}1\\0\\1\end{pmatrix}=\begin{pmatrix}-1\\0\\1\end{pmatrix} \text{ and } A\begin{pmatrix}0\\0\\1\end{pmatrix}=\begin{pmatrix}1\\1\\2\end{pmatrix}.$$

If $X = (x_1, x_2, x_3)^T$ and I is an identity matrix of order 3, then the system $(A-2I)X = \begin{pmatrix} 4 \\ 1 \\ 1 \end{pmatrix}$ has

(A) no solution
(B) infinitely many solutions
(C) an unique solution
(D) exactly two solutions

[Main 2022 25th June 1st Shift]

4. Let $A = \begin{bmatrix} 0 & -2 \\ 2 & 0 \end{bmatrix}$. If M and N are two matrices given by $M = \sum_{k=1}^{10} A^{2k}$ and $N = \sum_{k=1}^{10} A^{2k-1}$ then MN^2 is

(A) a non-identity symmetric matrix
(B) a skew-symmetric matrix
(C) neither symmetric nor skew-symmetric matrix
(D) an identity matrix

[Main 2022 25th June 1st Shift]

5. Let A be a 3×3 matrix having entries from the set $\{-1, 0, 1\}$. The number of all such matrices A having sum of all entries equal to 5, is _______

[Main 2022 25th June 1st Shift]

6. Let $A = \begin{pmatrix} 2 & -2 \\ 1 & -1 \end{pmatrix}$ and $B = \begin{pmatrix} -1 & 2 \\ -1 & 2 \end{pmatrix}$. Then the number of elements in the set $\{(n, m) : n, m \in \{1, 2, \ldots, 10\}$ and $nA^n + mB^m = I\}$ is _______ **[Main 2022 25th June 2nd Shift]**

7. Let A be a 3×3 invertible matrix. If $|\text{adj}(24A) = \text{adj}(3\,\text{adj}(2A))|$, then $|A|^2$ is equal to:

(A) 6^6 (B) 2^{12} (C) 2^6 (D) 1

[Main 2022 26th June 1st Shift]

8. Let $X = \begin{bmatrix} 0 & 1 & 0 \\ 0 & 0 & 1 \\ 0 & 0 & 0 \end{bmatrix}$, $Y = \alpha I + \beta X + \gamma X^2$ and $Z = \alpha^2 I - \alpha\beta X + (\beta^2 - \alpha\gamma)X^2, \alpha, \beta, \gamma \in R.$ If $Y^{-1} = \begin{bmatrix} \frac{1}{5} & \frac{-2}{5} & \frac{1}{5} \\ 0 & \frac{1}{5} & \frac{-2}{5} \\ 0 & 0 & \frac{1}{5} \end{bmatrix}$, then $(\alpha - \beta + \gamma)^2$ is equal to _______.

[Main 2022 26th June 2nd Shift]

9. The positive value of the determinant of the matrix A, whose $\text{Adj}(\text{Adj}(A)) = \begin{pmatrix} 14 & 28 & -14 \\ -14 & 14 & 28 \\ 28 & -14 & 14 \end{pmatrix}$, is _______.

[Main 2022 27th June 1st Shift]

10. Let A and B be two 3×3 matrices such that $AB = I$ and $|A| = \frac{1}{8}$ then $|\text{adj}(B\,\text{adj}(2A))|$ is equal to

(A) 16 (B) 32 (C) 64 (D) 128

[Main 2022 27th June 2nd Shift]

11. Let A be a matrix of order 2×2, whose entries are from the set $\{0, 1, 2, 3, 4, 5\}$. If the sum of all the entries of A is a prime number $p, 2 < p < 8$, then the number of such matrices A is:

[Main 2022 27th June 2nd Shift]

12. Let A be a matrix of order 3×3 and det (A) = 2. The det (det (A) adj (5 adj (A^3))) is equal to _______.

(A) 512×10^6 (B) 256×10^6
(C) 1024×10^6 (D) 256×10^{11}

[Main 2022 28th June 1st Shift]

13. Let $A = \begin{pmatrix} 1+i & 1 \\ -i & 0 \end{pmatrix}$ where $i = \sqrt{-1}$. Then, then number of elements in the set $\{n \in \{1, 2, \ldots, 1000\} : A^n = A\}$ is

[Main 2022 28th June 2nd Shift]

14. The probability that a randomly chosen 2×2 matrix with all the entries from the set of first 10 primes, is singular, is equal to:

(A) $\frac{133}{10^4}$ (B) $\frac{18}{10^3}$

(C) $\frac{19}{10^3}$ (D) $\frac{271}{10^4}$

[Main 2022 29th June 1st Shift]

15. Let $A=\left[a_{ij}\right]$ be a square matrix of order 3 such that $a_{ij}=2^{j-i}$, for all $i, j=1,2,3$. Then, the matrix $A^2+A^3+\ldots+A^{10}$ is equal to

(A) $\left(\frac{3^{10}-3}{2}\right)A$ (B) $\left(\frac{3^{10}-1}{2}\right)A$

(C) $\left(\frac{3^{10}+1}{2}\right)A$ (D) $\left(\frac{3^{10}+3}{2}\right)A$

[Main 2022 29th June 1st Shift]

16. Let $A=\begin{pmatrix}2 & -1\\ 0 & 2\end{pmatrix}$. If $B=I-{}^5C_1(\text{adj } A)+{}^5C_2(\text{adj } A)^2-\ldots-{}^5C_5$, then the sum of all elements of the matrix B is:

(A) –5 (B) –6 (C) –7 (D) –8

[Main 2022 29th June 2nd Shift]

17. Let $M=\begin{bmatrix}0 & -\alpha\\ \alpha & 0\end{bmatrix}$, where α is a non-zero real number an $N=\sum_{k=1}^{49} M^{2k}$. If $\left(I-M^2\right)N=-2I$, then the positive integral value of α is ______.

[Main 2022 29th June 2nd Shift]

18. Let $A=\begin{bmatrix}2 & -1 & -1\\ 1 & 0 & -1\\ 1 & -1 & 0\end{bmatrix}$ and

$B=\begin{bmatrix}9^2 & -10^2 & 11^2\\ 12^2 & 13^2 & -14^2\\ -15^2 & 16^2 & 17^2\end{bmatrix}$,

then the value of $A'BA$ is:

(A) 1224 (B) 1042

(C) 540 (D) 539

[Main 2022 25th July 1st Shift]

19. Let $A=\begin{bmatrix}1 & a & a\\ 0 & 1 & b\\ 0 & 0 & 1\end{bmatrix}, a, b\in R$. If for some $n\in N$, $A^n=\begin{bmatrix}1 & 48 & 2160\\ 0 & 1 & 96\\ 0 & 0 & 1\end{bmatrix}$ then $n+a+b$ is equal to ______.

[Main 2022 25th July 2nd Shift]

20. Let A be a 2×2 matrix with $\det(A)=-1$ and $\det\left((A+I)(\text{Adj}(A)+I)\right)=4$. Then the sum of the diagonal elements of A can be:

(A) –1 (B) 2 (C) 1 (D) $-\sqrt{2}$

[Main 2022 26th July 1st Shift]

21. Let $A=\begin{bmatrix}1\\ 1\\ 1\end{bmatrix}$ and $B=\begin{bmatrix}9^2 & -10^2 & 11^2\\ 12^2 & 13^2 & -14^2\\ -15^2 & 16^2 & 17^2\end{bmatrix}$,

then the value of $A'BA$ is:

(A) 1224 (B) 1042 (C) 540 (D) 539

[Main 2022 26th July 2nd Shift]

22. The number of matrices $A=\begin{bmatrix}a & b\\ c & d\end{bmatrix}$, where $a, b, c, d\in\{-1,0,1,2,3,\ldots,10\}$, such that $A=A^{-1}$, is ______.

[Main 2022 26th July 2nd Shift]

23. Let $A=\begin{pmatrix}1 & 2\\ -2 & -5\end{pmatrix}$. Let $\alpha, \beta\in R$ be such that $\alpha A^2+\beta A=2I$. Then $\alpha+\beta$ is equal to

(A) –10 (B) –6 (C) 6 (D) 10

[Main 2022 27th July 1st Shift]

24. Let S be the set containing all 3×3 matrices with entries from $\{-1,0,1\}$. The total number of matrices $A\in S$ such that the sum of all the diagonal elements of A^TA is 6 is ______.

[Main 2022 27th July 1st Shift]

25. Let $A=\begin{pmatrix}4 & -2\\ \alpha & \beta\end{pmatrix}$

If $A^2+\gamma A+18I=O$, then $\det(A)$ is equal to ______

(A) –18 (B) 18 (C) –50 (D) 50

[Main 2022 27th July 2nd Shift]

26. Consider a matrix

$$A=\begin{bmatrix}\alpha & \beta & \gamma \\ \alpha^2 & \beta^2 & \gamma^2 \\ \beta+\gamma & \gamma+\alpha & \alpha+\beta\end{bmatrix},$$

where α, β, γ are three distinct natural numbers. If

$$\frac{\det(\operatorname{adj}(\operatorname{adj}(\operatorname{adj}(\operatorname{adj} A))))}{(\alpha-\beta)^{16}(\beta-\gamma)^{16}(\gamma-\alpha)^{16}}=2^{32}\times 3^{16},$$

then the number of such 3-tuples (α, β, γ) is ______.

[Main 2022 27th July 2nd Shift]

27. Let the matrix $A=\begin{bmatrix}0 & 1 & 0\\ 0 & 0 & 1\\ 1 & 0 & 0\end{bmatrix}$ and the matrix $B_0 = A^{49}+2A^{98}$. If $B_n = \operatorname{Adj}(B_{n-1})$ for all $n \ge 1$, then $\det(B_4)$ is equal to

(A) 3^{28} (B) 3^{30} (C) 3^{32} (D) 3^{36}

[Main 2022 28th July 1st Shift]

28. Let $A=\begin{bmatrix}1 & -1\\ 2 & \alpha\end{bmatrix}$ and $B=\begin{bmatrix}\beta & 1\\ 1 & 0\end{bmatrix}$, $\alpha, \beta \in R$. Let α_1 be the value of α which satisfies $(A+B)^2 = A^2+\begin{bmatrix}2 & 2\\ 2 & 2\end{bmatrix}$ and α_2 be the value of α which satisfies $(A+B)^2 = B^2$. Then $|\alpha_1-\alpha_2|$ is equal to ______.

[Main 2022 28th July 1st Shift]

29. Let A and B be any two 3×3 symmetric and skew symmetric matrices respectively. Then which of the following is NOT true?

(A) $A^4 - B^4$ is a symmetric matrix
(B) $AB-BA$ is a symmetric matrix
(C) $B^5 - A^5$ is a skew-symmetric matrix
(D) $AB+BA$ is a skew-symmetric matrix

[Main 2022 28th July 2nd Shift]

30. Let A and B be two 3×3 non-zero real matrices such that AB is a zero matrix. Then

(A) the system of linear equations $AX = 0$ has a unique solution
(B) the system of linear equations $AX = 0$ has infinitely many solutions
(C) B is an invertible matrix
(D) $\operatorname{adj}(A)$ is an invertible matrix

[Main 2022 29th July 1st Shift]

31. The number of matrices of order 3×3, whose entries are either 0 or 1 and the sum of all the entries is a prime number, is ______. **[Main 2022 29th July 1st Shift]**

32. Which of the following matrices can NOT be obtained from the matrix $\begin{bmatrix}-1 & 2\\ 1 & -1\end{bmatrix}$ by a single elementary row operation?

(A) $\begin{bmatrix}0 & 1\\ 1 & -1\end{bmatrix}$ (B) $\begin{bmatrix}1 & -1\\ -1 & 2\end{bmatrix}$

(C) $\begin{bmatrix}-1 & 2\\ -2 & 7\end{bmatrix}$ (D) $\begin{bmatrix}-1 & 2\\ -1 & 3\end{bmatrix}$

[Main 2022 29th July 2nd Shift]

33. Let $x=\begin{bmatrix}1\\ 1\\ 1\end{bmatrix}$ and $A=\begin{bmatrix}-1 & 2 & 3\\ 0 & 1 & 6\\ 0 & 0 & -1\end{bmatrix}$. For $k \in N$, if $X'A^kX = 33$, then k is equal to:

[Main 2022 29th July 2nd Shift]

34. If A and B are two non-zero $n\times n$ matrices much that $A^2+B = A^2B$, then

(A) $AB = I$ (B) $A^2B = I$
(C) $A^2 = I$ or $B = I$ (D) $A^2B = BA^2$

[Main 2023 24th Jan 1st Shift]

35. The number of square matrices of order 5 with entries from the set $\{0, 1\}$, such that the sum of all the elements in each row is 1 and the sum of all the elements in each column is also 1, is

(A) 225 (B) 120 (C) 150 (D) 125

[Main 2023 24th Jan 2nd Shift]

36. Let A be a 3×3 matrix such that $|\operatorname{adj}(\operatorname{adj}(\operatorname{adj} A))| = 12^4$. Then $|A^{-1}\operatorname{adj} A|$ is equal to

(A) $2\sqrt{3}$ (B) $\sqrt{6}$ (C) 12 (D) 1

[Main 2023 24th Jan 2nd Shift]

37. Let $x, y, z > 1$ and

$$A=\begin{bmatrix}1 & \log_x y & \log_x z\\ \log_y x & 2 & \log_y z\\ \log_z x & \log_z y & 3\end{bmatrix}.$$

Then $|\operatorname{adj}(\operatorname{adj} A^2)|$ is equal to

(A) 6^4 (B) 2^8 (C) 4^8 (D) 2^4

[Main 2023 25th Jan 1st Shift]

38. Let $A=\begin{bmatrix}\frac{1}{\sqrt{10}} & \frac{3}{\sqrt{10}}\\ \frac{-3}{\sqrt{10}} & \frac{1}{\sqrt{10}}\end{bmatrix}$ and $B=\begin{bmatrix}1 & -i\\ 0 & 1\end{bmatrix}$, where $i=\sqrt{-1}$. If $M=A^TBA$, then the inverse of the matrix $AM^{2023}A^T$ is

(A) $\begin{bmatrix}1 & -2023i\\ 0 & 1\end{bmatrix}$ (B) $\begin{bmatrix}1 & 0\\ -2023i & 1\end{bmatrix}$

(C) $\begin{bmatrix}1 & 0\\ 2023i & 1\end{bmatrix}$ (D) $\begin{bmatrix}1 & 2023i\\ 0 & 1\end{bmatrix}$

[Main 2023 25th Jan 2nd Shift]

39. Let A, B, C be 3×3 matrices such that A is symmetric and B and C are skew-symmetric. Consider the statements

(S1) $A^{13}B^{26}-B^{26}A^{13}$ is symmetric

(S2) $A^{26}C^{13}-C^{13}A^{26}$ is symmetric

Then,

(A) Only S2 is true

(B) Only S1 is true

(C) Both S1 and S2 are false

(D) Both S1 and S2 are true

[Main 2023 25th Jan 2nd Shift]

40. Let α and β be real numbers. Consider a 3×3 matrix A such that $A^2=3A+\alpha I$. If $A^4=21A+\beta I$, then

(A) $\alpha=1$ (B) $\alpha=4$

(C) $\beta=8$ (D) $\beta=-8$

[Main 2023 29th Jan 1st Shift]

41. The set of all values of $t\in R$, for which the matrix

$$\begin{bmatrix}e^t & e^{-t}(\sin t-2\cos t) & e^{-t}(-2\sin t-\cos t)\\ e^t & e^{-t}(2\sin t+\cos t) & e^{-t}(\sin t-2\cos t)\\ e^t & e^{-t}\cos t & e^{-t}\sin t\end{bmatrix}$$

is invertible, is

(A) $\left\{(2k+1)\frac{\pi}{2}, k\in Z\right\}$

(B) $\left\{k\pi+\frac{\pi}{4}, k\in Z\right\}$

(C) $\{k\pi, k\in Z\}$

(D) $\mathbb{R}$

[Main 2023 29th Jan 2nd Shift]

42. Let A be a symmetric matrix such that $|A|=2$ and $\begin{bmatrix}2 & 1\\ 3 & \frac{3}{2}\end{bmatrix}A=\begin{bmatrix}1 & 2\\ \alpha & \beta\end{bmatrix}$. If the sum of the diagonal elements of A is s, then $\frac{\beta s}{\alpha^2}$ is equal to ______.

[Main 2023 29th Jan 2nd Shift]

43. Let

$\begin{bmatrix}m & n\\ p & q\end{bmatrix}$, $d=|A|\neq 0\,|A-d(\text{Adj}\,A)|=0$.

Then

(A) $(1+d)^2=(m+q)^2$

(B) $1+d^2=(m+q)^2$

(C) $(1+d)^2=m^2+q^2$

(D) $1+d^2=m^2+q^2$

[Main 2023 30th Jan 1st Shift]

44. If P is a 3×3 real matrix such that $P^T=aP+(a-1)I$, where $a>1$, then

(A) P is singular matrix

(B) $|\text{Adj }P|>1$

(C) $|\text{Adj }P|=\frac{1}{2}$

(D) $|\text{Adj }P|=1$

[Main 2023 30th Jan 2nd Shift]

45. Let $A=\begin{pmatrix}0 & 4 & -1\\ 0 & 12 & -3\end{pmatrix}$. Then the sum of the diagonal elements of the matrix $(A+I)^{11}$ is equal to

(A) 6144 (B) 4094 (C) 4097 (D) 2050

[Main 2023 31st Jan 1st Shift]

46. Let $A=[a_{ij}]$, $a_{ij}\in Z\cap[0,4]$, $1\leq i, j\leq 2$.

The number of matrices A such that the sum of all entries is a prime number $p\in(2,13)$ is ______.

[Main 2023 31st Jan 2nd Shift]

47. Let A be a $n\times n$ matrix such that $|A|=2$. If the determinant of the matrix $\text{Adj}(2\cdot\text{Adj}(2A^{-1}))$ is 2^{84}, then n is equal to ______.

[Main 2023 31st Jan 2nd Shift]

48. If $A=\frac{1}{2}\begin{bmatrix}1 & \sqrt{3}\\ -\sqrt{3} & 1\end{bmatrix}$, then:

(A) $A^{30}-A^{25}=2I$
(B) $A^{30}+A^{25}+A=I$
(C) $A^{30}+A^{25}-A=I$
(D) $A^{30}=A^{25}$

[Main 2023 01st Feb 2nd Shift]

49. Let $A=\left[a_{ij}\right]_{2\times 2}$, where $ij\neq 0$ for all i, j and $A^2=I$, Let a be the sum of all diagonal elements of A and $b=|A|$ Then $3a^2+4b^2$ is equal to

(A) 4 (B) 14 (C) 7 (D) 3

[Main 2023 6th April 1st Shift]

50. Let P be a square matrix such that $P^2=I-P$. For $\alpha,\beta,\gamma,\delta\in\mathbb{N}$, if $P^{\alpha}+P^{\beta}=\gamma\ell-29P$ and $P^{\alpha}-P^{\beta}=\delta\ell-13P$, then $\alpha+\beta+\gamma-\delta$ is equal to

(A) 18 (B) 40 (C) 22 (D) 24

[Main 2023 6th April 2nd Shift]

51. Let $\begin{bmatrix}2 & 1 & 0\\ 1 & 2 & -1\\ 0 & -1 & 2\end{bmatrix}$.

If $\left|\text{adj}(\text{adj}(\text{adj } 2A))\right|=(16)^n$, then n is equal to

(A) 8 (B) 10 (C) 9 (D) 12

[Main 2023 8th April 1st Shift]

52. Let $P=\begin{bmatrix}\frac{\sqrt{3}}{2} & \frac{1}{2}\\ -\frac{1}{2} & \frac{\sqrt{3}}{2}\end{bmatrix}$, $A=\begin{bmatrix}1 & 1\\ 0 & 1\end{bmatrix}$ and $Q=PAP^T$. If $P^TQ^{2007}P=\begin{bmatrix}a & b\\ c & d\end{bmatrix}$ then $2a+b-3c-4d$ is equal to

(A) 2004 (B) 2005 (C) 2007 (D) 2006

[Main 2023 8th April 1st Shift]

53. If $A=\begin{bmatrix}1 & 5\\ \lambda & 10\end{bmatrix}$, $A^{-1}=\alpha A+\beta I$ and $\alpha+\beta=-2$, then $4\alpha^2+\beta^2+\lambda^2$ is equal to

(A) 12 (B) 19 (C) 14 (D) 10

[Main 2023 8th April 2nd Shift]

54. If A is a 3×3 matrix and $|A|=2$, then $\left|3\,\text{adj}\left(|3A|A^2\right)\right|$ is equal to

(A) $3^{12}\cdot 6^{11}$ (B) $3^{12}\cdot 6^{10}$
(C) $3^{10}\cdot 6^{11}$ (D) $3^{11}\cdot 6^{10}$

[Main 2023 10th April 1st Shift]

55. If $A=\frac{1}{5!\,6!\,7!}\begin{bmatrix}5! & 6! & 7!\\ 6! & 7! & 8!\\ 7! & 8! & 9!\end{bmatrix}$, then $\left|\text{adj}(\text{adj}(2A))\right|$ is equal to

(A) 2^{20} (B) 2^8 (C) 2^{12} (D) 2^{16}

[Main 2023 10th April 2nd Shift]

56. Let $A=\begin{bmatrix}0 & 1 & 2\\ a & 0 & 3\\ 1 & c & 0\end{bmatrix}$, where $a,c\in R$. If $A^3=A$ and the positive value of a belongs to the interval $(n-1,n]$, where $n\in\mathbb{N}$, then n is equal to ______.

[Main 2023 11th April 1st Shift]

57. Let $A=\begin{bmatrix}1 & \frac{1}{51}\\ 0 & 1\end{bmatrix}$. If $B=\begin{bmatrix}1 & 2\\ -1 & -1\end{bmatrix}A\begin{bmatrix}-1 & -2\\ 1 & 1\end{bmatrix}$, then the sum of all the elements of the matrix $\sum_{n=1}^{50}B^n$ is equal to

(A) 75 (B) 125 (C) 50 (D) 100

[Main 2023 12th April 1st Shift]

58. The number of symmetric matrices of order 3, with all the entries from the set $\{0,1,2,3,4,5,6,7,8,9\}$ is

(A) 6^{10} (B) 10^6 (C) 9^{10} (D) 10^9

[Main 2023 13th April 1st Shift]

59. Let $B=\begin{bmatrix}1 & 3 & \alpha\\ 1 & 2 & 3\\ \alpha & \alpha & 4\end{bmatrix}$, $\alpha>2$ be the adjoint of matrix A and $|A|=2$.

Then $\begin{bmatrix}\alpha & -2\alpha & \alpha\end{bmatrix}B\begin{bmatrix}\alpha\\ -2\alpha\\ \alpha\end{bmatrix}$ is equal to

(A) 0 (B) 16 (C) -16 (D) 32

[Main 2023 13th April 1st Shift]

60. Let for $A=\begin{bmatrix}1 & 2 & 3\\ \alpha & 3 & 1\\ 1 & 1 & 2\end{bmatrix}$, $|A|=2$.

If $\left|2\,\text{adj}\left(2\,\text{adj}(2A)\right)\right| = 32^n$, then $3n+\alpha$ is equal to

(A) 9 (B) 11 (C) 12 (D) 10

[Main 2023 13th April 2nd Shift]

61. Let the determinant of a square matrix A of order m be $m-n$, where m and n satisfy $4m+n=22$ and $17m+4n=93$.
If $\det\left(n\,\text{adj}\left(\text{adj}(mA)\right)\right) = 3^a 5^b 6^c$, then $a+b+c$ is equal to

(A) 84 (B) 96 (C) 101 (D) 109

[Main 2023 15th April 1st Shift]

62. If $A = \begin{bmatrix} \sqrt{2} & 1 \\ -1 & \sqrt{2} \end{bmatrix}, B = \begin{bmatrix} 1 & 0 \\ 1 & 1 \end{bmatrix}, C = ABA^T$ and $X = A^T C^2 A$, then $\det X$ is equal to:

(A) 243 (B) 729 (C) 27 (D) 891

[Main 2024 1st Feb 1st Shift]

63. Let $A = I_2 - MM^T$, where M is a real matrix of order 2×1 such that the relation $M^T M = I_1$ holds. If λ is a real number such that the relation $AX = \lambda X$ holds for some non-zero real matrix X of order 2×1, then the sum of squares of all possible values of λ is equal to ______.

[Main 2024 1st Feb 2nd Shift]

64. Consider the matrix

$$f(x) = \begin{bmatrix} \cos x & -\sin x & 0 \\ \sin x & \cos x & 0 \\ 0 & 0 & 1 \end{bmatrix}.$$

Given below are two statements:

Statement I: $f(-x)$ is the inverse of the matrix $f(x)$.

Statement II: $f(x)f(y) = f(x+y)$

In the light of the above statements, choose the correct answer from the options given below

(A) Statement I is false but Statement II is true.

(B) Both Statement I and Statement II are false.

(C) Statement I is true but Statement II is false.

(D) Both Statement I and Statement II are true.

[Main 2024 27th Jan 1st Shift]

65. Let A be a 2×2 real matrix and I be the identity matrix of order 2. If the roots of the equation $|A - xI| = 0$ be -1 and 3, then the sum of the diagonal elements of the matrix A^2 is **[Main 2024 27th Jan 2nd Shift]**

66. Let $A = \begin{bmatrix} 1 & 0 & 0 \\ 0 & \alpha & \beta \\ 0 & \beta & \alpha \end{bmatrix}$ and $|2A|^3 = 2^{21}$ where $\alpha, \beta \in Z$, then a value of α is

(A) 3 (B) 5 (C) 17 (D) 9

[Main 2024 29th Jan 1st Shift]

67. Let A be a square matrix such that $AA^T = I$. Then $\frac{1}{2}A\left[\left(A+A^T\right)^2 + \left(A-A^T\right)^2\right]$ is equal to

(A) $A^2 + I$ (B) $A^3 + U = I$

(C) $A^2 + A^T$ (D) $A^3 + A^T$

[Main 2024 29th Jan 1st Shift]

68. Let $A = \begin{bmatrix} 2 & 1 & 2 \\ 6 & 2 & 11 \\ 3 & 3 & 2 \end{bmatrix}$ and $P = \begin{bmatrix} 1 & 2 & 0 \\ 5 & 0 & 2 \\ 7 & 1 & 5 \end{bmatrix}$. The sum of the prime factors of $\left|P^{-1}AP - 2I\right|$ is equal to

(A) 26 (B) 27 (C) 66 (D) 23

[Main 2024 29th Jan 2nd Shift]

69. Let $R = \begin{pmatrix} x & 0 & 0 \\ 0 & y & 0 \\ 0 & 0 & z \end{pmatrix}$ be a non-zero 3×3 matrix, where

$x\sin\theta = y\sin\left(\theta + \frac{2\pi}{3}\right) = z\sin\left(\theta + \frac{4\pi}{3}\right)$
$\neq 0, \theta \in (0, 2\pi)$. For a square matrix M, let trace (M) denote the sum of all the diagonal entries of M. Then, among the statements

(I) Trace $(R) = 0$

(II) If trace $\left(\text{adj}\left(\text{adj}(R)\right)\right) = 0$, then R has exactly one non-zero entry.

(A) Both (I) and (II) are true.

(B) Neither (I) nor (II) is true.

(C) Only (II) is true.

(D) Only (I) is true.

[Main 2024 30th Jan 2nd Shift]

70. Let A be a 3×3 matrix and $\det(A) = 2$. If $n = \det\Big(\underbrace{\text{adj}\left(\text{adj}\left(\ldots(\text{adj}\,A)\right)\right)}_{2024\text{ - times}}\Big)$. Then the remainder when n is divided by 9 is equal to ______. **[Main 2024 31st Jan 2nd Shift]**

6.7 Answer Key

6.7.1 Know Your Concepts

1. (0) 2. $5\sqrt{3}\left(\sqrt{6}-5\right)$ 3. (2) 4. (1) 5. (–9,2,7) 6. (17)
7. (121) 8. (10) 9. (0) 10. (Infinite solutions) 11. ($\lambda = -2$)
12. (4) 13. (0) 15. (1) 16. (12) 17. (2) 18. (2) 19. (2)
20. (–1) 21. (3) 22. (4) 23. (0) 24. $\begin{bmatrix} -1 & -2 \\ -1 & -1 \end{bmatrix}$ 25. $\theta = 2n\pi, n \in I$
26. (200) 27. $\dfrac{-9 \pm \sqrt{53}}{2}$ 28. $\dfrac{5\pi}{12}$ 29. (4)

6.7.2 JEE Main: Sample Questions

Single Answer Type

1. (B) 2. (C) 3. (C) 4. (B) 5. (B) 6. (C) 7. (D) 8. (C)
9. (B) 10. (C) 11. (B) 12. (C) 13. (B) 14. (A) 15. (C) 16. (D)
17. (D) 18. (B) 19. (B) 20. (B) 21. (B) 22. (B) 23. (C) 24. (D)
25. (B) 26. (C) 27. (B) 28. (A) 29. (D) 30. (C) 31. (B) 32. (C)
33. (D) 34. (B) 35. (C) 36. (B) 37. (D) 38. (D) 39. (D) 40. (B)
41. (D) 42. (A) 43. (A) 44. (C) 45. (D) 46. (D) 47. (B) 48. (B)
49. (B) 50. (C) 51. (C) 52. (B) 53. (C) 54. (C) 55. (B) 56. (C)
57. (D) 58. (A)

Integer Answer Type

1. (1) 2. (1) 3. (2) 4. (–1) 5. (–8) 6. (5) 7. (0) 8. (0)
9. (2) 10. (1) 11. (3) 12. (5) 13. (7) 14. (3) 15. (0) 16. (1)
17. (1) 18. (0) 19. (2)

6.7.3 JEE Advanced: Sample Questions

Single Answer Type

1. (C) 2. (B) 3. (A) 4. (A) 5. (D) 6. (D) 7. (A) 8. (D)
9. (C) 10. (D) 11. (A) 12. (D) 13. (A) 14. (C) 15. (A) 16. (C)
17. (A) 18. (C) 19. (D) 20. (B) 21. (D) 22. (B) 23. (B) 24. (C)
25. (B) 26. (C) 27. (C) 28. (B) 29. (B) 30. (C) 31. (D) 32. (C)
33. (D) 34. (D) 35. (B) 36. (A) 37. (A) 38. (B) 39. (C) 40. (C)
41. (D)

Multiple Answer Type

1. (A, B, C, D) 2. (B, C, D) 3. (A, B, C) 4. (A, B)
5. (A, B, C, D) 6. (A, C) 7. (A, B, C, D) 8. (C, D)
9. (A, B, C) 10. (C, D) 11. (A, B, C) 12. (A, C, D)
13. (A, B, D) 14. (C, D) 15. (B, D) 16. (A, B, C, D)
17. (A, D) 18. (A, B, D) 19. (A, C, D) 20. (B, C)
21. (A, C) 22. (A, B, C, D) 23. (A, C, D) 24. (A, B)

Comprehension Based Questions

1. (C) 2. (C) 3. (B) 4. (B) 5. (C) 6. (D) 7. (A) 8. (B)
9. (B) 10. (C) 11. (C) 12. (C)

Match Matrix

1. (A – R; B – S; C – Q; D – P) 2. (A – R; B – S; C – Q; D – P)
3. (A – S; B – P; C – P, R; D – Q) 4. (A – P; B – T; C – Q, D – R)

Subjective Answer Type (Hard Nut)

1. (4) **2.** (0) **3.** (1) **4.** (0) **5.** (–4) **6.** (2) **7.** (3)

8. (1) **9.** (π) **10.** $-\left(\frac{\pi}{4}+\frac{2}{3}\right)$ **11.** (5) **12.** (2) **13.** (3) **14.** (0)

15. (0) **17.** (0) **19.** $\frac{-3\pi}{4}$ **20.** $=\frac{n}{6}\left[2n^3+n+3\right]$ **23.** –1 or 5 **24.** (2)

25. (1) **26.** (4) **27.** (2) **28.** (1) **29.** (7) **30.** $\frac{3}{2}$ **31.** $\cos^{-1}\left(\frac{7}{\sqrt{202}}\right)$

32. $\left(\frac{-4\sqrt{2}}{3},\frac{2}{3},2\sqrt{2}\right),\left(\frac{4\sqrt{2}}{3},\frac{2}{3},-2\sqrt{2}\right),(3,3,-1)$ **33.** Neither symmetric nor skew symmetric

34. $\begin{bmatrix}-4 & 7 & -7\\ 3 & -5 & 5\end{bmatrix}$ **35.** $[a\neq-3, b\in R];\left[a=-3, b\neq\frac{1}{3}\right];\left[a=-3, b=\frac{1}{3}\right]$ **36.** 2 or 4 **37.** $f(a)=\frac{1}{4}$

6.7.4 JEE Main: Archive

Determinants

1. (B)	**2.** (C)	**3.** (D)	**4.** (C)	**5.** (C)	**6.** (D)	**7.** (B)	**8.** (B)
9. (58)	**10.** (B)	**11.** (B)	**12.** (C)	**13.** (D)	**14.** (4)	**15.** (C)	**16.** (C)
17. (D)	**18.** (B)	**19.** (D)	**20.** (C)	**21.** (D)	**22.** (D)	**23.** (A)	**24.** (C)
25. (D)	**26.** (A)	**27.** (B)	**28.** (24)	**29.** (D)	**30.** (A)	**31.** (D)	**32.** (6)
33. (B)	**34.** (B)	**35.** (B)	**36.** (B)	**37.** (B)	**38.** (28)	**39.** (B)	**40.** (113)
41. (B)	**42.** (A)	**43.** (D)	**44.** (D)	**45.** (C)	**46.** (A)		

Matrices

1. (B)	**2.** (100)	**3.** (B)	**4.** (A)	**5.** (414)	**6.** (1)	**7.** (C)	**8.** (100)
9. (14)	**10.** (C)	**11.** (180)	**12.** (A)	**13.** (25)	**14.** (C)	**15.** (A)	**16.** (C)
17. (1)	**18.** (17)	**19.** (24)	**20.** (B)	**21.** (D)	**22.** (50)	**23.** (D)	**24.** (5376)
25. (B)	**26.** (42)	**27.** (C)	**28.** (2)	**29.** (C)	**30.** (B)	**31.** (282)	**32.** (C)
33. (10)	**34.** (D)	**35.** (B)	**36.** (A)	**37.** (B)	**38.** (D)	**39.** (A)	**40.** (D)
41. (D)	**42.** (5)	**43.** (A)	**44.** (D)	**45.** (C)	**46.** (196)	**47.** (5)	**48.** (C)
49. (A)	**50.** (D)	**51.** (B)	**52.** (B)	**53.** (C)	**54.** (D)	**55.** (D)	**56.** (2)
57. (D)	**58.** (B)	**59.** (C)	**60.** (B)	**61.** (B)	**62.** (B)	**63.** (2)	**64.** (D)
65. (10)	**66.** (B)	**67.** (D)	**68.** (A)	**69.** (B)	**70.** (7)		

6.8 Solutions

6.8.1 Know Your Concepts

1. Let $\Delta=\begin{vmatrix}265 & 240 & 219\\ 240 & 225 & 198\\ 219 & 198 & 181\end{vmatrix}$

Operating C_1-C_2 and C_2-C_3

$$\Delta=\begin{vmatrix}25 & 21 & 219\\ 15 & 27 & 198\\ 21 & 17 & 181\end{vmatrix}$$

Operating C_1-C_2 and C_3-10C_2

$$\Delta=\begin{vmatrix}4 & 21 & 9\\ -12 & 27 & -72\\ 4 & 17 & 11\end{vmatrix}$$

Operating R_1-R_3 and R_2+3R_3

$$\Delta=\begin{vmatrix}0 & 4 & -2\\ 0 & 78 & -39\\ 4 & 17 & 11\end{vmatrix}=0$$

[As the first two rows are proportional]

2. Let $\Delta=\begin{vmatrix}\sqrt{13}+\sqrt{3} & 2\sqrt{5} & \sqrt{5}\\ \sqrt{15}+\sqrt{26} & 5 & \sqrt{10}\\ 3+\sqrt{65} & \sqrt{15} & 5\end{vmatrix}$

$$\Rightarrow \Delta = \sqrt{5}\times\sqrt{5}\begin{vmatrix} \sqrt{13}+\sqrt{3} & 2 & 1 \\ \sqrt{15}+\sqrt{26} & \sqrt{5} & \sqrt{2} \\ 3+\sqrt{65} & \sqrt{3} & \sqrt{5} \end{vmatrix}$$

$$= 5\left[\begin{vmatrix} \sqrt{13} & 2 & 1 \\ \sqrt{26} & \sqrt{5} & \sqrt{2} \\ \sqrt{65} & \sqrt{3} & \sqrt{5} \end{vmatrix} + \begin{vmatrix} \sqrt{3} & 2 & 1 \\ \sqrt{15} & \sqrt{5} & \sqrt{2} \\ 3 & \sqrt{3} & \sqrt{5} \end{vmatrix}\right]$$

$$=5\left[\sqrt{13}\begin{vmatrix} 1 & 2 & 1 \\ \sqrt{2} & \sqrt{5} & \sqrt{2} \\ \sqrt{5} & \sqrt{3} & \sqrt{5} \end{vmatrix} + \sqrt{3}\begin{vmatrix} 1 & 2 & 1 \\ \sqrt{5} & \sqrt{5} & \sqrt{2} \\ \sqrt{3} & \sqrt{3} & \sqrt{5} \end{vmatrix}\right]$$

$$= 5\left[0+\sqrt{3}\begin{vmatrix} -1 & 2 & 1 \\ 0 & \sqrt{5} & \sqrt{2} \\ 0 & \sqrt{3} & \sqrt{5} \end{vmatrix}\right] [C_1 - C_2]$$

$$= -5\sqrt{3}\left(5-\sqrt{6}\right) = 5\sqrt{3}\left(\sqrt{6}-5\right)$$

3. Let, $\Delta = \begin{vmatrix} a+b+2c & a & b \\ c & b+c+2a & b \\ c & a & c+a+2b \end{vmatrix}$

$C_1 \to C_1 + C_2 + C_3$

$$\Delta = \begin{vmatrix} 2(a+b+c) & a & b \\ 2(a+b+c) & b+c+2a & b \\ 2(a+b+c) & a & c+a+2b \end{vmatrix}$$

$$= 2(a+b+c)\begin{vmatrix} 1 & a & b \\ 1 & b+c+2a & b \\ 1 & a & c+a+2b \end{vmatrix}$$

$R_2 - R_1,\ R_3 - R_1$

$$= 2(a+b+c)\begin{vmatrix} 1 & a & b \\ 0 & a+b+c & 0 \\ 1 & a & c+a+2b \end{vmatrix}$$

$$= 2(a+b+c)^2\begin{vmatrix} 1 & a & b \\ 0 & 1 & 0 \\ 0 & 0 & c+a \end{vmatrix}$$

$$= 2(a+b+c)^3 \Rightarrow K = 2$$

4. $D_1 = \begin{vmatrix} 1 & 1 & 1 \\ x^2 & y^2 & z^2 \\ x & y & z \end{vmatrix}$

operating $C_1 - C_2$ and $C_2 - C_3$

$$= \begin{vmatrix} 0 & 0 & 1 \\ (x+y)(x-y) & (y+z)(y-z) & z^2 \\ x-y & y-z & z \end{vmatrix}$$

$$= (x-y)(y-z)\begin{vmatrix} 0 & 0 & 1 \\ x+y & y+z & z^2 \\ 1 & 1 & z \end{vmatrix}$$

$$= (x-y)(y-z)(x-z)$$

$$D_2 = \begin{vmatrix} 1 & 1 & 1 \\ yz & zx & xy \\ x & y & z \end{vmatrix}$$

operating $C_1 - C_2$ and $C_2 - C_3$

$$= \begin{vmatrix} 0 & 0 & 1 \\ -z(x-y) & -x(y-z) & xy \\ x-y & y-z & z \end{vmatrix}$$

$$= (x-y)(y-z)\begin{vmatrix} 0 & 0 & 1 \\ -z & -x & xy \\ 1 & 1 & z \end{vmatrix}$$

$$= (x-y)(y-z)\begin{vmatrix} 0 & 0 & 1 \\ -z & -x & xy \\ 1 & 1 & z \end{vmatrix}$$

$$= (x-y)(y-z)(x-z) \Rightarrow \frac{D_1}{D_2} = 1$$

5. $\begin{vmatrix} x & 3 & 7 \\ 2 & x & 2 \\ 7 & 6 & x \end{vmatrix} = 0$

operating $C_1 - C_3$

$$\begin{vmatrix} x-7 & 3 & 7 \\ 0 & x & 2 \\ -(x-7) & 6 & x \end{vmatrix} = 0 \Rightarrow (x-7)\begin{vmatrix} 1 & 3 & 7 \\ 0 & x & 2 \\ -1 & 6 & x \end{vmatrix} = 0$$

$$\Rightarrow (x-7)\left\{1\left(x^2-12\right)-1(6-7x)\right\} = 0$$

$$\Rightarrow (x-7)\left(x^2+7x-18\right) = 0$$

$$\Rightarrow x = -9, 2, 7$$

6. $\begin{vmatrix} x^3+4x & x+3 & x-2 \\ x-2 & 5x & x-1 \\ x-3 & x+2 & 4x \end{vmatrix}$

$= ax^5+bx^4+cx^3+dx^2+ex+f$

Put $x=0$

$f = \begin{vmatrix} 0 & 3 & -2 \\ -2 & 0 & -1 \\ -3 & 2 & 0 \end{vmatrix} = 0-3(0-3)-2(-4-0)$

$= 9+8 = 17$

7. We know that if k is the value of a determinant of third order, then the value of determinant formed by its co-factor is k^2.
Hence 121.

8. Let, $\Delta = \begin{vmatrix} 1 & 3\cos\theta & 1 \\ \sin\theta & 1 & 3\cos\theta \\ 1 & \sin\theta & 1 \end{vmatrix}$

Operating $C_1 - C_3$

$\Delta = \begin{vmatrix} 0 & 3\cos\theta & 1 \\ \sin\theta - 3\cos\theta & 1 & 3\cos\theta \\ 0 & \sin\theta & 1 \end{vmatrix}$

$= (3\cos\theta - \sin\theta)^2$

$\Rightarrow$ Max. value $= 10$

9. $\alpha+\beta+\gamma = 0$

$\Delta = \begin{vmatrix} \alpha & \beta & \gamma \\ \beta & \gamma & \alpha \\ \gamma & \alpha & \beta \end{vmatrix}$

$R_1 \to R_1 + R_2 + R_3$

$= \begin{vmatrix} \alpha+\beta+\gamma & \alpha+\beta+\gamma & \alpha+\beta+\gamma \\ \beta & \gamma & \alpha \\ \gamma & \alpha & \beta \end{vmatrix}$

$= 0$

10. $\Delta = \begin{vmatrix} 1 & 2 & 3 \\ 2 & 3 & 4 \\ 3 & 4 & 5 \end{vmatrix}$

Operating $R_1 - R_2$ and $R_2 - R_3$

$\Delta = 0$

$\Delta_x = \begin{vmatrix} 4 & 2 & 3 \\ 5 & 3 & 4 \\ 6 & 4 & 5 \end{vmatrix}$

operating $R_1 - R_2$ and $R_2 - R_3$

$\Delta_x = 0$

Similarly, $\Delta_y = 0$ and $\Delta_z = 0$

Hence infinite solutions.

11. For a non-trivial solution,

$\Delta = 0$

$\Rightarrow \Delta = \begin{vmatrix} 2 & 5 & 8 \\ 1 & 4 & 7 \\ 6 & 4 & -\lambda \end{vmatrix} = 0 \Rightarrow \lambda = -2$

12. Let $\Delta = \begin{vmatrix} a^2 & b^2 & c^2 \\ (a+1)^2 & (b+1)^2 & (c+1)^2 \\ (a-1)^2 & (b-1)^2 & (c-1)^2 \end{vmatrix}$

$R_2 - R_1$ and $R_3 - R_1$

$\Delta = \begin{vmatrix} a^2 & b^2 & c^2 \\ 2a+1 & 2b+1 & 2c+1 \\ -2a+1 & -2b+1 & -2c+1 \end{vmatrix}$

$R_2 - R_3$

$\Delta = \begin{vmatrix} a^2 & b^2 & c^2 \\ 4a & 4b & 4c \\ -2a+1 & -2b+1 & -2c+1 \end{vmatrix}$

$= 4\begin{vmatrix} a^2 & b^2 & c^2 \\ a & b & c \\ -2a+1 & -2b+1 & -2c+1 \end{vmatrix}$

$R_3 + 2R_2$

$= 4\begin{vmatrix} a^2 & b^2 & c^2 \\ a & b & c \\ 1 & 1 & 1 \end{vmatrix} \Rightarrow K = 4$

13. Since, $\dfrac{a_n+1}{a_n} = \dfrac{a_{n+2}}{a_{n+1}} = \dfrac{a_{n+4}}{a_{n+3}} = \ldots = (r)$ say

$C_2 - C_1$ and $C_3 - C_2 = 0$

14. Let, $\Delta = \begin{vmatrix} a-b-c & 2a & 2a \\ 2b & b-c-a & 2b \\ 2c & 2c & c-a-b \end{vmatrix}$

Operating $R_1 \to R_1 + R_2 + R_3$

$$\Rightarrow \Delta = \begin{vmatrix} a+b+c & a+b+c & a+b+c \\ 2b & b-c-a & 2b \\ 2c & 2c & c-a-b \end{vmatrix}$$

$C_1 - C_2$ and $C_2 - C_3$

$$= (a+b+c)\begin{vmatrix} 0 & 0 & 1 \\ a+b+c & -a-b-c & 2b \\ 0 & +a+b+c & c-a-b \end{vmatrix}$$

$$= (a+b+c)^3 \begin{vmatrix} 0 & 0 & 1 \\ 1 & -1 & 2b \\ 0 & +1 & c-a-b \end{vmatrix}$$

$$= (a+b+c)^3 \{0-0+1(1-0)\}$$

$$= (a+b+c)^3$$

15. Multiplying C_1, C_2, C_3 respectively by l, m, n

$$\Delta = \frac{1}{lmn}\begin{vmatrix} \ell(1+\ell^2) & \ell m^2 & \ell n^2 \\ \ell^2 m & m(1+m^2) & mn^2 \\ \ell^2 n & m^2 n & n(1+n^2) \end{vmatrix}$$

$$= \begin{vmatrix} 1+\ell^2 & m^2 & n^2 \\ \ell^2 & 1+m^2 & n^2 \\ \ell^2 & m^2 & 1+n^2 \end{vmatrix}$$

$C_1 + (C_2 + C_3)$

$$= \begin{vmatrix} 1 & m^2 & n^2 \\ 1 & 1+m^2 & n^2 \\ 1 & m^2 & 1+n^2 \end{vmatrix} \qquad [\ell^2 + m^2 + n^2 = 0]$$

$R_2 - R_1$ and $R_3 - R_1$

$$\begin{vmatrix} 1 & m^2 & n^2 \\ 0 & 1 & 0 \\ 0 & 0 & 1 \end{vmatrix} = 1$$

16. Let, $\Delta = \begin{vmatrix} {}^xC_1 & {}^xC_2 & {}^xC_3 \\ {}^yC_1 & {}^yC_2 & {}^yC_3 \\ {}^zC_1 & {}^zC_2 & {}^zC_3 \end{vmatrix}$

$$\Rightarrow \Delta = \begin{vmatrix} x & \frac{x(x-1)}{2} & \frac{x(x-1)(x-2)}{6} \\ y & \frac{y(y-1)}{2} & \frac{y(y-1)(y-2)}{6} \\ z & \frac{z(z-1)}{2} & \frac{z(z-1)(z-2)}{6} \end{vmatrix}$$

$$= \frac{1}{12}xyz\begin{vmatrix} 1 & x-1 & x^2-3x+2 \\ 1 & y-1 & y^2-3y+2 \\ 1 & z-1 & z^2-3z+2 \end{vmatrix}$$

$C_2 \to C_2 + C_1$

$$= \frac{1}{12}xyz\begin{vmatrix} 1 & x & x^2-3x+2 \\ 1 & y & y^2-3y+2 \\ 1 & z & z^2-3z+2 \end{vmatrix}$$

$C_3 \to C_3 + 3C_2 - 2C_1$

$$= \frac{1}{12}xyz(x-y)(y-z)(z-x)$$

$$\Rightarrow \alpha\beta = 12$$

17. $A = -\begin{vmatrix} \ell & m & 2n \\ d & e & 2f \\ a & b & 2c \end{vmatrix} = \begin{vmatrix} d & e & 2f \\ \ell & m & 2n \\ a & b & 2c \end{vmatrix}$

$$= -\begin{vmatrix} 2f & e & d \\ 2n & m & \ell \\ 2c & b & a \end{vmatrix} = \begin{vmatrix} 2f & d & e \\ 2n & \ell & m \\ 2c & a & b \end{vmatrix}$$

$$= 2\cdot\frac{1}{2}\begin{vmatrix} f & 2d & e \\ n & 2\ell & m \\ c & 2a & b \end{vmatrix}$$

$$= \frac{1}{2}\begin{vmatrix} f & 2d & e \\ 2n & 4\ell & 2m \\ c & 2a & b \end{vmatrix} = \frac{B}{2}$$

$$\Rightarrow \frac{B}{A} = 2$$

18. Let, $\Delta = \begin{vmatrix} \tan A & 1 & 1 \\ 1 & \tan B & 1 \\ 1 & 1 & \tan C \end{vmatrix}$

$$\Rightarrow \Delta = \begin{vmatrix} \tan A & 1 & 0 \\ 1 & \tan\beta & 1-\tan\beta \\ 1 & 1 & \tan C - 1 \end{vmatrix}$$

$$\Rightarrow \Delta = \tan A(\tan B \tan C - 1) - 1(\tan C - 1 - 1 + \tan B)$$

$$= \tan A \tan B \tan C - \tan A - \tan B - \tan C + 2$$

$$= 2[\tan A + \tan B + \tan C = \tan A \tan B \tan C]$$

19. Differentiating and putting $x = 0$

$$\therefore f'(0) = \begin{vmatrix} 0 & 1 & 0 \\ 0 & 1 & 0 \\ 0 & 0 & 0 \end{vmatrix} + \begin{vmatrix} 1 & 0 & -1 \\ 1 & 0 & 1 \\ 0 & 0 & 0 \end{vmatrix}$$

$$+ \begin{vmatrix} 1 & 0 & -1 \\ 0 & 1 & 0 \\ 2 & 0 & 0 \end{vmatrix} = 2$$

20. $\Delta = \begin{vmatrix} p+a & b & c \\ a & q+b & c \\ a & b & r+c \end{vmatrix} = 0$

$R_1 - R_2$ and $R_2 - R_3$

$$\Delta = \begin{vmatrix} p & -q & 0 \\ 0 & q & -r \\ a & b & r+c \end{vmatrix} = 0$$

$$\Rightarrow p(qr + qc + br) + q(0 + ar) = 0$$

$$\Rightarrow pqr + pqc + prb + qra = 0$$

$$\Rightarrow 1 + \frac{c}{r} + \frac{b}{q} + \frac{a}{p} = 0$$

$$\Rightarrow \frac{a}{p} + \frac{b}{q} + \frac{c}{r} = -1$$

21. $t_r(A) = t_r(B)$

$$\Rightarrow x^2 + y^2 + z^2 = 2x + 2y + 2z - 3$$

$$\Rightarrow (x^2 - 2x + 1) + (y^2 - 2y + 1) + (z^2 - 2z + 1) = 0$$

$$\Rightarrow (x-1)^2 + (y-1)^2 + (z-1)^2 = 0$$

$$\Rightarrow x = 1, y = 1, z = 1$$

$$\therefore x + y + z = 3$$

22. We know that in a square matrix of order n

$$|\text{adj}(A)| = |A|^{n-1}$$

$$\Rightarrow |\text{adj}(\text{adj } A)| = |\text{adj}(A)|^{n-1} = |A|^{(n-1)^2}$$

$$\therefore n^2 - 2n + 1 = 9 \Rightarrow n^2 - 2n - 8 = 0$$

$\Rightarrow n = 4$, as $n = -2$ is not possible

23. A is an idempotent matrix, hence

$$A^2 = A \Rightarrow A^n = A$$

So that $AB = A(1 - A) = A - A^2 = A - A = 0$

24. $I + A = \begin{bmatrix} 1 & 0 \\ 0 & 1 \end{bmatrix} + \begin{bmatrix} 1 & 2 \\ 1 & 1 \end{bmatrix} = \begin{bmatrix} 2 & 2 \\ 1 & 2 \end{bmatrix}$

$$I - A = \begin{bmatrix} 1 & 0 \\ 0 & 1 \end{bmatrix} - \begin{bmatrix} 1 & 2 \\ 1 & 1 \end{bmatrix} = \begin{bmatrix} 0 & -2 \\ -1 & 0 \end{bmatrix}$$

$$|I - A| = \begin{vmatrix} 0 & -2 \\ -1 & 0 \end{vmatrix} = -2 \neq 0$$

Hence, found.

25. $[1 \;\; x \;\; 1] \begin{bmatrix} 1 & 3 & 2 \\ 0 & 5 & 1 \\ 0 & 3 & 2 \end{bmatrix} \begin{bmatrix} 1 \\ 1 \\ x \end{bmatrix} = 0$

$$\Rightarrow [1 \quad 5x+6 \quad x+4] \begin{bmatrix} 1 \\ 1 \\ x \end{bmatrix} = 0$$

$$\Rightarrow [1 + 5x + 6 + x^2 + 4x] = 0$$

$$\Rightarrow x^2 + 9x + 7 = 0$$

$$\Rightarrow x = \frac{-9 \pm \sqrt{53}}{2}$$

26. $A = \begin{bmatrix} 1 & 2 & a \\ 0 & 1 & 4 \\ 0 & 0 & 1 \end{bmatrix} \therefore A^2 = \begin{bmatrix} 1 & 2 & a \\ 0 & 1 & 4 \\ 0 & 0 & 1 \end{bmatrix} \begin{bmatrix} 1 & 2 & a \\ 0 & 1 & 4 \\ 0 & 0 & 1 \end{bmatrix}$

$$= \begin{bmatrix} 1 & 4 & 2a+8 \\ 0 & 1 & 8 \\ 0 & 0 & 1 \end{bmatrix}$$

$$A^3 = A^2 \cdot A = \begin{bmatrix} 1 & 6 & 3a+24 \\ 0 & 1 & 12 \\ 0 & 0 & 1 \end{bmatrix}$$

Generalising the result,

$$A^n = \begin{bmatrix} 1 & 2n & na+8\sum_{K=0}^{n-1} K \\ 0 & 1 & 4n \\ 0 & 0 & 1 \end{bmatrix} = \begin{bmatrix} 1 & 18 & 2007 \\ 0 & 1 & 36 \\ 0 & 0 & 1 \end{bmatrix}$$

$\Rightarrow 2n = 18 \Rightarrow n = 9$

$\therefore 9a + \sum_{K=0}^{8} K = 2007$

$\Rightarrow 9a + 8\left(\frac{8\times 9}{2}\right) = 2007 \Rightarrow 9a = 2007 - 288$

$\Rightarrow a = 191$

$\therefore n + a = 200$

27. We have, $\begin{vmatrix} a & b & c \\ b & c & a \\ c & a & b \end{vmatrix} = 0$

$\Rightarrow -\frac{1}{2}(a+b+c)\left[(a-b)^2 + (b-c)^2 + (c-a)^2\right] = 0$

$\Rightarrow$ Either $a+b+c = 0$ or $a = b = c$

If $a+b+c = 0$, then we must have

$\cos\theta + \cos 3\theta + \cos 2\theta = 0$ and

$\sin\theta + \sin 3\theta - \sin 2\theta = 0$

$\Rightarrow \cos\theta(2\cos\theta + 1) = 0$ and

$\sin 2\theta(2\cos\theta - 1) = 0$

The above equations do not hold simultaneously.

Therefore, the only possibility is $a = b = c$

$\Rightarrow e^{i\theta} = e^{-2i\theta} = e^{3i\theta}$ which is satisfied only when,

$e^{i\theta} = 1 \Rightarrow \cos\theta = 1, \sin\theta = 0$

$\Rightarrow \theta = 2n\pi, n \in I$

28. Let, $A = \begin{bmatrix} 0 & 2\sin\beta & \tan\gamma \\ \cos\alpha & \sin\beta & -\tan\gamma \\ \cos\alpha & -\sin\beta & \tan\gamma \end{bmatrix}$

As A is orthogonal

$\Rightarrow AA^T = I = A^T A$

$$\Rightarrow \begin{bmatrix} 0 & 2\sin\beta & \tan\gamma \\ \cos\alpha & \sin\beta & -\tan\gamma \\ \cos\alpha & -\sin\beta & \tan\gamma \end{bmatrix} \begin{bmatrix} 0 & \cos\alpha & \cos\alpha \\ 2\sin\beta & \sin\beta & -\sin\beta \\ \tan\gamma & -\tan\gamma & \tan\gamma \end{bmatrix} = I$$

$$\Rightarrow \begin{bmatrix} 4\sin^2\beta + \tan^2\gamma & 2\sin^2\beta - \tan^2\gamma & -2\sin^2\beta + \tan^2\gamma \\ 2\sin^2\beta - \tan^2\gamma & \cos^2\alpha + \sin^2\beta + \tan^2\gamma & \cos^2\alpha - \sin^2\beta - \tan^2\gamma \\ -2\sin^2\beta + \tan^2\gamma & \cos^2\alpha - \sin^2\beta - \tan^2\gamma & \cos^2\alpha + \sin^2\beta + \tan^2\gamma \end{bmatrix} = I$$

$\Rightarrow 4\sin^2\beta + \tan^2\gamma = \cos^2\alpha + \sin^2\beta + \tan^2\gamma = 1$

and $2\sin^2\beta - \tan^2\gamma = -2\sin^2\beta + \tan^2\gamma$

$= \cos^2\alpha - \sin^2\beta - \tan^2\gamma = 0$

$\Rightarrow 2\sin^2\beta = \tan^2\gamma; \cos^2\alpha = \frac{3}{2}\tan^2\gamma; 3\tan^2\gamma = 1$

$\Rightarrow \gamma = \frac{\pi}{6}, \alpha = \frac{\pi}{4} \quad \therefore \alpha + \gamma = \frac{5\pi}{12}$

29. 4 [Do it yourself]

30. $A = \begin{bmatrix} 1+a & 1 & 1 \\ 1 & 1+b & 1 \\ 1 & 1 & 1+c \end{bmatrix}$

$$|A| = abc \begin{vmatrix} 1+\frac{1}{a} & \frac{1}{b} & \frac{1}{c} \\ \frac{1}{a} & 1+\frac{1}{b} & \frac{1}{c} \\ \frac{1}{a} & \frac{1}{b} & 1+\frac{1}{c} \end{vmatrix}$$

$C_1 \to C_1 + C_2 + C_3$

$$|A| = abc \begin{vmatrix} 1 & \frac{1}{b} & \frac{1}{c} \\ 1 & 1+\frac{1}{b} & \frac{1}{c} \\ 1 & \frac{1}{b} & 1+\frac{1}{c} \end{vmatrix} \quad \left[\because \frac{1}{a}+\frac{1}{b}+\frac{1}{c} = 0\right]$$

Applying $R_2 \to R_2 - R_1$ and $R_3 \to R_3 - R_1$

$$\therefore |A| = abc\begin{vmatrix} 1 & \frac{1}{b} & \frac{1}{c} \\ 0 & 1 & 0 \\ 0 & 0 & 1 \end{vmatrix} = abc \neq 0$$

$\Rightarrow A$ is non-singular

6.8.2 JEE Main: Sample Questions

Single Answer Type

1. **(B)**

$$a_{ij} = \lim_{x\to 0} \frac{\sin(ix)}{\tan(jx)} = \frac{i}{j}$$

$$A = \begin{bmatrix} 1 & \frac{1}{2} & \frac{1}{3} \\ 2 & 1 & \frac{2}{3} \\ 3 & \frac{3}{2} & 1 \end{bmatrix} \Rightarrow A^2 = \begin{bmatrix} 3 & \frac{3}{2} & 1 \\ 6 & 3 & 2 \\ 9 & \frac{9}{2} & 3 \end{bmatrix}$$

$\Rightarrow A^2 = 3A$

2. **(C)**

$T_r(A) = \alpha^2 + \beta^2 + \gamma^2$

$T_r(B) = 2\alpha + 2\beta + 2\gamma - 3$

$\Rightarrow \alpha^2 + \beta^2 + \gamma^2 = 2\alpha + 2\beta + 2\gamma - 3$

$\Rightarrow (\alpha - 1)^2 + (\beta - 1)^2 + (\gamma - 1)^2 = 0$

$\Rightarrow \alpha = \beta = \gamma = 1$

$\Rightarrow \frac{1}{\alpha} + \frac{1}{\beta} + \frac{1}{\gamma} = 3$

3. **(C)**

$$\Delta = \begin{vmatrix} \tan\alpha & \cot\alpha & 8\cos 2\alpha \\ \cot\alpha & 8\cos 2\alpha & \tan\alpha \\ 8\cos 2\alpha & \tan\alpha & \cot\alpha \end{vmatrix} = 0$$

$$\Rightarrow \begin{vmatrix} \sin^2\alpha & \cos^2\alpha & 2\sin 4\alpha \\ \cos^2\alpha & 2\sin 4\alpha & \sin^2\alpha \\ 2\sin 4\alpha & \sin^2\alpha & \cos^2\alpha \end{vmatrix} = 0$$

$\Rightarrow 1 + 2\sin 4\alpha = 0$

4. **(B)**

$$\begin{vmatrix} 1-p & 2 & 1 \\ -3 & p & -2 \\ 2 & -2 & 1+p \end{vmatrix} = 0 \Rightarrow p = 0, 3, -3$$

5. **(B)**

Let $M = \begin{bmatrix} a & b \\ c & d \end{bmatrix}$

$\Rightarrow -\text{tr. }(M^2) + (\text{tr. }(M))^2 = 2 \det M$

6. **(C)**

Has non-trivial solutions.

$\therefore \Delta = 0$

$$\Delta = \begin{vmatrix} p & p+1 & p-1 \\ p+1 & p & p+2 \\ p-1 & P+2 & p \end{vmatrix} = 0$$

$R_2 \to R_2 - R_1;\ R_3 \to R_3 - R_1$

$$\Delta = \begin{vmatrix} p & p+1 & p-1 \\ 1 & -1 & 3 \\ -1 & 1 & 1 \end{vmatrix} = 0$$

$R_3 \to R_3 + R_2$

$$\Delta = \begin{vmatrix} p & p+1 & p-1 \\ 1 & -1 & 3 \\ 01 & 0 & 4 \end{vmatrix} = 0 \Rightarrow p = \frac{-1}{2}$$

$\therefore$ Exactly one real value of p.

7. **(D)**

$(a-1)\,x + 0y + z = \alpha$ (1)

$x + (b-1)\,y + 0z = \beta$ (2)

$0x + y + (c-1)z = \gamma$ (3)

For no unique solution $D = 0$

$$\begin{vmatrix} (a-1) & 0 & 1 \\ 1 & (b-1) & 0 \\ 0 & 1 & (c-1) \end{vmatrix} = 0$$

$(a-1)(b-1)(c-1) + 1 = 0$

$\therefore a = 2;\ b = 2;\ c = 0$

Hence, $|a + b + c| = 4$

8. **(C)** [Do it yourself]

9. **(B)**

$$M = \begin{pmatrix} 0 & i \\ i & 0 \end{pmatrix},\ M^2 = MM = \begin{pmatrix} -1 & 0 \\ 0 & -1 \end{pmatrix}$$

$$M^3 = \begin{pmatrix} 0 & -i \\ -i & 0 \end{pmatrix},\ M^4 = \begin{pmatrix} 1 & 0 \\ 0 & 1 \end{pmatrix} = I,$$

$M^{4k} = I\ \forall\ k \in N$

So, $I + M + M^2 + M^3 = \begin{bmatrix} 0 & 0 \\ 0 & 0 \end{bmatrix}$

$\therefore I + M + M^2 + M^3 + M^4 + M^5 + \ldots + M^{2010}$

$= (I + M + M^2 + M^3) + M^4(I + M + M^2 + M^3) + \ldots + M^{2008}(I + M + M^2)$

$$= I + M + M^2 = \begin{pmatrix} 1 & 0 \\ 0 & 1 \end{pmatrix} + \begin{pmatrix} 0 & i \\ i & 0 \end{pmatrix} + \begin{pmatrix} -1 & 0 \\ 0 & -1 \end{pmatrix}$$

$$= \begin{pmatrix} 0 & i \\ i & 0 \end{pmatrix} = M \text{ (itself)}$$

10. (C)

$$AB = \begin{bmatrix} -x & -y & 7z \\ 0 & y & 14z \\ x & -y & 7z \end{bmatrix}$$

Equating coefficients $x^2 = 4$; $y^2 = 9$ and $z^2 = 1$
$\Rightarrow x = \pm 2$, $y = \pm 3$ and $z = \pm 1$
Number of solutions $= 2 \times 2 \times 2 = 8$

11. (B)

$$P = \begin{bmatrix} 1 & 0 & 0 \\ 4 & 1 & 0 \\ 16 & 4 & 1 \end{bmatrix} \Rightarrow P^2 = \begin{bmatrix} 1 & 0 & 0 \\ 8 & 1 & 0 \\ 16+32 & 8 & 1 \end{bmatrix}$$

$$\text{so, } P^3 = \begin{bmatrix} 1 & 0 & 0 \\ 12 & 1 & 0 \\ 16+32+48 & 12 & 1 \end{bmatrix}$$

(from the symmetry)

$$\therefore P^{50} = \begin{bmatrix} 1 & 0 & 0 \\ 200 & 1 & 0 \\ \frac{16.50.51}{2} & 200 & 1 \end{bmatrix}$$

As, $P^{50} - Q = I \Rightarrow q_{31} = \frac{16.50.51}{2}$

$q_{32} = 200$ and $q_{21} = 200$

$$\therefore \frac{q_{31} + q_{32}}{q_{21}} = \frac{16.50.51}{2.200} + 1 = 102 + 1 = 103$$

12. (C)

Put $\Delta = 0$

$$\Rightarrow \begin{vmatrix} a^3 & (a+1)^3 & (a+2)^3 \\ a & a+1 & a+2 \\ 1 & 1 & 1 \end{vmatrix} = 0$$

Apply $C_2 \to C_2 - C_1$

$C_3 \to C_3 - C_2$

and then expanding along R_3,

We get $a = -1$

13. (B)

Here, $\Delta_1 = \Delta_0^2$

$\Delta_2 = \Delta_1^2 = (\Delta_0^2)^2 = \Delta_0^4; \Delta_3 = \Delta_2^2 = (\Delta_0^4)^2 = \Delta_0^8$

$\Delta_4 = \Delta_3^2 = (\Delta_0^8)^2 = \Delta_0^{16} \Rightarrow \Delta_n = \Delta_0^{2^h}$

14. (A)

Here, points P, Q, R are distinct so $u \neq v \neq w$.

$$\text{We have } \begin{vmatrix} 3u^2 & 2u^3 & 1 \\ 3v^2 & 2v^3 & 1 \\ 3w^2 & 2w^3 & 1 \end{vmatrix} = 0$$

$$\Rightarrow \begin{vmatrix} u^2 & u^3 & 1 \\ v^2 & v^3 & 1 \\ w^2 & w^3 & 1 \end{vmatrix} = 0$$

Apply, $R_1 \to R_1 - R_2$ and $R_2 \to R_2 - R_3$

$$\Rightarrow \begin{vmatrix} u^2 - v^2 & u^3 - v^3 & 0 \\ v^2 - w^2 & v^3 - w^3 & 0 \\ w^2 & w^3 & 1 \end{vmatrix} = 0$$

$$\Rightarrow \begin{vmatrix} v+u & u^2 + v^2 + uv & 0 \\ v+w & v^2 + w^2 + vw & 0 \\ w^2 & w^3 & 1 \end{vmatrix} = 0$$

Apply $R_1 \to R_1 - R_2$

$$\Rightarrow \begin{vmatrix} u-w & (u^2 - w^2) + v(u-w) & 0 \\ v+w & v^2 + w^2 + vw & 0 \\ w^2 & w^3 & 1 \end{vmatrix} = 0$$

$$\Rightarrow \begin{vmatrix} 1 & u+w+v & 0 \\ v+w & v^2 + w^2 + vw & 0 \\ w^2 & w^3 & 1 \end{vmatrix} = 0$$

$\therefore$ Expanding along C_3, we get

$$[(v^2 + w^2 + vw) - (v+w)(\overline{w+v+u})] = 0$$

$$\Rightarrow (v^2 + w^2 + vw) = (v+w)^2 + u(v+w)$$

$$\Rightarrow v^2 + w^2 + vw = v^2 + w^2 + 2vw + uv + uw$$

$$\Rightarrow \boxed{uv + vw + wu = 0}$$

15. **(C)**

We have $B = \begin{vmatrix} p+x & q+y & r+z \\ a+x & b+y & c+z \\ a+p & b+q & c+r \end{vmatrix}$

Apply $R_1 \to R_1 + R_2 + R_3$, we get

$$B = 2\begin{vmatrix} a+p+x & b+q+y & c+r+z \\ a+x & b+y & c+z \\ a+p & b+q & c+r \end{vmatrix}$$

Apply, $R_2 \to R_2 - R_1$ and $R_3 \to R_3 - R_1$, we get

$$= 2\begin{vmatrix} a+p+x & b+q+y & c+r+z \\ -p & -q & -r \\ -x & -y & -z \end{vmatrix}$$

Apply $R_1 \to R_1 + R_2 + R_3$

$$\therefore B = 2\begin{vmatrix} a & b & c \\ -p & -q & -r \\ -x & -y & -z \end{vmatrix} = 2\begin{vmatrix} a & b & c \\ p & q & r \\ x & y & z \end{vmatrix} = 2\text{det. } A$$

$\therefore$ det. $B = 2(6) = 12$

16. **(D)**

For non-trivial solutions,

$$\begin{vmatrix} \sin\theta & -\cos\theta & \lambda+1 \\ \cos\theta & \sin\theta & -\lambda \\ \lambda & \lambda+1 & \cos\theta \end{vmatrix} = 0$$

$\Rightarrow 2\cos\theta(\lambda^2 + \lambda + 1) = 0$ (always positive)

$\therefore \theta = (2n+1)\frac{\pi}{2}, n \in I$ and $\lambda \in R$

17. **(D)**

For non-trivial solutions,

$$\begin{vmatrix} 1 & -\cos\theta & \cos 2\theta \\ -\cos\theta & 1 & -\cos\theta \\ \cos 2\theta & -\cos\theta & 1 \end{vmatrix} = 0$$

Apply, $C_1 \to C_1 - C_3$

$$\Rightarrow \begin{vmatrix} 2\sin^2\theta & -\cos\theta & \cos 2\theta \\ 0 & 1 & -\cos\theta \\ -2\sin^2\theta & -\cos\theta & 1 \end{vmatrix} = 0$$

$$\Rightarrow 2\sin^2\theta \begin{vmatrix} 1 & -\cos\theta & \cos 2\theta \\ 0 & 1 & -\cos\theta \\ -1 & -\cos\theta & 1 \end{vmatrix} = 0$$

(Taking out 2 $\sin^2\theta$ as common from the above)

Now, expanding along C_1, we get

$2\sin^2\theta[1(1-\cos^2\theta) - 1(\cos^2\theta - \cos 2\theta)] = 0$

$\Rightarrow 2\sin^2\theta[(\sin^2\theta - \cos^2\theta) + \cos 2\theta] = 0$

$\Rightarrow 2\sin^2\theta[-\cos 2\theta + \cos 2\theta] = 0 \; \forall \; \theta \in R$

18. **(B)**

We have $\Delta = \begin{vmatrix} 3 & -7 & 5 \\ 3 & 1 & 5 \\ 2 & 3 & 5 \end{vmatrix} = 5\begin{vmatrix} 3 & -7 & 1 \\ 3 & 1 & 1 \\ 2 & 3 & 1 \end{vmatrix}$

Apply, $R_1 \to R_1 - R_2$ and $R_2 \to R_2 - R_3$, we get

$$\Delta = 5\begin{vmatrix} 0 & -8 & 0 \\ 1 & -2 & 0 \\ 2 & 3 & 1 \end{vmatrix}$$

Now, expanding along C_3, we get

$\Delta = 5(0(-2) - 1(-8)) = 40 \neq 0$

$\Rightarrow$ System is consistent with a unique non-trivial solution.

19. **(B)**

We have

$$\Delta = \begin{vmatrix} 1 & -2 & 1 \\ 2 & 1 & -2 \\ 1 & 3 & -3 \end{vmatrix}$$

$= 1(-3+6) + 2(-6+2) + 1(6-1)$

$\Rightarrow \Delta = 3 - 8 + 5 = 0$

$\therefore$ For atleast one solution, $\Delta_1 = \Delta_2 = \Delta_3 = 0$

Now, $\Delta_1 = 0 \Rightarrow \begin{vmatrix} a & -2 & 1 \\ b & 1 & -2 \\ c & 3 & -3 \end{vmatrix} = 0$

$\Rightarrow a(3) - b(3) + c(3) = 0$

$\Rightarrow a - b + c = 0$

(Expanding along C_1)

Similarly, $\Delta_2 = 0 \Rightarrow \begin{vmatrix} 1 & a & 1 \\ 2 & b & -2 \\ 1 & c & -3 \end{vmatrix} = 0$

(expanding along R_1)

$\Rightarrow 1(-3b + 2c) - a(-6+2) + 1(2c - b) = 0$

$\Rightarrow 4a - 4b + 4c = 0 \Rightarrow \boxed{a - b + c = 0}$

Also, $\Delta_3 = 0 \Rightarrow \begin{vmatrix} 1 & -2 & a \\ 2 & 1 & b \\ 1 & 3 & c \end{vmatrix} = 0$

$\Rightarrow 1(c-3b)+2(2c-b)+a(5)=0$

$\Rightarrow 5a-5b+5c=0 \Rightarrow \boxed{a-b+c=0}$

(Expanding along R_1)

20. **(B)**

Let $x17 = kl = 100x + 10 + 7$

$3y6 = km = 300 + 10y + 6$

$12z = kn = 100 + 20 + z$

Now, we have $\Delta = \begin{vmatrix} x & 3 & 1 \\ 7 & 6 & z \\ 1 & y & 2 \end{vmatrix}$

Apply, $R_2 \to R_2 + 100R_1 + 10R_3$

we get $\Delta = \begin{vmatrix} x & 3 & 1 \\ k\ell & km & kn \\ 1 & y & 2 \end{vmatrix}$

$\therefore$ Take out k as common from R_2, we get

$\Delta = k\begin{vmatrix} x & 3 & 1 \\ \ell & m & n \\ 1 & y & 2 \end{vmatrix} \Rightarrow \Delta$ is divisible by k.

21. **(B)**

We have $N = \begin{vmatrix} 28 & 25 & 38 \\ 42 & 38 & 65 \\ 56 & 47 & 83 \end{vmatrix}$

Apply, $C_3 \to C_3 - C_1$ and $C_1 \to C_1 - C_2$

$\Rightarrow N = \begin{vmatrix} 3 & 25 & 10 \\ 4 & 38 & 23 \\ 9 & 47 & 27 \end{vmatrix}$

Apply, $C_2 \to C_2 - 2C_3$, we get

$N = \begin{vmatrix} 3 & 5 & 10 \\ 4 & -8 & 23 \\ 9 & -7 & 27 \end{vmatrix}$

Apply, $C_3 \to C_3 - 3C_1$ and $C_2 \to C_2 + C_1$

$\Rightarrow N = \begin{vmatrix} 3 & 8 & 1 \\ 4 & -4 & 11 \\ 9 & 2 & 0 \end{vmatrix}$

Now, expanding along C_3, we get

$N = 1(8 + 36) - 11(6 - 72) + 0$

$= 44 + 11 \times 66 = 22\,(2 + 33)$

$= 22 \times 35 = 2 \times 11 \times 5 \times 7$

$\therefore N = 770 = 2 \times 5 \times 7 \times 11$

Hence, number of ways $= 2^{n-1} = 2^4 = 2^3 = 8$

22. **(B)**

For non-zero solutions, $\begin{vmatrix} 1 & 2a & a \\ 1 & 3b & b \\ 1 & 4c & c \end{vmatrix} = 0$

Apply, $R_1 \to R_1 - R_2$ and $R_2 \to R_2 - R_3$, we get

$\begin{vmatrix} 0 & 2a-3b & a-b \\ 0 & 3b-4c & b-c \\ 1 & 4c & c \end{vmatrix} = 0$

Now expanding along R_1 we get

$\Rightarrow (2a-3b)(b-c)-(a-b)(3b-4c)=0$

$\Rightarrow (2ab-2ac-3b^2+3bc)-$

$(3ab-4ac-3b^2+4bc)=0$

$\Rightarrow -ab+2ac-bc=0 \Rightarrow 2ac = b(a+c)$

$\Rightarrow \frac{2}{b} = \frac{a+c}{ac} \Rightarrow \frac{2}{b} = \frac{1}{c}+\frac{1}{a}$ or $b = \frac{2}{\frac{1}{a}+\frac{1}{c}}$

$\Rightarrow a, b, c$ are in H.P.

23. **(C)**

We have, $x^3 + ax^2 + bx + c = 0$ with roots α, β, γ

$\therefore \alpha+\beta+\gamma = -a;\ \alpha\beta+\beta\gamma+\gamma\alpha = b$

Since the given system of equations (in u, v and w) has non-trivial solutions,

So $\begin{vmatrix} \alpha & \beta & \gamma \\ \beta & \gamma & \alpha \\ \gamma & \alpha & \beta \end{vmatrix} = 0$

$\Rightarrow \alpha^3+\beta^3+\gamma^3-3\alpha\beta\gamma = 0$

$\Rightarrow (\alpha+\beta+\gamma)(\alpha^2+\beta^2+\gamma^2-\alpha\beta-\beta\gamma-\gamma\alpha)$

$= 0$

$\Rightarrow (\alpha+\beta+\gamma)((\alpha+\beta+\gamma)^2 -$

$3(\alpha\beta+\beta\gamma+\gamma\alpha)) = 0$

$\Rightarrow -a(a^2-3b)=0$

As, $a \neq 0$ (given) $\Rightarrow a^2 = 3b$

24. (D)

Let $\Delta_1 = \begin{vmatrix} ap^2 & 2ap & 1 \\ aq^2 & 2aq & 1 \\ ar^2 & 2ar & 1 \end{vmatrix} = (a)(2a)\begin{vmatrix} p^2 & p & 1 \\ q^2 & q & 1 \\ r^2 & r & 1 \end{vmatrix}$

Apply, $R_1 \to R_1 - R_2$ and $R_2 \to R_2 - R_3$, we get

$$\Delta_1 = 2a^2 \begin{vmatrix} p^2-q^2 & p-q & 0 \\ q^2-r^2 & q-r & 0 \\ r^2 & r & 1 \end{vmatrix}$$

(expanding along C_3)

$$= 2a^2 \begin{vmatrix} p^2-q^2 & p-q \\ q^2-r^2 & q-r \end{vmatrix}$$

$$= 2a^2(p-q)(q-r)\begin{vmatrix} p+q & 1 \\ q+r & 1 \end{vmatrix}$$

$$= 2a^2(p-q)(q-r)(p-r)$$

$\therefore$ $\boxed{\Delta_1 = -2a^2(p-q)(q-r)(r-p)}$ (1)

Now, $\Delta_2 = \begin{vmatrix} apq & a(p+q) & 1 \\ aqr & a(q+r) & 1 \\ arp & a(r+p) & 1 \end{vmatrix}$

$$= (a)(a)\begin{vmatrix} pq & p+q & 1 \\ qr & q+r & 1 \\ rp & r+p & 1 \end{vmatrix}$$

Apply, $R_1 \to R_1 - R_2$, $R_2 \to R_2 - R_3$

We get $\Delta_2 = a^2 \begin{vmatrix} q(p-r) & (p-r) & 0 \\ r(q-p) & (q-p) & 0 \\ rp & r+p & 1 \end{vmatrix}$

(expanding along C_3)

We get $\Delta_2 = a^2 \begin{vmatrix} q(p-r) & (p-r) \\ r(q-p) & (q-p) \end{vmatrix}$

$$= a^2(p-r)(q-p)\begin{vmatrix} q & 1 \\ r & 1 \end{vmatrix}$$

$$= a^2(p-r)(q-p)(q-r)$$

$\therefore$ $\boxed{\Delta_2 = a^2(p-q)(q-r)(r-p)}$ (2)

$\therefore$ From (1) and (2), we get

$\Delta_1 = -2\Delta_2 \Rightarrow \Delta_1 + 2\Delta_2 = 0$

25. (B)

Let

$$F(x) = \begin{vmatrix} 1 & 1+\sin x & 1+\sin x+\cos x \\ 2 & 3+2\sin x & 4+3\sin x+2\cos x \\ 3 & 6+3\sin x & 10+6\sin x+3\cos x \end{vmatrix}$$

Apply, $R_2 \to R_2 - 2R_1$ and $R_3 \to R_3 - 3R_1$, we get

$$F(x) = \begin{vmatrix} 1 & 1+\sin x & 1+\sin x+\cos x \\ 0 & 1 & 2+\sin x \\ 0 & 3 & 7+3\sin x \end{vmatrix}$$

(Expanding along C_1), we get

$F(x) = 1(7+3\sin x - 6 - 3\sin x) = 1$

$\therefore F'\left(\frac{\pi}{2}\right) = 0$

26. (C)

Clearly, $\Delta_2 = {\Delta_1}^2$, $\Delta_3 = {\Delta_2}^2 = {\Delta_1}^4$

$\therefore {\Delta_1}^2 = \frac{\Delta_3}{\Delta_2}$

27. (B)

We have $\begin{bmatrix} 1 & 1 \\ 0 & 1 \end{bmatrix}\begin{bmatrix} 1 & 2 \\ 0 & 1 \end{bmatrix} = \begin{bmatrix} 1 & 3 \\ 0 & 1 \end{bmatrix}$

Now, $\begin{bmatrix} 1 & 3 \\ 0 & 1 \end{bmatrix}\begin{bmatrix} 1 & 3 \\ 0 & 1 \end{bmatrix} = \begin{bmatrix} 1 & 4 \\ 0 & 1 \end{bmatrix}$

$\therefore \begin{bmatrix} 1 & 1 \\ 0 & 1 \end{bmatrix}\begin{bmatrix} 1 & 2 \\ 0 & 1 \end{bmatrix}\begin{bmatrix} 1 & 3 \\ 0 & 1 \end{bmatrix}\cdots\begin{bmatrix} 1 & n \\ 0 & 1 \end{bmatrix}$

$$= \begin{bmatrix} 1 & \frac{n(n+1)}{2} \\ 0 & 1 \end{bmatrix}$$

But, we are given $\begin{bmatrix} 1 & \frac{n(n+1)}{2} \\ 0 & 1 \end{bmatrix} = \begin{bmatrix} 1 & 378 \\ 0 & 1 \end{bmatrix}$

$\therefore \frac{n(n+1)}{2} = 378 \Rightarrow n = 27$

28. **(A)**
We have $AB = I_3$

$\Rightarrow \begin{bmatrix} 1 & 2 & x \\ 0 & 1 & 0 \\ 0 & 0 & 1 \end{bmatrix}\begin{bmatrix} 1 & -2 & y \\ 0 & 1 & 0 \\ 0 & 0 & 1 \end{bmatrix} = \begin{bmatrix} 1 & 0 & 0 \\ 0 & 1 & 0 \\ 0 & 0 & 1 \end{bmatrix}$

$\Rightarrow \begin{bmatrix} 1 & 0 & x+y \\ 0 & 1 & 0 \\ 0 & 0 & 1 \end{bmatrix} = \begin{bmatrix} 1 & 0 & 0 \\ 0 & 1 & 0 \\ 0 & 0 & 1 \end{bmatrix}$

$\therefore$ On comparing, we get $x + y = 0$

29. **(D)**
We have, $A + 2X = B \Rightarrow X = \frac{B - A}{2}$

$\Rightarrow X = \frac{1}{2}\begin{bmatrix} -5 & 1 \\ 5 & 7 \end{bmatrix} = \begin{bmatrix} -\frac{5}{2} & \frac{1}{2} \\ \frac{5}{2} & \frac{7}{2} \end{bmatrix}$

30. **(C)**
M-I: 'Cayley Hamilton theorem'
That is, every square matrix satisfies its characteristic equation. That is, $|A - \lambda I| = 0$

Now, $A - \lambda I = \begin{bmatrix} a & b \\ c & d \end{bmatrix} - \lambda \begin{bmatrix} 1 & 0 \\ 0 & 1 \end{bmatrix}$

$= \begin{bmatrix} a-\lambda & b \\ c & d-\lambda \end{bmatrix}$

$\therefore |A - \lambda I| = 0 \Rightarrow \begin{vmatrix} a-\lambda & b \\ c & d-\lambda \end{vmatrix} = 0$

$\Rightarrow (a-\lambda)(d-\lambda) - bc = 0$

$\Rightarrow \lambda^2 + ad - \lambda(a+d) - bc = 0$

$\therefore$ On comparing, we get $k = (ad - bc)$

M-II: We have,

$A^2 = \begin{bmatrix} a & b \\ c & d \end{bmatrix}\begin{bmatrix} a & b \\ c & d \end{bmatrix} = \begin{bmatrix} a^2+bc & ab+bd \\ ac+cd & bc+d^2 \end{bmatrix}$

Also,

$(a+d)A = (a+d)\begin{bmatrix} a & b \\ c & d \end{bmatrix} = \begin{bmatrix} a(a+d) & b(a+d) \\ c(a+d) & d(a+d) \end{bmatrix}$

$kI_2 = k\begin{bmatrix} 1 & 0 \\ 0 & 1 \end{bmatrix} = \begin{bmatrix} k & 0 \\ 0 & k \end{bmatrix}$

$\therefore A^2 - (a+d)A + kI_2 = 0$

$\Rightarrow \begin{bmatrix} bc-ad & 0 \\ 0 & bc-ad \end{bmatrix} + \begin{bmatrix} k & 0 \\ 0 & k \end{bmatrix} = 0$

$\therefore k = (ad - bc)$

31. **(B)**

Given $[x \quad -5 \quad -1]_{1\times3}\begin{bmatrix} 1 & 0 & 2 \\ 0 & 2 & 1 \\ 2 & 0 & 3 \end{bmatrix}_{3\times3}\begin{bmatrix} x \\ 4 \\ 1 \end{bmatrix} = 0$

$\Rightarrow [x-2-10 \quad 2x-8]_{1\times3}\begin{bmatrix} x \\ 4 \\ 1 \end{bmatrix}_{3\times1} = 0$

$\Rightarrow [x(x-2) + (-4)\times10 + (2x-8)\times1] = [0]$

$\Rightarrow x^2 - 2x - 40 + 2x - 8 = 0$

$\Rightarrow x^2 = 48 \Rightarrow x = \pm 4\sqrt{3}$

32. **(C)**
Tr. (A) + 2 Tr. $(B) = -1$ (1)
2 Tr. (A) – Tr. $(B) = 3$ (2)
$\therefore$ On solving (1) and (2), we get
Tr. $(A) = 1$ and Tr. $(B) = -1$
Hence, Tr. (A) – Tr. (B)
$= 1 - (-1) = 2$

33. **(D)**

Let $X = \begin{pmatrix} a & b \\ c & d \end{pmatrix}$

Now,

$X^2 = \begin{pmatrix} a & b \\ c & d \end{pmatrix}\begin{pmatrix} a & b \\ c & d \end{pmatrix} = \begin{pmatrix} a^2+bc & ab+bd \\ ac+cd & bc+d^2 \end{pmatrix}$

$\equiv \begin{pmatrix} 1 & 0 \\ 0 & 1 \end{pmatrix}$

$\therefore$ We must have
$a^2 + bc = 1$ (1)
$b(a + d) = 0$ (2)
$c(a + d) = 0$ (3)
$c + d^2 = 1$ (4)
So, from (1), $bc = 1 - a^2$
and from (4), $bc = 1 - d^2$
$\therefore 1 - a^2 = 1 - d^2$
$\Rightarrow a^2 - d^2 = 0 \Rightarrow (a-d)(a+d) = 0$
$\therefore a = d$ or $a = -d$
Case-I: When $a = d$
$\therefore$ from (2), $2ab = 0$ and from (3),
$2ac = 0$

Now, if $a = 0$, then $d = 0$ and from (1),

$$c = \frac{1}{b}$$

$$\therefore X = \begin{pmatrix} 0 & b \\ \frac{1}{b} & 0 \end{pmatrix}$$

$\Rightarrow$ Infinite number of matrices are possible with $b \in R - \{0\}$

Example: If $b = 2$ then $X = \begin{pmatrix} 0 & 2 \\ \frac{1}{2} & 0 \end{pmatrix}$

Now, $X^2 = XX = \begin{pmatrix} 0 & 2 \\ \frac{1}{2} & 0 \end{pmatrix}\begin{pmatrix} 0 & 2 \\ \frac{1}{2} & 0 \end{pmatrix} = \begin{pmatrix} 1 & 0 \\ 0 & 1 \end{pmatrix}$

Verified and so on.

Case-II: When $a = -d$

then from (1) and (4) we get $d^2 = 1 - bc$

or $bc = 1 - d^2$ $\quad \therefore X = \begin{pmatrix} -\alpha & \beta \\ \gamma & \alpha \end{pmatrix}$

where $\beta\gamma = 1 - \alpha^2$

34. **(B)**

We have

$$A - \lambda I = \begin{bmatrix} 1 & 3 \\ 2 & 2 \end{bmatrix} - \lambda \begin{bmatrix} 1 & 0 \\ 0 & 1 \end{bmatrix} = \begin{bmatrix} 1-\lambda & 3 \\ 2 & 2-\lambda \end{bmatrix}$$

As $|A - \lambda I| = 0 \Rightarrow \begin{vmatrix} 1-\lambda & 3 \\ 2 & 2-\lambda \end{vmatrix} = 0$

$\Rightarrow (1 - \lambda)(2 - \lambda) - 6 = 0$

$\Rightarrow \boxed{\lambda^2 - 3\lambda - 4 = 0}$

35. **(C)**

$$|A| = \begin{vmatrix} 1 & \sin\theta & 1 \\ -\sin\theta & 1 & \sin\theta \\ -1 & -\sin\theta & 1 \end{vmatrix}$$

$\therefore$ Expanding along R_1, we get

$= 1(1+\sin^2\theta) - \sin\theta(-\sin\theta + \sin\theta) + 1\cdot(1+\sin^2\theta)$

$\Rightarrow |A| = 2(1+\sin^2\theta)$

As, $\sin^2\theta \in [0,1]$ $\quad \therefore |A| \in [2,4]$

Apply $R_1 \to R_1 + R_3$ we get

$$|A| = \begin{vmatrix} 0 & 0 & 2 \\ -\sin\theta & 1 & \sin\theta \\ -1 & -\sin\theta & 1 \end{vmatrix}$$

Now, expanding along R_1, we get

$|A| = 2(\sin^2\theta + 1) \in [2,4]$

36. **(B)**

If matrix A is singular, so

$$\begin{vmatrix} 3-x & 2 & 2 \\ 2 & 4-x & 1 \\ -2 & -4 & -1-x \end{vmatrix} = 0$$

Apply $R_1 \to R_1 + R_3$

$R_2 \to R_2 + R_3$, we get $\begin{vmatrix} 1-x & -2 & 1-x \\ 0 & -x & -x \\ -2 & -4 & -1-x \end{vmatrix} = 0$

Apply $C_3 \to C_3 - C_2$, we get

$$\begin{vmatrix} 1-x & -2 & 3-x \\ 0 & -x & 0 \\ -2 & -4 & 3-x \end{vmatrix} = 0$$

$$\Rightarrow -(3-x)\begin{vmatrix} 1-x & +2 & 1 \\ 0 & +x & 0 \\ -2 & +4 & 1 \end{vmatrix} = 0$$

Expanding along R_1, we get

$\Rightarrow (x-3)[(1-x)\cdot(x-0) - 2(0-0) + 1(0+2x)]$

$= 0$

$\Rightarrow (x-3)(x - x^2 + 2x) = 0$

$\Rightarrow (3x - x^2)(x-3) = 0$

$\Rightarrow -x(x-3)^2 = 0$

$\therefore x = 0$ or 3

37. **(D)**

We have $A = \begin{bmatrix} 1 & \tan x \\ -\tan x & 1 \end{bmatrix}$

$\therefore$ det. $A = \begin{vmatrix} 1 & \tan x \\ -\tan x & 1 \end{vmatrix}$

$= (1 + \tan^2 x) = \sec^2 x$

$\Rightarrow$ det. $A^T = \begin{vmatrix} 1 & -\tan x \\ \tan x & 1 \end{vmatrix} = \sec^2 x$

Now,

$f(x) = \text{det.}(A^T A^{-1}) = \text{det.}(A^T)\det(A^{-1})$

$= \dfrac{\det(A^T)}{\det(A)} = 1$

Hence, $f(x) = 1$. Now, verify alternatives.

38. (D)

Let $A=\begin{bmatrix} a & b \\ c & d \end{bmatrix}$

Now, $A\begin{bmatrix} 1 \\ -1 \end{bmatrix}=\begin{bmatrix} -1 \\ 2 \end{bmatrix} \Rightarrow \begin{bmatrix} a & b \\ c & d \end{bmatrix}\begin{bmatrix} 1 \\ -1 \end{bmatrix}=\begin{bmatrix} -1 \\ 2 \end{bmatrix}$

$\Rightarrow \begin{bmatrix} a-b \\ c-d \end{bmatrix}=\begin{bmatrix} -1 \\ 2 \end{bmatrix}$

$\therefore a-b=-1$ (1)

$c-d=2$ (2)

Similarly, $A^2\begin{bmatrix} 1 \\ -1 \end{bmatrix}=\begin{bmatrix} 1 \\ 0 \end{bmatrix}$

$\Rightarrow AA\begin{bmatrix} 1 \\ -1 \end{bmatrix}=\begin{bmatrix} 1 \\ 0 \end{bmatrix}$

$\Rightarrow A\begin{bmatrix} -1 \\ 2 \end{bmatrix}=\begin{bmatrix} 1 \\ 0 \end{bmatrix} \Rightarrow \begin{bmatrix} a & b \\ c & d \end{bmatrix}\begin{bmatrix} -1 \\ 2 \end{bmatrix}=\begin{bmatrix} 1 \\ 0 \end{bmatrix}$

$\Rightarrow \begin{bmatrix} -a+2b \\ -c+2d \end{bmatrix}=\begin{bmatrix} 1 \\ 0 \end{bmatrix}$

$\therefore -a+2b=1$ (3)

$-c+2d=0$ (4)

$\therefore$ We get

$a=-1, b=0, c=4, d=2$

$\Rightarrow (a+b+c+d)=5$

39. (D)

Let $A=\begin{pmatrix} a_{11} & a_{12} & a_{13} \\ a_{21} & a_{22} & a_{23} \\ a_{31} & a_{32} & a_{33} \end{pmatrix}$

Given, $a_{11}=a_{22}=a_{33}=(a+b)$;

$a_{12}=a_{23}=ab$; $a_{21}=a_{32}=1$

and all other elements are zero.

$\therefore |A| = \text{det. } A = \begin{vmatrix} a+b & ab & 0 \\ 1 & a+b & ab \\ 0 & 1 & a+b \end{vmatrix}$

Expanding along R_1, we get

$=[(a+b)((a+b)^2-ab)-ab(a+b)]$

$=(a+b)\,((a+b)^2-2ab)$

$\therefore$ det. $A=(a+b)\,(a^2+b^2)$

40. (B)

We have

Det. $(D_2)=\begin{vmatrix} 0 & 1 \\ 2 & 0 \end{vmatrix}=0-2=-2$

(As, $f(2)=2 \Rightarrow$ unity as the element of 1st row and 2nd column)

Similarly, Det. $(D_3)=\begin{vmatrix} 0 & 3 & 1 \\ 3 & 0 & 3 \\ 3 & 3 & 0 \end{vmatrix}$

(Expanding along R_1)

(As, $f(3)=3 \Rightarrow$ unity as element 1st row and 3rd column)

$=0(0-9)-3(0-9)+1(9-0)$

$=0+27+9=36$

$\Rightarrow$ Det. (D_2) + Det. $(D_3)=-2+36=34$

41. (D)

We have $A=\begin{bmatrix} 1 & 2r-1 \\ 0 & 1 \end{bmatrix}$

Now, $\prod_{r=1}^{50}\begin{bmatrix} 1 & 2r-1 \\ 0 & 1 \end{bmatrix}$

$=\begin{bmatrix} 1 & 1 \\ 0 & 1 \end{bmatrix}\begin{bmatrix} 1 & 3 \\ 0 & 1 \end{bmatrix}\begin{bmatrix} 1 & 5 \\ 0 & 1 \end{bmatrix}\begin{bmatrix} 1 & 7 \\ 0 & 1 \end{bmatrix}\cdots\begin{bmatrix} 1 & 99 \\ 0 & 1 \end{bmatrix}$

$=\begin{bmatrix} 1 & \underbrace{1+3+5+\ldots+97+99}_{\text{A.P.}} \\ 0 & 1 \end{bmatrix}$

$=\begin{bmatrix} 1 & \frac{50}{2}(1+99) \\ 0 & 1 \end{bmatrix}=\begin{bmatrix} 1 & 2500 \\ 0 & 1 \end{bmatrix}$

42. (A)

As A is an involutory matrix,

So, $A^2=I \Rightarrow A=A^{-1}$

Now, $A^2=\left(\frac{A}{2}\right)(2A)=I$

$\Rightarrow \left(\frac{A}{2}\right)^{-1}=2A$

43. (A)

We have

$BC=\begin{bmatrix} 3 & 4 \\ 2 & 3 \end{bmatrix}\begin{bmatrix} 3 & -4 \\ -2 & 3 \end{bmatrix}=\begin{bmatrix} 1 & 0 \\ 0 & 1 \end{bmatrix}=I$

$\therefore t_r(A)+t_r\left(\frac{ABC}{2}\right)+t_r\left(\frac{A(BC)^2}{4}\right)+$

$t_r\left(\frac{A(BC)^3}{8}\right)+\ldots$

$= t_r(A) + \frac{1}{2}t_r(A) + \frac{1}{4}t_r(A) + \ldots$ (Infinite G.P.)

$= \frac{t_r(A)}{1-\frac{1}{2}} = 2t_r(A) = 2(2+1) = 6$

44. **(C)**

We have, $A^2 = A$ (1)

Now, $(I - 0.4A)^{-1} = (I - \alpha A)$

$\therefore$ on pre-multiplying with $(I - 0.4A)$, we get

$I = (I - 0.4A)(I - \alpha A)$

$= I - I\alpha A - 0.4A + 0.4\alpha A$

[using (1) and $I^2 = I$]

$= I - (0.4 + \alpha)A + 0.4\alpha A$

Hence, $0.4 + \alpha = 0.4\alpha$

$\Rightarrow 0.6\alpha = -0.4 \Rightarrow \alpha = \frac{-4}{6} = \frac{-2}{3}$

45. **(D)**

We have $A = \begin{bmatrix} 3 & 1 \\ -6 & -2 \end{bmatrix}$

Now,

$A^2 = AA = \begin{bmatrix} 3 & 1 \\ -6 & -2 \end{bmatrix}\begin{bmatrix} 3 & 1 \\ -6 & -2 \end{bmatrix} = \begin{bmatrix} 3 & 1 \\ -6 & -2 \end{bmatrix}$

$\Rightarrow A^2 = A$

$\therefore A$ is an idempotent matrix.

So, $A^n = A, \forall\, n \geq 2$ and $n \in N$

$\therefore (I + A)^{99} = {}^{99}C_0 I + {}^{99}C_1 A + {}^{99}C_2 A^2 + \ldots + {}^{99}C_{98}A^{98} + {}^{99}C_{99}A^{99}$

as, $A = A^2 = A^3 = A^4 = \ldots = A^{99}$

$\Rightarrow (I + A)^{99} = I + ({}^{99}C_1 + {}^{99}C_2 + \ldots + {}^{99}C_{99})A$

$= I + (2^{99} - 1)A$

46. **(D)**

We have $A^2 = A$

Also, $A = \begin{pmatrix} d_1 & 0 & 0 \\ 0 & d_2 & 0 \\ 0 & 0 & d_3 \end{pmatrix} \Rightarrow d_1, d_2, d_3$ be the diagonal elements of matrix A of order 3.

Now,

$d_1^2 = d_1 \Rightarrow d_1 = 0$ or 1

$d_2^2 = d_2 \Rightarrow d_2 = 0$ or 1 and

$d_3^2 = d_3 \Rightarrow d_3 = 0$ or 1

$\therefore$ Number of ways in which diagonal elements can be filled up is 2^3, but all diagonal elements can't be zero.

$\therefore$ Number of required matrices $= 2^3 - 1 = 8 - 1 = 7$

47. **(B)**

Given, $AB = I$

So,

$\begin{bmatrix} 2 & 0 & 7 \\ 0 & 1 & 0 \\ 1 & -2 & 1 \end{bmatrix}\begin{bmatrix} -\alpha & 14\alpha & 7\alpha \\ 0 & 1 & 0 \\ \alpha & -4\alpha & -2\alpha \end{bmatrix} = \begin{bmatrix} 1 & 0 & 0 \\ 0 & 1 & 0 \\ 0 & 0 & 1 \end{bmatrix}$

$\Rightarrow \begin{bmatrix} 5\alpha & 0 & 0 \\ 0 & 1 & 0 \\ 0 & 10\alpha - 2 & 5\alpha \end{bmatrix} = \begin{bmatrix} 1 & 0 & 0 \\ 0 & 1 & 0 \\ 0 & 0 & 1 \end{bmatrix}$

$\therefore$ We get, $5\alpha = 1 \Rightarrow \alpha = \frac{1}{5}$

Now, $t_r(B) = (-\alpha) + 1 + (-2\alpha)$

$= 1 - 3\alpha = 1 - \frac{3}{5} = \frac{2}{5}$

48. **(B)**

We have $\det(2A^9 B^{-1}) = -2$

$\Rightarrow \frac{2^2 |A|^9}{|B|} = -2 \Rightarrow 4(3\alpha - 4)^9 = -2(2)$

$\Rightarrow (3\alpha - 4)^9 = -1$

Which is possible when

$(3\alpha - 4) = -1 \Rightarrow 3\alpha = 3 \Rightarrow \alpha = 1$

So, only one real value of α exists.

49. **(B)**

We know that

A adj. $A = |A|\, I_n$ (1)

$\therefore$ (AB) (adj. AB) $= |AB|\, I_n$ (2)

Now, consider

(AB) (adj. $B \cdot$ adj. A)

$= A\,(B$ adj. $B)$ adj. A (using associativity)

$= A\,|B|\, I_n$ adj. A

$= |B|\, A$ adj. A

$= |B|\, |A|\, I_n$

$\therefore (AB)(\text{adj. } B \cdot \text{adj. } A) = |AB|\, I_n$ (3)

$\therefore$ From (2) and (3), we get

(AB) (adj. AB) $= (AB)$ (adj. $B \cdot$ adj. A)

Now, on premultiplying by $(AB)^{-1}$, we get

$(AB)^{-1}(A - B)(\text{adj. } AB) = (AB)^{-1}(AB)$

(adj. $B \cdot$ adj. A)

$\Rightarrow \boxed{\text{adj.}(AB) = (\text{adj. } B)\cdot(\text{adj. } A)}$ Proved

50. **(C)**

If A^{-1} exists, then det. $A \neq 0$

Now,

$$\text{det. } A = \begin{vmatrix} x+\lambda & x & x \\ x & x+\lambda & x \\ x & x & x+\lambda \end{vmatrix}$$

Apply, $C_1 \to C_1 + C_2 + C_3$ and then take out $(2x+\lambda)$ as common from C_1,

$$\text{det. } A = (3x+\lambda)\begin{vmatrix} 1 & x & x \\ 1 & x+\lambda & x \\ 1 & x & x+\lambda \end{vmatrix}$$

Apply row transformations,

$$\Rightarrow \text{det. } A = (3x+\lambda)\begin{vmatrix} 1 & x & x \\ 0 & \lambda & 0 \\ 0 & 0 & \lambda \end{vmatrix}$$

(expanding along C_1)

$\therefore$ det. $A = \lambda^2(3x+\lambda)$

Thus, A^{-1} will exist if $\lambda \neq 0$ and $(3x+\lambda) \neq 0$.

51. **(C)**

For square matrix $A, |A| \neq 0$ (Given)

(A) If A is a diagonal matrix

Then, $A^{-1} = \dfrac{\text{adj. } A}{|A|}$

must also be a diagonal matrix.

$$\text{e.g., } A = \begin{pmatrix} 1 & 0 & 0 \\ 0 & 2 & 0 \\ 0 & 0 & 3 \end{pmatrix} \Rightarrow |A| = 6$$

$$\therefore A^{-1} = \frac{\begin{pmatrix} 6 & 0 & 0 \\ 0 & 3 & 0 \\ 0 & 0 & 2 \end{pmatrix}}{6} = \begin{pmatrix} 1 & 0 & 0 \\ 0 & 2 & 0 \\ 0 & 0 & 3 \end{pmatrix} = A$$

$\Rightarrow A^{-1}$ is also a diagonal matrix.

$\Rightarrow$ option (A) is correct.

(B) If A is symmetric, then

A^{-1} is also symmetric matrix.

because, If $A^T = A$ (Given)

Now, let $B = A^{-1}$

$\therefore B^T = (A^{-1})^T = (A^T)^{-1} = A^{-1} = B$

$\therefore B = A^{-1}$ is also a symmetrix matrix.

$\therefore$ Option (B) is correct.

(C) If $A^{-1} = A$

$\therefore$ On premultiplying by A, we get

$AA^{-1} = A^2 \Rightarrow A^2 = I$

$\Rightarrow A$ is an involutory matrix

$\therefore$ Option (C) is incorrect.

(D) If $A^{-1} = A$

$\therefore$ On pre-multiplying by A, we get

$AA^{-1} = A^2 \Rightarrow A^2 = I$

$\Rightarrow A$ is an involutory matrix.

$\therefore$ Option (D) is correct.

52. **(B)**

(A) If a system of n simultaneous linear equations has a unique solution, then coefficient matrix must be non-singular.

$\Rightarrow$ option (A) is incorrect.

(B) If system of n simultaneous linear equations has a unique solution, then coefficient matrix must be non-singular

$\Rightarrow$ option (B) is correct.

(C) Since, $A^{-1} = \dfrac{\text{adj. } A}{|A|}$ (suppose $A_{n \times n}$)

and $|A^{-1}| = \left|\dfrac{\text{adj. } A}{|A|}\right| = \dfrac{|A|^{n-1}}{|A|^n} = \dfrac{1}{|A|}$

If A^{-1} exists, then adj. A is a non-singular matrix.

$\Rightarrow$ (adj. $A)^{-1}$ must exist.

$\Rightarrow$ Option (C) is incorrect.

(D) We have

$F(x)F(y) =$

$$\begin{bmatrix} \cos x & -\sin x & 0 \\ \sin x & \cos x & 0 \\ 0 & 0 & 0 \end{bmatrix}\begin{bmatrix} \cos y & -\sin y & 0 \\ \sin y & \cos y & 0 \\ 0 & 0 & 0 \end{bmatrix}$$

$$= \begin{bmatrix} \cos(x+y) & -\sin(x+y) & 0 \\ \sin(x+y) & \cos(x+y) & 0 \\ 0 & 0 & 0 \end{bmatrix}$$

$= F(x+y)$

$\Rightarrow$ Option (D) is incorrect.

53. **(C)**

$$\text{We have } A = \begin{pmatrix} 3 & -3 & 4 \\ 2 & -3 & 4 \\ 0 & -1 & 1 \end{pmatrix}$$

$\Rightarrow |A| = 3(-3+4)+3(2-0)+4(-2-0)$

$= 3+6-8 = 1$

Now, $A^2 = AA = \begin{pmatrix} 3 & -3 & 4 \\ 2 & -3 & 4 \\ 0 & -1 & 1 \end{pmatrix}\begin{pmatrix} 3 & -3 & 4 \\ 2 & -3 & 4 \\ 0 & -1 & 1 \end{pmatrix}$

$= \begin{pmatrix} 3 & -4 & 4 \\ 0 & -1 & 0 \\ -2 & 2 & -3 \end{pmatrix}$

$\therefore A^4 = A^2A^2 = \begin{pmatrix} 3 & -4 & 4 \\ 0 & -1 & 0 \\ -2 & 2 & -3 \end{pmatrix}\begin{pmatrix} 3 & -4 & 4 \\ 0 & -1 & 0 \\ -2 & 2 & -3 \end{pmatrix}$

$= \begin{pmatrix} 1 & 0 & 0 \\ 0 & 1 & 0 \\ 0 & 0 & 1 \end{pmatrix} = I$ (on multiplying by A^{-1})

$\Rightarrow A^4 = I \quad \therefore A^3 = A^{-1}$

54. **(C)**

Given, $A = \begin{pmatrix} 0 & 2b & c \\ a & b & c \\ a & -b & c \end{pmatrix}$

The rows of matrix A are unit orthogonal vectors.

Hence, $\vec{R}_1 \cdot \vec{R}_2 = 0 \Rightarrow 2b^2 = c^2$

$\vec{R}_2 \cdot \vec{R}_3 = 0 \Rightarrow a^2 - b^2 - c^2 = 0$

$\vec{R}_3 \cdot \vec{R}_3 = 1 \Rightarrow a^2 + b^2 + c^2 = 1$ (note this step)

$\therefore a^2 = \frac{1}{2}, b^2 = \frac{1}{6}, c^2 = \frac{1}{3} \Rightarrow |abc| = \frac{1}{6}$

55. **(B)**

Clearly, $A = \begin{pmatrix} 3 & 0 & 0 \\ 0 & 3 & 0 \\ 0 & 0 & 3 \end{pmatrix} \Rightarrow 3^3 = 27 = |A|$

$\therefore$ det. (adj. (adj. A)) $= |A|^4 = (27)^4$

$= (25+2)^4 = 25\lambda + 16$

where λ is some positive integer.

$\therefore \left\{\frac{\text{det(adj. (adj. } A))}{5}\right\} = \left\{\frac{25\lambda+16}{5}\right\}$

$= \left\{\frac{16}{5}\right\} = \left\{3+\frac{1}{5}\right\} = \frac{1}{5}$

56. **(C)**

We have, $S = ABCD = A(BCD) = AA^T$ (1)

Now, $S^3 = SSS = (ABCD)(ABCD)(ABCD)$

$= (ABC)(DAB)(CDA)(BCD)$

$= D^T C^T B^T A^T = (BCD)^T A^T = (A^T)^T A^T$

$= AA^T$ (2)

$\therefore$ From (1) and (2), we get

$S = S^3 \Rightarrow$ S-1 is correct

On multiplying both the sides by S,

we get $\boxed{S^2 = S^4} \Rightarrow$ S-2 is correct.

57. **(D)**

Given, $|A| \neq 0$ and $A = AB - BA$ (given) (1)

From (1), we get

$BA + A = AB$

$\Rightarrow BA + IA = AB \Rightarrow (B+I)A = AB$

$\Rightarrow |(B+I)A| = |AB| \Rightarrow |B+I||A| = |A||B|$

As, $|A| \neq 0$

$\Rightarrow$ det. $(B+I) =$ det. B (2)

Similarly, from (1), we get

$AB - A = BA$

$\Rightarrow AB - AI = BA \Rightarrow A(B-I) = BA$

$\Rightarrow |A(B-I)| = |BA| \Rightarrow |A||B-I| = |B||A|$

As, $|A| \neq 0$

$\Rightarrow$ det. $(B-I) =$ det. B (3)

$\therefore$ From (2) and (3), we get

$\boxed{\text{det.}(B+I) = \text{det.}(B-I)}$

(D) none of these

58. **(A)**

We have

$$A = \begin{pmatrix} 0 & \sin\alpha & \sin\alpha\cdot\sin\beta \\ -\sin\alpha & 0 & \cos\alpha\cdot\cos\beta \\ -\sin\alpha\cdot\sin\beta & -\cos\alpha\cdot\cos\beta & 0 \end{pmatrix}$$

As matrix A is skew-symmetric matrix of odd order

$\Rightarrow |A| = 0 \Rightarrow A$ is a singular matrix

$\therefore A^{-1}$ does not exist.

$\Rightarrow |A|$ is independent of α and β.

We have

$A = \begin{pmatrix} 4 & 6 & -1 \\ 3 & 0 & 2 \\ 1 & -2 & 5 \end{pmatrix}, B = \begin{pmatrix} 2 & 4 \\ 0 & 1 \\ -1 & 2 \end{pmatrix}, B = \begin{pmatrix} 3 \\ 1 \\ 2 \end{pmatrix}$

Now,

$(AB)^T C = (A_{3\times3}B_{3\times2})^T C_{3\times1} = (AB)^T{}_{3\times2} C_{3\times1}$

$= (2\times3)(3\times1) = 2\times1$ (exists)

Similarly, $C^T AB$ exists and $A^T A B B^T C$ exists

but $C^T C(AB)^T$ will not exists

$\Rightarrow$ (i) $(AB)^T C$ (iii) $C^T AB$ and (iv) $A^T ABB^T C$ exists.

$\Rightarrow$ exactly 3 are defined.

Integer Answer Type

1. (1)

$$\lambda = \frac{t^2 - 3t + 4}{t^2 + 3t + 4}$$

$$\Rightarrow (\lambda - 1)t^2 + 3(\lambda + 1)t + 4(\lambda - 1) = 0$$

Since, t is real

$$\Rightarrow 9(\lambda + 1)^2 - 16(\lambda - 1)^2 \geq 0$$

$$\Rightarrow \frac{1}{7} \leq \lambda \leq 7$$

$$\text{Now, } \Delta = \begin{vmatrix} 3 & -1 & 4 \\ 1 & 2 & -3 \\ 6 & 5 & \lambda \end{vmatrix} = 7(\lambda + 5) \neq 0$$

Hence, unique solution.

2. (1)

$$\Delta = \begin{vmatrix} \operatorname{cosec}\alpha & 1 & 0 \\ 1 & 2\operatorname{cosec}\alpha & 1 \\ 0 & 1 & 2\operatorname{cosec}\alpha \end{vmatrix}$$

Operate $C_2 \to C_2 - \sin\alpha C_1$

$$= \begin{vmatrix} \operatorname{cosec}\alpha & 0 & 0 \\ 1 & 2\operatorname{cosec}\alpha - \sin\alpha & 1 \\ 0 & 1 & 2\operatorname{cosec}\alpha \end{vmatrix}$$

$$= 4\operatorname{cosec}^3\alpha - 3\operatorname{cosec}\alpha$$

$$= \frac{1}{2}\left[\left(\tan^3\frac{\alpha}{2} + \cot^3\frac{\alpha}{2}\right)\right]$$

If $0 < \alpha < \pi \Rightarrow 0 < \frac{\alpha}{2} < \frac{\pi}{2}$

$$\Rightarrow \tan\frac{\alpha}{2} > 0 \Rightarrow \tan^3\frac{\alpha}{2} > 0$$

$$\therefore \tan^3\frac{\alpha}{2} + \cot^3\frac{\alpha}{2} \geq 2$$

$\therefore \Delta \geq 1 \Rightarrow$ least value of $\Delta = 1$

3. (2)

Let, $\Delta_1 + \Delta_2 > 33$ (1)

$$\Delta_1 = \begin{vmatrix} x & 2 & 2 \\ 3 & x & 2 \\ 3 & 3 & x \end{vmatrix} = x^3 - 18x + 30$$

And $\Delta_2 = 21x^2 - x^3$

From (1)

$$x^3 - 18x + 30 + 21x^2 - x^3 > 33$$

$$\Rightarrow (7x + 1)(x - 1) > 0$$

$$\Rightarrow x < -\frac{1}{7} \text{ or } x > 1$$

$\Rightarrow$ Least positive integral value = 2

4. (–1)

$$\Delta = \begin{vmatrix} \alpha & \alpha + \beta & \alpha + \beta + \gamma \\ 3\alpha & 4\alpha + 3\beta & 5\alpha + 4\beta + 3\gamma \\ 6\alpha & 9\alpha + 6\beta & 11\alpha + 9\beta + 6\gamma \end{vmatrix}$$

$R_2 - 3R_1$, $R_3 - 6R_1$

$$= \begin{vmatrix} \alpha & \alpha + \beta & \alpha + \beta + \gamma \\ 0 & \alpha & 2\alpha + \beta \\ 0 & 0 & -\alpha \end{vmatrix} = -\alpha^3 = -1$$

5. (–8)

Since, $2 < e < 3$, $3 < \pi < 4$ and $3 < \pi^2 - 6 < 4$, the given determinant reduces to

$$\begin{vmatrix} 2 & 3 & 3 \\ 3 & 3 & 2 \\ 3 & 2 & 3 \end{vmatrix} \qquad R_1 \to R_1 + R_2 + R_3$$

$$= 8\begin{vmatrix} 1 & 1 & 1 \\ 3 & 3 & 2 \\ 3 & 2 & 3 \end{vmatrix} = -8$$

6. (5)

Possible values are –2, – 1, 0, 1, 2

That is, $\begin{vmatrix} 1 & 0 \\ -1 & 0 \end{vmatrix} = 1$, $\begin{vmatrix} 1 & -1 \\ 0 & 0 \end{vmatrix} = 0$, $\begin{vmatrix} 0 & 1 \\ 1 & -1 \end{vmatrix} = -1$,

$\begin{vmatrix} 1 & -1 \\ 1 & 1 \end{vmatrix} = 2$, $\begin{vmatrix} -1 & 1 \\ 1 & 1 \end{vmatrix} = -2$

7. (0)

Let,

$$\begin{vmatrix} (1+x)^a & (1+2x)^b & 1 \\ 1 & (1+x)^a & (1+2x)^b \\ (1+2x)^b & 1 & (1+x)^a \end{vmatrix}$$

$$= A + Bx + Cx^2 + \ldots$$

Put $x = 0$, then $A = 0$

Differentiating both sides with respect to x and then put $x = 0$

$$\begin{vmatrix} a & 2b & 0 \\ 1 & 1 & 1 \\ 1 & 1 & 1 \end{vmatrix} + \begin{vmatrix} 1 & 1 & 1 \\ 0 & a & 2b \\ 1 & 1 & 1 \end{vmatrix} + \begin{vmatrix} 1 & 1 & 1 \\ 1 & 1 & 1 \\ 2b & 0 & 0 \end{vmatrix} = B$$

$\Rightarrow B = 0$

$\Rightarrow$ Coefficient of $x = 0$

8. **(0)**

$\alpha+\beta+\gamma-0$

Applying $C_1 \to C_1 + C_2 + C_3$, then

$$\begin{vmatrix} \alpha+\beta+\gamma & \beta & \gamma \\ \alpha+\beta+\gamma & \gamma & \alpha \\ \alpha+\beta+\gamma & \alpha & \beta \end{vmatrix} = \begin{vmatrix} 0 & \beta & \gamma \\ 0 & \gamma & \alpha \\ 0 & \alpha & \beta \end{vmatrix} = 0$$

9. **(2)**

$$\Delta(x) = \begin{vmatrix} x-2 & (x-1)^2 & x^3 \\ (x-1) & x^2 & (x+1)^3 \\ x & (x+1)^2 & (x+2)^3 \end{vmatrix}$$

$R_2 - R_1$ and $R_3 - R_1$, then

$$\Delta(x) = \begin{vmatrix} x-2 & (x-1)^2 & x^3 \\ 1 & 2x-1 & 3x^2+3x+1 \\ 2 & 4x & 6x^2+12x+8 \end{vmatrix}$$

$R_3 - 2R_2$

$$\Delta(x) = \begin{vmatrix} x-2 & (x-1)^2 & x^3 \\ 1 & 2x-1 & 3x^2+3x+1 \\ 0 & 2 & 6x+6 \end{vmatrix}$$

Expand with respect to C_1 and equating the coefficient of $-x = 2$

10. **(1)**

[Do it yourself]

11. **(3)**

$A^3 - A = 0$

$\Rightarrow b + \frac{1}{2} = 0, \frac{b-a}{2} = -1 \Rightarrow b = -\frac{1}{2}, a = \frac{3}{2}$

$b - a = 0, b + \frac{1}{2} = 1 \Rightarrow b = \frac{1}{2}, a = \frac{1}{2}$

$b - a = 0, b + \frac{1}{2} = -1 \Rightarrow b = -\frac{3}{2}, a = -\frac{3}{2}$

Hence $n(p) = 3$

12. **(5)**

$A\begin{bmatrix} 1 \\ -1 \end{bmatrix} = \begin{bmatrix} -1 \\ 2 \end{bmatrix} \Rightarrow \begin{bmatrix} a & b \\ c & d \end{bmatrix}\begin{bmatrix} 1 \\ -1 \end{bmatrix} = \begin{bmatrix} -1 \\ 2 \end{bmatrix}$

$\Rightarrow a - b = -1, c - d = 2$

$A^2\begin{bmatrix} 1 \\ -1 \end{bmatrix} = A\begin{bmatrix} -1 \\ 2 \end{bmatrix} = \begin{bmatrix} 1 \\ 0 \end{bmatrix}$

$\Rightarrow \begin{bmatrix} a & b \\ c & d \end{bmatrix}\begin{bmatrix} -1 \\ 2 \end{bmatrix}\begin{bmatrix} 1 \\ 0 \end{bmatrix}$

$\Rightarrow -a + 2b = 1, -c + 2d = 0$

$\Rightarrow d = 2, c = 4, a = -1, b = 0$

$\Rightarrow a + b + c + d = 5$

13. **(7)**

$$\begin{bmatrix} a_n \\ b_n \end{bmatrix} = 2\sqrt{2}\begin{bmatrix} \cos\frac{\pi}{12} & -\sin\frac{\pi}{12} \\ \sin\frac{\pi}{12} & \cos\frac{\pi}{12} \end{bmatrix}\begin{bmatrix} a_{n-1} \\ b_{n-1} \end{bmatrix}$$

$$= \left(2\sqrt{2}\right)^2\begin{bmatrix} \cos\frac{\pi}{12} & -\sin\frac{\pi}{12} \\ \sin\frac{\pi}{12} & \cos\frac{\pi}{12} \end{bmatrix}^2\begin{bmatrix} a_{n-2} \\ b_{n-2} \end{bmatrix}$$

$$= \left(2\sqrt{2}\right)^{n-1}\begin{bmatrix} \cos\frac{\pi}{12} & -\sin\frac{\pi}{12} \\ \sin\frac{\pi}{12} & \cos\frac{\pi}{12} \end{bmatrix}^{n-1}\begin{bmatrix} a_1 \\ b_1 \end{bmatrix}$$

Since, $\begin{bmatrix} \cos\theta & -\sin\theta \\ \sin\theta & \cos\theta \end{bmatrix}^n = \begin{bmatrix} \cos n\theta & -\sin n\theta \\ \sin n\theta & \cos n\theta \end{bmatrix}$

$$\Rightarrow \begin{bmatrix} a_{22} \\ b_{22} \end{bmatrix} = \left(2\sqrt{2}\right)^{21}\begin{bmatrix} \frac{1}{\sqrt{2}} & \frac{1}{\sqrt{2}} \\ -\frac{1}{\sqrt{2}} & \frac{1}{\sqrt{2}} \end{bmatrix}\begin{bmatrix} 1 \\ 1 \end{bmatrix}$$

$$= 2^{31}\begin{bmatrix} 1 & 1 \\ -1 & 1 \end{bmatrix}\begin{bmatrix} 1 \\ 1 \end{bmatrix}$$

$$= 2^{32}\begin{bmatrix} 1 \\ 0 \end{bmatrix}$$

$\Rightarrow a_{22} = 2^{32} = 2^{(2)^5}$

$\Rightarrow a + b = 7$

14. **(3)**

$(\text{adj } A)\text{adj}(\text{adj } A) = |\text{adj } A| I = |A|^2 \cdot I$

$\Rightarrow (A \text{ adj } A)\text{adj}(\text{adj } A) = |A|^2 A$

$\Rightarrow (|A|I)\text{adj}(\text{adj}A) = |A|^2 A$

$\Rightarrow \text{adj}(\text{adj}A) = |A|A = 3A$

$\Rightarrow \lambda = 3$

15. (0)

$(A^3 + B^3 + C^3)(A - B)$

$= A^4 - A^3B + B^3A - B^4 + C^3A - C^3B = 0$

$\Rightarrow \det((A^3 + B^3 + C^3)(A - B)) = 0$

$\Rightarrow \det(A^3 + B^3 + C^3) = 0$ as $\det(A - B) \neq 0$

$\Rightarrow \det(A^3 + B^3 + C^3) = 0$ as $\det(A - B) \neq 0$

$\Rightarrow A^3 + B^3 + C^3 = 0$

16. (1)

[Do it yourself]

17. (1)

[Do it yourself]

18. (0)

$A^T_{(1\times3)}B_{3\times3}A_{3\times1} = (A^TBA)_{1\times1} = [0]$

19. (2)

$A = I + B$, where $B = \begin{bmatrix} 0 & 1 \\ 0 & 0 \end{bmatrix}$, $B^2 = 0$

$A^n = (I + B)^n = I + {}^nC_1B = \begin{bmatrix} 1 & n \\ 0 & 1 \end{bmatrix}$

$A + A^2 + A^3 + A^4 + A^5 = \begin{bmatrix} 5 & \sum_{r=1}^{5} r \\ 0 & 5 \end{bmatrix} = \begin{bmatrix} 5 & 15 \\ 0 & 5 \end{bmatrix}$

$\therefore |A + A^2 + A^3 + A^4 + A^5| = 25$

6.8.3 JEE Advanced: Sample Questions

Single Answer Type

1. (C)

$$\Delta = \begin{vmatrix} \operatorname{cosec}\alpha & 1 & 0 \\ 1 & 2\operatorname{cosec}\alpha & 1 \\ 0 & 1 & 2\operatorname{cosec}\alpha \end{vmatrix}$$

$= 4\operatorname{cosec}^3\alpha - 3\operatorname{cosec}\alpha$

$\Rightarrow$ min value = 1

2. (B)

The degree of the determinant

$= n + n + 2 + n + 3 = 3n + 5$

and the degree of the expression of RHS $= 2$

$\Rightarrow 3n + 5 = 2 \Rightarrow n = -1$

3. (A)

Equating the coefficient of x^4

$4 - p = 0 \Rightarrow p = 4$

Put, $x = 0$, we get $\begin{vmatrix} 1 & -1 & 2 \\ 0 & -3 & 4 \\ -1 & 0 & -2 \end{vmatrix} - t = 0$

$\Rightarrow t = 4$

Since, the sum of the coefficients of the given equation is 0

$\Rightarrow$ One root is 1

$\therefore$ It's one root is $1 = \dfrac{p}{t}$

4. (A)

Given

$(1 + x + x^2)^n = \alpha_0 + \alpha_1x + \alpha_2x^2 + \ldots + \alpha_{2n}x^{2n}$ (1)

Replace x by $\dfrac{1}{x}$

$(x^2 + x + 1)^n$

$= \alpha_0x^{2n} + \alpha_1x^{2n-1} + \alpha_2x^{2n-2} + \ldots + \alpha_{2n}$ (2)

From (1) and (2)

$\alpha_{n-1} = \alpha_{n+1}, \alpha_{n-3} = \alpha_{n+3}, \alpha_{n-7} = \alpha_{n+7}$

Hence, $p = 0$

5. (D)

As $x, y, z > 1$, $\log_x y = \dfrac{\ln y}{\ln x} > 0$

Similarly, $\log_y z$, $\log_z x$ are also positive

$\Rightarrow \log_x y + \log_y z + \log_z x > 0$

The given determinant is a symmetric determinant and its value will be zero only if

$\log_x y = \log_y z = \log_z x$

(since, $\log_x y + \log_y z + \log_z x \neq 0$)

Let $\log_x y = a$, $\log_y z = b$, $\log_z x = c$

Then, $a = b = c$

Also,

$\log_x y \cdot \log_y z \cdot \log_z x$

$= \dfrac{\log y}{\log x} \cdot \dfrac{\log z}{\log y} \cdot \dfrac{\log x}{\log z} = 1$

$\Rightarrow abc = 1 \Rightarrow a^3 = 1$

$\Rightarrow a = 1 = b = c \qquad [a = b = c]$

$\Rightarrow \log y = \log x = \log z \Rightarrow x = y = z$

6. **(D)**

$C_1 \to C_1\alpha + C_2$

$$= \begin{vmatrix} a\alpha + b & b & a\alpha + b \\ b\alpha + c & c & b\alpha + c \\ a\alpha^2 + 2b\alpha + c & b\alpha + c & 0 \end{vmatrix}$$

$$= \begin{vmatrix} a\alpha + b & b & a\alpha + b \\ b\alpha + c & c & b\alpha + c \\ 0 & b\alpha + c & 0 \end{vmatrix}$$

$= 0$

7. **(A)**

$$\text{Area} = \frac{1}{2}\begin{vmatrix} a & x & 1 \\ a+d & xr & 1 \\ a+2d & xr^2 & 1 \end{vmatrix}$$

$$= \frac{1}{2}\begin{vmatrix} a & x & 1 \\ a & xr & 1 \\ a & xr^2 & 1 \end{vmatrix} + \frac{1}{2}\begin{vmatrix} 0 & x & 1 \\ d & xr & 1 \\ 2d & xr^2 & 1 \end{vmatrix}$$

$$= 0 + \frac{1}{2}\begin{vmatrix} 0 & x & 1 \\ d & xr & 1 \\ 2d & xr^2 & 1 \end{vmatrix}$$

Which clearly shows that the area is independent of a.

8. **(D)**

$[\sin^2\theta] = 0, \sin^2\theta \neq 1$

$= 1, \sin^2\theta = 1$

If $\sin^2\theta \neq 1 \Rightarrow D = 2\sin\theta\cos\theta - 2i - 1$

$\text{Re}(D) = 2\sin\theta\cos\theta - 1$

$\Rightarrow -2 \le \text{Re}(D) \le 0$

$\Rightarrow -\dfrac{3\pi}{4} \le \arg D \le \dfrac{-\pi}{2}$

If $\sin^2\theta = 1, \sin\theta = \pm 1, \cos\theta = 0$

$\text{Arg}(D) = \arg(1 - 2i)$ or $\arg(-1 - 2i)$

9. **(C)**

α, β, γ are the roots of the equation

$px^3 + qx^2 - rx - r = 0$

$\therefore \alpha + \beta + \gamma = -\dfrac{q}{p}$

$\alpha\beta + \alpha\gamma + \beta\gamma = -\dfrac{r}{p}$

$\alpha\beta\gamma = +\dfrac{r}{p}$

$$\text{Now,} \begin{vmatrix} 1+\alpha & 1 & 1 \\ 1 & 1+\beta & 1 \\ 1 & 1 & 1+\gamma \end{vmatrix}$$

$= \alpha\beta + \beta\gamma + \alpha\gamma + \alpha\beta\gamma = -\dfrac{r}{p} + \dfrac{r}{p} = 0$

10. **(D)**

$C_1 \to C_1 \cos D + C_2 \sin D$

$\therefore \Delta = 0$

Hence Δ is independent of A, B, C and D all.

11. **(A)**

From the first two equations, we get

$x = z + 2$ and $y = 4 - 2z$

So, the third equation

$\Rightarrow (z+2) + 2(4 - 2z) + az = b$

$\Rightarrow (a - 3)z = b - 10$

This will be true for all real z if $a = 3$ and $b = 10$

12. **(D)**

Since,

$$f(x) = \begin{vmatrix} x^3 & \sin x & \tan x \\ \cos x & x^8 & 1 \\ e^{x^2} & \sec x & 2 \end{vmatrix}$$

Since, $f(-x) = -f(x)$

$\Rightarrow f(x)$ is an odd function.

$\Rightarrow \displaystyle\int_{-\frac{\pi}{2}}^{\frac{\pi}{2}} f(x)\,dx = 0$

13. **(A)**

We know that

$$\left(a^3+b^3+c^3-3abc\right)^2=\begin{vmatrix} a & c & b \\ b & a & c \\ c & b & a \end{vmatrix}\begin{vmatrix} a & c & b \\ b & a & c \\ c & b & a \end{vmatrix}$$

$$=\begin{vmatrix} A & B & B \\ B & A & B \\ B & B & A \end{vmatrix}=\begin{vmatrix} B & A & B \\ B & B & A \\ A & B & B \end{vmatrix}$$

14. **(C)**

Applying $C_1 \to C_1 + C_2$, we get

$$\begin{vmatrix} 2 & \cos^2 x & \sin 2x \\ 2 & 1+\cos^2 x & \sin 2x \\ 1 & \cos^2 x & 1+\sin 2x \end{vmatrix}$$

Applying $R_2 \to R_2 - R_1$ and $R_3 \to R_3 - R_1$, we get

$$\begin{vmatrix} 2 & \cos^2 x & \sin 2x \\ 0 & 1 & 0 \\ -1 & 0 & 1 \end{vmatrix}=2+\sin 2x$$

The maximum value of $\sin 2x$ is 1, and min. value of $\sin 2x$ is (-1)

Therefore $\alpha = 3, \beta = 1$

$\Rightarrow \alpha^{2n} - \beta^{2n}$ is always an even integer for $n \in N$.

15. **(A)**

$$R_2 \to R_2 - \frac{(R_1+R_3)}{2}$$

$$\Rightarrow \begin{vmatrix} x+3 & x+4 & x+\lambda \\ 0 & 0 & \mu-\dfrac{(\lambda+\delta)}{2} \\ x+5 & x+6 & x+\delta \end{vmatrix}=0$$

But λ, μ, δ are in A.P.

$$\therefore \mu = \frac{\lambda+\delta}{2}$$

$$\therefore \begin{vmatrix} x+3 & x+4 & x+\lambda \\ 0 & 0 & 0 \\ x+5 & x+6 & x+\delta \end{vmatrix}=0 \ \forall\, x \in R$$

$\therefore$ An identity in x.

16. **(C)**

$$\text{The determinant} = \begin{vmatrix} n-1 & n^2-n & n^2 \\ n & n^2-1 & n^2+n \\ n-1 & n^2+1 & n^2 \end{vmatrix}$$

$[R_3 \to R_3 - R_1]$

$$=\begin{vmatrix} n-1 & n^2-n & n^2 \\ n & n^2-1 & n^2+n \\ 0 & n+1 & 0 \end{vmatrix}$$

$$=-(n+1)\left\{n\left(n^2-1\right)-n^3\right\}$$

$$=n(n+1)=72=8\times 9$$

$$\Rightarrow n=8$$

17. **(A)**

$$\Delta=\begin{vmatrix} 100x_1+10y_1+z_1 & y_1 & z_1 \\ 100x_2+10y_2+z_2 & y_2 & z_2 \\ 100x_3+10y_3+z_3 & y_3 & z_3 \end{vmatrix}$$

$$=\begin{vmatrix} 2A & y_1 & z_1 \\ 2B & y_2 & z_2 \\ 2C & y_3 & z_3 \end{vmatrix}, \text{ where } A, B, C \in I$$

$$=2\begin{vmatrix} A & y_1 & z_1 \\ B & y_2 & z_2 \\ C & y_3 & z_3 \end{vmatrix},$$

which is divisible by 2 but not necessarily by 4 or 8.

18. **(C)**

We have,

$$f(x)=(x-1)(x-1)(x-2)\begin{vmatrix} 1 & 2 & 3 \\ x-1 & x-2 & x-3 \\ x & x & x \end{vmatrix}$$

$R_3 \to R_3 - R_2$

$$=(x-1)^2(x-2)\begin{vmatrix} 1 & 2 & 3 \\ x-1 & x-2 & x-3 \\ 1 & 2 & 3 \end{vmatrix}$$

$$=0$$

$$\Rightarrow f(49)=0$$

19. **(D)**

The given system of equations is consistent.

$$\Rightarrow \Delta = \begin{vmatrix} 2 & 1 & -5 \\ 1 & -2 & 1 \\ 2 & -14 & -a \end{vmatrix} = 0$$

$$\Rightarrow 2(2a+14)-1(-a-2)-5(14+4)=0$$

$$\Rightarrow 4a+28+a+2-90=0$$

$$\Rightarrow 5a-60=0 \Rightarrow a=12$$

20. **(B)**

$C_1 \to C_1 - C_3$

$$\Rightarrow \begin{vmatrix} 1-\alpha^2 & \alpha & \alpha^2 \\ 0 & \cos pa & \cos(p-d)a \\ 0 & \sin pa & \sin(p-d)a \end{vmatrix} = 0$$

$$\Rightarrow (1-\alpha^2)$$

$$\{\cos pa \sin(p-d)a - \sin pa \cos(p-d)a\} = 0$$

$$\Rightarrow (1-\alpha^2)\{\sin(p-d-p)a\} = 0$$

$$\Rightarrow (1-\alpha^2)\sin(-da) = 0$$

$\Rightarrow$ Not dependent on p.

21. **(D)**

$R_1 \to R_1 - R_2$ and $R_2 \to R_2 - R_3$

$$\therefore \Delta = \begin{vmatrix} +x & -y & 0 \\ 0 & y & -z \\ a-x & b-y & c \end{vmatrix} = 0$$

$$\Rightarrow x(cy+bz-yz)+y(0+z(a-x))=0$$

$$\Rightarrow cxy+bzx-xyz+ayz-xyz=0$$

$$\Rightarrow ayz+bzx+cxy=2xyz$$

Dividing by xyz

$$\therefore \frac{a}{x}+\frac{b}{y}+\frac{c}{z}=2$$

22. **(B)**

$$\Delta = \begin{vmatrix} 1 & 1 & 1 \\ 2 & 3 & 1 \\ 1 & 2 & 0 \end{vmatrix}$$

$$\Rightarrow \Delta = 1(0-2)-1(0-1)+1(4-3)$$

$$= -2+1+1=0$$

Also, $\Delta_x = \Delta_y = \Delta_z = 0$

$\Rightarrow$ Infinite solutions.

23. **(B)**

$$\text{Since, } \begin{bmatrix} 1 & x & 1 \end{bmatrix} \begin{bmatrix} 1 & 3 & 2 \\ 0 & 5 & 1 \\ 0 & 3 & 2 \end{bmatrix} \begin{bmatrix} x \\ 1 \\ -2 \end{bmatrix} = 0$$

$$\Rightarrow \begin{bmatrix} 1 & 5x+6 & x+4 \end{bmatrix} \begin{bmatrix} x \\ 1 \\ -2 \end{bmatrix} = 0$$

$$\Rightarrow x+5x+6-2x-8=0 \Rightarrow x=\frac{1}{2}$$

24. **(C)**

Since the diagonal elements of a skew symmetric matrix are always zero,

$\Rightarrow$ trace of $A = 0$

25. **(B)**

$$A^2 = \begin{bmatrix} -5 & -8 & 0 \\ 3 & 5 & 0 \\ 1 & 2 & -1 \end{bmatrix} \begin{bmatrix} -5 & -8 & 0 \\ 3 & 5 & 0 \\ 1 & 2 & -1 \end{bmatrix}$$

$$= \begin{bmatrix} 1 & 0 & 0 \\ 0 & 1 & 0 \\ 0 & 0 & 1 \end{bmatrix} = I$$

$\Rightarrow$ The matrix A is involutory.

26. **(C)**

$A \text{ adj } A = |A| I$

$$|A| = xyz - (8x+4y+3z)+28$$

$$= 6-20+28 = 68$$

$\Rightarrow (\text{adj } A)^{-1}$ always exists when ever $(A)^{-1}$ exists

$$\therefore A \text{ adj}(A) = |A| I = 68 \begin{bmatrix} 1 & 0 & 0 \\ 0 & 1 & 0 \\ 0 & 0 & 1 \end{bmatrix}$$

$$= \begin{bmatrix} 68 & 0 & 0 \\ 0 & 68 & 0 \\ 0 & 0 & 68 \end{bmatrix}$$

27. **(C)**

Since the given equations has non-zero solutions,

$$\Rightarrow \begin{vmatrix} a^3 & (a+1)^3 & (a+2)^3 \\ a & a+1 & a+2 \\ 1 & 1 & 1 \end{vmatrix} = 0$$

$C_2 - C_1$ and $C_3 - C_2$

$$\begin{vmatrix} a^3 & (a+1)^3 - a^3 & (a+2)^3 - (a+1)^3 \\ a & 1 & 1 \\ 1 & 0 & 0 \end{vmatrix} = 0$$

Expanding along R_3 we get

$-6(a+1) = 0 \Rightarrow a = -1$

28. **(B)**

$$|A| = \begin{vmatrix} a & 0 & 0 \\ 0 & a & 0 \\ 0 & 0 & a \end{vmatrix} = a^3$$

$$|\text{adj } A| = |A|^2 = (a^3)^2 = a^6$$

29. **(B)**

A is an idempotent matrix. Hence $A^2 = A$

$\Rightarrow A^n = A$ so that

$AB = A(I - A) = A - A^2 = A - A = 0$

30. **(C)**

We have $\begin{bmatrix} \omega^2 & -\omega^2 \\ \omega & \omega \end{bmatrix}^{-1} \begin{bmatrix} \omega & \omega^2 \\ -\omega & \omega^2 \end{bmatrix}^{-1}$

$$= \left(\begin{bmatrix} \omega & \omega^2 \\ -\omega & \omega^2 \end{bmatrix} \begin{bmatrix} \omega^2 & -\omega^2 \\ \omega & \omega \end{bmatrix} \right)^{-1} = \begin{bmatrix} 2 & 0 \\ 0 & 2 \end{bmatrix}^{-1}$$

$$= \frac{1}{4} \begin{bmatrix} 2 & 0 \\ 0 & 2 \end{bmatrix}$$

31. **(D)**

Using $AA^T = I$

32. **(C)**

$(XY)^T = Y^T X^T$

$= ((AB)^T - (BA)^T)((AB)^T + (BA)^T)$

$= (B^T A^T - A^T B^T)(B^T A^T + A^T B^T)$

$= (BA - AB)(BA + AB)$

$= -YX$

33. **(D)**

Find a, b, c, d

Clearly $\begin{bmatrix} a & b \\ c & d \end{bmatrix}$ is nilpotent

34. **(D)**

$$A^{-1} = \frac{\text{adj } A}{|A|}$$

$$= \frac{1}{2A^2 + A - 20} \begin{bmatrix} A & 5 \\ 4 & 2A+1 \end{bmatrix}$$

$$\therefore \frac{1}{2A^2 + A - 20} \begin{bmatrix} A & 5 \\ 4 & 2A+1 \end{bmatrix} \begin{bmatrix} A-5 & B \\ 2A-2 & C \end{bmatrix}$$

$$= \begin{bmatrix} 14 & D \\ E & F \end{bmatrix}$$

$$\Rightarrow \frac{A^2 + 5A - 10}{2A^2 + A - 20} = 14$$

$\Rightarrow 3A^2 + A - 30 = 0$

$\Rightarrow A = 3, -\frac{10}{3} \Rightarrow$ The absolute value of the difference of these two solutions

$= 3 + \frac{10}{3} = \frac{19}{3}$

35. **(B)**

Let $A = \begin{bmatrix} 0 & a \\ -a & 0 \end{bmatrix}$

$$BC = \begin{bmatrix} 1 & 4 \\ 2 & 9 \end{bmatrix} \begin{bmatrix} 9 & -4 \\ -2 & 1 \end{bmatrix} = \begin{bmatrix} 1 & 0 \\ 0 & 1 \end{bmatrix} = I$$

$\therefore B^2C^2 = (BC)^2 = I^2 = I$

Similarly, $B^2C^2 = B^3C^3 = \ldots = B^nC^n = I$

Let,

$$D = A^3(BC) + A^5(B^2C^2) + A^7(B^3C^3) + \ldots + A^{2n+1}(B^nC^n)$$

$= A^3 + A^5 + A^7 + \ldots + A^{2n+1}$

$= A(A^2 + A^4 + A^6 + \ldots + A^{2n})$

Now, $A^2 = A \cdot A = \begin{bmatrix} 0 & a \\ -a & 0 \end{bmatrix} \begin{bmatrix} 0 & a \\ -a & 0 \end{bmatrix}$

$$= \begin{bmatrix} -a^2 & 0 \\ 0 & -a^2 \end{bmatrix} = -a^2 I$$

$\therefore D = A\left(-a^2 + a^4 - a^6 + \ldots + (-1)^n a^{2n}\right)$

Hence D is skew symmetric.

36. **(A)**

$AA^{-1} = I$

$$\Rightarrow \begin{bmatrix} 0 & 1 & 2 \\ 1 & 2 & 3 \\ 3 & a & 1 \end{bmatrix} \begin{bmatrix} \frac{1}{2} & -\frac{1}{2} & \frac{1}{2} \\ -4 & 3 & 6 \\ \frac{5}{2} & -\frac{3}{2} & \frac{1}{2} \end{bmatrix} = \begin{bmatrix} 1 & 0 & 0 \\ 0 & 1 & 0 \\ 0 & 0 & 1 \end{bmatrix}$$

$$\Rightarrow \begin{bmatrix} 1 & 0 & b+1 \\ 0 & 1 & 2(b+1) \\ 4(1-a) & 3(a-1) & ab+2 \end{bmatrix} = \begin{bmatrix} 1 & 0 & 0 \\ 0 & 1 & 0 \\ 0 & 0 & 1 \end{bmatrix}$$

$\Rightarrow b+1=0, ab+2=1, a-1=0$

$\Rightarrow a=1, b=-1$

37. **(A)**

38. **(B)**

39. **(C)**

40. **(C)**

41. **(D)**

Multiple Answer Type

1. **(A, B, C, D)**

$$\Delta_r = \begin{vmatrix} 2^{r-1} & \frac{1}{r(r+1)} & \sin r\theta \\ x & y & z \\ 2^n - 1 & \frac{n}{n+1} & \frac{\sin\left(\frac{n+1}{2}\right)\theta \sin\frac{n}{2}\theta}{\sin\frac{\theta}{2}} \end{vmatrix}$$

$$\sum_{r=1}^{n} 2^{r-1} = 2^n - 1$$

$$\sum_{r=1}^{n} \frac{1}{r(r+1)} = \frac{n}{n+1} \text{ and}$$

$$\sum_{r=1}^{n} \sin r\theta = \frac{\sin\left(\frac{n+1}{2}\right)\theta \sin\frac{n}{2}\theta}{\sin\frac{\theta}{2}}$$

$$\therefore \sum_{r=1}^{n} \Delta_r = 0$$

2. **(B, C, D)**

$$\Delta = \begin{vmatrix} x^2 & (y+z)^2 & yz \\ y^2 & (z+x)^2 & zx \\ z^2 & (x+y)^2 & xy \end{vmatrix}$$

Clearly factor of Δ are $(x+y+z); (x-y)$ and $\left(x^2+y^2+z^2\right)$

3. **(A, B, C)**

$$\Delta = \begin{vmatrix} a & b & c \\ b & c & a \\ c & a & b \end{vmatrix}$$

$R_1 \to R_1 + (R_2 + R_3)$

$$= (a+b+c)\begin{vmatrix} 1 & 1 & 1 \\ b & c & a \\ c & a & b \end{vmatrix}$$

$C_1 \to C_1 - C_2, C_2 \to C_2 - C_3$

$$= (a+b+c)\begin{vmatrix} 0 & 0 & 1 \\ b-c & c-a & 1 \\ c-a & a-b & 1 \end{vmatrix}$$

$$= (a+b+c)\left\{(b-c)(a-b)-(c-a)^2\right\}$$

$$= (a+b+c)\left\{ab-ac-b^2+bc-c^2-a^2+2ac\right\}$$

$$= -(a+b+c)\left\{f(w)\cdot f\left(w^2\right)\right\}$$

4. **(A, B)**

$$\text{Let, } \Delta = \begin{vmatrix} A & 6 & 8 \\ 8 & B & 6 \\ 8 & 8 & C \end{vmatrix}$$

Operating $R_3 \to 100R_1 + 10R_2 + R_3$

$$= \begin{vmatrix} A & 6 & 8 \\ 8 & B & 6 \\ 72n_1 & 72n_2 & 72n_3 \end{vmatrix}$$

where n_1, n_2, n_3 are integer

$$= 72\begin{vmatrix} A & 6 & 8 \\ 8 & B & 6 \\ n_1 & n_2 & n_3 \end{vmatrix}$$

$\Rightarrow$ Divisible by 72 and 12

5. **(A, B, C, D)**

Operating $R_1 \to R_1 + R_2 + R_3$

6. **(A, C)**

Since, $\left(AA^T\right)^T = \left(A^T\right)^T \cdot A^T$

$= AA^T$

$\Rightarrow AA^T$ is symmetric

Also, $\left(A^T A\right)^T = A^T \left(A^T\right)^T$

$= A^T \cdot A$

$\Rightarrow A^T A$ is symmetric

7. **(A, B, C, D)**
{Do it yourself}

8. **(C, D)**

$$A_r = \begin{bmatrix} C_{r-1}^2 & 0 \\ 0 & C_r^2 \end{bmatrix}$$

$$\Rightarrow \sum_{r=1}^{n} A_r = \begin{bmatrix} \sum_0^{n-1}(C_r)^2 & 0 \\ 0 & \sum_1^{n}(C_r)^2 \end{bmatrix}$$

$$= \begin{bmatrix} {}^{2n}C_n - 1 & 0 \\ 0 & {}^{2n}C_n - 1 \end{bmatrix} = \begin{bmatrix} K_1 & 0 \\ 0 & K_2 \end{bmatrix}$$

$\Rightarrow K_1 = K_2 = {}^{2n}C_n - 1$

9. **(A, B, C)**

Since, $|A| = \begin{vmatrix} 3 & -3 & 4 \\ 2 & -3 & 4 \\ 0 & -1 & 1 \end{vmatrix} = 1$

$\because \text{adj}(\text{adj } A) = |A|^{3-2} A = A$

and $|\text{adj}(\text{adj } A)| = |A| = 1$

Also, $|\text{adj } A| = |A|^{3-1} = |A|^2 = 1^2 = 1$

10. **(C, D)**
[Do it yourself]

11. **(A, B, C)**

$A^2 + A + I = 0$

$\Rightarrow (A - I)\left(A^2 + A + I\right) = 0 \qquad [A - I \neq 0]$

$\Rightarrow A^3 = I \Rightarrow A^{3K} = I, A^{3K+1} = A, A^{3K+2} = A^2$

$A^4 + A^9 = A + I = -\left(A^{-1}\right)$

$A^6 = I \Rightarrow A^{-1} = A^5$

$A^7 + A^8 = A + A^2 = -I$

12. **(A, C, D)**

$$A = \begin{bmatrix} 2-\pi & 3-\pi & 4-\pi \\ 3-\pi & 4-\pi & 5-2\pi \\ 4-\pi & 5-2\pi & 6-2\pi \end{bmatrix}$$

$\Rightarrow A^T = A$

$\Rightarrow (\text{adj } A)^T = \text{adj}(A)$

$\Rightarrow \left((\text{adj } A)^T\right)^3 = (\text{adj } A)^3$

$\Rightarrow \left((\text{adj } A)^3\right)^T = (\text{adj } A)^3$

$\text{Tr}(A) = 12 - 4\pi$

13. **(A, B, D)**

$$A = \begin{bmatrix} 1 & 0 \\ 0 & 1 \end{bmatrix} + \begin{bmatrix} 0 & 1 \\ -1 & 0 \end{bmatrix} = I + B$$

$B^2 = -I, (I + B)^n$

$= \left({}^nC_0 - {}^nC_2 + {}^nC_4 - {}^nC_6 + \ldots\right)I$

$+ \left({}^nC_1 - {}^nC_3 + {}^nC_5 - {}^nC_7\right)B$

$$\Rightarrow A^n = \left(2^{n/2}\cos\frac{n\pi}{4}\right)I + \left(2^{n/2}\sin\frac{n\pi}{4}\right)B$$

$$A^n = \begin{bmatrix} 2^{n/2}\cos\frac{n\pi}{4} & 2^{n/2}\sin\frac{n\pi}{4} \\ -2^{n/2}\sin\frac{n\pi}{4} & 2^{n/2}\cos\frac{n\pi}{4} \end{bmatrix}$$

14. **(C, D)**

15. **(B, D)**

16. **(A, B, C, D)**

17. **(A, D)**

18. **(A, B, D)**

19. **(A, C, D)**

20. **(B, C)**

21. **(A, C)**

22. **(A, B, C, D)**

23. **(A, C, D)**

24. **(A, B)**

Comprehension Based Questions

1. **(C)**

p, q, r are the entries of first row and C_{21}, C_{22}, C_{23} are cofactors of second row

$\therefore p \cdot C_{21} + q \cdot C_{22} + r \cdot C_{23} = 0$

2. **(C)**
x, y, z are the entries of second row and C_{21}, C_{22}, C_{23} are cofactors of second row
$\therefore x \cdot C_{21} + y.C_{22} + z \cdot C_{23} = 0$

3. **(B)**
$q \cdot M_{12} - y \cdot M_{22} + m \cdot M_{32}$
$= -q \cdot C_{12} - y \cdot C_{22} - m \cdot C_{32}$
$= -(q \cdot C_{12} + y \cdot C_{22} + m \cdot C_{32}) = -\Delta$
[$\because q, y, m$ are entries of second column and C_{12}, C_{22}, C_{32} are cofactors of second column]

4. **(B)**

$$\Delta = \begin{vmatrix} 1 & 1 & 1 \\ a^2 & b^2 & c^2 \\ bc & ca & ab \end{vmatrix} = \frac{abc}{abc}\begin{vmatrix} a & b & c \\ a^3 & b^3 & c^3 \\ 1 & 1 & 1 \end{vmatrix}$$

$$= \begin{vmatrix} 1 & 1 & 1 \\ a & b & c \\ a^3 & b^3 & c^3 \end{vmatrix}$$

$$= (a-b)(b-c)(c-a)(a+b+c)$$

5. **(C)**

$$\Delta = \begin{vmatrix} a & b & c \\ a^2 & b^2 & c^2 \\ bc & ca & ab \end{vmatrix} = \begin{vmatrix} a^3 & b^2 & c^2 \\ a^3 & b^3 & c^3 \\ 1 & 1 & 1 \end{vmatrix}$$

$$= \begin{vmatrix} 1 & 1 & 1 \\ a^2 & b^2 & c^2 \\ a^3 & b^3 & c^3 \end{vmatrix}$$

$$= (a-b)(b-c)(c-a)(ab+bc+ca)$$

6. **(D)**
$\Delta = abc(a-b)(b-c)(c-a)$

7. **(A)**
Case I
All three diagonal elements are 1
Number of matrices $= {}^3C_1 = 3$
Case II
Two diagonal elements are zero and one element is $1 = {}^3C_1 \cdot {}^3C_1 = 9$
$\Rightarrow$ No. of matrices = 12

8. **(B)**

$$A\begin{bmatrix} x \\ y \\ z \end{bmatrix} = \begin{bmatrix} 1 \\ 0 \\ 0 \end{bmatrix}$$

For unique solution
$\det(A) \neq 0$
Case I

$$\det(A) = \begin{vmatrix} 1 & a & b \\ a & 1 & c \\ b & c & 1 \end{vmatrix} = 2abc \neq 0$$

Here a, b, c is selected from 1, 0, 0
(No case possible)
Case II

$$\det(A) = \begin{vmatrix} 1 & a & b \\ a & 0 & c \\ b & c & 0 \end{vmatrix} = 2abc - c^2 \neq 0$$

Hence a, b, c are selected from 1, 1, 0
(2 cases are possible)
Similarly, $2abc - b^2 \neq 0$
and $2abc - a^2 \neq 0$
Hence there are exactly 6 matrices for unique solution.

9. **(B)**
More than 2 matrices

10. **(C)**
$AB = A \Rightarrow |AB| = |A|$
$\Rightarrow |A| = 0$ or $|B| = 1$
$BA = B \Rightarrow |BA| = |B|$
$\Rightarrow |A| = 1$ or $B = 0$

11. **(C)**
$AB = A,\ BA = B$
$\Rightarrow (AB)A = A^2$ or $A(BA) = A^2$
or AB^2 or $A = A^2$
Similarly, $B^2 = B$
$(A+B)^2 = A^2 + B^2 + AB + BA$
$= 2(A+B)$
$(A+B)^3 = 2^2(A+B)$
$\Rightarrow (A+B)^7 = 2^6(A+B) = 64(A+B)$

12. (C)

$$(A+I)^5 = A^5 + 5A^4 + 10A^3 + 10A^2 + 5A + I$$
$$= I + 5A + 10A + 10A + 5A + A$$
$$\left[A^2 = A \Rightarrow A^3 = A^4 = A^5 = A\right]$$
$$= I + 31A$$

Match Matrix

1. (A – R; B – S; C – Q; D – P)

$$f(x) = \begin{vmatrix} \sec^2 x & 1 & 1 \\ \cos^2 x & \cos^2 x & \operatorname{cosec}^2 x \\ 1 & \cos^2 x & \cot^2 x \end{vmatrix}$$

Operate $C_2 \to C_2 - C_1 \cos^2 x$

$$f(x) = \begin{vmatrix} \sec^2 x & 0 & 1 \\ \cos^2 x & \cos^2 x - \cos^4 x & \operatorname{cosec}^2 x \\ 1 & 0 & \cot^2 x \end{vmatrix}$$

$$= \sec^2 x\left\{\left(\cos^2 x - \cos^4 x\right)\cot^2 x - 0\right\} + 1\left(0 - \left(\cos^2 x - \cot^4 x\right)\right)$$

$$= \left(1 - \cos^2 x\right)\cot^2 x - \cos^2 x \sin^2 x$$

$$= \cos^2 x\left(1 - \sin^2 x\right) = \cos^4 x$$

$$\Rightarrow f(x) = \cos^4 x$$

$$\therefore \int_0^{\pi/4} f(x)dx = \int_0^{\pi/4} \cos^4 x dx = \frac{3\pi}{32} + \frac{1}{4}$$

$$\therefore \int_0^{\pi/4} f(x)dx - \frac{1}{4} = \frac{3\pi}{32}$$

Max. value of $f(x) = 1$

Min. value of $f(x) = 0$

Period of $f(x) = \pi$

2. (A – R; B – S; C – Q; D – P)

$$a_k = \int_0^{\pi/2} (k + \sin\theta)^2 \cos\theta d\theta$$

$$= \left[\frac{(k+\sin\theta)^3}{3}\right]_0^{\pi/2} = \frac{(k+1)^3 - k^3}{3}$$

$$= \frac{(k+1-k)\left((k+1)^2 + k(k+1) + k^2\right)}{3}$$

$$= \frac{3k^2 + 3k + 1}{3}$$

$$\text{Let, } \Delta = \begin{vmatrix} a_k & k & k^2 \\ a_{k+1} & k+1 & (k+1)^2 \\ a_{k+2} & k+2 & (k+2)^2 \end{vmatrix}$$

$$\Rightarrow \Delta = \begin{vmatrix} \frac{(K+1)^3 - K^2}{3} & K & K^2 \\ \frac{(K+2)^3 - (K+1)^3}{3} & K+1 & (K+1)^2 \\ \frac{(K+3)^3 - (K+2)^3}{3} & K+2 & (K+2)^2 \end{vmatrix}$$

$$= \frac{1}{3}\begin{vmatrix} 1 & k & k^2 \\ 1 & k+1 & (k+1)^2 \\ 1 & k+2 & (k+2)^2 \end{vmatrix}$$

$$= \frac{1}{3}\begin{vmatrix} 1 & k & k^2 \\ 0 & 1 & (2k+1) \\ 0 & 1 & (2k+3) \end{vmatrix} = \frac{2}{3}$$

$\therefore p = 2, q = 3$

$|p - q| = 1,\ p^2 + q^2 = 13$

3. (A – S; B – P; C – P, R; D – Q)

(B) As A is idempotent,

$A^2 = A^3 = A^4 = \ldots = A$

$(A+I)^n$

$= I + {}^nC_1A + {}^nC_2A^2 + \ldots + {}^nC_nA^n$

$= I + \left({}^nC_1 + {}^nC_2 + \ldots + {}^nC_n\right)A$

$\Rightarrow 2^n - 1 = 127$

$\Rightarrow n = 7$

(C) Since matrix A is skew symmetric and $|A| = \left|-A^T\right| = (-1)^n|A|$

$\Rightarrow |A|\left(1 - (-1)^n\right) = 0$ if n is odd

$\Rightarrow A$ is singular

(D) $(I - A)\left(I + A + A^2 + \ldots + A^7\right)$

$= I - A^8 = I$ if $A^8 = 0$

$\therefore \left|A + A^2 + A^3 + A^4 + A^5\right| = 25$

4. (A – P; B – T; C – Q, D – R)

$$A=\begin{bmatrix}0&1&1\\1&0&1\\1&1&0\end{bmatrix}\Rightarrow|A|=2$$

$$|A-I|=\begin{vmatrix}-1&1&1\\1&-1&1\\1&1&-1\end{vmatrix}$$

$$A+I=\begin{bmatrix}1&1&1\\1&1&1\\1&1&1\end{bmatrix}\Rightarrow|A+I|=0$$

$(A+I)^2=3(A+I)\Rightarrow A^2=A+2I$

Similarly, $A^3=3A+2I,\ A^4=5A+6I$

$A=I+2A^{-1}\Rightarrow 2A^{-1}=A-I$

$A^3-A^2+2A=4A$

$\Rightarrow|A^3-A^2+2A|=4^3|A|=128$

$|A^4-4A-7I|=|A-I|=4$

$A^4-8A^{-1}-A^2=8I$

Subjective Answer Type (Hard Nut)

1. The first determinant

$$=\begin{vmatrix}(a-b)^2&a^2&b^2\\1&1&1\\c^2&a^2&(c-a)^2\end{vmatrix}$$

$$=\begin{vmatrix}(a-b)^2-c^2&0&b^2-(c-a)^2\\1&1&1\\c^2&a^2&(c-a)^2\end{vmatrix}$$

$[R_1-R_3]$

$$=(b+c-a)\begin{vmatrix}-(a+c-b)&0&(b+a-c)\\1&1&1\\c^2&a^2&(c-a)^2\end{vmatrix}$$

$$=(2s-2a)\begin{vmatrix}-(2s-2b)&0&(2s-2c)\\1&1&1\\c^2&a^2&(c-a)^2\end{vmatrix}$$

$$=4(s-a)(s-b)\{a^2-(c-a)^2\}+4(s-a)(s-c)(a^2-c^2)$$

Similarly operating 2nd and 3rd determinant and adding

$=16s(s-a)(s-b)(s-c)=16\Delta^2$

$\Rightarrow\lambda=4$

2. Since, $\dfrac{4x^2}{1+4x^2}=y$ (1)

$\dfrac{4y^2}{1+4y^2}=z$ (2)

and, $\dfrac{4z^2}{1+4z^2}=x$ (3)

From (1), $\dfrac{y}{x}=\dfrac{4x}{1+4x^2}$ (since, $1+4x^2\ge 4x$)

$\Rightarrow y\le x$, similarly, $z\le y$ and $x\le z$

$\therefore x=y=z$

Hence $\begin{vmatrix}x&y&z\\y&z&x\\z&x&y\end{vmatrix}=0$

3. Since the elements in the first determinant is the cofactor of the corresponding elements of the second determinant,

$$\text{L.H.S}=\begin{vmatrix}\lambda&c&-b\\-c&\lambda&a\\b&-a&\lambda\end{vmatrix}^3$$

$=\left[\lambda(\lambda^2+a^2+b^2+c^2)\right]^3=(1+a^2+b^2+c^2)^3$

$\Rightarrow\lambda=1$

4. Since the given equation has a non-trivial solution,

$$\Rightarrow\begin{vmatrix}-a^2&ab&ac\\ab&-b^2&bc\\ac&bc&-c^2\end{vmatrix}=0\Rightarrow\begin{vmatrix}0&c&b\\c&0&a\\b&a&0\end{vmatrix}^2=0$$

$$\Rightarrow\begin{vmatrix}0&c&b\\c&0&a\\b&a&0\end{vmatrix}-0$$

5. $\because A+B+C=\pi$

Let

$$\Delta=e^{iA}\cdot e^{iB}\cdot e^{iC}\begin{vmatrix}e^{iA}&e^{-i(A+C)}&e^{-i(A+B)}\\e^{-i(B+C)}&e^{iB}&e^{-i(A+B)}\\e^{-i(B+C)}&e^{-i(A+C)}&e^{iC}\end{vmatrix}$$

$$\Rightarrow \Delta = e^{i\pi}\begin{vmatrix} e^{iA} & e^{-i(\pi-B)} & e^{-i(\pi-C)} \\ e^{-i(\pi-A)} & e^{iB} & e^{-i(\pi-C)} \\ e^{-i(\pi-A)} & e^{-i(\pi-B)} & e^{iC} \end{vmatrix}$$

$$\Rightarrow \Delta = -\begin{vmatrix} e^{iA} & -e^{iB} & -e^{iC} \\ -e^{iA} & e^{iB} & -e^{iC} \\ -e^{iA} & -e^{iB} & e^{iC} \end{vmatrix}$$

$R_1 + R_2$ and $R_2 + R_3$

$$\Delta = -\begin{vmatrix} 0 & 0 & -2e^{iC} \\ -2e^{iA} & 0 & 0 \\ -e^{iA} & -e^{iB} & e^{iC} \end{vmatrix}$$

$$= -\left\{-2e^{iC}\left(2e^{iA}e^{iB}\right)\right\}$$

$$= 4e^{i(A+B+C)} = 4e^{i\pi} = -4$$

6. $\begin{vmatrix} p & b & c \\ a & q & c \\ a & b & r \end{vmatrix} = 0$

Operating $R_1 - R_2$ and $R_2 - R_3$

$$\Rightarrow \begin{vmatrix} p-a & b-q & 0 \\ 0 & q-b & c-r \\ a & b & r \end{vmatrix} = 0$$

$$\Rightarrow (p-a)(qr-br-bc+br)- (b-q)(-a(c-r)) = 0$$

$$\Rightarrow r(p-a)(q-b)-b(p-a)(c-r)+ a(b-q)(c-r) = 0$$

$$\Rightarrow +r(p-a)(q-b)+b(p-a)(r-c)+ a(q-b)(r-c) = 0$$

Dividing by $(p-a)(q-b)(r-c)$

$$\frac{r}{r-c}+\frac{b}{q-b}+\frac{a}{p-a} = 0$$

$$\Rightarrow \frac{r}{r-c}+\frac{b}{q-b}+1+\frac{a}{p-a}+1 = 2$$

$$\Rightarrow \frac{p}{p-a}+\frac{q}{q-b}+\frac{r}{r-c} = 2$$

7. $\begin{vmatrix} 1 & 0 & 0 \\ {}^nC_1 & 3 & 3 \\ {}^nC_2 & 3(n+1) & 3(n+4) \end{vmatrix}$

$[C_2 - C_1 \text{ and } C_3 - C_2]$

$$= 3^3\begin{vmatrix} 1 & 0 & 0 \\ {}^nC_1 & 1 & 1 \\ {}^nC_2 & n+1 & n+4 \end{vmatrix} = 3^2\left[1(n+4-n-1)\right]$$

$$= 3^3$$

$$\Rightarrow n = 3$$

8. Applying $C_1 \to C_1 - C_3\sin\theta$ and $C_2 \to C_2 + C_3\cos\theta$

$$\Rightarrow f(\theta) = \begin{vmatrix} 1 & 0 & -\sin\theta \\ 0 & 1 & \cos\theta \\ \sin\theta & -\cos\theta & 0 \end{vmatrix}$$

Applying $R_3 \to R_3 - R_1\sin\theta + R_2\cos\theta$

$$\Rightarrow f(\theta) = \begin{vmatrix} 1 & 0 & -\sin\theta \\ 0 & 1 & \cos\theta \\ 0 & 0 & 1 \end{vmatrix} = 1$$

$f(\theta) = 1$ for all $\theta, \Rightarrow f\left(\frac{\pi}{6}\right) = 1$

9. Since, $f(x) = \begin{vmatrix} 2\cos^2 x & \sin 2x & -\sin x \\ \sin 2x & 2\sin^2 x & \cos x \\ \sin x & -\cos x & 0 \end{vmatrix}$

$\left.\begin{matrix} C_1 \to C_1 + 2\sin xC_3 \\ C_2 \to C_2 - 2\cos xC_3 \end{matrix}\right\}$, further operating

$C_1 \to C_1 - 4\sin xC_3;\ C_2 \to C_2 + 4\cos xC_3$

$$f(x) = \begin{vmatrix} 2 & 0 & -\sin x \\ 0 & 2 & \cos x \\ \sin x & -\cos x & 0 \end{vmatrix}$$

$$= 2\sin^2 x + 2\cos^2 x = 2$$

$$\Rightarrow \int_0^{\frac{\pi}{2}}\left(f(x)+f'(x)\right)dx = 2\int_0^{\frac{\pi}{2}} dx = \pi$$

10. Applying

$R_1 \to R_1 - \sec xR_3,\ R_2 \to R_2 - \cos^2 xR_3$

$\therefore F(x) =$

$$\begin{vmatrix} 0 & 0 & \sec^2 x + \cot x \operatorname{cosec} x \\ 0 & \cos^2 x - \cos^4 x & \operatorname{cosec}^2 x \\ 1 & \cos^2 x & \cos^2 x \end{vmatrix}$$

$\Rightarrow f(x) =$

$$\left(\sec^2 x + \cot x \operatorname{cosec} x\right)\left(-\cos^2 x + \cos^4 x\right)$$

$$= -\sin^2 x - \cos^3 x$$

$$\therefore \int_0^{\frac{\pi}{2}} f(x)\,dx = \int_0^{\frac{\pi}{2}} \left(-\sin^2 x - \cos^3 x\right)dx$$

$$= -\frac{1}{2}\cdot\frac{\pi}{2} - \frac{2}{3} = -\left(\frac{\pi}{4} + \frac{2}{3}\right)$$

11. $f(x) =$

$$\begin{vmatrix} x^2-4x+6 & 2x^2+4x+10 & 3x^2-2x+16 \\ x-2 & 2x+2 & 3x-1 \\ 1 & 2 & 3 \end{vmatrix}$$

$$f'(x) = \begin{vmatrix} 2x-4 & 4x+4 & 6x-2 \\ x-2 & 2x+2 & 3x-1 \\ 1 & 2 & 3 \end{vmatrix}$$

$$+\begin{vmatrix} x^2-4x+6 & 2x^2+4x+10 & 3x^2-2x+16 \\ 1 & 2 & 3 \\ 1 & 2 & 3 \end{vmatrix}$$

$$+\begin{vmatrix} - & - & - \\ - & - & - \\ 0 & 0 & 0 \end{vmatrix}$$

$$\Rightarrow f'(x) = 0 \Rightarrow f(x) = \cos t$$

$$f(0) = \begin{vmatrix} 6 & 10 & 16 \\ -2 & 2 & -1 \\ 1 & 2 & 3 \end{vmatrix}$$

$$= 6(6+2) - 10(-6+1) + 16(-4-2)$$

$$\Rightarrow f(0) = 48 + 50 - 96 = 2$$

$$\Rightarrow f(x) = 2 \Rightarrow [f(x)] = 2$$

Now, $\int_{-2}^{2} x^2[f(x)]\,dx = 2\times 2\int_0^2 x^2 dx$

$$= 4\left[\frac{x^3}{3}\right]_0^2$$

$$= \frac{32}{3}$$

$$\therefore \left\{\frac{32}{3}\right\} = \frac{2}{3} = \frac{\alpha}{\beta}$$

$$\Rightarrow \alpha + \beta = 5$$

12. $\begin{vmatrix} 3^2+K & 4^2 & 3^2+3+K \\ 4^2+K & 5^2 & 4^2+4+K \\ 5^2+K & 6^2 & 5^2+5+K \end{vmatrix} = 0$

$C_3 \to C_3 - C_1$

$$\Rightarrow \begin{vmatrix} 3^2+K & 4^2 & 3 \\ 4^2+K & 5^2 & 4 \\ 5^2+K & 6^2 & 5 \end{vmatrix} = 0 \Rightarrow \begin{vmatrix} 3^2+K & 4^2 & 3 \\ 7 & 9 & 1 \\ 9 & 11 & 1 \end{vmatrix} = 0$$

$[R_2 - R_1; R_3 - R_2]$

$$\Rightarrow K = 1$$

$$\therefore \sqrt{2^K\sqrt{2^K\sqrt{2^K}}\ \ldots} = 2^{\frac{1}{2}}\cdot 2^{\frac{1}{4}}\cdot 2^{\frac{1}{8}}\ldots$$

$$= 2^{\frac{1}{2}/\frac{1}{2}} = 2$$

13. $C_1 \to C_1 - C_2, C_2 \to C_2 - C_3$

$$\therefore \begin{vmatrix} a & 0 & 1 \\ -b & 2b & 1 \\ 0 & -2c & 1+3c \end{vmatrix} = 0$$

$$\Rightarrow a(b+3bc+c) + 1(bc) = 0$$

$$\Rightarrow ab + bc + ac = -3abc$$

$$\Rightarrow \left|\frac{1}{a} + \frac{1}{b} + \frac{1}{c}\right| = 3$$

14. Let, $f(x) = \begin{vmatrix} (1+x)^{22} & (1+x)^{44} & (1+x)^{66} \\ (1+x)^{33} & (1+x)^{66} & (1+x)^{99} \\ (1+x)^{44} & (1+x)^{88} & (1+x)^{144} \end{vmatrix}$

$f(x)$ is a polynomial

$$\Rightarrow f(x) = A_0 + A_1x + A_2x^2 + \ldots + A_{232}x^{232}$$

Thus, coefficient of $x = f'(0)$

Differentiating both sides with respect to x and then put $x = 0$ in both sides

$$f'(0) = \begin{vmatrix} 22 & 44 & 66 \\ 1 & 1 & 1 \\ 1 & 1 & 1 \end{vmatrix} + \begin{vmatrix} 1 & 1 & 1 \\ 33 & 66 & 99 \\ 1 & 1 & 1 \end{vmatrix} + \begin{vmatrix} 1 & 1 & 1 \\ 1 & 1 & 1 \\ 44 & 88 & 144 \end{vmatrix}$$

$$\Rightarrow f'(0) = 0 + 0 + 0 \Rightarrow f'(0) = 0$$

Hence, coefficient of x in the given determinant = 0

15. $(1+x+x^2)^n = a_0 + a_1x + a_2x^2 + \ldots + a_{2n}x^{2n}$ (1)

Replace x by $\frac{1}{x}$

$$(x^2+x+1)^n$$

$= a_0x^{2n} + a_1x^{2n-1} + a_2x^{2n-2} + \ldots + a_{2n}$ (2)

Equating coefficients of like terms from (1) and (2)

$a_0 = a_{2n}, a_1 = a_{2n-1}, \ldots, a_r = a_{2n-r}$

$\Rightarrow a_{n-1} = a_{n+1}, a_{n-3} = a_{n+3}, a_{n-7} = a_{n+7}$

Therefore, value of determinant = 0

16. [Do it yourself]

17. Operating $R_1 \to R_1 - R_2, R_2 \to R_2 - R_3$

So,

$$\Delta = \begin{vmatrix} \sin(A+B)\sin(A-B) & \cot A - \cot B & 0 \\ \sin(B+C)\sin(B-C) & \cot B - \cot C & 0 \\ \sin^2 C & \cot C & 1 \end{vmatrix}$$

$$= \begin{vmatrix} \sin C \sin(A-B) & \dfrac{\sin(B-A)}{\sin A \sin B} & 0 \\ \sin A \sin(B-C) & \dfrac{\sin(C-B)}{\sin B \sin C} & 0 \\ \sin^2 C & \cot C & 1 \end{vmatrix}$$

Expanding along 3rd column

$$\Delta = \frac{\sin(A-B)\sin(C-B)}{\sin B} - \frac{\sin(B-C)\sin(B-A)}{\sin B}$$

$$= \frac{\sin(A-B)}{\sin B}[-\sin(B-C) + \sin(B-C)] = 0$$

18. From the given conditions

$\frac{y-x}{b-a} \neq \pm 1, \frac{z-y}{c-b} \neq \pm 1, \frac{z-x}{c-a} \neq \pm 1$

$\Rightarrow x+a \neq y+b \neq z+c$

The determinant is a symmetric one. The determinant will be equal to zero if

$x+a+y+b+z+c=0$

But $a+b+c=0$ (given)

$\Rightarrow x+y+z=0$

$\Rightarrow x+z = 2\left(\frac{-y}{2}\right) \Rightarrow x, -\frac{y}{2}, z$ are in A.P.

19. Clearly $\alpha = -i$ where $i^2 = -1$

$$\text{So, } \Delta(\alpha) = \alpha^n \cdot \frac{1}{\alpha^n}\begin{vmatrix} 1 & 1 & 1 \\ 1 & \alpha & \alpha^3 \\ \frac{1}{\alpha} & 1 & 0 \end{vmatrix}$$

$$= \begin{vmatrix} 1 & 1 & 1 \\ 1 & -i & i \\ i & 1 & 0 \end{vmatrix} = 1(-i) - 1(-i^2) + 1(1+i^2)$$

$= -1 - i$

$\Rightarrow$ arg is $\frac{-3\pi}{4}$

20. Let, $M_n = (a_{ij})$ where $i, j = 1, 2, 3, \ldots, n$

We first find out a_{11} for the nth matrix, which is the nth term of the series 1, 2, 6, 15

Let, $S = 1+2+6+15+\ldots+t_{n-1}+t_n$ (1)

$\Rightarrow S = 1+2+6+\ldots+t_{n-1}+t_n$ (2)

Subtracting,

$0 = 1+1+4+9+\ldots+(t_n - t_{n-1}) - t_n$

$\Rightarrow t_n = 1 + (1+4+9+\ldots \text{to } (n-1) \text{ terms})$

$= 1 + (1^2+2^2+3^2+4^2+\ldots+(n-1)^2)$

$= 1 + \frac{(n-1)n(2n-1)}{6}$

The consecutive distance between the elements of the diagonal of the nth matrix is $(n+1)$

$\Rightarrow$ First term $= 1 + \frac{(n-1)n(2n-1)}{6}$ and

common difference $= (n+1)$

Hence the sum of the elements of the diagonal in M_n

$$= \frac{n}{2}\left[2\left(1 + \frac{n(n-1)(2n-1)}{6}\right) + (n-1)(n+1)\right]$$

$$= \frac{n}{6}\left[2n^3 + n + 3\right]$$

21. It is given that

$X=(A+I)^{-1}(A-I)$

So, $X^T=\left(A^T-I\right)\left(A^T+I\right)^{-1}$ (1)

As, $\left(A^T-I\right)\left(A^T+I\right)$ commutes, so

$\left(A^T+I\right)\left(A^T-I\right)=\left(A^T-I\right)\left(A^T+I\right)$

$\Rightarrow\left(A^T+1\right)^{-1}\left(A^T+I\right)\left(A^T-I\right)\left(A^T+I\right)^{-1}$

$=\left(A^T+I\right)^{-1}+\left(A^T-I\right)\left(A^T+I\right)\left(A^T+I\right)^{-1}$

$\Rightarrow\left(A^T-I\right)\left(A^T+I\right)^{-1}=\left(A^T+I\right)^{-1}\left(A^T-I\right)$ (2)

From (1) and (2)

$X^T=\left(A^T+I\right)^{-1}\left(A^T-I\right)$

$=\left(A^T(I+A)\right)^{-1}\left(A^T(I-A)\right)$

[As A is an orthogonal matrix]

$=(A+I)^{-1}\left(A^T\right)^{-1}A^T(I-A)$

$=(-1)^n(A+I)^{-1}(A-I)$

$=-(A+I)^{-1}(A-I)$ [As n is an odd integer]

Hence, X is a skew symmetric matrix.

22. Let $A=\begin{bmatrix}a & p\\ b & q\\ c & r\end{bmatrix}\Rightarrow A'=\begin{bmatrix}a & b & c\\ p & q & r\end{bmatrix}$

$$\Rightarrow AA'=\begin{bmatrix}a^2+p^2 & ab+pq & ac+pr\\ ab+pq & b^2+q^2 & bc+qr\\ ac+pr & bc+rq & c^2+r^2\end{bmatrix}$$

$$\Rightarrow|AA'|=\begin{vmatrix}a & p & 0\\ b & q & 0\\ c & r & 0\end{vmatrix}\begin{vmatrix}a & b & c\\ p & q & r\\ 0 & 0 & 0\end{vmatrix}$$

Hence, AA' is singular.

23. $t_r(A)=t_r(B)-2$

$\Rightarrow\alpha^2+\beta^4+2\gamma^2=2\beta^2+\gamma^2-2\gamma+2\alpha-1-2$

$\Rightarrow\left(\alpha^2-2\alpha+1\right)+\left(\beta^4-2\beta^2+1\right)+\left(\gamma^2+2\gamma+1\right)=0$

$\Rightarrow(\alpha-1)^2+\left(\beta^2-1\right)^2+(\gamma+1)^2=0$

$\Rightarrow\alpha=1,\beta=\pm1,\gamma=-1$

Now, $|c|=\alpha\beta\gamma-2\alpha-2\beta+4$

$\Rightarrow|c|=-1$ or 5

24. **(2)** [Do it yourself]

25. **(1)** [Do it yourself]

26. **(4)** [Do it yourself]

27. **(2)** [Do it yourself]

28. **(1)** [Do it yourself]

29. $AB=\begin{bmatrix}3ax^2 & 3bx^2 & 3cx^2\\ a & b & c\\ 6ax & 6bx & 6cx\end{bmatrix}$

Since, $t_r(AB)=t_r(C)$

$\Rightarrow 3ax^2+b+6cx$

$=(x+2)^2+2x+5x\ \forall\ x\in R$ (identify)

$\Rightarrow 3ax^2+bcx+b=6x^2+6x+4$

$\Rightarrow a=2, c=1, b=4$

30. Since, $A^2=\begin{bmatrix}-1 & 3 & 5\\ 1 & -3 & -5\\ -1 & 3 & 5\end{bmatrix}\begin{bmatrix}-1 & 3 & 5\\ 1 & -3 & -5\\ -1 & 3 & 5\end{bmatrix}$

$=\begin{bmatrix}-1 & 3 & 5\\ 1 & -3 & -5\\ -1 & 3 & 5\end{bmatrix}$

$=A$

$\therefore A^2=A^3=A^4=A^5=\ldots=A$

But it is given $A^x=A$

$\Rightarrow x=2,3,4,5,\ldots$ $[x\neq1]$

$$\therefore\prod\frac{\left(x^3+1\right)}{\left(x^3-1\right)}=\prod\frac{(x+1)}{x-1}\prod\left(\frac{x^2-x+1}{x^2+x+1}\right)$$

Putting $x=2,3,4,5,\ldots$

$$\prod\left(\frac{x^3+1}{x^3-1}\right)=\lim_{n\to\infty}\prod_{x=2}^{n}\left(\frac{x+1}{x-1}\right)\prod_{x=2}^{n}\left(\frac{x^2-x+1}{x^2+x+1}\right)$$

$$= \lim_{n\to\infty}\left[\frac{3\cdot4\cdot5\ldots(n-1)n(n+1)}{1\cdot2\cdot3\ldots(n-3)(n-2)(n-1)}\right]\times$$

$$\lim_{n\to\infty}\left[\frac{3\cdot7\ldots\left(n^2-n+1\right)}{7\cdot13\ldots\left(n^2-n+1\right)\left(n^2+n+1\right)}\right]$$

$$=\frac{3}{2}\lim_{n\to\infty}\frac{\left(1+\frac{1}{n}\right)}{\left(1+\frac{1}{n}+\frac{1}{n^2}\right)}=\frac{3}{2}$$

31. Since, $AX=\lambda x \Rightarrow (A-\lambda I)X=0$

As $X\neq 0$

$\therefore \det(A-\lambda I)=0$

$$\Rightarrow \begin{vmatrix} 3-\lambda & 2 & 2 \\ 2 & 4-\lambda & 1 \\ -2 & -4 & -1-\lambda \end{vmatrix}=0$$

$$\Rightarrow -\lambda(3-\lambda)^2=0$$

$$\Rightarrow \lambda=0,3$$

Clearly $\lambda=0, \mu=3$

For,

$$\lambda=0, AX=0\Rightarrow \begin{bmatrix} 3 & 2 & 2 \\ 2 & 4 & 1 \\ -2 & -4 & -1 \end{bmatrix}\begin{bmatrix} x \\ y \\ z \end{bmatrix}=\begin{bmatrix} 0 \\ 0 \\ 0 \end{bmatrix}$$

$$\Rightarrow 3x+2y+2z=0 \text{ and } 2x+4y+z=0$$

$$\Rightarrow \frac{x}{-6}=\frac{y}{1}=\frac{z}{8}$$

So, $X=\begin{bmatrix} -6 \\ 1 \\ 8 \end{bmatrix}$

For, $\mu=3, (A-3I)Y=0$

$$\Rightarrow \begin{bmatrix} 0 & 2 & 2 \\ 2 & 1 & 1 \\ -2 & -4 & -4 \end{bmatrix}\begin{bmatrix} \alpha \\ \beta \\ \gamma \end{bmatrix}=\begin{bmatrix} 0 \\ 0 \\ 0 \end{bmatrix}$$

$$\Rightarrow Y=\begin{bmatrix} 0 \\ -1 \\ 1 \end{bmatrix}$$

If θ is the angle between X and Y

$$\cos\theta=\frac{0\cdot(-6)+(-1)\cdot1+1\cdot8}{\sqrt{0+1+1}\sqrt{36+1+64}}=\frac{7}{\sqrt{202}}$$

$$\Rightarrow \theta=\cos^{-1}\left(\frac{7}{\sqrt{202}}\right)$$

32. $\left(\frac{-4\sqrt{2}}{3},\frac{2}{3},2\sqrt{2}\right),\left(\frac{4\sqrt{2}}{3},\frac{2}{3},-2\sqrt{2}\right),$

$(3,3,-1)$

[Do it yourself]

33. Since, A is symmetric

$\therefore c=8, b=-1$ and $a=2$ (1)

and B is skew symmetric

$\therefore d=e=f=0$ and

$2b+c=6, a=2, b-a=-3$ (2)

From (1) and (2)

$a=2, b=-1, c=8, d=0, e=0, f=0$

$$\therefore A=\begin{bmatrix} 3 & 2 & -1 \\ 2 & 5 & 8 \\ -1 & 8 & 2 \end{bmatrix} \text{ and } B=\begin{bmatrix} 0 & 3 & 2 \\ -3 & 0 & -6 \\ -2 & 6 & 0 \end{bmatrix}$$

$$\Rightarrow AB=\begin{bmatrix} -4 & 3 & -6 \\ -31 & 54 & -26 \\ -28 & 9 & -50 \end{bmatrix}$$

Which is neither symmetric nor skew symmetric.

34. $BPA=\begin{bmatrix} 1 & 0 & 1 \\ 0 & 1 & 0 \end{bmatrix}$

$\Rightarrow P=B^{-1}\begin{bmatrix} 1 & 0 & 1 \\ 0 & 1 & 0 \end{bmatrix}A^{-1}$ (1)

$B=\begin{bmatrix} 2 & 3 \\ 3 & 4 \end{bmatrix}, B^{-1}=\begin{bmatrix} -4 & 3 \\ 3 & -2 \end{bmatrix}$ (2)

and $A=\begin{bmatrix} 1 & 1 & 1 \\ 2 & 4 & 1 \\ 2 & 3 & 1 \end{bmatrix}$

$|A|=-1\neq 0$

$|A|=-1\neq 0$

$$A^{-1}=\begin{bmatrix}-1 & -2 & 3\\ 0 & 1 & -1\\ 2 & 1 & -2\end{bmatrix} \qquad (3)$$

$$\therefore P=\begin{bmatrix}-4 & 3\\ 3 & -2\end{bmatrix}\begin{bmatrix}1 & 0 & 1\\ 0 & 1 & 0\end{bmatrix}\begin{bmatrix}-1 & -2 & 3\\ 0 & 1 & -1\\ 2 & 1 & -2\end{bmatrix}$$

$$\Rightarrow P=\begin{bmatrix}-4 & 7 & -7\\ 3 & -5 & 5\end{bmatrix}$$

35. $[a\neq -3, b\in R]$; $\left[a=-3, b\neq \frac{1}{3}\right]$;

$\left[a=-3, b=\frac{1}{3}\right]$

[Do it Yourself]

36. $A^T A=I$

$\Rightarrow |A^T A|=|I| \Rightarrow |A^T||A|=I$

$\Rightarrow |A||A|=1$

$\Rightarrow |A|=\pm 1$

$$\Rightarrow \begin{vmatrix}a & b & c\\ b & c & a\\ c & a & b\end{vmatrix}=\pm 1$$

$\Rightarrow a^3+b^3+c^3=3\pm 1=2$ or 4

37. $\begin{bmatrix}a & b\\ c & 1-a\end{bmatrix}$ is an idempotent matrix

$$\Rightarrow \begin{bmatrix}a & b\\ c & 1-a\end{bmatrix}^2=\begin{bmatrix}a & b\\ c & 1-a\end{bmatrix}$$

$$\Rightarrow \begin{bmatrix}a^2+bc & b\\ c & bc+(1-a)^2\end{bmatrix}=\begin{bmatrix}a & b\\ c & 1-a\end{bmatrix}$$

$\Rightarrow a^2+bc=a \Rightarrow a-a^2=bc=\frac{1}{4}$

$\Rightarrow f(a)=\frac{1}{4}$

6.8.4 JEE Main: Archive

Determinants

1. **(B)**

$x+y+z=\alpha$

$\alpha x+2\alpha y+3z=-1$

$x+3\alpha y+5z=4$

Has an inconsistent solution.

$$D=\begin{vmatrix}1 & 1 & 1\\ \alpha & 2\alpha & 3\\ 1 & 3\alpha & 5\end{vmatrix}=0$$

$\Rightarrow (\alpha-1)^2=0$

$\alpha=1$

For $\alpha=1$

$$D_1=\begin{vmatrix}1 & 1 & 1\\ -1 & 2 & 3\\ 4 & 3 & 5\end{vmatrix}$$

$=(10-9)-(-5-12)+(-3-8)$

$=1+17-11\neq 0$

For $\alpha=1$ the system of equations has an inconsistent solution.

2. **(C)**

$$\Delta=\begin{vmatrix}1 & 1 & \alpha\\ 3 & 1 & 1\\ 1 & 0 & 2\end{vmatrix}=-(\alpha+3)$$

$$\Delta_1=\begin{vmatrix}2 & 1 & \alpha\\ 4 & 1 & 1\\ 1 & 0 & 2\end{vmatrix}=-(3+\alpha)$$

$$\Delta_2=\begin{vmatrix}1 & 2 & \alpha\\ 3 & 4 & 1\\ 1 & 1 & 2\end{vmatrix}=-(\alpha+3)$$

$$\Delta_3=\begin{vmatrix}1 & 1 & 2\\ 3 & 1 & 4\\ 1 & 0 & 1\end{vmatrix}=0$$

$\alpha\neq -3, x=1, y=1, z=0,$

Now points $(\alpha,1),(1,\alpha)$ and $(1,-1)$ are collinear.

$$\begin{vmatrix} \alpha & 1 & 1 \\ 1 & \alpha & 1 \\ 1 & -1 & 1 \end{vmatrix} = 0$$

$\Rightarrow \alpha(\alpha+1)-1(1-1)+1(-1-\alpha)=0$

$\alpha^2+\alpha-1-\alpha=0$

$\alpha=\pm 1$

3. (D)

$$\Delta = \begin{vmatrix} -k & 3 & -14 \\ -15 & 4 & -k \\ -4 & 1 & 3 \end{vmatrix} = 121-k^2$$

$\Delta \neq 0 \quad k \in R-\{11,-11\}$ (Unique solution)

If $k=11$

$$\Delta_z = \begin{vmatrix} -11 & 3 & 25 \\ -15 & 4 & 3 \\ -4 & 1 & 4 \end{vmatrix} \neq 0$$

No solution

If $k=-11$

$$\Delta_z = \begin{vmatrix} 11 & 3 & 25 \\ -15 & 4 & 3 \\ -4 & 1 & 4 \end{vmatrix} \neq 0$$

No solution

4. (C)

$$\begin{vmatrix} 3 & -2 & 1 \\ 5 & -8 & 9 \\ 2 & 1 & a \end{vmatrix} = 0$$

$3(-8a-9)+2(5a-18)+1(21)=0$

$\Rightarrow a=-3$

$$\text{Also, } \Delta_2 = \begin{vmatrix} 3 & -2 & b \\ 5 & 8 & 3 \\ 2 & 1 & -1 \end{vmatrix}^{\frac{1}{3}}$$

If $b=\dfrac{1}{3}$

$\Delta_2=0$

So b must be equal to

$=-\dfrac{1}{3}$

5. (C)

For infinitely many solutions,

$\Delta - 0 - \Delta_x = \Delta_y = \Delta_z$

$$\Delta = \begin{vmatrix} \alpha & 1 & 1 \\ 1 & 2 & 3 \\ 1 & 3 & 5 \end{vmatrix} = 0$$

$\Rightarrow \alpha(10-9)-1(5-3)+1(3-2)=0$

$\Rightarrow \alpha-2+1=0$

$\Rightarrow \alpha=1$

$$\Delta_x = \begin{vmatrix} 5 & 1 & 1 \\ 4 & 2 & 3 \\ \beta & 3 & 5 \end{vmatrix} = 0$$

$\Rightarrow 5(10-9)-1(20-3\beta)+1(12-2\beta)$

$\Rightarrow 5-20+3\beta+12-2\beta$

$\Rightarrow -3+\beta=0$

$\Rightarrow \beta=3$

6. (D)

$$\Delta = \begin{vmatrix} 1 & 2 & 1 \\ 2 & 3 & -1 \\ -2 & 1 & 2 \end{vmatrix}$$

$=(6+y)-2((2\alpha-\alpha)+1(\alpha+3\alpha))$

$=7-2\alpha+4\alpha$

$=7+2\alpha$

$\Delta=0 \Rightarrow \alpha=-\dfrac{7}{2}$

$$\Delta_1 = \begin{vmatrix} 2 & 2 & 1 \\ \alpha & 3 & -1 \\ -\alpha & 1 & 2 \end{vmatrix}$$

$=14+2\alpha$

$\alpha=-x_2=7$

$\Delta_1 \neq 0$

7. (B)

$a=\alpha-i\beta; \alpha\in R; \beta \in R$

$4ix+(1=i)y=0$ and

$8\left(\cos\dfrac{2\pi}{3}+i\sin\dfrac{2\pi}{3}\right)x+\bar{a}y=0$

$$\begin{vmatrix} 4i & 1+i \\ 8e^{i2\pi/3} & \bar{a} \end{vmatrix} = 0$$

$\Rightarrow 4i\bar{a}-(1+i)8e^{i2\pi/3}=0$

$$\Rightarrow 4i(\alpha+i\beta)-8(1+i)\left(\frac{-1+i\sqrt{3}}{2}\right)=0$$

$$\Rightarrow i\alpha-\beta+1+\sqrt{3}+i\left(1-\sqrt{3}\right)=0$$

$$\Rightarrow \beta=\sqrt{3}+1$$

$$\alpha=\sqrt{3}-1$$

So, $\frac{\alpha}{\beta}=\frac{\sqrt{3}-1}{\sqrt{3}+1}=2-\sqrt{3}$

8. (B)

$$\begin{vmatrix} 2 & 3 & -1 \\ 1 & 1 & 1 \\ 1 & -1 & |\lambda| \end{vmatrix}=0$$

$\Rightarrow |\lambda|=7\Rightarrow \pm 7$ (1)

System:

$2x+3y-z=-2$ (2)

$x+y+z=4$ (3)

$x-y+|\lambda|z=4\lambda-4$ (4)

Eliminating y from equations (2) and (3) we get

$x+4z=14$ (5)

$(3)+(4)\Rightarrow x+\left(\frac{|\lambda|+1}{2}\right)z=2\lambda$ (6)

Clearly for $\lambda=-7$, system is inconsistent.

9. (58)

$2x-3y=\gamma+5$

$\alpha x+5y=\beta+1$

Infinitely many solutions,

$$\frac{\alpha}{2}=\frac{5}{-3}=\frac{\beta+1}{\gamma+5}$$

$$\alpha=\frac{-10}{3},\ 5\gamma+25=-3\beta-3$$

$9\alpha=-30,\ 3\beta+5\gamma=-28$

Now, $9\alpha+3\beta+5\gamma=-58$

$|9\alpha+3\beta+5\gamma|=58$

10. (B)

$$\begin{vmatrix} 2 & 1 & -1 \\ 1 & -3 & 2 \\ 1 & 4 & \delta \end{vmatrix}=0$$

$\Rightarrow \delta=-3$

And $\begin{vmatrix} 7 & 1 & -1 \\ 1 & -3 & 2 \\ k & 4 & -3 \end{vmatrix}=0\Rightarrow k=6$

$\Rightarrow \delta+K=3$

Alternate

$2x+y-z=7$ (1)

$x-3y+2z=1$ (2)

$x+4y+\delta z=k$ (3)

Equations (2) and (3)

We get $2x+y+(2+\delta)z=1+K$ (4)

For infinite solutions,

From equations (1) and (4)

$2+\delta=-1\Rightarrow \boxed{\delta=-3}$

$1+k=7\Rightarrow \boxed{k=6}$

$\delta+k=3$

11. (B)

The system of equations has no solution.

$$D=\begin{vmatrix} 3\sin 3\theta & -1 & 1 \\ 3\cos 2\theta & 4 & 3 \\ 6 & 7 & 7 \end{vmatrix}=0$$

$21\sin 3\theta+42\cos 2\theta-42=0$

$\sin 3\theta+2\cos 2\theta-2=0$

Number of solutions is 7 in $(0, 4\pi)$

12. (C)

13. (D)

$$D=\begin{vmatrix} 8 & 1 & 4 \\ 1 & 1 & 1 \\ \lambda & -3 & 0 \end{vmatrix}=0\Rightarrow \lambda=4$$

Also, $D_1=D_2=D_3=0$

So $\mu=-2$

Point $\left(4,-2,\frac{-1}{2}\right)$

Distance from plane $=\frac{10}{3}$

14. (4)

$$\Delta=\begin{vmatrix} P! & (P+1)! & (P+2)! \\ (P+1)! & (P+2)! & (P+3)! \\ (P+2)! & (P+3)! & (P+4)! \end{vmatrix}$$

$\Delta = P!(P+1)!(P+2)!$

$$\begin{vmatrix} 1 & 1 & 1 \\ P+1 & P+2 & P+3 \\ (P+2)(P+1) & (P+3)(P+2) & (P+4)(P+3) \end{vmatrix}$$

$\Delta = 2P!(P+1)!(P+2)!$

Which is divisible by P^{α} and $(P+2)^{\beta}$

$\therefore \alpha = 3, \beta = 1$

15. **(C)**

$x+y+z=6$ (1)

$2x+5y+\alpha z=\beta$ (2)

$x+2y+3z=14$ (3)

On solving

$x=z-2 \Rightarrow y=8-2z$ in (2)

$2(z-2)+5(8-2z)+\alpha z=\beta$

$(\alpha-8)z=\beta-36$

for having infinite solutions

$\alpha-8=0$ and $\beta-36=0$

$\alpha=8, \beta=36$ $(\alpha+\beta=44)$

16. **(C)**

$x+2y+3z=3$ (1)

$4x+3y-4z=4$ (2)

$8x+4y-\lambda z=9+\mu$ (3)

$(1)\times 4-(2) \Rightarrow 5y+16z=8$ (4)

$(2)\times 2-(3) \Rightarrow 2y+(\lambda-8)z=-1-\mu$ (5)

$(4)\times 2-(3)\times 5 \Rightarrow 72-5\lambda=0 \Rightarrow \lambda=\frac{72}{5}$

$21+5\mu=0 \Rightarrow \mu=\frac{-21}{5}$

$\Rightarrow (\lambda,\mu) \equiv \left(\frac{72}{5}, -\frac{-21}{5}\right)$

17. **(D)**

$$\Delta = \begin{vmatrix} a & 2a & -3a \\ 2a+1 & 2a+3 & a+1 \\ 3a+5 & a+5 & a+2 \end{vmatrix}$$

$= a(15a^2+31a+36)=0 \Rightarrow a=0$

$\Delta \neq 0$ for all $a \in R-\{0\}$

Hence $S_1 = R-\{0\}$ $S_2=\Phi$

18. **(B)**

$$D = \begin{vmatrix} \alpha & 2 & 1 \\ 2\alpha & 3 & 1 \\ 3 & \alpha & 2 \end{vmatrix} = 0 \Rightarrow \alpha = -1, 3$$

$$D_x = \begin{vmatrix} 2 & 1 & 1 \\ 3 & 1 & 1 \\ \alpha & 2 & \beta \end{vmatrix} = 0 \Rightarrow \beta = 2$$

$$D_y = \begin{vmatrix} \alpha & 1 & 1 \\ 2\alpha & 1 & 1 \\ 3 & 2 & \beta \end{vmatrix} = 0$$

$$D_z = \begin{vmatrix} \alpha & 2 & 1 \\ 2\alpha & 3 & 1 \\ 3 & \alpha & \beta \end{vmatrix} = 0$$

$\beta=2, \alpha=-1$

$\alpha=-1, \beta=2$ Infinite solutions

19. **(D)**

$$\begin{vmatrix} 1 & 1 & k \\ 2 & 3 & -1 \\ 3 & 4 & 2 \end{vmatrix} = 0$$

$\Rightarrow 1(10)-1(7)+k(-1)-0$

$\Rightarrow k=3$

For $k=3$, 2nd system is

$4x+5y=7$ (1)

and $7x+8y=10$ (2)

Clearly, they have a unique solution

$(2)-(1) \Rightarrow 3x+3y=3$

$\Rightarrow x+y=1$

20. **(C)**

$$\begin{vmatrix} 1 & -1 & 1 \\ 2 & 2 & \alpha \\ 3 & -1 & 4 \end{vmatrix} = 0;$$

$8+\alpha-2(-4+1)+3(-\alpha-2)=0$

$8+\alpha+6-3\alpha-6=0$

$\alpha=4$

21. **(D)**

By equation 1 and $3y+2z=8$

$y=8-2z$

And $x=-2+z$

Now putting this in equation 2

$\alpha(z-2)+\beta(-2z+8)+7z=3$

$\Rightarrow(\alpha-2\beta+7)z=2\alpha-8\beta+3$

So, equations have a unique solution if

$\alpha-2\beta+7\neq0$

Equations have no solution if

$\alpha-2\beta+7=0$ and $2\alpha-8\beta+3\neq0$

And, equations have infinite solutions if

$\alpha-2\beta+7=0$ and $2\alpha-8\beta+3=0$

22. **(D)**

$$\begin{vmatrix}\lambda & 1 & 1\\ 1 & \lambda & 1\\ 1 & 1 & \lambda\end{vmatrix}=0\ (\lambda+2)\begin{vmatrix}1 & 1 & 1\\ 1 & \lambda & 1\\ 1 & 1 & \lambda\end{vmatrix}=0$$

$$(\lambda+2)\left[(\lambda^2-1)-1(\lambda-1)+(1-\lambda)\right]=0$$

$$(\lambda+2)\left[(\lambda^2-2\lambda+1)=0\right]$$

$$(\lambda+2)(\lambda-1)^2-0\rightarrow\lambda-\ \ 2,\lambda=1$$

at $\lambda=1$, the system has infinite solutions, for inconsistent $\lambda=-2$

$$\sum\left(|-2|^2+|-2|\right)=6$$

23. **(A)**

$$\begin{vmatrix}\alpha & 1 & 1\\ 1 & \alpha & 1\\ 1 & 1 & \alpha\end{vmatrix}=0$$

$$\alpha(\alpha^2-1)-1(\alpha-1)+1(1-\alpha)=0$$

$$\alpha^3-3\alpha+2=0$$

$$\alpha^2(\alpha-1)+\alpha(\alpha-1)-2(\alpha-1)=0$$

$$(\alpha-1)(\alpha^2+\alpha-2)=0\ \ \alpha=1,\alpha=-2,1$$

For $\alpha=1,\beta=1$

$$\left.\begin{matrix}x+y+z=1\\ x+y+z=b\end{matrix}\right\}\text{infinite solution}$$

For $\alpha=2,\beta=1$

$\Delta=4$

$$\Delta_1=\begin{vmatrix}1 & 1 & 1\\ 1 & 2 & 1\\ 1 & 1 & 2\end{vmatrix}=3-1-1=1\Rightarrow x=\frac{1}{4}$$

$$\Delta_2=\begin{vmatrix}2 & 1 & 1\\ 1 & 1 & 1\\ 1 & 1 & 2\end{vmatrix}=2-1=1\Rightarrow y=\frac{1}{4}$$

$$\Delta_3=\begin{vmatrix}2 & 1 & 1\\ 1 & 2 & 1\\ 1 & 1 & 1\end{vmatrix}=2-1=1\Rightarrow z=\frac{1}{4}$$

For $\alpha=2\Rightarrow$ unique solution

24. **(C)**

Given system of equations is

$x+y+az=b$ (1)

$2x+5y+2z=6$ (2)

$x+2y+3z=3$ (3)

It is given that the system of equations has infinite solutions.

$\Rightarrow\Delta=0$

$$\Rightarrow\begin{vmatrix}1 & 1 & a\\ 2 & 5 & 2\\ 1 & 2 & 3\end{vmatrix}=0$$

$$\Rightarrow1(15-4)-1(6-2)+a(4-5)=0$$

$\Rightarrow\alpha=7$

Also, we have $\Rightarrow\Delta_1=0$

$$\Rightarrow\begin{vmatrix}b & 1 & a\\ 6 & 5 & 2\\ 3 & 2 & 3\end{vmatrix}=0$$

$$\Rightarrow b(15-4)-1(18-6)+7(12-15)=0$$

$\Rightarrow b=3$

$\therefore a=7,b=3\ \therefore 2a+3b=14+9=23$

Hence, this is the correct option.

25. **(D)**

Given, the system of equations

$x+y+z=6;\ x+2y+\alpha z=10$

$x+3y+5z=\beta$

Now finding,

$$\Delta=\begin{vmatrix}1 & 1 & 1\\ 1 & 2 & \alpha\\ 1 & 3 & 5\end{vmatrix}$$

$$\Rightarrow\Delta=1(10-3\alpha)-(5-\alpha)+(3-2)$$

$$\Rightarrow\Delta=6-2\alpha$$

For unique solution $\Rightarrow \alpha \neq 3$

But in option (D), system has unique solution for $\alpha = 3$ which is completely wrong.

26. **(A)**

Given,

S is the set of all values of $\theta \in [-\pi, \pi]$ for which the system of linear equations

$x + y + \sqrt{3} = 0$

$-x + (\tan\theta)y + \sqrt{7}z = 0\ x + y + (\tan\theta)z = 0$

has non-trival solutions,

So, by condition of non-trival solution we get,

$$\begin{vmatrix} 1 & 1 & \sqrt{3} \\ -1 & \tan\theta & \sqrt{7} \\ 1 & 1 & \tan\theta \end{vmatrix} = 0$$

$$\Rightarrow \left(\tan^2\theta - \sqrt{7}\right) - \left(-\tan\theta - \sqrt{7}\right) + \sqrt{3}(-1 - \tan\theta) = 0$$

$$\Rightarrow \tan^2\theta + \left(1 - \sqrt{3}\right)\tan\theta - \sqrt{3} = 0$$

$$\Rightarrow (\tan\theta + 1)\left(\tan\theta - \sqrt{3}\right) = 0$$

$$\Rightarrow \tan\theta = \sqrt{3}, -1$$

$$\Rightarrow \theta = \frac{\pi}{3}, \frac{-2\pi}{3}, \frac{-\pi}{4}, \frac{3\pi}{4}$$

Hence,

$$\frac{120}{\pi}\sum\theta = 120\left(\frac{1}{3} - \frac{2}{3} - \frac{1}{4} + \frac{3}{4}\right) = 20$$

27. **(B)**

Given,

For the system of linear equations

$2x - y + 3z = 5;\ 3x + 2y - z = 7$

$4x + 5y + \alpha z = \beta,$

Now finding,

$$D = \begin{vmatrix} 2 & -1 & 3 \\ 3 & 2 & -1 \\ 4 & 5 & \alpha \end{vmatrix}$$

$$\Rightarrow D = 2(2\alpha + 5) + (3\alpha + 4) + 3(7)$$

$$\Rightarrow D = 7\alpha + 35$$

Now we know that, for unique solution $D \neq 0 \Rightarrow \alpha \neq -5$

So, in option (B) $\alpha = -6$ corresponds to infinitely many solutions, which is completely wrong; hence, this is the answer.

28. **(24)**

Given, system of equations,

$6\lambda x - 3y + 3z = 4\lambda^2$

$2x + 6\lambda y + 4z = 1$

$3x + 2y + 3\lambda z = \lambda$

For no solution $\Delta = 0$,

$$\text{So,} \begin{vmatrix} 6\lambda & -3 & 3 \\ 2 & 6\lambda & 4 \\ 3 & 2 & 3\lambda \end{vmatrix} = 0$$

$$\Rightarrow 9\lambda^3 - 7\lambda - 2 = 0$$

$$\Rightarrow (\lambda - 1)(3\lambda + 1)(3\lambda + 2) = 0$$

$$\text{Hence, } \lambda \in \left\{1, \frac{-1}{3}, \frac{-2}{3}\right\}$$

Now putting the value of λ in $12\sum_{\lambda \in S}|\lambda|$ we get,

$$12\sum_{\lambda \in S}|\lambda| = 12 \times \left(1 + \frac{1}{3} + \frac{2}{3}\right) = 24$$

29. **(D)**

Given,

A is a 2×2 matrix with real entries such that $A' = \alpha A + 1,$

$$\text{Now let } A = \begin{bmatrix} a & b \\ c & d \end{bmatrix}$$

Now putting the value in $A' = \alpha A + I$ we get,

$$\Rightarrow \begin{bmatrix} a & c \\ b & d \end{bmatrix} = \begin{bmatrix} \alpha a & \alpha b \\ \alpha c & \alpha d + 1 \end{bmatrix}$$

Equating both sides we get,

$$a = \alpha a + 1 \Rightarrow a = \frac{1}{1-\alpha} \tag{1}$$

$$b = \alpha c \tag{2}$$

$$c = \alpha b \tag{3}$$

From equations (2) and (3) we get,

$c = 0$ or $\alpha = \pm 1$

(not possible as mentioned in question)

$\therefore c = 0 \ \Rightarrow c = 0, b = 0$

$$\text{And } d = \alpha d + 1 \Rightarrow d = \frac{1}{1-\alpha}$$

Also given, $\left|A^2 - A\right| = 4$

$$\Rightarrow |A||A - I| = 4$$

$$\Rightarrow \left(\frac{1}{1-\alpha}\right)^2\left(\frac{1}{1-\alpha} - 1\right)^2 - 4$$

Now let $\frac{1}{1-\alpha}=t$

So, $t^2(t-1)^2=4$

$\Rightarrow t(t-1)=\pm 2$

$\Rightarrow t=2$ or $t=-1$

$\Rightarrow \alpha=\frac{1}{2},2$

So, sum of all possible values will be

$\frac{1}{2}+2=\frac{5}{2}.$

30. **(A)**

Given, system of linear equations,

$7x+11y+\alpha z=13$

$5x+4y+7z=\beta$

$175x+194y+57z=361$

We know that, for infinite solutions

$D=D_1=D_2=D_3=0.$

Solving $D=0$ we get,

$$\begin{vmatrix} 7 & 11 & \alpha \\ 5 & 4 & 7 \\ 175 & 194 & 57 \end{vmatrix}=0$$

Now applying operation $R_3\to R_3-25R_1$ we get,

$$\begin{vmatrix} 7 & 11 & \alpha \\ 5 & 4 & 7 \\ 0 & -81 & 57-25\alpha \end{vmatrix}=0$$

$\Rightarrow 81(49-5\alpha)+(57-25\alpha)(-27)=0$

$\Rightarrow 270\alpha=-2430\Rightarrow\alpha=-9$

Solving $D_1=0$ we get,

$$\begin{vmatrix} 13 & 11 & -9 \\ \beta & 4 & 7 \\ 361 & 194 & 57 \end{vmatrix}=0$$

$\Rightarrow 13(4\times57-7\times194)-$

$\beta(11\times57+9\times194)+361(11\times7+9\times4)=0$

$\Rightarrow 13(-1130)-\beta(2373)+361(113)=0$

$\Rightarrow 13(-10)-\beta(21)+361=0$

$\Rightarrow 21\beta=231\ \Rightarrow\beta=11$

$\therefore \alpha+\beta+2=4$

31. **(D)**

Given that,

$$\begin{vmatrix} x+1 & x & x \\ x & x+\lambda & x \\ x & x & x+\lambda^2 \end{vmatrix}=\frac{9}{8}(103x+81)$$

Put $x=0$ as $x\in R$

$$\Rightarrow\begin{vmatrix} 1 & 0 & 0 \\ 0 & \lambda & 0 \\ 0 & 0 & \lambda^2 \end{vmatrix}=\frac{9}{8}(103(0)+81)$$

On expanding the determinant we get,

$\Rightarrow \lambda^3=\frac{9}{8}\times81\ \Rightarrow\lambda^3=\frac{9^3}{2^3}$

$\Rightarrow\lambda=\frac{9}{2}$ and $\frac{\lambda}{3}=\frac{9}{6}$

The required quadratic equation is

$x^2-\left(\frac{9}{2}+\frac{9}{6}\right)x+\frac{9}{2}\times\frac{9}{6}=0$

$\Rightarrow x^2-(6)x+\frac{27}{4}=0\Rightarrow 4x^2-24x+27=0$

Hence, the required quadratic equation is $4x^2-24x+27=0$

32. **(6)**

$$\text{Given that } D_k=\begin{vmatrix} 1 & 2k & 2k-1 \\ n & n^2+n+2 & n^2 \\ n & n^2+n & n^2+n+2 \end{vmatrix}$$

$$\Rightarrow\sum_{k=1}^{n}D_k=\begin{vmatrix} \sum 1 & 2\sum k & 2\sum k-\sum 1 \\ n & n^2+n+2 & n^2 \\ n & 2\frac{n(n+1)}{2} & 2\left(\frac{n(n+1)}{2}\right)-n \end{vmatrix}$$

$$\Rightarrow\sum_{k=1}^{n}D_k=\begin{vmatrix} n & 2\frac{n(n+1)}{2} & 2\left(\frac{n(n+1)}{2}\right)-n \\ n & n^2+n+2 & n^2 \\ n & n(n+1) & n^2 \end{vmatrix}$$

$$\Rightarrow\sum_{k=1}^{n}D_k=\begin{vmatrix} n & n(n+1) & n^2 \\ n & n^2+n+2 & n^2 \\ n & n^2+n & n^2+n+2 \end{vmatrix}$$

$R_1\to R_1-R_2$ and $R_2\to R_2-R_3$

$$\Rightarrow \sum_{k=1}^{n} D_k = \begin{vmatrix} 0 & -2 & 0 \\ 0 & 2 & -n-2 \\ n & n^2+n & n^2+n+2 \end{vmatrix}$$

Expand the determinant along R_1

$$\Rightarrow \sum_{k=1}^{n} D_k = 2\left((-n)(-n-2)\right) = 96$$

$\Rightarrow n^2 + 2n = 48 \;\Rightarrow n = 6, -8$

So, $n = 6$

Hence, this is the required answer.

33. (B)

Given, system of linear equations,

$2x + 4y + 2az = b;\ x + 2y + 3z = 4$

$x + 2y + 3z = 4$

Now finding,

$$\Delta = \begin{vmatrix} 2 & 4 & 2a \\ 1 & 2 & 3 \\ 2 & -5 & 2 \end{vmatrix}$$

$\Delta = -18(a-3)$

Hence, for $a = 3$, $\Delta = 0$

Now finding,

$$\Delta_3 = \begin{vmatrix} b & 2 & 4 \\ 4 & 1 & 2 \\ 8 & 2 & -5 \end{vmatrix}$$

$\Delta = -18(a-3)$

Hence, for $a = 3, \Delta = 0$

Now finding,

$$\Delta_3 = \begin{vmatrix} b & 2 & 4 \\ 4 & 1 & 2 \\ 8 & 2 & -5 \end{vmatrix}$$

$\Rightarrow \Delta_3 = 9(b-8)$

So, for $b = 8, \Delta_3 = 0$

Solving options we get,

If $a = 3$ and $b = 8$, we'll have infinite solutions

For $a = 8, b = 8, \Delta \neq 0, \Delta_3 = 0 \Rightarrow$ unique solution

For $a = b = 6, \Delta \neq 0, \Delta_3 \neq 0 \Rightarrow$ unique solution

$\therefore$ For $a = 3, b = 6$, we'll have no solution

34. (B)

Given system of equations.

$2x + y - z = 5$

$2x - 5y + \lambda z = \mu$

$x + 2y - 5z = 7$

For infinite solutions, $\Delta = 0 = \Delta_1 = \Delta_2 = \Delta_3$

$$\Rightarrow \Delta = \begin{vmatrix} 2 & 1 & -1 \\ 2 & -5 & \lambda \\ 1 & 2 & -5 \end{vmatrix} = 0$$

$\Rightarrow 51 - 3\lambda = 0 \;\Rightarrow \lambda = 17$

$$\Rightarrow \Delta_3 = \begin{vmatrix} 5 & 2 & 1 \\ \mu & 2 & -5 \\ 7 & 1 & 2 \end{vmatrix} = 0$$

$\Rightarrow\ -3(\mu + 13) = 0$

$\Rightarrow \mu = -13$

Now $(\lambda + \mu)^2 + (\lambda - \mu)^2$

$= (17 + 13)^2 + (17 - 13)^2$

$= 900 + 16 = 916$

35. (B)

Given, system of linear equations,

$-x + 2y - 9z = 7$

$-x + 3y + 7z = 9$

$-2x + y + 5z = 8$

$-3x + y + 13z = \lambda$

Solving equations (1), (2) and (3) we get,

$x = -3,\ y = 2,\ z = 0$

Now, substituting in equation (4) we get,

$3 \times 3 + 2 = \lambda \Rightarrow \lambda = 11$

Finding the distance of point (–3, 2, 0) from the plane $2x - 2y + z = 11$ we get

$$d = \left|\frac{-6-4-11}{\sqrt{2^2+2^2+1}}\right| = 7$$

36. (B)

Using family of planes

$2x + 3y - z - 5$

$= k_1(x + \alpha y + 3z + 4) + k_2(3x - y + \beta z - 7)$

$2 = k_1 + 3k_2,\ 3 = k_1\alpha - k_2,$

$-1 = 3k_1 + \beta k_2, -5 = 4k_1 - 7k_2$

On solving we get

$$k_2 = \frac{13}{19}, k_1 = \frac{-1}{19}, \alpha = -70, \beta = \frac{-16}{13}$$

$$13\alpha\beta = 13(-70)\left(\frac{-16}{13}\right) = 1120$$

37. **(B)**

$x + 2y + 3z = 5$

$2x + 3y + z = 9$

$4x + 3y + \lambda z = \mu$

For infinite following $\Delta = \Delta_1 = \Delta_2 = \Delta_3 = 0$

$$\Delta = \begin{vmatrix} 1 & 2 & 3 \\ 2 & 3 & 1 \\ 4 & 3 & \lambda \end{vmatrix} = 0 \Rightarrow \lambda = -13$$

$$\Delta_1 = \begin{vmatrix} 5 & 2 & 3 \\ 9 & 3 & 1 \\ \mu & 3 & -13 \end{vmatrix} = 0 \Rightarrow \mu = 15$$

$$\Delta_2 = \begin{vmatrix} 1 & 5 & 3 \\ 2 & 9 & 1 \\ 4 & 15 & -13 \end{vmatrix} = 0$$

$$\Delta_3 = \begin{vmatrix} 1 & 2 & 5 \\ 2 & 3 & 9 \\ 4 & 3 & 15 \end{vmatrix} = 0$$

for $\lambda = -13, \mu = 15$

the system of equations has infinite solutions, hence $\lambda + 2\mu = 17$.

38. **(28)**

$$A = \begin{bmatrix} 2 & 0 & 1 \\ 1 & 1 & 0 \\ 1 & 0 & 1 \end{bmatrix} B = [B_1, B_2, B_3]$$

$$B_1 = \begin{bmatrix} x_1 \\ y_1 \\ z_1 \end{bmatrix}, B_2 = \begin{bmatrix} x_2 \\ y_2 \\ z_2 \end{bmatrix}, B_3 - \begin{bmatrix} x_3 \\ y_3 \\ z_3 \end{bmatrix}$$

$$AB_1 = \begin{bmatrix} 2 & 0 & 1 \\ 1 & 1 & 0 \\ 1 & 0 & 1 \end{bmatrix}\begin{bmatrix} x_1 \\ y_1 \\ z_1 \end{bmatrix} = \begin{bmatrix} 1 \\ 0 \\ 0 \end{bmatrix}$$

$x_1 = 1, y_1 = -1, z_1 = -1$

$$AB_2 = \begin{bmatrix} 2 & 0 & 1 \\ 1 & 1 & 0 \\ 1 & 0 & 1 \end{bmatrix}\begin{bmatrix} x_2 \\ y_2 \\ z_2 \end{bmatrix} = \begin{bmatrix} 2 \\ 3 \\ 0 \end{bmatrix}$$

$x_2 = 2, y_2 = 1, z_2 = -2$

$$AB_3 = \begin{bmatrix} 2 & 0 & 1 \\ 1 & 1 & 0 \\ 1 & 0 & 1 \end{bmatrix}\begin{bmatrix} x_3 \\ y_3 \\ z_3 \end{bmatrix} = \begin{bmatrix} 3 \\ 2 \\ 1 \end{bmatrix}$$

$x_3 = 2, y_3 = 0, z_3 = -1$

$$B = \begin{bmatrix} 1 & 2 & 2 \\ -1 & 1 & 0 \\ -1 & -2 & -1 \end{bmatrix}$$

$\alpha = |B| = 3$

$\beta = 1$

$\alpha^3 + \beta^3 = 27 + 1 = 28$

39. **(B)**

$$\begin{vmatrix} 1 & \frac{3}{2} & \alpha + \frac{3}{2} \\ 1 & \frac{1}{3} & \alpha + \frac{1}{3} \\ 2\alpha + 3 & 3\alpha + 1 & 0 \end{vmatrix} = 0$$

$$\Rightarrow (2\alpha + 3)\left\{\frac{7\alpha}{6}\right\} - (3\alpha + 1)\left\{\frac{-7}{6}\right\} = 0$$

$$\Rightarrow (2\alpha + 3)\cdot\frac{7\alpha}{6} + (3\alpha + 1)\cdot\frac{7}{6} = 0$$

$$\Rightarrow 2\alpha^2 + 3\alpha + 3\alpha + 1 = 0$$

$$\Rightarrow 2\alpha^2 + 6\alpha + 1 = 0$$

$$\Rightarrow \alpha = \frac{-3 + \sqrt{7}}{2}, \frac{-3 - \sqrt{7}}{2}$$

Hence option (B) is correct.

40. **(113)**

$\because a, b, c$ and in A.P.

$\Rightarrow 2b = a + c \Rightarrow a - 2b + c = 0 \therefore ax + by + c$ passes through fixed point (1, –2)

$\therefore P = (1, -2)$

For infinite solution,

$D = D1 = D2 = D3 = 0$

$$D : \begin{vmatrix} 1 & 1 & 1 \\ 2 & 5 & \alpha \\ 1 & 2 & 3 \end{vmatrix} = 0 \Rightarrow \alpha = 8$$

$$D_1:\begin{vmatrix}6 & 1 & 1\\ \beta & 5 & \alpha\\ 4 & 2 & 3\end{vmatrix}=0\Rightarrow\beta=6$$

$\therefore Q=(8,6)$

$\therefore PQ^2=113$

41. **(B)**

$x+y+z=4\mu,$

$x+2y+2\lambda z=10\mu,\ x+3y+4\lambda$

${}^2z=\mu^2+15$

$$\Delta=\begin{vmatrix}1 & 1 & 1\\ 1 & 2 & 2\lambda\\ 1 & 3 & 4\lambda^2\end{vmatrix}=(2\lambda-1)^2$$

For a unique solution

$\Delta\neq 0,\ 2\lambda-1\neq 0,\left(\lambda\neq\frac{1}{2}\right)$

Let $\Delta=0,\ \lambda=\frac{1}{2}$

$$\Delta_y=0,\ \Delta_x=\Delta_z=\begin{vmatrix}4\mu & 1 & 1\\ 10\mu & 2 & 1\\ \mu^2+15 & 3 & 1\end{vmatrix}$$

$=(\mu-15)(\mu-1)$

For infinite solutions $\lambda=\frac{1}{2},\ \mu=1$ or 15

42. **(A)**

$$\begin{vmatrix}2\cos^4 x & 2\sin^4 x & 3+\sin^2 2x\\ 3+2\cos^4 x & 2\sin^4 x & \sin^2 2x\\ 2\cos^4 x & 3+2\sin^2 4x & \sin^2 2x\end{vmatrix}$$

$R_2\to R_2-R_1,\ R_3\to R_3-R_1$

$$\begin{vmatrix}2\cos^4 x & 2\sin^4 x & 3+\sin^2 2x\\ 3 & 0 & -3\\ 0 & 3 & -3\end{vmatrix}$$

$f(x)=45\ \ f'(x)=0$

43. **(D)**

$$\begin{vmatrix}1 & 1 & 1\\ 1 & 2 & \lambda^2\\ 1 & 3 & \lambda\end{vmatrix}=0$$

$\Rightarrow 2\lambda^2-\lambda-1=0\quad \lambda=1,-\frac{1}{2}$

$$\begin{vmatrix}1 & 1 & 5\\ 2 & \lambda^2 & 9\\ 3 & \lambda & \mu\end{vmatrix}=0\Rightarrow\mu=13$$

Infinite solution $\lambda=1$ and $\mu=13$

For unique solution $\lambda\neq 1$

For no solution $\lambda=1$ and $\mu\neq 13$

If $\lambda\neq 1$ and $\mu\neq 13$

Considering the case when

$\lambda=-\frac{1}{2}$ and $\mu\neq 13$

this will generate no solution case

44. **(D)**

$$D=\begin{vmatrix}1 & -2 & 1\\ 2 & \alpha & 3\\ 3 & -1 & \beta\end{vmatrix}$$

$=1(\alpha\beta+3)+2(2\beta-9)+1(-2-3\alpha)$

$=\alpha\beta+3+4\beta-18-2-3\alpha$

For infinite solutions $D=0,\ D_1=0,\ D_2=0$ and $D_3=0$

$D=0$

$\alpha\beta-3\alpha+4\beta=17\ldots$

$$D_1=\begin{vmatrix}-4 & -2 & 1\\ 5 & \alpha & 3\\ 3 & -1 & \beta\end{vmatrix}=0$$

$$D_2=\begin{vmatrix}1 & -4 & 1\\ 2 & 5 & 3\\ 3 & 3 & \beta\end{vmatrix}=0$$

$\Rightarrow 1(5\beta-9)+4(2\beta-9)+1(6-15)=0$

$13\beta-9-36-9=0$

$13\beta=54,\ \beta=\frac{54}{13}$ put in (1)

$\frac{54}{13}\alpha-3\alpha+4\left(\frac{54}{13}\right)=17$

$54\alpha-39\alpha+216=221$

$15\alpha=5\qquad\alpha=\frac{1}{3}$

Now, $12\alpha+13\beta=12\cdot\frac{1}{3}+13\cdot\frac{54}{13}$

$=4+54=58$

45. **(C)**

$$f(0)=\begin{vmatrix}0 & 1 & 1\\2 & 0 & 6\\0 & 4 & -2\end{vmatrix}=12$$

$$f'(x)=\begin{vmatrix}3x^2 & 4x & 3\\3x^2+2 & 2x & x^3+6\\x^3-x & 4 & x^2-2\end{vmatrix}+$$

$$\begin{vmatrix}x^3 & 2x^2+1 & 1+3x\\6x & 2 & 3x^2\\x^3-x & 4 & x^2-2\end{vmatrix}+$$

$$\begin{vmatrix}x^3 & 2x^2+1 & 1+3x\\3x^2+2 & 2x & x^3+6\\3x^2-1 & 0 & 2x\end{vmatrix}$$

$$\therefore f'(0)=\begin{vmatrix}0 & 0 & 3\\2 & 0 & 6\\0 & 4 & -2\end{vmatrix}+\begin{vmatrix}0 & 1 & 1\\0 & 2 & 0\\0 & 4 & -2\end{vmatrix}+\begin{vmatrix}0 & 1 & 1\\2 & 0 & 6\\-1 & 0 & 0\end{vmatrix}$$

$$=24-6=18 \therefore 2f(0)+f'(0)=42$$

46. **(A)**

$$\text{Let } A=\begin{bmatrix}x_1 & y_1 & z_1\\x_2 & y_2 & z_2\\x_3 & y_3 & z_3\end{bmatrix}$$

$$\text{Given } A\begin{bmatrix}1\\0\\1\end{bmatrix}=\begin{bmatrix}2\\0\\2\end{bmatrix} \quad (1)$$

$$\therefore \begin{bmatrix}x_1+z_1\\x_2+z_2\\x_3+z_3\end{bmatrix}=\begin{bmatrix}2\\0\\2\end{bmatrix}$$

$\therefore x_1+z_1=2$ (2)

$x_2+z_2=0$ (3)

$x_3+z_3=0$ (4)

$$\text{Given } A\begin{bmatrix}-1\\0\\1\end{bmatrix}=\begin{bmatrix}-4\\0\\4\end{bmatrix}$$

$$\therefore \begin{bmatrix}-x_1+z_1\\-x_2+z_2\\-x_3+z_3\end{bmatrix}=\begin{bmatrix}4\\0\\4\end{bmatrix}$$

$\Rightarrow -x_1+z_1=-4$ (5)

$-x_2+x_2=0$ (6)

$-x_3+z_3=4$

$$\text{Given } A\begin{bmatrix}0\\1\\0\end{bmatrix}=\begin{bmatrix}0\\2\\0\end{bmatrix}$$

$$\therefore \begin{bmatrix}y_1\\y_2\\y_3\end{bmatrix}=\begin{bmatrix}0\\2\\0\end{bmatrix}$$

$\therefore y_1=0, y_2=2, y_3=0$

$\therefore$ from (2), (3), (4), (5), (6) and (7)

$x_1=3x, x_2=0, x_3=-1$

$y_1=0, y_2=2, y_3=0$

$z_1=-1, z_2=0, z_3=3$

$$\therefore A=\begin{bmatrix}3 & 0 & -1\\0 & 2 & 0\\-1 & 0 & 3\end{bmatrix}$$

$$\therefore \text{Now } (A-31)\begin{bmatrix}x\\y\\z\end{bmatrix}=\begin{bmatrix}-1\\2\\3\end{bmatrix}$$

$$\therefore \begin{bmatrix}0 & 0 & -1\\0 & -1 & 0\\-1 & 0 & 0\end{bmatrix}\begin{bmatrix}x\\y\\z\end{bmatrix}=\begin{bmatrix}-1\\2\\3\end{bmatrix}$$

$$\begin{bmatrix}-z\\-y\\-x\end{bmatrix}=\begin{bmatrix}1\\2\\3\end{bmatrix}$$

$[z=-1], [y=-2], [x=-3]$

Matrices

1. **(B)**

$S=\{\sqrt{n:1}:1\le n\le 50 \text{ and } n \text{ is odd}\}$

$=\{\sqrt{1}, \sqrt{3}, \sqrt{5}\ldots\sqrt{49}\}$, 25 terms

$|A|=1+a^2$

$$\sum_{a\in S}\det(\text{adj } A)=\sum_{a\in S}|A|^2=\sum(1+a^2)^2$$

$$=22100=100\lambda$$

$\lambda=221$

2. **(100)**

$$A=\begin{bmatrix}-1 & a\\ 0 & b\end{bmatrix}$$

$$A^2=\begin{bmatrix}-1 & a\\ 0 & b\end{bmatrix}\begin{bmatrix}-1 & a\\ 0 & b\end{bmatrix}$$

$$=\begin{bmatrix}1 & -a+ab\\ 0 & b^2\end{bmatrix}$$

$$\therefore T_n=\left\{A\in S;\, A^{n(n+1)}=I\right\}$$

$\therefore$ b must be equal to 1

$\therefore$ In this case A^2 will become an identity matrix and a can take any value from 1 to 100

$\therefore$ Total number of common elements will be 100.

3. **(B)**

$$A+\begin{bmatrix}a_1 & b_1 & c_1\\ a_2 & b_2 & c_2\\ a_3 & b_3 & c_3\end{bmatrix} \quad A\begin{bmatrix}0\\0\\1\end{bmatrix}=\begin{bmatrix}c_1\\c_2\\c_3\end{bmatrix}=\begin{bmatrix}1\\1\\2\end{bmatrix}$$

$$\Rightarrow c_1=1, c_2=1, c_3=2$$

$$A\begin{bmatrix}1\\0\\1\end{bmatrix}=\begin{bmatrix}c_1+a_1\\c_2+a_2\\c_3+a_3\end{bmatrix}=\begin{bmatrix}-1\\0\\1\end{bmatrix}$$

$$\Rightarrow a_1=-2, a_2=-1, a_3=-1$$

$$A\begin{bmatrix}1\\1\\0\end{bmatrix}=\begin{bmatrix}a_1+b_1\\a_2+b_2\\a_3+b_3\end{bmatrix}=\begin{bmatrix}1\\1\\0\end{bmatrix}$$

$$\Rightarrow b_1=3, b_2=2, b_3=1$$

$$\Rightarrow A=\begin{bmatrix}-2 & 3 & 1\\ -1 & 2 & 1\\ -1 & 1 & 2\end{bmatrix}$$

$$\Rightarrow A-2I=\begin{bmatrix}-4 & 3 & 1\\ -1 & 0 & 1\\ -1 & 1 & 0\end{bmatrix}$$

$$|A-2I|=0$$

$$\text{Now, }\begin{bmatrix}-4 & 3 & 1\\ -1 & 0 & 1\\ -1 & 1 & 0\end{bmatrix}\begin{bmatrix}x_1\\x_2\\x_3\end{bmatrix}=\begin{bmatrix}4\\1\\1\end{bmatrix}$$

$$-4x_1+3x_2+x_3=4 \qquad (1)$$

$$-x_1+x_3=1 \qquad (2)$$

$$-x_1+x_2=1 \qquad (3)$$

$$(1)-[(2)+3(3)]$$

$0=0\Rightarrow$ infinite solutions

4. **(A)**

$$A=\begin{bmatrix}0 & -2\\ 2 & 0\end{bmatrix}$$

$$A^2=\begin{bmatrix}0 & -2\\ 2 & 0\end{bmatrix}\begin{bmatrix}0 & -2\\ 2 & 0\end{bmatrix}=\begin{bmatrix}-4 & 0\\ 0 & -4\end{bmatrix}=-4I$$

$$A^3=-4A$$

$$A^4=(-4I)(-4I)=(-4)^2I$$

$$A^5=(-4I)(-4I)=(-4)^2I$$

$$M=\sum_{k=1}^{10}A^{2k}=A^2+A^4+\ldots+A^{20}$$

$$=\left[-4+(-4)^2+(-4)^3+\ldots+(-4)^{20}\right]I$$

$$=-4\lambda I$$

$\Rightarrow M$ is a symmetric matrix

$$N=\sum_{k=1}^{10}A^{2k-1}=A+A^3+\cdots+A^{19}$$

$$=A\left[1+(-4)+(-4)^2+\ldots+(-4)^9\right]$$

$=\lambda A\Rightarrow$ skew symmetric

$\Rightarrow N^2$ is a symmetric matrix

$\Rightarrow MN^2$ is a non-identity symmetric matrix

5. **(414)**

Case-I: $1\to 7$ times

and $-1\to 2$ times

Number of possible matrices $=\dfrac{9!}{7!2!}=36$

Case-II: $1\to$ times,

$-1\to 1$ times

and $0\to 2$ times

Number of possible matrices $=\dfrac{9!}{6!2!}=252$

Case-III: $1\to 5$ times,

and $0\to 4$ times

Number of possible matrices $= \dfrac{9!}{5!4!} = 126$

Hence, the total number of all such matrices $A = 414$

6. (1)

$A^2 = A$ and $B^2 = B$

Therefore equation $nA^n + mB^m = I$ becomes

$nA + mB = I$, which gives $m = n = 1$

Only one set is possible.

7. (C)

$|\text{adj}\,(24A)| = |\text{adj}\,3\,(\text{adj}\,2A)|$

$\Rightarrow |24a|^2 = (3\,\text{adj}\,(2A))^2$

$\Rightarrow \left(24^3|A|\right)^2 = \left(3^3|\text{adj}\,(2A)|\right)^2 = 3^6\left(|2A|^2\right)^2$

$\Rightarrow 24^6|A|^2 = \left(24^3|A|\right)^2 = 3^6 \times 2^{12}|A|^4$

$\Rightarrow |A|^2 = \dfrac{24^6}{3^6 \times 2^{12}} = 64$

8. (100)

$$X = \begin{bmatrix} 0 & 1 & 0 \\ 0 & 0 & 1 \\ 0 & 0 & 0 \end{bmatrix}, X^2 = \begin{bmatrix} 0 & 0 & 1 \\ 0 & 0 & 0 \\ 0 & 0 & 0 \end{bmatrix}$$

$$Y = \begin{bmatrix} \alpha & \beta & \gamma \\ 0 & \alpha & \beta \\ 0 & 0 & \alpha \end{bmatrix}, Z = \begin{bmatrix} \alpha^2 & -\alpha\beta & \beta^2 - \alpha\gamma \\ 0 & \alpha^2 & -\alpha\beta \\ 0 & 0 & \alpha^2 \end{bmatrix}$$

$Y \cdot Y^{-1} = I$

$$\begin{bmatrix} \alpha & \beta & \gamma \\ 0 & \alpha & \beta \\ 0 & 0 & \alpha \end{bmatrix} \begin{bmatrix} \frac{1}{5} & \frac{-2}{5} & \frac{1}{5} \\ 0 & \frac{1}{5} & \frac{-2}{5} \\ 0 & 0 & \frac{1}{5} \end{bmatrix} = \begin{bmatrix} 1 & 0 & 0 \\ 0 & 1 & 0 \\ 0 & 0 & 1 \end{bmatrix}$$

$\dfrac{\alpha}{5} = 1 \Rightarrow \alpha = 5$

$-\dfrac{2}{5}\alpha + \dfrac{\beta}{5} = 0 \Rightarrow \beta = 10$

$\dfrac{\alpha}{5} - \dfrac{2\beta}{5} + \dfrac{\gamma}{5} = 0 \Rightarrow \gamma = 15$

$\Rightarrow (\alpha - \beta + \gamma)^2 = (5 - 10 + 15)^2 = 100$

9. (14)

$$\text{Adj}\,(\text{Adj}\,A) = \begin{bmatrix} 14 & 18 & -14 \\ -14 & 14 & 28 \\ 28 & -14 & 14 \end{bmatrix}$$

$$|\text{Adj}\,(\text{Adj}\,A)| = \begin{bmatrix} 14 & 28 & -14 \\ -14 & 14 & 28 \\ 28 & -14 & 14 \end{bmatrix}$$

$$= 14 \times 14 \times 14 \begin{vmatrix} 1 & 2 & -1 \\ -1 & 1 & 2 \\ 2 & -1 & 1 \end{vmatrix}$$

$= (14)^3[3 - 2(-5) - 1(-1)] = (14)^3[14] = (14)^4$

$|A|^4 = (14)^4 \Rightarrow |A| = 14$

10. (C)

$AB = i$

$|\text{adj}(B\,\text{adj}(2A))| = |B\,\text{adj}(2A)|^2$

$= |B|^2\,|\text{adj}\,(2A)|^2$

$= |B|^2\left(|2A|^2\right)^2 = |B|^2\left(2^6|A|^2\right)^2$

$|A| = \dfrac{1}{8}$ and $|AB| = 1 \Rightarrow |A||B| = 1$

$\Rightarrow \dfrac{1}{8}|B| = 1 \;\Rightarrow |B| = 8$

Required value = 64

11. (180)

Let $A = \begin{bmatrix} a & b \\ c & d \end{bmatrix}$; $a, b, c, d \in \{0, 1, 2, 3, 4, 5\}$

$a + b + c + d = p,\ p \in \{3, 5, 7\}$

Case (i)

$a + b + c + d = 3;\ a, b, c, d \in \{0, 1, 2, 3\}$

No. of ways $= {}^{3+4-1}C_{4-1} = {}^6C_3 = 56$ (1)

Case (ii)

$a + b + c + d = 5;\ a, b, c, d \in \{0, 1, 2, 3, 4, 5\}$

No. of ways $= {}^{5+4-1}C_{4-1} = {}^8C_3 = 56$ (2)

Case (iii)

$a + b + c + d = 7$

No. of ways = total ways when

$a, b, c, d \in \{0, 1, 2, 3, 4, 5, 6, 7\} -$

Total ways when $a, b, c, d \notin \{6, 7\}$

No. of ways $= {}^{7+4-1}C_{4-1}\left(\frac{\lfloor 4}{\lfloor 3}+\frac{\lfloor 4}{\lfloor 2}\right)$

$= {}^{10}C_3 - 16 = 104$ (3)

Hence total number of ways = 180

12. (A)

$\left|(\det(A))\text{adj}\left(5\,\text{adj}(A)\right)\right|$

$= \left|2\,\text{adj}\left(5\,\text{adj}\left(A^3\right)\right)\right|$

$= 2^3\left|\text{adj}\left(5\,\text{adj}\left(A^3\right)\right)\right|$

$= 2^3 \cdot \left|5\,\text{adj}\left(A^3\right)\right|^2$

$= 2^3\left(5^3 \cdot \left|\text{adj}\left(A^3\right)\right|\right)^2$

$= 2^3 \cdot 5^6 \cdot \left|\text{adj}\,A^3\right|^2$

$= 2^3 \cdot 5^6\left(\left(|A|^3\right)^2\right)^2$

$= 2^3 \cdot 5^6 \cdot 2^{12} = 2^{15} \times 5^6$

$= 2^9 \times 10^6 = 512 \times 10^6$

13. (25)

$$A = \begin{bmatrix} 1+i & 1 \\ -i & 0 \end{bmatrix}$$

$$A^2 = \begin{bmatrix} 1+i & 1 \\ -i & 0 \end{bmatrix}\begin{bmatrix} 1+i & 1 \\ -i & 0 \end{bmatrix}$$

$$A^2 = \begin{bmatrix} i & 1+i \\ -i+1 & -i \end{bmatrix}$$

$$A^4 = \begin{bmatrix} i & 1+i \\ -i+1 & -i \end{bmatrix}\begin{bmatrix} i & 1+i \\ -i+1 & -i \end{bmatrix}$$

$$A^4 = \begin{bmatrix} 1 & 0 \\ 0 & 1 \end{bmatrix} = I$$

$A^{4n+1} = A$

$n = 1, 5, 9, \ldots, 97$

$\Rightarrow$ Total elements in the set is 25.

14. (C)

Let matrix A be singular, then $|A| = 0$

Number of singular matrices = All entries are same + only two prime number are used in the matrix

$= 10 + 10 \times 9 \times 2 = 190$

Required probability $= \frac{190}{10^4} = \frac{19}{10^3}$

15. (A)

$$A = \begin{pmatrix} 1 & 2 & 2^2 \\ \frac{1}{2} & 1 & 2 \\ \frac{1}{2^2} & \frac{1}{2} & 1 \end{pmatrix}$$

$A^2 = 3A$

$A^3 = 3^2 A$

$A^2 + A^3 + \ldots A^{10}$

$= 3A + 3^2 A + \ldots + 3^9 A = \frac{3\left(3^9 - 1\right)}{3-1}A$

$= \frac{3^{10} - 3}{2}A$

16. (C)

$$B = (I - \text{adj}\,A)^5 = \begin{bmatrix} -1 & -1 \\ 0 & -1 \end{bmatrix}^5 = \begin{bmatrix} -1 & -5 \\ 0 & -1 \end{bmatrix}$$

Sum of its all elements = –7

17. (1)

$$M = \begin{bmatrix} 0 & -\alpha \\ \alpha & 0 \end{bmatrix};\ M^2 = \begin{bmatrix} -\alpha^2 & 0 \\ 0 & -\alpha^2 \end{bmatrix} = -\alpha^2 I$$

$N = M^2 + M^4 + \ldots + M^{98}$

$= \left[-\alpha^2 + \alpha^4 - \alpha^6 + \ldots\right]I$

$= -\alpha^2 \frac{\left(1 - \left(-\alpha^2\right)^{49}\right)}{1+\alpha^2} \cdot I$

$I - M^2 = \left(1+\alpha^2\right)I$

$\left(I - M^2\right)N = -\alpha^2\left(\alpha^{98} + 1\right) = -2$

$\alpha = 1$

18. (17)

$$A = \begin{bmatrix} 2 & -1 & -1 \\ 1 & 0 & -1 \\ 1 & -1 & 0 \end{bmatrix} \Rightarrow A^2 = A \Rightarrow A^n = A.$$

$\forall\, n \in \{1, 2, \ldots, 100\}$

Now, $B = A - I = \begin{bmatrix} 1 & -1 & -1 \\ 1 & -1 & -1 \\ 1 & -1 & -1 \end{bmatrix}$

$B^2 = -B$

$\Rightarrow B^3 = -B^2 = B$

$\Rightarrow B^5 = B$

$\Rightarrow B^{99} = B$

Also, $\omega^{3k} = 1$

So, $n =$ common of $\{1, 3, 5, \ldots, 99\}$ and

$\{3, 6, 9, \ldots, 99\} = 17$

19. **(24)**

20. **(B)**

Let $A = \begin{bmatrix} a & b \\ c & d \end{bmatrix}$; $ad - bc = -1$

$|A + I||\text{adj } A + I| = 4$

$\Rightarrow ad - bc + a + d + 1 = 2$ or -2

$a + d = 2$ or -2

21. **(D)**

$$A'BA = [1 \ \ 1 \ \ 1]\begin{bmatrix} 9^2 & -10^2 & 11^2 \\ 12^2 & 13^2 & -14^2 \\ -15^2 & 16^2 & 17^2 \end{bmatrix}\begin{bmatrix} 1 \\ 1 \\ 1 \end{bmatrix}$$

$$= \left[9^2 + 12^2 - 15^2 \quad -10^2 + 13^2 + 16^2 \quad 11^2 - 14^2 + 17^2\right]\begin{bmatrix} 1 \\ 1 \\ 1 \end{bmatrix}$$

$$= \left[9^2 + 12^2 - 15^2 - 10^2 + 13^2 + 16^2 + 11^2 - 14^2 + 17^2\right]$$

$= [539]$

22. **(50)**

$A = \begin{bmatrix} a & b \\ c & d \end{bmatrix}$

Given $A = A^{-1}$

$\therefore A^2 = A \cdot A^{-1} = I$

$$\begin{bmatrix} a & b \\ c & d \end{bmatrix}\begin{bmatrix} a & b \\ c & d \end{bmatrix} = \begin{bmatrix} 1 & 0 \\ 0 & 1 \end{bmatrix}$$

$$\Rightarrow \begin{bmatrix} a^2 + bc & ab + bd \\ ac + cd & bc + d^2 \end{bmatrix} = \begin{bmatrix} 1 & 0 \\ 0 & 1 \end{bmatrix}$$

$\therefore a + bc = 1$ (1)

$ab + bd = 0$ (2)

$ac + cd = 0$ (3)

$bc + d^2 = 1$ (4)

(1) – (4) gives

$a^2 - d^2 = 0$

$\Rightarrow (a + d) = 0$ or $a - d = 0$

Case-I

$a + d = 0 \Rightarrow (a, d) = (-1, 1), (0, 0), (1, -1)$

$(a)\, (a, d) = (-1, 1)$

$\therefore$ from equation (1)

$1 + bc = 1 \Rightarrow bc = 0$

$b = 0 \quad c = 12$ possibilities

$c = 0 \quad b = 12$ possibilities

but $(0, 0)$ is repeated

$\therefore 2 \times 12 = 24$

$24 - 1$ (repeated) $= 23$ pairs

$(b)\, (a, d) = (1, -1) \Rightarrow bc = 0 \rightarrow 23$ pairs

$(c)\, (a, d) = (0, 0) \Rightarrow bc = 1$

$\Rightarrow (b, c) = (1, 1)$ and $(-1, -1)$, 2 pairs

Case-II

$a = d$

from (2) and (3)

$a \neq 0$ then $b = c = 0$

$a^2 = -1$

$1 = d$

$(a, d) = (1, 1), (-1, -1) \rightarrow 2$ pairs

$\therefore$ Total $= 23 + 23 + 2 + 2$

$= 50$ pairs

23. **(D)**

Characteristic equation of matrix A

$|A - \lambda I| = 0$

$$\begin{vmatrix} 1 - \lambda & 2 \\ -2 & -5 - \lambda \end{vmatrix} = 0$$

$\Rightarrow \lambda^2 + 4\lambda = 1$

$\Rightarrow A^2 + 4A = I$

$\Rightarrow 2A^2 + 8A = 2I$ (1)

Given that $\alpha A^2 + \beta A = 2I$ (2)

Comparing equations (1) and (2) we get

$\alpha = 2, \beta = 8$

$\therefore \alpha + \beta = 10$

24. **(5376)**

$\text{Tr}\left(AA^T\right) = 6$

$$AA^T = \begin{bmatrix} a & d & g \\ b & e & h \\ c & f & i \end{bmatrix}\begin{bmatrix} a & b & c \\ d & e & f \\ g & h & i \end{bmatrix}$$

Now given

$a^2 + d^2 + g^2 + b^2 + e^2 + h^2 + c^2 + f^2 + i^2 = 6$

$= {}^9C_3 \times 2^6 = 5376$

25. **(B)**

The characteristic equation for A is

$|A - \lambda I| = 0$

$$\Rightarrow \begin{vmatrix} 4\lambda & -2 \\ \alpha & \beta - \lambda \end{vmatrix} = 0$$

$\Rightarrow (4 - \lambda)(\beta - \lambda) + 2\alpha = 0$

$\Rightarrow \lambda^2 - (\beta + 4)\lambda + 4\beta + 2\alpha = 0$

Put $\lambda = A$

$A^2 - (\beta + 4)A + (4\beta + 2\alpha)I = 0$

On comparison

$-9(\beta + 4) = \gamma$ and $4\beta + 2\alpha = 18$

and $|A| = 4\beta + 2\alpha = 18$

26. **(42)**

$$A = \begin{bmatrix} \alpha & \beta & \gamma \\ \alpha^2 & \beta^2 & \gamma^2 \\ \beta + \gamma & \gamma + \alpha & \alpha + \beta \end{bmatrix}$$

$R_3 \to R_3 + R_1$

$$\Rightarrow |A| = |\alpha + \beta + \gamma| \begin{vmatrix} \alpha & \beta & \gamma \\ \alpha^2 & \beta^2 & \gamma^2 \\ 1 & 1 & 1 \end{vmatrix}$$

$\Rightarrow |A| = (\alpha + \beta + \gamma)(\alpha - \beta)(\beta - \gamma)(\gamma - \alpha)$

$\because |\text{adj } A| = |A|^{n-1}$

$|\text{adj}\,(\text{adj } A)| = |A|^{(n-1)^2}$

$\left|\text{adj}\left(\text{adj}\left(\text{adj}\,(\text{adj } A)\right)\right)\right|$

$= |A|^{(n-1)^4} = |A|^{2^4} = |A|^{16}$

$\therefore (\alpha + \beta + \gamma)^{16} = 2^{32} \cdot 3^{16}$

$\Rightarrow (\alpha + \beta + \gamma)^{16} = \left(2^2 \cdot 3\right)^{16} = (12)^{16}$

$\Rightarrow \alpha + \beta + \gamma = 12$

$\because \alpha, \beta, \gamma \in N$

$(\alpha - 1) + (\beta - 1) + (\gamma - 1) = 9$

Number of all tuples $(\alpha, \beta, \gamma) = {}^{11}C_2 = 55$

1 case for $\alpha = \beta = \gamma$

and 12 case when any two of these are equal

So, No. of distinct tuples (α, β, γ)

$= 55 - 13 = 42$

27. **(C)**

$$A^2 = \begin{bmatrix} 0 & 1 & 0 \\ 0 & 0 & 1 \\ 1 & 0 & 0 \end{bmatrix}\begin{bmatrix} 0 & 1 & 0 \\ 0 & 0 & 1 \\ 1 & 0 & 0 \end{bmatrix}$$

$$= \begin{bmatrix} 0 & 0 & 1 \\ 1 & 0 & 0 \\ 0 & 1 & 0 \end{bmatrix}$$

$a \leftrightarrow R_2$

$$-\begin{bmatrix} 1 & 0 & 0 \\ 0 & 0 & 1 \\ 0 & 1 & 0 \end{bmatrix}$$

$R_2 \leftrightarrow R_3$

$$\begin{bmatrix} 1 & 0 & 0 \\ 0 & 1 & 0 \\ 0 & 0 & 1 \end{bmatrix} = I$$

$B_0 = A^{49} + 2A^{98}$

$= A + 2I$

$B = \text{Adj}\,(B - 1)$

$B_4 = \text{Adj}\left(\text{Adj}\left(\text{Adj}\,(\text{Adj } B_0)\right)\right)$

$$= |B_0|^{(n-1)^4}$$

$$= |B_0|^{16}$$

$$B_0 = \begin{bmatrix} 0 & 1 & 0 \\ 0 & 0 & 1 \\ 1 & 0 & 0 \end{bmatrix} + \begin{bmatrix} 2 & 0 & 0 \\ 0 & 2 & 0 \\ 0 & 0 & 2 \end{bmatrix}$$

$$= \begin{bmatrix} 2 & 1 & 0 \\ 0 & 2 & 1 \\ 1 & 0 & 2 \end{bmatrix}$$

$$= 2(4-0) - 1(0-1)$$

$$= 9$$

$$B_4 (9)^{16} = (3)^{32}$$

28. **(2)**

$$A + B = \begin{bmatrix} \beta+1 & 0 \\ 3 & \alpha \end{bmatrix}$$

$$(A+B)^2 = \begin{bmatrix} \beta+1 & 0 \\ 3 & \alpha \end{bmatrix} \begin{bmatrix} \beta+1 & 0 \\ 3 & \alpha \end{bmatrix}$$

$$= \begin{bmatrix} (\beta+1)^2 & 0 \\ 3(\beta+1)+3\alpha & \alpha^2 \end{bmatrix}$$

$$A^2 = \begin{bmatrix} 1 & -1 \\ 2 & \alpha \end{bmatrix} \begin{bmatrix} 1 & -1 \\ 2 & \alpha \end{bmatrix}$$

$$= \begin{bmatrix} -1 & -1-\alpha \\ 2+2\alpha & \alpha^2 - 2 \end{bmatrix}$$

$$\therefore \begin{bmatrix} 1 & -\alpha+1 \\ 2\alpha+4 & \alpha^2 \end{bmatrix} = \begin{bmatrix} (\beta+1)^2 & 0 \\ 3(\alpha+\beta+1) & \alpha^2 \end{bmatrix}$$

$$\boxed{\alpha = 1} = \alpha_1$$

$$B^2 = \begin{bmatrix} \beta & 1 \\ 1 & 0 \end{bmatrix} \begin{bmatrix} \beta & 1 \\ 1 & 0 \end{bmatrix}$$

$$= \begin{bmatrix} \beta^2+1 & \beta \\ \beta & 1 \end{bmatrix} = \begin{bmatrix} (\beta+1)^2 & 0 \\ 3(\beta+1)+3\alpha & \alpha^2 \end{bmatrix}$$

$$\therefore \beta = 0, \alpha = -1 = \alpha_2$$

$$|\alpha_1 - \alpha_2| = |1-(-1)| = 2$$

29. **(C)**

Given that $A^T = A, B^T = -B$

(A) $C = A^4 - B^4$

$$C^T = \left(A^4 - B^4\right) = \left(A^4\right)^T - \left(B^4\right)^T$$

$$= A^4 - B^4 = C$$

(B) $C = AB - BA$

$$C^T = (AB - BA)^T = (AB)^T - (BA)^T$$

$$= B^T A^T - A^T B^T = -BA + AB = C$$

(C) $C = B^5 - A^5$

$$C^T = \left(B^5 - A^5\right)^T = \left(B^5\right)^T - \left(A^5\right)^T$$

$$= -B^5 - A^5$$

(D) $C = AB + BA$

$$C^T = (AB + BA)^T = (AB)^T + (BA)^T$$

$$= -BA - AB = -C$$

$\therefore$ Option (C) is not true.

30. **(B)**

$AB = 0 \Rightarrow |AB| = 0$

$|A|\,|B| = 0$

$|A| = 0 \quad |B| = 0$

If $|A| \neq 0, B = 0$ (not possible)

If $|B| \neq 0, A = 0$ (not possible)

Hence $|A| = |B| = 0$

$\Rightarrow AX = 0$ has infinitely many solutions

31. **(282)**

$$A = \begin{bmatrix} a_{11} & a_{12} & a_{13} \\ a_{21} & a_{22} & a_{23} \\ a_{31} & a_{32} & a_{33} \end{bmatrix} a_{ij} \in \{0,1\}$$

$$\sum a_{ij} = 2, 3, 5, 7$$

Total matrix $= {}^9C_2 + {}^9C_3 + {}^9C_5 + {}^9C_7$

$= 282$

32. **(C)**

$$A = \begin{bmatrix} -1 & 2 \\ 1 & -1 \end{bmatrix}$$

(1) $R_1 \to R_1 + R_2; \begin{bmatrix} 0 & 1 \\ 1 & -1 \end{bmatrix}$ possible

(2) $R_1 \leftrightarrow R_2; \begin{bmatrix} 1 & -1 \\ -1 & 2 \end{bmatrix}$ possible

(3) Option is not possible

(4) $R_2 \to R_2 + 2R_1; \begin{bmatrix} -1 & 2 \\ -1 & 3 \end{bmatrix}$ possible

33. **(10)**

$$X = \begin{bmatrix} 1 \\ 1 \\ 1 \end{bmatrix}; A = \begin{bmatrix} -1 & 2 & 3 \\ 0 & 1 & 6 \\ 0 & 0 & -1 \end{bmatrix}$$

$X^T A^k X = 33$

$$[1 \ 1 \ 1]\begin{vmatrix} -1 & 2 & 3 \\ 0 & 1 & 6 \\ 0 & 0 & -1 \end{vmatrix}\begin{bmatrix} 1 \\ 1 \\ 1 \end{bmatrix} = 33$$

$$[1 \ 1 \ 1]\begin{vmatrix} -1 & 2 & 3 \\ 0 & 1 & 6 \\ 0 & 0 & -1 \end{vmatrix}\begin{bmatrix} 1 \\ 1 \\ 1 \end{bmatrix} = 33$$

As

$$A^2 = \begin{bmatrix} -1 & 2 & 3 \\ 0 & 1 & 6 \\ 0 & 0 & -1 \end{bmatrix}\begin{bmatrix} -1 & 2 & 3 \\ 0 & 1 & 6 \\ 0 & 0 & -1 \end{bmatrix} = \begin{bmatrix} 1 & 0 & 6 \\ 0 & 1 & 0 \\ 0 & 0 & 1 \end{bmatrix}$$

$$A^4 = \begin{bmatrix} 1 & 0 & 6 \\ 0 & 1 & 0 \\ 0 & 0 & 1 \end{bmatrix}\begin{bmatrix} 1 & 0 & 6 \\ 0 & 1 & 0 \\ 0 & 0 & 1 \end{bmatrix} = \begin{bmatrix} 1 & 0 & 12 \\ 0 & 1 & 0 \\ 0 & 0 & 1 \end{bmatrix}$$

$$A^8 = \begin{bmatrix} 0 & 0 & 24 \\ 0 & 1 & 0 \\ 0 & 0 & 1 \end{bmatrix}$$

$$A^{10} = \begin{bmatrix} 1 & 0 & 6 \\ 0 & 1 & 0 \\ 0 & 0 & 1 \end{bmatrix}\begin{bmatrix} 1 & 0 & 24 \\ 0 & 1 & 0 \\ 0 & 0 & 1 \end{bmatrix} = \begin{bmatrix} 1 & 0 & 30 \\ 0 & 1 & 0 \\ 0 & 0 & 1 \end{bmatrix}$$

For $K \to$ Even $A^K = \begin{bmatrix} 1 & 0 & 3K \\ 0 & 1 & 0 \\ 0 & 0 & 1 \end{bmatrix}$

$X^T A^K X = 33$ (This is not correct)

$$[1 \ 1 \ 1]\begin{bmatrix} 1 & 0 & 3K \\ 0 & 1 & 0 \\ 0 & 0 & 1 \end{bmatrix}\begin{bmatrix} 1 \\ 1 \\ 1 \end{bmatrix}$$

$$= [1-1 \quad 3K+1]\begin{bmatrix} 1 \\ 1 \\ 1 \end{bmatrix} = [3K+3]$$

$\therefore 3K + 3 = 3 \qquad \therefore K = 10$

But it should be dropped as 33 is not a matrix.

If K is odd,

$X^T A^K X = 33$

$X^T A A^{K-1} X = 33$

$$[1 \ 1 \ 1]\begin{bmatrix} -1 & 2 & 3 \\ 0 & 1 & 6 \\ 0 & 0 & -1 \end{bmatrix}\begin{bmatrix} 1 & 0 & 3K-3 \\ 0 & 1 & 0 \\ 0 & 0 & 1 \end{bmatrix}\begin{bmatrix} 1 \\ 1 \\ 1 \end{bmatrix} = 33$$

$$[-1 \ 3 \ 8]\begin{bmatrix} 3K-2 \\ 1 \\ 1 \end{bmatrix} = [33]$$

$[-3K + 13] = [33]$

$K = \dfrac{20}{3}$ (not possible)

34. **(D)**

$A^2 + B = A^2 B$

$(A^2 - I)(B - I) = I \qquad (1)$

$A^2 + B = A^2 B$

$A^2 (B - I) = B$

$A^2 = B(B - I)^{-1}$

$A^2 = B(A^2 - I)$

$A^2 = BA^2 - B \quad A^2 + B = BA^2$

$A^2 B = BA^2$

35. **(B)**

In each row and each column, exactly one is to be placed.

$\therefore$ No. of such materials

$= 5 \times 4 \times 3 \times 2 \times 1 = 120$

Alternate:

$$\begin{bmatrix} 0 & 0 & 1 & 0 & 0 \\ 0 & 1 & 0 & 0 & 0 \\ 0 & 0 & 0 & 1 & 0 \\ 0 & 0 & 0 & 0 & 1 \\ 1 & 0 & 0 & 0 & 0 \end{bmatrix}\begin{matrix} \to 5 \text{ ways} \\ \to 4 \text{ ways} \\ \to 3 \text{ ways} \\ \to 2 \text{ ways} \\ \to 1 \text{ ways} \end{matrix}$$

Step 1: Select any 1 place for 1's in row 1. Automatically some column will get filled with 0's.
Step 2: From next now select 1 place for 1's. Automatically some column will get filled with 0's.
$\Rightarrow$ Each time one less place will be available for putting 1's.
Repeat step 2 till last row.
Required ways $= 5\times4\times3\times2\times1 = 120$
$3\sqrt{3}$

36. **(A)**
Given $|\text{adj}(\text{adj})(\text{adj. } A)| = 12^4$

$\Rightarrow |A|^{(n-1)^3} = 12^4$
Given $n = 3$
$\Rightarrow |A|^8 = 12^4$
$\Rightarrow |A|^2 = 12$
$|A| = 2\sqrt{3}$

We are asked
$|A^{-1}\cdot \text{adj } A|$
$= |A^{-1}|\cdot|\text{adj } A|$
$= \dfrac{1}{|A|}\cdot|A|^{3-1}$
$= |A| = 2\sqrt{3}$

37. **(B)**

$$|A| = \frac{1}{\log x\cdot\log y\cdot\log z}\begin{vmatrix}\log x & \log y & \log z\\ \log x & 2\log y & \log z\\ \log x & \log y & 3\log z\end{vmatrix}$$

$= 2$
$\Rightarrow |\text{adj}(\text{adj } A^2)| = |A^2|^4 = 2^8$

38. **(D)**

$$AA^T = \begin{bmatrix}\frac{1}{\sqrt{10}} & \frac{3}{\sqrt{10}}\\ \frac{-3}{\sqrt{10}} & \frac{1}{\sqrt{10}}\end{bmatrix}\begin{bmatrix}\frac{1}{\sqrt{10}} & \frac{-3}{\sqrt{10}}\\ \frac{3}{\sqrt{10}} & \frac{1}{\sqrt{10}}\end{bmatrix} = \begin{bmatrix}1 & 0\\ 0 & 1\end{bmatrix}$$

$$B^2 = \begin{bmatrix}1 & -i\\ 0 & 1\end{bmatrix}\begin{bmatrix}1 & -i\\ 0 & 1\end{bmatrix} = \begin{bmatrix}1 & -2i\\ 0 & 1\end{bmatrix}$$

$$B^3 = \begin{bmatrix}1 & -3i\\ 0 & 1\end{bmatrix}$$

$\vdots$

$$B^{2023} = \begin{bmatrix}1 & -2023i\\ 0 & 1\end{bmatrix}$$

$M = A^T BA$
$M^2 = M\cdot M = A^T BAA^T BA = A^T B^2 A$
$M^3 = M^2\cdot M = A^T B^2 AA^T BA = A^T B^3 A$
$\vdots$

$M^{2023} = \ldots A^T B^{2023} A$

$AM^{2023}A^T = \underbrace{AA^T}\, B^{2023}\, \underbrace{AA^T} = B^{2023}$

$$= \begin{bmatrix}1 & -2023i\\ 0 & 1\end{bmatrix}$$

Inverse of $(AM^{2023}A^T)$ is $\begin{bmatrix}1 & 2023i\\ 0 & 1\end{bmatrix}$

39. **(A)**
Given, $A^T = A, B^T = -B, C^T = -C$
Let $M = A^{13}B^{26} - B^{26}A^{13}$
Then, $M^T = (A^{13}B^{26} - B^{26}A^{13})^T$
$= (A^{13}B^{26})^T - (B^{26}A^{13})^T$
$= (B^T)^{26}(A^T)^{13} - (A^T)^{13}(B^T)^{26}$
$= B^{26}A^{13} - A^{13}B^{26} = -M$
Hence, M is skew symmetric
Let, $N = A^{26}C^{13} - C^{13}A^{26}$
then, $N^T = (A^{26}C^{13})^T - (C^{13}A^{26})^T$
$= -(C)^{13}(A)^{26} + A^{26}C^{13} = N$
Hence, N is symmetric.
$\therefore$ Only S2 is true.

40. **(D)**
$A^2 = 3A + \alpha I \quad A^3 = 3A^2 + \alpha A$
$A^3 = 3(3A + \alpha I) + \alpha A;\ A^3 = 9A + \alpha A + 3\alpha I$
$A^4 = (9+\alpha)A^2 + 3\alpha A$
$= (9+\alpha)(3A + \alpha I) + 3\alpha A$
$= A(27 + 6\alpha) + \alpha(9+\alpha)$
$\Rightarrow 27 + 6\alpha = 21 \Rightarrow \alpha = -1$

41. **(D)**

If its invertible, then determinant value $\neq 0$

So,

$$\begin{vmatrix} e^t & e^{-t}(\sin t - 2\cos t) & e^{-t}(-2\sin t - \cos t) \\ e^t & e^{-t}(2\sin t + \cos t) & e^{-t}(\sin t - 2\cos t) \\ e^t & e^{-t}\cos t & e^{-t}\sin t \end{vmatrix} \neq 0$$

$$\Rightarrow e^t \cdot e^{-t} \cdot e^{-t} \begin{vmatrix} 1 & \sin t - 2\cos t & -2\sin t - \cos t \\ 1 & 2\sin t + \cos t & \sin t - 2\cos t \\ 1 & \cos t & \sin t \end{vmatrix} \neq 0$$

Applying, $R_1 \to R_1 - R_2$ then $R_2 \to R_2 - R_3$

We get

$$e^{-t} \begin{vmatrix} 0 & -\sin t - \cos t & -3\sin t + \cos t \\ 0 & 2\sin t & -2\cos t \\ 1 & \cos t & \sin t \end{vmatrix} \neq 0$$

By expanding we have,

$$e^{-t} \times 1\left(2\sin t\cos t + 6\cos^2 t + 6\sin^2 t - 2\sin t\cos t\right) \neq 0$$

$$\Rightarrow e^{-t} \times 6 \neq 0$$

for $\forall\, t \in \mathbb{R}$

42. **(5)**

$$\begin{bmatrix} 2 & 1 \\ 3 & \frac{3}{2} \end{bmatrix} \begin{bmatrix} a & b \\ b & c \end{bmatrix} = \begin{bmatrix} 1 & 2 \\ \alpha & \beta \end{bmatrix}$$

Now $ac - b^2 = 2$ and $2a + b = 1$

and $2b + c = 2$

Solving all the above equations we get

$$\frac{1-b}{2} \times \left(\frac{2-2b}{1}\right) - b^2 = 2$$

$$\Rightarrow (1-b)^2 - b^2 = 2$$

$$\Rightarrow 1 - 2b = 2$$

$$\Rightarrow b = -\frac{1}{2} \text{ and } a = \frac{3}{4} \text{ and } c = 3$$

Hence $\alpha = 3a + \frac{3b}{2} = \frac{9}{4} - \frac{3}{4} = \frac{3}{2}$

and $\beta = 3b + \frac{3c}{2} = -\frac{3}{2} + \frac{9}{2} = 3$

also $s = a + c = \frac{15}{4}$

$$\therefore \frac{\beta s}{\alpha^2} = \frac{3 \times 15}{4 \times \frac{9}{4}} = 5$$

43. **(A)**

$$A = \begin{bmatrix} m & n \\ p & q \end{bmatrix}, |A - d\,(\text{adj } A)| = 0$$

$$\Rightarrow |A - d\,(\text{adj } A)| = \left| \begin{bmatrix} m & n \\ p & q \end{bmatrix} - d \begin{bmatrix} q & -n \\ -p & m \end{bmatrix} \right|$$

$$= \begin{vmatrix} m - qd & n(1+d) \\ p(1+d) & q - md \end{vmatrix} = 0$$

$$\Rightarrow (m - qd)(q - md) - np(1+d)^2 = 0$$

$$\Rightarrow mq - m^2 d - q^2 d + mqd^2 - np(1+d)^2 = 0$$

$$(mq - np) + d^2(mq - np) - d\left(m^2 + q^2 + 2np\right) = 0$$

$$\Rightarrow d + d^3 - d\left((m+q)^2 - 2d\right) = 0$$

$$\Rightarrow 1 + d^2 = (m+q)^2 - 2d$$

$$\Rightarrow (1+d)^2 = (m+q)^2$$

$\therefore$ Option (A) is correct.

44. **(D)**

$$P^T = aP + (a-1)I$$

$$\Rightarrow P = aP^T + (a-1)I$$

$$\Rightarrow P^T - P = a\left(P - P^T\right)$$

$$\Rightarrow P = P^T, \quad \text{as } a \neq -1$$

Now, $P = aP + (a-1)I$

$$\Rightarrow P = -I \Rightarrow |P| = 1$$

$$\Rightarrow |\text{Adj } P| = 1$$

45. **(C)**

$$A^2 = \begin{bmatrix} 1 & 0 & 0 \\ 0 & 4 & -1 \\ 0 & 12 & -3 \end{bmatrix} \begin{bmatrix} 1 & 0 & 0 \\ 0 & 4 & -1 \\ 0 & 12 & -3 \end{bmatrix}$$

$$= \begin{bmatrix} 1 & 0 & 0 \\ 0 & 4 & -1 \\ 0 & 12 & -3 \end{bmatrix} = A$$

$\Rightarrow A^3 = A^4 = \ldots = A$

$(A+I)^{11}$

$= {}^{11}C_0 A^{11} + {}^{11}C_1 A^{10} + \ldots {}^{11}C_{10} A + {}^{11}C_{11} I$

$= \left({}^{11}C_0 + {}^{11}C_1 + \ldots {}^{11}C_{10}\right) A + I$

$= \left(2^{11} - 1\right) A + I = 2047A + I$

$\therefore$ Sum of diagonal elements

$= 2047(1+4-3)+3 = 4094+3 = 4097$

46. (196)

As given $a+b+c+d = 3$ or 5 or 7 or 11

if sum = 3

$\left(1+x+x^2+\ldots+x^4\right)^4 \to x^3$

$\left(1-x^5\right)^4 (1-x)^{-4} \to x^3$

$\therefore {}^{4+3-1}C_3 = {}^6C_3 = 20$

If sum = 5

$\left(1-4x^5\right)(1-x)^{-4} \to x^5$

$\Rightarrow {}^{4+5-1}C_5 - 4x^4\ {}^{4+0-1}C_0 = {}^8C_5 - 4 = 52$

If sum = 7

$\left(1-4x^5\right)(1-x)^{-4} \to x^7$

$\Rightarrow {}^{4+5-1}C_4 - {}^{4\cdot 4+0-1}C_0 = {}^8C_5 - 4 = 52$

If sum = 11

$\left(1-4x^5+6x^{10}\right)(1-x)^{-4} \to x^{11}$

$\Rightarrow {}^{4+11-1}C_{11} - 4\cdot {}^{4+6-4}C_6 + 6\cdot {}^{4+1-1}C_1$

$= {}^{14}C_{11} - 4\cdot {}^9C_6 + 6\cdot 4 = 364 - 336 + 24 = 52$

$\therefore$ Total matrices = 20 + 52 + 80 + 52 = 204

47. (5)

$\left|\text{Adj}\left(2\ \text{Adj}\left(2A^{-1}\right)\right)\right|$

$= \left|2\ \text{Adj}\left(\text{Adj}\left(2A^{-1}\right)\right)\right|^{n-1}$

$2^{n(n-1)} \left|\text{Adj}\left(2A^{-1}\right)\right|^{n-1}$

$= 2^{n(n-1)} \left|\left(2A^{-1}\right)\right|^{(n-1)(n-1)}$

$= 2^{n(n-1)} 2^{n(n-1)(n-1)} \left|A^{-1}\right|^{(n-1)(n-1)}$

$= 2^{n(n-1)+n(n-1)(n-1)} \dfrac{1}{|A|^{(n-1)^2}}$

$= \dfrac{2^{n(n-1)+n(n-1)(n-1)}}{2^{(n-1)^2}}$

$= 2^{n(n-1)+n(n+1)^2-(n-1)^2}$

$= 2^{(n-1)\left(n^2-n+1\right)}$

Now, $2^{(n-1)\left(n^2-n+1\right)}$

$2^{(n-1)\left(n^2-n+1\right)} = 2^{84}$

So, $n = 5$

48. (C)

$A = \dfrac{1}{2}\begin{bmatrix} 1 & \sqrt{3} \\ -\sqrt{3} & 1 \end{bmatrix}$

$A = \begin{bmatrix} \cos 60° & \sin 60° \\ -\sin 60° & \cos 60° \end{bmatrix}$

If $A = \begin{bmatrix} \cos\alpha & \sin\alpha \\ -\sin\alpha & \cos\alpha \end{bmatrix}$

Here $\alpha = \dfrac{\pi}{3}$

$A^2 = \begin{bmatrix} \cos\alpha & \sin\alpha \\ -\sin\alpha & \cos\alpha \end{bmatrix}\begin{bmatrix} \cos\alpha & \sin\alpha \\ -\sin\alpha & \cos\alpha \end{bmatrix}$

$= \begin{bmatrix} \cos 2\alpha & \sin 2\alpha \\ -\sin 2\alpha & \cos 2\alpha \end{bmatrix}$

$A^{30} = \begin{bmatrix} \cos 30\alpha & \sin 30\alpha \\ -\sin 30\alpha & \cos 30\alpha \end{bmatrix}$

$A^{30} = \begin{bmatrix} 1 & 0 \\ 0 & 1 \end{bmatrix} = I$

$A^{25} = \begin{bmatrix} \cos 25\alpha & \sin 25\alpha \\ -\sin 25\alpha & \cos 25\alpha \end{bmatrix} = \begin{bmatrix} \dfrac{1}{2} & \dfrac{\sqrt{3}}{2} \\ \dfrac{-\sqrt{3}}{2} & \dfrac{1}{2} \end{bmatrix}$

$A^{25} = A$

$A^{25} - A = 0$

49. (A)

Let, $A = \begin{bmatrix} p & q \\ r & s \end{bmatrix}$

$A^2 = \begin{bmatrix} p & q \\ r & s \end{bmatrix}\begin{bmatrix} p & q \\ r & s \end{bmatrix} = \begin{bmatrix} p^2+qr & pq+qs \\ rp+rs & qr+s^2 \end{bmatrix}$

Given, $A^2 = I$

$$\Rightarrow \begin{bmatrix} p^2+qr & pq+qs \\ rp+rs & qr+s^2 \end{bmatrix} = \begin{bmatrix} 1 & 0 \\ 0 & 1 \end{bmatrix}$$

On comparing both sides we get,

$\Rightarrow p^2+qr=1, q(p+s)=0$

And $r(p+s)=0, qr+s^2=1$

Now, on solving the above relation, we get,

$q \neq 0 \Rightarrow p+s=0 \Rightarrow a=0$

And

$b=|A|=ps-qr=-p^2-qr=-1 \ (\because s=-p)$

$\therefore 3a^2+4b^2=4$

50. **(D)**

Given,

P is a square matrix such that $P^2 = I - P$,

Solving $P^2 = I - P$ we get,

$\Rightarrow P^4 = (I-P)(I-P)$

$\Rightarrow P^4 = I + P^2 - 2P$

$\Rightarrow P^4 = I + I - P - 2P = 2I - 3P$

Now solving, $P^4 \cdot P^2 = P^6 = 2I - 5P + 3P^2$

$\Rightarrow P^6 = 2I - 5P + 3(I-P) = 5I - 8P$ (1)

Similarly, $P^8 = 13I - 21P$ (2)

Now adding (2) + (1) we get,

$P^8 + P^6 = 18I - 29P$

Now subtracting (2) – (1) we get,

$P^8 - P^6 = 8I - 13P$

Comparing with $P^\alpha - P^\beta = \delta\ell - 13P$ and $P^\alpha + P^\beta = \gamma\ell - 29P$ we get,

$\alpha = 8, \beta = 6, \gamma = 18, \delta = 8$

Hence, $\alpha+\beta+\gamma-\delta = 8+6+18-8 = 24$

51. **(B)**

Given that $A = \begin{bmatrix} 2 & 1 & 0 \\ 1 & 2 & -1 \\ 0 & -1 & 2 \end{bmatrix}$ and

$|\text{adj}(\text{adj}(\text{adj } 2A))| = 16^n$.

We know that $|\text{adj } A| = |A|^{n-1}$ and

$|\text{adj}(\text{adj } A)| = |A|^{(n-1)^2}$

Similarly, $|\text{adj}(\text{adj}(\text{adj } A))| = |A|^{(n-1)^3}$

where n is the order of the square matrix.

Let us find $|A|$.

$$|A| = 2(2\times2-(-1)(-1)) - 1(1\times2-0(-1)) + 0(1\times(-1)-0\times2)$$

$= 2(3)-(2)+0 = 4$

$\Rightarrow |2A|^{(n-1)^3} = |2A|^{(3-1)^3}$

$= |2A|^8 = (2^8)^3 |A|^8$

$\Rightarrow 2^{24} \times 4^8 = 16^n$

$\Rightarrow 16^6 \times 16^4 = 16^n \Rightarrow 16^{10} = 16^n$

$\Rightarrow n = 10$

Therefore, the value of n is 10.

52. **(B)**

Given,

$$P = \begin{bmatrix} \frac{\sqrt{3}}{2} & \frac{1}{2} \\ -\frac{1}{2} & \frac{\sqrt{3}}{2} \end{bmatrix}$$

$$\therefore PP^T = \begin{bmatrix} \frac{\sqrt{3}}{2} & \frac{1}{2} \\ -\frac{1}{2} & \frac{\sqrt{3}}{2} \end{bmatrix} \begin{bmatrix} \frac{\sqrt{3}}{2} & -\frac{1}{2} \\ \frac{1}{2} & \frac{\sqrt{3}}{2} \end{bmatrix} = \begin{bmatrix} 1 & 0 \\ 0 & 1 \end{bmatrix}$$

$\Rightarrow PP^T = I$

Now,

$P^T (PAP^T)^{2007} P$

$$= P^T \underbrace{(PAP^T)(PAP^T)(PAP^T)\ldots(PAP^T)}_{2007 \text{ times}} P$$

$\Rightarrow P^T (PAP^T)^{2007} P = A^{2007}$

Now,

$$A = \begin{bmatrix} 1 & 1 \\ 0 & 1 \end{bmatrix}$$

$$A^2 = \begin{bmatrix} 1 & 2 \\ 0 & 1 \end{bmatrix}$$

$$A^3 = \begin{bmatrix} 0 & 3 \\ 0 & 1 \end{bmatrix}$$

$$A^{2007}=\begin{bmatrix}1 & 2007\\0 & 1\end{bmatrix}$$

So,

$$\Rightarrow P^T\left(PAP^T\right)^{2007}P=\begin{bmatrix}1 & 2007\\0 & 1\end{bmatrix}$$

$$\Rightarrow P^TQP=\begin{bmatrix}1 & 2007\\0 & 1\end{bmatrix}=\begin{bmatrix}a & b\\c & d\end{bmatrix}$$

So,

$a=1, b=2007, c=0, d=1$

$2a+b+3c-4d=2+2007-4=2005$

53. **(C)**

Given,

$$A=\begin{bmatrix}1 & 5\\\lambda & 10\end{bmatrix}, A^{-1}=\alpha A+\beta I \text{ and}$$

$\alpha+\beta=-2,$

Now solving by using the characteristic equation, we get

$$|A-k|=0\Rightarrow\begin{vmatrix}1-k & 5\\\lambda & 10-k\end{vmatrix}=0$$

$$\Rightarrow k^2-11k+10-5\lambda=0$$

$$\Rightarrow A^2-11A+(10-5\lambda)I=0$$

{by putting $k = A$}

$$\Rightarrow A^{-1}=\frac{1}{10-5\lambda}(-A+11I)$$

On comparing with $A^{-1}=\alpha A+\beta I$ we get,

$$\Rightarrow \alpha=-\frac{1}{10-5\lambda}, \beta=\frac{11}{10-5\lambda}$$

And given,

$$\alpha+\beta=-2\Rightarrow\lambda=3, \alpha=\frac{1}{5}, \beta=\frac{-11}{5}$$

Hence, the value of $4\alpha^2+\beta^2+\lambda^2=14$

54. **(D)**

We need to find the value of $\left|3\text{ adj}\left(|3A|A^2\right)\right|$

We know that $|kA|=k^n|A|$ where n is the order of the matrix and k is a constant.

$$\Rightarrow |3A|=3^3(2)$$

$$\Rightarrow \left|3\text{ adj}\left(|3A|A^2\right)\right|=3^3\left|\text{adj}\left(\left(3^3\cdot 2\right)A^2\right)\right|$$

We know that $\text{adj}(kA)=k^{n-1}\text{ adj}(A)$

$$=3^3\left|\left(3^3\cdot 2\right)^2\text{ adj}\left(A^2\right)\right|$$

Now we know that $|\text{adj } A|=|A|^{n-1}$

$$=3^3\left(\left(3^3\cdot 2\right)^2\right)^3\left|\text{adj}\left(A^2\right)\right|$$

$$=3^3\left(3^3\cdot 2\right)^6\left|A^2\right|^{3-1}$$

$$=3^3\left(3^3\cdot 2\right)^6(2)^4$$

$$=2^{10}\cdot 3^{21}$$

Hence, the required answer is

$2^{10}\cdot 3^{21}=6^{10}\cdot 3^{11}$

55. **(D)**

Given,

$$A=\frac{1}{5!6!7!}\begin{bmatrix}5! & 6! & 7!\\6! & 7! & 8!\\7! & 8! & 9!\end{bmatrix}$$

$$\Rightarrow |A|=\frac{1}{5!6!7!}\begin{vmatrix}5! & 6! & 7!\\6! & 7! & 8!\\7! & 8! & 9!\end{vmatrix}$$

$$\Rightarrow |A|=\frac{1}{5!6!7!}5!\times 6!\times 7!\begin{vmatrix}1 & 6 & 42\\1 & 7 & 56\\1 & 8 & 72\end{vmatrix}$$

$$\Rightarrow |A|=2$$

Using the property of adjoint of matrix we get,

$$|\text{adj}(\text{adj}(A))|=|A|^{(n-1)^2} \text{ and } |kA|=k^n|A|,$$

where n is the order of square matrix,

Using the above two formulae we get,

$$|\text{adj}(\text{adj}(2A))|=|2A|^4=2^{12}|A|^4$$

$$\Rightarrow |\text{adj}(\text{adj}(2A))|=2^{12}\cdot 2^4$$

$$\Rightarrow |\text{adj}(\text{adj}(2A))|=2^{16}$$

56. **(2)**

Given,

$$A=\begin{bmatrix}0 & 1 & 2\\ a & 0 & 3\\ 1 & c & 0\end{bmatrix}$$

$$\Rightarrow A^2=\begin{bmatrix}0 & 1 & 2\\ a & 0 & 3\\ 1 & c & 0\end{bmatrix}\begin{bmatrix}0 & 1 & 2\\ a & 0 & 3\\ 1 & c & 0\end{bmatrix}$$

$$\Rightarrow A^2=\begin{bmatrix}a+2 & 2c & 3\\ 3 & a+3c & 2a\\ ca & 1 & 2+3c\end{bmatrix}$$

Now finding,

$$A^3=\begin{bmatrix}a+2 & 2c & 3\\ 3 & a+3c & 2a\\ ca & 1 & 2+3c\end{bmatrix}\begin{bmatrix}0 & 1 & 2\\ a & 0 & 3\\ 1 & c & 0\end{bmatrix}$$

$$\Rightarrow A^3=\begin{bmatrix}2ac+3 & a+2+3c & 2a+4+6c\\ a^2+3ca+2a & 3+2ac & 6+3a+9c\\ a+2+3c & ca+2c+3c^2 & 2ca+3\end{bmatrix}$$

Equating $A^3=A$ and comparing both sides we get,

$$2ca+3=0\Rightarrow c=-\frac{3}{2a}$$

And $a+2+3c=1$

$$\Rightarrow a+2+3\left(-\frac{3}{2a}\right)=1$$

$$\Rightarrow a+1-\frac{9}{2a}=0$$

$$\Rightarrow 2a^2+2a-9=0$$

$$\Rightarrow a=\frac{-2+\sqrt{4+4\times 9\times 2}}{4}$$

$$=\frac{-2+\sqrt{76}}{4}\approx\frac{6.7}{4}\approx 1.4$$

Hence, $a\in(1,2]$

So, on comparing with $a\in(n-1,n]$ we get $n=2$

57. **(D)**

Given,

$$A=\begin{bmatrix}1 & \frac{1}{51}\\ 0 & 1\end{bmatrix},$$

$$B=\begin{bmatrix}1 & 2\\ -1 & -1\end{bmatrix}A\begin{bmatrix}-1 & -2\\ 1 & 1\end{bmatrix}$$

So, $B=MAN$

Now let $M=\begin{bmatrix}1 & 2\\ -1 & -1\end{bmatrix}$ and $N=\begin{bmatrix}-1 & -2\\ 1 & 1\end{bmatrix}$

Solving,

$$\underset{(M)}{\begin{bmatrix}1 & 2\\ -1 & -1\end{bmatrix}}\underset{(N)}{\begin{bmatrix}-1 & -2\\ 1 & 1\end{bmatrix}}=\begin{bmatrix}1 & 0\\ 0 & 1\end{bmatrix}$$

$\therefore MN=I=NM$

Now using, $B=MAN$ we get

$$B^n=(MAN)^n=(MAN)(MAN)\ldots(MAN)$$

$$\Rightarrow B^n=MA\underbrace{NM}_{I}A\underbrace{NM}_{I}\ldots\underbrace{NM}_{I}AN$$

$$\Rightarrow B^n=M\underbrace{AAAAA\ldots A}_{n\text{ times}}N=MA^nN$$

Using $A=\begin{bmatrix}1 & \frac{1}{51}\\ 0 & 1\end{bmatrix}$

$$\Rightarrow A=\begin{bmatrix}1 & 0\\ 0 & 1\end{bmatrix}+\begin{bmatrix}0 & \frac{1}{51}\\ 0 & 0\end{bmatrix}=I+E$$

Finding, $E^2=\begin{bmatrix}0 & \frac{1}{51}\\ 0 & 0\end{bmatrix}\begin{bmatrix}0 & \frac{1}{51}\\ 0 & 0\end{bmatrix}=\begin{bmatrix}0 & 0\\ 0 & 0\end{bmatrix}$

$\Rightarrow E^2=0,$

so all higher powers will also be zero,

$$\therefore A^n=(I+E)^n$$

$$=I+nE+\underbrace{{}^nC_2E^2}_{0}+\underbrace{{}^nC_3E^3}_{0}+\underbrace{{}^nC_4E^4}_{0}\ldots$$

$$\Rightarrow A^n=I+nE$$

$$\Rightarrow A^n=\begin{bmatrix}1 & \frac{n}{51}\\ 0 & 1\end{bmatrix}$$

So,

$$B^n = MA^nN = \begin{bmatrix} 1 & 2 \\ -1 & -1 \end{bmatrix}\begin{bmatrix} 1 & \frac{n}{51} \\ 0 & 1 \end{bmatrix}\begin{bmatrix} -1 & -2 \\ 1 & 1 \end{bmatrix}$$

$$\Rightarrow B^n = \begin{bmatrix} 1 & \frac{n}{51}+2 \\ -1 & \frac{-n}{51}-1 \end{bmatrix}\begin{bmatrix} -1 & -2 \\ 1 & 1 \end{bmatrix}$$

$$\Rightarrow B^n = \begin{bmatrix} 1+\frac{n}{51} & \frac{n}{51} \\ \frac{-n}{51} & 1-\frac{n}{51} \end{bmatrix}$$

$$\sum_{n=1}^{50} B^n = \begin{bmatrix} 50+\frac{50\cdot 51}{2\cdot 51} & \frac{50\cdot 51}{2\cdot 51} \\ \frac{-50\cdot 51}{2\cdot 51} & 50-\frac{50\cdot 51}{2\cdot 51} \end{bmatrix}$$

$$= \begin{bmatrix} 75 & 25 \\ -25 & 25 \end{bmatrix}$$

$\therefore$ Sum = 100

58. **(B)**

Given that $A = \{0, 1, 2, 3, \ldots, 8, 9\}$

We need to find the number of symmetric matrices.

We know that a matrix is symmetric if $A^T = A$.

$$A = \begin{bmatrix} a_{11} & a_{12} & a_{13} \\ a_{21} & a_{22} & a_{23} \\ a_{31} & a_{32} & a_{33} \end{bmatrix}$$

Now, a_{11}, a_{22}, a_{33} can have any digit from 0 to 9

Number of ways this can be done is $10\times 10\times 10 = 10^3$

Now $a_{12} = a_{21}$. So number of ways this can be done is 10×1

Similarly, $a_{13} = a_{31}$,

So number of ways this can be done is 10×1

Also, $a_{23} - a_{32}$.

So number of ways this can be done is 10×1

$\therefore$ Total symmetric matrix

$= 10^3\times 10\times 10\times 10 = 10^6$

Hence this is the correct option.

59. **(C)**

We have been given that

$$B = \begin{bmatrix} 1 & 3 & \alpha \\ 1 & 2 & 3 \\ \alpha & \alpha & 4 \end{bmatrix}, \alpha > 2$$

And $\text{adj}(A) = B, |A| = 2$

$\Rightarrow |\text{adj}(A)| = |B|$

We know that $|\text{adj } A| = |A|^{n-1}$

$\Rightarrow 2^2 = (8-3\alpha) - 3(4-3\alpha) + \alpha(-\alpha)$

$\Rightarrow \alpha^2 - 6\alpha + 8 = 0$

$\Rightarrow (\alpha-4)(\alpha-2) = 0$

$\alpha = 4, 2$ but $\alpha > 2$ so $\boxed{\alpha = 4}$

Now,

$$[\alpha \quad -2\alpha \quad \alpha] B \begin{bmatrix} \alpha \\ -2\alpha \\ \alpha \end{bmatrix}$$

$$= [4 \quad -8 \quad 4]\begin{bmatrix} 1 & 3 & 4 \\ 1 & 2 & 3 \\ 4 & 4 & 4 \end{bmatrix}\begin{bmatrix} 4 \\ -8 \\ 4 \end{bmatrix}$$

$$[12 \quad 12 \quad 8]\begin{bmatrix} 4 \\ -8 \\ 4 \end{bmatrix} = 48 - 96 + 32 = -16$$

Hence this is the correct option.

60. **(B)**

Given that $A = \begin{bmatrix} 1 & 2 & 3 \\ \alpha & 3 & 1 \\ 1 & 1 & 2 \end{bmatrix}$

$\Rightarrow |A| = 2$

$\Rightarrow 1(6-1) - \alpha - 7 = 2$

$\Rightarrow -\alpha - 2 = 2$

$\Rightarrow \alpha = -4$

$\Rightarrow |2\text{ adj}(2\text{ adj}(2A))| = 32^n$

We know that

$|kA| = k^n |A|$, $\text{adj}(\text{adj}(A)) = |A|^{(n-1)^2}$ and

$\text{adj}(kA) = k^{n-1}\,\text{adj}(A)$.

$2^3 |\text{adj}(2\,\text{adj}(2A))|$

$= 2^3 |2^{3-1}\,\text{adj}(\text{adj}(2A))|$

$= 2^3 \cdot (4)^3 |\text{adj}(\text{adj}(2A))|$

$= 2^9 |2A|^{(2)^2}$

$= 2^9 \cdot |2A|^4$

$= 2^9 \cdot 2^{12} |A|^4$

$= 2^{21} \cdot 2^4 = 2^{25} = (32)^n = (2)^{5n}$

$\therefore n = 5$

$\Rightarrow 3n + \alpha = 15 - 4 = 11$

Hence this is the correct option.

61. **(B)**

We have been given that

$|A| = m - n$, where $4m + n = 22$ (1)

and $17m + 4n = 93$ (2)

Solving (1) and (2)

$m = 5, n = 2$

Order = 5

$\Rightarrow |A| = 3$

We know that

$|\text{adj}\,A| = |A|^{\omega-1}\,|\text{adj}(\text{adj}\,A)| = |A|^{(\omega-1)^2}$

where $\omega \times \omega$ is the order of the matrix.

$\therefore \det(n\,\text{adj}(\text{adj}(mA)))$

$= |2\,\text{adj}(\text{adj}(5A))|$

$= 2^5 |5A|^{16}$

$= 2^5\, 5^{80}\, |A|^{16} = 2^5 \cdot 3^{16} \cdot 5^{80}$

$= 3^{11}\, 5^{80}\, 6^5$

So, $a + b + c = 96$

Hence this is the required option.

62. **(B)**

$$A = \begin{bmatrix} \sqrt{2} & 1 \\ -1 & \sqrt{2} \end{bmatrix} \Rightarrow \det(A) = 3$$

$$B = \begin{bmatrix} 1 & 0 \\ 1 & 1 \end{bmatrix} \Rightarrow \det(B) = 1$$

Now

$C = ABA^T \Rightarrow \det(C) = (\det(A))^2\, x \det(B)$

$|C| = 9$

Now $|X| = |A^T C^2 A|$

$= |A^T||C|^2|A|$

$= |A|^2 |C|^2$

$= 9 \times 81 = 729$

63. **(2)**

$A = I_2 - 2MM^T$

$A^2 = (I_2 - 2MM^T)(I_2 - 2MM^T)$

$= I_2 - 2MM^T - 2MM^T + 4MM^T MM^T$

$= I_2 - 4MM^T + 4MM^T = I_2$

$AX = \lambda X$

$A^2 X = \lambda AX$

$X = \lambda(\lambda X)$

$X = \lambda^2 X$

$X(\lambda^2 - 1) = 0$

$\lambda^2 = 1$

$\lambda = \pm 1$

Sum of the square of all possible values = 2

64. **(D)**

$$f(-x) = \begin{bmatrix} \cos x & \sin x & 0 \\ -\sin x & \cos x & 0 \\ 0 & 0 & 1 \end{bmatrix}$$

$$f(x) \cdot f(-x) = \begin{bmatrix} 1 & 0 & 0 \\ 0 & 1 & 0 \\ 0 & 0 & 1 \end{bmatrix} = I$$

Hence statement I is correct

Now, checking statement II,

$$f(y) = \begin{bmatrix} \cos y & -\sin y & 0 \\ \sin y & \cos y & 0 \\ 0 & 0 & 1 \end{bmatrix}$$

$$f(x) \cdot f(y) = \begin{bmatrix} \cos(x+y) & -\sin(x+y) & 0 \\ \sin y & \cos y & 0 \\ 0 & 0 & 1 \end{bmatrix}$$

$f(x)\cdot f(y)=f(x+y)$

Hence statement II is also correct.

65. (10)

$|A-xI|=0$

Roots are -1 and 3

Sum of roots $=\text{tr}(A)=2$

Product of roots $=|A|=-3$

Let $A=\begin{bmatrix} a & b \\ c & d \end{bmatrix}$

We have $a+d=2$

$ad-bc=-3$

$$A^2=\begin{bmatrix} a & b \\ c & d \end{bmatrix}\times\begin{bmatrix} a & b \\ c & d \end{bmatrix}=\begin{bmatrix} a^2+bc & ab+bd \\ ac+ad & bc+d^2 \end{bmatrix}$$

We need $a^2+bc+bc+d^2$

$=a^2+2bc+d^2$

$=(a+d)^2-2ad+2bc$

$=4-2(ad-bc)=4-2(-3)$

$=4+6=10$

66. (B)

$|A|=\alpha^2-\beta^2$

$|2A|^3=2^{21}\Rightarrow |A|=2^4$

$\alpha^2-\beta^2=16$

$(\alpha+\beta)(\alpha-\beta)=16\Rightarrow \alpha=4$ or 5

67. (D)

$AA^T=I=A^TA$

On solving the given expression, we get

$$\frac{1}{2}A\left[A^2+\left(A^T\right)^2+2AA^T+A^2+\left(A^T\right)^2-2AAA^T\right]$$

$$=A\left[A^2+\left(A^T\right)^2\right]=A^3+A^T$$

68. (A)

$$|P^{-1}AP-2I|=|P^{-1}AP-2P^{-1}P|$$

$$=|P^{-1}(A-2I)P|$$

$$=\left|P^{-1}|A-2I|P\right|$$

$|A-2I|$

$$=\begin{vmatrix} 0 & 1 & 2 \\ 6 & 0 & 11 \\ 3 & 3 & 0 \end{vmatrix}=69$$

So, prime factors of 69 are 3 and 23

So, sum = 26

69. (B)

$$1\cdot x\sin\theta=y\sin\left(\theta+\frac{2\pi}{3}\right)=z\sin\left(\theta+\frac{4\pi}{3}\right)\neq 0$$

$\Rightarrow x, y, z\neq 0$

Also,

$$\sin\theta+\sin\left(\theta+\frac{2\pi}{3}\right)+\sin\left(\theta+\frac{4\pi}{3}\right)=0\ \forall\ \theta\in R$$

$$\Rightarrow \frac{1}{x}+\frac{1}{y}+\frac{1}{z}=0$$

$\Rightarrow xy+yz+zx=0$

(i) Trace $(R)=x+y+z$

If $x+y+z=0$ and $xy+yz+zx=0$

$\Rightarrow x=y=z=0$

Statement (i) is False

(ii) $\text{Adj}(\text{Adj}(R))=|R|R$

Trace $(\text{Adj}(\text{Adj}(R)))$

$=xyz(x+y+z)\neq 0$

Statement (ii) is also False

70. (7)

$|A|=2$

$$\underbrace{\text{adj}(\text{adj}((\text{adj}............(a))))}_{2024\text{ - times}}=|A|^{(n-1)^{2024}}$$

$=|A|^{2024}=2^{2024}$

$2^{2024}=(2^2)2^{2022}=4(8)^{674}=4(9-1)^{674}$

$\Rightarrow 2^{2024}\equiv 4(\text{mod }9)$

$\Rightarrow 2^{2024}\equiv 9m+4,\ m\leftarrow$ even

$2^{9m+4}\equiv 16\cdot\left(2^3\right)^{3m}\equiv 16(\text{mod }9)$

$\equiv 7$

6.9 Summary

- Let us consider the system of three homogeneous linear equations in three variables x, y, z.
  - $a_1x+b_1y+c_1z=0$ (1)
  - $a_2x+b_2y+c_2z=0$ (2)
  - $a_3x+b_3y+c_3z=0$ (3)

  Solving (2) and (3) by cross multiplication and then putting the value of x, y, z in (1), we get
  - $a_1(b_2c_3-b_3c_2)+b_1(c_2a_3-c_3a_2)+c_1(a_2b_3-a_3b_2)=0$
  - This can be written as $\begin{vmatrix} a_1 & b_1 & c_1 \\ a_2 & b_2 & c_2 \\ a_3 & b_3 & c_3 \end{vmatrix}=0$
- The expression $\begin{vmatrix} a_1 & b_1 & c_1 \\ a_2 & b_2 & c_2 \\ a_3 & b_3 & c_3 \end{vmatrix}$ is called determinant of order 3.
- If (x_1, y_1), (x_2, y_2) and (x_3, y_3) are the vertices of a triangle, then area of triangle is defined as $\left|\frac{1}{2}\begin{vmatrix} x_1 & y_1 & 1 \\ x_2 & y_2 & 1 \\ x_3 & y_3 & 1 \end{vmatrix}\right|$
- If $\vec{a}=a_1\vec{i}+a_2\vec{j}+a_3\vec{k}, \vec{b}=b_1\vec{i}+b_2\vec{j}+b_3\vec{k}$; then their cross product is evaluated as $\vec{a}\times\vec{b}=\begin{vmatrix} \vec{i} & \vec{j} & \vec{k} \\ a_1 & a_2 & a_3 \\ b_1 & b_2 & b_3 \end{vmatrix}$
- Various properties of determinant are explained in theory which we use to solve the determinant.
- If the value of a determinant is Δ, then the value of the determinant formed by the corresponding element of determinant will be Δ^2.
- Cramer's rules are explained to solve the system of linear equation and homogenous equations.
  - If $\Delta \neq 0$, then the system of equations will have unique finite solutions and so equations are consistent.
  - If $\Delta=0$ and at least one of $\Delta_1, \Delta_2, \Delta_3$ is non-zero, then the system is inconsistent, that is, it has no solution.
  - If $\Delta=0$ and $\Delta_1, \Delta_2, \Delta_3=0$ then the system will have an infinite number of solutions that is, the equations are consistent.
- In the case of homogeneous equations,
  - if $\Delta\neq 0$, the solution is called trivial solution.
  - if $\Delta=0$, then solution is called non-trivial solution.
- A matrix is a rectangular arrangement of (mn) numbers containing m rows and n columns.
- Matrix A of order $(m\times n)$ is written as

$$A=\begin{bmatrix} a_{11} & a_{12} & \cdots\cdots & a_{1n} \\ a_{21} & a_{22} & \cdots\cdots & a_{2n} \\ \cdots & \cdots\cdots & & \cdots \\ \cdots & \cdots\cdots & & \cdots \\ a_{m1} & a_{m2} & \cdots\cdots & a_{mn} \end{bmatrix}$$

- Various types of matrices have been explained in the theory.
- Any two matrices can be added or subtracted if they have same number of rows and columns.
- Any two matrices A and B can be multiplied if the number of columns in A is equal to the number of rows in B.
- If A is a non-singular matrix, then their inverse is defined as $A^{-1}=\frac{(\text{adj } A)}{|A|}$.
- Properties of matrices are already explained in the theory.
- We can find the solution of system of linear equations by the matrix method.
- Reflection of matrices has been explained in the theory.

CHAPTER 7

Probability

OBJECTIVES

7.1 Introduction
7.2 Probability of Occurrence of an Event
7.3 Odds in Favour and Odds Against an Event
7.4 Conditional Probability
7.5 Theorem of Total Probability
7.6 Bayes' Theorem (Inverse Probability Theorem)
7.7 Binomial Theorem on Probability
7.8 Probability Distribution of a Random Variable
7.9 Know Your Concepts
7.10 JEE Main: Sample Questions
7.11 JEE Advanced: Sample Questions
7.12 JEE Main: Archive
7.13 Answer Key
7.14 Solutions
7.15 Summary

7.1 Introduction

In our day-to-day lives, we often make guesses and use statements such as, 'it will most probably rain today', 'it is almost certain that Ramesh will compete in the IIT'. Whenever we use such statements, we have an intuition that enables us to claim that one event is more likely to happen than another. In probability theory, we measure the degree of certainty or uncertainty of such events in terms of numbers lying between 0 and 1.

7.1.1 Random Experiment

An experiment whose all-possible outcomes are known in advance but the outcome of any specific performance cannot be predicted before the completion of the experiment, is known as a random experiment. If we throw a die randomly then it cannot be said in advance that the number which will appear is 1 or 2 or 3 or 4 or 5 or 6, so we can say that the throwing of a die is a random experiment.

7.1.2 Sample Space

The set of all possible outcomes of a random experiment is called the *sample space* and is represented by S. Each outcome is a point of the sample space and it is called the *sample point*.

Discrete Sample Space: A sample space having finite number of sample points is called a *discrete sample space*.

Event: The subset of a sample space is called an *event* and is denoted by E.

Simple Event: Every singleton subset of a sample space is called a *simple event*.

Mixed Event: A subset of the sample space which has more than one element is called a *mixed event*.

Equally Likely Event: Those events, the chances of whose happening is neither lesser nor greater than others, are said to be *equally likely events*.

Mutually Exclusive Events: Any two events A and B are said to be mutually exclusive if $A \cap B = \phi$.

Exhaustive Set of Events: The set of all possible events of a random experiment is called an *exhaustive set of events* of the experiment.

7.2 Probability of Occurrence of an Event

7.2.1 Mathematical or A Priori or Classical Definition of Probability

Let S be the sample space, then the probability of occurrence of an event E is denoted by $P(E)$ and is defined as

$$P(E)=\frac{n(E)}{n(S)}=\frac{\text{number of elements in } E}{\text{number of elements in } S}$$

7.2.2 Statistical or Empirical Definition of Probability

Let a random experiment be repeated n times under similar conditions and the event E occurs r times in n trials, then

$$P(E)=\lim_{n\to\infty}\frac{r}{n}$$

$P(E)$ is generally denoted by p.

7.2.2.1 Probability of occurrence of an event is a number lying between 0 and 1

Let S be the sample space and E be an event

Then $\phi\subseteq E\subseteq S$

$$\Rightarrow n(\phi)\le n(E)\le n(S)$$

$$\Rightarrow 0\le n(E)\le n(S)$$

$$\frac{0}{n(S)}\le\frac{n(E)}{n(S)}\le\frac{n(S)}{n(S)}$$

$$\Rightarrow 0\le P(E)\le 1$$

7.2.3 Complement of an Event

Let S be the sample space for a random experiment and E be an event.

Then, the complement of the event E is denoted by E' or E^c and is defined as the set of all those elements of S which does not belong to E.

Thus E' occurs if and only if E does not occur.

If E is any event and E' be the complement of the event E, then $P(E) + P(E') = 1$

Let S be the sample space, then

$$n(E)+n(E')=n(S)$$

$$\Rightarrow\frac{n(E)}{n(S)}+\frac{n(E')}{n(S)}=1$$

$$\Rightarrow P(E)+P(E')=1$$

7.3 Odds in Favour and Odds Against an Event

Let S be the sample space and E be an event. Let E' denote the complement of event E, then

(i) Odds in favour of an event E

$$=\frac{n(E)}{n(E')}=\frac{\text{Number of cases favourable to event } E}{\text{Number of cases against } E}$$

$$=\frac{P(E)}{P(E')}$$

(ii) Odds against an event E

$$=\frac{n(E')}{n(E)}=\frac{\text{Number of cases against the event } E}{\text{Number of cases favourable to event } E}$$

$$=\frac{P(E')}{P(E)}$$

Note:

(I) Pack of 52 playing cards

(i) **Four suits**

- Spades, Hearts, Diamonds and Clubs.
- Spades and clubs are black faced cards.
- Hearts and diamonds are red faced cards.
- Each suit consists of 13 cards.
- The aces, kings, queens, jacks are called face cards or honours cards.

(ii) **Game of Bridge:** It is played by 4 players, each player is given 13 cards.

(iii) **Game of Whist:** It is played by two pairs of persons.

(II) If A and B are mutually exclusive events, then

$$P(A\cup B)=P(A)+P(B)$$

(III) $P(A\cap B')=P(A)-P(A\cap B)$

$$P(A'\cap B)=P(B)-P(A\cap B)$$

(IV) P(Exactly one of A, B occurs)

$$=P(A\cap B')+P(A'\cap B)$$

$$=P(A)+P(B)-2P(A\cap B)$$

$$=P(A\cup B)-P(A\cap B)$$

$$=P(A'\cup B')-P(A'\cap B')$$

(V) If A, B, C are any three events of the sample space S, then

$$P(A\cup B\cup C)=P(A)+P(B)+P(C)-P(A\cap B)-P(A\cap C)-P(B\cap C)+P(A\cap B\cap C)$$

- P(Exactly one of A, B, C occur) $=P(A)+P(B)+P(C)-2P(A\cap B)-2P(A\cap C)-2P(B\cap C)+3P(A\cap B\cap C)$
- P(Exactly two of A, B, C occur) $=P(A\cap B)+P(B\cap C)+P(A\cap C)-3P(A\cap B\cap C)$
- P(at least two of A, B, C occur) $=P(A\cap B)+P(B\cap C)+P(A\cap C)-2P(A\cap B\cap C)$

7.3.1 Independent Event

Two or more events are said to be independent if occurrence or non-occurrence of any of them does not influence the occurrence or non-occurrence of the other events.

7.4 Conditional Probability

The probability of occurrence of an event B when it is known that some event A has occurred is called a conditional probability and is denoted by $P\left(\frac{B}{A}\right)$.

We read $P\left(\frac{B}{A}\right)$ as 'B occurs given that A occurs'.

Let S be the sample space, and A and B be any two events. In case of occurrence of event A when B has already occurred, B works as a sample space and $A\cap B$ works as the event, therefore,

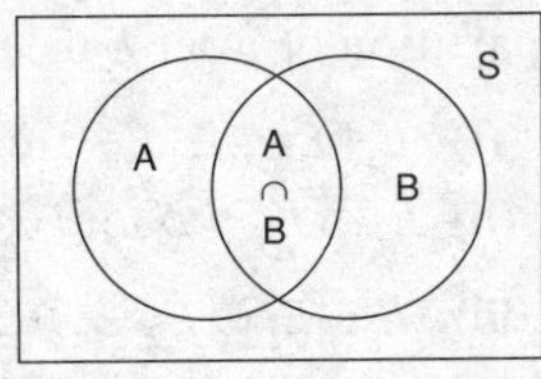

$$P\left(\frac{A}{B}\right)=\frac{n(A\cap B)}{n(B)}=\frac{\frac{n(A\cap B)}{n(S)}}{\frac{n(B)}{n(S)}}=\frac{P(A\cap B)}{P(B)}$$

Thus, $P\left(\frac{A}{B}\right)=\frac{P(A\cap B)}{P(B)},0<P(B)\le 1$

Similarly, $P\left(\frac{B}{A}\right)=\frac{P(A\cap B)}{P(A)},0<P(A)\le 1$

$$\therefore P(A\cap B)=P(B)P\left(\frac{A}{B}\right)$$

If A and B are independent events, then the probability of occurrence of event A is not affected by the occurrence or non-occurrence of event B, therefore

$$P(A\cap B)=P(A)P(B)\left[P\left(\frac{A}{B}\right)=P(A)\right]$$

Example

When a die is thrown, sample space,

$$S=\{1,2,3,4,5,6\}$$

Let $B=$ The event of occurrence of a number 4 and $A=$ The event of occurrence of a number greater than 3.

We have to find $P\left(\frac{B}{A}\right)$, that is, the probability of getting a '4' given that a number greater than 3 has occurred.

Clearly $A=\{4,5,6\},B=\{4\}\Rightarrow P\left(\frac{B}{A}\right)=\frac{1}{3}$

Also, $P(A\cap B)=\frac{1}{6}$ and $P(A)=\frac{3}{6}=\frac{1}{2}$

$$\Rightarrow P\left(\frac{B}{A}\right)=\frac{P(A\cap B)}{P(A)}=\frac{\frac{1}{6}}{\frac{1}{2}}=\frac{1}{3}$$

Note:

(i) Mutually exclusive events will never be independent and independent events will never be mutually exclusive.

Proof

Let A and B be two mutually exclusive events.

Then $P(A\cup B)=P(A)+P(B)$ (7.1)

Now,

$P(A\cap B)=P(A)+P(B)-P(A\cup B)=0$ (7.2)

But $P(A)>0,P(B)>0$ $[\because A\ne\phi,B\ne\phi]$

$\therefore P(A)P(B) > 0$ (7.3)

$\therefore P(A \cap B) \neq P(A)P(B)$

$\Rightarrow$ A and B are not independent events.

Again, if A and B are independent events then

$P(A \cap B) = P(A)P(B)$

Now, $P(A \cup B) = P(A) + P(B) - P(A)P(B)$

$\neq P(A) + P(B)$

$\Rightarrow$ A and B are not mutually exclusive events.

(ii) If A and B are two independent events, then

$$P(A \cup B) = 1 - P(A')P(B')$$

(iii) If A and B are independent events, then

- A and B' are independent events
- A' and B are independent events
- A' and B' are independent events

(iv) If A and B are two events such that $B \neq \phi$, then

$$P\left(\frac{A}{B}\right) + P\left(\frac{A'}{B}\right) = 1$$

(v) If A and B are two events such that $A \neq \phi$, then

$$P(B) = P(A)P\left(\frac{B}{A}\right) + P(A')P\left(\frac{B}{A'}\right).$$

7.5 Theorem of Total Probability

Let $E_1, E_2, \ldots, E_n$ be n mutually exclusive and exhaustive events associated with a random experiment. If E is any event which occurs together with one of the events $E_1, E_2, \ldots, E_n$ $(E \subseteq E_1 \cup E_2 \cup \ldots \cup E_n)$, then

$$P(E) = P(E_1)P\left(\frac{E}{E_1}\right) + P(E_2)P\left(\frac{E}{E_2}\right) + \ldots + P(E_n)P\left(\frac{E}{E_n}\right)$$

Proof: Let $E_1, E_2, \ldots, E_n$ be n mutually exclusive and exhaustive events, therefore,

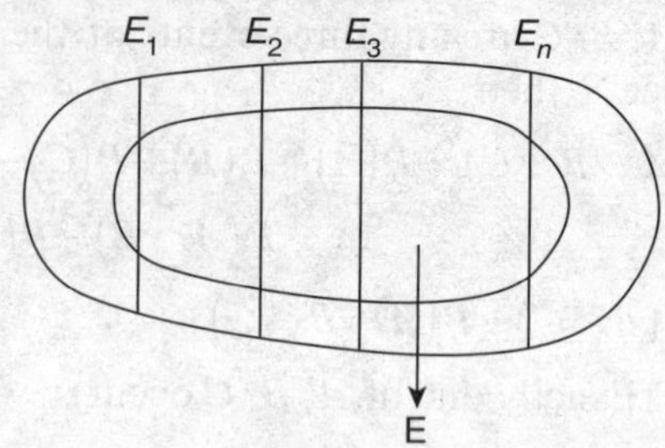

$E_i \cap E_j = \phi$ for $i \neq j$

and $S = E_1 \cup E_2 \cup \ldots \cup E_n$

Since, $E_1, E_2, \ldots, E_n$ are disjoint

$\therefore$ $E \cap E_1, E \cap E_2, \ldots, E \cap E_n$ are also disjoint

Now, $E = E \cap S$

$= E \cap (E_1 \cup E_2 \cup \ldots \cup E_n)$

$= (E \cap E_1) \cup (E \cap E_2) \cup \ldots (E \cap E_n)$

$\Rightarrow P(E) = P(E \cap E_1) + P(E \cap E_2) + \ldots + P(E \cap E_n)$

$\Rightarrow P(E) = P(E_1)P\left(\frac{E}{E_1}\right) + P(E_2)P\left(\frac{E}{E_2}\right) + \ldots + P(E_n)P\left(\frac{E}{E_n}\right)$

[As $E \cap E_1, E \cap E_2, \ldots, E \cap E_n$ are mutually exclusive]

7.5.1 Partition of a Set

A family of sets $E_1, E_2, \ldots, E_n$ is said to form a partition of a set E if

(i) $E_1, E_2, \ldots, E_n$ are non-empty

(ii) $E_i \cap E_j = \phi$ for $i \neq j$

(iii) $E = E_1 \cup E_2 \cup \ldots \cup E_n$.

7.6 Bayes' Theorem (Inverse Probability Theorem)

If $E_1, E_2, E_3, \ldots, E_n$ be n mutually exclusive and exhaustive events and E is an event which occurs together with $E_1, E_2, \ldots, E_n$, that is,

$(E_1, E_2, \ldots, E_n$ form a partition of the sample space S and E be any event$)$, then

$$P\left(\frac{E_K}{E}\right) = \frac{P(E_K)P\left(\frac{E}{E_K}\right)}{P(E_1)P\left(\frac{E}{E_1}\right) + P(E_2)P\left(\frac{E}{E_2}\right) + \ldots P(E_n)P\left(\frac{E}{E_n}\right)}$$

Proof

$E_1, E_2, \ldots, E_n$ form a partition of S, therefore

(i) $E_1, E_2, \ldots, E_n$ are non-empty

(ii) $E_i \cap E_j = \phi$ for $i \neq j$

(iii) $S = E_1 \cup E_2 \cup \ldots \cup E_n$

Now, $\quad E = E \cap S = E \cap (E_1 \cup E_2 \cup \ldots \cup E_n)$

$$= (E \cap E_1) \cup (E \cap E_2) \cup \ldots (E \cap E_n) \quad (7.4)$$

Since, $E_1, E_2, \ldots E_n$ are disjoint sets

$\therefore\ E \cap E_1,\ E \cap E_2, \ldots, E \cap E_n$ are also disjoint

$\therefore$ from (i)

$$P(E) = P(E \cap E_1) + P(E \cap E_2) + \ldots + P(E \cap E_n) \quad (7.5)$$

Now, $\quad P\left(\frac{E_K}{E}\right) = \frac{P(E_K \cap E)}{P(E)}$

$$= \frac{P(E_K)P\left(\frac{E}{E_K}\right)}{P(E \cap E_1) + P(E \cap E_2) + \ldots + P(E \cap E_n)}$$

$$\Rightarrow P\left(\frac{E_K}{E}\right) = \frac{P(E_K)P\left(\frac{E}{E_K}\right)}{P(E_1)P\left(\frac{E}{E_1}\right) + P(E_2)P\left(\frac{E}{E_2}\right) + \ldots + P(E_n)P\left(\frac{E}{E_n}\right)}$$

Note:

If $P(E_1) = P(E_2) = \ldots = P(E_n)$, then by Bayes's theorem

$$P\left(\frac{E_K}{E}\right) = \frac{P\left(\frac{E}{E_K}\right)}{P\left(\frac{E}{E_1}\right) + P\left(\frac{E}{E_2}\right) + \ldots + P\left(\frac{E}{E_n}\right)}$$

7.6.1 Special Case of Bayes's Theorem

If $E_1, E_2, \ldots, E_n$ form a partition of an event E, then

$$P\left(\frac{E_K}{E}\right) = \frac{P(E_k)}{P(E_1) + P(E_2) + \ldots + P(E_n)}$$

Proof:

Since $E_1, E_2, \ldots, E_n$ form a partition of E, therefore

(i) $E_1, E_2, \ldots, E_n$ are non-empty

(ii) No two of $E_1, E_2, \ldots, E_n$ have any common element

(iii) $E = E_1 \cup E_2 \cup \ldots \cup E_n$

Clearly $E \cap E_1, E \cap E_2, \ldots, E \cap E_n$ are non-empty, pair-wise disjoint and

$$E = (E \cap E_1) \cup (E \cap E_2) \cup \ldots \cup (E \cap E_n)$$

Hence by the addition theorem of probability,

$$P(E) = P(E \cap E_1) + P(E \cap E_2) + \ldots P(E \cap E_n) \quad (7.6)$$

Now, $\quad P\left(\frac{E_K}{E}\right) = \frac{P(E_K \cap E)}{P(E)}$

$$= \frac{P(E_K \cap E)}{P(E \cap E_1) + P(E \cap E_2) + \ldots P(E \cap E_n)}$$

$$= \frac{P(E_K)P\left(\frac{E}{E_K}\right)}{P(E_1)P\left(\frac{E}{E_1}\right) + P(E_2)P\left(\frac{E}{E_2}\right) + \ldots + P(E_n)P\left(\frac{E}{E_n}\right)}$$

$$= \frac{P(E_K)}{P(E_1) + P(E_2) + \ldots P(E_n)}$$

[since $E_1, E_2, \ldots, E_n$ are subsets of E

$\therefore P\left(\frac{E}{E_i}\right) = 1,\ i = 1, 2, \ldots, n$]

7.7 Binomial Theorem on Probability

Let E be an event and

Let, p = Probability of occurrence of event E in one trial

$q = 1 - p$ = Probability of non-occurrence of event E in one trial

That is, $p + q = 1$

Also let X = number of successes, that is, number of times event E occurs in n trials.

Then the probability of occurrence of event E exactly r times in n trials is denoted by $P(X = r)$ or $P(r)$ and is defined as $P(X = r)$ or $P(r) = {}^nC_r p^r q^{n-r}$

$= (r+1)^{\text{th}}$ term in the expansion of $(q+p)^n$

Proof

Probability of occurrence of event E in one trial $= p$

Probability of non-occurrence of event E in one trial $= q = 1 - p$

Event E occurs exactly r times in n trials

$\Rightarrow$ It does not occur $(n-r)$ times in n trials.

Now, r trials in which event E occurs can be selected out of n trials in ${}^nC_r$ ways.

Also, the n trials are independent.

Therefore, the probability of occurrence of event E exactly r times out of n trials is given by $P(X = r) = {}^nC_r\,(p.p \ldots r \text{ times})(qq \ldots (n-r) \text{ times})$

$= {}^nC_r p^r q^{n-r}$

7.7.1 Random Variable

A *random variable* is a real valued function whose domain is the sample space of a random experiment.

7.8 Probability Distribution of a Random Variable

If the values of a random variable together with the corresponding probabilities are given, then this description is called the *probability distribution of a random variable*.

Example:

Let X denote the number of heads occurred when two coins are tossed.

$P(X = 0)$ = Probability of occurrence of zero heads

$$= P(TT) = \frac{1}{2} \cdot \frac{1}{2} - \frac{1}{4}$$

$P(X = 1)$ = Probability of occurrence of one head

$$= P(TH) + P(HT) = \frac{1}{2} \cdot \frac{1}{2} + \frac{1}{2} \cdot \frac{1}{2} = \frac{1}{2}$$

$P(X = 2)$ = Probability of occurrence of two heads

$$= P(HH) = \frac{1}{2} \cdot \frac{1}{2} = \frac{1}{4}$$

Thus, the probability distribution is

X	0	1	2
P(X)	$\frac{1}{4}$	$\frac{1}{2}$	$\frac{1}{4}$

That is, $X : (0, 1, 2)$

$$P(X) : \left(\frac{1}{4}, \frac{1}{2}, \frac{1}{4}\right)$$

7.8.1 Binomial Distribution

Let p = Probability of occurrence of event E in one trial and $q = 1 - p$ = Probability of non-occurrence of event E in one trial.

Clearly $p + q = 1$

Let X denote the number of successes in n trials.

Then $P(X = r)$ = Probability of r successes

$= {}^nC_r p^r q^{n-r}$

Therefore, the probability distribution of the random variable X is as given below.

X	0	1	2	…	r	…	n
P(X)	q^n	${}^nC_1 pq^{n-1}$	${}^nC_2 p^2 q^{n-2}$	…	${}^nC_r p^r q^{n-r}$	…	p^n

Note:

(i) Mean of the binomial distribution $= np$

(ii) Variance of the binomial distribution $= npq$

SOLVED EXAMPLES

Example 7.1 If three coins are tossed randomly then find the probability of getting at least one head.

Solution Since,

$$S = \{(H, H, H), (H, H, T), (H, T, H), (T, H, H), (H, T, T), (T, H, T), (T, T, H), (T, T, T)\}$$

$\therefore \quad n(S)=2^3=8$

Let E denote the event of getting at least one head

$$\therefore E=\{(H,H,H),(H,H,T),(H,T,H),(T,H,H),(H,T,T),(T,H,T),(T,T,H)\}$$

$\therefore \quad n(E)=7$

$\therefore \quad P(E)=\dfrac{n(E)}{n(S)}=\dfrac{7}{8}$

Example 7.2 If three fair and unbiased dice are rolled on the Ludo board at once, find the probability that the sum of the numbers is 10.

Solution Since, $n(S)=6^3=216$

If E denotes the event that the sum of all three dice appeared is 10, then

$$E \to (1,3,6) \to 3! = 6$$
$$(1,4,5) \to 3! = 6$$
$$(2,2,6) \to \frac{3!}{2!} = 3$$
$$(2,3,5) \to 3! = 6$$
$$(2,4,4) \to \frac{3!}{2!} = 3$$
$$(3,3,4) \to \frac{3!}{2!} = 3$$

$\therefore \quad n(E)=27$

$\therefore \quad P(E)=\dfrac{n(E)}{n(S)}=\dfrac{27}{216}=\dfrac{1}{8}$

Example 7.3 A special die with numbers 1, –1, 2, –2, 0 and 3 is thrown thrice. What is the probability that the total is (i) zero (ii) 6?

Solution Clearly $n(S)=216$

(i) If E denotes the event that the sum of the upper faced value is zero, then

$$E \to 1,-1,0 \to 3! = 6$$
$$2,-2,0 \to 3! = 6$$
$$-1,-2,3 \to 3! = 6$$
$$1,1,-2 \to \frac{3!}{2!} = 3$$
$$-1,-1,2 \to \frac{3!}{2!} = 3$$
$$0,0,0 \to 1$$

$\therefore n(E)=25$

$\therefore P(E)=\dfrac{25}{216}$

(ii) If E_1 denotes the event that the sum of upper faced value = 6, then

$$E_1 \to 1,2,3, \to 3! = 6$$
$$0,3,3 \to \frac{3!}{2!} = 3$$
$$2,2,2 \to 1$$

$\therefore P(E_1)=\dfrac{10}{216}=\dfrac{5}{108}$

Example 7.4 What are the odds in favour of throwing at least 8 in a single throw with two dice?

Solution Let S be the sample space and E the event of occurrence of a total of at least 8, then

$$S=\{1,2,3,4,5,6\}\times\{1,2,3,4,5,6\}$$

$$\therefore n(S)=36$$

Throwing at least 8 with two dice means throwing 8 or 9 or 10 or 11 or 12.

$$\therefore E=\begin{Bmatrix}(6,2)(5,3),(4,4),(3,5),(2,6),(6,3),\\(5,4),(4,5),(3,6),(6,4),(5,5),(4,6),\\(6,5),(5,6),(6,6)\end{Bmatrix}$$

$\therefore \quad n(E)=15$

$\therefore \quad n(E')=36-15=21$

$\therefore$ odds in favour of $E=\dfrac{n(E)}{n(E')}=\dfrac{15}{21}=\dfrac{5}{7}$

Example 7.5 In a hand at bridge what is the chance that the 4 kings are held by a specified player?

Solution We know that in a bridge game, 13 cards distributed to each of the 4 players.

Let S = The sample space

and E = The event that the specified player gets 4 kings and 9 other cards out of remaining 48 cards.

$$\therefore \quad n(S) = {}^{52}C_{13}$$

$$\text{and} \quad n(E) = {}^{4}C_4 \times {}^{48}C_9 = {}^{48}C_9$$

$$\therefore \quad P(E) = \frac{n(E)}{n(S)} = \frac{{}^{48}C_9}{{}^{52}C_{13}}$$

$$= \frac{(48!)}{(9)!(39!)} \times \frac{(13)!(39)!}{(52)!} = \frac{11}{4165}$$

Example 7.6 What is the probability of getting 9 cards of the same suit in one hand in a game of bridge?

Solution Let S denote the sample space and E denote the event that a particular player can get 9 cards out of thirteen of one suit.

$$\therefore \quad n(E) = {}^{13}C_9 \times {}^{39}C_4 \times 4$$

[There are four suits in a set of cards]

$$\text{and} \quad n(S) = {}^{52}C_{13}$$

$$\therefore \quad P(E) = \frac{{}^{13}C_9 \times {}^{39}C_4 \times 4}{{}^{52}C_{13}}$$

Example 7.7 Two dice are thrown simultaneously; what is the probability obtaining a total score less than 11?

Solution Since, $n(S) = 36$

Let E denote the event that total score is less than 11, then E' be the event of obtaining a total score greater than or equal to 11.

$$\therefore \quad E' = \{(5,6),(6,5),(6,6)\}$$

$$\therefore \quad n(E') = 3$$

$$\therefore \quad P(E') = \frac{n(E')}{n(S)} = \frac{3}{36} = \frac{1}{12}$$

$$P(E) = 1 - P(E')$$

$$= 1 - \frac{1}{12} = \frac{11}{12}$$

Example 7.8 Two cards are drawn at random from a pack of cards. Find the probability that both the cards are of red colour or they are Queens.

Solution Let S be the sample space.

E_1 = The event that the two cards drawn are red

E_2 = The event that the two cards drawn are Queens

Number of red cards = 26

and number of Queens = 4

Now, $n(S) = {}^{52}C_2, n(E_1) = {}^{26}C_2, n(E_2) = {}^{4}C_2$

$E_1 \cup E_2$ = The event that the two cards drawn are both red or both Queens

$E_1 \cap E_2$ = The event that the two cards drawn are Queens of red colour.

Required probability

$$P(E_1 \cup E_2) = P(E_1) + P(E_2) - P(E_1 \cap E_2)$$

$$= \frac{{}^{26}C_2}{{}^{52}C_2} + \frac{{}^{4}C_2}{{}^{52}C_2} - \frac{{}^{2}C_2}{{}^{52}C_2}$$

$$= \frac{26 \times 25}{52 \times 51} + \frac{4 \times 3}{52 \times 51} - \frac{1 \times 2}{52 \times 51}$$

$$= \frac{285}{2652}$$

Example 7.9 If A and B are two candidates seeking admission in IIT, the probability that A is selected is 0.5 and the probability that both A and B are selected is at most 0.3. Is it possible that the probability of B getting selected is 0.9.

Solution Let E_1 denote the event that A is selected and E_2 denote the event that B is selected.

It is given that

$$P(E_1) = 0.5, P(E_1 \cap E_2) \le 0.3$$

Since, by addition theorem of probability

$$P(E_1 \cup E_2) = P(E_1) + P(E_2) - P(E_1 \cap E_2)$$

$$\ge 0.5 + P(E_2) - 0.3$$

[Maximum value of $P(E_1 \cap E_2) = 0.3$]

$$\Rightarrow P(E_1 \cup E_2) \ge 0.2 + P(E_2)$$

$$\therefore P(E_2) \le P(E_1 \cup E_2) - 0.2$$

$$\Rightarrow P(E_2) \le 1 - 0.2 \qquad [P(E_1 \cup E_2) \le 1]$$

$$\Rightarrow P(E_2) \le 0.8$$

Hence $P(E_2) \ne 0.9$

Example 7.10 A bag contains 5 red and 8 black balls and another bag contains 3 red and 6 black balls. A ball is drawn from each bag. What is the probability that one is red and the other is black?

Solution Let E_1 denote the event of drawing a red ball from Bag I and E_2 denote the event of drawing a black ball from Bag II, E_3 denote the drawing of a black ball from Bag I and E_4 denote the drawing of a red ball from Bag II.

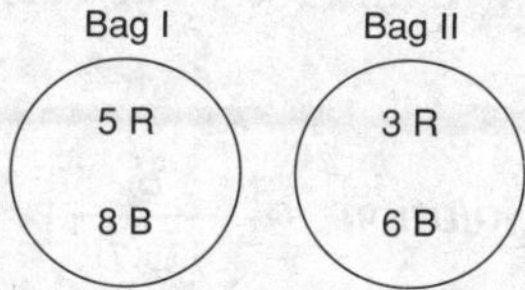

Let $A = E_1 \cap E_2$ and $B = E_3 \cap E_4$

$$\therefore \quad P(A) = P(E_1 \cap E_2) = P(E_1) \cdot P(E_2)$$

$[E_1 \text{ and } E_2 \text{ are independent}]$

$$= \frac{5}{13} \cdot \frac{6}{9} = \frac{10}{39}$$

Also, $P(B) = P(E_3 \cap E_4) = P(E_3) \cdot P(E_4)$

$[E_3 \text{ and } E_4 \text{ are independent}]$

$$= \frac{8}{13} \cdot \frac{3}{9} = \frac{8}{39}$$

Since, $P(A \cup B) = P(A) + P(B)$

[A and B are mutually exclusive]

$$= \frac{10}{39} + \frac{8}{39} = \frac{6}{13}$$

Example 7.11 A speaks the truth in 75% cases and B in 80% of the cases; in what percentage of cases are they likely to contradict each other in stating the same fact?

Solution Let E_1 denote the event that A speaks the truth, E_2 that B speaks the truth, E_1' denote that A speaks a lie and E_2' that B speaks a lie.

Let, $C = E_1 \cap E_2'$ = the event that A speaks the truth and B speaks a lie and $D = E_1' \cap E_2$ = the event that A speaks a lie and B speaks the truth.

Let E denote the event that A and B contradict in stating the same fact.

Then, $E = C \cup D$

Since A and B are independent $\Rightarrow E_1$ and E_2 are independent, E_1 and E_2' are independent, E_1' and E_2 are independent.

Now, $P(E_1) = \frac{75}{100} = \frac{3}{4}, P(E_2) = \frac{80}{100} = \frac{4}{5}$

$$P(E_1') = 1 - \frac{3}{4} = \frac{1}{4}, P(E_2')1 - \frac{4}{5} = \frac{1}{5}$$

$$P(C) = P(E_1 \cap E_2') = P(E_1)P(E_2')$$

$$= \frac{3}{4} \cdot \frac{1}{5} = \frac{3}{20}$$

$$P(D) = P(E_1' \cap E_2) = P(E_1')P(E_2)$$

$$= \frac{1}{4} \cdot \frac{4}{5} = \frac{4}{20}$$

Since C and D are mutually exclusive events, therefore

$$P(E) = P(C \cup D) = P(C) + P(D) = \frac{3}{20} + \frac{4}{20} = \frac{7}{20}$$

Example 7.12 A student appears for test I, II and III. The student is successful if he passes either in tests I and II or tests I and III. The probability of the student passing in test I, II, III are p, q and $\frac{1}{2}$ respectively. If the probability that the student is successful is $\frac{1}{2}$, then find the relation between p and q.

Solution Let E_1, E_2 and E_3 denote the events that the student is successful in test I, II and III respectively, then

P (The student is successful)

$$= P(E_1 \cap E_2 \cap E_3') + P(E_1 \cap E_2' \cap E_3) + P(E_1 \cap E_2 \cap E_3)$$

$$= P(E_1)P(E_2)P(E_3') + P(E_1)P(E_2')P(E_3) + P(E_1)P(E_2)P(E_3)$$

$[E_1, E_2, E_3 \text{ are independent}]$

$$= pq\left(1 - \frac{1}{2}\right) + P(1-q)\frac{1}{2} + pq\frac{1}{2}$$

$$\Rightarrow \frac{1}{2} = pq + \frac{p}{2} - \frac{pq}{2} \Rightarrow \frac{1}{2} = \frac{1}{2}(pq + p) \Rightarrow pq + p = 1$$

Example 7.13 The probability that A, B, C solve a problem is $\frac{1}{2}, \frac{1}{3}$ and $\frac{1}{4}$. If the probability that the problem will be solved is λ and that the problem is solved by only one of them is μ, then find $\lambda + \mu$.

Solution $\lambda = 1-$ Problem will not be solved

$$= 1 - P(A' \cap B' \cap C')$$

$$= 1 - P(A')P(B')P(C')$$

$$= 1 - \left(1-\frac{1}{2}\right)\left(1-\frac{1}{3}\right)\left(1-\frac{1}{4}\right)$$

$$= 1 - \frac{1}{2} \times \frac{2}{3} \times \frac{3}{4} = 1 - \frac{1}{4} = \frac{3}{4}$$

$$\mu = P(A \cap B' \cap C') + P(A' \cap B \cap C') + P(A' \cap B' \cap C)$$

$$= \frac{1}{2}\left(1-\frac{1}{3}\right)\left(1-\frac{1}{4}\right) + \left(1-\frac{1}{2}\right)\frac{1}{3}\left(1-\frac{1}{4}\right) + \left(1-\frac{1}{2}\right)\left(1-\frac{1}{3}\right)\frac{1}{4}$$

$$= \frac{1}{2} \times \frac{2}{3} \times \frac{3}{4} + \frac{1}{2} \times \frac{1}{3} \times \frac{3}{4} + \frac{1}{2} \times \frac{2}{3} \times \frac{1}{4}$$

$$= \frac{1}{4} + \frac{1}{8} + \frac{1}{12} = \frac{11}{24}$$

$$\therefore \qquad \lambda + \mu = \frac{29}{24}$$

Example 7.14 A letter is selected from each of the words MATHEMATICS and STATISTICS. Find the probability that the same letters are selected.

Solution The number of ways of choosing a letter from the word MATHEMATICS is 11 and from STATISTICS is 10.

A T I C S (same letter)

MATHEMATICS $(2A, 2T, 1T, 1C, 1S)$

STATISTICS $(1A, 3T, 2I, 1C, 3S)$

Hence the probability of selecting the same letter is

$$\frac{2}{11} \times \frac{1}{10} + \frac{2}{11} \times \frac{3}{10} + \frac{1}{11} \times \frac{2}{10} + \frac{1}{11} \times \frac{1}{10} + \frac{1}{11} \times \frac{3}{10}$$

$$= \frac{14}{110} = \frac{7}{55}$$

Example 7.15 If two events A and B are such that $P(A') = 0.3, P(B) = 0.4$ and $P(A \cap B') = 0.5$, then find the value of $P\left(\frac{B}{(A \cup B')}\right)$.

Solution $P\left(\frac{B}{A \cup B'}\right) = \frac{P(B \cap (A \cup B'))}{P(A \cup B')}$

$$= \frac{P(A \cap B)}{P(A) + P(B') - P(A \cap B')}$$

$$= \frac{P(A) - P(A \cap B')}{P(A) + P(B') - P(A \cap B')}$$

$$= \frac{0.7 - 0.5}{0.7 + 0.6 - 0.5} = \frac{0.2}{0.8} = \frac{1}{4}$$

Example 7.16 The first twelve letters of the English alphabet are written down at random. What is the probability that there are exactly 4 letters between A and B?

Solution Here, $n(S) = 12!$

If A is first place then B will be placed in 6th place.

Similarly A can be placed at 2nd, 3rd, 4th, 5th, 6th and 7th place, then B must be placed at 7th, 8th, 9th, 10th, 11th and 12th place respectively.

Also, A and B can be interchanged by 2! ways. Therefore, A and B can be placed at the required place by $2! \times 7 = 14$ ways.

The remaining 10 letters can be arranged by 10! ways.

Thus, required probability $= \frac{14 \times 10!}{12!} = \frac{7}{66}$

Example 7.17 Out of 21 tickets consecutively numbered, three are drawn at random. Find the probability that the numbers on them are in A.P.

Solution Since, 3 tickets out of 21 tickets can be selected by ${}^{21}C_3$ ways.

$$\therefore \qquad n(S) = {}^{21}C_3 = \frac{21!}{3!18!} = \frac{19 \cdot 20 \cdot 21}{2 \times 3}$$

$$= 1330$$

Now the three tickets will be in A.P. if we choose two numbers from 1 to 21, which are either both even or both odd. This can be done in ${}^{11}C_2 + {}^{10}C_2$ ways.

Hence, required probability $= \dfrac{{}^{11}C_2 + {}^{10}C_2}{1330}$

$$= \frac{\frac{11!}{2!9!} + \frac{10!}{2!8!}}{1330} = \frac{\frac{10\times 11}{2} + \frac{9\times 10}{2}}{1330}$$

$$= \frac{55+45}{1330} = \frac{100}{1330} = \frac{10}{133}$$

Example 7.18 A bag contains 5 black and 3 red balls. A ball is taken out of the bag and is not returned to it. Then,

(i) What is the probability of drawing a red ball in the second draw if the ball in the first draw is red?

(ii) What is the probability of drawing a black ball in the second draw if the ball in the first draw is red?

Solution Let A denote the event of drawing a red ball in the first draw, B denote the event of drawing a red ball in second draw and C denote the event of drawing a black ball in the second draw.

Since $A \cap B$ = The event of drawing a red ball in both draws

$\therefore \quad n(A \cap B) = {}^3C_1 \times {}^2C_1 = 6$

$n(A)$ = Number of ways of drawing red ball in both the draws + number of ways of drawing a red ball in the first draw and a black ball in the second draw.

$$= {}^3C_1 \cdot {}^2C_1 + {}^3C_1 \cdot {}^5C_1 = 21$$

(i) $P\left(\dfrac{B}{A}\right) = \dfrac{n(A \cap B)}{n(A)} = \dfrac{6}{21} = \dfrac{2}{7}$

(ii) $P\left(\dfrac{C}{A}\right) = \dfrac{n(A \cap C)}{n(A)} = \dfrac{{}^3C_1 \cdot {}^5C_1}{21} = \dfrac{15}{21} = \dfrac{5}{7}$

Example 7.19 The probability of event A is $\dfrac{3}{4}$. The probability of event B, given that event A occurs is $\dfrac{1}{4}$. The probability of event A, given that event B occurs is $\dfrac{2}{3}$, then find the probability that neither event occurs.

Solution Since, $P(A) = \dfrac{3}{4}$

$$P\left(\frac{B}{A}\right) = \frac{1}{4}$$

$$P\left(\frac{A}{B}\right) = \frac{2}{3}$$

Now, $\quad P\left(\dfrac{B}{A}\right) = \dfrac{P(B \cap A)}{P(A)}$

$$\Rightarrow \frac{1}{4} = \frac{P(B \cap A)}{\frac{3}{4}} \Rightarrow P(B \cap A) = \frac{1}{4} \times \frac{3}{4} = \frac{3}{16}$$

Also, $\quad P\left(\dfrac{A}{B}\right) = \dfrac{P(A \cap B)}{P(B)}$

$$\Rightarrow P(B) = \frac{\frac{3}{16}}{\frac{2}{3}} = \frac{3}{16} \times \frac{3}{2} = \frac{9}{32}$$

Now, $\quad P(A \cup B) = P(A) + P(B) - P(A \cap B)$

$$= \frac{3}{4} + \frac{9}{32} - \frac{3}{16} = \frac{27}{32}$$

$$\therefore \quad P(A' \cap B') = 1 - P(A \cup B) = 1 - \frac{27}{32} = \frac{5}{32}$$

Example 7.20 A bag X contains 3 white and 2 black balls; another bag Y contains 2 white and 4 black balls. A bag and a ball out of it is picked at random. What is the probability that the ball is white?

Solution Let E_1 denote the event of selecting bag X, E_2 denote the event of selecting bag Y and E the event of drawing a white ball.

Let $A = E_1 \cap E$ and $B = E_2 \cap E$

Then, $P(E) = P(A \cup B)$

$$= P(A) + P(B)$$

[Since A and B are mutually exclusive]

$$= P(E_1 \cap E) + P(E_2 \cap E)$$

$$= P(E_1)P\left(\frac{E}{E_1}\right) + P(E_2)P\left(\frac{E}{E_2}\right)$$

$$= \frac{1}{2} \cdot \frac{3}{5} + \frac{1}{2} \cdot \frac{2}{6} = \frac{7}{15}$$

Example 7.21 A man is known to speak the truth 3 out of 4 times. He throws a die and reports that it is a six. Find the probability that it is actually a six.

Solution Suppose that E denote the event that man reports occurrence of 6.

E_1 denote the event of occurrence of 6 when a die is thrown and E_2 denote the event of non-occurrence of 6 when a die is thrown.

$P\left(\dfrac{E}{E_1}\right)$ = Probability that man reports 6 when 6 has actually occurred

= Probability that the man speaks the truth $= \dfrac{3}{4}$

$P\left(\dfrac{E}{E_2}\right)$ = Probability that man reports 6 when 6 has not actually occurred

= Probability that the man tells the lie $= \dfrac{1}{4}$

Now using Bayes theorem

$$P\left(\frac{E_1}{E}\right) = \frac{P(E_1)P\left(\frac{E}{E_1}\right)}{P(E_1)P\left(\frac{E}{E_1}\right) + P(E_2)P\left(\frac{E}{E_2}\right)}$$

$$= \frac{\frac{1}{6}\cdot\frac{3}{4}}{\frac{1}{6}\cdot\frac{3}{4} + \frac{5}{6}\cdot\frac{1}{4}} = \frac{3}{8}$$

Example 7.22 Bag X contains 2 white and 3 red balls and bag Y contains 4 white and 5 red balls. One ball is drawn at random from one of the bags and is found to be red. Find the probability that is was from bag Y.

Solution Let E denote the event that one ball drawn is red.

Also let E_1 denote the event that bag Y is selected and E_2 denote the event that bag X is selected.

Now, $P\left(\dfrac{E}{E_1}\right)$ = Probability of drawing a red ball when one ball is drawn from bag $Y = \dfrac{5}{9}$ and $P\left(\dfrac{E}{E_2}\right) = \dfrac{3}{5}$

$$P(E_1) = P(E_2) = \frac{1}{2}$$

Now as it is known that the ball drawn is red, using Bayes's theorem

$$P\left(\frac{E_1}{E}\right) = \frac{P(E_1)P\left(\frac{E}{E_1}\right)}{P(E_1)P\left(\frac{E}{E_1}\right) + P(E_2)P\left(\frac{E}{E_2}\right)}$$

$$= \frac{\frac{1}{2}\cdot\frac{5}{9}}{\frac{1}{2}\cdot\frac{5}{9} + \frac{1}{2}\cdot\frac{3}{5}} = \frac{25}{52}$$

Example 7.23 A pack of cards is counted keeping them face downwards. It is found that one card is missing. One card is drawn and is found to be red. Find the probability that the missing card is red.

Solution Let E denote the event of drawing a red card when one card is drawn out of 51 cards (excluding the missing card). Let E_1 denote the event that the missing card is red and E_2 denote the event that the missing card is black.

Now, $$P(E_1) = \frac{^{26}C_1}{^{52}C_1} = \frac{26}{52} = \frac{1}{2}$$

$$P(E_2) = \frac{1}{2}$$

$P\left(\dfrac{E}{E_1}\right)$ = Probability of drawing a red card when the missing card is red $= \dfrac{25}{51}$

$P\left(\dfrac{E}{E_2}\right)$ = Probability of drawing a red card when the missing card is black $= \dfrac{26}{51}$

Now using Bayes theorem

$$P\left(\frac{E_1}{E}\right) = \frac{P(E_1)P\left(\frac{E}{E_1}\right)}{P(E_1)P\left(\frac{E}{E_1}\right) + P(E_2)P\left(\frac{E}{E_2}\right)}$$

$$= \frac{\frac{1}{2}\cdot\frac{25}{51}}{\frac{1}{2}\cdot\frac{25}{51} + \frac{1}{2}\cdot\frac{26}{51}} = \frac{25}{51}$$

Example 7.24 Rahul and Akhilesh are weak students in Mathematics and their chances of solving a problem correctly are $\dfrac{1}{8}$ and $\dfrac{1}{12}$ respectively. They are given a question and they obtain the same answer. If the probability of a common mistake is $\dfrac{1}{1001}$. What is the probability that the answer was correct?

Solution The probability that both Rahul and Akhilesh get the correct answer is $\frac{1}{8}\times\frac{1}{12}=\frac{1}{96}$ and the probability that both get the wrong answer is $\left(1-\frac{1}{8}\right)\left(1-\frac{1}{12}\right)=\frac{77}{96}$

Let E_1 denote the event of both getting the correct answer and E_2 denote the event of both getting the same answer. Let E be the event of both getting the same answer.

We have $\quad P(E_1)=\frac{1}{96}, P(E_2)=\frac{77}{96}$

$$P\left(\frac{E}{E_1}\right)=1 \text{ and } P\left(\frac{E}{E_2}\right)=\frac{1}{1001}$$

Since, $\quad P(E_1)P\left(\frac{E}{E_1}\right)=P(E)P\left(\frac{E_1}{E}\right)$

Now, $\quad P\left(\frac{E_1}{E}\right)=\dfrac{P(E_1)P\left(\frac{E}{E_1}\right)}{P(E_1)P\left(\frac{E}{E_1}\right)+P(E_2)P\left(\frac{E}{E_2}\right)}$

$$=\frac{\frac{1}{96}\times 1}{\frac{1}{96}\times 1+\frac{77}{96}\times\frac{1}{1001}}=\frac{13}{14}$$

Example 7.25 An urn contains five balls. Two balls are drawn and are found to be white. What is the probability that all the balls are white?

Solution Let E denote the event of drawing two white balls when two balls are drawn out of 5 balls in the urn, E_1 denote the event that the urn contains 5 white balls, E_2 denote the event that the urn contains 4 white balls, E_3 denote the event that the urn contains 3 white balls, and E_4 denote the event that urn contains 2 white balls.

$$P(E_1)=P(E_2)=P(E_3)=P(E_4)=\frac{1}{4}$$

$P\left(\frac{E}{E_1}\right)$ = drawing 2 white balls when urn contains 5 white balls and 2 balls are drawn from the urn $=\frac{{}^5C_2}{{}^5C_2}=1$

$P\left(\frac{E}{E_2}\right)$ = drawing 2 white balls when the urn contains 4 white balls and 2 balls are drawn from the urn $=\frac{{}^4C_2}{{}^5C_2}=\frac{6}{10}=\frac{3}{5}$

$$P\left(\frac{E}{E_3}\right)=\frac{{}^3C_2}{{}^5C_2}=\frac{3}{10} \text{ and } P\left(\frac{E}{E_4}\right)=\frac{{}^2C_2}{{}^5C_2}=\frac{1}{10}$$

$$P\left(\frac{E_1}{E}\right)=\frac{P(E_1)P\left(\frac{E}{E_1}\right)}{P(E_1)P\left(\frac{E}{E_1}\right)+P(E_2)P\left(\frac{E}{E_2}\right)+P(E_3)P\left(\frac{E}{E_3}\right)+P(E_4)P\left(\frac{E}{E_4}\right)}$$

$$=\frac{\frac{1}{4}.1}{\frac{1}{4}\cdot 1+\frac{1}{4}\cdot\frac{3}{5}+\frac{1}{4}\cdot\frac{3}{10}+\frac{1}{4}\cdot\frac{1}{10}}=\frac{1}{2}$$

Example 7.26 A man takes a step forward with a probability 0.4 and a step backward with a probability 0.6. Find the probability that at the end of eleven steps, he is one step away from the starting point.

Solution Suppose p denotes the probability that the man takes a step forward and q denotes that he takes a step backward.

$\therefore \quad p=0.4$ and $q=0.6$

Clearly $p+q=1$

Let E_1 denote the event that out of 11 steps the man takes 5 steps forward and 6 steps backward, E_2 denote the event that out of 11 steps the man takes 5 steps forward and 6 steps backward and E denote the event that at the end of 11 steps the man is one step away from the starting point.

Clearly $E=E_1\cup E_2$ and E_1 and E_2 are mutually exclusive.

$$\therefore \quad P(E)=P(E_1)+P(E_2)$$

$$={}^{11}C_6p^6q^5+{}^{11}C_5p^5q^6$$

$$={}^{11}C_5(0.4)^6(0.6)^5+{}^{11}C_5(0.4)^5(0.6)^6$$

$$\left[{}^{11}C_6={}^{11}C_5\right]$$

$= {}^{11}C_5\,(0.4)^5\,(0.6)^5\,(0.4+0.6)$

$= {}^{11}C_5\,(0.24)^5$

Example 7.27 Find the minimum number of tosses of a pair of dice so that the probability of getting the sum of the digits on the dice is equal to 7 and at least one toss is greater than 0.95

$$\left(\log_{10} 2 = 0.3010, \log_{10}^{3} = 0.4771\right).$$

Solution Let S be the sample space and E denote the event of occurrences of a total of 7 when a pair of dice is tossed.

Clearly $n(S) = 36$

and $E = \{(1,6),(2,5),(3,4),(4,3),(5,2),(6,1)\}$

$\Rightarrow n(E) = 6$

$\therefore P(E) = \frac{6}{36} = \frac{1}{6}$

Let the occurrence of event E be called a success and X denote the number of successes.

$\therefore \quad p = \frac{1}{6}$ and $q = 1 - \frac{1}{6} = \frac{5}{6}$

Also let n denote the number of tosses.

Probability of at least one success

$= 1 -$ Probability of no success $= 1 - P(x = 0)$

$$= 1 - {}^{n}C_0 p^0 q^n = 1 - \left(\frac{5}{6}\right)^n$$

From question

$$1 - \left(\frac{5}{6}\right)^n > 0.95$$

$$\Rightarrow \left(\frac{5}{6}\right)^n < (1 - 0.95) = 0.05 = \frac{1}{20}$$

Taking log on both sides on the base 10

$\therefore n[\log_{10} 5 - \log_{10} 6] < \log_{10} 1 - \log_{10} 20$

$\Rightarrow n[\log_{10} 10 - \log_{10} 2 - \log_{10} 2 - \log_{10} 3] < - \log_{10} 2 - \log_{10} 10$

$\Rightarrow n[1 - 2\log_{10} 2 - \log_{10} 3] < -1 - \log_{10} 2$

$\Rightarrow -0.791n < -1.3010$

$$\Rightarrow n > \frac{1.3010}{0.0791} = 16.44$$

Therefore, the least number of trials = 17.

Example 7.28 If A and B are independent events and C is an event such that exactly one of A or B occurs. Prove that

$$P(C) \ge P(A \cup B) \cdot P(A' \cap B').$$

Solution Since C is an event such that exactly one of A or B occurs, therefore,

$$P(C) = P(A \cap B') + P(A' \cap B)$$
$$= P(A) \cdot P(B') + P(A') \cdot P(B)$$

[Since A and B are independent $\Rightarrow A'$ and B are independent and A and B' are independent]

Now, $\quad P(A \cup B) \cdot P(A' \cap B')$

$= \{P(A) + P(B) - P(A \cap B)\}\, P(A')P(B')$

$\le [P(A) + P(B)]\, P(A')P(B')$

$= P(A)P(A')P(B') + P(B)P(A')P(B')$

$\le P(A)P(B') + P(B)P(A')$

$[\because P(A') \le 1$ and $P(B') \le 1]$

$= P(C)$

Thus, $\quad P(C) \ge P(A \cup B) \cdot P(A' \cap B')$

Example 7.29 Five ordinary dice are rolled at random and the sum of the numbers shown on them is 16. What is the probability that the numbers shown on each is any one from 2, 3, 4 or 5?

Solution Suppose that the integers x_1, x_2, x_3, x_4 and x_5 are shown on the dice, then

$$x_1 + x_2 + x_3 + x_4 + x_5 = 16$$

where $1 \le x_i \le 6\ (i = 1, 2, 3, 4, 5)$

The number of total solutions of this equation

$=$ coefficient of x^{16} in $\left(x^1 + x^2 + x^3 + x^4 + x^5 + x^6\right)^5$

$=$ coefficient of x^{16} in $x^5\left(1 + x + x^2 + x^3 + x^4 + x^5\right)^5$

$=$ coefficient of x^{11} in $\left(1 + x + x^2 + x^3 + x^4 + x^5\right)^5$

$=$ coefficient of x^{11} in $\left(\frac{1 - x^6}{1 - x}\right)^5$

$=$ coefficient of x^{11} in $\left(1 - x^6\right)^5 (1 - x)^{-5}$

$= \text{coefficient of } x^{11} \text{ in}$

$$(1-5x^6+\ldots)(1+{}^5C_1x+{}^6C_2x^2+\ldots+{}^9C_5x^5 +\ldots+{}^{15}C_{11}x^{11}+\ldots)$$

$$= {}^{15}C_{11}-5.{}^9C_5$$

$$= {}^{15}C_4-5.{}^9C_4 = 735$$

If S be the sample space, then $n(S) = 735$

Let E be the occurrence of the event, then $n(E)$ = the number of integral solutions of $x_1+x_2+x_3+x_4+x_5=16$ where (i = 1, 2, 3, 4, 5)

$= \text{coefficient of } x^{16} \text{ in } (x^2+x^3+x^4+x^5)^5$

$= \text{coefficient of } x^{16} \text{ in } x^{10}(1+x+x^2+x^3)^5$

$= \text{coefficient of } x^6 \text{ in } (1+x+x^2+x^3)^5$

$= \text{coefficient of } x^6 \text{ in } \left(\frac{1-x^4}{1-x}\right)^5$

$= \text{coefficient of } x^6 \text{ in}$

$$(1-5x^4+\ldots)(1+{}^5C_1x+{}^6C_2x^2+\ldots+{}^{10}C_6x^6+\ldots)$$

$$= {}^{10}C_6-5.{}^6C_2 = 135$$

∴ The required probability

$$P(E)=\frac{n(E)}{n(S)}=\frac{135}{735}=\frac{9}{49}$$

Example 7.30 If A and B are two events such that $P(A\cup B)'=\frac{1}{6}$, $P(A\cap B)=\frac{1}{4}$ and $P(A')=\frac{1}{4}$ where A' denotes the complement of event A, then prove that A and B are independent events but they are not equally likely.

Solution Since, $P(A\cup B)'=\frac{1}{6}$

$$\Rightarrow 1-P(A\cup B)=\frac{1}{6}$$

$$\Rightarrow 1-P(A)-P(B)+P(A\cap B)=\frac{1}{6}$$

$$\Rightarrow P(A')-P(B)+P(A\cap B)=\frac{1}{6}$$

$$\Rightarrow \frac{1}{4}-P(B)+\frac{1}{4}=\frac{1}{6}$$

$$\Rightarrow P(B)=\frac{1}{2}-\frac{1}{6}\Rightarrow P(B)=\frac{1}{3}$$

and $\qquad P(A)=1-P(A')$

$$=1-\frac{1}{4}=\frac{3}{4}$$

⇒ Events A and B are not equally likely.

Also, $\quad P(A)\cdot P(B)=\frac{3}{4}\cdot\frac{1}{3}=\frac{1}{4}=P(A\cap B)$

Hence A and B are independent events.

Example 7.31 Two cards are lost in a pack of 52 playing cards. If a card is drawn from the remaining cards, find the probability that it is a spade.

Solution Let E_1 denote the event that the lost cards are spade, E_2 that one spade and one non-spade, and E_3 that both are non-spade, then

$$P(E_1)=\frac{13}{52}\cdot\frac{12}{51},\quad P(E_2)=\frac{13}{52}\cdot\frac{39}{51}$$

$$P(E_3)=\frac{39}{52}\cdot\frac{38}{51}$$

If E denotes the event that the card drawn is a spade,

$$P\left(\frac{E}{E_1}\right)=\frac{11}{50}, P\left(\frac{E}{E_2}\right)=\frac{12}{50}, P\left(\frac{E}{E_3}\right)=\frac{13}{50}$$

Now,

$$P(E)=P(E_1)P\left(\frac{E}{E_1}\right)+P(E_2)P\left(\frac{E}{E_2}\right)+P(E_3)P\left(\frac{E}{E_3}\right)$$

$$=\frac{13}{52}\cdot\frac{12}{51}\cdot\frac{11}{50}+\frac{13}{52}\cdot\frac{39}{51}\cdot\frac{12}{50}\cdot 2+\frac{39}{52}\cdot\frac{38}{51}\cdot\frac{13}{50}$$

$$=\frac{1}{4}\left[\frac{12\cdot 11+39\cdot 24+39\cdot 38}{50\cdot 51}\right]=\frac{1}{4}\cdot\frac{2550}{2550}=\frac{1}{4}$$

Example 7.32 If A and B are independent events such that $P(A)>\frac{1}{2}, P(A\cap B')=\frac{3}{25}$ and $P(A'\cap B)=\frac{8}{25}$. Find $P(A)$ and $P(B)$.

Solution Let $P(A)=a, P(B)=b$

$$P(A\cap B')=\frac{3}{25}\Rightarrow P(A)P(B')=\frac{3}{25}$$

$$\Rightarrow a(1-b)=\frac{3}{25} \quad (1)$$

and $$P(A'\cap B)=\frac{8}{25}$$

$$\Rightarrow b(1-a)=\frac{8}{25} \quad (2)$$

(2) – (1), we get

$$b-a=\frac{1}{5}\Rightarrow b=a+\frac{1}{5}$$

Now, from (1)

$$a\left(1-a-\frac{1}{5}\right)=\frac{3}{25}$$

$$\Rightarrow a\left(\frac{4}{5}-a\right)=\frac{3}{25}$$

$$\Rightarrow \frac{4}{5}a-a^2=\frac{3}{25}$$

$$\Rightarrow a^2-\frac{4}{5}a+\frac{3}{25}=0$$

$$\Rightarrow a=\frac{3}{5},\frac{1}{5}$$

As $$a>\frac{1}{2}\Rightarrow a=\frac{3}{5}$$

$$b=\frac{3}{5}+\frac{1}{5}=\frac{4}{5}$$

$\therefore$ $$P(A)=\frac{3}{5}, P(B)=\frac{4}{5}$$

Example 7.33 A, B, C in order cut a pack of cards, replacing them after each cut, on condition that the first who cuts a spade shall win a prize, find their respective chances.

Solution There are 13 cards of spades in a complete pack of cards. Therefore, the chance that any of them will cut a spade is $\frac{13}{52}$, that is, $\frac{1}{4}$ and the chance that it is not a spade is $1-\frac{1}{4}=\frac{3}{4}$.

Now $A's$ respective chance of success

$$=\frac{1}{4}+\left(\frac{3}{4}\right)^3\times\frac{1}{4}+\left(\frac{3}{4}\right)^6\times\frac{1}{4}+\ldots\text{to }\infty$$

$$=\frac{1}{4}\cdot\frac{1}{1-\frac{27}{64}}=\frac{16}{37}$$

Similarly $B's$ respective chance of success

$$=\frac{3}{4}\times\frac{1}{4}+\left(\frac{3}{4}\right)^4\times\frac{1}{4}+\left(\frac{3}{4}\right)^7\times\frac{1}{4}+\ldots\text{to }\infty$$

$$=\frac{3}{16}\times\frac{1}{1-\frac{27}{64}}=\frac{12}{37}$$

and $C's$ respectively chance of success

$$=1-\frac{16}{37}-\frac{12}{37}=\frac{9}{37}$$

Hence their respective chances are

$$\frac{16}{37},\frac{12}{37},\frac{9}{37}$$

Example 7.34 If 10 things are distributed among three persons, find the probability that a particular person gets more than 5 things.

Solution Since 10 things can be distributed among three persons in 3^{10} ways.

The number of ways in which a man may have all 10 things is 1.

The number of ways in which he may have 9 things is ${}^{10}C_9\times 2=20$

Similarly, he may have 8 things in ${}^{10}C_8\times 2^2$ ways = 180

He may have 7 things in ${}^{10}C_7\times 2^3$ ways = 960 ways

and he may have 6 things in ${}^{10}C_6\times 2^4=3360$ ways

Therefore, the chance of a man having more than 5 things

$$=\frac{1+20+180+960+3360}{3^{10}}=\frac{4521}{3^{10}}$$

Example 7.35 A doctor is to visit a patient from the past experience, it is known that the probabilities that he will come by train, bus, scooter or by other means of transport are respectively $\frac{3}{10}, \frac{1}{5}, \frac{1}{10}$ and $\frac{2}{5}$. The probabilities that he will be late are $\frac{1}{4}, \frac{1}{3}$ and $\frac{1}{12}$, if he comes by train, bus and scooter respectively, but if he comes by other means of transport, then he will not be late. When he arrives, he is late. What is the probability that he comes by train?

Solution Let E_1, E_2, E_3, E_4 be the events that the doctor comes by train, bus, scooter and other means of transport respectively. Also let E be the event that the doctor visits the patient late.

Then, $P(E_1) = \frac{3}{10}, P(E_2) = \frac{1}{5}, P(E_3) = \frac{1}{10}$

and $\quad P(E_4) = \frac{2}{5}$

Now, $P\left(\frac{E}{E_1}\right)$ = Probability that the doctor arriving late comes by train $= \frac{1}{4}$

Similarly, $P\left(\frac{E}{E_2}\right) = \frac{1}{3}, P\left(\frac{E}{E_3}\right) = \frac{1}{12}$

and $P\left(\frac{E}{E_4}\right) = 0$, since he is not late if he comes by other means of transport.

Using Bayes' theorem

$P\left(\frac{E_1}{E}\right)$ = Probability that the doctor arriving late comes by train

$$= \frac{P(E_1)P\left(\frac{E}{E_1}\right)}{P(E_1)P\left(\frac{E}{E_1}\right) + P(E_2)P\left(\frac{E}{E_2}\right) + P(E_3)P\left(\frac{E}{E_3}\right) + P(E_4)P\left(\frac{E}{E_4}\right)}$$

$$= \frac{\frac{3}{10}\times\frac{1}{4}}{\frac{3}{10}\times\frac{1}{4}+\frac{1}{5}\times\frac{1}{3}+\frac{1}{10}\times\frac{1}{12}+\frac{2}{5}\times 0} = \frac{3}{40}\times\frac{120}{18} = \frac{1}{2}$$

7.9 Know Your Concepts

1. If a number with two digits is formed with the digits 2, 3, 5, 7, 9 without repetition of digits, what is the probability that the number formed is 35?
2. What is the probability that four *S's* come consecutively in the word 'MISSISSIPI'?
3. In a throw of two dice, find the chance of getting a total of 3 or 6 or 12.
4. 4 boys and 3 girls are seated in a row, find the odds in favour and odds against all the girls sitting together.
5. A bag contains 7 black and 4 white balls. Two balls are drawn simultaneously from the bag. Find the probability that one is a white ball and other a black ball.
6. *A* can solve 75% of the problems in a book on mathematics and *B* can solve 70%. What is the probability that either *A* or *B* can solve problem chosen at random?
7. Consider the sample space '*S*' representing the adults in a small town who have completed the requirements for a college degree. They have been categorised according to sex and employment as under:

	Employed	**Unemployed**
Male	460	40
Female	140	260

   An employed person is selected at random. Find the probability that chosen one is male.
8. A pair of fair dice is rolled till we get a sum of 4 or 8. What is the probability that 4 comes before 8?
9. A problem on mathematics is given to three students whose chances of solving it are $\frac{1}{2}, \frac{1}{3}$ and $\frac{1}{4}$ respectively. Find the probability that the problem will be solved.
10. There is a 30% chance that it rains on any particular day. What is the probability that there is at least one rainy day within a 7-days period?

11. A man throws a die and tosses a coin alternately, starting with the coin. Find the probability that he gets 5 or 6 on the die before he gets a head on the coin.

12. X and Y are two independent events. The probability that both X and Y occur is $\frac{1}{6}$ and the probability neither of them occurs is $\frac{1}{3}$. Find the probabilities of each of their occurrences, separately.

13. A word of 6 letters is formed from a set of 16 different letters of the English alphabet (with replacement). Find out the probability that exactly two letters are repeated.

14. If 5 balls are drawn at random from a bag containing 8 white and 4 red balls. What is the probability that two of them are red and 3 are white?

15. A university has to select an examiner from a list of 50 persons. 20 of them are women and 30 are men, 10 of them knowing Hindi and 40 not knowing Hindi, 15 of them are teachers and the remaining 35 are not teachers. What is the probability that the university selects a Hindi-knowing, woman teacher?

16. A and B stand in a ring with 10 other persons. If the arrangement of the persons is at random, find the chance that there are just 3 persons between A and B.

17. The sum of two natural numbers is 20. Find the chance that their product will be less than 50.

18. If $a \in [-20, 0]$, then find the probability that the graph of the function

$y = 16x^2 + 8(a+5)x - 7a - 5 = 0,$

is strictly above the x-axis.

19. If $\frac{1-3p}{2}, \frac{1+4p}{3}, \frac{1+p}{6}$ are the probabilities of three mutually exclusive and exhaustive events, then find the set of all values of p.

20. If a box contains 20 cards, 10 of these cards have the letter I printed on them. The other 10 have the letter T printed on them. If we choose 3 cards at random and keep them in the same order, find the probability of making the word IIT.

7.10 JEE Main: Sample Questions

Single Answer Type

1. A number with six digits is written down at random. The probability that the sum of digits of the number is even is

(A) $\frac{1}{2}$ (B) $\frac{3}{8}$

(C) $\frac{3}{7}$ (D) none of these

2. Two small squares on a chess board are chosen at random. The probability that they have a common side is,

(A) $\frac{1}{3}$ (B) $\frac{1}{9}$

(C) $\frac{1}{18}$ (D) none of these

3. A bag contains 17 tickets numbered 1 to 17. A ticket is drawn and replaced, then one more ticket is drawn and replaced. The probability that the first number drawn is even and the second is odd is

(A) $\frac{81}{289}$ (B) $\frac{72}{289}$

(C) $\frac{64}{289}$ (D) none of these

4. A and B play a game of tennis. The situation of the game is as follows; if one scores two consecutive points after a deuce he wins; if loss of a point is followed by win of a point, it is deuce. The chance of a server to win a point is $\frac{2}{3}$. The game is at deuce and A is serving. Probability that A will win the match is, (serves are changed after each game)

(A) $\frac{3}{5}$ (B) $\frac{2}{5}$ (C) $\frac{1}{2}$ (D) $\frac{4}{5}$

5. A die is thrown three times and the sum of three numbers obtained is 15. The probability of the first throw being 4 is

(A) $\frac{1}{18}$ (B) $\frac{1}{5}$ (C) $\frac{4}{5}$ (D) $\frac{17}{18}$

6. Fifteen coupons are numbered 1, 2, 3, ..., 15. Seven coupons are selected at random one at a time with replacement. The probability that the largest number appearing on the selected coupon is 9, is

(A) $\left(\frac{9}{16}\right)^6$ (B) $\left(\frac{8}{15}\right)^7$

(C) $\left(\frac{3}{5}\right)^7$ (D) none of these

7. A fair die is thrown until a score of less than 5 points is obtained. The probability of obtaining not less than 2 points on the last throw is

(A) $\frac{3}{4}$ (B) $\frac{5}{6}$ (C) $\frac{4}{5}$ (D) $\frac{1}{3}$

8. A fair coin is tossed a fixed number of times. If the probability of getting 7 heads is equal to getting 9 heads, then the probability of getting 2 heads is,

(A) $\frac{15}{2^8}$ (B) $\frac{2}{15}$

(C) $\frac{15}{2^{13}}$ (D) none of these

9. A fair die is tossed eight times. The probability that on the eighth throw a third six is observed is,

(A) ${}^8C_3\frac{5^5}{6^8}$ (B) $\frac{{}^7C_2.5^5}{6^8}$

(C) $\frac{{}^7C_2.5^5}{6^7}$ (D) none of these

10. If the papers of 4 students can be checked by any one of the 7 teachers, then the probability that all the 4 papers are checked by exactly 2 teachers is,

(A) $\frac{2}{7}$ (B) $\frac{32}{343}$

(C) $\frac{12}{49}$ (D) none of these

11. A and B each throw a die. The probability that A's throw is not greater than B's is

(A) $\frac{1}{6}$ (B) $\frac{5}{12}$ (C) $\frac{1}{2}$ (D) $\frac{7}{12}$

12. A speaks the truth in 70% cases and B speaks the truth in 80% cases. The probability that they will contradict each other in describing a single event is

(A) 0.38 (B) 0.56 (C) 0.62 (D) 0.94

13. If A and B are two events, the probability that exactly one of them occurs is given by

(A) $P(A) + P(B) - 2P(A \cap B)$

(B) $P(A \cap B') - P(A' \cap B)$

(C) $P(A \cup B) + P(A \cap B)$

(D) $P(A') + P(B') + 2P(A' \cap B')$

14. If A and B are independent events such that $P(A) = 0.3$, $P(A \cup \bar{B}) = 0.8$, then $P(B)$ equals to

(A) $\frac{1}{7}$ (B) $\frac{2}{7}$ (C) $\frac{3}{4}$ (D) $\frac{1}{4}$

15. For three events A, B and C, P(exactly one of the events A or B occurs) = P(exactly one of the events B or C occurs) = P(exactly one of the events C or A occurs) = p and P(all the three events occur simultaneously) $= p^2$, where $0 < p < \frac{1}{2}$. Then the probability of at least one of the three events A, B and C occurring is

(A) $\frac{3p+2p^2}{2}$ (B) $\frac{p+3p^2}{2}$

(C) $\frac{3p+p^2}{2}$ (D) $\frac{3p+2p^2}{4}$

16. A four-digit number is formed from numbers 1, 2, 3, 5 with no repetitions. The probability that the number is divisible by 5 is

(A) $\frac{3}{4}$ (B) $\frac{1}{4}$

(C) $\frac{1}{8}$ (D) none of these

17. Three natural numbers are taken at random from the set $A = \{x : 1 \le x \le 100, x \in N\}$. then the probability that the A.M. of the numbers taken is 25 will be

(A) $\frac{{}^{77}C_2}{{}^{100}C_3}$ (B) $\frac{{}^{25}C_2}{{}^{100}C_3}$

(C) $\frac{{}^{74}C_2}{{}^{100}C_3}$ (D) none of these

18. A bag contains 5 white and 3 black balls. Two balls are drawn at random. The probability that the drawn balls are of different colours, is equal to

(A) $\frac{2}{7}$ (B) $\frac{15}{28}$ (C) $\frac{3}{7}$ (D) $\frac{4}{7}$

19. The probability that at least one of the events A and B happens is $\frac{3}{5}$. Probability of their simultaneous happening is $\frac{1}{5}$. The value of $P(\bar{A}) + P(\bar{B})$ is equal to

(A) $\frac{2}{5}$ (B) $\frac{4}{5}$ (C) $\frac{6}{5}$ (D) $\frac{7}{5}$

20. One number is selected at random from the first two hundred positive integers. The probability that it is divisible by 6 or 8, is equal to

(A) $\frac{29}{100}$ (B) $\frac{33}{100}$ (C) $\frac{43}{100}$ (D) $\frac{1}{4}$

21. Two fair and ordinary dice are rolled simultaneously. The probability of getting the sum of outcomes of the dice as a multiple of 4, is equal to

(A) $\frac{5}{9}$ (B) $\frac{1}{4}$ (C) $\frac{1}{9}$ (D) $\frac{1}{3}$

22. The probabilities of occurrence of the events A and B are $\frac{1}{4}$ and $\frac{1}{2}$ respectively. If the probability of occurrence of A and B simultaneously is $\frac{7}{50}$, then probability of neither A nor B happening, is equal to

(A) $\frac{39}{100}$ (B) $\frac{1}{4}$

(C) $\frac{11}{100}$ (D) None of these

23. A five-digit number (having all different digits) is formed using the digits 1, 2, 3, 4, 5, 6, 7, 8 and 9. The probability that the formed number either begins or ends with an odd digit, is equal to

(A) $\frac{5}{6}$ (B) $\frac{1}{6}$ (C) $\frac{1}{3}$ (D) $\frac{2}{3}$

24. The probabilities of solving a problem by students A, B and C independently are $\frac{1}{2}, \frac{1}{3}$, and $\frac{1}{4}$ respectively. If they start solving the given problem. independently, then the probability that at least two of them will solve the problem successfully, is equal to

(A) $\frac{5}{24}$ (B) $\frac{9}{24}$ (C) $\frac{7}{24}$ (D) $\frac{11}{24}$

25. The probability that the persons P_1 and P_2 will die in a year are p and q respectively. The probability that at the end of the year only of them will be alive, is equal to

(A) $p + q - pq$ (B) $p + q - p^2 - q^2$

(C) $p + q - 2pq$ (D) $pq(p + q - pq)$

26. Two fair and ordinary dice are rolled simultaneously 4 times. The probability that both dice will show the same outcome exactly twice, is equal to

(A) $\frac{25}{216}$ (B) $\frac{25}{36}$ (C) $\frac{25}{108}$ (D) $\frac{25}{72}$

27. Three numbers are chosen at random without replacement from the set of integers $\{1, 2, 3, \ldots, 10\}$. The probability that the minimum of the chosen numbers is 3 or the maximum of the chosen numbers is 7, is equal to

(A) $\frac{23}{120}$ (B) $\frac{13}{120}$ (C) $\frac{13}{60}$ (D) $\frac{11}{40}$

28. Three integers are selected simultaneously from the set of integers $\{1, 2, 3, 4, \ldots, 50\}$. The probability that the selected numbers are consecutive, is equal to

(A) $\frac{9}{(25)(49)}$ (B) $\frac{6}{(25)(49)}$

(C) $\frac{3}{(25)(49)}$ (D) None of these

29. Two numbers are selected simultaneously from the set $\{6, 7, 8, 9, \ldots, 39\}$. If the sum of selected numbers is even then the probability that both the selected numbers are odd, is equal to

(A) $\frac{11}{51}$ (B) $\frac{40}{51}$ (C) $\frac{51}{91}$ (D) $\frac{40}{91}$

30. A bag contains 4 red, 5 white and 6 black balls. Three balls are selected from the bag simultaneously. The probability that one of the colours will be missing in the selected balls, is equal to

(A) $\frac{301}{455}$ (B) $\frac{366}{455}$

(C) $\frac{261}{455}$ (D) none of these

31. Three children are selected at random from a group of 6 boys and 4 girls. It is known that in this group exactly one girl and one boy belong to the same parent. The probability that the selected group of children have no blood relations, is equal to

(A) $\frac{1}{15}$ (B) $\frac{13}{15}$ (C) $\frac{14}{15}$ (D) $\frac{2}{15}$

32. In a certain city 40% of the people have brown hair, 25% have brown eyes and 15% have brown eyes as well as brown hair. If a person selected at random has brown hair, the probability that they have also brown eyes, is equal to

(A) $\frac{2}{5}$ (B) $\frac{1}{4}$ (C) $\frac{1}{2}$ (D) $\frac{3}{8}$

33. A pair of fair and ordinary dice is rolled simultaneously. It is found that they show different outcomes. The probability that the sum of the outcomes will be either 6 or 10, is equal to

(A) $\frac{1}{5}$ (B) $\frac{1}{6}$ (C) $\frac{1}{3}$ (D) $\frac{2}{3}$

34. In a bag there are 15 red and 5 white balls. Two balls are drawn in succession, without replacement. The first drawn ball is found to be red. The probability that second ball is also red, is equal to

(A) $\frac{3}{10}$ (B) $\frac{7}{10}$ (C) $\frac{5}{19}$ (D) $\frac{14}{19}$

35. Two fair dice are rolled simultaneously. One of the dice shows four. The probability of other die showing six, is equal to

(A) $\frac{2}{11}$ (B) $\frac{1}{18}$ (C) $\frac{1}{6}$ (D) $\frac{1}{36}$

36. Two fair dice are rolled simultaneously. It is found that one of them shows odd prime numbers. The probability that remaining die also shows an odd prime number, is equal to

(A) $\frac{1}{5}$ (B) $\frac{2}{5}$

(C) $\frac{3}{5}$ (D) none of these

37. Let 'A', 'B' and 'C' be three independent events with $P(A)=\frac{1}{3}$, $P(B)=\frac{1}{2}$ and $P(C)=\frac{1}{4}$. The probability of exactly 2 of these events occurring, is equal to

(A) $\frac{1}{4}$ (B) $\frac{7}{24}$ (C) $\frac{3}{4}$ (D) $\frac{17}{24}$

38. Three different dice are rolled simultaneously, three times. The probability that all of them show different numbers only two times, is equal to

(A) $\frac{1}{3}$ (B) $\frac{2}{3}$ (C) $\frac{100}{243}$ (D) $\frac{143}{243}$

39. Two subsets A and B of a set S consisting of 'n' elements are constructed randomly. The probability that $A \cap B = \phi$ and $A \cup B = S$ is equal to

(A) $1-\left(\frac{3}{4}\right)^n$ (B) $\left(\frac{3}{4}\right)^n$ (C) $\frac{1}{2^n}$ (D) $\frac{1}{3^n}$

40. A fair coin is tossed 'n' times. The probability that a head will turn up an even number of times, is equal to

(A) $\frac{n-1}{2n}$ (B) $\frac{1}{2}$

(C) $\frac{n+1}{2n}$ (D) $\frac{2^{n-1}-1}{2^n}$

41. A and B each throw a fair die. The probability that A's throw is not greater than B's throw, is equal to

(A) $\frac{1}{3}$ (B) $\frac{2}{3}$ (C) $\frac{7}{12}$ (D) $\frac{5}{12}$

42. A and B play a game where each is asked to select a number from 1 to 25. If the numbers selected by A and B match, both of them win a prize. The probability that they win their third prize on the 5th game is equal to

(A) $\frac{6.(24)^2}{(25)^5}$ (B) $\frac{6.(21)^2}{(25)^5}$

(C) $\frac{(24)^2}{(25)^2}$ (D) $\frac{(21)^2}{(25)^5}$

43. A and B are two events such that $P(A)=\frac{1}{2}$, $P(B)=\frac{2}{3}$. Then which of the following is not correct?

(A) $P(A \cup B) \geq \frac{2}{3}$ (B) $P(A \cap B') \leq \frac{1}{3}$

(C) $0 \leq P(A \cap B) \leq \frac{1}{2}$ (D) none of these

44. A natural number 'n' is selected at random from the set of first 100 natural numbers. The probability that $n+\frac{100}{n} \leq 50$ is equal to

(A) $\frac{9}{10}$ (B) $\frac{39}{50}$
(C) $\frac{9}{20}$ (D) none of these

45. A bag contains 4 tickets numbered 1, 2, 3, 4 and another bag contains 6 tickets numbered 2, 4, 6, 7, 8, 9. One bag is chosen and a ticket is drawn. The probability that the ticket bears the number 4, is equal to

(A) $\frac{5}{12}$ (B) $\frac{5}{24}$ (C) $\frac{7}{12}$ (D) $\frac{19}{24}$

46. A bag contains four tickets marked with numbers 112, 121, 211, 222. One ticket is drawn at random from the bag. Let E_i ($i = 1, 2, 3$) denote the event that i^{th} digit on the selected ticket is 2, then which of the following is not correct?

(A) E_1 and E_2 are independent
(B) E_2 and E_3 are independent
(C) E_3 and E_1 are independent
(D) E_1, E_2, E_3 are independent

47. A real estate man has eight master keys to open several new homes. Only one master key will open any given home. If 40% of these homes are usually left unlocked, the probability that the real estate man can get into a specific home, if it is given that he selected 3 keys randomly before leaving his office, is equal to

(A) $\frac{5}{8}$ (B) $\frac{3}{8}$ (C) $\frac{3}{4}$ (D) $\frac{1}{4}$

48. Two numbers are chosen at random (without replacement) from the set $\{1, 2, 3, \ldots, 5n\}$. the probability that $n_1^4 - n_2^4$ is divisible by 5, is equal to

(A) $\frac{n-1}{5n-1}$ (B) $\frac{4(4n-1)}{5(5n-1)}$
(C) $\frac{17n-5}{5(5n-1)}$ (D) $\frac{8n}{5(5n-1)}$

49. Two persons are selected at random from n persons seated in a row ($n \geq 3$). The probability that the selected persons are not seated consecutively, is equal to

(A) $\frac{n-2}{n}$ (B) $\frac{n-1}{n}$ (C) $\frac{n+2}{n+3}$ (D) $\frac{n-2}{n-1}$

50. Three numbers are selected simultaneously from the set $\{1, 2, 3, \ldots, 25\}$. The probability that the product of selected numbers is divisible by 4, is equal to

(A) $\frac{1}{115}$ (B) $\frac{98}{115}$
(C) $\frac{773}{1150}$ (D) $\frac{963}{1150}$

51. A man alternatively tosses a fair coin and rolls a fair ordinary die. He starts with the coin. The probability that he gets a tail on the coin before getting 5 or 6 on the die, is equal to

(A) $\frac{3}{4}$ (B) $\frac{1}{2}$ (C) $\frac{1}{3}$ (D) $\frac{2}{3}$

52. Three ordinary and fair dice are rolled simultaneously. The probability of the sum of outcomes being at least equal to 8, is equal to

(A) $\frac{81}{108}$ (B) $\frac{27}{216}$ (C) $\frac{81}{216}$ (D) $\frac{181}{216}$

53. Two squares are chosen from the squares of an ordinary chess board. It is given that the selected squares do not belong to the same row or column. The probability that the chosen squares are of the same colour, is equal to

(A) $\frac{25}{49}$ (B) $\frac{32}{49}$ (C) $\frac{25}{64}$ (D) $\frac{1}{2}$

54. A committee consists of 9 experts taken from three institutions A, B and C, of which 2 are from A, 3 from B and 4 from C. If three experts resign from the committee, then the probability of exactly two of the resigned experts being from the same institution, is equal to

(A) $\frac{4}{7}$ (B) $\frac{25}{84}$ (C) $\frac{55}{84}$ (D) $\frac{37}{84}$

55. The sum of two natural numbers n_1 and n_2 is known to be equal to 100. The probability that their product being greater than 1600, is equal to

(A) $\frac{20}{33}$ (B) $\frac{58}{99}$ (C) $\frac{13}{33}$ (D) $\frac{59}{99}$

56. A person while dialing a telephone number, forgets the last three digits of the number but remembers that exactly two of them are same. He dials the number randomly.

The probability that he dialed the correct number, is equal to

(A) $\frac{1}{135}$ (B) $\frac{1}{27}$ (C) $\frac{1}{54}$ (D) $\frac{1}{270}$

57. A bag B_1 has 3 white balls and 2 red balls. Another bag B_2 has 4 white and 6 red balls. A ball is drawn randomly from bag B_1 without seeing its colour and is put in bag B_2. Now a ball is drawn from bag B_2. The probability of both the drawn balls being of the same colour, is

(A) $\frac{41}{55}$ (B) $\frac{31}{55}$

(C) $\frac{29}{55}$ (D) none of these

58. A bag contains 'W' white balls and 'R' red balls. Two players P_1 and P_2 alternately draw a ball from the bag, replacing the ball each time after the draw, till one of them draws a white ball and wins the game. 'P_1' beings the game. The probability of P_2 being the winner, is equal to

(A) $\frac{W^2}{(W+2R)R}$ (B) $\frac{R}{(2W+R)}$

(C) $\frac{R^2}{(2W+R)W}$ (D) $\frac{R}{(W+2R)}$

Integer Answer Type

1. The integers m and n are chosen at random between 1 and 100. If the probability that a number of the form $7^m + 7^n$ is divisible by 5 is λ, then find the value of 12λ.

2. For a post, three persons A, B and C appear in the interview. The probability of A being selected is twice that of B and the probability of B being selected is thrice that of C. If the odds in favour of B to be selected is $\lambda_1 : \lambda_2$ then find the value of $\left[\frac{\lambda_2}{\lambda_1}\right]$ where $[\cdot]$ denotes the GIF.

3. Three houses are available in a locality. Three persons apply for the houses. Each applies for one house without consulting others. If the probability that all the three apply for the same house is $\frac{K_1}{K_2}$, then find the value of $K_2 - K_1$.

4. Find the least number of dice that must be thrown, so that there is a better than even chance of obtaining a six.

5. A natural number x is chosen at random from the first one hundred natural numbers. If the probability that $\frac{(x-20)(x-40)}{(x-30)} < 0$ is $\frac{\lambda_1}{\lambda_2}$, then find the value of $\frac{|\lambda_1 - \lambda_2|}{2}$.

6. If a head means one and a tail means two, then coefficients of the quadratic equation $ax^2 + bx + c = 0$ are chosen by tossing three fair coins. If the probability that the roots of the equations are imaginary is $\frac{K_1}{K_2}$. Find the value of $\frac{K_1 + K_2}{3}$.

7. A fair coin is tossed 99 times. Let X be the number of times a head occurs. If $P(X = r)$ is maximum when r is λ, then find the value of $\left[\sqrt{\lambda}\right]$, where [·] denotes the Greatest Integer Function.

8. An unbiased cubic die marked with 1, 2, 2, 3, 3, 3 is rolled 3 times. If the probability of getting a total score of 4 or 6 is $\frac{\lambda_1}{\lambda_2}$ find the value of $\sqrt{\lambda_1}$.

9. The letters of the word 'REGULATION' are arranged at random. If the probability that there are exactly 4 letters between R and E is $\frac{\lambda_1}{\lambda_2}$. Find the value of $|\lambda_1 - \lambda_2|$.

10. 7 boys and 8 girls have to sit on 15 chairs arranged in a row numbered from 1 to 15. If the probability that the end seats are occupied by the boys and between any two boys, an even number of girls occupy seats is ${}^9C_4 \frac{\lambda_1 ! \lambda_2 !}{\lambda_3 !}$, then find the value of $\frac{\lambda_1 + \lambda_2}{\lambda_3}$.

11. A die is thrown three times. The chance that the highest number shown on the die is 4 is p, then what is the value of $[1/p]$ where [.] represents greatest integer function?

12. Suppose A and B are two events with $P(A)=0.5$ and $P(A\cup B)=0.8$. Let $P(B)=p$ if A and B are mutually exclusive and $P(B)=q$ if A and B are independent events, then what is the value of q/p?

13. Five different games are to be distributed among four children randomly. The probability that each child get at least one game is p, then the value of $[1/p]$ is, where [.] represents the greatest integer function.

14. Thirty-two players ranked 1 to 32 are playing in a knockout tournament. Assume that in every match between any two players the better ranked player wins, the probability that ranked 1 and ranked 2 players are winner and runner up respectively is p, then what is the value of $[2/p]$ where [.] represents the greatest integer function.

15. An urn contains three red balls and n white balls. Mr A draws two balls together from the urn. The probability that they have the same colour is $\frac{1}{2}$. Mr B draws one ball from the urn, notes its colour and replaces it. He then draws a second ball from the urn and finds that both balls have the same colour, the probability of the same is $\frac{5}{8}$. What is the possible value of n?

16. A drawer contains a mixture of red and blue socks, at most 17 in all. It so happens that when two socks are selected randomly without replacement, there is a probability of exactly $\frac{1}{2}$ that both are red or both are blue. What is the largest possible number of red socks in the drawer that is consistent with this data?

17. Two different numbers are taken from the set (0, 1, 2, 3, 4, 5, 6, 7, 8, 9, 10). The probability that their sum and positive difference are both multiples of 4 is $x/55$, then what does x equal?

18. There are two red, two blue, two white and certain number (greater than 0) of green socks in a drawer. If two socks are taken at random from the drawer without replacement, the probability that they are of the same colour is $\frac{1}{5}$, then what is the number of green socks?

19. If the probability that the product of the outcome of three rolls of a fair die is a prime number is p, then what is the value of $1/(4p)$?

20. If the probability of a six-digit number N whose six digits are 1, 2, 3, 4, 5, 6 written in random order is divisible by 6 is p, then what is the value of $1/p$?

7.11 JEE Advanced: Sample Questions

Single Answer Type

1. Find the probability of getting the sum as a prime number when two dice are thrown together.

(A) $\frac{7}{18}$ (B) $\frac{11}{36}$ (C) $\frac{5}{12}$ (D) $\frac{7}{12}$

2. If A, B, C are any three events such that $P(A)=0.3$, $P(B)=0.4$, $P(C)=0.8$ $P(A\cap B)=0.08$, $P(A\cap C)=0.28$, $P(A\cap B\cap C)=0.09$. If $P(A\cup B\cup C)\geq 0.75$, then the range of $P(B\cap C)$ is

(A) $[0.23, 0.48]$ (B) $[0.23, 0.47]$

(C) $[0.22, 0.48]$ (D) none of these

3. Three friends A, B and C play the game 'Pahle Hum Pahle Tum' in which they throw a die one after the other and the one who will get a composite number first will be announced as winner, if A started the game followed by B and then C then what is the ratio of their winning probabilities?

(A) 10 : 5 : 4 (B) 9 : 6 : 4

(C) 8 : 6 : 5 (D) 9 : 4 : 5

4. A pair of fair dice is rolled together till a sum of either 5 or 7 is obtained. The probability that 5 comes before 7 is

(A) $\frac{1}{5}$ (B) $\frac{2}{5}$ (C) $\frac{3}{5}$ (D) $\frac{4}{5}$

5. In a bulb manufacturing company, it is known that 10% of the bulbs are defective, what is the probability that out of 10 randomly selected bulbs, not more than 10% are defective?

(A) $\frac{15(9^9)}{10^9}$ (B) $\frac{9(9^9)}{10^{10}}$

(C) $\frac{9(9^9)}{10^9}$ (D) $\frac{19(9^9)}{10^{10}}$

6. India and Pakistan are playing a best four of 7 match series. What is the probability that the tournament ends up in 6 matches, assuming that no match ends up in a draw?

(A) $\frac{5}{32}$ (B) $\frac{5}{48}$ (C) $\frac{5}{16}$ (D) $\frac{5}{24}$

7. Two cards are drawn at random from a well shuffled pack of cards. What is the probability that both are numbered cards or both are black cards?

(A) $\frac{101}{1326}$ (B) $\frac{401}{663}$ (C) $\frac{362}{663}$ (D) $\frac{432}{663}$

8. An ordinary deck of 52 playing cards is randomly divided into 4 groups of 13 cards each. What is the probability that each group has exactly 1 jack?

(A) $\frac{39\times 26}{51\times 50}$ (B) $\frac{39\times 26\times 13}{51\times 50\times 49}$

(C) $\frac{39\times 26\times 11}{51\times 50\times 49}$ (D) none of these

9. Three of the six vertices of a regular hexagon are chosen at random. The probability that the triangle with these three vertices is equilateral, equals

(A) $\frac{1}{2}$ (B) $\frac{1}{5}$ (C) $\frac{1}{10}$ (D) $\frac{1}{20}$

10. A bag contains n white and n black balls. Pairs of balls are drawn without replacement until the bag is empty. If the number of ways in which each pair consists of one black and one white ball is 576, then n is equal to

(A) 4 (B) 5 (C) 24 (D) 12

11. A die is thrown n times (n being odd). The probability that an even face turns odd number of times, is

(A) $\frac{n}{2n+1}$ (B) less than $\frac{1}{2}$

(C) $\frac{1}{2}$ (D) greater than $\frac{1}{2}$

12. If A and B are two events such that $P(A)=0.2$ and $P(A\cup B)=0.7$. If A and B are independent events, then $P(B')$ equals

(A) $\frac{2}{7}$ (B) $\frac{7}{9}$

(C) $\frac{3}{8}$ (D) none of these

13. Let A, B, C be three events such that A and B are independent and $P(C)=0$, then events A, B, C are

(A) independent

(B) pair wise independent but not totally independent

(C) $P(A)=P(B)=P(C)$

(D) none of these

14. If two events A and B are such that $P(A)>0$ and $P(B)\neq 1$, then $P\left(\frac{A'}{B'}\right)$ is equal to

(A) $1-P\left(\frac{A}{B}\right)$ (B) $1-P\left(\frac{A'}{B}\right)$

(C) $\frac{1-P(A\cup B)}{P(B)}$ (D) $\frac{P(A)}{P(B)}$

15. If 7 digits out of 1 to 9 are arranged in the spaces of number 1 2 6 3 _ _ _ _ _ _ _ 6, what is the probability that the number is divisible by 9?

(A) $\frac{1}{9}$ (B) $\frac{1}{16}$ (C) $\frac{1}{36}$ (D) 1

16. Two subsets A and B of a set containing n elements are chosen at random. The probability that $A\subseteq B$ is

(A) $\frac{1}{2}$ (B) $\frac{2^n}{n!}$ (C) $\left(\frac{2}{3}\right)^n$ (D) $\left(\frac{3}{4}\right)^n$

17. Three smallest squares are chosen randomly on a chess board, then the probability that these squares have exactly two corners, but no side common is

(A) $\frac{80}{^{64}C_3}$ (B) $\frac{72}{^{64}C_3}$

(C) $\frac{6^3}{^{64}C_2}$ (D) none of these

18. If the letters of the word BANANA are arranged randomly, then the probability that the word thus formed does not contain the pattern BAN is

(A) $\frac{1}{5}$ (B) $\frac{4}{5}$

(C) $\frac{2}{5}$ (D) none of these

19. A bag contains 100 balls numbered from 1 to 100. Four balls are removed at random without replacement. The probability that the number on the last ball is smaller than the number on first ball, is equal to

(A) $\frac{1}{4}$ (B) $\frac{3}{4}$ (C) $\frac{2}{5}$ (D) $\frac{1}{2}$

20. The probability that an archer hits the target when it is windy is equal to $\frac{2}{5}$, when it is not windy his probability of hitting the target is $\frac{7}{10}$. On any shot the probability of gust of wind is $\frac{3}{10}$. The probability that there is no gust of wind on the occasion when he missed the target, is equal to

(A) $\frac{5}{13}$ (B) $\frac{18}{39}$ (C) $\frac{7}{13}$ (D) $\frac{23}{39}$

21. Three numbers are selected from the set {1, 2, 3, ..., 23, 24}, without replacement. The probability that the formed numbers from an A.P. is equal to

(A) $\frac{11}{23}$ (B) $\frac{12}{23}$

(C) $\frac{3}{46}$ (D) none of these

22. *A*, *B* and *C* are contesting the election for the post of secretary of a club which does not allow ladies to become members. The probabilities of *A*, *B* and *C* winning the election are $\frac{1}{3}, \frac{2}{9}$ and $\frac{4}{9}$ respectively. The probability of introducing the clause of admitting lady members to the club by *A*, *B* and *C* are 0.6, 0.7 and 0.5 respectively. The probability that ladies will be taken as members in the club after the election is

(A) $\frac{26}{45}$ (B) $\frac{28}{45}$

(C) $\frac{27}{45}$ (D) none of these

23. A card is drawn from a pack. The card is replaced and the pack is reshuffled. If this is done six times, the probability that 2 hearts, 2 diamonds and 2 black cards are drawn is

(A) $90.\left(\frac{1}{4}\right)^6$ (B) $\frac{45}{2}\left(\frac{3}{4}\right)^4$

(C) $\frac{90}{2^{10}}$ (D) none of these

24. Ram's gardener is not dependable. The probability that he will forget to water the rose bush is $\frac{2}{3}$. The rose bush is in questionable condition. If watered, the probability of its withering is $\frac{1}{2}$, if not watered, the probability of its withering is $\frac{3}{4}$. Ram went out of station and upon returning, he finds that the rose bush has withered. The probability that the gardener did not water the bush is

(A) $\frac{1}{2}$ (B) $\frac{1}{3}$

(C) $\frac{3}{4}$ (D) none of these

25. Ten pairs of shoes are in a closet. Four shoes are selected at random. The probability that there will be at least one pair among the four shoes selected is

(A) $\frac{90}{323}$ (B) $\frac{99}{323}$

(C) $\frac{47}{323}$ (D) none of these

Multiple Answer Type

1. If A and B are two events, the probability that exactly one of them occurs is
(A) $P(A)+P(B)-2(A\cap B)$
(B) $P(A\cup B)-P(A\cap B)$
(C) $P(A')+P(B')-2P(A'\cap B')$
(D) $P(A\cap B')+P(A'\cap B)$

2. For two events A and B, if
$P(A)=P\left(\frac{A}{B}\right)=\frac{1}{4}$ and $P\left(\frac{B}{A}\right)=\frac{1}{2}$, then
(A) A and B are independent
(B) A and B are mutually exclusive
(C) $P\left(\frac{A'}{B}\right)=\frac{3}{4}$
(D) $P\left(\frac{B'}{A'}\right)-\frac{1}{2}$

3. Four whole numbers are randomly selected and are multiplied. Let us define
E_1 : Last digit in the product is 1, 3, 7, 9
E_2 : Last digit in the product is 2, 4, 6, 7, then
(A) $P(E_1)=\left(\frac{4}{10}\right)^4$ (B) $P(E_2)=\left(\frac{4}{10}\right)^4$
(C) $P(E_2)=\frac{8^4-4^4}{10^4}$ (D) $P(E_2)>P(E_1)$

4. Two real numbers, x and y are selected at random. Given that $0\le x\le 1, 0\le y\le 1$. Let A be the event that $y^2\le x$; B be the event that $x^2\le y$, then
(A) $P(A\cap B)=\frac{1}{3}$
(B) A and B are exhaustive events
(C) A and B are mutually exclusive
(D) A and B are independent events

5. Two fair dice are thrown independently. Three events A, B and C are defined as follows.
A: odd face on first die
B: odd face on second die
C: sum of the numbers on two dice is odd
Then
(A) the events A, B and C are mutually exclusive
(B) the events A, B and C are mutually independent
(C) the events A, B and C are pair wise independent
(D) $P(A\cap B)=P(B\cap C)=P(C\cap A)$

6. If E and F are independent events such that $0<P(E)<1$ and $0<P(F)<1$, then
(A) E and F are mutually exclusive
(B) E and F' are independent
(C) E' and F' are independent
(D) $P\left(\frac{E}{F}\right)+P\left(\frac{E'}{F}\right)=1$

7. For any two events A and B in a sample space
(A) $P\left(\frac{A}{B}\right)\ge\frac{P(A)+P(B)-1}{P(B)}$, $P(B)\ne 0$ is always true
(B) $P(A\cap B')=P(A)-P(A\cap B)$ does not hold
(C) $P(A\cup B)=1-P(A')P(B')$, if A and B are independent
(D) $P(A\cup B)=1-P(A')P(B')$, if A and B are disjoint

8. If A and B are two events such that $P(A)=\frac{5}{8}$ and $P(B)=\frac{3}{4}$, then
(A) $P(A'\cup B)\ge\frac{1}{2}$
(B) $\frac{1}{8}\le P(A'\cap B)\le\frac{3}{8}$
(C) $\frac{3}{8}\le P(A\cap B)\le\frac{5}{8}$
(D) $P(A'\cup B)\ge\frac{3}{4}$

9. If E' and F' are the complementary events of the events E and F respectively and if $0<P(F)<1$, then

(A) $P\left(\frac{E}{F}\right)+P\left(\frac{E'}{F}\right)=1$

(B) $P\left(\frac{E}{F}\right)+P\left(\frac{E}{F'}\right)=1$

(C) $P\left(\frac{E'}{F}\right)+P\left(\frac{E}{F'}\right)=1$

(D) $P\left(\frac{E}{F'}\right)+P\left(\frac{E'}{F'}\right)=1$

10. A bag contains four tickets with numbers 112, 121, 211, 222. One ticket is drawn at random from the bag. Let $E_i\,(i=1,2,3)$ denote the event that i^{th} digit on the drawn ticket is 2, then

(A) E_1, E_2, E_3 are pair wise independent

(B) E_1, E_2' are independent

(C) E_2' and E_3' are not independent

(D) E_1, E_2, E_3 are mutually independent

11. A fair die is rolled four times. Find the probability that each number is no smaller than the preceding number

(A) $\frac{7}{12}$ (B) $<\frac{1}{3}$ (C) $>\frac{1}{3}$ (D) $\frac{11}{72}$

12. The probability that a student passes at least in one of the three examinations *A*, *B*, *C* is 0.75; the probability that he passes in at least two of the exams is 0.5 and the probability he passes exactly two of the exams is 0.4. If α, β, γ are respectively the probabilities of the student passing in *A*, *B*, *C*, then

(A) $\alpha+\beta+\gamma=\frac{17}{20}$ (B) $\alpha+\beta+\gamma=\frac{20}{27}$

(C) $\alpha\beta\gamma=\frac{1}{10}$ (D) $\alpha^2+\beta^2+\gamma^2=1$

13. A number is chosen at random from the set of integers $\{1, 2, 3, 4, \ldots, n\}$. Let *A* denote the event that the number chosen is divisible by 4, *B* denote the event that the number chosen is divisible by 5 and *C* denote the event that the number chosen is divisible by 7, then

(A) A, B, C are always mutually independent

(B) *A* and *B* are always independent

(C) *B* and *C* are independent if *n* is of the form $35K$ (*K* positive integer)

(D) *A* and *C* are independent if *n* is of the form 28λ (λ is positive integer)

14. If *M* and *N* are any two events, then which one of the following represents the probability of the occurrence of exactly one of them?

(A) $P(M) + P(N) - 2P(M \cap N)$

(B) $P(M) + P(N) - P(M \cap N)$

(C) $P(\bar{M})+P(\bar{N})-2P(\bar{M}\cap\bar{N})$

(D) $P(M\cap\bar{N})+P(\bar{M}\cap N)$

15. The probabilities of events, $A \cap B$, *A*, *B* and $A \cup B$ are respectively in A.P. with the second term equal to the common difference. Therefore, *A* and *B* are:

(A) mutually exclusive

(B) independent

(C) such that one of them must occur

(D) such that one is twice as likely as the other

16. In a single throw with 3 ordinary unbiased dice, the chance of

(A) throwing a "4 – 5 – 6" is equal to $\frac{1}{216}$

(B) throwing a total of 11 is $\frac{1}{8}$

(C) all the 3 dice showing up with different faces is $\frac{4}{9}$

(D) all the 3 dice showing up alike faces is $\frac{1}{36}$

17. Let $0 < P(A) < 1$, $0 < P(B) < 1$ and $P(A \cup B) = P(A) + P(B) - P(A)\cdot P(B)$, then:

(A) $P(B/A) = P(B) - P(A)$

(B) $P(A^C \cup B^C) = P(A^C) + P(B^C)$

(C) $P((A \cup B)^C) = P(A^C)\cdot P(B^C)$

(D) $P(A/B) = P(A)$

18. If $\bar{E}$ and $\bar{F}$ are the complementary events of events *E* and *F* respectively and if $0 < p(F) < 1$, then:

(A) $P(E/F) + P(\bar{E}/F) = 1$

(B) $P(E/F) + P(E/\bar{F}) = 1$

(C) $P(\bar{E}/F) + P(E/\bar{F}) = 1$

(D) $P(E/\bar{F})\, P(\bar{E}/\bar{F}) = 1$

19. Two numbers are chosen from {1, 2, 3, 4, 5, 6, 7, 8} one after another without replacement. Then the probability that

(A) the smaller value of two is less than 3 is $\frac{13}{28}$

(B) the bigger value of two is more than 5 is $\frac{9}{14}$

(C) product of two number is even is $\frac{11}{14}$

(D) none of these

Comprehension Based Questions

Paragraph for Question No. 1 to 3

Three vertices are chosen at random from the vertices of a regular hexagon then

1. The probability that the triangle with these vertices is an equilateral triangle is equal to

(A) $\frac{1}{10}$ (B) $\frac{3}{5}$ (C) $\frac{9}{10}$ (D) $\frac{3}{10}$

2. The probability that the triangle has exactly one side common with the side of the hexagon is

(A) $\frac{1}{10}$ (B) $\frac{3}{5}$ (C) $\frac{3}{10}$ (D) $\frac{9}{10}$

3. The probability that the triangle has exactly two sides common with the side of the hexagon is

(A) $\frac{1}{10}$ (B) $\frac{3}{5}$ (C) $\frac{3}{10}$ (D) $\frac{9}{10}$

Paragraph for Question No. 4 to 6

5 Indian and 5 Russian couples meet at a party and shake hands.

4. The probability that no wife shakes hands with her husband is

(A) $\frac{3}{5}$ (B) $\frac{18}{19}$ (C) $\frac{23}{38}$ (D) $\frac{27}{38}$

5. The probability that no Indian wife shakes hands with a male is

(A) $\frac{14}{19}$ (B) $\frac{9}{19}$ (C) $\frac{27}{38}$ (D) $\frac{27}{38}$

6. The probability that no wife shakes hands with her husband and no Indian wife shakes hands with a male is

(A) $\frac{23}{38}$ (B) $\frac{27}{38}$ (C) $\frac{9}{19}$ (D) $\frac{3}{5}$

Paragraph for Question No. 7 to 9

Different words are being formed by arranging the letters of the word "SUCCESS".

7. The probability of choosing a word in which the two *C* are together but no two S are together is

(A) $\frac{2}{35}$ (B) $\frac{2}{7}$ (C) $\frac{24}{105}$ (D) 1

8. The probability of word in which no two *C* and no two *S* are together is

(A) $\frac{2}{35}$ (B) $\frac{8}{35}$ (C) $\frac{24}{105}$ (D) 1

9. The probability of selecting a word that begins and ends with *S* is

(A) $\frac{1}{7}$ (B) $\frac{2}{7}$ (C) $\frac{2}{35}$ (D) $\frac{2}{5}$

Paragraph for Question No. 10 to 12

A bag contains 3 white, 5 black and 2 red balls. Two persons draw balls in turns without replacement. The first person to draw a white ball wins the game. If a red ball is drawn, the game is a tie, then

10. The probability that person who begins the game is the winner is equal to

(A) $\frac{67}{168}$ (B) $\frac{333}{840}$ (C) $\frac{73}{210}$ (D) $\frac{83}{210}$

11. The probability that the second person is the winner is equal to

(A) $\frac{43}{210}$ (C) $\frac{87}{420}$ (C) $\frac{77}{420}$ (D) $\frac{11}{70}$

12. The probability that the game is a tie is equal to

(A) $\frac{1}{5}$ (B) $\frac{2}{5}$ (C) $\frac{3}{5}$ (D) $\frac{4}{5}$

Paragraph for Question No. 13 to 15

A fair die is tossed repeatedly until a six obtained. Let *X* denote the number of tosses required.

13. The probability that $X = 3$ equals

(A) $\frac{25}{216}$ (B) $\frac{25}{36}$ (C) $\frac{5}{36}$ (D) $\frac{125}{216}$

14. The probability that $X \geq 3$ equals

(A) $\frac{125}{216}$ (B) $\frac{25}{36}$ (C) $\frac{5}{36}$ (D) $\frac{5}{216}$

15. The conditional probability that $X \geq 6$ given $X > 3$ equals

(A) $\frac{125}{216}$ (B) $\frac{25}{216}$ (C) $\frac{25}{36}$ (D) $\frac{5}{36}$

Match Matrix

1. A box contains 25 tickets, each with a different number from 1 to 25. Four tickets are drawn at random and without replacement. Let a, b, c, d be the number of the tickets drawn.

Column I	Column II
(A) The probability that $\min(a, b, c, d) < 10$ is equal to	(P) $\frac{189}{2530}$
(B) The probability that the second smallest number chosen is 10 is equal to	(Q) $\frac{286}{575}$
(C) The probability that the sum of $a+b+c+d$ is odd is equal to	(R) $\frac{207}{230}$
(D) The probability that the product of $abcd$ is even is equal to	(S) $\frac{217}{230}$
	(T) $\frac{1083}{1265}$

2. A bag contains some white and some black balls, all combinations being equally likely. The total number of balls in the bag is 12. Four balls are drawn at random from the bag at random without replacement. Now match the entries from the following two columns.

Column I	Column II
(A) Probability that all the four balls are black is equal to	(P) $\frac{14}{32}$
(B) If the bag contains 10 black and 2 white balls, then the probability that all four balls are black is equal to	(Q) $\frac{1}{5}$
(C) If all the four balls are black, then the probability that the bag contains 10 black balls is equal to	(R) $\frac{70}{429}$
(D) Probability that two balls are black and two are white	(S) $\frac{13}{165}$

(Continued)

3. Suppose that E_1 and E_2 are two events of a random experiment such that

$$P(E_1) = \frac{1}{4},\ P\left(\frac{E_2}{E_1}\right) = \frac{1}{2} \text{ and } P\left(\frac{E_1}{E_2}\right) = \frac{1}{4},$$

observe the lists given below

Column I	Column II
(A) $P(E_2)$	(P) $1/4$
(B) $P(E_1 \cup E_2)$	(Q) $5/8$
(C) $P(\bar{E}_1 / \bar{E}_2)$	(R) $1/8$
(D) $P(E_1 / \bar{E}_2)$	(S) $1/2$
	(T) $3/8$
	(U) $3/4$

4.

Column I	Column II
(A) The probability of a bomb hitting a bridge is $\frac{1}{2}$. Two direct hits are needed to destroy it. The number of bombs required so that the probability of the bridge being destroyed is greater than 0.9 can be	(P) 4

(Continued)

Column I	Column II
(B) A bag contains 2 red, 3 white and 5 black balls, a ball is drawn its colour is noted and replaced. The number of times, a ball can be drawn so that the probability of getting a red ball for the first time is at least $\frac{1}{2}$	(Q) 6
(C) A drawer contains a mixture of red socks and blue socks, at most 17 in all. It so happens that when two socks are selected randomly without replacement, there is a probability of exactly $\frac{1}{2}$ that both are red or both are blue. Then number of red socks in the drawer can be	(R) 7
(D) There are two red, two blue, two white and certain number (greater than 0) of green socks in a drawer. If two socks are taken at random from the drawer without replacement, the probability that they are of the same colour is $\frac{1}{5}$, then the number of green socks are	(S) 10

5. An urn contains r red balls and b black balls

Column I	Column II
(A) If the probability of getting two red balls in first two draws (without replacement) is $\frac{1}{2}$, then value of r can be	(P) 10

(*Continued*)

Column I	Column II
(B) If the probability of getting two red balls in first two draws (without replacement) is $\frac{1}{2}$ and b is an even number, then r can be	(Q) 3
(C) If the probability of getting exactly two red balls in four draws (with replacement) from the urn is $\frac{3}{8}$ and $b=10$, then r can be	(R) 8
(D) If the probability of getting exactly n red balls in $2n$ draw (with replacement) is equal to probability of getting exactly n black balls in $2n$ draws (with replacement), then the ratio r/b can be	(S) 2

Subjective Answer Type (Hard Nut)

1. In a multiple-choice question, there are four alternative answers, of which one or more are correct. A candidate will get marks only if he ticks all the correct answers. The candidate decides to tick at random. He is allowed up to three chances to answer the question. Find the probability that he will get marks for the question.

2. Let an ordered pair A be defined as $A(x, y)$ where $x \in$ prime number, such that $x < 10$ and $y \in N$ and $y \leq 10$. Find the probability that the ordered pair A satisfies the relation $x^2 - 3y^2 = 1$.

3. If four cards are drawn at random from a pack of 52 cards, then find the probability of getting exactly one pair of cards.

4. Find the probability that a factor of 97020 is divisible by 45.

5. A person goes to office either by car, scooter, bus or train, probability of which being $\frac{1}{7}, \frac{3}{7}, \frac{2}{7}$ and $\frac{1}{7}$ respectively. Probability that he reaches office late, if he takes car, scooter, bus or train is $\frac{2}{9}, \frac{1}{9}, \frac{4}{9}$ and $\frac{1}{9}$ respectively. Given that he reached office in time, then what is the probability that he travelled by a car.

6. Five different digits from the set of numbers $\{1, 2, 3, 4, 5, 6, 7\}$ are written in random order. If the probability that the 5 digit number thus formed is divisible by 9 is $\frac{x}{y}$ (x and y are co-prime numbers), then find the value of $y-10x$.

7. Let A, B, C be 3 independent events such that $P(A)=\frac{1}{3}, P(B)=\frac{1}{2}, P(C)=\frac{1}{4}$, then find the probability of exactly 2 events occurring out of 3 events.

8. There are m stations between two cities A and B. A train is to stop at three of these m stations. What is the probability that no two of these three stations are consecutive?

9. A letter is known to have come either from TATANAGAR or CALCUTTA. On the envelope, just two consecutive letters TA are visible. What is the probability that the letter came from CALCUTTA?

10. A bag contains 10 balls of white as well as black. A person draws 6 balls and finds that 3 balls are white and the other three are black. Find the probability that the number of white balls in the bag is the same as that of black balls.

11. A bag contains 'm' number of white and 'n' number of black balls. Two players A and B alternately draw a ball from the bag, replacing the ball each time after the draw till one of them draws a white ball and wins the game. If A begins the game and the probability of A winning the game is three times that of B, show that $m : n = 2 : 1$.

12. 5 girls and 10 boys sit at random in a row having 15 chairs numbered as 1 to 15. Find the probability that end seats are occupied by the girls and between any two girls, there always lie an odd number of boys.

13. Seven digits from the numbers 1, 2, 3, 4, 5, 6, 7, 8 and 9 are written in random order. Find the probability that this seven-digit number is divisible by 9.

14. Three critics review a book, odds in favour of the book are 5 : 2, 4 : 3 and 3 : 4 respectively for the three critics. Find the probability that the majority is in favour of the book.

15. Six different balls are put in three different boxes, no box being empty. Find the probability of putting balls in the boxes in equal numbers.

16. There are 8 coloured balls and correspondingly 8 coloured bags (same as the balls). The balls are placed in the bags, each one in one bag. Find the probability that 5 of the balls are placed in the respective coloured bags.

17. A and B are two factories producing 20% and 30% defective goods. From a sample one item is taken and it is found to be defective. Find the probability that it was produced by factory B.

18. A point is chosen from 1st quadrant randomly such that $x, y \in [0, 4]$, find the probability that it satisfies $[x]+[y]=3$.

19. A number is chosen at random from the numbers 10 to 99. By seeing the number, a man will laugh if the product of the digits is 12. If he chooses three numbers with replacement then find the probability that he will laugh at least once.

20. There are 10 different objects 1, 2, 3, …, 10 arranged at random in 10 places marked 1, 2, 3, …, 10. Then find the probability that exactly five of these objects occupy placcs corresponding to their number.

7.12 JEE Main: Archive

1. Bag A contains 2 white, 1 black and 3 red balls and bag B contains 3 black, 2 red and n white balls. One bag is chosen at random and 2 balls drawn from it at random, are found to be 1 red and 1 black. If the probability that both balls come from bag A is $\frac{6}{11}$, then n is equal to ______

(A) 13 (B) 6 (C) 4 (D) 3

[Main 2022 24th June 1st Shift]

2. If a random variable X follows the binomial distribution B (33, p) such that $3P(X=0)=P(X=1)$, then the value of $\frac{P(X=15)}{P(X=18)}-\frac{P(X=16)}{P(X=17)}$ is equal to

(A) 1320 (B) 1088 (C) $\frac{120}{1331}$ (D) $\frac{1088}{1089}$

[Main 2022 24th June 1st Shift]

3. A random variable X has the following probability distribution:

X	0	1	2	3	4
$P(X)$	k	$2k$	$4k$	$6k$	86

The value of $P(1<X<4 \mid X\leq 2)$ is equal to:

(A) $\frac{4}{7}$ (B) $\frac{2}{3}$ (C) $\frac{3}{7}$ (D) $\frac{4}{5}$

[Main 2022 24th June 2nd Shift]

4. In an examination, there are 10 true–false type questions. Out of 10, a student can guess the answer of 4 questions correctly with probability $\frac{3}{4}$ and the remaining 6 questions correctly with probability $\frac{1}{4}$. If the probability that the student guesses the answers of exactly 8 questions correctly out of 10 is $\frac{27k}{4^{10}}$, then k is equal to ______.

[Main 2022 24th June 2nd Shift]

5. Let E_1 and E_2 be two events such that the conditional probabilities $P(E_1 \mid E_2)=\frac{1}{2}$. $P(E_2 \mid E_1)=\frac{3}{4}$ and $P(E_1\cap E_2)=\frac{1}{8}$. Then,

(A) $P(E_1\cap E_2)=P(E_1)\cdot P(E_2)$

(B) $P(E_1'\cap E_2')=P(E_1')\cdot P(E_2)$

(C) $P(E_1\cap E_2')=P(E_1)\cdot P(E_2)$

(D) $P(E_1'\cap E_2)=P(E_1)\cdot P(E_2)$

[Main 2022 25th June 1st Shift]

6. A biased die is marked with numbers 2, 4, 8, 16, 32, 32 on its faces and the probability of getting a face with mark n is $\frac{1}{n}$. If the die is thrown thrice, then the probability, that the sum of the numbers obtained is 48, is

(A) $\frac{7}{2^{11}}$ (B) $\frac{7}{2^{12}}$ (C) $\frac{3}{2^{10}}$ (D) $\frac{13}{2^{12}}$

[Main 2022 25th June 2nd Shift]

7. Let a biased coin be tossed 5 times. If the probability of getting 4 heads is equal to the probability of getting 5 heads, then the probability of getting at most two heads is:

(A) $\frac{275}{6^5}$ (B) $\frac{36}{5^4}$ (C) $\frac{181}{5^5}$ (D) $\frac{46}{6^4}$

[Main 2022 26th June 1st Shift]

8. If the probability that a randomly chosen 6-digit number formed by using digits 1 and 8 only is a multiple of 21 is p, then $96p$ is equal to ______.

[Main 2022 26th June 2nd Shift]

9. Five numbers x_1, x_2, x_3, x_4, x_5 are randomly selected from the numbers $1, 2, 3, \ldots, 18$ and are arranged in the increasing order $(x_1<x_2<x_3<x_4<x_5)$. The probability that $x_2=7$ and $x_4=11$ is:

(A) $\frac{1}{136}$ (B) $\frac{1}{72}$ (C) $\frac{1}{68}$ (D) $\frac{1}{34}$

[Main 2022 27th June 1st Shift]

10. Let X be a random variable having binomial distribution $B(7, p)$. If $P(X=3)=5P(X=4)$, then the sum of the mean and the variance of X is:

(A) $\frac{105}{16}$ (B) $\frac{7}{16}$ (C) $\frac{77}{36}$ (D) $\frac{49}{16}$

[Main 2022 27th June 1st Shift]

11. If a point $A(x, y)$ lies in the region bounded by the y-axis, straight lines $2y+x=6$ and $5x-6y=30$, then the probability that $y<1$ is:

(A) $\frac{1}{6}$ (B) $\frac{5}{6}$ (C) $\frac{2}{3}$ (D) $\frac{6}{7}$

[Main 2022 27th June 2nd Shift]

12. Let $S=\{E, E_2, \ldots E_8\}$ be a sample space of random experiment such that $P(E_n)=\frac{n}{36}$ for every $n=1, 2, \ldots, 8$. Then the number of elements in the set $\left\{A \subset S : P(A) \geq \frac{4}{5}\right\}$ is ________.

[Main 2022 27th June 2nd Shift]

13. The probability, that in a randomly selected 3-digit number at least two digits are odd, is

(A) $\frac{19}{36}$ (B) $\frac{15}{36}$ (C) $\frac{13}{36}$ (D) $\frac{23}{36}$

[Main 2022 28th June 1st Shift]

14. If the sum and the product of mean and variance of a binomial distribution are 24 and 128 respectively, then the probability of one or two successes is:

(A) $\frac{33}{2^{32}}$ (B) $\frac{33}{2^{29}}$ (C) $\frac{33}{2^{28}}$ (D) $\frac{33}{2^{27}}$

[Main 2022 25th July 1st Shift]

15. If A and B are two events such that $P(A)=\frac{1}{3}, P(B)=\frac{1}{5}$ and $P(A \cup B)=\frac{1}{2}$, then $P(A \mid B')+P(B \mid A')$ is equal to

(A) $\frac{3}{4}$ (B) $\frac{5}{8}$ (C) $\frac{5}{4}$ (D) $\frac{7}{8}$

[Main 2022 25th July 2nd Shift]

16. The mean and variance of a binomial distribution are α and $\frac{\alpha}{3}$ respectively. If $P(X=1)=\frac{4}{243}$, then $P(X=4 \text{ or } 5)$ is equal to:

(A) $\frac{5}{9}$ (B) $\frac{64}{81}$ (C) $\frac{16}{27}$ (D) $\frac{145}{243}$

[Main 2022 26th July 1st Shift]

17. Let E_1, E_2, E_3 be three mutually exclusive events such that

$P(E_1)=\frac{2+3p}{6}, P(E_2)=\frac{2-p}{8}$

and $P(E_3)=\frac{1-p}{2}$. If the maximum and minimum values of p are p_1 and p_2, then (p_1+p_2). is equal to:

(A) $\frac{2}{3}$ (B) $\frac{5}{3}$ (C) $\frac{5}{4}$ (D) 1

[Main 2022 26th July 1st Shift]

18. Let X be a binomially distributed random variable with mean 4 and variance $\frac{4}{3}$. Then $54P(X \leq 2)$ is equal to

(A) $\frac{73}{27}$ (B) $\frac{146}{27}$ (C) $\frac{146}{81}$ (D) $\frac{126}{81}$

[Main 2022 26th July 2nd Shift]

19. Let X have a binomial distribution $B(n, p)$ such that the sum and the product of the mean and variance of X are 24 and 128 respectively. If $P(X>n-3)=\frac{k}{2^n}$, then k is equal to

(A) 528 (B) 529 (C) 629 (D) 630

[Main 2022 27th July 2nd Shift]

20. A six faced die is biased such that $3 \times P$ (a prime number) = $6 \times P$ (a composite number) = $2 \times P(1)$. Let X be a random variable that counts the number of times one gets a perfect square on some throws of this die. If the die is thrown twice, then the mean of X is

(A) $\frac{3}{11}$ (B) $\frac{5}{11}$ (C) $\frac{7}{11}$ (D) $\frac{8}{11}$

[Main 2022 27th July 2nd Shift]

21. Out of 60% female and 40% male candidates appearing in an exam, 60% candidates qualify it. The number of females qualifying the exam is twice the number of males qualifying it. A candidate is randomly chosen from the qualified candidates. The probability that the chosen candidate is female, is:

(A) $\frac{2}{3}$ (B) $\frac{11}{16}$ (C) $\frac{23}{32}$ (D) $\frac{13}{16}$

[Main 2022 28th July 1st Shift]

22. Let A and B be two events such that $P(B|A)=\frac{2}{5}, P(A|B)=\frac{1}{7}$ and $P(A\cap B)=\frac{1}{9}$. Consider

$(S1)\,P(A'\cup B)=\frac{5}{6}$,

$(S2)\,P(A'\cap B')=\frac{1}{18}$. Then

(A) both (S1) and (S2) are true
(B) both (S1) and (S2) are false
(C) only (S1) is true
(D) only (S2) is true

[Main 2022 28th July 2nd Shift]

23. A bag contains 4 white and 6 black balls. Three balls are drawn at random from the bag. Let X be the number of white balls, among the drawn balls. If σ^2 is the variance of X, then $100\sigma^2$ is equal to

[Main 2022 28th July 2nd Shift]

24. Let $S=\{1, 2, 3, \ldots, 2022\}$.

Then the probability, that a randomly chosen number n from the set S such that HCF (n, 2022) = 1, is

(A) $\frac{128}{1011}$ (B) $\frac{166}{1011}$ (C) $\frac{127}{337}$ (D) $\frac{112}{337}$

[Main 2022 29th July 1st Shift]

25. Bag I contains 3 red, 4 black and 3 white balls and Bag II contains 2 red, 5 black and 2 white balls. One ball is transferred from Bag I to Bag II and then a ball is draw from Bag II. The ball so drawn is found to be black in colour. Then the probability, that the transferred ball is red, is:

(A) $\frac{4}{9}$ (B) $\frac{5}{18}$ (C) $\frac{1}{6}$ (D) $\frac{3}{10}$

[Main 2022 29th July 2nd Shift]

26. The sum and product of the mean and variance of a binomial distribution are 82.5 and 1350 respectively. They the number of trials in the binomial distribution is:

[Main 2022 29th July 2nd Shift]

27. Let N denote the number that turns up when a fair die is rolled. If the probability that the system of equations

$x+y+z=1$
$2x+Ny+2z=2$
$3x+3y+Nz=3$

has unique solution is $\frac{k}{6}$, then the sum of value of k and all possible values of N is

(A) 18 (B) 19 (C) 20 (D) 21

[Main 2023 24th Jan 1st Shift]

28. Let Ω be the sample space and $A\subseteq\Omega$ be an event. Given below are two statements:

(S1): If $P(A)=0$, then $A=\phi$

(S2): If $P(A)=$, then $A=\Omega$

Then

(A) only (S1) is true
(B) only (S2) is true
(C) both (S1) and (S2) are true
(D) both (S1) and (S2) are false

[Main 2023 24th Jan 1st Shift]

29. Let M be the maximum value of the product of two positive integers when their sum is 66. Let the sample space

$$S=\left\{x\in Z : x(66-x)\geq\frac{5}{9}M\right\}$$

and the event

$A=\{x\in S : x \text{ is a multiple of } 3\}$.

Then $P(A)$ is equal to

(A) $\frac{15}{44}$ (B) $\frac{1}{3}$ (C) $\frac{1}{5}$ (D) $\frac{7}{22}$

[Main 2023 25th Jan 1st Shift]

30. Let x and y be distinct integers where $1 \le x \le 25$ and $1 \le y \le 25$. Then, the number of ways of choosing x and y, such that $x + y$ is divisible by 5, is ______

[Main 2023 25th Jan 1st Shift]

31. Let N be the sum of the numbers appeared when two fair dice are rolled and let the probability that $N-2, \sqrt{3N}, N+2$ are in geometric progression be $\frac{k}{48}$. Then the value of k is

(A) 2 (B) 4 (C) 16 (D) 8

[Main 2023 25th Jan 2nd Shift]

32. 25% of the population are smokers. A smoker has 27 times more chances to develop lung cancer then a non-smoker. A person is diagnosed with lung cancer and the probability that this person is a smoker is $\frac{k}{10}$. Then the value of k is ______.

[Main 2023 25th Jan 2nd Shift]

33. Fifteen football players of a club-team are given 15 T-shirts with their names written on the backside. If the players pick up the T-shirts randomly, then the probability that at least 3 players pick the correct T-shirt is

(A) $\frac{5}{24}$ (B) $\frac{2}{15}$ (C) $\frac{1}{6}$ (D) $\frac{5}{36}$

[Main 2023 29th Jan 1st Shift]

34. Let $S = \{w_1, w_2, \ldots\}$ be the sample space associated to a random experiment. Let $P(w_n) = \frac{P(w_{n-1})}{2}, n \ge 2$.

Let $A = \{2k + 3\ell; k, \ell \in N\}$ and $B = \{w_n; n \in A\}$. Then $P(B)$ is equal to

(A) $\frac{3}{32}$ (B) $\frac{3}{64}$ (C) $\frac{1}{16}$ (D) $\frac{1}{32}$

[Main 2023 29th Jan 2nd Shift]

35. If an unbiased die, marked with –2, –1, 0, 1, 2, 3 on its faces, is thrown five times, then the probability that the product of the outcomes is positive, is:

(A) $\frac{881}{2592}$ (B) $\frac{521}{2592}$ (C) $\frac{440}{2592}$ (D) $\frac{27}{288}$

[Main 2023 30th Jan 1st Shift]

36. A bag contains six balls of different colours. Two balls are drawn in succession with replacement. The probability that both the balls are of the same colour is p. The next four balls are drawn in succession with replacement and the probability that exactly three balls are of the same colours is q. If $p : q = m : n$, where m and n are co-prime, then $m + n$ is equal to

[Main 2023 30th Jan 2nd Shift]

37. A bag contains 6 balls. Two balls are drawn from it at random and both are found to be black. The probability that the bag contains at least 5 black balls is

(A) $\frac{5}{7}$ (B) $\frac{2}{7}$ (C) $\frac{3}{7}$ (D) $\frac{5}{6}$

[Main 2023 31st Jan 1st Shift]

38. Let A be the event that the absolute difference between two randomly chosen real numbers in the sample space $[0, 60]$ is less than or equal to a. If $P(A) = \frac{11}{36}$, then a is equal to ______.

[Main 2023 31st Jan 2nd Shift]

39. In a binomial distribution $B(n, p)$, the sum and product of the mean and variance are 5 and 6 respectively, then find $6(n + p - q)$ is equal to:

(A) 51 (B) 52 (C) 53 (D) 50

[Main 2023 01st Feb 1st Shift]

40. Two dice are thrown independently. Let A be the event that the number appeared on the 1st die is less than the number appeared on the 2nd die, B be the event that the number appeared on the 1st die is even and that on the second die is odd, and C be the event that the number appeared on the 1st die is odd and that on the 2nd is even. Then

(A) the number of favourable cases of the event $(A \cup B) \cap C$ is 6

(B) A and B are mutually exclusive

(C) The number of favourable cases of the events A, B and C are 15, 6 and 6 respectively

(D) B and C are independent

[Main 2023 01st Feb 2nd Shift]

41. A pair of dice is thrown 5 times. For each throw, a total of 5 is considered a success.

If the probability of at least 4 successes is $\frac{k}{3^{11}}$, then k is equal to

(A) 82 (B) 75 (C) 164 (D) 123

[Main 2023 06th Apr 1st Shift]

42. Three dice are rolled. If the probability of getting different numbers on the three dice is $\frac{p}{q}$, where p and q are co-prime, then $q-p$ is equal to

(A) 2 (B) 1 (C) 3 (D) 4

[Main 2023 06th Apr 2nd Shift]

43. In a bolt factory, machines A, B and C manufacture respectively 20%, 30% and 50% of the total bolts. Of their output 3, 4 and 2 percent are respectively defective bolts. A bolt is drawn at random from the product. If the bolt drawn is found to be defective, then the probability that it is manufactured by the machine C is

(A) $\frac{5}{14}$ (B) $\frac{9}{28}$ (C) $\frac{3}{7}$ (D) $\frac{2}{7}$

[Main 2023 08th Apr 1st Shift]

44. If the probability that the random variable X takes values x is given by

$P(X=x)=k(x+1)3^{-x}, x=0,1,2,3\ldots,$

where k is a constant, then $P(X \geq 2)$ is equal to

(A) $\frac{7}{27}$ (B) $\frac{7}{18}$ (C) $\frac{11}{18}$ (D) $\frac{20}{27}$

[Main 2023 08th Apr 2nd Shift]

45. Let N denote the sum of the numbers obtained when two dice are rolled. If the probability that $2^N < N!$ is $\frac{m}{n}$ where m and n are coprime, then $4m-3n$ is equal to

(A) 6 (B) 12 (C) 10 (D) 8

[Main 2023 10th Apr 1st Shift]

46. Let a die be rolled n times. Let the probability of getting odd numbers seven times be equal to the probability of getting odd numbers nine times. If the probability of getting even numbers twice is $\frac{k}{2^{15}}$, then k is equal to

(A) 60 (B) 15 (C) 90 (D) 30

[Main 2023 10th Apr 2nd Shift]

47. Let

$S=\{M=[a_{ij}], a_{ij} \in \{0,1,2\}, \{1 \leq i, j \leq 2\}\}$

be a sample space and

$A\{M \in S : M \text{ is invertible}\}$

be an even. Then $P(A)$ is equal to

(A) $\frac{16}{27}$ (B) $\frac{47}{81}$ (C) $\frac{49}{81}$ (D) $\frac{50}{81}$

[Main 2023 11th Apr 1st Shift]

48. Let the probability of getting a head for a biased coin be $\frac{1}{4}$. It is tossed repeatedly until a head appears. Let N be the number of tosses required. If the probability that the equation $64x^2+5Nx+1=0$ has no real root is $\frac{p}{q}$, where p and q are co-prime, then $q-p$ is equal to ________.

[Main 2023 11th Apr 2nd Shift]

49. Two dice A and B are rolled. Let the numbers obtained on A and B be α and β respectively. If the variance of $\alpha-\beta$ is $\frac{p}{q}$, where p and q are co-prime, then the sum of the positive divisors of p is equal to

(A) 72 (B) 36 (C) 48 (D) 31

[Main 2023 12th Apr 1st Shift]

50. If a fair $n(n>1)$ faces die is rolled repeatedly until a number less than n appears, if the mean of the number of tosses required is $\frac{n}{9}$, then n is equal to ________.

[Main 2023 12th Apr 1st Shift]

51. A coin is biased so that the head is 3 times as likely to occur as the tail. This coin is tossed until a head or three tails occur. If X denotes the number of tosses of the coin, then the mean of X is

(A) $\frac{38}{16}$ (B) $\frac{15}{16}$ (C) $\frac{21}{16}$ (D) $\frac{81}{64}$

[Main 2023 13th Apr 1st Shift]

52. The random variable X follows binomial distribution $B(n, p)$, for which the difference of the mean and the variance is 1.

If $2P(X=2)=3P(X=1)$, then $n^2P(X>1)$ is equal to

(A) 15 (B) 11 (C) 12 (D) 16

[Main 2023 13th Apr 2nd Shift]

53. A bag contains 6 white and 4 black balls. A die is rolled once and a number of balls equal to the number obtained on the die are drawn from the bag at random. The probability that all the balls drawn are white is

(A) $\frac{1}{4}$ (B) $\frac{11}{50}$ (C) $\frac{1}{5}$ (D) $\frac{9}{50}$

[Main 2023 15th Apr 1st Shift]

54. A bag contains 8 balls, whose colours are either white or black. 4 balls are drawn at random without replacement and it was found that 2 balls are white and other 2 balls are black. The probability that the bag contains equal number of white and black balls is:

(A) $\frac{2}{5}$ (B) $\frac{2}{7}$ (C) $\frac{1}{7}$ (D) $\frac{1}{5}$

[Main 2024 01st Feb 1st Shift]

55. Let the probability that Ajay will not appear in JEE exam be $p=\frac{2}{7}$, while the probability that both Ajay and Vijay will appear in the exam be $q=\frac{1}{5}$. Then the probability that Ajay will appear in the exam and Vijay will not appear is:

(A) $\frac{9}{35}$ (B) $\frac{18}{35}$ (C) $\frac{24}{35}$ (D) $\frac{3}{35}$

[Main 2024 01st Feb 2nd Shift]

56. A fair die is tossed repeatedly until a six is obtained. Let X denote the number of tosses required and let

$a=P(X=3), b=P(X\geq 3)$

and $c=P(X\geq 6\,|\,X>3)$. Then $\frac{b+c}{a}$ is equal to ________.

[Main 2024 27th Jan 1st Shift]

57. An urn contains 6 white and 9 black balls. Two successive draws of 4 balls are made without replacement. The probability that the first draw gives all white balls and the second draw gives all black balls, is:

(A) $\frac{5}{256}$ (B) $\frac{5}{715}$ (C) $\frac{3}{715}$ (D) $\frac{3}{256}$

[Main 2024 27th Jan 2nd Shift]

58. A fair die is thrown until 2 appears. Then the probability that 2 appears in an even number of throws is

(A) $\frac{5}{6}$ (B) $\frac{1}{6}$ (C) $\frac{5}{11}$ (D) $\frac{6}{11}$

[Main 2024 29th Jan 1st Shift]

59. An integer is chosen at random from the integers 1, 2, 3, …, 50. The probability that the chosen integer is a multiple of at least one of 4, 6 and 7 is

(A) $\frac{8}{25}$ (B) $\frac{21}{50}$ (C) $\frac{9}{50}$ (D) $\frac{14}{25}$

[Main 2024 29th Jan 2nd Shift]

60. Two integers x and y are chosen with replacement from the set $\{0,1,2,3,\ldots,10\}$. Then the probability that $|x-y|>5$ is:

(A) $\frac{30}{121}$ (B) $\frac{62}{121}$ (C) $\frac{60}{121}$ (D) $\frac{31}{121}$

[Main 2024 30th Jan 1st Shift]

61. Bag A contains 3 white, 7 red balls and bag B contains 3 white, 2 red balls. One bag is selected at random and a ball is drawn from it. The probability of drawing the ball from the bag A if the ball drawn in white, is:

(A) $\frac{1}{4}$ (B) $\frac{1}{9}$ (C) $\frac{1}{3}$ (D) $\frac{3}{10}$

[Main 2024 30th Jan 2nd Shift]

62. Two marbles are drawn in succession from a box containing 10 red, 30 white, 20 blue and 15 orange marbles, with replacement being made after each drawing. Then the probability that first drawn marble is red and the second drawn marble is white, is

(A) $\frac{2}{25}$ (B) $\frac{4}{25}$ (C) $\frac{2}{3}$ (D) $\frac{4}{75}$

[Main 2024 31st Jan 1st Shift]

63. Three rotten apples are accidently mixed with fifteen good apples. Assuming the random variable x to be the number of rotten apples in a draw of two apples, the variance of x is

(A) $\frac{37}{153}$ (B) $\frac{57}{153}$ (C) $\frac{47}{153}$ (D) $\frac{40}{153}$

[Main 2024 31st Jan 1st Shift]

64. A coin is based so that a head is twice as likely to occur as a tail. If the coin is tossed 3 times, then the probability of getting two tails and one head is

(A) $\frac{2}{9}$ (B) $\frac{1}{9}$ (C) $\frac{2}{27}$ (D) $\frac{1}{27}$

[Main 2024 31st Jan 2nd Shift]

65. Three urns A, B and C contain 7 red, 5 black, 5 red, 7 black and 6 red, 6 black balls, respectively. One of the urns is selected at random and a ball is drawn from it. If the ball drawn is black, then the probability that it is drawn from urn A is:

(A) $\frac{5}{18}$ (B) $\frac{5}{16}$ (C) $\frac{4}{17}$ (D) $\frac{7}{18}$

[Main 2024 04 Apr 1st Shift]

66. In a tournament, a team plays 10 matches with probabilities of winning and losing each match as $\frac{1}{3}$ and $\frac{2}{3}$ respectively. Let x be the number of matches that the team wins, and y be the number of matches that team loses. If the probability $P(|x-y|\le 2)$ is p, then $3^9 p$ equals _______.

[Main 2024 04 Apr 2nd Shift]

67. The coefficients a, b, c in the quadratic equation $ax^2+bx+c=0$ are chosen from the set $\{1, 2, 3, 4, 5, 6, 7, 8\}$. The probability of this equation having repeated roots is:

(A) $\frac{1}{128}$ (B) $\frac{1}{64}$ (C) $\frac{3}{256}$ (D) $\frac{3}{128}$

[Main 2024 05 Apr 1st Shift]

68. The coefficients a, b, c in the quadratic equation $ax^2+bx+c=0$ are from the set $\{1, 2, 3, 4, 5, 6\}$. If the probability of this equation having one real root bigger than the other is p, then $216p$ equals:

(A) 57 (B) 76 (C) 38 (D) 19

[Main 2024 05 Apr 2nd Shift]

69. Let the mean and the standard deviation of the probability distribution

X_i	α	1	0	-3
$P(X)$	$\frac{1}{3}$	K	$\frac{1}{6}$	$\frac{1}{4}$

be μ and σ, respectively. If $\sigma-\mu=2$, then $\alpha+\mu$ is equal to _______.

[Main 2024 05 Apr 2nd Shift]

70. A company has two plants A and B to manufacture motorcycles. 60% of the motorcycles are manufactured at plant A and the remaining are manufactured at plant B. 80% of the motorcycles manufactured at plant A are rated of the standard quality, while 90% of the motorcycles manufactured at plant B are rated of the standard quality. A motorcycle picked up randomly from the total production is found to be of the standard quality. If p is the probability that it was manufactured at plant B, then $126p$ is

(A) 54 (B) 66 (C) 64 (D) 56

[Main 2024 06 Apr 1st Shift]

71. If three letters can be posted to any one of the 5 different addresses, then the probability that the three letters are posted to exactly two addresses is:

(A) $\frac{18}{25}$ (B) $\frac{12}{25}$ (C) $\frac{6}{25}$ (D) $\frac{4}{25}$

[Main 2024 06 Apr 2nd Shift]

72. From a lot of 12 items containing 3 defectives, a sample of 5 items is drawn at random. Let the random variable X denote the number of defective items in the sample. Let items in the sample be drawn one by one without replacement. If variance of X is $\frac{m}{n}$, where $\text{GCD}(m,n)=1$, then $n-m$ is equal to _______.

[Main 2024 06 Apr 2nd Shift]

73. Let the sum of two positive integers be 24. If the probability that their product is not less than $\frac{3}{4}$ times their greatest possible product is $\frac{m}{n}$, where $gcd(m,n)=1$, then $n-m$ equals

(A) 10 (B) 9 (C) 11 (D) 8

[Main 2024 08 Apr 1st Shift]

74. Three balls are drawn at random from a bag containing 5 blue and 4 yellow balls. Let the random variables X and Y respectively denote the number of blue and yellow balls.

If $\bar{X}$ and $\bar{Y}$ are the means respectively, then $7\bar{X}+4\bar{Y}$ is equal to_______.

[Main 2024 08 Apr 1st Shift]

75. There are three bags X, Y and Z. Bag X contains 5 one-rupee coins and 4 five-rupee coins, Bag Y contains 4 one-rupee coins and 5 five-rupee coins and Bag Z contains 3 one-rupee coins and 6 five-rupee coins. A bag is selected at random and a coin drawn from it at random is found to be a one-rupee coin. Then the probability, that it came from bag Y, is:

(A) $\frac{1}{4}$ (B) $\frac{1}{2}$ (C) $\frac{5}{12}$ (D) $\frac{1}{3}$

[Main 2024 08 Apr 2nd Shift]

76. Let a, b and c denote the outcome of three independent rolls of a fair tetrahedral die, whose four faces are marked 1, 2, 3, 4. If the probability that $ax^2+bx+c=0$ has all real roots is $\frac{m}{n}$, $\gcd(m,n)=1$, then $m+n$ is equal to _______.

[Main 2024 09 Apr 1st Shift]

77. If an unbiased dice is rolled thrice, then the probability of getting a greater number in the ith roll than the number obtained in the $(i-1)^{th}$ roll, $i=2,3$, is equal to

(A) $\frac{3}{54}$ (B) $\frac{2}{54}$ (C) $\frac{1}{54}$ (D) $\frac{5}{54}$

[Main 2024 9th Apr 2nd Shift]

7.13 Answer Key

7.13.1 Know Your Concepts

1. $\frac{1}{20}$ **2.** $\frac{4}{165}$ **3.** $\frac{2}{9}$ **4.** (1:6, 6:1) **5.** $\frac{28}{55}$ **6.** $\frac{37}{40}$ **7.** $\frac{23}{30}$

8. $\frac{3}{8}$ **9.** $\frac{3}{4}$ **10.** $1-(0.7)^7$ **11.** $\frac{1}{2}$ **12.** $\frac{1}{2},\frac{1}{3}$ **13.** $\frac{120[18080]}{16^6}$

14. $\frac{14}{33}$ **15.** $\frac{3}{125}$ **16.** $\frac{2}{11}$ **17.** $\frac{4}{19}$ **18.** $\frac{13}{20}$ **19.** $\left[-\frac{1}{4},\frac{1}{3}\right]$ **20.** $\frac{5}{38}$

7.13.2 JEE Main: Sample Questions

Single Answer Type

1. (A)	**2.** (C)	**3.** (B)	**4.** (C)	**5.** (B)	**6.** (D)	**7.** (A)	**8.** (C)
9. (B)	**10.** (B)	**11.** (D)	**12.** (A)	**13.** (A)	**14.** (B)	**15.** (A)	**16.** (B)
17. (C)	**18.** (B)	**19.** (C)	**20.** (A)	**21.** (B)	**22.** (A)	**23.** (A)	**24.** (C)
25. (C)	**26.** (A)	**27.** (D)	**28.** (C)	**29.** (D)	**30.** (A)	**31.** (C)	**32.** (D)
33. (A)	**34.** (D)	**35.** (A)	**36.** (A)	**37.** (A)	**38.** (C)	**39.** (C)	**40.** (B)
41. (C)	**42.** (A)	**43.** (D)	**44.** (C)	**45.** (B)	**46.** (D)	**47.** (A)	**48.** (C)
49. (A)	**50.** (C)	**51.** (A)	**52.** (D)	**53.** (A)	**54.** (C)	**55.** (D)	**56.** (D)
57. (C)	**58.** (D)						

Integer Answer Type

1. (3)	**2.** (2)	**3.** (8)	**4.** (4)	**5.** (9)	**6.** (5)	**7.** (7)	**8.** (5)
9. (8)	**10.** (1)	**11.** (5)	**12.** (2)	**13.** (4)	**14.** (3)	**15.** (1)	**16.** (7)
17. (6)	**18.** (4)	**19.** (6)	**20.** (2)				

7.13.3 JEE Advanced: Sample Questions

Single Answer Type

1. (C)	**2.** (A)	**3.** (B)	**4.** (B)	**5.** (D)	**6.** (C)	**7.** (B)	**8.** (C)
9. (C)	**10.** (A)	**11.** (C)	**12.** (C)	**13.** (A)	**14.** (C)	**15.** (D)	**16.** (D)
17. (C)	**18.** (B)	**19.** (D)	**20.** (C)	**21.** (C)	**22.** (A)	**23.** (C)	**24.** (C)
25. (B)							

Multiple Answer Type

1. (A, B, C, D) **2.** (A, C, D) **3.** (A, C, D) **4.** (A, B)
5. (A, C) **6.** (B, C, D) **7.** (A, C) **8.** (A, B, C, D)
9. (A, D) **10.** (A, B) **11.** (A, B) **12.** (A, C)
13. (C, D) **14.** (A, C, D) **15.** (A, D) **16.** (B, D)
17. (C, D) **18.** (A, D) **19.** (A, B, C)

Comprehension Based Questions

1. (A) **2.** (B) **3.** (C) **4.** (B) **5.** (A) **6.** (B) **7.** (A) **8.** (B)
9. (A) **10.** (D) **11.** (A) **12.** (B) **13.** (A) **14.** (B) **15.** (D)

Match Matrix

1. (A – T; B – P; C – Q; D – S) **2.** (A – Q; B – P; C – R; D – Q)
3. (A – R; B – Q; C – U; D – P) **4.** (A – R, S; B – P, Q, R, S; C – P, Q, R, S; D – P)
5. (A – Q, R; B – R; C – P; D – P, Q, R, S)

Subjective Answer Type (Hard Nut)

1. $\frac{1}{5}$ **2.** $\frac{1}{20}$ **3.** $\frac{{}^{13}C_3\times{}^{3}C_1\times{}^{4}C_2\times{}^{4}C_1\times{}^{4}C_1+{}^{13}C_2\times{}^{2}C_1\times{}^{4}C_3\times{}^{4}C_1}{{}^{52}C_4}$ **4.** $\frac{1}{6}$

5. $\frac{1}{7}$ **6.** (1) **7.** $\frac{1}{4}$ **8.** $\frac{(m-3)(m-4)}{(m-1)(m-2)}$ **9.** $\frac{9}{11}$ **10.** $\frac{10}{33}$ **12.** $\frac{20\times10!\times5!}{15!}$

13. $\frac{1}{9}$ **14.** $\frac{209}{343}$ **15.** $\frac{1}{6}$ **16.** $\frac{1}{360}$ **17.** $\frac{3}{5}$ **18.** $\frac{1}{4}$ **19.** $\frac{43}{45}$ **20.** $\frac{11}{3600}$

7.13.4 JEE Main: Archive

1. (C) **2.** (A) **3.** (A) **4.** (479) **5.** (C) **6.** (D) **7.** (D) **8.** (33)
9. (C) **10.** (C) **11.** (B) **12.** (19) **13.** (A) **14.** (C) **15.** (B) **16.** (C)
17. (D) **18.** (B) **19.** (B) **20.** (D) **21.** (A) **22.** (A) **23.** (56) **24.** (D)
25. (B) **26.** (96) **27.** (C) **28.** (D) **29.** (B) **30.** (120) **31.** (B) **32.** (9)
33. (C) **34.** (B) **35.** (B) **36.** (14) **37.** (A) **38.** (10) **39.** (B) **40.** (A)
41. (D) **42.** (D) **43.** (A) **44.** (A) **45.** (D) **46.** (A) **47.** (D) **48.** (27)
49. (C) **50.** (10) **51.** (C) **52.** (B) **53.** (C) **54.** (B) **55.** (B) **56.** (12)
57. (C) **58.** (C) **59.** (B) **60.** (A) **61.** (C) **62.** (D) **63.** (D) **64.** (A)
65. (A) **66.** (8288) **67.** (B) **68.** (C) **69.** (5) **70.** (A) **71.** (B) **72.** (71)
73. (A) **74.** (17) **75.** (D) **76.** (19) **77.** (D)

7.14 Solutions

7.14.1 Know Your Concepts

1. $n(E)=1$

$n(S)={}^{5}P_2=20$

$P(E)=\frac{1}{20}$

2. $n(S)=\frac{11!}{4!4!2!}=5\cdot7\cdot9\cdot10\cdot11$

$n(E)=\frac{8!}{4!2!}=4\cdot5\cdot6\cdot7$

$P(E)=\frac{n(E)}{n(S)}=\frac{4\cdot5\cdot6\cdot7}{5\cdot7\cdot9\cdot10\cdot11}=\frac{4}{165}$

3. If E_1, E_2 and E_3 denote the event of total of 3 or 6 or 12.

$E_1=\begin{cases}1, & 2\\ 2, & 1\end{cases}\Rightarrow n(E_1)=2!=2$

$E_2=\begin{cases}1, & 5\\ 2, & 4\\ 3, & 3\end{cases}\Rightarrow n(E_2)=2!+2!+1=5$

$E_3=\{6, 6=n(E_3)=1\ \ n(S)=6^2=36$

$P(E_1\cup E_2\cup E_3)=\frac{8}{36}=\frac{2}{9}$

4. Odds in favour—1:6; odds against—6:1
[Do it Yourself]

5. Two balls can be drawn out of 11 balls by ${}^{11}C_2$ ways. Probability to select a white ball and a black ball

$$= \frac{{}^4C_1 \times {}^7C_1}{{}^{11}C_2} = \frac{4\times7\times2}{11\times10} = \frac{28}{55}$$

6. $P(A) = \frac{75}{100} = \frac{3}{4}, P(B) = \frac{70}{100} = \frac{7}{10}$

$P(A \cap B) = P(A)P(B) = \frac{21}{40}$

$P(A \cup B) = \frac{3}{4} + \frac{7}{10} - \frac{21}{40} = \frac{30+28-21}{40} = \frac{37}{40}$

7. $\frac{23}{30}$

[Do it yourself]

8. $n(S) = 36$

If E denotes the event that the sum is 4

$E = \{(1,3),(2,2),(3,1)\}$

$n(E) = 3 \Rightarrow P(E) = \frac{3}{36} = \frac{1}{12}$

If E_1 denotes the event that the sum is 8

$E_1 = \{(2,6),(3,5),(4,4),(5,3),(6,2)\}$

$n(E_1) = \frac{5}{36}$

$P(\bar{E}\bar{E}_1) = \frac{28}{36}$, because there are 8 cases in favour of getting a sum of 4 or 8. 4 can come before 8 in the following cases.

$$P(E) + P(\bar{E}\bar{E}_1)P(E) + P(\bar{E}\bar{E}_1)P(\bar{E}\bar{E}_1)P(E) + \ldots$$

$$= \frac{1}{12} + \frac{28}{36}\times\frac{1}{12} + \left(\frac{28}{36}\right)^2 \times \frac{1}{12} + \ldots$$

$$= \frac{1}{12}\cdot\frac{1}{1-\frac{28}{36}} = \frac{1}{12}\cdot\frac{36}{8} = \frac{3}{8}$$

9. Let E_1 denote the event of the first student solving the problem, E_2 the second student solving the problem, E_3 the third student solving the problem.

Let $E = (E_1 \cup E_2 \cup E_3)$ denote the event of the problem being solved by at least one student, that is, the event that the problem will be solved.

$P(E_1) = \frac{1}{2}, P(E_2) = \frac{1}{3}, P(E_3) = \frac{1}{4}$

$P(E_1') = 1 - \frac{1}{2} = \frac{1}{2}; P(E_2') = 1 - \frac{1}{3} = \frac{2}{3}$

and $P(E_3') = 1 - \frac{1}{4} = \frac{3}{4}$

Since, E_1, E_2, E_3 are independent events

$\therefore P(E) = P(E_1 \cup E_2 \cup E_3)$

$= 1 - P(E_1')P(E_2')P(E_3')$

$= 1 - \frac{1}{2}\times\frac{2}{3}\times\frac{3}{4} = 1 - \frac{1}{4} = \frac{3}{4}$

10. P (at least one rainy day)

$= 1 - P$ (no rainy day) $= 1 - (0.7)^7$

11. The probability of getting a head in a single toss of a coin is $p = \frac{1}{2}$. The probability of getting 5 or 6 in a single throw of a die is $q = \frac{2}{6} = \frac{1}{3}$. Therefore, the required probability is

$$q + (1-q)(1-p)q + (1-q)(1-p)(1-q)(1-p)q + \ldots$$

$$= \frac{q}{1-(1-q)(1-p)} = \frac{\frac{1}{3}}{1-\frac{2}{3}\times\frac{1}{2}} = \frac{1}{2}$$

12. Let $P(X) = a, P(Y) = b$

$P(X \cap Y) = \frac{1}{6} \Rightarrow P(X)\cdot P(Y) = \frac{1}{6}$

$\Rightarrow ab = \frac{1}{6}$

Again, $P(X' \cap Y') = \frac{1}{3}$

$\Rightarrow P(X \cup Y)' = \frac{1}{3}$

$P(X \cup Y) = 1 - P(X \cup Y)'$

$= 1 - \frac{1}{3} = \frac{2}{3}$

$\therefore P(X) + P(Y) - P(X \cap Y) = \frac{2}{3}$

$\Rightarrow a+b-\frac{1}{6}=\frac{2}{3} \quad \Rightarrow a+b=\frac{5}{6}$

$\therefore a+\frac{1}{6a}=\frac{5}{6} \Rightarrow 6a^2-5a+1=0$

$\Rightarrow a=\frac{1}{2},\frac{1}{3} \quad \therefore b=\frac{1}{3},\frac{1}{2}$

Thus, $P(X)=\frac{1}{2}, P(Y)=\frac{1}{3}$ or

$P(X)=\frac{1}{3}, P(Y)=\frac{1}{2}$

13. Since replacement is possible, the total number of ways of choosing 6 letters is 16^6. We can choose 2 letters (which will repeat) from 16 in ${}^{16}C_2$ ways, let these letters be A and B. We have the following cases.

Case I: A and B both come two times and the rest two from the remaining 14.

So, the number of ways in this case is

${}^{14}C_2 \times \frac{6!}{2!2!}=16380$

Case II: A appears 3 times and B appears 2 times or A 2 times and B 3 times and one from remaining 14. The number of ways in this case is

$2\times {}^{14}C_1 \times \frac{6!}{3!2!}=1680$

Case III: A and B both appear 3 times.

Number of ways in this case is $\frac{6!}{3!3!}=20$

Total number of favorable cases

$=120[16380+1680+20]=120[18080]$

$\therefore$ Required probability $=\frac{120[18080]}{16^6}$

14. Required probability $=\frac{{}^4C_2\times{}^8C_3}{{}^{12}C_5}=\frac{14}{33}$

15. We have to select a woman who knows Hindi and she should be a teacher.

$\therefore$ Probability to select a woman $=\frac{20}{50}$

Probability to select a Hindi knowing candidate $=\frac{10}{50}$

Probability to select a teacher $=\frac{15}{50}$

Since the events are independent, the probability that the university selects a Hindi knowing woman teacher

$=\frac{20}{50}\times\frac{15}{50}\times\frac{10}{50}=\frac{3}{125}$

16. In case of ring or round table, we fix the position of one, so 12 persons can stand to form a ring by 11! ways and 3 persons between A and B can be selected by ${}^{10}C_3$ ways. A and B can interchange their position by 2! ways. Also 3 persons between A and B can stand by 3! ways and the other 7 in 7! ways.

Thus, the number of favourable ways

$=2!3!7!\,{}^{10}C_3$

Hence, required probability

$=\frac{2!3!7!\,{}^{10}C_3}{(11!)}=\frac{2}{11}$

17. Let x and y be two natural numbers. Then $x+y=20$, we have to find out $P(xy<50)$

It is given that

$x+y=20,\ y<\frac{50}{x}$

$\therefore x+\frac{50}{x}>20$

$\Rightarrow (x-10)^2>50$

$\Rightarrow 10-5\sqrt{2}<x<10+5\sqrt{2}$

We have to find the probability that

$(x-10)^2>50$

The number of ways in which x can take values is 19, since $x+y=20$

Clearly, $E=\{1,2,18,19\}, n(E)=4$

$n(s)=19$

Hence, required probability $=\frac{4}{19}$

18. It is given that the graph of the function $y=16x^2+8(a+5)x-7a-5$ is strictly above the x-axis.

Therefore, $y>0\ \forall\ x\in R$

$\Rightarrow 7a+5<0\ [c>0]$

$\Rightarrow a<-\frac{5}{7}$

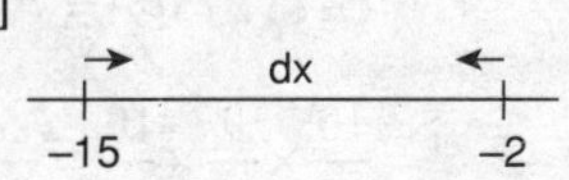

Also, $D<0 \Rightarrow 64(a+5)^2+64(7a+5)<0$

$\Rightarrow (a+15)(a+2)<0$

$\therefore$ Required probability $= \dfrac{\int_{-15}^{-2} dx}{\int_{-20}^{0} dx}$

$= \dfrac{13}{20}$

19. Let A, B and C be three mutually exclusive and exhaustive events, where

$P(A)=\dfrac{1-3p}{2}$, $P(B)=\dfrac{1+4p}{3}$

$P(C)=\dfrac{1+p}{6}$

Since events A, B, C are mutually exclusive

$P(A\cup B\cup C)=P(A)+P(B)+P(C)$

$=\dfrac{1-3p}{2}+\dfrac{1+4p}{3}+\dfrac{1+p}{6}=1$

From question, A, B, C are exhaustive events

$\therefore P(A\cup B\cup C)=1 \Rightarrow 1=1$, true for all values of p

Since, $0\le P(A)\le 1$

$\therefore 0\le \dfrac{1-3p}{2}\le 1 \Rightarrow -\dfrac{1}{3}\le p\le \dfrac{1}{3}$

Again, $0\le P(B)\le 1 \Rightarrow -\dfrac{1}{4}\le p\le \dfrac{1}{2}$

Also, $0\le \dfrac{1+p}{6}\le 1 \Rightarrow -1\le p\le 5$

Hence, $-\dfrac{1}{4}\le p\le \dfrac{1}{3}$

$\Rightarrow p\in\left[-\dfrac{1}{4},\dfrac{1}{3}\right]$

20. Let A: denote the event of getting a card with mark I in first the draw,

B: denote the event with mark I in the second draw,

C: denote the event with mark T in the third draw.

Required probability is

$P(A\cap B\cap C)=P(A)P\left(\dfrac{B}{A}\right)P\left(\dfrac{C}{A\cap B}\right)$

$-\dfrac{10}{20}\times\dfrac{9}{19}\times\dfrac{10}{18}=\dfrac{5}{38}$

7.14.2 JEE Main: Sample Questions

Single Answer Type

1. **(A)**

As we are considering all possible six-digit numbers, half of these would have the sum of their digits to be even and half of these would have sum of their digits to be odd.

Hence the required probability would be $\dfrac{1}{2}$.

2. **(C)**

There are 64 small squares on a chessboard.

$\Rightarrow$ Total number of ways to choose two squares $= {}^{64}C_2 = 32.63$

For favourable ways we must choose two consecutive small squares for any row or any columns

$\Rightarrow$ Number of favorable ways = (7.8)2

$\Rightarrow$ Required probability $=\dfrac{7.8.2}{32.63}=\dfrac{1}{18}$

3. **(B)**

Let A be the event that the first drawn number is even and B be event that the second drawn number is odd.

Total number of even numbers $=\left[\dfrac{17}{2}\right]=8$

$\Rightarrow P(A)=\dfrac{8}{47}$ and $P(B)=\dfrac{9}{17}\Rightarrow P(A\cap B)$

$=\dfrac{72}{289}$

4. **(C)**

Let us assume that 'A' wins after n deuces, $n\in(0,\infty)$

Probability of a deuce $=\dfrac{2}{3}.\dfrac{2}{3}+\dfrac{1}{3}.\dfrac{1}{3}=\dfrac{5}{9}$

(A wins his serve then B wins his serve or A loses his serve then B also loses his serve)

Now probability of 'A' winning the game

$=\sum_{n=0}^{\infty}(5/9)^n.\left(\dfrac{2}{3}\right)\dfrac{1}{3}=\dfrac{1}{1-(5/9)}.\dfrac{2}{9}=\dfrac{1}{2}$

5. **(B)**

If the first throw is four, then the sum of numbers appearing on the last two throws must be equal to eleven. That means last two throws are (6, 5) or (5, 6).

Now there are 10 ways to get the sum as 15.

[(5, 5, 5) (4, 5, 6) (3, 6, 6)]

$\Rightarrow$ Required probability $=\dfrac{2}{10}=\dfrac{1}{5}$

6. **(D)**
Total ways = 15^7
For favourable ways, we must have 7 coupons numbered from 1 to 9 so that '9' is selected at least once. Thus, the total number of favorable ways are, $9^7 - 8^7$

$\Rightarrow$ Required probability $= \dfrac{9^7 - 8^7}{15^7}$

7. **(A)**
A score less than 5 means the occurrence of 1, 2, 3, or 4. Now, on the last throw we should not obtain a score less then 2, that is, 1. Clearly the favourable outcomes are 2, 3 or 4.

Thus, the required probability $= \dfrac{3}{4}$.

8. **(C)**
Let the coin be tossed 'n' times and X be the random variable representing the number of heads appearing in 'n' trials.

$$P(X=7) = P(X=9) \Rightarrow {}^nC_7\left(\frac{1}{2}\right)^7\left(\frac{1}{2}\right)^{n-7}$$

$$= {}^nC_9\left(\frac{1}{2}\right)^{n-9} \times \left(\frac{1}{2}\right)^9$$

$\Rightarrow {}^nC_7 = {}^nC_9 \Rightarrow n = 16$

Now $P(X=2) = {}^{16}C_2\left(\dfrac{1}{2}\right)^2\left(\dfrac{1}{2}\right)^{14}$

$$= \frac{{}^{16}C_2}{2^{16}} = \frac{16.15}{2^{17}} = \frac{15}{2^{13}}.$$

9. **(B)**
The third six occurs on the 8th trial. It means that in the first 7 trials we must get exactly 2 sixes and the 8th trial must result in a six.

$\Rightarrow$ Required probability $= {}^7C_2 \cdot \left(\dfrac{1}{6}\right)^2 \cdot \left(\dfrac{5}{6}\right)^5 \cdot$

$$\left(\frac{1}{6}\right) = \frac{{}^7C_2.5^5}{6^8}$$

10. **(B)**
Total number of ways in which 4 papers can be distributed among 7 teachers = 7^4
Now exactly 2 teachers out of 7 can be chosen in ${}^7C_2$ ways and the total number of ways in which 4 papers can be given to these 2 teachers (each one getting at least one) = $(2^4 - 2) = 14$

$\Rightarrow$ Total number of ways in which exactly 2 teachers check all four papers = ${}^7C_2 \cdot 14$ = 21.14

$\Rightarrow$ Required probability $= \dfrac{21.14}{7^4} = \dfrac{3.2}{7^2} = \dfrac{6}{49}$.

11. **(D)**
$A \le B$
$P(A > B)$

B draw	A draw	Total numbers
1	2, 3, …, 6	5
2	3, 4, …, 6	4
3	4, 5, 6	3
4	5, 6	2
5	6	1
		Total = 15

$P(A > B) = \dfrac{15}{30}$

$P(A \le B) = 1 - \dfrac{15}{30} = \dfrac{7}{12}$.

12. **(A)**

$P(A) = \dfrac{7}{10}$, $P(B) = \dfrac{8}{10}$

$P(C) = \dfrac{7}{10}\left(1 - \dfrac{8}{10}\right) + \left(1 - \dfrac{7}{10}\right)\dfrac{8}{10}$

$= \dfrac{14}{100} + \dfrac{24}{100} = \dfrac{38}{100} = 0.38$

13. **(A)**
$P(A\bar{B} \cup \bar{A}B) = P(A) - P(AB) + P(B) - P(AB)$
$= P(A) + P(B) - 2P(AB)$

14. **(B)**
$P(A) + P(\bar{B}) - P(A)\,P(\bar{B}) = 0.8$
$0.3 + (1 - x) - 0.3(1 - x) = 0.8$
$(1 - x)\,(0.7) = 0.5$
$1 - x = \dfrac{5}{7} \Rightarrow x = 1 - \dfrac{5}{7} = \dfrac{2}{7}$

15. **(A)**
Given $P(A) + P(B) - 2P(AB) = P$
$P(B) + P(C) - 2P(BC) = P$
$P(C) + P(A) - 2P(AC) = P$
$2[P(A) + P(B) + P(C) - P(AB) + P(BC) + P(CA)] = \dfrac{3P}{2}$
and $P(ABC) = P^2$
Hence $P(A \cup B \cup C) = \dfrac{3P}{2} + P^2 = \dfrac{3P + 2P^2}{2}$.

16. **(B)**

Required probability $= \frac{3!}{4!} = \frac{1}{4}$

17. **(C)**

$x_1 + x_2 + x_3 = 75$

coefficient x^{75} in $(x^1 + x^2 + \ldots + x^{100})^3 = {}^{74}C_2$

Hence probability $= \frac{{}^{74}C_2}{{}^{100}C_3}$.

18. **(B)**

Drawn balls should be such that one of them is white and another is black.

Total ways of drawing two balls $= {}^8C_2 = 28$

Total number of favourable ways $= {}^5C_1 \cdot {}^3C_1 = 15$

Thus, required probability $= \frac{15}{28}$

19. **(C)**

$P(A \cup B) = \frac{3}{5}, P A \cap B) = \frac{1}{5}$

Now, $P(A \cup B) = P(A) + P(B) - P(A \cap B)$

$= 2 - P(A \cap B) - P(\bar{A}) - P(\bar{B})$

$\Rightarrow \quad P(\bar{A}) + P(\bar{B}) = 2 - P(A \cap B) - P(A \cup B)$

$= 2 - \frac{1}{5} - \frac{3}{5} = \frac{6}{5}$

20. **(A)**

Total number of integers that are divisible by six = 38

Total number of integers that are divisible by eight = 28

Total number of integers that are divisible by six as well as by eight, that is, by 24 = 8

Thus, total number of integers that are divisible by 6 or 8 = 38 + 28 –8 = 58

Thus, required probability $= \frac{58}{200} = \frac{29}{100}$

21. **(B)**

In the case sum can be 4 or 8 or 12.

Total ways of getting the sum as 4

= 3 (namely (2, 2), (1, 3), (3, 1))

Total ways of getting the sum as 8

= 5 (namely (2, 6), (6, 2), (3, 5), (5, 3), (4, 4))

Total ways of getting the sum as 12 = 1 (namely (6, 6))

Thus, total ways of getting the sum as a multiple of four = 3 + 5 + 1 = 9

Thus, required probability $= \frac{9}{36} = \frac{1}{4}$

22. **(A)**

$P(A) = \frac{1}{4}, P(B) = \frac{1}{2}, P(A \cap B) = \frac{7}{50}$

Now, $P(\bar{A} \cap \bar{B}) = P(\bar{A}) + P(\bar{B}) - P(\bar{A} \cup \bar{B})$

$= 1 - P(A) + 1 - P(B) - (1 - P(A \cap B))$

$= 1 - \frac{1}{4} - \frac{1}{2} + \frac{7}{50} = \frac{39}{100}$

23. **(A)**

Total formed numbers that begin with an odd digit $= {}^5C_1 \cdot {}^8P_4 = 5(8)\,(7)\,(6)\,(5)$

Total formed numbers that end with an odd digit $= {}^5C_1 \cdot {}^8P_4 = (8)\,(7)\,(6)\,(5)$

Total formed numbers that begin with an odd digit and also end with an odd digit

$$= {}^5C_2 \cdot 2! \cdot {}^7P_3 = 5 \cdot (4)\,(7)\,(6)\,(5)$$

Thus, total formed numbers that begin with an odd digit or end with an odd digit is equal to $5 \cdot 7 \cdot 6 \cdot 60$

Total formed numbers $= {}^9P_5 = 9 \cdot 8 \cdot 7 \cdot 6 \cdot 5$

Thus, required probability $= \frac{5}{6}$

24. **(C)**

Probability that *A* and *B* can solve but '*C*' is unable to solve

$= \frac{1}{2} \cdot \frac{1}{3} \cdot \frac{3}{4} = \frac{1}{8}$

Probability that *A* and *C* can solve but '*B*' is unable to solve

$= \frac{1}{2} \cdot \frac{2}{3} \cdot \frac{1}{4} = \frac{1}{12}$

Probability that *B* and *C* can solve but '*A*' is unable to solve

$= \frac{1}{2} \cdot \frac{1}{3} \cdot \frac{1}{4} = \frac{1}{24}$

Probability that all of them can solve the problem

$= \frac{1}{2} \cdot \frac{1}{3} \cdot \frac{1}{4} = \frac{1}{24}$

Thus, required probability that at least two of them can solve the problem

$= \frac{7}{24}$

25. **(C)**

Let 'A' denote the event that 'P_1' dies in a year and 'B' denote the event that 'P_2' dies in a year. Thus, required probability

$= P(\overline{A} \cap B) + P(A \cap \overline{B})$

$= P(A) + P(B) - 2(A \cap B)$

$= p + q - 2pq$

26. **(A)**

Probability of showing equal outcomes in any specific roll of dice is $\frac{6}{36}$, that is, $\frac{1}{6}$.

Thus, required probability

= Getting exactly 2 successes in 4 trials.

$= {}^4C_2 . \left(\frac{1}{6}\right)^2 . \left(\frac{5}{6}\right)^2 = \frac{25}{216}$

27. **(D)**

Let

A: Minimum of the chosen number is 3.

B: Maximum of the chosen number is 7.

$$P(A) = \frac{{}^7C_2}{{}^{10}C_3} = \frac{21}{120}$$

$$P(B) = \frac{{}^6C_2}{{}^{10}C_3} = \frac{15}{120}$$

$$P(A \cap B) = \frac{{}^3C_1}{{}^{10}C_3} = \frac{3}{120}$$

Thus, required probability

$= P(A) + P(B) - P(A \cap B) = \frac{11}{40}$

28. **(C)**

Possible set of three consecutive numbers are $\{1, 2, 3\}$, $\{2, 3, 4\}$, $\{3, 4, 5\}$, ..., $\{48, 49, 50\}$, which are 48 in number.

Thus, required probability $= \frac{{}^{48}C_1}{{}^{50}C_3} = \frac{3}{(25)(49)}$

29. **(D)**

Total number of even numbers in the set is 18, and total number of odd numbers is 16.

A: Sum of selected numbers is even.

B: Selected numbers are odd.

$$P(A) = \frac{{}^{18}C_2 + {}^{16}C_2}{{}^{34}C_2}, \ P(A \cap B) = \frac{{}^{16}C_2}{{}^{34}C_2}$$

$$P(B | A) = \frac{P(A \cap B)}{P(A)} = \frac{{}^{16}C_2}{{}^{16}C_2 + {}^{18}C_2} = \frac{40}{91}$$

30. **(A)**

(i) Selected balls can be W W B or W B B (when red colour is missing).
Corresponding probability

$$= \frac{{}^5C_2 . {}^6C_1 + {}^5C_1 . {}^6C_2}{{}^{15}C_3} = \frac{135}{{}^{15}C_3}$$

(ii) Selected balls can be R R W or R W W (when black colour is missing).
Corresponding probability

$$= \frac{{}^4C_2 . {}^5C_1 + {}^4C_1 . {}^5C_2}{{}^{15}C_3} = \frac{70}{{}^{15}C_3}$$

(iii) Selected balls can be RBB, RRB (when white colour is missing).
Corresponding probability

$$= \frac{{}^4C_1 . {}^6C_2 + {}^4C_2 . {}^6C_1}{{}^{15}C_3} = \frac{96}{{}^{15}C_3}$$

Thus, required probability $= \frac{301}{455}$

31. **(C)**

Probability that the boy and his sister are both selected

$$= \frac{{}^8C_1}{{}^{10}C_3} = \frac{1}{15}$$

Thus, required probability $= 1 - \frac{1}{15} = \frac{14}{15}$

32. **(D)**

E_1: Person has brown hair.

E_2: Person has brown hair.

$$P(E_1) = \frac{40}{100} = \frac{2}{5}$$

$$P(E_2) = \frac{25}{100} = \frac{1}{4}$$

$$P(E_1 \cap E_2) = \frac{15}{100} = \frac{3}{20}$$

$$P(E_2 / E_1) = \frac{P(E_1 \cap E_2)}{P(E_1)}$$

$$= \frac{3/20}{2/5} = \frac{3}{8}$$

33. **(A)**

E_1: Event that dice show different outcomes.

E_2: Outcome is either 6 or 10.

$P(E_1)=\frac{5}{6}, P(E_1 \cap E_2)=\frac{4+2}{36}=\frac{1}{6}.$

As 6 can occur in four ways, namely, (2, 4), (4, 2) (1, 5), (5, 1) and ten can occur in 2 ways, namely, (6, 4) and (4, 6), the required probability,

$$P(E_1 / E_2)=\frac{P(E_1 \cap E_2)}{P(E_1)}=\frac{1/6}{5/6}=\frac{1}{5}$$

34. **(D)**

Total number of ways of selecting the second ball

$= {}^{19}C_1 = 19.$

Total number of ways of selecting the second ball suitably

$= {}^{14}C_1 = 14.$

Thus, required probability $\frac{14}{19}$.

35. **(A)**

Following equally likely outcomes may occur when one of the dice show four; (4, 1), (1, 4) (4, 2), (2, 4) (4, 3), (3, 4) (4, 4), (4, 4) (4, 5), (5, 4) (4, 6), (6, 4).

Out of these eleven outcomes exactly 2 outcomes favour the cause of the second die showing a six.

Thus, required probability $=\frac{2}{11}$.

36. **(A)**

3 and 5 are the only odd prime numbers, among the possible outcomes. The following are the outcomes when one of them show odd prime numbers,

(3, 1), (1, 3) (3, 2), (2, 3) (3, 3), (3, 4) (4, 3), (3, 5) (5, 3), (3, 6) (6, 3), (5, 1), (1, 5), (5, 2), (2, 5), (5, 4), (4, 5), (5, 5), (5, 6), (6, 5).

Out of these 20 equally likely outcomes exactly 4 favour the presence of odd prime numbers on both dice.

Thus, required probability $=\frac{4}{20}=\frac{1}{5}$.

37. **(A)**

Probability of exactly 2 of the three events happening

$= P(A \cap B)+P(A \cap C)+P(B \cap C)- 3P(A \cap B \cap C)$

$=\frac{1}{3}.\frac{1}{2}+\frac{1}{3}.\frac{1}{4}+\frac{1}{2}.\frac{1}{4}-3.\frac{1}{3}.\frac{1}{2}.\frac{1}{4}=\frac{1}{4}$

38. **(C)**

Probability of showing different numbers on a single trial

$=\frac{({}^6C_3)(3!)}{6^3}=\frac{5}{9}.$

Thus, required probability

$={}^3C_2.\left(\frac{5}{9}\right)^2.\frac{4}{9}=\frac{100}{243}.$

39. **(C)**

Let 'A' have 'r' elements, then 'B' has all the remaining $(n-r)$ elements and none of the elements that are already present in A.

Thus, total number of favourable ways

$=\sum_{r=0}^{n} {}^nC_r = 2^n$

Hence, required probability $==\frac{2^n}{4^n}=\frac{1}{2^n}.$

40. **(B)**

Total outcomes $= 2^n$.

Total number of favourable outcomes

$= {}^nC_0 + {}^nC_2 + {}^nC_4 + \ldots {}^nC_{\left[\frac{n}{2}\right]} = 2^{n-1}$

Thus, required probability $=\frac{2^{n-1}}{2^n}=\frac{1}{2}.$

41. **(C)**

Let A's throw be r, $r = 1, 2, 3, \ldots 6$, then B's outcome can be $(r, r+1, \ldots 6)$,

Thus, the required probability

$=\sum_{r=1}^{6}\frac{7-r}{36}=\sum_{r=1}^{6}\frac{r}{36}=\frac{7}{12}.$

42. **(A)**

Probability of winning the prize in a single game

$=\frac{25}{25^2}=\frac{1}{25}.$

In this case, first 4 games must result in exactly two prizes and the 5th game must result in a prize.

Thus, required probability

$= {}^4C_2\left(\frac{1}{25}\right)^2.\left(\frac{24}{25}\right)^2.\frac{1}{25} = \frac{6.24^2}{(25)^5}$

43. (D)

$P(A \cup B) \geq \text{maximum, } \{P(A), P(B)\} = \frac{2}{3}$

$P(A \cap B') \leq \text{minimum, } \{P(A), P(B)\} = \frac{1}{3}$

$P(A \cap B) \geq \text{Maxi } \{0, 1-(P(A)+P(B)))\} = 0$

and $P(A \cap B) \leq \text{minimum } \{P(A), P(B)\} = \frac{1}{2}$

Thus, all the statements are correct.

44. (C)

$n+\frac{100}{n} \leq 50$

$\Rightarrow n^2 - 50n + 100 \leq 0$

$\Rightarrow 25 - 5\sqrt{21} \leq n \leq 25 + 5\sqrt{21}$

$\Rightarrow n = 3, 4, 5 \ldots, 47.$

Thus, favourable number of ways $= (47 - 3 + 1) = 45.$

Thus, required probability $= \frac{45}{100} = \frac{9}{20}.$

45. (B)

E_1: First bag is chosen, $P(E_1) = \frac{1}{2}$

E_2: Second bag is chosen, $P(E_2) = \frac{1}{2}$

A: Drawn number is 4.

Now, $P(A) = P(E_1) \cdot P(A/E_1) + P(E_2) \cdot P(A/E_2)$

$= \frac{1}{2}.\frac{1}{4} + \frac{1}{2}.\frac{1}{6} = \frac{5}{24}$

46. (D)

$P(E_1) = \frac{2}{4} = \frac{1}{2}, P(E_2) = \frac{2}{4} = \frac{1}{2}, P(E_3) = \frac{2}{4} = \frac{1}{2}$

$P(E_1 \cap E_2) = \frac{1}{4} = P(E_1).(E_2)$

$\Rightarrow$ 'E_1' and 'E_2' are independent.

$P(E_1 \cap E_3) = \frac{1}{4} = P(E_1).(E_3)$

$\Rightarrow$ 'E_1' and 'E_3' are independent.

$P(E_2 \cap E_3) = \frac{1}{4} = P(E_2).(E_3)$

$\Rightarrow$ 'E_2' and 'E_3' are independent.

$P(E_1 \cap E_2 \cap E_3) = \frac{1}{4} \neq P(E_1).P(E_2).(E_3)$

$\Rightarrow E_1, E_2$ and E_3 are not independent.

47. (A)

E_1: Specific home is locked.

E_2: Specific home is unlocked.

A: Real estate man get into the home.

$P(E_2) = \frac{40}{100} = \frac{2}{5}, P(E_1) = \frac{3}{5}$

$P(A) = P(E_1) \cdot P(A/E_1) + P(E_2) \cdot P(A/E_2)$

$= \frac{3}{5}.\frac{{}^7C_2}{{}^8C_3} + \frac{2}{5}.1 = \frac{25}{40} = \frac{5}{8}.$

48. (C)

$n_1^4 - n_2^4 = (n_1^2 + n_2^2)(n_1 + n_2)(n_1 - n_2)$

We can segregate the numbers as multiples of five

(5λ), $5\lambda + 1$, $5\lambda + 2$, $5\lambda + 3$, $5\lambda + 4$, as follows:

$R_1 \rightarrow 1 \quad 6 \ldots 5n-4 \longrightarrow 5\lambda+1$

$R_2 \rightarrow 2 \quad 7 \ldots 5n-3 \longrightarrow 5\lambda+2$

$R_3 \rightarrow 3 \quad 8 \ldots 5n-2 \longrightarrow 5\lambda+3$

$R_4 \rightarrow 4 \quad 9 \ldots 5n-1 \longrightarrow 5\lambda+4$

$R_5 \rightarrow 5 \quad 10 \ldots 5n- \longrightarrow 5\lambda$

We can select either both the numbers from R_5, or any two numbers from the first four rows. (As square of any number that is not a multiple of three will be in the form of $5\lambda' + 1$ or $5\lambda' - 1$)

Thus, required probability

$= \frac{{}^nC_2 + {}^{4n}C_2}{{}^{5n}C_2} = \frac{n(n-1) + 4n(4n-1)}{5n(5n-1)}$

$= \frac{17n-5}{5(5n-1)}$

49. (A)

Total ways of selecting two persons

$= {}^nC_2 = \frac{n(n-1)}{2}$

Total ways of selecting two consecutively seated persons

$= {}^{n-1}C_1 = (n-1)$

Thus, required probability

$$= 1 - \frac{(n-1)\,2}{n\,(n-1)} = \frac{n-2}{n}$$

50. **(C)**

Product can't be divisibly by 4 in the following mutually exclusive cases;

(i) All the selected numbers are odd.
corresponding ways ${}^{13}C_3 = 286$

(ii) Two of the selected numbers are odd and another is a multiple of 2 only.
Corresponding ways $= {}^{13}C_2 . {}^{6}C_1 = 468$
Thus, required probability

$$= 1 - \frac{(286+468)}{{}^{25}C_3} = \frac{773}{1150}$$

51. **(A)**

Probability of getting 5 or 6 on a specific roll of dice

$$= \frac{2}{6} = \frac{1}{3}$$

and, probability of getting a tail $= \frac{1}{2}$.

The desired outcome can happen, in general, on $(2r + 1)^{th}$ trial. That means the first $2r$ trials should result neither in a tail nor in 5 or 6, and the $(2r + 1)^{th}$ trial must result in a tail.

If the corresponding probability is p_r then

$$p_r = \left(\frac{1}{2}\right)^r . \left(\frac{2}{3}\right)^r . \frac{1}{2} = \frac{1}{2} . \left(\frac{1}{3}\right)^r$$

Thus, required probability

$$= \sum_{r=0}^{\infty} p_r = \frac{1}{2} \sum_{r=0}^{\infty} \left(\frac{1}{3}\right)^r = \frac{3}{4}.$$

52. **(D)**

Let the outcome on the dice be n_1, n_2 and n_3. We first try to find the number of ways in which sum of outcomes is at the most equal to seven.

For this, $n_1 + n_2 + n_3 \leq 7$, where $n_i \in [1, 6]$.

$\Rightarrow (n_1 - 1) + (n_2 - 1) + (n_3 - 1) + y_4 = 4$

That is, $y_1 + y_2 + y_3 + y_4 = 4$, where y_1, y_2, y_3, y_4 are non-negative integers. This can happen in ${}^{4+4-1}C_4$, that is, ${}^{7}C_4$ ways. Thus, the required probability

$$= 1 - \frac{{}^{7}C_4}{6^3} = \frac{181}{216}$$

53. **(A)**

Let the first chosen square be white. In this case, the total number of squares that don't belong to the row or column associated with the selected square is 49. Out of these 25 are white and 24 are black.

Thus, the corresponding probability

$$= \frac{1}{2} . \frac{25}{49}$$

Similarly, if the first square is black, then the corresponding probability

$$= \frac{1}{2} . \frac{25}{49}$$

Thus, the required probability

$$= \frac{1}{2}\left(\frac{25}{49} + \frac{25}{49}\right) = \frac{25}{49}$$

54. **(C)**

These are three mutually exclusive cases:

(i) 2 are from A and another is from either B or C.
Corresponding probability

$$= \frac{{}^{2}C_2 . ({}^{3}C_1 + {}^{4}C_1)}{{}^{9}C_3} = \frac{7}{84}$$

(ii) 2 are from B and another is either from A or C.
Corresponding probability

$$= \frac{{}^{3}C_2 . ({}^{2}C_1 + {}^{4}C_1)}{{}^{9}C_3} = \frac{18}{84}$$

(iii) 2 are from C and another is either from A or B.
Corresponding probability

$$= \frac{{}^{4}C_2 . ({}^{2}C_1 + {}^{3}C_1)}{{}^{9}C_3} = \frac{30}{84}$$

Thus, required probability =

$$\frac{7+18+30}{84} = \frac{55}{84}$$

55. (D)

Total number of ways in which $n_1 + n_2 = 100$ is equal to 99.

Now, $n_1 . n_2 > 1600$

$\Rightarrow n_1(100 - n_1) > 1600$

$\Rightarrow n_1^2 - 100n + 1600 < 0$

$\Rightarrow (n_1 - 80)(n_1 - 20) < 0$

$\Rightarrow 20 < n_1 < 80$

$\Rightarrow 21 \le n_1 \le 79$

Thus, number of favourable ways $= 79 - 21 + 1 = 59$

Hence, required probability $= \frac{59}{99}$.

56. (D)

Total number of ways of dialing the last three digits such that exactly 2 of them are the same

$= {}^{10}C_2 . 2 . \frac{3!}{2!} = 270$

Thus, required probability $= \frac{1}{270}$.

57. (C)

E_1: Ball drawn from B_1 is white.

E_2: Ball drawn from B_1 is red.

A: Both the drawn balls are of same colour.

$P(A) = P(E_1) \cdot P(A/E_1) + P(E_2) \cdot P(A/E_2)$

$= \frac{3}{5} \cdot \frac{5}{11} + \frac{2}{5} \cdot \frac{7}{11} = \frac{29}{55}$

58. (D)

Probability of drawing a white ball at any draw

$= \frac{W}{W+R}$.

Now, P_2 can be the winner in general on $2r^{th}$, $r \ge 1$, draw. That means in the first $(2r - 1)$ draws no player draws a white ball and $2r^{th}$ draw results in a white ball for P_2.

If the corresponding probability is p_r, then

$$p_r = \left(\frac{R}{W+R}\right)^r \cdot \left(\frac{R}{W+R}\right)^{r-1} \cdot \left(\frac{W}{W+R}\right)$$

$$= \left(\frac{R}{W+R}\right)^{2r} \cdot \frac{W}{R}$$

Hence, required probability

$$= \sum_{r=1}^{\infty} p_r = \frac{W}{R} \sum_{r=1}^{\infty} \left(\frac{R}{W+R}\right)^{2r} = \frac{R}{(W+2R)}$$

Integer Answer Type

1. (3)

The sum $7^m + 7^n$ will be divisible by 5 when m, n are odd or are even. Two integers can be

selected (with replacement) out of 100 integers $= (100)^2 \Rightarrow n(S) = (100)^2$

$n(E) = 2\left[{}^{25}C_1 \times {}^{25}C_1 + {}^{25}C_1 \times {}^{25}C_1\right]$

$$P(E) = \frac{2\left[(25)^2 + (25)^2\right]}{(100)^2} = \frac{1}{4}$$

$\therefore 12\lambda = 3$

2. (2)

Let $P(C) = K \Rightarrow P(B) = 3K$ and $P(A) = 6K$

Since, $10K = 1 \Rightarrow K = \frac{1}{10}$

and $P(B) = \frac{3}{10}$

odds in favour of $B = \frac{3}{7}$

$\Rightarrow \lambda_1 = 3, \lambda_2 = 7$

$\therefore \left[\frac{\lambda_2}{\lambda_1}\right] = 2$

3. (8)

Let the three persons be A, B, C and the three houses by H_1, H_2, H_3.

For H_1: Probability of 'A' applying = Probability of 'B' applying = Probability of 'C' applying $= \frac{1}{3}$ each

Similarly, for H_2 and H_3

Probability of all the three persons applying for $H_1 = \frac{1}{27}$. Similarly for H_2 and H_3.

$\therefore$ Required probability $= 3 \times \frac{1}{27} = \frac{1}{9} = \frac{K_1}{K_2}$

$\Rightarrow K_2 - K_1 = 8$

4. (4)
{Do it yourself}

5. (9)
$n(S) = 100$

Now, $\frac{(x-20)(x-40)}{(x-30)} < 0$

$\Rightarrow x \in (-\infty, 20) \cup (30, 40)$

$\Rightarrow E = \{1, 2, 3, \ldots, 19, 31, 32, \ldots, 39\}$

$n(E) = 28$

Required probability $= \frac{28}{100} = \frac{7}{25}$

$\Rightarrow \lambda_1 = 7, \lambda_2 = 25$

$\Rightarrow \left|\frac{\lambda_1 - \lambda_2}{2}\right| = 9$

6. (5)
$b^2 - 4ac < 0$

For $b = 1$ any a and c which can be chosen in 4 ways

For $b = 2$, either $a = 1, c = 2$ or $a = 2, c = 1$ or $a = 2, c = 2$

$\Rightarrow$ Required probability $= \frac{7}{8}$

$\Rightarrow K_1 = 7, K_2 = 8$

$\therefore \frac{K_1 + K_2}{3} = 5$

7. (7)
A fair coin is tossed 99 times

$P(\text{Head} = \text{Success}) = P(\text{tail} = \text{failure}) = \frac{1}{2}$

$P(X = r \text{ that is } r \text{ head}) = {}^{99}C_r \left(\frac{1}{2}\right)^{99}$

which is maximum when $r = 49, 50$

$\Rightarrow \left[\sqrt{\lambda}\right] = 7$

8. (5)
The cubic die is unbiased but marking is different.

So, $P(1) = \frac{1}{6}, P(2) = \frac{2}{6}, P(3) = \frac{1}{2}$

Now in 3 throws a sum of 4 is possible as $1 + 1 + 2$ which can happen in 3 ways.

$\Rightarrow P(\text{total sum} = 4) = 3 \cdot \frac{1}{6} \cdot \frac{1}{6} \cdot \frac{1}{3} = \frac{1}{36}$

Similarly, to get a total of 6 in three throws is possible as $1 + 2 + 3$ or $2 + 2 + 2$

$\Rightarrow P(\text{total sum} = 6) = 6\left(\frac{1}{6}\right)\left(\frac{1}{3}\right)\left(\frac{1}{2}\right) + \left(\frac{1}{3}\right)^3$

$= \frac{1}{6} + \frac{1}{27} = \frac{11}{54}$

So, total probability (sum 4 or 6)

$= \frac{1}{36} + \frac{11}{54} = \frac{25}{108}$

$\therefore \lambda_1 = 25 \Rightarrow \sqrt{\lambda_1} = 5$

9. (8)
Since all letters are distinct

$\Rightarrow n(S) = 10!$

When there are exactly 4 letters between R and E,
the required number of arrangements
$= {}^8C_4 \times 4! \times 2 \times 51$

$P(E) = \frac{2 \times 5 \times 81}{101} = \frac{1}{9}$

$\Rightarrow |\lambda_1 - \lambda_2| = 8$

10. (1)
Number of ways in which 2 end seats are occupied by boys $= {}^7C_2 \times 2!$

Now boys can occupy seats only after an even number of girls.

Already 1 and 15 are occupied by boys.

Hence after 1, a boy can occupy seat No. 4, 6, 8, 10, 12, 14, 2 in ${}^7C_5 \times 5!$ ways

Seats left for girls = 3, 5, 7, 9, 11, 13

Total No. of ways in which girls can sit $= {}^8P_4$

Required probability =

$\frac{{}^9C_4 7! 8!}{15!} \Rightarrow \frac{\lambda_1 + \lambda_2}{\lambda_3} = 1$

11. (5)
Highest number in three throws 4

$\Rightarrow$ At least one of the throws must be equal to 4

Number of ways when three blocks are filled from $\{1, 2, 3, 4\} = 4^3$

$\therefore$ number of ways when filled from $\{1, 2, 3\} = 3^3$

$\therefore$ required number of ways $= 4^3 - 3^3$

$\therefore$ Probability $p = \frac{4^3 - 3^3}{6^3} = \frac{37}{216}$

12. **(2)**
When A and B are mutually exclusive, $P(A \cap B) = 0$

$\therefore P(A \cup B) = P(A) + P(B)$ (1)

$\Rightarrow 0.8 = 0.5 + p \Rightarrow p = 0.3$ (2)

$P(A \cup B) = P(A) + P(B)$

$= P(A) + P(B) - P(A \cap B)$

$= P(A) + P(B) - P(A)P(B)$

$\Rightarrow 0.8 = 0.5 + q - (0.5)q$

$\Rightarrow 0.3 = q/2$

$\Rightarrow q = 0.6$

$\Rightarrow p/q = 2$ (3)

13. **(4)**
Total ways of distribution $= n(S) = 4^5$
Total ways of distribution so that each child get at least one game

$n(E) = 4^5 - {}^4C_1 3^5 + {}^4C_2 2^5 - {}^4C_3$

$= 1024 - 4 \times 243 + 6 \times 32 - 4 = 240$

Required probability $p = \frac{n(E)}{n(S)} = \frac{240}{4^5} = \frac{15}{64}$

14. **(3)**
For ranked 1 and 2 players to be winners and runners up, respectively, they should not be paired against each other in any round

$\Rightarrow p = \frac{30}{31} \times \frac{14}{15} \times \frac{6}{7} \times \frac{2}{3} = \frac{16}{31}$

15. **(1)**
There are n white balls in the turn
$\Rightarrow$ Probability of Mr A to draw two balls of same colour is

$\frac{{}^3C_2 + {}^nC_2}{{}^{n+3}C_2} = \frac{1}{2}$ (given)

$\Rightarrow \frac{6 + n(n-1)}{(n+3)(n+2)} = \frac{1}{2}$

$\Rightarrow n^2 - 7n + 6 = 0$

$\Rightarrow n = 1$ or 6 (1)

Also required probability for Mr B according to the question is

$\frac{3}{n+3}\frac{3}{n+3} + \frac{n}{n+3}\frac{n}{n+3} = \frac{5}{8}$ (given)

Solving, we get $n^2 - 10n + 9 = 0$, $n = 1$ or 9 (2)

From (1) and (2), $n = 1$

16. **(7)**
Let there be x red socks and y blue socks.
Let $x > y$

Then, $\frac{{}^xC_2 + {}^yC_2}{{}^{x+y}C_2} = \frac{1}{2}$

$\Rightarrow \frac{x(x-1) + y(y-1)}{(x+y)(x+y-1)} = \frac{1}{2}$

$\Rightarrow 2x^2 - 2x + 2y^2 - 2y = x^2 + 2xy + y^2 - x - y$

$\Rightarrow x^2 - 2xy + y^2 = x + y$

$\Rightarrow (x - y)^2 = x + y$

$\Rightarrow |x - y| = (x + y)^{1/2}$

Since $x + y \le 17$, $x - y \le \sqrt{17}$

As $x - y$ must be an integer $\Rightarrow x - y = 4$

$\therefore x + y = 16$

Solving, we get $x = 7$

17. **(6)**
Let the two numbers be 'a' and 'b'
According to the question $a + b = 4p$ and $a - b = 4q$ where $p, q \in I$

$\Rightarrow 2a = 4(p+q)$ and $2b = 4(p-q)$

$\Rightarrow a = 2I_1$ and $b = 2I_2$

Hence, both a and b must be even
Also, if $(a-b)$ is a multiple of 4 then $(a+b)$ will also be a multiple of 4
Hence, $n(S) = {}^{11}C_2$

$n(A) = (0, 4), (0, 8), (2, 6), (2, 10), (4, 8), (6, 10) = 6$

$P(A) = \frac{6}{{}^{11}C_2} = \frac{6}{55}$

18. **(4)**
Let the number of green socks be $x > 0$
E : Two socks drawn are of the same colour

$\Rightarrow P(E) = P(RR \text{ or } BB \text{ or } WW \text{ or } GG)$

$= \frac{3}{{}^{6+x}C_2} + \frac{{}^xC_2}{{}^{6+x}C_2}$

$$=\frac{6}{(x+6)(x+5)}+\frac{x(x-1)}{(x+6)(x+5)}=\frac{1}{5}\text{(given)}$$

$$\Rightarrow 5\left(x^2-x+6\right)=x^2+11x+30$$

$$\Rightarrow 4x^2-16x=0$$

$$\Rightarrow x=4$$

19. **(6)**

Total number of cases $n(S)=6^3=216$

Product is prime only when two outcomes are 1 and the third is prime, that is, 2, 3, 5.
If it is 2, 1, 1, then the number of cases is 3.
Similarly, 3 cases for 3, 1, 1 and 5, 1, 1 each.
Hence, favourable cases = 9

Hence, required probability $p=\frac{1}{24}$

$$\Rightarrow \frac{1}{4p}=6$$

20. **(2)**

Total number of cases = $n(S)=6!$

Sum of the given digits is 1 + 2 + 3 + 4 + 5 + 6 = 21, which is divisible by 3
Now we have to form the number which is divisible by 6, then we have to ensure that the digit in unit place is even.

$\Rightarrow$ Favourable cases = $n(A)=3.5!$

Hence, $P(A)=\frac{3.5!}{6!}=\frac{1}{2}$

7.14.3 JEE Advanced: Sample Questions

Single Answer Type

1. **(C)**

$n(S)=36$

$$E=\begin{Bmatrix}(1,1),(1,2),(2,1),(1,4),(4,1),\\(1,6),(6,1),(2,3),(3,2),(2,5),\\(5,2),(3,4),(4,3),(5,6),(6,5)\end{Bmatrix}$$

$$n(E)=15\Rightarrow P(E)=\frac{15}{36}=\frac{5}{12}$$

2. **(A)**

Since,

$P(A\cup B\cup C)=P(A)+P(B)+P(C)-P(A\cap B)-P(B\cap C)-P(C\cap A)+P(A\cap B\cap C)$

$$\Rightarrow P(A\cup B\cup C)=0.3+0.4+0.8-(0.08+0.28+P(B\cap C))+0.09$$

$$=1.23-P(B\cap C)$$

$$\Rightarrow P(B\cap C)=1.23-P(A\cup B\cup C)$$

We know that $0\le P(A\cup B\cup C)\le 1$

$$\Rightarrow 0.23\le P(B\cap C)\le 0.48$$

3. **(B)**

Probability of getting a composite number is $\frac{2}{6}=\frac{1}{3}$

Probability that A will win the game is

$$\frac{1}{3}+\left(\frac{2}{3}\right)\left(\frac{2}{3}\right)\left(\frac{2}{3}\right)\left(\frac{1}{3}\right)+\ldots$$

$$=\frac{\frac{1}{3}}{1-\frac{8}{27}}=\frac{9}{19}$$

Probability that B will win the game is

$$\left(\frac{2}{3}\right)\left(\frac{1}{3}\right)+\left(\frac{2}{3}\right)\left(\frac{2}{3}\right)\left(\frac{2}{3}\right)\left(\frac{1}{3}\right)+\ldots$$

$$=\frac{\frac{2}{9}}{1-\frac{8}{27}}=\frac{6}{19}$$

Probability that C will win the game

$$=\left(\frac{2}{3}\right)\left(\frac{2}{3}\right)\left(\frac{1}{3}\right)+\left(\frac{2}{3}\right)\left(\frac{2}{3}\right)\left(\frac{2}{3}\right)\left(\frac{2}{3}\right)\left(\frac{2}{3}\right)\left(\frac{1}{3}\right)+\ldots$$

$$=\frac{4}{19}$$

Solving required ratio is $9:6:4$

4. **(B)**

Sum of 5 can be given by (1, 4), (4, 1), (3, 2) and (2, 3) $\Rightarrow$ 4 ways

Sum of 7 can be given by

(1, 6), (6, 1), (2, 5), (5, 2), (4, 3) and (3, 4)

$\Rightarrow$ 6 ways

The number of ways of throwing neither 5 nor 7 is

$36-(4+6)=26$

Now, Probability (a sum of 5 in a single throw)

$$=\frac{4}{36}=\frac{1}{9}$$

$P(\text{neither 5 nor 7}) = \dfrac{26}{36} = \dfrac{13}{18}$

Required probability

$= \dfrac{1}{9} + \left(\dfrac{13}{18}\right)\left(\dfrac{1}{9}\right) + \left(\dfrac{13}{18}\right)^2\left(\dfrac{1}{9}\right) + \ldots = \dfrac{2}{5}$

5. (D)

Total number of experiments $= n = 10$

Number of favorable cases $= r = 0$ or 1

(Here, 'favourable case' is getting defective bulbs)

Probability of getting favourable case

$= P(F) = \dfrac{1}{10}$

Probability of not getting favourable case

$= P(F') = \dfrac{9}{10}$

Required probability is

$^{10}C_0\left(\dfrac{1}{10}\right)^0\left(\dfrac{9}{10}\right)^{10} + {^{10}C_1}\left(\dfrac{1}{10}\right)^1\left(\dfrac{9}{10}\right)^9$

$= \dfrac{19(9^9)}{10^{10}}$

6. (C)

Let us assume that India won the tournament $\Rightarrow$ India wins total 4 matches out of 6 and Pakistan won only 2 matches, with the condition that India won the last match.

So, if we see at the result of the first five matches, then India won 3 and Pakistan won 2 matches then 6th match is won by India.

So, required probability

$= \left({^5C_3}\right)\left({^2C_2}\right)\left(\dfrac{1}{2^6}\right) = \dfrac{5}{32}$

We will get a similar value if Pakistan won the tournament.

So, required probability

$= \dfrac{5}{32} + \dfrac{5}{32} = \dfrac{5}{16}$

7. (B)

Probability that both the cards are numbered is

$P(N) = \dfrac{^{36}C_2}{^{52}C_2} = \dfrac{36 \times 35}{52 \times 51}$

Probability that both the cards are black is

$P(B) = \dfrac{^{26}C_2}{^{52}C_2} = \dfrac{26 \times 25}{52 \times 51}$

Probability that both the cards are numbered and they are black is

$P(N \cap B) = \dfrac{^{18}C_2}{^{52}C_2} = \dfrac{18 \times 17}{52 \times 51}$

So, required probability is

$P(N \cup B) = P(N) + P(B) - P(N \cap B)$

$= \dfrac{1604}{2652} = \dfrac{401}{663}$

8. (C)

Probability that the first group has only one Jack

$= \dfrac{\left({^4C_1}\right)\left({^{48}C_{12}}\right)}{^{52}C_{13}} = \dfrac{39 \times 38 \times 37}{51 \times 50 \times 49}$

Similarly, probability that the second group has a Jack is

$\dfrac{^3C_1 \times {^{36}C_{12}}}{^{39}C_{13}}$

Similarly, probability that the third group has a Jack is $\dfrac{13}{25}$

Then, the probability that the remaining fourth group also has a Jack

$= \dfrac{39 \times 26 \times 13}{51 \times 50 \times 49}$

9. (C)

Let $ABCDEF$ be the hexagon. Three vertices of a triangle can be selected in $^6C_3 = 20$ ways.

The only possible equilateral triangles are ACE, BDF. Hence the required probability

$= \dfrac{2}{20} = \dfrac{1}{10}$

10. (A)

$\left({^nC_1} \times {^nC_1}\right)\left({^{n-1}C_1} \times {^{n-1}C_1}\right)\ldots$

$\left({^2C_1} \times {^2C_1}\right)\left({^1C_1} \times {^1C_1}\right) = 576$

$\Rightarrow n^2(n-1)^2 \ldots 2^2 \cdot 1^2 = 576$

$\Rightarrow (n!)^2 = (24)^2 \Rightarrow n! = 24 = 4!$

$\Rightarrow n = 4$

11. **(C)**

The required probability

$$= {}^nC_1\left(\frac{1}{2}\right)\left(\frac{1}{2}\right)^{n-1} + {}^nC_3\left(\frac{1}{2}\right)^3\left(\frac{1}{2}\right)^{n-3} + \ldots + {}^nC_n\left(\frac{1}{2}\right)^n$$

$$= \left(\frac{1}{2}\right)^n\left({}^nC_1 + {}^nC_3 + \ldots + {}^nC_n\right)$$

$$= \frac{1}{2^n}\cdot 2^{n-1} = \frac{1}{2}$$

12. **(C)**

$$P(A\cup B) = P(A) + P(B) - P(A)P(B)$$

$$\Rightarrow 0.7 = 0.2 + P(B) - 0.2P(B)$$

$$\Rightarrow P(B) = \frac{0.5}{0.8} = \frac{5}{8} \Rightarrow P(B') = 1 - \frac{5}{8} = \frac{3}{8}$$

13. **(A)**

$$P(C) = 0 \Rightarrow C = \phi$$

$$\Rightarrow A\cap C = B\cap C = A\cap B\cap C = \phi$$

$$\Rightarrow P(A\cap B\cap C) = 0 \Rightarrow P(A)P(B), P(C)$$

and $P(A\cap C) = P(B\cap C) = 0$

$$\Rightarrow P(A)P(C) = P(B)P(C)$$

Hence the events are independent.

14. **(C)**

$$P\left(\frac{A'}{B'}\right) = \frac{P(A'\cap B')}{P(B')} = \frac{P(A\cup B)'}{P(B')}$$

$$= \frac{1 - P(A\cup B)}{P(B')}$$

15. **(D)**

Sum of all numbers between 1 to 9

$$= \frac{10\times 9}{2} = 45$$

and $1+2+6+3+6 = 18$

Total is $18 + 45 = 63$ which is divisible by 9, so whatever the arrangement the number will always be divisible by 9, so probability is 1.

16. **(D)**

The number of favourable cases is

$${}^nC_0\times 2^n + {}^nC_1\times 2^{n-1} + \ldots + {}^nC_n 2^0 = 3^n$$

Total number of cases is $2^n\times 2^n = 4^n$

Thus, required probability $= \left(\frac{3}{4}\right)^n$

17. **(C)**

If the shaded square is chosen we can choose 2 squares out of 4 others at the corner in ${}^4C_2$ ways. Such shaded squares are 6^2.

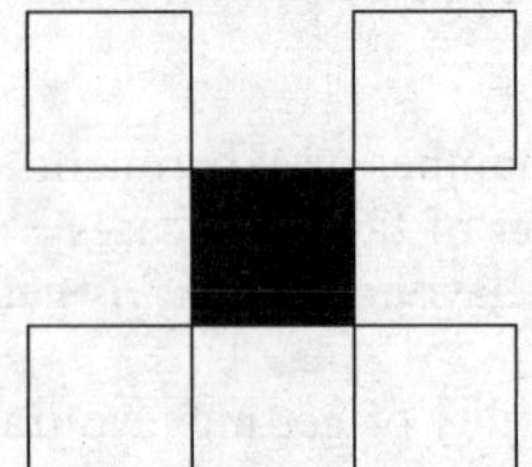

Hence the required probability $= \frac{{}^4C_2\cdot 6^2}{{}^{64}C_2}$

18. **(B)**

Total number of words formed $= \frac{6!}{3!2!} = 60$

The number of words containing the pattern BAN $= \frac{4!}{2!} = 12$

So, the required probability $= \frac{60-12}{60} = \frac{4}{5}$

19. **(D)**

If n_1, n_2, n_3 and n_4 be the numbers on the drawn balls (in that order) than, we must have

$n_1 > n_4$.

Total number of order relations between n_1, n_2, n_3 and n_4 is 4! That is, 24.

Now total number of favourable order relations have to counted.

(i) If 'n_1' is the largest, then favourable order relations are $3! = 6$.

(ii) If n_1 is the second largest, then favourable order relations are $2(2!) = 4$.

(iii) In n_1 is the third largest, then favourable order relations are $2! = 2$.

Thus, required probability $= \frac{12}{24} = \frac{1}{2}$.

Alternate solution

Clearly n_1 and n_4 are unequal. Thus there are only two cases

$n_1 < n_4$ or $n_1 > n_4$.

Thus, required probability $= \frac{1}{2}$

20. **(C)**

E_1: There is a gust of wind.

E_2: There is no gust of wind.

A: Archer misses the target.

$P(A) = P(E_1)\cdot P(A/E_1) + P(E_2)\cdot P(A/E_2)$

$= \frac{3}{10}.\frac{3}{5}+\frac{7}{10}.\frac{3}{10}$

$= \frac{39}{100}$

Now, required probability

$= P(E_2 / A) = \frac{P(E_2 \cap A)}{P(A)}$

$= \frac{P(E_2).\, P(A / E_2)}{P(A)}$

$= \frac{\frac{7}{10}.\frac{3}{10}}{\frac{39}{100}} = \frac{7}{13}$

21. **(C)**

Let the selected numbers be n_1, n_2 and n_3.

We must have

$2n_2 = n_1 + n_3$

Thus $n_1 + n_2$ must be even. That means n_1 and n_3 both must have same nature.

Thus, required probability $= \frac{2.\,^{12}C_2}{^{24}C_3} = \frac{3}{46}$

22. **(A)**

Let E_A = the event of A becoming secretary

Similarly, E_B = the event of admitting lady members

$\therefore P(E_A) = \frac{1}{3}, P(E_B) = \frac{2}{9}, P(E_C) = \frac{4}{9}$

Clearly $E_{A,}$ E_B, E_C are mutually exclusive and exhaustive.

Also, $P\left(\frac{E_1}{E_A}\right) = 0.6$

$P\left(\frac{E_1}{E_B}\right) = 0.7, P\left(\frac{E_1}{E_C}\right) = 0.5$

$\therefore$ the required probability,

$= P(E_A)P\cdot\left(\frac{E_1}{E_A}\right) + P(E_B)P\cdot\left(\frac{E_1}{E_B}\right) + P(E)\cdot P\left(\frac{E_1}{E_C}\right)$

$= \frac{1}{3}.\frac{3}{5}+\frac{2}{9}.\frac{7}{10}+\frac{4}{9}.\frac{1}{2}$

$= \frac{1}{5}+\frac{7}{45}+\frac{2}{9} = \frac{9+7+10}{45} = \frac{26}{45}$

23. **(C)**

The probability of getting a heart in one draw $= \frac{13}{52} = \frac{1}{4}$

Similarly, for a diamond, the probability $= \frac{1}{4}$

The probability of getting a black card in one draw $= \frac{26}{52} = \frac{1}{2}$

$\therefore$ the required probability

$= {}^6C_2 \times {}^4C_2 \times \left(\frac{1}{4}\right)^2 \times \left(\frac{1}{4}\right)^2 \times \left(\frac{1}{2}\right)^2$

$= 15 \times 6 \times \frac{1}{2^{10}} = \frac{90}{2^{10}}$

24. **(C)**

Let A = the event that the rose bush has withered

Let A_1 = the event that the gardener did not water

A_2 = the event that the gardener watered

By Baye's theorem

$P\left(\frac{A_1}{A}\right)$

$= \frac{P(A_1).P(A/A_1)}{P(A_1).P(A/A_1)+P(A_2).P(A/A_2)}$ ...(1)

Given $P(A_1) = \frac{2}{3}$

$\therefore P(A_2) = \frac{1}{3}$

$P(A/A_1) = \frac{3}{4}, P(A/A_2) = \frac{1}{2}$

$\therefore$ from (1) $P(A_1(A) = \frac{\frac{2}{3}.\frac{3}{4}}{\frac{2}{2}.\frac{3}{4}+\frac{1}{3}.\frac{1}{2}} = \frac{6}{6+2} = \frac{3}{4}$

25. **(B)**

If E be the event that no pair is selected, then

$P(E) = \frac{20}{20}.\frac{18}{19}.\frac{16}{18}.\frac{14}{17} = \frac{224}{323}$

$\therefore$ reqd. prob. $= 1 - P(E) = 1 - \frac{224}{323} = \frac{99}{323}$.

Multiple Answer Type

1. **(A, B, C, D)**

P(exactly one of A, B occurs)

$= P\{(A\cap B')\cup(A'\cap B)\}$

$= P(A\cap B')+P(A'\cap B)$

$= P(A)-P(A\cap B)+P(B)-P(A\cap B)$

$= P(A)+P(B)-2P(A\cap B)$

$= P(A\cup B)-P(A\cap B)$

Also, P(exactly one of A, B occurs)

$= \{1-P(A'\cap B')\}-\{1-P(A'\cup B')\}$

$= P(A'\cup B')-P(A'\cap B')$

$= P(A')+P(B')-2P(A'\cap B')$

Hence, (A), (B), (C), (D)

2. **(A, C, D)**

We have, $P(A)=P\left(\frac{A}{B}\right)=\frac{P(A\cap B)}{P(B)}$

$\Rightarrow P(A\cap B)=P(A)P(B)$

$\Rightarrow$ A and B are independent

Since,

$P(A\cap B)=P(A)P\left(\frac{B}{A}\right)=\left(\frac{1}{4}\right)\left(\frac{1}{2}\right)=\frac{1}{8}\neq 0$

A and B cannot be mutually exclusive

As A and B are independent

$P\left(\frac{B'}{A'}\right)=P(B')$

$\Rightarrow P\left(\frac{B'}{A'}\right)=1-P(B)=\frac{1}{2}$

Also, $P\left(\frac{A'}{B}\right)=\frac{P(A'\cap B)}{P(B)}$

$=\frac{P(B)-P(A\cap B)}{P(B)}=1-P\left(\frac{A}{B}\right)$

$=1-\frac{1}{4}=\frac{3}{4}$

Hence (A), (C), (D)

3. **(A, C, D)**

{Do it yourself}

4. **(A, B)**

$$P(A)=\frac{\int_0^1\sqrt{x}dx}{\int_0^1 dx}=\frac{2}{3}$$

$$P(A\cap B)=\frac{\int_0^1\sqrt{x}dx-\int_0^1 x^2dx}{\int_0^1 dx}=\frac{2}{3}-\frac{1}{3}=\frac{1}{3}$$

Also, $P(B)=\frac{1-\int_0^1 x^2dx}{\int_0^1 dx}=\frac{2}{3}$

5. **(A, C)**

{Do it yourself}

6. **(B, C, D)**

Since, E and F are independent events, therefore, $P(E\cap F)=P(E)P(F)\neq 0$

$\Rightarrow$ E and F are not mutually exclusive

Now, $P(E\cap F')=P(E)-P(E\cap F)$

$=P(E)-P(E)P(F)$

$=P(E)(1-P(F))$

$=P(E)P(F')$

and, $P(E'\cap F')=P(E\cup F)'$

$=1-P(E\cup F)$

$=1-\{1-P(E')P(F')\}$

$=P(E')P(F')$

Finally, $P\left(\frac{E}{F}\right)+P\left(\frac{E'}{F}\right)$

$=\frac{P(E\cap F)}{P(F)}+\frac{P(E'\cap F)}{P(F)}$

$=\frac{P(E\cap F)+P(E'\cap F)}{P(F)}=\frac{P(F)}{P(F)}=1$

7. **(A, C)**

$P\left(\frac{A}{B}\right)=\frac{P(A\cap B)}{P(B)}$

$P(A \cap B) \geq P(A) + P(B)$

$1 \geq P(A) + P(B) - P(A \cap B)$

$1 \geq P(A \cup B)$

8. **(A, B, C, D)**

$P(A \cup B) \geq \max\{P(A), P(B)\} = \frac{3}{4}$

and $P(A \cap B) \leq \min\{P(A), P(B)\} = \frac{5}{8}$

Now, $P(A \cap B) = P(A) + P(B) - P(A \cup B)$

$= P(A) + P(B) - 1 = \frac{3}{8}$

$P(A \cap B') = P(A) - P(A \cap B)$

$P(A' \cap B) = P(B) - P(A \cap B)$

9. **(A, D)**

$P\left(\frac{E}{F}\right) = \frac{P(E \cap F)}{P(F)}$

$P\left(\frac{E'}{F}\right) = \frac{P(E' \cap F)}{P(F)}$

$\therefore P\left(\frac{E}{F}\right) + P\left(\frac{E'}{F}\right) = \frac{P(E \cap F) + P(E' \cap F)}{P(F)}$

$= \frac{P(F)}{P(F)} = 1$

$\Rightarrow$ 'A' is correct

Similarly, 'D' is correct.

10. **(A, B)**

$P(E_i) = \frac{2}{4} = \frac{1}{2}$ for $i = 1, 2, 3$

Also for $i \neq j$,

$P(E_i \cap E_j) = \frac{1}{4} = P(E_i)P(E_j)$

Therefore E_i and E_j are independent for $i \neq j$

Also,

$P(E_1 \cap E_2 \cap E_3) = \frac{1}{4} \neq P(E_1)P(E_2)P(E_3)$

$\Rightarrow E_1, E_2, E_3$ are not independent

11. **(A, B)**

[Do it yourself]

12. **(A, C)**

{Do it yourself}

13. **(C, D)**

{Do it yourself}

14. **(A, C, D)**

{Do it yourself}

15. **(A, D)**

{Do it yourself}

16. **(B, D)**

{Do it yourself}

17. **(C, D)**

{Do it yourself}

18. **(A, D)**

{Do it yourself}

19. **(A, B, C)**

{Do it yourself}

Comprehension Based Questions

1. **(A)**

$n(S)$ = no. of total triangle $= {}^6C_3 = 20$

Equilateral triangle is $A_1 A_3 A_5$ and $A_2 A_4 A_6$

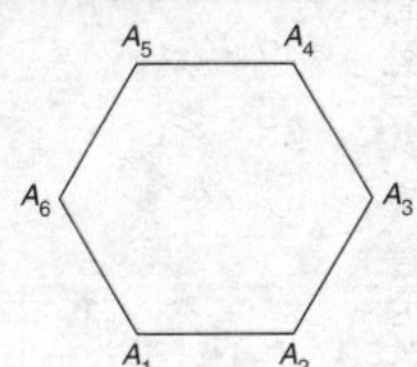

$\therefore P(E) = \frac{2}{20} = \frac{1}{10}$

2. **(B)**

Let $A_1 A_2$ be the common side

No. of triangle having $A_1 A_2 = 2$

$(\Delta A_1 A_2 A_4$ or $\Delta A_1 A_2 A_5)$

Total no. of triangles having one side common $= 6 \times 2 = 12$

$P(E) = \frac{12}{20} = \frac{3}{5}$

3. **(C)**

Number of triangles having two sides common = 6

$P(E) = \frac{6}{20} = \frac{3}{10}$

4. **(B)**
Total number of possible hand shakes = $^{20}C_2 = 190$
No. of hand shakes = hand shakes between Indian male and Russian male + hand shakes between women and men excluding husband of the women.
$= {}^{10}C_2 + \frac{1}{2}\cdot 10\times 9 = 90$
$\Rightarrow P$ (no wife shakes hands with her husband) $= \frac{90}{190} = \frac{9}{19}$.

5. **(A)**
No. of hands shakes when no Indian wife shakes hands with a male
= No. of hand shakes between Indian women and Russian women + hand shakes between Indian men and Russian men
$= {}^{15}C_2 + {}^{5}C_2 = 105 + 10 = 115$
Required probability $= \frac{115}{190} = \frac{23}{38}$

6. **(B)**
Number of possible shakes hands = $^{10}C_2 + {}^{10}C_2 + {}^{10}C_2 = 135$
$P(E) = \frac{135}{190} = \frac{27}{38}$

7. **(A)**
SUCCESS
– U – CC – E –, If 'S' is placed at the place
– $n(E) = {}^4C^3 \times 3!$
$n(S) = \frac{7!}{3!2!}$
$P(E) = \frac{^4C_3\times 3!}{\frac{7!}{3!\times 2!}} = \frac{4\times 3!\times 3!\times 2!}{7!} = \frac{2}{35}$

8. **(B)**
Let us first find the words in which no two S are together.
The remaining letters C, C, U, E can be arranged in $\frac{4!}{2!} = 12$ ways.
These are five favourable places for three SSS × C × C × U × E ×
Thus the total number of ways = $12 \times {}^5C_3 = 120$
Hence the number of words having CC and SSS are separated = 120 – 24 = 96.
Consequently required probability
$= \frac{96}{7\times 6\times 10} = \frac{8}{35}$

9. (A)
SUCCESS
I, C, C, E, S can be arranged by $\frac{5!}{2!}$
$n(E) = \frac{5!}{2!}$
$n(S) = \frac{\frac{5!}{2!}}{\frac{7!}{3!2!}} = \frac{5!\times 3!}{7!}$
$= \frac{6}{7\times 6} = \frac{1}{7}$

10. **(D)**
$P = P$(W or BBW or BBBBW)
$= \frac{3}{10} + \frac{5}{10}\times\frac{4}{9}\times\frac{3}{8} + \frac{5}{10}\times\frac{4}{9}\times\frac{3}{8}\times\frac{2}{7}\times\frac{3}{6}$
$= \frac{83}{210}$

11. **(A)**
$P = P$(BW or BBBW or BBBBBW)
$= \frac{5}{10}\times\frac{3}{9} + \frac{5}{10}\times\frac{4}{9}\times\frac{3}{8}\times\frac{3}{7} + \frac{5}{10}\times\frac{4}{9}\times\frac{3}{8}\times\frac{2}{7}\times\frac{1}{6}\times\frac{3}{5}$
$= \frac{43}{210}$

12. **(B)**
$P = P$(R or BR or BBR or BBBR or BBBBR or BBBBBR)
$= \frac{2}{10} + \frac{5}{10}\times\frac{2}{9} + \frac{5}{10}\times\frac{4}{9}\times\frac{2}{8} + \frac{5}{10}\times\frac{4}{9}\times\frac{3}{8}\times\frac{2}{7}$
$+ \frac{5}{10}\times\frac{4}{9}\times\frac{3}{8}\times\frac{2}{7}\times\frac{2}{6} + \frac{5}{10}\times\frac{4}{9}\times\frac{3}{8}\times\frac{2}{7}\times\frac{1}{6}\times\frac{2}{5}$
$= \frac{2}{5}$

13. **(A)**
$P(X = 3) = \left(\frac{5}{6}\right)\left(\frac{5}{6}\right)\left(\frac{1}{6}\right) = \frac{25}{216}$

14. **(B)**
Given that $P(X \le 2) = \frac{1}{6} + \frac{5}{6}\times\frac{1}{6} = \frac{11}{36}$
Hence the required probability $= 1 - \frac{11}{36} = \frac{25}{36}$

15. **(D)**

For $x \geq 6$, the probability is

$$\frac{5^5}{6^6}+\frac{5^5}{6^7}+\ldots \text{ to } \infty$$

$$=\frac{5^5}{6^6}\left(\frac{1}{1-\frac{5}{6}}\right)=\left(\frac{5}{6}\right)^5$$

For $X > 3$

$$\frac{5^3}{6^4}+\frac{5^4}{6^5}+\frac{5^5}{6^6}+\ldots\text{to } \infty=\left(\frac{5}{6}\right)^3$$

Hence the conditional probability is

$$\frac{\left(\frac{5}{6}\right)^6}{\left(\frac{5}{6}\right)^3}=\frac{25}{36}$$

Match Matrix

1. **(A – T; B – P; C – Q; D – S)**

(A) $P = 1 - P$(all numbers are ≥ 10)

$$=1-\frac{{}^{16}C_4}{{}^{25}C_4}=1-\frac{182}{1265}=\frac{1083}{1265}$$

(B) $P=\dfrac{{}^{9}C_1\times 1\times {}^{15}C_2}{{}^{25}C_4}=\dfrac{189}{2530}$

(C) $P = P$ (1 even, 3 odd) + P(3 even, 1 odd)

$$=\frac{{}^{13}C_3\times {}^{12}C_1+{}^{13}C_1\times {}^{12}C_3}{{}^{25}C_4}=\frac{286}{575}$$

(D) $P = 1 - P$ ($abcd$ is odd)

$$=1-\frac{{}^{13}C_4}{{}^{25}C_4}=\frac{217}{230}$$

2. **(A – Q; B – P; C – R; D – Q)**

(A) $P=\dfrac{1}{13}\left(\dfrac{{}^{4}C_4+{}^{5}C_4+\ldots+{}^{12}C_4}{{}^{12}C_4}\right)=\dfrac{1}{5}$

(B) $P=\dfrac{{}^{10}C_4}{{}^{12}C_4}=\dfrac{14}{33}$

(C) $P=\dfrac{1}{13}\left(\dfrac{{}^{2}C_2\,{}^{10}C_2+{}^{3}C_2\,{}^{9}C_2+{}^{4}C_2\,{}^{8}C_2+\ldots+{}^{10}C_2\,{}^{2}C_2}{{}^{12}C_4}\right)=\dfrac{1}{5}$

(D) $P=\dfrac{{}^{10}C_4}{{}^{4}C_4+{}^{5}C_4+\ldots+{}^{12}C_4}$

$$=\frac{{}^{10}C_4}{{}^{13}C_5}=\frac{70}{429}$$

3. **(A – R; B – Q; C – U; D – P)**

(A) Given $P\left(\dfrac{E_2}{E_1}\right)=\dfrac{1}{2}$

$$\Rightarrow \frac{P(E_2\cap E_1)}{P(E_1)O}=\frac{1}{2}\Rightarrow P(E_2\cap E_1)=\frac{1}{8}$$

Also, $P\left(\dfrac{E_1}{E_2}\right)=\dfrac{1}{4}$

$$\Rightarrow \frac{P(E_1\cap E_2)}{P(E_2)}=\frac{1}{4}\Rightarrow \frac{1}{8P(E_2)}=\frac{1}{4}$$

$$\Rightarrow P(E_2)=\frac{1}{2}$$

(B) $P(E_1\cup E_2)=P(E_1)+P(E_2)-P(E_1\cap E_2)$

$$=\frac{1}{4}+\frac{1}{2}-\frac{1}{8}=\frac{5}{8}$$

(C) $P\left(\dfrac{\overline{E_1}}{\overline{E_2}}\right)=\dfrac{P(\overline{E_1}\cap\overline{E_2})}{P(\overline{E_2})}$

$$=\frac{1-P(E_1\cup E_2)}{1-P(E_2)}$$

$$=\frac{1-\frac{5}{8}}{1-\frac{1}{2}}=\frac{3}{4}$$

(D) $P\left(\dfrac{E_1}{\overline{E_2}}\right)=\dfrac{P(E_1\cap\overline{E_2})}{P(\overline{E_2})}$

$$=\frac{P(E_1)-P(E_1\cap E_2)}{1-P(E_2)}$$

$$=\frac{\frac{1}{4}-\frac{1}{8}}{1-\frac{1}{2}}=\frac{\frac{1}{8}}{\frac{1}{2}}=\frac{1}{4}$$

4. (A – R, S; B – P, Q, R, S; C – P, Q, R, S; D – P)

1. $P(\text{success}) = 1/2$; $P(\text{failure}) = 1/2$

   Suppose 'n' bombs are to be dropped. Let E be the event that the bridge is destroyed. Then,

   $P(E) = 1 - P(0 \text{ or } 1 \text{ succes})$

   $$= 1 - \left(\left(\frac{1}{2}\right)^n + \frac{1}{2}\left(\frac{1}{2}\right)^{n-1}\right)$$

   $$= 1 - \left(\frac{1}{2^n} + \frac{n}{2^n}\right) \geq 0.9$$

   $$\Rightarrow \frac{1}{10} \geq \frac{n+1}{2^n} \text{ or } \frac{2^n}{10(n+1)} \geq 1$$

2. The bag contains 2 red, 3 white and 5 black balls. Hence

   $P(S) = 1/5$; $P(F) = 4/5$; Let E be the event of getting a red ball

   $$P(E) = P(S \text{ or } FS \text{ or } FFS \text{ or} \ldots) \geq \frac{1}{2}$$

   $$\therefore P(F)^n \leq \frac{1}{2}; \left(\frac{4}{5}\right)^n \leq \frac{1}{2}$$

   The value of n consistent is 4

3. Let there be x red socks and y blue socks and $x > y$.

   Then $\dfrac{{}^xC_2 + {}^yC_2}{{}^{x+y}C_2} = \dfrac{1}{2}$

   or $\dfrac{x(x-1) + y(y-1)}{(x+y)(x+y-1)} = \dfrac{1}{2}$

   Multiplying both sides by $2(x+y)(x+y-1)$ and expanding, we get

   $$2x^2 - 2x + 2y^2 - 2y = x^2 + 2xy + y^2 - x - y$$

   Rearranging, we have

   $x^2 - 2xy + y^2 = x + y$

   $$\Rightarrow (x-y)^2 = x + y \Rightarrow |x - y| = x + y$$

   Now, $x + y \leq 17$

   $x - y \leq \sqrt{17}$

   As $x - y$ must be an integer, so $x - y = 4$

   $\therefore x + y = 16$

   Adding both together and dividing by 2 yields $x \leq 10$

4. Let the number of green socks be $x > 0$. Let E: be the event that two socks drawn are of the same colour

   $P(E) = P(RR \text{ or } BB \text{ or } WW \text{ or } GG)$

   $$= \frac{{}^3C_2}{{}^{6+x}C_2} + \frac{{}^xC_2}{{}^{6+x}C_2}$$

   $$= \frac{6}{(x+6)(x+5)} + \frac{x(x-1)}{(x+6)(x+5)} = \frac{1}{5}$$

   $$\Rightarrow 5(x^2 - x + 6) = x^2 + 11x + 30$$

   $$\Rightarrow 4x^2 - 16x = 0 \Rightarrow x = 4$$

5. (A – QR; B – R; C – P; D – P, Q, R, S)

1. $= \frac{1}{2} 2r(r-1) = (r-b)(r+b-1)$

   $= 2r(r-1) = (r+b)(r+b-1)$

   $\Rightarrow 2r^2 - 2r = r^2 + (2b-1)r + b^2 - 1$

   $\Rightarrow r^2 - (1+2b)r + 1 - b^2 = 0$

   $\Rightarrow b^2 + 2br + r - r^2 - 1 = 0$

   $$\Rightarrow b = \frac{-2r \pm \sqrt{4r^2 - 4(r - r^2 - 1)}}{2}$$

   $$= -r \pm \sqrt{2r^2 - r + 1}$$

   Since b is integer, possible values of r are 3 and 8

2. $$\left(\frac{r}{r+10}\right)^2 \left(\frac{10}{r+10}\right)^2 = \frac{3}{8}$$

3. $$\left(\frac{r}{r+10}\right)^2 \left(\frac{10}{r+10}\right)^2 = \frac{1}{16}$$

   $\Rightarrow r = 10$

4. Probability of getting exactly n red balls in $2n$ draws is always equal to probability of getting exactly n black balls in $2n$ draws for any value of r and b, hence the ratio r/b can be 10, 3, 8, 2.

Subjective Answer Type (Hard Nut)

1. Total number of ways of answering the question

$= {}^4C_1 + {}^4C_2 + {}^4C_3 + {}^4C_4 = 15$

out of these only one combination is correct. Probability of answering the question correctly at the first trial $= \dfrac{1}{15}$

Probability of answering the question correctly at the second trial

$= \frac{14}{15} \cdot \frac{1}{14} = \frac{1}{15}$

(with the assumption that the student doesn't repeat the mistake of the first trial)

Similarly, probability of answering the question correctly at the third trial

$= \frac{14}{15} \cdot \frac{13}{14} \cdot \frac{1}{13} = \frac{1}{15}$

$\therefore$ Probability of required event

$= \frac{1}{15} + \frac{1}{15} + \frac{1}{15} = \frac{1}{5}$

2. Since, x is prime and $x < 10$

$\Rightarrow x = 2, 3, 5, 7$

$\Rightarrow$ Total number of $A(x, y)$ pairs

$= n(A) = 4 \times 10 = 40$

and $x^2 = 3y^2 + 1$

For the above condition to be satisfied only two such pairs (2, 1) and (7, 4) are possible

$\Rightarrow P(A) = \frac{2}{40} = \frac{1}{20}$

3. We can get exactly one pair of two ways

(i) If we get two cards of the same front along with two other cards of different front.

Number of ways =

${}^{13}C_3 \times {}^{3}C_1 \times {}^{4}C_2 \times {}^{4}C_1 \times {}^{4}C_1$

(ii) If we get three cards of the same front and one card of another front

Number of ways =

${}^{13}C_2 \times {}^{2}C_1 \times {}^{4}C_3 \times {}^{4}C_1$

So, required probability

$$= \frac{{}^{13}C_3 \times {}^{3}C_1 \times {}^{4}C_2 \times {}^{4}C_1 \times {}^{4}C_1 + {}^{13}C_2 \times {}^{2}C_1 \times {}^{4}C_3 \times {}^{4}C_1}{{}^{52}C_4}$$

4. $97020 = 2^2 \times 3^2 \times 5 \times 7^2 \times 11$

Number of factors $= 3 \times 3 \times 2 \times 2 \times 3 = 108$

Number of factors divisible by 45 =

$3 \times 1 \times 1 \times 2 \times 3 = 18$

Hence, the required probability $= \frac{18}{108} = \frac{1}{6}$

5. Let C, S, B, T be the events of the person going by car, scooter, bus or train respectively

Given that

$P(C) = \frac{1}{7}, P(S) = \frac{3}{7}, P(B) = \frac{2}{7}, P(T) = \frac{1}{7}$

Let L be the event of the person reaching the office in time

$\Rightarrow P\left(\frac{\bar{L}}{C}\right) = \frac{7}{9}, P\left(\frac{\bar{L}}{S}\right) = \frac{8}{9}, P\left(\frac{\bar{L}}{B}\right)$

$= \frac{5}{9}, P\left(\frac{\bar{L}}{T}\right) = \frac{8}{9}$

$$\Rightarrow P\left(\frac{C}{\bar{L}}\right) = \frac{\frac{1}{7} \times \frac{7}{9}}{\frac{1}{7} \times \frac{7}{9} + \frac{3}{7} \times \frac{8}{9} + \frac{2}{7} \times \frac{5}{9} + \frac{8}{9} \times \frac{1}{7}} = \frac{1}{7}$$

6. If the sum of the digits in a number is divisibly by 9 then the number itself is divisible by 9.

The least and the greatest sum of five digits chosen from the set is 15 and 25 respectively

$\Rightarrow$ Either number comprises $\{1, 2, 4, 5, 6\}$ or $\{1, 2, 3, 5, 7\}$

$\Rightarrow$ Favourable cases $= 2 \times 5!$

Total number of cases $= {}^{7}P_5 = 42 \times 60$

$\Rightarrow$ Required probability $= \frac{5! \times 2!}{42 \times 60}$

$= \frac{2}{21} = \frac{x}{y}$

$\Rightarrow y - 10x = 1$

7. P(exactly two of A, B, C occur)

$\Rightarrow P(A \cap B) + P(B \cap C) + P(C \cap A) - 3P(A \cap B \cap C)$

$\Rightarrow P(A)P(B) + P(B)P(C) + P(C)P(A) - 3P(A)P(B)P(C)$

$= \frac{1}{3} \cdot \frac{1}{2} + \frac{1}{2} \cdot \frac{1}{4} + \frac{1}{4} \cdot \frac{1}{3} - 3 \cdot \frac{1}{3} \cdot \frac{1}{2} \cdot \frac{1}{4} = \frac{1}{4}$

8. $\frac{(m-3)(m-4)}{(m-1)(m-2)}$

[Do it yourself]

9. Let E_1 and E_2 denote the events that letter came from TATANAGAR and CULCUTTA respectively. Let E be the event that the two consecutive letters visible on the envelope are TA

$\therefore P(E_1)=\frac{1}{2}, P(E_2)=\frac{1}{2}, P\left(\frac{E}{E_1}\right)=\frac{2}{8}=\frac{1}{4}$

(Here consider TA as single letter)

$\therefore$ Favourable ways $= {}^2C_1 = 2$

Also, when one TA is considered as a single letter, then total letters become 8.

So total ways $= {}^8C_1 = 8$. Similarly,

$$P\left(\frac{E}{E_2}\right)=\frac{1}{7}$$

Therefore, the required probability (by Baye's theorem)

$$P\left(\frac{E_2}{E}\right)=\frac{P(E_2)P\left(\frac{E}{E_2}\right)}{P(E_1)P\left(\frac{E}{E_1}\right)+P(E_2)P\left(\frac{E}{E_2}\right)}$$

$$=\frac{\left(\frac{1}{2}\right)\left(\frac{1}{7}\right)}{\left(\frac{1}{2}\right)\left(\frac{1}{4}\right)+\left(\frac{1}{2}\right)\left(\frac{1}{7}\right)}$$

$$=\frac{4}{11}$$

10. Let $E_{m,n}$ be the event representing a bag containing m white and n black balls.

$S = E_{3,7}; E_{4,6}; E_{5,5}; E_{6,4}; E_{7,3}$

Let A be event representing drawing of 3 white and 3 black balls.

$P(E_{3,7}) = P(E_{4,6}) = P(E_{5,5}) = P(E_{6,4})$

$= P(E_{7,3}) = \frac{1}{5}$

$$P\left(\frac{E_{5,5}}{A}\right)$$

$$=\frac{P(E_{5,5})P\left(\frac{A}{E_{5,5}}\right)}{\left[\begin{array}{l}P(E_3,7)P\left(\frac{A}{E_{3,7}}\right)+P(E_{4,6})\\ P\left(\frac{A}{E_{4,6}}\right)+P(E_{5,5})P\left(\frac{A}{E_{5,5}}\right)+P(E_{6,4})\\ P\left(\frac{A}{E_{6,4}}\right)+P(E_{7,3})P\left(\frac{A}{E_{7,3}}\right)\end{array}\right]}$$

$$=\frac{{}^5C_3\times{}^5C_3}{{}^5C_3\times{}^7C_3+{}^4C_3\times{}^6C_3+{}^5C_3\times{}^5C_3+{}^6C_3\times{}^4C_3+{}^7C_3\times{}^3C_3}$$

$$=\frac{100}{330}=\frac{10}{33}$$

11. Let E_1 denote the event of drawing a white ball at any draw and E_2 that for a black ball and let E be the event for A winning the game.

$\therefore P(E_1)=\frac{m}{m+n}$ and $P(E_2)=\frac{n}{m+n}$

$P(E) = P(E_1 \text{ or } E_2E_2E_1 \text{ or } E_2E_2E_2E_2E_1 \text{ or} \ldots)$

$= P(E_1)+P(E_2)P(E_2)P(E_1)+(P(E_2))^4 P(E_1)$

$$=\frac{P(E_1)}{1-\{P(E_2)\}^2}=\frac{\frac{m}{m+n}}{1-\left(\frac{n}{m+n}\right)^2}$$

$$P(E)=\frac{m(m+n)}{m^2+2mn}=\frac{m+n}{m+2n}$$

Then $P(E')=1-P(E)=1-\frac{m+n}{m+2n}$

$$=\frac{n}{m+2n}$$

From question

$P(E)=3P(E')$

$$\Rightarrow \frac{m+n}{m+2n}=\frac{3n}{m+2n}$$

$$\Rightarrow m=2n \Rightarrow \frac{m}{n}=2:1$$

12. There are four gaps in between the girls where the boys can sit. Let the number of boys in these gaps be $2a+1, 2b+1, 2c+1, 2d+1$, then

$2a+1+2b+1+2c+1+2d+1=10$

$\Rightarrow a+b+c+d=3$

$=$ coefficient of x^3 in $(1-x)^{-4}$

$= {}^6C_3 = 20$

Thus, boys and girls can sit in $20\times 10!\times 5!$ ways

Total ways $=15!$

Hence the required probability $= \dfrac{20\times 10!\times 5!}{15!}$

13. Let $a_1, a_2, a_3, a_4, a_5, a_6, a_7$ be the seven digits and the remaining two be a_8 and a_9

Let

$a_1+a_2+a_3+a_4+a_5+a_6+a_7=9K, K\in I$

Also, $a_1+a_2+\ldots+a_9=1+2+3+\ldots+9$

$=\dfrac{9\times 10}{2}=45$

$\therefore a_8+a_9=45-9K$

Since, $a_1+a_2+\ldots+a_9$ and $a_1+a_2+\ldots+a_7$ are divisible by 9 if and only if a_8+a_9 is divisibly by 9. Let S be the sample space and E be the event that the sum of the digits a_8 and a_9 is divisible by 9.

Since, $a_8+a_9=45-9K$

Max. value of $a_8+a_9=17$ and min. value = 3

$3\le 45-9K\le 17$

$\Rightarrow \dfrac{28}{9}\le K\le \dfrac{42}{9}$

$\Rightarrow K=4$ $\quad [\because K \text{ is positive integer}]$

$\therefore a_8+a_9=45-36=9$

Possible pair of (a_8, a_9) can be

$\{(1,8),(2,7),(3,6),(4,5)\}=E$

$n(E)=4$ and $n(S)={}^9C_2=36$

$\therefore P(E)=\dfrac{4}{36}=\dfrac{1}{9}$

14. Let the critics be E_1, E_2 and E_3

Since, $P(E_1)=\dfrac{5}{5+2}$, $P(E_2)=\dfrac{4}{4+3}$ and $P(E_3)=\dfrac{3}{3+4}$

Clearly, the event of majority being in favour

= The event of at least two critics being in favour

$\therefore$ The required probability

$=P(E_1E_2\bar{E}_3)+P(\bar{E}_1E_2E_3)+P(E_1\bar{E}_2E_3)+P(E_1E_2E_3)$

$=\dfrac{1}{7^3}(80+24+45+60)=\dfrac{209}{343}$

15. Putting 6 different balls into three different boxes (where no boxes remain empty) = 540

Number of groups of (2 each) from 6 numbers $={}^6C_2=15$ ways

So, three groups are formed in 15 possible ways

The number of ways to put these into the boxes $=15\times 3!=90$ ways

Required probability $=\dfrac{90}{540}=\dfrac{1}{6}$

16. 8 different coloured balls and 8 bags of respective colour balls can be put in the bags (one each) in 8! ways

When 5 balls are in their respective bags

$\Rightarrow$ Possible ways $={}^8C_5$

Now 3 balls go into de-arrangement (in 2 ways)

$\Rightarrow P(E)=\dfrac{{}^8C_5\times 2}{8!}=\dfrac{1}{360}$

17. Let E be the event that the item is defective. The item produced has equal chances of being produced by either factory.

Let E_1 = The item produced from factory A

E_2 = from factory B

$$P\left(\frac{E_2}{E}\right)=\frac{P(E_2)P\left(\frac{E}{E_2}\right)}{P(E_1)P\left(\frac{E}{E_1}\right)+P(E_2)P\left(\frac{E}{E_2}\right)}$$

$$=\frac{\frac{1}{2}\left(\frac{3}{10}\right)}{\frac{1}{2}\left(\frac{2}{10}\right)+\frac{1}{2}\left(\frac{3}{10}\right)}=\frac{3}{5}$$

18. $0\le x, y\le 4\Rightarrow n(S)=16$ sq. units

Clearly, $n(E)=4$

$\therefore P(E)=\dfrac{n(E)}{n(S)}=\dfrac{1}{4}$

19. Numbers marked from 10 to 99 = 90 numbers
These are two-digit numbers. A product of 12 is obtained where (2, 6), (6, 2), (3, 4), (4, 3) are obtained
∴ In one attempt probability of success
$= \frac{4}{90} = \frac{2}{45}$
Probability of at least one laughter (success) in three trials
= 1- probability of failure in the three trials
$= \frac{43}{45}$

20. $\frac{11}{3600}$
{Do it yourself}

7.14.4 JEE Main: Archive

1. **(C)**
E_1 = denotes selection for 1st bag
E_2 = denotes selection for 2nd bag
$P(E_1) = \frac{1}{2}, P(E_2) = \frac{1}{2}$
A = selected balls are 1 red and 1 black
$P\left(\frac{A}{E_1}\right) = \frac{{}^3C_1 \times {}^1C_1}{{}^6C_2} = \frac{1}{5}$
$P\left(\frac{A}{E_1}\right) = \frac{{}^3C_1 \times {}^2C_1}{{}^{(n+5)}C_2} = \frac{12}{(n+5)(n+4)}$
$$P\left(\frac{E_1}{A}\right) = \frac{P(E_1) \times P\left(\frac{A}{E_1}\right)}{P(E_1) \times P\left(\frac{A}{E_1}\right) + P(E_2) \times P\left(\frac{A}{E_2}\right)}$$
$$= \frac{\frac{1}{10}}{\frac{1}{10} + \frac{6}{(n+5)(n+4)}} = \frac{6}{11}$$
$\Rightarrow n = 4$

2. **(A)**
$n = 33$, let probability of success is p and $q = 1 - p$
$3p(x = 0) = p(x = 1)$
$3 \cdot {}^{33}C_0 (q)^{33} = {}^{33}C_1 \, pq^{32}$
$p = \frac{1}{12}, q = \frac{11}{12}, \frac{q}{p} = 11$
$$\frac{p(x=15)}{p(x=18)} - \frac{p(x=16)}{p(x=17)}$$
$$\frac{{}^{33}C_{15} \, p^{15} q^{18}}{{}^{33}C_{18} \, p^{17} q^{15}} - \frac{{}^{33}C_{16} \, p^{16} \, q^{17}}{{}^{33}C_{17} \, p^{17} \, q^{16}}$$
$$= \left(\frac{q}{p}\right)^3 - \left(\frac{q}{p}\right)$$
$= (11)^3 - 11 = 1320$

3. **(A)**
$$P\left(\frac{1 < x < 4}{x \le 2}\right) = \frac{P(1 < x < 4 \cap x \le 2)}{P(x \le 2)}$$
$$= \frac{P(1 < x \le 2)}{P(x \le 2)} = \frac{P(x = 2)}{P(x \le 2)}$$
$$= \frac{4k}{k + 2k + 4k} = \frac{4}{7}$$

4. **(479)**
$A = \{1, 2, 3, 4\} : P(A) = \frac{3}{4} \rightarrow$ Correct
$B = \{5, 6, 7, 8, 9, 10\}; P(B) = \frac{1}{4} \rightarrow$ Correct
8 Correct Ans.:
$(4, 4) : {}^4C_4 \left(\frac{3}{4}\right)^4 \cdot {}^6C_4 \cdot \left(\frac{1}{4}\right)^4 \cdot \left(\frac{3}{2}\right)^2$
$(3, 5) : {}^4C_3 \left(\frac{3}{4}\right)^3 \cdot \left(\frac{1}{4}\right)^1 \cdot {}^6C_5 \left(\frac{1}{4}\right)^5 \cdot \left(\frac{3}{4}\right)$
$(2, 6) : {}^4C_2 \left(\frac{3}{4}\right)^2 \left(\frac{1}{4}\right)^2 \cdot {}^6C_6 \left(\frac{1}{4}\right)^6$
Total
$= \frac{1}{4^{10}} \left[3^4 \times 15 \times 3^2 + 4 \times 3^3 \times 6 \times 3 + 6 \times 3^2\right]$
$= \frac{27}{4^{10}} [2.7 + 15 + 72 + 2]$
$\Rightarrow K = 479$

5. **(C)**
(A) $P(E_1) \cdot P(E_2) = \frac{1}{6} \cdot \frac{1}{4}$
$= \frac{1}{24} \neq P(E_1 \cap E_2)$

(B) $P(E_1' \cap E_2') = 1 - P(E_1 \cup E_2)$

$= 1 - (P(E_1) + P(E_2) - P(E_1 \cap E_2))$

$= 1 - \left(\frac{1}{6} + \frac{1}{4} - \frac{1}{8}\right) = \frac{17}{24}$

$P(E_1')P(E_2) = \frac{5}{6} \times \frac{1}{4} = \frac{5}{24}$

(C) $P(E_1 \cap E_2') = P(E_1) - P(E_1 \cap E_2)$

$= \frac{1}{6} - \frac{1}{8} = \frac{1}{24}$

(D) $P(E_1' \cap E_2) = P(E_2) - P(E_1 \cap E_2)$

$= \frac{1}{4} - \frac{1}{8} = \frac{1}{8}$

6. **(D)**

$P(n) = \frac{1}{n}$

$P(2) = \frac{1}{2}\ P(8) = \frac{1}{8}$

$P(4) = \frac{1}{4}\ P(16) = \frac{1}{16}$

$P(32) = \frac{2}{32}$

Possible cases

16, 16, 16 and 32, 8, 8

$\text{Probability} = \frac{1}{16^3} + \frac{2}{32} \times \frac{1}{8} \times \frac{1}{8} \times 3 = \frac{13}{16^3}$

7. **(D)**

$P(H) = x, P(T) = 1 - x$

$P(4H\ 1T) = P(5H)$

${}^5C_1(x)^4(1-x)^1 = {}^5C_5x^5$

$5(1-x) = x$

$6x = 5 = 0\ x = \frac{5}{6}$

$P(\text{atmost } 2H)$

$= P(OH, 5T) + P(1H, 4T) + P(2H, 3T)$

$= {}^5C_0\left(\frac{1}{6}\right)^5 + {}^5C_1\frac{5}{6}\cdot\left(\frac{1}{6}\right)^4 + {}^5C_2\left(\frac{5}{6}\right)^3\left(\frac{1}{6}\right)^3$

$= \frac{1}{6^5}(1 + 25 + 250) = \frac{276}{6^5}$

$= \frac{46}{6^4}$

8. **(33)**

$2 \times 2 \times 2 \times 2 \times 2 \times 2 = 64$

Divisible by 21 when divided by 3

Case-I: All $1 \rightarrow$ (1)

Case-II: All $8 \rightarrow$ (2)

Case-II: 3 ones and 3 eights

$\frac{6!}{3! \times 3!} = 20$

Required probability $\therefore p = \frac{22}{64}$

$96p = 96 \times \frac{22}{64} = 33$

9. **(C)**

No. of ways to select and arrange

x_1, x_2, x_3, x_4, x_5

from 1, 2, 3, …, 18

$n(S) = {}^{18}C_5$

$x_1\,(x_2)\,x_3\,(x_4)\,x_5$

7 1 1

$n(E) = {}^6C_1 \times {}^3C_1 \times {}^7C_1$

$P(E) = \frac{6 \times 3 \times 7}{{}^{18}C_5}$

$\frac{1}{17 \times 4} = \frac{1}{68}$

10. **(C)**

$B(7, p)$

$n = 7, p = p$

given

$P(x=3) = 5P(x=4)$

${}^7C_3 \times p^3(1-p)^4 = 5 \cdot {}^7C_4 p^4(1-p)^3$

$\frac{{}^7C_3}{5 \times {}^7C_4} = \frac{p}{1-p}$

$1 - p = 5p$

$6p = 1$

$p = \frac{1}{6} \Rightarrow q = \frac{5}{6}$

$n = 7$

$\text{Mean} = np = 7 \times \frac{1}{6} = \frac{7}{6}$

$\text{Var} = npq = 7 \times \frac{1}{6} \times \frac{5}{6} = \frac{35}{36}$

Sum

$= \frac{7}{6} + \frac{35}{36}$

$= \frac{42+35}{36} = \frac{77}{36}$

11. **(B)**

$\text{Required probability} = \frac{\text{ar}(ADEC)}{\text{ar}(ABC)}$

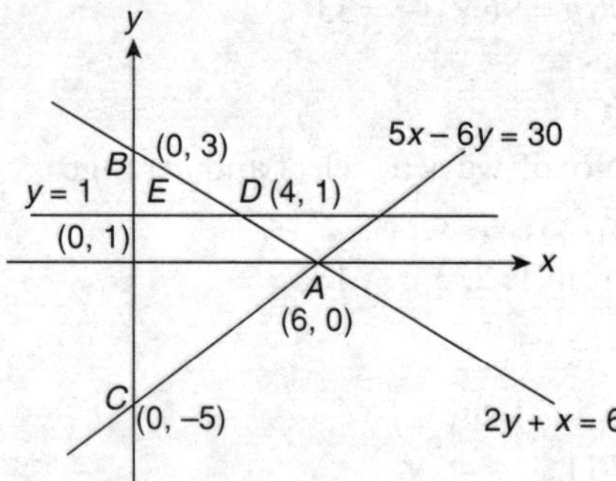

$= 1 - \frac{\text{ar}(BDE)}{\text{ar}(ABC)}$

$= 1 - \frac{\frac{1}{2} \times 2 \times 4}{\frac{1}{2} \times 8 \times 6} = 1 - \frac{1}{6} = \frac{5}{6}$

12. **(19)**

$P(A') < \frac{1}{5} = \frac{36}{180}$

5 times the sum of missing number should be less than 36

If 1 digit is missing = 7

If 2 digits are missing = 9

If 3 digits are missing = 2

If 0 digits are missing = 1

Alternate:

A is subset of S hence

A can have elements:

Type 1: $\{\}$

Type 2: $\{E_1\}, \{E_2\}, \{E_8\}$

Type 3: $\{E_1, E_2\}, \{E_1, E_3\} \ldots \{E_1, E_8\}$

:

:

Type 6:

$\{E_1, E_2, \ldots, E_5\}, \ldots \{E_4, E_5, E_6, E_7, E_8\}$

Type 7: $\{E_1, E_2 \ldots, E_6\}, \ldots \{E_3, E_4, \ldots, E_8\}$

Type 8: $\{E_1, E_2, \ldots, E_7\}\{E_2, E_3, \ldots, E_8\}$

Type 9: $\{E_1, E_2, \ldots, E_8\}$

As $P(A) \geq \frac{4}{5}$;

Note: Type 1 to Type 4 elements can not be in set A as maximum probability of type 4 elements.

$\{E_5, E_6, E_7, E_8\}$ is

$\frac{5}{36} + \frac{6}{36} + \frac{7}{36} + \frac{8}{36} = \frac{13}{18} < \frac{4}{5}$

Now for Type 5 acceptable elements let's call probability as P_5

$P_5 = \frac{n_1 + n_2 + n_3 + n_4 + n_5}{36} \leq \frac{4}{5}$

$\Rightarrow n_1 + n_2 + n_3 + n_4 + n_5 \geq 28.8$

Hence, 2 possible ways

$\{E_5, E_6, E_7, E_8, E_3 \text{ or } E_4\}$

$P_6 = n_1 + n_2 + n_3 + n_4 + n_5 + n_6 \geq 28.8$

$\Rightarrow$ 9 possible ways

$P_7 \Rightarrow n_1 + n_2 + \ldots + n_7 \geq 288$

$\Rightarrow$ 7 possible ways

$P_8 \Rightarrow n_1 + n_2 + \ldots + n_8 \geq 28.8$

$\Rightarrow$ 1 possible way

Total = 19

13. **(A)**

At least two digits are odd

= exactly two digits are odd + exactly three 3 digits are odd. For exactly three digits are odd.

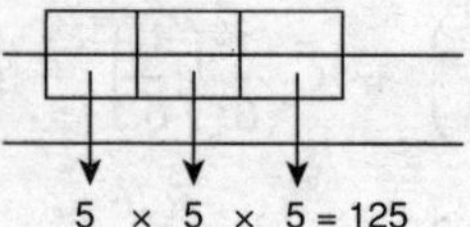

For exactly two digits odd:

If 0 is used then: $2 \times 5 \times 5 = 50$

If 0 is not used then: ${}^3C_1 \times 4 \times 5 \times 5 = 300$

Required Probability $= \frac{475}{900} = \frac{19}{36}$

14. **(C)**

$np + npq = 24$ (1)

$np \cdot npq = 128$ (2)

Solving (1) and (2):

We get $p = \frac{1}{2}, q = \frac{1}{2}, n = 32$

Now,

$P(X = 1) + P(X = 2)$

$= {}^{32}C_1 pq^{31} + {}^{32}C_2 p^2 q^{30}$

$= \frac{33}{2^{28}}$

15. **(B)**

16. **(C)**

$np = \alpha$ (1)

$npq = \frac{\alpha}{3}$ (2)

From (1) and (2)

$q = \frac{1}{3}$ and $p = \frac{2}{3}$

${}^nC_1\, q^{\,n-1}\, p^1 = \frac{4}{243}$

$\frac{n}{3^n} = \frac{2}{243}$

$n = 6$

$P(4 \text{ or } 5) = {}^nC_4 \left(\frac{2}{3}\right)^4 \left(\frac{1}{2}\right)^2 + {}^6C_5 \left(\frac{2}{3}\right)^5 \cdot \left(\frac{1}{3}\right)^0$

$= \frac{16}{27}$

17. **(D)**

$0 \le P(E_i) \le 1$ for $i = 1, 2, 3$

$\Rightarrow -\frac{2}{3} \le p \le 1$

E_1 and E_2 and E_3 are mutually exclusive

$P(E_1) + P(E_2) + P(E_3) \le 1$

$\Rightarrow \frac{2}{3} \le p \le 1$

$p_1 = 1,\ p_2 = \frac{2}{3}$

$p_1 + p_2 = \frac{5}{3}$

18. **(B)**

$np = 4$

$npq = \frac{4}{3}$

$n = 6,\ p = \frac{2}{3},\ q = \frac{1}{3}$

$54\big(P(X = 2) + P(X = 1) + P(X = 0)\big)$

$$54\left({}^6C_2\left(\frac{2}{3}\right)^2\left(\frac{1}{3}\right)^4 + {}^6C_1\left(\frac{2}{3}\right)^1\left(\frac{1}{3}\right)^5 + {}^6C_0\left(\frac{2}{3}\right)^0\left(\frac{1}{3}\right)^6\right)$$

$= \frac{146}{27}$

19. **(B)**

Let α = mean and β = Variance $(\alpha > \beta)$

So, $\alpha + \beta = 24,\ \alpha\beta = 128$

$\Rightarrow \alpha = 16$ and $\beta = 8$

$\Rightarrow np = 16\ \ npq = 8 \Rightarrow q = \frac{1}{2}$

$\therefore p = \frac{1}{2},\ n = 32$

$p(x > n - 3) = \frac{1}{2^n}\left({}^nC_{n-2} + {}^nC_{n-1} + {}^nC_n\right)$

$\therefore k = {}^{32}C_{30} + {}^{32}C_{31} + {}^{32}C_{32} = \frac{32 \times 31}{2} + 32 + 1$

$= 496 + 33 = 529$

20. **(D)**

Let

So, P (a prime number) $= 2k$,

P (a composite number) $= k$,

and $P(1) = 3k$

and $3 \times 2k + 2 \times k + 3k = 1$

$\Rightarrow k = \frac{1}{11}$

P (Success) $= P(1 \text{ or } 4) = 3k + k = \frac{4}{11}$

Number of trials, $n = 2$

$\therefore$ mean $= np = 2\times\frac{4}{11} = \frac{8}{11}$

21. (A)

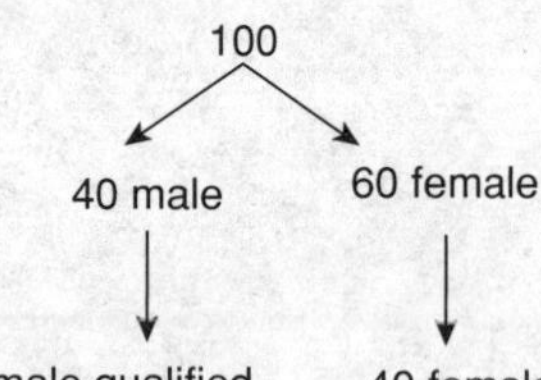

Probability that chosen candidate is female $= \frac{40}{60} = \frac{2}{3}$

22. (A)

$P(A|B) = \frac{1}{7} \Rightarrow \frac{P(A\cap B)}{P(B)} = \frac{1}{7}$

$\Rightarrow P(B) = \frac{7}{9}$

$P(B|A) = \frac{2}{5} \Rightarrow \frac{P(A\cap B)}{P(A)} = \frac{2}{5}$

$\Rightarrow P(A) = \frac{5}{18}$

Now, $P(A'\cup B) = 1 - P(A\cup B) + P(B)$

$= 1 - P(A) + P(A\cap B) = \frac{5}{6}$

$P(A'\cap B') = 1 - P(A\cup B)$

$= 1 - P(A) - P(B) + P(A\cap B) = \frac{1}{18}$

$\Rightarrow$ Both (S1) and (S2) are true.

23. (56)

X	0	1	2	3
P(X)	$\frac{1}{6}$	$\frac{1}{2}$	$\frac{3}{10}$	$\frac{1}{30}$

$\sigma^2 = \sum X^2 P(X) - \left(\sum XP(X)\right)^2 = \frac{56}{100}$

$100\sigma^2 = 56$

24. (D)

Total number of elements = 2022

$2022 = 2\times 3\times 337$

$\text{HCF}(n, 2022) = 1$

is feasible when the value of 'n' and 2022 has no common factor.

A = Number which are divisible by 2 from $\{1, 2, 3, \ldots, 2022\}$

$n(A) = 1011$

B = Number which are divisible by 3 by 3 from $(1, 2, 3 \ldots 2022)$

$n(B) = 674$

$A\cap B$ = Number which are divisibly by 6 from $\{1, 2, 3, \ldots 2022\}$

$6, 12, 18, \ldots, 2022$

$\boxed{337 = n(A\cap B)}$

$n(A\cup B) = n(A) + n(B) - n(A\cap B)$

$= 1011 + 674 - 337$

$= 1348$

C = Number which divisible by 337 from $\{1, \ldots, 1022\}$

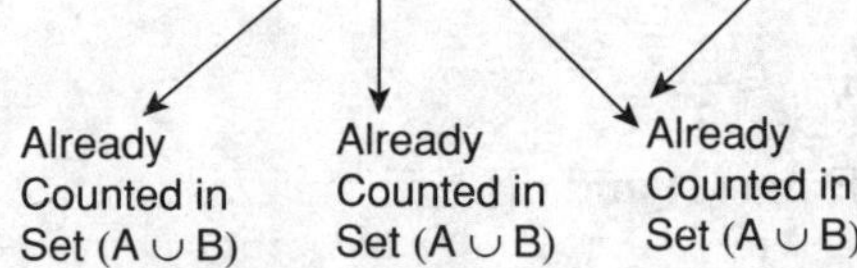

Total elements which are divisible by 2 or 3 or 337

$= 1348 + 2 = 1350$

Favourable cases = Element which are neither divisible by 2, 3 or 337

$= 2022 - 1350$

$= 672$

Required probability $= \frac{672}{2022} = \frac{112}{337}$

25. (B)

$3R$	$2R$
$4B$	$5B$
$3W$	$2W$

A: Drawn ball from boy II is black

B: Red ball transferred

$P\left(\frac{B}{A}\right) = \frac{P(A\cap B)}{P(A)}$

$$= \frac{\frac{3}{9}\times\frac{5}{10}}{\frac{3}{9}\times\frac{5}{10}+\frac{4}{9}\times\frac{6}{10}+\frac{3}{9}\times\frac{5}{10}}$$

$$= \frac{15}{15+24+15} = \frac{15}{54} = \frac{5}{18}$$

26. (96)

Let, mean $= np$

and variance $= v = npq,\ p+q=1$

$\text{Sum} = m+v = \frac{165}{2}$

$\text{Product} = mv = 1350$

On solving,

$m = np = 60$ and

$v = npq = \frac{45}{2} \therefore q = \frac{3}{8} \therefore P = \frac{5}{8}$

Hence $n = 96$

27. (C)

$x+y+z=1$

$2x+Ny+2z=2$

$3x+3y+Nz=3$

$$\Delta = \begin{vmatrix} 1 & 1 & 1 \\ 2 & N & 2 \\ 3 & 3 & N \end{vmatrix}$$

$= (N-2)(N-3)$

For unique solution $\Delta \neq 0$

So $N \neq 2, 3$

$\Rightarrow P$ (system has unique solution) $= \frac{4}{6}$

So $k = 4$

Therefore sum $= 4+1+4+5+6 = 20$

28. (D)

Ω =sample space

A = be an event

If $P(A) = 0 \Rightarrow A = \phi$

If $P(A) = 1 \Rightarrow A = \Omega$

Then both statements are true.

29. (B)

$M = 33 \times 33$

$x(66-x) \geq \frac{5}{9} \times 33 \times 33$

$11 \leq x \leq 55$

$A: \{12, 15, 18, \ldots, 54\}$

$P(A) = \frac{15}{45} = \frac{1}{3}$

30. (120)

$x + y = 5\lambda$

Cases:

x	y	Number of ways
5λ	5λ	20
$5\lambda+1$	$5\lambda+4$	25
$5\lambda+2$	$5\lambda+3$	25
$5\lambda+3$	$5\lambda+2$	25
$5\lambda+4$	$5\lambda+1$	25

31. (B)

$n(s) = 36$

Given: $N-2, \sqrt{3N}, N+2$ are in G.P.

$3N = (N-2)(N+2)$

$3N = N^2 - 4$

$\Rightarrow N^2 - 3N - 4 = 0$

$(N-4)(N+1) = 0 \Rightarrow \boxed{N=4}$ or $N = -1$ rejected

$(\text{Sum} = 4) \equiv \{(1,3), (3,1), (2,2)\}$

$n(A) = 3$

$P(A) = \frac{3}{36} = \frac{1}{12} = \frac{4}{48} \Rightarrow \boxed{k=4}$

32. (9)

E_1 : Smokers

$P(E_1) = \frac{1}{4}$

E_2 : non-smokers

$P(E_2) = \frac{3}{4}$

E : diagnosed with lung cancer

$P\left(\frac{E}{E_1}\right) = \frac{27}{28}$

$P\left(\frac{E}{E_2}\right) = \frac{1}{28}$

$$P\left(\frac{E_1}{E}\right)=\frac{P(E_1)P(E/E_1)}{P(E)}$$

$$=\frac{\frac{1}{4}\times\frac{27}{28}}{\frac{1}{4}\times\frac{27}{28}+\frac{3}{4}\times\frac{1}{28}}=\frac{27}{30}=\frac{9}{10}$$

$K=9$

33. (C)

Required probability =

$$1-\frac{D_{(15)}+{}^{15}C_1\cdot D_{(14)}+{}^{15}C_2 D_{(13)}}{15!}$$

Taking $D_{(15)}$ as $\frac{15!}{e}$ $D_{(14)}$ as $\frac{14!}{e}$ $D_{(13)}$ as $\frac{13!}{e}$

We get, $1-\left(\frac{\frac{15!}{e}+15\cdot\frac{14!}{e}+\frac{15\times 14}{2}\times\frac{13!}{e}}{15!}\right)$

$$=1-\left(\frac{1}{e}+\frac{1}{e}+\frac{1}{2e}\right)=1-\frac{5}{2e}\approx 0.08$$

34. (B)

Let $P(w_1)=\lambda$ then $P(w_2)=\frac{\lambda}{2}\ \ldots$

$$P(w_n)=\frac{\lambda}{2^{n-1}}$$

As $\sum_{k=1}^{\infty}P(w_k)=1\Rightarrow\frac{\lambda}{1-\frac{1}{2}}=1\Rightarrow\lambda=\frac{1}{2}$

So, $P(w_n)=\frac{1}{2^n}$

$A=\{2k+3\ell;\,k,\ell\in N\}=\{5,7,8,9,10,\ldots\}$

$B=\{w_n:n\in A\}$

$B=\{w_5,w_7,w_8,w_9,w_{10},w_{11},\ldots\}$

$A=N-\{1,2,3,4,6\}$

$\therefore P(B)=1-[P(w_1)+P(w_2)+P(w_3)+P(w_4)+P(w_6)]$

$$=1-\left[\frac{1}{2}+\frac{1}{4}+\frac{1}{8}+\frac{1}{16}+\frac{1}{64}\right]$$

$$=1-\frac{32+16+8+4+1}{64}=\frac{3}{64}$$

35. (B)

Either all outcomes are positive or any two are negative.

Now, $p=P$ (positive) $=\frac{3}{6}=\frac{1}{2}$

$q=p$ (negative) $=\frac{2}{6}=\frac{1}{3}$

Required probability

$$={}^5C_5\left(\frac{1}{2}\right)^5+{}^5C_2\left(\frac{1}{3}\right)^2\left(\frac{1}{2}\right)^3+{}^5C_4\left(\frac{1}{3}\right)^4\left(\frac{1}{2}\right)^1$$

$$=\frac{521}{2592}$$

$\therefore$ Option (B) is correct.

36. (14)

$$p=\frac{{}^6C_1}{6\times 6}=\frac{1}{6}$$

$$q=\frac{{}^6C_1\times{}^5C_1\times 4}{6\times 6\times 6\times 6}=\frac{5}{54}$$

$\therefore p:q=9:5\Rightarrow m+n=14$

37. (A)

$$\frac{{}^5C_2+{}^6C_2}{{}^2C_2+{}^3C_2+{}^4C_2+{}^5C_2+{}^8C_2}=\frac{10+15}{1+3+6+10+15}$$

$$=\frac{25}{35}=\frac{5}{7}$$

38. (10)

$|x-y|<a\Rightarrow -a<x-y<a$

$\Rightarrow x-y<a$ and $x-y>-a$

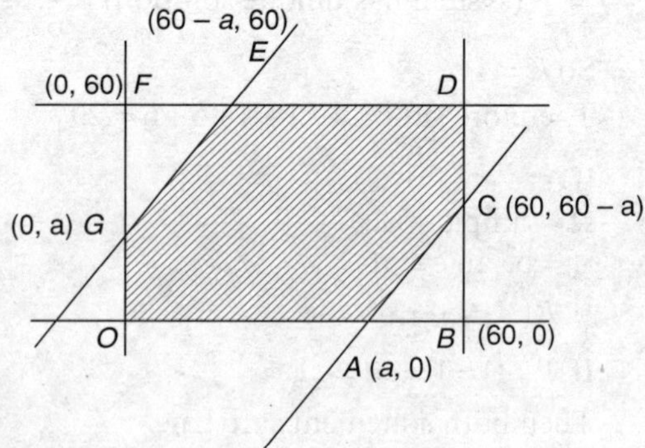

$$P(A)=\frac{\text{ar}\,(OACDEG)}{(OBDF)}$$

$$=\frac{\text{ar}\,(OBDF)-\text{ar}\,(ABC)-\text{ar}\,(EFG)}{\text{ar}\,(OBDF)}$$

$$\Rightarrow \frac{11}{36} = \frac{(60)^2 - \frac{1}{2}(60-a)^2 - \frac{1}{2}(60-a)^2}{3600}$$

$\Rightarrow 1100 = 3600 - (60-a)^2$

$\Rightarrow (60-a)^2 = 2500 \Rightarrow 60 - a = 50$

$\Rightarrow a = 10$

39. **(B)**

$np + npq = 5, np \cdot npq = 6$

$np(1+q) = 5, n^2p^2q = 6$

$n^2p^2(1+q)^2 = 25, n^2p^2q = 6$

$\frac{6}{q}(1+q)^2 = 25$

$6q^2 + 12q + 6 = 25q$

$6q^2 - 13q + 6 = 0$

$6q^2 - 9q - 4q + 6 = 0$

$(3q-2)(2q-3) = 0$

$q = \frac{2}{3}, \frac{3}{2}, q = \frac{2}{3}$ is accepted

$p = \frac{1}{3} \Rightarrow n \cdot \frac{1}{3} + n \cdot \frac{1}{3} \cdot \frac{2}{3} = 5$

$\frac{3n + 2n}{9} = 5$

$n = 9$

So $6(n + p - q) = 6\left(9 + \frac{1}{3} - \frac{2}{3}\right) = 52$

40. **(A)**

A: no. on 1st die < no. on 2nd die

A: no. on 1st die = even and no. of 2nd die = odd

C: no. on 1st die = odd and no. on 2nd die = even

$n(A) = 5 + 4 + 3 + 2 + 1 = 15$

$n(B) = 9$

$n(C) = 9$

$n((A \cup B) \cap C) = (A \cap C) \cup (B \cap C)$

$= (3 + 2 + 1) + 0 = 6$

41. **(D)**

Given that,

A pair of dice is thrown 5 times.

Required sample space is

$\{(1,1), (1,2), (1,3) \ldots (6,6)\}$

Total number of observations = 36

Now, a sum of 5 is observed in

$\{(1,4), (4,1), (2,3), (3,2)\}$

No. of favourable outcomes = 4

$P(\text{sucess})\, p = \frac{4}{36} = \frac{1}{9}$

$P(\text{failure})\, q = 1 - \frac{1}{9} = \frac{8}{9}$

We know that for a binomial distribution, $B(n, p)$

$P(X = k) = {}^nC_k p^k q^{n-k}$ where $k \geq 0$ and $p + q = 1$

$\therefore$ Required probability

$\Rightarrow P(X \geq 4) = P(X = 4) + P(X = 5)$

$= {}^5C_4 \left(\frac{1}{9}\right)^4 \frac{8}{9} + {}^5C_5 \left(\frac{1}{9}\right)^5$

$= 5 \cdot \frac{8}{9^5} + \frac{1}{9^5} = \frac{41}{9^5}$

$= \frac{41 \times 3}{3^{10} \times 3} = \frac{123}{3^{11}}$

$\Rightarrow k = 123$

Hence, this is the correct option.

42. **(D)**

Given that three dice are thrown.

The total number of outcomes when three dice are thrown together is $6^3 = 6 \times 6 \times 6$

The number of outcomes such that all the outcomes are different is $6 \times 5 \times 4$.

For example, if the outcome in dice 1 is '6'

43. **(A)**

Let $X \equiv$ Event that product is defective, then

$P\left(\frac{X}{A}\right) = \frac{3}{100}$

$P\left(\frac{X}{B}\right) = \frac{4}{100}$

$P\left(\frac{X}{C}\right) = \frac{2}{100}$

Now,

$$P\left(\frac{C}{X}\right) = \frac{P(C)P\left(\frac{X}{C}\right)}{P(A)P\left(\frac{X}{A}\right) + P(B)P\left(\frac{X}{B}\right) + P(C)P\left(\frac{X}{C}\right)}$$

$$\Rightarrow P\left(\frac{C}{X}\right)=\frac{\frac{5}{100}\times\frac{2}{100}}{\frac{20}{100}\times\frac{3}{100}+\frac{30}{100}\times\frac{4}{100}+\frac{50}{100}\times\frac{2}{100}}$$

$$\Rightarrow P\left(\frac{C}{X}\right)=\frac{100}{60+120+100}$$

$$\Rightarrow P\left(\frac{C}{X}\right)=\frac{5}{14}$$

Hence this the correct option.

44. **(A)**

Given,

The probability that the random variable X takes values x is given by $P(X=X)=k(x+1)$ 3^{-z}, $x = 0, 1, 2, 3, \ldots$, where k is a constant,

Now we know that,

$$P(X=0)+P(x=1)+P(x=2)+\ldots=1$$

$$\Rightarrow \frac{k}{3^0}+\frac{2k}{3^1}+\frac{3k}{3^2}+\ldots=1$$

$$\Rightarrow k\left(1+\frac{2}{3}+\frac{3}{3^2}+\ldots\right)=1$$

Now finding,

$$S=1+\frac{2}{3}+\frac{3}{3^2}+\frac{4}{3^3}+ \qquad (1)$$

$$\Rightarrow \frac{S}{3}=\frac{1}{3}+\frac{2}{3^2}+\frac{3}{3^2}+ \qquad (2)$$

Now on subtracting the above two equations, we get,

$$\Rightarrow \frac{2S}{3}=1+\frac{1}{3}+\frac{1}{3^2}+\ldots$$

$$\Rightarrow S=\frac{9}{4}$$

Hence, $k\left(1+\frac{2}{3}+\frac{3}{3^2}+\ldots\right)=1$

$$\Rightarrow k=\frac{4}{9}$$

Now finding,

$$P(X\ge 2)=P(2)+P(3)+\ldots$$

$$\Rightarrow P(X\ge 2)=1-P(0)-P(1)$$

$$\Rightarrow P(X\ge 2)=1-\left(\frac{k}{1}+\frac{2k}{3}\right)=1-\frac{20}{27}=\frac{7}{27}$$

45. **(D)**

Given that, $2^N < N!$

N is the sum of numbers of two dice.

$\Rightarrow 2\le N\le 12$

Let us check the given condition $2^N < N!$

$N = 1$ (not possible) $\rightarrow 0$

$N = 2$ (not possible) $\rightarrow 1$

$N = 3$ (not possible) $\rightarrow 2$

$N = 4$ (possible)

We need to find the probability for $N \ge 4$

$$\Rightarrow P(N\ge 4)=1-P(N=2)-P(N=3)$$

$$=1-\frac{1}{36}-\frac{2}{36}$$

$$=\frac{11}{12}=\frac{m}{n}$$

Now let us find $4m-3n$.

$= 4\times 11-3\times 12 = 8$

Hence, the required answer is 8

46. **(A)**

Given that P(getting odd 7 times) = P(getting odd 9 times)

This is the case of binomial distribution with n successes and probability of success $= p = \frac{3}{6}$ and failure is $q = 1-p = \frac{1}{2}$.

We know that $P(X=k)={}^nC_k\,(p)^k\,(q)^{n-k}$

Now $P(X=7)=P(X=9)$

$$\Rightarrow {}^nC_7\left(\frac{1}{2}\right)^7\left(\frac{1}{2}\right)^{n-7}={}^nC_9\left(\frac{1}{2}\right)^9\left(\frac{1}{2}\right)^{n-9}$$

$$\Rightarrow {}^nC_7={}^nC_9$$

$$\Rightarrow {}^nC_{n-7}={}^nC_9$$

$$\Rightarrow n=16$$

$$\Rightarrow P(2\text{ times even})={}^{16}C_2\left(\frac{1}{2}\right)^{14}\left(\frac{1}{2}\right)^2$$

$$=\frac{15\times 4}{2^{15}}=\frac{60}{2^{15}}=\frac{k}{2^{15}}$$

$$\Rightarrow k=60$$

Hence this the required option.

47. **(D)**

Given,

$$S=\left\{M=\left[a_{ij}\right], a_{ij}\in\{0,1,2\}, \{1\le i, j\le 2\}\right\}$$

So, let $M = \begin{bmatrix} a & b \\ c & d \end{bmatrix}$ where $a, b, c, d \in \{0, 1, 2\}$

So, total sample space will be $3^4 = 81$

Now we will find $P(\bar{A})$ or possibilities for $|M| = 0$

So, for $|M| = 0$ cases will be,

1. $\begin{bmatrix} 1 & 1 \\ 1 & 1 \end{bmatrix}$ or $\begin{bmatrix} 0 & 0 \\ 0 & 0 \end{bmatrix}$ or $\begin{bmatrix} 2 & 2 \\ 2 & 2 \end{bmatrix} \rightarrow$ Total matrix = 3
2. Two 2's and two 0's $\rightarrow$ Total matrix = 4
3. Two 1's and Two 2's $\rightarrow$ Total matrix = 4
4. Two 1's and Two 2's $\rightarrow$ Total matrix = 4
5. One 1 and three 0's $\rightarrow$ Total matrix = 4
6. One 2 and three 0's $\rightarrow$ Total matrix = 4
7. One 1 and one 2 and two 0's $\rightarrow$ Total matrix = 8

So, total cases for which $|M| = 0$ is 31

So, probability will be $P(\bar{A}) = \frac{31}{81}$

Hence, $P(A) = 1 - \frac{31}{81} = \frac{50}{81}$

48. **(27)**

Given a quadratic equation

$64x^2 + 5nx + 1 = 0$ has no real roots, so

$D < 0$

$\Rightarrow 25N^2 - 4 \times 64 < 0$

$\Rightarrow N < \frac{16}{5} \Rightarrow N = 1, 2, 3$

And, for a biased coin, the probability of getting head is $P = \frac{1}{4}$, and probability of getting tail will be $Q = \frac{3}{4}$

So, required probability is

$\frac{1}{4} + \frac{3}{4} \cdot \frac{1}{4} + \left(\frac{3}{4}\right)^2 \cdot \frac{1}{4} + \left(\frac{3}{4}\right)^3 \cdot \frac{1}{4} = \frac{p}{q}$

$\Rightarrow \frac{\frac{1}{4}\left(1 - \left(\frac{3}{4}\right)^3\right)}{1 - \left(\frac{3}{4}\right)} = \frac{p}{q}$

$\Rightarrow 1 - \left(\frac{3}{4}\right)^3 = \frac{p}{q}$

$\Rightarrow 1 - \left(\frac{3}{4}\right)^3 = \frac{p}{q}$

$\Rightarrow 1 - \frac{27}{64} = \frac{p}{q}$

$\Rightarrow \frac{p}{q} = \frac{37}{64}$

$q - p = 27$

49. **(C)**

Given that variance of $\alpha - \beta$ is $\frac{p}{q}$

$\alpha \in \{1, 2, 3, 4, 5, 6\}$

$\beta \in \{1, 2, 3, 4, 5, 6\}$

$(\alpha - \beta) = 0$ (6 case)

$(\alpha - \beta) = -1$ (5 case)

$(\alpha - \beta) = -2$ (4 case)

$(\alpha - \beta) = -3$ (3 case)

$(\alpha - \beta) = -4$ (2 case)

$(\alpha - \beta) = -5$ (1 case)

$(\alpha - \beta) = 1$(5 case)

$(\alpha - \beta) = 2$ (4 case)

$(\alpha - \beta) = 3$ (3 case)

$(\alpha - \beta) = 4$ (2 case)

$(\alpha - \beta) = 5$ (1 case)

Mean = 0

We know that variance $= \frac{\sum_{i=1}^{n} f_i (x_i - \bar{x})^2}{\sum_{i=1}^{n} f_i}$

Variance

$$= \sigma^2 = \frac{0^2 \times 6 + 2 \times 1^2 + 5 + 2 \times 2^2 \times 4 + 2 \times 3^2 \times 3 + 2 \times 4^2 \times 2 \times 5^2 \times 1}{36}$$

$= \frac{2}{36} \times (5 + 16 + 27 + 32 + 25) = \frac{105}{18} = \frac{35}{6}$

$\therefore p = 35$

Sum of divisors of $p = 1 + 5 + 7 + 35 = 48$

Option (C) is correct.

50. **(10)**
Given,
A fair $n(n>1)$ faces die is rolled repeatedly until a number less than n appears,

x_i	1	2	3	4	...
p_i	$\frac{n-1}{n}$	$\frac{1}{n}\left(\frac{n-1}{n}\right)$	$\frac{1}{n^2}\cdot\frac{n-1}{n}$	$\frac{1}{n^3}\cdot\left(\frac{n-1}{n}\right)$	...

Now we know that, mean is given by,
Mean

$$=\sum_{i=1}^{\infty} p_i x_i = 1\cdot\frac{n-1}{n}+\frac{2}{n}\cdot\left(\frac{n-1}{n}\right)+\frac{3}{n^2}\left(\frac{n-1}{n}\right)+\ldots$$

$$\Rightarrow \frac{n}{9}=\left(1-\frac{1}{n}\right)S \quad (1)$$

Where, $S=1+\frac{2}{n}+\frac{3}{n^2}+\frac{4}{n^3}+\ldots \quad (2)$

$$\frac{1}{n}S=\frac{1}{n}+\frac{2}{n^2}+\frac{3}{n^3}+\ldots \quad (3)$$

Now subtracting equation (2) – (3) we get,

$$\left(1-\frac{1}{n}\right)S=\frac{1}{1-\frac{1}{n}}$$

$$\Rightarrow \left(1-\frac{1}{n}\right)S=\frac{1}{1-\frac{1}{n}}$$

Now putting the value of S in equation (1) we get,

$$\Rightarrow \frac{n}{9}=\left(1-\frac{1}{n}\right)\times\frac{1}{\left(1-\frac{1}{n}\right)^2}=\frac{n}{n-1}$$

$$\Rightarrow n=10$$

51. **(C)**
It is given that $P(H)=3P(T)$

$$\therefore P(H)=\frac{3}{4}, P(T)=\frac{1}{4}$$

Since the coin is tossed till either 1H or 3T occurs. So, the process will end in a maximum of 3 throws.

X_i	1	2	3
$P(X_i)$	$\frac{3}{4}$	$\frac{1}{4}\cdot\frac{3}{4}$	$\frac{1}{4}\times\frac{1}{4}\times\frac{1}{4}+\frac{1}{4}\times\frac{1}{4}\times\frac{3}{4}$

Now we know that,

Mean is given by $\sum_{i=1}^{n} X_i\ P(X_i)$

$$\text{Mean}=1\times\frac{3}{4}+2\times\frac{3}{16}+3\times\frac{1}{16}=\frac{21}{16}$$

Hence, this is the correct option.

52. **(B)**
Given that the difference between the mean and variance is 1.

$\Rightarrow np-npq=1$

$\Rightarrow np(1-q)=1$

$\Rightarrow np^2=1$

Also given that,

$\Rightarrow 2P(X=2)=3P(X=1)$

$\Rightarrow 2\cdot{}^nC_2 p^2 q^{n-2}=3\cdot{}^nC_1 p\cdot q^{n-1}$

$\Rightarrow 2\cdot\frac{n(n-1)}{2}\cdot p=3\cdot n\cdot q$

$\Rightarrow (n-1)p=3(1-p)$

$\Rightarrow \left(\frac{1}{p^2}-1\right)p=3(1-p)$

$\Rightarrow \frac{(1-p)(1+p)}{p}=3(1-p)$

$\Rightarrow 1+p=3p$

$\Rightarrow p=\frac{1}{2}$

$\therefore n=4$

Now,

$$n^2P(x>1)=n^2(1-P(x=1)-P(x=0))$$

$$=16\left(1-{}^4C_1\cdot\left(\frac{1}{2}\right)^4-\left(\frac{1}{2}\right)^4\right)=11$$

Hence this is the correct option.

53. **(C)**
Given
Bag has 6 white and 4 black balls.
The given die is rolled and the number

denotes the number of balls drawn from the bag,
So, the probability that all drawn balls are white will be,

$$=\frac{1}{6}\left[\frac{{}^{6}C_1}{{}^{10}C_1}+\frac{{}^{6}C_2}{{}^{10}C_2}+\frac{{}^{6}C_3}{{}^{10}C_3}+\frac{{}^{6}C_4}{{}^{10}C_4}+\frac{{}^{6}C_5}{{}^{10}C_5}+\frac{{}^{6}C_6}{{}^{10}C_6}\right]$$

$$=\frac{1}{6}\left[\frac{6}{10}+\frac{15}{45}+\frac{20}{120}+\frac{15}{210}+\frac{6}{252}+\frac{1}{210}\right]$$

$$=\frac{504}{2520}=\frac{1}{5}$$

54. **(B)**

$$P(4W\ 4B/2W\ 2B)=\frac{P(4W\ 4B)\times P(2W\ 2B/4W\ 4B)}{P(2W\ 6B)\times P(2W\ 2B/2W\ 6B)+P(3W\ 5B)\times P(2W\ 2B/3W\ 5B)+\ldots+P(6W\ 2B)\times P(2W\ 2B/6W\ 2B)}$$

$$=\frac{\frac{1}{5}\times\frac{{}^{4}C_2\times{}^{4}C_2}{{}^{8}C_4}}{\frac{1}{5}\times\frac{{}^{2}C_2\times{}^{6}C_2}{{}^{8}C_4}+\frac{1}{5}\times\frac{{}^{3}C_2\times{}^{5}C_2}{{}^{8}C_4}+\ldots+\frac{1}{5}\times\frac{{}^{6}C_2\times{}^{2}C_2}{{}^{8}C_4}}$$

$$=\frac{2}{7}$$

55. **(B)**

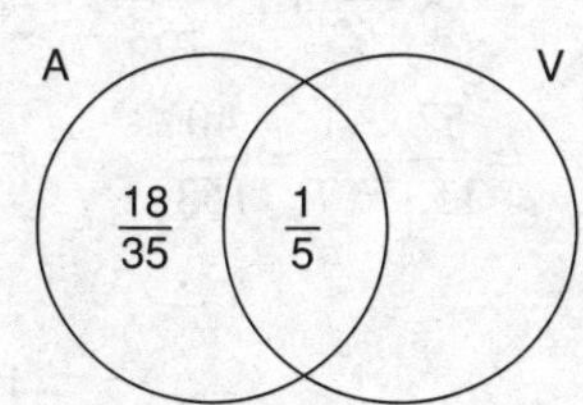

$$P(\bar{A})=\frac{2}{7}=p$$

$$P(A\cap V)=\frac{1}{5}=q$$

$$P(A)=\frac{5}{7}$$

Ans. $P(A\cap\bar{V})=\frac{18}{35}$

56. **(12)**

$$a=P(X=3)=\frac{5}{6}\times\frac{5}{6}\times\frac{1}{6}=\frac{25}{216}$$

$$b=P(X\geq 3)=\frac{5}{6}\times\frac{5}{6}\times\frac{1}{6}+\left(\frac{5}{6}\right)^3\cdot\frac{1}{6}+\left(\frac{5}{6}\right)^4\cdot\frac{1}{6}+\ldots$$

$$=\frac{\frac{25}{216}}{1-\frac{5}{6}}=\frac{25}{216}\times\frac{6}{1}=\frac{25}{36}$$

$$P(X\geq 6)=\left(\frac{5}{6}\right)^5\cdot\frac{1}{6}+\left(\frac{5}{6}\right)^6\cdot\frac{1}{6}+\ldots$$

$$=\frac{\left(\frac{5}{6}\right)^5\cdot\frac{1}{6}}{1-\frac{5}{6}}=\left(\frac{5}{6}\right)^5$$

$$c=\frac{\left(\frac{5}{6}\right)^5}{\left(\frac{5}{6}\right)^3}=\frac{25}{36}$$

$$\frac{b+c}{a}=\frac{\left(\frac{5}{6}\right)^2+\left(\frac{5}{6}\right)^2}{\left(\frac{5}{6}\right)^2\cdot\frac{1}{6}}=12$$

57. **(C)**

$$\frac{{}^{6}C_4}{{}^{15}C_4}\times\frac{{}^{9}C_4}{{}^{11}C_4}=\frac{3}{715}$$

Hence option (C) is correct.

58. **(C)**

Required probability =

$$\frac{5}{6}\times\frac{1}{6}+\left(\frac{5}{6}\right)^3\times\frac{1}{6}+\left(\frac{5}{6}\right)^5\times\frac{1}{6}+\ldots$$

$$=\frac{1}{6}\times\frac{\frac{5}{6}}{1-\frac{25}{36}}=\frac{5}{11}$$

59. **(B)**

Given set = {1, 2, 3, …, 50}

$P(A)$ = Probability that number is multiple of 4

$P(B)$ = Probability that number is a multiple of 6

$P(C)$ = Probability that number is a multiple of 7

Now,

$P(A)=\frac{12}{50}, P(B)=\frac{8}{50}, P(C)=\frac{7}{50}$

again

$P(A\cap B)=\frac{4}{50}, P(B\cap C)$

$=\frac{1}{50}, P(A\cap C)=\frac{1}{50}$

$P(A\cap B\cap C)=0$

Thus

$P(A\cup B\cup C)=\frac{12}{50}+\frac{8}{50}+\frac{7}{50}-\frac{4}{50}-\frac{1}{50}-\frac{1}{50}+0$

$=\frac{21}{50}$

60. **(A)**

If $x=0, y=6,7,8,9,10$

If $x=1, y=7,8,9,10$

If $x=2, y=8,9,10$

If $x=3, y=9,10$

If $x=4, y=10$

If $x=5$, y = no possible value

Total possible ways $=(5+4+3+2+1)\times 2$

$=30$

Required probability $=\frac{30}{11\times 11}=\frac{30}{121}$

61. **(C)**

E_1 : A is selected	A	B
	3 W	3 W
	7 R	2 R

E_2 : B is selected

E : white ball is drawn

$$P\left(\frac{E_1}{E}\right)=\frac{P(E)\cdot P\left(\frac{E}{E_1}\right)}{P(E_1)\cdot P\left(\frac{E}{E_1}\right)+P(E_2)\cdot P\left(\frac{E}{E_2}\right)}$$

$$=\frac{\frac{1}{2}\times\frac{3}{10}}{\frac{1}{2}\times\frac{3}{10}+\frac{1}{2}\times\frac{3}{5}}$$

$$=\frac{3}{3+6}=\frac{1}{3}$$

62. **(D)**

Probability of drawing first red and then white

$=\frac{10}{75}\times\frac{30}{75}=\frac{4}{75}$

63. **(D)**

3 bad apples, 15 good apples.

Let X be the number of bad apples

Then $P(X=0)=\frac{^{15}C_2}{^{18}C_2}=\frac{105}{153}$

$P(X=1)=\frac{^{3}C_1\times {^{15}C_1}}{^{18}C_2}=\frac{45}{153}$

$P(X=2)=\frac{^{3}C_2}{^{18}C_2}=\frac{3}{153}$

$E(X)=0\times\frac{105}{153}+1\times\frac{45}{153}+2\times\frac{3}{153}=\frac{51}{153}$

$=\frac{1}{3}$

$\text{Var}(X)=E(X^2)-(E(X))^2$

$=0\times\frac{105}{153}+1\times\frac{45}{153}+4\times\frac{3}{153}-\left(\frac{1}{3}\right)^2$

$=\frac{57}{153}-\frac{1}{9}=\frac{40}{153}$

64. **(A)**

Let the probability of tails be $\frac{1}{3}$

$\Rightarrow$ Probability of getting a head $=\frac{2}{3}$

$\therefore$ Probability of getting 2 tails and 1 head

$=\left(\frac{1}{3}\times\frac{2}{3}\times\frac{1}{3}\right)\times 3$

$=\frac{2}{27}\times 3=\frac{2}{9}$

65. **(A)**

$A\ B\ C$

$7R, 5B\ 5R, 7B\ 6R, 6B$

$$P(B)=\frac{1}{3}\cdot\frac{5}{12}+\frac{1}{3}\cdot\frac{7}{12}+\frac{1}{3}\cdot\frac{6}{12}$$

$$\text{required probability}=\frac{\frac{1}{3}\cdot\frac{5}{12}}{\frac{1}{3}\cdot\left[\frac{5}{12}+\frac{7}{12}+\frac{6}{12}\right]}=\frac{5}{18}$$

66. **(8288)**

$P(W)=\frac{1}{3}, P(L)=\frac{2}{3}$

x = number of matches that team wins

y = number of matches that team loses

$|x-y|\le 2$ and $x+y=10$

$|x-y|=0\,1\,2\ x, \quad y\in N$

Case I: $-I:|x-y|=0\Rightarrow x=y$

$\because x+y=10\Rightarrow x=5=y$

$$P(|x-y|-0)={}^{10}C_5\left(\frac{1}{3}\right)^5\left(\frac{2}{3}\right)^5$$

Case II: $|x-y|=1\Rightarrow x-y=\pm 1$

$x=y+1$	$x=y-1$
$\because x+y=10$	$\because x+y=10$
$2y=9$	$2y=11$
Not possible	Not possible

Case III: $|x-y|=2\Rightarrow x-y=\pm 2$

$x-y=2$ OR $x-y=-2$

$\because x+y=10 \because x+y=10$

$x=6, y=4\ x=4, y=6$

$$P(|x-y|=2)={}^{10}C_6\left(\frac{1}{3}\right)^6\left(\frac{2}{3}\right)^4+{}^{10}C_4\left(\frac{1}{3}\right)^4\left(\frac{2}{3}\right)^6$$

$$p={}^{10}C_5\frac{2^5}{3^{10}}+{}^{10}C_6\frac{2^4}{3^{10}}+{}^{10}C_4\frac{2^6}{3^{10}}$$

$$3^9p=\frac{1}{3}\left({}^{10}C_5 2^5+{}^{10}C_6 2^4+{}^{10}C_4 2^6\right)$$

$=8288$

67. **(B)**

$ax^2+bx+c=0$

$a, b, c\in\{1, 2, 3, 4, 5, 6, 7, 8\}$

Repeated roots $D=0$

$\Rightarrow b^2-4ac=0\Rightarrow b^2=4ac$

$$\text{Prob}=\frac{8}{8\times 8\times 8}=\frac{1}{64}$$

$\Rightarrow(a, b, c)$

$(1, 2, 1); (2, 4, 2); (1, 4, 4); (4, 4, 1); (3, 6, 3);$

$(2, 8, 8); (8, 8, 2); (4, 8, 4)$

68. **(C)**

$D>0$

$b^2>4ac$

$b=3:(a, c)=(1,1)(1,2)(2,1)$

$b=4:(a, c)=(1,1)(1,2)(2,1)(1,3)(3,1)$

$b=5:(a, c)=(1,1)(1,2)(2,1)(1,3)(3,1)$

$(1,4)(4,1)$

$(1,5)(5,1)(1,6)(6,1)(2,3)(3,2)(2,2)$

$b=6:(a, c)=(1,1)(1,2)(2,1)(1,3)(3,1)$

$(1,4)(4,1)$

$(1,5)(5,1)(1,6)(6,1)(2,3)(3,2)(2,4)$

$(4,2)(2,2)$

fav. cases = 38

$$\text{Probability: }\frac{38}{6\times 6\times 6}$$

69. **(5)**

$$\frac{1}{3}+k+\frac{1}{6}+\frac{1}{4}=1\Rightarrow k=\frac{1}{4}$$

$$\mu=\frac{\alpha}{3}+\frac{1}{4}-\frac{3}{4}$$

$$\mu=\frac{\alpha}{3}-\frac{1}{2}$$

$$\alpha=\sqrt{\left(\alpha^2\frac{1}{3}+\frac{1}{4}+9\frac{1}{4}\right)-\left(\frac{\alpha}{3}-\frac{1}{2}\right)^2}$$

$$\alpha=\sqrt{\frac{2\alpha^2}{9}+\frac{\alpha}{3}+\frac{9}{4}}$$

$\alpha=\mu+2$

$$\sigma^2=(\mu+2)^2\Rightarrow\frac{2\alpha^2}{9}+\frac{\alpha}{3}+\frac{9}{4}=\frac{\alpha^2}{9}+\frac{9}{4}+\alpha$$

$$\frac{\alpha^2}{9}-\frac{2\alpha}{3}=0$$

$\alpha = 0$, (reject) or $\alpha = 6$

$(\because x = 0$ is already given$)$

$\Rightarrow \sigma + \mu = 2\mu + 2$

$= 5$

70. (A)

	A	B
Manufactured	60%	40%
Standard quality	80%	90%

P(Manufactured at B/found standard quality) = ?

A: Found S. Q

B: Manufacture B

C: Manufacture A

$$P(E_1) = \frac{40}{100}$$

$$P(E_2) = \frac{60}{100}$$

$$P\left(\frac{A}{E_1}\right) = \frac{90}{100}$$

$$P\left(\frac{A}{E_2}\right) = \frac{80}{100}$$

$$\because P\left(\frac{E_1}{A}\right) = \frac{P\left(\frac{A}{E_1}\right)P(E_1)}{P\left(\frac{A}{E_1}\right)P(E_1) + P\left(\frac{A}{E_2}\right)P(E_2)} = \frac{3}{7}$$

$\therefore 126P = 54$

71. (B)

Total method $= 5^3$

Favourable $= {}^5C_2\left(2^3 - 2\right) = 60$

$$\text{Probability} = \frac{60}{125} = \frac{12}{25}$$

72. (71)

$$a = 1 - \frac{{}^3C_5}{{}^{12}C_5}$$

$$b = 3 \cdot \frac{{}^9C_4}{{}^{12}C_5}$$

$$c = 3 \cdot \frac{{}^9C_3}{{}^{12}C_5}$$

$$d = 1 \cdot \frac{{}^9C_2}{{}^{12}C_5}$$

$u = 0 \cdot a + 1 \cdot b + 2 \cdot c + 3 \cdot d = 1.25$

$\sigma^2 = 0 \cdot a + 1 \cdot b + 4 \cdot c + 9 \cdot d - u^2$

$$\sigma^2 = \frac{105}{176}$$

Ans. $176 - 105 = 71$

73. (A)

$x + y = 24$, $x, y \in N$

$AM > GM \Rightarrow xy \le 144$

$xy \ge 108$

Favourable pairs of (x, y) are

$(13,11), (12,12), (14,10), (15,9), (16,8),$
$(17,7), (18,6), (6,18), (7,17), (8,16), (9,15)$
$(10,14), (11,13)$

That is, 13 cases

Total choices for $x + y = 24$ is 23

$$\text{Probability} = \frac{13}{23} = \frac{m}{n}$$

$n - m = 10$

74. (17)

Blue Balls	0	1	2	3	4	5
Prob.	$\frac{{}^5C_0 \cdot {}^4C_1}{{}^9C_3}$	$\frac{{}^5C_1 \cdot {}^4C_2}{{}^9C_3}$	$\frac{{}^5C_2 \cdot {}^4C_1}{{}^9C_3}$	$\frac{{}^5C_3 \cdot {}^4C_0}{{}^9C_3}$	0	0

$$7\bar{x} = \frac{{}^5C_1\,{}^4C_2 + {}^5C_2 \cdot {}^4C_1 \times 2 + {}^5C_3 \cdot {}^4C_0 \times 3}{{}^9C_3} \times 7$$

$$\frac{30 + 80 + 30}{84} \times 7$$

$$= \frac{140}{12} = \frac{70}{6} = \frac{35}{3}$$

Yellow	**0**	**1**	**2**	**3**	**4**
		${}^5C_2\,{}^4C_1$	${}^5C_1\,{}^4C_2$	${}^5C_0\,{}^4C_3$	0

$$4\bar{y} = \frac{40 + 60 + 12}{84} \times 4 = \frac{112}{21} = \frac{16}{3}$$

75. **(D)**

$X\ Y\ Z$

5 one and 4 five 4 one and 5 five 3 one and 6 five

$$P=\frac{\frac{4}{9}}{\frac{5}{9}+\frac{4}{9}+\frac{3}{9}}=\frac{4}{12}=\frac{1}{3}$$

76. **(19)**

$a, b, c \in \{1, 2, 3, 4\}$

Tetrahedral dice

$ax^2+bx+c=0$

has all real roots

$\Rightarrow D \geq 0$

$\Rightarrow b^2-4ac \geq 0$

Let $b=1 \Rightarrow 1-4ac \geq 0$ (Not feasible)

$b=2 \Rightarrow 4-4ac \geq 0$

$1 \geq ac \Rightarrow a=1, c=1$

$b=3 \Rightarrow 9-4ac \geq 0$

$\frac{9}{4} \geq ac$

$\Rightarrow a=1, c=1$

$\Rightarrow a=1, c=2$

$\Rightarrow a=2, c=1$

$b=4 \Rightarrow 16-4ac \geq 0$

$4 \geq ac$

$\Rightarrow a=1, c=1$

$\Rightarrow a=1, c=2 \Rightarrow a=2, c=1$

$\Rightarrow a=1, c=3 \Rightarrow a=3, c=1$

$\Rightarrow a=1, c=4 \Rightarrow a=4, c=1$

$\Rightarrow a=2, c=2$

1 2 3 4

$$\text{Probability}=\frac{12}{(4)(4)(4)}=\frac{3}{16}=\frac{m}{m}$$

$m+n=19$

77. **(D)**

Favourable cases $= {}^6C_3$

Total out comes $= 6^3$

Probability of getting a greater number than the previous one

$$=\frac{{}^6C_3}{r^3}=\frac{20}{216}=\frac{5}{54}$$

7.15 Summary

In this chapter, we studied about the axiomatic approach of probability. The main features of this chapter are as follows.

- **Random Experiment:** An experiment whose all possible outcomes are known in advance but the outcome of any specific performance cannot be predicted before the completion of the experiment.
- **Sample Space:** The set of all possible outcomes of a random experiment.
- **Discrete Sample Space:** A sample space having a finite number of sample points.
- **Event:** A subset of a sample space.
- **Equally Likely Event:** Those events, the chances of whose happening is neither lesser nor greater than others.
- **Mutually Exclusive Events:** Any two events A and B are said to be mutually exclusive if $A \cap B = \phi$.
- **Exhaustive Set of Events:** The set of all possible events of a random experiment.
- **Probability of an Event:** If S denotes sample space and E denotes event, then

$$P(E)=\frac{n(E)}{n(S)}=\frac{\text{number of elements in } E}{\text{number of elements in } S}$$

- *I)* $0 \leq P(E) \leq 1$ *II)* $P(E)+P(E')=1$, where E′ denote complement of the event E

Odds in favour of $E=\frac{n(E)}{n(E')}=\frac{P(E)}{P(E')}$

Odds in against of $E=\frac{n(E')}{n(E)}=\frac{P(E')}{P(E)}$

- If A and B are any two events, then $P(A \cup B)=P(A)+P(B)-P(A \cap B)$
  - If A and B are mutually exclusive, then $P(A \cup B)=P(A)+P(B)$
- **Independent Event:** Two or more events are said to be independent if occurrence or non-occurrence of any of them does not influence the occurrence or non-occurrence of the other events.

- **Conditional Probability:**
  - $P\left(\frac{A}{B}\right)=\frac{P(A\cap B)}{P(B)}, 0<P(B)\le 1$
  - $P\left(\frac{B}{A}\right)=\frac{P(A\cap B)}{P(A)}, 0<P(A)\le 1$

  If A and B are independent events, then $P(A\cap B)=P(A)P(B)$
- Mutually exclusive events will never be independent and independent events will never be mutually exclusive.
- **Total Probability Rule:** If $E_1, E_2, \ldots, E_n$ are n mutually exclusive and exhaustive events associated with a random experiment, if E be any event which occurs together with one of the events $E_1, E_2, \ldots, E_n$, then

$$P(E)=P(E_1)P\left(\frac{E}{E_1}\right)+P(E_2)P\left(\frac{E}{E_2}\right)+\ldots+P(E_n)P\left(\frac{E}{E_n}\right)$$

- **Baye's Theorem:**

$$P\left(\frac{E_K}{E}\right)=\frac{P(E_K)P\left(\frac{E}{E_K}\right)}{P(E_1)P\left(\frac{E}{E_1}\right)+P(E_2)P\left(\frac{E}{E_2}\right)+\ldots+P(E_n)P\left(\frac{E}{E_n}\right)}$$

- **Binomial Theorem on Probability:** $P(X=r)$ or $P(r)={}^nC_r p^r q^{n-r}$
- **Probability Distribution:** If the values of a random variable together with the corresponding probabilities are given

  Mean of a Binomial Distribution $= np$

  Variance of Binomial Distribution $= npq$

CHAPTER 8

Statistics

OBJECTIVES

8.1 Introduction
8.2 Measures of Central Value
8.3 Dispersion
8.4 Short Cut Method
8.5 Step Deviation Method for Finding S.D.
8.6 Know Your Concepts
8.7 JEE Main: Sample Questions
8.8 JEE Main: Archive
8.9 Answer Key
8.10 Solutions
8.11 Summary

8.1 Introduction

Statistics is the science that deals with collection, classification and tabulation of numerical facts as the basis for explanation, description and comparison of phenomena. In other words, statistics is the science that deals with the collection, analysis and interpretation of numerical data.

8.2 Measures of Central Value

8.2.1 Mean

The mean is the most common measure of central tendency. The mean is computed by summing all the scores in the distribution and dividing that sum by the total number of scores.

(I) Arithmetic mean of individual series (ungrouped data): If $x_1, x_2, \ldots, x_n$ are n observations, then the arithmetic mean $\bar{x}$ is given by

$$\bar{x} = \frac{x_1 + x_2 + \ldots + x_n}{n} = \frac{1}{n}\sum_{i=1}^{n} x_i.$$

(II) Arithmetic mean for discrete frequency distribution: If $f_1, f_2, \ldots, f_n$ are the frequencies of the random variables $x_1, x_2, \ldots, x_n$ respectively, then the arithmetic mean $\bar{x}$ is given by

$$\bar{x} = \frac{f_1x_1 + f_2x_2 + \ldots + f_nx_n}{f_1 + f_2 + \ldots + f_n}$$

$$= \frac{1}{N}\sum_{i=1}^{n} f_ix_i, \text{ where } N = \sum_{i=1}^{n} f_i$$

We can write this as $\bar{x} = \frac{1}{N}\sum_{i=1}^{n} f_i(x_i - O)$

That is, the deviations are measured from the origin.

(III) Arithmetic mean for grouped or continuous frequency distribution: Arithmetic mean

$$\bar{x} = A + \frac{1}{N}\sum_{i=1}^{n} f_i(x_i - A),$$ where A = assumed mean,

f_i = frequency and $x_i - A$ = deviation of each item from the assumed mean.

Alternatively, $\bar{x} = A + \frac{1}{N}\sum_{i=1}^{n} f_id_i$ where $d_i = x_i - A$.

Properties of A.M.

(I) The algebraic sum of the deviations of a set of values from their arithmetic mean is zero.

That is, $\sum_{i=1}^{n} f_i(x_i - \bar{x}) = 0$

Since $\sum_{i=1}^{n} f_i(x_i - \bar{x}) = \sum_{i=1}^{n} f_ix_i - \bar{x}\sum_{i=1}^{n} f_i$

$= N\bar{x} - \bar{x}N = 0$

(II) The sum of the squares of the deviations of a set of values is minimum about their mean $\bar{x}$.

(III) Combined arithmetic mean: If $\bar{x}_i (i = 1, 2, \ldots, k)$ are the means of K-components series of size $n_i\,(i = 1, 2, \ldots, k)$ respectively, then the mean $\bar{x}$

of the composite series obtained on combining the component series is given by the formula

$$\bar{x} = \frac{n_1\bar{x}_1 + n_2\bar{x}_2 + \ldots + n_k\bar{x}_k}{n_1 + n_2 + \ldots + n_k} = \frac{\sum_{i=1}^{k} n_i\bar{x}_i}{\sum_{i=1}^{k} n_i}$$

8.2.2 Median

The median is the value of the variate which divides the total frequency into two equal halves. The median is also referred to as the score at the 50th percentile in the distribution.

Calculation of Median

(i) Individual series: If the data is raw, arrange in ascending or descending order. Let n be the number of observations.

If n is odd, Median = Value of $\left(\frac{n+1}{2}\right)^{\text{th}}$ term

If n is even, Median $= \frac{1}{2}$ [Value of $\left(\frac{n}{2}\right)^{\text{th}}$ term + Value of $\left(\frac{n}{2}+1\right)^{\text{th}}$ term]

(ii) Discrete Series: In this case, we first find the cumulative frequencies of the variable arranged in ascending or descending order and the median is given by the $\left(\frac{n}{2}+1\right)^{\text{th}}$ observation, where n is the cumulative frequency.

(iii) For grouped or continuous distributions: In this case we use the formula

$$\text{Median} = l + \frac{\left(\frac{N}{2} - c\right)}{f} \times i$$

where l = lower limit of the median class, f = frequency of the median class, N = the sum of all frequencies, i = the width of the median class and c = the cumulative frequency of the class preceding the median class.

(iv) Quartile: Since the median divides a distribution into two equal parts, similarly the quartiles, quintiles, deciles and percentiles divide the distribution respectively into 4, 5, 10 and 100 equal parts. The first quartile is denoted by Q_1 and the third quartile by Q_3; $Q_3 - Q_1$ is called the inter-quartile range. The formulae for calculating quartiles, deciles and percentiles are

$$Q_J = l + \frac{\left(\frac{JN}{4} - c\right)}{f} \times i, \; J = 1, 2, 3$$

$$D_J = l + \frac{\left(\frac{JN}{10} - c\right)}{f} \times i, \; J = 1, 2, 3, \ldots, 9$$

$$P_K = l + \frac{\left(\frac{KN}{100} - c\right)}{f} \times i, \quad K = 1, 2, 3, \ldots, 99$$

8.2.3 Mode

It is the value of the variate in the frequency distribution for which the frequency is maximum. The formula for calculating the mode is

$$\text{Mode} = l + \left[\frac{f_1 - f_0}{2f_1 - f_0 - f_2}\right] \times i$$

where l = The lower limit of the model class, f_1 = the frequency of the model class, f_0 = the frequency of the class preceding the model class, f_2 = the frequency of the class succeeding the model class and i = the size of model class.

Note:

(i) For a moderately symmetrical distribution, the relation connecting mean, median and mode is

$$\text{Mean} - \text{Mode} = 3(\text{Mean} - \text{Median})$$

(ii) For a symmetrical distribution Mean, Median and Mode coincide.

8.3 Dispersion

It is a measure of the variations. It measures the spread of the observed values of a distribution around the centre value. The measures of dispersion commonly used are:

(i) Range

(ii) Quartile deviation or semi-interquartile range

(iii) Mean deviation

(iv) Standard deviation

Measures of dispersions are also called averages of the second kind.

(i) Range: It is the difference between the greatest and least values of the variables.

(ii) Quartile Deviation: The difference between the upper and lower quartiles, that is, $Q_3 - Q_1$ is called the interquartile range and $Q = \frac{Q_3 - Q_1}{2}$ is called the quartile deviation or semi-interquartile range. The coefficient of quartile deviation is $\frac{Q_3 - Q_1}{Q_3 + Q_1}$. It is independent of the unit of measurement.

(iii) Mean Deviation: Mean deviation of a distribution is the arithmetic mean of the absolute deviations of the terms of the distribution from its statistical mean (A.M. or Median or Mode).

(a) **Mean deviation of a simple (individual) distribution:** If $x_1, x_2, \ldots, x_n$ are n variables and K is the A.M. or median or mode, then the Mean Deviation (M.D.) about K is given by,

$$\delta_K = \frac{|x_1 - K| + |x_2 - K| + \ldots + |x_n - K|}{n}$$

$$= \frac{\sum_{i=1}^{n} |x_i - K|}{n} = \frac{1}{n}\sum_{i=1}^{n} |d_i|,$$

where $d_i = x_i - k$.

(b) **Mean deviation of ungrouped (unclassified) frequency distribution (discrete frequency distribution):** Let $x_1, x_2, \ldots, x_n$ be the n observations having frequencies $f_1, f_2, \ldots, f_n$ respectively and K be the A.M. or median or mode. Then, the mean deviation about K is given by

$$\text{M.D.} = \frac{|x_1 - K| f_1 + |x_2 - K| f_2 + \ldots + |x_n - K| f_n}{f_1 + f_2 + \ldots + f_n}$$

$$= \frac{\sum_{i=1}^{n} f_i |d_i|}{N}, \text{ where } d_i = x_i - K \text{ and}$$

$$N = \sum_{i=1}^{n} f_i$$

(c) **Mean deviation of grouped (classified) frequency distribution:** Mean deviation (M.D.) about K

$$= \frac{\sum_{i=1}^{n} |x_i - K| f_i}{\sum_{i=1}^{n} f_i}$$

where x_i = mid value (class mark) of the ith class, f_i = frequency of the ith class and K = mean or median or mode.

Note:

(a) Mean deviation about the median is least.

(b) Mean deviation of individual observations about K can also be found by using the formula

$$\text{M.D.} = \frac{1}{n}\left[S_1 - S_2 - K(n_1 - n_2)\right]$$

where K = central value or average, n = number of observations in the distribution, S_1 = sum of the values of observations greater than or equal to K, S_2 = sum of values of observations less than K, n_1 = number of observations greater than or equal to K and n_2 = number of observations less than K.

(iv) Variance and Standard Deviation Variance: The variance of a variate is the arithmetic mean of the squares of all deviations of the values of the variate x from the arithmetic mean of the observations and is denoted by σ^2 or $\text{var}(x)$.

Thus, if $x_1, x_2, \ldots, x_n$ are the n given values of the variate and $\bar{x}$ is their mean, then variance, $\sigma^2 = \frac{1}{n}\sum_{i=1}^{n} (x_i - \bar{x})^2$.

8.3.1 Standard Deviation

The standard deviation $(s \text{ or } \sigma)$ is defined as the positive square root of the variance.

$$\text{Standard deviation } (\sigma) = \sqrt{\frac{1}{n}\sum_{i=1}^{n} (x_i - \bar{x})^2}$$

Standard Deviation in Different Cases

(a) When the distribution is simple: Let $x_1, x_2, \ldots, x_n$ be n values of a variable x, then by definition

Variance, $\sigma^2 = \frac{1}{n}\sum_{i=1}^{n}(x_i - \bar{x})^2$, where

$\bar{x}$ = A.M. of $x_1, x_2, \ldots, x_n$.

$$\text{or } \sigma^2 = \frac{1}{n}\left[\sum_{i=1}^{n}\left(x_i^2 - 2x_i\bar{x} + \bar{x}^2\right)\right]$$

$$= \frac{1}{n}\left[\sum_{i=1}^{n} x_i^2 - 2\bar{x}\sum_{i=1}^{n} x_i + \sum_{i=1}^{n}\bar{x}^2\right]$$

$$= \frac{1}{n}\left[\sum_{i=1}^{n} x_i^2 - 2\bar{x}\cdot n\bar{x} + n\bar{x}^2\right] \left[\because \bar{x} = \frac{\sum_{i=1}^{n} x_i}{n}\right]$$

$$= \frac{1}{n}\left[\sum_{i=1}^{n} x_i^2 - n\bar{x}^2\right]$$

$$\therefore \sigma^2 = \frac{1}{n}\sum_{i=1}^{n} x_i^2 - \bar{x}^2$$

$$\text{and S.D., } \sigma = \sqrt{\frac{1}{n}\sum_{i=1}^{n} x_i^2 - \bar{x}^2}$$

(b) When the distribution is ungrouped (unclassified) frequency distribution (discrete frequency distribution): Let $x_1, x_2, \ldots, x_n$ be the n observations having frequency $f_1, f_2, \ldots, f_n$ respectively. Then the S.D. σ is given by

$$\sigma^2 = \frac{1}{n}\sum_{i=1}^{n}(x_i - \bar{x})^2 f_i \text{, where } \bar{x} = \text{A.M.}$$

$$= \frac{1}{N}\left[\sum_{i=1}^{n}\left(x_i^2 - 2x_i\bar{x} + \bar{x}^2\right) f_i\right]$$

$$= \frac{\sum_{i=1}^{n} x_i^2 f_i}{N} - 2\bar{x}\frac{\sum_{i=1}^{n} x_i f_i}{N} + \bar{x}^2\frac{\sum_{i=1}^{n} f_i}{N}$$

$$= \frac{\sum_{i=1}^{n} x_i^2 f_i}{N} - 2\bar{x}\cdot\bar{x} + \bar{x}^2$$

$$\left[\frac{\sum_{i=1}^{n} x_i f_i}{N} = \bar{x} \text{ and} \sum_{i=1}^{n} f_i = N\right]$$

$$\therefore \sigma = \sqrt{\frac{\sum_{i=1}^{n} x_i^2 f_i}{N} - \bar{x}^2}$$

(c) When distribution is grouped (classified) frequency distribution (continuous frequency distribution): Let x_i = class mark (mid value) of the ith class, f_i = frequency of x_i = frequency of ith class, $N = \sum_{i=1}^{n} f_i$ = total frequency and $\bar{x}$ = A.M. of given data.

$$\text{Then, S.D. } \sigma = \sqrt{\frac{\sum_{i=1}^{n} x_i^2 f_i}{N} - \bar{x}^2}$$

8.4 Short Cut Method

(a) For individual series: Let $x_1, x_2, \ldots, x_n$ be the n observations and $\bar{x}$ be their arithmetic mean.

Let A = Assumed mean and $d_i = x_i - A$

$$\text{then S.D. } \sigma = \sqrt{\frac{\sum_{i=1}^{n} d_i^2}{n} - \left(\frac{\sum_{i=1}^{n} d_i}{n}\right)^2}$$

(b) For ungrouped frequency distribution: Let $x_1, x_2, \ldots, x_n$ be the n observations having frequencies $f_i, f_2, \ldots, f_n$ respectively and $\bar{x}$ be their A.M.

Let $N = \sum_{i=1}^{n} f_i$ and $d_i = x_i - A$, where A is the assumed mean.

$$\text{Then, S.D. } \sigma = \sqrt{\frac{\sum f_i d_i^2}{N} - \left(\frac{\sum f_i d_i}{N}\right)^2}$$

(c) For grouped frequency distribution:

$$\text{S.D. } \sigma = \sqrt{\frac{\sum f_i d_i^2}{N} - \left(\frac{\sum f_i d_i}{N}\right)^2}$$

where x_i = class mark (mid value) of the ith class, f_i = frequency of x_i, $d_i = x_i - A$, A being the assumed mean and $N = \sum_{i=1}^{n} f_i$.

8.5 Step Deviation Method for Finding S.D.

This method is used when deviations of observations from assumed mean are large numbers and class intervals are equal.

$$\text{S.D.} \quad (\sigma) = h\sqrt{\frac{\sum f_i u_i^2}{N} - \left(\frac{\sum f_i u_i}{N}\right)^2}$$

where x_i = class mark (mid value) of the ith class, f_i = frequency of the ith class, $N = \sum_{i=1}^{n} f_i$, A = assumed mean, h = class interval and $u_i = \frac{x_i - A}{h}$.

Note:
If $x_1, x_2, \ldots, x_n$ are the n observations having A.M. $\bar{x}$ and S.D. σ, then

$$x_1^2 + x_2^2 + \ldots + x_n^2 = n\left(\sigma^2 + \bar{x}^2\right).$$

SOLVED EXAMPLES

Example 8.1 The following are the marks obtained by 10 students in an examination. Find their mean deviation from (i) mean and (ii) mode.

50, 40, 22, 35, 45, 60, 50, 70, 50, 78

Solution:

(i) Mean deviation about the mean: Let $\bar{x}$ be the mean of the given data,

$$\bar{x} = \frac{50 + 40 + 22 + 35 + 45 + 60 + 50 + 70 + 78}{10}$$

$$\Rightarrow \bar{x} = \frac{500}{10} = 50$$

Calculation of M.D.

x_i	$\lvert d_i \rvert = \lvert x_i - \bar{x} \rvert = \lvert x_i - 50 \rvert$
50	0
40	10
22	28
35	15
45	5
60	10
50	0
70	20
50	0
78	28
Total	116

M.D.

$$= \frac{1}{n}\sum \lvert d_i \rvert = \frac{1}{10}\sum \lvert x_i - 50 \rvert = \frac{116}{10} = 11.6$$

(ii) Mean deviation from mode: The mode of the given distribution is 50 (because 50 occurs the maximum number of times).

Calculation for mean deviation.

x_i	$\lvert d_i \rvert = \lvert x_i - 50 \rvert$
50	0
40	10
22	28
35	15
45	5
60	10
50	0
70	20
50	0
78	28
Total	116

$$\text{M.D.} = \frac{1}{n}\sum d_i = \frac{1}{10}(116) = 11.6$$

Example 8.2 The following are the marks obtained by 9 students in mathematics. Find their mean deviation.

49, 68, 21, 32, 54, 38, 41, 66, 59

Solution: As it is not given about which statistical mean we have to compute, we will compute it about the median.

Arranging the observations in ascending order of magnitudes, we get

21, 32, 38, 41, 49, 54, 59, 66, 68

Here, the number of observations = 9 (odd number)

$\therefore \quad \text{Median} = \dfrac{9+1}{2}\text{th term} = 5^{\text{th}}\text{ term} = 49$

Calculation of mean deviation

x_i	$\lvert d_i \rvert = \lvert x_i - 49 \rvert$
21	28
32	17
38	11
41	8
49	0
54	5
59	10
66	17
68	19
Total	115

$\therefore \quad \text{M.D.} = \dfrac{1}{n}\sum d_i = \dfrac{115}{9} = 12.78$

Example 8.3 Calculate the mean deviation and the coefficient of mean deviation of the following from the A.M.

x_i	10	11	12	13	14
f_i	3	12	18	12	3

Solution: We know that

$$\text{Mean } \overline{x} = \frac{\sum f_i x_i}{\sum f_i} = \frac{30+132+216+156+42}{3+12+18+12+3}$$

$$\Rightarrow \overline{x} = \frac{576}{48} = 12$$

Now,

x_i	f_i	$f_i x_i$	$\lvert x_i - 12 \rvert$	$f_i \lvert x_i - 12 \rvert$
10	3	30	2	6
11	12	132	1	12
12	18	216	0	0
13	12	156	1	12
14	3	42	2	6

$$N = \sum f_i = 48$$

$$\sum f_i \lvert x_i - 12 \rvert = 36$$

$$\text{M.D.} = \frac{\sum f_i \lvert x_i - 12 \rvert}{N} = \frac{36}{48} = 0.75$$

$$\text{Coefficient of M.D.} = \frac{\text{Mean deviation}}{\text{Mean}}$$

$$= \frac{0.75}{12} = 0.0625$$

Example 8.4 Find the mean deviation from the median of the following frequency distribution.

Age (in years)	10	12	15	18	21	23
Frequency	3	5	4	10	8	4

Solution: Calculation of mean deviation from the median.

Age (in years) x_i	Frequency f_i	C.F.	$f_i x_i$	$\lvert x_i - 18 \rvert$	$f_i (x_i - 18)$
10	3	3	30	8	24
12	5	8	60	6	30
15	4	12	60	3	12
18	10	22	180	0	0
21	8	30	168	3	24
23	4	34	92	5	20
	34				110

Here, $N = \sum f_i = 34, \frac{N}{2} = 17$ and $\frac{N}{2} + 1 = 18$

∴ Median

$$= \frac{(\text{Value of 17th term}) + (\text{Value of 18th term})}{2}$$

$$= \frac{18+18}{2} = 18 \text{ years}$$

∴ Mean deviation from the median

$$= \frac{\sum f_i |x_i - 18|}{N} = \frac{110}{34}$$

$$= 3.23 \text{ years}$$

Example 8.5 Find the mean deviation of the following distribution from the median.

Class	10–20	20–30	30–40	40–50	50–60	60–70	70–80
Frequency	4	6	10	20	10	6	4

Solution: Calculation of mean deviation about the median.

Class	**x_i (Class mark)**	**f_i**	**Cumulative frequency**	**$\lvert x_i - 45\rvert$**	**$f_i \lvert x_i - 45\rvert$**
10–20	15	4	4	30	120
20–30	25	6	10	20	120
30–40	35	10	20	10	100
40–50	45	20	40	0	0
50–60	55	10	50	10	100
60–70	65	6	56	20	120
70–80	75	4	60	30	120
Total	$N = \sum f_i = 60$				$\sum f_i \lvert x_i - 45\rvert = 680$

Here, $N = 60 \Rightarrow \frac{N}{2} = 30$ and the cumulative frequency just greater than $\frac{N}{2}$ is 40 so the corresponding class is 40–50.

Thus, 40–50 is the median class.

Now, median $M = l + \frac{\frac{N}{2} - c}{f} \times i$

Here, $l = 40, c = 20, i = 10, f = 20$

$$\therefore \quad M = 40 + \frac{30 - 20}{20} \times 10 = 45$$

Hence, mean deviation from the median

$$= \frac{\sum f_i |x_i - 45|}{\sum f_i} = \frac{680}{60} = 11.33$$

Example 8.6 Find the variance and standard deviation of the following data.

5, 9, 10, 12, 8, 13, 6

Solution: Here, n = number of observations = 7

And $\bar{x} = \frac{\sum x_i}{n} = \frac{5+9+10+12+8+13+6}{7}$

$$= \frac{63}{7} = 9$$

$$\sum x_i^2 = 5^2 + 9^2 + 10^2 + 12^2 + 8^2 + 13^2 + 6^2 = 619$$

We know that variance $(\sigma^2) = \frac{1}{n} \sum x_i^2 - \bar{x}^2$

$$\Rightarrow \sigma^2 = \frac{619}{7} - 9^2 = 7.43$$

And $\sigma = \sqrt{7.43} = 2.72$

Thus, the variance = 7.43 and S.D. = 2.72

Example 8.7 The marks obtained by 10 students in an examination are as follows:

40, 62, 58, 73, 86, 83, 50, 54, 71, 75.

Calculate the standard deviation.

Solution: Let the assumed mean be 62

x_i	$d_i = x_i - 62$	d_i^2
40	–22	484
50	–12	144
54	–8	64
58	–4	16
62	0	0
71	9	81
73	11	121
75	13	169
83	21	441
86	24	576
	$\sum d_i = 32$	$\sum d_i^2 = 2096$

Now, $\sigma^2 = \frac{1}{n}\sum d_i^2 - \left(\frac{\sum d_i}{n}\right)^2$

$\therefore \quad \sigma = \sqrt{\frac{2096}{10} - \left(\frac{32}{10}\right)^2} = \sqrt{209.6 - 10.24}$

$= \sqrt{199.36}$

$\sigma = 14.12$

Thus, the standard deviation is 14.12.

Example 8.8 The mean and variance of 5 observations are 4.4 and 8.24 respectively. If 3 of the observations are 1, 2 and 6, find the remaining two observations.

Solution: Suppose that the remaining two observations are a and b.

Then, mean $= 4.4 = \frac{1+2+6+a+b}{5}$

$\Rightarrow a+b+9 = 22$

$\Rightarrow a+b = 13$

Variance $= 8.24 = \frac{1}{5}\left(1^2+2^2+6^2+a^2+b^2\right) - (4.4)^2$

$\left[\sigma^2 = \frac{1}{n}\sum x_i^2 - \bar{x}^2\right]$

$\Rightarrow 41.2 = 41 + a^2 + b^2 - 96.80$

$\Rightarrow a^2 + b^2 = 97$

Now, $(a+b)^2 + (a-b)^2 = 2\left(a^2+b^2\right)$

$\Rightarrow 169 + (a-b)^2 = 2 \times 97$

$\Rightarrow (a-b)^2 = 25 \Rightarrow a-b = \pm 5$

If $a-b=5$ and $a+b=13 \Rightarrow a=9, b=4$

If $a-b=-5$ and $a+b=13 \Rightarrow a=4, b=9$

Hence the remaining two observations are 9 and 4.

Example 8.9 The following table gives the frequency distribution of marks obtained by a batch of 20 students. Find their mean and standard deviation.

Marks obtained	5	15	25	35	45
Number of students	5	4	6	3	2

Solution: Since the marks distribution is in A.P., we can use the step deviation method.

Here, h = c.d. of A.P. = 10

Let the assumed mean be 25.

x_i	f_i	$u_i = \frac{x_i - 25}{10}$	$f_i u_i$	$f_i u_i^2$
5	5	–2	–10	20
15	4	–1	–4	4
25	6	0	0	0
35	3	1	3	3
45	2	2	4	8
	$\sum f_i = N = 20$		$\sum f_i u_i = -7$	$\sum f_i u_i^2 = 35$

Actual mean, $\bar{x} = a + h \times \frac{\sum f_i u_i}{N}$

$= 25 + 10 \times \frac{(-7)}{20}$

$= 25 - 3.5 = 21.5$

And standard deviation,

$$\sigma = h\sqrt{\frac{\sum f_i u_i^2}{N} - \left(\frac{\sum f_i u_i}{N}\right)^2}$$

$= 10\sqrt{\frac{35}{20} - \left(-\frac{7}{20}\right)^2}$

$= 10\sqrt{1.75 - (0.35)^2} = 10\sqrt{1.6275}$

$10 \times 1.27 = 12.7$

Example 8.10 Calculate the mean, variance and standard deviation for the following distribution.

Classes	30–40	40–50	50–60	60–70	70–80	80–90	90–100
Frequencies	3	7	12	15	8	3	2

Solution:

Classes	**Mid-value** (x_i)	f_i	$f_i x_i$	$(x_i - \bar{x})^2$	$f_i (x_i - \bar{x})^2$
30–40	35	3	105	729	2187
40–50	45	7	315	289	2023
50–60	55	12	660	49	588
60–70	65	15	975	9	135
70–80	75	8	600	169	1352
80–90	85	3	255	529	1587
90–100	95	2	190	1089	2178
		$\sum f_i = N = 50$	$\sum f_i x_i = 3100$		$\sum (x_i - \bar{x})^2 = 10050$

Mean, $\bar{x} = \frac{\sum f_i x_i}{N} = \frac{3100}{50} = 62$

Now, Variance $\sigma^2 = \frac{\sum f_i (x_i - \bar{x})^2}{N}$

$= \frac{10050}{50} = 201$

And standard deviation, $\sigma = \sqrt{201} = 14.17$

Example 8.11 Calculate the mean and standard deviation for the following data.

Wages up to (in Rs)	10	20	30	40	50	60	70
No. of workers	9	26	58	81	121	139	140

Solution: Consider the following table

Class interval	Mid value (x_i)	Frequency (f_i)	$u_i = \frac{x_i - A}{h}$	$f_i u_i$	$f_i u_i^2$
0–10	5	9	−3	−27	81
10–20	15	17	−2	−34	68
20–30	25	32	−1	−32	32
30–40	35	23	0	0	0
40–50	45	40	1	40	40
50–60	55	18	2	36	72
60–70	65	1	3	3	9
		$N = \sum f_i = 140$		$\sum f_i u_i = -14$	$\sum f_i u_i^2 = 302$

Here, assumed mean, $A = 35$ and $h = 10$

Now, actual mean, $\bar{x} = A + \frac{\sum f_i u_i}{\sum f_i} \times h$

$$= 35 + \frac{(-14)}{140} \times 10 = 34$$

And standard deviation,

$$\sigma = h\sqrt{\frac{\sum f_i u_i^2}{N} - \left(\frac{\sum f_i u_i}{N}\right)^2}$$

$$= 10\sqrt{\frac{302}{140} - \left(\frac{-14}{140}\right)^2}$$

$$= 10\sqrt{2.15 - 0.01} = 10 \times 0.46 = 14.60$$

Example 8.12 Coefficient of variation of two distributions are 60% and 70% and their standard deviations are 21 and 16, respectively. What are their arithmetic means?

Solution: For first distribution

Standard deviation, $\sigma_1 = 21$

Coefficient of variation = 60

Let $\bar{x}$ be the mean

$\because$ coefficient of variation $= \frac{\sigma_1}{\bar{x}} \times 100$

$\therefore$ $\frac{\sigma_1}{\bar{x}} \times 100 = 60 \Rightarrow \bar{x} = \frac{21 \times 10}{6} = 35$

For second distribution

$\sigma_2 = 16$, coefficient of variation = 70

Let $\bar{y}$ be the mean

Then, $70 = \frac{\sigma_2}{\bar{y}} \times 100 = \frac{16}{\bar{y}} \times 100 \Rightarrow \bar{y} = \frac{160}{7}$

$\Rightarrow \bar{y} = 22.85$

8.6 Know Your Concepts

1. Find the mean deviation from the mean for the following data.

   (i) 4, 7, 8, 9, 10, 12, 13, 17

   (ii) 4.5, 4.75, 5, 5.25, 5.5, 6.25, 6.5, 7.75, 8.5

2. The lengths (in cm) of 10 rods in a shop are given below.

   42.0, 52.3, 55.2, 72.9, 52.8, 79.0, 32.5, 15.2, 27.9, 30.2

   (i) Find the mean deviation from the median.

   (ii) Find the mean deviation from the mean.

3. Find the mean deviation from median for the following set of observations:

x_i	15	21	27	30	35
f_i	3	5	6	7	8

4. The height distribution of 100 children is as follows:

Height (cm)	95–105	105–110	115–125	125–135	135–145	145–155
No. of students	9	13	26	30	12	10

Calculate the mean deviation from the mean height.

5. Find the mean deviation from the median of the given distribution.

Class	0–6	6–12	12–18	18–24	24–30
Frequency	8	10	12	9	5

6. The mean and standard deviation of 20 observations are found to be 10 and 2 respectively. On rechecking, it was found that an observation, 8, was incorrect. Calculate the correct mean and standard deviation in each of the following cases.

(i) If the wrong item is omitted,

(ii) If it is replaced by 12.

7. Mean and standard deviation of 100 items are 50 and 4 respectively. Find the sum of all the items and also the sum of the squares of the items.

8. Find the mean and S.D. of the following distribution.

Marks	10	20	30	40	50	60	70
No. of students	1	5	12	22	17	9	4

9. The marks obtained by 200 pupils of a class in their annual examination are given in the following table. Find the mean and S.D. of these marks.

Marks	1–10	11–20	21–30	31–40	41–50	51–60	61–70	71–80	81–90	91–100
No. of students	2	3	10	19	30	47	54	28	5	2

10. Let $\bar{x}$ and $\bar{y}$ be the arithmetic means for the variables x and y respectively and σ_x, σ_y their standard deviations. If the variables x and y are related by $my = nx + k$, where m, n, k are constants, show that

(i) $m\bar{y} = n\bar{x} + k$

(ii) $m^2\sigma_y^2 = n^2\sigma_x^2$

11. Find the mean, variance and S.D. using short-cut method.

Class interval	10–20	20–30	30–40	40–50	50–60	60–70	70–80	80–90	90–100
Frequency	3	4	7	7	15	9	6	6	3

12. If for a distribution $\sum(x-5) = 3, \sum(x-5)^2 = 43$ and total number of items $N = 18$, find the mean and standard deviation.

8.7 JEE Main: Sample Questions

Single Answer Type

1. In the frequency distribution of the discrete data given below, the frequency K against value 0 is missing

Variable x:	0	1	2	3	4	5
Frequency f:	K	20	40	40	20	4

If the mean is 2.5, then the missing frequency K will be

(A) 0 (B) 1 (C) 3 (D) 4

2. The variance of α, β and γ is 9, the variance of $5\alpha, 5\beta$ and 5γ is

(A) 45 (B) $\frac{9}{5}$ (C) $\frac{5}{9}$ (D) 225

3. If data is less scattered about its mean then which of the following is not always true?
(i) Coefficient of variance is small
(ii) S.D. is large
(iii) Variance is small
(iv) Mode is large
(A) only (i) (B) only (iv)
(C) (i) and (iv) (D) (ii) and (iv)

4. If the values $1, \frac{1}{2}, \frac{1}{3}, \frac{1}{4}, \frac{1}{5}, \ldots, \frac{1}{n}$ occur at frequencies $1, 2, 3, 4, 5, \ldots, n$ in a distribution then mean is

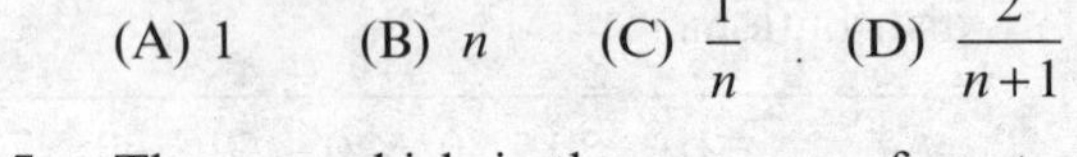

(A) 1 (B) n (C) $\frac{1}{n}$ (D) $\frac{2}{n+1}$

5. The one which is the measure of central tendency is
(A) mode (B) mean deviation
(C) S.D. (D) coefficient of correlation

6. The mean deviation from the mean of the A.P. $a, a+d, a+2d, \ldots, a+2nd$ is
(A) $n(n+1)d$ (B) $\frac{n(n+1)d}{(2n+1)}$
(C) $\frac{n(n+1)d}{2n}$ (D) $\frac{n(n-1)d}{2n+1}$

7. For two data sets each of size 5, the variances are given to be 4 and 5 and the corresponding means are given to be 2 and 4 respectively. The variance of the combined data set is
(A) $\frac{5}{2}$ (B) $\frac{11}{2}$ (C) 6 (D) $\frac{13}{2}$

8. If the mean deviation of number $1, 1+d, 1+2d, \ldots, 1+100d$ from their mean is 255, then the d is equal to
(A) 10.0 (B) 20.0 (C) 10.1 (D) 20.2

9. The standard deviation of n observations $x_1, x_2, \ldots, x_n$ is 2. If $\sum_{i=1}^{n} x_i = 20$ and $\sum_{i=1}^{n} x_i^2 = 100$, then n is
(A) 10 or 20 (B) 5 or 10
(C) 5 or 20 (D) 5 or 15

10. Suppose, a population A has 100 observations 101, 102, …, 200 and another population B has 100 observations 151, 152, …, 250. If V_A and V_B represent the variances of the two population respectively, then $\frac{V_A}{V_B}$ is
(A) $\frac{9}{4}$ (B) $\frac{4}{9}$ (C) $\frac{2}{3}$ (D) 1

11. The average marks of boys in a class is 52 and that of girls is 42. The average marks of boys and girls combined is 50. The percentage of boys in the class is
(A) 40 (B) 20 (C) 80 (D) 60

12. The mean of the numbers $a, b, 8, 5, 10$ is 6 and the variance is 6.80. Then which one of the following are the possible values of a and b?
(A) $a = 0, b = 7$ (B) $a = 5, b = 2$
(C) $a = 1, b = 6$ (D) $a = 3, b = 4$

13. In a class of 100 students, there are 70 boys whose average marks are 750. If the average marks of the complete class are 720, then the average marks of the girls are
(A) 700 (B) 650 (C) 690 (D) 680

14. The S.D. of the numbers 31, 32, 33, …, 46, 47 is
(A) $\sqrt{\frac{17}{12}}$ (B) $\sqrt{\frac{47^2-1}{12}}$
(C) $2\sqrt{6}$ (D) $4\sqrt{3}$

15. If the S.D. of a variable x is σ, then the S.D. of $\frac{ax+b}{c}$ (a, b, c are constants) is
(A) $\frac{a}{c}\sigma$ (B) $\left|\frac{a}{c}\right|\sigma$ (C) $\left|\frac{c}{a}\right|\sigma$ (D) $\frac{c}{a}\sigma$

16. If the mean of n observations $1^2, 2^2, 3^2, \ldots, n^2$ is $\frac{46n}{11}$, then n is equal to
(A) 11 (B) 12 (C) 23 (D) 22

17. The S.D. of the wages of 85 employees is Rs 15.40. After one year, each of them is given an increment of Rs 25. The S.D. of the new wages (in Rs) is
(A) 15.40 (B) 40.40 (C) 20.40 (D) 10.40

18. The mean deviation about the median of the following data:

Class	16–20	21–25	26–30	31–35	36–40	41–45	46–50	51–55
Frequency	5	6	12	14	26	12	16	9

(A) 26 (B) 9.35
(C) 7.35 (D) none of these

8.8 JEE Main: Archive

1. If the mean deviation about the mean of the numbers 1, 2, 3, …, n, where n is odd, is $\frac{5(n+1)}{n}$, then n is equal to ______.
[Main 2022 25th June 2nd Shift]

2. The mean of the numbers a, b, 8, 5, 10 is 6 and their variance is 6.8. If M is the mean deviation of the numbers about the mean, then $25M$ is equal to:
(A) 60 (B) 55 (C) 50 (D) 45
[Main 2022 26th June 1st Shift]

3. The mean and standard deviation of 50 observations are 15 and 2 respectively. It was found that one incorrect observation was taken such that the sum of correct and incorrect observations is 70. If the correct mean is 16, then the correct variance is equal to
(A) 10 (B) 36 (C) 43 (D) 60
[Main 2022 26th June 2nd Shift]

4. The mean and variance of the data 4, 5, 6, 6, 7, 8, x, y where $x < y$ are 6, and $\frac{9}{4}$ respectively. Then $x^4 + y^2$ is equal to
(A) 162 (B) 320 (C) 674 (D) 420
[Main 2022 27th June 2nd Shift]

5. The mean and standard deviation of 15 observations are found to be 8 and 3 respectively. On rechecking it was found that, in the observations, 20 was misread as 5. Then, the correct variance is equal to ______.
[Main 2022 28th June 1st Shift]

6. Let the mean and the variance of 5 observations x_1, x_2, x_3, x_4, x_5 be $\frac{24}{5}$ and $\frac{194}{25}$ respectively. If the mean and variance of the first 4 observation are $\frac{7}{2}$ and a respectively, then $(4a + x_5)$ is equal to
(A) 13 (B) 15 (C) 17 (D) 18
[Main 2022 29th June 1st Shift]

7. The number of values of $a \in N$ such that the variance of 3, 7, $12a$, 43 (a is a natural number) is:
(A) 0 (B) 2
(C) 5 (D) infinite
[Main 2022 29th June 2nd Shift]

8. Let the six numbers $a_1, a_2, a_3, a_4, a_5, a_6$ be in A.P. and $a_1 + a_3 = 10$. If the mean of these six numbers is $\frac{19}{2}$ and their variance is σ^2, then $8\sigma^2$ is equal to
(A) 220 (B) 210 (C) 200 (D) 105
[Main 2023 24th Jan 2nd Shift]

9. The mean and variance of the marks obtained by the students in a test are 10 and 4 respectively. Later, the marks of one of the students is increased from 8 to 12. If the new mean of the marks is 10.2, then their new variance is equal to:
(A) 4.04 (B) 4.08 (C) 3.96 (D) 3.92
[Main 2023 25th Jan 1st Shift]

10. Three rotten apples are mixed accidently with seven good apples and four apples are drawn one by one without replacement. Let the random variable X denote the number of rotten apples. If μ and σ^2 represent mean and variance of X, respectively, then $10(\mu^2 + \sigma^2)$ is equal to
(A) 20 (B) 250 (C) 25 (D) 30
[Main 2023 29th Jan 1st Shift]

11. Let $X = \{11, 12, 13, \ldots, 40, 41\}$ and $Y = \{61, 62, 63, \ldots, 90, 91\}$ be two sets of observations. If $\bar{x}$ and $\bar{y}$ are their respective

means and σ^2 is the variance of all the observations in $X \cup Y$, then $\left|\overline{x}+y-\sigma^2\right|$ is equal to ______.

[Main 2023 29th Jan 2nd Shift]

12. The mean and variance of 7 observations are 8 and 16 respectively. If one observation 14 is omitted a and b are respectively mean and variance of remaining 6 observation, then $a+3b-5$ is equal to

[Main 2023 30th Jan 1st Shift]

13. Let S be the set of all values of a_1 for which the mean deviation about the mean of 100 consecutive positive integers $a_1, a_2, a_3, \ldots, a_{100}$ is 25. Then S is

(A) ϕ (B) $\{99\}$ (C) $\mathbb{N}$ (D) $\{9\}$

[Main 2023 30th Jan 2nd Shift]

14. If the variance of the following frequency distribution is 3, then find the value of α.

x_i	2	3	4	5	6	7	8
Frequency f_i	3	6	16	α	9	5	6

[Main 2023 31st Jan 1st Shift]

15. Let the mean and standard deviation of marks of class A of 100 students be respectively 40 and $\alpha(>0)$, and the mean and standard deviation of marks of class B of n students be respectively 55 and $30-\alpha$. If the mean and variance of the marks of the combined class of $100+n$ students are respectively 50 and 350, then the sum of variances of classes A and B is

(A) 500 (B) 650 (C) 450 (D) 900

[Main 2023 31st Jan 2nd Shift]

16. The mean and variance of 5 observations are 5 and 8 respectively. If 3 observations are 1, 3, 5, then the sum of cubes of the remaning two observations is

(A) 1072 (B) 1792 (C) 1216 (D) 1456

[Main 2023 01st Feb 1st Shift]

17. Let $9=x_1<x_2<\ldots<x_7$ be in an A.P. with a common difference d. If the standard deviation of $x_1, x_2, \ldots x_7$ is 4 and the mean is $\overline{x}$, then $\overline{x}+x_6$ is equal to

(A) $18\left(1+\frac{1}{\sqrt{3}}\right)$ (B) 34

(C) $2\left(9+\frac{8}{\sqrt{7}}\right)$ (D) 25

[Main 2023 01st Feb 2nd Shift]

18. The mean and variance of a set of 15 numbers are 12 and 14 respectively. The mean and variance of another set of 15 numbers are 14 and σ^2 respectively. If the variance of all the 30 numbers in the two sets is 13, then σ^2 is equal to

(A) 10 (B) 11 (C) 9 (D) 12

[Main 2023 6th April 1st Shift]

19. If the mean and variance of the frequency distribution

x_i	2	4	6	8	10	12	14	16
f_i	4	4	α	15	8	β	4	5

are 9 and 15.08 respectively, then the value of $\alpha^2+\beta^2-\alpha\beta$ is ______.

[Main 2023 6th April 2nd Shift]

20. Let the mean and variance of 8 numbers $x, y, 10, 12, 6, 12, 4, 8$ be 9 and 9.25 respectively. If $x>y$, then $3x-2y$ is equal to ______.

[Main 2023 8th April 1st Shift]

21. Let the mean and variance of 12 observations be $\frac{9}{2}$ and 4 respectively. Later on, it was observed that two observations were considered as 9 and 10 instead of 7 and 14 respectively. If the correct variance is $\frac{m}{n}$, where m and n are coprime, then $m+n$ are coprime, then $m+n$ is equal to

(A) 315 (B) 316 (C) 314 (D) 317

[Main 2023 8th April 2nd Shift]

22. If the mean of the frequency distribution

Class	0–10	10–20	20–30	30–40	40–50
Frequency	4	4	α	15	8

is 28, then its variance is ______.

[Main 2023 10th April 1st Shift]

23. Let μ be the mean and σ be the standard deviation of the distribution

x_i	0	1	2	3	4	5
f_i	$k+2$	$2k$	k^2-1	k^2-1	k^2+1	$k-3$

where $\Sigma f_i = 62$. If $[x]$ denotes the greatest integer $\le x$, then $\left[\mu^2+\sigma^2\right]$ is equal to

(A) 9 (B) 8 (C) 7 (D) 6

[Main 2023 10th April 2nd Shift]

24. Let sets A and B have 5 elements each. Let the mean of the elements in sets A and B be 5 and 8 respectively and the variance of the elements in sets A and B be 12 and 20 respectively. A new set C of 10 elements is formed by subtracting 3 from each element of A and adding 2 to each element of B. Then the sum of the mean and variance of the elements of C is

(A) 40 (B) 32 (C) 38 (D) 36

[Main 2023 11th April 1st Shift]

25. Let the mean of 6 observations 1, 2, 4, 5, x and y be 5 and their variance be 10. Then their mean deviation about the mean is equal to

(A) $\frac{7}{3}$ (B) 3 (C) $\frac{8}{3}$ (D) $\frac{10}{3}$

[Main 2023 11th April 2nd Shift]

26. Let the positive numbers a_1, a_2, a_3, a_4 and a_5 be in a G.P. Let their mean and variance be $\frac{31}{10}$ and $\frac{m}{n}$ respectively, where m and n are co-prime. If the mean of their reciprocals is $\frac{31}{10}$ and $a_3+a_4+a_5=14$, then $m+n$ is equal to________.

[Main 2023 12th April 1st Shift]

27. Let the mean of the data

x	1	3	5	7	9
Frequency (f)	4	24	28	α	8

be 5. If m and σ^2 are respectively the mean deviation about the mean and the variance of the data, then $\frac{3\alpha}{m+\sigma^2}$ is equal to ______.

[Main 2023 13th April 1st Shift]

28. The mean and standard deviation of the marks of 10 students were found to be 50 and 12 respectively. Later, it was observed that two marks 20 and 25 were wrongly read as 45 and 50 respectively. Then the correct variance is ______.

[Main 2023 13th April 2nd Shift]

29. The mean and standard deviation of 10 observations are 20 and 8 respectively. Later on, it was observed that one observation was recorded as 50 instead of 40. Then the correct variance is ______.

(A) 11 (B) 13 (C) 12 (D) 14

[Main 2023 15th April 1st Shift]

30. Let the median and the mean deviation about the median of 7 observations 170, 125, 230, 190, 210, a, b be 170 and $\frac{205}{7}$ respectively. Then the mean deviation about the mean of these 7 observations is:

(A) 31 (B) 28 (C) 30 (D) 32

[Main 2024 01st Feb 1st Shift]

31. Consider 10 observations $x_1, x_2, \ldots, x_{10}$ such that $\sum_{i=1}^{10}(x_i-\alpha)=2$ and $\sum_{i=1}^{10}(x_i-\beta)^2=40$, where α, β are positive integers. Let the mean and the variance of the observations be $\frac{6}{5}$ and $\frac{84}{25}$ respectively. The $\frac{\beta}{\alpha}$ is equal to

(A) 2 (B) $\frac{3}{2}$ (C) $\frac{5}{2}$ (D) 1

[Main 2024 01st Feb 2nd Shift]

32. Let $a_1, a_2, \ldots, a_{10}$ be 10 observations such that $\sum_{k=1}^{10} a_k = 50$ and $\sum_{\forall k<j} a_k \cdot a_j = 1100$.

Then the standard deviation of $a_1, a_2, \ldots, a_{10}$ is equal to

(A) 5 (B) $\sqrt{5}$ (C) 10 (D) $\sqrt{115}$

[Main 2024 27th Jan 1st Shift]

33. The mean and standard deviation of 15 observations were found to be 12 and 3 respectively. On rechecking it was found that an observation was read as 10 in place of 12. If μ and σ^2 denote the mean and variance of the correct observations respectively, then $15(\mu + \mu^2 + \sigma^2)$ is equal to ________.

[Main 2024 27th Jan 2nd Shift]

34. If the mean and variance of the data $65, 68, 58, 44, 48, 45, 60, \alpha, \beta, 60$ where $\alpha > \beta$ are 56 and 66.2 respectively, then $\alpha^2 + \beta^2$ is equal to

[Main 2024 29th Jan 1st Shift]

35. If the mean and variance of five observations are $\frac{24}{5}$ and $\frac{194}{25}$ respectively and the mean of first four observations is $\frac{7}{2}$, then the variance of the first four observations in equal to

(A) $\frac{4}{5}$ (B) $\frac{77}{12}$ (C) $\frac{5}{4}$ (D) $\frac{105}{4}$

[Main 2024 29th Jan 2nd Shift]

36. Let M denote the median of the following frequency distribution.

Class	0–4	4–8	8–12	12–16	16–20
Frequency	3	9	10	8	6

Then $20M$ is equal to:

(A) 416 (B) 104 (C) 52 (D) 208

[Main 2024 30th Jan 1st Shift]

37. The variance σ^2 of the data

x_i	0	1	5	6	10	12	17
f_i	3	2	3	2	6	3	3

Is ________.

[Main 2024 30th Jan 2nd Shift]

38. Let the mean and the variance of 6 observation $a, b, 68, 44, 48, 60$ be 55 and 194, respectively if $a > b$, then $a + 3b$ is

(A) 200 (B) 190 (C) 180 (D) 210

[Main 2024 31st Jan 2nd Shift]

39. Let $\alpha, \beta \in R$. Let the mean and the variance of 6 observations $-3, 4, 7, -6, \alpha, \beta$ be 2 and 23, respectively. Then mean deviation about the mean of these 6 observations is

(A) $\frac{13}{3}$ (B) $\frac{16}{3}$ (C) $\frac{11}{3}$ (D) $\frac{14}{3}$

[Main 2024 04th Apr 1st Shift]

40. If the mean of the following probability dristribution of a random variable X:

X	0	2	4	6	8
P(X)	a	$2a$	$a+b$	$2b$	$3b$

is $\frac{46}{9}$, then the variance of the distribution is

(A) $\frac{173}{27}$ (B) $\frac{566}{81}$ (C) $\frac{151}{27}$ (D) $\frac{581}{81}$

[Main 2024 04th Apr 1st Shift]

41. From a lot of 10 items, which include 3 defective items, a sample of 5 items is drawn at random. Let the random variable X denote the number of defective items in the sample. If the variance of X is σ^2, then $96\,\sigma^2$ is equal to ________.

[Main 2024 05th Apr 1st Shift]

42. The mean and standard deviation of 20 observations are found to be 10 and 2 respectively. On rechecking, it was found that an observation by mistake was taken as 8 instead of 12. The correct standard deviation is

(A) 1.8 (B) 1.94 (C) $\sqrt{3.96}$ (D) $\sqrt{3.86}$

[Main 2024 06th Apr 1st Shift]

43. Let $a, b, c \in N$ and $a < b < c$. Let the mean, the mean deviation about the mean and the variance of the 5 observations $9, 25, a, b, c$ be 18, 4 and $\frac{136}{5}$, respectively. Then $2a + b - c$ is equal to ________.

[Main 2024 08th Apr 2nd Shift]

44. The frequency distribution of the age of students in a class of 40 students is given below.

Age	15	16	17	18	19	20
No. of students	5	8	5	12	x	y

If the mean deviation about the median is 1.25, then $4x+5y$ is equal to:

(A) 46 (B) 43 (C) 44 (D) 47

[Main 2024 09th Apr 1st Shift]

45. If the variance of the frequency distribution

x	c	$2c$	$3c$	$4c$	$5c$	$6c$
f	2	1	1	1	1	1

is 160, then the value of $c \in N$ is

(A) 7 (B) 8 (C) 5 (D) 6

[Main 2024 09th Apr 2nd Shift]

8.9 Answer Key

8.9.1 Know Your Concepts

1. (i) 3 (ii) 1.1 **2.** (i) 16.44 (ii) 16.44 **3.** (5.1) **4.** (11.288) **5.** (6.32)
6. (i) Correct Mean : 10.10, Correct S.D. : 1.997 (ii) Correct Mean : 10.20, Correct S.D. : 1.9899)
7. (5000, 251600) **8.** (13.27) **9.** (Mean = 54.85, S.D. = 16.37)
11. (Mean = 56, Variance = 422.33, S.D. = 20.55) **12.** (Mean = 5.17, S.D. = 1.52)

8.9.2 JEE Main: Sample Questions

Single Answer Type

1. (D) **2.** (D) **3.** (D) **4.** (D) **5.** (A) **6.** (B) **7.** (B) **8.** (C)
9. (C) **10.** (D) **11.** (C) **12.** (D) **13.** (B) **14.** (C) **15.** (B) **16.** (A)
17. (A) **18.** (C)

8.9.3 JEE Main: Archive

1. (21) **2.** (A) **3.** (C) **4.** (B) **5.** (17) **6.** (B) **7.** (A) **8.** (B)
9. (C) **10.** (A) **11.** (603) **12.** (37) **13.** (C) **14.** (5) **15.** (A) **16.** (A)
17. (B) **18.** (A) **19.** (25) **20.** (25) **21.** (D) **22.** (151) **23.** (B) **24.** (C)
25. (C) **26.** (211) **27.** (8) **28.** (269) **29.** (B) **30.** (C) **31.** (A) **32.** (B)
33. (2521) **34.** (6344) **35.** (C) **36.** (D) **37.** (29) **38.** (C) **39.** (A) **40.** (B)
41. (56) **42.** (C) **43.** (33) **44.** (C) **45.** (A)

8.10 Solutions

8.10.1 Know Your Concepts

1. (i) Let $\bar{x}$ be the mean of the given data. Then

$$\bar{x} = \frac{4+7+8+9+10+12+13+17}{8}$$

$$\Rightarrow \bar{x} = \frac{80}{8} = 10$$

Mean deviation

x_i	$\lvert d_i\rvert = \lvert x_i - \bar{x}\rvert = \lvert x_i = 10\rvert$
4	6
7	3
8	2
9	1
10	0
12	2
13	3
17	7
Total	24

$$\therefore M.D. = \frac{1}{n}\sum \lvert d_i\rvert = \frac{1}{8}(24) = 3$$

(ii) 1.1

[Do it yourself]

2. (i) First we arrange the length of the rods in ascending order

15.2, 27.9, 30.2, 32.5, 42.0, 52.3, 52.8, 55.2, 72.9, 79.0

Median

$$= \frac{\text{Value of } \left(\frac{10}{2}\right)^{\text{th}} \text{ observation } + \text{ Value of } \left(\frac{10}{2}+1\right)^{\text{th}} \text{ observation}}{2}$$

$$= \frac{42.0 + 52.3}{2} = \frac{94.3}{2} = 47.2$$

$$|d_i| = |x_i - 47.2| = 164.4$$

$$\text{M.D.} = \frac{1}{10} \times 164.4$$

$$= 16.44$$

(ii) 16.44
[Do it yourself]

3. Calculation of mean deviation from the median

x_i	f_i	c.f.	$f_i x_i$	$\|x_i - 30\|$	$f_i\|x_i - 30\|$
15	3	3	45	15	45
21	5	8	105	9	45
27	6	14	162	3	18
30	7	21	210	0	0
35	8	29	280	5	40
Total	$N = 29$				

Here $N = \sum f_i = 29 \therefore \frac{N+1}{2} = 15$

Median = 30
Mean deviation from the median

$$= \frac{\sum f_i |x_i - 30|}{N} = \frac{148}{29} = 5.1$$

4. We construct the following table

Height in cm	x_i	f_i	$f_i x_i$	$\|x_i - k\|$	$f_i\|x_i - k\|$
95–105	100	9	900	25.3	227.7
105–115	110	13	1430	15.3	198.9
115–125	120	26	3120	5.3	137.8
125–135	130	30	3900	4.7	141.0
135–145	140	12	1680	14.7	176.4
145–155	150	10	1500	24.7	247.0
Total		$\sum f_i = 100$	12530		1128.8

$$k = \frac{\sum f_i x_i}{\sum f_i} = \frac{12530}{100} = 125.3$$

$$\text{M.D. from mean } k = \frac{\sum f_i |x_i - k|}{\sum f_i}$$

$$= \frac{1128.8}{100} = 11.288$$

5. Calculation of mean deviation about the median.

Class	x_i	f_i	c.f.	$\|x_i - 14\|$	$f_i\|x_i - 14\|$
0–6	3	8	8	11	88
6–12	9	10	18	5	50
12–18	15	12	30	1	12
18–24	21	9	39	7	63
24–30	27	5	44	13	65
		$N = \sum f_i = 44$			$\sum f_i\|x_i - 14\| = 278$

Here $N = 44$, so, $\frac{N}{2} = 22$ and the c.f.

Just greater than $\frac{N}{2}$ is 30.

Therefore $12 - 18$ is the median class

$$\text{Median, } M = \ell + \frac{\frac{N}{2} - c}{f} \times h$$

Here, $\ell = 12, \frac{N}{2} = 22$

$c = 18, f = 12, h = 6$

$$\therefore M = 12 + \frac{22 - 18}{12} \times 6$$

$$= 12 + \frac{4}{12} \times 6 = 14$$

$\therefore$ Mean deviation from median

$$= \frac{\sum f_i |x_i - 14|}{\sum f_i} = \frac{278}{44} = 6.32$$

6. Since, $n = 20, \bar{x} = 10$ and $\sigma = 2$

$\bar{x} = \frac{1}{n}\sum x_i \Rightarrow \sum x_i = n\bar{x} = 20 \times 10 = 200$

$\therefore$ Incorrect $\sum x_i = 200$

and, $\sigma = 2 \Rightarrow \sigma^2 = 4$

Now,

$\frac{1}{n}\sum x_i^2 - \bar{x}^2 = 4 \Rightarrow \frac{1}{20}\sum x_i^2 - 100 = 4$

or, $\sum x_i^2 = 2080$

$\therefore$ Incorrect $\sum x_i^2 = 2080$

(i) If we omit the wrong item, 8 from the observations, then 19 observations are left correct $\sum x_i + 8 =$ incorrect $\sum x_i$

$\Rightarrow$ Correct $\sum x_i = 200 - 8 = 192$

$\therefore$ Correct mean $= \frac{192}{19} = 10.10$

and

correct $\sum x_i^2 + 8^2 =$ incorrect $\sum x_i^2$

$\therefore$ Correct $\sum x_i^2 = 2080 - 64 = 2016$

$\therefore$ Correct variance $= \frac{1}{19}$ (Correct $\sum x_i^2$) – (correct mean)^2

$= \frac{2016}{19} - \left(\frac{192}{19}\right)^2 = \frac{1440}{361}$

$\therefore$ Correct standard deviation

$= \sqrt{\frac{1440}{361}} = 1.997$

(ii) If we replace the wrong item by 12

correct $\sum x_i =$ Incorrect

$\sum x_i - 8 + 12$

$= 200 + 4 = 204$

Correct mean $= \frac{204}{20} = 10.2$

Correct $\sum x_i^2 =$ incorrect

$\sum x_i^2 - 8^2 + 12^2$

$= 2080 - 8^2 + 12^2 = 2160$

Correct variance $= \frac{1}{20}$ (correct $\sum x_i^2$) – (correct mean)^2

$= \frac{2160}{20} - \left(\frac{204}{20}\right)^2 = \frac{1584}{400}$

$\therefore$ Correct standard deviation =

$\sqrt{\frac{1584}{400}} = 1.9899$

7. It is given that

$\bar{x} = 50, n = 100$ and $\sigma = 4$

$\Rightarrow 50 = \frac{\sum x_i}{100} \Rightarrow \sum x_i = 5000$

Now, $\sigma^2 = \frac{\sum x_i^2}{n} - (\bar{x})^2$

$\Rightarrow 16 = \frac{\sum x_i^2}{100} - (50)^2 \Rightarrow \frac{\sum x_i^2}{100} = 2516$

$\therefore \sum x_i^2 = 251600$

8. Calculation of mean and standard deviation

x_i	f_i	d_i $x_i - 40$	di^2	$f_i d_i$	$f_i d_i^2$
10	1	–30	900	–30	900
20	5	–20	400	–100	2000
30	12	–10	100	–120	1200
40	22	0	0	0	0
50	17	10	100	170	1700
60	9	20	400	180	3600
70	4	30	900	120	3600
	$\sum f_i = N$ $= 70$			$\sum f_i d_i$ $= 220$	$\sum f_i d_i^2$ $= 13000$

Here assumed mean = 40

Actual mean $\bar{x} = 40 + \frac{220}{70} = 40 + 3.1 = 43.1$

Standard deviation

$$\sigma = \sqrt{\frac{1}{N}\sum f_i d_i^2 - \left(\frac{\sum f_i d_i}{N}\right)^2}$$

$$= \sqrt{\frac{13000}{70} - \left(\frac{220}{70}\right)^2}$$

$$= \sqrt{\frac{1300}{7} - \frac{484}{49}} = \frac{1}{7}\sqrt{9100 - 484} = 13.27$$

9. 54.85; 16.37

[Do it yourself]

10. Let number of observations for variable $x = p$

$y = ax + b$ where $a = \frac{n}{m}, b = \frac{k}{m}$

$\therefore$ Number of observations for variable $y = p$

Now, $\bar{y} = \frac{\sum y}{p} = \frac{\sum (ax+b)}{p}$

$= a\frac{\sum x}{p} + \frac{bp}{p} = a\bar{x} + b$

$\Rightarrow \bar{y} = \frac{n}{m}\bar{x} + \frac{k}{m} \Rightarrow m\bar{y} = n\bar{x} + k$

Again, $\sigma_x^2 = \frac{\sum (x-\bar{x})^2}{p}$

$\sigma_y^2 = \frac{\sum (y-\bar{y})^2}{p} = \frac{\sum \left[(ax+b)-(a\bar{x}+b)\right]^2}{p}$

$= \frac{\sum \left[a(x-\bar{x})\right]^2}{p} = a^2\sigma_x^2$

$\Rightarrow m^2\sigma_y^2 = n^2\sigma_x^2$

11. 56, 422.33, 20.55

[Do it yourself]

12. Standard deviation

$= \sqrt{\frac{\sum x_i^2}{n} - \left(\frac{\sum x_i}{n}\right)^2}$

$= \sqrt{\frac{43}{18} - \left(\frac{3}{18}\right)^2} = \sqrt{\frac{86-1}{36}}$

$= 1.52$

8.10.2 JEE Main: Sample Questions

Single Answer Type

1. **(D)**

Since, mean $(\bar{x}) = 2.5$ and discrete data

Variable *x*:	0	1	2	3	4	5
Frequency *f*:	K	20	40	40	20	4

We know that

$\sum f = K + 20 + 40 + 40 + 20 + 4 = 124 + K$

Also,

$\sum xf = (0 \times K) + (1 \times 20) + (2 \times 40) + (3 \times 40) + (4 \times 20) + (5 \times 4)$

$= 0 + 20 + 80 + 120 + 80 + 20$

$= 320$

Therefore, mean $(\bar{x}) = \frac{\sum xf}{\sum f} = \frac{320}{124+K}$

or, $2.5 = \frac{320}{124+K}$

$\Rightarrow K = 128 - 124 = 4$

2. **(D)**

Since, data are multiplied by 5, then variance should be 25 times.

So, new variance $= 25 \times 9 = 225$

3. **(D)**

4. **(D)**

Mean

$= \frac{1\cdot 1 + \frac{1}{2}\cdot 2 + \frac{1}{3}\cdot 3 + \frac{1}{4}\cdot 4 + \frac{1}{5}\cdot 5 + \ldots + \frac{1}{n}\cdot n}{1+2+3+\ldots+n}$

$= \frac{n}{\frac{n(n+1)}{2}} = \frac{2}{n+1}$

5. **(A)**

Except mode, all are a measure of dispersion. Hence (A).

6. **(B)**

The mean of the series

$a, a+d, a+2d, \ldots, a+2\text{ nd}$ is

$\bar{x} = \frac{1}{(2n+1)}[a + a + d + a + 2d + \ldots + a + 2nd]$

$= \frac{1}{(2n+1)}\left[\frac{2n+1}{2}(a + a + 2nd)\right] = a + nd$

Therefore, mean deviation from mean

$= \frac{1}{(2n+1)}\sum_{r=0}^{2n} \left|(a+rd) - (a+nd)\right|$

$$= \frac{1}{2n+1}\sum_{r=0}^{2n} |r-n| d$$

$$= \frac{1}{2n+1} 2d\,(1+2\ldots+n) = \frac{n(n+1)}{(2n+1)} d$$

7. **(B)**

Since $\sigma_x^2 = 4$ and $\sigma_y^2 = 5$

Also, $\bar{x} = 2$ and $\bar{y} = 4$

Now, $\dfrac{\sum x_i}{5} = 2 \Rightarrow \sum x_i = 10$

and, $\dfrac{\sum y_i}{5} = 4 \Rightarrow \sum y_i = 20$

$$\Rightarrow \sigma x^2 = \left(\frac{1}{5}\sum x_i^2\right) - (\bar{x})^2$$

$$= \left(\frac{1}{5}\sum x_i^2\right) - (2)^2$$

and, $\sigma_y^2 = \frac{1}{5}\left(\sum y_i^2\right) - 16$

$\therefore \sum x_i^2 = 40$ and $\sum y_i^2 = 105$

$$\sigma_z^2 = \frac{1}{10}\left(\sum x_i^2 + \sum y_i^2\right) - \left(\frac{\bar{x}+\bar{y}}{2}\right)^2$$

$$= \frac{1}{10}(40+105) - 9 = \frac{145-90}{10} = \frac{55}{10} = \frac{11}{2}$$

8. **(C)**

The mean of the series

$1, 1+d, 1+2d, \ldots, 1+100d$ is

$$\bar{x} = \frac{1+(1+d)+(1+2d)+\ldots+(1+100d)}{101}$$

$$= \frac{101+101\times 50\times d}{101} = 1+50d$$

Therefore, mean deviation from mean

$$= \frac{1}{101}\sum_{r=0}^{100} |(1+rd)-(1+50d)|$$

$$= \frac{2d}{101}\left(\frac{50\times 51}{2}\right)$$

$$\Rightarrow 255 = \frac{50\times 51\times d}{101} \Rightarrow d = \frac{255\times 101}{25\times 51} = 10.1$$

9. **(C)**

We know

$$\text{S.D.} = \sqrt{\frac{\sum x_i^2}{n} - \left(\frac{\sum x_i}{n}\right)^2}$$

$$\Rightarrow 2 = \sqrt{\frac{100}{n} - \frac{400}{n^2}} \Rightarrow 4 = \frac{100}{n} - \frac{400}{n^2}$$

$$\Rightarrow n^2 - 25n + 100 = 0 \Rightarrow n = 5 \text{ or } 20$$

10. **(D)**

Since variance is independent of change of origin, the variance of the observations 101, 102, …, 200 is the same as the variance of the observations 151, 152, …, 250

$$\therefore V_A = V_B \Rightarrow \frac{V_A}{V_B} = 1$$

11. **(C)**

Let the number of boys and girls be x and y

$\therefore 52x + 42y = 50(x+y) \Rightarrow x = 4y$

$\therefore$ Total number of students in the class = $x + y = 4y + y = 5y$

$\therefore$ Required percentage of boys

$$= \frac{4y}{5y}\times 100\% = 80\%$$

12. **(D)**

From given condition

$$6.80 = \frac{\left[(6-a)^2 + (6-b)^2 + (6-8)^2 + (6-5)^2 + (6-10)^2\right]}{5}$$

$$\Rightarrow 34 = (6-a)^2 + (6-b)^2 + 4 + 1 + 16$$

$$\Rightarrow (6-a)^2 + (6-b)^2 = 13 = 9 + 4 = 3^2 + 2^2$$

$$\Rightarrow a = 3, b = 4$$

13. **(B)**

Since the total number of students = 100 and the number of boys = 70

Number of girls = $(100 - 70) = 30$

Now the total marks of 100 students = $100 \times 720 = 72000$

and total marks of 70 boys = $70 \times 750 = 52500$

$\Rightarrow$ Total marks of 30 girls

= $72000 - 52500 - 19500$

$\therefore$ Average marks of 30 girls = $\dfrac{19500}{30} = 650$

14. **(C)**

Let the assumed mean be $\bar{x} = 39$

x_i	$d_i = x_i - \bar{x}$	di^2
31	−8	64
32	−7	49
33	−6	36
34	−5	25
35	−4	16
36	−3	9
37	−2	4
38	−1	1
39	0	0
40	1	1
41	2	4
42	3	9
43	4	16
44	5	25
45	6	36
46	7	49
47	8	64
	$\sum d_i = 0$	$\sum d_i^2 = 408$

$$\text{S.D.} = \sqrt{\frac{\sum d_i^2}{17} - \left(\frac{\sum d_i}{17}\right)^2}$$

$$= \sqrt{\frac{408}{17}} = \sqrt{24} = 2\sqrt{6}$$

15. **(B)**

Since the S.D. of a variable x is σ

Clearly S.D. of $\frac{ax+b}{c}$ is $\left|\frac{a}{c}\right|\sigma$

16. **(A)**

Since

$$\frac{1^2+2^2+3^2+\ldots+n^2}{n} = \frac{46n}{11}$$

$$\Rightarrow \frac{(n+1)(2n+1)}{6} = \frac{46n}{11} \Rightarrow 22n^2 - 243n + 11 = 0$$

$$(n-11)(22n-1) = 0 \Rightarrow n = 11$$

17. **(A)**

Remains same.

18. **(C)**

8.10.3 JEE Main: Archive

1. **(21)**

Mean deviation about mean of first n natural numbers is $\frac{n^2-1}{4n}$

$\therefore n = 21$

2. **(A)**

$$\sigma^2 = \frac{\sum_{i=1}^{5}(x_i - \bar{x})^2}{n}$$

Mean = 6

$$\frac{a+b+8+5+10}{5} = 6$$

$a + b = 7\ b = 7 - a$

$$6.8 = \frac{(a-6)^2 + (b-6)^2 + (8-6)^2 + (5-6)^2 + (10-6)^2}{5}$$

$$34 = (a-6)^2 + (7-a-6)^2 + 4 + 1 + 18$$

$a^2 - 7a + 12 = 0 \Rightarrow a = 4$ or $a = 3$

$a = 4\ a = 3$

$b = 3\ b = 4$

$$M = \frac{\sum_{i=1}^{5}|x_i - x|}{n}$$

$$M = \frac{|a-6| + |b-6| + |8-6| + |5-6| + |10-6|}{5}$$

where $a = 3, b = 4$

when $a = 4, b = 3$

$$M = \frac{3+2+2+1+4}{5}\quad M = \frac{2+3+2+1+7}{5}$$

$$M = \frac{12}{5}\ M = \frac{12}{5}\ 25M = 25 \times \frac{12}{5} = 60$$

3. **(C)**

No. of observations: −50

mean $(\bar{x}) = 15$

Standard deviation $(\sigma) = 2$

Let incorrect observation is x_1 and correct observation is (x_1')

Given $x_1 + x_1' = 70$

$$\bar{x} = \frac{x_1 + x_2 + \ldots + x_{50}}{50} = 15 \text{ (given)}$$

$$\Rightarrow x_1 + x_2 + \ldots + x_{50} = 750 \quad (1)$$

Now

Mean of correct observation is 16

$$\frac{x_1' + x_2 + \ldots + x_{50}}{50} = 16$$

$$x_1' + x_2 + x_3 + \ldots x_{50} = 16 \times 50 \quad (2)$$

Eq. (2) – Eq. (1)

$$\Rightarrow x_1' - x_1 = 16 \times 50 - 15 \times 50$$

$x_1' - x_1 = 50$ and $x_1 + x_1' = 70$

$x_1' = 60 \quad x_1 = 10$

$$\Rightarrow 4 = \frac{x_1^2 + x_2^2 + \ldots + x_{50}^2}{50} - 15^2 \quad (3)$$

$$\Rightarrow \sigma^2 = \frac{x_1'^2 + x_2^2 + \ldots x_{50}^2}{50} - 16^2 \quad (4)$$

From (3)

$$\Rightarrow 4 = \frac{(10)^2}{50} + \frac{x_2^2 + x_3^2 + \ldots + x_{50}^2}{50} - 225$$

$$\Rightarrow 4 = 2 - 225 + \frac{\left(x_2^2 + x_3^2 + \ldots + x_{50}^2\right)}{50}$$

$$\Rightarrow 227 = \frac{\left(x_2^2 + x_3^2 + \ldots x_{50}^2\right)}{50}$$

From (4)

$$\sigma^2 = \frac{(60)^2}{50} + \left(\frac{x_2^2 + x_3^2 + \ldots + x_{50}^2}{50}\right) - (16)^2$$

$$\sigma^2 = \frac{60 \times 60}{50} + 227 - 256$$

$$\sigma^2 = 72 + 227 - 256 \qquad \sigma^2 = 43$$

4. **(B)**

$$\text{Mean } \bar{x} = \frac{4+5+6+6+7+8+x+y}{8} = 6$$

$$\Rightarrow x + y = 48 - 36 = 12$$

Variance

$$= \frac{1}{8}\left(16+25+36+36+49+64+x^2+y^2\right)$$

$$-36 = \frac{9}{4}$$

$$\Rightarrow x^2 + y^2 = 80$$

$$\therefore x = 4; y = 8$$

$$x^4 + y^2 = 256 + 64 = 320$$

5. **(17)**

We have

$$\text{Variance} = \frac{\sum_{r=1}^{15} x_r^2}{15} \left(\frac{\sum_{r=1}^{15} x_r}{15}\right)^2$$

Now, as per information given in equation

$$\frac{\sum x_r^2}{15} - 8^2 = 3^2 \Rightarrow \sum x_r^2 = \log 5$$

Now, the new

$$\sum x_r^2 = \log 5 - 5^2 + 20^2 = 1470$$

And, new $\sum x_r = (15 \times 8) - 5 + (20) = 135$

$$\therefore \text{Variance} = \frac{1470}{15} - \left(\frac{135}{15}\right)^2 = 98 - 81 = 17$$

6. **(B)**

$$\bar{x} = \frac{\sum x_i}{5} = \frac{24}{5} \Rightarrow \sum x_i = 24$$

$$\sigma^2 = \frac{\sum x_i^2}{5} - \left(\frac{24}{5}\right)^2 = \frac{194}{25}$$

$$\Rightarrow \sum x_i^2 = 154$$

$$x_1 + x_2 + x_3 + x_4 = 14 \Rightarrow x_5 = 10$$

$$\sigma^2 = \frac{x_1^2 + x_2^2 + x_3^2 + x_4^2}{4} - \frac{49}{4} = a$$

$$x_1^2 + x_2^2 + x_3^2 + x_4^2 = 4a + 49$$

$$x_5^2 = 154 - 4a - 49$$

$$\Rightarrow 100 = 105 - 4a \Rightarrow 4a = 5$$

$$4a + x_5 = 15$$

7. **(A)**

Mean = 13

Variance

$$= \frac{9+49+144+a^2+(43-a)^2}{5} - 13^2 \in N$$

$$\Rightarrow \frac{2a^2 - a + 1}{5} \in N$$

$\Rightarrow 2a^2 - a + 1 - 5n = 0$ must have solutions as natural numbers its $D = 40n - 7$ always has 3 at the units place

$\Rightarrow D$ can't be perfect square

So, a can't be integer

8. **(B)**

$a_1 + a_3 = 10 = a_1 + d \Rightarrow 5$

$a_1 + a_2 + a_3 + a_4 + a_5 + a_6 = 57$

$\Rightarrow \frac{6}{2}[a_1 + a_6] = 57 \Rightarrow a_1 + a_6 = 19$

$\Rightarrow 2a_1 + 5d = 19$ and $a_1 + d = 5$

$\Rightarrow a_1 = 2, d = 3$

Numbers: 2, 5, 8, 11, 14, 17

Variance $= \sigma^2 =$ mean of squares – square of mean

$$= \frac{2^2 + 5^2 + 8^2 + (11)^2 + (14)^2 + (17)^2}{6} - \left(\frac{19}{2}\right)^2$$

$$= \frac{699}{6} - \frac{361}{4} = \frac{105}{4} \quad 8\sigma^2 = 210$$

9. **(C)**

$$\sum_{i=1}^{n} x_i = 10\,n$$

$$\sum_{i=1}^{n} x_i - 8 + 12 = (10.2)n \quad \therefore n = 20$$

$$\text{Now } \frac{\sum_{i=1}^{20} x_i^2}{20} - (10)^2 = 4 \Rightarrow \sum_{i=1}^{20} x_i^2 = 2080$$

$$\frac{\sum_{i=1}^{20} x_i^2 - 8^2 + 12^2}{20} - (10.2)^2$$

$$= 108 - 104.04 = 3.96$$

10. **(A)**

X	P(X)	XP(X)	$x^2P(x)$
0	$\frac{1}{6}$	0	0
1	$\frac{1}{2}$	$\frac{1}{2}$	$\frac{1}{2}$
2	$\frac{3}{10}$	$\frac{6}{10}$	$\frac{12}{10}$
3	$\frac{1}{30}$	$\frac{1}{10}$	$\frac{9}{30}$

$$\sum xP(x) = \frac{6}{2} = \mu$$

$$\sigma^2 = \sum x^2 P(x) - \mu^2$$

$$\sigma^2 + \mu^2 = 0 + \frac{1}{2} + \frac{12}{10} + \frac{9}{30} = 2$$

$$10\left(\sigma^2 + \mu^2\right) = 20$$

11. **(603)**

$$\bar{x} = \frac{\sum_{i=11}^{41} i}{31} = \frac{11+41}{2} = 26 \text{ (31 elements)}$$

$$\bar{y} = \frac{\sum_{j=61}^{91} j}{31} = \frac{61+91}{2} = 76 \text{ (31 elements)}$$

$$\text{Combined mean, } \mu = \frac{31+26+31\times 76}{31+31}$$

$$= \frac{26+76}{2} = 51$$

$$\sigma^2 = \frac{1}{61}\times\left(\sum_{i=1}^{31}(x_i-\mu)^2 + \sum_{i=1}^{31}(y_i-\mu)^2\right) = 705$$

Since, $x_i \in X$ are in A.P. with 31 elements and common difference 1, same is $y_i \in y$, when written in increasing order.

$$\therefore \sum_{i=1}^{31}(x_i-\mu)^2 = \sum_{i=1}^{31}(y_i-\mu)^2$$

$$= 10^2 + 11^2 + \ldots + 40^2$$

$$= \frac{40\times 41\times 81}{6} - \frac{9\times 10\times 19}{6} = 21855$$

$$\therefore \left|\bar{x} + y - \sigma^2\right| = |26 + 76 - 705| = 603$$

12. **(37)**

$$\frac{x_1 + x_2 + \ldots + x_7}{7} = 8$$

$$\frac{x_1 + x_2 + x_3 \ldots + x_6 + 14}{7} = 8$$

$$\Rightarrow x_1 + x_2 + \ldots + x_6 = 42$$

$$\therefore \frac{x_1 + x_2 \ldots + x_6}{6} = \frac{42}{6} = 7 = a$$

$$\frac{\sum x_i^2}{7} - 8^2 = 16$$

$\sum x_i^2 = 560$

$\Rightarrow x_1^2 + x_2^2 + \ldots + x_6^2 = 364$

$b = \dfrac{x_1^2 + x_2^2 + \ldots + x_6^2}{6} - 7^2 = \dfrac{364}{6} - 49$

$b = \dfrac{70}{6}$

$a + 3b - 5 = 7 + 3 \times \dfrac{70}{6} - 5 = 37$

13. **(C)**

Let a_1 be any natural number

$a_1, a_1+1, a_1+2, \ldots, a_1+99$ are values of $a_i'S$

$\bar{x} = \dfrac{a_1 + (a_1+1) + (a_1+2) + \ldots + a_1 + 99}{100}$

$= a_1 + \dfrac{99}{2}$

Mean deviation about mean $= \dfrac{\sum_{i=1}^{100} |x_i - \bar{x}|}{100}$

$= \dfrac{2\left(\frac{99}{2} + \frac{97}{2} + \frac{95}{2} + \ldots + \frac{1}{2}\right)}{100}$

$= \dfrac{1+3+\ldots+99}{100} = \dfrac{\frac{50}{2}[1+99]}{100} = 25$

14. **(5)**

x_i	f_i	$d = x_i - 5$	$f_i d_i^2$	$f_i d_i$
2	3	−3	27	−9
3	6	−2	24	−12
4	16	−1	16	−16
5	α	0	0	0
6	9	1	9	9
7	5	2	20	10
8	6	3	54	18

$\sigma_x^2 = \sigma_d^2 = \dfrac{\sum f_i d_i^2}{\sum f_i} - \left(\dfrac{\sum f_i d_i}{\sum f_i}\right)^2$

$= \dfrac{150}{45+\alpha} - 0 = 3 \Rightarrow 150 = 135 + 3\alpha$

$\Rightarrow 3\alpha = 15 \Rightarrow \alpha = 5$

15. **(A)**

A	*B*	*A + B*
$\bar{x}_1 = 40$	$\bar{x}_2 = 55$	$\bar{x} = 50$
$\sigma_1 = \alpha$	$\sigma_2 = 30 - \alpha$	$\sigma^2 = 350$
$n_1 = 100$	$n_2 = n$	$100 + n$

$\bar{x} = \dfrac{100 \times 40 + 55n}{100 + n}$

$5000 + 50n = 4000 + 55n$

$1000 = 5n$

$n = 200$

$\sigma_1^2 = \dfrac{\sum x_i^2}{100} - 40^2 \quad \sigma_2^2 = \dfrac{\sum x_j^2}{100} - 55^2$

$350 = \sigma^2 = \dfrac{\sum x_i^2 + \sum x_j^2}{300} - (\bar{x})^2$

$2850 \times 3 = \alpha^2 + 2(30-\alpha)^2 + 1600 + 6050$

$8550 = \alpha^2 + 2(30-\alpha)^2 + 7650$

$\alpha^2 + 2(30-\alpha)^2 = 900$

$\alpha^2 - 40\alpha + 300 = 0 \quad \alpha = 10, 30$

$\sigma_1^2 + \sigma_2^2 = 10^2 + 20^2 = 500$

16. **(A)**

$\dfrac{1+3+5+a+b}{5} = 5$

$a + b = 16$ (1)

$\sigma^2 = \dfrac{\sum x_1^2}{5} - \left(\dfrac{\sum x}{5}\right)^2$

$8 = \dfrac{1^2 + 3^2 + 5^2 + a^2 + b^2}{5} - 25$

$a^2 + b^2 = 130$ (2)

by (1), (2)

$a = 7, b = 9$

17. **(B)**

$9 = x_1 < x_2 < \ldots < x_7$

$9, 9+d, 9+2d, \ldots 9+6d$

$\bar{x}_{\text{new}} = \dfrac{21d}{7} = 3d$

$$16=\frac{1}{7}\left(0^2+1^2+\ldots+6^2\right)d^2-9d^2$$

$$=\frac{1}{7}\left(\frac{6\times7\times13}{6}\right)d^2-9d^2\ 16=4d^2$$

$d^2=4 \quad d=2$

$\bar{x}+x_6=6+9+10+9$

18. **(A)**

Given that,

For set 1:

Mean = 12 and Variance = 14

For set 2:

Mean = 14 and Variance = σ^2

Now, For set 1 we have,

$$\Rightarrow \frac{\sum_{i=1}^{i=15} x_i}{15}=12$$

$$\Rightarrow \sum_{i=1}^{i=15} x_i=15\times12$$

Also, $$\frac{\sum_{i=1}^{i=15} x_i^{\ 2}}{15}-\left(\frac{\sum_{i=1}^{i=15} x_i}{15}\right)^2=14$$

$$\Rightarrow \frac{\sum_{i=1}^{i=15} x_i^{\ 2}}{15}-12^2=14$$

$$\Rightarrow \sum_{i=1}^{i=15} x_i^{\ 2}=(14+144)\times15$$

Now for set 2:

$$\Rightarrow \sum_{i=1}^{i=15} x_i^{\ 2}=(14+144)\times15 \Rightarrow \frac{\sum_{i=1}^{i=15} y_i}{15}=14$$

and $$\Rightarrow \sum_{i=1}^{i=15} y_i=15\times14$$

Also, $$\frac{\sum_{i=1}^{i=15} y_i^{\ 2}}{15}-\left(\frac{\sum_{i=1}^{i=15} y_i}{15}\right)^2=\sigma^2$$

$$\Rightarrow \frac{\sum_{i=1}^{i=15} y_i^2}{15}-14^2=\sigma^2$$

$$\Rightarrow \sum_{i=1}^{i=15} y_i^{\ 2}=\left(\sigma^2+196\right)\times15$$

Now,

The combined variance will be

$$\Rightarrow 13=\frac{\left(\sum_{i=1}^{i=15} x_i^{\ 2}\right)\times15+\left(\sum_{i=1}^{i=15} y_i^{\ 2}\right)\times15}{15+15}-\left(\frac{\sum_{i=1}^{i=15} x_i+\sum_{i=1}^{i=15} y_i}{15+15}\right)^2$$

$$\Rightarrow 13=\frac{(14+144)\times15+\left(\sigma^2+196\right)\times15}{30}-13^2$$

$\therefore \sigma^2=10$

Hence, $\sigma^2=10$

19. **(25)**

The mean and variance of the frequency distribution

x_i	2	4	6	8	10	12	14	16
f_i	4	4	α	15	8	β	4	5

are 9 and 15.08 respectively,

Now mean is given by,

Mean

$$=\frac{8+16+120+80+56+80+6\alpha+12\beta}{40+\alpha+\beta}=9$$

$$\Rightarrow 360+9\alpha+9\beta=360+6\alpha+12\beta$$

$$\Rightarrow 3\alpha-3\beta=0$$

$$\Rightarrow \alpha=\beta\ldots \qquad (1)$$

Now variance is given by,

$$15.08=\frac{16+64+36\alpha+960+800+144\alpha+784+1280}{40+2\alpha}-9^2$$

$$\Rightarrow 96.08=\frac{3904+180\alpha}{40+2\alpha}$$

$$\Rightarrow (40+2\alpha)(96.08)=3904+180\alpha$$

$$\Rightarrow 3843.20+(192.16)\alpha=3904+180\alpha$$

$$\Rightarrow (12.16)\alpha=60.80 \Rightarrow \alpha=5=\beta$$

Hence, $\alpha^2+\beta^2-\alpha\beta=25+25-25=25$

20. (25)

Given, the data $x, y, 10, 12, 4, 6, 8, 12$

So,

$$\text{Mean} = \frac{x+y+10+12+4+6+8+12}{8}$$

$$\Rightarrow 9 = \frac{x+y+52}{8}$$

$\Rightarrow x+y+52=72 \Rightarrow x+y=20$

And,

$$\text{Variance} = \left(\frac{\sum x_i^2}{n}\right) - \left(\frac{\sum x_i}{n}\right)^2$$

$$\Rightarrow 9.25 = \left(\frac{x^2+y^2+100+144+16+36+64+144}{8}\right) - (9)^2$$

$\Rightarrow x^2+y^2=218$

$\Rightarrow (x+y)^2-2xy=218$

$\Rightarrow 20^2-2xy=218$

$\Rightarrow 2xy=182 \Rightarrow xy=91$

Now solving $x+y=20$ & $xy=91$ we get,

$x=13, y=7 \Rightarrow 3x-2y=25$

21. (D)

Given,

The mean and variance of 12 observations be $\frac{9}{2}$ and 4 respectively,

So, mean will be,

$$\bar{x} = \frac{x_1+x_2+\ldots+9+10+\ldots+x_{12}}{12} = \frac{9}{2}$$

$\Rightarrow x_1+x_2\ldots+x_{12}+19=54$

Now removing the observations 9 and 10 and adding the observations 7 and 14,

So, the new total sum of the observations will be,

$x_1+x_2+\ldots+x_{12}+7+14=54-19+7+14$

Now, the new mean will be,

$$\frac{x_1+x_2+\ldots+x_{12}+7+14}{12} = \frac{56}{12}$$

$$\Rightarrow \bar{x}_{\text{new}} = \frac{14}{3}$$

Using the formula of the variance we get,

$$\frac{x_1^2+x_2^2+\ldots+x_{12}^2+9^2+10^2}{12} - \left(\frac{9}{2}\right)^2 = 4$$

$\Rightarrow x_1^2+x_2^2+\ldots+x_{12}^2+9^2+10^2-81\times 3=4\times 12$

$\Rightarrow x_1^2+x_2^2+\ldots+x_{12}^2+9^2+10^2=291$

Now removing 9^2 and 10^2 and adding 7^2 and 14^2 we get,

$\Rightarrow x_1^2+x_2^2+\ldots+x_{12}^2+7^2+14^2=355$

$$\text{New variance} = \frac{\sum x_i^2}{N} - (\bar{x})^2$$

$$= \frac{355}{12} - \left(\frac{14}{3}\right)^2$$

$$= \frac{281}{36}$$

Now on comparing with $\frac{m}{n} = \frac{281}{36}$ we get,

$\Rightarrow m=281$ and $n=36$

$\therefore m+n=317$

22. (151)

Given that the mean for the frequency distribution is 28.

We know that Mean $(\bar{x}) = \frac{\sum f_i x_i}{\sum f_i}$

Now for the class interval 0–10 we get $x_1 = \frac{0+10}{2} = 5.$

Similarly we can calculate x_i for other class intervals.

Now we know that mean = 28

$$\Rightarrow \frac{2\times 5+3\times 15+x\times 25+5\times 35+4\times 45}{14+x} = 28$$

$\Rightarrow x=6$

Also we know that variance

$$(\sigma^2) = \frac{\sum f_i(\bar{x}-x_i)^2}{\sum f_i}$$

$$\sigma^2 = \frac{2\times(28-5)^2+3\times(28-15)^2+6\times(28-25)^2+5\times(28-35)^2+4\times(28-45)^2}{20}$$

$$\sigma^2 = \frac{2\times(23)^2 + 3\times(13)^2 + 6\times(3)^2 + 5\times(7)^2 + 4\times(17)^2}{20}$$

$\sigma^2 = 151$

Hence this is the required answer.

23. (B)

Given,

x_i	0	1	2	3	4	5
f_i	$k+2$	$2k$	k^2-1	k^2-1	k^2+1	$k-3$

Now we know that,

$$\text{Mean} = \frac{\sum X_i f_i}{\sum f_i}$$

$$\Rightarrow \mu = \frac{2k+2k^2-2+3k^2-3+4k^2+4+5k-15}{62}$$

$$\Rightarrow \mu = \frac{9k^2+7k-16}{62} \quad (1)$$

And $\sum f_i = 62$

$\Rightarrow 3k^2+4k-2 = 62$

$\Rightarrow 3k^2+4k-64 = 0$

$\Rightarrow (3k+16)(k-4) = 0$

$\Rightarrow k = 4$

Now putting the value of k in above Eq. (1) we get,

$$\mu = \frac{156}{62}$$

Now finding variance we get,

$$\sigma^2 = \frac{\sum (X_i)^2 f_i}{\sum f_i} - (\text{ Mean })^2$$

$$\Rightarrow \sigma^2 = \frac{1\cdot(2k)+4(k^2-1)+9(k^2-1)+16(k^2+1)+25(k-3)}{62} - \left(\frac{156}{62}\right)^2$$

$$\Rightarrow \sigma^2 = \frac{29k^2+27k-72}{62} - \mu^2$$

Now putting the value of k in above equation we get,

$$\Rightarrow \sigma^2 = \frac{500}{62} - \mu^2$$

$$\Rightarrow \sigma^2 + \mu^2 = \frac{500}{62}$$

$$\Rightarrow \left[\sigma^2 + \mu^2\right] = 8$$

24. (C)

Let the elements in set A be $\{x_1, x_2, x_3, x_4, x_5\}$

Now given mean is = 5

Now subtracting 3 from each term we get,

New mean as

$(x_1 - 3\ x_2 - 3\ldots, x_5 - 3) = 5 - 3 = 2$

So, sum of elements will be $2\times 5 = 10$

Also given variance,

$V\text{ ar }(X) = 12$

Now we know that subtracting 3 from each term will not change the variance,

So, $\text{Var}(x_1-3, x_2-3, \ldots x_5-3) = 12$

$$\Rightarrow \frac{\sum (x_i-3)^2}{5} - 4 = 12$$

$$\Rightarrow \sum (x_i-3)^2 = 80$$

Now let elements in set B be $\{y_1, y_2 \ldots y_5\}$

Given mean $(y_1, y_2 \ldots y_5) = 8$

Now adding 2 to each element we get,

New mean $(y_1+2, y_2+2, \ldots y_5+2) = 10$

So, sum of elements will be $10\times 5 = 50\ d$

Also given $\text{Var}(y_1, y_2 \ldots y_5) = 20$

Similarly new variance,

$\text{Var}(y_1+2, y_2+2\ldots y_5+2) = 20$

$$\Rightarrow \frac{\sum (y_i+2)^2}{5} - 100 = 20$$

$$\Rightarrow \sum (y_i+2)^2 = 120\times 5$$

Now finding, combined mean we get,

$$\frac{\sum_{i=1}^{5} (x_i-3) + \sum (y_i+2)}{10} = \frac{10+50}{10} = 6$$

And combined variance

$$=\frac{\sum (x_i-3)^2+\sum (y_i+2)^2}{10}-6^2$$

$$=\frac{80+120\times 5}{10}-36=32$$

Hence, the sum of combined mean and variance will be $32+6=38$

25. **(C)**

Consider the data 1, 2, 4, 5, x, y

So,

$$\text{Mean}=\frac{1+2+4+5+x+y}{6}$$

$$\Rightarrow 5=\frac{x+y+12}{6}$$

$$\Rightarrow x+y=18$$

And,

$$\text{Variance}=\left(\frac{\sum x_i^2}{n}\right)-\left(\frac{\sum x_i}{n}\right)^2$$

$$\Rightarrow 10=\left(\frac{1^2+2^2+4^2+5^2+x^2+y^2}{6}\right)-(5)^2$$

$$\Rightarrow x^2+y^2=164$$

We have $x+y=18$ and $x^2+y^2=164$.

Solving the above equations we get,

$x=10,\ y=8$,

Now mean deviation about mean will be

$$=\frac{\sum_{i=1}^{6}|x_i-\bar{x}|}{6}$$

$$=\frac{|1-5|+|2-5|+|4-5|+|5-5|+|10-5|+|8-5|}{6}$$

$$=\frac{16}{6}=\frac{8}{3}$$

Therefore, the required answer is $\frac{8}{3}$.

26. **(211)**

5 positive numbers $a_1, a_2, \ldots, a_5$ are in geometric progression. So, let the numbers be $\frac{a}{r^2}, \frac{a}{r}, a, ar, ar^2$

According to question,

$$\frac{\frac{a}{r^2}+\frac{a}{r}+a+ar+ar^2}{5}=\frac{31}{10}$$

$$\text{And, } \frac{\frac{r^2}{a}+\frac{r}{a}+\frac{1}{a}+\frac{1}{ar}+\frac{1}{ar^2}}{5}=\frac{31}{40}$$

$$\Rightarrow \frac{\frac{1}{a}\left(r^2+r+1+\frac{1}{r}+\frac{1}{r^2}\right)}{5}=\frac{31}{40}$$

By (1) and (2), we get

$$a^2=4\Rightarrow a=2$$

So,

$$\left(r^2+r+1+\frac{1}{r}+\frac{1}{r^2}\right)=\frac{31}{4}$$

$$\Rightarrow \left(r+\frac{1}{r}\right)^2+r+\frac{1}{r}=\frac{27}{4}+2$$

$$\Rightarrow t^2+t=\frac{35}{4}$$

where, $t=r+\frac{1}{r}$

$$4t^2+4t-35=0$$

$$\Rightarrow t=\frac{5}{2}\Rightarrow r=2$$

So, numbers are $\frac{1}{2}$, 1, 2, 4, 8

Variance is

$$=\frac{\frac{1}{4}+1+4+16+64}{5}-\left(\frac{\frac{1}{2}+1+2+4+8}{5}\right)^2$$

$$=\frac{186}{25}=\frac{m}{n}\quad m+n=211$$

Hence this is the required option.

27. **(8)**

Given,

The mean of the below data is 5

x	1	3	5	7	9
Frequency (f)	4	24	28	α	8

Now finding mean we get,

$$\frac{4+72+28\times5+7\alpha+72}{64+\alpha}=5$$

$\Rightarrow \alpha=16$

So, the total number of observations will be, $\sum f_i=80$

Now finding mean deviation we get,

$$\text{M.D.}=\frac{\sum f_i|x_i-5|}{\sum f_i}$$

$$=\frac{4\times4+24\times2+0+16\times2+8\times4}{80}=\frac{8}{5}$$

So, $m=\frac{8}{5}$

Now finding, variance we get,

$$\sigma^2=\frac{\sum f_i|x_i-5|^2}{\sum f_i}$$

$$\Rightarrow \sigma^2=\frac{4\times16+24\times4+0+16\times4+8\times16}{80}$$

$$\Rightarrow \sigma^2=\frac{22}{5}$$

Hence, the value of $\frac{3\alpha}{m+\sigma^2}=\frac{3\times16}{\frac{8}{5}+\frac{22}{5}}=8$

28. (269)

Let the observations be

$x_1, x_2, x_3, \ldots, x_8, 45, 50$,

Mean = 50

$$\Rightarrow \frac{x_1+x_2+x_3+\ldots+x_8+45+50}{10}=50$$

$$\Rightarrow x_1+x_2+x_3+\ldots+x_8=405 \quad (1)$$

Hence, new mean

$$(\bar{X})_{\text{new}}=\frac{405+20+25}{10}=45$$

Now,

$$\text{S.D.}=\sqrt{\frac{\sum_{i=1}^{8}x_i^2+45^2+50^2}{10}-(50)^2}$$

$$\Rightarrow 12=\sqrt{\frac{\sum_{i=1}^{8}x_i^2+4525}{10}-2500}$$

$$\Rightarrow \sum_{i=1}^{8}x_i^2=21915$$

Now,

$$(\text{Variance})_{\text{new}}=\frac{\sum_{i=1}^{8}x_i^2+20^2+25^2}{10}-(45)^2$$

$$\Rightarrow (\text{Variance})_{\text{new}}=\frac{21915+20^2+25^2}{10}-(45)^2$$

$$\Rightarrow (\text{Variance})_{\text{new}}=2294-2025$$

$$\Rightarrow (\text{Variance})_{\text{new}}=269$$

Hence this is the correct answer

29. (B)

Incorrect mean is

$$\frac{\sum_{i=1}^{10}x_i}{10}=20$$

$$\Rightarrow \sum_{i=1}^{10}x_i=200\ldots \quad (1)$$

So, corrected mean is

$$\frac{\sum_{i=1}^{10}x_i-50+40}{10}=\frac{200-10}{10}=19$$

Incorrect variance is

$$\frac{\sum_{i=1}^{10}x_i^2}{10}-400=64$$

$$\Rightarrow \sum_{i=1}^{10}x_i^2=4640$$

Now, corrected variance is

$$=\frac{\sum_{i=1}^{9}x_i^2+40^2-50^2}{10}-(19)^2$$

$= \frac{4640-900}{10} - (19)^2$

$= 374 - 361$

$= 13$

30. **(C)**

$\text{Median} = 170 \Rightarrow 125, a, b, 170, 190, 210, 230$

Mean deviation about

Median =

$$\frac{0+45+60+20+40+170-a+170-b}{7} = \frac{205}{7}$$

$\Rightarrow a + b = 300$

Mean

$$= \frac{170+125+230+190+210+a+b}{7} = 175$$

Mean deviation

About mean =

$$\frac{50+175-a+175-b+5+15+35+55}{7} = 30$$

31. **(A)**

$x_1, x_2 \dots x_{10}$

$$\sum_{i=1}^{10}(x_i - \alpha) = 2 \Rightarrow \sum_{i=1}^{10} x_i - 10\alpha = 2$$

$$\text{Mean } \mu = \frac{6}{5} = \frac{\sum x_i}{10}$$

$\therefore \sum x_i = 12$

$10\alpha + 2 = 12 \therefore \alpha = 1$

$$\text{Now } \sum_{i=1}^{10}(x_i - \beta)^2 = 40$$

Let $y_i = x_i - \beta$

$$\therefore \sigma_y^2 = \frac{1}{10}\sum y_i^2 - (\bar{y})^2$$

$$\sigma_x^2 = \frac{1}{10}\sum (x_i - \beta)^2 - \left(\frac{\sum_{i=1}^{10}(x_i - \beta)}{10}\right)^2$$

$$\frac{84}{25} = 4 - \left(\frac{12-10\beta}{10}\right)^2$$

$$\therefore \left(\frac{6-5\beta}{5}\right)^2 = 4 - \frac{84}{25} = \frac{16}{25}$$

$6 - 5\beta = \pm 4 \Rightarrow \beta = \frac{2}{5}$ (not possible) or

$\beta = 2$

Hence $\frac{\beta}{\alpha} = 2$

32. **(B)**

$$\sum_{k=1}^{10} a_k = 50$$

$$a_1 + a_2 + \ldots + a_{10} = 50 \qquad (1)$$

$$\sum_{\forall k<j} a_k a_j = 1100 \qquad (2)$$

If $a_1 + a_2 + \ldots + a_{10} = 50$

$(a_1 + a_2 + \ldots + a_{10})^2 = 2500$

$$\Rightarrow \sum_{i=1}^{10} a_i^2 + 2\sum_{k<j} a_k a_j = 2500$$

$$\Rightarrow \sum_{i=1}^{10} a_i^2 = 2500 - 2(1100)$$

$$\sum_{i=1}^{10} a_i^2 = 300$$

Standard deviation 'σ'

$$= \sqrt{\frac{\sum a_i^2}{10} - \left(\frac{\sum a_i}{10}\right)^2} = \sqrt{\frac{300}{10} - \left(\frac{50}{10}\right)^2}$$

$= \sqrt{30 - 25} = \sqrt{5}$

33. **(2521)**

Let the incorrect mean be μ' and standard deviation be σ'

We have

$$\mu' = \frac{\sum x_i}{15} = 12 \Rightarrow \sum x_i = 180$$

As per given information correct

$\sum x_i = 180 - 10 + 12$

$\Rightarrow \mu$ (correct mean) $= \frac{182}{15}$

Also

$$\sigma' = \sqrt{\frac{\sum x_i^2}{15} - 144} = 3 \Rightarrow \sum x_i^2 = 2295$$

$$\sigma^2 \text{ (correct variance)} = \frac{2339}{15} - \frac{182\times182}{15\times15}$$

Required value

$$= 15\left(\mu + \mu^2 + \sigma^2\right)$$

$$= 15\left(\frac{182}{15} + \frac{182\times182}{15\times15} + \frac{2339}{15} - \frac{182\times182}{15\times15}\right)$$

$$= 15\left(\frac{182}{15} + \frac{2339}{15}\right) = 2521$$

34. (6344)

$\bar{x} = 56$

$\sigma^2 = 66.2$

$$\Rightarrow \frac{\alpha^2 + \beta^2 + 25678}{10} - (56)^2 = 66.2$$

$$\therefore \sigma^2 + \beta^2 = 6344$$

35. (C)

$$\bar{X} = \frac{24}{5}; \sigma^2 = \frac{194}{25}$$

Let first four observation be x_1, x_2, x_3, x_4

$$\text{here, } \frac{x_1 + x_2 + x_3 + x_4 + x_5}{5} = \frac{24}{5} \quad (1)$$

$$\text{Also, } \frac{x_1 + x_2 + x_3 + x_4}{4} = \frac{7}{2}$$

$$\Rightarrow x_1 + x_2 + x_3 + x_4 = 14$$

Now from Eq. (1)

$x_5 = 10$

$$\text{Now, } \sigma^2 = \frac{194}{25}$$

$$\frac{x_1^2 + x_2^2 + x_3^2 + x_4^2 + x_5^2}{5} - \frac{576}{25} = \frac{194}{25}$$

$$\Rightarrow x_1^2 + x_2^2 + x_3^2 + x_4^2 = 54$$

Now, variance of first 4 observations

$$\text{Var} = \frac{\sum_{i=1}^{4} x_i^2}{4} - \left(\frac{\sum_{i=1}^{4} x_i}{4}\right)^2 = \frac{54}{4} - \frac{49}{4} = \frac{5}{4}$$

36. (D)

Class	Frequency	Cumulative Frequency
0–4	3	3
4–8	9	12
8–12	10	22
12–16	8	30
16–20	6	36

$$M = 1 + \left(\frac{\frac{N}{2} - C}{f}\right)h$$

$$M = 8 + \frac{18-12}{10} \times 4$$

$M = 10.4$

$20M = 208$

37. (29)

x_i	f_i	$f_i x_i$	$f_i x_i^2$
0	3	0	0
1	2	2	2
5	3	15	75
6	2	12	72
10	6	60	600
12	3	36	432
		$\sum f_i x_i = 176$	$\sum f_i x_i^2 = 2048$

$$\therefore \sum f_i x_i = 176$$

$$\text{So } \bar{x} = \frac{\sum f_i x_i}{\sum f_i} = \frac{176}{22} = 8$$

$$\text{for } \sigma^2 = \frac{1}{N}\sum f_i x_i^2 - (\bar{x})^2$$

$$= \frac{1}{22} \times 2048 - (8)^2$$

$= 93.090964$

$= 29.0909$

38. (C)

$a, b, 68, 44, 48, 60$

Mean = 55

Variance = 194 $a > b$

$$\frac{a+b+68+44+48+60}{6}=55$$

$\Rightarrow 220+a+b=330$

$\therefore a+b=110$ (1)

Also $\sum \frac{(x_i-\bar{x})^2}{n}=194$

$\Rightarrow (a-55)^2+(b-55)^2+(68-55)^2+(44-55)^2$

$+(48-55)^2+(60-55)^2=194\times 6$

$\Rightarrow (a-55)^2+(b-55)^2+169+121+$
$49+25=1164$

$\Rightarrow (a-55)^2+(b-55)=1164-364=800$

$a^2+3025-110a+b^2+3025-110b=800$

$\Rightarrow a^2+b^2=800-6050+12100$

$a^2+b^2=6850$ (2)

Solve (1) and (2):

$a=75, b=35$

$\therefore a+3b=75+3(35)=75+105=180$

39. **(A)**

$\frac{\sum x_i}{6}=2$ and $\frac{\sum x_i^2}{N}-\mu^2=23$

$\alpha+\beta=10$

$\alpha^2+\beta^2=52$

Solving we get $\alpha=4, \beta=6$

$$\frac{\sum |x_i-\bar{x}|}{6}=\frac{5+2+5+8+2+4}{6}=\frac{13}{3}$$

40. **(B)**

$\sum P_i=1$

$a+2a+a+b+2b+3b=1$

$4a+6b=1$ (1)

$E(x)=\text{mean}=\frac{46}{9}$

$\sum P_iX_i=\frac{46}{9}\Rightarrow 4a+4a+4b+12b+24b=\frac{46}{9}$

$8a+40b=\frac{46}{9}$

$4a+20b=\frac{23}{9}$ (2)

Subtract (1) from (2) we get

$b=\frac{1}{9}$ and $a=\frac{1}{12}$

$\text{Variance}=E\left(x_i^2\right)-E\left(x_i\right)^2$

$E\left(x_i\right)^2=0^2\times 9^2+2^2\times 2a+4^2(a+b)+$
$6^2(2b)+8^2(3b)$

$=24a+280b$

Put $a=\frac{1}{12}, b=\frac{1}{9}$

$E\left(x_i^2\right)=2+\frac{280}{9}=\frac{298}{9}$

$\therefore \sigma^2=E\left(x_i^2\right)-E\left(x_i\right)^2$

$=\frac{298}{9}-\left(\frac{46}{9}\right)^2$

$\sigma^2=\frac{298}{9}-\frac{2116}{81}$

$=\frac{566}{81}$

41. **(56)**

X = denotes number of defective

x	0	1	2	3
$P(x)$	$\frac{7}{15}$	$\frac{5}{12}$	$\frac{5}{12}$	$\frac{1}{12}$
x_1^2	0	1	4	9
$P_ix_1^2$	0	$\frac{5}{12}$	$\frac{20}{12}$	$\frac{9}{12}$
p_ix_i	0	$\frac{5}{12}$	$\frac{10}{12}$	$\frac{3}{12}$

$\mu=\sum p_ix_i=\frac{18}{12}$

$\sum p_ix_1^2=\frac{34}{12}$

$\sigma^2=\sum p_ix_1^2-(\mu)^2$

$$= \frac{34}{12} - \left(\frac{18}{12}\right)^2 = \frac{17}{6} - \frac{9}{4}$$

$$\frac{34-27}{12} = \frac{7}{12}$$

$$96\sigma^2 = 96 \times \frac{7}{12} = 56$$

42. **(C)**

Mean $(\overline{x}) = 10$

$$\Rightarrow \frac{\sum x_i}{20} = 10$$

$$\sum x_i = 10 \times 20 = 200$$

If 8 is replaced by 12, then

$$\sum x_i = 200 - 8 + 12 = 204$$

$$\therefore \text{Correct mean } (\overline{x}) = \frac{\sum x_i}{20}$$

$$= \frac{204}{20} = 10.2$$

$\because$ Standard deviation = 2

$\therefore$ Variance = $(\text{S.D.})^2 = 2^2 = 4$

$$\Rightarrow \frac{\sum x_i^2}{20} - \left(\frac{\sum x_i}{20}\right)^2 = 4$$

$$\Rightarrow \frac{\sum x_i^2}{20} - (10)^2 = 4$$

$$\Rightarrow \frac{\sum x_i^2}{20} = 104$$

$$\Rightarrow \sum x_i^2 = 2080$$

Now, replaced '8' observations by '12'

Then, $\sum x_i^2 = 2080 - 8^2 + 12^2 = 2160$

$\therefore$ Variance of removing observations

$$\Rightarrow \frac{\sum x_i^2}{20} - \left(\frac{\sum x_i}{20}\right)^2$$

$$\Rightarrow \frac{2160}{20} - (10.2)^2$$

$$\Rightarrow 108 - 104.04 = 3.96$$

Correct standard deviation = $\sqrt{3.96}$

43. **(33)**

$a, b, c \in N \; a < b < c$

$$\overline{x} = \text{mean} = \frac{9+25+a+b+c}{5} = 18$$

$$a+b+c = 56$$

$$\text{Mean deviation} = \frac{\sum |x_i - \overline{x}|}{n} = 4$$

$$= 9+7+|18-a|+|18-b|+|18-c| = 20$$

$$= |18-a|+|18-a|+|18-c| = 4$$

$$\text{Variance} = \frac{\sum |x_i - \overline{x}|^2}{n} = \frac{136}{5}$$

$$= 81+49+|18-a|^2+|18-b|^2+|18-c|^2 = 136$$

$$= (18-a)^2+(18-b)^2+(18-c)^2 = 6$$

Possible values

$(18-a)^2 = 1, (18-b)^2 = 1\,(18-c)^2 = 4 a < b < c$

So $18-a = 1 18-b = -1 18-c = -2$

$a = 17 \; b = 19 \; c = 20$

$a+b+c = 56$

$2a+b-c34 = 19-20 = 33$

44. **(C)**

$x + y = 10$ (1)

Median = 18 = M

$$\text{M.D.} = \frac{\sum f_i |x_i - M|}{\sum f_i}$$

$$1.25 = \frac{36+x+2y}{40}$$

$x + 2y = 14$ (2)

by (1) and (2)

$x = 6, y = 4$

$\Rightarrow 4x + 5y = 24 + 20 = 44$

Age (x_i)	f	$\lvert x_i - M\rvert$	$f_i \lvert x_i - M\rvert$
15	5	3	15
16	8	2	16

Age (x_i)	f	$\lvert x_i - M\rvert$	$f_i\lvert x_i - M\rvert$
17	5	1	5
18	12	0	0
19	x	1	x
20	y	2	$2y$

45. **(A)**

x	C	$2C$	$3C$	$4C$	$5C$	$6C$
f	2	1	1	1	1	1

$$\bar{x} = \frac{(2+2+3+4+5+6)C}{7} = \frac{22C}{7}$$

$$\text{Var}(x) = \frac{c^2\left(2+2^2+3^2+4^2+5^2+6^2\right)}{7} - \left(\frac{22c}{7}\right)^2$$

$$= \frac{92c^2}{7} - c^2 \times \frac{484}{49}$$

$$= \frac{(644-484)c^2}{49} = \frac{160c^2}{49}$$

$$= 160 = \frac{160\times c^2}{49} \Rightarrow c = 7$$

8.12 Summary

- **Measures of central value**
  - Arithmetic mean of individual series (ungrouped data)

    $$\bar{x} = \frac{x_1 + x_2 + \ldots + x_n}{n}$$

  - Arithmetic mean for discrete frequency distribution

    $$\bar{x} = \frac{1}{N}\sum_{i=1}^{n} f_i x_i, \text{ where } N = \sum_{i=1}^{n} f_i$$

  - Arithmetic mean for grouped or continuous frequency distribution

    $$\bar{x} = A + \frac{1}{N}\sum_{i=1}^{n} f_i\,(x_i - A)$$

    where A = assumed mean, f_i = frequency and $x_i - A$ =deviation of each item from the assumed mean.

- **Median:** If n be the number of observations,
  - then Median = Value of $\left(\frac{n+1}{2}\right)^{\text{th}}$ term, if n is odd
  - and Median = $\frac{1}{2}$ [Value of $\left(\frac{n}{2}\right)^{\text{th}}$ term + Value of $\left(\frac{n}{2}+1\right)^{\text{th}}$ term], if n is even
- If the series is discrete, we first find the cumulative frequencies of the variable and median $= \left(\frac{n}{2}+1\right)^{\text{th}}$ observation; where n is the cumulative frequency
- For grouped or continuous distributions

  $$\text{Median} = l + \frac{\left(\frac{N}{2} - c\right)}{f} \times i$$

  where l = lower limit of the median class, f = frequency of the median class, N = the sum of all frequencies, i = the width of the median class, c = the cumulative frequency of the class preceding the median class.

- **Quartile:** $Q_J = l + \dfrac{\left(\frac{JN}{4} - c\right)}{f} \times i,\ J = 1, 2, 3$
- $\text{Mode} = l + \left[\dfrac{f_1 - f_0}{2f_1 - f_0 - f_2}\right] \times i$

  where l = the lower limit of the model class, f_1 = the frequency of the model class, f_0 = the frequency of the class preceding the model class, f_2 = the frequency of the class succeeding the model class, i = the size of model class.

- The measures of dispersion commonly used are
  - Range
  - Quartile Deviation
  - Mean Deviation
  - Standard Deviation

- Range is the difference between the greatest and least values of the variables

$Q = \dfrac{Q_3 - Q_1}{2}$ is called quartile deviation

- Mean deviation of a simple distribution

$= \dfrac{\sum_{i=1}^{n} |x_i - K|}{n}$, where K is the A.M. or median or mode

- Mean deviation of ungrouped frequency distribution $= \dfrac{\sum_{i=1}^{n} f_i |d_i|}{N}$, where $d_i = x_i - K$ and $N = \sum_{i=1}^{n} f_i$

- Mean deviation of grouped (Classified) frequency distribution

$$= \frac{\sum_{i=1}^{n} |x_i - K| f_i}{\sum_{i=1}^{n} f_i}$$

where x_i = mid value (Class mark) of the ith class, f_i = frequency of the ith class, K = mean or median or mode.

Also, M.D. = $= \dfrac{1}{n}\left[S_1 - S_2 - K(n_1 - n_2)\right]$

- Variance
  - $\sigma^2 = \dfrac{1}{n}\sum_{i=1}^{n} (x_i - \bar{x})^2$
- Standard deviation, $\sigma = \sqrt{\dfrac{1}{n}\sum_{i=1}^{n} (x_i - \bar{x})^2}$
- When the distribution is simple
  - $\sigma^2 = \dfrac{1}{n}\sum_{i=1}^{n} x_i^2 - \bar{x}^2$
  - $\sigma = \sqrt{\dfrac{1}{n}\sum_{i=1}^{n} x_i^2 - \bar{x}^2}$
- When the distribution is ungrouped
  - $\sigma = \sqrt{\dfrac{\sum_{i=1}^{n} x_i^2 f_i}{N} - \bar{x}^2}$
- When the distribution is grouped
  - $\sigma = \sqrt{\dfrac{\sum_{i=1}^{n} x_i^2 f_i}{N} - \bar{x}^2}$
- **Short Cut Method**
  - S.D., $\sigma = \sqrt{\dfrac{\sum_{i=1}^{n} d_i^2}{n} - \left(\dfrac{\sum_{i=1}^{n} d_i}{n}\right)^2}$

For ungrouped frequency distribution

- S.D., $\sigma = \sqrt{\dfrac{\sum f_i d_i^2}{N} - \left(\dfrac{\sum f_i d_i}{N}\right)^2}$

For grouped frequency distribution

- S.D. $\sigma = \sqrt{\dfrac{\sum f_i d_i^2}{N} - \left(\dfrac{\sum f_i d_i}{N}\right)^2}$

- **Step Deviation Method for Finding S.D.**

S.D. $(\sigma) = h\sqrt{\dfrac{\sum f_i u_i^2}{N} - \left(\dfrac{\sum f_i u_i}{N}\right)^2}$

where x_i = Class mark (mid value) of the ith class, f_i = Frequency of the ith class, $N = \sum_{i=1}^{n} f_i$, A = assumed mean, h = class interval and $u_i = \dfrac{x_i - A}{h}$.

CHAPTER 9

Set Theory and Real Number System

OBJECTIVES

9.1 Set Theory: Introduction
9.2 Types of Sets
9.3 Set Operations
9.4 Cardinal Number of a Set
9.5 Real Number System
9.6 Concept of Intervals
9.7 Wavy Curve Method
9.8 Absolute Value
9.9 Greatest Integer and Fractional Part
9.10 Know Your Concepts
9.11 JEE Main: Sample Questions
9.12 Answer Key
9.13 Summary

The concept of sets forms the basis of modern mathematics. Sets are widely used in different branches of mathematics. Sets are used in the study of relations and functions which are very useful in modern mathematics.

9.1 Sets

Any collection of well defined and distinct objects is called a *set*. Sets are usually denoted by capital letters A, B, C and so on, and their elements by small letters a, b, c and so on. Let $A = \{a, b, c\}$. As a is a member of A, we write $a \in A$ (the symbol $\in$ denotes 'belongs to') and d is not a member of A, so we write $d \notin A$ ($\notin$ denotes 'does not belong to').

Note:

(i) The collection of good cricket players of India is not a set since the term 'Good player' is vague and is not well defined.

(ii) * The set of all natural numbers is denoted by N

** The set of integers by Z

*** The set of rational numbers by Q

**** The set of real numbers by R

***** The set of positive integers by Z^+

****** The set of positive rational numbers by Q^+

******* The set of positive real numbers by R^+

9.1.1 Description of Sets

I. Set Listing Method/Tabular Form/Roster Form

In this method, a set is described by listing all the elements, separated by commas within brackets. The set of all natural numbers in the roster form can be written as $\{1, 2, 3, \ldots\}$.

II. Set Builder Form

If $A = \{1, 2, 3, 4\}$, then in set builder form, we can write,

$$A = \{x : x \in N; x < 5\}$$

9.2 Types of Sets

(i) Finite set: A set having a finite number of elements is called a *finite set*.

(ii) Infinite set: A set having an infinite number of elements is called an *infinite set*.

(iii) Null set (or empty set or void set): A set having no elements is called an *empty set* and is denoted by ϕ.

Let $A = \{x : x \in N, 3 < x < 4\} \Rightarrow A = \phi$

Note:

(i) The set $\{0\}$ is not an empty set as it contains the element '0'.

(ii) The set $\{\phi\}$ is not an empty set. It is a set containing one element ϕ.

Subset

If A and B are any two sets such that every element of set A is also an elements of set B, we say that A is a subset of B; we write this as $A \subseteq B$.

If $A = \{1, 2, 4\}$ and $B = \{1, 2, 3, 4\}$
$\Rightarrow A \subseteq B$

Note:
If $A = \{1, 2, 4\}$, the total number of subsets is

$$\phi, \{1\}, \{2\}, \{4\}, \{1,2\}, \{1,4\}, \{2,4\}, \{1,2,4\}$$

That is, if the set A has 3 elements, the total number of subsets $= 2^3$.

Generalising this, if a set A has n elements, then the total number of subsets $= 2^n$.

Equality of Two Sets

Any two sets A and B are said to be equal if $A \subseteq B$ and $B \subseteq A$.

Proper Subset

If A and B are two sets such that every element of set A is also an element of set B, and there exists at least one element in B which is not in A, then A is called a *proper subset* of B, we write this as '$A \subset B$'.

Example: $N \subset Z \subset Q \subset R$.

Super Set of a Set

A set A is said to be a *super set* of B if B is a subset of A, we write this as $A \supseteq B$.

Power Set

The set of all the subsets of a set A is called a *power set* of A and is denoted by $P(A)$.

Example: Let $A = \{2,3\}$

$$P(A) = \{\phi, \{2\}, \{3\}, \{2,3\}\}$$

Universal Set

A set which is the superset of any considered set is called *universal set* and is denoted by U.

Example: Let $A = \{1,2,3\}$, $B = \{4,5,6\}$ and $U = \{1,2,3,4,5,6,7\}$. Clearly U is a universal set.

The Venn diagram is shown below.

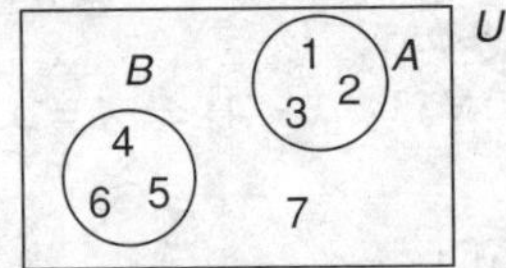

9.3 Set Operations

Union of Two Sets

If A and B are any two sets, then their union is denoted by $A \cup B$ and is defined as the set of all those elements which either belongs to A or to B, that is, $A \cup B = \{x : x \in A \text{ or } x \in B\}$.

Example: Let $A = \{1,2,3,4\}$, $B = \{2,4,5,6\}$

$$A \cup B = \{1,2,3,4,5,6\}.$$

The Venn diagram is shown below.

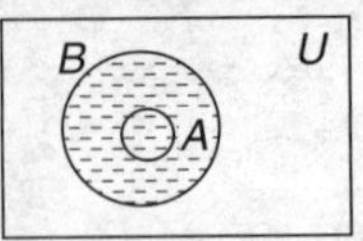

$A \cup B$ when $A \subseteq B$

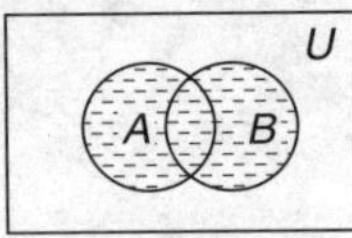

$A \cup B$ when neither $A \subseteq B$ or $B \subseteq A$

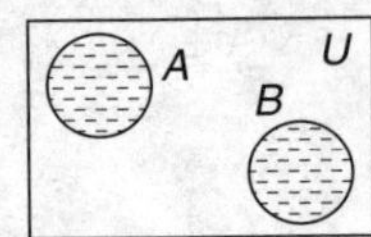

$A \cup B$ when A and B are disjoint sets

Intersection of Two Sets

If A and B are any two sets, then their intersection is denoted by $A \cap B$ and is defined as the set of all those elements which belong to A and B both.

That is, $A \cap B = \{x : x \in A \text{ and } x \in B\}$.

Example: Let $A = \{1,2,3,4\}$ and $B = \{2,4,5,6\}$

$$A \cap B = \{2,4\}$$

The Venn diagram is shown below.

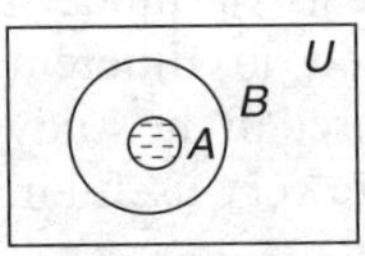

$A \cap B$ when $A \subseteq B$, $A \cap B = A$

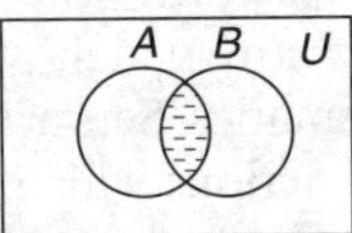

$A \cap B$ when neither $A \subseteq B$ or $B \subseteq A$

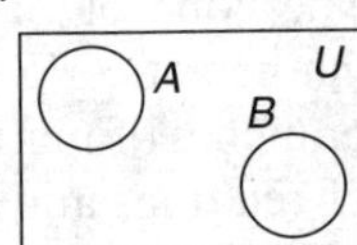

$A \cup B = \phi$

Disjoint Sets

Two sets A and B are said to be disjoint sets if they have no common element, that is, $A \cap B = \phi$.

Example: Let $A = \{1,2,3\}$ and $B = \{4,5\}$
Then $A \cap B = \phi$

The Venn diagram is shown below.

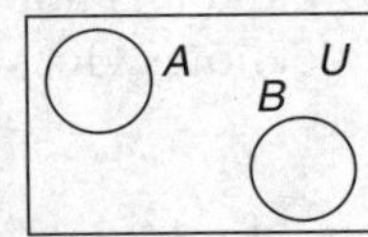

Pair Wise Disjoint Sets

A family of sets $A_1, A_2, \ldots, A_n$ are said to be a pair wise disjoint family of sets, if no two members of this family of sets have a common element. Thus, the family of sets $A_1, A_2, \ldots, A_n$ are said to be pair wise disjoint sets if $A_i \cap A_j = \phi$ for $i \neq j$.

Difference of Two Sets

If A and B are any two sets, then their difference is denoted by '$A - B$' and is defined as the set of all those elements of A which does not belong to B.

Example: Let $A = \{1,2,3,4,5\}$ $B = \{2,4,6,7\}$

$$A - B = \{1,3,5\} \text{ and } B - A = \{6,7\}$$

The Venn diagram is shown below.

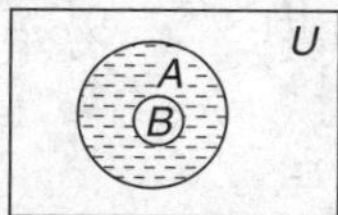

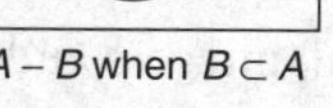
$A-B$ when $B \subset A$

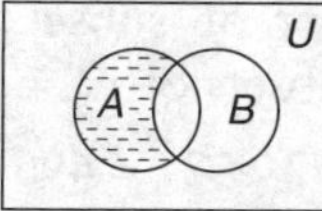

$A-B$ when neither $A \subset B$ not $B \subset A$

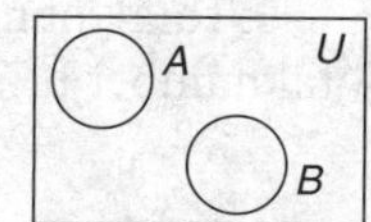

$A-B$ when A and B are disjoint sets. Clearly $A-B=A$

Symmetric Difference

If A and B are two sets, then the symmetric difference of sets A and B is denoted by $A\Delta B$ and is defined as $A\Delta B=(A-B)\cup(B-A)$.

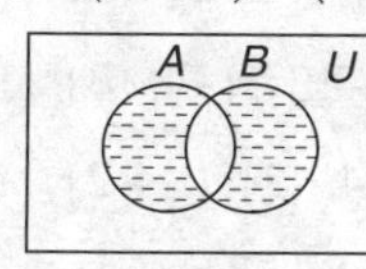

$A \Delta B$

Complement of a Set

The complement of a set A is the set of all those elements of the universal set U which are not elements of A. It is denoted by A' or A^c.

$$A'=\{x : x\in U \text{ and } x\notin A\}$$

Example: Let $U=\{1,2,3,4,5,6,7,8,9\}$

and $\quad A=\{2,4,6,8\}$

$$A'=U-A=\{1,3,5,7,9\}$$

The Venn diagram is shown below.

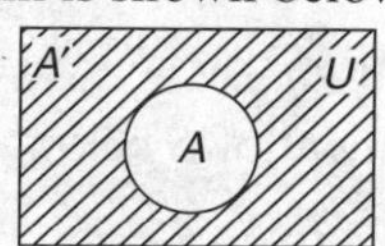

Cartesian Product of Two Sets

If A and B are any two sets then their Cartesian product is denoted by $A\times B$ and is defined as $A\times B=\{(a,b) : a\in A, b\in B\}$.

9.3.1 Laws of Operations

1. Idempotent laws: For any set A
   (i) $A\cap A=A$ (ii) $A\cup A=A$
2. Identities laws: For any set A
   (i) $A\cap U=A$ (ii) $A\cup\phi=A$
3. Commutative laws: For any two set A and B
   (i) $A\cup B=B\cup A$ (ii) $A\cap B=B\cap A$
4. Associative laws: For any three sets A, B and C
   (i) $A\cup(B\cup C)=(A\cup B)\cup C$
   (ii) $A\cap(B\cap C)=(A\cap B)\cap C$
5. Distributive laws: For any three sets A, B and C
   (i) $A\cup(B\cap C)=(A\cup B)\cap(A\cup C)$
   (ii) $A\cap(B\cup C)=(A\cap B)\cup(A\cap C)$
6. DeMorgan's law: For any two sets A and B
   (i) $(A\cup B)'=A'\cap B'$
   (ii) $(A\cap B)'=A'\cup B'$

Other Form of DeMorgan's Law

If A, B, C are any three sets, then

(i) $A-(B\cup C)=(A-B)\cap(A-C)$

(ii) $A-(B\cap C)=(A-B)\cup(A-C)$

Some Important Theorems

(i) For any two sets A and B, $A-B=A\cap B'$

(ii) For any two sets A and B
$A\cup B=B\Leftrightarrow A\subseteq B$
$A\cap B=A\Leftrightarrow A\subseteq B$

(iii) For any two sets A and B
$(A-B)\cup B=A\cup B$

(iv) If $A\cap B'=\phi$, then $A\subseteq B$

(v) If A and B are any two sets, then
$(A-B)\cup B=A$ if and only if $B\subseteq A$

(vi) If $A\cup B=C$ and $A\cap B=\phi$, then $A=C-B$

9.4 Cardinal Number of a Set

The number of distinct elements in a finite set is called the cardinal number of a set. The cardinal number of a set A is denoted by $n(A)$.

Some Theorems on Number of Elements of Sets

(I) If A and B are finite sets, then
$n(A\cup B)=n(A)+n(B)-n(A\cap B)$

(II) If A and B are disjoint sets, that is, $A\cap B=\phi$
then $n(A\cup B)=n(A)+n(B)$

(III) $n(A)=n(A-B)+n(A\cap B)$

(IV) $n(B)=n(B-A)+n(A\cap B)$

(V) $n(A\cup B)=n(A-B)+n(B-A)+n(A\cap B)$

(VI) If A, B, C are any three finite sets, then

$$n(A\cup B\cup C)=n(A)+n(B)+n(C)-n(A\cap B)-n(B\cap C)-n(A\cap C)+n(A\cap B\cap C)$$

SOLVED EXAMPLES

Example 9.1 Write the set $A=\left\{\frac{1}{2},\frac{2}{3},\frac{3}{4},\frac{4}{5},\frac{5}{6},\frac{6}{7}\right\}$ in set builder form.

Solution Since in every element of set A, the numerator is a natural number from 1 to 6 and the denominator is one more than the numerator, in set builder form we can write this as,

$$A=\left\{x: x=\frac{n}{n+1}, n\in N, 1\le n\le 6\right\}.$$

Example 9.2 Match each of the sets on the left in the roster from with the same set on the right given in the set builder form.

(i) $\{5,10,15,\ldots\}$	(A) $\{x: x$ is a letter of the word COLLEGE$\}$
(ii) $\{C,O,L,E,G\}$	(B) $\{x: x$ is an odd natural number less than $10\}$
(iii) $\{1,3,5,7,9\}$	(C) $\{x: x=5n, n\in N\}$
(iv) $\{2,3\}$	(D) $\{x: x$ is a prime number divisor of $12\}$

Solution

(i) Matches with (C), As the set of all natural numbers which are multiple of 5
(ii) Matches with (A)
(iii) Matches with (B)
(iv) Matches with D, as the prime divisors of 12 are 2 and 3.

Example 9.3 If $A=\{x: x=4^n-3n-1 \text{ and } n\in N\}$ and $B=\{y: y=9(n-1) \text{ and } n\in N\}$,
Prove that $A\subset B$.

Solution Since $x=4^n-3n-1$

$$=(1+3)^n-3n-1$$

$$=\left(1+{}^nC_1\cdot 3+{}^nC_2\cdot 3^2+{}^nC_3\cdot 3^3+\ldots + {}^nC_n\cdot 3^n-3n-1\right)$$

$$=9\left({}^nC_2+{}^nC_3\cdot 3+\ldots+{}^nC_n\cdot 3^{n-2}\right), n\in N$$

And $y=9(n-1), n\in N$

Taking, $n=1, 2, 3, \ldots$

$$y=0, 9, 18, \ldots$$

A contains 0 and some positive integral multiple of 9, while B contains 0 and all positive integral multiples of 9. Hence $A\subset B$.

Example 9.4 Find the total number of subsets of $\{1, 2, 3, \ldots, 9\}$ containing at least one odd number.

Solution Required number of subsets

$=$ Total number of subsets of $\{1, 2, \ldots, 9\}$ $-$ Total number of subsets of $\{2,4,6,8\}$

$$=2^9-2^4=496$$

Example 9.5 Let z denote the set of all integers and

$$A=\{(a,b): a^2+3b^2=28, a, b\in z\}$$

and $\quad B=\{(a,b): a>b, a, b\in z\}$,

then find the number of elements in $(A\cap B)$.

Solution Since $A=\{(a,b); a^2+b^2=28, a, b\in z\}$

$$=\{(5,1),(-5,-1),(5,-1),(-5,1),(1,3),(-1,3),(-1,-3),(1,-3),(4,2),(4,-2),(-4,2),(-4,-2)\}$$

And $B=\{(a,b): a>b, a, b\in z\}$

$$\therefore \quad A\cap B=\{(5,1),(5,-1),(-1,-3),(1,-3),(4,2),(4,-2)\}$$

$\therefore$ Number of elements in $A\cap B=6$

Example 9.6 Suppose $A_1, A_2, \ldots, A_{30}$ are thirty sets each with five elements and $B_1, B_2, \ldots, B_n$ are n sets each with three elements. Let $\bigcup_{i=1}^{30} A_i=\bigcup_{j=1}^{n} B_j=S$.

Assume that each element of S belongs to exactly 10 of A_i's and exactly nine of B_j's. Find n.

Solution Let $n(S)=p$

Now each element of S belongs to exactly 10 of A_i's, so, $\sum_{i=1}^{30} n(A_i)=10p$

But each A_i contains 5 elements

So, $\sum_{i=1}^{30} n(A_i)=5\times 30=150$

$\Rightarrow 10p=150\Rightarrow p=15$

Similarly, $\sum_{j=1}^{n} n(B_j)=9p$ and $\sum_{j=1}^{n} n(B_j)=3n$

$\Rightarrow 9p=3n\Rightarrow p=\dfrac{n}{3}$

$\Rightarrow \dfrac{n}{3}=15 \Rightarrow n=45$

Example 9.7 A survey shows that 63% of Americans like cheese where as 76% like apples. If x% of the Americans like both cheese and apples, then find x.

Solution It is given that $n(C)=63$, $n(A)=76$ and $n(C\cap A)=x$

Since $n(C\cup A)=n(C)+n(A)-n(C\cap A)$

$\Rightarrow 100=63+76-x \Rightarrow x=39$

Also, $n(C\cap A)\le n(C)\Rightarrow x\le 63$

Thus, $39\le x\le 63$

Example 9.8 A market research group conducted a survey of 1000 consumers and reported that 720 consumers liked product A and 450 consumers liked product B. What is the least number that must have liked both products?

Solution Let S = Set of consumers questioned

X = Set of consumers who liked product A

Y = Set of consumers who liked product B

We have to find the minimum value of $n(X\cap Y)$

From question,

$$n(S)=1000,\ n(X)=720,\ n(Y)=450$$

Now, $n(X\cap Y)=n(X)+n(Y)-n(X\cup Y)$

$$=720+450-n(X\cup Y)$$

$$=1170-n(X\cup Y) \quad (1)$$

$n(X\cap Y)$ will be minimum when $n(x\cup Y)$ is maximum.

But, $X\cup Y\subseteq S$

$\therefore \quad n(X\cup Y)\subseteq n(S)$

Maximum value of $n(X\cup Y)=n(S)=1000$

Hence from (1), the least value of

$$n(x\cap Y)=1170-1000=170$$

Example 9.9 Let A and B be two sets containing 23 and 60 elements respectively. Find the maximum and minimum number of elements in $A\cup B$.

Solution From question, $n(A)=23$ and $n(B)=60$

We know that,

$$n(A\cup B)=n(A)+n(B)-n(A\cap B)$$

$$\Rightarrow n(A\cup B)=23+60-n(A\cap B)$$

$$=83-n(A\cap B)$$

Now, $n(A\cup B)$ will be maximum for a minimum value of $n(A\cap B)$ which is zero, when A and B become disjoint sets $(A\cap B=\phi)$.

$\Rightarrow n(A\cup B)_{max}=83$

Similarly, $n(A\cup B)$ is minimum when $n(A\cap B)$ is maximum (when $A\subset B$)

$\Rightarrow n(A\cup B)_{min.}=83-(n(A\cap B))_{max.}$

$$=83-23=60$$

9.5 Real Number System

I. Natural Numbers (*N*) and Whole Numbers (*W*)

The numbers which are used for counting are known as *natural numbers* (set of positive integers), that is, $\{1, 2, 3, 4, 5, \ldots\}$. If 0 is included in the set of natural numbers, then it is called *whole numbers* (non-negative integers), that is, $W=\{0, 1, 2, \ldots\}$.

II. Integers (Z or I)

When negative of natural numbers are included in the set of whole numbers then the set of integers are formed, that is, $\{\ldots, -3, -2, -1, 0, 1, 2, 3, \ldots\}$.

III. Prime Numbers

If an integer (other than 1) is not divisible by any other integer except one and itself, then it is called a prime number, that is, 2, 3, 5, 7, 11, 13, ….

Note:

If the difference of two prime numbers is 2, then they are called *twin primes*, that is, $\{5, 7\}$, $(11, 13)$.

IV. Co-prime Numbers

Two integers are said to be relatively prime or co-prime if they have no common factor, that is, 4 and 15, 3 and 16, and so on.

V. Perfect Number

If the sum of all the divisors of a number n is equal to $2n$, then the number n is called a perfect number, that is, 6 is a perfect number because the sum of all divisors = 1 + 2 + 3 + 6 = 12 = 2(6).

Note:

(i) If x and y are co-prime then LCM of x and $y=x\cdot y$.

(ii) LCM of a rational and an irrational number does not exist.

(iii) LCM of $\frac{a}{b}$ and $\frac{c}{d}=\frac{\text{LCM}(a \text{ and } c)}{\text{HCF}(b \text{ and } d)}$.

(iv) HCF of $\frac{a}{b}$ and $\frac{c}{d}=\frac{\text{HCF}(a \text{ and } c)}{\text{LCM}(b \text{ and } d)}$.

(v) If $\frac{a}{b}=\frac{c}{d}=\frac{e}{f}=\ldots=\lambda$ (say), then

$$\lambda=\frac{a+c+e+\ldots}{b+d+f+\ldots}=\frac{\text{Sum of numerators}}{\text{Sum of denominators}}$$

(vi) A rational number in decimal form is either recurring or terminating.

(vii) An irrational number always give its absolute value.

(viii) Real Number = Rational Number + Irrational Number

9.6 Concept of Intervals

Open Interval

If a real number x is such that $a < x < b$, that is, the end points are not included, then we say that x lies in the open interval and is denoted by (a,b) or $]a,b[$.

Closed Interval

If a real number x is such that $a \le x \le b$, that is, the end points are included, then we say that x lies in the closed interval and is denoted by $[a,b]$.

Open–closed interval: $(a,b] = \{x : a < x \le b\}$

Closed–open interval: $[a,b) = \{x : a \le x < b\}$

The infinite intervals are defined as follows

(i) $(a,\infty) \Rightarrow x > a$ **(ii)** $[a,\infty) \Rightarrow x \ge a$

(iii) $(-\infty,b) \Rightarrow x < b$ **(iv)** $(-\infty,b] \Rightarrow x \le b$

Important Inequality

(i) $a \le b \Rightarrow$ either $a < b$ or $a = b$

(ii) $a < b$ and $b < c \Rightarrow a < c$

(iii) $a < b \Rightarrow a + c < b + c, c \in R$

(iv) $a < b \Rightarrow -a > -b$

(v) $a < b \Rightarrow ma < mb$ if $m > 0$ and $ma > mb$ if $m < 0$.

(vi) $0 < a < b \Rightarrow a^r < b^r$ if $r > 0$ and $a^r > b^r$ if $r < 0$

(vii) $a + \frac{1}{a} \ge 2$ if $a > 0$ and equality holds for $a = 1$

(viii) $a + \frac{1}{a} \le -2$ if $a < 0$ and equality holds for $a = -1$

(ix) $\log_a x > \log_a y \Rightarrow \begin{cases} x > y & \text{if} \quad a > 1 \\ x < y & \text{if} \quad 0 < a < 1 \end{cases}$

9.7 Wavy Curve Method

The method of intervals (or wavy curve) is used for solving inequalities of the form

$$f(x) = \frac{(x-a_1)^{n_1}(x-a_2)^{n_2}\ldots(x-a_k)^{n_k}}{(x-b_1)^{m_1}(x-b_2)^{m_2}\ldots(x-b_p)^{m_p}} > 0$$

$$(< 0, \le 0 \text{ or } \ge 0)$$

where $n_1, n_2, \ldots, n_k;\ m_1, m_2, \ldots, m_p$ are natural numbers. $a_1, a_2, \ldots, a_k;\ b_1, b_2, \ldots, b_p$ are any real numbers such that $a_i \ne b_j$, where $i = 1, 2, 3, \ldots, k$ and $j = 1, 2, 3, \ldots, p$.

(i) All zeros of the function $f(x)$ contained on the left hand side of the inequality should be marked on the number line with black circles.

(ii) All points of discontinuities of the function $f(x)$ contained on the left hand side of the inequality should be marked on the number line with white circles.

(iii) Check the value of $f(x)$ for any real number greater than the right most marked number on the number line.

(iv) From right to left, beginning above the number line (in case the value of $f(x)$ is positive otherwise from below the number line), a wavy curve should be drawn to pass through all the marked points so that when it passes through a simple point, the curve intersects the number line, and, when passing through a double point, the curve remains located on one side of the number line.

(v) The appropriate intervals are chosen in accordance with the sign of inequality (the function $f(x)$ is positive whenever the curve is situated above the number line, it is negative if the curve is found below the number line). Their union represents the solution of inequality.

Note:
Points of discontinuity will never be included in the answers.

9.8 Absolute value

If $x \in R$, then the magnitude of x is called its absolute value and is denoted by $|x|$ and is defined as, $|x| = \begin{cases} x, & x \ge 0 \\ -x, & x < 0 \end{cases}$

$x = 0$ can be included either with positive values of x or with negative values of x.

$|x|$ in fact represents the distance of number x from the origin measured along the number line.

We can also define $|x| = \sqrt{x^2}$.

If x and a are real numbers, then

$$|x - a| = (x - a) \text{ if } x \ge a$$
$$= -(x - a) \text{ if } x < a$$

Properties of Modulus (Absolute Value)

(i) $|x| > a$ where $a \in R^+ \Rightarrow x < -a$ or $x > a$
$\Rightarrow x \in (-\infty, -a) \cup (a, \infty)$

(ii) $|x| < a$ where
$a \in R^+ \Rightarrow -a < x < a \Rightarrow x \in (-a, a)$

(iii) $|xy| = |x||y|$

(iv) $\left|\dfrac{x}{y}\right| = \dfrac{|x|}{|y|}, y \neq 0$

(v) $|x+y| \leq |x| + |y|$; equality sign holds if $xy \geq 0$

(vi) $|x-y| \geq ||x| - |y||$

(vii) $||x| - |y|| \leq |x \pm y| \leq |x| + |y|$

Note:

(i) $|x| > -a$ where $a \in R^+$, this is true for all real values of x.

(ii) $|x| < -a$ where $a \in R^+$, not possible.

9.9 Greatest Integer and Fractional Part

If x is any real number, then the greatest integer of x is denoted by $[x]$ and is defined as $[x] \leq x$.

If $x = 2.69 \Rightarrow 2 < x < 3 \Rightarrow [x] = 2$

If $x = -3.63 \Rightarrow -4 < x < -3 \Rightarrow [x] = -4$

Points to Remember

(i) $a \leq x < a+1 \Leftrightarrow [x] = a$

(ii) $[x] > a \Rightarrow [x] \geq a+1 \Rightarrow x \geq a+1$

(iii) $[x] < a \Rightarrow [x] \leq a-1 \Rightarrow x < a$

(iv) $a_1 \leq [x] \leq a_2 \Rightarrow a_1 \leq x < a_2 + 1 \Rightarrow x \in [a_1, a_2 + 1]$
i.e. $-1 \leq [x] \leq 4 \Rightarrow -1 \leq x < 5 \Rightarrow x \in [-1, 5)$

Fractional Part

The difference between the number x and its integral value '$[x]$' is called the fractional part of x and is symbolically denoted by $\{x\}$.

i.e. $\{x\} = x - [x]$

we know that $x - 1 < [x] \leq x$

$\Rightarrow -x \leq -[x] < 1 - x$

$\Rightarrow 0 \leq x - [x] < 1$

$\Rightarrow 0 \leq \{x\} < 1$

Note:

(i) $[[x]] = [x]$ **(ii)** $[\{x\}] = 0$ **(iii)** $\{[x]\} = 0$

SOLVED EXAMPLES

Example 9.10 If $\dfrac{(x-1)(2-x)}{(x-3)} \geq 0$, then find the interval in which x lies.

Solution The given expression is $\dfrac{(x-1)(2-x)}{(x-3)}$.

We find the value of x by putting the numerator and denominator equal to zero.

$$(x-1)(2-x) = 0 \Rightarrow x = 1, 2$$

And $(x-3) = 0 \Rightarrow x = 3$

Plot them on the number line

1 2 3

As the coefficient of the highest power of x in the numerator and denominator are of opposite signs, a wavy curve will come from below (always from right side)

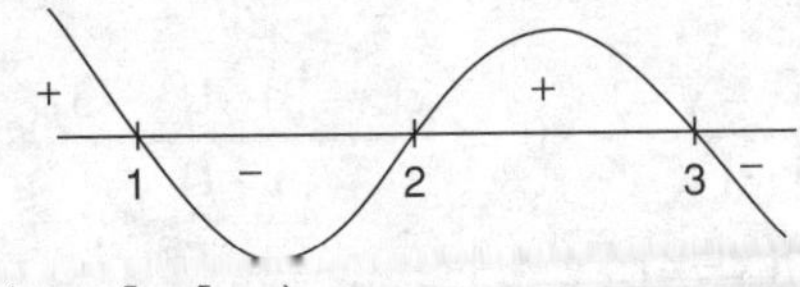

$\Rightarrow x \in (-\infty, 1] \cup [2, 3)$

Example 9.11 Find the interval in which x lies

$$\frac{(x-2)^2(1-x)(x-3)^3(x-4)^2}{(x+1)} \geq 0.$$

Solution We evaluate the value of x by putting the numerator and denominator equal to zero.

$$(x-2)^2(1-x)(x-3)^3(x-4)^2 = 0$$

$\Rightarrow x = 2, 1, 3, 4$

where 2 and 4 are repeated an even number of times.

And $x + 1 = 0 \Rightarrow x = -1$

As the coefficient of the highest power of x in the numerator and denominator are of opposite signs, the curve will come from below.

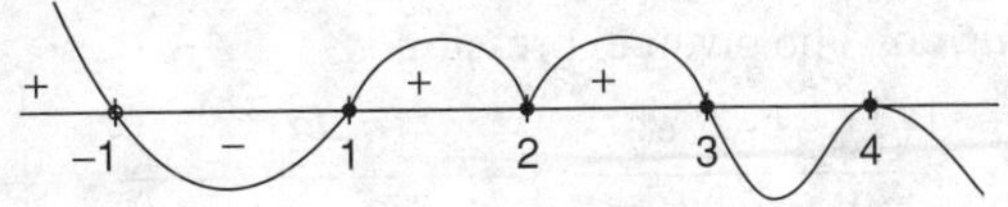

$\Rightarrow x \in (-\infty, -1) \cup [1, 2] \cup [2, 3] \cup \{4\}$

$\Rightarrow x \in (-\infty, -1) \cup [1, 3] \cup \{4\}$

Example 9.12 Solve for x

$$-15 < \frac{3(x-2)}{5} \leq 0$$

Solution $\frac{3(x-2)}{5} \le 0 \Rightarrow x-2 \le 0$

$\Rightarrow x \le 2$

$\Rightarrow x \in (-\infty, 2]$ (1)

Again, $\frac{3(x-2)}{5} > -15 \Rightarrow x-2 > \frac{-15\times 5}{3}$

$\Rightarrow x-2 > -25 \Rightarrow x > -23$

$\Rightarrow x \in (-23, \infty)$ (2)

From (1) and (2)

$$x \in (-23, 2]$$

Example 9.13 Solve for x

$$\frac{x^2-3x+4}{x+1} > 1 \text{ for all } x \in R$$

Solution $\frac{x^2-3x+4}{x+1} - 1 > 0$

$\Rightarrow \frac{x^2-3x+4-x-1}{x+1} > 0 \Rightarrow \frac{x^2-4x+3}{x+1} > 0$

$\Rightarrow \frac{x^2-3x-x+3}{x+1} > 0 \Rightarrow \frac{x(x-3)-1(x-3)}{(x+1)} > 0$

$\Rightarrow \frac{(x-1)(x-3)}{(x+1)} > 0$

Find the value of x by putting the numerator and denominator equal to zero, we get x = 1, 3, –1. As the coefficeint of the highest power of x in the numerator and denominator are of the same sign, the curve will come from above.

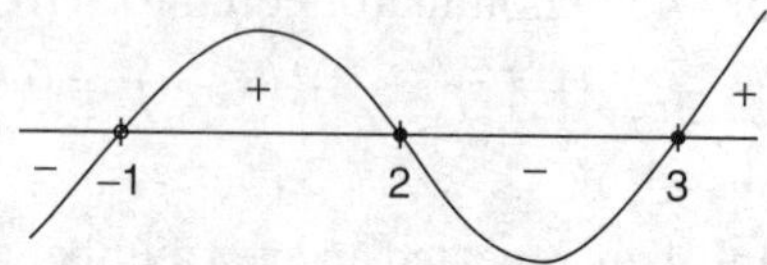

$\Rightarrow x \in (-1, 1) \cup (3, \infty)$

Example 9.14 Solve for x

$$\frac{x^2-6x+9}{5-4x-x^2} \ge 0$$

Solution The given expression is

$$\frac{x^2-6x+9}{-(x^2+4x-5)} = \frac{(x-3)^2}{-(x^2+5x-x-5)}$$

$$= \frac{(x-3)^2}{-\{x(x+5)-1(x+5)\}}$$

$$= \frac{(x-3)^2}{-(x-1)(x+5)}$$

Put $(x-3)^2 = 0 \Rightarrow x = 3, 3$

$(x-1)(x+5) = 0 \Rightarrow x = 1, -5$

$\Rightarrow x \in (-5, 1) \cup \{3\}$

Example 9.15 Find the number of natural numbers satisfying the inequality

$$\frac{(x-2)(x+2)^2(x-4)^2}{x(x+1)(x-3)} \le 0.$$

Solution The given expression is

$$\frac{(x-2)(x+2)^2(x-4)^2}{x(x+1)(x-3)}$$

Putting numerator and denominator = 0, we get

$x = 2, -2$ (two times), 4(two times)

$x = 0, -1, 3$

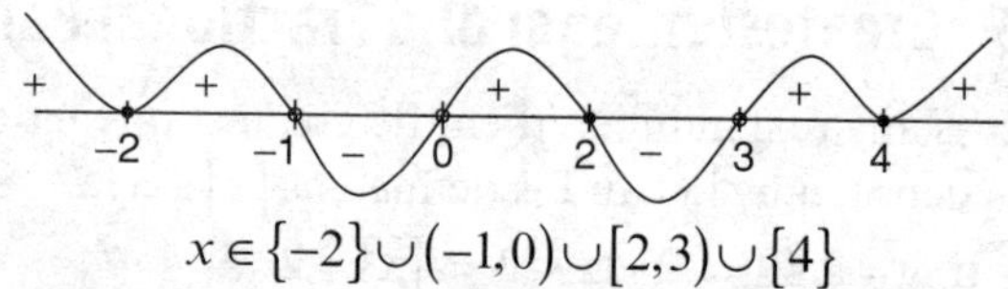

$$x \in \{-2\} \cup (-1, 0) \cup [2, 3) \cup \{4\}$$

$\Rightarrow$ There are only two natural numbers 2 and 4 which satisfy the given inequality.

Example 9.16 Solve for x

$$\frac{|x+3|+x}{x+2} > 1$$

Solution Since, $|x+3| = (x+3), x \ge -3$

$= -(x+3), x < -3$

Case I: When $x \ge -3$

$$\frac{x+3+x}{x+2} > 1 \Rightarrow \frac{2x+3}{x+2} - 1 > 0$$

$\Rightarrow \frac{2x+3-x-2}{x+2} > 0$

$\Rightarrow \frac{x+1}{x+2} > 0$

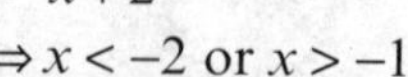

$\Rightarrow x < -2$ or $x > -1$

As condition is $x \ge -3$

Hence, $x \in [-3, -2) \cup (-1, \infty)$

Case II: $x < -3$

$$\frac{-x-3+x}{x+2} > 1 \Rightarrow \frac{-3}{x+2} - 1 > 0$$

$$\Rightarrow \frac{-3-x-2}{x+2} > 0 \Rightarrow \frac{-(x+5)}{(x+2)} > 0$$

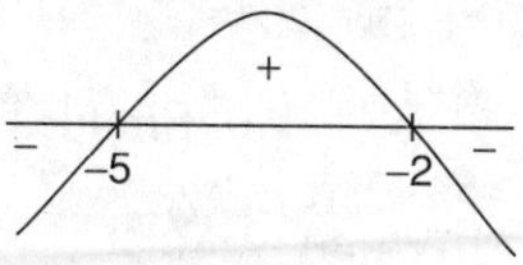

$$x \in (-5,-2)$$

From Case I and II,

$$x \in (-5,-2) \cup (-1,\infty)$$

Example 9.17 Find the set of real values of x satisfying $||x-1|-1| \le 1$.

Solution Since $||x-1|-1| \le 1$

$\Rightarrow -1 \le |x-1|-1 \le 1$

$\Rightarrow 0 \le |x-1| \le 2$

Since $|x-1| \ge 0, \forall x \in R$

Now, $|x-1| \le 2$

$\Rightarrow -2 \le x-1 \le 2$

$\Rightarrow -1 \le x \le 3$

Thus $x \in [-1,3]$.

Example 9.18 Find the set of real values of x satisfying $|x-1| \le 3$ and $|x-1| \ge 1$.

Solution Since $|x-1| \le 3$

$\Rightarrow -3 \le x-1 \le 3$

$\Rightarrow -2 \le x \le 4 \Rightarrow x \in [-2,4]$ (1)

Again, $|x-1| \ge 1 \Rightarrow x-1 \le -1$ or $x-1 \ge 1$

$\Rightarrow x \le 0$ or $x \ge 2$

$\Rightarrow x \in (-\infty,0] \cup [2,\infty)$ (2)

From (1) and (2)

$$x \in [-2,0] \cup [2,4]$$

Example 9.19 Solve for x

$$|x^2+6x+6| = |x^2+4x+9| + |2x-3|$$

Solution Since

$$|x^2+6x+6| = |(x^2+4x+9)+(2x-3)|$$

We know that

$$|a+b| = |a|+|b| \text{ if } ab \ge 0$$

$\Rightarrow (x^2+4x+9)(2x-3) \ge 0$

Now, $x^2+4x+9 > 0$ for all real values of x

As we know $ax^2+bx+c > 0$ for all real values of x if $a > 0$ and $D < 0$

$$\therefore \quad 2x-3 \ge 0 \Rightarrow x \ge \frac{3}{2}$$

Hence, $x \in \left[\frac{3}{2},\infty\right)$

Example 9.20 Solve for x

$$\left|\frac{x^2-3x-1}{x^2+x+1}\right| < 3$$

Solution $\left|\frac{x^2-3x-1}{x^2+x+1}\right| < 3$

$$\Rightarrow -3 < \frac{x^2-3x-1}{x^2+x+1} < 3$$

As $x^2+x+1 > 0$, for all real values of x.

Now, $\frac{x^2-3x-1}{x^2+x+1} < 3$

$\Rightarrow x^2-3x-1 < 3x^2+3x+3$

$\Rightarrow 2x^2+6x+4 > 0 \Rightarrow x^2+3x+2 > 0$

$\Rightarrow x^2+2x+x+2 > 0$

$\Rightarrow x(x+2)+1(x+2) > 0$

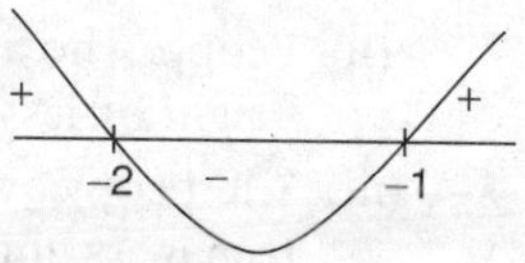

$\Rightarrow (x+1)(x+2) > 0$

$\Rightarrow x \in (-\infty,-2) \cup (-1,\infty)$

Again, $x^2-3x-1 > -3x^2-3x-3$

$\Rightarrow 4x^2+2 > 0$ which is true for all $x \in R$

Hence, $x \in (-\infty,-2) \cup (-1,\infty)$

Example 9.21 Solve for x

$$|x-1| + |x-2| \ge 3$$

Solution Since, $|x-1| = (x-1), x \ge 1$

$= -(x-1), x < 1$

Also, $|x-2| = (x-2), x \ge 2$

$= -(x-2), x < 2$

Case I: When $x < 1$

$$-(x-1)-(x-2) \ge 3$$

$\Rightarrow -x+1-x+2 \ge 3$

$\Rightarrow -2x+3 \ge 3 \Rightarrow -2x \ge 0 \Rightarrow x \le 0$

$\Rightarrow x \in (-\infty,0]$

Case II: when $1 \le x < 2$

$$x-1-x+2 \ge 3 \Rightarrow 1 \ge 3. \text{ No solution}$$

Case III: $x \ge 2$

$$x-1+x-2 \ge 3 \Rightarrow 2x \ge 6 \Rightarrow x \ge 3$$

$\Rightarrow x \in [3,\infty)$

Hence, $x \in (-\infty,0] \cup [3,\infty)$

9.10 Know Your Concepts

1. (i) $A-(B-C)=(A-B)\cup(A\cap C)$
   (ii) $A-B=(A\cup B)-B=A-(A\cap B)$
   Verify these equalities if
   $A=\{1,2,3,4,5,6,7\}$, $B=\{3,5,6,7,9,11\}$ and
   $C=\{2,5,6,9,20\}$.
2. Let U be the set of all people and
   $M=\{\text{Males}\}$, $S=\{\text{College students}\}$,
   $T=\{\text{Teenagers}\}$,
   $W=\{\text{People having height more than five feet}\}$.
   Express each of the following in the notation of set theory.
   (i) College students having a height of more than five feet.
   (ii) People who are not teenagers and have a height of less than five feet.
   (iii) All people who are neither male nor teenagers nor college students.
3. If $A=\{x: x \text{ is a digit in the number } 3591\}$
   $B=\{x: x\in N, x<10\}$.
   Find $A\cup B$, $A\cap B$, $A-B$ and $B-A$.
4. If $A=\{(x,y): x,y\in R, x^2+y^2=1\}$ and
   $B=\{(x,0): x\in R, -1\le x\le 1\}$, then find $A\cap B$.
5. In a town of 10,000 families it was found that 40% families buy newspaper A, 20% families buy newspaper B and 10% families buy newspaper C, 5% families buy A and B, 3% buy B and C and 4% buy A and C. If 2% families buy all the three newspapers, find the number of families which buy
   (i) A only
   (ii) B only
   (iii) none of A, B and C
6. Find the solution set of $\left|\dfrac{x^2-1}{x^2+x+1}\right|<1$.
7. Solve the following inequalities.
   (i) $\dfrac{5-4x}{3x^2-x-4}<4$
   (ii) $\left(\dfrac{3}{2}\right)^{1+x^2}>\left(\dfrac{2}{3}\right)^{3x-1}$
   (iii) $\dfrac{2^{x-1}-1}{2^{x+1}+1}<2$
   (iv) $2^{x+2}-2^{x+1}+2^{x-1}-2^{x-2}\le 9$
   (v) $\dfrac{2x-1}{2x^3+3x^2+x}>0$. Find least x, $x\in N$.
   (vi) $\dfrac{x^2-13x+40}{\sqrt{19x-x^2-78}}\le 0$
8. Solve the following for real x.
   (i) $|x^2-5x+2|=4$
   (ii) $|x+2|=|x-2|$
   (iii) $|x^2+x-1|=2x-1$
   (iv) $|2x+3|-|2x-5|=6$
   (v) Find the number of solutions of $|x|+|x-1|=5$
9. Find the set of odd integer values of x from the set of integers
   $A=\{x: x\in I \text{ and } |x-6|\le 8\}$,
   satisfying $\dfrac{(2x-1)(x-1)^2(x-2)^3}{(x-4)^4}>0$.
10. Find the complete set of reals which satisfy
    $$\frac{4}{1+x}+\frac{2}{1-x}<1.$$
11. Find the set of values of x satisfying the equality
    $$\left|\frac{x+1}{x}\right|+|x+1|=\frac{(x+1)^2}{|x|}$$
12. Find the complete set of reals which satisfy
    $$\frac{|x-3|}{x^2-5x+6}\ge 2.$$
13. Find the total number of real solutions of the equation $x^2+x+|x|+1\le 0$.
14. Find the number of integer values of x satisfying $\dfrac{x^2+6x-7}{|x+4|}<0$.
15. Solve for x, $|x+3|+|x-3|<2x$.
16. Solve for x, $|x+1|+|x-1|>2x$.
17. Solve the inequality $\left|1-\dfrac{|x|}{1+|x|}\right|\ge\dfrac{1}{2}$.

18. Solve the inequality

(i) $(x^2-5x+6)(x^2+3x+4)<0$

(ii) $\dfrac{(x+1)^4(x-2)(x-3)^3(x-4)^2}{x^3-36x}\geq 0$

9.11 JEE Main: Sample Questions

Single Answer Type

1. Let $A=\{x: x\in R, |x|<1\}$, $B=\{x: x\in R, |x-1|\geq 1\}$ and $A\cup B=R-D$, then set D is

(A) $\{x:1<x\leq 2\}$ (B) $\{x:1\leq x<2\}$
(C) $\{x:1\leq x\leq 2\}$ (D) none of these

2. If $aN=\{ax: x\in N\}$ and $bN\cap CN=dN$, where $b, c\in N$ are relatively prime, then

(A) $d=bc$ (B) $c=bd$
(C) $b=cd$ (D) none of these

3. Consider the set of all determinants of order 3 with entries 0 or 1 only. Let B be the subset of A consisting of all determinants with value 1. Let C be the subset of the set of all determinants with value – 1, then

(A) C is empty
(B) B has as many elements as C
(C) $A=B\cup C$
(D) B has twice as many elements as C

4. Two finite sets have m and n element. The total number of subsets of the first set is 56 more than the total number of subsets of the second set. The values of m and n are

(A) 7, 6 (B) 6, 3
(C) 5, 1 (D) 8, 7

5. Let $A=\{(x,y):|x-3|<1 \text{ and } |y-3|<1\}$ and $B=\{(x,y):4x^2+9y^2-32x-54y+109\leq 0\}$, then

(A) $A\subset B$ (B) $B\subset A$
(C) $A=B$ (D) none of these

6. Let A, B, C be subsets of the universal set U. If $n(U)=692, n(B)=230, n(C)=370$, $n(B\cap C)=20, n(A\cap B'\cap C')=10$, then value of $n(A'\cap B'\cap C')$ is equal to

(A) 127 (B) 172
(C) 182 (D) none of these

7. For any two sets A and B, the value of $(A-B)\cup(B-A)\cup(A\cup B)$ is equal to

(A) $A\cup B$ (B) $A\cap B$
(C) ϕ (D) none of these

8. If $A=\{(x,y): x^2+y^2=25\}$ and $B=\{(x,y): x^2+9y^2=144\}$ then $A\cap B$ contains

(A) 3 points (B) 4 points
(C) infinite points (D) none of these

9. In a city, 20 per cent of the population travels by car, 50 per cent of the population travels by bus and 10 per cent travels by both car and bus. Then, the person travelling by car or bus is

(A) 60% (B) 70%
(C) 80% (D) none of these

10. If $n(A)=1000, n(B)=500, n(A\cap B)\geq 1$ and $n(A\cup B)=p$, then

(A) $1000\leq p\leq 1497$ (B) $1000\leq p\leq 1498$
(C) $1000\leq p\leq 1499$ (D) none of these

11. Let A and B be two sets containing 2 and 4 elements respectively. The number of subsets of $A\times B$ having 3 or more elements is

(A) 211 (B) 256
(C) 220 (D) 219

12. If $f(x)+2f\left(\dfrac{1}{x}\right)=3x, x\neq 0$ and $S=\{x\in R: f(x)=f(-x)\}$. Then S

(A) is an empty set
(B) contains exactly one element
(C) contains exactly two elements
(D) contains more than two elements

13. Two sets A and B as under
$A=\{(a,b)\in R\times R:|a-5|<1 \text{ and } |b-5|<1\}$;
$B=\{(a,b)\in R\times R: 4(a-6)^2+9(b-5)^2\leq 36\}$,
then

(A) neither $A\subset B$ nor $B\subset A$
(B) $B\subset A$
(C) $A\subset B$
(D) $A\cap B=\phi$

9.12 Answer Key

9.12.1 Know Your Concepts

1. True

2. (i) $S \cap W$ (ii) $(U-T)\cap(U-W)$ or $T' \cap W'$ (iii) $U-(M\cup T\cup S)$, that is, $(M\cup T\cup S)'$

3. $A\cup B = B$; $A\cap B = A$; $A-B=\phi$; $B-A=\{2,4,6,7,8\}$

4. $\{(-1,0),(1,0)\}$ **5.** (i) 3300, (ii) 1400, (iii) 4000 **6.** $x\in\left[-2,-\frac{1}{2}\right]\cup(0,\infty)$

7. (i) $\left(-\infty,-\frac{\sqrt{7}}{2}\right]\cup\left(-1,\frac{\sqrt{7}}{2}\right]\cup\left(\frac{4}{3},\infty\right)$ (ii) $(-\infty,-3)\cup(0,\infty)$ (iii) $x\in R$ (iv) $(-\infty,2)$ (v) 1 (vi) $(6,8]$

8. (i) $\left\{\frac{5\pm\sqrt{33}}{2},2,3\right\}$ (ii) 0 (iii) $\left\{1,\frac{\sqrt{17}-3}{2}\right\}$ (iv) 2 (v) 2

9. $\{-1,3,5,9,11,13\}$ **10.** $(-\infty,-1)\cup(1,\infty)$ **11.** $x\in(0,\infty)\cup\{-1\}$ **12.** $x\in\left[\frac{3}{2},2\right)$

13. 0 **14.** 6 **15.** No solution **16.** $(-\infty,1)$ **17.** $x\in[-1,1]$

18. (i) (2, 3) (ii) $(-6,0]\cup[2,3]\cup(6,\infty)\cup\{4\}$

9.12.2 JEE Main: Sample Questions

Single Answer Type

1. B **2.** A **3.** B **4.** B **5.** A **6.** B **7.** A **8.** B
9. A **10.** C **11.** D **12.** C **13.** C

9.13 Summary

- Any collection of well defined and distinct objects is called a set.
- A set can be described in either Roster Form or Set Builder Form.
- Union, intersection, disjoint sets have been explained with Venn diagrams.
- Theorems on number of elements of sets
  - $n(A\cup B)=n(A)+n(B)-n(A\cap B)$
  - $n(A\cup B)=n(A)+n(B)$ if $A\cap B=\phi$
  - $n(A\cup B\cup C)=n(A)+n(B)+n(C)-n(A\cap B)-n(B\cap C)-n(A\cap C)+n(A\cap B\cap C)$
- Concepts covered in chapter
  - Different types of numbers
  - Interval and inequality
  - Wavy curve method
  - Absolute value and its property
  - Greatest integer function and fractional part